Ken Albrecht

Forage Quality, Evaluation, and Utilization

Editor

George C. Fahey, Jr.

Associate Editors

Michael Collins, David R. Mertens, and Lowell E. Moser

Based on the National Conference on Forage Quality, Evaluation, and Utilization held at the University of Nebraska, Lincoln, on 13-15 April 1994

Steering Committee

Terry J. Klopfenstein, co-chair
James R. Forwood, co-chair(deceased)

Robert F Barnes
Michael Collins
Barbara Emil
George C. Fahey, Jr.
David R. Mertens
C. Jerry Nelson
Craig A. Roberts
C. Alan Rotz
Monty Rouquette
Dan Undersander
Walt Wedin

American Society of Agronomy, Inc.
Crop Science Society of America, Inc.
Soil Science Society of America, Inc.
Madison, Wisconsin, USA

1994

Cover Artwork: Courtesy of the University of Nebraska-Lincoln

Copyright © 1994 by the American Society of Agronomy, Inc.
Crop Science Society of America, Inc.
Soil Science Society of America, Inc.

ALL RIGHTS RESERVED UNDER THE U.S. COPYRIGHT ACT OF 1976 (P.L. 94-553)

Any and all uses beyond the limitations of the "fair use" provision of the law require written permission from the publisher(s) and/or the author(s); not applicable to contributions prepared by officers or employees of the U.S. Government as part of their official duties.

American Society of Agronomy, Inc.
Crop Science Society of America, Inc.
Soil Science Society of America, Inc.
677 South Segoe Road, Madison, WI 53711 USA

Library of Congress Cataloging-in-Publication Date

National Conference on Forage Quality, Evaluation, and Utilization (1994: University of Nebraska)
Forage quality, evaluation, and utilization / editor-in-chief, George C. Fahey, Jr. ; associate editors, Lowell E. Moser, David R. Mertens, and Michael Collins.
    p.   cm.
"Based on the National Conference on Forage Quality, Evaluation, and Utilization held at the University of Nebraska, Lincoln, on 13–15 April 1994."
Includes bibliographical references.
ISBN 0-89118-119-9
1. Forage plants—Quality—Congresses. 2. Forage plants—Utilization—Congresses. 3. Feeds—Quality—Congresses. 4. Feed utilization efficiency—Congresses. 5. Ruminants—Feeding and feeds—Congresses. I. Fahey, George C. II. American Society of Agronomy. III. Crop Science Society of America. IV. Soil Science Society of America. V. Title.
SF94.6.N37 1994
636.2'0855—dc20
                                                  93-46888
                                                       CIP

Printed in the United States of America

CONTENTS

Page

PREFACE vii
CONTRIBUTORS ix

SECTION I. OVERVIEW OF FORAGE SCIENCE

1 **Milestones in Forage Research(1969-1994)**

R. L. Reid   1

2 **The Impact of Forage Quality and Supplementation Regimen on Ruminant Animal Intake and Performance**

J. A. Paterson, R. L. Belyea, J. P. Bowman, M. S. Kerley, and J. E. Williams   59

3 **Plant Factors Affecting Forage Quality**

C. J. Nelson and L. E. Moser   115

4 **Plant Environment and Quality**

D. R. Buxton and S. L. Fales   155

SECTION II. IDENTIFICATION AND QUANTITATIVE MEASUREMENT OF FORAGE QUALITY COMPONENTS

5 **Quantifying Forage Protein Quality**

G. A. Broderick   200

6 **Carbohydrates and Forage Quality**

K. J. Moore and R. D. Hatfield   229

7 **Minerals in Forages**

J. W. Spears   281

8 Fungal Endophytes, Other Fungi, and Their Metabolites as Extrinsic Factors of Grass Quality

    C. W. Bacon   318

9 Intrinsic Chemical Factors in Forage Quality

    L. Bush and H. Burton   367

10 The Application of Near Infrared Reflectance Spectroscopy (NIRS) to Forage Analysis

    J. S. Shenk and M. O. Westerhaus   406

SECTION III. INTAKE AS A CRITICAL ELEMENT OF FORAGE QUALITY

11 Regulation of Forage Intake

    D. R. Mertens   450

12 Measurement of Forage Intake

    J. C. Burns, K. R. Pond, and D. S. Fisher   494

13 Prediction of Intake as an Element of Forage Quality

    D. J. Minson and J. R. Wilson   533

SECTION IV. ROLE OF DIGESTION AND METABOLISM IN DETERMINING FORAGE QUALITY

14 Processes of Digestion and Factors Influencing Digestion of Forage-Based Diets by Ruminants

    N. R. Merchen and L. D. Bourquin   564

15 Measurement of In Vivo Forage Digestion by Ruminants

    R. C. Cochran and M. L. Galyean   613

16 Estimation of Digestibility of Forages by Laboratory Methods

W. P. Weiss   644

17 Methodology for Estimating Digestion and Passage Kinetics of Forages

W. C. Ellis, J. H. Matis, T. M. Hill, and M. R. Murphy   682

SECTION V. INTEGRATING CONCEPTS AFFECTING CHANGES IN FORAGE QUALITY

18 Modeling Forage Quality Changes in the Growing Crop

G. W. Fick, P. W. Wilkens, and J. H. Cherney   757

19 Foraging Behavior in Grazing Animals and its Impact on Plant Communities

J. Hodgson, D. A. Clark, and R. J. Mitchell   796

20 Changes in Forage Quality During Harvest and Storage

C. A. Rotz and R. E. Muck   828

21 Assessing Forage Quality Using Integrated Models of Intake and Digestion by Ruminants

A. W. Illius and M. S. Allen   869

SECTION VI. IMPROVING FORAGE QUALITY AND EVALUATION

22 Alteration of Plants Via Genetics and Plant Breeding

K. P. Vogel and D. A. Sleper   891

23 Modification of Forage Quality After Harvest

L. L. Berger, G. C. Fahey, Jr., L. D. Bourquin, and E. C. Titgemeyer   922

24 Forage Quality Indices: Development and Application

J. E. Moore   967

PREFACE

The chapters in this text were prepared in conjunction with the National Conference on Forage Quality, Evaluation, and Utilization held at the University of Nebraska, Lincoln, from April 13-15, 1994. The format of the book generally follows the symposium agenda. All chapters were reviewed by the Editor-in-Chief, the three Associate Editors, and two outside reviewers. Chapters were revised in accordance with important points made by these individuals.

This is the 25th anniversary of the first National Conference on Forage Quality Evaluation and Utilization held at the same location on September 3-4, 1969. The Proceedings of that Conference were published by the Nebraska Center for Continuing Education in 1970. The Conference was considered a milestone in the field of forage science and the proceedings have proven to be an extremely valuable reference for scientists in a variety of disciplines. Impetus for the 25th anniversary Conference came about as a result of the enthusiasm of two individuals - Dr. Terry J. Klopfenstein and Dr. James R. Forwood. Both served as co-chairs of the Conference Steering Committee until Dr. Forwood's untimely death in January, 1992.

The Editors of the book and the Steering Committee for the Conference would like to dedicate these proceedings to the memory of Jim Forwood. His enthusiasm and encouragement to summarize the state of the knowledge of forage quality and utilization was a major factor that stimulated planning for this Conference. Jim's own research activity centered around the forage/animal interface. He was keenly interested in developing research techniques and improved grazing systems that met livestock nutritional needs and plant requirements. He was equally comfortable interacting with plant or animal scientists. Jim was a native Nebraskan and received the B.S. degree from the University of Nebraska in Natural Resources and Wildlife Management. After graduation, Jim became interested in forages and range when he worked on the USDA-ARS project on pasture weed control. He finished a Master's degree at the University of Nebraska and went to Kansas State University to further his education in Range Management. After receiving a Ph.D., he joined the USDA-ARS Animal Physiology and Nutrition Unit at the University of Missouri as a pasture agronomist. In 1989, he transferred to the USDA-ARS range program at Fort Collins, Colorado, where he worked until his death. His interests included range management, forage management, animal science, and wildlife management, but most of all, he was interested in how animals utilized the forage available to them. We are deeply indebted to him for his leadership which resulted in forage and animal scientists from many places coming together to take part in this Conference. We wish he could have participated in this milestone event.

The variety of disciplines, industries, and associations involved in the field of forage quality research makes it a rich and diverse field of study, one not dominated by any single professional group. This Conference and the monograph resulting from the Conference provide a multidisciplinary look at this field. The varied perspectives on a shared interest--the future of forage science--strengthen the Conference and the future of the field.

The information presented in this book provides an historical foundation as well as a review of state-of-the-art developments in forage science. Twenty-five years of progress in forage quality, evaluation, and utilization are detailed along with "cutting edge" research in this field. New directions for future research and development are explored.

The text revolves around six major sections. These include: 1) Overview of Forage Science; 2) Identification and Quantitative Measurement of Forage Quality Components; 3) Intake as a Critical Element of Forage Quality; 4) Role of Digestion and Metabolism in Determining Forage Quality; 5) Integrating Concepts Affecting Changes in Forage Quality; and 6) Improving Forage Quality and Evaluation. Between three and six chapters contribute to the development of each major section topic. Nearly half of the chapters are co-authored by scientists at different institutions. Thus, different perspectives on a particular subject are brought out, even within the same chapter. In addition, chapter authors have attempted to define those areas where more information is needed and where new research efforts should concentrate.

Finally, the Steering Committee and editors of the book owe a debt of gratitude to all authors, reviewers, and sponsors of the Conference. Without their outstanding effort, including their moral and financial support, this Conference would not have been possible. A special thanks goes to the American Society of Agronomy, the Crop Science Society of America, and the Soil Science Society of America for publishing the book.

George C. Fahey, Jr.
Editor-in-Chief

David R. Mertens
Associate Editor

Michael Collins
Associate Editor

Lowell E. Moser
Associate Editor

CONTRIBUTORS

M. S. Allen	Assistant Professor, Department of Animal Science, Michigan State University, East Lansing, MI 48824
C. W. Bacon	Toxicology and Mycotoxin Research Unit, Richard B. Russell Agricultural Research Center, USDA-ARS, Athens, GA 30613
R. L. Belyea	Professor, Department of Animal Sciences, University of Missouri, Columbia, MO 65201
L. L. Berger	Professor, Department of Animal Sciences, University of Illinois, Urbana, IL 61801
L. D. Bourquin	Postdoctoral Research Associate, Department of Food Science and Human Nutrition, Michigan State University, East Lansing, MI 48824
J. P. Bowman	Assistant Professor, Department of Animal and Range Sciences, Montana State University, Bozeman, MT 59717
G. A. Broderick	Research Dairy Scientist, U.S. Dairy Forage Research Center, USDA-ARS, Madison, WI 53706
J. C. Burns	Lead Scientist, USDA-ARS and Departments of Crop Science and Animal Science, North Carolina State University, Raleigh, NC 27695
H. Burton	Professor, Department of Agronomy, University of Kentucky, Lexington, KY 40546-0091
L. Bush	Professor, Department of Agronomy, University of Kentucky, Lexington, KY 40546-0091
D. R. Buxton	Field Crops Research Unit and Cluster Scientist, U.S. Dairy Forage Research Center, USDA-ARS, Department of Agronomy, Iowa State University, Ames, IA 50011

J. H. Cherney	Associate Professor, Department of Soil, Crop and Atmospheric Sciences, Cornell University, Ithaca, NY 14853
D. A. Clark	Dairying Research Corporation, Hamilton, New Zealand
R. C. Cochran	Associate Professor, Department of Animal Sciences and Industry, Kansas State University, Manhattan, KS 66506
W. C. Ellis	Professor, Department of Animal Science, Texas A&M University, College Station, TX 77843-2471
G. C. Fahey, Jr.	Professor, Department of Animal Sciences, University of Illinois, Urbana, IL 61801
S. L. Fales	Associate Professor, Department of Agronomy, The Pennsylvania State University, University Park, PA 16802
G. W. Fick	Professor, Department of Soil, Crop and Atmospheric Sciences, Cornell University, Ithaca, NY 14853
D. S. Fisher	USDA-ARS and Department of Crop Science, North Carolina State University, Raleigh, NC 27695
M. L. Galyean	Professor, Clayton Livestock Research Center, New Mexico State University, Clayton, NM 88415
R. D. Hatfield	Research Plant Physiologist, U.S. Dairy Forage Research Center, USDA-ARS, Madison, WI 53706
T. M. Hill	Assistant Professor, Department of Animal Science, University of Maine, Orono, ME 04469
J. Hodgson	Massey University, Palmerston North, New Zealand
A. W. Illius	Institute of Cell, Animal and Population Biology, University of Edinburgh, Edinburgh EH9 3JT, United Kingdom
M. S. Kerley	Associate Professor, Department of Animal Sciences, University of Missouri, Columbia, MO 65201
J. H. Matis	Professor, Department of Statistics, Texas A&M University, College Station, TX 77843-2471

N. R. Merchen	Professor, Department of Animal Sciences, University of Illinois, Urbana, IL 61801
D. R. Mertens	Research Dairy Scientist, U.S. Dairy Forage Research Center, USDA-ARS, Madison, WI 53706
D. J. Minson	Division of Tropical Crops and Pastures, CSIRO, Cunningham Laboratory, St. Lucia, Brisbane, Queensland 4067, Australia
R. J. Mitchell	Agricultural Research, Palmerston North, New Zealand
J. E. Moore	Professor, Department of Animal Science, University of Florida, Gainesville, FL 32611-0900
K. J. Moore	Professor, Department of Agronomy, Iowa State University, Ames, IA 50011
L. E. Moser	Professor, Department of Agronomy, University of Nebraska, Lincoln, NE 68506-0915
R. E. Muck	Agricultural Engineer, USDA-ARS, U.S. Dairy Forage Research Center, University of Wisconsin, Madison, WI 53706
M. R. Murphy	Associate Professor, Department of Animal Sciences, University of Illinois, Urbana, IL 61801
C. J. Nelson	Professor, Department of Agronomy, University of Missouri, Columbia, MO 65211
J. A. Paterson	Professor and Head, Department of Animal and Range Sciences, Montana State University, Bozeman, MT 59717
K. R. Pond	Associate Professor, Department of Animal Science, North Carolina State University, Raleigh, NC 27695
R. L. Reid	Professor, Division of Animal and Veterinary Sciences, West Virginia University, Morgantown, WV 26506-6108
C. A. Rotz	Agricultural Engineer, USDA-ARS, U.S. Dairy Forage Research Center, Michigan State University, East Lansing, MI 48824
J. S. Shenk	Professor, Department of Agronomy, The Pennsylvania State University, University Park, PA 16802

D. A. Sleper	Professor, Department of Agronomy, University of Missouri, Columbia, MO 65211
J. W. Spears	Professor, Department of Animal Science, North Carolina State University, Raleigh, NC 27695
E. C. Titgemeyer	Assistant Professor, Department of Animal Sciences and Industry, Kansas State University, Manhattan, KS 66506
K. P. Vogel	USDA-ARS, Department of Agronomy, University of Nebraska Lincoln, NE 68583-0937
W. P. Weiss	Associate Professor, Department of Dairy Science, Ohio Agricultural Research and Development Center, The Ohio State University, Wooster, OH 44691
M. O. Westerhaus	Department of Agronomy, The Pennsylvania State University, University Park, PA 16802
P. W. Wilkens	International Fertility Development Center, Muscle Shoals, AL 35662
J. E. Williams	Professor, Department of Animal Sciences, University of Missouri, Columbia, MO 65201
J. R. Wilson	Forage Physiologist, Division of Tropical Crops and Pastures, CSIRO, Cunningham Laboratory, St. Lucia, Brisbane, Queensland 4067, Australia

CHAPTER 1

MILESTONES IN FORAGE RESEARCH
(1969 - 1994)

R.L. Reid

INTRODUCTION

The purpose of this chapter will be to describe, briefly, the highlights in forage research which have occurred since the first conference held in Lincoln, Nebraska, in 1969. Use of the term "milestone" in this context may be somewhat misleading. The incidence of research milestones, considered as discrete and revolutionary events, may be about as frequent as the diminishing occurrence of physical milestones on the American landscape. More often, advances in research occur in small and incremental steps, sometimes with obscure or unknown origins, but generally as variations on a preceding theme. There are exceptions. In the area of feedstuff evaluation, development of the detergent system of analysis by Van Soest in the 1960's might reasonably be regarded as a milestone event associated with one individual. Perhaps the application of near infrared reflectance (NIR) analysis to forages (Norris et al., 1976) in the 1970's would be considered another, although the technique had been used for other purposes previously. Beyond this, definition of what constitutes a milestone tends to be largely in the eye of the particular beholder and might better be left to the judgment of a group of individuals with different backgrounds and experience. The organizing committee of this conference recognized this and proposed a number of subjects for discussion in this chapter. These included: methodologies for determining quality (detergent methods, NIRS); recognition of the importance of protein fractions in forages; advent of Coastal bermudagrass [*Cynodon dactylon* (L.) Pers.], endophyte-free tall fescue (*Festuca arundinacea* Schreb.) and improved switchgrass (*Panicum virgatum L.)*; preservatives and inoculants; forage packaging systems; brown midrib plants; advances in understanding of the kinetics of forage digestion and passage.

These are specific areas of research in which much progress has been made in the last 25 years and most of them will be reviewed in detail in other chapters. I will attempt to present these subjects in a fairly broad context, and add some topics which I believe are of particular interest or significance. As a working and flexible definition of a milestone, I propose: a) an original event, concept, or general area of research leading to a significant advance in understanding and(or) application; b) a key review/conference/book(s) summarizing a research field and acting as a base for further study and development.

Div. of Animal and Vet. Sciences, West Virginia Univ., Morgantown, WV 26506-6108

A Brief Recapitulation

"What's past is prologue"
                                          The Tempest

In order to understand the significance of changes which have occurred in forage quality and utilization research since 1969, a short summary of the state of knowledge at that time is in order. Probably no better or more succinct account can be found than that of Mott and Moore (Figures 1 and 2) in their perspective on forage evaluation methods at the 1969 conference. Forage quality (Figure 1) was defined as some product of digestibility and intake of the diet, with intake taking precedence as the more important of the two aspects of quality.

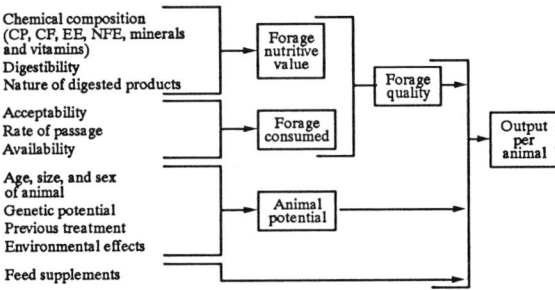

Figure 1. Forage quality and output per animal. Mott and Moore (1969).

At much the same time, Raymond (1968) defined quality in the equation:

Nutrient intake = Intake of feed DM x digestibility of feed DM
                  x efficiency of utilization of digested nutrients

The involvement of efficiency as a significant factor in forage evaluation was not generally recognized or measured and relatively few determinations had been made of forage consumption, although a number of regional projects were initiated in the 1960's to measure intake under different conditions and to define plant and animal factors affecting it (Richards, 1969). The prevailing view was that ruminants increased energy intake as energy digestibility of the ration increased until some "requirement" was met (Conrad et al., 1964; Conrad, 1966); with low quality roughages, intake was limited by physical capacity of the tract. Waldo (1969) concluded that fiber mass restricted intake of forage by retention in the reticulorumen and that reduction of the mass occurred by passage and by digestion, with physical breakdown of particles to a critical size required for transfer to the lower tract. Riewe and Lippke (1969) described factors affecting the digestibility of forages (level of feeding, kind and level of concentrate, method of forage processing, frequency of feeding, and ruminant species) and discussed forage composition effects in relation to the summative system developed by Van

Soest (1967, 1969).
The application of laboratory analytical methods to forage and pasture evaluation under different research and management conditions (Figure 2) was reviewed also by Mott and Moore (1969).

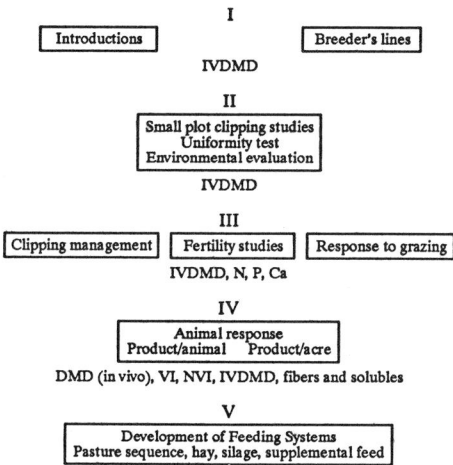

Figure 2. Scheme for forage evaluation. Mott and Moore (1969).

In vitro procedures were tested extensively in the NC-64 regional project, based on the two-stage Tilley and Terry (1963) technique, and methods and variables studied in this collaborative approach were discussed by Johnson (1969) and Barnes (1969). Van Soest (1969) concluded that "the difference in R^2 between the Tilley in vitro system and any group of chemical tests in accounting for variance in in vivo digestibility is a measure of the proportional effect of unassayed limiting factors active in the substrate." Use of a cellulolytic enzyme digestion instead of rumen fluid, with good results, was reported by Jarrige et al. (1970) at the 11th International Grassland Congress in Australia. Van Soest (1969) reviewed the concept of nutritive entities and emphasized that "we have not progressed sufficiently far in our understanding of the basic factors influencing the nutritive availability and qualities of forage that we can invariably make an intelligent choice of laboratory chemical methods to be routinely used in the accurate evaluation of commercial forages." The application of solubility measurements, forage lignins, and nylon bag measurements to the estimation of nutritive quality was discussed by Donefer (1969), Allinson (1969), and Lowrey (1969), respectively. On the basis of information available at that time, Mott and Moore (1969) proposed use of in vitro procedures for the initial steps in the sequence of agronomic testing of forages (Figure 2), followed by involvement of the animal in estimation of intake and pasture evaluation.

Pasture and range research at the time of the 1969 conference was conducted primarily in the western states using well-established methods of estimating the chemical and botanical composition of the grazing animal's diet (Theurer, 1969; Raleigh, 1969). Matches (1969), in reviewing pasture research techniques, stated that "Defining the plant-animal interrelationships influencing output per animal and per pasture should be our goal in pasture research". At much the same time, in Australia, McDonald (1968) concluded that "The interaction between grazing animal and grazed plant is extremely complex and will be much influenced by season of year, rate of plant growth, species of plants and animals involved, stocking rates, and management practices. The defoliation of plants by animals has radical effects on the growth of the plants, and that neglected subject merits much research." Partly because of the complexity of the problem, and partly due to the shift away from pasture use in the United States in the 1950's and 1960's (Reid and Klopfenstein, 1983), relatively little attention was directed, at least in the humid areas of the country, to study of pasture-animal interactions. For many of the same reasons, breeding research on the selection of higher quality pasture and forage plants was limited, with one or two notable exceptions. Burton (1969), in Georgia, reviewed criteria and techniques used in plant breeding programs and described the success of a selection approach leading to the development of Coastcross-1 bermudagrass, based on nylon bag measurements of digestibility.

Topics which were not discussed specifically at the 1969 conference included: effects of forage conservation procedures and additives on plant composition and quality; role of crop residues in feeding systems; the significance of minerals and secondary plant metabolites in forage plant digestion, metabolism, and animal health; digestive kinetics and nutrient availability at different points in the ruminant alimentary tract; the development of modelling and systems approaches to the study of forage digestion and to pasture and forage management and feeding. These subjects will be considered in detail in presentations at the present conference.

MAJOR DEVELOPMENTS IN THE PERIOD 1969 - 1994

Progress will be evaluated within the following general areas of research:

(1) Definition and understanding of forage quality.
(2) Advances in understanding of plant composition and structure.
(3) Laboratory methods for forage analysis and prediction of quality.
(4) Advances in knowledge of digestive kinetics and application to forage evaluation.
(5) Progress in breeding better forage plants.
(6) Forage conservation methods.
(7) The plant-animal interface and pasture evaluation.
(8) Development of modelling and systems approaches.

Definition of Forage Quality

"Now good digestion wait on appetite,
And health on both!"

Macbeth

The major milestone in definition of the nutritional quality of forages for ruminants - significance of the intake factor - had been established well before the 1969 conference (Crampton, 1957; Crampton et al., 1960; Blaxter, 1962). While no advance of comparable magnitude has, arguably, been made since that time, there have been a number of interesting developments in understanding and quantification of components of the basic Raymond (1968) equation.

The importance of level of intake of the diet on the performance of ruminant animals has been attested in a large body of literature during the last 25 years and will be discussed in detail in several papers at this conference. The literature may be summarized, simply, by the statement of Forbes (1986) that "The composition and availability of the diet are major factors influencing voluntary intake. While the concentration of protein, the balance of amino acids and deficiency or excess of minerals can all affect intake, the major parameter of a food which determines the amount eaten is the concentration of available energy." This viewpoint has not changed appreciably and would apply equally to stall-fed and grazing animals. Forbes (1986) cited the opinion of Moir in 1970 that "Voluntary intake can only be fully understood and manipulated when the control factors and their integration mechanisms are known". Several good reviews on the subject of intake regulatory mechanisms and factors affecting intake have appeared since 1969, e.g., Baile and Forbes (1974); Bines (1976); Weston (1982); Van Soest (1982); Forbes (1986, 1993); Weston and Poppi (1987); Garnsworthy and Cole (1990). A model of feed transfer and intake regulation proposed by Weston (1982) is shown in Figure 3. This has the merit of linking factors influencing digesta load to energy metabolism and the state of energy deficit in the animal.

In a recent critique of accepted theories of intake regulation, Ketelaars and Tolkamp (1992a), Tolkamp and Ketelaars (1992), and Ketelaars and Tolkamp (1992b) questioned the concept of rumen fill as a major factor controlling intake in ruminants and proposed that feed intake behavior is directed towards maximizing efficiency of oxygen utilization by the animal, on a costs and benefits basis. This is an interesting hypothesis and could explain a number of anomalies in intake data and theory.

The significance of intake in the productivity of farm animals in the United States was recognized in a National Research Council (1987) publication; "Predicting Feed Intake of Food-Producing Animals". This contained reviews of control mechanisms in the main domestic animal species and, for each species, developed prediction equations and adjustment factors for intake estimation and diet formulation. In France, forage 'fill values' (related to cell wall content) were developed for sheep and cattle (Jarrige, 1989). Forbes' (1986) book "The Voluntary Food Intake of Farm Animals" provides an excellent account of the more recent methods of measuring food intake in animals and of the many single-factor and multiple-factor theories proposed for intake regulation.

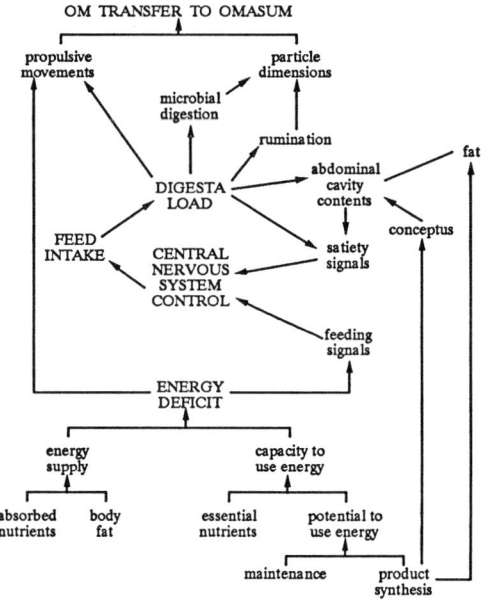

Figure 3. Regulation of feed transfer and intake. Weston (1982).

Digestibility estimates are still the most frequently used animal measures of forage quality, as they have been for the last century, and no major changes in technique or interpretation of digestibility data have been introduced, although digestibility is now frequently determined at ad libitum rather than, or in addition to, the maintenance level of feeding. A comprehensive review of methods for conducting digestibility trials, and of factors affecting digestibility, was published by Schneider and Flatt (1975). In terms of the quality equation, Leng (1991) stated that "To optimize ruminant production from forages, the need is to optimize digestibility and intake, maximize microbial growth efficiency and adjust the nutrients so that they closely correspond to the quantities and balances required for the productive function", and concluded that "The past 30 years of research has emphasized that it is the efficiency of feed utilization which largely determines the level of production from ruminants on forage-based diets." It is in this area, that of defining the nutrient requirements for optimizing microbial growth and efficiency, that much progress has been made and reference will be made to these developments in subsequent sections.

Environmental Effects on Quality

In the period prior to the 1969 conference, relatively little research had been conducted on the influence of environment on forage quality, although the work of J.T. Reid (Reid et al., 1959; Reid, 1973) at Cornell University clearly showed the marked effect of time of cutting, latitude, and altitude on the digestibility and intake of forage crops grown in temperate regions. The effect of climate in depressing the digestibility of tropical vs temperate forages was indicated in a landmark paper by Minson and McLeod (1970) at the International Grassland Congress in Australia. Frequency distribution plots of published digestibility data for large populations of forages showed that, on average, temperate grasses had digestible dry matter (DMD) coefficients 12.8 percentage units higher than tropical grasses, and the authors demonstrated that digestibility decreased by approximately one percentage unit for each 1°C rise in growing temperature. In a series of papers, Deinum in the Netherlands (Deinum, 1971; Deinum and Dirven, 1974, 1975) and Wilson in Queensland (Wilson and Ford, 1971, 1973; Wilson et al., 1976, 1991a; Wilson and Minson, 1980; Wilson, 1982) examined the effects of temperature, light, and humidity on the composition and quality of forages. Possible consequences of such changes on gut digestion and animal response are outlined in Figure 4, adapted from Van Soest (1983). The subject is well reviewed by Van Soest (1982), Wilson (1982), Norton (1982), and Jones and Wilson (1987).

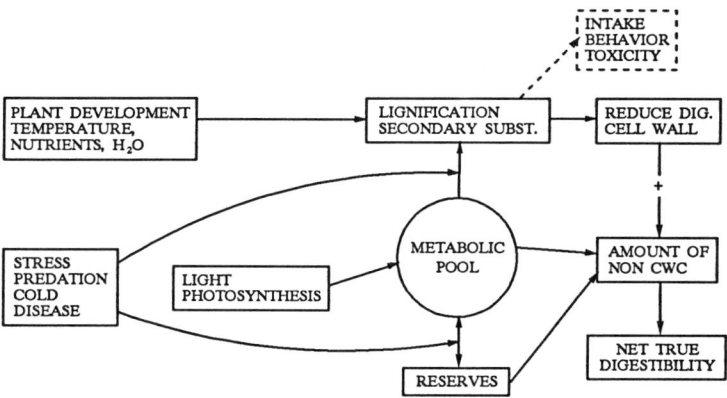

Figure 4. Environmental effects on plant quality. Adapted from Van Soest (1983).

Implications of climatic effects on the quality of different classes of forages - warm-season and cool-season annual and perennial grasses, and legumes - grown in the southern United States are discussed by Riewe (1981), and the need for considering possible forage x environmental interactions in designing forage evaluation experiments is stated by Cherney and Volenec (1992). Climate also affects the grazing behavior and performance of ruminant animals, and effects upon intake are reviewed by Weston (1982) and Young (1987). The influence of environment on forage composition and quality will be discussed in detail in later papers at this conference (Chapters 4 and 18).

Plant Composition and Structure and Relationship to Quality

"More composition and fierce quality".

King Lear

The chemistry of forage plants was reviewed at length in a series of publications in the early 1970's (Butler and Bailey, 1973), and this series remains a vade mecum for forage chemists and those interested in forage utilization. Van Soest (1969, 1975), however, pointed out that chemical and nutritive criteria do not necessarily correspond and indicated the importance of physico-chemical properties of plants in determining their susceptibility to microbial breakdown and rate of passage through the alimentary tract. Minson (1982) discussed the effects of chemical and physical properties of forages upon intake by ruminant animals. The significance of physical structures in plants in relation to their digestive utilization was emphasized by development of scanning (SEM) and transmission (TEM) electron microscopic techniques (Akin et al., 1973; Akin, 1979) and by increased interest in the use of C4 grasses in grazing and feeding systems (Akin, 1982, 1986, 1989; Akin et al., 1983, 1984). Visual depiction of leaf and stem anatomy of forage plants, and of their sequential degradation by ruminal microorganisms, showed clear differences in the proportions of tissue types in temperate (C3) and tropical (C4) grasses, accounting at least in part for differences in their rates of degradation and digestibility characteristics. A comparison of the digestive kinetics of isolated parenchyma and sclerenchyma cell walls from a C3 and C4 grass by Grabber et al. (1992) indicated that the anatomical arrangement of sclerenchyma was a major factor influencing cell digestion. Wilson (1991b) stated that, in terms of rates of breakdown to small particles, extent of digestion and retention times of DM in the rumen, large differences occurred between classes of forage, immature vs mature, legume vs grass, tropical vs temperate grass, leaf vs stem, and species vs species. Much of the more recent work on the chemistry and organization of forage cell walls, and the effects of structure on ruminal degradation, was presented at a conference in Wisconsin in 1991 (Jung et al., 1993). This line of research obviously merits further development; some implications will be discussed by Minson and Wilson (Chapter 13).

Nitrogen Fractions

The period since 1969 has seen significant advances in understanding of the utilization of N compounds in forages in two areas: a) potential losses of N by conservation procedures, particularly those involving heating during storage (Goering et al., 1972, 1973). Van Soest and Mason (1991) reviewed the effects of the Maillard reaction on the production of artifact lignin polymers and such compounds as 4-methyl-imidazole in high temperature-NH_3 treated forages; b) more generally, profound effects of the form of N in feeds on ruminal degradability, efficiency of microbial protein synthesis, and availability of amino acids at the animal tissue level. The evolution of present protein systems for ruminants is outside the scope of this conference and is discussed in other bulletins or books, e.g., ARC (1980, 1984), Ørskov (1982), National Research Council (1985), and Jarrige (1989). Less attention has been directed to the question of N availability in forage crops and pastures (Beever and Siddons, 1986), although Beever and Thomson (1977) noted differences among forms of forage (grass, silage, hay) and plant species in percentages of N escaping ruminal degradation, and similar observations with tropical forages were made by Aii and Stobbs (1980). Recognition of the generally wasteful nature of the process of N utilization in grazing ruminants (Ulyatt et al., 1975; Kemp et al., 1979) was confirmed by observations, in Australian trials, that dairy cows grazing high-N tropical and temperate grasses increased milk production in response to supplementation with protected protein (Stobbs et al., 1977; Minson, 1981a). Similar growth responses were obtained in steers grazing smooth bromegrass (*Bromus inermis* Leyss.) pastures in Nebraska trials (Anderson et al., 1988). Reasons for the observed responses of grazing animals to supplementary protein have not always been clear (Beever and Siddons, 1986) and may reside, in part, from increases in intake and(or) effects on the animal's endocrine system and partitioning of nutrients. Leng (1991, 1993) suggested that the production responses of heat-stressed animals in tropical environments to supplementary bypass protein result from improved efficiency of utilization of the low-quality feed.

Forage conservation has marked effects on both the amount and solubility/degradability characteristics of N compounds in forages (McDonald, 1982; Minson, 1990; Reid, 1994). Changes in protein solubility in hays and artificially-dried forages depend on the time and temperature of drying, and ammoniation of heated feeds can increase levels of acid detergent insoluble N (ADIN) and 4-methyl-imidazole (Mason et al., 1990; Perdok and Leng, 1987). Treatment of perennial ryegrass (*Lolium perenne* L.) with anhydrous NH_3 was found to increase concentrations of total N, neutral detergent and ADIN, water insoluble N, soluble NAN and 4-methyl-imidazole and lysinoalanine, but decreased water soluble carbohydrate (WSC) and cellulase-undegraded cell walls (Mason et al., 1989).

Ensiling effects on N forms and content in forage crops have been widely investigated and reviewed in recent years (McDonald, 1981; Woolford, 1984; Thomas and Thomas, 1985). Using the nylon bag incubation technique, Tamminga et al. (1991) showed that the proportion of N that escaped degradation in the rumen was much lower in silages than in hays and, in silage, was positively influenced by DM content and date of harvesting, and negatively by crude protein concentration of the forage. Waldo (1985) considered that while wilting legumes to approximately 350 g Kg^{-1} DM before ensiling reduced N breakdown, it did not result in silage containing adequate bypass protein for high-producing ruminants. He suggested the possibility of selecting legumes with a higher concentration of tannins to reduce ruminal N degradation. Alternative stratagems have been to use formic and inorganic acid and aldehyde treatments to "fix" soluble plant proteins; Waldo (1985) reported that use of formic acid and formaldehyde increased the proportion of ruminally undegraded N and improved N retention in steers.

The general principle which emerges from the many studies which have been conducted in recent years to manipulate the ruminal availability of N in fresh and conserved forages is that stated by Van Vuuren et al. (1990) as "synchronization of available energy and N to the rumen biota." Further progress seems attainable in the areas of selection of forage species differing in anatomical characteristics or containing compounds, such as tannins, which regulate rates or extent of degradation of N and carbohydrate components, use of additives in the conservation process, and, as suggested by Beever and Siddons (1986), the strategic use of energy supplements to balance degraded N relative to degraded OM supply.

Secondary Compounds

While most attention has centered, justifiably, on N and carbohydrate fractions in forages in relation to efficiency of utilization, much interesting research has been conducted since the 1969 conference on secondary components such as minerals and "antiquality" components, and this will be the subject of several papers in these proceedings. Factors influencing the concentration and availability of minerals in forages were discussed in reviews by Fleming (1973) and Butler and Jones (1973) in the Butler and Bailey (1973) volumes on herbage chemistry and biochemistry. Interest was, however, largely confined to the function of minerals in metabolic disorders of ruminant animals. A series of studies in France (Durand and Kawashima, 1980; Durand and Komisarczuk, 1988; Komisarczuk-Bony and Durand, 1991) demonstrated the essential function of minerals in optimal amounts for ruminal microbial function, and these authors concluded that, as for N, mineral supply must be balanced with available energy to optimize microbial protein synthesis and carbohydrate digestion. Several animal trials, mainly performed in Australia with forages grown on soils deficient in specific nutrients, support this additional role of minerals in forage utilization and the evidence is presented in detail by Minson (1990) in his recent book "Forage in Ruminant Nutrition."

The pathway to several classes of the secondary organic components of forages is outlined by Burns (1985) in Figure 5 and the subject of toxins in plants is treated extensively in a number of recent publications (Hegarty, 1982; Barry and Blaney, 1987; Cheeke, 1989 a,b,c; D'Mello et al., 1991).

Perhaps the main new idea which has emerged in relation to these "antiquality" components is their role in plants as defense mechanisms against herbivore predation (Rosenthal and Janzen, 1979; Rosenthal and Berenbaum, 1991, 1992). The principles of 'chemical ecology' are not new; Feeny (1992) refers to the observation of Theophrastus in 300 BC that: "All trees, it may be said, have worms, but some less, as fig and apple, some more, as pear. Speaking generally, those least liable to be worm-eaten are those which have a bitter acrid juice."

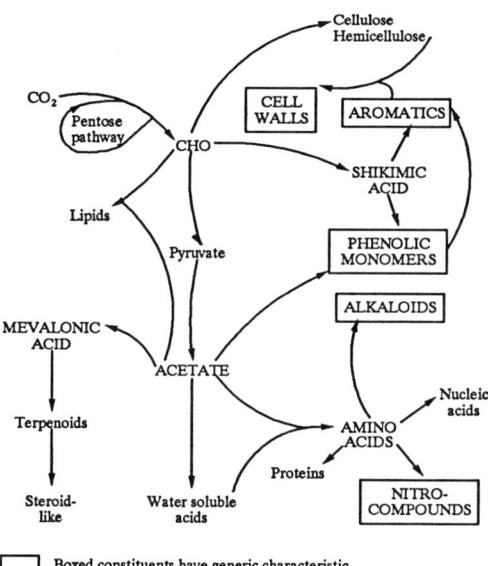

Figure 5. Pathways to secondary plant compounds. Burns (1985).

Bryant et al. (1992) described plant chemical defenses as based on either toxins or digestion inhibitors, with the latter generally claimed as the more important. These authors, however, suggested that the role of compounds such as tannins in decreasing ruminal digestion is overstated and that toxicity may be the major factor inhibiting intake. The role of tannins in forage plants is of particular interest in relation to protein utilization and has been examined in a series of studies in New Zealand (Barry and Blaney, 1987). The defensive function of tannins in cervid nutrition is discussed by Robbins et al. (1987).

In general terms, Bryant et al. (1992) commented that: "Adaptation by plants to resource limitation and disturbance has resulted in a continuum of chemical defenses against mammalian herbivory that range from low in graminoids of eutrophic grasslands to high in woody species characteristic of unproductive habitats." Hagerman and Butler (1992) described the nature of behavioral and physiological adaptations in herbivorous animals to the presence of secondary compounds in the diet, and Provenza and Pfister (1991) have reviewed the various stratagems used by ruminants in detecting and minimizing the ingestion of phytotoxins. Relatively little work has, however, been done to define the effect of environmental variables on concentrations of secondary constituents in forage plants (Provenza and Pfister, 1991; Reid and Jung, 1991).

These concepts offer a new and interesting departure from the traditional way of looking at the function of secondary compounds in plants. They have obvious implications in the nutrition and behavior of free-grazing ruminant animals, wild and domesticated, and in such specific problems as the continuing tall fescue debate (see later section).

Forage Analysis and Prediction of Quality

"Good reasons must, of force, give place to better."

Julius Caesar

In the decade prior to the 1969 conference, a series of papers by Van Soest (Van Soest, 1965, 1967, 1968, 1969) defined the general theory of the detergent fiber system of analysis and its potential application to forage evaluation, and led to publication of a manual of procedures by Goering and Van Soest in 1970 which is now used internationally. The techniques were updated in 1991 (Van Soest et al., 1991). The system has been reviewed in detail in Van Soest's book "Nutritional Ecology of the Ruminant" (1982), which stands as a major contribution in the areas of ruminant nutrition and forage quality and evaluation. Possibly the major limitation of the detergent analysis is that it has, on occasion, been put to uses by investigators which its architect never intended. The subject will be discussed again at this conference and requires no elaboration at this point.

In 1969 and in 1973, Barnes reviewed factors which affect the in vitro fermentation procedure and its application to forage evaluation. In brief, the two-stage in vitro method (Tilley and Terry, 1963) was considered superior to other laboratory methods in the prediction of in vivo digestibility, but was difficult to standardize. The use of short term incubations was unreliable in the prediction of intake. Nylon bag digestion measurements were useful in examining rate phenomena but no technique adequately estimated rate of passage. It was concluded (Barnes, 1973) that a summative approach, such as in vitro estimation of rate and extent of digestion and a physical measurement of rate of passage, might be successful in predicting herbage intake. But some estimate of the efficiency of utilization would also be required.

The widespread use of in vitro techniques in the 1970's was described by Marten (1981), who also remarked on the uncertainty of the fungal enzyme (cellulase, hemicellulase) methods being tested at that time. Minson (1981b), in a companion paper, commented that, with introduction of an active broad spectrum cellulase from *Trichoderma reesei*, the pepsin-cellulase method was replacing the rumen fluid-pepsin technique in Australia. A useful updating and comparison of cellulase procedures is given by workers in Missouri (Bughrara and Sleper, 1986; Bughrara et al., 1992a,b), who concluded that: "Digestibility estimates obtained utilizing prepared cellulase solutions are often faster, less expensive, and more precise than estimates obtained by rumen fermentation procedures." These workers (Bughrara et al., 1992b) used cellulase solutions to develop regression equations for estimation of IVDMD of tall fescue by NIR spectroscopy.

Development of an effective continuous culture apparatus by Hoover et al. (1976 a,b) incorporated, as described by the authors, solid feed inputs at variable rates, differing output rates for solid and liquid digesta, homogeneous mixing of fermenter contents, and better maintenance of protozoa at turnover rates of more than one volume per day. The technique therefore permits examination of such variables as diet composition and particle size on fermentation responses like fiber digestion and metabolite production under controlled turnover conditions.

The milestone analytical event of the 1970's was undoubtedly the application of the NIR technique to forage evaluation. The original paper (Norris et al., 1976) developed and tested prediction equations for protein, NDF, ADF, lignin, IVDMD, in vivo digestibility, intake, and intake of digestible energy. The correlation coefficients obtained were better than those obtained using other laboratory procedures at that time and, by and large, the reliable performance of NIR in the prediction of a diversity of feed, pasture, and fecal characteristics has been maintained to the present.

The principle of the technique, as defined by Norris (1985), is that "each of the major chemical components of a sample has near infrared absorption properties which can be used to differentiate one component from the others. The summation of these absorption properties, combined with the radiation scattering properties of the sample, determines the diffuse reflectance of the sample.--The compositional information can be extracted by proper treatment of the reflectance data." Murray (1986), in reviewing European experience with NIR, claimed that "By probing feedstuffs with NIR diffuse reflectance, using fast scanning monochromators, we can feel the molecular fabric of foods." Norris summarized the advantages of NIR analysis as speed, simplicity of sample preparation, multiplicity of analysis with one operation, and nonconsumption of the sample. Disadvantages are instrumentation requirements, dependence on calibration, complexity in the choice of data treatment, and lack of sensitivity for minor constituents. Some applications of the technique were presented by Shenk et al. (1979). In the United States, NIR research in the 1970's and 1980's was coordinated in a national network project (Templeton et al., 1981), resulting in a number of workshops and culminating in the publication of USDA Agriculture Handbook No. 643 (Marten et al., 1985, revised in 1989). This contains papers on all aspects of NIR technology, in addition to reviews on alternative methods of forage analysis (Barton, 1985; Coleman and Windham, 1985). The subject will be updated at this conference by Shenk and Westerhaus (Chapter 10).

Recently, techniques of nuclear magnetic resonance (NMR) scanning and microspectrophotometry have been applied to the analysis of cell wall and other components of feeds (Barton, 1985, 1991; Hartley et al., 1990). Ultraviolet fluorescence microscopy has been used to determine phenolic compounds in cell walls (Akin et al., 1990) and to study the accumulation of lignin in cell walls of straw fractions of differing ruminal degradabilities (Goto et al., 1992). These methods and the development of light microscopic, SEM, and TEM analyses, possibly in combination with immunocytochemical procedures (Grenet, 1991), appear to offer powerful new tools for the examination of cell wall structures at the cellular level and for their differential breakdown by ruminal bacteria, protozoa, and fungi.

With application of the concept of protein degradability to feed evaluation arose the need for laboratory methods to estimate solubility/degradability (Ørskov and McDonald, 1979). Efforts in this area have resulted in a plethora of chemical, enzymatic, and in situ techniques which have been described by one worker, charitably, as "cumbersome". The dimensions of the problem, and progress in resolving it, are described in a series of papers by Cornell workers (e.g., Pichard and Van Soest, 1977; Sniffen, 1980; Van Soest, 1982; Roe and Sniffen, 1990), and French evaluation methods are reviewed by Aufrère et al. (1991). The current situation will be defined by Broderick (Chapter 5).

In the area of mineral analysis, introduction of inductively coupled plasma-atomic emission spectroscopy (ICP-AES) provided a rapid and accurate means of analysis of a number of macro- and microelements in forages, and NIR methods also have been applied to minerals. The more difficult analytical assessment of the availability of minerals in forages was attempted in an Australian study (Playne et al., 1978) using the nylon bag approach. Significant differences were found in rates of ruminal solubilization of a number of minerals from tropical forages and the technique has been the forerunner of several investigations of the release rates of minerals in the alimentary tract.

Digestive Kinetics and Application to Forage Evaluation

"Is it not strange that sheep's guts should hale souls out of men's bodies?"
Much Ado About Nothing

Hungate (1966), in his book "The Rumen and its Microbes", commented that the ruminant alimentary tract was not "a simple tube through which food and food residues are propelled, ingestion of fresh food at one end causing discharge of spent digesta at the other. Rather it is a regulated system in which the digesta are variously treated in the different parts". He defined characteristics of a continuous fermentation system (turnover time, turnover rate) and described the use of internal and external markers to estimate extent of digestion and rates of passage of feed components in different sections of the alimentary tract. Increased ruminal turnover rate reduced digestibility, due to lesser retention time, but increased microbial yield to the host. Hungate recognized that the rumen was not exactly a continuous fermentation system, but felt that it provided a reasonable model for reference purposes.

At the first conference, Waldo (1969) concluded that the factor distinguishing forage from concentrate feeds was the fiber mass, which restricted intake by extending retention in the rumen and whose disappearance occurred by both chemical digestion and passage, with reduction of particle size required for passage. Particle size, and its rate of reduction, had significant effects on intake. Smaller particles passed more rapidly than large particles and had a faster rate of digestion; however, the increase in rate of digestion did not match that of passage, so that the extent of digestion often declined.

Much of the theory and many of the methods currently used in determining digestive kinetics were, therefore, in place at the time of the 1969 conference. Major advances have occurred in the development of appropriate models to describe breakdown and passage phenomena, in improving marker methodology, and in using kinetic data to explain observed differences in forage digestibility and intake relating to differences in forage photosynthetic type (C3 vs C4 grasses, legumes), species and cultivar, method of forage utilization (grazing, conservation), plant growth stage and morphology, and environmental effects, and to account for animal differences caused by physiological factors and the effects of climate. A comprehensive account of the quantitative aspects of digestive kinetics and metabolism in ruminants is found in a recent book edited by Forbes and France (1993).

Van Soest (1982) reviewed various models for the breakdown of feed particles and for the calculation of passage data and rates of digestion, and this subject will be further discussed by Ellis and Murphy (Chapter 17). Methods for the estimation of particle size distributions in the diet and in the digesta were described by Faichney (1986). Ulyatt et al. (1986) concluded that there were two particle pools in the rumen, with a well regulated threshold which was not affected by dietary or animal variables. In sheep, the critical particle size for escape from the reticulorumen was 1 to 2 mm and, in cattle, 2 to 4 mm. Particle size reduction took place during the processes of initial eating (chewing, mastication), rumination, ruminal microbial breakdown, and rumen contraction. Chewing during eating was shown to be highly efficient in reducing large particles, cattle were less efficient than sheep in reducing particle size, fresh diets and high quality diets were chewed more effectively than dry or low quality diets, and animals differed considerably in their efficiency of chewing. Chewing during the rumination process was found also to be an effective mechanism for particle size reduction and was thought to have the major role of preparing cell wall material for reticuloruminal clearance. In comparison, microbial digestion and the effect of ruminal contractions had relatively little effect on particle size reduction (Ulyatt, 1983; Ulyatt et al., 1986). Recent studies (des Bordes and Welch, 1984; Sutherland, 1988; Wattiaux et al., 1992) show that changes in functional specific gravity during fermentation may also influence selective retention of fiber particles in the rumen.

Wilson et al. (1989a,b) ascribed some of the differences in intake between tropical and temperate grasses to different effects of chewing and digestion by cattle on particle breakdown. Chewing while eating reduced particle size more in length than in width, while digestion reduced particle width but not length. Reduction in width was faster in ryegrass than in green panic (*Panicum maximum* var. *trichoglume*) and was related to differences in anatomical structure of the leaf. However, it is still not clear from these and other studies (Pond et al., 1984) why the greatest reduction in particle size during chewing appears to occur in C4 grasses with high cell wall and lignin concentrations.

The ruminal digestion of forages, or of forage components such as fiber, N, or minerals, is now generally measured by in vitro or in sacco procedures. Ørskov (1991) concluded that three critical characteristics of fibrous feeds, i.e., the soluble fraction, the insoluble but degradable component, and the rate of degradation of the insoluble material, could be described by an exponential equation:

$$p = a + b (1 - e^{-ct}),$$

where p is degradation at time t and a, b, and c are characteristics of the feed; a represents immediately soluble material, b the insoluble but degradable fraction, and c the rate constant of fermentation of b. This equation was originally developed for dietary protein (Ørskov and McDonald, 1979) and is used in the calculation of degradability values for a range of feedstuffs in the UK Tables of Nutritive Value and Chemical Composition (Givens and Moss, 1990). An example of values for fresh grass is given in Table 1.

TABLE 1. Inclusion of degradability values in feed tables.[1]

Fresh grass, all species

Nitrogen	Mean	SD	Min	Max	n
a(%)*	20.5	12.2	0.96	46.0	51
b(%)	71.4	11.9	44.2	91.3	51
c(h^{-1})	0.13	0.06	0.04	0.47	51
degradability (%) at outflow rate					
.02h^{-1}	81.5	6.8	61.0	93.0	51
.05h^{-1}	70.9	8.4	47.0	87.0	51
.08h^{-1}	63.5	9.3	41.0	82.0	51

[1]Givens and Moss (1990).
*p = a + b (1 - e^{-ct})

The incorporation of a lag phase, in which no disappearance occurred, was proposed for fibrous feeds by Mertens (1977). The equation is:

$$\text{Digestibility} = (e^{-k_p L})(a)(k_d) / (k_d + k_p),$$

where L is the discrete lag time, a the potentially digestible fiber fraction, kd the first-order rate constant for digestion and kp the first-order rate constant for passage. This equation has subsequently been modified (Mertens and Ely, 1979; Allen and Mertens, 1988), but Mertens (1987a) concluded that digestibility is always determined by the fraction of the forage which is potentially digestible and the first-order rate constants for digestion and passage. Mertens (1987a) described forage factors which influenced these parameters, criticized the wide diversity of mathematical approaches used to describe kinetic characteristics, and considered that the most serious error in kinetic measurements was the failure to estimate the indigestible residue by extended fermentations.

Passage rates of digesta are calculated from estimates of rumen volume and turnover time, or by marker recovery from the rumen or feces (Van Soest, 1982; Owens and Goetsch, 1986; Galyean, 1987). Ellis et al. (1982) described types of flow markers as: 1) microbial, 2) fluid or solute, and 3) particulate. Ruminal volume or fill is affected by a number of factors (animal size, breed, physiological condition, diet, and level of intake), and fill has a major effect on forage intake. Rumen volume increases with intake and, as commented by Van Soest (1982), the stretch factor tends to offset increases in passage rates, so that increases in k_p values with higher consumption may be less than expected. France et al. (1991) described the kinetic properties of total and individual ruminal pool volumes in steady and non-steady state conditions and the use of single and dual-marker techniques to measure pool volumes and outflow rates.

Fractional passage rates from the rumen, as reviewed by Owens and Goetsch (1986), depend both on rumen volume and rates of flow. Liquid passage rates are regulated by osmotic, neural, and hormonal factors, while particle passage rates are controlled by position in the rumen and, possibly, particle size. Microbial passage rates depend on fractional rates of passage of liquid and solid particles and on the extent of association between feed particles and bacteria. These authors concluded: a) that marker methods to estimate passage rates and ruminal volume are questionable from both a mathematical (lag and mixing times) and chemical (marker migration) standpoint, and are difficult to verify; b) that an increase in passage rate may be either of advantage or disadvantage to the animal, "depending on the need to clear the rumen to increase feed intake, the post-ruminal digestibility of nutrients escaping digestion in the rumen, and the requirements of the host animal for microbial products (protein, B-vitamins)". Ulyatt et al. (1986) reached the somewhat pessimistic conclusion that "Our understanding of the physiology of passage regulation is poor and remains a major scientific challenge if we wish to manipulate the ruminoreticulum to increase clearance and improve feed utilization".

From an applied viewpoint, Ørskov (1991) suggested that it may be feasible to increase digestibility and intake of fibrous diets by ruminants by increasing the potential degradability of the insoluble but fermentable fraction of the diet and by increasing degradation rates. Methods to increase the potentially degradable part of low-quality forage will be discussed in a later section. Possibilities for increasing degradation rates may lie in plant genetic selection, adjustment of the ruminal environment, selection of animals for higher ruminal volume and increased retention times, various kinds of chemical treatment, and, possibly, selection of ruminal bacteria for enhanced fiber degradation rates.

Breeding Better Forage Plants

"He that hath learned no wit by nature nor art
May complain of good breeding"

As You Like It

In reviewing genetic variation in forage species, Cooper (1973) made the general point that a breeding program for nutritional quality should first consider the purpose for which the forage is to be used - whether as a "complete and balanced diet for livestock, or primarily as an energy source, to be supplemented with protein and/or minerals from elsewhere." The forage breeder has then a choice between selection within established varieties, introducing the required characteristics from less adapted or productive varieties [(e.g., high digestibility from Kenya 56/14 into Coastal bermudagrass (Burton et al., 1967)], or production of induced mutants. Cooper concluded that breeding for quality would be valuable in two kinds of livestock systems: 1) where one component had a major effect on animal performance, e.g., estrogens in subterranean clover (*Trifolium subterraneum* L.) and alkaloids in *Phalaris tuberosa*; 2) where further improvement of digestibility, intake, and efficiency in intensive production systems utilizing well adapted species/varieties with high DM production was justified.

Advances in the breeding of legumes (Bray, 1982) and grasses (Hacker, 1982) were discussed at an Australian conference in 1981 (Hacker, 1982) and at a later CSIRO-DSIR Workshop (Ulyatt and Black, 1984). Screening and selection programs against negative quality factors (estrogenic isoflavones in subterranean clover, coumarin in sweet clover (*Melilotus* Mill.), bloat in temperate legumes, and mimosine in *Leucaena leucocephala*) were reasonably effective, but little success was reported in locating tannin-containing genotypes of *Trifolium* and *Medicago* species. Bray (1982) suggested that, in the legumes, the highest priority should be directed to reducing the levels of antiquality compounds rather than selecting for positive attributes such as digestibility. Hacker (1982), in reviewing quality evaluations of grass species, concluded that in most studies there was a range of at least ten digestibility units, with about half the variation being genetic, and more variation in stem than in leaf digestibility. Generally, digestibility was strongly negatively correlated with lignin, negatively with other cell wall components, and positively with N concentration, although the relationships were often influenced by environmental factors. In a recent study, Humphreys (1989) found that there was more genetic variation for water soluble carbohydrate (WSC) than for other components in perennial ryegrass, that WSC was correlated positively with digestibility, and that it was feasible to select for ryegrass varieties with high WSC without sacrificing DM production.

In the United States, selection among bermudagrass hybrids based on a nylon bag digestibility procedure (Lowrey, 1969) led to identification of the Coastcross-1 hybrid which, as described by Burton (1969), was 12% more digestible than Coastal bermudagrass and supported significantly better performance of dairy heifers and of steers in grazing trials. Burton (1969) noted that a simple way to improve quality was to increase leaf percentage, and showed higher digestibility of dwarf than of tall forms of Tift 23 pearlmillet [*Pennisetum americanum* (L.) Leeke]; steers grazing the dwarf form made 20% better ADG, with no reduction in liveweight gains per ha. An alternative approach to improve quality was to breed later-maturing cultivars of millet (Burton et al., 1968).

Similarly, Nebraska workers (Anderson et al., 1988) have shown that a switchgrass strain selected for high IVDMD gave better ADG and liveweight gains per ha in grazing yearling cattle than a low IVDMD strain or the control cultivar

Pathfinder. Subsequent studies in Nebraska (Gabrielsen et al., 1990; Fritz et al., 1991;Moore et al.,1993) indicate that the higher IVDMD results from differences in cell wall composition and digestibility among strains rather than to differences in concentration of cell-soluble material.

The effects of quantity and composition of lignin on the digestibility of forage plants are well documented (Allinson, 1969; Van Soest, 1982; Jung, 1989). The lower lignin content of brown midrib mutants (bmr) of corn (*Zea mays* L.), a trait which also has been introduced into sorghum [*Sorghum bicolor* (L.) Moench.], sudangrass, and millet, results in higher digestibility and intake of the bmr genotypes although, as commented in a review by Cherney et al. (1991), animal performance is not always improved. Responses in liveweight gains are observed fairly consistently when bmr forages constitute the only or principal source of feed. These authors project that geneticists will incorporate the bmr trait into diverse sorghum and pearlmillet material and, with isolation and characterization of the gene, may succeed in transferring it to other species.

Selection of forage grasses for greater intake is a more difficult proposition. Hacker(1982) suggested breeding for characteristics such as leaf strength and flexibility, and leaf density within the sward. In terms of efficiency of utilization, Hacker commented that: "At this stage the breeder can only wait for the nutritionist to postulate a selection criterion." Some progress in this line is evident in recent years (Ulyatt, 1981; Black, 1987). A consensus view of a group of grassland research workers on criteria for improving the feeding value of grasses and legumes for liveweight gain and wool production is summarized in Table 2 (Wheeler and Corbett, 1989).

TABLE 2. Ranking of quality criteria for breeding grasses and legumes.[1]

	Grasses		Legumes	
Rank	For gain	For wool	For gain	For wool
1	High digestibility	High digestibilty	High digestibility	High digestibility
2	Easy comminution	Easy comminution	Easy comminution	High S-amino acids
3	High NSC*	High S-amino acids	Appropriate tannins	Easy comminution
4	High CP	High NSC	High NSC	Appropriate tannins
5	Adequate minerals	High CP	Adequate minerals	High CP
6	High palatability	Adequate minerals	High CP	High NSC
7	High S-amino acids	High palatability	High S-amino acids	High palatability
8	High lipid	Appropriate tannins constituents	Low anti-quality	Adequate minerals
9	Low anti-quality constituents	Low anti-quality constituents	High palatability	Low anti-quality constituents
10	Appropriate tannins	High lipid	High lipid	High lipid
11	Erect growth	Erect growth	Erect growth	Erect growth

[1]Wheeler and Corbett (1989).
*Non-structural carbohydrate.

The responses are interesting in that, while there is general agreement on the primary importance of digestibility and comminution properties (influencing the rate of passage and intake), greater attention is directed to the significance of plant components (non-structural carbohydrate, "appropriate" tannins, S-amino acids) considered to regulate the efficiency of ruminal metabolism and nutrient conversion to end products. The authors concluded that what is required from the breeding viewpoint is: 1) a high rate of breakdown of non-protein organic matter, promoting rapid outflow from the rumen and increased intake; and 2) a reduction in protein solubility to reduce pre-duodenal N losses and increase amino acid supplies to the small intestine. The potential for selecting for such variation is seen in a study by Broderick and Buxton (1991) showing differences in alfalfa (*Medicago sativa* L.) lines for protein degradation rates and net protein escape.

Improved Tall Fescue?

The dimensions of tall fescue research in the United States during the last 25 years will be presented by Bacon (Chapter 8) at this conference and will be considered only briefly here. The economic losses ascribed to fescue pastures infected by the endophytic fungus *Acremonium coenophialum* Morgan-Jones and Gams are high; survey results by Hoveland (1990) indicated possible total annual losses due to lower calving rates and reductions in calf weaning weights of approximately $600 million. The research effort devoted to the problem has been proportionately large and has revealed a fascinating and complex picture of the nature of the plant-endophyte-animal relationship in fescue toxicosis (Bacon et al., 1977,1986; Siegel et al., 1987a; Fribourg et al., 1991). A similar picture has been described in New Zealand for perennial ryegrass (Siegel et al., 1987a) and annual ryegrass (*Lolium multiflorum* Lam.)(Latch et al., 1988). These relationships appear to be good examples of the concepts of chemical ecology mentioned earlier and discussed by Siegel et al.(1987a) with reference to fungal endophytes. Development of endophyte-free cultivars of tall fescue in a number of states has resulted in significantly improved performance of grazing cattle and higher calving rates and weaning weights. However, endophyte-free stands were found to be less tolerant of over-grazing and more susceptible to environmental stress (Fribourg et al.,1991; Bouton et al., 1993). In contrast, possession of the endophyte appears to confer advantages in terms of resistance to insect predation [e.g., Argentine stem weevil (*Listronotus bonariensis* Kuschel) on endophyte-free perennial ryegrass] and greater tolerance of environmental stress in turf grasses.

The apparent paradox and opportunity for forage breeders is expressed by Siegel et al. (1987b): "The production of improved endophyte-infected cultivars may prove to be as auspicious as the introduction of endophyte-free grasses if endophyte biotypes can be isolated (naturally occurring) or produced (genetically engineered) which do not synthesize chemicals toxic to animals. The modified endophyte(s) can be introduced into the plant by either maternal line selection or by artificial inoculation at the seedling stage. This presumes that the chemicals thought responsible for animal toxicosis (i.e., ergot alkaloids) are different from those toxic to insects (i.e., peramine and loline alkaloids) and that improved infected cultivars produce only the latter compounds." Fletcher et al. (1990) in

New Zealand described the isolation of a novel strain of endophyte which protected its host from Argentine stem weevil but did not produce lolitrem B. They concluded that release of cultivars containing novel endophytes would permit the positive effects of endophytes to dominate the negative effects.

These are interesting research possibilities. Presumably the breeding approach will need to be balanced against the practical difficulties of replacing the very large acreage of endophyte-infected fescue in the United States, and with the many existing livestock management systems which have been developed to use the infected grass effectively. A debate at the Southern Pasture and Forage Crop Improvement Conference in Alabama in 1992 on whether livestock producers should use endophyte-free fescue showed that opinions were very much divided on the question.

Forage Conservation

"Good hay, sweet hay, hath no fellow"
A Midsummer Night's Dream

Early and comprehensive accounts of factors influencing the composition of grassland products, and their digestibilities, are found in books by Watson (1951) and Schneider (1947), and were updated by Sullivan (1973) and McDonald and Whittenbury (1973). Sullivan (1973) concluded that even the best hays lost some quality in drying and summarized normal DM losses from field drying as: 4 to 15% from respiration during wilting and drying; 2 to 5% from leaf shattering in grass hays and 3 to 35% for legume hays; 5 to 14% from leaching by rain. Field losses of TDN could range from 25% (without rain) to 42% with rain damage. McDonald and Whittenbury (1973) classified ensiling losses into four types: field; respiration (aerobic); fermentation; effluent losses. In the field, wilting resulted in proteolysis and loss of WSC. Under aerobic conditions in the silo, respiration continued, with loss of fermentable carbohydrate and production of heat. Losses associated with the activity of homo- and heterofermentative lactic acid bacteria were low, but increased in clostridial fermentations. Effluent losses depended primarily on the DM content of the ensiled crop, with little effluent produced at approximately 300 g Kg^{-1} DM. Silage additives were characterized as: a) stimulants; and b) inhibitors. Stimulants were mainly carbohydrates (sugars, starch, molasses), with variable results from bacterial inoculants. Inhibitors included mineral acids and formic acid, formic acid salts, and sodium metabisulphite. Some preliminary trials were in progress at that time to study the effects of formaldehyde on silage fermentation.

Since the early 1970's, major developments have occurred mainly in processing technology (development of hay and silage-packaging equipment) and in the testing of methods to prevent nutrient losses and(or) enhance quality at different stages in conservation. Active interest has developed in the uses of low quality roughages such as crop residues in livestock feeding systems, and in improvement of their nutritive quality for cattle and sheep. These subjects will be discussed in detail by Rotz and Muck (Chapter 20) and by Berger et al. (Chapter 23), and only a few highlights will be mentioned here.

Hays and Other Dried Forages

Jarrige et al. (1982) concluded that the major developments in hay making were in equipment which increased the loss of water by mechanical conditioning in the early stages of drying, and mechanized bale handling systems. To these might be added processes for chemical conditioning at harvest, using compounds such as potassium carbonate and formic acid (Harris and Tulberg, 1980; Jones, 1991), and methods for the chemical preservation of wet hay. Preservatives such as propionic acid and ammonia applied at baling to harvested forage at high moisture content inhibit microbial activity, reduce heating, and frequently improve quality. Ammoniation has been shown to increase both total N content and IVDMD and intake of treated hays (Knapp et al., 1975). Because of potential toxicity problems of ammonia to operators, and animal feeding disorders associated with formation of 4-methyl-imidazole (Perdok and Leng, 1987), Kentucky workers (Henning et al., 1990) recommended use of urea for preservation of moist hay. Added urea is converted by plant ureases to NH_3 and has comparable effects in reducing bale temperature and increasing N concentrations and IVDMD and cell wall digestibility.

Effects of dehydration of forages on feeding value were reviewed by Thomson and Beever (1980). Dehydration processes frequently resulted in: a) reductions in organic matter and gross energy digestibility, of variable magnitude; b) a beneficial shift in site of digestion in the alimentary tract, particularly for N, caused by a decreased ruminal degradability and increased amino acid supply to the small intestine; and c) possibly a higher efficiency of conversion of ruminally degraded energy into VFA energy. Comparisons of the nutritive quality of dried with fresh forages led Jarrige et al. (1982) to conclude that the various conservation losses resulted in a higher percentage of cell walls, a decrease in organic matter and N digestibility and, generally, a reduced DM intake resulting from a lower ruminal breakdown rate of the forage.

A good review of the possible DM and quality losses involved in field drying of forages is provided by Macdonald and Clark (1987), and a summary of their projections of losses under different scenarios is given in Table 3. It is obvious that there is still considerable room for improvement in hay making operations.

Silage

In terms of the technology of silage making and feeding, Jarrige et al. (1982) summarized the main advances since the 1960's as: a) development of the flail, double-chop and precision-chop harvesters; b) development of additive applicators; c) improved polythene coverings; and d) methods for mechanical cutting and feeding of silage. More recently, a variety of new methods for bagging and wrapping silage has been introduced. Fenlon et al. (1989) showed comparative spoilage losses of 102 vs 215 g Kg^{-1} DM for wrapped and bagged bales, respectively, and a Canadian study (Nicholson et al., 1991) demonstrated that bagged silage contained less NPN (468 vs 585 g Kg^{-1} of total N) than big bale silage and supported better weight gains and feed efficiency in beef calves, with no differences in digestibility by sheep.

Much of the more recent research on silage quality has centered on effects of

TABLE 3. Estimates of total dry matter losses in haymaking and use, assuming optimistic, "realistic", and pessimistic scenarios.[1]

Source of loss	Optimistic		Realistic		Pessimistic	
	Loss (%)	Left (%)	Loss (%)	Left (%)	Loss (%)	Left (%)
Relative to fresh cut material		100		100		100
Cutting/conditioning	5	95	10	90	20	80
Respiration	5	90	10	81	15	68
Raking/tedding	5	86	10	73	20	54
Rainfall/weathering	0	86	10	66	15	46
Baling	5	81	10	59	20	37
Storage	5	77	10-20*	53-47	30	18
Handling/transport/feeding	5	74	10	48-43	30	18
Forage dry matter potentially consumed		74		48-43		18

[1]Macdonald and Clark (1987).
*Assuming either indoor or good outdoor storage.

wilting and on the role of additives in enhancing quality. The function of additives, as summarized by McDonald (1981) is: a) to stimulate lactic acid fermentation; b) to inhibit microbial growth (mineral acids, formic acid, formaldehyde, SO_2, etc.); c) to inhibit aerobic breakdown (NH_3, propionic acid); and d) to provide nutrients in the form of N compounds or minerals.

McDonald and Whittenbury (1973) noted that half of the protein in fresh herbage may be degraded during silage making. The process is reduced by wilting through inhibition of the activity of microorganisms, especially clostridia (Thomas and Thomas, 1985). Waldo (1985) commented that "Extensive protein degradation will cause low intakes, low production responses, low or negative N retentions, and decreased protein deposition by ruminants consuming such silages". He concluded that good but variable legume silage could be made, without additives, by wilting to approximately 350 g Kg^{-1} DM before ensiling. Extensive studies on wilting of grass silage have been conducted in Europe, and results have been somewhat inconsistent. Wilting tends to lower digestibility (Marsh, 1979; Thomas and Thomas, 1985), while several studies have shown some increase in intake of wilted silage without, however, commensurate increases in milk production by dairy cows (Castle and Watson, 1984; Gordon, 1989).

In view of the inconsistency of response to wilting, use of additives has been widely evaluated in the United States and in western Europe. Protein degradation in silage is inhibited by formic acid, formaldehyde, or a combination of the two (McDonald, 1982; Waldo, 1985). Formaldehyde-treated silage fed to young cattle was shown to increase protein supply to the small intestine by 33%, due to increased passage of undegraded dietary protein (UDP) from the rumen (Thomson and Beever, 1980). However, O'Kiely and Flynn (1988) in Ireland noted no benefit in the performance of cattle from silage treated with a combination of formaldehyde and formic acid over formic acid alone.

Additives designed to stimulate fermentation include both nutritive and non-nutritive compounds, with much work being conducted on bacterial and enzyme preparations. Pitt and Leibensperger (1987) reviewed the effectiveness of inoculants (dried or inactive lactic acid bacteria) and reported that of seven studies in the 1980's, one showed consistently positive results, three positive effects some of the time, and three no effect. They concluded that level of inoculation was very important in determining effectiveness, with inoculation concentrations of 10^5 bacteria/g needed to give consistent responses. In studies in Northern Ireland with cattle fed grass silages, Gordon (1989) found that inoculation with *Lactobacillus plantarum* was beneficial in terms of digestibility, intake and animal performance. When fed to dairy cows, intake and milk production on inoculant-treated silage were higher than on formic acid and control silages. In reviewing recent work with bacterial inoculants, mainly in Europe, Mayne and Steen (1990) concluded that improvements in animal performance were positively correlated with WSC concentration in the herbage at ensiling and that increases in digestibility might be due to "a better match of energy and N supply to the rumen". The apparently improved response in recent compared with earlier trials may result, in part, from what Mayne and Steen (1990) describe as a "new generation of inoculant additives" becoming available in the late 1980's.

An alternative approach has been to use cellulase and hemicellulase additives to degrade the cell walls of ensiled forage and increase the supply of WSC to microorganisms, with the possibility of further enhanced breakdown within the rumen. A number of studies have shown reductions in structural carbohydrates and increases in lactic acid content of treated silages, and Finnish work (Huhtanen et al., 1985) indicated increases in digestibility and N retention. Results of recent studies on effects of cellulase-hemicellulase additives on digestibility and intake have been somewhat conflicting (Chamberlain and Robertson, 1989; Jacobs and McAllan, 1991) and it is probably too early to evaluate the potential of this technique in improving silage quality.

Ammoniation of cereal and forage crops at ensiling improves aerobic stability, reduces proteolysis, and increases concentrations of total and insoluble N (Johnson et al., 1982), and was shown to increase N retention by heifers fed alfalfa silage (Glenn, 1990). Much the same compositional and quality effects result from treatment with urea.

Low Quality Roughage

Major advances in both the use and improvement of the feeding quality of fibrous plant residues, mainly cereal straws and corn stover, were made in the 1970's and 1980's. The subject is reviewed by a number of writers (e.g., Klopfenstein and Owen, 1981; Wilkins, 1982; Reid and Klopfenstein, 1983; Sundstøl, 1988, 1991; Campling et al., 1990) and will be further discussed by Berger et al. (Chapter 23) at this conference.

Ward (1980) described the potential of crop residues for feeding beef cattle in the United States in the 1980's. The main nutritive constraints were a low energy digestibility (TDN 350 to 550 g Kg^{-1}) and protein content (40 to 80 g Kg^{-1}). It was proposed that improved utilization could result from better management practices (harvest time, genetic selection) and chemical treatment [NaOH, $Ca(OH)_2$, NH_3]. Probably the most adopted practice is that of ammoniation, which is effective not only in increasing N concentration in the treated roughage but in improving digestibility and intake. Corah (1990) reviewed the results of several trials on ammoniation in the United States and Canada and reported a two-fold increase in N content and increases in intake ranging from 16 to 70% for straws, corn cobs and stover. The use of alternative methods for the harvesting, storage, treatment, processing, and feeding of straw to ruminants is outlined by Sundstøl (1991), and a summary of effects of different treatments on feeding quality is given in Table 4. Treatment with strong alkalis like NaOH was most effective in increasing OM digestibility.

TABLE 4. Effects of straw treatments on organic matter digestibility in sheep and dry matter intake by bulls.[1]

Method	Increase in OMD relative to untreated, %	Increase in DMI relative to untreated, %
Urea treatment	13-18	45
Ammonia treatment	18-29	19
NaOH, wet treatment	43	
NaOH, dry treatment	23-31	
chopped		18
pelleted		92
Physical treatment		
ground	0	7
ground and pelleted	-4	37

[1]Sundstøl (1991).

Another promising approach is use of alkaline hydrogen peroxide, which degrades lignin and increases breakdown of cell wall carbohydrates by ruminal microorganisms (Kerley et al., 1985; Willms et al., 1991). Amjed et al. (1992) reported that peroxide treatment of crop residues increased IVDMD, extent of NDF digestion, and digestibility of monosaccharides. Similarly, studies in Israel (Ben-Ghedalia and Miron, 1981; Miron and Ben-Ghedalia, 1992) showed that treatment with SO_2 gas converted straws into "highly degradable and fermentable materials", with a major effect on solubilization of the hemicellulose component of the cell wall.

In reviewing the overall potential benefits and limitations of different methods of forage processing, Wilkins (1982) concluded that although processing might improve feeding value, in many cases "these treatments do not provide outstanding advantages in terms of the use either of land or support energy". A cost-benefits study on uses of straw in the United Kingdom showed that the most profitable option for using wheat was to sell the grain and burn the straw (Campling et al., 1990). In the Great Plains region of the United States, the grazing of corn stover and milo stubble fields after the harvesting of grain represents probably the most economical use of crop residues, and systems for utilizing this resource by beef cattle are presented by a number of writers (Ward, 1978; Reid and Klopfenstein, 1983; Klopfenstein, 1987; Nelson et al., 1989; Fernandez-Rivera and Klopfenstein, 1989).

The Plant-Animal Interface and Pasture Evaluation

"Good pasture makes fat sheep"

As You Like It

Until recent years, relatively little attention was paid to the production, management, and utilization of high quality pasture as a major source of nutrients for ruminant livestock in the humid regions of the United States and Canada. However, research on rangeland in the western states has been a continuing process. Richards (1969) described the objectives of western regional research in the 1960's as determining the energy, protein, and phosphorus requirements of beef cattle and sheep on range, measuring utilization of range plant nutrients, and testing methods for increasing nutrient utilization to meet livestock requirements. Progress in these areas was described by Theurer (1969) and Raleigh (1969) at the first conference.

The nature of research techniques used in the United States to assess pasture quality in the period 1975 to 1980 was surveyed by Mochrie (1981). A general conclusion was that "The heavy reliance on visual and clipping methods and on animal performance by gain alone indicates that more critical assessment of forage yields and quality, and animal consumption by techniques sophisticated enough to relate to economic animal production, are needed". Concurrently, Robards (1981) commented on Australian methods that "Weighing of animals at regular intervals is the most common animal measurement in pasture evaluation experiments --- The accurate and dependable measurement of intake by grazing animals remains a major challenge."

During the last 10 to 15 years, there has been increased interest in pasture utilization and research in the United States, resulting in part from changing economics, shifts in consumer attitudes towards the quality of their food, potential environmental problems, and the development of highly sophisticated electronic and remote sensing equipment for the measurement of animal behavior and plant resources. Changes in management systems on a regional basis were reviewed by Reid and Klopfenstein (1983). The development of improved pasture systems for the production of forage-fed beef in the south was discussed in a 1975 conference (Stuedemann et al., 1977). In the northeast and north-central states, there is an active interest in use of intensified grazing methods for dairy and beef cattle and sheep. In the western states, two successful Grazing Livestock Nutrition Conferences were held in Wyoming and Colorado in 1987 and 1991 (Judkins et al., 1987; McCollum and Judkins, 1991).

Greater dependence on pasture as the principal source of nutrients for livestock in Europe, Australia, and New Zealand led to the development of innovative research procedures for studying pasture-animal interactions in the 1970's and 1980's and this work will be discussed by Hodgson et al. (Chapter 19). A key event, for attendants at the 11th International Grassland Congress in Queensland, was demonstration of procedures for measurement of bite size, bite rate, and grazing time by cattle grazing tropical pastures, and of the significance of sward structure in determining these characteristics (Stobbs, 1970, 1973a,b, 1974). Intake could be measured as the product of grazing time, rate of biting, and intake per bite. The main determinant of intake, as defined by Stobbs (1973a), was bite size, and it was calculated that herbage intake by cattle grazing tropical pastures would be restricted when average bite size was less than 0.3 g OM.

These concepts were developed by Hodgson (1982, 1990), who concluded that "The vertical distribution of foliage exerts the major influence on ingestive behavior in sown temperate swards, whereas in tropical swards variables associated with leaf density and leaf/stem ratio are of dominant importance." In New Zealand and Scottish studies on temperate pastures (Hodgson, 1985; Sheath et al., 1987), sheep adjusted bite size, rate of biting, and grazing time in relation to the amount of green leaf on offer. Sheath et al. (1987) found that herbage intake increased in a curvilinear fashion with pasture allowance or herbage mass, and lambs on legume pastures performed better than those on grass, particularly at low allowance. The importance of the distribution of leaf, stem, and dead material in determining the quality of selected herbage was seen in North Carolina studies (Fisher et al., 1988) with warm-season and cool-season pasture grasses. Recent studies with constructed swards or 'cage patches' (Burlison et al., 1991; Laca et al., 1992) suggest that both sward height and bulk density have additive and independent effects in determining bite weights of herbage. A model incorporating some of these ideas is shown in Figure 6 (Arnold, 1985).

Techniques for measuring the ingestive behavior of grazing animals were reviewed by Forbes (1988), who commented that research on the effects of herbage height or mass on grazing behavior had already led in some countries to systems of pasture management based on this criterion. A number of papers on plant-animal interface research were presented at the 15th International Grassland Congress in Japan in 1985 and are published in extended form by Horn et al. (1987). Comprehensive reviews on factors influencing the grazing behavior of ruminant animals were published by Arnold and Dudzinski (1978) and Arnold (1981), and they considered the evidence relating selection or 'palatability' behavior to chemical composition of the diet. This subject, as it applies to behavioral responses of ruminants to poisonous plants on rangeland, is examined in a number of interesting studies by workers in Utah (Provenza et al., 1988; Provenza and Pfister, 1991). Computerized technology for recording and integrating grazing behavior activities is described by Coulter and O'Sullivan (1988) and Luginbuhl et al. (1991).

Use of indicator techniques, frequently in combination with determinations of in vitro digestibility on esophageal samples, has been the traditional approach to estimating intake by grazing animals. The procedures as applied to pasture and range have been reviewed, and their limitations discussed, by many writers (e.g., Kotb and Luckey, 1972; Corbett, 1978; Holechek et al., 1982). The more recent, and extensive, literature on the application of marker methods to estimation of digestibility and intake by grazing ruminants is evaluated by Cochran et al. (1987) and Pond et al. (1987). Among the more promising techniques being examined at the present time are use of the long chain N-alkanes in a dual-marker system as developed by Mayes et al. (1986, 1988), for the estimation of intake, and the use of controlled release Cr_2O_3 capsules for determining fecal output (Laby et al., 1984; Parker et al., 1989, 1991; Furnival et al., 1990). A recent report (Vulich and Hanrahan, 1992) showed that "the n-alkane technique can provide herbage intake estimates of high accuracy (bias of 1 to 5%) and high precision (rsd's of 2 to 7%), with the correlation between actual intake and that estimated through the n-alkane technique being of the order of 0.93-0.98."

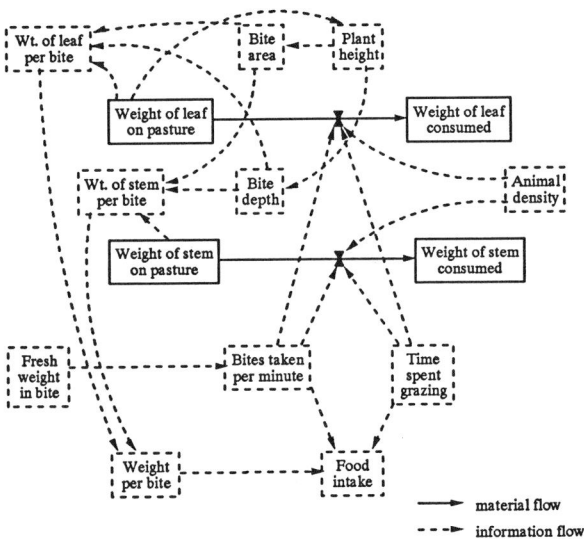

Figure 6. Grazing systems dynamics. Arnold (1985).

Milne (1982-83) commented that herbage intake was still a "crude measure" of nutrient intake by the grazing animal, since it does not indicate the nature of absorbed energy substrates or measure microbial protein and the amount of protein potentially available for absorption, and this is especially important with supplement feeding. There have been notable advances in the development of equipment for infusing markers and sampling digesta in grazing animals, enabling the calculation of nutrient flow and microbial protein synthesis. These techniques, and calorimetric methods for the measurement of energy exchange and retention, are described by Corbett (1981). The application of such technology in the future, as seen by Milne (1982-83), may lie in the construction of grazing models which can be used to predict the outcome of management decisions relating to stocking rates or use of supplementary feed for animals on pasture.

Two problems which continue to plague pasture investigators are the use of appropriate designs for grazing trials and the confusion which attends grazing terminology. These subjects are addressed in recent publications by the Crop Science Society of America (Marten, 1989) and the Forage and Grazing Terminology Committee (Allen, 1991). The first contains a number of reviews relating to statistical design and the interpretation of grazing experiments. The second is a consensus report by individuals from societies and agencies in the United States involved with grazing lands, with input from Australia and New Zealand, which proposes "clear definitions of terms used in the grazing of animals." The report has been accepted by the majority of societies concerned with the management of pastures and range in the United States, and its development at an international level would be a worthwhile objective for the future.

Modelling and Systems Approaches

"Hast any philosophy in thee, shepherd?"

As You Like It

One of the main features distinguishing research papers presented at this and the 1969 conference is the greatly increased use of modelling techniques. This difference results, in part, from the rapid advances in computer technology which have been made during the last 25 years. As stated by Baldwin et al. (1976): "Computer modelling techniques provide means for not only integration of information and evaluation of data and concepts in a quantitative fashion, but also for evaluation of alternative hypotheses for probable adequacy, estimation of system parameters not directly measurable, and identification of critical experiments". Baldwin et al. (1976) reviewed the development and use of: a) balance models to describe nutrient conversion and ruminal fermentation patterns, and b) dynamic models, as formulated by Waldo et al. (1972) for estimation of ruminal cellulose digestion and passage, and by Ulyatt et al. (1975) to incorporate particle pools and lower tract digestion. Baldwin et al. (1977), Beever (1978), Mertens and Ely (1979), and Murphy et al. (1986) subsequently developed dynamic models to describe ruminal digestion and passage which incorporated several chemical components of the diet and physical inputs of particle size. A general model of plant and animal factors regulating nutrient supply (Figure 7) was proposed by Faichney and Black (1979). Black et al. (1980-81) further outlined a computer program to predict ruminal digestion of diet components, fermentation products, microbial yields and the outflow of protein and other material from the rumen. Modelling approaches to many aspects of ruminant metabolism are described in the book "Quantitative Aspects of Ruminant Digestion and Metabolism", edited by Forbes and France (1993).

Following demonstration of the significance of NDF level of the diet as a principal factor influencing intake, Mertens (1985, 1987b) developed mathematical models to predict intake and digestibility of diets by lactating cows. Intake was considered to be mediated by physical, physiological, and psychogenic factors and equations for physical and physiological regulation were solved simultaneously to calculate maximal DM intake and the NDF content of the diet associated with maximal intake. Limitations to the model were defined by Williams et al. (1989) as: a) assumption of a constant gut capacity, b) simplification in assessing the contribution of body tissue reserves to energy requirements, and c) failure to account for differences in NDF quality as compared with quantity. The role of the animal, and specifically the effect of body size on intake and digestibility of forage, was examined in simulation models by Illius and Gordon (1991) and the use of integrated models involving interactions of plant and animal will be discussed by Illius and Allen (Chapter 21).

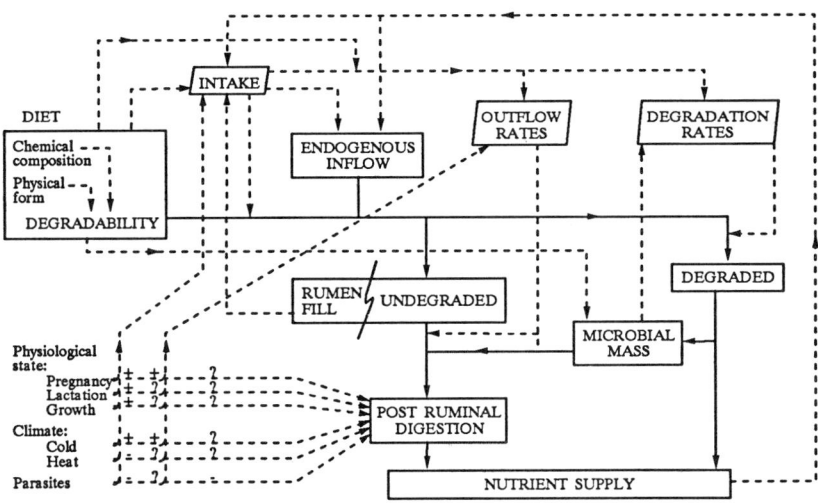

Figure 7. A model of forage and animal factors regulating nutrient supply to the ruminant. Faichney and Black (1979).

Modelling approaches to the characteristics of pasture swards, and their utilization by grazing animals, have been applied in a number of studies. A general philosophy of systems in grazing trials was outlined by Morley and Spedding (1968). Christian (1981) stated that "The comparative scarcity of fully-functional models may be largely attributed to the uncertainties of weather effects, to the heterogeneity of pasture growth and composition, and to the difficulties of relating animal intake to pasture characteristics." Models of rangeland ecosystems to relate changes in vegetative cover and livestock production to climatic conditions, and to formulate optimal grazing management practices, were developed by Van Dyne et al. (1977) and Redetzke and Van Dyne (1979). Plant factors influencing livestock performance on pasture include sward height, density, stage of maturity, concentrations of N, cell wall and lignin (Seman et al., 1991), while animal factors include animal species, body weight, sex, age, production level, bite rate, grazing time, and ambient temperature. In a rangeland model, Demment and Greenwood (1988) related compensatory ingestive and processing behaviors in the animal to environmental changes, to achieve a high rate of energy digestion per unit of time. The optimal solution of the model maintained high intake by sacrificing diet quality and a high passage rate by increasing mastication and rumination effort on poor quality forage.

Christian (1981) commented that, in the early 1980's, agricultural modelling was something of an "amateur science" with results "not always remarkable for either programming methods or scientific insights", and stressed the fact that simulation had yet to prove itself in general farming environments. Current modelling approaches have become increasingly complex and sophisticated, as seen, for example, in the analysis of alternative prediction models for silage quality and intake (Pitt et al., 1985; Rook and Gill, 1990; Rook et al., 1990a,b), and development of the Cornell Net Carbohydrate and Protein System for prediction of the nutrient requirements and performance of cattle (Russell et al., 1992; Sniffen et al., 1992; Fox et al., 1992). Owens (1987) considers that "With complex models, direct involvement by the reader is necessary to visualize application," and this may be a problem which is difficult to resolve. Modelling has undoubtedly become an integral part of the research approach to improving forage and pasture utilization in the last 25 years. It will continue to enlarge our understanding of the complex factors operating in grassland ecosystems, in fermentation reactions and ruminant physiology, and in the animal's response to digested nutrients in a variety of feeding situations. It is hoped that simplified models, or model components, will be developed to effect further improvements in feeding or grazing management at the farm level.

CONCLUSIONS

"He draweth out the thread of his verbosity finer
than the staple of his argument"

Love's Labour Lost

Summaries and extrapolations are best left brief. It will be evident from this review, and the literature cited, that the study of forage quality and utilization is very much of a hybrid enterprise, drawing from both plant and animal sciences, with elements of chemistry, biochemistry, microbiology, toxicology, ecology, ethology, and a variety of other disciplines. The proliferation of new concepts and techniques within these disciplines during the last 25 years, and their application to the areas of forage evaluation and ruminant nutrition, has produced an enormous amount of new literature which is difficult to evaluate either completely or equitably. I regret the inevitable oversights, inaccuracies, and possible misinterpretations in this review.

What of the future? Individual authors will discuss some of the research possibilities and objectives within their own areas of interest. The potential of genetic engineering techniques applied to forage plants appears to be great, although it is fair to say that much of this potential remains to be realized in the field. In discussions at the 17th International Grassland Congress in New Zealand, Peacock (1993) concluded that genetic transfer systems were already in place for several temperate and tropical legumes and a few species of grasses, with the general objectives of pest and pathogen protection and the improvement of nutritional quality. Changes in nutritive value could be achieved both by gene addition and down-modulation of existing genes. Similarly, the possibility of applying genetic engineering techniques to ruminal microorganisms is being actively investigated, with the objectives of enhancing plant cell wall degrading capacity and(or) modifying ruminal detoxifying and metabolic activity (Russell and Wilson, 1988; Flores, 1989; Gilbert and Hazlewood, 1991). The difficulties of this approach are considerable and it will be of interest to see how engineered microorganisms cope with the hurly-burly of the ruminal environment.

In a more general framework, I am reminded of Christian's (1981) description of a systems model as composed of both a biological and a managerial section. The latter "involves decision-making procedures based on assessment of the biological situation on the one hand, and the relevant economic position on the other". Perhaps unfortunately, application of the economic component at the international level in recent years has resulted in the elimination or restructuring of productive research institutes and programs in pasture and forage research in many countries. Emphasis in the developed countries has shifted from production studies to environmental concerns. In the United States, it is difficult to predict whether forage research will continue to develop in the pattern in which it has been described at this conference, or will turn in another direction. What will be the consequences for American agricultural and land use policy of the present radical changes in international, political, and economic systems and food demand? How will public perceptions of environmental issues and of the nutritional quality and safety of their diet affect livestock production methods? These matters await resolution. What will be the agenda of a third Forage Quality, Evaluation, and Utilization Conference in 2019?

REFERENCES

Aii, T., and T.H. Stobbs. 1980. Solubility of the protein of tropical pasture species and the rate of its digestion in the rumen. Anim. Feed Sci. Tech. 5:183-192.

Akin, D.E. 1979. Microscopic evaluation of forage digestion by rumen microorganisms - a review. J. Anim. Sci. 48:701-710.

Akin, D.E. 1982. Microbial breakdown of feed in the digestive tract. p. 201-223. *In* J.B. Hacker (ed.) Nutritional limits to animal production from pastures. Commonwealth Agric. Bureaux, Farnham Royal, UK.

Akin, D.E. 1986. Interaction of ruminal bacteria and fungi with southern forages. J. Anim. Sci. 62:962-977.

Akin, D.E. 1989. Histological and physical factors affecting digestibility of forages. Agron. J. 81:17-25.

Akin, D.E., F.E. Barton II, and S.W. Coleman. 1983. Structural factors affecting leaf degradation of old world bluestem and weeping love grass. J. Anim. Sci. 56:1434-1446.

Akin, D.E., R.H. Brown, and L.L. Rigsby. 1984. Digestion of stem tissues in *Panicum* species. Crop Sci. 24:769-773.

Akin, D.E., H.E. Amos, F.E. Barton II, and D. Burdick. 1973. Rumen microbial degradation of grass tissue revealed by scanning electron microscopy. Agron. J. 65:825-828.

Akin, D.E., N. Ames-Gottfried, R.D. Hartley, R.G. Fulcher, and L.L. Rigsby. 1990. Microspectrophotometry of phenolic compounds in Bermudagrass cell walls in relation to rumen microbial digestion. Crop Sci. 30:396-401.

Allen, M.S., and D.R. Mertens. 1988. Evaluating constraints on fiber digestion by rumen microbes. J. Nutr. 118:261-270.

Allen, V.G. (ed.) 1991. Terminology for grazing lands and grazing animals. Forage and grazing terminology committee, Pocahontas Press, Inc., Blacksburg, VA.

Allinson, D.W. 1969. Forage lignins and their relationship to nutritive value. p. S1 to S9. *In* R.F. Barnes, D.C. Clanton, C.H. Gordon, T.J. Klopfenstein, and D.R. Waldo (ed.) Proc. Natl. Conf. Forage Quality Evaluation and Utilization, Nebraska Center Cont. Educ., Lincoln, NE.

Amjed, M., H.G. Jung, and J.D. Donker. 1992. Effect of alkaline hydrogen peroxide treatment on cell wall composition and digestion kinetics of sugarcane residues and wheat straw. J. Anim. Sci. 70:2877-2884.

Anderson, B.E., J.K. Ward, K.P. Vogel, M.G. Ward, H.J. Gorz, and F.A. Haskins. 1988. Forage quality and performance of yearlings grazing switchgrass strains selected for differing digestibility. J. Anim. Sci. 66:2239-2244.

Anderson, S.J., T.J. Klopfenstein, and V.A. Wilkersen. 1988. Escape protein supplementation of yearling steers grazing smooth brome pastures. J. Anim. Sci. 66:237-242.

ARC. 1980. The nutrient requirements of ruminant livestock. CAB International, Wallingford, Oxon.

ARC. 1984. The nutrient requirements of ruminant livestock. Suppl. No. 1. CAB, Farnham Royal, UK.

Arnold, G.W. 1981. Grazing behavior. p. 79-104. In F.H.W. Morley (ed.) Grazing animals, World Anim. Sci. B1. Elsevier Sci. Publ. Co., Amsterdam.

Arnold, G.W. 1985. Ingestive behavior. p. 183-200. In A.F. Fraser (ed.) Ethology of farm animals, World Animal Sci. A5. Elsevier, Amsterdam.

Arnold, G.W., and M.L. Dudzinski. 1978. The ethology of free-ranging domestic animals. Elsevier Sci. Publ. Co., Amsterdam.

Aufrère, J., D. Graviou, C. Demarquilly, R. Vérité, B. Michalet-Doreau, and P. Chapoutot. 1991. Predicting in situ degradability of feed proteins in the rumen by two laboratory methods (solubility and enzymatic degradation). Anim. Feed Sci. Tech. 33:97-116.

Bacon, C.W., J.K. Porter, J.D. Robbins, and E.S. Luttrell. 1977. *Epichloe typhina* from toxic tall fescue grasses. Appl. Environ. Microbiol. 34:576-581.

Bacon, C.W., P.C. Lyons, J.K. Porter, and J.D. Robbins. 1986. Ergot toxicity from endophyte-infected grasses: A review. Agron. J. 78:106-116.

Baile, C.A., and J.M. Forbes. 1974. Control of feed intake and regulation of energy balance in ruminants. Physiol. Rev. 54:160-214.

Baldwin, R.L., L.J. Koong, and M.J. Ulyatt. 1977. A dynamic model of ruminant digestion for evaluation of factors affecting nutritive value. Agric. Syst. 2:255-288.

Baldwin, R.L., L.J. Koong, M. Ulyatt, and N. Smith. 1976. Towards a synthesis. p. 175-181. In Reviews in rural science, no. II, From plant to animal protein. Univ. of New England Publ. Unit, Armidale, NSW.

Barnes, R.F. 1969. Collaborative research with the two stage *in vitro* technique. p. N-1 to N-18. In R.F. Barnes, D.C. Clanton, C.H. Gordon, T.J. Klopfenstein, and D.R. Waldo (ed.) Proc. Natl. Conf. Forage Quality Evaluation and Utilization, Nebraska Center Cont. Educ., Lincoln, NE.

Barnes, R.F. 1973. Laboratory methods of evaluating feeding value of herbage. p. 179-214. *In* G.W. Butler and R.W. Bailey (ed.) Chemistry and biochemistry of herbage. Vol. 3. Academic Press, London and New York.

Barry, T.N., and B.J. Blaney. 1987. Secondary compounds of forages. p. 91-119. *In* J.B. Hacker and J.H. Ternouth (ed.) The nutrition of herbivores. Academic Press, Australia.

Barton, F.E. II. 1985. Considerations of chemical analyses. p. 68-82. *In* G.C. Marten, J.S. Shenk, and F.E. Barton II (ed.) Near infrared reflectance spectroscopy (NIRS): Analysis of forage quality. USDA-ARS Agric. Handbook No. 643.

Barton, F.E. II. 1991. New methods for the structural and compositional analysis of cell walls for quality determination. Anim. Feed Sci. Tech. 32:1-11.

Beever, D.E. 1978. Prediction of end products of digestion. p. 9.1. *In* D.F. Osbourn, D.E. Beever, and D.J. Thomson (ed.) Ruminant digestion and feed evaluation. Grassl. Res. Inst., Hurley, England.

Beever, D.E., and D.J. Thomson. 1977. The potential of protected protein in ruminant nutrition. p. 66-82. *In* W. Haresign and D. Lewis (ed.) Recent advances in animal nutrition. Butterworths, London-Boston.

Beever, D.E., and R.C. Siddons. 1986. Digestion and metabolism in the grazing ruminant. p. 479-497. *In* L.P. Milligan, W.L. Grovum, and A. Dobson (ed.) Control of digestion and metabolism. Reston Books, Prentice-Hall, Englewood Cliffs, NJ.

Ben-Ghedalia, D., and J. Miron. 1981. Effect of sodium hydroxide, ozone and sulfur dioxide on the composition and in vitro digestibility of wheat straw. J. Sci. Food Agric. 32:224-228.

Bines, J.A. 1976. Factors influencing voluntary food intake in cattle. p. 287-305. *In* H. Swan and W.H. Broster (ed.) Principles of cattle production. Butterworths, London-Boston.

Black, J.L. 1987. Nutritional criteria for forage nutritive value. p. 29-43. *In* K.J. Hutchinson (ed.) Improving the nutritive value of forage. Aust. Agric. Council SCA Tech. Rep. No. 20.

Black, J.L., D.E. Beever, G.J. Faichney, B.R. Howarth, and N. Mc C. Graham. 1980-81. Simulation of the effects of rumen function on the flow of nutrients from the stomach of sheep: Pt. 1 - Description of a computer program. Agric. Syst. 6:195-219.

Blaxter, K.L. 1962. The energy metabolism of ruminants. C.C. Thomas, Publ., Springfield, IL.

Bouton, J.H., R.W. Gates, D.P. Belesky, and M. Owsley. 1993. Yield and persistence of tall fescue in the southeastern Coastal Plain after removal of its endophyte. Agron. J. 85:52-55.

Bray, R.A. 1982. Selecting and breeding better legumes. p. 287-303. *In* J.B. Hacker (ed.) Nutritional limits to animal production from pastures. Commonwealth Agric. Bureaux, Farnham Royal, UK.

Broderick, G.A., and D.R. Buxton. 1991. Genetic variation in alfalfa for ruminal protein degradability. Can. J. Plant Sci. 71:755-760.

Bryant, J.P., P.B. Reichardt, T.P. Clausen, F.D. Provenza, and P.J. Kuropat. 1992. Woody plant-mammal interactions. p. 344-370. *In* G.A. Rosenthal and M.R. Berenbaum (ed.) Herbivores: Their interactions with secondary plant metabolites, Vol. II. Academic Press, Inc., San Diego, CA.

Bughrara, S.S., and D.A. Sleper. 1986. Digestion of several temperate forage species by a prepared cellulase solution. Agron. J. 78:94-98.

Bughrara, S.S., D.A. Sleper, and P.R. Beuselinck. 1992a. Comparison of cellulase solutions for use in digesting forage samples. Agron. J. 84:631-636.

Bughrara, S.S., D.A. Sleper, and C.A. Roberts. 1992b. Cellulase solution concentrations used to develop spectral equations for predicting tall fescue digestibility. Agron. J. 84:636-638.

Burlison, A.J., J. Hodgson, and A.W. Illius. 1991. Sward canopy structure and the bite dimensions and bite weight of grazing sheep. Grass Forage Sci. 46:29-38.

Burns, J.C. 1985. Antiquality factors in temperate forage legumes in the United States. p. 260-267. *In* R.F. Barnes, P.R. Ball, R.W. Brougham, G.C. Marten, and D.J. Minson (ed.) Forage legumes for energy-efficient animal production. Proc. Trilateral Workshop, Palmerston North, New Zealand, 1984. USDA-ARS, Springfield, VA.

Burton, G.W. 1969. The plant breeder and forage evaluation. p. P-1 to P-8. *In* R.F. Barnes, D.C. Clanton, C.H. Gordon, T.J. Klopfenstein, and D.R. Waldo (ed.) Proc. Natl. Conf. Forage Quality Evaluation and Utilization, Nebraska Center Cont. Educ., Lincoln, NE.

Burton, G.W., R.H. Hart, and R.S. Lowrey. 1967. Improving forage quality in bermudagrass by breeding. Crop Sci. 7:329-332.

Burton, G.W., J.B. Gunnells, and R.S. Lowrey. 1968. Yield and quality of early and late-maturing, near-isogenic populations of pearl millet. Crop Sci. 8:431-434.

Butler, G.W., and R.W. Bailey (ed.). 1973. Chemistry and biochemistry of herbage. Vols. 1, 2, 3. Academic Press, London and New York.

Butler, G.W., and D.I.H. Jones. 1973. Mineral biochemistry of herbage. p. 127-162. *In* G.W. Butler and R.W. Bailey (ed.) Chemistry and biochemistry of herbage. Vol. 2. Academic Press, London and New York.

Campling, R.C., J.W.W. Ng'ambi, and O.A.S. Al-Saghier. 1990. Cereal straw in ruminant diets. Outlook Agric. 19:37-41.

Castle, M.E., and J.N. Watson. 1984. Silage and milk production: A comparison between unwilted and wilted grass silages. Grass Forage Sci. 39:187-193.

Chamberlain, D.G., and S. Robertson. 1989. The effects of various enzyme mixtures as silage additives on feed intake and milk production in dairy cows. p. 187-189. *In* C.S. Mayne (ed.) Silage for milk production. Occ. Symp. No. 23, Br. Grassl. Soc.

Cheeke, P.R. (ed.). 1989a. Toxicants of plant origin. Vol. I. Alkaloids. CRC Press, Boca Raton, FL.

Cheeke, P.R. (ed.). 1989b. Toxicants of plant origin. Vol. II. Glycosides. CRC Press, Boca Raton, FL.

Cheeke, P.R. (ed.). 1989c. Toxicants of plant origin. Vol. III. Proteins and amino acids. CRC Press, Boca Raton, FL.

Cherney, J.H., and J.J. Volenec. 1992. Forage evaluation as influenced by environmental replication: A review. Crop Sci. 32:841-846.

Cherney, J.H., D.J.R. Cherney, D.E. Akin, and J.D. Axtell. 1991. Potential of brown-midrib, low-lignin mutants for improving forage quality. Adv. Agron. 46:157-198.

Christian, K.R. 1981. Simulation of grazing systems. p. 361-377. *In* F.H.W. Morley (ed.) Grazing animals. World Anim. Sci. B1. Elsevier Sci. Publ. Co., Amsterdam-Oxford-New York.

Cochran, R.C., E.S. Vanzant, K.A. Jacques, M.L. Galyean, D.C. Adams, and J.D. Wallace. 1987. Internal markers. p. 39-48. *In* M.B. Judkins, D.C. Clanton, M.K. Peterson, and J.D. Wallace (ed.) Proc. Grazing Livestock Nutr. Conf., Univ. of Wyoming, Laramie.

Coleman, S.W., and W.R. Windham. 1985. In vivo and in vitro measurements of forage quality. p. 83-95. *In* G.C. Marten, J.S. Shenk, and F.E. Barton II (ed.) Near infrared reflectance spectroscopy (NIRS): Analysis of forage quality. USDA-ARS Agric. Handbook No. 643.

Conrad, H.R. 1966. Symposium on factors influencing the voluntary intake of herbage by ruminants: Physiological and physical factors limiting feed intake. J. Anim. Sci. 25:227-235.

Conrad, H.R., A.D. Pratt, and J.W. Hibbs. 1964. Regulation of feed intake in dairy cows. I. Change in importance of physical and physiological factors with increasing digestibility. J. Dairy Sci. 47:54-62.

Cooper, J.P. 1973. Genetic variation in herbage constituents. p. 379-417. *In* G.W. Butler and R.W. Bailey (ed.) Chemistry and biochemistry of herbage. Vol. 2. Academic Press, London and New York.

Corah, L.R. 1990. Ammoniation of low quality forages. Agri-Practice 11:35-38.

Corbett, J.L. 1978. Measuring animal performance. p. 163-213. *In* L.'t Mannetje (ed.) Measurement of grassland vegetation and animal production. Commonwealth Agric. Bureaux, Farnham Royal, UK.

Corbett, J.L. 1981. Determination of the utilization of energy and nutrients by grazing animals. p. 383-399. *In* J.L. Wheeler and R.D. Mochrie (ed.) Forage evaluation: Concepts and techniques. AFGC/CSIRO, Griffin Press, Ltd., Netley, S. Australia.

Coulter, B.S., and M. O'Sullivan. 1988. A remote monitor for animal behavior using radio telemetry. p. 450-454. *In* Proc. 12th Gen. Meeting of European Grassl. Fed., Dublin, Ireland. Wicklow Press Ltd.

Crampton, E.W. 1957. Interrelations between digestible nutrient and energy content, voluntary dry matter intake, and the overall feeding value of forages. J. Anim. Sci. 15:546-552.

Crampton, E.W., E. Donefer, and L.E. Lloyd. 1960. A Nutritive Value Index for forages. J. Anim. Sci. 19:538-544.

Deinum, B. 1971. Climate, nitrogen and grass. 3. Some effects of light intensity on nitrogen metabolism. Neth. J. Agric. Sci. 19:184-188.

Deinum, B., and J.G.P. Dirven. 1974. A model for the description of the effects of different environmental factors on the nutritive value of forages. p. 338-346. *In* V.G. Iglovikov and A. P. Movsisyants (ed.) Proc. 12th Int. Grassl. Cong., Moscow, USSR.

Deinum, B., and J.G.P. Dirven. 1975. Climate, nitrogen and grass. 6. Comparison of yield and chemical composition of some temperate and tropical grass species grown at different temperatures. Neth. J. Agric. Sci. 23:69-82.

Demment, M.W., and G.B. Greenwood. 1988. Forage ingestion: Effects of sward characteristics and body size. J. Anim. Sci. 66:2380-2392.

des Bordes, C.K., and J.G. Welch. 1984. Influence of specific gravity on rumination and passage of indigestible particles. J. Anim. Sci. 59:470-475.

D'Mello, J.P.F., C.M. Duffus, and J.H. Duffus (ed.). 1991. Toxic substances in crop plants. Royal Soc. Chem., Cambridge, UK.

Donefer, E. 1969. Forage solubility measurements in relation to nutritive value. p. Q 1 to Q 7. *In* R.F. Barnes, D.C. Clanton, C.H. Gordon, T.J. Klopfenstein, and D.R. Waldo (ed.) Proc. Natl. Conf. Forage Quality Evaluation and Utilization, Nebraska Center Cont. Educ., Lincoln, NE.

Durand, M., and R. Kawashima. 1980. Influence of minerals in rumen microbial digestion. p. 375-408. *In* Y. Ruckebusch and P. Thivend (ed.) Digestive physiology and metabolism in ruminants. MTP, Lancaster, England.

Durand, M., and S. Komisarczuk. 1988. Influence of major minerals on rumen microbiota. J. Nutr. 118:249-260.

Ellis, W.C., C. Lascano, R. Teeter, and F.N. Owens. 1982. Solute and particulate flow markers. p. 37-55. *In* F.N. Owens (ed.) Protein requirements for cattle: Symposium. Oklahoma Agric. Exp. Sta. MP-109, Stillwater, OK.

Faichney, G.J. 1986. The kinetics of particulate matter in the rumen. p. 173-195. *In* L.P. Milligan, W.L. Grovum, and A. Dobson (ed.) Control of digestion and metabolism in ruminants. Reston Books, Prentice-Hall, Englewood Cliffs, NJ.

Faichney, G.J., and J.L. Black. 1979. Factors affecting rumen function and the supply of nutrients. p. 179-192. *In* J.L. Black and P.J. Reis (ed.) Physiological and environmental limitations to wool growth. Univ. of New England Publ. Unit, Armidale, Australia.

Feeny, P. 1992. The evolution of chemical ecology: Contributions from the study of herbivorous insects. p. 1-44. *In* G.A. Rosenthal and M.R. Berenbaum (ed.) Herbivores: Their interactions with secondary plant metabolites. Vol. II. Academic Press, Inc., San Diego, CA.

Fenlon, D.R., J. Wilson, and J.R. Weddell. 1989. The relationship between spoilage and *Listeria monocytogenes* contamination in bagged and wrapped big bale silage. Grass Forage Sci. 44:97-100.

Fernandez-Rivera, S., and T.J. Klopfenstein. 1989. Yield and quality components of corn crop residues and utilization of these residues by grazing cattle. J. Anim. Sci. 67:597-605.

Fisher, D.S., J.C. Burns, and K.R. Pond. 1988. The grazing ruminants diet from pasture. p. 29-30. North Carolina State Univ. ANS Rept. No. 243.

Fleming, G.A. 1973. Mineral composition of herbage. p. 529-566. In G.W. Butler and R.W. Bailey (ed.) Chemistry and biochemistry of herbage. Vol. 1. Academic Press, London and New York.

Fletcher, L.R., J.H. Hoglund, and B.L. Sutherland. 1990. The impact of *Acremonium* endophytes in New Zealand, past, present, and future. Proc. New Zealand Grassl. Assoc. 52:227-235.

Flores, D.A. 1989. Application of recombinant DNA to rumen microbes for the improvement of low quality feed utilization. J. Biotechnol. 10:95-112.

Forbes, J.M. 1986. The voluntary intake of farm animals. Butterworths, London.

Forbes, J.M. 1993. Voluntary feed intake. p. 479-494. In J.M. Forbes and J. France (ed.) Quantitative aspects of ruminant digestion and metabolism. C.A.B. International, Wallingford, Oxon.

Forbes, J.M., and J. France (ed.). 1993. Quantitative aspects of ruminant digestion and metabolism. C.A.B. International, Wallingford, Oxon.

Forbes, T.D.A. 1988. Researching the plant-animal interface: The investigation of ingestive behavior in grazing animals. J. Anim. Sci. 66:2369-2379.

Fox, D.G., C.J. Sniffen, J.D. O'Connor, J.B. Russell, and P.J. Van Soest. 1992. A net carbohydrate and protein system for evaluating cattle diets: III. Cattle requirements and diet adequacy. J. Anim. Sci. 70:3578-3596.

France, J., R.C. Siddons, M.S. Dhanoa, and J.H.M. Thornley. 1991. A unifying mathematical analysis of methods to estimate rumen volume using digesta markers and intraruminal sampling. J. Theor. Biol. 150:145-155.

Fribourg, H.A., C.S. Hoveland, and K.D. Gwinn. 1991. Tall fescue and the fungal endophyte - a review of current knowledge. p. 30-38. Tennessee Farm and Home Science No. 160, Univ. of Tennessee Agric. Exp. Sta., Knoxville, TN.

Fritz, J.O., K.J. Moore, and K.P. Vogel. 1991. Ammonia-labile bonds in high- and low-digestibility strains of switchgrass. Crop Sci. 31:1566-1570.

Furnival, E.P., J.L. Corbett, and M.W. Inskip, 1990. Evaluation of controlled release devices for administration of chromium sesquioxide using fistulated grazing sheep. Aust. J. Agric. Res. 41:969-975.

Gabrielsen, B.C., K.P. Vogel, B.E. Andersen, and J.K. Ward. 1990. Alkali-labile cell-wall phenolics and forage quality in switchgrass selected for differing digestibility. Crop Sci. 30:1313-1320.

Galyean, M.L. 1987. Factors influencing digesta flow in grazing ruminants. p. 77-89. *In* M.B. Judkins, D.C. Clanton, M.K. Peterson, and J.D. Wallace (ed.) Proc. Grazing Livestock Nutr. Conf., Univ. of Wyoming, Laramie.

Garnsworthy, P.C., and D.J.A. Cole. 1990. The importance of intake in feed evaluation. p. 147-160. *In* J. Wiseman and D.J.A. Cole (ed.) Feedstuff evaluation. Butterworths, London-Boston.

Gilbert, H.J., and G.P. Hazlewood. 1991. Genetic modification of fibre digestion. Proc. Nutr. Soc. 50:173-186.

Givens, D.I., and A.R. Moss (ed.). 1990. UK tables of nutritive value and chemical composition of feedingstuffs. ADAS Feed Evaluation Unit, Stratford-on-Avon, UK.

Glenn, B.P. 1990. Effects of dry matter concentration and ammonia treatment of alfalfa silage on digestion and metabolism by heifers. J. Dairy Sci. 73:1081-1090.

Goering, H.K., and P.J. Van Soest. 1970. Forage fiber analyses (apparatus, reagents, procedures and some applications). Agric. Handbook No. 379. ARS, USDA, Washington, DC.

Goering, H.K., P.J. Van Soest, and R.W. Hemken. 1973. Relative susceptibility of forages to heat damage as affected by moisture, temperature and pH. J. Dairy Sci. 56:137-143.

Goering, H.K., C.H. Gordon, R.W. Hemken, D.R. Waldo, P.J. Van Soest, and L.W. Smith. 1972. Analytical estimates of nitrogen digestibility in heat-damaged forages. J. Dairy Sci. 56:1275-1280.

Gordon, F.J. 1989. An evaluation through lactating cattle of a bacterial inoculant as an additive for grass silage. Grass Forage Sci. 44:169-179.

Goto, M., K. Takabe, O. Morita, and I. Abe. 1992. Ultraviolet microscopy of lignins in specific cell walls of barley straw fractions with different rumen degradability. Anim. Feed Sci. Tech. 36:229-237.

Grabber, J.H., G.A. Jung, S.M. Abrams, and D.B. Howard. 1992. Digestion kinetics of parenchyma and sclerenchyma cell walls isolated from orchardgrass and switchgrass. Crop Sci. 32:806-810.

Grenet, E. 1991. Electron microscopy as a method for investigating cell wall degradation in the rumen. Anim. Feed. Sci. Tech. 32:27-33.

Hacker, J.B. 1982. Selecting and breeding better quality grasses. p. 305-326. *In* J.B. Hacker (ed.) Nutritional limits to animal production from pastures. Commonwealth Agric. Bureaux, Farnham Royal, UK.

Hagerman, A.E., and L.G. Butler. 1992. Tannins and lignins. p. 355-388. *In* G.A. Rosenthal and M.R. Berenbaum (ed.) Herbivores: Their interactions with secondary plant metabolites, Vol. I. Academic Press, Inc., San Diego, CA.

Harris, C.E., and J.N. Tullberg. 1980. Pathways of water loss from legumes and grasses cut for conservation. Grass Forage Sci. 35:1-11.

Hartley, R.D., D.E. Akin, D.S. Himmelsbach, and D.C. Beach. 1990. Microspectrophotometry of Bermudagrass cell walls in relation to lignification and cell wall biodegradability. J. Sci. Food Agric. 50:179-189.

Hegarty, M.P. 1982. Deleterious factors in forages affecting animal production. p. 133-150. *In* J.B. Hacker (ed.) Nutritional limits to animal production from pastures. Commonwealth Agric. Bureaux, Farnham Royal, UK.

Henning, J.C., C.T. Dougherty, J. O'Leary, and M. Collins. 1990. Urea for preservation of moist hay. Anim. Feed Sci. Tech. 31:193-204.

Hodgson, J. 1982. Influence of sward characteristics on diet selection and herbage intake by the grazing animal. p. 153-166. *In* J.B. Hacker (ed.) Nutritional limits to animal production from pastures. Commonwealth Agric. Bureaux, Farnham Royal, UK.

Hodgson, J. 1985. The control of herbage intake in the grazing ruminant. Proc. Nutr. Soc. 44:339-346.

Hodgson, J. 1990. Grazing management. Science into practice. Longman Handbooks in Agriculture, Harlow, Essex.

Holechek, J.L., M. Vavra, and R.D. Pieper. 1982. Methods for determining the nutritive quality of range ruminant diets: A review. J. Anim. Sci. 54:363-376.

Hoover, W.H., B.A. Crooker, and C.J. Sniffen. 1976a. Effects of differential solid-liquid removal rates on protozoa numbers in continuous cultures of rumen contents. J. Anim. Sci. 43:528-534.

Hoover, W.H., P.H. Knowlton, M.D. Stern, and C.J. Sniffen. 1976b. Effects of differential solid-liquid removal rates on fermentation parameters in continuous cultures of rumen contents. J. Anim. Sci. 43:535-542.

Horn, F.P., J. Hodgson, J.J. Mott, and R.W. Brougham (ed.). 1987. Grazing-lands research at the plant-animal interface. Winrock International, Morrilton, AR.

Hoveland, C.S. 1990. Importance and economic significance of the *Acremonium* endophytes to performance of animals and grass plants. *In* R.E. Joost and S.S. Quisenberry (ed.) Int. Symp. on *Acremonium*/grass interactions, Prog. and Abstr. Louisiana Agric. Exp. Sta., Baton Rouge, LA.

Huhtanen, P., K. Hissa, S. Saaskkola, and E. Poutiainen. 1985. Enzymes as silage additives. Effect on fermentation quality, digestibility in sheep, degradability in sacco and performance in growing cattle. J. Agric. Sci., Finland 57:284-292.

Humphreys, M.O. 1989. Water-soluble carbohydrates in perennial ryegrass breeding. III. Relationships with herbage production, digestibility and crude protein content. Grass Forage Sci. 44:423-430.

Hungate, R.E. 1966. The rumen and its microbes. Academic Press, New York and London.

Illius, A.W., and I.J. Gordon. 1991. Prediction of intake and digestion in ruminants by a model of rumen kinetics integrating animal size and plant characteristics. J. Agric. Sci. (Camb.) 116:145-157.

Jacobs, J.L., and A.B. McAllan. 1991. Enzymes as silage additives. 1. Silage quality, digestion, digestibility and performance in growing cattle. Grass Forage Sci. 46:63-73.

Jarrige, R. (ed.). 1989. Ruminant nutrition. Recommended allowances and feed tables. INRA. John Libbey, Eurotext, London, Paris.

Jarrige, R., P. Thivend, and C. Demarquilly. 1970. Development of a cellulolytic enzyme digestion for predicting the nutritive value of forages. p. 762-766. *In* M.J.T. Norman (ed.) Proc. 11th Int. Grassl. Cong., Surfers Paradise, Australia. Univ. of Queensland Press, St. Lucia.

Jarrige, R., C. Demarquilly, and J.P. Dulphy. 1982. Forage conservation. p. 364-378. *In* J.B. Hacker (ed.) Nutritional limits to animal production from pastures. Commonwealth Agric. Bureaux, Farnham Royal, UK.

Johnson, C.O.L.E., J.T. Huber, and W.G. Bergen. 1982. Influence of ammonia treatment and time of ensiling on proteolysis in corn silage. J. Dairy Sci. 65:1740-1747.

Johnson, R.R. 1969. The development and application of in vitro rumen fermentation methods for forage evaluation. p. M-1 to M-11. *In* R.F. Barnes, D.C. Clanton, C.H. Gordon, T.J. Klopfenstein, and D.R. Waldo (ed.) Proc. Natl. Conf. Forage Quality Evaluation and Utilization. Nebraska Center Cont. Educ., Lincoln, NE.

Jones, L. 1991. Laboratory studies on the effect of potassium carbonate solution on the drying of cut forage. Grass Forage Sci. 46:153-158.

Jones, D.I.H., and A.D. Wilson. 1987. Nutritive quality of forage. p. 65-89. *In* J.B. Hacker and J.H. Ternouth (ed.) The nutrition of herbivores. Academic Press, Australia.

Judkins, M.B., D.C. Clanton, M.K. Petersen, and J.D. Wallace (ed.). 1987. Proc. Grazing Livestock Nutr. Conf., Univ. of Wyoming, Laramie.

Jung, H.G. 1989. Forage lignins and their effects on fiber digestibility. Agron. J. 81:33-38.

Jung, H.G., D.R. Buxton, R.D. Hatfield, and J. Ralph (ed.). 1993. Forage cell wall structure and digestibility. ASA, CSSA, and SSSA, Madison, WI.

Kemp, A., O.J. Hemkes, and T. van Steenbergen. 1979. The crude protein production of grassland and the utilization by milking cows. Neth. J. Agric. Sci. 27:36-47.

Kerley, M.S., G.C. Fahey, Jr., L.L. Berger, J.M. Gould, and F.L. Baker. 1985. Alkaline hydrogen peroxide treatment unlocks energy in agricultural by-products. Science 230:820-822.

Ketelaars, J.J.M.H., and B.J. Tolkamp. 1992a. Toward a new theory of feed intake regulation in ruminants. 1. Causes of differences in voluntary feed intake: Critique of current views. Livest. Prod. Sci. 30:269-296.

Ketelaars, J.J.M.H., and B.J. Tolkamp. 1992b. Toward a new theory of feed intake regulation in ruminants. 3. Optimum feed intake: In search of a physiological background. Livest. Prod. Sci. 31:235-258.

Klopfenstein, T.J. 1987. Using crop residues in a total cattle management program. Proc. Washington Beef Research Info. Day, p. 26.

Klopfenstein, T.J., and F.G. Owen. 1981. Value and potential use of crop residues and by-products in dairy rations. J. Dairy Sci. 64:1250-1268.

Knapp, W.R., D.A. Holt, and V.L. Lechtenberg. 1975. Hay preservation and quality improvement by anhydrous ammonia treatment. Agron. J. 67:766-769.

Komisarczuk-Bony, S., and M. Durand. 1991. Nutrient requirements of rumen microbes. p. 133-141. In Y.W. Ho, H.K. Wang, N. Abdullah, and Z.A. Tajuddin (ed.) Recent advances on the nutrition of herbivores. Malaysian Soc. Anim. Prod., Kuala Lumpur.

Kotb, A.R., and T.D. Luckey. 1972. Markers in nutrition. Nutr. Abst. Rev. 42:813-845.

Laby, R.H., G.A. Graham, S.R. Edwards, and B. Kautzner. 1984. A controlled release intraruminal device for the administration of fecal dry-matter markers to the grazing ruminant. Can. J. Anim. Sci. 64(Suppl.):337-338.

Laca, E.A., E.D. Ungar, N. Seligman, and M.W. Demment. 1992. Effects of sward height and bulk density on bite dimensions of cattle grazing homogeneous swards. Grass Forage Sci. 47:91-102.

Latch, G.C.M., M.J. Christensen, and R.E. Hickson. 1988. Endophytes of annual and hybrid ryegrasses. N.Z. J. Agric. Res. 31:57-63.

Leng, R.A. 1991. Optimizing herbivore nutrition. p. 269-281. In Y.W. Ho, H.K. Wong, N. Abdullah, and Z.A. Tajuddin (ed.) Recent advances on the nutrition of herbivores. Malaysian Soc. Anim. Prod., Kuala Lumpur.

Leng, R.A. 1993. Overcoming low productivity of ruminants in tropical developing countries. J. Anim. Sci. 71 (Suppl. 1):284 (Abst.)

Lowrey, R.S. 1969. The nylon bag technique for the estimation of forage quality. p. 0-1 to 0-12. In R.F. Barnes, D.C. Clanton, C.H. Gordon, T.J. Klopfenstein, and D.R. Waldo (ed.) Proc. Natl. Conf. Forage Quality Evaluation and Utilization, Nebraska Center Cont. Educ., Lincoln, NE.

Luginbuhl, J.M., K.R. Pond, J.C. Burns, D.S. Fisher, and J.C. Russ. 1991. A computer interface system to monitor the ingestive and ruminating behavior of grazing ruminants. p. 177. In F.T. McCollum III and M.B. Judkins (ed.) Proc. 2nd Grazing Livestock Nutr. Conf., Steamboat Springs, CO. Oklahoma State Univ. Agric. Exp. Sta. MP-133, Stillwater, OK.

Macdonald, A.D., and E.A. Clark. 1987. Water and quality loss during field drying of hay. Adv. Agron. 41:407-437.

Marsh, R. 1979. The effects of wilting on fermentation in the silo and on nutritive value of silage. Grass Forage Sci. 34:1-10.

Marten, G.C. 1981. Chemical, in vitro, and nylon bag procedures for evaluating forage in the USA. p. 39-55. *In* J.L. Wheeler and R.D. Mochrie (ed.) Forage evaluation: Concepts and techniques. AFGC/CSIRO, Griffin Press, Ltd., Netley, S. Australia.

Marten, G.C. (ed.) 1989. Grazing research: Design, methodology, and analysis. CSSA Spec. Publ. No. 16. CSSA, Madison, WI.

Marten, G.C., J.S. Shenk, and F.E. Barton II (ed.). 1985 (revised in 1989). Near infrared reflectance spectroscopy (NIRS): Analysis of forage quality. USDA Agric. Handbook 643. U.S. Gov. Print. Office, Washington, DC.

Mason, V.C., J.E. Cook, M.S. Dhanoa, and R.D. Hartley. 1989. Chemical composition and nutritive value to sheep of untreated and oven-ammoniated ryegrass hays prepared from crops of different maturities. Anim. Feed Sci. Tech. 26:207-220.

Mason, V.C., J.E. Cook, M.S. Dhanoa, C.J. Hoadley, and R.D. Hartley. 1990. Chemical composition, digestibility in vitro and biodegradability of grass hays oven-treated with different amounts of ammonia. Anim. Feed Sci. Tech. 29:237-249.

Matches, A.G. 1969. Pasture research methods. p. I-1 to I-29. *In* R.F. Barnes, D.C. Clanton, C.H. Gordon, T.J. Klopfenstein, and D.R. Waldo (ed.) Proc. Natl. Conf. Forage Quality Evaluation and Utilization, Nebraska Center Cont. Educ., Lincoln, NE.

Mayes, R.W., C.S. Lamb, and P.M. Colgrove. 1986. The use of dosed and herbage n-alkanes as markers for the determination of herbage intake. J. Agric. Sci. (Camb.) 107:161-170.

Mayes, R.W., C.S. Lamb, and P.M. Colgrove. 1988. Digestion and metabolism of dosed even-chain and herbage odd-chain n-alkanes in sheep. p. 159-163. *In* Proc. 12th Gen. Meeting of European Grassl. Fed., Dublin, Ireland. Wicklow Press Ltd.

Mayne, C.S., and R.W.J. Steen. 1990. Recent research on silage additives for milk and beef production. p. 30-42. Agric. Res. Inst. N. Ireland, 63rd Ann. Rept. 1989-90.

McCollum, F.T., and M.B. Judkins (ed.). 1991. Proc. 2nd Grazing Livestock Nutr. Conf., Steamboat Springs, CO. Oklahoma State Univ. Agric. Exp. Sta. MP-133. Stillwater, OK.

McDonald, I.W. 1968. The nutrition of grazing ruminants. Nutr. Abst. Rev. 38:381-400.

McDonald, P. 1981. The biochemistry of silage. John Wiley & Sons, New York.

McDonald, P. 1982. The effect of conservation processes on the nitrogenous components of forages. p. 41-49. In D.J. Thomson, D.E. Beever, and R.G. Gunn (ed.) Forage protein in ruminant animal production. Occ. Publ. No. 6, Br. Soc. Anim. Prod.

McDonald, P, and R. Whittenbury. 1973. The ensilage process. p. 33-60. In G.W. Butler and R.W. Bailey (ed.) Chemistry and biochemistry of herbage. Vol. 3. Academic Press, London and New York.

Mertens, D.R. 1977. Dietary fiber components: Relationship to the rate and extent of ruminal digestion. Fed. Proc. 36:187-192.

Mertens, D.R. 1985. Factors influencing feed intake in lactating cows: From theory to application using neutral detergent fiber. Proc. Georgia Nutr. Conf., p. 1-18.

Mertens, D.R. 1987a. Chemical and physical factors affecting forage degradation. p. 69-75. In M.B. Judkins, D.C. Clanton, M.K. Peterson, and J.D. Wallace (ed.) Proc. Grazing Livestock Nutr. Conf., Univ. of Wyoming, Laramie.

Mertens, D.R. 1987b. Predicting intake and digestibility using mathematical models of ruminal function. J. Anim. Sci. 64:1548-1558.

Mertens, D.R., and L.O. Ely. 1979. A dynamic model of fiber digestion and passage in the ruminant for evaluating forage quality. J. Anim. Sci. 49:1085-1095.

Milne, J.A. 1982-83. Evolution of research on the grazing ruminant. p. 181-192. In M.M. Alcock (ed.) The hill farming research organization biennial report, 1982-83, Penicuik, Midlothian, Scotland.

Minson, D.J. 1981a. The effects of feeding protected and unprotected casein on the milk production of cows grazing ryegrass. J. Agric. Sci. (Camb.) 96:233-241.

Minson, D.J. 1981b. An Australian view of laboratory techniques for forage evaluation. p. 57-73. In J.L. Wheeler and R.D. Mochrie (ed.) Forage evaluation: Concepts and techniques. AFGC/CSIRO, Griffin Press, Ltd., Netley, S. Australia.

Minson, D.J. 1982. Effects of chemical and physical composition of herbage eaten upon intake. p. 167-182. In J.B. Hacker (ed.) Nutritional limits to animal production from pastures. Commonwealth Agric. Bureaux, Farnham Royal, UK.

Minson, D.J. 1990. Forage in ruminant nutrition. Academic Press, Inc., San Diego, CA.

Minson, D.J., and M.N. McLeod. 1970. The digestibility of temperate and tropical grasses. p. 719-722. *In* M.J.T. Norman (ed.) Proc. 11th Int. Grassl. Cong., Surfers Paradise, Australia. Univ. of Queensland Press, St. Lucia.

Miron, J., and D. Ben-Ghedalia. 1992. The effect of sulphur dioxide application levels on the biodegradation of wheat straw carbohydrates by rumen microorganisms and by *Trichoderma viride* cellulase. Bioresource Tech. 41:139-144.

Mochrie, R.D. 1981. Survey of techniques used in grazing trials in U.S. from 1975 to 1980. p. 449-459. *In* J.L. Wheeler and R.D. Mochrie (ed.) Forage evaluation: Concepts and techniques. AFGC/CSIRO, Griffin Press, Ltd., Netley, S. Australia.

Moore, K.J., K.P. Vogel, and A.A. Hopkins. 1993. Improving the digestibility of warm-season perennial grasses. p. 34. *In* Proc. 17th Int. Grassl. Cong., Palmerston North, New Zealand. In press.

Morley, F.H.W., and C.R.W. Spedding. 1968. Agricultural systems and grazing experiments. Herbage Abst. 38:279-287.

Mott, G.O., and J.E. Moore. 1969. Forage evaluation techniques in perspective. p. L-1 to L-7. *In* R.F. Barnes, D.C. Clanton, C.H. Gordon, T.J. Klopfenstein, and D.R. Waldo (ed.) Proc. Natl. Conf. Forage Quality Evaluation and Utilization, Nebraska Center Cont. Educ., Lincoln, NE.

Murphy, M.R., R.L. Baldwin, and M.J. Ulyatt. 1986. An update of a dynamic model of ruminant digestion. J. Anim. Sci. 62:1412-1422.

Murray, I. 1986. Near infrared reflectance analysis of forages. p. 141-156. *In* W. Haresign and D.J.A. Cole (ed.) Recent advances in animal nutrition - 1986. Butterworths, London.

National Research Council. 1985. Ruminant nitrogen usage. National Academy Press, Washington, DC.

National Research Council. 1987. Predicting feed intake of food-producing animals. National Academy Press, Washington, DC.

Nelson, M.L., C.T. Gaskins, and J.R. Males. 1989. Grazing corn harvest residue in range beef cattle production systems. p. 1-11. Washington State Univ. Res. Bull. XB1008.

Nicholson, J.W.G., R.E. McQueen, E. Charmley, and R.S. Bush. 1991. Forage conservation in round bales or silage bags: Effect on ensiling characteristics and animal performance. Can. J. Anim. Sci. 71:1167-1180.

Norris, K.H. 1985. Definition of NIRS analysis. p. 6. *In* G.C. Marten, J.S. Shenk, and F.E. Barton II (ed.) Near infrared reflectance spectroscopy (NIRS): Analysis of forage quality. USDA Agric. Handbook 643. U.S. Gov. Print. Office, Washington, DC.

Norris, K.H., R.F. Barnes, J.E. Moore, and J.S. Shenk. 1976. Predicting forage quality by infrared reflectance spectroscopy. J. Anim. Sci. 43:889-897.

Norton, B.W. 1982. Differences between species in forage quality. p. 89-110. *In* J.B. Hacker (ed.) Nutritional limits to animal production from pastures. Commonwealth Agric. Bureaux, Farnham Royal, UK.

O'Kiely, P., and A.V. Flynn. 1988. Comparison of formic acid and a mixture of formic acid and formaldehyde as preservatives for silage for beef cattle. Irish J. Agric. Res. 27:111-122.

Ørskov, E.R. 1982. Protein nutrition in ruminants. Academic Press, Inc., New York.

Ørskov, E.R. 1991. Manipulation of fibre digestion in the rumen. Proc. Nutr. Soc. 50:187-196.

Ørskov, E.R., and I. McDonald. 1979. The estimation of protein degradability in the rumen from incubation measurements weighted according to rate of passage. J. Agric. Sci. (Camb.) 92:499-503.

Owens, F.N. 1987. New techniques for studying digestion and absorption of nutrients by ruminants. Fed. Proc. 46:283-289.

Owens, F.N., and A.L. Goetsch. 1986. Digesta passage and microbial protein synthesis. p. 196-223. *In* L.P. Milligan, W.L. Grovum, and A. Dobson (ed.) Control of digestion and metabolism in ruminants. Reston Books, Prentice-Hall, Englewood Cliffs, NJ.

Parker, W.J., S.N. McCutcheon, and D.H. Carr. 1989. Effect of herbage type and level of intake on the release of chromic oxide from intraruminal controlled release capsules in sheep. N.Z. J. Agric. Res. 32:546-557.

Parker, W.J., S.N. McCutcheon, and G.A. Wickham. 1991. Effect of administration and ruminal presence of chromic oxide controlled release capsules on herbage intake of sheep. N.Z. J. Agric. Res. 34:193-200.

Peacock, W.J. 1993. High technology for our grasslands. *In* Proc. 17th Int. Grassl. Cong., Palmerston North, New Zealand. In press.

Perdok, H.B., and R.A. Leng. 1987. Hyperexcitability in cattle fed ammoniated roughages. Anim. Feed Sci. Tech. 17:121-143.

Pichard, G., and P.J. Van Soest. 1977. Protein solubility of ruminant feeds. p. 91-98. Proc. Cornell Nutr. Conf., Ithaca, NY.

Pitt, R.E., R.E. Muck, and R.Y. Leibensperger. 1985. A quantitative model of the ensilage process in lactate silages. Grass Forage Sci. 40:279-303.

Pitt, R.E., and R.Y. Leibensperger. 1987. The effectiveness of silage inoculants: A systems approach. Agric. Systems 25:27-49.

Playne, M.J., M.G. Echeverria, and R.G. Megarrity. 1978. Release of nitrogen, sulfur, phosphorus, calcium, magnesium and sodium from four tropical hays during their digestion in nylon bags in the rumen. J. Sci. Food Agric. 29:520-526.

Pond, K.R., W.C. Ellis, and D.E. Akin. 1984. Ingestive mastication and fragmentation of forage. J. Anim. Sci. 58:1567-1574.

Pond, K.R., J.C. Burns, and D.S. Fisher. 1987. External markers - Use and methodology in grazing studies. p. 49-53. *In* M.B. Judkins, D.C. Clanton, M.K. Petersen, and J.D. Wallace (ed.) Proc. Grazing Livestock Nutr. Conf., Univ. of Wyoming, Laramie.

Provenza, F.D., and J.A. Pfister. 1991. Influence of plant toxins on food ingestion by herbivores. p. 199-206. *In* Y.W. Ho, H.K. Wong, N. Abdullah, and Z.A. Tajuddin (ed.) Recent advances on the nutrition of herbivores. Malaysian Soc. Anim. Prod., Kuala Lumpur.

Provenza, F.D., D.F. Balph, J.D. Olsen, D.D. Dwyer, M.H. Ralphs, and J.A. Pfister. 1988. Toward understanding the behavioral responses of livestock to poisonous plants. p. 407-424. *In* L.F. James, M.H. Ralphs, and D.B. Neilsen (ed.) The ecology and economic impact of poisonous plants on livestock production. Westview Press, Boulder, CO.

Raleigh, R.J. 1969. Application of chemical and botanical analysis of range forage to range livestock management. p. K-1 to K-13. *In* R.F. Barnes, D.C. Clanton, C.H. Gordon, T.J. Klopfenstein, and D.R. Waldo (ed.) Proc. Natl. Conf. Forage Quality Evaluation and Utilization, Nebraska Center Cont. Educ., Lincoln, NE.

Raymond, W.F. 1968. Components in the nutritive value of forages. p. 47-62. *In* C.M. Harrison, M. Stelly, and S.A. Breth (ed.) Forage economics-quality. ASA, CSSA, and SSSA, Madison, WI.

Redetzke, K.A., and G.M. Van Dyne. 1979. Data-based, empirical, dynamic matrix modeling of rangeland grazing systems. p. 157-172. *In* N. French (ed.) Perspectives in grassland ecology. Springer-Verlag, New York.

Reid, J.T. 1973. Quality hay. p. 532-548. *In* M.E. Heath, D.S. Metcalfe, and R.F. Barnes (ed.) Forages, 3rd ed., Iowa State Univ. Press, Ames.

Reid, J.T., W.K. Kennedy, K.L. Turk, S.T. Slack, G.W. Trimberger, and R.P Murphy. 1959. Effect of growth stage, chemical composition, and physical properties upon the nutritive value of forages. J. Dairy Sci. 42:567-571.

Reid, R.L. 1994. Nitrogen components of forages and feedstuffs. *In* J.M. Asplund (ed.) Principles of amino acid nutrition for ruminants. CRC Press, Boca Raton, FL. In press.

Reid, R.L., and T.J. Klopfenstein. 1983. Forage and crop residues: Quality evaluation and systems of utilization. J. Anim. Sci. 57 (Suppl. 2):534-562.

Reid, R.L., and G.A. Jung. 1991. Plant/soil interactions in nutrition of the grazing animal. p. 48-63. *In* F.T. McCollum III and M.B. Judkins (ed.) Proc. 2nd Grazing Livestock Nutr. Conf., Steamboat Springs, CO. Oklahoma State Univ. Agric. Exp. Sta. MP-133, Stillwater, OK.

Richards, C.R. 1969. Regional research and forage evaluation. p. W-1 to W-10. *In* R.F. Barnes, D.C. Clanton, C.H. Gordon, T.J. Klopfenstein, and D.R. Waldo (ed.) Proc. Natl. Conf. Forage Quality Evaluation and Utilization, Nebraska Center Cont. Educ., Lincoln, NE.

Riewe, M.E. 1981. Expected animal response to certain grazing strategies. p. 341-355. *In* J.L. Wheeler and R.D. Mochrie (ed.) Forage evaluation: Concepts and techniques. AFGC/CSIRO, Griffin Press, Ltd., Netley, S. Australia.

Riewe, M.E., and H. Lippke. 1969. Considerations in determining the digestibility of harvested forages. p. F-1 to F-17. *In* R.F. Barnes, D.C. Clanton, C.H. Gordon, T.J. Klopfenstein, and D.R. Waldo (ed.) Proc. Natl. Conf. Forage Quality Evaluation and Utilization, Nebraska Center Cont. Educ., Lincoln, NE.

Robards, G.E. 1981. Techniques used in practice in forage evaluation in Australia. p. 461-472. *In* J.L. Wheeler and R.D. Mochrie (ed.) Forage evaluation: Concepts and techniques. AFGC/CSIRO, Griffin Press, Ltd., Netley, S. Australia.

Robbins, C.T., T.A. Hanley, A.E. Hagerman, O. Hjeljord, D.L. Baker, C.C. Schwartz, and W.W. Mautz. 1987. Role of tannins in defending plants against ruminants: Reduction in protein availability. Ecology 68:98-107.

Roe, M.B., and C.J. Sniffen. 1990. Techniques for measuring protein fractions in feedstuffs. p. 81-88. Proc. Cornell Nutr. Conf., Ithaca, NY.

Rook, A.J., and M. Gill. 1990. Prediction of the voluntary intake of grass silages by beef cattle. I. Linear regression analyses. Anim. Prod. 50:425-438.

Rook, A.J., M.S. Dhanoa, and M. Gill. 1990a. Prediction of the voluntary intake of grass silages by beef cattle. 2. Principal component and ridge regression analyses. Anim. Prod. 50:439-454.

Rook, A.J., M.S. Dhanoa, and M. Gill. 1990b. Prediction of the voluntary intake of grass silages by beef cattle. Anim. Prod. 50:455-466.

Rosenthal, G.A., and D.H. Janzen (ed.). 1979. Herbivores: Their interactions with secondary plant metabolites. Academic Press, New York.

Rosenthal, G.A., and M.R. Berenbaum (ed.). 1991. Herbivores: Their interactions with secondary plant metabolites. Vol. I. The chemical participants. 2nd ed. Academic Press, Inc., San Diego, CA.

Rosenthal, G.A., and M.R. Berenbaum (ed.). 1992. Herbivores: Their interactions with secondary plant metabolites. Vol. II. Ecological and evolutionary processes. 2nd ed. Academic Press, Inc., San Diego, CA.

Russell, J.B., and D.B. Wilson. 1988. Potential opportunities and problems for genetically altered rumen microorganisms. J. Nutr. 118:271-279.

Russell, J.B., J.D. O'Connor, D.G. Fox, P.J. Van Soest, and C.J. Sniffen. 1992. A net carbohydrate and protein system for evaluating cattle diets. I. Ruminal fermentation. J. Anim. Sci. 70:3551-3561.

Schneider, B.H. 1947. Feeds of the world. Their digestibility and composition. West Virginia Univ. Agric. Exp. Sta., Morgantown. Jarrett Printing Co., Charleston, WV.

Schneider, B.H., and W.P. Flatt. 1975. The evaluation of feeds through digestibility experiments. The Univ. of Georgia Press, Athens, GA.

Seman, D.H., M.H. Frere, J.A. Stuedemann, and S.R. Wilkinson. 1991. Simulating the influence of stocking rate, sward height and density on steer productivity and grazing behavior. Agric. Syst. 37:165-181.

Sheath, G.W., P.V. Rattray, and D.C. Smeaton. 1987. Influence of pasture quantity and quality on intake and production of sheep. p. 33-43. In F.P. Horn, J. Hodgson, J.J. Mott, and R.W. Brougham (ed.) Grazing-lands research at the plant-animal interface. Winrock International, Morrilton, AR.

Shenk, J.S., M.O Westerhaus, and M.R. Hoover. 1979. Analysis of forage by infrared reflectance. J. Dairy Sci. 62:807-812.

Siegel, M.R., G.C.M. Latch, and M.C. Johnson. 1987a. Fungal endophytes of grasses. Ann. Rev. Phytopathol. 25:293-315.

Siegel, M.R., L.P. Bush, and D.D. Dahlman. 1987b. What am I losing by removing fungal endophyte from tall fescue? p. 41-44. Proc. 43rd Southern Pasture and Forage Improvement Conf., Clemson, SC.

Sniffen, C.J. 1980. Dynamics of protein solubility and degradability in ruminant rations. Proc. Distillers Feed Conf., 35: 69-75.

Sniffen, C.J., J.D. O'Connor, P.J. Van Soest, D.G. Fox, and J.B. Russell. 1992. A net carbohydrate and protein system for evaluating cattle diets: Carbohydrate and protein availability. J. Anim. Sci. 70:3562-3577.

Stobbs, T.H. 1970. Automatic measurement of grazing time by dairy cows on tropical grass and legume pastures. Trop. Grassl. 4:237-244.

Stobbs, T.H. 1973a. The effect of plant structure on the intake of tropical pastures. I. Variation in bite size of grazing cattle. Aust. J. Agric. Res. 24:809-819.

Stobbs, T.H. 1973b. The effect of plant structure on the intake of tropical pastures. II. Differences in sward structure, nutritive value, and bite size of animals grazing *Setaria anceps* and *Chloris gayana* at various stages of growth. Aust. J. Agric. Res. 24:821-829.

Stobbs, T.H. 1974. Rate of biting by Jersey cows as influenced by the yield and maturity of pasture swards. Trop. Grassl. 8:81-86.

Stobbs, T.H., D.J. Minson, and M.N. McLeod. 1977. The response of dairy cows grazing a nitrogen fertilized grass pasture to a supplement of protected protein. J. Agric. Sci. (Camb.) 89:137-141.

Stuedemann, J.A., D.L. Huffman, J.C. Purcell, and O.L. Walker. 1977. Forage-fed beef: Production and marketing alternatives in the south. Southern Coop. Series Bull. 220.

Sullivan, J.T. 1973. Drying and storing herbage as hay. p. 1-31. *In* G.W. Butler and R.W. Bailey (ed.) Chemistry and biochemistry of herbage. Vol. 3. Academic Press, London and New York.

Sundstøl, F. 1988. Straw and other fibrous by-products. Livestock Prod. Sci. 19:136-158.

Sundstøl, F. 1991. Large scale utilization of straw for ruminant production systems. p. 55-60. *In* Y.W. Ho, H.K. Wong, N. Abdullah, and Z.A. Tajuddin (ed.) Recent advances on the nutrition of herbivores. Malaysian Soc. Anim. Prod., Kuala Lumpur.

Sutherland, T.M. 1988. Particle separation in the forestomachs of sheep. p. 43-73. *In* A. Dobson and M.J. Dobson (ed.) Aspects of digestive physiology in ruminants. Comstock Publ. Assoc., Ithaca, NY.

Tamminga, S., R. Ketalaar, and A.M. van Vuuren. 1991. Degradation of nitrogenous compounds in conserved forages in the rumen of dairy cows. Grass Forage Sci. 46:427-435.

Templeton, Jr., W.C., J.S. Shenk, K.H. Norris, G.W. Fissel, G.C. Marten, J.H. Elgin, Jr., and M.O. Westerhaus. 1981. Forage analysis with near-infrared reflectance spectroscopy: Status and outline of national research project. p. 528-531. *In* Proc. 14th Int. Grassl. Cong., J.A. Smith and V.W. Hays (ed.) Westview Press, Boulder, CO.

Theurer, C.B. 1969. Determination of botanical and chemical composition of the grazing animal's diet. p. J-1 to J-20. *In* R.F. Barnes, D.C. Clanton, C.H. Gordon, T.J. Klopfenstein, and D.R. Waldo (ed.) Proc. Natl. Conf. Forage Quality Evaluation and Utilization, Nebraska Center Cont. Educ., Lincoln, NE.

Thomas, C., and P.C. Thomas. 1985. Factors affecting the nutritive value of grass silages. p. 223-256. *In* W. Haresign and D.J.A. Cole (ed.) Recent advances in animal nutrition - 1985. Butterworths, London.

Thomson, D.J., and D.E. Beever. 1980. The effect of conservation on the digestion of forages by ruminants. p. 291-308. *In* Y. Ruckebusch and P. Thivend (ed.) Digestive physiology and metabolism in ruminants. Avi Publ. Co., Inc., Westport, CT.

Tilley, J.M.A., and R.A. Terry. 1963. A two-stage technique for *in vitro* digestion of forage crops. J. Br. Grassl. Soc. 18:104-111.

Tolkamp, B.J., and J.J.M.H. Ketelaars. 1992. Toward a new theory of feed intake regulation in ruminants. 2. Costs and benefits of feed consumption: An optimization approach. Livest. Prod. Sci. 30:297-317.

Ulyatt, M.J. 1981. The feeding value of herbage: Can it be improved? N.Z. J. Agric. Sci. 15:200-205.

Ulyatt, M.J. 1983. Fiber in human and animal nutrition. p. 103-107. *In* G. Wallace and L. Bell (ed.) The Royal Society of New Zealand Bull. 20, Wellington, New Zealand.

Ulyatt, M.J., and J.L. Black. 1984. Plant breeding and feeding value of pastures and forages. CSIRO-DSIR Workshop Summary, DSIR, Palmerston North, New Zealand.

Ulyatt, M.J., J.C. Macrae, R.T.J. Clarke, and P.D. Pearce. 1975. Quantitative digestion of fresh herbage by sheep. IV. Protein synthesis in the stomach. J. Agric. Sci. (Camb.) 84:453-458.

Ulyatt, M.J., D.W. Dellow, A. John, C.S.W. Reid, and G.C. Waghorn. 1986. Contribution of chewing during eating and rumination to the clearance of digesta from the ruminoreticulum. p. 498-515. In L.P. Milligan, W.L. Grovum, and A. Dobson (ed.) Control of digestion and metabolism in ruminants. Reston Books, Prentice-Hall, Englewood Cliffs, NJ.

Van Dyne, G.M., L.A. Joyce, and B.K. Williams. 1977. Models and the formulation and testing of hypotheses in grazingland ecosystems management. p. 41-84. In M. Holdgate and M. Woodman (ed.) Ecosystems restoration - Principles and case studies. Plenum Press, London.

Van Soest, P.J. 1965. Symposium on factors influencing the voluntary intake of herbage by ruminants: Voluntary intake in relation to chemical composition and digestibility. J. Anim. Sci. 24:834-843.

Van Soest, P.J. 1967. Development of a comprehensive system of feed analysis and its application to forages. J. Anim. Sci. 26:119-128.

Van Soest, P.J. 1968. Structural and chemical characteristics which limit the nutritive value of forages. p. 63-76. In C.M. Harrison (ed.) Forage economics-quality. ASA Spec. Publ. 13, ASA, Madison, WI.

Van Soest, P.J. 1969. The chemical basis for the nutritive evaluation of forages. p. U-1 to U-19. In R.F. Barnes, D.C. Clanton, C.H. Gordon, T.J. Klopfenstein, and D.R. Waldo (ed.) Proc. Natl. Conf. Forage Quality Evaluation and Utilization, Nebraska Center Cont. Educ., Lincoln, NE.

Van Soest, P.J. 1975. Physico-chemical aspects of fiber digestion. p. 351-365. In I.W. McDonald and A.C.I. Warner (ed.) Digestion and metabolism in the ruminant. Univ. of New England Publ. Unit, New South Wales.

Van Soest, P.J. 1982. Nutritional ecology of the ruminant. O & B Books, Inc., Corvallis, OR.

Van Soest, P.J. 1983. Methods of forage analysis: What do they measure? Univ. Arkansas Agric. Exp. Sta. Special Report 113.

Van Soest, P.J., and V.C. Mason. 1991. The influence of the Maillard reaction upon the nutritive value of fibrous feeds. Anim. Feed Sci. Tech. 32:45-53.

Van Soest, P.J., J.B. Robertson, and B.A. Lewis. 1991. Methods for dietary fiber, neutral detergent fiber, and nonstarch polysaccharides in relation to animal nutrition. J. Dairy Sci. 74:3583-3597.

Vulich, S.A., and J.P. Hanrahan. 1992. Estimation of intake by the grazing animal with special reference to the n-alkane technique. p. 164. Abstr., 43rd Ann. Meeting of E.A.A.P., Madrid, Spain.

Van Vuuren, A.M., S. Tamminga, and R.S. Ketelaar. 1990. Ruminal availability of nitrogen and carbohydrates from fresh and preserved herbage in dairy cows. Neth. J. Agric. Sci. 38:499-512.

Waldo, D.R. 1969. Factors influencing the voluntary intake of forages. p. E-1 to E-21. In R.F. Barnes, D.C. Clanton, C.H. Gordon, T.J. Klopfenstein, and D.R. Waldo (ed.) Proc. Natl. Conf. Forage Quality Evaluation and Utilization, Nebraska Center Cont. Educ., Lincoln, NE.

Waldo, D.R. 1985. Nutritional value of legumes preserved as silage. p. 220-224. In R.F. Barnes, P.R. Ball, R.W. Brougham, G.C. Marten, and D.J. Minson (ed.) Forage legumes for energy-efficient animal production. Proc. Trilateral Workshop, Palmerston North, New Zealand, 1984. USDA-ARS, Springfield, VA.

Waldo, D.R., L.W. Smith, and E.L. Cox. 1972. Model of cellulose disappearance from the rumen. J. Dairy Sci. 55:125-129.

Ward, G.M. 1980. Energy, land and feed constraints on beef production in the 80's. J. Anim. Sci. 51:1051-1064.

Ward, J.K. 1978. Utilization of corn and grain sorghum residues in beef cow forage systems. J. Anim. Sci. 46:831-840.

Watson, S.J. 1951. Grassland and grassland products. Arnold & Co., London.

Wattiaux, M.A., L.D. Satter, and D.R. Mertens. 1992. Effect of microbial fermentation on functional specific gravity of small forage particles. J. Anim. Sci. 70:1262-1270.

Weston, R.H. 1982. Animal factors affecting feed intake. p. 183-198. In J.B. Hacker (ed.) Nutritional limits to animal production from pastures. Commonwealth Agric. Bureaux, Farnham Royal, UK.

Weston, R.H., and D.P. Poppi. 1987. Comparative aspects of food intake. p. 133-161. *In* J.B. Hacker and J.H. Ternouth (ed.) The nutrition of herbivores. Academic Press, Inc., Orlando, FL.

Wheeler, J.L., and J.L. Corbett. 1989. Criteria for breeding forages of improved feeding value: Results of a Delphi survey. Grass Forage Sci. 44:77-83.

Wilkins, R.J. 1982. Improving forage quality by processing. p. 389-408. *In* J.B. Hacker (ed.) Nutritional limits to animal production from pastures. Commonwealth Agric. Bureaux, Farnham Royal, UK.

Williams, C.B., P.B. Oltenacu, and C.J. Sniffen. 1989. Application of neutral detergent fiber in modeling feed intake, lactation response, and body weight changes in dairy cattle. J. Dairy Sci. 72:652-663.

Willms, C.L., L.L. Berger, N.R. Merchen, and G.C. Fahey, Jr. 1991. Utilization of alkaline hydrogen peroxide-treated wheat straw in cattle growing and finishing diets. J. Anim. Sci. 69:3917-3924.

Wilson, J.R. 1982. Environmental and nutritional factors affecting herbage quality. p. 111-131. *In* J.B. Hacker (ed.) Nutritional limits to animal production from pastures. Commonwealth Agric. Bureaux, Farnham Royal, UK.

Wilson, J.R. 1991b. Plant structures: Their digestive and physical breakdown. p. 207-216. *In* Y.W. Ho, H.K. Wong, N.Abdullah, and Z.A. Tajuddin (ed.) Recent advances on the nutrition of herbivores. Malaysian Soc. Anim. Prod., Kuala Lumpur.

Wilson, J.R., and C.W. Ford. 1971. Temperature influences on the growth, digestibility, and carbohydrate composition of two tropical grasses, *Panicum maximum* var. *trichoglume* and *Setaria sphacelata*, and two cultivars of the temperate grass, *Lolium perenne*. Aust. J. Agric. Res. 22:563-571.

Wilson, J.R., and C.W. Ford. 1973. Temperature influences on the in vitro digestibility and soluble carbohydrate accumulation of tropical and temperate grasses. Aust. J. Agric. Res. 24:187-198.

Wilson, J.R., and D.J. Minson. 1980. Prospects for improving the digestibility and intake of tropical grasses. Trop. Grassl. 14:253-259.

Wilson, J.R., A.O. Taylor, and G.R. Dolby. 1976. Temperature and atmospheric humidity effects on cell wall content and dry matter digestibility of some tropical and temperate grasses. N.Z. J. Agric. Res. 19:41-46.

Wilson, J.R., M.N. McLeod, and D.J. Minson. 1989a. Particle size reduction of the leaves of a tropical and a temperate grass by cattle. I. Effect of chewing during eating and varying times of digestion. Grass Forage Sci. 44:55-63.

Wilson, J.R., D.E. Akin, M.N. McLeod, and D.J. Minson. 1989b. Particle size reduction of the leaves of a tropical and temperate grass by cattle. II. Relation of anatomical structure to the process of leaf breakdown through chewing and digestion. Grass Forage Sci. 44:65-75.

Wilson, J.R., B. Deinum, and F.M. Engels. 1991a. Temperature effects on anatomy and digestibility of leaf and stem of tropical and temperate forage species. Neth. J. Agric. Sci. 39:31-48.

Woolford, M.K. 1984. The silage fermentation. Marcel Dekker, New York.

Young, B.A. 1987. The effect of climate upon food intake. p. 163-190. *In* J.B. Hacker and J.H. Ternouth (ed.) The nutrition of herbivores. Academic Press Inc., Orlando, FL.

CHAPTER 2

THE IMPACT OF FORAGE QUALITY AND SUPPLEMENTATION REGIMEN ON RUMINANT ANIMAL INTAKE AND PERFORMANCE

J.A. Paterson, R.L. Belyea, J.P. Bowman, M.S. Kerley, and J.E. Williams

INTRODUCTION

The livestock producer's primary goal in forage management is to maintain forage quality at a level that will support desired levels of gain or milk production. Forage quality can be defined as a function of both forage intake and digestibility. Traditional methods used to estimate forage quality have included the measurement of plant cell walls and crude protein (CP) concentration, in vitro or in vivo estimates of digestibility and, ultimately, animal productivity (growth or milk production). Cell wall content generally is regarded as the most important factor affecting forage utilization because it comprises the major fraction of forage dry matter (DM) and is correlated with forage intake and digestibility.

The cell wall, composed chiefly of the structural polysaccharides cellulose and hemicelluloses, is degraded by the ruminal microflora. The microflora's ability to degrade and ferment these structural polysaccharides determines the digestible energy (DE) obtained from the forage. Because nutritional quality is assessed as the DE and protein concentration in the forage, microbial fermentation has important implications on forage digestibility. One challenge is to determine how ruminal microbial output resulting from forage digestion can be measured to predict animal performance.

J. A. Paterson and J. P. Bowman, Dep. of Animal & Range Sciences, Montana State University, 119 Linfield Hall, Bozeman, MT 59717; R. L. Belyea, M.S. Kerley, and J. E. Williams, Dep. of Animal Sciences, University of Missouri, Animal Science Research Center, Columbia, MO 65201.

The rate and extent of forage digestibility will determine the clearance rate of forage from the rumen. Rate of removal from the rumen and rate of digestion affect forage intake and, subsequently, milk production and growth by the ruminant. For example, as forage intake increases from 1.5-2.5% of the animal's body weight (BW), daily gain would be expected to increase from 0.10 kg d^{-1} - 0.77 kg d^{-1} [assuming a total digestible nutrient (TDN) content of 630 g kg^{-1} DM]. Likewise, for a steer consuming forage containing 500 g TDN kg^{-1} DM, daily gain would be projected to be 0.40 kg d^{-1}. However, if TDN content was increased to 700 g kg^{-1} DM, daily gain would be expected to increase to 1.0 kg d^{-1}.

In many production settings, supplementary nutrients are necessary to obtain acceptable levels of performance from forage-fed animals. A second challenge is to consistently predict the impact supplementation will have on animal performance. One supplementation strategy would be to maximize forage use by maximizing forage intake and digestion, but the supplement should not supply nutrients in excess of the animal's requirements.

Forage utilization also can be affected by controllable factors. Such factors include physical form, treatment with chemicals, enzymes, or microbiologicals, and grazing management.

EFFECTS OF GRAZING MANAGEMENT ON ANIMAL PERFORMANCE

Ruminant productivity is the ultimate measure of forage quality. Weight gain, when used as an index of forage quality, requires long-term studies and should relate animal performance to the amount of available forage as well as forage composition and digestibility throughout the grazing season. For example, by using Coastal bermudagrass, it was shown that there was a linear relationship between available forage consumed and rate of gain of steers until maximal forage intake was reached (Peterson et al., 1965). Using six warm-season perennial grasses, Duble et al. (1972) reported high correlations between forage quality measurements and animal gain. These experiments concluded that daily gain of yearling crossbred heifers was correlated ($r^2 = 0.85$) to available forage when regressions were computed within the following ranges of in vitro DM digestibility: 1) greater than 600 g kg^{-1} DM, 2) between 500 and 600 g kg^{-1} DM, and 3) less than 500 g kg^{-1} DM. Daily gain increased within each digestibility grouping as available forage increased up to a point above which daily gain finally plateaued. As warm-season perennial grasses increased in maturity and decreased in digestibility, additional available forage was required to maximize animal gains. The kilograms of available forage ha^{-1} at which daily gain did not increase were 500 kg for greater than 600 g in vitro DM disappearance (IVDMD) kg^{-1} DM, 1000 kg for 500-600 g IVDMD kg^{-1} DM, and 1250 kg for less than 500 g IVDMD kg^{-1} DM. Above these levels of available forage, it was assumed that forage digestibility was the primary factor influencing animal performance; below these levels, forage availability was the dominant factor.

Using seven sorghum x sudan hays and five bermudagrass hays harvested at different locations and maturities, weight gain of steers and laboratory analysis revealed a decline in forage quality with advancing maturity within each forage species and location (Lippke, 1980). When forage CP was greater than 60 g CP kg^{-1} DM, weight gain of steers was a function of DE intake (r^2=0.98). Intake was a better indicator of weight gain (r^2 =0.93) than digestibility (r^2 =0.77).

Missouri researchers (Hedrick et al., 1982) conducted a 3 yr study to evaluate the effects of interseeding legumes on forage quality. Steers grazed either N-fertilized tall fescue, tall fescue interseeded with red clover, or tall fescue interseeded with birdsfoot trefoil. Steers which grazed fescue-red clover or fescue-birdsfoot trefoil pastures gained 16 to 28 kg more during the summer than steers which grazed only tall fescue. Significant correlations were observed between steer average daily gain (ADG) and forage CP, IVDMD, acid detergent fiber (ADF), neutral detergent fiber (NDF), Mg, and P. Components having the highest correlations with gain were IVDMD (r^2 =0.73) followed by NDF (r^2=0.71), ADF (r^2=0.69), and CP (r^2=0.56). Also, concentrations of ADF, NDF, and CP were significantly correlated with forage IVDMD. It was concluded that as much as 80% of the variation in ADG was accounted for by forage quality components.

Other researchers have attempted to describe the relationships between animal performance, available forage, and digestibility. Guerrero et al. (1984) studied the relationship between weight gain, available forage, and DM digestibility when steers grazed bermudagrass. Forage was classified according to the following IVDMD: 1) high (> 600 g IVDMD kg^{-1} DM), medium (530-600 g kg^{-1} DM), or low (<530 g kg^{-1} DM). Within each level of digestibility, ADG of steers was fitted to an asymptotic function of available forage. Results suggested that within each digestibility class, increases in ADG occurred with increasing amounts of available forage until a point was reached at which available forage no longer limited animal production. On high, medium, and low digestibility forages, approximate asymptotic ADG values were 0.94, 0.74, and 0.31 kg d^{-1}, while 68, 83, and 89 g forage kg^{-1} BW were needed to produce the asymptotic gain values. For steers to maintain BW on high, medium, and low digestibility forages, 14, 18, and 43 g of forage DM kg^{-1} BW, respectively, were required. The ability to predict performance of cattle grazing forages based on laboratory analyses would benefit nutritionists who formulate supplements and also would benefit producers who rely on forages to feed cattle and sheep.

To determine the relationship between forage quality and ADG, data from three studies (Duble et al., 1972; Lippke, 1980; Hedrick et al., 1982) with 10 forage species and three cutting dates were analyzed for the correlation between net energy of gain (NEg) and ADG. The NEg of the forage species was calculated from percent IVDMD, which then was used to predict ADG of steers and heifers (NRC, 1984). The predicted ADG was correlated highly with NEg ($r^2=0.89$) while actual ADG had a low correlation to NEg ($r^2=0.16$, Figure 1). Forage ADF content also has been shown to be highly correlated with net energy of lactation (NE_L) (Mertens, 1987) as well as ADG (Hedrick et al., 1982). In comparing the 10 forage species and different cutting dates, the ADF content of forages was correlated highly ($r^2=0.80$) with actual ADG of steers and heifers, while the predicted ADG and ADF content of forage had a low correlation ($r^2=0.06$; Figure 2). Although collected on a limited number of grazed forages, the ADF content of forage appears to be a potential predictor of animal performance.

The use of warm-season grasses during mid-summer has been one method for maintaining high quality forage when used in conjunction with cool-season grasses. Warm-season grasses (e.g., switchgrass) have enhanced weight gains of grazing livestock during the summer (Krueger and Curtis, 1979; Burns et al., 1984). Warm-season grasses can produce approximately 70% of their total yield during mid-summer and have provided 212 cow grazing d ha^{-1} or about 60% of the annual grazing time (Jung et al., 1978). Rountree et al. (1974) showed that carrying capacity of switchgrass and Caucasian bluestem was two to three times higher than that of tall fescue from mid-June through September. Daily gain of weaned steer calves was 0.5 kg d^{-1} greater grazing Caucasian bluestem than tall fescue during early and mid-summer (Williams et al., 1993, unpublished data).

A limitation in the utilization of warm-season grasses by cattle is understanding grazing management. Grazing of these grasses so that they remain in a vegetative state can maintain animal performance into the summer when cool-season grass quality typically declines. The technique of a cool-season grass complementing a warm-season grass worked successfully with tall fescue and Caucasian bluestem (Table 1). Cattle which grazed tall fescue (176 d) followed by Caucasian bluestem (55 d) gained an additional 56 kg compared to those grazing tall fescue alone. This type of grazing system extended the quality of summer forage and decreased the costs of gain compared to spring and summer grazing of tall fescue only. These results also suggested that grazing annual rye (which was drilled into a fescue sod pasture during September) in the spring and early summer resulted in the fastest rates of gain and lowest cost of gain of the four systems evaluated.

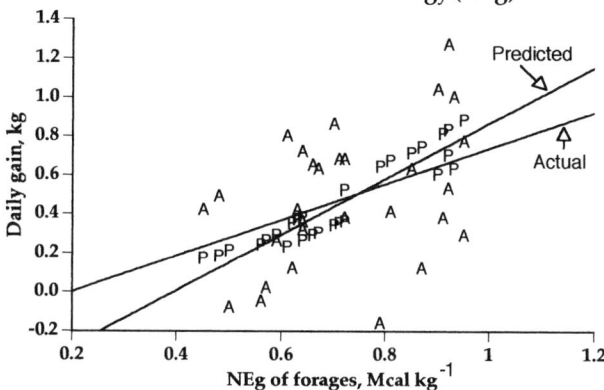

Figure 1. Relationship between actual and predicted ADG of cattle grazing various forages with increasing concentration of Net Energy (NEg).

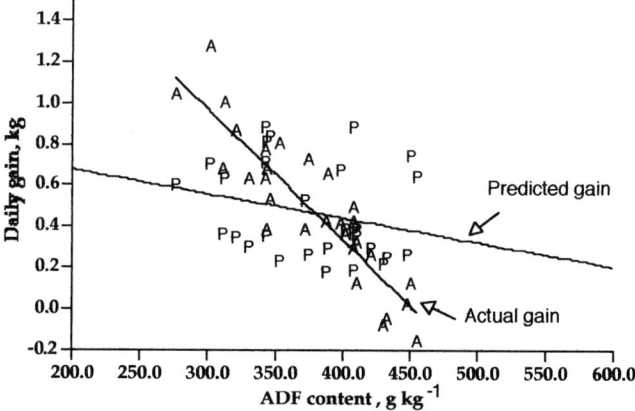

Figure 2. Relationship between actual and predicted ADG of cattle grazing various forages with increasing ADF content

Table 1. Effects of forage system on performance of steers[a].

Item	Forage system			
	Tall fescue 176 d	Tall fescue 213 d	Tall fescue 176 d Caucasian bluestem 55 d[b]	Annual rye 176 d
Initial weight, kg	218.00	218.00	217.00	217.00
Total gain, kg	56.00	80.00	136.00	167.00
Daily gain, kg	0.32	0.35	0.59	0.95
Cost/kg gain, $	1.74	1.37	0.78	0.74

[a]Williams et al., 1993 (unpublished).
[b]Steers grazed tall fescue pastures for 176 d followed by Caucasian bluestem for an additional 55 d.

Most studies have shown that grass-legume pastures are more productive than pure grass pastures fertilized with moderate rates of N. Spooner and McGuire (1979) concluded that greater animal performance could be expected if legumes were included in tall fescue pastures. Likewise, fescue-legume mixtures resulted in greater yields of DM and a more uniform distribution of forage production over the grazing season compared to tall fescue pastures without legumes (Matches, 1979). Incorporating legumes in grass pastures also extended the grazing season (Hoveland, 1960). Studies showed that orchardgrass pastures containing ladino clover supported 2.5-4 steers ha^{-1} from April to mid-August and produced 500 kg beef ha^{-1} in Tennessee (Fribourg et al., 1979; McLaren et al., 1983). Tall fescue interseeded with legumes supported more animals over a longer grazing season than tall fescue alone (High et al., 1965). Koch (1987) renovated pastures by sod-seeding with alfalfa and compared the animal performance to that from N-fertilized pastures. When heifers were fed silages harvested from these pastures, consistent increases in N intake, digestible protein and digestible DM by heifers fed grass-alfalfa silage vs N-fertilized grass silage were observed. Raymond (1969) also reported that legumes resulted in increased DM intake compared to grasses, even though digestibilities of the forages were similar.

Holloway et al. (1979) measured performance of cow-calf pairs and changes in forage digestibility when cows grazed either tall fescue-legume or tall fescue pastures only. Cows grazing fescue-legume pastures consumed more digestible DM than cows grazing fescue only. Pasture quality differences resulted in large advantages for the tall fescue-legume pastures during the summer. Cows grazing fescue-legume pastures gained 18 kg more and weaned heavier calves than those grazing fescue pastures without legumes. Although milk from cows grazing fescue-legume pastures contained more butterfat, the amount of milk produced was similar to those grazing fescue pastures without legumes.

Vicini et al.(1982 a,b) compared the effects of three different forage species and creep grazing on intake and digestibility by cows and calves over a 2 yr period. The three forage systems compared were: 1) Kentucky bluegrass and white clover; 2) Kentucky 31 tall fescue plus Kentucky bluegrass/white clover (bluegrass and white clover were available only to the calves as a forage creep); and 3) Kentucky 31 tall fescue. Mean forage intake by calves over the entire grazing season was 13% greater from the Kentucky bluegrass/white clover pasture than the fescue-creep and fescue pastures, even though cows had lower levels of intake when grazing bluegrass/white clover pasture. It was concluded that when high quality forage was incorporated into a grazing system as creep pasture, forage intake by calves could be equal to or greater than that of cows and calves grazing the same quality pasture. Milk production by cows and 205-d adjusted weights of calves which grazed the bluegrass-white clover pastures were greater than the other treatments. Incorporation of high quality forages into a pasture system and availability to calves through creep grazing increased cow milk production and calf weight gain.

EFFECTS OF FERTILIZATION ON FORAGE QUALITY AND UTILIZATION

Effects of fertilization on forage quality have been examined by numerous researchers. When precipitation was not limiting, fertilization of pastures increased forage yield and quality. Nitrogen fertilization has been shown to increase protein digestibility by sheep fed first and second cutting orchardgrass hay (Reid et al., 1966), but acceptability declined with increasing N fertilization. Minson and Milford (1967) found that application of N increased DM intake of pangola and digitgrass by sheep. However, fertilization of bermudagrass with three levels of N (67, 202, 326 kg ha^{-1}) had no effect on quantity of forage consumed by cows or calves (Horn et al., 1979). Calves tended to select less digestible forages with higher CP as level of N fertilization increased. Cows apparently selected forage without regard to level of N fertilization. Forage intake was positively correlated with IVDMD but negatively correlated with lignin content. Decker et al. (1971) also observed no increase in DM intake of bermudagrass fertilized with up to 672 kg N ha^{-1}.

Puoli et al. (1991) found that addition of 75 kg N from urea ha^{-1} increased DM intake of switchgrass and big bluestem by cattle, but DM and NDF digestibilities were not affected by N fertilization. Minson (1982) reported variable responses in the DM digestibility of tropical grasses fertilized with N. Puoli et al. (1991) also observed that fertilization decreased ruminal turnover time and increased in situ DM degradation rates of switchgrass and big bluestem. Since warm-season grasses can be limiting in CP, application of N appears to be one method for improving utilization and increasing animal performance.

A 3 yr study evaluated daily gain and total seasonal gain of beef cattle grazing fertilized, mixed grass-legume pastures (Raguse et al., 1988). Nitrogen was applied alone (at rates of 45 or 90 kg ha^{-1}) and at these rates with S and P added. The ADG of steers grazing mixed grass-legume pastures fertilized with N, P, and S were greater than those for steers grazing unfertilized or N-fertilized pasture treatments. The increased weight gain was greater in the first year than in the second or third years. The greater forage production from those pastures during the first year compared to other years accounted for most of the increase in ADG. The in vitro organic matter disappearance (IVOMD) and N concentrations were higher for fertilized pastures in March. Cattle grazing the fertilized forage selected a diet which was more digestible and higher in N during winter and early spring compared to the unfertilized pasture. Jones (1967) concluded that N fertilization made its greatest contribution to forage production during the winter.

Fertilization of forages with S in geographical areas in which soil S is deficient has been shown to increase forage yield and animal performance (Jones et al., 1982). The improved forage quality associated with fertilization may be partially due to an increase in forage S which corrects a microbial S deficiency in the rumen. Spears et al. (1985) applied 0 or 132 kg S ha^{-1} as gypsum to orchardgrass and Kenhy tall fescue. Sulfur fertilization of orchardgrass did not affect hay intake, but DM, NDF, and ADF digestibilities were increased, while N digestibility was decreased. Dry matter and fiber digestibilities of tall fescue were not changed by S fertilization, but N digestibility was increased. Sulfur retention was not affected. The failure of S fertilization to increase fiber digestibility of tall fescue may have been due to the higher S content in fescue hay compared to that in orchardgrass hay. Puoli et al.(1991) sprayed sodium sulfate and urea onto switchgrass hay and conducted a digestibility study. Addition of S had no effect on DM and NDF digestibilities, while urea addition increased DM and NDF intakes and DM and NDF digestibilities. Supplemental S increased apparent S digestibility but retention was not changed. It was concluded that S concentration of warm-season grasses was adequate for efficient digestion and animal production.

Corn silage also may contain insufficient S levels to meet animal needs for an optimal N:S ratio of 15. Buttery et al. (1986) examined the effect of S fertilization on corn silage composition and ensiling properties as well as digestibility of forage. Addition of S in the form of ammonium sulfate at the rate of 67.2 kg ha^{-1} lowered the N:S ratio before and after ensiling, increased DM and NDF digestibilities, and increased N retention by lambs. Although there were no effects on intake, S fertilization was more effective in increasing N retention than S supplementation.

Fertilization of forages that are deficient in N and S is one practical method for improving forage quality and performance of grazing animals. The application of N and S to forages deficient in these elements increases digestibility but results in variable effects on forage consumption. With forages such as warm-season grasses, N fertilization would be the preferred method to increase digestion and N retention.

IMPACT OF FORAGE QUALITY ON MILK PRODUCTION

The relationship between forage quality and milk production has been recognized since the first part of the 20th century. Morrison (1950) emphasized the importance of cutting forages early because of higher digestibility and nutrient content, increased intake, and better animal performance. Studies conducted in the 1950's quantified effectors of forage quality, such as date of harvest, species of forage, harvest method, and fertility. A Northeast regional project (Reid et al., 1959) showed that as forage harvest was delayed from early to late stages, TDN declined about 0.5 percentage units d^{-1} and DM intake by lactating cows declined from 250 g kg^{-1} BW for early cut forage to 150 g kg^{-1} BW for late cut forage. Quality of forage was about the same whether harvested as hay or silage. Differences between grasses and legumes were less important than harvest stage effects. Aftermath forages had lower quality than first cutting forage harvested at the same maturity stage. Reid et al. (1959) pointed out that TDN values of high and low quality forages were disproportionate, compared to DM intake and milk production responses. The precision of TDN as a measure of energy available for milk production was questioned (Kleiber, 1959; Reid et al., 1959).

Since 1960, there have been various studies comparing the feeding value of hay vs silage, grasses vs legumes, and legumes vs corn silage. Brown et al. (1963) fed lactating cows diets containing (g kg^{-1} DM) 1000 g grass silage, 750 g silage:250 g grass hay, 500 g of each, 250 g silage:750 g hay and 1000 g hay. Forage DM intake and milk yield increased with increasing amounts of hay in the diet. Cows fed silage responded better (34 percentage unit increase in milk yield) to grain supplementation than did cows fed hay (22 percentage unit increase). Merrill and Slack (1965) reviewed experiments in which legume and grass forage (hays or silages) were fed to lactating cows. Cows fed low (300 g DM kg^{-1} forage or less) and medium DM (300-400 g DM kg^{-1}) silages had milk yields similar to those fed hays. However, cows fed low and medium DM silages had 17 and 14% lower intakes and lost 0.1 kg BW d^{-1} compared to cows fed hay which gained 0.2-0.3 kg^{-1} d. Cows fed high DM silages (400-650 g kg^{-1} DM) had feed intakes and milk yields 102 and 104% of hay-fed cows and gained similar amounts of weight (0.3 kg d^{-1}).

Thomas et al. (1969) fed lactating cows direct-cut alfalfa forage or alfalfa hay. Forage DM intake was 23% higher for hay-fed cows but milk yields were 8% lower. The DM of alfalfa silage contained more energy and(or) was more efficiently used than the DM from hay. Hibbs et al. (1979) compared intake and production of dairy cows fed grass and legume silages. Cows consumed more alfalfa silage and produced more milk compared to cows fed grass silages. Alfalfa had a faster rate of digestion which allowed cows to eat more DM than cows fed grass silage. Larsen and Tenpas (1983) found that cows ate 3% more alfalfa silage than red clover silage and produced 2% more milk. It was concluded that although alfalfa silage was slightly better in quality than red clover silage, differences in production were small.

Broderick (1985) fed lactating cows diets containing alfalfa silage or corn silage (trial 1) or alfalfa silage, alfalfa hay, and corn silage (trial 2). In both trials, DM intake was similar across diets, but cows fed higher levels of corn silage produced less milk compared to those fed the other diets. The NDF digestibility of alfalfa was 85% compared to 70% for corn silage. Cows fed corn silage diets had higher milk protein content than those fed alfalfa diets. Broderick (1985) concluded that high quality alfalfa silage was comparable to corn silage for milk production. DePeters and Smith (1986) fed either early cut (240 g CP kg^{-1} DM and 290 g ADF kg^{-1} DM) or late cut (180 g CP kg^{-1} and 380 g ADF kg^{-1} DM) alfalfa hay to lactating cows at moderate (50:50) or high (70:30) forage to concentrate ratios. There were no differences among treatments in DM intake or milk yield. Cows fed the early cut alfalfa with the higher concentrate level gained BW while the other groups lost weight, suggesting that only the early cut forage provided adequate dietary energy.

Nelson and Satter (1990) harvested alfalfa at early, mid, and late stages of maturity as silage in year one or as hay and silage in year two. Protein content of forages at each of the three maturity stages was similar across years and forage types. The ADF content was lower in the second year than in the first year. Forages were fed to lactating cows in a 55:45 forage to concentrate ratio. In year one, cows were in high, medium, or low production groups while in year two there was no grouping. In year one, only the high producers responded to maturity stage. Milk yields were 30, 29, and 26 kg d^{-1} for the early, mid, and late harvested silages, respectively. In year two, maturity stage had no effect on milk yield, but cows fed the silage diets consumed 3.0 kg d^{-1} more DM and produced 1.5 kg more milk than those fed hay. Dry matter digestibility decreased from 68-63% and milk production decreased 0.15-0.20 kg d^{-1} as maturity increased from one stage to the next. Retention time of particles in the rumen was 18 h for late cut vs 15 h for early cut forages, and was greater for hays (21-24 h) than for silages (15 h). Cows spent 13 h d^{-1} eating and ruminating the hays compared with 12 h d^{-1} for silages. Increased production responses to alfalfa silage were attributed to shorter ruminal retention times and greater digestible DM intakes compared with alfalfa hay.

Llamas-Llamas and Combs (1990) harvested alfalfa hay at early vegetative (270 g CP and 260 g ADF kg^{-1} DM), late bud (210 g CP and 400 g ADF kg^{-1} DM), and full bloom stages (190 g CP and 390 g ADF kg^{-1}

DM). These were fed at 68:32, 53:47, and 45:55 forage to concentrate ratios, respectively. Cows fed the early cut hay-low concentrate diet had greater forage and total DM intakes than the other two diets, while milk yields were similar among diets. The NDF of the early vegetative alfalfa was 15-20% more digestible than the other hays. The higher digestibility was attributed to a greater soluble DM content (370 g kg^{-1} DM) and faster rate of digestion (0.10 h^{-1}) of the early harvested alfalfa than the other hays (250 g soluble matter kg^{-1} DM and 0.06 h^{-1}). It was concluded that high producing cows could consume large amounts of high quality forage with minimal concentrate and yet maintain acceptable milk yields. Kaiser and Combs (1989) fed cows alfalfa hays harvested at increasing maturities and fed at different forage to concentrate ratios. Cows in early lactation consuming low NDF and ADF alfalfa hay produced between 32-45 kg milk d^{-1} when fed diets containing 400 g or less concentrate kg^{-1} DM.

Alhadhrami and Huber (1992) fed lactating cows alfalfa hays containing increasing quantities of ADF and NDF. Diets were supplemented with 500 or 650 g concentrate kg^{-1} DM. Increased fiber (decreased quality) led to reduced DM intakes and milk yields. Peak intake and milk yield occurred with the alfalfa hay containing 280 g ADF and 380 g NDF kg^{-1} DM. Shaver et al. (1988) fed lactating cows diets containing forage (prebloom, midbloom, or full bloom alfalfa, or mature bromegrass) plus concentrate. Digestion and turnover kinetics indicated that the higher milk yield observed for prebloom alfalfa was due to more potentially digestible NDF (250 g kg^{-1} DM), less NDF residue, and a 25-30% faster rate of digestion. Peak intakes and milk production occurred for cows fed the early bloom forage alfalfa. Intake of the lower quality diets was limited by ruminal fill. Thomas et al. (1985) reported that cows in early lactation fed red clover silage had 22% higher forage intakes, produced 8% more milk, and had greater weight gains than cows fed ryegrass silage. Castle et al. (1983) observed that increasing the proportion of white clover silage to grass silage resulted in increased feed intake and milk yield by dairy cows.

Fescue forage infected with an endophytic fungus (*Acremonium*) and containing ergopeptides will not support high levels of milk production. Strahan et al. (1987) fed lactating cows infected or uninfected fescue hay or a control hay (alfalfa or orchardgrass). Forage DM intake was between 9.1 and 11.7 kg d^{-1} for uninfected or low infected tall fescue compared with 7.1 kg d^{-1} for heavily infected tall fescue. Cows fed the infected tall fescue produced less milk and lost more BW compared with cows fed alfalfa or orchardgrass hays. Baxter et al. (1986) fed two varieties of tall fescue and orchardgrass forage to lactating Jerseys. Both fescue varieties had low levels of endophyte-infection but one variety had high levels of perloline. Cows ate less fescue than orchardgrass and produced less milk.

Teller et al. (1992) showed that wilted grass silage resulted in increased intake and milk yield, compared to direct-cut grass silage. Digestibility of wilted silage was not different from direct cut silage, but cows consuming the wilted silage had greater ruminal microbial protein synthesis and higher total ruminal VFA. Weiss and Shockey (1991) fed lactating cows early cut, high quality orchardgrass silage and alfalfa silage supplemented with 200, 400, or 600 g concentrate kg^{-1} diet DM. Increasing concentrate from 200 to 400 g kg^{-1} diet DM increased milk yield 6 kg d^{-1} for the orchardgrass diet and 3 kg d^{-1} for the alfalfa diet. Increasing concentrate to 600 g kg^{-1} diet DM resulted in no additional increase in milk production for either diet. Fiber digestibility was higher for the orchardgrass silage diet than alfalfa silage diet even though DM digestibilities were similar between the two forages. High quality orchardgrass silage was similar to alfalfa in terms of DM intake, net energy, and milk production. Feeding more than 400 g concentrate kg^{-1} diet DM did not improve feeding value of silage-based diets. Vinet et al.(1980) found that cows fed only early cut timothy hay without concentrate produced about 15% more milk than those fed late cut timothy plus concentrate. Steacy et al. (1983) fed dairy cows early, mid, and late cut grass-legume hay plus concentrate. Hays contained 330, 360, and 390 g ADF kg^{-1} DM. Hay DM intake decreased about 1.0 kg d^{-1} for each increase in ADF concentration.

Coppock and Merrill (1967) showed that corn silage was a higher yielding forage than perennials. As DM content of the corn plant increased from early harvest to late harvest, energy content (TDN) was relatively unchanged, but when fed to lactating cows, DM intake increased from 19.5-23.1 g kg^{-1} BW. Although all corn silage diets were considered to be feasible and economical, it was suggested that feeding perennial forages, particularly legumes, would supplement the characteristically low protein content of corn silage. Coppock (1969) suggested that because corn silage is a high energy feed, mature cows are prone to obesity if excessive amounts are fed during late lactation and the dry period. Overconditioned cows had lower DM intakes and milk production and increased incidences of ketosis in subsequent lactations.

Thomas et al. (1970) fed lactating cows either corn silage or alfalfa hay as forage sources for three lactations. Cows fed corn silage consumed less DM, produced similar amounts of milk, and were more efficient than cows fed hay (less feed/milk). Trimberger et al. (1972) reported that cows fed only corn silage for three lactations suffered losses due to displaced abomasums, mastitis, ketosis, and other problems.

Hemken and Vandersall (1967) fed diets containing corn silage or corn silage plus hay for three lactations. Cows fed only corn silage had intake and production responses similar to those fed corn silage plus hay. The importance of adequately supplementing corn silage-based diets with protein, vitamins, and minerals was stressed. Holter et al. (1973) fed lactating cows diets containing either corn silage or corn silage plus increasing amounts of hay. Increasing hay increased forage DM intake 24%, but did not change total DM intake. Milk yield was increased by moderate hay intake. Milk fat and milk protein content also were increased by feeding hay. Cows fed only corn silage had fewer incidences of mastitis, ketosis, and services per conception than cows fed corn silage plus hay.

Belyea et al. (1975a,b) fed lactating cows diets containing only corn silage or 50% corn silage and 50% alfalfa silage or 50% corn silage and 50% alfalfa hay for three lactations. Feed intakes, milk yields, and body weight changes were not different among treatments.

Holter et al. (1975) fed diets containing concentrates plus corn silage or corn silage plus increasing amounts of haycrop silage. Dry matter intakes, milk yields, or digestibilities of energy, protein, or fiber were not affected by dietary treatments and there were no associative effects between corn silage and haycrop silage. Grieve et al. (1980) fed corn silage, corn silage plus haycrop silage, or corn silage plus chopped hay to dairy heifers from birth through three lactations. Cows fed corn silage plus hay consumed 10% more feed and produced 10% more milk than those fed only corn silage. Differences were greater during the first lactation than during later lactations. Cows fed corn silage produced more milk kg^{-1} feed compared to the other two groups.

Low N content is a nutritional limitation of corn silage. Protein supplementation is essential for optimal milk production and increases diet cost. Huber and Thomas (1971) evaluated the use of non-protein N for supplementing corn silage in a series of experiments. Cows fed corn silage ensiled with urea had higher milk production and better persistency than animals fed corn silage alone. Cows fed corn silage diets in which one-half the supplemental N was from urea and one-half from natural protein had 10-15% higher DM intakes and milk production than cows fed diets supplemented with only urea or natural protein. Huber and Santanna (1972) showed that corn silage ensiled with ammonia supported more gain of heifers compared with those fed untreated silage. Lactating cows fed an unsupplemented control diet (13.3 g N kg^{-1} DM) had lower DM intakes compared with cows fed corn silage plus a natural protein supplement or corn silage ensiled with either urea or ammonia. Intakes and milk production were similar for cows fed either diet. Mixing non-protein N with forage at ensiling time was preferred to other methods of addition. Huber et al. (1973) ensiled corn forage with increasing DM content with either urea or ammonia additions. Ammoniated corn silage contained 10-12 g insoluble N kg^{-1} DM compared with 6-9 for urea-treated and untreated silage. Cows fed ammoniated corn silage produced more milk d^{-1} than controls or those fed urea-treated silage. Conversion of high DM silage (420 g kg^{-1} DM or higher) into milk was 5-10% greater than silages with DM between 300-430 g kg^{-1} DM. Treating corn silage with ammonia lowered feed costs 8-14%.

Huber et al. (1980) fed Holstein cows corn silage treated with either ammonia or urea at time of ensiling. The concentrate fraction of the diet contained either soybean meal or urea as N supplements. Adding urea to the concentrate portion of the diet reduced DM intake compared with cows fed concentrate with no urea (but supplemented with soybean meal). Cows fed corn silage treated with urea had lower DM intakes than cows fed corn silage treated with ammonia. Milk production was 1 kg d^{-1} lower for the cows fed urea-treated corn silage plus urea in the concentrate. Cows fed ammonia-treated corn silage and soybean meal in the concentrate (no urea) had the highest forage and total DM intakes and had slightly higher milk production. Persistency was lower in later lactation for cows fed urea-treated corn silage plus high urea concentrate (78%) compared to other diets (87-89% persistency). Erfle et al. (1978) found that supplementing corn silage with high levels of urea reduced palatability and lactation persistency. Ruminal microbial numbers were reduced when natural protein supplement was replaced by urea. Sauer et al. (1979) fed lactating cows diets containing corn silage and cereal grains supplemented with soybean meal and(or) urea. Including urea in a blended diet was as effective as including a natural protein supplement.

Owen (1967) summarized research for sorghum silages. As maturity of sorghum increased from early to late, intake of silage DM increased 25%, but there was no effect on milk yield. Louisiana researchers (Morgan et al., 1987) compared forage sorghum to triticale, corn, ryegrass, and grain sorghum silages for lactating cows. Dry matter intake of sorghum silage was 18 kg d^{-1} compared with 23.6, 25, and 26 kg d^{-1} for cows fed triticale, corn, and ryegrass silages, respectively. Cows fed forage sorghum had lower milk protein concentrations (3 g kg^{-1}) compared with those fed the triticale and corn silages (avg. 3.2 g kg^{-1}).

Burgess et al. (1973) fed corn, barley, wheat, and oat silages to lactating cows. Cows fed corn silage had 10% lower DM intakes and 20% higher milk production than cows fed the small grain silages, for which milk production was similar. Snyder et al. (1979) showed that cows fed barley silage wilted to 500 g kg^{-1} DM and supplemented with protein had greater DM intake (21 kg d^{-1}) and produced more milk (25 kg d^{-1}) than those fed lower DM silage (19 and 23 kg d^{-1}, respectively).

Burgess et al. (1989) fed lactating cows concentrate plus either chopped wheat silage, chopped barley silage, or timothy silage. Cows fed the wheat silage had lower DM intakes and produced less milk than cows fed the other two diets. Baxter et al. (1980) fed diets containing corn silage, wheat silage, corn silage plus alfalfa silage, or corn silage plus wheat silage to lactating Jerseys. Digestibility of the corn silage plus wheat silage diet was similar to the corn silage plus alfalfa diet (690-710 g kg^{-1}). Milk production was higher for cows fed corn silage plus alfalfa silage. Wheat silage was not as palatable as alfalfa because of ensiling difficulties.

Thomas et al. (1981) reported that early cut, primary growth perennial ryegrass silage had higher digestibility than late cut silage and cows fed the early cut silage produced 3 kg more milk d^{-1} than those fed the late cut silage. DePeters et al. (1989) fed lactating cows increasing proportions of a winter cereal silage mixture (85% oats, wheat, and barley and 15% vetch) and decreasing proportions of alfalfa hay. Intake was reduced 1.5 kg d^{-1} with the higher levels of silage, but milk yield was similar among treatments. Messman et al. (1992) compared the feeding value of pea-triticale silage, pearl millet, and corn silage-alfalfa hay. Dry matter intakes were lower for the cows fed pearl millet (19.5 kg d^{-1}) than the other two diets (23 kg d^{-1}). Digestibility of the pea-triticale diet was greater than the other two diets, while milk production was similar among diets. Milk fat content from cows fed the triticale silage was higher than for cows fed the other diets. Cows fed pearl millet lost 0.02 kg BW d^{-1} while the other groups gained 0.22-0.35 kg BW d^{-1}. It was concluded that pearl millet and pea-triticale silage had adequate nutrients for cows in mid-lactation. Acosta et al. (1991) harvested barley silage at boot stage, soft dough-15 cm stubble height, and soft dough-25 cm stubble height. Boot stage silage had higher DM digestibility than soft dough stages (750 vs 620 g kg^{-1} DM). When fed to lactating cows, DM intakes were not different between silages (16-17 kg d^{-1}), but cows consuming boot stage silage produced slightly more milk (27 kg milk d^{-1}) compared with cows consuming dough stage silages (26 kg milk d^{-1}).

Stage of harvest is often used as an indicator of forage quality and harvest schedule. However, this can be a very empirical measure. For example, alfalfa harvested at boot stage in the heat of summer often will have higher fiber levels (particularly lignin) than alfalfa harvested at boot stage during the spring. This greatly affects DM intake and milk production. A more meaningful measure of forage quality may be potential availability of digestible nutrients, which is a function of both intake and digestibility. Unfortunately, in practical conditions, intake and digestibility of forages usually are unknown. The NDF and ADF concentrations of forages provide useful information about quality. High quality forages are characterized by relatively low concentrations of both NDF and ADF. The proportion of digestible NDF is relatively uniform across harvest stage within a species (i.e., about 15 and 25 percentage units for legumes and for grasses, respectively). Therefore, effects of NDF on intake are manifested primarily through effects on indigestible NDF (residue). For example, high quality alfalfa with 400 g kg^{-1} NDF and 150 g kg^{-1} potentially digestible NDF will have about 250 g indigestible NDF kg^{-1} DM. Lower quality alfalfa with 600 g kg^{-1} NDF will have 450 g indigestible NDF kg^{-1} DM. Rate of digestion is an important factor affecting intake; high rates of digestion result in minimal ruminal residue and rapid turnover. High quality forages achieve digestion endpoints quickly, minimize fill limitations, and allow maximum intake. A mature dairy cow (636 kg BW) should be able to consume about 16-18 kg d^{-1} of higher quality alfalfa (DM basis) compared with 11-13 kg d^{-1} of lower quality alfalfa.

Data from recent studies provide generalizations and insight about quality of forage and effects of different factors on quality (Table 2). First, high quality alfalfa is characterized by 280-300 g ADF kg^{-1} DM and 400-420 g NDF kg^{-1} DM. For grasses, the corresponding values are 300-320 g ADF kg^{-1} DM and 500-550 g NDF kg^{-1} DM. When fiber concentrations exceed these values, intake and milk production generally decline. Second, with high quality forage, differences in intake and milk production due to forage species appear to be small. As date of harvest advanced, there was an increase in fiber concentration and differences in intake and milk production due to forage species began to emerge. Third, diets based on high quality forages will support high levels of milk production with moderate amounts of concentrate (500-600 g concentrate kg^{-1} diet DM). This minimizes digestion problems which can occur with high levels of concentrate. Feeding minimal concentrate also can reduce feed costs.

A review of several experiments showed that intake of dietary NDF was relatively constant when compared across experiments, even though differences in dietary conditions, cell wall content, and DM intakes varied (Table 2). This is an important practical concept for diet evaluation and formulation. Average NDF intake was 10-13 g kg^{-1} of BW in 37 of 46 treatment groups. Mertens (1987) suggested that forage NDF intake was limited to about 11-12 g kg^{-1} of BW. Since forage usually provides at least 80% of total NDF intake, much of the data in Table 2 are consistent with Mertens suggestions. However, in several treatment groups, NDF intakes were between 14 and 15 g kg^{-1} BW. In those instances, cows were consuming forages with ADF and NDF in diets of less than 300 and 400 g kg^{-1}, respectively. This suggests that high quality forages have rapidly digested fiber, resulting in forage NDF intakes which exceed expectations. As dynamic, multiple variable models are developed and refined (Williams et al., 1989), intake and production responses will be more accurately estimated.

Two contributions from the USDA Beltsville Laboratory had major impacts on the understanding of forage utilization by lactating cows. One was quantification of the depressions in energy digestibility with increasing levels of intake (Moe and Tyrrell, 1975; Tyrrell and Moe, 1975). Depressions were exacerbated with increasing grain, such that diets containing high grain and medium quality forage had about the same net energy as high quality forage with limited or no grain supplementation. A second contribution was development of a net energy system, replacing TDN as a measure of energy (Moe et al., 1972). Concurrently, development and adoption of the detergent fiber analysis scheme provided a more accurate approach for determining the relative contribution of cell wall and non-cell wall components to digestibility (Van Soest, 1982). The detergent system provided a method to systematically estimate the net energy values of feeds for which there were no direct animal measurements (Van Soest, 1973; Van Soest et al., 1984).

Table 2. Intake and production responses of dairy cows to forages of different qualities.

Reference	Diet	Forage[1] ADF	NDF	Daily Intake Total kg	Forage kg	NDF[2]	Milk Yield[3]
Weiss and Shockey (1991)	800AS[4]	330	400	21.3	17.0	13	24
	600AS	330	400	22.4	13.0	12	27
	400AS	330	400	23.2	9.3	11	28
	800GS	330	530	17.1	13.7	14	21
	600GS	330	530	20.5	12.3	15	27
	400GS	330	530	21.8	8.7	12	27
Alhadhrami and Huber (1992)	500F[5]	260	360	24	12.0	11	30
	500F	280	380	26	13.0	13	31
	500F	320	400	25	12.5	14	28
	500F	380	490	23	11.5	13	28
	350F	260	360	24	8.4	11	29
	350F	280	380	23	8.1	11	30
	350F	320	400	26	9.1	14	29
	350F	380	490	23	8.1	13	30
Broderick (1985) Trial 1	600AS[6]	340	430	20.8	12.3	10	26
	600CS	240	550	20.7	12.4	12	26
	790CS	240	550	20.0	15.8	14	24
Trial 2	630AS[6]	390	480	24.1	15.1	13	30
	600AH	410	550	24.0	14.4	14	29
	600CS	230	460	23.1	13.9	12	30
	760CS	230	460	23.9	18.2	14	28
Llamas-Llamas Combs (1990)	EVAH[7]	250	360	26.1	17.3	12	34
	LBAH	400	520	24.2	11.9	12	33
	FBAH	390	520	24.8	10.0	10	34
Nelson and Satter (1990)	EAS[8]	340	380	20.7	12.4	8	30
	LAS	390	450	21.9	13.1	10	28
	EAH	320	420	23.8	14.3	10	30
	LAH	390	480	23.3	14.0	11	31
	EAS[8]	310	380	20.2	12.0	10	34

Table 2. Continued

Nelson and Satter (1992)							
Trial 1	LAS	350	450	19.0	11.4	11	32
	EAH	330	440	18.0	10.8	10	31
	LAH	370	490	18.5	11.1	11	32
Trial 2	EAS[8]	310	380	21.7	13.0	9	38
	LAS	350	450	22.6	13.6	12	37
	EAH	330	440	22.6	13.6	11	34
	LAH	370	490	21.9	13.1	12	36
DePeters and Smith (1986)	500FEC[9]	290	400	19.5	9.8	11	35
	700FEC	290	400	18.3	5.5	10	31
	500FNC	380	460	18.6	9.3	11	31
	700FNC	380	460	19.5	5.9	12	31
Nelson and Satter (1990)	ECAS[10]	360	400	24.9	13.7	12	30
	MCAS	440	520	25.3	13.9	15	29
High producers	LCAS	480	550	23.6	13.0	15	26
Low producers	ECAS	360	400	19.8	10.9	10	17
	MCAS	440	520	20.3	11.2	12	18
	LCAS	480	550	18.5	10.2	12	16

1 Forage ADF and NDF concentrations, g kg^{-1} DM.
2 NDF intake, g kg^{-1} BW.
3 Milk yield (uncorrected), kg d^{-1}.
4 g kg^{-1} diet DM from alfalfa silage (AS) or grass silage (GS).
5 Forage, g kg^{-1} DM of diet.
6 Alfalfa silage (AS), alfalfa hay (AH), corn silage (CS), g kg^{-1} DM of diet.
7 Early vegetative (EV), late bud (LB), or full bloom (FB) alfalfa hay (AH).
8 Early (E) or late (L) cut alfalfa hay (AH) or silage (AS).
9 Forage, g kg^{-1} diet DM, early cut (EC) or normal cut (NC).
10 Early (EC), mid (MC), or late (LC) cut alfalfa silage (AS).

INTERACTION OF FORAGE QUALITY AND SOURCE OF SUPPLEMENTATION ON INTAKE AND PERFORMANCE OF THE RUMINANT

Supplementation of forage-based diets has been practiced in an attempt to economically improve animal productivity on both cool- and warm-season grass pastures. Lusby (1990) summarized the reasons producers consider supplementing nutrients to ruminants grazing or fed forage-based diets. These include: correction of a nutrient deficiency in the forage; increasing the carrying capacity of the pasture or stretching forage supplies; providing a carrier for growth-promoting additives; aiding in the prevention or treatment of potential health problems; and enhancement of cattle management.

Summaries by Allden (1981), Freer (1981), Minson (1982), Doyle (1987), Horn and McCollum (1987), and Petersen (1987) provide additional insights into the effects of nutrient supplementation on forage intake. This review focuses on the influence of protein and(or) energy supplementation on forage intake by cattle and sheep. It is assumed that enhancement of forage intake and(or) digestibility through supplementation will increase animal productivity. Table 3 presents a summary of selected papers demonstrating how supplemental protein and(or) energy influenced forage intake when forages contained increasing levels of CP.

Non-protein N and true protein supplementation have been advocated with the idea of improving the profitability of ruminants by enhancing the utilization of grazed or harvested forages (Petersen, 1987). With low protein forages and crop residues (< 70 g CP kg^{-1} DM), a major response to protein supplementation has been due to satisfying minimal ruminal microbial requirements for N, possibly specific amino acids and(or) carbon chains. Results presented in Figure 3 suggest that non-protein N and(or) protein supplementation increased forage intake. The majority of the studies indicated that protein supplementation caused greater intake responses with lower rather than higher quality forages. The correlation coefficient of forage intake and of forage CP was less than 0.20 for both unsupplemented and supplemented forages. However, when results from supplementation of low CP forages were compared (Fick et al., 1973; Hennessy et al., 1983; McCollum and Galyean, 1985; DelCurto et al., 1990a; Hannah et al., 1991), it clearly showed that providing additional CP to ruminants consuming low CP forages stimulated forage intake. As the level of CP in the forage increased, the magnitude of the intake response either declined or was not evident (Krysl et al., 1989; DelCurto et al., 1990a; Forcherio et al., 1992). As forage CP increased, production responses due to additional CP supplementation may not have been due to changes in forage intake, but rather may have been due to changes in either forage digestibility or metabolic efficiencies of nutrient utilization, including effects of ruminal degradable vs undegradable protein (Penning et al., 1988; Donaldson et al., 1991; Veira et al., 1991).

Table 3. Summary of the influence of protein and/or energy supplementation and forage quality on intake by cattle and sheep.

Investigator(s)	Animals (wt)	Forage and description	Supplement	Forage intake	Comments
Forages containing less than 70 g CP kg^{-1} DM					
Hannah et al., 1991	Holstein steers (424 kg)	Dormant bluestem *Andropogon spp.*	1) None 2) 1.8 kg containing 12.8% CP 3) 1.8 kg containing 27.1% CP 4) 2.7 kg of dehydrated alfalfa	%BW 0.74 0.78 1.22 1.07	Increased forage intake with 27% CP.
Freeman et al., 1992	Steers (357 kg)	Prairie hay (*Agropyron smithii*)	1) None 2) 0.60 kg containing 43% CP 3) 1.20 kg containing 22% CP 4) 0.60 kg containing 22% CP	1.61 1.64 1.63 1.62	No changes in intakes.
Kartchner, 1980	Cows (458 kg)	*Trial 1* Native fall-winter range *Bouteloua gracilis Agropyron, Buchloe*	1) None 2) 0.75 kg cottonseed meal 3) barley at isocaloric levels	kg d^{-1} 8.6 8.2 7.8	No advantages of supplementation.
	Cows (497 kg)	*Trial 2* Native fall-winter range	1) None 2) 0.70 kg soybean meal 3) barley at isocaloric levels	6.8 8.0 6.3	Protein increased intake.
DelCurto et al., 1990b	Steers (242 kg)	Bluestem hay *Andropogon gerardii Sorgastrum nutans*	1) None 2) 0.99 kg containing 12% CP 3) 0.99 kg containing 28% CP 4) 0.99 kg containing 41% CP	% BW 0.87 0.85 1.36 1.21	Moderate levels increased forage intake.

Table 3. Continued.

Investigator(s)	Animals (wt)	Forage and Description	Supplement	Forage Intake	Comments
Sanson et al., 1990	Steers (270-550 kg)	Meadow hay *Andropogon gerardii* *Calamovilfa longifolia* *Panicum virgatum*	1) None 2) 1.98 g of cornstarch/kg BW 3) 3.96 g of cornstarch/kg BW	2.1 2.0 1.6	Hay DM intake decreased quadratically with increasing levels of corn.
Lamb and Eadie, 1979	Mature sheep	Oat straw Hay (6.3% CP; 70% NDF)	1) None 2) 235 g rolled barley 3) 470 g rolled barley 4) 705 g rolled barley	Oat straw / Hay g kg^{-1} BW$^{0.75}$ 45.6 / 63.2 50.9 / 57.8 47.4 / 55.0 45.4 / 51.5	Intake of straw increased with lowest level of barley; hay intake decreased linearly.
Henning et al., 1980	Wethers	Maize straw	1) None 2) 7.8% of DM from corn grain 3) 15.6% of DM from corn grain 4) 23.5% of DM from corn grain 5) 31.3% of DM from corn grain	kg d^{-1} 1.12 1.25 1.10 0.93 0.83	Linear decrease in digestibility with increasing corn. Decline not due to changes in ruminal pH.

Table 3. Continued.

Investigator(s)	Animals (wt)	Forage and Description	Supplement	Forage Intake	Comments
Mulholland et al., 1976	Wethers (31 kg)	Ground pelleted oat straw	1) None 2) 5% starch 3) 10% starch 4) 15% starch 5) 20% starch 6) 30% starch 7) 40% starch	1.36 1.21 1.24 1.23 1.19 1.13 0.76	Cellulose digestibility decreased 18 percentage units with 30% addition; 10% had minimal effects.
Crabtree and Williams, 1971	Sheep	Straw Hay	g/d of concentrate 1) None 2) 100 3) 200	$g\ d^{-1}$ Straw Hay 242 451 259 400 163 296	Linear decline in hay intake. Straw intake increased with 100 g addition, then declined.
Crabtree and Williams, 1971	Sheep	Hay	% of diet SBM Barley 0 0 0 50 25 0 25 50	287 g/d 316 g/d 412 g/d 298 g/d	SBM increased hay intake. SBM+barley decreased hay intake.
Martin and Hibberd, 1990	Heifers and cows	Native grass hay *Andropogon gerardii* *Schizachyrium scoparium* *Panicum virgatum* *Sorghastrum nutans*	1) None 2) 1 kg soybean hulls 3) 2 kg soybean hulls 4) 3 kg soybean hulls	$kg\ d^{-1}$ 9.70 10.1 9.83 9.07	At highest level, hulls only decreased forage OM intake by 0.64 kg.

Table 3. Continued.

Investigator(s)	Animals (wt)	Forage and Description	Supplement	Forage Intake	Comments
				% BW	
Sunvold et al., 1991	Steers (374 kg)	Dormant bluestem range *Andropogon* spp.	1) None 2) SBM/sorghum (0.32% BW) 3) Wheat midds (0.39% BW) 4) Wheat midds (0.77% BW)	0.87 1.07 0.99 1.15	Supplements increased forage intake and indigestible ADF passage.
				% BW	
McCollum and Galyean, 1985	Steers (214 kg)	Prairie hay *Andropogon* spp.	1) None 2) 800 g cottonseed meal	1.69 2.15	Supplementation increased dilution rates.
DelCurto et al., 1990c	Steers (259 kg)	Dormant range forage *Andropogon* spp. *Sorghastrum*	1) None 2) SBM/sorghum 3) Alfalfa hay 4) Dehydrated alfalfa	0.49 1.07 1.05 1.21	Supplementation increased forage intakes two-fold.
Brandyberry et al., 1991	Steers (316-400 kg)	Summer bluestem range *Andropogon* spp. *Sorghastrum*	1) Salt limiting SBM/sorghum 2) Same supplement hand-fed 3) Same supplement no salt	1.68 1.56 1.69	No effects due to supplementation type or season of year.

Table 3. Continued.

Investigator(s)	Animals (wt)	Forage and Description	Supplement	Forage Intake	Comments
Wagner, 1989	Cows	Winter range	1) 1.27 kg containing 15% CP 2) 1.27 kg containing 40% CP	kg d^{-1} 9.35 9.72	Cows receiving 15% CP lost twice as much weight as 40% CP supplement.
Turner, 1983	Cows	Winter range	1) None 2) 0.91 kg containing 15% CP 3) 0.91 kg containing 30% CP 4) 1.81 kg containing 15% CP	% BW 1.11 1.20 1.40 1.70	
Hennessy et al., 1983	Steers (142 kg)	Pasture hay	Protein / Sorghum grain None / None 0.60 kg / None 1.20 kg / None None / 0.60 kg None / 1.12 kg 0.60 kg / 0.60 kg 0.60 kg / 1.12 kg 1.2 kg / 0.60 kg 1.2 kg / 1.12 kg	2.0 2.6 2.9 1.4 1.6 2.6 2.5 2.2 2.2	Greater response due to protein alone.

Table 3. Continued.

Investigator(s)	Animals (wt)	Forage and Description	Supplement	Forage Intake	Comments
Forero et al., 1980	Cows (428 kg)	Warm season grass hay	1) 1.2 kg of 15% CP (natural) 2) 1.2 kg of 40% CP (natural) 3) 1.2 kg of 40% CP (urea) 4) 2.4 kg of 20% CP (urea)	% BW 1.6 2.2 2.0 1.9	
Coffey et al., 1989	Pregnant and nonpregnant ewes	Timothy hay *Phleum praetense* L.	Alfalfa hay or SBM	Lactating 3.99% BW Nonpregnant 2.70% BW	Pregnancy increased particulate flow rates.
Hennessy and Williamson, 1990	Cattle (172 kg)	Native pasture hay *Axonopus affinus*	1) None 2) Urea+maize 3) Urea+maize 4) Urea+maize+protected casein 5) Protected casein	kg d^{-1} 2.6 3.5 3.5 3.7 3.2	Increase in forage intakes due to CP.
Fick et al., 1973	Sheep (44 kg)	Pangola grass hay *Digitaria decumbens*	g N/d g energy 0 0 0 50 0 100 0 200 10 0 10 50 10 100 10 200	% BW 3.44 3.87 3.61 3.03 4.35 4.43 4.44 4.30	N supplementation increased forage intake more than energy supplementation.

Table 3. Continued.

Investigator(s)	Animals (wt)	Forage and Description	Supplement	Forage Intake	Comments
Doyle et al., 1988	Lambs (25 kg)	Oat hay	1) None 2) 175 g oats/sunflower meal 3) 350 g oats/sunflower meal 4) 505 g oats/sunflower meal	$g\,d^{-1}$ 464 437 252 145	Hay intake not reduced by lowest level of oats; substitution at higher levels; 92 g/100 g of supplement.
Hunt et al, 1988	Steers (390 kg)	Wheat straw	1) None 2) 25% alfalfa hay 3) 50% alfalfa hay 4) 75% alfalfa hay 5) 100% alfalfa hay	Diet NDF intake, $kg\,d^{-1}$ 3.9 4.5 4.5 4.6 3.9	Quadratic straw intake response. Intakes above 25% addition were due to physical changes in ruminal environment.
Zorilla-Rios et al., 1991	Steers (308 kg)	Wheat straw	1) None 2) 150 g/d SBM 3) 500 g/d SBM	3.3 3.5 3.6	SBM increased intake and steer gains.
Chase and Hibberd, 1989	Cows (367 kg)	Grass hay *Andropogon* spp. *Panicum virgatum* *Sorghastrum nutans*	1) 1.4 kg maize 2) 2.05 kg maize Supplements fed daily or twice the amount on alternate days	1.39 kg maize — Daily 2.4% BW, Alternate 2.3% BW 2.05 kg maize — Daily 2.1% BW, Alternate 2.1% BW	

Table 3. Continued.

Investigator(s)	Animals (wt)	Forage and Description	Supplement	Forage Intake	Comments
				% BW	
Chase et al., 1988	Heifers (425 kg)	Grass hay *Andropogon* spp. *Panicum virgatum* *Sorghastrum nutans*	1) NaCl 2) NaHCO$_3$ 3) NH$_4$Cl 4) NH$_4$HCO$_3$	1.7 1.6 1.2 2.3	Buffer alone did not increase OM intakes. Buffered NH$_3$ increased intake.
Chase and Hibberd, 1987	Cows (395 kg)	Grass hay *Andropogon* spp. *Panicum virgatum* *Sorghastrum nutans*	1) None 2) 1 kg corn grain 3) 2 kg corn grain 4) 3 kg corn grain	2.3 2.1 1.7 1.3	Hay intake decreased by corn. Digestible OM intake improved with 1 kg of corn grain.
				kg d^{-1}	
Hennessy and Williamson, 1990	Steers/Heifers (152 kg)	Pasture hay	1) None 2) 30 g urea/d 3) 60 g urea/d 4) 90 g urea/d 5) 120 g urea/d	2.6 3.5 3.5 3.7 3.2	Urea supplementation increased intake. Urea increased daily gains of cattle.

Table 3. Continued.

Investigator(s)	Animals (wt)	Forage and Description	Supplement		Forage Intake	Comments
			N supplement	Energy supplement	kg d^{-1}	
Lee et al., 1987	Steers (227 kg)	Pasture hay *Capillipedium specigium* *Axonopus affinis*	None None Urea Urea Protein Protein	None 570 g/d corn None 570 g/d corn None 570 g/d corn	1.98 2.45 4.32 3.97 4.40 3.91	Urea and protein increased hay intake. Significant interaction between N and corn.
Hunter and Siebert, 1980	Steers (290-490 kg)	Spear grass *Heteropogon contortus*	1) None 2) Urea + S 3) Cottonseed meal		1.48 1.88 2.28	Results demonstrate limitations of urea supplements for poor quality forage.
Forages containing between 80-120 g CP kg^{-1} DM						
Cordes et al., 1988	Cows (454 kg)	Timothy grass hay *Phleum pratense*	1) None 2) Corn grain-urea 3) Dry corn gluten feed		3.6 3.6 4.2	Corn-urea depressed digestibility by 12%. Gluten feed did not change hay digestibility.
					% BW	
Pordomingo et al., 1991	Steers (507 kg)	Bluegramma rangeland *Bouteloua gracilis* *Buchloe dactyloides* *Hilaria mutica*	1) None 2) Corn at 0.2% BW 3) Corn at 0.4% BW 4) Corn at 0.6% BW		2.76 2.95 2.26 1.98	0.2% corn tended to increase OM intake while higher levels depressed intake.
Cochran et al., 1990	Heifers (543 kg)	Bluestem range *Andropogon* spp.	1) None 2) Monensin bolus		2.9 3.0	No effect on forage intake. Minimal effect on digestibility.

Table 3. Continued.

Investigator(s)	Animals (wt)	Forage and Description	Supplement	Forage Intake	Comments
Langlands, 1969	Sheep	*Phalaris/Trifolium*	1) None 2) 100 g wheat 3) 200 g wheat 4) 300 g wheat 5) 400 g wheat	kg d^{-1} 0.83 0.72 0.77 0.68 0.56	Forage intake decreased with increasing levels of wheat. At high stocking rates, the decline in intake was less than at lower stocking rates.
Hodge and Bogdanovic, 1983	Lambs	Pasture hay	1) None 2) 250 g oats or peas 3) 500 g oats or peas 4) Ad lib oats or peas	g d^{-1} Oats Peas 221 221 84 213 43 215 39 50	Peas had less of depressing effect than did oats.
Hall et al., 1990	Cows and steers (450 and 254 kg)	Bermuda grass hay *Cynodan dacylon*	1) None 2) Ground corn 3) Poured fat 4) Corn + fat 5) Corn mixed with fat	kg d^{-1} 5.9 5.5 6.5 5.2 5.1	Intake influenced by supplement type.

Table 3. Continued.

Investigator(s)	Animals (wt)	Forage and Description	Supplement	Forage Intake	Comments
Hannah et al., 1989	Steers (302 kg)	Tall fescue pasture *Festuca arundinacea*	June 1) None 2) Ground corn, 1% BW 3) Whole corn, 1% BW August 1) None 2) Ground corn, 1% BW 3) Corn gluten feed, 1% BW	% BW 1.84 1.74 1.44 2.20 1.70 1.89	Starch digestibility lower for whole corn compared with pelleted corn. Corn depressed forage digestion more than corn gluten feed.
Forcherio et al., 1992	Lactating cows (471 kg)	Tall fescue pasture *Festuca arundinacea*	1) None 2) Blood meal + corn 3) Blood meal + soybean hulls 4) Soybean meal + corn 5) Soybean meal + soybean hulls	1.6 1.6 1.5 1.5 1.8	No increase in forage intake due to either source of supplemental CP or energy.
Cremin et al., 1991	Nursing calves (197 kg)	Tall fescue pasture *Festuca arundinacea*	1) None 2) 0.6 kg containing 13% CP 3) 0.6 kg containing 35% CP 4) 1.62 kg containing 13% CP	kg d^{-1} 1.85 1.65 1.60 1.14	Milk OM intake not affected by creep. Forage OM intake negatively correlated with level of creep.
Krysl et al., 1989	Steers (334 kg)	Blue gramma rangeland *Bouteloua gracilis*	1) None 2) 0.5 kg soybean meal 3) 0.5 kg steam flaked sorghum	7.7 8.7 8.2	Minimal effects of supplement on forage intakes.

Table 3. Continued.

Investigator(s)	Animals (wt)	Forage and Description	Supplement	Forage Intake (%BW)	Comments
Vanzant et al., 1990	Steers (270 kg)	Tall grass prairie 80% NDF *Trial 2* Tall grass prairie 74% NDF *Andropogon spp. Sorghastrum mutans*	*Trial 1* 1) None 2) 0.45 kg sorghum grain 3) 0.91 kg sorghum grain 4) 1.82 kg sorghum grain *Trial 2* 1) None 2) Corn at 0.37% BW 3) Wheat at 0.37% BW 4) Sorghum at 0.37% BW	Trial 1 2.12 2.20 2.20 2.23 Trial 2 2.45 2.50 2.40 2.49	No effects of supplement on intake, NDF digestibility or indigestible ADF fill. No effect on forage intake, digestibility.
DelCurto et al., 1990b	Steers (330 kg)	Dormant range forage 8.3% CP; 81% NDF *Andropogon spp. Sorghastrum mutans*	1) 1.48 kg containing 13% CP 2) 1.46 kg containing 25% CP 3) 1.44 kg containing 39% CP	0.87 1.31 0.99	Quadratic change in intake.

Forages Containing More Than 130 g CP kg⁻¹ DM.

Investigator(s)	Animals (wt)	Forage and Description	Supplement	Forage Intake (%BW)	Comments
Lake et al., 1974	Twin cattle (290 kg)	Irrigated pasture 38% ADF	1) None 2) 1.36 kg corn	2.90 2.57	Forage DM intake decreased by the amount of DM supplied by supplement.

Table 3. Continued.

Investigator(s)	Animals (wt)	Forage and Description	Supplement	Forage Intake	Comments
Lee et al., 1985	Cows (397 kg)	Chopped native hay	Cottonseed meal + meat+fish meal (43% CP) -- g kg^{-1} BW$^{0.75}$ -- 1) None 2) 5.3 3) 10.5 4) 15.8 5) 21.0	 1.2 1.4 1.8 2.1 2.0	Increase in forage intake due to CP.
Stakelum, 1986	Cows (520 kg)	Pasture	1) None 2) 3.3 kg barley	2.5 2.3	No increase in intake.
Penning et al., 1988	Ewes	Ryegrass (fresh or grazed) *Lolium perenne*	1) None 2) Barley and corn starch 3) Barley and soybean meal 4) Barley+soybean meal+fish meal 5) Barley and fish meal	2.00 1.71 1.75 1.70 1.80	Forage intakes depressed by supplementation. Milk yields were increased by supplement.

Table 3. Continued.

Investigator(s)	Animals (wt)	Forage and Description	Supplement	Forage Intake	Comments
Orr and Treacher, 1989	Pregnant ewes	*Lolium perenne L.*	1) None 2) 450 g barley based; 16% CP 3) 900 g barley based; 24% CP	1.22 1.11 1.00	
Lagassee et al., 1990	Steers (224 kg)	Bermuda grass hay *Cynodon dactylon*	1) None 2) 15% alfalfa hay (*Medicago sativa*) 3) 30% alfalfa hay	2.43 2.31 1.99	Digestible OM intake decreased between 15 and 24% with alfalfa.
		Orchardgrass hay *Dactylis glomerata*	1) None 2) 15% alfalfa hay 3) 30% alfalfa hay	2.98 2.55	Only 30% inclusion increased intake.
Veira et al., 1991	Steers (278 kg)	Grass silages *Phleum pratense Festuca rubra Poa compresa Agropyron repens Lotus corniculatus*	1) None 2) Rumen protected amino acids (8.2 g lysine; 2.6 g methionine)	2.24 2.27	No effect on feed intake. Amino acids increased gain 16%.
Donaldson et al., 1991	Steers (300 kg)	Annual ryegrass *Lolium multiflorum L.*	1) 1.5 kg corn grain 2) 1.5 kg low escape CP (0.13 kg) 3) 1.5 kg high escape CP (0.25 kg)	kg d^{-1} 7.3 10.5 10.3	Quadratic response for intake.

Table 3. Continued.

Investigator(s)	Animals (wt)	Forage and Description	Supplement	Forage Intake	Comments
				g d^{-1}	
Orskov and **Fraser**, 1975	Lambs	Chopped grass hay	-- g kg^{-1} BW$^{0.75}$ -- 1) None 2) 25 g whole barley 3) 25 g pelleted barley 4) 50 g whole barley 5) 50 g pelleted barley	991 790 820 642 472	At high intakes, pelleted barley reduced forage intake more than whole barley. Pelleting reduced ADF digestibility more than whole.

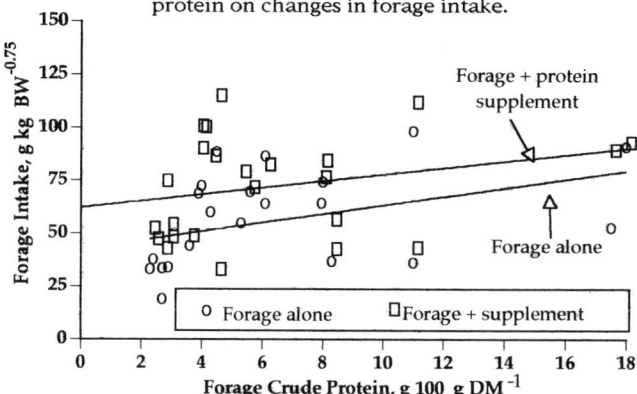

Figure 3. Effect of supplying protein supplements to cattle and sheep grazing forage with increasing levels of crude protein on changes in forage intake.

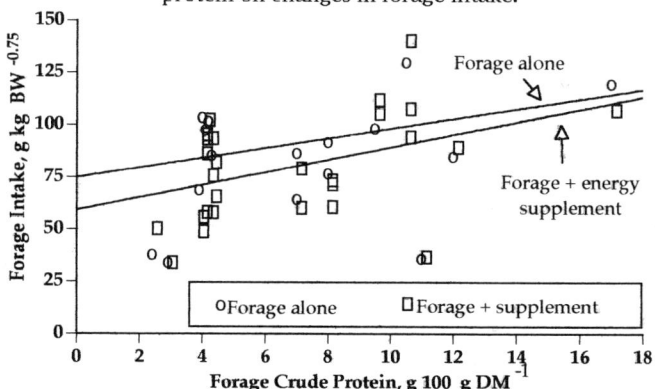

Figure 4. Effect of supplying energy supplements to cattle and sheep grazing forage with increasing levels of crude protein on changes in forage intake.

Protein supplements containing greater than 200 g CP kg DM^{-1} resulted in greater forage intake responses than did supplements containing less than 200 g CP kg DM^{-1} (Forero et al.,1980; Orr and Treacher, 1989; Wagner, 1989; DelCurto et al., 1990a; Hannah et al., 1991). Supplementation of low CP-containing forages has been shown to increase forage intake, but the increase in consumption still may not be great enough for the animal to satisfy its requirements for maintenance or desired growth.

Ellis (1990) outlined the desired information necessary to formulate supplements for grazing ruminants. Information required included both animal and ruminal microbial requirements, nutrient content of the forage in terms of degradable and undegradable protein and available energy, forage intake, and interactions between the forage and supplement. Ellis cited data of Egan (1977) which suggested that voluntary intakes of different types of forage may not be maximized until levels of truly absorbed protein exceeded 200 g kg^{-1} of the digested OM. Traditionally, we have not considered that protein supplementation of forages containing high (greater than 200 g CP kg DM^{-1}) protein would enhance animal performance. However, Horn (1990) showed that cottonseed meal supplementation of steers grazing wheat pasture increased animal performance by 15% compared to non-supplemented animals and by 4% compared with energy-supplemented animals. The efficiency with which the supplement was converted to additional gain was better for the cottonseed meal supplement compared with the energy supplement.

Unlike protein supplementation, energy supplementation has had either a minimal or negative influence on forage intake depending on the quantity of supplement fed (Figure 4). Forage intake reductions appear to be primarily related to form and source of supplemental energy (whole vs processed; starch vs rapidly digestible fiber). Results of Martin and Hibberd (1990), Forcherio et al. (1992), and Sunvold et al. (1991) showed that when rapidly digestible fiber-based supplements were fed, reductions in forage consumption were not as evident when compared to starch-based supplements (Langlands, 1969; Orskov and Fraser, 1975; Mulholland et al., 1976; Kartchner, 1981).

Horn and McCollum (1987) regressed the substitution effects of starch-based supplements on forage intake (substitution ratio defined as the unit change in forage intake per unit increase in concentrate intake) as forage digestibility increased. As forage digestibility increased, the substitution ratio increased and was greater for sheep than for cattle. However, not all starch-based supplements consistently decrease forage intake. Pordimingo et al. (1991) indicated a stimulatory effect of small quantities (0.2% BW) of corn-based supplements on forage intake. Depressing effects on forage intake were measured in this study when higher levels of supplemental corn were offered. Earlier work by Lamb and Eadie (1979), Henning et al. (1980), and Vanzant et al. (1990) also showed that forage intakes were not negatively affected by small quantities of starch-based supplements. Clearly, as level of starch increased in the diet, negative effects on forage intake were demonstrated, a response possibly due to an imbalance of N:energy in the diet and(or) rumen.

Work by Grant and Mertens (1992) indicated that low ruminal pH could decrease fiber digestion rate and increase lag times, even though Henning et al. (1980) could not show a clear relationship between changes in ruminal pH (<6.2) and voluntary forage intakes.

Physical characteristics of the energy supplement also may play a role in changing either forage intake or digestibility. Orskov and Fraser (1975) and Hannah et al. (1989) suggested that ground or whole cereal grains both could negatively influence forage intake and digestibility, but to different extents.

EFFECTS OF PHYSICAL FORM OF THE FORAGE ON ANIMAL PRODUCTION

The effects of physical form on forage quality have been studied extensively. Moore (1964) summarized the effects of wafering vs grinding and(or) pelleting of forages. In general, grinding or pelleting resulted in faster eating rates, less salivation, lower ruminal pH, faster rates of cellulose digestion, and increased proportion of propionic acid in the rumen. Pelleting or grinding increased rate of passage and depressed ruminal digestion. In lactating cows, grinding and(or) pelleting of forage generally increased intake 10-30% and milk yield 10-20%. Wafered forage often did not elicit the magnitude of response observed for ground and pelleted forage because wafered forage was not reduced in particle size as extensively as ground and pelleted forage. Waldo (1973) showed that particle size reduction of forages resulted in decreased digestion and increased passage rate from the rumen. The depression was greater with an increased level of intake. Cell walls were more adversely affected than cell solubles. Particle size reduction depressed digestion of grasses more than legumes.

There have been numerous studies in which coarsely chopped grass hay (Adams, 1988), coarsely chopped alfalfa hay (Belyea et al., 1985, 1988; Shaver et al., 1986), and coarsely chopped alfalfa silage (Brouk and Belyea, 1992) were compared to unprocessed forage. Compared to grinding and pelleting, coarsely chopped forage elicited few adverse effects on DM intake, milk production, milk composition, DM digestibility, or ruminal fermentation. It might be concluded from these studies that coarsely chopped hays or silages were similar to long forage in providing fiber that is capable of sustaining rumination. Either could be used as an adequate source of fiber in blended diets, eliminating the need to feed long hay.

The effects of forage quality and physical form on chewing activity have been extensively investigated. Welch and Smith (1970) compared the rumination activity of cows fed forages differing in maturity. Chewing time increased from 258 min meal^{-1} for early cut orchardgrass hay to 449 min meal^{-1} for late cut hay. Chewing activity was highly correlated with dietary NDF content. Santini et al. (1983) found that intake of forage, when adjusted for particle length, was positively related to chewing activity. Total chewing time increased from 554 min d^{-1} for shorter length forage to 720 min d^{-1} for longer length forage. An adjustment of intake to account for differences in particle length was proposed to account for differences in intake level. Sudweeks et al. (1981) summarized studies in which the chewing times for different forages were measured. Long, fibrous forages or crop residues elicited total chewing times (eating and ruminating activities) of 70-100 min kg DM^{-1} compared with coarsely chopped forages which had chewing times of 40-70 min kg DM^{-1}. Total chewing times were 20-40 min kg DM^{-1} for ground and(or) pelleted forages. The minimum amount of total chewing time for effective rumination and appropriate ruminal fermentation appeared to be about 35-45 min kg DM^{-1}.

Beauchemin and Buchanan-Smith (1989) fed lactating cows diets containing corn and alfalfa silage in proportions that resulted in dietary NDF concentrations of 260, 300, or 340 g kg DM^{-1}. Increasing the dietary NDF increased eating time from 214-260 min d^{-1}, ruminating time from 344-414 min d^{-1}, and total chewing time from 558-678 min d^{-1}. Increasing the percentage of NDF in the diet increased chewing activity per kg DM intake but not per kilogram NDF intake. Increased NDF in the diet resulted in an increased number of meals per day, duration of rumination periods, and number of boli d^{-1}. Beauchemin and Buchanan-Smith (1990) fed lactating cows diets composed of high moisture shelled corn plus either alfalfa silage or alfalfa silage plus long hay (fed prior to concentrate or together). Adding hay to the diet increased DM and NDF intakes and increased milk yield. Hay did not affect most of the measures of chewing activity during eating but increased rumination time from 274-328 min d^{-1}, number of boli d^{-1} from 297-380 and number of rumination periods d^{-1}. The number of chews kg NDF intake^{-1} was not changed by feeding hay. Woodford and Murphy (1988) found that increasing alfalfa pellets and decreasing alfalfa silage in a diet containing 60% concentrate decreased total chewing time (eating and ruminating) from 650 min d^{-1} to 380 min d^{-1}.

Woodford et al. (1986) fed cows 280, 360, 450, and 530 g long alfalfa hay kg^{-1} diet DM. Forage DM intake increased as the percentage of dietary forage increased but total DM intake was not changed. Total chewing time increased from 629-766 min d^{-1}, with similar increases in both eating and rumination activities. Cows were fed alfalfa hay chopped to mean particle lengths of 0.26, 0.46, 0.64, and 0.90 cm. Physical form did not affect intake but longer particle lengths increased total chewing time. It was concluded that diets should contain a minimum of 270 g NDF kg^{-1} and 180 g ADF kg^{-1} DM and that a minimum forage particle size should be at least 0.64 cm for optimal milk fat percentage and ruminal fermentation. Shaver et al. (1986) fed prebloom alfalfa as long, chopped, or pelleted forage to dairy cows in early or mid-lactation or to non-lactating cows. Physical form did not affect DM intake or milk yield. Total chewing activity was not different when cows were fed long and chopped alfalfa, but was significantly reduced for cows fed pelleted alfalfa. Rumination was depressed in cows fed pelleted hay, compared with cows fed the other two diets. Reducing the particle size of early cut alfalfa did not greatly affect ruminal fill or turnover. Digestion of fiber in the pelleted alfalfa diet was depressed, possibly due to low ruminal pH.

In summary, physical form affects forage utilization. Coarsely chopped or long forage appears to provide adequate effective fiber for chewing. Finely chopped or ground forage apparently does not elicit adequate rumination which can, in turn, adversely affect the ruminal environment and digestion. Dietary NDF reflects rumination activity more adequately than does dietary ADF. For diets containing forages of adequate particle length, it is important to maintain a minimum dietary NDF and ADF percentage.

MEASUREMENT OF FORAGE QUALITY-FUTURE IMPROVEMENTS

A truly comprehensive measure of forage quality should not only predict the quantity of digestible energy and protein consumed, but should also measure the synchrony between energy and protein fermentation in the rumen. Two major voids exist in the literature: (1) fermentation patterns of different forage carbohydrates, and (2) proteolytic degradation patterns of forage proteins. Our future goal should be to optimize the nutritional value of forage-based diets for ruminants. To achieve this goal, it is necessary to focus on maximizing microbial efficiency. Once achieved, greater animal productivity should result through increases in forage intake and (or) digestion.

Continued improvement in grazing systems for ruminants is necessary. Research has clearly shown the benefits derived when legumes have been incorporated into improved pastures. A specific challenge will be to maintain forage digestibility at as high a level as possible during mid summer when cool-season grasses decline in digestibility and result in lower intake and performance. Increased usage of warm-season grasses with cool-season grasses may be one alternative for maintaining higher digestibility and ruminant performance during the summer months. Furthermore, a clearer understanding of the relationships between forage availability, digestibility and stocking density can help to elucidate subsequent responses in animal productivity.

Supplementation of forages has been and needs to continue to be an active field of study. Protein supplementation of low protein forages generally improves forage intake and animal performance. However, energy supplementation strategies are still inadequate because we do not yet completely understand such interacting factors as dietary protein levels, ruminal N requirements, effects of supplement on ruminal pH, amount and frequency of supplementation, physical form of supplement and whether the supplement is based on rapidly digestible starch or fiber. A major challenge for the forage scientist continues to be the accurate measurement of forage intake by the grazing animal. Intake studies which incorporate a measurement of performance will be helpful in evaluating supplementation programs and will provide useful information for more accurately balancing ruminant diets.

REFERENCES

Acosta, Y.M., C.C. Stallings, C.E. Polan, and C.N. Miller. 1991. Evaluation of barley silage harvested at boot and soft dough stages. J. Dairy Sci. 74:167-176.

Adams, M.W., R.L. Belyea, and F.A. Martz. 1988. Effects of variety and particle size upon utilization of fescue hays by lactating dairy cows and dairy heifers. J. Dairy Sci. 71:1275-1282.

Alhadhrami, G., and J.T. Huber. 1992. Effects of alfalfa hay of varying fiber fed at 35 or 50% of diet on lactation and nutrient utilization by dry cows. J. Dairy Sci. 75:3091-3099.

Allden, W.G. 1981. Energy and protein supplements for grazing livestock. p. 289-309. *In* F.H. Morley (ed.) Grazing animals. Elsevier Scientific Publ. Co., New York, NY.

Baxter, H.D., M.J. Montgomery, and J.R. Owen. 1980. Digestibility and feeding value of corn silage fed with boot stage wheat silage and alfalfa silage. J. Dairy Sci. 63:255-261.

Baxter, H.D., J.R. Owen, R.C. Buckner, R.W. Hemken, M.R. Siegel, L.P. Bush, and M.J. Montgomery. 1986. Comparison of low alkaloid tall fescue and orchardgrass for lactating cows. J. Dairy Sci. 69:1329-1336.

Beauchemin, K.A., and J.G. Buchanan-Smith. 1989. Effects of dietary neutral detergent fiber concentration and supplementary long hay on chewing activities and milk production of dairy cows. J. Dairy Sci. 72:2288-2300.

Beauchemin, K.A., and J.G. Buchanan-Smith. 1990. Effects of fiber source and method of feeding on chewing activities, digestive function and productivity of dairy cows. J. Dairy Sci. 73:749-762.

Belyea, R.L., C.E. Coppock, W.G. Merrill, and S.T. Slack. 1975a. Effects of silage based diets on feed intake, milk production, and body weight of dairy cows. J. Dairy Sci. 58:1328-1335.

Belyea, R.L., C.E. Coppock, and G.B. Lake. 1975b. Effects of silage diets on health, reproduction, and blood metabolites of dairy cattle. J. Dairy Sci. 58:1336-1346.

Belyea, R.L., F.A. Martz, and G.A. Mbagaya. 1989. Effect of particle size of alfalfa hay on intake, digestibility, milk yield, and ruminal cell wall of dairy cattle. J. Dairy Sci. 72:958-963.

Belyea, R.L., P.J. Marin, and H.T. Sedgwick. 1985. Utilization of chopped and long alfalfa by dairy heifers. J. Dairy Sci. 68:1297-1301.

Brandyberry, S.D., R.C. Cochran, E.S. Vanzant, T. DelCurto, and L.R. Corah. 1991. Influence of supplementation method on forage use and grazing behavior by beef cattle grazing bluestem range. J. Anim. Sci. 69:4128-4136.

Broderick, G.A. 1985. Alfalfa silage or hay versus corn silage as the sole forage for lactating dairy cows. J. Dairy Sci. 68:3262-3271.

Brouk, M.J., and R.L. Belyea. 1993. Chewing activity and digestive responses of cows fed alfalfa forages. J. Dairy Sci. 76:175-182.

Brown, L.D., D. Hillman, C.A. Lassiter, and C.F. Huffman. 1963. Grass silage vs. hay for lactating dairy cows. J. Dairy Sci. 46:407-410.

Burgess, P.W., J.W.G. Nicholson, and E.A. Grant. 1973. Yield and nutritive value of corn, barley, wheat, and forage oats as silage for lactating dairy cows. Can. J. Anim. Sci. 53:245-250.

Burgess, P.W., G.C. Misener, R.E. McQueen, and J.W.G. Nicholson. 1989. Evaluation of barley and wheat head-chop silages for dairy cows. Can. J. Anim. Sci. 69:947-954.

Burns, J.C., R.D. Mochrie, and D.H. Timothy. 1984. Steer performance from two perennial Pennisetum species, switchgrass and a fescue-'Coastal' bermudagrass system. Agron. J. 76:775-780.

Buttrey, S.A., V.G. Allen, J.P. Fontenot, and R.B. Reneau. 1986. Effect of sulfur fertilization on chemical composition, ensiling characteristics and utilization of corn silage by lambs. J. Anim. Sci. 63:1236-1245.

Castle, M.E., D. Reid, and J.N. Watson. 1983. Silage and milk production: Studies with diets containing white clover silage. Grass Forage Sci. 38:193-200.

Chase, Jr., C.C., and C.A. Hibberd. 1989. Effect of level and frequency of maize supplementation on the utilization of low-quality grass hay by beef cows. Anim. Feed Sci. Technol. 24:129-139.

Chase, Jr., C.C., C.A. Hibberd, and F.N. Owens. 1988. Buffer and ammonia additions to corn supplemented native grass hay diets for beef heifers. J. Anim. Sci. 66:1790-1799.

Chase, Jr., C.C., and C.A. Hibberd. 1987. Utilization of low quality native grass hay by beef cows fed increasing quantities of corn grain. J. Anim. Sci. 65:557-566.

Cochran, R.C., E.S. Vanzant, J.G. Riley, and C. Owensby. 1990. Influence of intraruminal monensin administration on performance and forage use in beef cattle grazing early-summer bluestem range. J. Prod. Agric. 3:88-92.

Coffey, K.P., J.A. Paterson, C.S. Saul, L.S. Coffey, K.E. Turner, and J.G. Bowman. 1989. The influence of pregnancy and source of supplemental protein on intake, digestive kinetics and amino acid absorption by ewes. J. Anim. Sci. 67:1805-1814.

Coppock, C.E., and W.G. Merrill. 1967. Corn silage in combination with hay-crop silage. Dairy Herd Manage. 4:100.

Coppock, C. E. 1969. Problems associated with all corn silage feeding. J. Dairy Sci. 52:848-858.

Cordes, C.S., K.E. Turner, J.A. Paterson, J.G.P. Bowman, and J.R. Forwood. 1988. Corn gluten feed supplementation of grass hay diets for beef cows and yearling heifers. J. Anim. Sci. 66:522-531.

Crabtree, J.R., and G.L. Williams. 1971. The voluntary intake and utilization of roughage-concentrate diets by sheep. Anim. Prod. 13:83-92.

Cremin, Jr., J.D., D.B. Faulkner, N.R. Merchen, G.C. Fahey, Jr., R.L. Fernando, and C.L. Willms. 1991. Digestion criteria in nursing beef calves supplemented with limited levels of protein and energy. J. Anim. Sci. 69:1322-1331.

Decker, A.M., R.W. Hemken, J.R. Miller, N.A. Clark, and A.V. Okorie. 1971. Nitrogen fertilization, harvest management, and utilization of Midland bermudagrass (*Cynodon dactylon* L.). p. 71-74. Maryland Agric. Exp. Stn. Bull. 487, College Park.

DelCurto, T., R.C. Cochran, D.L. Harmon, A.A. Beharka, K.A. Jacques, G. Towne, and E.W. Vanzant. 1990a. Supplementation of dormant tallgrass-prairie forage: I. Influence of varying supplemental protein and(or) energy levels on forage utilization characteristics of beef steers in confinement. J. Anim. Sci. 68:515-531.

DelCurto, T., R.C. Cochran, L.R. Corah, A.A. Beharka, E.S. Vanzant, and D.E. Johnson. 1990b. Supplementation of dormant tallgrass-prairie forage: II. Performance and forage utilization characteristics in grazing beef cattle receiving supplements of different protein concentrations. J. Anim. Sci. 68:532-542.

DelCurto, T., R.C. Cochran, T.G. Nagaraja, L.R.Corah, A.A. Beharka, and E.S. Vanzant. 1990c. Comparison of soybean meal/sorghum grain, alfalfa hay and dehydrated alfalfa pellets as supplemental protein sources for beef cattle consuming dormant tallgrass-prairie forage. J. Anim. Sci. 68:2901-2915.

DePeters, E. J., and N. E. Smith. 1986. Forage quality and concentrate for cows in early lactation. J. Dairy Sci. 69:135-141.

DePeters, E.J., J.F. Medrano, and D.L. Bath. 1989. A nutritional evaluation of mixed winter wheat cereals with vetch utilized as silage or hay. J. Dairy Sci. 72:3247-3254.

Donaldson, R.S., M.A. McCann, H.E. Amos, and C.S. Hoveland. 1991. Protein and fiber digestion by steers grazing winter annuals and supplemented with ruminal escape protein. J. Anim. Sci. 69:3067-3071.

Doyle, P.T. 1987. Supplements other than forages. p. 429-464. *In* J.B. Hacker and J.H. Ternouth (ed.) The nutrition of herbivores. Academic Press, New York.

Doyle, P.T., H. Dove, M. Freer, F.J. Hart, R.M. Dixon, and A.R. Egan. 1988. Effects of a concentrate supplement on the intake and digestion of a low-quality forage by lambs. J. Agric. Sci. (Camb.) 111:503-511.

Duble, R.L., J.A. Lancaster, and E.C. Holt. 1972. Forage characteristics limiting animal performance on warm-season perennial grasses. Agron. J. 63:795-798.

Egan, A.R. 1977. Nutritional status and intake regulation in sheep. VIII. Relationships between voluntary intake of herbage by sheep and the protein/energy ratio in the digestion products. J. Agric. Res. 28:907-915.

Ellis, W.C. 1990. Nutritional principles involved in supplementing grazing cattle. Southern Pasture and Forage Crop Improvement Conf., p. 43-48.

Erfle, J.D., S. Mahadevan, and F.D. Sauer. 1978. Urea as a supplemental nitrogen source for lactating cows. Can. J. Anim. Sci. 58:77-86.

Fick, K.R., C.B. Ammerman, C.H. McGowan, P.E. Loggins, and J.A. Cornell. 1973. Influence of supplemental energy and biuret nitrogen on the utilization of low quality roughage by sheep. J. Anim. Sci. 36:137-143.

Forero, O., F.N. Owens, and K.S. Lusby. 1980. Evaluation of slow-release urea for winter supplementation of lactating range cows. J. Anim. Sci. 50:532-538.

Forcherio, J.C., J.A. Paterson, and M.S. Kerley. 1992. Effect of source of supplemental energy and level of undegradable protein on forage intake and performance of cow-calf pairs grazing endophyte-infected tall fescue pasture. J. Anim. Sci. 70 (Suppl. 1):188 (Abstr.).

Freeman, A.S., M.L. Galyean, and J.S. Caton. 1992. Effects of supplemental protein percentage and feeding level on intake, ruminal fermentation, and digesta passage in beef steers fed prairie hay. J. Anim. Sci. 70:1562-1572.

Freer, M. 1981. The control of food intake by grazing animals. p. 105-124. *In* F.W. Morley (ed.) Grazing animals. Elsevier Scientific Publ. Co., New York.

Fribourg, H.A., J.B. McLaren, K.M. Barth, J.M. Bryan, and J.T. Connell. 1979. Productivity and quality of bermudagrass and orchardgrass-ladino clover pastures for beef steers. Agron. J. 71:315-320.

Grant, R.J., and D.R. Mertens. 1992. Influence of buffer pH and raw corn starch addition on in vitro fiber digestion kinetics. J. Dairy Sci. 75:2762-2768.

Grieve, D.G., J.B. Stone, G.K. Macleod, and R.A. Curtis. 1980. All silage forage programs for dairy cattle. II. Performance through three lactations. J. Dairy Sci. 63:594-600.

Guerrero, J.N., B.E. Conrad, E.C. Holt, and H. Wu. 1984. Prediction of animal performance on bermudagrass pasture from available forage. Agron. J. 76:577-580.

Hall, K.L., A.L. Goetsch, K.M. Landis, L.A. Forster, Jr., and A.C. Brake. 1990. Effects of a fat and ground maize supplement on feed intake and digestion by cattle consuming bermudagrass hay (*Cynodon dactylon*). Anim. Feed Sci. Technol. 30:275-288.

Hannah, S.M., M.T. Rhodes, J.A. Paterson, M.S. Kerley, J.E. Williams, and K.E. Turner. 1989. Influence of energy supplementation on forage intake, digestibility and grazing time by cattle grazing tall fescue. Nutr. Rep. Int. 40:1153-1157.

Hannah, S.M., R.C. Cochran, E.S. Vanzant, and D.L. Harmon. 1991. Influence of protein supplementation on site and extent of digestion, forage intake, and nutrient flow characteristics in steers consuming dormant bluestem-range forage. J. Anim. Sci. 69:2624-2633.

Hedrick, H.B., J.A. Paterson, A.G. Matches, J.D. Thomas, N.G. Krouse, R.E. Morrow, and W.C. Stringer. 1982. The production, characteristics and utilization of forage-fed beef. Univ. of Missouri Agric. Exp. Stn. Bull. 1043.

Hemken, R.W., and J.W. Vandersall. 1967. Feasibility of an all silage forage program. J. Dairy Sci. 50:417-421.

Hennessy, D.W., P.J. Williamson, J.V. Nolan, T.J. Kempton, and R.A. Leng. 1983. The roles of energy- or protein-rich supplements in the subtropics for young cattle consuming basal diets that are low in digestible energy or protein. J. Agric. Sci. (Camb.) 100:657-666.

Hennessy, D.W., and P.J. Williamson. 1990. Feed intake and live weight of cattle on subtropical native pasture hays. I. The effect of urea. Aust. J. Agric. Res. 41:1169-1177.

Henning, P.A., Y. Van Der Linden, M.E. Mattheyse, W.K. Nauhaus, H.M. Schwartz, and F.M.C. Gilchrist. 1980. Factors affecting the intake and digestion of roughage by sheep fed maize straw supplemented with maize grain. J. Agric. Sci. (Camb.) 94:565-573.

Hibbs, J.W., H.R. Conrad, and R.W. Van Keuren. 1979. Effects of different forages and methods of feeding concentrate on dry matter intake and milk production in dairy cows fed complete diets. Research Circular 248, Ohio Agricultural Research and Development Center, Wooster.

High, Jr., J.W., L.M. Safley, O.H. Long, H.R. Duncan, and T.W. High, Jr. 1965. Combinations of orchardgrass, fescue, and ladino clover pastures for producing yearling steers. Tennessee Agric. Exp. Stn. Bull. 388.

Hodge, R.W., and B. Bogdanovic. 1983. Feeding hay supplemented with peas or low protein oats to crossbred lambs born in the spring. Aust. J. Exp. Agric. Anim. Husb. 23:19-23.

Holloway, J.W., W.T. Butts, Jr., J.D. Beaty, J.T. Hopper, and N.S. Hall. 1979. Forage intake and performance of lactating beef cows grazing high or low quality pastures. J. Anim. Sci. 48:692-700.

Holter, J.B., W.E. Urban, W.S. Kennett, and C.J. Sniffen. 1973. Corn silage with and without grass hay for lactating dairy cows. J. Dairy Sci. 56:915-922.

Holter, J.B., W. Johns III, and W.E. Urban. 1975. No associative effects between corn and haycrop silages fed to lactating cows. J. Dairy Sci. 58:1865-1870.

Horn, F.P., J.P. Telford, J.E. McCroskey, D.F. Stephens, J.V. Whiteman, and Robert Totusek. 1979. Relationship of animal performance and dry matter intake to chemical constituents of grazed forage. J. Anim. Sci. 49:1051-1058.

Horn, G.W., and F.T. McCollum. 1987. Energy supplementation of grazing ruminants. p. 125-130. *In* M. B. Judkins, D.C. Clanton, M.K. Petersen, and J.D. Wallace (ed.) Proc. Grazing Livestock Nutr. Conf., Univ. of Wyoming, Laramie.

Horn, G.W. 1990. Energy and protein supplementation of growing cattle on wheat pasture. Southern Pasture and Forage Crop Improvement Conf., p. 54-59.

Hoveland, C. 1960. Bermudagrass for forage. Alabama Agric. Exp. Stn. Bull. 328.

Huber, J.T., H.F. Bucholtz, and R.L. Boman. 1980. Ammonia versus urea-treated silages with varying urea in concentrate. J. Dairy Sci. 63:76-81.

Huber, J.T., R.E. Lichtenwalner, and J.W. Thomas. 1973. Factors affecting response of lactating cows to ammonia-treated corn silages. J. Dairy Sci. 56:1283-1289.

Huber, J. T., and O. P. Santana. 1972. Ammonia-treated corn silage for dairy cattle. J. Dairy Sci. 55:489-493.

Huber, J.T., and J.W. Thomas. 1971. Urea-treated corn silage in low protein rations for lactating cows. J. Dairy Sci. 54:224-230.

Hunt, C.W., T.J. Klopfenstein, and R.A. Britton. 1988. Effect of alfalfa addition to wheat straw diets on intake and digestion in beef cattle. Nutr. Rep. Int. 38:1249-1257.

Hunter, R.A., and B.D. Siebert. 1980. The utilization of spear grass (*Heteropogon contortus*). IV. The nature and flow of digesta in cattle fed spear grass alone and with protein or nitrogen or sulfur. Aust. J. Agric. Res. 31:1037-1047.

Jones, M.B. 1967. Forage and nitrogen production by subclover-grass and nitrogen-fertilized California grassland. Agron. J. 59:209-214.

Jones, M.B., V.V. Rendig, D.T. Torell, and T.S. Inouye. 1982. Forage quality for sheep and chemical composition associated with sulfur fertilization on a sulfur deficient site. Agron. J. 74:775-780.

Jung, G.A., C.F. Gross, R.E. Kocher, L.R. Burdett, and W.C. Sharp. 1978. Warm-season range grasses extend beef cattle forage. Pennsylvania Agric. Exp. Stn. Sci. Agric. 25:6.

Kaiser, R.M., and D.K. Combs. 1989. Utilization of three maturities of alfalfa by dairy cows fed rations that contain similar concentrations of fiber. J. Dairy Sci. 72:2301-2307.

Kartchner, R.J. 1980. Effects of protein and energy supplementation of cows grazing native winter range forage on intake and digestibility. J. Anim. Sci. 51:432-438.

Kleiber, M. 1959. Symposium on forage evaluation: II. Progress in feed evaluation. Agron. J. 51:217-219.

Koch, D.W., J.B. Holter, D.M. Coates, and J.R. Mitchell. 1987. Animal evaluation of forages following several methods of field renovation. Agron. J. 79:1044-1048.

Krueger, C.R., and D.C. Curtis. 1979. Evaluation of big bluestem, indiangrass, sideoats grama and switchgrass pastures with yearling steers. Agron. J. 71:480-484.

Krysl, L.J., M.E. Branine, A.U. Cheema, M.A. Funk, and M.L. Galyean. 1989. Influence of soybean meal and sorghum grain supplementation on intake, digesta kinetics, ruminal fermentation, site and extent of digestion and microbial protein synthesis in beef steers grazing blue grama rangeland. J. Anim. Sci. 67:3040-3051.

Lagasse, M.P., A.L. Goetsch, K.M. Landis, and L.A. Forster, Jr. 1990. Effects of supplemental alfalfa hay on feed intake and digestion by Holstein steers consuming high-quality bermudagrass or orchardgrass hay. J. Anim. Sci. 68:2839-2847.

Lake, R.P., D.C. Clanton, and J.F. Karn. 1974. Intake, digestibility and nitrogen utilization of steers consuming irrigated pasture as influenced by limited energy supplementation. J. Anim. Sci. 38:1291-1297.

Lamb, C.S., and J. Eadie. 1979. The effect of barley supplements on the voluntary intake and digestion of low quality roughages by sheep. J. Agric. Sci. (Camb.) 92:235-241.

Langlands, J.P. 1969. The feed intake of sheep supplemented with varying quantities of wheat while grazing pastures differing in herbage availability. Aust. J. Agric. Res. 20:919-924.

Larsen, H.J., and G.H. Tenpas. 1983. Red clover haylage vs. alfalfa haylage in the dairy cow diet. Univ. of Wisconsin, Marshfield Exp. Stat. Rep. No. MSH-521-6003-83-1.

Lee, G.J., D.W. Hennessy, P.J. Williamson, J.V. Nolan, T.J. Kempton, and R.A. Leng. 1985. Responses to protein meal supplements by lactating beef cattle given a low-quality pasture hay. Aust. J. Agric. Res. 36:729-741.

Lee, G.J., D.W. Hennessy, J.V. Nolan, and R.A. Leng. 1987. Responses to nitrogen and maize supplements by young cattle offered a low-quality pasture hay. Aust. J. Agric. Res. 38:195-207.

Lippke, H. 1980. Forage characteristics related to intake, digestibility and gain by ruminants. J. Anim. Sci. 50:952-961.

Llamas-Llamas, G., and D.K. Combs. 1990. Effect of alfalfa maturity on fiber utilization by high producing dairy cows. J. Dairy Sci. 73:1069-1080.

Lusby, K.S. 1990. Supplementation of cattle on rangeland. Proc. Southern Pasture and Forage Crop Improvement Conf., p. 64-71.

Martin, S.K., and C.A. Hibberd. 1990. Intake and digestibility of low-quality native grass hay by beef cows supplemented with graded levels of soybean hulls. J. Anim. Sci. 68:4319-4325.

Matches, A.G. 1979. Management of tall fescue. p. 171-199. *In* R.C. Buckner and L.P. Bush (ed.) Tall fescue, 1st edition. No. 20, Agronomy Series. ASA, CSSA, SSSA, Madison, WI.

McCollum, F.T., and M.L. Galyean. 1985. Influence of cottonseed meal supplementation on voluntary intake, rumen fermentation and rate of passage of prairie hay in beef steers. J. Anim. Sci. 60:570-577.

McLaren, J.B., R.J. Carlisle, and H.A. Fribourg. 1983. Bermudagrass, tall fescue, and orchardgrass pasture combinations with clover or N fertilization for grazing steers. I. Forage growth and consumption and animal performance. Agron. J. 75:587-592.

Merrill, W.G., and S.T. Slack. 1965. Feeding value of perennial forages for dairy cows - A review. Animal Science Mimeograph Series No. 3, Department of Animal Science, New York State College of Agriculture, Cornell University, Ithaca, NY.

Mertens, D.A. 1987. Predicting intake and digestibility using mathematical models of rumen function. J. Anim. Sci. 64:1548-1558.

Messman, M.A., W.P. Weiss, P.R. Henderburg, and W.L. Shockey. 1992. Evaluation of pearl millet and field plot silages for midlactating dairy cows. J. Dairy Sci. 75:2769-2775.

Minson, D.J., and R. Milford. 1967. The voluntary intake and digestibility of diets containing different proportions of legume and mature Pangola grass. Aust.J. Exp. Agric. Anim. Husb. 7:546-551.

Minson, D.J. 1982. Effects of chemical and physical composition of herbage eaten upon intake. p 167-182. In J. B. Hacker (ed.) Nutritional limits to animal production from pastures. Commonwealth Agricultural Bureaux, Slough, UK.

Moe, P.W., and H.F. Tyrrell. 1975. Efficiency of conversion of digested energy to milk. J. Dairy Sci. 58:602-610.

Moe, P.W., W.P. Flatt, and H.F. Tyrrell. 1972. Net energy value of feeds for lactation. J. Dairy Sci. 55:945-958.

Moore, L.A. 1964. Symposium on forage utilization: Nutritive value of forage as affected by physical form. Part I. General principles involved with ruminants and effect of feeding pelleted or wafered forage to dairy cattle. J. Anim. Sci. 23:230-238.

Morgan, E.B., B.D. Nelson, T.F. Brown, M.E. McCormick, and A. Saxton. 1987. Comparison of alfalfa, ryegrass, and triticale haylages and corn and sorghum silages with lactating Holstein cows. Annu. Prog. Rep., Southeast Louisiana Agric. Exp. Stn., Franlinton. p. 95-106.

Morrison, F.B. 1950. Feeds and feeding. 21st Ed. Morrison Publ. Co., Ithaca, NY.

Mulholland, J.G., J.B. Coombe, and W.R. McManus. 1976. Effect of starch on the utilization by sheep of a straw diet supplemented with urea and minerals. Aust. J. Agric. Res. 27:139-153.

Nelson, W.F., and L.D. Satter. 1990. Effect of stage of maturity and method of preservation of alfalfa on production by lactating dairy cows. J. Dairy Sci. 73:1800-1811.

Nelson, W.F., and L.D. Satter. 1992. Impact of alfalfa maturity and preservation method on milk production by cows in early lactation. J. Dairy Sci. 75:1562-1570.

NRC. 1984. Nutrient requirements of beef cattle, 6th revised edition. p. 77-82. National Academy Press, Washington, DC.

Orr, R.J., and T.T. Treacher. 1989. The effect of concentrate level on the intake of grass silages by ewes in late pregnancy. Anim. Prod. 48:109-120.

Orskov, E.R., and C. Fraser. 1975. The effects of processing of barley-based supplements on rumen pH, rate of digestion and voluntary intake of dried grass in sheep. Br. J. Nutr. 4:493-500.

Owens, F.G. 1967. Factors affecting nutritive value of corn and sorghum silage. J. Dairy Sci. 50:404-416.

Penning, P.D., R.J. Orr, and T.T. Treacher. 1988. Responses of lactating ewes, offered fresh herbage indoors and when grazing, to supplements containing differing protein concentrations. Anim. Prod. 46:403-415.

Petersen, R.G., H.L. Lucas, and G.O. Mott. 1965. Relationship between rate of stocking and per animal and per acre performance on pasture. Agron. J. 57:27-30.

Petersen, M.K. 1987. Nitrogen supplementation of grazing livestock. p. 115-121. *In* M.B. Judkins, D.C. Clanton, M.K. Petersen, and J.D. Wallace (ed.) Proc. Grazing Livestock Nutrition Conf., Univ. of Wyoming, Laramie.

Pordomingo, A.J., J.D. Wallace, A.S. Freeman, and M.L. Galyean. 1991. Supplemental corn grain for steers grazing native rangeland during summer. J. Anim. Sci. 69:1678-1687.

Puoli, J.R., G.A. Jung, and R.L. Reid. 1991. Effects of nitrogen and sulfur on digestion and nutritive quality of warm-season grass hays for cattle and sheep. J. Anim. Sci. 69:843-852.

Raguse, C.A., K.L. Taggard, J.L. Hull, J.G. Morris, M.R. George, and L. C. Larsen. 1988. Conversion of fertilized annual range forage to beef cattle live weight gain. Agron. J. 80:591-596.

Raymond, W.F. 1969. The nutritive value of forage crops. Adv. Agron. 21:1-107.

Reid, R.L., G.A. Jung, and S.J. Murray. 1966. Nitrogen fertilization in relation to the palatability and nutritive value of orchardgrass. J. Anim. Sci. 25:636-645.

Reid, J.T., W.K. Kennedy, K.L. Turk, S.T. Slack, G.W. Trimberger, and R.P. Murphy. 1959. Symposium on forage evaluation: What is forage quality from the animal standpoint? Agron. J. 51:213-216.

Rountree, B.H., A.G. Matches, and F.A. Martz. 1974. Season too long for your grass pasture? Crop Soils 26:7-10.

Sanson, D.W., D.C. Clanton, and I.G. Rush. 1990. Intake and digestion of low-quality meadow hay by steers and performance of cows on native range when fed protein supplements containing various levels of corn. J. Anim. Sci. 68:595-603.

Santini, F.J., A.R. Hardie, N.A. Jorgensen, and M.F. Finner. 1983. Proposed use of adjusted take based on forage particle length for calculation of roughage indexes. J. Dairy Sci. 66:811-820.

Sauer, F.D., J.D. Erfle, S. Mahadevan, and J.R. Lessard. 1979. Urea in corn silage as a supplemental nitrogen source for lactating cows. Can. J. Anim. Sci. 59:403-410.

Shaver, R.D., A.J. Nytes, L.D. Satter, and N.A. Jorgensen. 1986. Influence of level of feed intake and forage physical form on digestion and passage of prebloom alfalfa hay in dairy cows. J. Dairy Sci. 69:1545-1559.

Shaver, R.D., L.D. Satter, and N.A. Jorgensen. 1988. Impact of forage fiber content on digestion and digesta passage in lactating cows. J. Dairy Sci. 71:1556-1565.

Snyder, T.J., C.E. Polan, and C.N. Miller. 1979. Effects of dry matter content of barley silage on nutrient preservation and animal response. J. Dairy Sci. 62:297-303.

Spears, J.W., J.C. Burns, and Patricia A. Hatch. 1985. Sulfur fertilization of cool season grasses and effect on utilization of minerals, nitrogen and fiber by steers. J. Dairy Sci. 68:347-355.

Spooner, A.E., and W.S. McGuire. 1979. Tall fescue pasture for growing and finishing animals. p. 233-246. *In* R.C. Buckner and L.P. Bush (ed.) Tall Fescue (1st ed.) MO-20 Agronomy Series, ASA, CSSA, SSSA, Madison, WI.

Stakelum, G. 1986. Herbage intake of grazing dairy cows: 1. Effect of autumn supplementation with concentrates and herbage allowance on herbage intake. Irish J. Agric. Res. 25:31-40.

Steacy, G.M., D.A. Christensen, M.I. Cochran, and G.M.J. Horton. 1983. An evaluation of three stages of maturity of hay fed with two concentrate levels for lactating dairy cows. Can. J. Anim. Sci. 63:623-629.

Strahan, S.R., R.W. Hemken, J.A. Jackson, Jr., R.C. Buckner, L.P. Bush, and M.R. Siegel. 1987. Performance of lactating dairy cows fed tall fescue forage. J. Dairy Sci. 70:1228-1234.

Sudweeks, E.M., L.O. Ely, D.R. Mertens, and L.R. Sisk. 1981. Assessing minimum amounts and form of roughages in ruminant diets: Roughage value index system. J. Anim. Sci. 53:1406-1411.

Sunvold, G.D., R.C. Cochran, and E.S. Vanzant. 1991. Evaluation of wheat middlings as a supplement for beef cattle consuming dormant bluestem-range forage. J. Anim. Sci. 69:3044-3054.

Teller, E., M. Vanbelle, M. Foulon, G. Collignon, and B. Matatu. 1992. Nitrogen metabolism in rumen and whole digestive tract of lactating dairy cows fed grass silage. J. Dairy Sci. 75:1296-1304.

Thomas, C., K. Aston, and S.R. Daley. 1985. Milk production from silage. 3. A comparison of red clover with grass silage. Anim. Prod. 41:23-31.

Thomas, C., S.R. Daley, K. Aston, and P.M. Hughes. 1981. Milk production from silage. 2. The influence of the digestibility of silage made from the primary growth of perennial ryegrass. Anim. Prod. 33:7-12.

Thomas, J.W., L.D. Brown, R.S. Emery, E.J. Benne, and J.T. Huber. 1969. Comparisons between alfalfa silage and hay. J. Dairy Sci. 52:195-204.

Thomas, J.W., L.D. Brown, and R.S. Emery. 1970. Corn silage compared to alfalfa hay for milking cows when fed various levels of grain. J. Dairy Sci. 53:342-350.

Trimberger, G.W., H.F. Tyrrell, D.A. Morrow, J.T. Reid, M.J. Wright, W.F. Shipe, W.G. Merrill, J.K. Loosli, C.E. Coppock, L.A. Moore, and C.H. Gordon. 1972. Effects of liberal concentrate feeding on health, reproductive efficiency, economy of milk production and other related responses of the dairy cow. New York Food and Life Sci. Bull. No. 8, Cornell University, Ithaca, NY.

Turner, M.E. 1983. The forage intake of supplemented cows grazing winter foothill rangelands of Montana. M.S. Thesis. Montana State Univ., Bozeman.

Tyrrell, H.F., and P.W. Moe. 1975. Effect of intake on digestive efficiency. J. Dairy Sci. 58:1151-1163.

Van Soest, P.J. 1982. Nutritional ecology of the ruminant. Durham and Downey Inc., Portland, OR.

Van Soest, P.J. 1973. Revised estimates of the net energy values of feeds. Proc. Cornell Nutr. Conf. Feed Manuf., Syracuse, NY. p. 11-23.

Van Soest, P.J., D.G. Fox, D.R. Mertens, and C.J. Sniffen. 1984. Discounts for net energy and protein-fourth revision. Proc. Cornell Nutr. Conf. Feed Manuf., Syracuse, NY. p. 121-136.

Vanzant, E.S., R.C. Cochran, K.A. Jacques, A.A. Beharka, T. Delcurto, and T.B. Avery. 1990. Influence of level of supplementation and type of grain in supplements on intake and utilization of harvested, early-growing-season, bluestem-range forage by beef steers. J. Anim. Sci. 68:1457-1468.

Veira, D.M., J.R. Seoane, and J.G. Proulx. 1991. Utilization of grass silage by growing cattle: Effect of a supplement containing ruminally protected amino acids. J. Anim. Sci. 69:4703-4709.

Vicini, J.L., E.C. Prigge, W.B. Bryan, and G.A. Varga. 1982a. Influence of forage species and creep grazing on a cow-calf system. I. Intake and digestibility of forages. J. Anim. Sci. 55:752-758.

Vicini, J.L., E.C. Prigge, W.B. Bryan, and G.A. Varga. 1982b. Influence of forage species and creep grazing on a cow-calf system. II. Calf production. J. Anim. Sci. 55:759-764.

Vinet, C., R. Bouchard, and G.J. St-Laurent. 1980. Effects of stage of maturity of timothy hay and concentrate supplementation on performance of lactating dairy cows. Can. J. Anim. Sci. 60:511-521.

Wagner, D. 1989. Strategies to improve performance of grazing cattle through protein supplementation. Agri-Practice 10:27-32.

Waldo, D.R. 1973. Effects of physical form of feed on digestibility. Proc. Maryland Nutr. Conf. Feed Manuf., College Park. p. 6-12.

Weiss, W.P., and W.L. Shockey. 1991. Value of orchardgrass and alfalfa silages fed with varying amounts of concentrates to dairy cows. J. Dairy Sci. 74:1933-1943.

Welch, J.G., and A.M. Smith. 1970. Forage quality and rumination in cattle. J. Dairy Sci. 53:797-800.

Williams, C.B., P.A. Oltenaeu, and C.J. Sniffen. 1989. Application of neutral detergent fiber in modeling feed intake, lactation response and body weight changes in dairy cattle. J. Dairy Sci. 72:652-663.

Woodford, J.A., N.A. Jorgensen, and G.P. Barrington. 1986. Impact of dietary fiber and physical form on performance of lactating cows. J. Dairy Sci. 69:1035-1047.

Woodford, S.T., and M.R. Murphy. 1988. Effect of forage physical form on chewing activity, dry matter intake and rumen function of dairy cows in early lactation. J. Dairy Sci. 71:674-686.

Zorrilla-Rios, J., G.W. Horn, W.A. Phillips, and R.W. McNew. 1991. Energy and protein supplementation of ammoniated wheat straw diets for growing steers. J. Anim. Sci. 69:1809-1819.

CHAPTER 3

PLANT FACTORS AFFECTING FORAGE QUALITY

C. J. Nelson and L. E. Moser

INTRODUCTION

The plant or, more generally, the mixture of plants growing in a field or ecosystem determines forage quality, depending on the growing and harvesting conditions. Although certain aspects of quality can be modified after harvest (Chapter 23), the plant species, growth stage, and condition at harvest generally dictate quality of the forage. A wide range of plants are used for forages. Each plant has a unique morphology and physiology which gives it specific adaptation, growth, and forage quality features. Although some generalizations can be made, it is extremely important to recognize the differences in quality that exist among groups of plants, individual species, and even cultivars. Further, these differences interact with stage of growth and the environment.

When communicating animal results from forage-based experiments, researchers should describe the forage species, cultivar, stage of maturity, and environmental conditions during growth (Chapter 4), and characterize the forage with appropriate quality tests. Animal performance will not be a direct function of plant factors which affect nutritive value if selectivity occurs, or if the amount of forage offered is not sufficient to maximize voluntary intake. In those cases, animal performance involves both the nutritive value of the selected material and the rate of voluntary intake. In addition to nutritional value, plant factors affecting selectivity and voluntary intake are covered in more detail in Chapters 11, 12, and 13.

In this chapter, our objective is to review the growth properties and morphological characteristics of plants used for forages in the broad sense, and to consider how physiological processes, anatomy, and morphology interact with biotic and abiotic factors to alter species distribution, growth rates, season of production, and forage quality. Forage quality will be considered mainly in terms of plant factors affecting availability and quality of the forage produced, emphasizing digestibility, with less detail on intake even though the two are often closely related (Chapter 13). Animal performance adds another level of complexity and is covered in detail in later chapters. With some species, and in certain environments, forages accumulate chemicals or antiquality factors that override desirable attributes associated with anatomical or morphological characters of the plant. These will be introduced, but dealt with more specifically in Chapters 8 and 9.

PLANT ADAPTATION AFFECTS FORAGE AVAILABILITY

More than any other group of agricultural plants, forages are distributed widely over the earth, but the forage plant resource, being comprised of several hundred species, sub-species, varieties, and ecotypes of grasses, legumes, forbs, and sedges, varies from location to location. The diversity is extremely large, allowing numerous species to be adapted to any one site or region. Further, forage

---

C. J. Nelson, Dep. of Agronomy, Univ. of Missouri, Columbia, MO 65211; L. E. Moser, Dep. of Agronomy, Univ. of Nebraska, Lincoln, NE 68506.

species differ markedly in their season of production, adaptation to stress environments, and ability to regrow and persist after grazing and cutting. Generally, decisions are made on what species to plant relative to adaptation and production, then the species or mixture of species is managed within that constraint to maximize forage quality.

Each individual species from within a plant type has its own set of adaptation features. For example, perennial ryegrass (*Lolium perenne* L.) is renowned worldwide for its high forage quality, but is adapted best to the northeast or northwest U.S., areas where temperatures are mild and rainfall is abundant during the growing season. In contrast, tall fescue (*Festuca arundinacea* Schreb.) also grows in similar areas, yet may not be used extensively because it is of lower quality. But tall fescue also is adapted to much harsher climates of Missouri and Kentucky in terms of drought and high summer temperatures than perennial ryegrass, and in these environments the higher quality grass is no longer a popular alternative. Similarly, tall fescue adaptation overlaps that of bermudagrass [*Cynodon dactylon* (L.) Pers.] in warm climates of the south, such as Alabama, Georgia, and Mississippi, even more so if tall fescue is infected with the endophytic fungus (*Acremonium coenophialum* Morgan-Jones and Gams) which improves its stress resistance. In addition, bermudagrass has the C_4 photosynthetic system making it better adapted than infected tall fescue to long, sustained periods of hot weather in the deep south provided there is sufficient rainfall.

Alternatively, C_4 prairie grasses such as switchgrass (*Panicum virgatum* L.) and big bluestem (*Andropogon gerardii* Vitman) overlap with tall fescue in the eastern regions of the central states of Kansas and Oklahoma. The C_4 photosynthetic system allows the tall, native, prairie grasses to be adapted to the much drier environments as one moves westward. As environmental conditions get even drier in the west, tall-grass species overlap with short-grass species that, in turn, are better adapted to low rainfall sites. Similarly, smooth bromegrass (*Bromus inermis* Leyss.) is popular in the northern Great Plains because of its higher nutritive value and greater winter hardiness than other cool-season species.

Collectively, these adaptation features among plants give a geographic mosaic of forage species that are utilized within the U.S. Often, many species are adapted similarly to specific sites within a given locality allowing management flexibility and alternatives in species selection. In turn, each species has its own unique growth and quality traits, creating in each microcosm a range of alternatives for the livestock producer who grows and uses his own forage, or for the producer who grows forages for sale.

REGIONAL DIFFERENCES AFFECT GRASSLAND PHILOSOPHIES

Nearly all cool-season forage species used in the eastern humid regions of the U.S. are introduced, except perhaps for reed canarygrass (*Phalaris arundinacea* L.) (Marten, 1985). Because only a limited germplasm was introduced originally, these species generally have a narrower range of adaptation in the new environment than in their native habitats where greater genetic diversity exists. Plant explorers, and now plant breeders and genetic engineers, are identifying and developing germplasm to extend the range of adaptation by improving resistance to environmental stress, especially to drought and cold, and to diseases and insects. To date, most emphasis on genetic improvement has been placed on introduced species.

Plants growing on the "edge of their range of adaptation" are more stressed environmentally than those growing in the "center", and thus are more vulnerable to additional stresses from biotic agents. In general, plants cultured in sub-optimal environments near the perimeter of the region of adaptation must be managed more conservatively to maintain stands and productivity, which may limit full exploitation of their potential for high forage quality. Many forage species used in the Great Plains and rangeland areas are native, but are being challenged ecologically and economically by introduced species, especially along edges of adaptation. In many environments, maintaining forage quality over a long grazing season involves use of both cool- and warm-season species.

The general philosophy of forage management also changes from region to region of the U.S., from more-intensive management systems in humid regions of the east and southeast that are based on monocultures or binary mixtures of introduced forage species, to extensively managed systems in the drier western states that are based on multiple species, usually native, growing in a diverse ecosystem. Thus, forage quality is equated with selecting and seeding desirable species and cultivars in the east, which are managed for productivity, persistence, and quality, expecting to rotate or renovate periodically. Conversely, in the west, emphasis is on ecosystem management and maintaining natural diversity, while enhancing the economic success of an introduced livestock industry.

The current international emphasis on sustainability of land resources and the environmental concerns of the public are rapidly causing agriculturalists in the east to think much more in terms of ecosystem management. Overall, management strategies are shifting from an emphasis on productivity to persistence of forage species. It is unknown whether managing more conservatively for persistence of desirable species, especially legumes, will be more or less compatible with producing high quality forage.

CLIMATE FACTORS AFFECTING FORAGES

In a global sense, forage quality is a result of species present, amount of forage available, and composition and texture of each species. The species present depend on their adaptation; thus, potential productivity, composition, and texture of the forage are limited by decisions made within the constraints of adapted species. Understanding the biological potential of a species at a given site will help evaluate its limitations in production, quality, and persistence, and give valuable insight into breeding and management strategies necessary to improve its performance and quality. The biological potential of adapted species depends on the climate.

Solar Radiation

Climatic factors alter natural plant distribution among a range of environments, indirectly altering forage quality through the physiological responses of plants which influence adaptation and forage availability. Solar radiation through photosynthesis is the primary driving force which sets the upper limit on productivity in a direct sense, but temperature and rainfall play major roles as modulators in determining the proportion of the potential productivity that is achieved at a given site (Snaydon, 1991). In addition, portions of the radiation spectrum modify the formative processes of growth, i.e. morphogenesis, which give the plant form (Casal et al., 1985; 1987). These morphogenic factors include tillering and branching, internode elongation, leaf expansion, and flowering of photoperiod-sensitive species (Briske, 1991). An understanding of these photo-morphogenic responses is just emerging.

Amount of radiation reaching crop canopies differs markedly among regions in the U.S., mostly due to latitude and differential cloud cover. On clear days, a maximum of about 2000 μmol photons $m^{-2} s^{-1}$ of photosynthetically active radiation at mid-day is quite similar throughout mid-latitudes including the U.S., but daylength and duration of the growing season have a marked effect on the total daily and annual accumulation (Figure 1). Actual length of the growing season is usually established by temperature or water stress. In some environments water stress limits growth at the time radiation is highest.

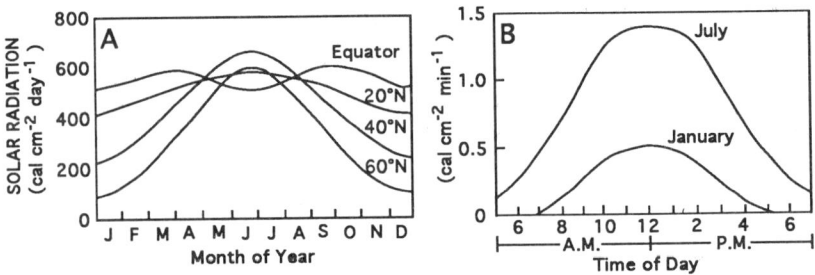

Figure 1. (A) Solar energy received at any given location is a function of latitude which alters solar angle and daylength. (B) Radiation per day at 40°N is also affected by season. Data are for cloudless days.

Photosynthesis of Grasses

Photosynthetically, C_3 and C_4 grasses respond similarly at low to moderate levels of radiation density. The major differences in photosynthesis are expressed at radiation densities above 1200 μmol m^{-2} s^{-1} (Figure 2), where the conversion efficiency of solar energy to fixed CO_2 for single leaves of C_4 species can be double (6%) that of C_3 species (<3%) (Cooper, 1970). Dry matter production of C_4 canopies, however, is not double that of C_3 canopies because most leaves in the lower canopy are shaded (Snaydon, 1991), and only a few leaves at the top of the canopy can reach the photosynthetic potential of the C_4 species (Nelson, 1988). Shaded leaves often operate at radiation densities of less than 1000 μmol m^{-2} s^{-1}, well below light saturation, where differences between C_3 and C_4 species are small. In addition, except for a few hours around solar noon, even the upper leaves of C_3 canopies are not light-saturated. Thus, under typical conditions the average conversion efficiency of 5 to 6% for leaves in the C_3 crop canopy is higher than that of a single leaf at full sun. Typical efficiencies for canopies of C_4 species may be slightly higher. Thus, with appropriate temperature and adequate water and minerals, maximum daily productivity of C_4 grasses in high radiation

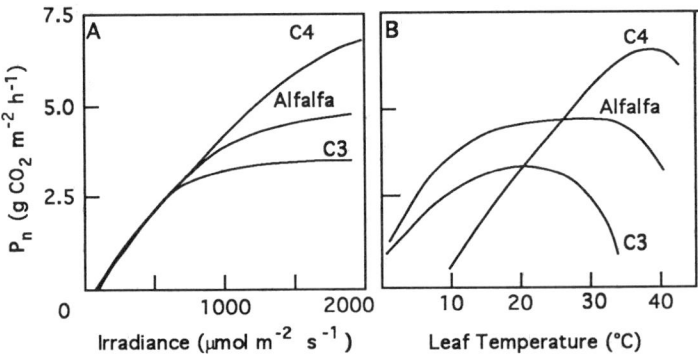

Figure 2. (A) At full sun (2000 μmol m^{-2} s^{-1} of photosynthetically active radiation) rates of net photosynthesis (Pn) for leaves of C_4 grasses are nearly double those for C_3 grasses, but they are similar at half-full sun. Legumes have C_3 photosynthesis and respond similar to C_3 grasses, except for alfalfa which has C_3 metabolism but an intermediate photosynthetic rate. (B) Temperature optima for photosynthesis are higher for C_4 than for C_3 grasses. Alfalfa has a broad temperature optimum which may help explain its adaptation to all states in the U.S. Adapted from several sources.

environments is often higher than for C_3 species (Cooper, 1970; Snaydon, 1991).

The C_4 grasses are more efficient in light conversion than C_3 species, but the major adaptation advantage is their better water-use efficiency (Brown and Simmons, 1979), drought resistance, and heat tolerance (Snaydon, 1991). These factors are related partly to the photosynthetic pathway, but also are associated with growth processes and resource acquisition. For example, nitrogen-use efficiency is higher for C_4 than for C_3 grasses (Brown, 1978), and higher temperature tolerance of C_4 plants is likely due to reduced photorespiration (Treharne and Nelson, 1975).

Since latitude is closely related to temperature, it is logical that C_4 grasses are more prevalent at lower latitudes and areas of lower rainfall (Figure 3). Thus, in terms of adaptation, C_4 species tend to be the base for forage and pasture systems in the southeast due to temperature, the Great Plains due to drought, and the southwest due to both temperature and drought.

While the division of physiological activity between mesophyll and bundle sheath cells in C_4 grasses is beneficial for CO_2 fixation and adaptation, the anatomical organization required for C_4 grasses causes forage quality to generally be lower than for C_3 grasses (Akin, 1989). In general, C_4 grasses have less mesophyll cell volume per unit leaf volume than do C_3 grasses, and the protein concentration in the mesophyll of C_4 grasses is considerably lower than in the mesophyll of C_3 grasses. Differences in the mesophyll content, which is nearly 100% digestible, contribute directly to differences in forage quality of leaves (Table 1). Further, digestion of bundle sheath cells of C_4 grasses is generally much slower than the adjacent mesophyll cells, which likely reduces the absolute digestibility of protein in the rumen to 50 to 60% of the total compared with 80% or more of the total in C_3 grasses. Currently, there are efforts to determine if some of the bundle sheath protein of C_4 grasses (largely Rubisco) may escape the rumen undigested, to be later made available to the animal in the lower gastrointestinal tract. That process would likely increase the efficiency of protein use.

Photosynthesis of Legumes and Other Dicots

All legumes appear to have the C_3 photosynthetic pathway. There are also other C_3 and C_4 dicot species that provide forage, including some C_4 weeds which

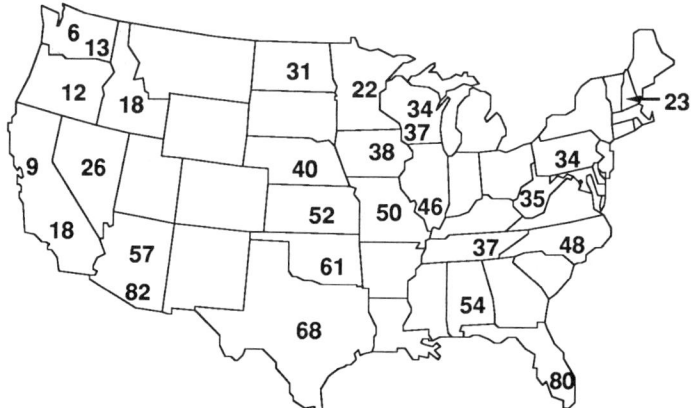

Figure 3. Percentage of C_4 species among grass floras for several areas of the U.S. No C_4 species were found on the arctic slopes of Alaska and in northern Manitoba. Twelve per cent of the species were C_4 in Alberta and the Gaspe Peninsula, Canada. Adapted from Teeri and Stowe (1976).

Table 1. Percentage of cross-sectional area comprised of different tissue types in leaf blades and stems of temperate (C_3) and tropical (C_4) grasses. (Adapted from Akin and Chesson, 1989).

Tissue	Temperate	Tropical
	------------% Area----------	
Leaf Blade	(n=6)	(n=9)
Total vascular	15±5	22±8
Lignified vascular	7±2	4±2
Parenchyma bundle sheath	6±2	15±7
Epidermis	23±6	35±10
Schlerenchyma	6±2	6±3
Mesophyll	57±5	38±9
Stem	(n=2)	(n=6)
Epidermis, xylem, and schlerenchyma ring	28±4	34±4
Parenchyma	52±3	55±6

have excellent forage quality (Marten and Anderson, 1975; Bosworth et al., 1980). These C_4 dicots, like C_4 grasses, have active photosynthesis and g.owth at high temperatures or under drought, and contribute to forage production in areas where adapted. In general, however, some weeds interfere with production of desirable species, making them undesirable in the pasture or field.

Even though legumes are C_3 plants, they can be separated into cool- and warm-season types based on their adaptation to temperature. For example, photosynthesis response of alfalfa (*Medicago sativa* L.) to light is intermediate between cool- and warm-season grasses (Figure 2), even though the leaf anatomy and enzymes involved with CO_2 fixation are clearly C_3. Alfalfa has a relatively high rate of photosynthesis among legumes and has a broad temperature range over which high rates are maintained, but the physiological or biochemical reasons for these responses are unknown. In that sense, alfalfa fits both a "cool-season" and "warm-season" legume category. Conversely, white clover (*Trifolium repens* L.) is adapted largely to cool, moist environments and typifies a cool-season legume. Sericea lespedeza [*Lespedeza cuneata* (Dum.-cours.) G. Don] and korean lespedeza [*Kummerowia stipulacea* (Maxim.) Makino] are adapted largely to warm climates and typify warm-season legumes.

Photosynthesis rates among cool- or warm-season legumes can differ genetically just as within the C_3 or C_4 groups of grasses. Rate of photosynthesis of sericea lespedeza is about half that of alfalfa (Brown and Radcliffe, 1986), although the upper temperature range and optimum (about 30°C) were similar to alfalfa, and very near that for several tropical legumes (Ludlow and Wilson, 1971). Photosynthesis of single leaves of birdsfoot trefoil (*Lotus corniculatus* L.) (Nelson and Smith, 1969), ladino clover (Wilfong et al., 1967), and red clover (*Trifolium pratense* L.) (Hesketh, 1963) is maximum at temperatures lower than for alfalfa, near 20-25°C. Their photosynthesis at sub-optimal temperatures is similar to cool-season forage grasses and is much higher than for true warm-season legumes.

Cool- and warm-season legumes also differ in nitrogen metabolism and transport forms from the nodules, warm-season legumes transport N as ureides whereas cool-season legumes transport amides. The response is thought to be related to carbohydrate conservation of ureide transporters allowing lower respiration and better adaptation to high temperatures. The significance of ureide metabolism on forage quality is unknown. Dark respiration is also a large component of the carbon budget and is involved in adaptation to temperature.

Growth Patterns, Radiation Interception, and Forage Quality

Legume forage plants tend to display their leaf area more horizontally than grasses, therefore reducing light penetration through the canopy (Figure 4). Similarly, the leaf area index (LAI), i.e. area of leaf blades per unit area of soil surface, required for maximizing radiation interception is lower for legumes, about 4-5, than for grasses, about 6-8. Thus, although quality would be higher, it is critical for maintenance of legume-grass mixtures that the legume not shade the grass excessively. In some cases forage quality is improved when plants develop in shade (Buxton and Casler, 1993), but not always (Allard et al., 1991).

Duration and density of radiation per day change markedly over the season in northern latitudes due to changes in solar angle. For example, radiation per day and potential productivity of cool-season forages are highest during June in the U.S. when days are longest (Figure 1). This corresponds well with productivity patterns of cool-season grasses in northern environments which have a long-day requirement for flowering. Typical growth curves of introduced cool-season grasses in the northern states are well known (Figure 5). As daily radiation increases during the growing season, the long-day plants are induced to flower, and stem growth contributes 50% or more of the total rate when it is active. Thus, the rapid growth of the reproductive culm occurs during long days of June, and before drought stress of July and August becomes a major limitation. But, the later vegetative growth of C_3 plants cannot fully utilize the high radiation of July and August, most likely because the vegetative canopy is changed markedly in architecture from that of the reproductive phase (Woledge, 1979). In southern environments, the higher temperature during spring causes earlier flowering of cool-season species, further shifting the time of maximal growth from the time of maximal radiation.

The transition from reproductive to vegetative growth involves initiating and developing new vegetative tillers from the reproductive culm, and then redeveloping a vegetative canopy during late spring and summer. Leaf elongation at the same temperature is slower on vegetative than reproductive tillers (Parsons and Robson, 1980). During summer and fall, the vegetative stem elongates very little on most cool-season grasses to separate the leaf blades spatially, and thus radiation does not penetrate these compact canopies very effectively (Figure 4). Further, photosynthetic rates of leaves on vegetative tillers during summer are

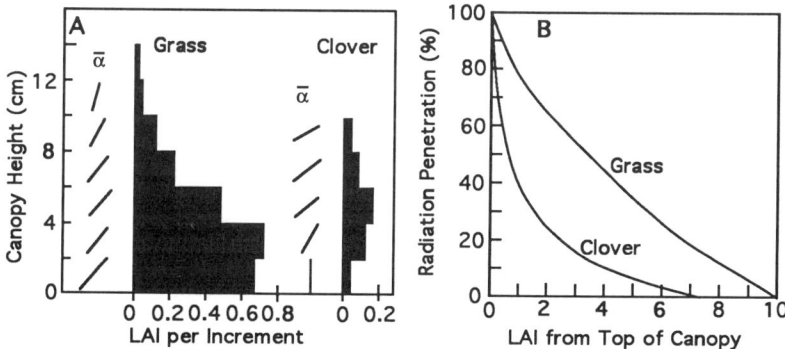

Figure 4. Perennial ryegrass and white clover differ in leaf angle ($\bar{\alpha}$) within the canopy (Panel A), with $\bar{\alpha}$ becoming more vertical at upper levels in grasses and more horizontal in clovers. This markedly alters the radiation penetration patterns in the canopy (Panel B), depending on whether the major component is grass or clover, and affects the LAI required for 90% light interception, i.e. 10% penetration. Adapted from Loomis and Williams (1969).

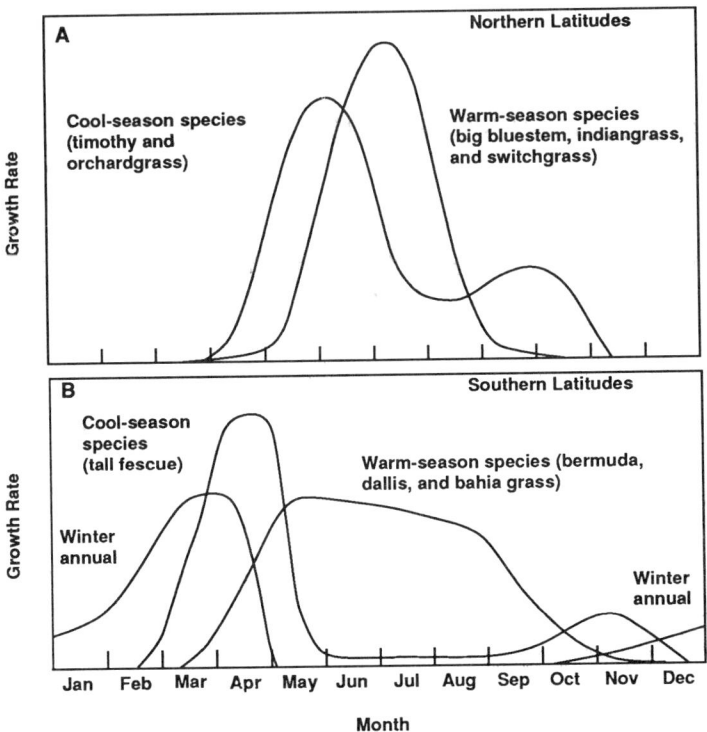

Figure 5. Growth distribution patterns for cool- and warm-season grasses in northern and southern latitudes of the U.S. Height and duration of responses will differ among species within each grouping. Note cool-season species in northern and southern latitudes differ in duration of low productivity during summer.

lower than those on reproductive tillers during spring. Leaves on compact vegetative tillers develop within a more-shaded environment which reduces the maximal photosynthetic rate up to 50% (Woledge, 1979), and respiration and photorespiration losses are higher due to high summer temperatures. Conversely, light effectively penetrates the extended reproductive canopy to illuminate leaves more uniformly in the higher efficiency range, and internode elongation is a strong sink for photosynthate which contributes directly to forage yield. For a few weeks during the early period of long days, the stems are elongating rapidly, the canopy shape is improved, and photosynthetic capacity and growth rate are maximized.

Temperature Effects on Forage Available

While solar radiation is a major factor determining potential productivity, temperature is the major factor that determines species diversity and the proportion of potential productivity that is actually achieved (McWilliam, 1978). Further, cold temperatures during winter tend to have a greater effect on species distribution than do high temperatures during summer. Generally, textbooks list optimal temperatures for growth, but the minimal and maximal temperatures are also critical because perennial forage plants are often growing outside of their optimal temperature range.

Cool-season or C_3 grasses have a minimal temperature for growth near 0°C, but growth is generally relatively slow below 5-7°C. Optimal average daily temperatures for growth are usually about 20°C; above 25°C, growth rate is reduced rapidly. There has been speculation that growth was carbohydrate-limited because the growth responses followed closely the temperature response of photosynthesis (Figure 2). Today, however, there is good evidence that photosynthesis is relatively faster than growth at low temperature, and carbohydrate (photosynthate) supply does not limit elongation growth of grass leaves (Nelson, 1992). Photosynthesis in low light can change density of the tissue, however, leading to lower specific leaf weight, i.e., mass per unit area (Schnyder and Nelson, 1989; Allard et al., 1991).

High temperatures and drought stress markedly reduce forage availability. Growth rates between the minimal and optimal temperatures are likely limited by metabolism associated with cell wall growth, but growth at temperatures above optimum is more likely to be limited by carbohydrate supply as photorespiration and dark respiration are much more active. For example, rates of dark respiration of tall fescue leaves are 40% higher at 30°C than at 27°C, and plants did not survive in moderate radiation at constant 34°C (Volenec and Nelson, 1984). Birdsfoot trefoil could not restore its carbohydrate in roots when regrown at 32/24°C (day/night) (Nelson and Smith, 1969). In addition, transpirational demand is generally higher at high temperatures leading to drought stress which reduces cell expansion and growth.

Temperature Effects on Forage Quality

Temperature also affects forage quality. Cell wall materials deposited at lower temperatures are less lignified and higher in digestibility. At high temperatures, lignin synthesis is preferentially increased, causing the forage produced to be lower in digestibility. Also, at low temperatures, storage carbohydrates tend to accumulate in leaf tissue of warm-season grasses and legumes, and especially in cool-season grasses. Some cool-season species may accumulate up to 30% of the leaf dry weight as total non-structural carbohydrate (TNC), mainly fructan, when plants are grown at 10 to 15°C (Table 2). The accumulated TNC is nearly 100% digestible which greatly improves forage quality. Fall growth of tall fescue is often higher in quality than at other times of the year, due largely to the higher concentration of TNC accumulated in the leaves (Brown et al., 1963). Warm-season species generally accumulate less TNC, especially at low temperatures, perhaps because they synthesize little fructan.

The environment in which a forage is growing interacts with the maturation process to affect forage quality. In controlled environment conditions, neutral detergent fiber (NDF) concentration increased in *Cynodon* spp., but decreased in

Table 2. Total non-structural carbohydrate (TNC) in leaves of 128 cool-season (C_3) and 57 warm-season (C_4) grasses grown at 10°/5°C (light/dark) or 25°/15°C. Components of TNC include fructan (Ftn), sucrose (Suc), glucose (Glu), fructose (Fru), and starch (Str). Data from Chatterton et al. (1989).

	TNC	Ftn	Suc	Glu	Fru	Str
Cool-season	mg kg^{-1}	---------------------mg kg^{-1}---------------------				
10°/5°C	312	115	58	29	24	86
25°/15°C	107	12	23	18	14	41
Warm-season						
10°/5°C	166	3	66	22	14	64
25°/15°C	92	4	20	13	8	47

Paspalum spp. with increasing temperature (Henderson and Robinson, 1982). Acid detergent fiber (ADF), cellulose, lignin, and silica concentrations all increased with increasing temperature, but hemicellulose concentrations decreased. Although ADF, cellulose, and silica levels decreased and lignin level increased with increasing radiation, temperature had a more profound effect on forage quality than did light flux. With crownvetch (*Coronilla varia* L.), Jung et al. (1981) found higher ADF and lignin levels with high radiation.

Since forage plant ontogeny is highly dependent on temperature, the accumulation of growing-degree days is closely related to forage quality. Buxton and Marten (1989) found declines in *in vitro* dry matter disappearance (IVDMD) and crude protein (CP) were closely related to cumulative growing degree days, actually more so than with the quantified morphological stage developed by Simon and Park (1983). Both NDF and ADF of alfalfa were better predicted from growing degree days for Iowa than from growth stages based on stem count or stem weight, but not for Texas (Sanderson, 1992b). In whole plant forage, growth stage based on weight accounted for 97% of the variation in NDF and IVDMD (Sanderson and Wedin, 1989).

Water Stress Effects on Availability and Quality

Water stress with minimal associated heat stress often improves forage quality (Vough and Marten, 1971). Wilson (1983) reported leaf and stem digestibility was at least as high for water-stressed warm-season grasses as for non-stressed plants. In many cases, water stress improved the forage quality due to an increased leaf:stem ratio and higher digestibility of both leaf and stem fractions. Halim et al. (1989) found water-stressed alfalfa was of higher quality (IVDMD) than alfalfa under normal water conditions, and the improved quality under stress was due to a delay in development. In their study, CP increased by 11% in the lower stem with water stress even though an adjustment was made for the change in plant maturity. Snaydon (1972) also reported that as the water supply to alfalfa decreased, the proportion of leaves increased and the IVDMD of both the leaf and stem fractions increased. However, phosphorus (P) level increased and whole-shoot nitrogen (N) content was unaffected.

Wilson (1983) found short periods of water stress had little effect on IVDMD of green panicgrass (*Panicum maximum* var. *trichoglume* Eyles), buffelgrass (*Cenchrus ciliaris* L.), and speargrass [*Heteropogon contortus* (L.) Beauv. ex Roem. & Schult.]. However, water-stressed plants had a slower decline in IVDMD with maturity, and had restricted stem development compared to unstressed plants. The effect of environmental factors on forage quality is discussed in detail in Chapter 4.

ONTOGENY EFFECTS ON AVAILABILITY AND QUALITY

Forage offered to animals is not a homogeneous product. It may be a single plant species or a mixture of plants. Additionally, it is comprised of leaf blades, leaf sheaths, stems and, in many cases, reproductive structures. The morphological development of forage plants interacting with environmental conditions, and each other in a mixture, will determine the amount and quality of each component. The ability livestock have to select among the components will determine diet quality.

Tiller size, architecture, and reproductive status of grasses are changed with ontogeny and affect the quality of forage harvested or on offer to grazing animals. Differentials in shoot ontogeny and quality evoke selective grazing by animals. Temperature is the major environmental factor that affects morphological development of tillers in the spring when water is available (Frank et al., 1985). The accumulation of heat units or growing degree days causes major growth and quality responses in forage plants. Since morphological development affects forage quality, quantifying the morphological stage of forage is necessary for descriptive and predictive purposes.

Fick and Mueller (1990) described a system for staging alfalfa shoots. Several systems have been developed for forage grasses (Simon and Park, 1983; Moore et al., 1991; Sanderson, 1992a). In contrast with most field crops, forage swards are comprised of a heterogeneous population of tillers which may be at different morphological stages and hence have different qualities. To characterize the population and predict forage quality of alfalfa, Fick and Mueller (1990) calculated weighted means based on shoot count (Mean Stage Count) or tiller weight (Mean Stage Weight). Moore et al. (1991) used a similar system to calculate weighted morphological indices for forage grasses. Detailed discussion of prediction of forage quality based on morphological indices is presented in Chapter 18.

Cool-season Grasses

Nearly all cool-season grasses have a vernalization or a daylength requirement to condition the tiller to flower. Vernalization requirements are species specific, but usually a period of 4 to 6 wk at temperatures less than 6°C is required during winter while growth is very slow or arrested. Once conditioned, either by vernalization, by short days during fall and winter, or both, flowering can be initiated during spring in response to lengthening of the day. During early floral initiation and development, the terminal stops initiating leaves and axillary buds as it differentiates into the inflorescence. As inflorescence development continues, the elongation of stem internodes pushes the terminal (inflorescence) to the top of the canopy to facilitate wind pollination.

During internode elongation and flowering, the terminal bud or growing point is elevated, making it subject to removal by grazing or cutting. Therefore, the plant must make a transition to new tillers (at least one must survive to replace the one that flowered) to regrow though the summer. The new tillers must also establish their own root system during the onset of potentially dry and hot weather and survive through the summer.

This growth strategy cannot take advantage of the highest radiation portion of the growing season. Except for a few grasses tested, most cool-season perennials do not have a true photoperiod-induced summer dormancy (Laude, 1953; 1964). Rather, the leaves grow slowly during summer in response to high temperature and lack of water (Horst and Nelson, 1979) in addition to having a lower photosynthetic potential of single leaves (Woledge, 1979). The leaf blades with short sheaths help shade the soil, however, and the vegetative canopy retains the terminal meristems near ground level where it is protected from close grazing.

These morphological features suggest survival is the natural summer strategy, and likely it is against drought, heat, and overgrazing. This also suggests that improving summer productivity of cool-season perennial grasses to exploit the high radiation environment may be difficult. Thus, developing complementary grazing systems using warm-season grasses for summer and cool-season forages for spring and fall has a sound physiological basis.

Many introduced cool-season grasses, such as tall fescue, still contain characteristics of their evolution in Mediterranean climates. In that environment, the vernalization requirement limits flowering to spring, and survival is enhanced by vegetative tillers that exist during summer when rainfall is very limited. Not all cool-season grasses react the same, however. For example, reed canarygrass and smooth bromegrass have extended vegetative culms during summer, perhaps because they evolved in more continental climates, and they also have rhizomes to help perennation. These morphological types would be able to use light more efficiently during summer due to the extended canopy, and the rhizomes are beneficial because the elevated terminal is more susceptible to close grazing. Reed canarygrass and smooth bromegrass are also relatively more drought tolerant than other cool-season perennial grasses, probably because they develop their root system to deeper depths (Gist and Smith, 1948).

Conversely, timothy *(Phleum pratense* L.) is unique in that it is a cool-season grass which has a very weak vernalization requirement. It likely evolved in more northern latitudes with a continental climate as it also is not very drought resistant. Since each regrowth can flower, timothy tends to be better adapted to infrequent cutting than to continuous stocking, and uses the high radiation during summer by producing reproductive canopies.

Warm-season Grasses

Native C_4 grasses are more drought hardy than the introduced C_3 grasses, and their growth patterns tend to match up closely with solar radiation in northern environments (compare Figures 1 and 5). In addition to having a more efficient photosynthetic system, big bluestem and indiangrass [*Sorghastrum nutans* (L.) Nash] generally flower 30 to 60 d later than C_3 grasses. Thus, their tall canopy, openly spaced leaves of reproductive growth, and heat and drought tolerance associated with C_4 photosynthesis allow them to exploit the high radiation of summer. In contrast, switchgrass flowers earlier than other C_4 perennials, but can flower more than once in central latitudes. Thus, it also can effectively use the high summer radiation.

Warm-season grasses are sensitive to close summer defoliation because they have very low concentrations of stored carbohydrate for regrowth (Forwood et al., 1988), and the culms elongate to elevate the terminal meristem and reproductive structure to alter regrowth patterns (Anderson et al., 1989). Further, the vegetative tillers that regrow after cutting or grazing must be protected from defoliation during early stages of growth.

In general, tropical grasses grown in the southern states do not have a vernalization requirement and usually are relatively insensitive to daylength or are quantitative short-day plants. Instead, they flower according to thermoperiod or plant size. Some tropical grasses, such as bermudagrass, are low growing with prostrate stolons and underground rhizomes that escape frequent and close grazing. Others, such as dallisgrass *(Paspalum dilatatum* Poir.), tiller profusely to withstand heavy grazing pressure. Thus, a range of morphological types exist within each adaptation category, giving the livestock producer and forage manager considerable flexibility in developing pasture and hay systems.

Legumes and Dicots

Legumes develop differently from grasses in that the terminal is always at the end of the elongated stem. The stem can be prostrate (stolon) with long petioles like ladino clover or vertical with short petioles like alfalfa or striate lespedeza *(Kummerowia striata* Thunb.). Legumes with prostrate stems generally are lower in productivity than those with upright stems, but can be grazed or cut closely because the terminal and axillary buds are protected near ground level, and more frequently because they retain leaf area near the soil surface. Conversely, animals generally graze the young tissue of upright legumes, which also removes the terminal bud, forcing them to regrow again from the base. Axillary buds along the stem of alfalfa tend to remain relatively dormant, generally forcing regrowth from the base, whereas other species will branch more profusely when the terminal is removed.

Growth patterns of red clover are intermediate between ladino clover and alfalfa, making it a popular legume for pasture in mixture with cool-season grasses. The stem grows upright, but leaves with long petioles are produced that extend well above the terminal, leaving the terminal low in the canopy until flowering is imminent. Thus, animals at low stocking rates graze the upper canopy, consume largely leaf blades and petioles of red clover, and leave the terminal intact to continue leaf production.

PLANT MATURITY AND FORAGE QUALITY

Plant maturity is the major factor affecting morphology and determining forage quality. The decline of forage quality with age results primarily from a decrease in the leaf:stem ratio (Ugherughe, 1986) and the decline in quality of the stem component. The proportion of leaf tissue for both big bluestem and switchgrass declines with maturity. At an immature stage in June, the leaf yields were twice those of stems, but at a mature stage in August, the stem yields were twice those of leaves (Griffin and Jung, 1983). At early head emergence, the CP, IVDMD, NDF, and lignin contents of leaves were 97, 604, 660, and 47 g kg^{-1}, respectively. Stems at the same morphological stage contained 43, 500, 753, and 72 g kg^{-1}, respectively.

Leaf blades accounted for 47% of the forage yield of switchgrass early in morphological development compared to only 26% late in the season (Twidwell et al., 1988). With several cool-season grasses, both lignin and hemicellulose-lignin concentrations increased with maturity more in stems than in leaves, and tetraploid cultivars of grasses had lower lignin content than diploid ones (Morrison, 1980). Albrecht et al. (1987) reported that the leaf:stem ratio of alfalfa decreased from 1.5 in the vegetative stage to around 0.5 when plants were mature, and Nordkvist and Aman (1986) reported that the stem fraction increased from 18.5 to 50.7% of the yield, and the leaf fraction declined from 72.9 to 18.4% as alfalfa matured.

Leaf Quality

Leaf tissue is nearly always the highest quality part of the forage (Hides et al., 1983). In grasses, leaves are composed of the lamina (blade) and the sheath. In a dicot plant, such as a legume, the leaves consist of the lamina (leaflets) and the petiole (leaf stalk). These leaf components differ in quality and should not be considered as a single entity. For example, Cherney et al. (1983) reported rates of cell wall digestion for leaf blades of oat (*Avena sativa* L.) and barley (*Hordeum vulgare* L.) were 10.7% h^{-1} compared to 7.9% h^{-1} for leaf sheaths. Since rate of cell wall digestion was 7.5% h^{-1} for stems, they concluded that sheaths should be considered as part of the stem when determining the quality of the leaf and stem components. Poppi et al. (1981) reported that cattle and sheep consumed 35 and 21% more of the leaf fraction than of the stem fraction of pangolagrass (*Digiteria decumbans* Stent.) and rhodesgrass (*Chloris gayana* Kunth). Both leaf and stem fractions were digested to the same extent, so the increased intake was attributed to the shorter time the leaf fraction was retained in the rumen compared to the stem fraction.

Minson (1990) reviewed a large number of studies and concluded leaf intake was higher than stem intake. Leaf sheaths of switchgrass were intermediate between the leaf blades and the stems in IVDMD, NDF, lignin, and extent of cell wall digestion, but leaf blades and sheaths had similar N concentrations (Twidwell et al., 1988). Leaflets of alfalfa, red clover, and birdsfoot trefoil were always high in digestibility, and the leaf blade cell wall component was more digestible in white clover (*Trifolium repens* L.) and alfalfa than in red clover (Wilman and Atimini, 1984). In general, they found petioles were less digestible than leaflets. In white clover, leaf petioles were more digestible than flower stalks.

Leaf blade fractions of legumes generally decline very little in digestibility and CP with maturity (Hides et al., 1983; Nordkvist and Aman, 1986). Albrecht et al. (1987) reported that alfalfa leaves had about 800 g kg^{-1} IVDMD over a wide range of growth stages. Cell wall content of the leaves increased by only 10% over the maturity range, so the decline in forage quality was attributed to a decrease in leaf:stem ratio and a decline in stem quality. In comparing late and early maturing clones of orchardgrass (*Dactylis glomerata* L.), Lentz and Buxton (1992) found 12% of the leaf fraction was indigestible compared to 49% of the stem and leaf sheath fraction. Digestion rates of leaves from the early and late clones were similar when harvested at similar dates.

Improvement of Leaf Quality

Selecting for leaf blade characteristics may genetically improve forage quality in some instances and not in others. Selecting on the basis of specific leaf weight and leaf expansion rate had little effect on the quality of perennial grasses (Carlson et al., 1983; Zarrough et al., 1983; Buxton and Marten, 1989). Sleper and Drolsom (1974) indicated that selecting for wide leaf blades in smooth bromegrass should increase yield and digestibility of first growth, perhaps due to change in the leaf:stem ratio.

An increase in leaf blade fraction was correlated with higher digestibility, but selection for total plant digestibility appeared to be as effective in reed canarygrass as selecting for leaf or stem digestibility (Christensen et al., 1984). Bruckner and Hanna (1990) concluded there was ample variation in rye (*Secale cereale* L.) and oat to allow improvement in IVDMD by selection. However, they found less variation in leaf IVDMD level in triticale (X *Triticosecale* Wittmack) and wheat [*Triticum aestivum* (L.) emend. Thell]. Nguyen et al. (1982) found little correlation between leaf tensile strength, IVDMD and fiber content of tall fescue, indicating that selecting on the basis of tensile strength would be of little value in quality improvement.

Stem Quality

Immature grass stems are generally high in quality (Minson, 1990), perhaps even higher than leaves on the same date. Stems decrease in quality faster than leaves of most forage plants, especially as the plants approach maturity. Concentration of NDF in reproductive stems of four species of cool-season grasses often ranged from 600 to 700 g kg^{-1} and approached 750 g kg^{-1} or more if plants were mature (Buxton, 1990). The lower digestibility of stems can be attributed to their anatomy. Leaf lamina are comprised of many thin-walled mesophyll cells whereas stems are composed of highly lignified xylem cells, a high concentration of vascular bundles, and other sclerenchyma cells.

The lignified ring in both grass and legume stems was the major factor providing structural strength and was the greatest limitation to the breakdown of stems in the rumen (Akin et al., 1990). In contrast, stem parenchyma was easily digested.

Forage yield is determined by yield per tiller or shoot and shoots per unit area. Often selecting for larger shoots or weight per shoot will have a positive effect on yield of reproductive (Sleper and Drolsom, 1974) and vegetative (Nelson et al., 1985) swards. However, Buxton (1990) found that NDF concentration was not affected by selecting for rapid growth rates or large tillers of several cool-season grasses. Sleper and Drolsom (1974) reported that selecting for an increase in culm diameter had a strong genetic association with improved quality and may offer a way to improve smooth bromegrass.

High plant density (population) often decreases stem size. Bolger and Meyer (1983) found CP and ADF levels were not affected by alfalfa plant density during the first production year. Thus, changes in number of tillers per plant, tiller density, or tiller weight apparently had little effect on forage quality. Volenec et al. (1987) also reported stem diameter decreased with increasing populations.

Stem material at 172 plants m^{-2} had 10 g kg^{-1} less lignin and 30 g kg^{-1} higher IVDMD than plants at a population of 11 plants m^{-2}. With increased plant density of alfalfa, stem diameter and lignin concentration decreased and IVDMD increased (Cherney et al., 1986).

Plant population influences forage quality differently for genotypes of grain sorghum [*Sorghum bicolor* (L.) Moench.]. As within-row plant spacing increased (lower plant population), plant height declined, leaf:stem ratio increased, and number of shoots per plant increased (Caravetta et al., 1990a). This resulted in a decrease of NDF and lignin as within-row spacing increased, and an increase in IVDMD and N concentrations in the forage (Caravetta et al., 1990b).

Improvement of Stem Quality

Since forage stems may be more highly variable in forage quality than the leaf lamina, significant improvement of forage grasses or legumes may be possible by selecting for stem quality, especially in C_4 grasses and legumes. Stems generally contribute to the forage yield in all cuttings of legumes and most tropical or warm-season grasses. Stems generally contribute only to the first growth of cool-season grasses. Significant differences also exist in quality of forage stems in C_3 grasses. For example, orchardgrass genotypes had a 24% variation in stem IVDMD and only a 7% variation in leaf IVDMD. 'Rebound' smooth bromegrass stems were 12% more digestible than stems of 'Barton' (Buxton and Marten, 1989). However, Barton leaves were 6% higher in IVDMD than Rebound leaves. Hence, variation in quality of plant parts is not uniform (Buxton and Marten, 1989), so merit exists for focusing on plant parts, especially stems, in forage quality improvement programs.

Buxton et al. (1987) sampled the lower six internodes and nodes of field-grown alfalfa accessions at a mature (early seed pod) stage. Some accessions had higher stem IVDMD or CP concentrations than the cultivars used as controls. They concluded that selection for improved nutritive value of stems would be a good way to improve overall quality of alfalfa forage. In other studies, Buxton and Hornstein (1986) compared cell wall concentration and components in alfalfa, red clover, birdsfoot trefoil, and white clover. The cell wall concentration was lowest for leaves and highest for lower stems within species. Since lower stems are of lowest quality and stems decline the most in quality with maturity, Buxton et al. (1985) concluded that plant breeders should focus attention on the lower plant canopy to improve overall quality of alfalfa or birdsfoot trefoil.

Kephart et al. (1989) determined the effects of divergent selection for alfalfa cell wall components on plant morphology. Lines selected for low-lignin had a higher leaf:stem ratio, especially in the canopy fraction above the sixth node. In the upper part of the canopy (above the sixth node), the leaf:stem ratio was 18 to 53% higher in lines selected for low lignin compared to those selected for high lignin. On a given calendar date, the low-lignin lines were less mature than high-lignin lines. In the stem bases, Kephart et al. (1990) found NDF was from 3 to 4% higher in high-lignin lines, and total-herbage NDF was 4 to 8% higher in high-lignin lines compared to those selected for low lignin. They concluded that lines selected for low herbage-lignin had altered plant morphology and components of stem cell walls.

The reproductive process is the main factor that alters the leaf:stem ratio with increasing maturity. In grasses, reproductive development changes the morphology of a tiller as it assumes a determinate growth habit and new leaves are no longer formed. Stem elongation and inflorescence development form lower quality stem material that dilutes the higher quality leaf material. Cereals often produce a significant amount of highly digestible grain which may partially offset the decline in leaf:stem ratio and stem quality. A highly digestible inflorescence resulted in barley forage being more digestible than oat forage (Cherney and Marten, 1982).

In general, leaf lamina of forage crops have the highest quality and are preferred by animals. For example, voluntary intake of the leaf fraction of a tropical legume *Lablab purpureus* was 79% higher for cattle and 61% higher for sheep compared to the stem fraction, even though both the leaf and stem fraction were digested to the same extent by cattle (Hendrickson et al., 1981). Voluntary intake of leaf tissue, averaged over five warm-season grasses, was 46% higher than intake of stem material even though leaf dry matter digestibility was slightly lower than stem digestibility (Laredo and Minson, 1973).

Stems may have high digestibility and nutritive value when they are young, but they decrease very rapidly with maturity (Terry and Tilley, 1964; Buxton et al., 1985). With warm-season grasses, leaf quality was not always greater than stem quality (Perry and Baltensperger, 1979). Grass inflorescences vary widely in digestibility. In their review, Hacker and Minson (1981) showed inflorescence digestibility of cool-season grasses ranged from about 400 to 800 g kg^{-1}. Digestibility and quality of the inflorescence decrease rapidly with maturity in forage grasses that do not produce a significant weight of digestible seed.

Regulating the Leaf:Stem Ratio

Leaf tissue of forage plants is generally of higher quality than stem tissue which has given rise to a great number of studies focused on management and breeding strategies to increase the leaf:stem ratio. Quality of leaves and stems of most grasses and legumes are nearly equal when young, but as the tissues age, quality of leaves decreases at a much slower rate. This is related to the mesophyll cells in leaves, which form a major part of leaf tissues, have a high CP content, and do not form secondary, highly lignified cell walls (Table 1). Conversely, epidermal and fiber cells in leaves form thick secondary cell walls, and stem tissue becomes strongly lignified as it gets older, especially the basal stem tissue. Thus, quality of forage is improved when plants are managed or bred to have a greater leaf:stem ratio.

Reducing Flowering Periods

Flowering alters the morphological development of the plant and the leaf:stem ratio because the terminal stops production of new leaves in all grasses and determinate dicots. In addition, stem growth continues after leaf growth stops causing a continued reduction in the leaf:stem ratio. Although leaf initiation from the terminal meristem of indeterminate species such as alfalfa continues as the flowers develop from axillary buds along the stem, the rate of leaf area production usually slows (Brown and Tanner, 1983), yet the stem continues to enlarge and mature. Thus, overall, the effect of flowering on indeterminate species is a gradual reduction in leaf:stem ratio, as leaf production slows, leaves in the lower canopy senesce, and the stem increases in weight. The quality of the forage decreases rapidly as the increase in stem weight, especially the basal portion, is associated with components like collenchyma cells and lignin which are only marginally digestible (Buxton and Casler, 1993).

High quality forage or pasture is most likely if plants are managed to minimize the proportion of time plants are in reproductive phases, i.e., grasses need to be cut early, in the boot stage even, and legumes need to be cut as soon as they begin to flower. These strategies capture the highest proportion of leaf production, and harvest the stem before its quality has decreased below acceptable levels. At the same time, however, cutting early may reduce yield and stand persistence, factors that need to be considered in the decision process.

The leaf:stem ratio is generally higher in C_3 than C_4 grasses, especially when comparing more upright species in both groups. Warm-season grasses generally have larger, more lignified stems that are not grazed or eaten as readily as those of cool-season grasses. Many warm-season grasses are bunchgrasses that do not tiller actively, and thus have the dry weight distributed proportionately higher in the canopy where a higher proportion of the above-ground biomass can be harvested. Coupled with the more active photosynthetic rate, the economic or harvestable yield of C_4 grasses is much greater, but it is not as densely packed in the canopy as with C_3 grasses to facilitate grazing. These C_4 species work very well for stored feeding systems when harvested early to keep the leaf:stem ratio high.

Many warm-season grasses such as bermudagrass or caucasian bluestem [*Bothriochloa caucasica* (Trin.) C.E. Hubbard] have relatively prostrate stems which display a large amount of biomass near the soil surface. The entire biomass, however, consists of stolons and short upright stems with many short leaves, so the herbage is still a mixture of stems and leaves. In contrast, most cool-season grasses can be managed to be in vegetative stages for high proportions of the growing season, by cutting the first growth early, just as the stem emerges, which allows sufficient time for the vegetative tillers to begin regrowth. Thus, it is important to think in terms of leaf factors such as enhanced expansion rates in improving quality of cool-season species, and stem factors such as lignification in warm-season grasses. In tall fescue, however, high rates of leaf area expansion are genetically associated with large tillers (Nelson et al., 1977).

Using Growth Regulators

Since maintaining a high proportion of leaves in the canopy is important in maintaining forage quality several growth regulators have been used to modify plant development. Mefluidide {N-[2,4-dimethyl-5-[[(trifluoromethyl) sulfonyl amino]phenyl]acetamide} has been applied to vegetative grasses in spring to reduce stem elongation of cool-season grasses, subdue seedhead formation, and maintain the vegetative state (Glenn et al., 1980; Allen et al., 1983; Wimmer et al., 1986; Sheaffer and Marten, 1986). An associated response is increased tiller formation which improves summer regrowth and leaf yield after flowering. Although absolute yields in spring are generally reduced, NDF and ADF remain low and CP remains high in mefluidide-treated compared to untreated grass (Allen et al., 1983). Sheaffer and Marten (1986) and Wimmer et al. (1986) also reported decreased NDF and increased CP with mefluidide treatment. Gains of both calves and cows were increased with mefluidide treatment despite lower forage yields, but not consistently (Wimmer et al., 1986).

The timing of application relative to time of initiation and development of the inflorescence is critical. Sheaffer and Marten (1986) reported year-to-year differences in the effectiveness of mefluidide. Mefluidide application to cool-season grasses at leafy stages before stems elongated consistently increased first-harvest leafiness and quality. Applying it at later stages of morphological development or in the fall was much less effective. Glenn et al. (1980) reported yield of tall fescue was reduced by mefluidide in the first harvest due to less culm and seedhead formation, but no difference in regrowth was noted between treated and untreated areas.

Roberts and Moore (1990) decreased seedhead production and increased IVDMD of tall fescue and CP by applying amidochlor {N-[(acetylamino)methyl] 2-chloro-N-(2,6-diethylphenyl) acetimide]} in one year but not in another. They concluded that amidochlor could be effective in suppressing seedhead development during spring if tiller initiation the previous fall was not interrupted by adverse environmental conditions. Forage treated with imazethapyr [5-ethyl-2-(4-isopropyl-4-methyl-5-oxo-2-imidazolin-2-yl) nicotinic acid] was 36 to 71 g kg^{-1} lower in NDF, 19 to 44 g kg^{-1} higher in IVDMD, and 9 to 20 g kg^{-1} higher in CP than untreated cool-season grasses (Fales et al., 1990). Yield was reduced by 0.8 to 2.8 Mg ha^{-1} depending on the rate of imazethapyr applied.

In the long term, growth regulators will have the most potential in cool-season pastures that are stocked intensively with high-producing animals for which quality is more critical than yield (Reynolds et al., 1993). Since most cool-season grasses flower only once per year, the timing can be worked out for the spring growth and does not need to be repeated during the year. Not all the area will need to be treated, only that part needing to be deferred, or that which is going to be grazed when stem elongation is normally active. Since forage production in spring is high, an alternative to chemical regulation is to harvest some of the pasture area, cutting when it is of high quality. Thus, other management alternatives will reduce the area needing to be chemically treated.

There have been attempts to use chemicals in mixtures of upright growing species like alfalfa-orchardgrass mixtures, with the objective being to delay flowering and reduce stem growth of the grass during spring. To date, these chemical procedures have not been agronomically or economically attractive. With the current national and international concerns about use of chemicals and the need for timely application, it appears regulation of stem growth using breeding and other cultural practices will be most feasible.

Breeding for Altered Leaf:Stem Ratio

There has been extensive discussion as to whether cultivars of alfalfa differ sufficiently in leaf:stem ratio or in digestibility of either component to allow genetic improvement. Some experiments have shown cultivar differences in quality when harvested on common dates, but it is difficult to determine if quality differences are confounded with maturity, which often differs among entries. Cultivars also differ in susceptibility to leaf diseases which affects loss of lower leaves, and rapidly changes the leaf:stem ratio and digestibility. These and several other factors need to be considered in cultivar assessment for quality.

Since alfalfa quality could be improved by increasing the proportion of leaves, multifoliolate alfalfa strains with higher leaf:stem ratios have been developed (Brick et al., 1976; Etzel et al., 1988; Ferguson and Murphy, 1973). Etzel et al. (1988) reported multifoliolate plants have rapid leaf expansion after defoliation that occurs on fewer, but larger shoots compared to trifoliolate plants. Volenec and Cherney (1990) compared alfalfa plants that averaged 4.4, 6.2, and 6.8 leaflets per leaf with trifoliolate plants which were selected for rapid or slow shoot elongation rate. Genotypes with an average of 4.4 and 6.2 leaflets per leaf produced larger stems, and although the multifoliolate strains had higher leaf:stem ratios, the multifoliolate trait was not consistently associated with higher quality of the forage.

Some proprietary cultivars of alfalfa have been selected for high CP and the associated higher digestibility. These have been tested in only limited situations.

Genetic progress has been made in breeding tall fescue for more rapid leaf elongation rate (Sleper and Nelson, 1989), resulting in enhanced yield, generally being expressed most during vegetative growth stages (Nelson et al., 1985). Plants with rapid leaf growth had wider leaves and fewer tillers than plants selected for slow leaf growth, but the ratio of vegetative to reproductive tillers was similar during spring growth. The expression of enhanced vegetative yield with rapid leaf growth occurred under both frequent and infrequent defoliation. Similar responses have been reported for perennial ryegrass populations selected for long or short leaves (Simonds et al., 1973).

Reducing Disease and Insect Effects

Diseases often alter the quality of forage, generally by affecting the leaves more than the stems. Edwards et al. (1981) reported early leaf senescence and reduced digestion when orchardgrass leaf tissue was infected with rust (*Puccinia graminis* Pers. F. sp. *dactylidis* Guyot et Massinot) compared to healthy tissue. It is unknown whether the disease complex physically limits ruminal microorganisms or chemically inhibits digestion of the tissue. Total non-structural carbohydrates (TNC), CP, and digestibility of leaves were reduced in orchardgrass infected with *Stagonospora arenaria* and in alfalfa infected with *Phoma medicaginis* (Mainer and Leath, 1978). In alfalfa, this reduction occurred with an 80% or greater level of disease severity while in orchardgrass CP was reduced at the 20%, and TNC was reduced at the 50% level of severity. Regrowth after cutting diseased alfalfa was higher in CP than it was from non-diseased alfalfa.

Wilson et al. (1991) found digestibility decreased as severity of rust (*Puccinia substriata* var. *indica*) increased on pearl millet [*Pennisetum glaucum* (L.) R. Br. K. Schum.]. The response was not linear; the initial infections caused the greatest reduction. Forage quality of susceptible and anthracnose-resistant plants of 'Saranac' alfalfa was similar, except for stem NDF, when both types were disease free (Lenssen et al., 1991b). Infection with anthracnose slowed shoot development, however, resulting in an increase in CP and digestibility of stems. Forage quality of leaf tissue was often lower in infected leaves of susceptible plants.

Insect infestations frequently affect yield, but also may alter forage quality. The potato leafhopper (*Empoasca fabae* Harris) reduces alfalfa production, but NDF was affected very little by feeding severity (Hutchins et al., 1989). Although CP of leaves was reduced by leaf hopper feeding, CP of stems was either increased or maintained, and overall herbage digestibility was not affected. They concluded that the visual chlorotic effect on alfalfa caused by leafhoppers was not necessarily indicative of low forage quality, but growth was slowed markedly.

Aphid (*Acyrthosiphon kondoi* Shinji) feeding on alfalfa did not affect true IVDMD or digestible NDF of stems, but lowered CP and true IVDMD in leaves (Lenssen et al., 1991a). Non-infested aphid-resistant alfalfa germplasm had higher leaf NDF and lower NDF and true IVDMD in the stems. With high populations of the three-cornered alfalfa hopper (*Spissistilus festinus* Say), CP was reduced and NDF was increased in alfalfa compared to non-infested plants (Moellenbeck and Quisenberry, 1991). Alfalfa weevil (*Hypera postica* Gyll.) feeds largely on leaf tissue, and can reduce the leaf:stem ratio and decrease quality of alfalfa.

Several grasses such as tall fescue and perennial ryegrass serve as a host for an endophytic fungus which grows between cells of the leaf sheath. The plant-fungus complex in tall fescue produces a toxin, most likely an alkaloid, which is known to interfere with rate of intake and the ruminal digestion process. The slower digestion reduces rate of passage out of the rumen which, in turn, limits forage intake. In perennial ryegrass, an endophytic fungus can cause ryegrass staggers.

Most emphasis is on genetic resistance to minimize disease and insect problems, but cultural practices also have helped. For example, the alfalfa weevil is currently controlled by an interacting suppression based on tolerant cultivars, increase in natural enemies for the larva and adult, and timely harvest of the first cutting. While these systems work well in northern states, the eggs laid in the fall overwinter in southern states and provide an early infestation, too early for the natural cycles of enemies, and damage in spring is extensive before alfalfa is ready to harvest. Until recently, this required two insecticide treatments in southern states for control, but management has reduced it to one application. Fall and winter management strategies are being evaluated in an effort to alter egg laying in the fall and survival of eggs over the winter.

CANOPY EFFECTS ON FORAGE QUALITY

Canopy architecture affects the physiological function of forage plants, the forage quality on offer to grazing animals, and grazing patterns. Upper leaves on a grass tiller are the youngest and lower leaves are the oldest. Leaf blades differ in anatomy depending on the level of insertion (location) of a leaf on a tiller (Wilson, 1976a). Cell wall content of leaf blades, sampled at different levels of insertion on the tiller of green panicgrass, but at the same stage of development, increased from 326 to 638 g kg^{-1} from low to high leaf insertion level, while the N content and the IVDMD decreased from 47 to 25 g kg^{-1}, and from 773 to 650 g kg^{-1}, respectively. Lignin also tended to increase with increase in insertion level. Sheaths responded similarly, but the gradients were not as great. Leaf blades and sheaths at high insertion levels had a greater proportion of sclerenchyma and vascular tissue, and smaller-sized mesophyll, bundle sheath, and epidermal cells than leaves of lower insertion levels (Wilson, 1976b).

Leaf blade anatomy at any insertion level did not change with age. Rather, lignification of the sheath sclerenchyma and thickening of the sheath cuticle increased significantly following ligule emergence. These changes paralleled a rather large decrease in forage quality, which would be expected in species that elongate the stem to cause sheath tissue to be included in the harvested portion. Other studies suggest grazing animals selectively avoid sheath tissue, especially in vegetative canopies where the whorl of sheaths discourages animals from grazing to short stubble heights (Hodgson, 1982).

Effects on Intake

L'Huillier et al. (1984) determined that diets of ewes grazing ryegrass and prairiegrass (*Bromus wildenowii* Kunth.) were comprised of 68 to 97% green grass leaves while dead material and pseudostems were rejected. Interestingly, defoliation of white clover in the mixture also was affected by the location of the green grass leaves. Forage intake by sheep was reduced in the pastures during summer because the surface canopy of ryegrass restricted grazing green leaf area of white clover which was located lower in the canopy.

Other stem factors also can affect animal performance indirectly. Standing stems in pastures have potential to irritate animals' eyes and cause pinkeye. Old standing stems and litter in plants such as little bluestem [*Schizachyrium scoparium* (Michx.) Nash] protect new growth from grazing; hence, animals are unable to effectively utilize new growth of such plants (Briske, 1991).

Parsons et al. (1991a) indicated that a difference in the patterns of leaf development between perennial ryegrass and white clover in a mixture affected their defoliation by sheep. Many white clover leaves escaped grazing because they expanded near the soil surface. Clover leaves often were eaten completely when they expanded near the top of the canopy. Grass leaves expand vertically above the whorl, and were grazed earlier, while they were still growing, but defoliation was only partial. Parsons et al. (1991b) concluded that mixtures are unstable, and if one component has no mechanism to escape grazing, the one with a physiological advantage will dominate. However, if each component can escape grazing to some extent, then stable mixtures can result because the species with the physiological advantage increases in abundance and is more subject to grazing.

With high grazing pressure, animal diet is more closely related to growth rates among plant species than to animal preference. High leaf density within a sward and low stem content favored larger bite size by grazing animals (Stobbs, 1973). But these factors change as the canopy changes (Penning et al., 1989). For example, when first exposed to the mixture, steers selected a diet high in aeschynomene (*Aeschynomene americana* L.) when grown in combination with limpograss [*Hemarthria altissma* (Poir.) Staph et C. E. Hubb.] (Sollenberger et al., 1987). However, in a 48-h grazing period, aeschynomene consumption was similar to its concentration in the grazed strata. This indicates a dynamic interface between animal and forage where canopy characteristics influence diet composition, and the resulting consumption alters the canopy. In turn, the changed canopy causes animals to respond by altering their selection.

In addition to composition of the tissue, the density of the canopy, i.e., mass per volume, and the way it is displayed affect intake, selectivity, and performance of grazing animals. Intake per bite is generally the major component affecting intake, and is strongly influenced by the mass of herbage available (Hodgson, 1981; Forbes and Coleman, 1993). Herbage intake by either cattle or sheep is usually increased as herbage mass increases. For example, in Scotland, intake of cool-season grass by sheep increased until about 2000 kg ha^{-1} of above-ground biomass was available, then remained constant (Hodgson, 1977). Conversely, in the southwestern U.S., bite size increased up to near 3400 kg ha^{-1} for old-world bluestem (Coleman, 1992), a C_4 species which responds similarly to tropical grasses (e.g., Stobbs, 1973). Coleman suggests herbage mass may be a poorer indicator of intake or bite size than is the leaf:stem ratio, a factor also needing consideration in spring growth of C_3 grasses. The live:dead ratio also should be considered as animals tend to select specific plant parts. Sheep are more effective than cattle in selecting higher quality diets from similar swards (Jamieson and Hodgson, 1979).

It also is well known that animal selectivity among plant parts in single-species pastures is more difficult than in pastures comprised of many species of diverse morphology. Likewise, as herbage availability decreases, selectivity generally decreases (Coleman, 1992), and intake will be more related to herbage growth rate. These principles interact with basic principles of growth and development described earlier in the chapter, whereby it is critical to have sufficient leaf area and a favorable display in order to intercept most of the solar radiation. It also is helpful to have at least one legume and one grass in the ecosystem to assist in N nutrition and to present an efficient canopy for intercepting radiation. Thus, the plant factors associated with producing high quality forage may not be consistent with the plant factors displaying high quality forage for a grazing animal. This is less of a problem for forages that are mechanically harvested, but it is still critical.

Relationships with Forage Production

Fertilization, especially N, increases leaf size of grasses (Nelson, 1992) and increases tillering (Brown and Ashley, 1974), but generally does not change the number of live leaves per tiller. In N-deficient situations, addition of N fertilizer often increases seedhead density per unit area. However, Jung et al. (1990) found that the response of reproductive stem density to N varied depending on the species and cultivar among the warm-season prairie grasses used in their study. Reproductive stem density was significantly increased with Plains bluestem (*Bothriochloa ischaemum* Keng.) and switchgrass, but not for little bluestem, big bluestem, or indiangrass.

Wilman and Pearse (1984) found perennial ryegrass maintained three live leaves over a wide range in N rates. Similar data were obtained by Davidson (1980) with vegetative plants of tall fescue. Thus, while leaf elongation rate and leaf size were enhanced by N (Volenec and Nelson, 1983), there was little reduction in rate of senescence. Thus, a large proportion of the herbage produced can be lost to death and decay, and does not contribute to animal production. If grass canopies fertilized with N are allowed to develop for a long time period, the high leaf area index produced will limit light penetration, increase death rate of young tillers, and increase length of the pseudostem. These features lead to 'bunchiness' of many grasses, high loss of older tissue, and longer pseudostems that discourage close grazing.

Tiller growth interacts with grazing dynamics. Frequent removal of the canopy by grazing often increases tillering by increasing the amount of light reaching the base of the plants (Crawley, 1983). The presence of a dense green canopy selectively screens out large amounts of red light, thus decreasing the red:far red ratio of radiation reaching the plant bases where new tillers arise. This altered ratio possibly interacts with the phytochrome system preventing basal buds from forming new tillers. Removal and modification of the canopy by grazing or cutting increases the red:far red ratio, possibly releasing dormant tillers (Briske, 1991). These light responses have been observed in ryegrass (Casal et al., 1985; 1987) and barley (Skinner and Simmons, 1993), but changing light quality at plant bases did not affect canopy architecture or tillering in little bluestem (Murphy and Briske, 1992).

Parsons and his co-workers (1983) sought to integrate the physiological condition of the sward, the grazing behavior (intake) of the ruminant, and animal performance. Their grazing trials also involved animal effects on the sward. Their early studies showed death and decay of older leaves occurred regularly in the perennial ryegrass canopy, especially under lenient grazing. For example, if the monoculture sward was maintained at a leaf area index (LAI) near 3.0, up to 40% of the above-ground dry matter was not utilized by grazing sheep, allowing losses due to leaf death and decay (Table 3).

When perennial ryegrass pastures were grazed to a LAI near 1.0, dry weight production of grass was reduced. However, the loss of dry matter to leaf death and decay was greatly reduced, and intake per land area was increased over pastures grazed to a LAI of 3.0 (Table 3). In addition, with closer defoliation, the average leaf developed in brighter light to achieve a higher photosynthesis potential (Woledge, 1979), and was younger in age than canopies maintained at a higher LAI. With less shade at the low LAI, tillering can be enhanced, providing more growing points per land area. The overall effect of close defoliation was that CO_2 uptake by the canopy was only marginally reduced, and even though intake per animal was reduced, intake per land area was increased. Thus, the overall efficiency of converting solar energy to animal intake was nearly doubled, largely because less was lost to death and decay. The problem, however, is that few species and grazing strategies have been evaluated in these intensive manners. The system gets much more complicated when legumes are added to the mixture (Parsons, 1991a; 1991b), seasonal factors such as reproductive growth and temperature are considered, and plant persistence is evaluated.

CHEMICAL CONSTITUENTS AFFECTING FORAGE VALUE

The chemical components of ingested forage affect the digestibility and nutritive value of the forage, and certain compounds interact with physiological reactions in the animal resulting in toxicity or animal stress. Intrinsic chemical factors are discussed in depth in Chapter 9 and extrinsic factors are discussed in Chapter 8. Basic chemical constituents comprising cell walls and cell contents will not be discussed here as they are discussed in detail elsewhere in this book. A recent text discusses cell wall structure in detail (Jung et al., 1993). This

Table 3. Harvest efficiency depends on canopy size. Photosynthesis was converted to organic matter (O.M.) allowing harvest efficiency to be calculated as the proportion of O.M. grazed relative to that produced. Data are average of three 2-week trials with sheep grazing at leaf area index (LAI) near 1.0 or 3.0. Adapted from Parsons and Leafe (1981).

Measurement	LAI=1.0	LAI=3.0
Gross Photosynthesis (kg O.M. ha^{-1} d^{-1})	208	299
Intake (kg O.M. ha^{-1} d^{-1})	52	38
Efficiency of harvest (%)	25	13

section will provide an overview of chemical constituents having an anti-metabolic effect on ruminant animals.

Many different plant toxicants are present in a wide array of forages which affect animal management and feeding decisions. From the plant side, the role of toxicants is often viewed as a method for plant protection from excessive herbivory which evolved under grazing. In addition, insect herbivory and disease infections might be deterred to some degree in plants with high levels of certain toxicants. Plant-derived toxicants have been referred to as "nature's pesticides" (Cheeke, 1989a). Likely, synthesis and accumulation of several of these chemical constituents evolved as plants gained adaptation to a broader range of habitats. Thus, their role in the entire ecosystem needs to be evaluated along with their specific effects on forage quality.

Glycosides

Glucosinolates, cardiac glycosides, cyanogenic glucosides, calcinogenic glycosides, carcinogenic glycosides, and some specific glycosides are reviewed in depth by Cheeke (1989b). Those most frequently encountered in forage-livestock programs are cyanogenic glucosides which are present in about 3000 species of higher plants located in 110 plant families. The source of HCN (prussic acid) has been identified in only 300 species (Poulton, 1990). Some common forage plants such as sorghums (*Sorghum* spp.), white clover, trefoils (*Lotus* spp.), and vetches (*Vicia* spp.) contain cyanogenic glucosides. Also, animals may encounter choke cherry (*Prunus virginiana* L.) that contains cyanogenic glucosides. Some poisonous plants such as bracken [*Pteridium aqualinum* (L.) Kuhn] and arrowgrass (*Triglochin* spp.) have this mechanism of toxicity (Tewe and Iyayi, 1989).

When tissue is mechanically disrupted, such as by mastication, a breakdown of the cyanogenic glucoside occurs releasing HCN and an aldehyde or ketone (Poulton, 1990). The HCN is rapidly absorbed from the digestive tract of the ruminant, and cyanide inhibits cytochrome oxidase in the respiratory electron transport chain (Tewe and Iyayi, 1989). The cyanogenic glucosides are in highest concentration in young leaves of plants like sorghum (Kingsbury, 1964). Therefore, the leaf:stem ratio and the size and age of the plant canopy affect the likelihood of HCN poisoning.

Alkaloids

Alkaloids occur widely, both in higher plants and lower plants such as bacteria and fungi (Willaman and Li, 1970). Many alkaloids serve as toxicity mechanisms in poisonous plants (Kingsbury, 1964). In addition to the poisonous plants, alkaloids are encountered in some grasses, primarily in the *Festuca*, *Lolium*, and *Phalaris* genera. Indole alkaloids occur in reed canarygrass, barley, other *Hordeum* spp., and giant reed (*Arundo donax* L.). Phalaris staggers is associated with alkaloids in hardinggrass (*Phalaris tuberosa* L.) or bulbous canarygrass (*Phalaris aquatica* L.). Reduced palatability of reed canarygrass occurs with high indole alkaloid content (Marten et al., 1976).

An association between an endophyte fungus (*Acremonium coenophialum* Morgan-Jones and Gams) and tall fescue produces "ergot type" alkaloids (ergopeptine related alkaloids) (Garner et al., 1993) and saturated pyrrolizidine alkaloids (lolines) (Bush et al., 1993). These alkaloids, individually or in combination, have been associated with a number of toxicoses in cattle such as fescue foot, hyperthermia, summer slump, fat necrosis, reduced gain, and reduced milk production (Schmidt and Osborn, 1993; Thompson and Stuedemann, 1993). Mares grazing infected tall fescue often are severely affected, having prolonged gestations, parturition problems, still-born foals, and re-breeding problems (Schmidt and Osborn, 1993).

The endophyte of tall fescue is seed-borne, allowing it to be excluded by establishment of stands using endophyte-free seed (Bacon and Siegel, 1988). Although this can alleviate the livestock toxicoses, the endophyte-free plants lose some of their insect, nematode, drought and grazing resistance (Hoveland, 1993). Current research is underway to retain the desirable attributes of the endophyte-tall fescue association for the plant while avoiding tall fescue toxicosis in the animal (Siegel, 1993).

A similar endophyte (*Acromonium lolii* Latch, Christensen et Samuels) infects perennial ryegrass and produces lolitrens (pyrrolizidine alkaloids) and peramines (indole terpenes) (Rowan, 1993). Sheep eating infected perennial ryegrass have reduced weight gains and experience "ryegrass staggers". A review by Joost and Quisenberry (1993) summarizes the state of knowledge of the endophyte-grass association.

Nitrates

Some common forage and weed plants accumulate nitrate (NO_3^-) to a level that leads to toxicity for ruminants. Nitrate is relatively nontoxic, but nitrite (NO_2^-), one of the intermediates in the reduction process, is extremely toxic. Nitrate is the most common form of N taken up by plants (Salisbury and Ross, 1985). Organic forms of nitrogen, including recycled N from organic matter and ammonium (NH_4^+), are converted to NO_3^- by nitrification reactions in the soil. Plants growing normally absorb NO_3^-, reduce it to the ammonium form, mostly in the leaves, and then incorporate it into amino acids and proteins. Nitrate reductase is the enzyme that catalyzes the reduction of NO_3^- to NO_2^-, a key regulatory step in the NO_3^- reduction process. Nitrite does not accumulate in plants as it is rapidly reduced to NH_4^+ (Salisbury and Ross, 1985).

Nitrate reductase is sensitive to environmental stress so drought, hail, 2,4-D applications, disease, shading, or other factors affecting the photosynthetic process in the leaves may reduce NO_3^- reductase activity (Wright and Davison, 1964). Once leaves are disrupted, NO_3^- can accumulate providing there is an ample supply in the soil. In high-NO_3^- plants, the lower internodes of the stem have the greatest concentration and are generally the toxic portion of the plant. Roots have less NO_3^- than the leaves, and the grain or fruit has none (Wright and Davison, 1964). If the stress-damaged forage is to be fed, avoiding the lower stalk makes the feed much safer for livestock.

When ruminants consume NO_3^-, there is a conversion to NO_2^- in the rumen. Nitrite is rapidly absorbed into the blood stream where it reacts with hemoglobin to form methemoglobin which cannot carry oxygen. Once 80 to 90% of the hemoglobin is converted to methemoglobin, the animal can die of asphyxiation (Britton, 1981). Other symptoms of NO_3^- poisoning may appear at 30 to 40% hemoglobin conversion.

Members of the Amaranthaceae (pigweed), Chenopodiaceae (goosefoot), Brassicaceae (mustard), Compositeae (sunflower), Solonaceae (nightshade), and Poaceae (grass) families are known to accumulate NO_3^- (Wright and Davison, 1964). In addition to this genetic tendency to accumulate NO_3^-, many plants in these families often occur in intensive cropping conditions where there may be high levels of NO_3^- in the soil. Soybean [*Glycine max* (L.) Merr.] also has been reported to accumulate NO_3^- (Wright and Davison, 1964), but generally NO_3^- accumulation does not occur in the legume family because much of the N used in the plant is symbiotically fixed in the NH_4^+ form and NO_3^- is not involved. However, if levels of NO_3^- in the soil are high, legumes can take it up and metabolize it.

Nitrate contamination also may occur in the water livestock consume. When calculating total NO_3^- intake by livestock, both plant and water sources need to be considered. Toxic concentrations of NO_3^- in forage have been reported to be around 2300 mg kg^{-1} dry weight (Hogg, 1981). Upon ensiling high-NO_3^- plants, the NO_3^- concentration is reduced by about 50% under the anaerobic conditions (Hogg, 1981). The products of the anaerobic reduction with ensiling are various oxides of N (NO, NO_2, N_2O_4) which collectively are referred to as silo gas (Wright and Davison, 1964). Silo gas can be very toxic to mammals.

Phenolics

Many phenolic compounds (aromatic rings) in forage plants affect forage quality and acceptance, and have a variety of anti-metabolic activities in animals (Cheeke, 1989c). Lignin is composed of low molecular weight phenolics (noncore lignin) and highly condensed phenolics (core lignin) which are discussed extensively elsewhere in this book. Lignin is the compound most highly negatively correlated with digestibility (Lapierre, 1993). The brown midrib mutants (low lignin) of forages are discussed in detail in Chapter 9.

High concentrations of plant estrogens such as coumestrans, isoflavones, and isoflavin, have been associated with reduced conception rates in livestock. Generally, this is temporary but permanent sterility occurs if ewes are exposed to high estrogenic plants for several grazing seasons (Adams, 1989). Plant estrogens occur in many different plants. Farnsworth et al. (1975) reported 145 species of plants that likely are estrogenic and 220 species of plants that contain isoflavonoid compounds. Plant estrogenic activity can be encountered with sheep grazing subterranean clover (*Trifolium subterraneum* L.), red clover, white clover, and alfalfa (Adams, 1989). Many other antifertility agents exist in plants that may affect estrus or stimulate the uterus (abortifacients) (Salunkhe et al., 1989), but they are not encountered in forages very frequently.

Tannins, sometimes referred to as plant polyphenols (Haslam, 1989), are one of the most common phenolics in plants. Tannins refer to a large group of compounds which generally are large molecules containing enough phenolic hydroxide groups to form effective cross links with protein (Kumar and Singh, 1984). Tannins reduce voluntary intake and digestibility of forages (Butler, 1989). Tannins also form a complex with protein which may be the major mechanism preventing bloat by tannin-containing legumes. Legumes containing tannins increase ruminal escape protein (Albrecht and Broderick, 1990). In addition to Chapter 9, details regarding the phenolics are reviewed in Cheeke (1989c) and polyphenols are reviewed in Haslam (1989).

Proteins (Bloat)

Bloat has been recognized as a livestock disorder since the beginning of agriculture. During the past 25 yr, the stable foam formation in the rumen that causes forage bloat has been tied to protein degradation in the rumen (Howarth, 1975). In immature, leafy legumes, plant tissue and cells are quickly broken down in the rumen, and cell contents containing protein are released. High levels of proteins released quickly in the rumen can cause stable foam formation and bloat. If the plant contains tannins, however, protein is bound (Jones and Lyttleton, 1971), and bloat is reduced.

Since speed of cell rupture of bloat-causing legumes is rapid, factors that delay degradation could reduce the likelihood of bloat (Lees et al., 1981). Non-bloat causing legumes had strong cell walls and sometimes strong tissue strength, while bloat-causing legumes (alfalfa and white clover) had weak cell walls and weak tissue strength, except red clover had moderately strong cell strength (Lees et al., 1981). In further work, Lees et al. (1982) confirmed tissue resistance to ruminal degradation and found that cicer milkvetch (*Astragalus cicer* L.) had a thick cuticle and a reticulate vein pattern in the leaflets which resisted leaf fragmentation.

When comparing alfalfa strains selected for slow and fast initial rates of digestion, Kudo et al. (1985) showed the initial rate of digestion was about 6% lower with the slow digesting strain compared to the mean of the slow and fast digesting strains. About a 25-30% reduction in rate would be needed during the first 6 to 8 h of digestion to obtain an alfalfa that might not cause bloat (Howarth et al., 1982). Bloat remains without a simple solution and likely will require an interdisciplinary approach from both the plant and animal standpoints. Reviews of bloat are given in Chapter 9, Howarth (1975), and Hall and Majak (1989).

Selenium

Selenium (Se) accumulation poses a toxicity problem in areas with soils having high levels of this element. Plants differ in their ability to accumulate Se. Selenium accumulators have from 1000 to 10000 mg Se kg^{-1} dry matter in their tissue and cause the most problems. Selenium absorbers have from 25 to 1000 mg kg^{-1} and non-accumulators have < 25 mg kg^{-1} in their tissue (Whanger, 1989). Plants in the *Astragalus* genus such as racemed poisonvetch (*Astragalus racemosus* Pursh) are the most recognized as Se accumulators. However, other plants in the same genus, such as cicer milkvetch, are non-accumulators.

Silica

Silica generally is associated with lowered cell wall digestibility, although it has been hard to show a consistent relationship. In certain grasses such as rice (*Oryza sativa* L.), the straw and hulls can be greatly reduced in digestibility, especially when dried, which is attributed to silica (Van Soest, 1993).

CONCLUSIONS

An emphasis on sustainable agricultural systems has refocused efforts onto forages and grasslands including plant factors that affect forage quality. In this perspective, forage quality is integrated more closely with yield and with stand persistence as the value of forages in a sustainable ecosystem context emerges. Livestock production, while an important component, will not be the sole use of forages in the future. How new roles for forages and grasslands will influence emphasis on quality over the next few years is not clear at present.

Today, a broader understanding has emerged relative to the role of plant factors in forage quality. A major factor has been the realization that some weeds are not of low forage quality, and can contribute to diversity of the plant population and to sward productivity and use. Several non-seeded plants in forage or grassland conditions may continue to be called weeds, but fill an ecological niche which, in concert with other issues, brings the perspectives of ecosystem management more strongly into forage management. Managing sustainable forage ecosystems necessitates filling all ecological niches with quality forage plants.

Important advances have been made in understanding the nature of antiquality factors and the significance of the leaf:stem ratio. The relationship of these factors with forage quality has been known for a long time, but only recently has technology and methodology been developed to properly measure individual characters with the speed and accuracy needed for routine field management evaluations and, especially, for breeding programs. For example, the development and application of fiber-solubility methods and near-infrared reflectance technology have been instrumental in pinpointing specific factors which influence forage quality.

The discovery and understanding of the effect of the fungal-endophyte relationships on tall fescue and perennial ryegrass quality and longevity have been profound. Forage breeders and managers know about the role of plant diseases and their effects on leaf:stem ratio, digestibility, and animal performance. Yet, the magnitude of the endophyte problem in tall fescue ushered in new perspectives about biological associations that are forcing broader investigations and interpretations of natural interrelationships within an ecosystem. This stimulus will likely help unravel other relationships involving herbivory and plant competition.

Role of Plant Breeding

Genetic resistance to diseases and insects in forages, especially alfalfa, has led to improved forage quality, and in some cases has improved persistence allowing stands to be managed more intensely for improved forage quality. Genetic improvement of plant materials will continue to be a major contributor to improved quality, especially in terms of disease resistance and overcoming antiquality factors. Regulation of the leaf:stem ratio and of quality factors of the stem also will continue to be areas of emphasis.

There has been a gradual shift worldwide as to how plant introductions fit into forage systems. There is a continued need to introduce and identify good herbaceous legumes and other high-quality species to complement the current primarily grass forage base. These efforts need to be continued and enhanced, especially in view of the rapidly decreasing areas of natural genetic diversity and the international emphasis on sustainability. There will continue to be environmental and cultural niches for more grasses and other plants that are introduced. Likely, however, there will be minimal efforts given these activities in the U.S. in the near future except for some legumes.

Antiquality factors will continue to be major deterrents of nutritive value, and more antiquality components will be discovered as researchers look more in-depth at ways to improve animal performance. Many of these are specific chemicals, and they may be manipulated genetically or offset by plant or animal management. The genetic advance in developing reed canarygrass with low alkaloid concentrations and overcoming mimosine problems by altered rumen microflora are but two recent examples. Altering mineral ratios to minimize grass tetany offers good potential. A major challenge in the future will be to modify the tall fescue-endophyte relationship to minimize the antiquality effects on livestock, but maintain aspects of the relationship that make tall fescue resistant to environmental stresses. Overcoming antiquality characters is likely more amenable to biotechnology techniques than are many other quality factors once limitations in gene transfer are overcome.

Understanding the roles of cellular constituents has moved forward rapidly during the past 25 years, especially regarding synthesis and regulation of cell wall and protein factors. Also, there is good insight regarding breakdown processes in the rumen. Collectively, this knowledge may be utilized in breeding efforts to alter rates of cell wall digestion or enhance protein escape from the rumen. Insertion of genes to provide rumen-escape proteins will likely be possible in the future which would improve the protein nutrition from forages, especially alfalfa. Methods useful in genetic selection for these plant factors need to be developed so hypotheses can be tested more easily.

Livestock bloat continues to be an enigma, yet some progress has been made in understanding the mechanism. Utilization of some legumes in pasture systems, most notably alfalfa and white clover, is still curtailed due to potential problems with bloat. The mechanisms and interrelationships between tannins, rumen-escape protein, and bloat need to be clarified.

The importance of the leaf:stem ratio has been established, and mechanisms for improving it are emerging. Managing forages and grasslands to maintain a high leaf:stem ratio is achievable, and valuable insight has been given to the components that affect stem quality, especially for legumes and several warm-season grasses, which should help in genetic improvement. Genetic improvement of cell wall quality may be a way to improve quality of warm-season grasses, and genetic improvement of leaf growth rates may be a way to improve quality of cool-season grasses.

Genetic improvement will need to continue to emphasize conventional methodologies in the near future. Once transformation and regeneration technologies are available for grasses and additional forage legumes, several characters will be able to be changed. Some quality characters likely will be easier to change as simple traits than would be complex traits such as yield or persistence. However, it probably will be several years before the regeneration technology is available, and by then only a few public research programs will be active in applying the technology. Thus, even progress using biotechnology will be slow. Except for alfalfa, and perhaps some annuals which have large seed markets, private industry is unlikely to give very much leadership or assistance in biotechnological improvement of forage quality.

Role of Management

Recognizing the importance of managing for reduced leaf senescence in forage canopies and understanding the role of light quality in tiller bud development of grasses and white clover are changing the management philosophy of many pastures. Previously, maintaining a high LAI to intercept radiation was emphasized. Now that is being changed in favor of managing the canopy to allow light penetration to tiller or regrowth sites, especially in tight sods like perennial ryegrass. Management principles based on light penetration have not been developed and applied to complex mixtures or grasses and legumes with different morphological development and environment interactions. The trade-offs on canopy management in terms of dry matter production, animal grazing behavior, and grass-legume compatibility need to be extensively evaluated.

Quality forage will receive more emphasis in the future. Dry matter yield improvement in forages has been disappointingly slow, and may not be the best long-term answer for livestock production as we move to lower inputs and longer duration pasture and forage systems. At the same time, society will expect ruminants to derive a higher proportion of their nutritional needs from forages which will require a high quality forage resource.

Interestingly, most projections of forage use for non-ruminant livestock, such as biomass for energy and sources of industrial raw materials, also have quality implications. Thus, new and relevant technologies regarding plant factors and forage quality may fit both ruminant nutrition and industrial applications. Long-term uses of grasslands as a waste repository may focus more emphasis on persistence and sustainability rather than quality, unless the biomass produced has utility for domestic ruminants, wildlife, or industrial uses.

The expected global climate may lead to higher temperatures, longer growing seasons, and more water stress in many locations. In addition, use of surface waters and aquifers such as the Ogallala for irrigation is being carefully scrutinized and monitored. These environmental factors will likely lead to land use changes, generally to favor forage and grassland use as a long-term land reserve or as a component of sustainable farming systems. We are also confident, the consumer will still be desirous of ruminant livestock products as the quality of these products is changed to fit healthy lifestyles. Likely, in the future we may be producing forages on more fragile and hostile landscapes as well as using forages as a component of short-term crop rotations in better environments. The long-term pastures and other forage production areas probably will have more diverse species present, some that will be used in only limited or specific conditions. Livestock producers will find it necessary to select specific forage species and cultivars and design management practices to effectively maintain and utilize specific forage mixtures.

REFERENCES

Adams, N. R. 1989. Phytoestrogens. p. 23-51. *In* P. R. Cheeke (ed.) Toxicants of plant origin. Vol. 4. Phenolics. CRC Press, Boca Raton, FL.

Akin, D. E. 1989. Histological and physical factors affecting digestibility of forages. Agron. J. 81:17-25.

Akin, D. E., and A. Chesson. 1989. Lignification as the major factor limiting forage feeding value especially in warm conditions. p. 1753-1760. *In* Proc. 16th Int. Grassl. Congr., Nice, France. 4-11 Oct. 1989. French Grassl. Soc.

Akin, D. E., L. L. Rigsby, C. E. Lyon, and W. R. Windham. 1990. Relationship of tissue digestion to textural strength in bermudagrass and alfalfa stems. Crop Sci. 30:990-993.

Albrecht, K. A., W. F. Wedin, and D. R. Buxton. 1987. Cell-wall composition and digestibility of alfalfa stems and leaves. Crop Sci. 27:735-741.

Albrecht, K. A., and G. A. Broderick. 1990. Degradation of forage legume protein by rumen microorganisms. p. 185. *In* Agronomy Abstracts. ASA, Madison, WI.

Allard, G., C. J. Nelson, and S. G. Pallardy. 1991. Shade effects on growth of tall fescue: I. Leaf anatomy and dry matter partitioning. Crop Sci. 31:163-167.

Allen, V. G., J. P. Fontenot, and W. H. McClure. 1983. Yield and quality of forage and performance of steers grazing orchardgrass treated with mefluidide. p. 9-13. *In* Proc. Forage and Grassl. Conf., Eau Claire, WI. Am. Forage and Grassl. Counc., Lexington, KY.

Anderson, B., A. G. Matches, and C. J. Nelson. 1989. Carbohydrate reserves and tillering of switchgrass following clipping. Agron. J. 81:13-16.

Bacon, C. W., and M. R. Siegel. 1988. Endophyte parasitism of tall fescue. J. Prod. Agric. 1:45-55.

Bolger, T. P., and D. W. Meyer. 1983. Influence of plant density on alfalfa yield and quality. p. 37-41. *In* Proc. Am. Forage and Grassl. Conf., Eau Claire, WI. Am. Forage and Grassl. Counc., Lexington, KY.

Bosworth, S. C., C. S. Hoveland, G. A. Buchanan, and W. B. Anthony. 1980. Forage quality of selected warm-season weed species. Agron. J. 72:1050-1054.

Brick, M. A., A. K. Dobrenz, and M. H. Schonhorst. 1976. Transmittance of the multifoliolate leaf characteristic in non-dormant alfalfa. Agron. J. 68:134-136.

Briske, D. D. 1991. Developmental morphology and physiology of grasses. p. 85-108. *In* R. K. Heitschmidt and J. W. Stuth (ed.) Grazing management: An ecological perspective. Timber Press, Portland, OR.

Britton, R. A. 1981. Nitrate toxicity in livestock. p. 9-11. *In* Living with nitrate. Nebraska Coop. Ext. Serv. Ext. Circ. EC 81-2400.

Brown, P. W., and C. B. Tanner. 1983. Alfalfa stem and leaf growth during water stress. Agron. J. 75:799-805.

Brown, R. H., R. E. Blaser, and J. P. Fontenot. 1963. Digestibility of fall grown Kentucky 31 fescue. Agron. J. 55:321-324.

Brown, R. H., and D. A. Ashley. 1974. Fertilizer effects on photosynthesis, organic reserves, and regrowth mechanisms of forages. p. 455-479. *In* D. A. Mays (ed.) Forage fertilization. ASA, Madison, WI.

Brown, R. H. 1978. A difference in N use efficiency in C_3 and C_4 plants and its implications in adaptation and evolution. Crop Sci. 18:93-98.

Brown, R. H., and R. E. Simmons. 1979. Photosynthesis of grass species differing in CO_2 fixation pathways. I. Water-use efficiency. Crop Sci. 19:375-379.

Brown, R. H., and D. E. Radcliffe. 1986. A comparison of apparent photosynthesis in sericea lespedeza and alfalfa. Crop Sci. 26:1208-1211.

Bruckner, P. L., and W. W. Hanna. 1990. In vitro digestibility of fresh leaves and stems of small grain species and genotypes. Crop Sci. 30:196-202.

Bush, L. P., F. F. Fannin, M. R. Siegel, D. L. Dahlman, and H. R. Burton. 1993. Chemistry, occurrence and biological effects of saturated pyrrolizidine alkaloids associated with endophyte-grass interactions. Agric. Ecosystems Environ. 44:81-102.

Butler, L. G. 1989. Sorghum polyphenols. p. 95-121. *In* P. R. Cheeke (ed.) Toxicants of plant origin. Vol. 4. Phenolics. CRC Press, Boca Raton, FL.

Buxton, D. R., J. S. Hornstein, W. F. Wedin, and G. C. Marten. 1985. Forage quality in stratified canopies of alfalfa, birdsfoot trefoil, and red clover. Crop Sci. 25:273-279.

Buxton, D. R., and J. S. Hornstein. 1986. Cell wall concentration and components in stratified canopies of alfalfa, birdsfoot trefoil, and red clover. Crop Sci. 26:180-184.

Buxton, D. R., J. S. Hornstein, and G. C. Marten. 1987. Genetic variation for forage quality of alfalfa stems. Can. J. Plant Sci. 67:1057-1067.

Buxton, D. R., and G. C. Marten. 1989. Forage quality of plant parts of perennial grasses and relationships to phenology. Crop Sci. 29:429-435.

Buxton, D. R. 1990. Cell wall components in divergent germplasms of four perennial forage grass species. Crop Sci. 30:402-408.

Buxton, D. R., and M. D. Casler. 1993. Environmental and genetic effects on cell wall composition and digestibility. p. 685-714. *In* H. G. Jung, D. R. Buxton, R. D. Hatfield, and J. Ralph (ed.) Forage cell wall structure and digestibility. ASA, Madison, WI.

Caravetta, G. J., J. H. Cherney, and K. D. Johnson. 1990a. Within-row spacing influences on diverse sorghum genotypes. I. Morphology. Agron. J. 82:206-210.

Caravetta, G. J., J. H. Cherney, and K. D. Johnson. 1990b. Within-row spacing influences on diverse sorghum genotypes. II. Dry matter yield and forage quality. Agron. J. 82:210-215.

Carlson, I. T., D. K. Christensen, and R. B. Pearce. 1983. Selection for specific leaf weight in reed canarygrass and its effects on the plant. p. 207-209. *In* J. A. Smith and V. W. Hays (ed.) Proc. 14th Int. Grassl. Congr., Lexington, KY. 15-24 June 1981. Westview Press, Boulder, CO.

Casal, J. J., V. A. Deregibus, and R. A. Sanchez. 1985. Variation in tiller dynamics and morphology in *Lolium multiflorum* Lam. vegetative and reproductive plants as affected by differences in red/far-red irradiation. Ann. Bot. 56:553-559.

Casal, J. J., R. A. Sanchez, and V. A. Deregibus. 1987. Tillering responses of *Lolium multiflorum* plants to changes of red/far-red ratio typical of sparse canopies. J. Exp. Bot. 38:1432-1439.

Chatterton, N. J., P. A. Harrison, J. H. Bennett, and K. H. Asay. 1989. Carbohydrate partitioning in 185 accessions of Gramineae grown under warm and cool temperatures. J. Plant Physiol. 134:169-179.

Cheeke, P. R. 1989a. Toxicants of plant origin. Vol. 1. Alkaloids. CRC Press, Boca Raton, FL.

Cheeke, P. R. 1989b. Toxicants of plant origin. Vol. 2. Glycosides. CRC Press, Boca Raton, FL.

Cheeke, P. R. 1989c. Toxicants of plant origin. Vol. 4. Phenolics. CRC Press, Boca Raton, FL.

Cherney, J. H., and G. C. Marten. 1982. Small grain forage potential. II. Interrelationships among biological, chemical, morphological, and anatomical determinants of quality. Crop Sci. 22:240-245.

Cherney, J. H., G. C. Marten, and R. D. Goodrich. 1983. Rate and extent of cell wall digestion of total forage and morphological components of oats and barley. Crop Sci. 23:213-216.

Cherney, J. H., J. J. Volenec, and K. D. Johnson. 1986. Forage quality of alfalfa as influenced by plant density. p. 127-131. *In* Proc. Am. Forage and Grassl. Conf., Athens, GA. Am. Forage and Grassl. Counc., Lexington, KY.

Christensen, D. W., D. D. Stuthman, and A. W. Hovin. 1984. Associations among morphological and digestibility characters in reed canarygrass. Crop Sci. 24:675-678.

Coleman, S. W. 1992. Plant-animal interface. J. Prod. Agric. 5:7-13.

Conn, E. E. 1980. Cyanogenic compounds. Annu. Rev. Plant Physiol. 31:433-451.

Cooper, J. P. 1970. Potential production and energy conversion in temperate and tropical grasses. Herb. Abstr. 40:1-15.

Crawley, M. J. 1983. Herbivory: The dynamics of animal-plant interactions.

Studies in Ecol. Vol. 10. Univ. of California Press, Berkeley, CA.

Davidson, D. J. 1980. Influence of nitrogen fertilization on yield and leaf senescence of two contrasting genotypes of tall fescue. M.S. thesis. University of Missouri, Columbia, MO.

Edwards, M. T., D. A. Sleper, and W. Q. Loegering. 1981. Histology of healthy and diseased orchardgrass leaves subjected to digestion in rumen fluid. Crop Sci. 21:341-343.

Etzel, M. G., J. J. Volenec, and J. J. Vorst. 1988. Leaf morphology, shoot growth, and gas exchange of multifoliolate alfalfa phenotypes. Crop Sci. 28:263-269.

Fales, S. L., R. R. Hill, and R. J. Hoover. 1990. Chemical regulation of growth and forage quality of cool-season grasses with the imazethapyr. Agron. J. 82:9-17.

Farnsworth, N. R., A. S. Bingel, G. A. Cordell, F. A. Crane, and H. H. S. Fong. 1975. Potential value of plants as sources of new anti-fertility agents. J. Pharm. Sci. 64:717-754.

Ferguson, J. E., and R. P. Murphy. 1973. Comparison of trifoliolate and multifoliolate phenotypes of alfalfa (*Medicago sativa* L.). Crop Sci. 13:463-465.

Fick, G. W., and S. C. Mueller. 1990. Alfalfa quality, maturity, and mean stage of development. Cornell Inf. Bull. No. 217.

Forwood, J. R., A. G. Matches, and C. J. Nelson. 1988. Forage yield, nonstructural carbohydrate levels, and quality trends of caucasian bluestem. Agron. J. 80:135-139.

Frank, A. B., J. D. Berdahl, and R. E. Barker. 1985. Morphological development and water use in clonal lines of four forage grasses. Crop Sci. 25:339-344.

Forbes, T. D. A., and S. W. Coleman. 1993. Forage intake and ingestive behavior of cattle grazing old world bluestems. Agron. J. 85:808-816.

Garner, G. B., G. E. Rottinghaus, C. N. Cornell, and H. Testereci. 1993. Chemistry of compounds associated with endophyte/grass interaction: Ergovaline- and ergopeptine-related alkaloids. Agric. Ecosystems Environ. 44:65-80.

Gist, G. R., and R. M. Smith. 1948. Root development of several common forage grasses to a depth of eighteen inches. J. Am. Soc. Agron. 40:1036-1042.

Glenn, S., C. E. Rieck, D. G. Ely, and L. P. Bush. 1980. Quality of tall fescue forage affected by mefluidide. J. Agric. Food Chem. 28:391-393.

Griffin, J. L., and G. A. Jung. 1983. Leaf and stem forage quality of big bluestem and switchgrass. Agron. J. 75:723-726.

Hacker, J. B., and D. J. Minson. 1981. The digestibility of plant parts. Herb. Abstr. 51:459-482.

Halim, R. A., D. R. Buxton, M. J. Hattendorf, and R. E. Carlson. 1989. Water stress effects on alfalfa forage quality after adjustment for maturity differences. Agron. J. 81:189-194.

Hall, J. W., and W. Majak. 1989. Plant and animal factors in legume bloat. p. 93-106. *In* P. R. Cheeke (ed.) Toxicants of plant origin. Vol. 3. Protein and amino acids. CRC Press, Boca Raton, FL.

Haslam, E. 1989. Plant polyphenols-vegetable tannins revisited.Cambridge Univ. Press, Cambridge, U.K.

Henderson, M. S., and D. L. Robinson. 1982. Environmental influences on fiber component concentrations of warm-season perennial grasses. Agron. J. 74:573-579.

Hendrickson, R. E., D. P. Poppi, and D. J. Minson. 1981. The voluntary intake, digestibility, and retention time by cattle and sheep of stem and leaf fractions of a tropical legume (*Lablab purpureus*). Aust. J. Agric. Res. 32:389-398.

Hesketh, J. D. 1963. Limitations to photosynthesis responsible for differences among species. Crop Sci. 3:493-496.

Hides, D. H., J. A. Lovatt, and M. V. Hayward. 1983. Influence of stage of maturity on the nutritive value of Italian ryegrasses. Grass Forage Sci. 38:33-38.

Hodgson, J. 1977. Factors limiting herbage intake by the grazing animal. p. 70-75. *In* B. Gilsenan (ed.) Proc. Int. Meet. on Anim. Prod. from Temperate Grassl. An Soras Taluntais, Dublin, Ireland.

Hodgson, J. 1981. Variations in the surface characteristics of the sward and the short-term rate of herbage intake by calves and lambs. Grass Forage Sci. 36:49-57.

Hodgson, J. 1982. Ingestive behaviour. p. 113-138. *In.* J. D. Leaver (ed.) Herbage intake handbook. Br. Grassl. Soc., Hurley, U.K.

Hogg, A. 1981. Nitrates and animal health. p. 11-12. *In* Living with nitrate. Nebraska Coop. Ext. Serv. Ext. Circ. EC 81-2400.

Horst, G. L., and C. J. Nelson. 1979. Compensatory growth of tall fescue following drought. Agron. J. 71:559-563.

Hoveland, C. S. 1993. Importance and economic significance of the *Acremonium* endophytes to performance of animals and grass plant. Agric. Ecosystems Environ. 44:3-12.

Howarth, R. E. 1975. A review of bloat in cattle. Can. Vet. J. 16:281-294.

Howarth, R. E., B. P. Goplen, S. A. Brandt, and K. -J. Cheng. 1982. Disruption of leaf tissues by rumen microorganisms: An approach to breeding bloat-safe forage legumes. Crop Sci. 22:564-568.

Hutchins, S. H., D. R. Buxton, and L. P. Pedigo. 1989. Forage quality of alfalfa as affected by potato leafhopper feeding. Crop Sci. 29:1541-1545.

Jamieson, W. S., and J. H. Hodgson. 1979. The effects of variation in sward characteristics upon the ingestive behaviour and herbage intake of calves and lambs under a continuous stocking management. Grass Forage Sci. 34:273-282.

Jones, W. T., and J. W. Lyttleton. 1971. Bloat in cattle. 34. A survey of legume forages that do and do not cause bloat. N.Z. J. Agric. Res. 14:101-107.

Joost, R., and S. Quisenberry. 1993. Introduction to acremonium/grass interactions. Agric. Ecosystems Environ. 44:1-2.

Jung, G. A., D. E. Brann, and G. W. Fissel. 1981. Environmental and plant growth stage effects on composition and digestibility of crownvetch stems and leaves at four locations in West Virginia. Agron. J. 73:122-128.

Jung, G. A., J. A. Shaffer, W. L. Stout, and M. Panciera. 1990. Warm-season grass diversity in yield, plant morphology, and nitrogen concentration and removal in northeastern USA. Agron. J. 82:21-26.

Jung, H. G., D. R. Buxton, R. D. Hatfield, and J. Ralph (ed.) 1993. Forage cell wall structure and digestibility. ASA-CSSA-SSSA, Madison, WI.

Kephart, K. D., D. R. Buxton, and R. R. Hill, Jr. 1989. Morphology of alfalfa selected for divergent herbage lignin concentration. Crop Sci. 29:778-782.

Kephart, K. D., D. R. Buxton, and R. R. Hill, Jr. 1990. Digestibility and cell wall components of alfalfa following selection for divergent herbage lignin concentration. Crop Sci. 30:207-212.

Kingsbury, J. M. 1964. Poisonous plants of the United States and Canada. Prentice Hall, Englewood Cliffs, NJ.

Kudo, H., K. -J. Cheng, M. R. Hanna, R. E. Howarth, B. P. Goplen, and J. W. Costerton. 1985. Ruminal digestion of alfalfa strains selected for slow and fast initial rates of digestion. Can. J. Anim. Sci. 65:157-161.

Kumar, R., and M. Singh. 1984. Tannins: Their adverse role in ruminant nutrition. J. Agric. Food Chem. 32:447-453.

Lapierre, C. 1993. Application of new methods for the investigation of lignin structure. p. 133-166. *In* H. G. Jung, D. R. Buxton, R. D. Hatfield, and J. Ralph (ed.) Forage cell wall structure and digestibility. ASA-CSSA-SSSA, Madison, WI.

Laredo, M. A., and D. J. Minson. 1973. The voluntary intake, digestibility, and retention time by sheep of leaf and stem fractions of five grasses. Aust. J. Agric. Res. 24:875-888.

Laude, H. M. 1953. The nature of summer dormancy in perennial grasses. Bot. Gaz. 114:284-292.

Laude, H. M. 1964. Plant response to high temperature. p. 15-31. *In* Forage plant physiology and soil-range relationships. ASA Spec. Pub. 5, ASA, Madison, WI.

Lees, G. L., R. E. Howarth, B. P. Goplen, and A. C. Fesser. 1981. Mechanized disruption of leaf tissues and cells in some bloat causing and bloat-safe legumes. Crop Sci. 21:444-448.

Lees, G. L., R. E. Howarth, and B. P. Goplen. 1982. Morphological characteristics of leaves from some legume forages: Relation to digestion and mechanical strength. Can. J. Bot. 60:2126-2132.

Lenssen, A. W., E. L. Sorenson, G. L. Posler, and S. L. Blodgett. 1991a. Depression of forage quality of alfalfa leaves and stems by *Acyrthosiphon kondoi* (Homoptera: Aphidacea). Environ. Entomol. 20:71-76.

Lenssen, A. W., E. L. Sorenson, G. L. Posler, and D. L. Stuteville. 1991b. Resistance to anthracnose protects forage quality of alfalfa. Crop Sci. 31:147-150.

Lentz, E. M., and D. R. Buxton. 1992. Digestion kinetics of orchardgrass as influenced by leaf morphology, fineness of grind and maturity group. Crop Sci. 32:482-486.

L'Huillier, P. J., D. P. Poppi, and T. J. Fraser. 1984. Influence of green leaf distribution on diet selection by sheep and the implications for animal performance. Proc. N.Z. Soc. Anim. Prod. 44:105-107.

Loomis, R. S., and W. A. Williams. 1969. Productivity and the morphology of crop stands: Patterns with leaves. p. 27-47. *In* J. D. Eastin, F. A. Haskins, C. Y. Sullivan, and C. H. M. van Bavel (ed.) Physiological aspects of crop yield. ASA, Madison, WI.

Ludlow, M. M., and G. L. Wilson. 1971. Photosynthesis of tropical pasture plants. I. Illuminance, carbon dioxide concentration, leaf temperature, and leaf-air vapor pressure difference. Aust. J. Biol. Sci. 24:449-470.

Mainer, A., and K. T. Leath. 1978. Foliar diseases alter carbohydrate and protein levels in leaves of alfalfa and orchardgrass. Phytopath. 68:1252-1255.

Marten, G. C., R. M. Jordan, and A. W. Hovin. 1976. Biological significance of reed canarygrass alkaloids and associated palatability variation to grazing sheep and cattle. Agron. J. 68:909-914.

Marten, G. C., and R. N. Anderson. 1975. Forage nutritive value and palatability of 12 common annual weeds. Crop Sci. 15:821-827.

Marten, G. C. 1985. Reed canarygrass. p. 207-216. *In* M. E. Heath, R. F Barnes, and D. E. Metcalfe (ed.) Forages. Iowa State Univ. Press, Ames, IA.

McWilliam, J. R. 1978. Responses of pasture plants to temperature. p. 17-34. *In* J. R. Wilson (ed.) Plant relations in pastures. CSIRO, East Melbourne, Australia.

Minson, D. J. 1990. Forage in ruminant nutrition. Academic Press Inc., San Diego, CA.

Moellenbeck, D. J., and S. S. Quisenberry. 1991. Effects of nymphal populations of threecornered alfalfa hopper (Homoptera: Membracidae) on Florida 77 alfalfa plants. J. Econ. Ent. 84:1889-1893.

Moore, K. J., L. E. Moser, K. P. Vogel, S. S. Waller, B. E. Johnson, and J. F. Pedersen. 1991. Describing and quantifying growth stages of perennial forage grasses. Agron. J. 83:1073-1077.

Morrison, I. M. 1980. Changes in the lignin and hemicellulose concentrations of ten varieties of temperate grasses with increasing maturity. Grass Forage Sci. 35:287-293.

Murphy, J. S., and D. D. Briske. 1992. Does light quality regulate tillering in the bunchgrass *Schizachyrium scoparium*. Soc. Range Manage. Abstr. no. 127.

Nelson, C. J., and Dale Smith. 1969. Growth of birdsfoot trefoil and alfalfa. IV. Carbohydrate reserve levels and growth analysis under two temperature regimes. Crop Sci. 9:589-591.

Nelson, C. J., K. H. Asay, and D. A. Sleper. 1977. Mechanisms of canopy development of tall fescue genotypes. Crop Sci. 17:476-478.

Nelson, C. J., D. A. Sleper, and J. H. Coutts. 1985. Field performance of tall fescue selected for leaf-area expansion rate. p. 320-322. *In* Proc. 15th Int. Grassl. Congr., Kyoto, Japan. 24-31 Aug., 1985. Jap. Soc. Grassl. Sci., Nishinasuno, Tochigi-ken, Japan.

Nelson, C. J. 1988. Genetic associations between photosynthetic characters and yield: Review of the evidence. Plant Physiol. Biochem. 26:543-554.

Nelson, C. J. 1992. Physiology of leaf growth of grasses. p. 175-179. *In* Proc. 14th Gen. Meet. European Grassl. Fed., Lahti, Finland. 8-11 June 1992.

Nguyen, H. T., D. A. Sleper, and A. G. Matches. 1982. Inheritance of forage quality and its relationship to leaf tensile strength in tall fescue. Crop Sci. 22:67-72.

Nordkvist, E., and P. Aman. 1986. Changes during growth in anatomical and chemical composition and in vitro degradability of lucerne. J. Sci. Food Agric. 37:1-7.

Parsons, A. J., and M. J. Robson. 1980. Seasonal changes in the physiology of S24 perennial ryegrass (*Lolium perenne* L.). 1. Response of leaf extension to temperature during the transition from vegetative to reproductive growth. Ann. Bot. 46:435-444.

Parsons, A. J., and E. L. Leafe. 1981. Photosynthesis and carbon balance of a grazed sward. p. 69-71. *In* C. E. Wright (ed.) Plant physiology and herbage production. Br. Grassl. Soc. Occ. Symp. no. 13, Hurley, U.K.

Parsons, A. J., E. L. Leafe, B. Collett, and W. Stiles. 1983. The physiology of grass production under grazing. I. Characteristics of leaf and canopy photosynthesis of continuously grazed swards. J. Appl. Ecol. 20:117-126.

Parsons, A. J., A. Harvey, and J. Woledge. 1991a. Plant-animal interactions in a continuously grazed mixture. I. Differences in the physiology of leaf expansion and the fate of leaves of grass and clover. J. Appl. Ecol. 28:619-634.

Parsons, A. J., A. Harvey, and I. R. Johnson. 1991b. Plant-animal interactions in continuously grazed mixtures. 2. The role of differences in physiology of plant growth and of selective grazing on the performance and stability of species in a mixture. J. Appl. Ecol. 28:635-658.

Penning, P. D., A. J. Parsons, G. E. Hopper, and R. J. Orr. 1989. Responses of ingestive behaviour by sheep to changes in sward structure. p. 791-792. *In* Proc. 16th Int. Grassl. Congr., Nice, France. 4-11 Oct. 1989. French Grassl. Soc.

Perry, L. J.,Jr., and D. D. Baltensperger. 1979. Leaf and stem yields and forage quality of three nitrogen fertilized warm season grasses. Agron. J. 71:355-358.

Poppi, D. P., D. J. Minson, and J. H. Ternouth. 1981. Studies of cattle and sheep eating leaf and stem fractions of grasses. I. The voluntary intake, digestibility, and retention time in the reticulo-rumen. Aust. J. Agric. Res. 32:99-108.

Poulton, J. E. 1990. Cyanogenesis in plants. Plant Physiol. 94:401-405.

Reynolds, J. H., W. A. Krueger, C. L. Walker, and J. C. Waller. 1993. Plant growth regulator effects on growth and forage quality of tall fescue. Agron. J. 85:545-548.

Roberts, C. A., and K. J. Moore. 1990. Chemical regulation of tall fescue reproductive development and quality with amidochlor. Agron. J. 82:523-526.

Rowan, D. D. 1993. Lolitrems, peramine, and paxilline: Mycotoxins of the ryegrass/endophyte interaction. Agric. Ecosystems Environ. 44:103-122.

Salisbury, F. B., and C. W. Ross. 1985. Plant physiology. Wadsworth Publishing Co., Belmont, CA.

Salunkhe, D. K., R. N. Adsule, and K. I. Bhonsle. 1989. Antifertility agents of plant origin. p. 53-81. *In* P. R. Cheeke (ed.) Toxicants of plant origin. Vol. 4. Phenolics. CRC Press, Boca Raton, FL.

Sanderson, M. A., and W. F. Wedin. 1989. Phenological stage and herbage quality relationships in temperate grasses and legumes. Agron. J. 81:864-869.

Sanderson, M. A. 1992a. Morphological development of switchgrass and kleingrass. Agron. J. 84:415-419.

Sanderson, M. A. 1992b. Predictors of alfalfa forage quality: Validation with field data. Crop Sci. 32:245-250.

Schmidt, S. P., and T. G. Osborn. 1993. Effects of endophyte-infested tall fescue on animal performance. Agric. Ecosystems Environ. 44:233-262.

Schnyder, H., and C. J. Nelson. 1989. Growth rates and assimilate partitioning in the elongation zone of tall fescue leaf blades at high and low irradiance. Plant Physiol. 90:1201-1206.

Sheaffer, C. C., and G. C. Marten. 1986. Effect of mefluidide on cool-season perennial grass forage yield and quality. Agron. J. 78:75-79.

Siegel, M. R. 1993. Acremonium endophytes: Our current state of knowledge and future directions for research. Agric. Ecosystems Environ. 44:301-321.

Simon, V., and B. H. Park. 1983. A descriptive scheme for stages of development in perennial forage grasses. p. 416-418. *In* J. A. Smith and V. W. Hays (ed.) Proc. 14th Int. Grassl. Congr. 15-24 June 1981. Westview Press, Boulder, CO.

Simonds, R. G., A. Davies, and A. Troughton. 1973. The effect of spacing on the regrowth of two genotypes of perennial ryegrass. J. Agric. Sci., Camb. 80:495-502.

Skinner, R. H., and S. R. Simmons. 1993. Modulation of leaf elongation, tiller appearance and tiller senescence in spring barley by far-red light. Plant, Cell, Environ. 16:555-562.

Sleper, D. A., and P. N. Drolsom. 1974. Analysis of several morphological traits and their associations with digestibility in *Bromus inermis* Leyss. Crop Sci. 14:34-36.

Sleper, D. A., and C. J. Nelson. 1989. Productivity of selected high and low leaf area expansion *Festuca arundinacea* strains. p. 379-380. *In* Proc. 16th Int. Grassl. Congr., Nice, France. 4-11 Oct. 1989. French Grassl. Soc.

Snaydon, R. W. 1972. The effect of total water supply and frequency of application upon lucerne. Aust. J. Agric. Res. 23:253-256.

Snaydon, R. W. 1991. The productivity of C_3 and C_4 plants: A reassessment. Functional Ecol. 5:321-330.

Sollenberger, L. E., J. E. Moore, K. H. Quesenberry, and P. T. Beede. 1987. Relationships between canopy botanical composition and diet selection in aeschynomene-limpograss pastures. Agron. J. 79:1049-1054.

Stobbs, T. H. 1973. The effect of plant structure on the intake of tropical pastures. II. Differences in sward structure, nutritive value, and bite size of animals grazing *Setaria anceps* and *Chloris gayana* at various stages of growth. Aust. J. Agric. Res. 24:821-829.

Teeri, J. A., and L. G. Stowe. 1976. Climatic patterns and the distribution of C_4 grasses in North America. Oecologia 23:1-12.

Terry, R. A., and J. M. A. Tilley. 1964. The digestibility of the leaves and stems of perennial ryegrass, cocksfoot, timothy, tall fescue, lucerne, and sainfoin as measured by an *in vitro* procedure. J. Br. Grassl. Soc. 19:363-372.

Tewe, O. O., and E. A. Iyayi. 1989. Cyanogenic glucosides. p. 43-60. *In* P. R. Cheeke (ed.) Toxicants of plant origin. Vol. 2. Glycosides. CRC Press, Boca Raton, FL.

Thompson, F. N., and J. A. Stuedemann. 1993. Pathophysiology of fescue toxicosis. Agric. Ecosystems Environ. 44:263-282.

Treharne, K. J., and C. J. Nelson. 1975. Effect of temperature on photosynthetic and photo-respiration activity in tall fescue. p. 61-69. *In* R. Marcelle (ed.) Environmental and biochemical control of photosynthesis. W. Junk Publishers, The Hague, Netherlands.

Twidwell, E. K., K. D. Johnson, J. H. Cherney, and J. J. Volenec. 1988. Forage quality and digestion kinetics of switchgrass herbage and morphological components. Crop Sci. 28:778-782.

Ugherughe, P. O. 1986. Relationship between digestibility of *Bromus inermis* plant parts. J. Agron. Crop Sci. 157:136-143.

Van Soest, P. J. 1993. Cell wall matric interactions and degradation--session synopsis. p. 377-395. *In* H. G. Jung, D. R. Buxton, R. D. Hatfield, and J. Ralph (ed.) Forage cell wall structure and digestibility. ASA-CSSA-SSSA, Madison, WI.

Volenec, J. J., J. H. Cherney, and K. D. Johnson. 1987. Yield components, plant morphology, and forage quality of alfalfa as influenced by plant population. Crop Sci. 27:321-326.

Volenec, J. J., and C. J. Nelson. 1983. Responses of tall fescue leaf meristems to nitrogen fertilization and harvest frequency. Crop Sci. 23:720-724.

Volenec, J. J, and C. J. Nelson. 1984. Influence of temperature on leaf dark respiration of diverse tall fescue genotypes. Crop Sci. 24:907-912.

Volenec, J. J., and J. H. Cherney. 1990. Yield components, morphology, and forage quality of multifoliolate alfalfa phenotypes. Crop Sci. 30:1234-1238.

Vough, L. R., and G. C. Marten. 1971. Influence of soil moisture and ambient temperature on yield and quality of alfalfa forage. Agron. J. 63:40-42.

Whanger, P. D. 1989. Selenocompounds in plants and their effects on animals. p. 141-167. *In* P. R. Cheeke (ed.) Toxicants of plant origin. Vol. 3. Protein and amino acids. CRC Press, Boca Raton, FL.

Wilfong, R. T., R. H. Brown, and R. E. Blaser. 1967. Relationships between leaf area index and apparent photosynthesis in alfalfa (*Medicago sativa* L.) and ladino clover (*Trifolium repens* L.). Crop Sci. 7:27-30.

Willaman, J. J., and H. Li. 1970. Alkaloid bearing plants and their contained alkaloids. Lloydia, The J. Nat. Prod. 33:1-286.

Wilman, D., and M. A. K. Atimimi. 1984. The in vitro digestibility and chemical composition of plant parts in white clover, red clover, and lucerne during primary growth. J. Sci. Food Agric. 35:133-138.

Wilman, D., and P. J. Pearse. 1984. Effects of applied nitrogen on grass yield, nitrogen content, tillers, and leaves in field swards. J. Agric. Sci., Camb. 103:201-211.

Wilson, J. R. 1976a. Variation of leaf characteristics with level of insertion on a grass tiller. I. Development rate, chemical composition, and dry matter digestibility. Aust. J. Agric. Res. 27:343-354.

Wilson, J. R. 1976b. Variation of leaf characteristics with level of insertion on a grass tiller. II. Anatomy. Aust. J. Agric. Res. 27:355-364.

Wilson, J. R. 1983. Effects of water stress on herbage quality. p. 470-472. *In* J. A. Smith and V. W. Hays (ed.) Proc. 14th Int. Grassl. Congr. 15-24 June, 1991. Westview Press, Boulder, CO.

Wilson, J. P., R. N. Gates, and W. W. Hanna. 1991. Effect of rust on yield and digestibility of pearl millet forage. Phytopath. 81:233-236.

Wimmer, S. K., J. K. Ward, B. E. Anderson, and S. S. Waller. 1986. Mefluidide effects on smooth brome composition and grazing cow-calf performance. J. Anim. Sci. 63:1054-1062.

Woledge, J. 1979. Effect of flowering on the photosynthetic capacity of ryegrass leaves grown with and without natural shading. Ann. Bot. 44:197-207.

Wright, M. J., and K. L. Davison. 1964. Nitrate accumulation in crops and nitrate poisoning in animals. Adv. Agron. 16:197-247.

Zarrough, K. M., C. J. Nelson, and J. H. Coutts. 1983. Relationship between tillering and forage yields of tall fescue. I. Yield. Crop Sci. 23:333-337.

CHAPTER 4

PLANT ENVIRONMENT AND QUALITY

Dwayne R. Buxton and Steven L. Fales

INTRODUCTION

No single factor impacts forage quality more than plant maturity, but plant environment modifies the impact of plant maturity. Plant environment includes those biotic and abiotic factors that influence growth and development of forages. Cumulative effects are integrated through physiological process and reflected in forage growth rate, developmental rate, yield, and herbage quality. Year-to-year, seasonal, and variations in environment related to geographical location alter herbage quality, even when forages are harvested at similar morphological stages. This makes prediction of forage quality difficult and may result in inconsistent performance of animals that consume the forage.

Plants rarely grow in ideal environments; instead they experience environmental fluctuations and stresses that modify morphology and rate of development, limit yield, and alter quality. Stress is caused when any environmental factor is not ideal for plant growth and development. It can be caused by numerous factors, but those that we will consider are temperature, water deficit, solar radiation, nutrient deficiency, and pests. Plant cell walls provide the first line of defense against many of these stresses. Secondary cell wall development, especially lignification, is an important aspect of protection. Lignification also restricts nutritive availability of cell walls to animals that consume them. Cell walls vary in digestibility, but usually are only partly available, whereas the cell contents within them are nearly completely digestible.

Most environmental stresses have a greater effect on forage yield than on digestibility or related quality factors. Some of these relationships were recently summarized by Buxton and Casler (1993). Plant environment often exerts its greatest influence over herbage quality by altering leaf/stem ratios, but it also causes other morphological modifications and changes in chemical composition of plant parts. Changes in plant morphology can alter herbage availability, especially by influencing intake of grazing animals by altering potential bite size. Canopy height is the most important pasture variable that effects animal bite size (Hodgson, 1981). In addition to altering leaf/stem ratios, plant environment influences senescence rates and amount of dead plant material. Animals generally

---

D. R. Buxton, Field Crops Res. Unit and cluster scientist of U.S. Dairy Forage Res. Ctr., Agric. Res. Ser., USDA, Dep. of Agronomy, Iowa State Univ., Ames, IA 50011; S. L. Fales, Dep. of Agronomy, Pennsylvania State Univ., University Park, PA 16802.

select young, green leaf tissues rather than stem and dead plant tissues. Many stresses slow plant growth and development, resulting in herbage quality being maintained at a higher level. Stresses that cause reductions in leaf/stem ratio decrease herbage quality because of the higher nutritive value of leaves.

TEMPERATURE

Because the nutritive value of forage is governed by the amount and availability of metabolic and anabolic products, including cell contents and cell walls, it follows that any factor that influences these products or the interrelationships among them also will affect forage quality. Temperature usually has a greater influence on forage quality than other environmental factors encountered by plants. Plant temperature is the result of complex interactions between plants and their environment and is influenced by radiation flux density, heat conduction, heat convection, latent heat, as well as morphological and anatomical features of plants. Furthermore, because of variations in canopy structure of plants, aspect (angle) of plant parts, and the resulting differences in radiation load, actual tissue temperature can vary widely within an individual plant at any given time.

General Effects on Forages

The effect of temperature on biochemical processes arises from the fact that temperature determines the kinetic energy of molecules, which in turn determines whether or not a reaction will occur. Energy barriers must be overcome before a reaction will occur. In other words, reactions require a minimum activation energy to proceed. The relationship between temperature and energy among molecules is predicted by the Boltzman energy distribution (Nobel, 1988).

The "law of mass action" states that the rate of a chemical reaction is proportional to the concentration of reactants. Thus, in the reaction, A + B ---> C + D, the velocity, v, can be expressed as:

$$v = k[A][B]$$

where k is the rate constant. In the late 1800s, the Swedish chemist Arrhenius observed that the rate constant and the absolute temperature (degrees Kelvin) at which reactions occur are related according to the following expression:

$$\ln k = A - E_a/RT$$

where T is absolute temperature, E_a represents the activation energy of the reaction, R is the gas constant, and A is an integration constant.

A related concept is that of Q_{10}, an empirical approach based on the observation that a given incremental temperature elevation increases a reaction rate by a constant factor. By definition:

$$Q_{10} = \text{(rate of a process at } T_K + 10°C)/\text{(rate of the process at } T_K)$$

where T_K is the reference temperature (K). The relationship also can be expressed as follows:

$$\ln k = \ln k_r + (T-T_R/10) \ln Q_{10}$$

where k is the rate constant, k_r is the rate constant at reference temperature T_R, and Q_{10} is the factor by which the rate constant increases for a given 10°C temperature increment (Johnson and Thornley, 1985). For a process with no energy barrier to surmount, $Q_{10} = 1.0$. Most important biochemical reactions, however, require significant activation energy and show Q_{10} values between 2 and 4 (Nobel, 1988).

As is the situation with simple chemical reactions, enzyme-catalyzed reactions generally increase with increasing temperature (within the range of temperatures at which the enzyme is stable and retains activity). Indeed, the function of enzymes, as is the situation with chemical catalysts, is to reduce the E_a required for a reaction to proceed. Most metabolic reactions in plants have Q_{10} values of approximately 2.0. However, Q_{10} varies among enzymes, depending on the E_a of a particular reaction as well as the thermal stability of the enzymes involved. Moreover, most plant species exhibit different temperature optima for various processes, with the rate of each reaction increasing as temperature is raised below the optimum and decreasing thereafter. Rising activity below the temperature optimum probably simply reflects the Boltzman energy distribution as increasing temperature causes more molecules to have sufficient kinetic energy to surmount energy barriers. Above the optimum temperature, however, increasing thermal motion may interfere with enzyme-substrate binding by changing the three-dimensional conformation of the enzymes.

Because reactions determining the synthesis of cell wall constituents are enzyme-mediated (Nishitani and Masuda, 1979), it is likely that temperature influences concentrations of various cell wall constituents and linkages among them. Other plant processes also are affected by temperature, including membrane phase changes (Lyons, 1973) and changes in viscosity and changes in rates of diffusion and translocation (Johnson and Thornley, 1985). Since each enzyme has a discrete Q_{10}, a change in temperature will result in a shift in carbon partitioning among alternate pathways. Unfortunately, there is virtually no information available concerning the direct effects of temperature on reactions leading to the formation of plant cell walls. Nevertheless, a body of circumstantial evidence, e.g., the generally positive relationship between temperature and concentrations of structural material (e.g., Fales, 1986; Akin et al., 1987; Wilson et al., 1991b), suggests the existence of temperature-dependent points of control for partitioning of photosynthate between cell walls and cell contents.

Effects on Plant Development

In the broadest sense, temperature (along with soil moisture) affects the quality of forage by determining which species will grow in a certain region. Temperature is the major determinant of geographical adaptation of plant species. This is particularly manifested in temperature extremes encountered over the ontogeny of plants. These extremes can cause plant death or severely weaken plants. Under field conditions, high-temperature stress frequently occurs concurrently with water stress making it difficult to separate the two effects.

Within a region or area, the primary effect of temperature on forage quality is to determine the rate of plant development and to influence the relative proportions of leaves and stems. A secondary temperature effect is to bring about differences in morphology or tissue type within leaves or stems. Because temperature seems to exert greater effects on digestibility than do other environmental variables, economic implications of changes in temperature should not be ignored in attempts to assess potential consequences of global warming. If global warming occurs, the length of vegetative stages will be reduced for many forages.

Most research on temperature as a factor in carbon partitioning has dwelled on the effects of temperature (primarily low, e.g., Pollack, 1990) on the carbohydrate status of metabolic sinks such as storage organs (Farrar, 1988). These studies have shown that temperature can alter sink metabolism by speeding or slowing individual reactions (particularly when diffusion is a rate-limiting part of sink growth), by changing rates of active transport across membranes, by affecting enzyme activity ("fine control"), and by changing the concentration of various enzymes ("coarse control") through modifications of gene expression. Because developing cell walls also represent a metabolic sink for photosynthate, temperature likely exerts its influence in a similar manner, an idea that is supported by a wealth of evidence strongly suggesting a greater conversion of photosynthate to structural components at high temperatures (Deinum and Dirven, 1975; Da Silva et al., 1987).

Plant temperature usually deviates from air temperature (Gates, 1968). Frequently, if air temperature is below 35°C, temperatures of tissues exposed to direct sunlight are above air temperature, especially if humidity is high. If air temperature is above 35°C, tissue temperature may be below air temperature, especially if humidity is low and adequate soil water is available. Within temperature regimens many forages are grown, temperature of leaves exposed to sunlight may be above air temperature. Cloudy weather can change the relationship between air temperature and leaf temperature in that leaf temperatures are more likely to be near air temperature. As a result, cloudy weather may cause degree days to be overestimated when used to predict plant response.

When harvested at a particular growth stage, highest yields are usually obtained when forages are grown at temperatures near the lower boundary of their optimal range (Fick et al., 1988). High growth temperatures decrease stem diameter and increase rate of maturation and lignification (Fick et al., 1988; Marten et al., 1988). When grown at temperatures above those optimal for growth, forages tend to be shorter at flowering and to bloom earlier than when grown under cooler temperatures. High growth temperatures also promote stem development over leaf development and consequently will lower leaf/stem ratios in herbage (Deinum, 1984). For example, alfalfa (*Medicago sativa* L.) grown at 18/10°C had a higher leaf/stem ratio at the late-bud stage than that grown at either 26/18 or 34/26°C (Walgenbach et al., 1981).

Optimal growth temperatures were established by Cooper and Tainton (1968) to be near 20°C for cool-season species and from 30 to 35°C for warm-season species. The marked effect of temperature on forage growth, development, and chemical composition was first demonstrated in studies with perennial ryegrass *(Lolium perenne* L.) by Alberda (1965). At temperatures below the optimum for growth, soluble sugars accumulated because of lower temperature sensitivity of photosynthesis compared to that of growth. At temperatures above the optimum for growth, soluble-sugar concentrations tended to decline, but the effect was not as consistent as was that for sugar accumulation.

Effects on Chemical Composition and Digestibility

The negative effect of elevated temperature on digestibility of forage grasses has been the subject of numerous studies over the past 30 yr. In the vast majority of flowering plants, environmental controls such as daylength and temperature modulate rate of development. The extent of the temperature effect within the range of temperatures normally experienced during the course of a growing season is illustrated in a study by Dirven and Deinum (1977) that showed an 80 g kg^{-1} decline in in vitro digestibility (IVDMD) of tall fescue (*Festuca arundinacea* Schreb.) when temperature was increased from 15/10°C to 25/20°C. Wilson and Minson (1980) summarized the results of several experiments relating temperature to digestibility and concluded that leaves of cool season grasses show an average 6.6 g kg^{-1} decline in IVDMD for each °C increase in growth temperature. Thorvaldsson (1992) found that the average decline in IVDMD of timothy (*Phloeum pratense* L.) was 0.6 g kg^{-1} d^{-1} for each °C increase in growth temperature. Changes in digestibility of this magnitude can have a significant impact on animal performance, particularly when forage represents a large proportion of the diet.

In early work at Wageningen, the Netherlands, Deinum et al. (1968) investigated the relative effects of temperature, solar radiation, and N nutrition on ryegrass digestibility under field conditions. Forage sampled in late spring showed 100 g kg^{-1} lower digestibility than that sampled in early spring, and this was attributed to increasing temperatures as well as increased stem growth and development. Samples taken in midsummer showed 70 g kg^{-1} lower digestibility than the early spring samples, interpreted to be a direct effect of increasing temperature. Of the three factors investigated, temperature consistently had the greatest negative association with degradability.

In other field studies, Minson and McLeod (1970) reported a significant negative correlation between mean daily temperature 1 mo before cutting and IVDMD of 28-d regrowths of several cool-season and warm-season grasses. Collins (1983) and Onstad and Fick (1983) reported alfalfa to be lower in digestibility and to have a faster rate of decline in digestibility during the summer harvest period than during the spring or fall. Vough and Marten (1971) attributed reductions in IVDMD of alfalfa grown in warm temperatures to a decrease in stem digestibility. Walgenbach et al. (1981) reported that alfalfa harvested at specific growth stages had greater crude protein concentrations in herbage when grown in warm rather than in cool temperatures. However, warm temperatures often decrease nonstructural carbohydrate concentrations as well as IVDMD (Vough and Marten, 1971).

Because ambient temperatures progressively rise during the first half or more of most growing seasons, ontogeny, maturity, and temperature are confounded

Table 1. Effect of growth temperature on chemical and digestibility characteristics of leaf blades and stem plus sheath of perennial forages (Wilson et al., 1991b).

Species	Plant part	Lignin 22°C	Lignin 32°C	Neutral detergent fiber 22°C	Neutral detergent fiber 32°C	NDF digestibility 22°C	NDF digestibility 32°C
		------- g kg^{-1} DM --------				-g kg^{-1} NDF-	
Bermudagrass	Leaf	13	22	373	514	752	620
	Stem	34	67	574	641	598	408
Switchgrass	Leaf	18	26	440	394	742	666
	Stem	42	68	638	648	647	421
Panicum laxum	Leaf	16	23	385	446	761	650
	Stem	30	50	511	550	548	416
Perennial ryegrass	Leaf	14	18	292	330	783	657
	Stem†	nd‡	nd	375	505	782	686
Alfalfa	Leaf	8	14	101	106	420	222
	Stem	86	99	420	419	360	299

†Leaf sheath only.
‡nd = not determined.

in field studies, and controlled-environment studies are required to isolate the effect of temperature. Deinum and Dirven (1974) proposed a conceptual model describing the relative effects of environmental variables on forage quality. They summarized research on the topic and observed that both inter- and intra-specific differences in digestibility seem to occur partly because of differences in leaf digestibility, but mainly as a result of differences in stem digestibility. The latter they attributed to variations in stem anatomy and morphology. These authors also observed that high temperatures always cause lower forage digestibility, but that the effect of increasing temperature is greater at lower temperatures and less near the optimal temperature for growth. The temperature effect also is more pronounced on older compared to younger tissue.

The depressed dry matter (DM) digestibility associated with elevated temperatures is most frequently attributed to higher concentrations of cell wall constituents (Ford et al., 1979), but little work has attempted to provide a mechanistic insight into this phenomenon. Elevated temperatures not only may increase the concentration of cell walls (neutral detergent fiber; NDF), they can also reduce cell wall digestibility (Table 1). It is likely that temperature plays a role in chemical composition and, thus, relative digestibility of cell wall constituents, but few studies have attempted to document this. Wilson (1982) demonstrated the negative effect of high temperature on the extent of cell wall degradation, and Allinson (1971) reported lower in vitro digestibility of cellulose in tall fescue grown at elevated temperatures. Yet, the question still remained as to what actually caused reduced DM digestibility, simply the mass of fiber (vs. cell contents) or some qualitative difference in the fiber itself.

Moir et al. (1977), studying the effect of elevated temperature on warm-season and cool-season grasses, found that increasing growth temperature from 18/10°C to 25/17°C resulted in an increase in both total cell wall concentration and the amount of cell wall digested. At temperatures above 25/17°C, total cell wall concentration continued to increase, but the amount digested did not. To determine the influence of temperature in increasing the total concentration of fiber vs. effects on digestibility of fiber, Fales (1986) sampled vegetative 28-d growths of tall fescue grown at 15/10, 20/18, and 30/27°C, and measured fractions of digestible and indigestible NDF, hemicellulose, cellulose, and lignin. Total concentrations of NDF, hemicellulose, and cellulose in DM increased up to the highest temperature imposed, and indigestible NDF (defined as that fraction of the total NDF remaining after 72 h of incubation in ruminal fluid) seemed to be more sensitive to temperature changes than was digestible NDF. No differences were observed in rate of in vitro NDF disappearance, and there was no apparent relationship between lignin concentration and fiber degradation. Fales (1986) concluded that high temperature had no effect on the rate of digestion of potentially digestible fiber, but reduced total fiber digestibility mainly by increasing the amount of indigestible fiber.

Change in cell wall concentration with temperature is often small in relation to the change in DM (Deinum and Dirven, 1976; Wilson et al., 1976). While it generally has been assumed that temperature effects on digestibility are mediated through lignin, negative correlations between elevated lignin concentrations and digestibility are not strong (Fales, 1986), and it is likely that an interaction of several factors is involved. The decrease in forage digestibility at high temperatures has been consistently associated with a substantial increase in the amount of indigestible cell wall (Moir et al., 1977; Fales, 1986). Akin et al. (1987) investigated the effects of increasing, then decreasing, temperature on leaf anatomy and 48 h in vitro tissue disappearance of tall fescue leaves. Under a temperature regimen of 20/18°C, epidermis, mesophyll, parenchyma bundle sheath, and phloem were completely removed during digestion, leaving a residue of vascular tissue, sclerenchyma, and cuticle (Figure 1). By contrast, at 30/27°C, only slight digestion of the mesophyll occurred, whereas the remainder of the leaf remained relatively intact. Histological reactions for lignin showed higher amounts of lignified vascular tissue at 30/27°C. Analyses for phenolic acids showed significant quadratic relationships between temperature and concentrations of p-coumaric and ferulic acids.

Other mechanisms by which elevated temperatures may cause reduced cell wall digestion also have been considered. Wilson et al. (1991b) hypothesized that

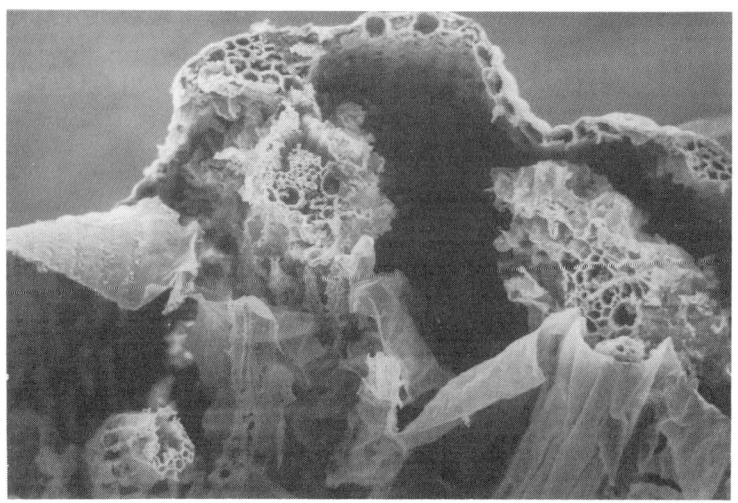

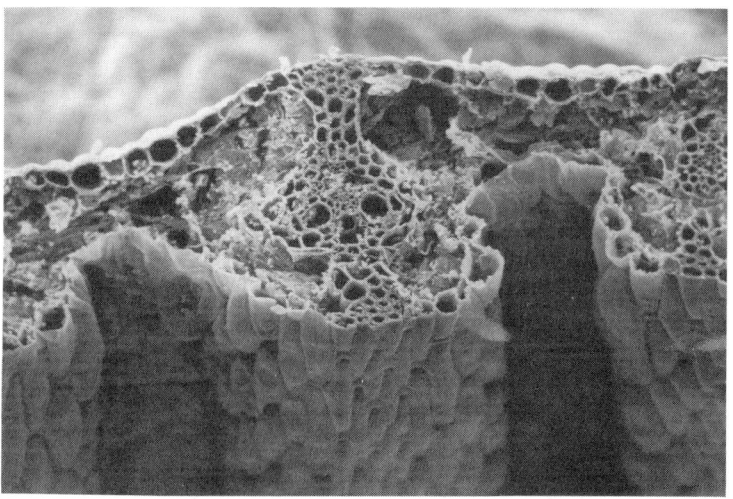

Fig. 1. Scanning electron micrograph of tall fescue leaves grown at 20/18 °C (top) and 30/27°C (bottom) incubated with rumen microorganisms for 48 h showing lower digestion of tissue grown at the higher temperatures (Akin et al., 1987).

temperature variation may alter digestion of cell walls by altering cell thickness. Thick cell walls may digest slower than thin cell walls as a result of the relative amount of surface area in relation to the amount of cell wall material. They found no consistent effect of temperature on thickness of cell walls when several forages were grown at 22/16 or 32/26°C. Instead, they concluded that temperature effects on leaf or stem anatomy were small and were not important factors in temperature-induced differences in forage digestibility. In contrast to Akin et al. (1987), they attribute the decrease in digestibility of forages grown at high temperature to increased lignin concentration with the greatest effect in walls of normally lignified cells such as vascular, sclerenchyma, and bundle sheath.

Cool-Season and Warm-Season Species

Teeri and Stowe (1976) and Hattersley (1983) reported that the geographic distribution of C_3 and C_4 species is largely determined by regional and seasonal temperatures, with C_4 types being more numerous in warm climates and seasons. Warm-season C_4 grasses have long been recognized for their relatively low digestibility and high concentrations of structural polysaccharides compared with C_3 grasses. Indeed, within *Poaceae*, the greatest differences in digestibility and in cell wall composition are between the temperate *(Festucoideae)* and the tropical *(Panicoideae)* subfamilies, the former exhibiting the C_3 photosynthetic pathway and the latter having the C_4 pathway. Warm-season grasses almost always have higher levels of cellulose and hemicelluloses than cool-season grasses (Bailey, 1973), and this fact has given rise to debate as to whether the differences are truly taxonomic or are caused by the different environmental conditions under which warm-season and cool-season species usually are grown.

Employing temperature-controlled growth chambers, Deinum and Dirven (1972) demonstrated that increasing the temperature during growth of a warm-season grass (*Brachiaria ruziziensis* R. Germ. and C. Evrard) increased DM production and tiller size, but had a negative relationship with tiller number, leaf/stem ratio, and concentration of organic nitrogen (N) in the DM. The authors also noted a positive relationship between temperature and crude fiber concentration in both leaf and stem tissue and postulated that temperature itself is a major factor contributing to the relatively poor quality of forages grown in warm climates.

In a study designed to test the relative responses of warm-season and cool-season grasses to temperature, Wilson and Ford (1971) compared the growth and quality characteristics of warm-season grasses with perennial ryegrass under temperature regimens ranging from 16 to 32°C. All species showed a temperature-related decrease in digestibility and an increase in cell wall concentration. Because the cell wall concentration of the warm-season species was consistently higher than that in the ryegrass over the range of temperatures tested, the authors concluded that warm-season grasses have intrinsic growth characteristics that promote higher levels of structural components and thus lower digestibility. This conclusion was supported by findings of Deinum and Dirven (1975, 1976) who showed that warm-season grasses have lower digestibility than cool-season grasses because of a lower leaf/stem ratio and because of the greater negative effect of temperature on stem digestibility than leaf digestibility.

Studies by Ford et al. (1979) showed that cool-season and warm-season grasses also respond differently to temperature with respect to individual cell wall constituents. Investigating 13 warm-season and 11 cool-season grass species at temperatures of 21/13, 27/19, and 32/24°C, they observed that NDF concentration rose with increasing temperature in the cool-season species, but decreased in the warm-season species. The main effect of higher temperatures on the warm-season grasses was to reduce cellulose concentrations, resulting in relatively higher proportions of hemicelluloses and lignin. Digestibility had a strong negative correlation with lignin concentration within the warm-season group but was not significantly related to lignin concentration in the cool-season group.

Grasses and Legumes

Differences in digestibility between grasses and legumes have been well-documented, with grasses normally showing higher cell wall concentrations and a more rapid accumulation of lignin, and thus a more rapid decline in digestibility with maturity (Waldo and Jorgensen, 1981; Buxton and Russell, 1988; Buxton and Brasche, 1991). Only a limited number of studies, however, have examined the differential response of grass and legume quality to temperature.

Faix (1974) studied the effects of both daylength and temperature on quality and morphological components of alfalfa, birdsfoot trefoil (*Lotus corniculatus* L.), and crownvetch (*Coronilla varia* L.). Leaf fiber concentration and digestibility changed little with temperature (17, 22, 27, or 32°C) or daylength, but stem fiber concentration was highest and DM digestibility was lowest for all species at 22°C. Growth rates and leaf/stem ratios were lowest at this temperature. The authors concluded that while stem quality in legumes responds to temperature, the effect could be masked by leaves. Thus, in general, the quality of legumes is determined primarily by the degree of leafiness. Wilson (1982), in a collection of data from several studies, concluded that both warm- and cool-season legumes decline in digestibility by a slower rate than cool-season grasses, approximately 3 g kg^{-1} per 1°C increase in growth temperature.

Ohlsson (1991) studied the response of two cultivars of red clover (*Trifolium pratense* L.) and two cultivars of timothy to temperature during 10 wk regrowth periods. He found that a temperature increase from 10 to 20°C advanced average stage of maturity by 21% in red clover and 9% in timothy. Concurrently, plant height increased 19% in red clover and 34% in timothy, tiller number of timothy decreased 11% and stem number of red clover decreased 22%, and leaf area increased by 18% in red clover and by 56% in timothy. Average N concentration was not affected by temperature, whereas digestibility at equivalent maturity stages was about 70 g kg^{-1} greater in red clover and 50 g kg^{-1} greater in timothy grown at 10°C rather than 20°C.

WATER DEFICIT

Water is a crucial component of plant cells as almost all metabolic processes depend on its presence. Moreover, adequate water is required for maintenance of turgor pressure, guard cell function, and solute diffusion in cells. Water provides the O_2 evolved during photosynthesis and the hydrogen used for CO_2 reduction. The amount of water present varies with cell type and physiological status.

Young, newly formed cells are composed mostly of water, whereas mature fiber cells are nearly devoid of water. On average, water concentration in herbage at anthesis may be near 750 g kg^{-1}, depending upon species and environmental conditions, and declines with advancing maturity.

Almost all water in forage plants comes from the soil through their root systems. Rosenberg et al. (1983) noted that plants function like water pumps moving water from the soil into the atmosphere in response to differences in water potential in soil, plant, and air. Only about 1% of the water that enters growing forage plants is retained as the vast majority is lost through transpiration. Water in excess of metabolic needs, however, serves important functions in solute movement from the roots to stems and leaves and in evaporative cooling of plants. The greatest resistance to water movement through plants normally occurs at the leaf:air interface. Stomata occupy only about 1% of the surface area of leaves, but the majority of water lost from living leaves passes through open stomata. Some water is lost through the cuticle as well.

General Effects on Forages

Both excess and deficiency of water can induce stresses in forages. Excess water, which can result in waterlogged soils, imposes stress because waterlogged soils are rapidly depleted of oxygen by microorganisms and root respiration, leaving forage root systems in an anoxic environment. Although anoxia can greatly reduce forage yield, there is little information as to its impact on forage quality.

For most forage-producing areas, drought is more common than excess soil water. In fact, one of the greatest concerns relative to climate change is even greater occurrence of drought in the future (Waggoner, 1993). Water-deficit stress is usually the major physical limitation to forage yield. When transpiration exceeds water absorption by roots, water deficits increase in plants and stress may occur, which adversely affects many enzymatic reactions and most physiological processes. Water deficits cause stomatal closure, reduce transpiration rates, and elevate foliage temperatures. Cell enlargement is particularly sensitive to water deficits. Cell division seems to be less sensitive than is cell enlargement (Levitt, 1980). Turgor pressure plays a crucial role in cell enlargement, providing the pressure necessary for cell wall expansion. As cell walls expand, turgor pressure decreases, which causes water potential inside cells to decrease. This creates a water potential difference between interior and exterior of cells, which moves more water into the cell. Solutes must continually accumulate inside growing cells for this process to continue. When water stress develops, turgor is lowered, which reduces cell enlargement.

The capacity of forages to sustain a positive or constant turgor as water potentials decrease is an important adaptation to water deficits. The most important physiological mechanism allowing plants to sustain turgor under water stress is osmoregulation, which is a lowering of osmotic potential. Decreased osmotic potential may result from condensing of cell sap during water loss and from an increase of solutes in cells under water stress (Turner, 1979). Solutes that increase in concentration include soluble sugars, free amino acids, K, chlorides, and organic acids (Turner and Jones, 1980). Under moderate to severe stress, the concentration of the amino acid, proline, increases more than other amino acids (Barker et al., 1993). Proline may serve as a storage pool of N as well as aid in drought tolerance by acting as a solute in osmoregulation (Stewart and Hanson, 1980). Photosynthetic rates are usually affected less by drought than are respiration rates and growth, causing a general increase in concentrations of nonstructural carbohydrates. Translocation of photosynthate, however, is relatively insensitive to water deficit (Setter, 1993). The effects vary depending on species, degree of stress, and plant developmental stage. Accumulation of nonstructural carbohydrates and N pools may facilitate rapid regrowth after water stress is relieved.

Effects on Plant Development

Leaf expansion is particularly sensitive to water deficits. Reduced cell growth results in smaller cells and less leaf area per plant (Jones et al., 1980). While reduced leaf area benefits water-stressed forages by reducing evaporative demand and by decreasing the amount of water lost through transpiration, it also lowers crop growth rates, especially during early stages of growth when interception of solar radiation is incomplete. Brouwer (1962) noted that shoots depend upon roots to furnish necessary water and essential minerals for growth. He hypothesized that limiting the activity of the root system relative to the activity of the shoot (such as when water or N are deficient) would result in increased levels of soluble carbohydrates. Because roots are in closer proximity to the limited water (or N), shoot growth would be restricted more than root growth. Several experiments have indeed shown that drought has a greater effect on shoot than on root growth resulting in increased root/shoot ratios (Brown and Tanner, 1983). This shift in root/shoot ratio also benefits plants because it improves their ability to extract water from the soil while decreasing water loss by decreasing leaf surface area. This shift in root/shoot ratio occurs even at mild water deficits sufficient to reduce shoot growth but not affect rates of photosynthesis.

The C_4 plants have a greater CO_2 uptake per unit of stomatal conductance than C_3 plants. This leads to much greater water use efficiency in C_4 than C_3 species (Tanner and Sinclair, 1983). Other plant species conserve water by having short periods of growth during which rate of development may be rapid (Blum, 1993). Additional water-conserving traits include inhibition of tillering and branching and hastening the death of established tillers (Turner, 1979). Furthermore, leaf area is also reduced by accelerating the rate of senescence of older leaves (Begg, 1980). Both N and soluble carbohydrates are mobilized and exported out of leaves as they senesce. Leaves comprise a large portion of forage yield and they represent an even higher proportion of total nutrients. Hence, their loss has an especially adverse effect on forage quality.

Bittman et al. (1988) reported that reproductive tillers of crested wheatgrass [*Agropyron desertorum* (Fesch. ex Link) Schult] and smooth bromegrass (*Bromus inermis* Leyss.) underwent greater leaf senescence during drought than vegetative tillers. Morphological adaptation of alfalfa to water stress through leaf loss was reported by Halim et al. (1989b). They found that if water stress occurred only at the vegetative or bud stages, leaf loss occurred, but alfalfa recovered by the early flower stage and returned to normal leaf/stem ratios with little loss in yield or change in plant maturity. If the stress occurred only at the flower stage, however, there was little time for recovery and return to normal leaf/stem ratios before harvest, and forage quality was lowered.

Droughts imposed throughout a growth cycle or for extended periods generally cause delayed plant maturity as well as reduced shoot length and increased leaf/stem ratio (Halim et al., 1989a; Peterson et al., 1992; Sheaffer et al., 1992). Wilson (1982) attributed effects of water stress that resulted in improved forage quality primarily to the influence of water stress in slowing growth and delaying further stem development. This in turn produces leafier plants of high digestibility. Halim et al. (1989a), however, found that the effect of water stress on most attributes of herbage quality were significant even after accounting for differences in plant maturity. The effect of water deficit on maturity may differ if the stress is applied only when plants are more mature. Late-applied stress sometimes hastens maturation of perennial forages (Wilson and Ng, 1975). Water stress may also hasten the maturity of annual forage plants.

Effects on Chemical Composition and Digestibility

High yield potential is usually negatively associated with many drought-adaptive traits among species and cultivars (Blum, 1993). Furthermore, xeromorphic features characteristic of plants grown in hot, dry environments such as thick cell walls (Cutler et al., 1977), thick cuticle, and highly lignified tissue (Levitt, 1980) are generally associated with low digestibility. Water stress, however, generally has a smaller effect on forage quality than on growth and development, and most of the effects on forage quality are positive, primarily because of the typical delay in maturity caused by water stress. For example, irrigating alfalfa to 65% of field capacity reduced yield by 49% compared to alfalfa irrigated to 112% (Halim et al., 1989a). This water-deficit stress treatment increased the leaf/stem ratio by 18%, but increased digestibility by only 8% for stems and less for leaves, and had no consistent effect on herbage N concentration. Inconsistent effects on total herbage N occurred because N concentration in stems increased up to 10%, whereas that in leaves decreased by up to 14%. Similar results have been reported for other forage legumes (Table 2) and generally for forage grasses (Table 3).

Other reports regarding the effect of drought on N concentration of herbage have been contradictory. In legumes, Gifford and Jensen (1967) and Walgenbach et al. (1981) reported increases in herbage N concentration as a result of drought, whereas Vough and Marten (1971) and Carter and Sheaffer (1983) found no effect. In grasses, Garwood et al. (1979) and Wilson (1983) reported inconsistent or no responses, and Bittman et al. (1988) reported reduced N concentration in droughted grasses. These inconsistent results may be a function of the degree to which water stress caused leaf senescence and changes in the leaf/stem ratio.

Table 2. Effect of drought under field conditions on chemical characteristics of leaves and stems of perennial legumes (Peterson et al., 1992).

Species	Plant part	Yield		Crude protein		Neutral detergent fiber	
		Control	Drought	Control	Drought	Control	Drought
		--t ha^{-1}--		---------g kg^{-1} DM----------			
Alfalfa	Leaf			275	250	273	266
	Stem			126	141	602	543
	Total	5.9	4.4	193	195	453	405
Birdsfoot trefoil	Leaf			308	293	216	165
	Stem			124	148	610	500
	Total	5.0	2.7	209	232	429	305
Cicer milkvetch	Leaf			291	247	232	205
	Stem			132	143	582	480
	Total	4.2	2.7	227	215	372	290
Red clover	Leaf			256	232	325	333
	Stem			117	137	524	466
	Total	6.6	4.0	205	207	398	367

Additionally, drought effects on N concentration may be determined by distribution of N in the soil profile in relation to location of the limited available soil water as suggested by Bittman et al. (1988). If both N and water are present in the same soil horizon, they may be taken up together and herbage N concentration may be unaffected or may be increased if N is more available than water. If sub-soil water is ample and most of the soil N is near the surface, however, growth may continue with reduced N uptake so that herbage N concentration declines.

There are reports of water stress having inconsistent effects (Garwood et al., 1979) or decreasing forage digestibility. Wilson (1983) reported that in the tropical legume siratro [*Macroptillium atropurpureum* (Mocino & Sesse ex DC.) Urb.], which has leaves capable of expanding under water stress, water-stressed leaves had lower digestibility than nonstressed leaves. Likewise, Pitman and Holt (1982) reported that water stress decreased the forage quality of three warm-season grasses, kleingrass (*Panicum coloratum* L.), green springletop [*Leptochloa dubia* (Kunth) Nees], and plains bristlegrass [*Setaria leucopila* (Scribn & Merr.) K. Schumi].

Cellulose concentration of alfalfa cell walls decreased and hemicellulose concentration increased with increasing stress in the work reported by Halim et

Table 3. Effect of drought under field conditions on chemical characteristics of leaves and stems of perennial grasses (Sheaffer et al., 1992).

Species	Plant part	Yield		Crude protein		Neutral detergent fiber	
		Control	Drought	Control	Drought	Control	Drought
		--t ha^{-1}--		----------g kg^{-1} DM----------			
Reed canarygrass	Leaf			186	178	555	561
	Stem			96	124	723	651
	Total	5.3	2.7	145	157	630	596
Orchardgrass	Leaf			154	153	558	546
	Stem			82	---	683	---
	Total	4.0	2.7	139	153	584	546
Smooth bromegrass	Leaf			189	176	561	560
	Stem			81	113	728	645
	Total	4.3	3.8	145	153	628	592
Timothy	Leaf			181	170	549	556
	Stem			94	149	694	556
	Total	4.7	2.7	147	165	606	559

al. (1989a). Conversely, Wilson (1983) reported that the forage legume siratro had a marked increase in cellulose and lignin concentrations and a decrease in hemicellulose concentration when grown under water stress compared to plants grown without water stress. While water stress may reduce cell wall concentration of leaves and stems, there is no evidence that it alters cell wall digestibility (Deetz et al., 1991).

Neutral detergent fiber concentration is the trait that seems to be most affected by water stress (Tables 2 and 3). The amount of C incorporated into cell walls is decreased during water stress. Much of the limited fixed carbon may be used to support higher levels of soluble sugars and ions during osmotic adjustment and may not be available for cell wall development.

SOLAR RADIATION

The first step in utilization of intercepted solar energy is its conversion into chemical energy by photosynthesis. During this process, photosynthesis initiates the flow of energy into ecosystems of the earth's biosphere. The carbon used is fixed from atmospheric CO_2, about 0.03% of the total gaseous composition. Photosynthesis occurs when green leaves are exposed to radiation in the visible range (e.g., radiation with wavelengths of 400 to 700 nm). Energy in these wavelengths represents approximately half of the total in solar radiation. Under ideal conditions, up to 7% of the energy in solar radiation can be stored in photosynthetic products of rapidly growing crops (Noble, 1988). For forages throughout the growing season, however, the average is much lower, perhaps less than 1%.

Other energy is absorbed by the plant's phytochrome system, which regulates many photomorphogenic aspects of plant growth and development such as stem elongation, leaf expansion, and flowering. Many of these phytochrome effects are induced by low levels of red and far-red radiation. Phytochrome, a photochromic protein, exists in two mutually photoconvertible forms. The red absorbing form of phytochrome (P_r) has a broad band of absorption with a peak near 660 nm. Upon absorption of red light, P_r is converted to the biologically active far-red absorbing form, P_{fr}, which has a broad band of absorption with a peak near 730 nm (Sanchez et al., 1993). Shifts in the red/far-red ratio of solar radiation determine which form of phytochrome predominates. Additionally, some morphological responses are a direct response of reduced irradiant flux density.

General Effects on Forages

Both the spectral quality and quantity of radiation received by plants is profoundly influenced by changes in canopy density (Ballare et al., 1991). Absorption and reflection of radiation by the upper portion of the plant canopy reduces radiation in the blue and red range and increases the proportion of radiation in the far-red waveband (Holmes, 1981). Wilson (1989) noted that red/far-red ratios of about 1:1 in full sunlight are reduced to about 0.4 to 0.7 under vegetative canopies transmitting less than 20% of solar radiation. These changes induce or accentuate morphogenetic responses such as promotion of stem and leaf elongation (Thompson and Harper, 1988; Ballare et al., 1991) or inhibition of tillering (Deregibus et al., 1985; Casal, 1988). Sanchez et al. (1993) noted that stem elongation of dicotyledonous species is one of the most obvious responses to increase in plant density mediated through changes in spectral quality and quantity. Ultraviolet and blue radiation have been reported to stimulate lignin deposition in plants by increasing phenylalanine and tyrosine ammonia-lyase activity (Guerra et al., 1985), but the response may be species specific (Jung and Russelle, 1991).

The potential amount of solar radiation received at the earth's surface is strongly influenced by the time of year and latitude (Rosenberg et al., 1983). The amount is surprisingly similar over a range of latitudes during late spring and early summer. The higher latitudes have long daylengths but the flux density of solar radiation is less than at lower latitudes. Conversely, lower latitudes have shorter daylengths but greater flux density. At high latitudes, photoperiod varies widely with season, with extended daylight in the summer. Variation in photoperiod plays important roles in induction of reproductive development in many forage species and affects forage quality through this indirect effect. Many cool-season forages tend to be long-day plants in that they flower during the long days of late spring after exposure to low temperatures during the previous winter.

Effect of Shading

Forages frequently experience shaded conditions. This may occur during periods of cloudiness; by neighboring plants, especially when grown in swards with mixtures differing in height; by companion crops commonly used during seedling establishment; or lower plant parts may be shaded by upper plant parts. Additionally, there is increased interest in incorporating forage systems within agroforestry programs (Uhl and Parker, 1986) where forages may be shaded by associated trees. Animals can graze pastures grown under plantation crops when trees are small before they reach full production (Wilson, 1989). Also, research is being conducted in which forage crops are grown in alleys between trees to enhance biomass yields for energy conversion (Colleti et al., 1992). Shading has both direct and indirect effects on forage quality in that it can alter the chemical composition of forages as well as alter morphological development and yield. Nitrogen concentration is usually greater in leaves and stem segments from the top of plant canopies than from the bottom. This has been attributed to decreased photosynthetically active radiation levels within the plant canopy and resulting in enhanced senescence rates (Lemaire et al., 1991).

Shade-adapted C_4 species are uncommon and the vast majority of native grasses under forests are C_3 species. Photosynthetic rates of individual leaves decrease in a curvilinear manner with increasing shading. Generally, a greater decrease in photosynthesis from intense shading occurs in C_4 than in C_3 species. Kephart (1987), for example, found that after adaptation to shaded conditions that removed 30% or 63% of incident solar radiation, photosynthetic rates were reduced 8 or 19%, respectively, for reed canarygrass (*Phalaris arundinacea* L.), a C_3 species, and 4 or 36% for switchgrass (*Panicum virgatum* L.), a C_4 species. Woledge and Parsons (1986) reported that swards of perennial ryegrass, a C_3 species, developed under 400 W m^{-2} photosynthetic active radiation had about four times the canopy photosynthetic rate as swards developed under 100 W m^{-2} photosynthetic active radiation. As photosynthetic active radiation levels decrease in shaded habitats, the rate of dark respiration also decreases (Pearce and Lee, 1969).

Pearce and Lee (1969) observed that alfalfa grown under high irradiance not only had greater photosynthetic rates but also had greater specific leaf weight (weight per unit leaf area) than plants grown under low irradiance, suggesting that leaves grown under low irradiance were thinner. Within 2 wk of being transferred to high irradiance, however, the low-irradiance plants had photosynthesis per unit leaf area and specific leaf weights similar to the high-irradiance plants. Prioul et al. (1980) obtained similar results with Italian ryegrass (*Lolium multiflorum* Lam.).

Decreased radiation load by shading generally improves water relations of forages by decreasing transpiration rates (Wong et al., 1985a, b). Kephart (1987) observed that the decrease in transpiration from shading of the C_3 and the C_4 species that he studied was similar in magnitude to the decrease for photosynthesis. As a result, shading had little influence on water-use efficiency. Shading also reduces air and soil temperatures, but the effect is relatively small. In studies by Eriksen and Whitney (1981, 1982), soil temperatures at 5 cm under 73% shade were 1 to 2°C lower than those under a grass or legume canopy in full sunlight. The literature also indicates that shading which removes half to two-thirds of solar radiation typically reduces air temperatures 1 to 3°C (Wong et al., 1985a; Barrios et al., 1986; Kephart et al., 1992). The greatest depression in air temperature likely occurs at midday. Shade traps escaping long-wave radiation, creating a warm-sky effect. A warm sky usually holds air and soil temperatures to higher levels during night.

Effects on Plant Development

Generally, growth rates and herbage yield of forages decrease with increasing shade (Walgenbach and Marten, 1981; Kephart et al., 1992). Under some circumstances, however, highest yields have been found in forages that received moderate shading rather than full sunlight (Eriksen and Whitney, 1981; Wong et al. 1985a; Samarakoon et al., 1990b). These incidents were under tropical conditions with low soil N where old grass stands may have tied up much of the available soil N. Wilson and Wong (1982) found that shade increased the availability of soil N for forage growth and Wilson (1989) speculated that this may be related to faster mineralization of N under shade because of a lower, more favorable temperature and because of improved soil water environment.

Morphological plasticity allows plants to adjust to environmental modification by changing their morphological form. This adaptation allows plants to maximize growth and to increase persistence under shaded conditions. Exposure to prolonged periods of shade causes most forages to modify proportioning of biomass among plant parts so that the potential for photosynthetic active radiation interception is maintained or increased and root growth is decreased (Cooper, 1967; Corre, 1983).

Shading also reduces tillering and stem production (Cooper and Tainton, 1968). In some species, shading also may induce stem, petiole, or leaf elongation so that leaves are lifted to higher elevations where they are more likely to be exposed to sunlight (Corre, 1983; Jones, 1985). Shaded grass leaves typically are longer, thinner, and sometimes narrower, than when grown in full sunlight (Allard et al., 1991; Kephart et al., 1992). Buxton and Lentz (1993) found that orchardgrass (*Dactylis glomerata* L.) leaves of vegetative tillers were 22% longer when plants were grown on 0.15-m centers, where interplant shading was greater, rather than on 0.60 m centers. Conversely, plant density had little effect on morphology of leaf blades from reproductive tillers.

The amount of diversity among species for morphological adjustment to shading is an issue not fully resolved. Wilson (1989) speculated that there are substantial differences among grass and legume species relative to morphological plasticity under shade. In support of this, Cooper (1967) observed that the leaf/stem ratio of birdsfoot trefoil declined with decreasing irradiance, whereas leaf/stem ratios of alfalfa and red clover increased. Kephart et al. (1992) reported that three C_3 [deertongue grass (*Panicum clandistinum* L.), reed canarygrass, and tall fescue] and two C_4 [switchgrass and big bluestem (*Andropogon gerardi* Vit.)] perennial forage grasses displayed similar responses for leaf area ratio (leaf area per plant weight) and specific leaf weight to shading. Mean leaf area ratio was 8.8 and 11.6 m^2 kg^{-1} under full sunlight and 63% shade, respectively. The leaf area ratio of the C_4 species increased with shading even though leaf blade mass per shoot was reduced by 21%. The leaf blade mass per shoot was unaffected by shading of the C_3 species. Shade did not affect the leaf/stem ratio, shoot length, or rate of morphological development for any of the grasses. Although most morphological responses to irradiance were similar for the C_3 and C_4 grasses, crop growth rate and herbage yield were affected more by shade in the C_4 than in the C_3 grasses.

The anatomical changes that occur in leaves as they become thinner from shading depends upon species because of differences in anatomy among species. Cool season, C_3 grass leaves have a loose arrangement of mesophyll cells and large internal air-spaces. There is a compact mesophyll layer associated with the abaxial epidermis (Esau, 1977; Prioul et al., 1980). Warm season, C_4 grasses, on the other hand, possess kranz anatomy, which is characterized by vascular bundles surrounded by one or two concentrical layers called bundle sheaths (Esau, 1977). The bundle sheath of C_4 leaves is surrounded by a concentric layer of mesophyll parenchyma cells (Esau, 1977; Jones, 1985). There are seldom more than the two mesophyll cells located between neighboring bundle sheaths, whereas in C_3 species, there are from four to as many as fifteen mesophyll cells between vascular bundles (Crookston and Moss, 1974).

Thinning of leaf blades may result both from reduction in cell size and from reductions in cell number. Cooper and Tainton (1968) working with perennial ryegrass and Prioul et al. (1980) working with Italian ryegrass reported that reduced leaf thickness from shading was caused by reduced cell size. Conversely, Allard et al. (1991) observed no change in mesophyll cell size of tall fescue as a consequence of adaptation to shade, even though shaded leaves had low specific leaf weight. Thus, reduced leaf thickness must have resulted from reduced cell number.

Dicotyledonous species, such as forage legumes, have a stratified anatomy with palisade cells over spongy mesophyll. Reduced leaf thickness from shading of these species is caused by fewer palisade and spongy mesophyll cells (Boardman, 1977). Unshaded leaves have a large proportion of palisade tissue, whereas those grown under shade often have reduced amounts of palisade and greater amounts of round and irregular shaped parenchyma. Nobel et al. (1975) found that the greatest effect of irradiance flux density on the palisade layers was in cell length rather than in cell diameter.

Effects on Chemical Composition and Digestibility

Shading typically has a smaller effect on forage quality than on forage morphology or yield. This is illustrated in work reported by Kephart et al. (1992) and Kephart and Buxton (1993). They found that imposing 63% shade on five perennial forage grasses reduced yield by 43% and specific leaf weight by 24%, but only reduced NDF concentration by 3%, cell wall lignin concentration by 4%, and increased digestibility of the herbage by 5%. Nitrogen concentration is much more responsive to shading than other quality characteristics and Kephart and Buxton (1993) found that 63% shade increased N concentration by 26%. The responses were generally greater for leaf blades than for stems. Increased concentration of nitrogenous compounds from shading is usually at the expense of soluble carbohydrates (Eriksen and Whitney, 1981; Walgenbach and Marten, 1981; Wilson and Wong, 1982). The N response for legumes, however, is generally much smaller than for grasses and N concentration has been reported to be similar across levels of irradiation in some studies (Wong and Wilson, 1980; Eriksen and Whitney, 1982; Wong et al., 1985b). This may occur because nodulation and N fixation are restricted under shade (Wilson, 1989). Walgenbach and Marten (1981) reported that concentrations of soluble nonprotein N of alfalfa increased, whereas soluble protein N decreased with shading. Shading also may increase silica concentration of herbage (Wilson, 1982), which tends to reduce herbage digestibility.

Cell wall components are deposited primarily in the order: hemicelluloses, cellulose, and lignin, although there is much overlap among these activities (Bidlack and Buxton, 1992). Baysdorfer and Bassham (1985) and Labhart et al. (1983) proposed that plant growth is limited by carbon availability during early stages of plant growth, but progressively becomes limited by size of the plant sink for carbon as plants continue to develop. Photosynthate availability, then, is normally high when the majority of secondary cell wall thickening and lignification occur. Shaded conditions may prolong processes associated with leaf development as Allard et al. (1991) found that leaf-area expansion continued for a longer time under shading. Kephart and Buxton (1993) hypothesized that shading may reduce photosynthate for secondary wall development. Most studies have indeed found that shading decreases the cell wall concentration of forages (Wilson and Wong, 1982; Kephart and Buxton, 1993). The hemicellulose fraction seems to be less sensitive to shading than are cellulose and lignin fractions (Henderson and Robinson, 1982a).

Reduced cell wall concentration from shading has been reflected in increased DM digestibility in some studies. Kephart and Buxton (1993) reported that digestibility of herbage was improved by about 5% with intense shading. Likewise, Samarakoon et al. (1990a,b) found that the digestible DM of grasses grown under shade was higher than that of herbage grown in full sunlight when fed to animals. Conversely, Wilson and Wong (1982) found a negative response of herbage digestibility to increased shading in spite of lowered cell wall concentration. Navarro-Chavira and McKersie (1983), Blair et al. (1983), and Struik (1983) also found that low irradiance reduced the digestibility of herbage. Hight et al. (1968) compared sun- and shade-grown temperate grass herbage and reported a large decline in intake and liveweight gain of sheep fed on shade-grown herbage. While there is conflict in the literature as to whether shading has a positive or negative effect on digestibility, the effects reported have all been relatively small.

Some of the differential responses reported for digestibility from shading could result from variation in species being investigated, influence of other environmental conditions, differing effects on leaf/stem ratio, and the length of time forages were exposed to shade. Henderson and Robinson (1982b) found species variation for digestibility among four C_4 grasses in response to low irradiance, which shifted with temperature. In their study, the temperature effect on herbage digestibility was consistently greater than the irradiance flux density effect. Wilson and Wong (1982) reported that leaf/stem ratio was increased in the upper canopy and reduced in the lower canopy from shading. Conversely, Deinum and Dirven (1972) and Kephart et al. (1992) found that the leaf/stem ratio was not influenced by low irradiance compared to high irradiance.

Under high radiation, cool-season perennial grasses may contain more than 40% nonstructural carbohydrates (Nosberger, 1993). Diurnal variation has been observed in concentrations of nonstructural carbohydrate, however. Water-soluble carbohydrate concentration in forages was shown by Holt and Hilst (1969) to follow a pattern of lowest values before sunrise and highest values in the afternoon. Lechtenberg et al. (1971) likewise found that IVDMD of alfalfa was about 16 g kg^{-1} greater in the late afternoon than in the early morning before sunrise. Starch concentration in leaves increased by 100 g kg^{-1} during the daylight, whereas that in stems did not change. These changes caused simultaneous shifts in the leaf/stem ratio from 1.1 to 1.5. Additionally, protein tends to be degraded at night through proteolysis followed by remobilization of N and protein synthesis during daylight. The result is fluctuation in protein concentration of 5 to 15% over a diurnal period (Seligman, 1993).

Photoperiod and Radiation Quality

Van Soest (1982) pointed out that at temperate latitudes, compositional changes of first-growth forages with increasing maturity include the influence of both warming temperature and lengthening photoperiod. During aftermath production, however, both temperature and photoperiod show little seasonal change as forages grow and mature. He speculated that this accounts for the closer association between lignification and digestibility in first growth forages than in aftermath production. Forage growth and maturity in tropical regions also occurs without marked seasonal changes in temperature and photoperiod. Thus, lignification and digestibility may not be as closely associated in forages grown in tropical regions either.

Both daylength and radiant flux density influence morphology, growth, flowering, and maturity (Thorvaldsson, 1987). Juan et al. (1993) reported that alfalfa grown under 13 h photoperiods had a higher leaf/stem ratio than alfalfa grown under 16 h photoperiods. Leaf and stem growth tend to be erect under long summer photoperiods and prostrate under the short days and cool temperatures of fall. When daylength exceeds the minimum photoperiod requirement of photoperiod-sensitive species, plant development changes from vegetative growth to reproductive development (Pulli, 1988). If the critical photoperiod is exceeded, reproductive development is enhanced. Heide (1985), for example, noted a three-fold increase in average timothy stem height as photoperiod increased from 8 to 24 h.

Long photoperiods generally result in high forage quality (Wilson, 1982; Deinum, 1984) because of greater photosynthetic activity, which in turn increases soluble sugars that dilute cell walls. Yield also increases and plant morphology is usually altered under long photoperiods. Shoot/root ratios generally increase and leaf/stem ratios decrease as daylengths are extended. The enhanced growth during long days often dilutes N in herbage causing N concentrations to be lowered.

Perennial forages adapted to high latitudes have high growth rates with a short life cycle to ensure that plants reach reproductive development before the end of the growing season (Heide, 1985). Despite a low leaf/stem ratio, at comparable growth stages, cultivars adapted to higher latitudes generally have higher forage quality than cultivars adapted to lower latitudes (Deinum et al., 1981). Perennial plants adapted to low latitudes have accelerated flowering and improved regrowth when they are grown in the long days of higher latitudes, but they may not survive the winters (Pulli, 1988). Conversely, when cultivars adapted to higher latitudes are grown in the short days of lower latitudes, flowering may be delayed, regrowth slowed, but probability of winter survival will be enhanced. Thus, moving perennial plants toward the poles can result in high annual production relative to locally adapted ecotypes but winter survival will be jeopardized. Moving plants toward the equator may result in higher quality forage than locally adapted ecotypes because of delayed maturity.

SOIL NUTRIENTS

The influence of amount and availability of mineral nutrients on forage production and quality has been reviewed extensively by Noller and Rhykerd (1974), Reid and Jung (1974) and, more recently, by Minson (1990). Fertilizer nutrients are applied to forage crops to increase yields by correcting deficiencies in the soil. In pastures, selected nutrients frequently are applied to manipulate botanical composition to maintain a balance of desired species. Application of fertilizer can have both direct and indirect effects on animals by inducing chemical, morphological, or physiological changes in plants. Specific effects, however, may be extremely difficult to diagnose because the utilization of a given mineral nutrient by an animal is governed by its absorption from the digestive system and this, in turn, is influenced by a complex interaction of various factors including concentration in the forage, forage intake, interactions with other minerals, and physiological status of animals. The nutrient requirements of animals vary with breed, maturity, and production level, and critical levels for growth of plants vary according to similar criteria. Plant environment will impact the nutrient concentration of forages. Belsky (1992) reported that forages growing under isolated savanna trees contained higher concentrations of N, P, K, Ca, B, and Cu, and lower concentrations of Mn, Zn, and Mo than plants not growing under the influence of the trees.

Soils deficient in one or more essential elements occur almost anywhere depending on the nutrient of concern. Most soils require N amendments to produce maximal yields for many plants, especially cereals and grasses that do not fix N. Phosphorus is added to most soils because of its low mobility and fixation and reactivity with so many soil fractions and chemicals. Phosphorus, S, and often Zn may be limiting on both alkaline and acid soils, whereas Ca, Mg, K, and often Mo may be limiting primarily on acid soils. Applying N generally results in increased growth rates and reduced nonstructural carbohydrate levels (Alberda, 1965; Colby et al., 1965).

Minerals in forage can be important in preventing livestock diseases as well as inhibiting or stimulating ruminal microbial activity (Hanna and Gates, 1990; Hoveland and Monson, 1980). Sodium deficiency for animals occurs in many warm-season grasses and legumes. Minson (1975) reported that mean Na levels ranged from 5.4 to 15.3 g kg^{-1} for *Panicum coloratum* Walt. and from 1.2 to 5.7 g kg^{-1} for *P. maximum* Jacq.

Nitrogen

Of all the nutrients, N has the greatest impact on plant growth, whether it is fixed and supplied by bacterial symbionts, as is the situation with legumes, or supplied by fertilization or mineralization of soil organic N, as is the situation with grasses. High levels of nitrogenous fertilizer have been used to greatly bolster grass yields under some intensively managed situations. High rates of fertilizer N, however, impose a threat of contamination of underground and surface water supplies because of its soluble nature and mobility with water movement. Absorbed primarily as nitrate (NO_3^-), but also as ammonium (NH_4^+), N is rapidly taken up by roots. Most of the absorbed N is used in the synthesis of protein in a process involving reduction of NO_3^- to NH_4^+ before incorporation into amino acids. The major portion of nonprotein N (NPN), about 10 to 30% of the total N, is comprised of amino acids. Normally, NO_3^- concentration is low in plants, but environmental conditions that restrict growth, such as drought or mineral deficiencies, can result in accumulation of toxic levels of NO_3^- for livestock.

Deficiency of NO_3^- has been reported to stimulate production of phenolics and lignin in wheat (Brown et al., 1984). Application of N fertilizer usually has little or no effect on extent of fiber digestion (Messman et al., 1991), but increases in rate of NDF digestion and intake by animals have been reported, especially if N concentration in the herbage was relatively low before N fertilization (Messman et al., 1991; Puoli et al., 1991). Increases in herbage digestibility also have been reported as a result of N fertilization of warm-season grasses, which normally have low N concentrations (George and Hall, 1983). This response occurs because forage N is limiting for ruminal microbial growth under low-N situations. Other times, N fertilization may modify herbage digestibility by altering the leaf/stem ratio. Under some conditions, N fertilization of tall fescue has been associated with reduced conception rates of cows grazing these pastures (Stricker et al., 1979).

Use of strategic applications of N to vary seasonal distribution of plant growth and to manipulate sward composition was reviewed in detail by Wedin (1974). Under favorable conditions of solar radiation, soil moisture, and temperature, application of N to grasses stimulates tiller development and increases leaf size. However, there seems to be little effect of N on number of leaves per tiller. Applied N usually increases the duration of green leaves (Rhykerd and Noller, 1974). Application of N fertilizer to grasses increases water concentration, reduces soluble carbohydrate concentration, and at higher rates results in progressively greater concentrations of N in the plant tissue (Noller and Rhykerd, 1974; Messman et al., 1991; Brink and Fairbrother, 1992). A resulting imbalance of protein and energy at high levels of N fertilization may be problematic, sometimes resulting in poor silage quality, poor performance of animals on pasture, animal reproduction problems, and other metabolic disorders (Reid and Jung, 1974). This does not occur to any appreciable degree in legumes.

By rapidly stimulating growth, N fertilization tends to cause lower concentrations of other minerals in forage, primarily through dilution. Whitehead et al. (1986) found S concentration of herbage to be more responsive to N fertilization than other minerals. The consequences of N fertilization on other aspects of forage quality of grasses is limited (Minson, 1990; Cox et al., 1993).

By virtue of increasing yield, N fertilization increases the carrying capacity of pastures. Thus, while individual animal performance as affected by N fertilization may be relatively small, performance per hectare can be increased, assuming that the management system is in place to effectively utilize the extra forage.

Phosphorus

In data collated by Minson (1990), concentrations of P measured in forages around the world ranged from 1 to 8 g kg^{-1}. Cool-season forages generally contain more P than warm-season forages, and legumes have higher P levels than grasses. Approximately 80% of the P content of animals resides in the skeleton, where it serves as a storage pool in maintaining the balance of P in blood plasma (Minson, 1990). The requirements for P are nearly equivalent for forage plants and the animals that consume them. Thus, if soil-available P is sufficient for vigorous growth of forage, P concentration in the forage should meet animal requirements. Where severe deficiencies occur, animal symptoms include poor intake and performance, breeding disorders, osteophagia (bone chewing), and rickets. Otherwise, concentrations of soil P have little effect on intake or DM digestibility (Minson, 1990), or relative concentrations of cell wall constituents (Reid and Jung, 1965). Because of the positive influence of P on legume proportion in mixed swards, however, P fertilization of pastures can indirectly increase animal performance.

Potassium

In contrast to P, K is required in substantially higher amounts by plants than by animals and limitations in soil K may limit yield but are unlikely to have negative effects on animal performance. Potassium concentration has shown little effect on voluntary intake or DM digestibility (Minson, 1990). More commonly, excess soil K interferes with the uptake of Ca, Mg, and Na in grazing situations, which can induce grass tetany, a potentially fatal metabolic disorder that causes considerable economic losses annually. Occurring primarily in the spring on cool-season grass pasture, the onset of grass tetany is associated with a rapid decrease in serum Mg. Tetany is difficult to predict from herbage Mg concentration, however, but is correlated with increasing herbage concentrations of both N and K. The topic was reviewed extensively by Grunes et al. (1970) and is covered in more detail in Chapter 7.

Calcium

Calcium concentrations in various forages worldwide range from 1 to 40 g kg^{-1} (Minson, 1990). The concentration of Ca in forages depends on the amount of exchangeable Ca in the soil and also is influenced by soil N and P. Most of the Ca in animals is present in bone, which acts as a reserve pool from which Ca can be drawn to maintain a uniform level in blood. Low concentrations of Ca in the blood of lactating animals causes parturient paresis (milk fever), a disease that is most prevalent where the prepartum diet is high in Ca, and it is brought on by the inability of animals to respond rapidly enough to the increased Ca demands of lactation. The ratio of Ca to P in the total feed is of some concern, since wide Ca/P ratios have been implicated in a variety of disorders including high incidence of parturient paresis, poor breeding performance, and suboptimal feed conversion efficiency. The Ca/P ratio in alfalfa may be as high as 8:1, much greater than the recommended ratio of 2:1 (Reid and Jung, 1974).

Deficiencies of Ca rarely occur where soils are limed. Overliming, on the other hand, has been implicated in micronutrient deficiencies and toxicities in livestock, because of pH effects on mineral availability to the plant (Reid and Jung, 1974). Calcium concentration is higher in legumes than in grasses, and is higher in cool-season species compared to warm-season species. Small increases in both voluntary intake and DM digestibility were observed when Ca was applied to Ca-deficient soil (Minson, 1990). This was attributed to "structural changes" in the plants.

Sulfur

Deficiencies of S in soil, most common in highly leached, sandy soils, cause a marked decrease in true protein concentration in forage, and can adversely impact ruminal microbial protein synthesis and fiber digestion (Bray and Hemsley, 1969). Bull (1971) reported that in vitro cellulose digestion was enhanced with increases in S concentration in corn (*Zea mays L.*) silage up to 2.3 g S kg^{-1} DM. Phenolic acids and lignin concentrations were reduced by application of S fertilizer to orchardgrass (Chestnut et al., 1986) and Italian ryegrass (Millard et al., 1987). Application of S fertilizer has had inconsistent results or no effect in other studies (Jung and Russelle, 1991; Puoli et al., 1991). Minson (1990) demonstrated an association between forage intake in sheep and S fertilization of S-deficient digitgrass (*Digiteria decumbens* Stent), according to the relationship:

$$I = 20.6 - 11.65\ S$$

where I was the intake (g kg^{-1} W$^{0.75}$), S was the S concentration (g kg^{-1}) in the grass, and W was the body weight of the animal. However, no response occurred when S concentration in the forage exceeded 1.8 g kg^{-1}.

PLANT PESTS

Pests markedly influence both yield and quality of forages. If infection is high, most diseases profoundly reduce yield and quality, whereas insects typically reduce yield more than quality. Invading weeds compete for soil water and nutrients, as well as for solar radiation. Thus, they usually reduce yield of forages, but their added biomass compensates for some of the loss. Many weeds are nutritious when immature and the overall effect on quality of forage herbage plus that of weeds may be relatively small.

Diseases

Diseased plants typically have lower digestibility and nonstructural carbohydrate concentrations than healthy plants, with variable differences in crude protein concentration (Braverman, 1986). Stem rust, caused by *Puccinia graminis* Pers., reduced digestibility of orchardgrass, presumably because of the increased resistance of leaf mesophyll tissue to degradation by ruminal microorganisms (Edwards et al., 1981). Digestibility of smooth bromegrass decreased 1.2 g kg^{-1} for each 1% increase in diseased area of leaves (Gross et al., 1975). Moreover, diseases often cause leaf loss. Willis et al. (1969) reported greater leaf retention, but no effect on crude protein concentration, when alfalfa plants grown in Kansas were sprayed weekly with a fungicide to reduce foliar disease compared to plants that were not sprayed. Similar results were reported by Campbell and Duthie (1990) for alfalfa grown in North Carolina.

Much is known about the biochemistry of host plant reaction to disease infection, which was recently reviewed by Nicholson and Hammerschmidt (1992). Disease resistance includes preformed barriers and antimicrobial compounds, mostly in external tissues, and a response phase. Plant surfaces are covered with waxes, cutin, and suberin, which restrict penetration of pathogens. In addition, external tissues contain phenolic compounds, alkaloids, and other compounds that inhibit development of fungi and bacteria (Kuc, 1993).

Spread of pathogens that pass these barriers may be contained by several reactions that occur in the host. These include hypersensitive cell death, rapid accumulation of phenols to isolate the pathogen, modification of cell walls by phenolic substances such as esterification of ferulic acid to the host cell walls, development of physical barriers such as apposition or papillae, and synthesis of specific antibiotics such as phytoalexins and other defense compounds. Lignin-like polymers accumulate as a rapid response to infection, which may present an additional barrier to the pathogen. Lignified tissues are more resistant to degradation by cell wall-digesting enzymes of fungal pathogens. All of these defense compounds can be produced by both resistant and susceptible plants. The speed, magnitude, and timing of response determine the degree of resistance (Kuc, 1993). During early stages of infection, phenolics increase more rapidly in resistant than in susceptible plants. However, at latter stages, phenolic concentrations frequently are higher in susceptible plants than in resistant plants (Goodman et al., 1986).

Not surprisingly, resistant plants possess increased enzyme activities of the general phenylpropanoid pathway and of the specific branch pathway of lignin biosynthesis, which can be detected at the time of hypersensitive cell death (Moerschbacher, 1989). During fungal attack, concentrations of ferulic, coumaric, and syringic acids in the cell wall increase (Southerton and Deverall, 1990). The general response of increased phenolics is not restricted to infection by fungal pathogens, however, as a similar response seems to occur as a result of infection by bacterial and virus pathogens (Estelle et al., 1992; Nicholson and Hammerschmidt, 1992). Other stresses that reduce the vigor of plant growth may predispose forages to pathogen susceptibility. Plant resistance to diseases through plant breeding can improve forage yield and nutritive value in the presence of disease organisms (Catherall, 1987; Karn et al., 1989). Plant disease resistance seems to have little influence on forage yield or quality when the pathogen is absent (Lenssen et al., 1991; Sherwood and Berg, 1991; Wilson et al., 1991a).

Insects

Welter (1993) emphasized that response of plants to insect feeding is mediated by environmental factors such as plant nutrient status. The response to insects is usually greater when soil nutrients are limiting and plant growth is less vigorous. Because of the relatively low value for forage crops compared to other crops and the relative high cost of control, losses to insects in forage crops often are not monitored and may be overlooked. In the USA, insect effects on alfalfa have been the subject of more investigations than for any other forage crop. Hutchins et al. (1990) recently reviewed the impact of insects on alfalfa growth and forage quality. They categorized insect pests broadly as either leaf-mass consumers or assimilate removers. Leaf-mass consumers damage plants mainly by consuming leaves and young developing buds. Defoliation by these insects can be divided into two physiologically distinct types: defoliation of the plant canopy and defoliation of regrowth buds after harvesting or grazing. Assimilate removers typically possess piercing-sucking mouthparts and extract plant juices or otherwise disrupt translocation functions of plants. The effects of both types of insects is often reflected in alterations to the leaf/stem ratio of plants.

Defoliating insects initially remove mostly leaf material, which slows subsequent stem development and maturation of plants while the leaf area is being reestablished. The result can be yield reductions with only minor effects on forage quality. Severe defoliation over an extended period can change the plant canopy from increased tillering or branching from removal of apical dominance and may lead to adverse effects on subsequent regrowth of plants and possibly to loss of stand. Primary insect defoliators of alfalfa in the USA include the alfalfa weevil (*Hypera postica* Gyllenhal) and the Egyptian alfalfa weevil (*H. brunneipennis* Boheman). Larvae of both species feed in stem terminals, where they skeletonize leaf tissue. The amount of damage to plants is influenced by plant maturity (more mature plants are less sensitive) and intensity of larval feeding. Yield declines linearly with larva density. Weevil feeding has only a moderate impact on alfalfa crude protein concentration and digestibility.

The second category of insects, assimilate removers, typically injure plants by extracting plant juices or otherwise disrupting translocation in plants. Aphids may have hormones associated with their saliva that changes photosynthate translocation pattern within plants to their benefit. The potato leafhopper (*Empoasca fabae* Harris), however, is probably the most destructive assimilate remover of alfalfa. Often it is the only insect that causes economic loss (Hutchins et al., 1990). The most pronounced effect of this insect is stunting of alfalfa and reduced forage yield. The herbage produced has a higher leaf/stem ratio than that from noninfected alfalfa and a delayed rate of maturation. Hutchins et al. (1989) found digestibility of alfalfa stems and leaves from stunted plants to be slightly higher than control plants that had not undergone potato leafhopper injury. Additionally, crude protein concentration of leaves was depressed while that in the stems was increased by potato leafhopper feeding. The overall effect on forage quality was small relative to the effect of the insect on alfalfa yield.

Cicer milkvetch (*Astragalus cicer* L.) and birdsfoot trefoil harbored lower populations of potato leafhopper, pea aphid (*Acyrthoaiphon pisum* Harris) and alfalfa weevil than did alfalfa, with cicer milkvetch having the fewest number of insects (Kephart et al., 1990). There is evidence, however, that the fiber of this promising legume is not fermented as readily as that of other forages (Weimer et al., 1993).

Weeds

Some invading weeds reduce forage quality, whereas others, depending upon the species and maturity, may have little effect. Toxic or unpalatable weeds reduce animal performance if present in sufficient quantity to reduce the overall acceptability of the animal diet (Cords, 1973; Marten and Anderson, 1975; Dutt et al., 1979). Despite these negative effects, many weeds are comparable to forage species in chemical composition and quality. Forwood et al. (1989) found that cattle grazing orchardgrass showed a preference for some associated weeds. Marten and Andersen (1975) and Temme et al. (1979) found that redroot pigweed (*Amaranthus retroflexus* L.), common lambsquarters (*Chenopodium album* L.), and common ragweed (*Ambrosia artemisiifolia* L.) had chemical traits and IVDMD similar to those of alfalfa in Minnesota and Wisconsin. Likewise, dandelion (*Taraxacum officinale* Weber), white cockle (*Lychnis alba* Mill.), and immature Jerusalem artichoke (*Helianthus tuberosus* L.) have forage quality equal or superior to many high quality forage species (Dutt et al., 1982; Marten et al., 1987). Moreover, quackgrass [*Agropyron repens* (L.) P. Beauv.] forage yield and quality are similar to those of commercial cultivars of cool-season grasses adapted to the North Central Region of the USA (Fawcett et al., 1978; Dutt et al., 1979; Sheaffer et al., 1990).

Conversely, giant foxtail (*Setaria faberii* Herrm.), Pennsylvania smartweed (*Polygonum pennsylvanicum* L.), yellow foxtail [*Setaria glauca* (L.) P. Beauv.], barnyardgrass [*Echinochloa crus-galli* (L.) P. Beauv.], and shepherd's purse [*Capsella bursa-pastoris* (L.) Medic], among other weed species, have relatively poor forage quality (Marten and Andersen, 1975; Temme et al., 1979; Bosworth et al., 1985). Dutt et al. (1982) reported that animal intake and IVDMD of alfalfa were reduced by infestation with yellow rocket (*Barbarea vulgaris* R. Br.). Cords (1973) found a negative correlation between crude protein concentration of alfalfa-weed hay and presence of the mature winter-annuals flixweed [*Descurainia sophia* (L.) Webb], downy bromegrass (*Bromus tectorum* L.), and wild barley (*Hordeum leporinum* Linl.) in Nevada. The rate of decline in IVDMD with time was generally greater for the weeds than for alfalfa. Likewise, Mueller et al. (1993) reported that infestation of johnsongrass ([*Sorghum halepense* (L.) Per.] into corn in North Carolina generally reduced corn density and forage yield. In vitro digestibility of the silage produced from the corn and associated weeds was reduced by 2.37 g kg^{-1} for each percentage unit of johnsongrass.

SUMMARY AND CONCLUSIONS

Herbage age and maturity generally have a larger influence on forage quality than environmental factors. Environmental factors, however, cause deviations in forage quality even when harvested at the same maturity. As animal production continues to improve through genetic gain, the need for high quality forage will increase. Producers control herbage maturity by selecting the harvesting or grazing date. As greater refinements in forage quality are required, the need to accurately predict forage quality before harvesting or grazing will become more critical. This will place increased emphasis on a better understanding of the effect of environmental factors on forage quality. Unfortunately, the mechanism by which environmental factors influence forage quality is not well understood, especially at the basic molecular level, and our understanding is not sufficient to predict the influence of environmental factors in a mechanistic manner or to be able to develop strategies to reduce the adverse effects of environment on forage quality.

Temperature usually has greater influence on forage quality than other environmental factors and is an area in particular where more information is needed. Although increasing temperature normally hastens maturity, the primary effect on digestibility may be through its effect on the leaf/stem ratio with high temperatures promoting stem over leaf growth. Digestibility of both leaves and stems is lowered by warm temperatures, however, because of resulting high cell wall and low nonstructural carbohydrate concentrations. High temperatures increase the concentration of the indigestible cell wall fraction in forages. Increasing temperatures may have a positive effect on forage quality by elevating the crude protein concentration.

Many other stresses slow plant development, which results in herbage quality being maintained at high levels for a longer time. This is illustrated by the effects of water deficit and insect damage on forages. Both types of stresses cause reduction in leaf mass. If the loss of leaf mass is not severe, these stresses may have little effect on forage quality or may actually improve forage quality relative to nonstressed forage harvested at the same time because of the slowing of plant maturity.

Shading of forages usually reduces yield and causes leaves to be thinner and grass leaves to be longer with small effects on most forage quality characteristics. Herbage grown under shaded conditions, however, usually has higher crude protein concentrations than unshaded herbage. The response is usually less in legumes than in grasses. Diurnal variation exists in forage quality with highest quality in the late evening as a result of photosynthate accumulation, which dilutes the cell wall concentration.

Soil nutrients only have small effects on forage quality. Nitrogen fertilization usually raises the crude protein concentration of nonlegume forages. Forage species with low N concentrations, such as warm-season grasses, may have improved digestibility because N fertilization may stimulate ruminal microbial activity. Additionally, application of S to S-deficient soils often stimulates digestibility. Foliar diseases probably have the greatest adverse effect on forage quality of the plant pests by reducing digestibility. Some weeds reduce forage quality but many have nutritive values similar to forages and have little impact on forage quality.

Plant stresses may become more frequent in the future if predictions for climate change come to pass. High temperature stress and drought in particular may become more common. The potentially adverse effects of climate change on forages through promotion of pests could be substantial. Until now, most stresses have only been considered relative to their independent effects on forages. Little is known about the interaction of stresses when more than one stress occurs at the same time such as is common under field conditions. These interactions likely will receive more attention in the future.

REFERENCES

Akin, D.E., S.L. Fales, L.L. Rigsby, and M.E. Snook. 1987. Temperature effects on leaf anatomy, phenolic acids, and tissue digestibility in tall fescue. Agron. J. 79:271-275.

Alberda, T. 1965. The influence of temperature, light intensity and nitrate concentration on dry matter production and chemical composition of *Lolium perenne* L. Neth. J. Agric. Sci. 13:335-360.

Allard, G., C.J. Nelson, and S.G. Pallardy. 1991. Shade effects on growth of tall fescue: I. Leaf anatomy and dry matter partitioning. Crop Sci. 31:163-167.

Allinson, D.W. 1971. Influence of photoperiod and thermoperiod on the IVDMD and cell wall components of tall fescue. Crop Sci. 11:456-458.

Bailey, R.W. 1973. Structural carbohydrates. p. 157-212. *In* G.W. Butler and R.W. Bailey (ed.) Chemistry and biochemistry of herbage. I. Academic Press, New York, NY.

Ballare, C.L., A.L. Scopel, and R.A. Sanchez. 1991. Photocontrol of stem elongation in plant neighbourhoods: Effects of photon fluence rate under natural conditions of radiation. Plant Cell Environ. 14:57-65.

Barker, D.J., C.Y. Sullivan, and L.E. Moser. 1993. Water deficit effects on osmotic potential, cell wall elasticity, and proline in five grasses. Agron. J. 85:270-275.

Barrios, E.P., F.J. Sundstrom, D. Babcock, and L. Leger. 1986. Quality and yield response of four warm-season lawngrasses to shade conditions. Agron. J. 78:270-273.

Baysdorfer, C., and J.A. Bassham. 1985. Photosynthate supply and utilization in alfalfa: A developmental shift from a source to a sink limitation of photosynthesis. Plant Physiol. 77:313-317.

Begg, J.E. 1980. Morphological adaptations of leaves to water stress. p. 33-42. *In* N.C. Turner and P.J. Kramer (ed.) Adaptation of plants to water and high temperature stress. John Wiley and Sons, New York, NY.

Belsky, A.J. 1992. Effects of trees on nutritional quality of understory gramineous forage in tropical savannas. Tropical Grassl. 26:12-20.

Bidlack, J. E., and D.R. Buxton. 1992. Content and deposition rates of cellulose, hemicellulose, and lignin during regrowth of forage grasses and legumes. Can. J. Plant Sci. 72:807-818.

Bittman, S., G.M. Simpson, and Z. Mir. 1988. Leaf senescence and seasonal decline in nutritional quality of three temperate forage grasses as influenced by drought. Crop Sci. 28:546-552.

Blair, R.M., R. Alcaniz, and A. Harrell. 1983. Shade intensity influences the nutrient quality and digestibility of southern deer browse leaves. J. Range Manage. 36:257-264.

Blum, A. 1993. Selection for sustained production in water-deficit environments. *In* D.R. Buxton, R.M. Shibles, R.A. Forsberg, B.L. Blad, K.H. Asay, G.M. Paulsen, and R.F. Wilson (ed.) International crop science I. CSSA, Madison, WI. In press.

Boardman, N.K. 1977. Comparative photosynthesis of sun and shade plants. Ann. Rev. Plant Physiol. 28:355-377.

Bosworth, S.C., C.S. Hoveland, and G.A. Buchanan. 1985. Forage quality of selected cool-season weed species. Weed Sci. 34:150-154.

Braverman, S.W. 1986. Disease resistance in cool-season forage, range and turf grasses. Bot. Rev. 52:1-112.

Bray, A.C., and J.A. Hemsley. 1969. Sulphur metabolism of sheep. IV. The effect of a varied dietary sulphur content on some body fluid sulphate levels and on the utilization of urea-supplemented roughage by sheep. Aust. J. Agric. Res. 20:759-773.

Brink, G.B., and T.E. Fairbrother. 1992. Bermudagrass-subterranean clover response to nitrogen application. J. Prod. Agric. 5:591-595.

Brouwer, R. 1962. Nutritive influences on the distribution of dry matter in the plant. Neth. J. Agric. Sci. 10:399-408.

Brown, P.W., and C.B. Tanner. 1983. Alfalfa stem and leaf growth during water stress. Agron. J. 75:799-804.

Brown, P.H., R.D. Graham, and D.J.D. Nicholas. 1984. The effects of manganese and nitrate supply on the levels of phenolics and lignin in young wheat plants. Plant Soil 81:437-440.

Bull, L.S. 1971. Corn and corn silage rations may be low in sulphur. Sulphur Inst. J. 7:7-9.

Buxton, D.R., and M.R. Brasche. 1991. Digestibility of structural carbohydrates in cool-season grass and legume forages. Crop Sci. 31:1338-1345.

Buxton, D.R., and M.D. Casler. 1993. Environmental and genetic effects on cell-wall composition and digestibility. p. 685-714. *In* H.G. Jung, D.R. Buxton, R.D. Hatfield, and J. Ralph (ed.) Forage cell wall structure and digestibility. ASA, CSSA, and SSSA, Madison, WI.

Buxton, D.R., and E. Lentz. 1993. Performance of morphologically diverse orchardgrass clones in spaced and solid plantings. Grass Forage Sci. In press.

Buxton, D.R., and J.R. Russell. 1988. Lignin constituents and cell wall digestibility of grass and legume stems. Crop Sci. 28:553-558.

Campbell, C.L., and J.A. Duthie. 1990. Impact of leaf spot diseases on yield and quality of alfalfa in North Carolina. Plant Disease 74:241-245.

Catherall, P.L. 1987. Selection of cocksfoot (*Dactylis glomerata*) with resistances to cocksfoot mottle virus. Tests Agrochem. Cult. 8:144-145.

Carter, P.R., and C.C. Sheaffer. 1983. Alfalfa response to soil water deficits: I. Growth, forage quality, yield, water use and water-use efficiency. Crop Sci. 23:669-675.

Casal, J.J. 1988. Light quality effects on the appearance of tillers of different orders in wheat (*Triticum aestivum*). Ann. Appl. Bio. 112:167-173.

Chestnut, A.B., G.C. Fahey, Jr., L.L. Berger, and J.W. Spears. 1986. Effects of sulfur fertilization on composition and digestion of phenolic compounds in tall fescue and orchardgrass. J. Anim. Sci. 63:1926-1934.

Colby, W.G., M. Drake, D.L. Field, and G. Kreowski. 1965. Seasonal pattern of fructosan in orchardgrass stubble as influenced by nitrogen and harvest management. Agron. J. 57:169-173.

Collins, M. 1983. Changes in composition of alfalfa, red clover, and birdsfoot trefoil during autumn. Agron. J. 75:287-291.

Colleti, J.P., R.C. Schultz, M.J. Thompson, D.R. Buxton, I.C. Anderson, L.C. Rule, and W.W. Simpkins. 1992. An agroforestry system for municipal biosolids treatment and biomass energy production. p. 321. *In* Agronomy abstracts. ASA, Madison, WI.

Cooper, C.S. 1967. Relative growth of alfalfa and birdsfoot trefoil seedlings under low light intensity. Crop Sci. 7:176-178.

Cooper, J.P., and N.M. Tainton. 1968. Light and temperature requirements for the growth of tropical and temperate grasses. Herbage Abstr. 38:167-176.

Cords, H.P. 1973. Weeds and alfalfa hay quality. Weed Sci. 21:400-401.

Corre, W.J. 1983. Growth and morphogenesis of sun and shade plants. I. The influence of light intensity. Acta Bot. Neerl. 32:49-62.

Cox, W.J., S. Kalonge, D.J.R. Cherney, and W.S. Reid. 1993. Growth, yield, and quality of forage maize under different nitrogen management practices. Agron. J. 85:341-347.

Crookston, R.K., and D.N. Moss. 1974. Interveinal distance for carbohydrate transport in leaves of C_3 and C_4 grasses. Crop Sci. 14:123-125.

Cutler, J. M., O. W. Rains, and R.S. Loomis. 1977. The importance of cell size in the water relations of plants. Physiol. Plant. 40:255-260.

Da Silva, J.H.S., W.L. Johnson, J.C. Burns, and C.E. Andrews. 1987. Growth and environment effects on anatomy and quality of temperate and subtropical forage grasses. Crop Sci. 27:1266-1273.

Deetz, D.A., H.G. Jung, and D.R. Buxton. 1991. Water-deficit effects on in vitro digestibility of cell wall neutral sugars from alfalfa stems harvested at three maturities. J. Dairy Sci. 74(Suppl.1):184.

Deinum, B. 1984. Chemical composition and nutritive value of herbage in relation to climate. p. 338-350. *In* H. Riley and A. Skjelvag (ed.) The impact of climate on grass production and quality. Proc. 10th General Mtg. European Grassl. Federation. 26-30 June 1984. As, Norway.

Deinum, B., J. de Beyer, P. H. Nordfeldet, A. Kornher, O. Ostgard, and G. van Bogaert. 1981. Quality of herbage at different latitudes. Neth. J. Agric. Sci. 29:141-150.

Deinum, B., and J.G.P. Dirven. 1972. Climate, nitrogen and grass. 5. Influence of age, light intensity, and temperature on the production and chemical composition of Congo grass. Neth. J. Agric. Sci. 20:125-132.

Deinum, B., and J.G.P. Dirvin. 1974. A model for the description of the effects of different environmental factors on the nutritive value of forages. p. 338-346. *In* V.G. Iglovikov and A.P. Moscsyants (ed.) Proc. 12th Int. Grassl. Congr., Moscow, U.S.S.R., 11-20 June 1974. Izd-vo MIR, Moscow.

Deinum, B., and J.G.P. Dirven. 1975. Climate, nitrogen and grass. 6. Comparison of yield and chemical composition of some temperate and tropical grass species grown at different temperatures. Neth. J. Agric. Sci. 23:69-82.

Deinum, B., and J.G.P. Dirven. 1976. Climate, nitrogen, and grass. 7. Comparison of chemical composition of *Brachiaria ruziziensis* and *Setaria sphaleceta* grown at different temperatures. Neth. J. Agric. Sci. 24:67-78.

Deinum, B., A.J.H. van Es, and P.J. Van Soest. 1968. Climate, nitrogen and grass. 2. The influence of light intensity, temperature and nitrogen on in vivo digestibility of grass and the prediction of these effects from some chemical procedures. Neth. J. Agric. Sci. 16:217-223.

Deregibus, V.A., R.Z. Sanchez, J.J. Casal, and J.M. Trlica. 1985. Tillering responses to enrichment of red light beneath the canopy in a humid natural grassland. J. Applied Bio. 22:199-206.

Dirven, J.G.P., and B. Deinum. 1977. The effect of temperature on the digestibility of grasses: An analysis. Forage Res. 3:1-17.

Dutt, T.E., R.G. Harvey, and R.S. Fawcett. 1982. Feed quality of hay containing perennial broadleaf weeds. Agron. J. 74:673-676.

Dutt, T.E., R.G. Harvey, R.S. Fawcett, N.A. Jorgensen, H.J. Larsen, and D.A. Schlough. 1979. Forage quality and animal performance as influenced by quackgrass control in alfalfa with pronamide. Weed Sci. 27:127-132.

Edwards, M.T., D.A. Sleper, and W.Q. Loegering. 1981. Histology of healthy and diseased orchardgrass leaves subjected to digestion in rumen fluid. Crop Sci. 21:341-343.

Eriksen, F.I., and A.S. Whitney. 1981. Effects of light intensity on growth of some tropical forage species. I. Interaction of light intensity and nitrogen fertilization on six forage grasses. Agron. J. 73:427-433.

Eriksen, F.I., and A.S. Whitney. 1982. Growth and N fixation of some tropical forage legumes as influenced by solar radiation regimes. Agron. J. 74:703-709.

Esau, K. 1977. Anatomy of seed plants. 2nd ed. John Wiley and Sons, Inc., New York, NY.

Estelle, J., B. Dumas, P. Geoffroy, N. Favet, D. Inze, M. Van Montagu, B. Fritig, and M. Legrand. 1992. Regulation of enzymes involved in lignin biosynthesis: Induction of O-methyltransferase mRNAs during the hypersensitive reaction to tobacco mosaic virus. Molecular Plant-Microbe Interact. 5:294-300.

Faix, J.J. 1974. The effect of temperature and daylength on the quality of morphological components of three legumes. Ph.D. diss. Cornell Univ., Ithaca, NY (Diss. Abstr. 35:202B).

Fales, S.L. 1986. Effects of temperature on fiber concentration, composition, and in vitro digestion kinetics of tall fescue. Agron. J. 78:963-966.

Farrar, J.F. 1988. Temperature and the partitioning and translocation of carbon. p. 203-235. *In* S.P. Long and F.I. Woodward (ed.) Plants and temperature. Soc. Exp. Biol., Cambridge, U.K.

Fawcett, R.S., R.G. Harvey, D.A. Schlough, and I.R. Block. 1978. Quackgrass control in established alfalfa with pronamide. Weed Sci. 26:193-198.

Fick, G.W., D.A. Holt, and D.G. Lugg. 1988. Environmental physiology and crop growth. p. 163-194. *In* A.A. Hanson, D.K. Barnes, and R.R. Hill, Jr. (ed.) Alfalfa and alfalfa improvement. Agronomy Monogr. 29. ASA, CSSA, and SSSA, Madison, WI.

Ford, C.W., I.M. Morrison, and J.R. Wilson. 1979. Temperature effects on lignin, hemicellulose and cellulose in tropical and temperate grasses. Aust. J. Agric. Res. 30:621-634.

Forwood, J.R., P. Stypinski, and J.A. Paterson. 1989. Forage selection by cattle grazing orchardgrass-legume pastures. Agron. J. 81:409-414.

Garwood, E.A., K.C. Tyson, and J. Sinclair. 1979. Use of water by six grass species. I. Dry-matter yields and response to irrigation. J. Agric. Sci., Camb. 93:13-24.

Gates, D.M. 1968. Transpiration and leaf temperature. Ann. Rev. Plant Physiol. 19:211-238.

George, J.R., and K.E. Hall. 1983. Herbage quality of three warm-season grasses with nitrogen fertilization. Iowa State J. Res. 58:247-259.

Gifford, R.O., and E.H. Jensen. 1967. Some effects of soil moisture regimes and bulk density on forage quality in the greenhouse. Agron. J. 59:75-77.

Goodman, R.N., Z. Kiraly, and K.R. Wood. 1986. The biochemistry and physiology of plant disease. Univ. of Missouri Press, Columbia, MO.

Gross, D.F., C.J. Mankin, and J.G. Ross. 1975. Effect of disease on in vitro digestibility of smooth bromegrass. Crop Sci. 15:273-275.

Grunes, D.L., P.R. Stout, and J.R. Brownell. 1970. Grass tetany of ruminants. Adv. Agron. 22:331-374.

Guerra, D., A.J. Anderson, and F.B. Salisbury. 1985. Reduced phenylalanine ammonia-lyase activities and lignin synthesis in wheat grown under low pressure sodium lamps. Plant Physiol. 78:126-130.

Halim, R.A., D.R. Buxton, J.J. Hattendorf, and R.E. Carlson. 1989a. Water-stress effects on alfalfa quality after adjustment for maturity differences. Agron. J. 81:189-194.

Halim, R.A., D.R. Buxton, J.J. Hattendorf, and R.E. Carlson. 1989b. Water-deficit effects on alfalfa at various growth stages. Agron. J. 81:765-770.

Hanna, W.W., and R. N. Gates. 1990. Plant breeding to improve forage utilization. p. 197-204. *In* D.E. Akin, L.G. Ljungdahl, J.R. Wilson, and P.J. Harris (ed.) Microbial and plant opportunities to improve lignocellulose utilization by ruminants. Proc. Trinational workshop microbial and plant opportunities to improve lignocellulose utilization by ruminants, Athens, GA, 30 Apr.- 4 May, 1990. Elsevier, New York, NY.

Hattersley, P.W. 1983. The distribution of C_3 and C_4 grasses in Australia in relation to climate. Oecologia 57:113-128.

Heide, O.M. 1985. Physiological aspects of climatic adaptation in plants with special reference to high-latitude environments. p. 1-22. *In* A. Kaurin, O. Junttial, and J. Nilsen (ed.) Plant production in the north. Norwegian Univ. Press, Oslo, Norway.

Henderson, M.S., and D.L. Robinson. 1982a. Environmental influences on fiber component concentrations of warm-season perennial grasses. Agron. J. 74:573-579.

Henderson, M.S. and D.L. Robinson. 1982b. Environmental influences on yield and *in vitro* true digestibility of warm-season perennial grasses and the relationships to fiber components. Agron. J. 74:943-946.

Hight, G.K., D.P. Sinclair, and R.J. Lancaster. 1968. Some effects of shading and of nitrogen fertilizer on the chemical composition of freeze-dried and oven-dried herbage and on the nutrition value of oven dried herbage fed to sheep. N. Z. J. Agric. Res. 11:286-302.

Hodgson, J. 1981. The influence of variation in the surface characteristics of the sward upon the short-term rate of herbage intake by calves and lambs. Grass Forage Sci. 36:49-57.

Holmes, M.G. 1981. Spectral distribution of radiation within plant canopies. p. 147-158. *In* H. Smith (ed.) Plants and the daylight spectrum. Academic Press, London, U.K.

Holt, D.A., and A.R. Hilst. 1969. Daily variation in carbohydrate content of selected forage crops. Agron. J. 61:239-242.

Hoveland, C.S., and W.G. Monson. 1980. Genetic and environmental effects on forage quality. p. 139-168. *In* C.S. Hoveland (ed.) Crop quality, storage, and utilization. ASA and CSSA, Madison, WI.

Hutchins, S.H., G.D. Buntin, and L.P. Pedigo. 1990. Impact of insect feeding on alfalfa regrowth: A review of physiological responses and economic consequences. Agron. J. 82:1035-1044.

Hutchins, S.H., D.R. Buxton, and L.P. Pedigo. 1989. Forage quality of alfalfa by potato leafhopper feeding. Crop Sci. 29:1541-1545.

Johnson, I.R., and J.H.M. Thornley. 1985. Temperature dependence of plant and crop processes. Ann. Bot. 55:1024-1032.

Jones, C.A. 1985. C_4 grasses and cereals. Growth, development, and stress response. John Wiley and Sons, New York, NY.

Jones, M.B., E. L. Leafe, and W. Stiles. 1980. Water stress in field-grown perennial ryegrass. II. Its effect on leaf water status, stomatal resistance and leaf morphology. Ann. Appl. Biol. 96:103-110.

Juan, N.A., C.C. Sheaffer, and D.K. Barnes. 1993. Temperature and photoperiod effects on multifoliolate expression and morphology of alfalfa. Crop Sci. 33:573-578.

Jung, H.G., and M.P. Russelle. 1991. Light source and nutrient regime effects on fiber composition and digestibility of forages. Crop Sci. 31:1065-1070.

Karn, J.F., J.M. Krupinsky, and J.D. Berdahl. 1989. Nutritive quality of four foliar disease resistant and susceptible strains of intermediate wheatgrass. Crop Sci. 29:436-439.

Kephart, K.D. 1987. Irradiance level effects on plant growth, nutritive quality, and energy exchange of C_3 and C_4 grasses. Ph.D. Diss., Iowa State Univ., Ames (Diss. Abstr. 87-16781).

Kephart, K.D., and D.R. Buxton. 1993. Forage quality responses of C_3 and C_4 perennial grasses to shade. Crop Sci. 33:831-837.

Kephart, K.D., D.R. Buxton, and S.E. Taylor. 1992. Growth of C_3 and C_4 perennial grasses in reduced irradiance. Crop Sci. 32:1033-1038.

Kephart, K.D., L.G. Higley, D.R. Buxton, and L.P. Pedigo. 1990. Cicer milkvetch forage yield, quality, and acceptability to insects. Agron. J. 82:477-483.

Kuc, J. 1993. Responses to plant infection and parasitic activity. *In* D.R. Buxton, R.M Shibles, R.A. Forsberg, B.L. Blad, and K.H. Asay, G.M. Paulsen, and R.F. Wilson (ed.) International crop science I. CSSA, Madison, WI. In press.

Labhart, C.H., J. Nosberger, and C.J. Nelson. 1983. Photosynthesis and degree of polymerization of fructosan during reproductive growth of meadow fescue at two temperatures and two photon flux densities. J. Exp. Bot. 34:1037-1046.

Lechtenberg, V.L., D.A. Holt, and H.W. Youngberg. 1971. Diurnal variation in nonstructural carbohydrates, *in vitro* digestibility, and leaf to stem ratio of alfalfa. Agron. J. 63:719-724.

Lemaire, G., B. Onillow, G. Gosse, M. Chartiea, and J.M. Allirand. 1991. Nitrogen distribution within a lucerne canopy during regrowth: Relation with light distribution. Ann. Bot. 68:483-488.

Lenssen, A.W., E.L. Sorensen, G.L. Posler, and D.L. Stuteville. 1991. Resistance to anthracnose protects forage quality of alfalfa. Crop Sci. 31:147-150.

Levitt, J. 1980. Response of plants to environmental stresses. Vol. II. 2nd ed. Academic Press, New York, NY.

Lyons, J.M. 1973. Chilling injury in plants. Ann. Rev. Plant Physiol. 24:445-466.

Marten, G.C., and R.N. Anderson. 1975. Forage nutritive value and palatability of 12 common annual weeds. Crop Sci. 15:821-827.

Marten, G.C., D.R. Buxton, and R.F Barnes. 1988. Feeding value (forage quality). p. 463-491. *In* A.A. Hansen, D.K. Barnes, and R.R. Hill, Jr. (ed.) Alfalfa and alfalfa improvement. Agronomy Monogr. 29. ASA, CSSA, and SSSA, Madison, WI.

Marten, G.C., C.C. Sheaffer, and D.L. Wyse. 1987. Forage nutritive value and palatability of perennial weeds. Agron. J. 79:980-986.

Messman, M.A., W.P. Weiss, and D.O. Erickson. 1991. Effects of nitrogen fertilization and maturity of bromegrass on in situ ruminal digestion kinetics of fiber. J. Anim. Sci. 69:1151-1161.

Millard, P., A.H. Gordon, A.J. Richardson, and A. Chesson. 1987. Reduced ruminal degradation of ryegrass caused by sulphur limitation. J. Sci. Food Agric. 40:305-314.

Minson, D.J. 1975. Pasture management and animal nutrition. Forage Res. 1:1-10.

Minson, D.J. 1990. Forage in ruminant nutrition. Academic Press, New York, NY.

Minson, D.J., and M.N. McLeod. 1970. The digestibility of temperate and tropical grasses. p. 719-722. *In* M.J.T. Norman (ed.) Proc. 11th Int. Grassl. Congr., Surfers Paradise, Queensland, Australia. 13-23 April 1970. University of Queensland Press, St. Lucia.

Moerschbacher, B.M. 1989. Lignin biosynthesis in stem rust infected wheat. p. 370-382. *In* N.G. Lewis and M.G. Paice (ed.) Plant cell wall polymers: Biogenesis and biodegradation. Proc. Symposium, Toronto, Ontario, Canada, 5-11 June 1988. American Chemical Society, Washington, DC.

Moir, K.W., J.R. Wilson, and G.W. Blight. 1977. The in vitro digested cell wall and fermentation characteristics of grasses as affected by temperature and humidity during their growth. J. Agric. Sci., Camb. 88:217-222.

Mueller, J.P., W.M. Lewis, J.T. Green, and J.C. Burns. 1993. Yield and quality of silage corn as altered by johnsongrass infestation. Agron. J. 85:49-52.

Navarro-Chavira, G., and B.D. McKersie. 1983. Growth, development and digestibility of guinea grass (*Panicum maximum* Jacq.) in two controlled environments differing in irradiance. Trop. Agric. 60:184-188.

Nicholson, R.L., and R. Hammerschmidt. 1992. Phenolic compounds and their role in disease resistance. Ann. Rev. Phytopathol. 30:369-389.

Nishitani, K., and Y. Masuda, 1979. Growth and cell wall changes in azuki bean epicotyls. I. Change in cell wall polysaccharides during intact growth. Plant Cell Physiol. 20:63-74.

Nobel, P.S. 1988. Principles underlying the prediction of temperature in plants with special reference to desert succulents. p. 1-23. *In*. S.P. Long and F.I. Woodward (ed.). Plants and temperature. Soc. Exp. Biol., Cambridge, U.K.

Nobel, P.S., L.J. Zaragoza, and W.K. Smith. 1975. Relation between mesophyll surface area, photosynthetic rate, and illumination level during development for leaves of *Plectranthus parviflorus* Hencke. Plant Physiol. 55:1067-1070.

Noller, C.H., and C.L. Rhykerd. 1974. Relationship of nitrogen fertilization and chemical composition of forage to animal health and performance. p. 363-394. *In* D.A. Mays (ed.) Forage fertilization. ASA, Madison. WI.

Nosberger, J. 1993. The physiological characteristics of successful, cool climate, hay and pasture species. *In* D.R. Buxton, R.M. Shibles, R.A. Forsberg, B.L. Blad, K.H. Asay, G.M. Paulsen, and R.F. Wilson (ed.) International crop science I. CSSA, Madison, WI. In press.

Ohlsson, Christer. 1991. Growth, development, and composition of temperate forage legumes and grasses in varying environments. Ph.D. Diss. Iowa State Univ., Ames (Diss. Abstr. 91-26231).

Onstad, D.W., and G.W. Fick. 1983. Predicting crude protein, in vitro true digestibility, and leaf proportion in alfalfa herbage. Crop Sci. 23:961-964.

Pearce, R.B., and D.R. Lee. 1969. Photosynthetic and morphological adaptation of alfalfa leaves to light intensity at different stages of maturity. Crop Sci. 9:791-794.

Peterson, P.R., C.C. Sheaffer, and M.H. Hall. 1992. Drought effects on perennial forage legume yield and quality. Agron. J. 84:774-779.

Pitman, W.D., and E.C. Holt. 1982. Environmental relationships with forage quality of warm-season perennial grasses. Crop Sci. 22:1012-1016.

Pollack, C.J. 1990. The response of plants to temperature change. J. Agric. Sci., Camb. 115:1-5.

Prioul, J.L., J. Brangeon, and A. Reyss. 1980. Interaction between external and internal conditions in the development of photosynthetic features in a grass leaf. I. Regional responses along a leaf during and after low-light or high-light acclimation. Plant Physiol. 66:762-769.

Pulli, S. 1988. Adaption of red clover to the longday environment. J. Agric. Sci., Finland 60:201-214.

Puoli, J.R., G.A. Jung, and R.L. Reid. 1991. Effects of nitrogen and sulfur on digestion and nutritive quality of warm-season grass hays for cattle and sheep. J. Anim. Sci. 69:843-852.

Reid, R.L., and G.A. Jung. 1965. Influence of fertilizer treatment on intake, digestibility, and palatability of tall fescue hay. J. Anim. Sci. 24:615-625.

Reid, R.L., and G.A. Jung. 1974. Effects of elements other than nitrogen on the nutritive value of forage p. 395-435. *In* D.A. Mays (ed.). Forage fertilization. ASA, Madison, WI.

Rhykerd, C.L., and C.H. Noller. 1974. The role of nitrogen in forage production. p. 416-424. *In* D.A. Mays (ed.). Forage fertilization. ASA, Madison, WI.

Rosenberg, N.J., B.L. Blad, and S.B. Verma. 1983. Microclimate: The biological environment. 2nd ed. Wiley Interscience, New York, NY.

Samarakoon, S. P., H.M. Shelton, and J.R. Wilson. 1990a. Voluntary feed intake by sheep and digestibility of shaded *Stenotaphrum secundatum* and *Pennisetum clandestinum* herbage. J. Agric. Sci., Camb. 114:143-150.

Samarakoon, S.P., J.R. Wilson, and H.M. Shelton. 1990b. Growth, morphology and nutritive quality of shaded *Stenotaphrum secundatum*, *Axonopus compressus* and *Pennisetum clandestinum*. J. Agric. Sci., Camb. 114:161-169.

Sanchez, R.A., J.J. Casal, C.L. Ballare, and A.L. Scopel. 1993. Plant responses to canopy density mediated by photomorphogenic precesses. *In* D.R. Buxton, R.M. Shibles, R.A. Forsberg, B.L. Blad, K.H. Asay, G.M. Paulsen, and R.F. Wilson (ed.) International crop science I. CSSA, Madison, WI. In press.

Seligman, N.G. 1993. Nitrogen redistribution in crop plants: Regulation and significance. *In* D.R. Buxton, R.M. Shibles, R.A. Forsberg, B.L. Blad, K.H. Asay, G.M. Paulsen, and R.F. Wilson (ed.) International crop science I. CSSA, Madison, WI. In press.

Setter, T.L. 1993. Assimilate allocation in response to water deficit stress. *In* D.R. Buxton, R.M. Shibles, R.A. Forsberg, B.L. Blad, K.H. Asay, G.M. Paulsen, and R.F. Wilson (ed.) International crop science I. CSSA, Madison, WI. In press.

Sheaffer, C.C., P.R. Peterson, M.H. Hall, and J.B. Stordahl. 1992. Drought effects on yield and quality of perennial grasses in the North Central United States. J. Prod. Agric. 5:556-561.

Sheaffer, C.C., D.L. Wyse, G.C. Marten, and P.H. Westra. 1990. The potential of quackgrass for forage production. J. Prod. Agric. 3:256-259.

Sherwood, R.T., and C.C. Berg. 1991. Anatomy and lignin content in relation to resistance of *Dactylis glomerata* to stagonospora leaf spot. Phytopathology 81:1401-1407.

Southerton, S.G., and B.J. Deveral. 1990. Changes in phenolic acid levels in wheat leaves expressing resistance to *Puccinia recondita* f. sp. *tritici*. Physiol. Mol. Plant Path. 37:437-450.

Stewart, C.R., and A.D. Hanson. 1980. Proline accumulation as a metabolic response to water stress. p. 173-189. *In* N.C. Turner and P.J. Kramer (ed.) Adaptation of plants to water and high temperature stress. John Wiley and Sons, New York, NY.

Stricker, J.A., A.G. Matches, G.B. Thompson, V.E. Jacobs, F.A. Martz, H.N. Wheaton, H.D. Currence, and G.F. Krause. 1979. Cow-calf production from tall fescue-ladino clover pastures with and without nitrogen fertilization on creep feeding: Spring calves. J. Anim. Sci. 48:13-25.

Struik, P.C. 1983. The effect of short and long shading, applied during different stages of growth, on development, productivity and quality of forage maize (*Zea mays*). Neth. J. Agric. Sci. 31:101-124.

Tanner, C.B., and T.R. Sinclair. 1983. Efficient water use in crop production: Research or re-search? p. 1-27. *In* H.M. Taylor, W.R. Jordon, and T.R. Sinclair (ed.) Limitations to efficient water use in crop plants. ASA, CSSA, and SSSA, Madison, WI.

Temme, D.G., R.G. Harvey, R.S. Fawcett, and A.W. Young. 1979. Effects of annual weed control on alfalfa forage quality. Agron. J. 71:51-54.

Teeri, J.A., and L.G. Stowe. 1976. Climatic patterns and the distribution of C_4 grasses in North America. Oecologia 23:1-12.

Thompson, L., and J.L. Harper. 1988. The effect of grasses on the quality of transmitted radiation and its influence on the growth of white clover *Trifolium repens*. Oecologia 75:343-347.

Thorvaldsson, G. 1987. The effects of weather on nutritional value of timothy in northern Sweden. Acta Agric. Scand. 37:305-319.

Thorvaldsson, G. 1992. The effects of temperature on digestibility of timothy (*Phleum pratense* L.) tested in growth chambers. Grass Forage Sci. 47:306-308.

Turner, N.C. 1979. Drought resistance and adaptation to water deficits in crop plants. p. 343-372. *In* H. Mussell and R.C. Staples (ed.) Stress physiology in crop plants. John Wiley and Sons, New York, NY.

Turner, N.C., and M.M. Jones. 1980. Turgor maintenance by osmotic adjustment: A review and evaluation. p. 87-103. *In* N.C. Turner and P.J. Kramer (ed.) Adaptation of plants to water and high temperature stress. John Wiley and Sons, New York, NY.

Uhl, C., and G. Parker. 1986. Viewpoint: Our steak in the jungle. Bioscience 36:642.

Van Soest, P.J. 1982. Nutritional ecology of the ruminant. O & B Books, Inc. Corvallis, OR.

Vough, L.R., and G.C. Marten. 1971. Influence of soil moisture and ambient temperature on yield and quality of alfalfa forage. Agron. J. 63:40-42.

Waggoner, P.E. 1993. Preparing for climate change. *In* D.R. Buxton, R.M. Shibles, R.A. Forsberg, B.L. Blad, K.H. Asay, G.M. Paulsen, and R.F. Wilson (ed.) International crop science I. CSSA, Madison, WI. In press.

Waldo, D.R., and N.A. Jorgensen. 1981. Forages for high animal production: Nutritional factors and effects of conservation. J. Dairy Sci. 64:1207-1229.

Walgenbach, R.P., and G.C. Marten. 1981. Release of soluble protein and nitrogen in alfalfa. II. Influence of shading. Crop Sci. 21:859-862.

Walgenbach, R.P., G.C. Marten, and G.R. Blake. 1981. Release of soluble protein and nitrogen in alfalfa. I. Influence of growth temperature and soil moisture. Crop Sci. 21:843-849.

Wedin, W.F. 1974. Fertilization of cool-season grasses. p. 95-118. In D.A. Mays (ed.). Forage fertilization. ASA, Madison, WI.

Weimer, P.J., R.D. Hatfield, and D.R. Buxton. 1993. Inhibition of ruminal cellulose fermentation by extracts of the perennial legume cicer milkvetch (*Astragalus cicer*). Appl. Environ. Microbiol. 59:405-409.

Welter, S.C. 1993. Responses of plants to insect injury: Eco-physiological insights. In D.R. Buxton, R.M. Shibles, R.A. Forsberg, B.L. Blad, K.H. Asay, G.M. Paulsen, and R.F. Wilson (ed.) International crop science I. CSSA, Madison, WI. In press.

Whitehead, D.C., K.M. Goulden, and R.D. Hartley. 1986. Fractions of nitrogen, sulphur, phosphorus, calcium and magnesium in the herbage of perennial ryegrass as influenced by fertilizer nitrogen. Anim. Feed Sci. Tech. 14:231-242.

Willis, W.G., D.L. Stuteville, and E.L. Sorensen. 1969. Effects of leaf and stem diseases on yield and quality of alfalfa forage. Crop Sci. 9:637-640.

Wilson, J.P., R.N. Gates, and W.W. Hanna. 1991a. Effect of rust on yield and digestibility of pearl millet forage. Phytopathology 81:233-236.

Wilson, J.R. 1982. Environmental and nutritional factors affecting herbage quality. p. 111-131. In J.B. Hacker (ed.) Nutritional limits to animal production from pastures. CAB, Farnham, U.K.

Wilson, J.R. 1983. Effects of water stress on *in vitro* dry matter digestibility and chemical composition of herbage of tropical pasture species. Aust. J. Agric. Res. 34:377-390.

Wilson, J.R. 1989. Ecophysiological constraints to production and nutritive quality of pastures under tree crops. p. 39-54. In Z. Ahmad Tajuddin (ed.) Proc. Int. Livestock-Tree Cropping Workshop, Serdang, Malaysia 5-9 December 1988. FAO, Rome, Italy.

Wilson, J.R., B. Deinum, and F.M. Engels. 1991b. Temperature effects on anatomy and digestibility of leaf and stem of tropical and temperate forage species. Neth. J. Agric. Sci. 39:31-48.

Wilson, J.R., and C.W. Ford. 1971. Temperature influence on the growth, digestibility, and carbohydrate composition of two tropical grasses, *Pannicum maximum* var *trichloglume* and *Setaria sphacellata*, and two cultivars of the temperate grass, *Lolium perenne*. Aust. J. Agric. Res. 22:563-571.

Wilson, J.R., and D.J. Minson. 1980. Prospects for improving the digestibility and intake of tropical grasses. Trop. Grassl. 14:253-257.

Wilson, J.R., and T.T. Ng. 1975. Influence of water stress on parameters associated with herbage quality of *Panicum maxium* var. *tricholume*. Aust. J. Agric. Res. 26:127-136.

Wilson, J.R., A.O. Taylor, and G.R. Dolby. 1976. Temperature and atmospheric humidity effects on cell wall content and dry matter digestibility of some tropical and temperate grasses. N. Z. J. Agric. Res. 19:41-46.

Wilson, J.R. and C.C. Wong. 1982. Effects of shade on some factors influencing nutritive quality of green panic and siratro pastures. Aust. J. Agric. Res. 33:937-950.

Woledge, J., and A.J. Parsons. 1986. The effect of temperature on the photosynthesis of ryegrass canopies. Ann. Bot. 57:487-497.

Wong, C.C., M.A. Mohd. Sharudin, and H. Rahim. 1985a. Shade tolerance potential of some tropical forages for integration with plantations. I. Grasses. MARDI Research Bull. 13:225-247.

Wong, C.C., H. Rahim, and M.A. Mohd. Sharudin. 1985b. Shade tolerance potential of some tropical forages for integration with plantations. II. Legumes. MARDI Res. Bull. 13:247-269.

Wong, C.C., and J.R. Wilson. 1980. Effects of shading on the growth and nitrogen content of green panic and siratro in pure and mixed swards defoliated at two frequencies. Aust. J. Agric. Res. 31:269-285.

CHAPTER 5

QUANTIFYING FORAGE PROTEIN QUALITY

G. A. Broderick

INTRODUCTION

The primary function of forage protein in the diet is to provide the ruminant with absorbed protein (AP) in the form of α-amino nitrogen (N). Protein requirements of ruminants are met from two major sources: microbially synthesized protein and dietary protein which escapes microbial degradation in the rumen. In this paper, microbial protein synthesized in the rumen and dietary protein escaping the rumen will be referred to, respectively, as bacterial crude protein (BCP) and undegraded intake protein (UIP) (NRC, 1989). Ideally, any method of assessing forage protein quality should describe the degree to which the forage contributes both BCP and UIP to meet the animal's AP requirements.

Protein nutrition of ruminants is complex. Determination of apparent crude protein (CP) digestibility has little value in ruminants because protein N disappears from the gastrointestinal tract only partly through absorption of α-amino N. Often, much of the apparent CP digestion is a result of ammonia absorption. Little is known about quantitative amino acid requirements of ruminants because microbial intervention prevents their convenient study. However, new methodology such as the total nutriture of productive ruminants by parenteral infusion (Storm and Ørskov, 1984) has allowed conduct of nutritional studies using amino acid mixtures of known composition. Until essential amino acid (EAA) needs have been quantified, information about EAA requirements and optimality of EAA patterns in dietary proteins must be inferred from requirements of nonruminant species and the amino acid patterns of proteins in the principal ruminant products, namely milk, tissue, and wool.

A comparison was made between the EAA and semi-essential amino acid (SEAA) compositions of milk protein, ruminal microbial proteins, and a number of feed proteins (Table 1). The geometric mean of EAA plus SEAA content in milk protein was similar or lower than all the listed feed proteins, except meat and bone meal, suggesting that most feed and microbial proteins would be adequate sources of EAA. Lysine and methionine were notable exceptions (Table 1). In these data, both bacterial and protozoal protein had contents of methionine and lysine which were similar to, or greater than, milk protein. Microbial protein content of cystine, a SEAA which spares the methionine requirement, also was similar to milk protein. However, relative to milk, the geometric mean of lysine and methionine content was lower for alfalfa (*Medicago sativa* L.) hay, corn (*Zea mays* L.) grain, soybean (*Glycine max* L.) meal, corn gluten meal, blood meal, and meat and bone meal. The geometric mean of lysine and methionine content in corn silage was greater than milk because of its relatively high methionine content, although the lysine content of corn

---

U. S. Dairy Forage Research Center, U. S. Department of Agriculture-Agricultural Research Service, 1925 Linden Dr. West, Madison, WI 53706.

Mention of a trademark or proprietary product in this paper does not constitute a guarantee or warranty of the product by the USDA or the Agricultural Research Service and does not imply its approval to the exclusion of other products that also may be suitable.

Table 1. Essential amino acid (EAA) and semi-essential amino acid (SEAA) composition of milk and selected feed proteins.[1]

Item	Cow's Milk	Bacterial Protein	Protozoal Protein	Alfalfa Hay	Corn Silage	Corn Grain	Solvent Soybean Meal	Corn Gluten Meal	Fish Meal (Menhaden)	Blood Meal	Meat & Bone Meal
CP (g kg^{-1} DM)	358	...	...	200	83	100	499	672	667	872	541
Amino acid (g kg^{-1} CP)											
ARG	34	61	57	46	117	54	68	34	61	41	69
HIS	26	26	24	20	25	25	24	23	24	50	19
ILE	65	74	82	44	30	39	45	42	47	11	33
LEU	99	88	98	73	112	112	73	169	73	125	61
LYS	75	99	119	51	52	24	60	17	77	79	57
MET	27	32	26	13	53	21	12	29	29	11	13
CYS	13	12	15	15	47	18	17	16	9	15	10
PHE	46	58	72	44	43	49	47	66	40	69	34
TYR	34	53	58	30	24	43	30	53	32	24	16
THR	47	63	58	38	43	39	37	37	41	45	33
TRP[2]	13	16	16	13	11	9	14	5	11	12	6
VAL	68	77	63	50	54	51	45	51	53	82	49
Geometric means[3]											
EAA+SEAA	34.5	46.2	47.3	31.4	42.5	33.5	33.6	32.6	34.7	34.4	26.0
LYS+MET	45.0	56.3	55.6	25.6	52.4	22.4	26.4	22.2	47.0	29.7	27.3

[1]Amino acid compositions for cow's milk (skimmed, dehydrated) and feed proteins are from NRC (Table 4; 1982); values for alfalfa hay are means of sun-cured hay and late vegetative hay. Amino acid compositions of bacterial and protozoal proteins are means from data of Weller (1957) and Purser and Buechler (1966).
[2]Contents of TRP in bacterial and protozoal protein computed from data of Holmes et al. (1953).
[3]Geometric mean is the antilog of the average log of concentration of each EAA and SEAA or LYS and MET.

silage was similar to alfalfa hay and soybean meal. Only the lysine and methionine content of fish meal was comparable to that in milk protein (Table 1). This comparison suggests that the EAA pattern of AP generally would be improved if feed protein were degraded in the rumen and converted to microbial protein. However, certain feeds would have complementary EAA patterns if their proteins escaped the rumen: alfalfa hay, soybean meal, and blood meal, with their relatively high lysine contents, would complement the relatively high methionine contents in corn grain and corn silage.

New methods are required to evaluate protein quality of forages because current approaches have proven quantitatively inadequate. Total CP analyses, based on Kjeldahl or other total N assays, usually are necessary for expressing data on a N basis. However, CP tells little about the contribution of amino acids from forage protein to the animal. Simple, uncomplicated interpretation of in situ or in vitro data for assessing ruminal protein degradation can be misleading; these methods will be discussed below.

In the future, speed of analysis often will be essential for forage protein evaluation systems but this need must be weighed in view of how the results are to be used. For example, rapid laboratory procedures for determining protein degradability would be valuable to farmers and nutritional consultants to permit timely decisions concerning supplement feeding, commodity purchasing, and so on. Speed of analysis also would be essential to screen large numbers of forage germplasm for differences in ruminal protein degradability. Plant breeders deal in thousands of samples and generally are willing to sacrifice relative accuracy to permit analysis of large numbers; knowing relative differences in forage protein degradability would be satisfactory for making genetic progress in plant breeding. Although relative degradabilities are useful, absolute values are required for diet formulation. Screening four replicates each of 22 samples of *Medicago* germplasm for ruminal protein degradability using an abbreviated version of our inhibitor in vitro (IIV) system (Broderick and Buxton, 1991) required about two months of laboratory work. A screening study with *Medicago* germplasm harvested from a single year yielded 4,400 samples each of leaves and stems (D.K. Barnes, personal communication).

Near infrared reflectance spectroscopy (NIRS) has application in those settings for rapid estimation of rate and extent of ruminal protein degradation. We used NIRS to estimate IIV protein degradation in roasted soybeans (G.F. Tremblay et al., unpublished data). The NIRS calibration equations developed from analyses of 121 samples for dry matter (DM), total N, and estimated protein escape gave means, standard errors of calibration, and coefficients of determination (r^2) of, respectively, 946 g kg^{-1}, 0.26, and 0.97 (DM), 6.71 g kg^{-1} DM, 0.05, and 0.99 (total N), and 236 g CP kg^{-1} DM, 0.98, and 0.90 (protein escape). Validation of these equations, conducted with analyses of 145 separate samples of roasted soybeans, yielded means, standard errors of validation, and r^2 of 945 g kg^{-1}, 0.63, and 0.86 (DM), 6.75 g kg^{-1} DM, 0.12, and 0.86 (total N), and 239 g CP kg^{-1} DM, 1.54, and 0.70 (protein escape). As yet, we have not attempted use of NIRS to estimate rate or extent of ruminal degradation of forage proteins. The NIRS approach would have particular advantage in settings where rapid turn around is required on large numbers of samples.

Future research will necessitate testing many other forage materials which may be useful for ruminant feeding. For example, several species containing condensed tannins, such as sainfoin (*Onobrychis viciifolia* Scop.; Reid et al., 1973; Egan and Ulyatt, 1980), birdsfoot trefoil (*Lotus corniculatus* L.; Barry and Reid, 1985; Waghorn et al., 1987) and "bigfoot" trefoil (*Lotus pedunculatus* Cav.; Barry and Reid, 1985; Barry et al., 1986), show promise of reduced ruminal protein degradation and improved protein utilization. However, most of these forages are not well adapted to growing conditions in North America; the plant genetic work necessary to develop adapted cultivars will involve continual screening to confirm that advantageous protein properties are maintained. Use of biotechnology to inject genes for resistant proteins (Spencer et al., 1988) or protein-protecting compounds such as tannins (Li et al., 1992) also will require rapid methodology to test forage materials at each developmental stage. Several silage treatments have been reported to improve protein utilization (e.g., Glenn et al., 1986; Charmley and Veira, 1990; Nagel and Broderick, 1992); heating alfalfa hay to increase acid detergent insoluble N (ADIN) levels to about 120 to 140 g kg^{-1} total N improved protein utilization in lactating cows (Broderick et al., 1993). Techniques must be available for routine analysis of the potential of such treatments if progress is to be made in improving forage protein value for ruminants.

What are the characteristics of an ideal forage protein evaluation system? To determine the contribution of AP due to consumption of a forage protein, it would be necessary to quantify: 1) Rates of ruminal degradation and passage of forage protein; 2) intestinal digestibility of the UIP; and 3) ruminal fermentable energy content. Knowing 1) and 2) allows determination of UIP (including a discount of UIP for unavailable protein) and computation of degraded intake protein (DIP). Forage fermentable energy, which is related to net energy of lactation and neutral detergent fiber (NDF) (Mertens, 1987), may be used to compute the contribution of AP from BCP synthesis (NRC, 1989). If proportionate loss of EAA were assumed during protein degradation, the contribution of EAA from UIP could be estimated from EAA composition of the original protein. For purposes of this review, we will concentrate on methodology for estimating forage UIP. Importance of DIP for BCP formation will not be discussed. Differential losses of individual EAA, particularly greater than average losses of the basic amino acids arginine and lysine (Craig and Broderick, 1984; Stern et al., 1985), may occur during protein degradation. However, because too few data are available, it is premature to consider disproportionate losses of individual EAA.

KINETICS OF RUMINAL DEGRADATION OF FORAGE PROTEIN

The model most commonly used to describe ruminal protein degradation is a first order disappearance model, without time-delay, which is similar to that applied to ruminal fiber digestion (Waldo et al., 1972; Mertens, 1987). With this model, total CP in feed is divided into fractions A, B, and C, which sum to unity. Fraction A is the proportion of total CP already degraded at 0-time (i.e., present in the feed as NPN), B is the fraction of total CP which is potentially digestible or degradable, and C is the fraction of total CP which is completely indigestible (ARC,1984; NRC, 1985). The proportion of total CP degraded in the rumen is determined by the fractional rates of degradation and passage (Broderick, 1978). When both degradation and passage account for removal, the rate of disappearance of potentially degradable protein, B, with respect to time, is expressed by the differential equation: $dB/dt = -k_d B - k_p B$, where k_d and k_p are fractional rates for ruminal degradation and passage, respectively. When only ruminal degradation accounts for disappearance, the equation becomes: $dB/dt = -k_d B$. Integration from zero to infinity yields, respectively, the amounts disappearing from the rumen due to degradation plus passage, $k_d + k_p$, and due to degradation only, k_d (Hungate, 1966). It follows that the proportion of total protein disappearance resulting from ruminal degradation is given by the ratio: $k_d / (k_d + k_p)$ (Mertens, 1987). Total protein degradation in this model is given by the equation:

$$\text{Degraded Protein} = A + B \times [k_d / (k_d + k_p)] \tag{1}$$

The fraction of total protein escaping undegraded is given by the complementary equation:

$$\text{Escaped Protein} = B \times [k_p / (k_d + k_p)] + C \qquad (2)$$

This basic model for describing protein degradation and escape is widely accepted and applied (Ørskov and McDonald, 1979; Van Soest et al., 1982; ARC, 1984; NRC, 1985).

Although fraction C will escape the rumen, it is indigestible and will not contribute to the animal's amino acid supply. When in situ methods are used to quantify rates and fractions, undegradable protein theoretically is the total CP remaining in situ at time approaching infinity. In practice, incubation times of 24 to 72 h are used. Often it is difficult to distinguish degradation rates obtained for slowly degraded proteins such as fish meal (ARC, 1984) and blood meal (Loerch et al., 1983) from rates approaching zero.

Ideally, fraction A is equivalent to forage NPN and can be quantified chemically as the proportion of buffer-soluble N which is not precipitated by protein denaturants such as trichloroacetic acid (TCA). Remaining CP can be apportioned between the other two CP fractions by determining digestible B or indigestible C, with the other fraction computed by difference. Indigestible N often has been estimated from ADIN (Goering et al., 1972; Yu, 1977; Thomas, 1982; Sniffen et al., 1992). Although ADIN present in unheated alfalfa had digestibility coefficients essentially equal to zero, additional ADIN produced by heating was reported to have apparent digestibilities of about 400 g kg^{-1} (Goering and Lindahl, 1975; Broderick et al., 1993). This suggested that forage ADIN may often be comprised of at least two fractions, one indigestible and one of low digestibility. Muscato et al. (1983) reported that an average 750 g kg^{-1} of forage ADIN was present as α-amino N. Krishnamoorthy et al. (1983) used *Streptomyces griseus* protease at high activity to solubilize potentially digestible CP (fraction B) and estimated indigestible CP (fraction C) by difference. Use of proteolytic enzymes, rather than chemical extractions, is a more logical approach for estimating digestible and indigestible CP.

Absence of a time-lag before protein degradation commences, such as occurs with fiber digestion, deserves comment. Time is required for colonization of feed particles and for elaboration of particulate-bound cellulases (Silva et al., 1987) before there is detectable fiber digestion. Both McDonald (1981) and Van Soest et al. (1982) incorporated time-lag parameters into their models for ruminal protein degradation. Laycock et al. (1985), using in vitro incubations with pure cultures of proteolytic rumen bacteria, observed a lag during protein degradation. This lag was due, however, to the time required for microbial growth in the in vitro system. Van Soest et al. (1982) suggested that a relative lag may be expected after feeding in meal-fed ruminants while microbial numbers increase following increased supply of fermentable substrate. However, in ruminants fed ad libitum, fermentable substrate would seldom be exhausted and microbial numbers only rarely would decline significantly. Thus, it seems unlikely that a time-lag such as that typical in fiber digestion would be observed with protein. Moreover, extensive in vitro research with the IIV system (Broderick and Craig, 1980; Broderick, 1978; 1987) did not reveal improved fit by incorporating a time-lag into models describing protein degradation. Time-lags have not been observed in protein degradation studies using in situ techniques. Omission of a time-lag from most models describing ruminal protein degradation seems warranted.

In the standard protein degradation model, fraction B is assumed to be a single fraction degraded at a single rate. Others have suggested that there are two or more fractions degraded at two or more rates. Broderick and Craig (1980) used a ruminal in vitro system and "curve-peeling" to identify one rapidly and one slowly degraded fraction in cottonseed meals which were heated to different extents.

Generally, heating decreased the proportion of the rapidly degraded protein fraction, without altering its degradation rate, and increased the slowly degraded fraction and decreased its degradation rate. Van Soest et al. (1982) used a similar approach with *Streptomyces griseus* protease to quantify two degradation fractions in soybean meal and brewer's grains and three degradation fractions in timothy (*Phleum pratense* L.) hay and wheat (*Triticum aestivum* L.) middlings. Although theoretically sound, using degradation kinetics to determine all potentially degradable fractions and their respective rates would be problematic. J. H. Matis (personal communication) stated that the need for statistical precision dictates five data points be obtained per regression parameter estimated. Hence, a curve-peeling approach to determine two intercepts (fractions) and two slopes (degradation rates) would require 20 data points. In vitro and in situ incubations rarely are conducted such that enough time-points are sampled to identify more than two fractions under Matis' restriction. However, Sniffen et al. (1992) used chemical extractions to identify three B fractions with different degradation rates.

It may be premature using current methods to attempt identification of more than one fraction and rate by regression procedures. Degradation curves obtained using a ruminal in vitro system by regressing log transformed data on time were tested for curvilinearity (Broderick, 1987). Of the 12 different proteins tested, only data from bovine serum albumin and meat meal showed significantly improved fit, versus linear regression, with non-linear regression. There was no improvement of fit ($P > 0.20$) for the other 10 proteins, suggesting most were degraded as a single fraction degraded at a single rate. Moreover, a Michaelis-Menten approach (described below) yielded a single overall degradation rate for protein mixtures which was not different from the proportionate mean of the individual rates, weighted for the amounts of each protein degradation fraction (Broderick and Clayton, 1992).

Inspection of equations 1 and 2 shows the importance of passage rate in determining ruminal degradation and escape. Mertens (1987) summarized the relationship of passage rate with extent of digestion, defined by the ratio: $k_d / (k_d + k_p)$ (Figure 1). At any degradation rate, as passage rate increases, extent of

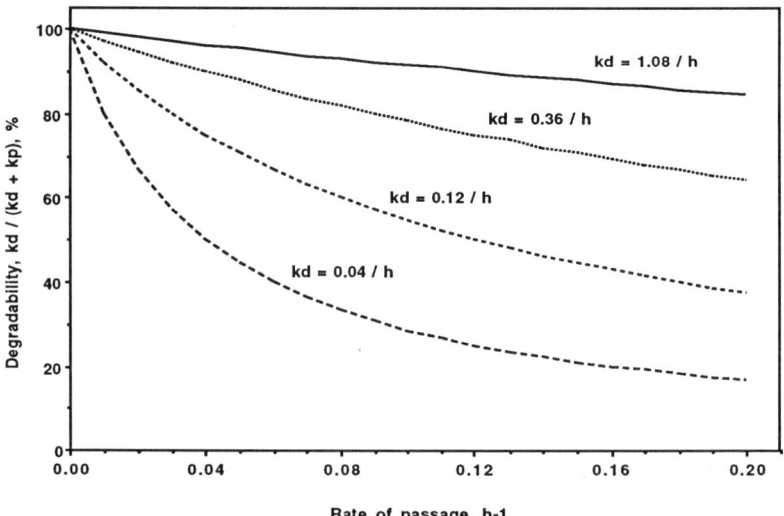

Fig. 1. The effect of rate of passage on extent of degradation at four different rates of degradation (k_d) (Mertens, 1987).

degradation decreases. However, the magnitude of this effect increases inversely with rate. Thus, the relative change in extent of degradation with rapidly degraded proteins, with k_d = 1.08 or 0.36 h^{-1} (typical of soluble proteins such as casein), as passage rate increases is quite small. The effect becomes much more important as degradation rate decreases, with k_d = 0.12 and k_d = 0.04 [degradation rates typical of solvent and expeller extracted soybean meals (Broderick, 1987)].

A two-compartment model incorporating differences in ruminal passage rates of liquid and particulate phases is described next. Proteins from forages could be separated physically into buffer-soluble and insoluble CP, and degradation and passage rates could be determined for each fraction. Techniques are available for measuring passage rates of both liquid and particulate phases (Hartnell and Satter, 1979; Owens and Hanson, 1992). In vitro methodology to be discussed later could then be used to arrive at degradation rates for these separate fractions. Equations for the two-compartment model for degraded and escaped protein are:

$$\text{Degraded Protein} = A + B_s \times [k_{ds} / (k_{ds} + k_{ps})] + B_i \times [k_{di} / (k_{di} + k_{pi})] \quad (3)$$

and

$$\text{Escaped Protein} = B_s \times [k_{ps} / (k_{ds} + k_{ps})] + B_i \times [k_{pi} / (k_{di} + k_{pi})] + C \quad (4)$$

where B_s and B_i are the soluble and insoluble degradable protein fractions, k_{ps} and k_{pi} are the passage rates for the soluble (liquid) and insoluble (particulate) fractions, k_{ds} and k_{di} are the degradation rates for the soluble and insoluble degradable protein fractions, and A and C are as defined earlier. Elimam and Ørskov studied the effect on ruminal passage of level of feed intake (1984a), particle size (1984b), and feeding frequency (1985) using ruminally cannulated cattle fed proteins labelled as chromium mordants. Later work was conducted with intact ruminants (Elimam and Ørskov, 1984c; Shaver et al., 1986). Owens and Goetsch (1986) and Owens and Hanson (1992) reviewed extensively the factors affecting liquid and particulate passage through the rumen.

Although the two-compartment model may seem a satisfying approach, there is an implied assumption that proteins vary in degradability according to solubility. Considerable effort has been made to classify feed proteins (Henderickx and Martin, 1963; Wohlt et al., 1973) and to formulate diets (Majdoub et al., 1978) based on protein solubility. However, it is now well known that soluble proteins may vary greatly in degradability. Mahadevan et al. (1980) reported that certain soluble proteins were less degradable than other insoluble proteins. Broderick et al. (1988) observed in vitro degradabilities for soluble proteins which ranged from 0.1 to 0.8 h^{-1}. Messman and Weiss (1993) reported that certain soluble proteins in fish meal and blood meal, identified by electrophoresis, persisted for as long as 20 h in ruminal in vitro incubations. Nevertheless, for this two-compartment approach to be satisfactory, it would only be necessary to assume that, within a protein source, degradabilities were related to solubility. Moreover, if overall degradation rates within the buffer-soluble and insoluble proteins were proportional to the weighted amounts of each degradation fraction (Broderick and Clayton, 1992), then this approach should produce a theoretically sound estimate of protein escape. It would be relatively easy to separate proteins into soluble and insoluble fractions, estimate the amount of NPN fraction (A) present with the soluble N, and the indigestible fraction (C) present with the insoluble N. There have been several comparisons of buffers for fractionating feed proteins into soluble and insoluble N (Crooker et al., 1978; Waldo and Goering, 1979; Craig and Broderick, 1981; Krishnamoorthy et al., 1983). An in vitro method, such as the IIV system (Broderick, 1987; Neutze et al., 1993), then could be used to determine degradation rates. These rates and fractions, along with observed or literature values for liquid and particulate passage, would allow computation of ruminal protein degradation and escape. This approach remains to be tested.

ESTIMATION OF RUMINAL PROTEIN DEGRADATION

In Vivo Methods

Direct in vivo measurement would appear to be the most desirable way to determine rates and extents of ruminal degradation of forage proteins. However, results obtained using this methodology have not been completely satisfactory. In vivo procedures require considerable investment of resources; each experiment involves use of several ruminants with cannulae in various sections of the gastrointestinal tract, plus measurement of the amount and composition of the digesta flow. Problems have occurred related to inaccuracies in: 1) Determination or alteration of flow due to marker, sampling, or cannulation effects; and 2) differentiation between protein of feed, microbial, and endogenous origin. There has been extensive research on marker and cannula techniques (Faichney,1986; Owens and Hansen, 1992), and many of the difficulties related to marker technology have been overcome. Accurate estimates of amino acid flow at the abomasum are available in the literature (e.g., Stern et al., 1985). However, errors in apportioning of the protein and amino acids flowing to the abomasum or small intestine between their feed and microbial origins has invalidated many of these estimates of ruminal protein escape.

The major problem with in vivo estimates of ruminal protein escape has been the accurate determination of microbial protein flow. Microbial protein has usually been distinguished from feed proteins using internal markers, such as diaminopimelic acid for bacteria (Hutton et al., 1971) and amino-ethylphosphonic acid for protozoa (Whitelaw et al., 1983); escaped protein is then estimated by difference. Many factors influence net synthesis of microbial protein in the rumen (Hobson and Wallace, 1982; Baldwin and Allison, 1983), and imprecise apportioning of N flows has yielded unusual and probably invalid results. Use of diaminopimelic acid, perhaps the most extensively applied bacterial marker, has been compromised by findings that it may not be degraded to the same extent as protoplasmic proteins when bacterial cells turn over in the rumen (Broderick and Merchen, 1992). Better techniques for quantifying microbial protein, such as nucleic acids or external markers (^{15}N or ^{35}S), used in conjunction with sufficient replication of measurements of protein flow at the abomasum or small intestine, should help improve accuracy of these determinations.

In Vitro Methods

In Vitro Systems Based On Isolated Proteolytic Enzymes

The nature of microbial proteolysis in the rumen will be discussed before considering the use of individual or mixtures of artificial enzymes to mimic ruminal protein degradation. Microbial proteolysis in the rumen is very complex. All of the enzymes which degrade protein to ammonia are of microbial rather than plant or host origin (Brock et al., 1982). Bacteria are the principal microorganisms involved in protein catabolism in the rumen. Ciliate protozoa and, to a lesser extent, anaerobic fungi also carry out protein degradation. Bacterial proteases occur mainly associated with the cell surface (Kopecny and Wallace, 1982); thus, the first step in the protein breakdown by bacteria is adsorption, either of soluble protein to the bacterial surface (Nugent and Mangan, 1981; Wallace, 1985) or of bacteria to insoluble protein. Insoluble proteins need not enter the soluble pool prior to attack by proteolytic bacteria. Proteolytic activity is about 75% particle-associated and 25% in the fluid phase (Brock et al., 1982), about the same distribution as microbial biomass (Craig et al., 1987). Numbers of proteolytic bacteria and the predominant proteolytic species present all vary with diet; large differences among animals fed the same diets also have been observed in the major proteolytic activities found in the rumen (Wallace and Cotta, 1988). Important contributions to protein degradation, particularly of solid phase and insoluble proteins, are made by ruminal protozoa. Four major classes of microbial proteases--serine, cysteine, aspartic, and metalloproteases--are found in the rumen (Wallace and Cotta, 1988). A multiplicity of protease inhibitors of plant, animal, or microbial origin caused only partial inhibition of bacterial (Brock et al., 1982) and protozoal (Forsberg et al., 1984) enzymes, and their effects on bacterial proteases were not additive (Brock et al., 1982). The microbiology of protein metabolism has been reviewed (Cotta and Hespell, 1986; Wallace and Cotta, 1988); factors influencing microbial proteolytic activity in the rumen and characterization of ruminal proteolytic activity also have been discussed (Broderick et al., 1991).

Proteases from commercial sources have been used to estimate protein degradability by measuring either residual N or solubilized digestion products. Poos-Floyd et al. (1985) compared results obtained with nine protein sources using five different proteases to ruminal escapes measured in vivo or calculated from growth trials. Correlations between in vivo protein escape and residual N remaining after 1 or 4 h in vitro digestion with either ficin or neutral protease from *Aspergillus oryzae* actually were greater than those obtained by in situ methods. *Streptomyces griseus* protease resulted in more rapid digestion but had a lower correlation to results from growth trials and in vivo escape (Poos-Floyd et al., 1985). However, the more similar the extent of enzymatic digestion was to in vivo degradation, the lower was its correlation with extent of in vivo degradation. Digestion by *S. griseus* protease previously was reported to result in satisfactory prediction of protein degradation (Krishnamoorthy et al., 1983), although the rate of digestion was much more rapid than in vivo. Laycock et al. (1985) studied the digestion of soybean meal by suspensions of a *Butyrivibrio* sp. and *Streptococcus bovis*, two proteolytic rumen bacteria, but found that degradation after 24-h incubation was only about half that observed in nylon bags incubated in situ.

Kopecny and Wallace (1982) isolated "coat proteases" from mixed ruminal bacteria to determine protein degradabilities. Canadian workers assessed ruminal protein degradability also using proteases isolated from a proteolytic rumen bacterium, *Bacteroides amylophilus* (Mahadevan et al., 1980), or from mixed ruminal organisms (Mahadevan et al., 1987). Mahadevan et al. (1987) found substantially different degradation rates with the proteases from mixed ruminal organisms and those from either *S. griseus* or *A. oryzae*. Rates of protein digestion with proteases from either mixed ruminal organisms and *S. griseus* are compared in Table 2. For example, digestion rates for blood meal and fish meal, two proteins known to be resistant to ruminal degradation, actually were more rapid than soybean meal in the presence of *S. griseus* protease. Kopecny et al. (1989) and Roe et al. (1991) also observed that degradabilities determined in vitro with, respectively, three and seven different commercial proteases were poorly correlated with in situ rates and extents of protein degradation. Since ruminal microbial proteases probably have specificities which differ from those of most commercial proteases or individual bacteria, it seems probable that simple use of commercial proteolytic enzymes at best give only empirical correlations with ruminal degradation. A mixture of four commercial proteases was elaborated in an attempt to match the proteolytic activity of mixed ruminal organisms (Luchini et al., 1993). Although this artificial mixture of commercial proteases gave satisfactory results for estimating ruminal in vitro rates for several proteins, it proved inaccurate for assessing proteins from heat-treated soybeans and alfalfa leaves and stems. Recently, Licitra et al. (1993) modified the *S. griseus* method by using a constant ratio of enzyme to substrate true protein, rather

Table 2. Rates of degradation of different proteins by protease isolated from mixed ruminal organisms and by *Streptomyces griseus* protease (Mahadevan et al., 1987).[1]

Protein source	Ruminal protease	*S. griseus* protease
	(mg protein degraded h^{-1})	
Soybean meal	1.06	0.70
Corn gluten meal	0.37	0.35
Blood meal	0.53	0.96
Fish meal	0.62	1.65

[1]Incubations conducted with 25 mg CP ml^{-1} medium with protease activities added to levels to hydrolyze 5.0 mg azo-casein h^{-1}. Rates determined from linear regression of amount degraded on time from 0 to 6 h.

than a constant ratio of enzyme to substrate CP. A number of different enzymes probably influence protein breakdown in the rumen, including cellulolytic and amylolytic as well as proteolytic activities. Assoumani et al. (1992) added amylases to a system based on *S. griseus*.

Proteases isolated from ruminal microbes show more promise than commercial enzymes for estimating ruminal protein degradability in vitro. Activities of commercial proteases from the single sources tested so far may have been too unlike the activity of mixed ruminal microbes. Moreover, using artificial substrates to match proteolytic activities to mixed ruminal microorganisms did not improve performance of two protease mixtures (Luchini et al., 1993). This suggests that intact ruminal microbes may be required to have the spatial and kinetic properties of normal protein attack. Use of free proteolytic enzymes from either source, commercial or microbial, may ultimately prove unsatisfactory for assessing ruminal degradability.

Ruminal In Vitro Systems

Quantifying rate and extent of protein degradation using in vitro incubations with mixed ruminal organisms is complicated by the fact that the microbes also utilize protein degradation products for growth, resulting in underestimation of degradation. In vitro ammonia production has been used extensively to assess protein degradability. However, an example of the potential flaw in this approach is illustrated in an early report which postulated existence of an inhibitor of deamination to account for apparently low in vitro ammonia production from concentrates (Warner, 1956). Problems related to use of ruminal in vitro ammonia production to estimate protein degradability have been reviewed (Broderick, 1982).

Borchers (1967) attempted to circumvent microbial uptake of protein degradation products by adding toluene, an inhibitor of amino acid deamination, to in vitro inocula and quantifying degradation from amino acid release. Broderick (1978) added hydrazine, an inhibitor of both deamination and ammonia uptake, to ruminal in vitro incubations containing limited amounts of protein and measured degradation from amino acid and ammonia production relative to blanks without added protein. Later evidence showing that direct utilization of amino acids was not prevented by hydrazine led to inclusion of chloramphenicol with hydrazine in the IIV method to inhibit completely the uptake of protein degradation products for microbial synthesis

(Broderick, 1987). Neither compound depressed proteolytic activity in short term incubations of 4 to 6 h in length.

The IIV method was used to estimate protein degradation rate by incubating limited substrate concentrations (i.e., under first-order conditions) with inhibited ruminal inoculum; extent of degradation was computed using the observed rates and an assumed ruminal passage rate (k_p), typically 0.06 h^{-1}, by equations (1) and (2) (Broderick, 1987). The IIV procedure successfully predicted relative differences in lactation performance of cows fed solvent and expeller soybean meal (Broderick, 1986; Broderick et al., 1990), and identified the optimal extent of heating required for protecting protein in soybeans (Faldet et al., 1991) and alfalfa hay (Broderick et al., 1993). However, the IIV procedure has several significant limitations: 1) Accumulation of ammonia and amino acids may make the system subject to end-product inhibition; 2) Degradation rates determined for feeds containing high levels of ammonia and free amino acids (e.g., grass and legume forage silages, especially alfalfa silage) are less accurate because breakdown of more slowly degraded residual protein must be computed from appearance of additional ammonia and amino acids in the presence of high backgrounds; and 3) Quantifying degradation rates from the gentle slopes obtained with slowly degraded proteins appears to be less accurate.

The tendency was for degradation rates to be underestimated using the limited substrate IIV procedure. A Michaelis-Menten approach based on the IIV system was developed (Broderick and Clayton, 1992) in which fractional degradation rate, k_d, was estimated as the tangent through the origin of the velocity versus [S] curve; the slope of this line can be determined from the ratio of the maximum velocity to the Michaelis constant (i.e., $k_d = V_{max} / K_m$; Mahler and Cordes, 1966). This approach overcame several limitations of the IIV procedure with limited substrate and may yield more accurate estimates of degradation rates. Nonlinear regression analysis of data from 2-h incubations in the integrated Michaelis-Menten equation yielded direct estimates of rate (Broderick and Clayton, 1992). Degradation rates estimated using the Michaelis-Menten method were more rapid than those obtained with a limited substrate approach; Michaelis-Menten rates were more consistent with in vivo estimates of rumen protein escape. For example, mean estimates of rates and escapes for protein in casein and alfalfa hay were, respectively, about 0.30 h^{-1} and 150 g kg^{-1} and 0.07 h^{-1} and 400 g kg^{-1} by the limited substrate method, and 0.75 h^{-1} and 80 g kg^{-1} and 0.20 h^{-1} and 250 g kg^{-1} by the Michaelis-Menten procedure. Literature values of ruminal escape for casein (100 g kg^{-1}; McDonald and Hall, 1954; Broderick, 1978) and alfalfa hay (280 g kg^{-1}; NRC, 1989) were more consistent with the Michaelis-Menten results.

Degradation rates and ruminal escapes were determined with the limited substrate and Michaelis-Menten procedures for a number of legume forages which were expected to differ in degradability due to variation in tannin concentrations (Albrecht and Broderick, 1990; Table 3). Some of the degradation rates obtained with the limited substrate approach were slightly negative, which probably means that these rates were essentially 0. However, it seemed unlikely that these forage proteins would be completely resistant to microbial degradation in the rumen. Degradation rates determined by the Michaelis-Menten method were more rapid: sainfoins and lespedezas (*Lespedeza cuneata* Dum.-Cours.), which resulted in rates of less than 0.01 h^{-1} and apparent escapes of 900 to 1060 g kg^{-1} by the limited substrate approach, yielded rates and escapes ranging from 0.03 to 0.05 h^{-1} and 530 to 610 g kg^{-1} by the Michaelis-Menten method. Tannin-containing forages had slower rates and greater escapes by either method. Determination of degradation rates in either the limited substrate or Michaelis-Menten approach requires numerous ammonia and amino acid analyses which would be an important practical limitation of wide application of either IIV methodology.

Neutze et al. (1993) used the limited substrate IIV system described above except they measured protein degradation as the net release of TCA-soluble N. The

Table 3. Tannin levels and ruminal inhibitor in vitro degradabilities of forage legume proteins determined by limited substrate or Michaelis-Menten procedures (Albrecht and Broderick, 1990).

Forage (cultivar)[1]	Tannin[2]	Limited Substrate		Michaelis-Menten	
		Rate, k_d	Escape[3]	Rate, k_d	Escape[3]
	(g TAE kg^{-1} DM)	(h^{-1})	(g kg^{-1} TN)	(h^{-1})	(g kg^{-1} TN)
Alfalfa (Magnum 3)	0	0.208	210	0.205	220
Cicer Milkvetch (Monarch)	0	0.195	220	0.266	180
BFT (Norcen)	Trace	0.178	240	0.266	180
BFT (High tannin)	5.8	0.173	240	0.249	190
BFT (Viking)	8.5	0.123	310	0.176	240
Red Clover (Marathon)	0	0.135	290	0.155	270
Sainfoin (Remont)	22.1	0.015	770	0.070	430
Lotus pedunculatus (Grasslands Maku)	28.6	0.023	690	0.094	380
Sainfoin (Nova)	27.0	-0.002	1020	0.045	540
Sainfoin (Eski)	29.5	-0.002	1000	0.046	540
Lespedeza (AU Lotan)	19.5	0.018	740	0.038	590
Lespedeza (AU Donnelly)	12.1	0.041	570	0.045	560
Lespedeza (I-76)	28.0	-0.007	1080	0.051	530
Lespedeza (Serala)	27.7	0.005	900	0.032	630
Lespedeza (Serala 76)	29.4	-0.006	1060	0.040	590
Lespedeza (Interstate)	31.1	-0.006	1050	0.034	610

[1]BFT = Birdsfoot trefoil (*Lotus corniculatus* L.); cicer milkvetch = *Atragalus cicer* L.; red clover = *Trifolium pratense* L.; lespedeza = sericea lespedeza.
[2]TAE = Tannic acid equivalents.
[3]Estimated ruminal escape, % = [B x (k_p / (k_p + k_d))], where it is assumed that ruminal passage rate, k_p = 0.06 h^{-1} (Broderick, 1987).

advantage of this approach is that extent of degradation can be measured by simple Kjeldahl or Dumas N assays rather than by dual ammonia and amino acid analyses. That end-product inhibition may have been reduced with this approach was indicated by a faster rate being observed for the rapidly degraded protein casein. Degradation rates reported for a number of soybean meals (Neutze et al., 1993) were of similar magnitude to rates determined from ammonia and amino acid release for several other samples of soybean meal (Broderick, 1987; Broderick and Clayton, 1992).

An alternative in vitro technique was described by Raab et al. (1983) in which starch or other fermentable carbohydrates were added in graded amounts to incubations of ruminal fluid with the protein being studied. Gas production and ammonia concentration were measured up to 24 h. Extrapolation of ammonia release to zero gas production was assumed to correct for ammonia (and amino acid) uptake for microbial growth, since zero gas production should represent zero growth where no N would be incorporated into microbial protein. Protein degradabilities at 24 h were then calculated. The method appears promising and might lend itself to analysis of the dynamics of protein degradation. To be successful, this procedure requires that the relationship between microbial protein synthesis and gas production be linear over the conditions of the incubations. The potential for partial uncoupling of fermentation from microbial growth suggests that a more direct estimate of microbial protein formation would be valuable.

A new in vitro procedure was developed to estimate rate and extent of protein degradation in uninhibited ruminal inoculum by using $^{15}NH_3$ to quantify microbial uptake of protein breakdown products for growth (Hristov and Broderick, 1993). Incubations were conducted for 6 h in stirrer flasks with buffered ruminal inoculum plus soluble carbohydrates and $(^{15}NH_4)_2SO_4$. Degradation rates were computed from net (i.e., protein-added minus blank) appearance of ammonia-N plus net synthesis of microbial CP (estimated from ^{15}N-enrichment of isolated microbial cells and total solids N). Escape also was computed using equation (2) assuming ruminal passage rate = 0.06 h^{-1}. Generally, degradation rates and ruminal escapes determined for seven proteins were of similar magnitude to those obtained for the same proteins using the limited substrate IIV method (Table 4). However, overall mean degradation rates averaged 28% greater than those previously estimated using the IIV system; rates obtained for two fish meals using the ^{15}N method were slower. This method is promising in that uninhibited ruminal organisms were used in incubations with "normal" microbial growth and without significant end-product accumulation. The procedure may prove useful with problematic forages such as legume and grass silages containing large amounts of NPN. An alternative approach would be to estimate microbial N using ^3H- or ^{14}C-labelled amino acids, since the radioisotopes are less costly and more easily measured than ^{15}N. This procedure using ^{15}N or radioisotopes has not yet been tested on hay or silage proteins.

Other in vitro techniques include labelling proteins under study with dyes (Mahadevan et al., 1979) or radioisotopes (Wallace, 1983) to follow their disappearance from incubations with mixed ruminal organisms or protease extracts. These methods worked well for soluble proteins, but proved unreliable with insoluble protein sources due to non-homogeneous labelling (Broderick et al., 1988). Therefore, their usefulness appears limited.

Solubility Methods

An ideal in vitro method would involve a simple physical or chemical laboratory technique to determine degradation characteristics. Solubility has long been used to estimate protein degradability (Chalmers and Synge, 1954; Henderickx and Martin, 1963; Wohlt et al., 1973). Ways of improving the solubility-degradability relationship, such as changes of solvent or pH, have been investigated

Table 4. Comparison of ruminal in vitro degradabilities for seven proteins determined from net microbial growth plus net ammonia release (Hristov and Broderick, 1993) or by the limited substrate inhibitor in vitro procedure.

Protein source	Net microbial growth plus ammonia[1]			Inhibitor in vitro[2]	
	Extent (6 h)	Rate	Escape	Rate	Escape
	(g kg^{-1} TN)	(h^{-1})	(g kg^{-1} TN)	(h^{-1})	(g kg^{-1} TN)
Casein	970	0.569	100	0.307	170
Solvent soybean meal	586	0.148	290	0.137	300
Expeller soybean meal	193	0.036	630	0.030	650
Low solubles fish meal	146	0.026	700	0.034	630
High solubles fish meal	314	0.063	490	0.066	460
Corn gluten meal	183	0.034	640	0.017	750
Roasted soybeans	256	0.050	550	0.045	560

[1] Degradation rate and escape computed from extent of CP degradation at 6 h estimated from amount of N detected in net microbial growth plus ammonia.

[2] Degradation rates and estimated escapes obtained previously for the same proteins using the inhibitor in vitro method (Broderick, 1987). Data are from Broderick (1992) (low solubles and high solubles fish meal), Broderick et al. (1990) (corn gluten meal), and Faldet and Satter (1991) (roasted soybeans). Values for casein, solvent soybean meal, and expeller soybean meal are means from incubations conducted in all three experiments.

(Crawford et al., 1978; Crooker et al., 1978; Craig and Broderick, 1981; Poos-Floyd et al., 1985). Provided that the protein contains little soluble, resistant protein, a positive relationship between CP solubility and degradation should be obtained. For example, solubility has been used to estimate degree of protection obtained by heat treating cottonseed meal (Sherrod and Tillman, 1964; Craig and Broderick, 1981) and soybeans (Hsu and Satter, 1991). However, protein solubility fell rapidly with initial heating, which reduced ruminal degradability only slightly, but solubility changed very little in the region where further heating rapidly decreased ruminal degradability (Craig and Broderick, 1981; Hsu and Satter, 1991). Moreover, the static nature of solubility determinations will not yield the dynamic constants required for a full description of ruminal degradation.

Sniffen et al. (1992), in the recently proposed Cornell Net Carbohydrate and Protein System, recommended the use of a series of chemical fractionations to identify five N components in feed CP. These are fraction A, which was TCA-soluble NPN; unavailable fraction C, which was ADIN; and three subfractions of degradable fraction B [TCA-precipitable N from buffer soluble CP (B1), neutral detergent insoluble N minus ADIN (B3), and the remainder of fraction B which was true protein of intermediate degradation rate (B2)]. Each of the three subfractions B were assigned different degradation rates and a ruminal passage rate characteristic for the feed of origin and level of intake. Separate passage rates for liquids (for B1) and solids (for B2 and B3) were not applied in this model. This ambitious effort at modelling ruminal N transactions, which includes proposed methodology for assessing UIP in various forage proteins, is very appealing; however, the model is some years and many experiments away from validation.

In Situ Methods

The technique gaining widest application for determining the degradation properties of proteins is the in situ method. In this procedure, the protein under study is added to artificial fiber bags (usually made of dacron or nylon) which then are suspended in the rumen and the time-course of DM and N loss from bags is used to estimate rate and extent of degradation (Mehrez and Ørskov, 1977). It is important that the pore size of the bag be standardized to minimize particle loss yet allow, as much as possible, unrestricted access by microorganisms. Some degree of compromise is required but it appears that a pore size of 40 to 60 µm is satisfactory (Mehrez and Ørskov, 1977; Lindberg and Varvikko, 1982). The quantity incubated and the bag size are also important. A sample size of about 3 to 5 g DM in a bag measuring 10 to 17 by 5 to 9 cm is recommended (Mehrez and Ørskov, 1977; Lindberg and Varvikko, 1982), a range of 16 to 30 mg sample DM cm^{-2} of surface area. Michalet-Doreau and Ould-Bah (1993) recommended a sample to bag surface area of 15 mg DM cm^{-2}. Several recent papers summarize standard in situ methodology in detail (ARC, 1984; Nocek, 1988; Michalet-Doreau and Ould-Bah, 1993).

Although used extensively, in situ techniques have been criticized for several reasons: 1) Restriction of microbial access to the protein (Meyer and Mackie, 1986); 2) Microbial N contamination within the bags (Nocek and Grant, 1987); and 3) Losses of soluble and particulate N which have not been degraded. Despite these limitations, in situ methods continue to be extensively applied because of ease of use, requiring only ruminally cannulated animals and Kjeldahl methodology, and because of the common misconception that in situ results will clearly reflect in vivo ruminal conditions. While it is true that factors which influence microbial activity also alter in situ protein degradation (Weakley et al., 1983), inherent errors in the in situ method must be borne in mind. Several workers have used internal (Nocek and Grant, 1987) and external markers (Varvikko, 1986) for microbial protein to correct for microbial contamination. The reason that in situ data have proven useful may be because the two major sources of error, underestimation of degradation (#2 above) and overestimation of the degraded fraction A (#3 above), tend to cancel each other out. The net result is that in situ rates yield apparently reliable estimates of escape when used with equation (2).

Microorganisms must associate with the substrate for degradation to occur, so estimates of N residues in the in situ technique include microbial as well as feed protein. This error, although small with most protein-rich feeds over short incubation times, becomes more problematic with longer incubations because of the low protein content of residual materials in the bag, particularly of the sort encountered with forages (Nocek and Grant, 1987). The low N, fibrous residues remaining after readily degraded fractions have disappeared will be colonized extensively (Varvikko, 1986).

Two further problems are also of major concern. One is that protein supplements often are ground to a particle size less than 1 mm before incubation. This improves reproducibility, but any influence of particle size on in vivo degradation rate is lost. Small particle size also will result in particle efflux from the bag without digestion. Michalet-Doreau and Ould-Bah (1993) recommended grinding samples through screen apertures of 1.5 to 3.0 mm to reduce the problem of particle efflux. The second concern is the assumption that, if the substrate disappears from the bag, it has been degraded. In view of the wide range of degradation rates observed for many soluble proteins (Nugent et al., 1983; Wallace, 1983; Broderick et al., 1988; Messman and Weiss, 1993), clearly this is not the case. Degradability end-points of proteins incubated in in situ bags with different pore sizes were found to be the same, despite much slower approaches to the end-points in bags of smaller pore size, suggesting that the small particles lost from the larger-pore bags were degraded eventually (Lindberg and Varvikko, 1982).

A comparison of ruminal degradabilities, estimated by the in situ and IIV methods, was made with seven protein concentrates (Broderick et al., 1988). These results are summarized in Table 5. Generally, in situ degradation rates were much slower than IIV rates (average 36% of IIV). However, the in situ method yielded proportions degraded at 0-time (fraction A) which averaged more than 2-fold greater than by IIV. Hence, ruminal protein escapes estimated by in situ did not differ as greatly from IIV as did the degradation rates (mean in situ ruminal escape was 142% of IIV). Thus, although in situ degradation rates appeared unrealistically low, the errors within the in situ method tended to correct for one another and degradabilities were of similar magnitude. Ruminal protein escapes computed from the IIV data were more similar to NRC (1989) values than those computed from in situ data: The mean (SD) ratio of estimated escapes for IIV:NRC and in situ:NRC were 100 (34)% and 140 (51)%, respectively. Until sufficient data are available from other, perhaps sounder, techniques, in situ results may be used so long as in situ estimates of rates and fractions are not considered accurate and emphasis is placed only on the extent of protein degradation or escape.

The mobile bag procedure, a variation on the in situ technique, has been used to estimate intestinal digestibility of UIP (Hvelplund, 1985; Rae and Smithard, 1985; DeBoer et al., 1987). In this approach, the protein under study is sealed in fiber bags as with a conventional in situ, except the bags are predigested in acid-pepsin (to simulate abomasal action) and inserted via cannulae into the proximal duodenum. Residual N or amino acids then are determined after recovery of the bags in the feces. The bags may be preincubated in the rumen to simulate more closely the actual UIP which will reach the small intestine. Although conceptually interesting, there has been a trend toward finding very high digestibilities with the mobile bag technique. Faldet et al. (1991) observed digestibilities of 980 to 1000 g kg^{-1} for soybean CP which had been heated from 0 to 3 h at 145°C; similar results were obtained by others (DeBoer et al., 1987). This suggests that the method may be insensitive to substantial differences in intestinal protein digestibility. Moreover, mere disappearance of protein from the small intestine does not mean the α-amino N can be utilized by the animal. In heat-damaged protein, for example, several amino acid derivatives are found that are readily absorbed but which do not function as amino acids in tissue protein synthesis (Hurrell and Finot, 1985). Also, some N disappearance which does not result in α-amino N absorption may occur due to microbial action in the large intestine.

SUGGESTED METHODOLOGY

Quantifying Degradation Fractions

Fraction A, the proportion of total CP present as NPN, may be determined simply as the buffer-extracted N from the forage sample which is not precipitated by a protein denaturant such as TCA at a final concentration of 50 to 100 g l^{-1} (Craig and Broderick, 1981; Krishnamoorthy et al., 1983; Sniffen et al., 1992). This method will solubilize ammonia, free amino acids, plus small and intermediate-sized peptides up to a MW of about 5,000 daltons. When fed to the animal, almost all of this α-amino N will either be degraded to ammonia or utilized by growing microbes (Wallace et al., 1990), although the magnitude of this conversion remains controversial (Chen et al., 1987). The TCA-extraction is to be preferred over in situ determination because of the presumption that all of the fraction A estimated by the in

Table 5. Kinetic estimates of protein degradation made using the inhibitor in vitro (IIV) and in situ methods (Broderick et al., 1988).

Protein source	IIV method			In situ method			Escape[1]		
	A	B	k_d	A	B	k_d	IIV	In situ	NRC
	(g kg^{-1} TN)		(h^{-1})	(g kg^{-1} TN)		(h^{-1})	-----(g kg^{-1} TN)-----		
Fish meal	180	820	0.042	310	690	0.014	480	560	600
Soybean meal	70	930	0.166	140	860	0.067	250	410	350
Linseed meal (*Linum usitatissimum* L.)	50	950	0.244	200	800	0.077	190	350	180
Sunflower meal (*Helianthus annuus* L.)	120	880	0.058	190	810	0.032	450	530	260
Rapeseed meal (*Brassica napus* L.)	140	860	0.124	180	820	0.036	280	510	280
Copra meal (*Cocos nucifera* L.)	80	920	0.050	240	760	0.017	500	590	630
Meat & bone meal	120	880	0.056	390	610	0.017	460	480	490

[1]Estimates of ruminal escape computed using equation (2) and assuming a ruminal passage rate, k_p, of 0.06h^{-1}. The NRC escape data are UIP values from Table 7-3 in (NRC, 1989).

situ method will be completely degraded, including intact, soluble proteins and feed particles which readily efflux from the bag during washing (see Table 5). Perhaps the best approach to estimating degradable and undegradable fractions B and C is to subject the whole forage protein sample to in vitro digestion with high activity protease preparations. The *S. griseus* method of Krishnamoorthy et al. (1983), where 6.6 units protease activity ml^{-1} was incubated 48 h with 10 mg feed DM ml^{-1}, appears to be appropriate for this purpose. All of NPN fraction A and all of degradable fraction B will be solubilized; fraction C will be the indigestible residue. Krishnamoorthy et al. (1983) found that this method gave estimates of unavailable CP which were similar but smaller than the ADIN assay. The mean (SD) for five concentrates was 70 (30)% as large as ADIN. Use of proteolytic enzymes, rather than chemical extractions, is a more logical approach for estimating the amount of CP which is biologically unavailable.

Should the investigator wish to apply a two-compartment model corresponding to soluble and insoluble fractions B_s and B_i [equations (3) and (4)], soluble B_s may be estimated as the TCA-insoluble N in the buffer extract from determination of fraction A (above). The CP solubilized by high activity *S. griseus* protease (Krishnamoorthy et al., 1983) from the buffer-insoluble residue corresponds to insoluble fraction B_i; CP not solubilized from the buffer-insoluble residue corresponds to unavailable fraction C. Degradation computations will require separate passage rates for the liquid and solid phases and may be estimated from the literature or determined as described earlier.

Quantifying Degradation Rates and Ruminal Escapes

Determining ruminal degradation rates for forage proteins is the most problematic phase of UIP estimation by in vitro kinetic methods. Since commercial enzyme preparations have not been shown to yield degradation rates similar to those obtained with mixed ruminal microbes, it is recommended that in vitro rates be determined with ruminal inoculum. Two possible approaches will be suggested to obtain rate estimates: 1) Using an abbreviated IIV approach, or 2) Using proteases isolated from mixed ruminal organisms. The abbreviated IIV method with forage proteins involves replicate assays at two time-points: a 0-h set to estimate fraction A and a 2-h set to compute rate. The rate estimate is given by the equation:

$$\text{Rate, h}^{-1} = [\ln \{[TN - (NH_3\text{-}N + TAA\text{-}N)_2] / TN\} - \ln \{[TN - (NH_3\text{-}N + TAA\text{-}N)_0] / TN\}] / 2\text{ h} \quad (5)$$

where TN is the total N added to each tube, and $(NH_3\text{-}N + TAA\text{-}N)_2$ and $(NH_3\text{-}N + TAA\text{-}N)_0$ are the amounts of N recovered as ammonia and total amino acids at 2 and 0 h, respectively. This procedure was used to estimate degradation rates and ruminal escapes in the screening of large numbers of samples of alfalfa germplasm (Broderick and Buxton, 1991) and of other legume forage species (Albrecht and Broderick, 1990). Generally, when working with forage samples which were freeze-dried or oven-dried at low temperature (55°C), fraction C will be small. Under these circumstances, fraction B is given by the equation:

$$B, \text{g kg}^{-1} = \{[TN - (NH_3\text{-}N + TAA\text{-}N)_0] / TN\} \times 1000 \quad (6)$$

If fraction C is not negligible, then the amount of N in fraction C is subtracted from the total N at both 0- and 2-h. Extent of degradation and escape may be computed using equations (1) and (2). If relatively few samples are being analyzed, then the complete IIV approach, including the taking of four to eight time-points over 4 h, is recommended as described (Broderick, 1987). The reader is reminded of the alternative IIV approach of Neutze et al. (1993) where protein degradation was measured as the net release of TCA-soluble N.

Application of the two-compartment model [equations (3) and (4)] to the abbreviated method would be analogous. Degradation rates would be estimated using equation (5) for both the buffer-soluble and buffer-insoluble fractions. Fraction A would be expected to be rather large when equation (5) is used for fraction B_s; fraction A would be almost nil in fraction B_i. Escape also would be computed for the two-compartment model as described above. It is suggested that an approach using the two-compartment model would be most appropriate for grass and legume silages which contain large amounts of NPN in fraction A and have more resistant fraction B.

Of the methods published by Kopecny and Wallace (1982) and Mahadevan et al. (1987) for isolating proteases from mixed ruminal microbes, the latter is probably more appropriate because it extracts protease from total ruminal inoculum (i.e., both bacteria and protozoa) rather than bacteria only. Mahadevan's freeze-dried YM-10 preparation (containing molecules with MW greater than 10,000 daltons) is preferred since it retains cellulase (and probably amylase) activity. Its use would be expected to yield more realistic degradation rates for forage proteins. For each batch of the freeze-dried extract, protease activity should be adjusted to yield, using the investigator's end-point for degradation, rates of about 0.54 and 0.14 h^{-1}, respectively, for the standard proteins casein and solvent-extracted soybean meal, corresponding to escapes of 100 and 300 g kg^{-1}. An approach exactly analogous to that described for the IIV method then may be applied using this system to estimate degradation rate and ruminal escape of forage proteins. Recently, Kohn and Allen

(1993) reported using an extraction procedure similar to Mahadevan et al. (1987), except without the YM-10 filtration, for preparing proteases from mixed ruminal microbes.

Comments on In Situ Estimates of Ruminal Protein Escape

The in situ method remains one of the most widely used for assessing degradability of forage proteins. However, the in situ technique should be applied only when the worker is cognizant of how the results may be misleading. The inherent errors in the method discussed above result in overestimation of the size of degraded fraction A, underestimation of the rate at which degradable fraction B is degraded, and probably overestimation of the size of undegradable fraction C. Although the biological significance of each parameter generated by the in situ model may be questionable, because these errors tend to correct one another, overall estimates of ruminal protein escape appear to be biologically reliable. Thus, emphasis should be given only to the in situ UIP estimate for forage proteins. In situ mathematical models (Ørskov and McDonald, 1979; McDonald, 1981) and methodology (Nocek, 1988; Michalet-Doreau and Ould-Bah, 1993) have been extensively reviewed elsewhere. Fraction C may be estimated using the enzymatic approach referred to earlier (Krishnamoorthy et al., 1983). Time-course data should be fitted by discounting residual CP for fraction C.

For quantifying ruminal in situ escape of forage proteins, 8 to 10 time-course samples over a period from 0 to 72 h would be appropriate. A later final time improves estimate of fraction C and obviates the need for an alternative method for its determination. Of course, the use of many time-points precludes study of large numbers of samples such as are generated in forage germplasm screening or routine assay of commercial feed samples. Collaborators in North Central Regional Project 189 (NC-189) have conducted an extensive study to identify sources of error when the in situ method is applied to forages (Wilkerson et al., 1990). For routine in situ assays, they have reduced sample numbers by first using a water wash followed by in situ incubation at a single time-point of 16 h (the mean retention time equivalent to a passage rate of 0.0625 h^{-1}) to estimate extent of degradation of forage proteins (Karges et al., 1992). The method includes a correction for microbial N contamination of residual N and uses ADIN to estimate fraction C. The residual N remaining after one mean retention time is assumed to be the CP which would escape the rumen at the equivalent passage rate. However, the incubation time required to yield a CP residue equal to the amount which would escape the rumen at a given passage rate is also a function of degradation rate (L. E. Armentano, personal communication). Thus, it would be more appropriate to use the data from the initial wash (0-h) and the single ruminal time (16 h) to do a two-point estimate of rate and escape as described for the abbreviated IIV approach. If it is assumed that escaped protein (EP), expressed as a percent of total N and corrected for microbial N and ADIN, is represented by in situ residual CP at 16 h, then the degradation rate is computed:

Degradation rate (k_d), h^{-1} = [ln (EP / 100) - ln (A / 100)] / 16   (7)

where fraction A also is expressed as a percent of total N. A comparison of the two approaches is summarized in Table 6 using in situ data obtained on esophageal extrusa samples from summer range (Karges et al., 1992). Over two seasons, the escaped CP values estimated by the NC-189 approach ranged from 180 to 280 g kg^{-1}, with a mean value of 209 g kg^{-1}. If it is assumed that A = 0, 100, or 200 g N kg^{-1} of total N in all samples, then the two-point estimate of escape yields means (at k_p = 1/16 or 0.0625 h^{-1}) of 390, 366, and 342 g kg^{-1}, respectively, corresponding to 1.82, 1.75, and 1.63 times more escaped CP than the reported values. This result

Table 6. Ruminal degradation estimates on protein in freeze-dried extrusa samples from native range grasses computed using data from a single 16 h in situ incubation time-point (Karges et al., 1992).

Year	Month	NC-189 method[1] CP	ADIN	UIP	A = 0[2] k_d	UIP	A = 100 g kg⁻¹ TN[2] k_d	UIP	A = 200 g kg⁻¹ TN[2] k_d	UIP
		(g kg⁻¹ DM)	(g kg⁻¹ TN)	(g kg⁻¹ TN)	(h⁻¹)	(g kg⁻¹ TN)	(h⁻¹)	(g kg⁻¹ TN)	(h⁻¹)	(g kg⁻¹ TN)
1988	June	141	12	211	0.097	391	0.091	367	0.083	343
	July	116	11	209	0.098	390	0.091	366	0.084	342
	August	104	14	279	0.080	439	0.073	415	0.066	390
	September	71	24	178	0.108	367	0.101	343	0.094	320
1989	June	100	16	207	0.098	388	0.092	364	0.084	340
	July	94	16	221	0.094	398	0.088	374	0.080	350
	August	91	21	187	0.105	374	0.098	350	0.091	326
	September	90	16	183	0.106	371	0.100	347	0.092	323
Mean CP Escape (g kg⁻¹ TN)				209		390		366		342
(SD)				(32)		(23)		(23)		(22)
Ratio =				1.00		1.86		1.75		1.63

[1] Estimates of ruminal protein escape (UIP) computed as residual N remaining after 16 h of in situ incubations, corrected for N present as ADIN and for assumed amounts of microbial N (2.9 g N kg⁻¹ residual DM) (Wilkerson et al., 1990).
[2] Estimates of degradation rate (k_d) and ruminal escape (UIP) computed assuming fraction A (NPN content) in each sample was equal to 0, 100, or 200 g kg⁻¹ total N (TN) using data from the initial wash (0-h) and the single ruminal time (16 h) to do a two-point estimate of rate and escape with equations (7) and (2), respectively. Rate of passage, k_p, is assumed equal to 1/16 h (0.0625 h⁻¹).

indicates that the NC-189 method will underestimate ruminal protein escape for forage proteins and probably should be altered to a two-point approach as described.

CONCLUSIONS

Although there has been extensive research on the problem of estimating UIP in ruminant feeds and great interest in making these determinations for forage proteins, no single method has yet evolved which will conveniently yield reliable UIP values. Eventually, feeding systems will be dynamic, incorporating degradation and passage rates for multiple fractions (e.g., Sniffen et al., 1992) and even EAA requirements and supplies. However, at present, there are no reliable biological methods for quantifying multiple fractions and estimating their degradation rates. Estimation of intestinal digestibility and EAA composition of UIP also is premature at this time. Until the tools are developed for accurate determination of the data required in multi-compartmental models of ruminant protein nutrition, we must rely on our established methods to make kinetic determinations of degradation and escape. For now, estimation of degradation rate and ruminal escape should be done using in vitro methods based on intact, mixed ruminal organisms, or their enzyme extracts. Commercial proteases do not give reliable results. Inherent errors with in situ methods result in overestimation of fraction A and underestimation of degradation rates; nevertheless, in situ estimates of overall protein escape appear to be reliable. In the future, NIRS methodology may be used to develop calibrations for ruminal escape of forage proteins.

REFERENCES

ARC. 1984. The Nutrient Requirements of Ruminant Livestock, Supplement No. 1. Commonwealth Agricultural Bureaux, Slough, England.

Albrecht, K.A., and G.A. Broderick. 1990. Degradation of forage legume protein by rumen microorganisms. p. 185. *In* Agronomy abstracts. ASA, Madison, WI.

Assoumani, M.B., F. Vedeau, L. Jacquot, and C.J. Sniffen. 1992. Refinement of an enzymatic method for estimating the theoretical degradability of proteins in feedstuffs for ruminants. Anim. Feed Sci. Tech. 39:357-368.

Baldwin, R.L., and M.J. Allison. 1983. Rumen metabolism. J. Anim. Sci. 57(Suppl. 2):461-477.

Barry, T.N., T.R. Manley, and S.J. Duncan. 1986. The role of condensed tannins in the nutritional value of *Lotus pedunculatus* for sheep. 4. Sites of carbohydrate and protein digestion as influenced by dietary reactive tannin concentrations. Br. J. Nutr. 55:123-137.

Barry, T.N., and C.S.W. Reid. 1985. Nutritional effects attributable to condensed tannins, cyanogenic glycosides and oestrogenic compounds in New Zealand forages. p. 251-259. *In* R.F. Barnes, P.R. Ball, R.W. Brougham, G.C. Marten, and D.J. Minson (ed.) Forage legumes for energy-efficient animal production. USDA-ARS, Washington, DC.

Borchers, R. 1967. Proteolytic activity of rumen fluid in vitro. J. Anim. Sci. 24:1033-1038.

Brock, F.M., C.W. Forsberg, and J.G. Buchanan-Smith. 1982. Proteolytic activity of rumen microorganisms and effects of proteinase inhibitors. Appl. Environ. Microbiol. 44:561-569.

Broderick, G.A. 1978. In vitro procedures for estimating rates of ruminal protein degradation and proportions of protein escaping the rumen undegraded. J. Nutr. 108:181-190.

Broderick, G.A. 1982. Estimation of protein degradation using in situ and in vitro methods. p. 72-80. In F. N. Owens (ed.) Protein requirements for cattle: Symposium. Oklahoma State Univ., Stillwater.

Broderick, G.A. 1986. Relative value of solvent and expeller soybean meal for lactating dairy cows. J. Dairy Sci. 69:2948-2958.

Broderick, G.A. 1987. Determination of protein degradation rates using a rumen in vitro system containing inhibitors of microbial nitrogen metabolism. J. Dairy Sci. 58:463-475.

Broderick, G. A. 1992. Relative value of fish meal versus solvent soybean meal for lactating dairy cows fed alfalfa silage as sole forage. J. Dairy Sci. 75:174-183.

Broderick, G.A., and D.R. Buxton. 1991. Genetic variation in alfalfa for ruminal protein degradability. Canadian J. Plant Sci. 71:755-760.

Broderick, G.A., and M.K. Clayton. 1992. Ruminal protein degradation rates estimated by nonlinear regression analysis of Michaelis-Menten in vitro data. Br. J. Nutr. 67:27-42.

Broderick, G.A., and W.M. Craig. 1980. Effect of heat treatment on ruminal degradation and escape, and intestinal digestibility of cottonseed meal protein. J. Nutr. 110:2381-2389.

Broderick, G.A., and N.R. Merchen. 1992. Markers for quantifying microbial protein synthesis in the rumen. J. Dairy Sci. 75:2618-2632.

Broderick, G.A., D.B. Ricker, and L.S. Driver. 1990. Expeller soybean meal and corn by-products versus solvent soybean meal for lactating dairy cows. J. Dairy Sci. 73:453-462.

Broderick, G.A., R.J. Wallace, and E.R. Ørskov. 1991. Control of rate and extent of protein degradation. p. 541-592. In T. Tsuda, Y. Sasaki, and R. Kawashima (ed.) Physiological Aspects of Digestion and Metabolism in Ruminants. Academic Press, Orlando, FL.

Broderick, G.A., R.J. Wallace, E.R. Ørskov, and L. Hansen. 1988. Comparison of estimates of ruminal protein degradation by in vitro and in situ methods. J. Anim. Sci. 66:1739-1745.

Broderick, G.A., J.H. Yang, and R.G. Koegel. 1993. Effect of steam heating alfalfa hay on utilization by lactating dairy cows. J. Dairy Sci. 76:165-174.

Chalmers, M.I., and R.L.M. Synge. 1954. The digestion of protein and nitrogenous compounds in ruminants. Adv. Protein Chem. 9:93-120.

Charmley, E., and D.M. Veira. 1990. Inhibition of proteolysis in alfalfa silages using heat at harvest: Effects on digestion in the rumen, voluntary intake and animal performance. J. Anim. Sci. 68:2042-2051.

Chen, G., C.J. Sniffen, and J.B. Russell. 1987. Concentration and estimated flow of peptides from the rumen of dairy cattle: Effects of protein quantity, protein solubility, and feeding frequency. J. Dairy Sci. 70:983-992.

Cotta, M.A., and R.B. Hespell. 1986. Protein and amino acid metabolism of rumen bacteria. p. 122-136. *In* L.P. Milligan, W.L. Grovum, and A. Dobson (ed.) Control of digestion and metabolism in ruminants. Prentice-Hall, Englewood Cliffs, NJ.

Craig, W.M., and G.A. Broderick. 1981. Comparison of nitrogen solubility in three solvents to in vitro protein degradation of heat-treated cottonseed meal. J. Dairy Sci. 64:769-774.

Craig, W.M., and G.A. Broderick. 1984. Amino acids released during protein degradation by rumen microbes. J. Anim. Sci. 58:436-443.

Craig, W.M., G.A. Broderick, and D.B. Ricker. 1987. Quantitation of microorganisms associated with the particulate phase of ruminal ingesta. J. Nutr. 117:56-62.

Crawford, R.J., W.H. Hoover, C.J. Sniffen, and B.A. Crooker. 1978. Degradation of feedstuff nitrogen in the rumen vs. nitrogen solubility in three solvents. J. Anim. Sci. 46:1768-1775.

Crooker, B.A., C.J. Sniffen, W.H. Hoover, and L.L. Johnson. 1978. Solvents for soluble nitrogen measurements in feedstuffs. J. Dairy Sci. 61:437-447.

DeBoer, G., J.J. Murphy, and J.J. Kennelly. 1987. Mobile nylon bag for estimating intestinal availability of rumen undegraded protein. J. Dairy Sci. 70:977-982.

Egan, A.R., and M.J. Ulyatt. 1980. Quantitative digestion of fresh herbage by sheep. VI. Utilization of nitrogen in five herbages. J. Agric. Sci., Camb. 94:447-456.

Elimam, M.E., and E.R. Ørskov. 1984a. Factors affecting the outflow of protein supplements from the rumen. 1. Feeding level. Anim. Prod. 38:45-51.

Elimam, M.E., and E.R. Ørskov. 1984b. Factors affecting the outflow of protein supplements from the rumen. 2. The composition and particle size of the basal diet. Anim. Prod. 39:201-206.

Elimam, M.E., and E.R. Ørskov. 1984c. Estimation of rates of outflow of protein supplements from the rumen by determining the rate of excretion of chromium-treated protein supplements in faeces. Anim. Prod. 39:77-80.

Elimam, M.E., and E.R. Ørskov. 1985. Factors affecting the fractional outflow of protein supplements from the rumen. 3. Effects of frequency of feeding, intake of water induced by the addition of sodium chloride, and the particle size of protein supplements. Anim. Prod. 40:309-313.

Faichney, G.J. 1986. The kinetics of particulate matter in the rumen. p. 173-195. *In* L.P. Milligan, W.L. Grovum, and A. Dobson (ed.) Control of digestion and metabolism in ruminants. Prentice-Hall, Englewood Cliffs, NJ.

Faldet, M. A., and L. D. Satter. 1991. Feeding heat-treated full fat soybeans to cows in early lactation. J. Dairy Sci. 74:3047-3054.

Faldet, M.A., V.L. Voss, G.A. Broderick, and L.D. Satter. 1991. Chemical, in vitro and in situ evaluation of heat treated soybean proteins. J. Dairy Sci. 74:2548-2554.

Forsberg, C.W., L.K.A. Lovelock, L. Krumholz, and J.G. Buchanan-Smith. 1984. Protease activities of rumen protozoa. Appl. Environ. Microbiol. 47:101-110.

Glenn, B.P., H.F. Tyrrell, D.R. Waldo, and H.K. Goering. 1986. Effects of diet nitrogen and forage nitrogen insolubility on performance of cows in early lactation. J. Dairy Sci. 69:2825-2836.

Goering, H.K., C.H. Gordon, R.W. Hemken, D.R. Waldo, P.J. Van Soest, and L.H. Smith. 1972. Analytical estimates of nitrogen digestibility in heat damaged forages. J. Dairy Sci. 55:1275-1280.

Goering, H.K., and I.L. Lindahl. 1975. Growth and metabolism of sheep fed rations containing alfalfa hay or dehydrated alfalfa. J. Dairy Sci. 58:759 (Abstr.).

Hartnell, G.F., and L.D. Satter. 1979. Determination of rumen fill, retention time and ruminal turnover rates of ingesta at different stages of lactation in dairy cows. J. Anim. Sci. 48:381-392.

Henderickx, H., and J. Martin. 1963. In vitro study of the nitrogen metabolism in the rumen. Compt. Rend. Rech., Inst. Rech. Sci. Ind. Agric. (Bruxelles) 31:7-61.

Hobson, P.N., and R.J. Wallace. 1982. Microbial ecology and activities in the rumen. Part II. *In* W.M. O'Leary (ed.) Critical reviews in microbiology. CRC Press, Boca Raton, FL.

Holmes, P., R.J. Moir, and E.J. Underwood. 1953. Ruminal flora studies in the sheep. V. The amino acid composition of rumen bacterial protein. Aust. J. Biol. Sci. 6:637-644.

Hristov, A., and G.A. Broderick. 1993. In vitro determination of ruminal protein degradability using ^{15}N-ammonia to correct for microbial nitrogen uptake. J. Anim. Sci. 71: submitted.

Hsu, J.T., and L.D. Satter. 1991. Protein dispersibility index combined with 420 nm absorbance to evaluate extent of heating of soybeans. J. Dairy Sci. 74(Suppl. 1):178 (Abstr.).

Hungate, R.E. 1966. The rumen and its microbes. p. 65-70. Academic Press, New York.

Hurrell, R.F., and R.A. Finot. 1985. Effects of food processing on protein digestibility and amino acid availability. p. 233-258. *In* J.W. Finley and D.T. Hopkins (ed.) Digestibility and amino acid availability in cereals and oilseeds. American Association of Cereal Chemists, St. Paul, MN.

Hutton, K., F.J. Bailey, and E.F. Annison. 1971. Measurement of the bacterial nitrogen entering the duodenum of the ruminant using diaminopimelic acid as a marker. Br. J. Nutr. 25:165-171.

Hvelplund, T. 1985. Digestibility of rumen microbial protein and undegraded dietary protein in the small intestine of sheep and by the in sacco procedure. Acta Agric. Scand. Suppl. 25:132.

Karges, K.K., T.J. Klopfenstein, V.A. Wilkerson, and D.C. Clanton. 1992. Effects of ruminally degradable and escape protein supplements on steers grazing summer native range. J. Anim. Sci. 70:1957-1964.

Kohn, R.A., and M.S. Allen. 1993. An in vitro method to measure protein degradation of feeds for ruminants using enzymes extracted from rumen contents. J. Dairy Sci. 76(Suppl. 1):175 (Abstr.).

Kopecny, J., B. Vencl, B. Vencl, J. Kyselova, and P. Brezina. 1989. Determination of rumen degradable protein with enzymes. Arch. Anim. Nutr. (Berlin) 7:635-645.

Kopecny, J., and R.J. Wallace. 1982. Cellular location and some properties of proteolytic enzymes of rumen bacteria. Appl. Environ. Microbiol. 43:1026-1033.

Krishnamoorthy, U., C.J. Sniffen, M.D.Stern, and P.J. Van Soest. 1983. Evaluation of a mathematical model of rumen digestion and an in vitro simulation of rumen proteolysis to estimate the rumen-undegraded nitrogen content of feedstuffs. Br. J. Nutr. 50:555-568.

Laycock, K.A., G.P.Hazlewood, and E.L. Miller. 1985. Potential use of proteolytic rumen bacteria for assessing feed protein degradability in vitro. Proc. Nutr. Soc. 44:54 (Abstr.).

Li, Y.-G., G.J. Tanner, A.C. Delves, and P.J. Larkin. 1992. Asymmetric intergeneric somatic hybrids between alfalfa and sainfoin by protoplast fusion. p. 40. *In* Abstr. First Int. Crop Sci. Congr. CSSA, Madison, WI.

Licitra, G., S. Carpino, P.J. Van Soest, and C.J. Sniffen. 1993. Improvement of the *Streptomyces griseus* method for degradable protein in ruminant feeds. J. Dairy Sci. 76(Suppl. 1):175 (Abstr.).

Lindberg, J.E., and T. Varvikko. 1982. The effect of bag pore size on the ruminal degradation of dry matter, nitrogenous compounds and cell walls in nylon bags. Swed. J. Agric. Res. 12:163-171.

Loerch, S.C., L.L. Berger, S.D. Plegge, and G.C. Fahey, Jr. 1983. Digestibility and rumen escape of soybean meal, blood meal, meat and bone meal and dehydrated alfalfa nitrogen. J. Anim. Sci. 57:1037-1047.

Luchini, N.D., G.A. Broderick, and D.K. Combs. 1993. Comparison of the proteolytic activity of commercial proteases with mixed rumen microorganisms. J. Dairy Sci. 76(Suppl. 1):175 (Abstr.).

Mahadevan, S., J.D. Erfle, and F.D. Sauer. 1979. A colorimetric method for the determination of proteolytic degradation of feed proteins by rumen microorganisms. J. Anim. Sci. 48:947-953.

Mahadevan, S., J.D. Erfle, and F.D. Sauer. 1980. Degradation of soluble and insoluble proteins by *Bacteroides amylophilus* protease and by rumen microorganisms. J. Anim. Sci. 50:723-728.

Mahadevan, S., J.D. Erfle, and F.D. Sauer. 1987. Preparation of protease from mixed rumen microorganisms and its use for the in vitro determination of the degradability of true protein in feedstuffs. Can. J. Anim. Sci. 67:55-64.

Mahler, H.R., and E.H. Cordes. 1966. Biological Chemistry. Harper and Row, New York.

Majdoub, A., G.T. Lane, and T.E. Aitchison. 1978. Milk production response to nitrogen solubility in dairy rations. J. Dairy Sci. 61:59-65.

McDonald, I.W., and R.J. Hall. 1954. The conversion of casein into microbial proteins in the rumen. Biochem. J. 67:400-405.

McDonald, I. 1981. A revised model for the estimation of protein degradability in the rumen. J. Agric. Sci., Camb. 96:251-258.

Mehrez, A.Z., and E.R. Ørskov. 1977. A study of the artificial fibre bag technique for determining the digestibility of feeds in the rumen. J. Agric. Sci., Camb. 88:645-650.

Mertens, D.R. 1987. Predicting intake and digestibility using mathematical models of ruminal function. J. Anim. Sci. 64:1548-1558.

Messman, M.A., and W.P. Weiss. 1993. Electrophoretic measurement of feed proteins in fluid and particulate phases of digesta. J. Dairy Sci. 76(Suppl. 1):176 (Abstr.).

Meyer, J.H.F., and R.I. Mackie. 1986. Microbiological evaluation of the intraruminal in sacculus digestion technique. Appl. Environ. Microbiol. 51:622-629.

Michalet-Doreau, B., and M.Y. Ould-Bah. 1993. In vitro and in sacco methods for the estimation of dietary nitrogen degradability in the rumen: A review. Anim. Feed Sci. Tech. 40:57-86.

Muscato, T.V., C.J. Sniffen, U. Krishnamoorthy, and P.J. Van Soest. 1983. Amino acid content of noncell and cell wall fractions in feedstuffs. J. Dairy Sci. 66:2198-2207.

Nagel, S.A., and G.A. Broderick. 1992. Effect of formic acid or formaldehyde treatment of alfalfa silage on nutrient utilization by dairy cows. J. Dairy Sci. 75:140-154.

Neutze, S.A., R.L. Smith, and W.A. Forbes. 1993. Application of an inhibitor in vitro method for estimating rumen degradation of feed protein. Anim. Feed Sci. Tech. 40:251-265.

Nocek, J.E., and A.L. Grant. 1987. Characterization of in situ nitrogen and fiber digestion and bacterial nitrogen contamination of hay crop forages preserved at different dry matter percentages. J. Anim. Sci. 64:552-564.

Nocek, J.E. 1988. In situ and other methods to estimate ruminal protein and energy digestibility: A review. J. Dairy Sci. 71:2051-2069.

NRC. 1982. United States-Canadian tables of feed composition. 3rd rev. National Academy Press, Washington, DC.

NRC. 1985. Ruminant nitrogen usage. National Academy Press, Washington, DC.

NRC. 1989. Nutrient requirements of domestic animals. No. 3. Nutrient requirements of dairy cattle. 6th rev. ed. National Academy Press, Washington, DC.

Nugent, J.H.A., W.T. Jones, D.J. Jordan, and J.L. Mangan. 1983. Rates of proteolysis in the rumen of the soluble proteins casein, Fraction I (18S) leaf protein, bovine serum albumin and bovine submaxillary mucoprotein. Br. J. Nutr. 50:357-368.

Nugent, J.H.A., and J.L. Mangan. 1981. Characteristics of the rumen proteolysis of fraction I (18S) leaf protein from lucerne (*Medicago sativa* L.). Br. J. Nutr. 46:39-58.

Ørskov, E.R., and I. McDonald. 1979. The estimation of protein degradability in the rumen from incubation measurements weighted according to rate of passage. J. Agric. Sci., Camb. 92:499-503.

Owens, F.N., and Goetsch, A. L. 1986. Digesta passage and microbial protein synthesis. p. 196-223. *In* L. P. Milligan, W. L. Grovum, and A. Dobson (ed.) Control of digestion and metabolism in ruminants. Prentice-Hall, Englewood Cliffs, NJ.

Owens, F.N., and C.F. Hanson. 1992. External and internal markers for appraising site and extent of digestion in ruminants. J. Dairy Sci. 75:2605-2617.

Poos-Floyd, M., T. Klopfenstein, and R.A. Britton. 1985. Evaluation of laboratory techniques for predicting ruminal protein degradation. J. Dairy Sci. 68:829-839.

Purser, D.B., and S.M. Buechler. 1966. Amino acid composition of rumen organisms. J. Dairy Sci. 49:81-84.

Raab, L., B. Cafantaris, T. Jilg, and K.H. Menke. 1983. Rumen protein degradation and biosynthesis. 1. A new method for determination of protein degradation in rumen fluid in vitro. Br. J. Nutr. 50:569-582.

Rae, R.C., and R.R. Smithard. 1985. Estimation of true nitrogen digestibility in cattle by a modified nylon bag technique. Proc. Nutr. Soc. 44:116 (Abstr.).

Reid, C.S.W., M.J. Ulyatt, and J.M. Wilson. 1973. Plant tannins, bloat and nutritive value. Proc. N. Z. Soc. Anim. Prod. 34:82-92.

Roe, M.B., L.E. Chase, and C.J. Sniffen. 1991. Comparison of in vitro techniques to the in situ technique for estimation of ruminal degradation of protein. J. Dairy Sci. 74:1632-1640.

Shaver, R.D., A.J. Nytes, L.D. Satter, and N.A. Jorgenson. 1986. Influence of amount of feed intake and forage physical form on digestion and passage of prebloom alfalfa hay in dairy cows. J. Dairy Sci. 69:1545-1559.

Sherrod, L.B., and A.D. Tillman. 1964. Further studies on the effects of different processing temperatures on the utilization of solvent-extracted cottonseed protein by sheep. J. Anim. Sci. 23:510-516.

Silva, A.T., R.J. Wallace, and E.R. Ørskov. 1987. Use of particle-bound microbial enzyme activity to predict the rate and extent of fibre degradation in the rumen. Br. J. Nutr. 57:407-415.

Sniffen, C.J., J.D. O'Conner, P.J. Van Soest, D.G. Fox, and J.B. Russell. 1992. A net carbohydrate and protein system for evaluating cattle diets: II. Carbohydrate and protein availability. J. Anim. Sci. 70:3562-3577.

Spencer, D., T.J.V. Higgins, M. Freer, H. Dove, and J.B.Coombe. 1988. Monitoring the fate of dietary proteins in rumen fluid using gel electrophoresis. Br. J. Nutr. 60:241-247.

Stern, M.D., K.A. Santos, and L.D. Satter. 1985. Protein degradation in rumen and amino acid absorption in small intestine of lactating dairy cattle fed heat-treated whole soybeans. J. Dairy Sci. 68:45-56.

Storm, E., and E.R. Ørskov. 1984. The nutritive value of rumen micro-organisms in ruminants. 4. The limiting amino acids of microbial protein in growing sheep determined by a new approach. Br. J. Nutr. 52:613-620.

Thomas, J.W., Y. Yu, T. Middleton, and C. Stallings. 1982. Estimations of protein damage. p. 81-98. *In* F. N. Owens (ed.) Protein requirements for cattle: symposium. Oklahoma State Univ., Stillwater.

Van Soest, P. J., C. J. Sniffen, D. R. Mertens, D. G. Fox, P. H. Robinson, and U. Krishnamoorthy. 1982. A net protein system for cattle: The rumen submodel for nitrogen. p. 265-279. *In* F. N. Owens (ed.) Protein requirements for cattle: symposium. Oklahoma State Univ., Stillwater.

Varvikko, T. 1986. Microbially corrected amino acid composition of rumen-undegraded feed protein and amino acid degradability in the rumen of feeds enclosed in nylon bags. Br. J. Nutr. 56:131-140.

Waghorn, G.C., M.J. Ulyatt, A. John, and M.T. Fisher. 1987. The effect of condensed tannins on the site of digestion of amino acids and other nutrients in sheep fed *Lotus corniculatus* L. Br. J. Nutr. 57:115-126.

Waldo, D.R., and H.K. Goering. 1979. Insolubility of proteins in ruminant feeds by four methods. J. Anim. Sci. 49:1560-1568.

Waldo, D.R., L.W. Smith, and E.L. Cox. 1972. Model of cellulose disappearance from the rumen. J. Dairy Sci. 55:125-129.

Wallace, R.J. 1983. Hydrolysis of ^{14}C-labelled proteins by rumen microorganisms and by proteolytic enzymes prepared from rumen bacteria. Br. J. Nutr. 50:345-355.

Wallace, R.J. 1985. Adsorption of soluble proteins to rumen bacteria and the role of adsorption in proteolysis. Br. J. Nutr. 53:399-408.

Wallace, R.J., and M.A. Cotta. 1988. Metabolism of nitrogen-containing compounds. p. 217-249. In P.N. Hobson (ed.) The rumen microbial ecosystem. Elsevier Applied Science, London.

Wallace, R.J., N. McKain, and C.J. Newbold. 1990. Metabolism of small peptides by mixed rumen bacteria and protozoa. Analysis of rates of hydrolysis and of intermediates which accumulate during the breakdown of alanine oligomers in vitro. J. Sci. Food Agric. 50:191-199.

Warner, A.C.I. 1956. Proteolysis by rumen micro-organisms. J. Gen. Microbiol. 14:749-762.

Weakley, D.C., M.D. Stern, and L.D. Satter. 1983. Factors affecting disappearance of feedstuffs from bags suspended in the rumen. J. Anim. Sci. 56:493-507.

Weller, R.A. 1957. The amino acid composition of hydrolysates of microbial preparations from the rumen of sheep. Aust. J. Biol. Sci. 10:384-389.

Wilkerson, V.A., S.M. Hannah, R.C. Cochran, and T.J. Klopfenstein. 1990. Rinsing procedures for in situ protein degradation as influenced by technician and method. J. Anim. Sci. 68(Suppl. 1):517 (Abstr.).

Whitelaw, F.G., L.A. Bruce, J.M. Eadie, and W.J. Shand. 1983. 2-Aminoethylphosphonic acid concentrations in some rumen ciliate protozoa. Appl. Environ. Microbiol. 46:951-959.

Wohlt, J.E., C.J. Sniffen, and W.H. Hoover. 1973. Measurement of protein solubility in common feedstuffs. J. Dairy Sci. 56:1052-1057.

Yu, Y. 1977. Effect of heating of forages on quantitative changes of acid-detergent insoluble nitrogen. J. Dairy Sci. 60:1813-1815.

CHAPTER 6

CARBOHYDRATES AND FORAGE QUALITY

K. J. Moore and R. D. Hatfield

INTRODUCTION

Carbohydrates are the most abundant class of compounds found in plants. They account for 50-80% of the dry biomass of forage species (Van Soest, 1982). From a plant perspective, carbohydrates play important roles in intermediary metabolism, energy transfer and storage, and plant structure. Photosynthetic energy is fixed in carbohydrates via the Calvin cycle and these carbohydrates serve as initial substrates for nearly all intermediary pathways in plants. Energy is translocated within plants as the disaccharide sucrose (Hawker, 1985), and stored in polymers such as starch and fructans (Manners, 1985; Pontis and Del Campillo, 1985). Carbohydrates constitute most of the plant cell wall and therefore play an important role in the structural integrity of individual cells, tissues, and organs (Hatfield, 1989).

Carbohydrates are extremely important from a nutritional perspective, providing the primary source of energy in ruminant diets. In ruminants, nearly all carbohydrate digestion (>90%) occurs within the rumen (Armstrong and Smithard, 1979; Sutton, 1979), although under certain circumstances, such as high rates of passage, a significant amount of carbohydrate digestion can occur in the small and large intestines (Hoover, 1978; Nocek and Tamminga, 1991). Simple sugars are rapidly fermented within the rumen to yield volatile fatty acids which are absorbed into the blood through the rumen wall (Morrison, 1979; Baldwin and Allison, 1983). Polysaccharides must be degraded to simple sugars before being utilized. Nonstructural polysaccharides, such as starch and fructans, are rapidly and generally completely degraded within the rumen (Nocek and Tamminga, 1991), while degradability of structural polysaccharides varies considerably. Pectins are rapidly and essentially completely degraded by ruminal microbes (Hatfield, 1989, 1993a). Cellulose and hemicellulosic polysaccharides are more slowly and incompletely degraded. Cellulose degradability of forages varies from 25-90% while hemicellulose digestibility varies from 45-90% (Pigden and Heaney, 1969). Degradation of β-glucans is intermediate to that of pectins and cellulose (Van Soest et al., 1991). This ability to degrade and utilize structural carbohydrates confers upon ruminants their unique ecological niche (Chesson and Forsberg, 1988).

K. J. Moore, Dep. of Agronomy, Iowa State Univ., Ames, IA 50011; R. D. Hatfield, USDA-ARS, U.S. Dairy Forage Research Center, Madison, WI 53706.

In addition to being a major source of energy in the diet of ruminants, carbohydrates have other nutritional roles. As components of dietary fiber, structural carbohydrates are important for normal rumen function. Fiber stimulates rumination and ensalivation, and the cation exchange properties of fiber are important in ruminal buffering capacity (Van Soest et al., 1991). Fiber also is involved in regulation of voluntary intake of forages (Mertens, 1987, 1993).

CHEMISTRY OF FORAGE CARBOHYDRATES

Classification / Terminology

The terminology used to describe the various carbohydrates of forages has often led to confusion in understanding their roles in plant growth and development and ruminant nutrition. In particular, the terms "fiber" and "plant cell wall" often are used interchangeably. These terms, however, are not synonymous and reflect different functional perspectives (Chesson and Forsberg, 1988; Van Soest, 1993).

Plants are unique among higher organisms in that they possess rigid cell walls (Bartnicki-Garcia, 1984). The walls of plant cells can be considered a composite consisting of cellulose fibrils embedded within a matrix of lignin and hemicellulosic polysaccharides (Monties, 1991). In addition, the intact cell wall contains fillers such as water, organic solvents, and phenolics which lend unique properties to the structure. The macromolecular composition of cell walls varies considerably among organs and tissues, and at the subcellular level. The primary cell wall is formed adjacent to the plasmalemma during cell elongation and consists almost entirely of polysaccharides (Albersheim, 1975). The secondary wall is formed during cell differentiation interior to the primary wall and varies greatly in composition depending on cell type. Individual cells are adjoined by the middle lamella which consists primarily of pectic substances which serve as an intercellular cementing agent (Varner and Lin, 1989).

The most obvious function of the cell wall is its role in morphogenesis. Cell walls form the structural framework of the plant architecture and provide mechanical and structural support for plant organs (Varner and Lin, 1989). In addition, walls play important roles in water balance, ion exchange, cell recognition, and protection from biotic stresses (Vian, 1982; Varner and Lin, 1989).

In contrast, fiber is a nutritional entity which is defined as much by its biological properties as its chemical composition (Van Soest et al., 1991). The concept of fiber, particularly in regard to forages, has traditionally referred to the complex of dietary nutrients which are relatively resistant to digestion and are slowly and only partially degraded by ruminants (Van Soest, 1982; Chesson and Forsberg, 1988). By this definition, cellulose, hemicelluloses, and lignin are the major components of fiber. The definition of fiber has more recently been expanded by nutritionists concerned with monogastric animals to include all compounds in the diet which are resistant to digestion by mammalian enzymes

(Van Soest et al., 1991). This definition includes pectins and β-glucans in addition to the components listed above.

The terms holocellulose and lignocellulose often are used in relation to forage quality. Holocellulose refers collectively to cellulose and hemicelluloses, (technically delignified cell walls), whereas lignocelluloses encompass lignin in addition to these structural polysaccharides (Pigden and Heaney, 1969; Bailey, 1973). Because of differences in the chemistry and microbial degradation of cellulose and hemicelluloses, it is not particularly useful to consider them collectively in studies of forage quality, so the term holocellulose is of limited value in this context (Baldwin and Allinson, 1983). The term lignocellulose often is used interchangeably with fiber, especially in areas of utilization not related to nutrition such as biofuels (Linden et al., 1980).

For the purposes of this chapter, we will attempt to avoid the terminology described above as much as possible, and use terms which refer to the role of carbohydrates in plants. Using this approach, carbohydrates can be classified as nonstructural and structural, with the nonstructural carbohydrates further classified as storage, transport, or metabolic.

Monosaccharides, Disaccharides, and Oligosaccharides

The term carbohydrate was given to describe molecules that were believed to be hydrates of carbon compounds because of their general formula of $C_n(H_2O)_n$. It has since been recognized that carbohydrates are polyhydroxyaldehydes, polyhydroxyketones, or molecules that can be hydrolyzed to produce these simpler molecules (Kennedy and White, 1983).

By definition, a monosaccharide is a carbohydrate that cannot be hydrolyzed to a simpler molecule whereas a disaccharide can be hydrolyzed to two monosaccharides, a trisaccharide to three monosaccharides, and so on. Such carbohydrates are referred to as oligosaccharides and may contain from 2-10 individual monosaccharides (Kandler and Hopf, 1980). There are no rigid rules that delineate oligosaccharides from polysaccharides other than polysaccharides usually contain several hundred monosaccharide units.

As all complex carbohydrates can be hydrolyzed (in principle) to monosaccharides, it is important to understand their classification. Monosaccharides are classified according to the functional groups they contain when in the open-chain form. An aldose contains an aldehyde group whereas a ketose contains a ketone group. Furthermore, a monosaccharide is referred to by the number of carbon atoms it contains: triose (3 C), tetrose (4 C), pentose (5 C), hexose (6 C), and so on. Combining concepts, an aldohexose is a six carbon monosaccharide that contains an aldehyde group (e.g., glucose) and a ketohexose is a six carbon monosaccharide that contains a ketone group (e.g., fructose). Although a wide range of sugars may be found in plants, particularly as key intermediates of metabolic pathways, pentoses and hexoses are the most abundant naturally occurring monosaccharides, with glucose the most prominent.

Most plant carbohydrates do not exist as an open chain linear molecule, but rather as a cyclic molecule created by the formation of a hemiacetal or

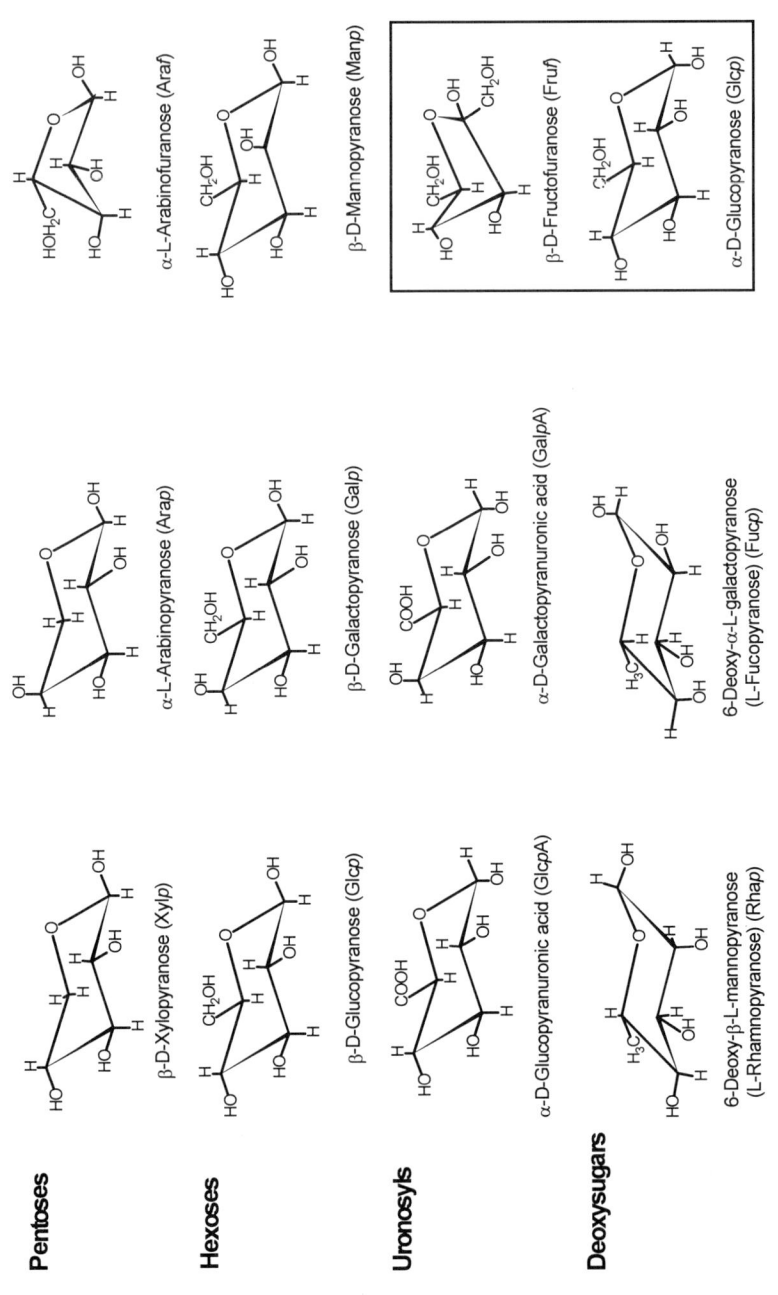

Figure 1. Common monosaccharides found in plants and their preferred conformation.

hemiketal. In the case of D-glucose, the six-membered ring hemiacetal is produced by the reaction between the aldehyde group (C_1) and the hydroxyl group of C_5. This ring structure provides an additional chiral center (C_1) resulting in two distinct isomers or anomers, α-D-glucose and β-D-glucose, differing only by their configuration about C_1 (Kennedy and White, 1983). As a hemiacetal, the glucose ring is readily opened by H_2O (as H^+) leading to isomerization. Therefore, either anomer (α or β) can be converted to an equilibrium mixture by opening and closing the hemiacetal ring. The same holds true for all other cyclic sugars (hemiacetals and hemiketals that are formed from a reaction between a hydroxyl group and a ketone group). If C_1 is involved in a glycosidic linkage forming an acetal (also referred to as a non-reducing sugar), the ring is not opened by H_2O. Therefore, glycosidically-linked monosaccharides that make up oligosaccharides and polysaccharides will have fixed α or β configurations.

In addition, individual monosaccharides assume preferential conformations that are the most stable shape or spatial arrangement of atoms in the molecule. For example, the preferred conformation of glucose is a 6-membered (pyranose) ring that minimizes torsional and van der Waals strain. The hydroxyl groups on the ring occupy equatorial or axial positions. X-ray analysis of β-D-glucose has shown the most stable conformation is to have all bulky groups in the equatorial positions (Kennedy and White, 1983). Monosaccharides commonly found in plants and their preferred conformations are illustrated in Figure 1.

As previously mentioned, oligosaccharides are carbohydrates that contain more than a single monosaccharide. A number of oligosaccharides function as transported forms of carbohydrate or as storage reserves among different plant tissues. The most prominent is sucrose, which is a non-reducing disaccharide composed of α glucose and fructose residues. Sucrose plays an important role in carbohydrate transport, is a glucose donor in starch synthesis, and is a storage molecule in some plants (Kandler and Hopf, 1980). Several other types of oligosaccharides have been isolated from a variety of plants (Table 1).

Polysaccharides

Non-Structural Polysaccharides. Starches and fructans are the major non-structural polysaccharides found in plants. Both are important as energy reserves and are stored in various plant organs, depending on species, until needed to provide energy for metabolic processes. Starch is the predominant storage polysaccharide in most higher plants. Fructans, however, occur in a number of forages and are the main storage polysaccharide in Festucoid grasses (Manners, 1985; Pontis and Del Campillo, 1985).

Starch polymers are composed of α-D-glucose units and occur in two forms in plants: amylose and amylopectin. Amylose is a linear molecule of α-D-glucose units linked 1→4. It has a relatively low molecular weight with typically less than 2,000 glucose residues per molecule. Amylose is soluble in water and takes on a helical conformation in solution. Amylopectin is a highly branched molecule with 1→4 linked α-D-glucose chains joined 1→6 at intervals along the backbone molecule. Amylopectin has a higher molecular weight than

Table 1. Common oligosaccharides that have been identified in plants.

Common Name	Composition	Function	Distribution
Sucrose	a-D-Glcp-(1→2)-β-D-Frucf	Storage/Transport	Ubiqutious
Raffinose	a-D-Galp-(1→6)-a-D-Glcp-(1→2)-β-D-Frucf	Storage/Transport	Widespread
Stachyose	a-D-Galp-(1→6)-a-D-Galp-(1→6)-a-D-Glcp-(1→2)-β-D-Frucf	Storage/Transport	Widespread
Verascose †	a-D-Galp-(1→6)-a-D-Galp-(1→6)-a-D-Galp-(1→6)-a-D-Glcp-(1→2)-β-D-Frucf	Storage/Transport	Limited
6-Kestose ‡	a-D-Glcp-(1→2)-β-D-Frucf-(6→2)-β-D-Frucf	Storage	Limited
1-Kestose ‡	a-D-Glcp-(1→2)-β-D-Frucf-(1→2)-β-D-Frucf	Storage	Limited
Neokestose ‡	β-D-Frucf-(2→6)-a-D-Glcp-(1→2)-β-D-Frucf	Storage	Limited

† Higher homologs with addition of Galp residues have been found in some plants.

‡ These oligosaccharides are typically found in conjunction with their higher homologs and polymeric members (fructans). In grasses, the predominant fructan is based on kestose with up to 50 β-D-Frucf residues bonded by β-(2→6) linkages.

amylose and generally ranges from 2,000 to 220,000 glucose residues per molecule (Smith, 1973; Preiss, 1988; Manners, 1985; Morrison and Karkalas, 1990).

Depending on species, starch is stored in amyloplasts within the cells of various plant organs including seeds, rhizomes, and tubers. Starch also is synthesized and stored in granules in chloroplasts when net photosynthesis exceeds sink demand. When sink demand exceeds energy available from photosynthesis, the starch then is metabolized to sucrose which is loaded into the phloem and transported to the developing tissue (Esau, 1977; Hawker, 1985; Morrison and Karkalas, 1990).

Fructans are unique polysaccharides composed almost entirely of fructose residues; furthermore, no bond of the fructofuranosyl ring is part of the polysaccharide backbone. These are one of a few natural polysaccharides that exist in the furanose form in plants. The unique structure of fructans results in their general high solubility in hot water, ease of hydrolysis in weak acid, and formation of non-reducing polymers. Three major types of fructans occur in plants: a) inulin type, those with β-(2→1)-D-fructofuranosyl units; b) phlein type,

Polysaccharide	Linkage	Linkage Sequence	Expected Conformation
Cellulose	*β*-D-Glc*p*(1→4)-*β*-D-Glc*p*	→Glc→Glc→Glc→Glc→Glc→	Extended Ribbon
Callose	*β*-D-Glc*p*(1→3)-*β*-D-Glc*p*	↗Glc↗Glc↗Glc↗	Coiled Springs
Mannans	*β*-D-Man*p*(1→4)-*β*-D-Man*p*	→Man→Man→Man→Man→	Extended Ribbon
Xylan	*β*-D-Xyl*p*(1→4)-*β*-D-Xyl*p*	→Xyl→Xyl→Xyl→Xyl→	Twisted Ribbon
Galacturonans	*α*-D-Gal*p*A(1→4)-*α*-D-Gal*p*A	→GalA→GalA→GalA→GalA→	Extended Ribbon
Galactan	*β*-D-Gal*p*(1→4)-*β*-D-Gal*p*	→Gal→Gal→Gal→Gal→Gal→	Extended Ribbon

Figure 2. Structural features of homopolysaccharides found in plants.

those with β-(6→2)-D-fructofuranosyl units; and c) the branched type that contains both kinds of glycosidic linkages (Pontis, 1990).

Structural Polysaccharides. There is a wide variety of polysaccharides found in plant walls that differ among cell types, tissues, organs, and plant species. Traditionally, these polysaccharides have been classified as pectic, hemicellulosic, and cellulosic fractions, depending upon the chemical extraction and fractionation procedures employed during isolation. Although such classifications are widely used and provide a general frame of reference, these should not be considered to represent discrete and finite groups of polysaccharides. Instead, the classification schemes reflect common solubilities rather than common compositions (Wilkie, 1985; Bacic et al., 1988). There is a great deal of overlap among the classes of polysaccharides that are extractable from walls with a given chemical treatment. The solubility of a given polysaccharide may differ from cell type to cell type or between developmental stages; therefore, fractionation without characterization is of limited value. Compositional and structural features of individual polysaccharides provide a better classification scheme.

It is not the objective of this chapter to provide an in depth review of structural polysaccharides, but rather to touch on some of their basic features. The major classes of polysaccharides, based upon composition and structural features, are given in Figures 2 and 3. In most cases, these are generalized features of a particular group. Individual structural features (bonding patterns of substituents, types of substituents, etc.) will vary depending upon the plant species, developmental stage, tissues, and cell types (Wilkie, 1979; Darvill et al., 1980; Selvendran, 1983; Bacic et al., 1988).

Polysaccharides are classified as homopolymers (Figure 2) or heteropolymers (Figure 3). There are few examples of true homopolymers, molecules composed of a single type of monosaccharide bonded by a single type of glycosidic linkage (Kennedy and White, 1983). Cellulose is the most abundant organic molecule in nature and belongs to the homopolymer group in that it contains only β-D-glucose linked through (1→4) glycosidic bonds (Figure 2). Callose is related to cellulose differing only in the type of glycosidic linkage, (1→3). It is found in all plants and appears to have a specialized role, frequently deposited in response to wounding. Although there are reports in the literature of homopolymers of xylan, galactan, and galacturonan, these are rare and reflect special cases. The majority of polysaccharides found in plants fall in the heteropolymer group. These are molecules that may contain a single type of monosaccharide linked by more than one type of glycosidic linkage (e.g, arabinans, (1→3),(1→4)-β-D-glucans; Figure 3) or contain more than a single monosaccharide with various glycosidic linkages (e.g., arabinogalactans, rhamnogalacturonans; Figure 3).

Biosynthesis of Carbohydrates

Carbohydrates are produced through the photosynthetic process of carbon fixation. Formation of individual types of sugars generally occurs through the

Heteropolymers	Backbone Linkages	Branch Linkages	Linkage Patterns
Arabinan	α-L-Araf(1→5)-α-L-Araf	α-L-Araf(1→3 or 2)-α-L-Araf	Ara→Ara→Ara→Ara→Ara ↑ Ara
Arabinogalactans			
Type I	β-D-Galp(1→4)-β-D-Galp	α-L-Araf(1→3 or 2)-β-D-Galp α-L-Araf(1→5)-α-L-Araf α-L-Araf(1→6)-β-D-Galp α-L-Araf(1→3)-β-D-Galp	Gal→Gal→Gal ↑ ↑ Ara Ara ↑ Ara
Type II	β-D-Galp(1→3)-β-D-Galp β-D-Galp(1→6)-β-D-Galp	**Minor substituents (R=)** α-D-GlcpA(1→3 or 6)-β-D-Galp L-Rhap(1→3 or 6)-β-D-Galp α-D-GlcpA(1→3 or 6)-β-D-Galp 4-O-CH$_3$-α-D-GlcpA(1→3 or 6)-β-D-Galp	Ara R ↓6 ↓6 →Gal→3Gal→3Gal→3Gal→ 6↑ 6↑ Gal 6↑ Ara→3Gal 6↑ R→3Gal ↑
Rhamnogalacturonans	α-D-Galp(1→2)-β-L-Rhap β-D-Galp(1→4)-α-D-GalpA	α-L-Araf(1→4)-β-L-Rhap β-D-Galp(1→4)-β-L-Rhap Acetyl = R' Galactans Arabinogalctans } = R Arabinans	Ara R ↓4 ↓4 →GalA→Rha→(GalA)→(Rha→GalA)→ ↑2or3 R' n=1 to70 n=to300

Xylans

Arabinoxylans	β-D-Xylp(1→4)-β-D-Xylp	α-L-Araf-(1→2or3)-β-D-Xylp	$\rightarrow Xyl \rightarrow Xyl \rightarrow (Xyl)_n \rightarrow Xyl \rightarrow$ with Ara↓3 branches, $n = 0$ to 7
Glucuronoarabinoxylans	"	α-D-GlcpA-(1→2)-β-D-Xylp	$\rightarrow (Xyl \rightarrow Xyl)_n \rightarrow Xyl \rightarrow Xyl \rightarrow$, Ara↓3, $n = 1$ to 7, GlcA↑2
Glucuronoxylans	"	α-D-GlcpA-(1→2)-β-D-Xylp	$\rightarrow Xyl \rightarrow (Xyl)_n \rightarrow Xyl \rightarrow Xyl \rightarrow$, GlcA↑2, $n = 10$ to 12

Xyloglucans

Dicot	β-D-Glcp-(1→4)-β-D-Glcp	α-D-Xylp-(1→6)-β-D-Glcp β-D-Galp-(1→2)-β-D-Xylp α-L-Fucp-(1→2)-β-D-Galp	Fuc↓2 Gal↓2 Xyl Xyl Xyl ↓6 ↓6 ↓6 →Glc→Glc→Glc→Glc→
Monocot	β-D-Glcp-(1→4)-β-D-Glcp	α-D-Xylp-(1→6)-β-D-Glcp β-D-Galp-(1→2)-β-D-Xylp Minor branch sugar	Xyl Xyl Xyl ↓6 ↓6 ↓6 →Glc→Glc→Glc→Glc→

Mixed linkage β-D-Glucans

	β-D-Glcp-(1→4)-β-D-Glcp
	β-D-Glcp-(1→3)-β-D-Glcp

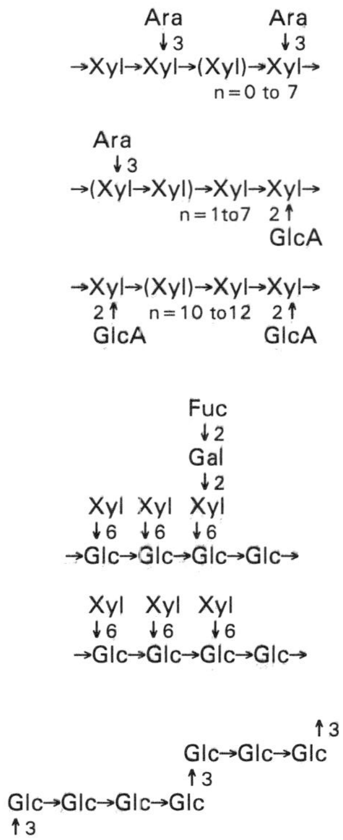

Glc→Glc→Glc↑3
Glc→Glc→Glc↑3
Glc→Glc→Glc

Figure 3. Structural features of heteropolysaccharides in plants.

action of epimerase, isomerase, oxidoreductase, and (or) decarboxylase enzymes from activated monosaccharides arising from the Calvin cycle or the breakdown of storage carbohydrates (Figure 4) (Feingold, 1982; Feingold and Barber, 1990).

Biosynthesis of oligosaccharides and polysaccharides requires activated sugars, in the form of nucleoside diphosphate monosaccharides (NDP-monosaccharides) (Delmer and Stone, 1988). The most predominant pathway is from interconversions of glucose derived directly from photosynthetic activity or from starch degradation (Figure 4). There are a few alternative pathways such as the conversion of inositol to glucuronic acid and scavenging pathways that can reclaim galacturonic acid and galactose through direct phosphorylation (Feingold and Avigad, 1980; Figure 4, dashed arrows).

Nonstructural Carbohydrates

Sucrose. Sucrose biosynthesis in plants can occur by two pathways. The principal reaction involves the conversion of UDP-Glc$p$ and Fru$f$-6-phosphate to sucrose-6-phosphate and is catalyzed by sucrose phosphate synthase. This reaction is coupled with sucrose phosphatase to yield free sucrose. The second pathway, which is considered to be of minor importance, is catalyzed by sucrose synthase and involves the conversion of UDP-Glc$p$ and Fru$f$ directly to sucrose (Avigad, 1982; Hawker, 1985).

Starch. The synthesis of starch can be summarized by the following equation:

$$\text{ADP-Glc}p + (\alpha\text{-D-glucan})_n \rightarrow \text{ADP} + (\alpha\text{-D-glucan})_{n+1}$$

The major enzyme, starch synthase, transfers glucose from ADP-Glc$p$ to the growing polymer of starch (α-D-glucan). Starch is not a homopolymer due to branching at C6 of some glucose residues in the glucan backbone, thus containing both α-(1$\rightarrow$4) and α-(1$\rightarrow$6) linkages. The formation of the branch point is carried out by a branching enzyme that cleaves an α-D-glucan chain and transfers the new reducing end to the C6 of another α-D-glucan chain. There appears to be multiple forms of both the starch synthase and branching enzymes which may be important in the particular type of starch that is formed and its ultimate use (Preiss, 1988).

Fructans. Fructan biosynthesis is initiated by the enzyme sucrose-sucrose transfructosylase which converts two sucrose molecules to glucose and fructosylsucrose, the initial homologue of the fructan series. Further chain elongation is catalyzed by a fructofuranosyltransferase which transfers terminal fructofuranosyl residues to the growing polymer. Fructosylsucrose can serve as both the donor and acceptor for this reaction while sucrose can serve only as an acceptor. Fructan biosynthesis occurs within the vacuole utilizing substrate synthesized within the cytoplasm and transported across the tonoplast (Meier and Reid, 1982; Pontis and Del Campillo, 1985).

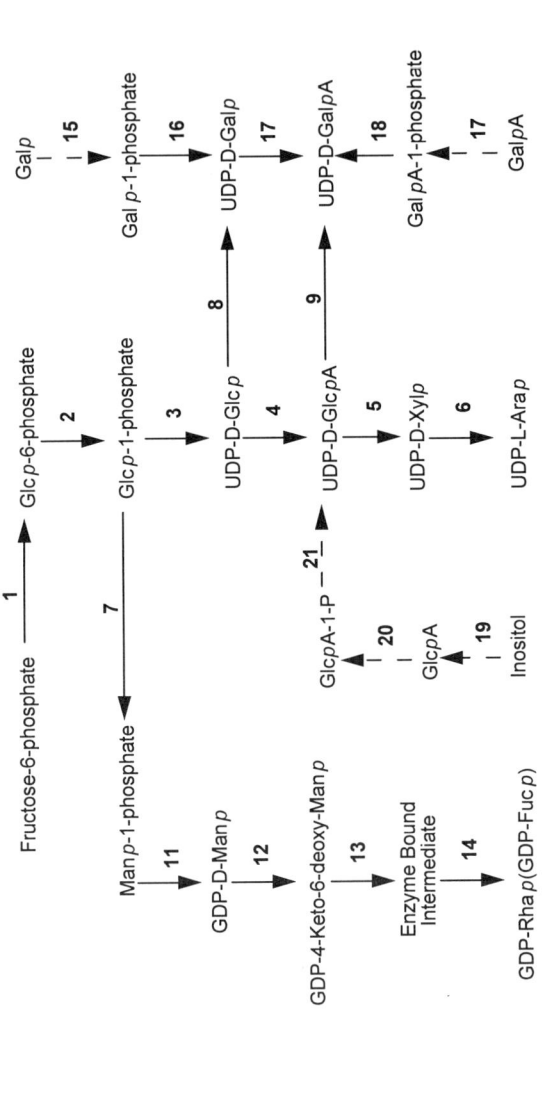

Figure 4. Formation of nucleotide monosaccharides. Dashed lines represent alternative and scavenging pathways.

Structural Polysaccharides

Biosynthesis of structural polysaccharides is more complicated and the details of the individual enzymes involved and their regulation are only now beginning to be unravelled. The more complicated the polysaccharide (types of monosaccharides or linkages), the more complex the biosynthetic apparatus must be.

For all wall polysaccharides except cellulose, synthesis (polymer formation) occurs within the endomembrane system of the cell. Polysaccharide synthesis is initiated in the endoplasmic reticulum and completed within the Golgi apparatus. A major requirement in wall synthesis is a transport system for the movement of materials to the outside of the cell. This is accomplished by the formation of vesicles from the outer fringes of the cisterna of individual dictyosomes that make up the whole Golgi apparatus (Delmer and Stone, 1988; Iiyama et al., 1993). As the vesicles fuse with the plasma membrane, their contents are deposited into the apoplastic space. Cellulose appears to be polymerized and simultaneously deposited in the wall matrix by a plasma membrane-cellulose-synthase complex (Delmer and Stone, 1988).

Delmer and Stone (1988) described the process of polysaccharide synthesis as consisting of three coordinated steps: chain initiation, chain elongation, and chain termination. It is important to note that our understanding of these individual steps, the enzymes involved, and their regulation, is far from complete. For matrix polysaccharides, it appears that synthases are membrane-bound complexes. The first step in the process, chain initiation, involves some type of initiator molecule to which the first sugar residue of a given polymer is attached. Several types of molecules have been suggested to fill this role including inositol, lipids, and proteins (Delmer and Stone, 1988; Iiyama et al., 1993). The type of chain initiation may be species-specific or, more likely, may be dependent upon the type of polysaccharide synthesized by a given complex.

Polymer chain elongation is somewhat better understood with evidence supporting a direct transfer of the activated sugar to the polymer (James et al., 1985; Delmer and Stone, 1988; Iiyama et al., 1993). It appears in plants that this process is carried out in a tailward growth fashion, namely, the new sugar residue is added to the non-reducing end of the polymer. The formation of even simple heteropolymers such as $(1\rightarrow3),(1\rightarrow4)$-$\beta$-D-glucans is a complicated and highly regulated process. The introduction of $(1\rightarrow3)$ linkages at regular intervals in the chain may be mediated by the binding of the transferase to the growing polymer; if the terminal sugar is linked $(1\rightarrow3)$, the successive Glc$p$ is attached by a $(1\rightarrow4)$ bond followed by 2 or 3 or more $(1\rightarrow4)$ additions. Stretches of $(1\rightarrow4)$ linkages causes a conformational shift in the synthase such that the addition of the next Glc$p$ is by a $(1\rightarrow3)$ linkage and the cycle repeats (Delmer and Stone, 1988; Iiyama et al., 1993). Alternatively, two specific transferases, one for $(1\rightarrow3)$ and one for $(1\rightarrow4)$ linkage formations that are coordinately regulated in some fashion, could produce the mixed-linked polymer.

Synthesis of branched polymers such as arabinoxylans or glucuronoarabinoxylans occur by the addition of monosaccharides to a main polymer backbone. The synthase complex must contain multiple transferases that are working simultaneously to produce the appropriately branched molecule. In some complexes, the activity of the transferase forming the backbone of the molecule is dependent upon the continued activity of a branching transferase. For example, soybean xyloglucan synthesis requires both UDP-Glc and UDP-Xyl for continued chain elongation because the addition of xylose side branches is required for extension of the β-glucan backbone (Hayashi, 1989). However, the addition of fucose and galactose onto the xylose branches seems to be an independent event that is not tightly associated with chain elongation. The addition of branch residues to homopolymer backbones during the synthesis of xyloglucans, glucuronoxylans, arabinoxylans, or any other branched polysaccharide requires a highly regulated mechanism to obtain the proper orientation of the branch residues. Stereospecificities of branch residues suggest the involvement of special binding proteins that hold the growing polysaccharide chain in the proper orientation to produce the correct molecular structures (Delmer and Stone, 1988; Iiyama et al., 1993).

To obtain a polymer of appropriate size, termination of the polysaccharide may be the result of physical disengagement of the polymer from the synthase complex. Alternatively, termination may be an active process that regulates the size of the molecule depending upon the physiological demands of the cell wall for a given stage of cell (wall) development (Delmer and Stone, 1988). It has been shown that the degree of polymerization (D.P.) of cellulose greatly increases with the onset of secondary wall synthesis (Delmer and Stone, 1988; Iiyama et al., 1993). Regulation of D.P. is independent of factors such as temperature that increase the rate of cellulose deposition. It is possible that for matrix polysaccharides synthesized in the endomembrane system, the rate of vesicle movement and fusion with the plasma membrane may control polymer length (Delmer and Stone, 1988).

Methods of Analysis

Extraction and Fractionation of Nonstructural Carbohydrates

There are a number of methods available for the extraction and fractionation of nonstructural carbohydrates from plant materials. The method used will depend upon the carbohydrates present in the sample and the degree of fractionation desired. Free sugars and fructans are soluble in water whereas starch is differentially soluble. Amylose is soluble and amylopectin insoluble in water (Smith, 1981). In addition, some structural carbohydrates such as pectins and β-glucans are partially soluble in water. Because concentrations of starch and nonstarch polysaccharides vary greatly among forages, extraction of carbohydrates with water without further fractionation is of little value. To avoid these problems, soluble sugars generally are extracted with ethanol solutions which will not solubilize the polysaccharides.

Free Sugars. Extraction with 80% ethanol is the most common method used to remove free sugars such as glucose, fructose, and sucrose. The extracts can then be analyzed in toto using nonspecific colorimetric or titrimetric assays (Smith, 1981; Sturgeon, 1990) or further fractionated using chromatographic techniques (Ericsson et al., 1978; Fales et al., 1982) or enzyme-coupled assays (Westhafer et al., 1982; Avigad, 1990). Some short-chain fructans also may be solubilized by 80% ethanol. When extracting free sugars from species that accumulate fructans, higher concentrations of ethanol should be used. Smith (1981) recommended ethanol concentrations of 85% for species which accumulate long-chain fructans and 90% for species which accumulate predominantly short-chain fructans.

Starch. Starch is generally extracted from forages using either acid or enzyme hydrolysis on residues which have been previously extracted with ethanol to remove soluble sugars (Smith, 1981). Glucose concentrations in the hydrolysate are determined using one of the assays mentioned above, and described in more detail below, and are converted to equivalent starch concentrations by multiplying by 0.9. Enzymatic hydrolysis is preferred to acid hydrolysis due to its specificity. Acid extraction can result in partial hydrolysis of structural polysaccharides or only partial hydrolysis of starch depending on the concentration of acid used (Smith, 1973). Intact starch may be extracted using dimethyl sulfoxide (70-90%) (Morrison and Karkalas, 1990). Amylose and amylopectin may be separated by selective precipitation from DMSO extracts or by using gel permeation chromatography (Morrison and Karkalas, 1990).

Fructans. Fructans are highly soluble and can be easily extracted with water from residues that have been previously extracted with ethanol (80-92%) to remove soluble sugars (Volenec and Nelson, 1984). Total concentration of fructans in the extract can be determined following hydrolysis with dilute acid (0.1 N H_2SO_4) by quantifying fructose using any of a number of colorimetric methods specific for ketoses (Pontis, 1990). Average chain length of fructans can be estimated by determining glucose concentration in the hydrolysate with an enzyme-coupled assay or by chromatography. Since each fructan polymer contains a single glucose residue, average D.P. can be estimated as the ratio of fructose to glucose (Volenec, 1986). The differential solubility of fructan polymers of various D.P. in ethanol solutions can be used to fractionate them by D.P. (Smith, 1973). However, chromatographic methods such as gel-permeation and thin-layer chromatography are less time-consuming and more commonly used today (Pontis, 1990).

Total Nonstructural Carbohydrates. As mentioned previously, utilization of nonstructural carbohydrates by ruminants is nearly complete. Therefore, in many studies of forage quality, extensive fractionation of nonstructural carbohydrates is not performed. Procedures for determining the concentration of total nonstructural carbohydrates (TNC) in forages have been developed which are generally used in studies of this nature (Smith, 1981). Total nonstructural carbohydrates can be extracted using acid or enzymes to hydrolyze starch as described above with the exception that these are applied to whole tissues. When using enzymes to extract TNC from fructan-accumulating species, it is necessary to hydrolyze the fructans with dilute acid prior to analysis for monosaccharides (Smith, 1981).

Extraction and Fractionation of Structural Carbohydrates

Cell Wall Isolation. For compositional analysis, wall preparation methods should have common features of inactivation of hydrolytic enzymes, particle size reduction, and removal of cytoplasmic contaminants (Aspinall, 1982; Masuda et al., 1982; Harris, 1983; Selvendran et al., 1985; Wilkie, 1985). The Van Soest detergent system (Van Soest et al., 1991) is the most widely used method for extraction of cell wall constituents in studies of forage quality (Figure 5). The detergent system, however, was not developed as a method for isolating cell walls per se, but rather as a method of partitioning forage dry matter into fractions based upon its bioavailability to ruminants (Van Soest, 1982). Consequently, the neutral detergent fiber (NDF) fraction which often is used synonomously with cell wall is actually a subfraction of the cell wall since the more soluble and nutritionally available wall polysaccharides are removed. The shortcomings of the detergent system as a method of isolating intact cell walls should not be taken as a criticism of the technique in general. For many research objectives, particularly those related to nutritional value, neutral detergent extraction may be the most appropriate method for isolating cell wall constituents.

The scheme illustrated in Figure 6 is a composite of several procedures commonly used to isolate plant cell walls. Options shown in the figure may be used depending upon the type of starting plant material and types of wall analysis to be performed on the isolated fractions. Detergents such as sodium dodecyl sulfate (SDS) and sodium deoxycholate (SDC) help remove cytoplasmic proteins but should be used with caution as they can solubilize portions of the wall matrix. This also will be true for treatments such as dimethyl sulfoxide (DMSO) that effectively removes starch from wall preparations (Selvendran et al., 1985) along with some matrix polysaccharides.

An alternative method of preparing plant samples is the Uppsala method developed by Theander and co-workers that is relatively rapid for preparing alcohol-insoluble residues(AIR) (Figure 7) (Theander, 1991). Parts of this method are contained within the scheme shown in Figure 6. For samples that contain significant levels of protein, the AIR also will contain elevated levels of proteins. However, if the analysis is concerned with carbohydrate composition, the protein contamination will not be a problem. The ease of sample preparation allows for handling several samples at one time.

No matter the choice of wall isolation method, removal of starch is necessary, particularly for samples containing leaf tissues. Enzyme methods work effectively as long as certain precautions are taken. Amylases from microbial and fungal sources are typically contaminated with wall hydrolytic enzymes. However, the heat stable α-amylase from *Bacillus licheniformis* and amyloglucosi-

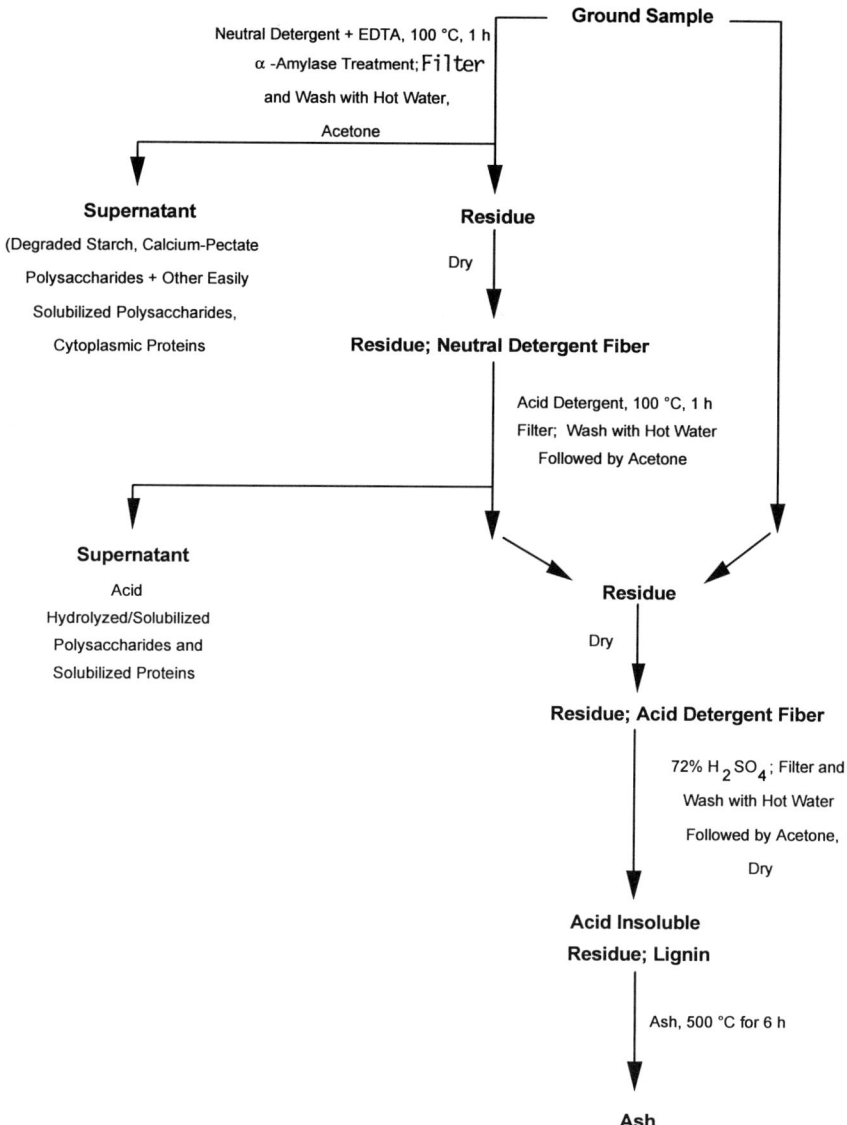

Figure 5. Schematic diagram of the detergent system of forage analysis.

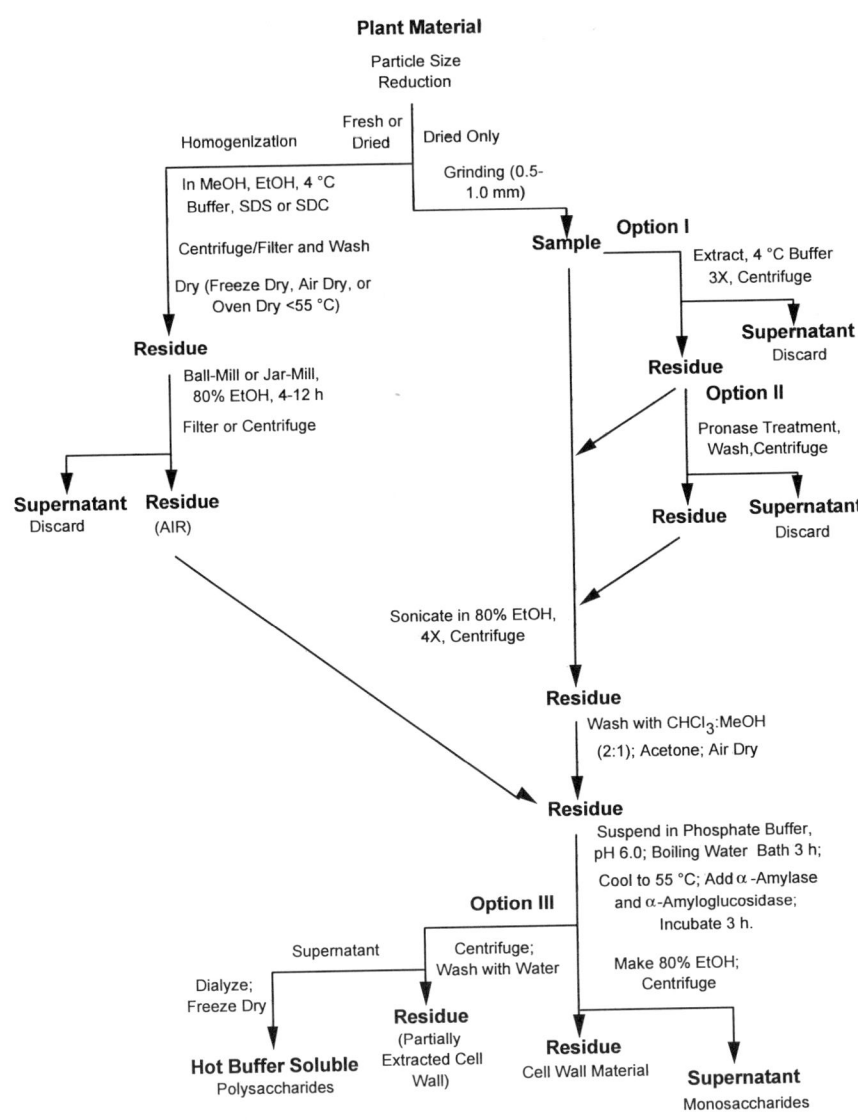

Figure 6. Isolation schemes for cell wall preparations.

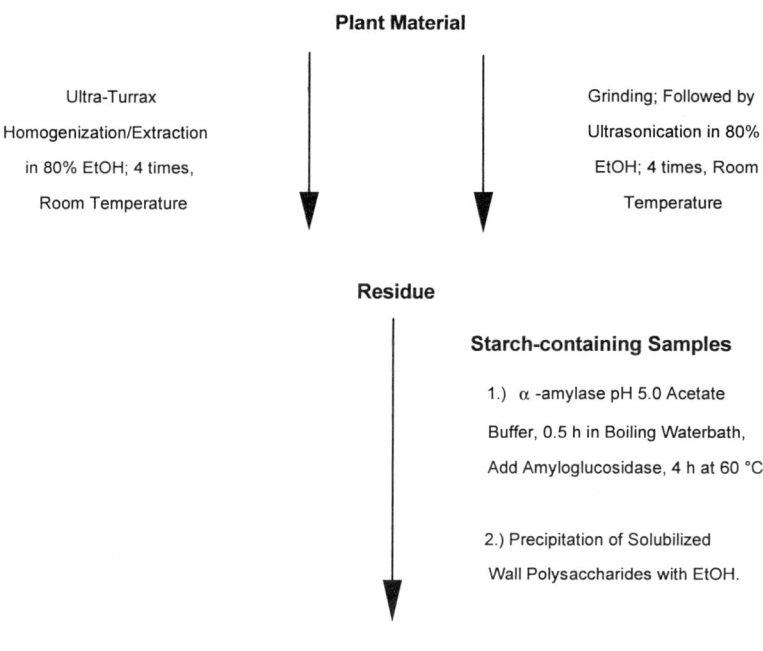

Figure 7. Brief outline of the Uppsala Method for preparing alcohol-insoluble residues (AIR).

dase from *Asergillus niger* are suitable for destarching walls as long as the incubation temperature is maintained at or above 55°C; incubation at this temperature inactivates all other hydrolytic enzymes while the amylolytic activity remains high. A safer, though more expensive alternative, is to use hog pancreas or human saliva amylases as they do not contain hydrolytic activities against wall polysaccharides.

Polysaccharide Isolation. Polysaccharide isolation is frequently based on a chemical extraction sequence that takes advantage of the different physicochemical properties of polysaccharides (Figure 8) (see reviews by Aspinall, 1982; Masuda et al., 1982; Selvendran et al., 1985; Wilkie, 1985). The progression from milder to harsher extraction conditions (hot water to 4 M KOH) effectively fractionates the wall into groups of polysaccharides with similar solubilities, but not necessarily similar compositions. For example, hot buffer extracts of bromegrass leaf walls contain at least three types of polysaccharides; (1→3),(1→4)-β-D-glucans, arabinoxylans, and rhamnogalacturonans (Hatfield, 1993b). Treatments such as acidic Na_2ClO_2 that are designed to remove lignin and render more of the wall matrix extractable solubilize a portion of the wall

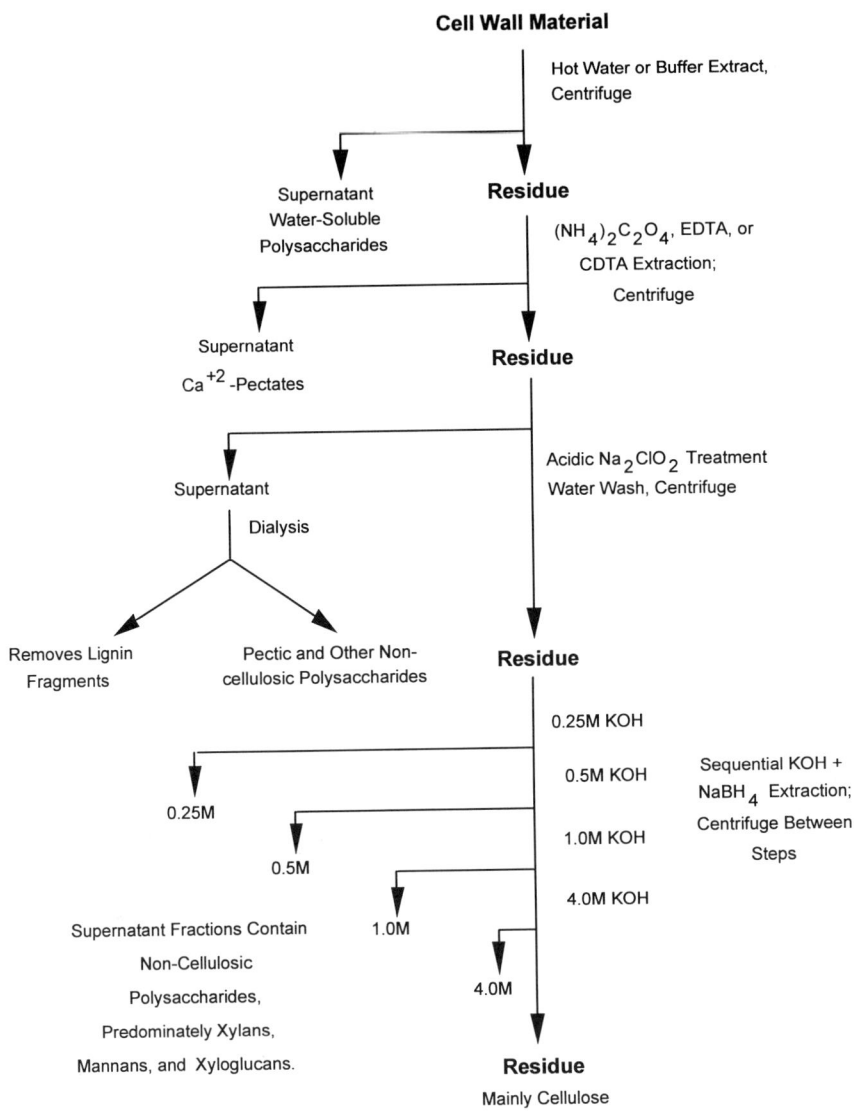

Figure 8. A typical wall fractionation scheme.

that should be reclaimed rather than discarded. For some forage samples, this can be a significant amount of the wall matrix (Hatfield, 1992). The addition of multiple steps in the KOH extraction, compared to the traditional 5% and 24% KOH extractions, does a reasonable job of fractionating polysaccharides that contain progressively less branching and, hence, lower solubilities (Carpita, 1984). A critical component of KOH extractions is the addition of $NaBH_4$ that prevents alkaline degradation of polysaccharides by reduction of the reducing terminal sugar.

A more selective method of wall fractionation is the use of highly purified hydrolases to specifically degrade a polysaccharide of interest. This technique has been applied to a number of wall samples using a range of hydrolases (Nishitani and Nevins, 1988, 1989; O'Neill et al., 1990). The key to the success of this approach is the purity of the enzyme. If the enzyme preparation is pure, the only material released will be from a specific type of polysaccharide (e.g., endopolygalacturonase releases oligosaccharides from polygalacturonans). The selection of the appropriate enzyme(s) is dependent upon the sample and the information sought. For example, some endoglucanases will hydrolyze both $(1\rightarrow 3),(1\rightarrow 4)$-$\beta$-D-glucans and xyloglucans (Streptomyces $(1\rightarrow 4)$-β-D-glucan endohydrolase; Barras et al., 1969) whereas, the $(1\rightarrow 3),(1\rightarrow 4)$-$\beta$-D-glucan endohydrolase from *Bacillus subtilis* is highly specific for the mixed linked β-glucan (Anderson and Stone, 1975). A disadvantage of enzyme fractionation is the loss of certain types of information; no estimate of the molecular mass of the polysaccharides is obtained, individual diversity of polysaccharides within a given group may be obscured, and not all of the polysaccharides of a given type will be released from the wall matrix by a single enzyme treatment. However, the fine structural detail of individual oligosaccharides can be determined and has led to detailed structural characterization of some complex wall polysaccharides (O'Neill, et al., 1990). Purified hydrolases also are useful for the structural characterization of isolated polysaccharide fractions (Hatfield, 1992; Hatfield et al., 1993).

Fractionation of polysaccharides. Some of the common means of fractionating polysaccharides isolated from plants are presented in Table 2. A detailed discussion of these various techniques will not be addressed here. Suffice it to say that the most frequently used methods include alcohol precipitation and anion exchange and gel filtration chromatography (the reader is encouraged to see more extensive reviews by Aspinall, 1982; Masuda et al., 1982; Wilkie, 1985). The fractionation of polysaccharide mixtures to homogeneity is rarely achieved in a single step, but rather the result of multiple techniques applied several times to a mixture. Judgment of homogeneity is based on the observation that additional fractionation procedures do not result in a compositional shift (monosaccharide and linkage composition).

An exception to the multiple step procedures is the use of affinity chromatography with a lectin, a carbohydrate binding protein, bound to a column matrix (Aspinall, 1982). Because lectins are highly specific for individual sugars of a given glycosidic configuration, only those molecules containing the specific

Table 2. Techniques used to fractionate polysaccharides from aqueous solutions.[†]

Technique	Reagents/Media	Polysaccharides Fractionated
Precipitation	Acetone	All types
	Ammonium sulfate	(1-3)(1-4)-β-D-glucans from cereal gums
	Barium hydroxide	Glucomannans, highly branched galactans
	Ethanol	All types
	Calcium salts	Acidic (e.g., Polygalacturonans)
	Copper reagents	Acidic and xylans, xyloglucans
	Fehling's solution	
	Cupric acetate	
	CTAB acidic	
	I_2/KI in $CaCl_2$	Highly branched xylans from linear xylans
	Lectins	Dependent upon specific lectin used
Chromatography		
Ion-exchange	DEAE	Separates acidic from neutral
Gel-filtration	2000kD to 100kD	Polysaccharides differing in molecular mass
	50kD to 5 kD	Polysaccharides from oligosaccharides
	DP 10 to DP 1	Separation of oligosaccharides
Hydroxyapatite		Glycoproteins
Affinity	Lectins	Dependent upon specific lectin used
	Antibodies	Antibody specific
Electrophoresis		Polygalacturonans
Enzyme Hydrolysis	Purified enzymes	Degrades a specific type of polysaccharide

[†] These techniques have been used in a wide range of applications, but the references listed are for reviews that may have compared several of these techniques. This is by no means a complete list of possible fractionation procedures but a general sampling (See reviews by Aspinall, 1982; Masuda et al., 1982; Selvendran et al., 1985; Wilkie, 1985).

sugar residue are retained on the column. This same principle also can be applied to the selective precipitation of polysaccharides by lectins although, generally, other polysaccharides are co-precipitated. Although the potential for highly selective purification of polysaccharides can be achieved with immobilized lectins, this technique has not been extensively used. The use of antibodies raised against

Table 3. Summary of colorimetric assays used for carbohydrate analysis.[†]

Assay	Sensitivity (μg/assay)	Wavelength (nm)	Comments
Hexose			
Phenol-sulfuric acid	10-200	490	Stable and easy
Anthrone	10-200	625	Variable response[‡]
Orcinol-sulfuric acid	10-100	420	Max 505 nm
Pentose			
Phenol-sulfuric acid	10-100	480	Hexoses abs[§]
Oricinol-HCl	4-40	660	Hexoses abs
Reducing Sugars			
Alkaline copper sulfate	5-80	500	Proteins can interfere
3,5-dinitrosalicylic acid	50-500	570	Lower sensitivity
2,2'-Bicinchoninate	1-5	520	Protein interference[¶]
PAHBAH[#]	1-100	410	Chromagen stability
Hexuronic acids			
Carbazole-sulfuric acid	5-150	525	Hexose interference
3,5-dimethylphenol	10-100	450	GalA from GlcA[††]
3-phenylphenol	1-200	520	Low hexose interference

[†] Most of the assays listed here have been modified since their original development. To avoid a large reference list, only select ones that describe the procedure with modifications will be given.

[‡] The anthrone method is sensitive but gives variable results depending upon the individual sugars present in the sample. Advantage has been taken of this observation in an attempt to develop sugar specific assays.

[§] There is a strong overlap in absorbance between pentoses and hexoses.

[¶] One of the most sensitive reducing sugar assays but suffers from interference from other compounds, particularly proteins, that can overestimate carbohydrate content.

[#] p-hydroxybenzoic acid hydrazide.

[††] Can be used to distinguish between galactopyranuronic acid (Ga/pA) and glucopyranuronic acid (Glc/pA) acids but not between Ga/pA and 4-O-methyl-Glc/pA.

specific polysaccharides or portions of a polysaccharide is another high affinity technique that may see increased exploitation in the future (Hoson, 1991). Disadvantages of this technique are the high cost, the need for a purified polysaccharide (or oligosaccharide), and the potential low antigencity of polysaccharides.

Table 4. Enzyme-coupled assays for monosaccharide analysis.

Assay	Sugars detected	Sensitivity mg ml^{-1}	Detection method/ Chromophore
Glucose oxidase	Glucose	0.5-50	H_2O_2/Peroxidase
Galactose oxidase	Galactose	0.5-50	H_2O_2/Peroxidase
Hexokinase-Glc 6-phosphate dehydrogenase	Glucose	0.5-60	NADPH
H-G6PD + PGI	Fructose	0.5-60	NADPH
H-G6PD + PMI	Mannose	0.5-60	NADPH
Galactose dehydrogenase	Galactose	0.5-50	NADH
Galactose dehydrogenase	Arabinose		NADH

† These assays can be quite specific and have a high degree of sensitivity.

Characterization of plant carbohydrates. Characterization of carbohydrates must include determining the amounts and types present in a given sample. Generally, it is not possible to accurately determine the amount of total carbohydrate in a sample by simple gravimetric methods because of contaminating molecules (e.g., proteins, ion-complexes) that are not easily removed. Measurement of total carbohydrate can be accurately determined using appropriate colorimetric assays (Table 3). Although a large number of assays are available, some may not be appropriate for particular samples because of the potential contaminants present. The phenol-sulfuric acid (Dubois et al., 1956) and the 3-phenyl phenol (Blumenkrantz and Asboe-Hansen, 1973) methods are a good combination for determining the total carbohydrate in a sample. Both methods are rapid, show low interference from non-carbohydrate components, and have good sensitivities and linear responses.

The next level of analysis is to determine the monosaccharide composition. Unfortunately, there are no simple colorimetric assays available to identify individual monosaccharides. There are a few enzyme-coupled assays that are specific for some sugars (Table 4) and these work well to identify the sugar of interest in complex mixtures. One such system is the glucose oxidase assay for determining the amount of Glc*p* in a sample (Table 4). The assay is based on the following reactions:

1) Glc*p* + H_2O + O_2 + glucose oxidase → Gluconate + H_2O_2 + glucose oxidase

2) H_2O_2 + *O*-dianisidine + peroxidase → H_2O + oxidized *O*-dianisidine (brown) + peroxidase

There is a 1:1 molar relationship between the amount of O-diansidine oxidized and Glc$p$ in the original sample. This coupled assay is a convenient method for determining free Glc$p$. It can be incorporated into the cell wall isolation scheme (Figure 6) that employs an α-amylase/amyloglucosidase treatment to determine the amount of starch in the original sample. There are a few other enzyme-based systems for determining the amount of a specific sugar of a sample (Table 4) as long as the sugar of interest is in the monosaccharide form. In some cases, the enzyme will react with the specific sugar in oligosaccharides or polysaccharides if it is in a non-reducing terminal position.

Frequently, the method of choice for compositional analysis of a sample is acid hydrolysis followed by gas liquid chromatography of the monosaccharides as alditol acetate derivatives (Blakeney et al., 1983; Selvendran et al., 1985; Wilkie, 1985). For most non-cellulosic polysaccharides, hydrolysis in 2M trifluroacetic acid (TFA) at 120°C for 2 h or 1M H_2SO_4 at 100°C for 2.5-3 h is sufficient to release all the monosaccharides. An advantage of TFA is easy removal of excess reagent by evaporation under a stream of nitrogen or dry filtered air. For cellulose and some recalcitrant non-cellulosic polysaccharides (linear xylans or mannans), it is necessary to solubilize the sample in 12M H_2SO_4 (1-4 h) followed by secondary hydrolysis in dilute acid (1M, 100°C 3-5 h). This latter procedure can be applied to the total wall sample to obtain a complete wall composition at least in terms of the neutral sugars present.

Likewise, high performance liquid chromatography (HPLC) techniques have been developed for the identification of monosaccharides, but generally these have suffered from poor resolution and low sensitivity. However, recent development of pulsed amperometric detectors (PAD) coupled with high resolution carbohydrate columns (based on anion exchange media) now rival GLC-flame ionization detector (FID) methods (Hotchkiss and Hicks, 1990). The advantage of HPLC is there is no need to derivatize the sample. However, to maintain optimal resolution and sensitivity, all ions must be removed. This is particularly true if H_2SO_4 is used for hydrolysis; samples must be passed through an anion exchange column to remove SO_4^{-2}. Trifluroacetic acid again has an advantage in that, depending upon the starting material, it may be possible to simply evaporate the TFA followed by dissolving the sample in deionized H_2O. The sample should still be passed through a micron filter to remove any particulates that would clog the HPLC column.

Quantitative identification of individual uronic acids is difficult. Total uronics can be determined by colorimetric analysis, as already discussed, but the identification of individual types of uronosyls (e.g., Gal$p$A, Glc$p$A, and 4-O-Methyl-Glc$p$A) cannot be completely accomplished by colorimetric assays (Scott, 1979; Table 3). Several factors must be considered for correct quantitative determination of uronic acids: incomplete hydrolysis, degradation losses through decarboxylation in acid solutions, and difficulties encountered in the formation of suitable derivatives for GLC analysis (Selvendran and DuPont, 1984; Selvendran et al., 1985). The latter problem can be overcome using new HPLC techniques that separate the different uronic acids. A major problem is the resistance to acid hydrolysis of uronosyl-sugar bonds (e.g.,

2-O-(4-O-methyl-α-D-Glc$p$A)-D-Xyl$p$, or any other uronic acid bonded to a neutral sugar). For soluble polysaccharides, uronosyls can be converted to the corresponding neutral sugars (e.g., Gal$p$A to Gal$p$; Glc$p$A to Glc$p$) using NaBH$_4$ to reduce the carbodiimide-activated carboxyl group (Taylor and Conrad, 1972). The resulting neutral polysaccharide, now containing only glycosidic linkages, is easily hydrolyzed and the neutral sugar composition determined as the alditol acetates. This procedure is limited to isolated polysaccharides and oligosaccharides.

Selvendran and DuPont (1984) suggested that polysaccharides be defined and classified according to the following properties: 1) monosaccharide composition; 2) positions of glycosidic linkages (e.g., 1→2, 1→3, 1→4, etc.); 3) monosaccharide ring form (furanose or pyranose); 4) configuration of glycosidic linkages (α or β); and 5) the degree of branching on the main chains. A major portion of this information can be obtained through methylation analysis. Briefly, the technique involves methylation of all free hydroxyls on a polysaccharide, followed by acid hydrolysis and conversion to partially methylated alditol acetates that are analyzed by capillary GLC (Harris et al., 1984; Selvendran et al., 1985; Carpita and Shea, 1989). Using this technique, it is possible to determine the monosaccharide composition, deduce the ring form (in many cases), and determine glycosidic linkages present in a polysaccharide. Reduction of the monosaccharides, with NaBH$_4$ after acid hydrolysis, labels the reducing end of the sugar and subsequent analysis with GLC-mass spectrometry (GLC-MS) allows identification of linkage patterns that might otherwise be indistinguishable by normal GLC-FID analysis (Carpita and Shea, 1989).

Other Types of Analysis. Analysis of carbohydrates has been achieved to varying degrees of success using direct spectroscopic methods. Most of the references to spectroscopy already mentioned have been limited to UV-visible spectra after the formation of a suitable chromophore (Table 3). Mass spectrometry is a powerful technique but is usually limited to monosaccharides or small oligosaccharides. Nevertheless, with the development of fast atom bombardment (FAB) techniques and coupling mass spectrometers with HPLC's, increased structural information can be obtained from even larger oligosaccharides. Infrared and Raman spectroscopy also provides useful information about polysaccharides but are generally of limited value in determining structural details (Perlin and Casu, 1982). A related technique, near infrared spectroscopy (NIR), has been gaining use for the analysis of forage samples. The technique represents a useful means of making relative comparisons of samples and provides an adequate means of quickly ranking samples. However, the spectral information cannot be used as an absolute measure of wall composition (monosaccharide components).

One of the most information-rich techniques for analysis of carbohydrates is nuclear magnetic resonance (NMR) spectrometry (Perlin and Casu, 1982). When a range of NMR experiments are performed on a sample, it is possible to determine the monosaccharide composition, glycosidic configurations (α or β), glycosidic linkages, ring form, and the relative abundance of each. Nuclear magnetic resonance has the added advantage of being non-destructive such that the sample can be subsequently subjected to other forms of analysis. The difficulty with NMR analysis is the need to solubilize the sample in an appropriate solvent, and its relative insensitivity in most cases, requiring on the order of several milligrams for high resolution analysis. Oligosaccharides can be analyzed effectively but polysaccharides are much more difficult, principally due to the complexity of the sample. As individual sugars that make up a complex polysaccharide may be similar, many signals may not completely resolve, resulting in a series of broad peaks that are difficult to interpret. Nevertheless, advances in NMR technology will undoubtedly lead to increased resolution and sensitivity of both liquid and solid state techniques and will find ever-increasing applications to the study of polysaccharides.

CARBOHYDRATES IN FORAGES

The concentrations and classes of carbohydrates isolated from forages vary greatly among genera, species and, to some extent, within species. Nearly all important cultivated and many native forages belong to either the grass family (*Poaceae*) or the legume family (*Fabaceae*), which are of the subclasses Monocotyledonae and Dicotyledonae, respectively. It is convenient to generalize about the carbohydrate composition of grasses and legumes, or monocots and dicots, but it should be realized that there is a large amount of variation in carbohydrate composition among forage species within these groupings.

There is a tendency to think of forages, particularly in regard to their nutritive value, as homogenous mixtures of nutrients. In fact, forages are a heterogenous complex consisting of different organs, tissues, cells, and subcellular components which vary in carbohydrate concentration and composition. The carbohydrate concentration of total herbage represents the relative contribution of each of these components and it is generally difficult to interpret changes in carbohydrate composition of total herbage because of this. The influence of environmental variables such as temperature and soil moisture on forage carbohydrate composition often is confounded with plant maturity and, therefore, morphology and anatomy.

The discussion of forage carbohydrates that follows is, by necessity, too brief to discuss all of the genetic and environmental influences on carbohydrate composition of forages in detail. These topics are discussed in more detail in Chapters 3, 4, 18, and 22.

Nonstructural Carbohydrates

The major roles of nonstructural carbohydrates in plants are in intermediary metabolism, energy transport, and energy storage. Each of these functions is dynamic, and concentrations of nonstructural carbohydrates vary greatly with the physiological status of the plant. Carbohydrates involved in intermediary metabolism are in a constant state of flux. Concentrations of transport and short-term storage carbohydrates vary diurnally, generally peaking in late afternoon (Lechtenberg et al., 1971, 1972). Some of the carbon fixed through photosynthesis is stored temporarily as sucrose or as starch within leaf cells. At night, these carbohydrates are mobilized and transported throughout the plant to meet the energy demand of cellular respiration. Longer-term carbohydrate storage occurs in specialized tissues within organs such as seeds, rhizomes, and stem bases. Concentrations of these storage carbohydrates tend to vary with seasonal cycles related to cold hardening, tillering, and regrowth following defoliation (Smith, 1973, 1981).

Nonstructural carbohydrates are rapidly fermented by ruminal microbes and can represent a substantial amount of the energy obtained by ruminants from consuming forages. However, quantitatively, they are less important than structural carbohydrates as a source of energy for ruminants consuming forage diets.

Nonstructural carbohydrates have a significant impact on forage quality in other ways. They are important substrates for reactions important in the postharvest physiology and preservation of forages. Respiratory enzymes continue to function for some time after forages are cut for hay and silage, converting nonstructural carbohydrates to energy and carbon dioxide. Under poor drying conditions, DM losses due to continued respiration of nonstructural carbohydrates can be significant, resulting in a loss of digestible energy (Moser, 1980). Nonstructural carbohydrates are the primary substrates for microorganisms that grow on poorly preserved forages during storage. In unfavorable storage conditions, microbial growth can result in significant losses in forage quality through molding and heating. Nonstructural carbohydrates are also substrates in the nonenzymatic browning reaction that occurs as the result of heating during storage (Van Soest and Mason, 1991). This reaction, also called the Maillard reaction, occurs between carbohydrates, amines, and amino acids in the presence of heat and water. The resulting polymer is similar to lignin in properties and thus the reaction renders the involved substrates partially unavailable to the ruminant animal (Van Soest, 1982). Silage preservation also is affected by concentrations of nonstructural carbohydrates. Proper ensiling is dependent upon the fermentation of nonstructural carbohydrates to lactate and other organic acids which eventually accumulate at high enough concentrations to inhibit further microbial activity. If initial concentrations of carbohydrates are too low to achieve stability, secondary fermentation reactions can occur which result in losses in silage quality (Moser, 1980; McDonald, 1981).

The monosaccharides, glucose and fructose, and the disaccharides, sucrose and maltose, account for most of the soluble sugars in forages. Smaller concentrations of arabinose, xylose, ribose, galactose, mannose, and myo-inositol are also generally present in extracts (McIlroy, 1967; Volenec and Nelson, 1984). Sucrose and other soluble sugars are generally present in higher concentrations in legumes than grasses and are usually higher in cool-season grasses than warm-season grasses (Table 5).

Starch is the predominant storage polysaccharide in legumes (Table 5). In most legume species, starch is accumulated in the roots although in some such as

Table 5. Typical concentrations of carbohydrates in temperate legumes, and cool and warm-season grasses (Adapted from Van Soest, 1982).

Category	Temperate Legumes	Cool-Season Grasses	Warm-Season Grasses
Nonstructural Carbohydrates	\- g kg^{-1} DM \-		
Soluble sugars	20 - 50	30 - 60	10 - 50
Starch	10 - 110	0 - 20	10 - 50
Fructans	-	30 - 100	-
Structural Carbohydrates			
Cellulose	200 - 350	150 - 450	220 - 400
Hemicelluloses	40 - 170	120 - 270	250 - 400
Pectin	40 - 120	10 - 20	10 - 20

white clover [*Trifolium repens* L.], it is stored in stolons. The leaves of legumes generally contain lower concentrations of free sugars than do stems, but generally have higher starch concentrations (Smith, 1973).

Grasses may be either starch or fructan accumulators depending on the species. Tropical and subtropical grasses typically accumulate starch (Table 5). Amylopectin is the main starch stored in most grasses and generally comprises 75% of the total starch concentration in grasses (McIlroy, 1967). Amylose, however, is the primary starch stored in stems of corn [*Zea mays* L.] and sorghum [*Sorghum bicolor* (L.) Moench.] (Bailey, 1973; Smith, 1981).

Species within the Festucoideae subfamily, which accounts for most temperate forage grasses, store fructans primarily of the phlein type (Table 5). Species that accumulate fructans in vegetative tissues generally store starch in their seeds (Pontis and Del Campillo, 1985). The D.P. of fructans varies with species. Species which accumulate long-chain fructans (D.P. > 90) include members of the Aveneae tribe and some species of the Festuceae tribe including orchardgrass [*Dactylis glomerata* L.] and kentucky bluegrass [*Poa pratensis* L.]. Other species of the Festuceae tribe, including smooth bromegrass [*Bromus inermis* Leyss.], tall fescue [*Festuca arundinacea* Schreb.], and perennial ryegrass [*Lolium perenne* L.], as well as members of the Hordeae tribe, accumulate short-chain fructans (D.P. < 60) (Smith, 1973, 1981; Meier and Reid, 1982).

Concentrations of fructans are generally highest in roots, rhizomes, and stem bases and usually low in leaves (Pontis and Del Campillo, 1985). This is in contrast to warm-season grasses which have higher concentrations of storage carbohydrates (starch) in leaves than in stems.

Structural Carbohydrates

There is considerable variation among plant species with respect to both the concentration and composition of structural carbohydrates (Table 5). Cellulose concentrations are typically higher in the walls of legumes than in grasses. This reflects a much lower concentration of hemicelluloses in legumes compared to grasses. Buxton et al. (1987) found that the ratio of hemicelluloses:cellulose ranged from 0.57 to 0.70 in mature temperate grasses and from 0.32 to 0.40 in mature legumes. On a DM basis, cellulose concentrations often appear similar between grasses and legumes, reflecting the lower concentration of cell walls in legumes (Table 5).

Warm-season perennial grasses generally have higher structural carbohydrate concentrations than cool-season grasses (Table 5). Windham et al. (1983) conducted a comparative study on the fiber composition of three warm and three cool-season grasses. Cellulose concentrations were higher in the warm-season grasses, averaging 319 g kg^{-1} DM compared to 251 g kg^{-1} DM for the cool-season grasses. Hemicellulose concentrations averaged 354 g kg^{-1} DM and 245 g kg^{-1} DM, respectively. There was no clear trend with respect to the ratio of hemicelluloses to cellulose between warm and cool-season grasses which ranged from 0.76 for bermudagrass [*Cynodon dactylon* (L.) Pers.] to 1.18 for orchardgrass. Cherney et al. (1988) determined structural carbohydrate concentrations of several cool-season and warm-season grasses in a study to evaluate their biomass fuel potential. Cellulose concentrations ranged from 254-285 g kg^{-1} DM for tall fescue and reed canarygrass [*Phalaris arundinacea* L.] and from 312-334 g kg^{-1} DM for switchgrass. However, the trend was reversed with annual grasses. Winter rye [*Secale cereale* L.] had higher cellulose concentrations (344-348 g kg^{-1} DM) than did sweet sorghum and a sorghum x sudangrass hybrid (244-285 g kg^{-1} DM). Hemicellulose concentrations followed similar trends.

Hemicelluloses consist of complex heteropolymers that vary considerably in primary composition, substitutions, and degree of branching (Figures 2 and 3). Arabinoxylans and xyloglucans are the predominant polysaccharides comprising hemicelluloses of monocots (Wilkie, 1979; Chesson and Forsberg, 1988). In dicots, arabinoxylans, xyloglucans, arabinans, and galactans are the major hemicellulosic polysaccharides (Chesson and Forsberg, 1988; Åman and Graham, 1990). Glucomannans and galactoglucomannans also have been isolated from dicot hemicelluloses, but quantitatively are of minor importance (Bailey, 1973). In both monocots and dicots, arabinoxylans are substituted with glucuronic acid and 4-O-methylglucuronic acid to varying degrees, forming glucuronoarabinoxylans. Arabinoxylans also may be substituted with galactose and acetyl groups. Xyloglucans of dicots often are substituted with galactose and fucose (Hatfield, 1989).

Hydrolysis of forage hemicelluloses yields the neutral monosaccharides glucose, xylose, arabinose, mannose, galactose, rhamnose, and fucose, and the uronic acids, galacturonic, glucuronic, and 4-*O*-methylglucuronic (Wilkie, 1979; Chesson and Forsberg, 1988; Åman and Graham, 1990). The relative proportions of each monosaccharide varies among species reflecting differences in polysaccharide structures.

Xylose and arabinose account for most of the neutral sugars isolated from hemicelluloses of grasses and legumes (Collings and Yokoyama, 1979; Wedig et al., 1987). Buxton et al. (1987) conducted comparative studies on structural neutral sugars isolated from stem bases of legumes and grasses. Glucose (predominantly of cellulose origin) and xylose accounted for 67% and 20% of the neutral sugars hydrolyzed from cell wall in legumes and 62% and 30%, respectively, in grasses. Arabinose concentrations were similar (approximately 4.4%) for grasses and legumes, but legumes had higher concentrations of galactose, mannose, and rhamnose.

There appears to be no clear distinction between the neutral sugar composition of structural carbohydrates of warm and cool-season grasses. Windham et al. (1983) found similar concentrations of hemicellulosic neutral sugars among three warm and three cool-season grasses when adjusted for recoveries.

UTILIZATION OF FORAGE CARBOHYDRATES

Nonstructural Carbohydrates

Utilization of nonstructural carbohydrates by ruminants is nearly complete. Sucrose and other soluble sugars are rapidly fermented within the rumen to yield volatile fatty acids. These acids are absorbed into the blood through the rumen wall and serve as the primary substrate for energy metabolism in the ruminant. Nonstructural polysaccharides must first be degraded to simple sugars before they can be fermented (Morrison, 1979).

Starch is hydrolyzed to maltotriose, maltose, and glucose by a variety of amylases. Amylases are present in saliva (nonruminant herbivores) and are produced by ruminal bacteria, protozoa, and fungi (Hoover and Stokes, 1991). Bacterial amylases account for most starch digestion. The predominant amylolytic microbes are *Bacteroides amylophilus, Succinivibrio dextrinosolvens, Selenomonas ruminantium,* and *Streptococcus bovis* (Morrison, 1979; Baldwin and Allinson, 1983).

Most starch digestion occurs within the rumen for livestock consuming forage diets (Nocek and Tamminga, 1991). However, ruminants have the capacity for significant postruminal digestion of starch (Armstrong and Smithard, 1979; Owens et al., 1986). Starch degradation occurs in the small intestine through the activities of pancreatic amylase and intestinal maltase and isomaltase (Nocek and Tamminga, 1991). Most starch escaping ruminal fermentation is digested within the small intestine. That starch which reaches the large intestine is readily fermented. Degradation of starch reaching the large intestine is generally in the range of 70 to 100% (Hoover, 1978).

Fructans are easily and rapidly degraded by ruminal microbes (McIlroy, 1967; Armstrong and Smithard, 1979). Specific fructanohydrolases have not been purified from ruminal microorganisms, but are presumed to exist. Invertases obtained from ruminal microbes have been demonstrated to completely degrade grass fructans by sequential hydrolysis of the terminal fructose residue (Chesson and Forsberg, 1988).

Structural Carbohydrates

As noted in the introduction, the degradability of cell wall polysaccharides varies considerably both within and between classes. Traditionally, studies on polysaccharide digestion have focused on isolated polymers. This reductionist approach has provided insight into the basic mechanisms by which cell wall polysaccharides are degraded, but is inadequate for studying the processes by which forage cell walls are degraded *in situ*. Chesson (1982) recognized that the degradation of cell wall polysaccharides is affected as much or more by interactions among cell wall polymers as by the individual properties of the polymers themselves. He advocated a holistic approach to studying cell wall degradation that takes into account these complex interactions. The discussion that follows will describe basic aspects of polysaccharide degradation by ruminal anaerobes and attempts to provide an overview of the interactions which occur among cell wall polymers with respect to their digestion.

Cellulose is degraded in the rumen by a complex of anaerobic microorganisms that include bacteria, protozoa, and fungi. Cellulolytic bacteria, of which *Ruminococcus flavefaciens, R. albus,* and *Fibrobacter succinogenes* are most important, are responsible for most cellulose digestion that occurs in the rumen (Bryant, 1973). Although ciliate protozoa and fungi having cellulolytic activity have been identified in ruminal microbial populations, their contribution to cellulose degradation is relatively minor (Dehority, 1993).

The cellulolytic bacteria adhere to the surface of the cell wall, placing the enzymes in close proximity to the substrate (Pell and Schofield, 1993; White et al., 1993). Cellulolysis is achieved by the action of several extracellular enzymes that are either bound to the surface of the organism or secreted into the surrounding medium. Three basic enzymatic activities are involved: endo-β-1,4-glucanase which randomly cleaves the polysaccharide into oligosaccharides; exo-β-1,4-glucanase which attacks the nonreducing end of the oligosaccharides, yielding cellobiose; and β-1,4-glucosidase, which hydrolyzes cellobiose to glucose (White et al., 1993).

The extent to which native cellulose is utilized by ruminal microorganisms is limited by its association with lignin and other cell wall constituents. Purified celluloses are, for the most part, completely degraded by ruminal microbes (Van Soest, 1973). There are, however, intrinsic factors which may limit the rate at which cellulose is digested. Crystallinity of cellulose has been suggested as a factor in reducing accessibility of cellulose to enzymatic attack (Kerley et al., 1988; Coughlin, 1992). Cellulose degradation has been shown to be inversely proportional to degree of crystallinity for purified substrates (Cowling, 1975). However, at present, there is little evidence to indicate that crystallinity is a rate-limiting factor in the degradation of native celluloses by ruminal microbes (Hatfield, 1993b).

Hemicellulose degradation in the rumen occurs in an analogous manner to that of cellulose, but involves a broader array of enzyme activities. The same cellulolytic bacteria listed above that are responsible for most cellulose degradation in the rumen are also the most important hemicellulolytic bacteria (Hespell, 1988). In addition to these, *Butyrivibrio fibrisolvens*, which has a relatively minor role in cellulose degradation, has a proportionally greater role in xylan degradation (White et al., 1993). Some ruminal fungi and protozoa also have hemicellulolytic activity, but their activity in hemicellulose degradation is relatively minor compared to ruminal bacteria (Dehority, 1993).

Isolated hemicelluloses are generally completely digestible by ruminal microorganisms. Degradation of hemicelluloses occurs through the activities of endo- and exo-glycanases which depolymerize and solubilize the main polysaccharide chains (White et al., 1993). Substituent groups and sidechains are removed from hemicellulosic polysaccharides and further degraded through the activities of a number of specific glycosidases (Dehority, 1993).

The degradability of native hemicelluloses in intact cell walls is variable and generally incomplete. The digestibility of hemicelluloses varies with its composition. Xylose residues are generally reported to be less digestible than arabinose in forage cell walls (Buxton et al., 1987). Consequently, the ratio of xylose:arabinose often has been negatively correlated with forage digestibility and has been suggested as an index of forage quality. The ratio of xylose to arabinose presumably reflects the degree of arabinose substitution along the xylan backbone of arabinoxylans (Albrecht et al., 1987).

Pectins are rapidly and completely degraded within the rumen through the activity of pectin lyases and pectin esterases (Chesson and Forsberg, 1988; White et al., 1993). The relative ease of pectin degradation is a function of its solubility and accessibility to pectinolytic enzymes. Many cellulolytic bacteria have pectinolytic activities, but cannot utilize the resulting oligogalaturonides and galacturonic acid to support their own growth. In addition to the predominant cellulolytics, *Lachnospira multiparus* is considered to be important as a pectinolytic organism (Dehority, 1993).

Phenolics and Carbohydrate Utilization

Cell Wall Phenolics

Digestion of polysaccharides from intact cell walls is limited by the presence of phenolic compounds within the cell wall matrix (Akin and Chesson, 1990; Eraso and Hartley, 1990; Hatfield, 1993b). These phenolics consist primarily of lignin and phenolic acids which are chemically bound to lignin or directly to cell wall polysaccharides. The importance of these compounds to utilization of structural carbohydrates requires their consideration in any discussion relative to cell wall polysaccharides and forage quality.

Lignin is the major non-polysaccharide component of plant cell walls. It is considered virtually indigestible (Van Soest, 1982) by ruminants and is implicated in limiting the digestion of cell wall polysaccharides (Jung, 1989). Lignin is the last major biopolymer to have evolved within the plant kingdom. Its most important function in plants is as a structural component to lend strength and rigidity to the cell wall (Monties, 1991), but also is important in limiting water loss by reducing permeability of the cell wall and in impeding disease microorganisms (Zeikus, 1980; Dean and Eriksson, 1992). These attributes which are desirable from the perspective of plant survival serve to limit the nutritional value of the plant for ruminants.

Forage lignins historically have been classified as core and non-core lignin. Core lignins consist of highly-condensed polymers formed by dehydrogenative polymerization of the hydroxycinnamyl alcohols (Table 6), *p*-coumaryl, coniferyl, and sinapyl alcohols (Sarkanen and Ludwig, 1971; Dean and Eriksson, 1992). Non-core lignin includes esterified and etherified phenolic acids which are either bound to core lignin or directly to cell wall polysaccharides (Jung and Deetz, 1993). Ferulic acid (FA) and *p*-coumaric acid *(p*-CA) are the predominant phenolics comprising non-core lignin. Non-core lignins are extractable with alkali or acid whereas core lignins are resistant to these reagents.

Lignin is formed as the result of a free radical reaction in which the monomeric precursors are condensed in a random arrangement. Because of extensive condensation during the polymerization of lignin, extraction of lignin with various solvents and reagents does not yield the immediate precursors, but rather a series of related compounds (Table 6). Alkali-extraction of forage lignins yields predominantly the cinnamic acid derivatives *p*-CA and FA. The double bond in the side-chain of cinnamic acid derivatives gives rise to the possibility of *cis*-trans isomerization, although trans isomers are thought to occur exclusively in native lignins (Jung and Fahey, 1983). Mild oxidation of forage lignins yields predominantly the benzaldehyde derivatives, *p*-hydroxybenzaldehyde, vanillin, and syringaldehyde. Together with the cinnamic acid derivatives, these compounds represent the most common lignin degradation products found in forages. Of minor importance are the benzoic acid derivatives, *p*-hydroxybenzoic acid, vanillic acid, syringic acid, and the acetophenone derivatives *p*-hydroxyacetophenone, acetovanillone, and acetosyringone (Table 6).

Lignins often are classified on the basis of their monomeric composition (Higuchi, 1981; Lewis and Yamamoto, 1990). The aromatic moieties of lignin precursors (monolignols) are referred to as *p*-hydroxyphenyl, guaiacyl, and syringyl based on their substitution at positions 3 and 5 (Table 6). Lignins which are formed predominantly from *p*-coumaryl alcohol are classified as *p*-hydroxyphenyl (H) lignins, those formed predominantly from coniferyl alcohol are classified as guaiacyl (G), and those formed predominantly from sinapyl alcohol are classified as syringyl (S). The monomeric composition of lignin is generally determined by quantifying the relative proportions of benzaldehyde derivatives released by nitrobenzene oxidation (Buxton and Russell, 1988).

Table 6. Classification, structure, and common names of lignin precursors and degradation products.

	p-Hydroxyphenyl	Guaiacyl	Syringyl
	$R_2 = H$ $R_3 = H$	$R_2 = OCH_3$ $R_3 = H$	$R_2 = OCH_3$ $R_3 = OCH_3$
	4-hydroxy-	3-methoxy-4-hydroxy-	3,5-dimethoxy-4-hydroxy-
Precursors:			
Cinnamyl alcohols $R_1 = -CH=CH-CH_2-OH$	p-coumaryl alcohol	coniferyl alcohol	sinapyl alcohol
Degradation Products:			
Cinnamic acids $R_1 = -CH=CH-COOH$	p-coumaric acid	ferulic acid	sinapic acid
Benzoic acids $R_1 = -COOH$	p-hydroxybenzoic acid	vanillic acid	syringic acid
Benzaldehydes $R_1 = -CHO$	p-hydroxybenz-aldehyde	vanillin	syringaldehyde
Acetophenones $R_1 = -COCH_3$	p-hydroxyaceto-phenone	acetovanillone	acetosyringone

Phenolics and Cell Wall Digestion

The mechanisms by which ruminal degradation of cellulose and hemicelluloses of forages is limited by phenolics are not fully understood. The close physical association of lignin with the cell wall polysaccharides and the existence of covalent bonds between lignin and cell wall polysaccharides are believed to be the major factors limiting the accessibility of the polysaccharides as substrate for the hydrolases secreted by ruminal microbes (Chesson, 1993). This steric interference of lignin with cell wall degrading enzymes is considered to be the primary mechanism by which lignin limits the digestibility of structural polysaccharides.

The negative relationship between lignin concentration and forage digestibility is well known (Jung and Vogel, 1986; Jung, 1989; Akin and Chesson, 1990). Lignification is generally negatively correlated with digestibility within a species over several growth stages (Reeves, 1987). However, the correlation does not always hold across species (Buxton and Russell, 1988) or within species at a single stage of development (Jung and Casler, 1991; Jung and Russelle, 1991). For example, the concentration of lignin is generally greater in legumes than in grasses at comparable stages of development. However, for a given lignin concentration, legumes are more digestible than grasses. Buxton and Russell (1988) studied the relationship between lignin and cell wall digestibility of smooth bromegrass, orchardgrass, alfalfa [*Medicago sativa* L.], birdsfoot trefoil [*Lotus corniculatus* L.], and red clover [*Trifolium pratense* L.]. They found that the lignin of grasses was 61% more inhibitory to cell wall digestion than the lignin of legumes.

Comparative studies with brown midrib (low lignin) mutants of sorghum, corn, and pearl millet [*Pennisetum glaucum* (L). R. Br.] and their normal isolines also have demonstrated the negative relationship between lignin concentration and forage digestibility. Brown midrib mutants are characterized by a dark brown leaf midrib and typically have lower lignin concentrations and are more digestible than their normal counterparts (Cherney et al., 1988, 1991; Moore et al., 1989; Wedig et al., 1989a,b). However, it should be pointed out that there are qualitative as well as quantitative differences between the lignins of normal and brown midrib genotypes (Akin et al., 1986; Fritz et al., 1990).

Delignification of forages generally results in improved digestibility of other cell wall constituents (Fahey et al., 1993). Oxidative reagents such as potassium permanganate, sodium chlorite, and hydrogen peroxide have been used to partially delignify forage cell walls, usually resulting in corresponding improvements in the digestibility of cell wall polysaccharides (Barton and Akin, 1977; Ford, 1978; Darcy and Belyea, 1980; Jung et al., 1992). In one experiment, partial delignification of wheat [*Triticum aestivum* L.] straw with peracetic acid resulted in approximately a one percentage unit increase in *in vitro* digestibility for each 0.4% decrease in lignin concentration (Streeter and Horn, 1982).

The hydroxycinnamic acids, p-CA and FA, are thought to play important roles in cell wall matrix interactions. Cinnamic acids may be esterified or etherified to cell wall polysaccharides (Bacic et al., 1988; Helm and Ralph, 1992, 1993). Because of their bifunctional nature, cinnamic acids are able to form bridges between cell wall components through formation of ester and ether linkages (Figure 9) (Iiyama et al., 1990; Hatfield, 1993a; Ralph and Helm, 1993). Lam et al. (1992a,b) verified the existence of cinnamic acid linkages in developing internodes of *Phalaris aquatica* L. They identified cinnamic acids that were either etherified, esterified, or both etherified and esterified to cell wall polysaccharides. They found evidence of ether and ester linkages for both FA and *p*-CA. Most of the etherified FA they found also was esterified to other wall polymers forming bridges, but *p*-CA was not involved in ester-ether bridging.

There is a great deal of circumstantial evidence to implicate phenolic acids in limiting cell wall digestion. There is generally a negative relationship between alkali-extractable phenolic concentration and forage digestibility (Hartley, 1972; Hartley and Jones, 1977; Jung, 1989; Fritz et al., 1990; Gabrielsen et al., 1990). Concentrations of *p*-CA increase with plant maturity, but FA concentration in cell walls remains relatively constant. Lam et al. (1992b) suggested that FA is primarily associated with primary wall development while deposition of *p*-CA occurs throughout wall formation including secondary wall formation.

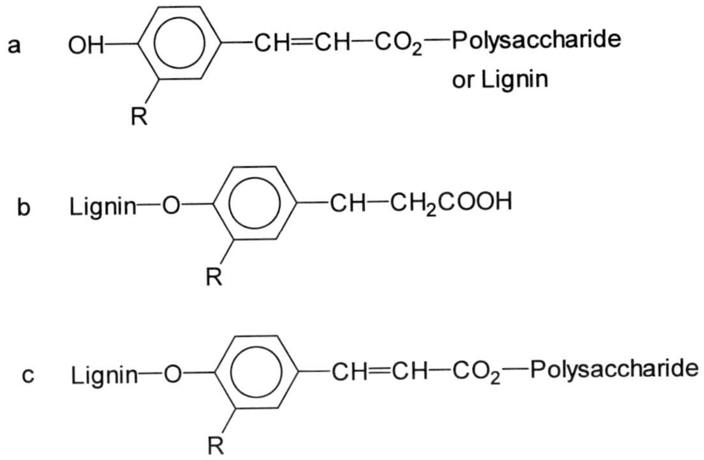

a OH—⟨C₆H₃R⟩—CH=CH—CO$_2$—Polysaccharide or Lignin

b Lignin—O—⟨C₆H₃R⟩—CH—CH$_2$COOH

c Lignin—O—⟨C₆H₃R⟩—CH=CH—CO$_2$—Polysaccharide

Figure 9. Linkages between cinnamic acids, lignin, and cell wall polysaccharides. a. Ester linkage with polysaccharide; b. Ether linkage with lignin; c. Ester-ether bridge between polysaccharide and lignin.

p-Coumaric acid and ferulic acid have been used by some researchers as an index of forage digestibility (Hartley, 1972; Burritt et al., 1984). Direct evidence for inhibition of cell wall digestion by cinnamic acids was obtained by Sawai et al. (1983) and Jung and Sahlu (1986). Sawai et al. (1983) esterified FA and p-CA to isolated forage fibers which were subsequently digested with a commercial cellulase. They observed a negative relationship between fiber degradability and amount of esterified hydroxycinnamic acids. Jung and Sahlu (1986) esterified cinnamic acids to isolated cellulose and measured cellulose degradation *in vitro*. They found that esterified cinnamic acids inhibited cellulose degradation more than free cinnamic acids and that this inhibition occurred at concentrations present in forages.

Phenolic dimers also have been isolated from forage lignins by alkali extraction (Hartley and Jones, 1976; Eraso and Hartley, 1990; Ford and Hartley, 1990; Hartley et al., 1990). Dehydrodiferulic acid and derivatives of truxillic and truxinic acids (cyclodimers of p-CA and FA) have been identified in alkali extracts of forages (Figure 10). Although phenolic dimers occur at relatively low concentrations, they may have a significant impact on cell wall degradability through their ability to form cross-linkages between hemicellulosic polysaccharides (Hartley and Ford, 1989; Lam et al., 1992b).

Other mechanisms for the inhibitory effects of phenolics on digestion which have been proposed include direct toxicity of phenolics to ruminal microbes, inhibition of microbial attachment to fiber particles during ruminal digestion, and the hydrophobic effect of lignin limiting water from the space adjacent to substrates (Akin and Chesson, 1990; Jung and Deetz, 1993). There

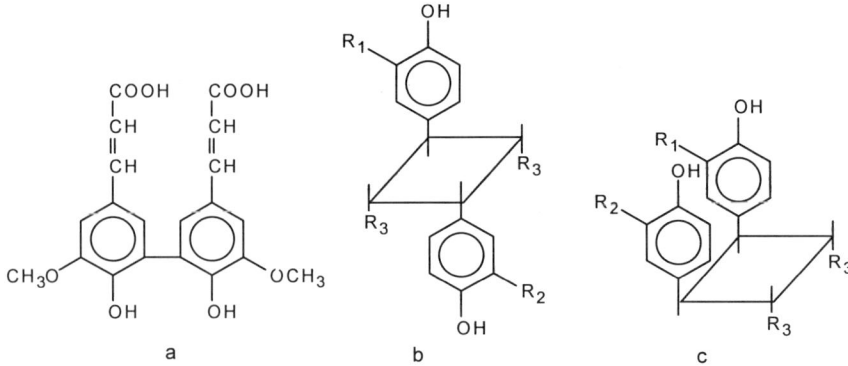

Figure 10. Phenolic dimers isolated from forage lignins. a. Diferulic acid; b. Truxillic acid derivatives; c. Truxinic acid derivatives.

is significant evidence that many of the phenolic compounds present in forages can have an inhibitory effect on ruminal fermentation (Akin, 1982; Jung and Sahlu, 1986; Varel and Jung, 1986). However, the concentrations at which such inhibition occurs are typically well above those that occur naturally in the rumen of livestock consuming forage diets. For this reason, the mechanism has largely been discounted as having a significant role in limiting the hydrolysis of cell wall polysaccharides by phenolics (Akin and Chesson, 1990; Jung, 1989). It has been postulated that hydrophobic lignin polymers present within the wall matrix may exclude water from space adjacent to enzyme binding sites, thereby preventing access to the enzymes. While there has been some evidence that such hydrophobic interactions occur, it is unlikely that they play a major role in limiting polysaccharide degradation (Jung and Deetz, 1993).

CONCLUSIONS

Much has been learned about the composition, structure, and biodegradability of forage carbohydrates in the last 25 years. This advancement in knowledge of forage carbohydrates has occurred on two fronts: applied studies on fiber composition and its degradation by ruminants carried out by agronomists and nutritionists, and more basic studies on the biosynthesis and structure of plant cell walls conducted by botanists, plant physiologists, and biochemists. From reading the Proceedings of the previous conference (Barnes, 1970), one can surmise that many of the principals believed that getting agronomists, plant breeders, and animal nutritionists together to discuss forage quality and evaluation was a major accomplishment of the previous decade. Perhaps a major achievement of the last 25 years is the crossover which has occurred between these more traditional disciplines involved with forage quality and the more basic disciplines of plant biochemistry, molecular biology, and microbiology.

In 1969, the Van Soest detergent system for extracting fiber constituents was relatively new and little was known about the detergent fiber composition of forages. Since that time, the detergent system has been widely employed in studies of forage quality. An extensive number of agronomic studies have characterized genetic and environmental influences on fiber composition of forages of economic importance (and many not so important) and nutritionists have characterized the relationships between fiber composition and forage intake, forage digestibility, and animal performance. The Van Soest system represented a paradigm shift with respect to forage quality evaluation. Prior to that time, the operative paradigm for forage quality evaluation was the proximate analysis system which was developed in the middle of the last century (Reid and Klopfenstein, 1983).

During this same period, botanists, plant physiologists, and biochemists made significant progress in elucidating the structure and biochemistry of the plant cell wall. Much of this progress was made possible by the development of improved instrumentation and methodologies such as electron microscopy, gas and liquid chromatography, mass spectroscopy, and nuclear magnetic resonance spectroscopy. In 1969, covalent bonds among cell wall constituents were only presumed to exist (Allinson, 1970). Now a number of these bonds have been characterized and their role in limiting forage digestibility is being investigated (Hatfield, 1993a). The challenge of the next decade and beyond is to ascertain the relevance of the increasing body of basic knowledge pertaining to cell wall structure and composition to the utilization of forages by ruminants. The nutritional implications of the chemical entities now being characterized by plant biochemists must be understood before the knowledge will be of any practical use in forage quality evaluation (Van Soest, 1993). This task will require collaborative research efforts on the part of plant biochemists, rumen microbiologists, and animal nutritionists.

REFERENCES

Akin, D. E. 1982. Forage cell wall degradation and p-coumaric, ferulic, and sinapic acids. Agron. J. 74:424-428.

Akin, D. E., and A. Chesson. 1990. Lignification as the major factor limiting forage feeding value especially in warm conditions. p. 1753-1760. *In* Proc. XVI Int. Grassland Congr., Vol. III. Nice, France, 4-11 October 1989, Association Française pour la Production Fourragère, Versailles, France.

Akin, D. E., W. W. Hanna, M. E. Snook, D. S. Himmelsbach, F. E. Barton, II, and W. R. Windham. 1986. Normal-23 and brown midrib-12 sorghum. II. Chemical variations and digestibility. Agron. J. 78:832-837.

Albersheim, P. 1975. The wall of growing plant cells. Sci. Am. 232:80-95.

Albrecht, K. A., W. F. Wedin, and D. R. Buxton. 1987. Cell-wall composition and digestibility of alfalfa stems and leaves. Crop Sci. 27:735-741.

Allinson, D.W. 1970. Forage lignins and their relationship to nutritive value. p. S1-S9. *In* R. F. Barnes, D. C. Clanton, C. H. Gordon, T. J. Klopfenstein, and D. R. Waldo (ed.) Proc. national conference on forage quality evaluation and utilization, Lincoln, NE, 3-4 Sept., 1969. Nebraska Center for Continuing Education, Lincoln, NE.

Aman, P., and H. Graham. 1990. Chemical evaluation of polysaccharides in animal feeds. p. 161-177. *In* J. Wiseman and D. J. A. Cole (ed.) Feedstuff evaluation. Butterworths, London.

Anderson, M. A., and B. A. Stone. 1975. A new substrate for investigating the specificity of β-glucan hydrolases. FEBS Letters 52:202-207.

Armstrong, D. G., and R. R. Smithard. 1979. The fate of carbohydrates in the small and large intestines of the ruminant. Proc. Nutr. Soc. 38:283-294.

Aspinall, G. O. 1982. Isolation and fractionation of polysaccharides. p. 19-34. *In* G. O. Aspinall (ed.) The polysaccharides, Vol. 1. Academic Press, New York.

Avigad, G. 1982. Sucrose and other disaccharides. p. 217-347. *In* F. A. Loewus and W. Tanner (ed.) Plant carbohydrates I: Intracellular carbohydrates. Springer-Verlag, New York.

Avigad, G. 1990. Disaccharides. p. 112-188. *In* P. M. Dey (ed.) Methods in plant biochemistry, Vol. 2. Academic Press, New York.

Bacic, A., P. J. Harris, and B. A. Stone. 1988. Structure and function of plant cell walls. p. 297-371. *In* J. Preiss (ed.) The biochemistry of plants, Vol. 14. Carbohydrates. Academic Press, New York.

Bailey, R. W. 1973. Structural carbohydrates. p. 157-211. *In* G. W. Butler and R. W. Bailey (ed.) Chemistry and biochemistry of herbage, Vol. 1. Academic Press, New York.

Baldwin, R. L., and M. J. Allison. 1983. Rumen metabolism. J. Anim. Sci. 57 (Suppl. 2):461-477.

Barras, D. R., A. E. Moore, and B. A. Stone. 1969. Enzyme substrate relationships among β-glucan hydrolases. Adv. Chem. Series 95:105-138.

Bartnicki-Garcia, S. 1984. Kingdoms with walls. p. 1-18. *In* W. M. Dugger and S. Bartnicki-Garcia (ed.) Structure, function, and biosynthesis of plant cell walls. Proc. 7th Ann. Sym. in Botany, Univ. of CA, Riverside, 12-14 January, 1984.

Barnes, R. F. 1970. Introduction and objectives of the conference on forage quality evaluation and utilization. p. A1-S5. *In* R. F. Barnes, D. C. Clanton, C. H. Gordon, T. J. Klopfenstein, and D. R. Waldo (ed.) Proc. national conference on forage quality evaluation and utilization, Lincoln, NE, 3-4 Sept., 1969. Nebraska Center for Continuing Education, Lincoln, NE.

Barton, F. E., II, and D. E. Akin. 1977. Digestibility of delignified forage cell walls. J. Agric. Food Chem. 25:1299-1303.

Blakeney, A. B., P. J. Harris, R. J. Henry, and B. A. Stone. 1983. A simple and rapid preparation of alditol acetates for monosaccharide analysis. Carbohydr. Res. 113:291-299.

Blumenkrantz N., and G. Asboe-Hansen. 1973. New method for quantitative determination of uronic acids. Anal. Biochem. 54:484-489.

Burritt, E. A., A. S. Bitner, J. C. Street, and M. J. Anderson. 1984. Correlations of phenolic acids and xylose content of cell wall with in vitro dry matter digestibility of three maturing grasses. J. Dairy Sci. 67:1209-1213.

Bryant, M. P. 1973. Nutritional requirements of the predominant rumen cellulolytic bacteria. Fed. Proc. 32:1809-1813.

Buxton, D. R., J. R. Russell, and W. F. Wedin. 1987. Structural neutral sugars in legume and grass stems in relation to digestibility. Crop Sci. 27:1279-1285.

Buxton, D. R., and J. R. Russell. 1988. Lignin constituents and cell-wall digestibility of grass and legume stems. Crop Sci. 28:553-558.

Carpita, N. C., and E. M. Shea. 1989. Linkage structure of carbohydrates by gas chromatography-mass spectrometry (GC-MS) of partially methylated alditol acetates. p. 157-216. *In* C. J. Biermann and G. D. McGinnis (ed.) Analysis of carbohydrates by GLC and MS. CRC Press, Boca Raton, FL.

Carpita, N. C. 1984. Fractionation of hemicelluloses from maize cell walls with increasing concentrations of alkali. Phytochem. 23:1089-1093.

Cherney, J. H., J. D. Axtell, M. M. Hassen, and K. S. Anliker. 1988. Forage quality characterization of a chemically induced brown-midrib mutant in pearl millet. Crop Sci. 28:783-787.

Cherney, J. H., D. J. R. Cherney, D. E. Akin, and J. D. Axtell. 1991. Potential of brown-midrib, low-lignin mutants for improving forage quality. Adv. Agron. 46:157-198.

Chesson, A. 1982. A holistic approach to plant cell wall structure and degradation. p. 85-90. *In* G. Wallace and L. Bell (ed.) Fibre in human and animal nutrition. The Royal Soc. of New Zealand, Wellington, NZ.

Chesson, A. 1993. Mechanistic models of forage cell wall degradation. p. 347-376. *In* H. G. Jung, D. R. Buxton, R. D. Hatfield, and J. Ralph (ed.) Forage cell wall structure and digestibility. ASA-CSSA-SSSA, Madison, WI.

Chesson, A., and C. W. Forsberg. 1988. Polysaccharide degradation by rumen microorganisms. p. 251-284. *In* P. N. Hobson (ed.) The rumen microbial system. Elsevier Applied Sci., New York.

Collings, G. F., and M. T. Yokoyama. 1979. Analysis of fiber components in feeds and forages using gas-liquid chromatography. J. Agric. Food Chem. 27:373-377.

Coughlan, M. P. 1991. Mechanisms of cellulose degradation by fungi and bacteria. Anim. Feed Sci. Technol. 32:77-100.

Coughlan, M. P. 1992. Enzymic hydrolysis of cellulose: An overview. Bioresource Technol. 39:107-115.

Cowling, E. B. 1975. Physical and chemical constraints in the hydrolysis of cellulose and lignocellulosic materials. Biotechnol. Bioeng. Symp. 5:163-181.

Darcy, B. K., and R. L. Belyea. 1980. Effect of delignification upon *in vitro* digestion of forage cellulose. J. Anim. Sci. 51:798-803.

Darvill, A., M. McNeil, P. Albersheim, and D. Delmer. 1980. The primary cell wall of flowering plants. p. 92-162. *In* N. E. Tolbert (ed.) The biochemistry of plants, Vol. 1. Academic Press, New York.

Dean, J. F. D., and K. E. Eriksson. 1992. Biotechnological modification of lignin structure and composition in forest trees. Holzforschung 46:135-147.

Dehority, B. A. 1993. Microbial ecology of cell wall fermentation. p. 425-453. *In* H. G. Jung, D. R. Buxton, R. D. Hatfield, and J. Ralph (ed.) Forage cell wall structure and digestibility. ASA-CSSA-SSSA, Madison, WI.

Delmer, D. P., and B. A. Stone. 1988. Biosynthesis of plant cell walls. p. 373-420. *In* J. Preiss (ed.) The biochemistry of plants, Vol. 14. Academic Press, New York.

Dubois, M., K. A. Giles, J. K. Hamilton, P. A. Rebers, and F. Smith. 1956. Colorimetric method for determination of sugars and related substances. Anal. Chem. 28:350-356.

Eraso, F., and R. D. Hartley. 1990. Monomeric and dimeric phenolic constituents of plant cell walls - possible factors influencing wall biodegradability. J. Sci. Food Agric. 51:163-170.

Esau, K. 1977. Anatomy of seed plants, 2nd ed. John Wiley and Sons, New York.

Ericsson, A., J. Hansen, and L. Dalgaard. 1978. A routine method for quantitative determination of soluble carbohydrates in small samples of plant material with gas-liquid chromatography. Anal. Biochem. 86:552-560.

Fahey, G. C., Jr., L. D. Bourquin, E. C. Titgemeyer, and D. G. Atwell. 1993. Postharvest treatment of fibrous feedstuffs to improve their nutritive value. p. 715-766. In H. G. Jung, D. R. Buxton, R. D. Hatfield, and J. Ralph (ed.) Forage cell wall structure and digestibility. ASA-CSSA-SSSA, Madison, WI.

Fales, S. L., D. A. Holt, V. L. Lechtenberg, K. Johnson, M. R. Ladisch, and A. Anderson. 1982. Fractionation of forage grass carbohydrates using liquid (water) chromatography. Agron. J. 74:1074-1077.

Feingold, D. S. 1982. Aldo (and keto) hexoses and uronic acids. p. 3-76. In F. A. Loewus and W. Tanner (ed.) Plant carbohydrates I: Intracellular carbohydrates. Springer-Verlag, New York.

Feingold, D. S., and G. Avigad. 1980. Sugar nucleotide transformations in plants. p. 101-170. In J. Preiss (ed.) The biochemistry of plants, Vol. 3. Academic Press, New York.

Feingold, D. S., and G. A. Barber. 1990. Nucleotide sugars. In P. M. Dey (ed.) Methods in plant biochemistry. Academic Press, New York.

Ford, C. W. 1978. Effect of particle size and delignification on the rate of digestion of hemicellulose and cellulose by cellulase in mature pangola grass stems. Aust. J. Agric. Res. 34:241-248.

Ford, C. W., and R. D. Hartley. 1990. Cyclodimers of *p*-coumaric and ferulic acids in the cell walls of tropical grasses. J. Sci. Food Agric. 50:29-43.

Fritz, J. O., K. J. Moore, and E. H. Jaster. 1990. Digestion kinetics and composition of cell walls isolated from morphological components of normal and brown midrib mutant sorghum x sudangrass hybrids. Crop Sci. 30:213-219.

Gabrielsen, B. C., K. P. Vogel, B. E. Anderson, and J. K. Ward. 1990. Alkali-labile cell-wall phenolics and forage quality in switchgrasses selected for differing digestibility. Crop Sci. 30:1313-1320.

Harris, P. J. 1983. Cell walls. p. 25-53. *In* J. L. Hall and A. L. Moore (ed.) Isolation of membranes and organelles from plant cells. Academic Press, London.

Harris, P. J., R. J. Henry, A. B. Blakeney, and B. A. Stone. 1984. An improved procedure for the methylation analysis of oligosaccharides and polysaccharides. Carbohydr. Res. 127:59-73.

Hartley, R. D. 1972. p-Coumaric and ferulic acid components of cell walls of ryegrass and their relationships with lignin and digestibility. J. Sci. Food Agric. 23:1347-1354.

Hartley, R. D., and C. W. Ford. 1989. Phenolic constituents of plant cell walls and wall biodegradability. p. 137-145. *In* N. G. Lewis and M. G. Paice (ed.) Plant cell wall polymers, biogenesis and biodegradation. ACS Symp. Ser. 399, ACS, Washington, DC.

Hartley, R. D., and E. C. Jones. 1976. Diferulic acid as a component of cell walls of *Lolium multiflorum*. Phytochem. 15:1157-1160.

Hartley, R. D., and E. C. Jones. 1977. Phenolic components and degradability of cell walls of grass and legume species. Phytochem. 16:1531-1534.

Hartley, R. D., W. H. Morrison, III, D. S. Himmelsbach, and W. S. Borneman. 1990. Cross-linking of cell wall phenolic arabinoxylans in graminaceous plants. Phytochem. 29:3705-3709.

Hatfield, R. D. 1989. Structural polysaccharides in forages and their degradability. Agron. J. 81:39-46.

Hatfield, R. D. 1992. Carbohydrate composition of alfalfa cell walls isolated from stem sections differing in maturity. J. Agric. Food Chem. 40:424-430.

Hatfield, R. D. 1993a. Cell wall polysaccharide interactions and degradability. p. 285-313. *In* H. G. Jung, D. R. Buxton, R. D. Hatfield, and J. Ralph (ed.) Forage cell wall structure and digestibility. ASA-CSSA-SSSA, Madison, WI.

Hatfield, R. D. 1993b. Characterization of a pectic fraction from smooth bromegrass cell walls using an endopolygalacturonase. J. Agric. Food Chem. 41:380-387.

Hatfield, R. D., J. Ralph, J. Grabber, and H. J. Jung. 1993. Structural characterization of isolated corn lignins. p. A-319. *In* Abstr. Keystone Symposia, The extracellular matrix of plants: Molecular, cellular and developmental biology. Santa Fe, NM, 9-15 Jan., 1993.

Hawker, J. S. 1985. Sucrose. p. 1-51. *In* P. M. Dey and R. A. Dixon (ed.) Biochemistry of storage carbohydrates in green plants. Academic Press, New York.

Hayashi, T. 1989. Xyloglucans in the primary cell wall. Annu. Rev. Plant Physiol. Plant Mol. Biol. 40:139-168.

Helm, R. F., and J. Ralph. 1992. Lignin-hydroxycinnamyl model compounds related to forage cell wall structure. 1. Ether-linked structures. J. Agric. Food Chem. 40:2167-2175.

Helm, R. F., and J. Ralph. 1993. Lignin-hydroxycinnamyl model compounds related to forage cell wall structure. 1. Ester-linked structures. J. Agric. Food Chem. 41:570-576.

Hespell, R. B. 1988. Microbial digestion of hemicelluloses in the rumen. Microbiol. Sci. 5:362-365.

Higuchi, T. 1981. Biosynthesis of lignin. p. 194-224. *In* W. Tanner and F. A. Loewus (ed.) Plant carbohydrates II: Extracellular carbohydrates. Encyclopedia of plant physiology, new series, Vol. 13B. Springer-Verlag, New York.

Hoover, W. H. 1978. Digestion and absorption in the hindgut of ruminants. J. Anim. Sci. 46:1789-1799.

Hoover, W. H., and S. R. Stokes. 1991. Balancing carbohydrates and proteins for optimum rumen microbial yield. J. Dairy Sci. 74:3630-3644.

Hoson, T. 1991. Structure and function of plant cell walls: Immunological approaches. Int. Rev. Cytol. 130:233-268.

Hotchkiss, A. T., and K. B. Hicks. 1990. Analysis of oligogalacturonic acids with 50 or fewer residues by high performance anion-exchange chromatography and pulsed amperometric detection. Anal. Biochem. 184:200-206.

Iiyama, K., T. B. T. Lam, and B. A. Stone. 1990. Phenolic acid bridges between polysaccharides and lignin in wheat internodes. Phytochem. 29:733-737.

Iiyama, K., T. B. T. Lam, P. J. Miekle, K. Ng, D. I. Rhodes, and B. A. Stone. 1993. Cell wall biosynthesis and its regulation. p. 621-683. *In* H. G. Jung, D. R. Buxton, R. D. Hatfield, and J. Ralph (ed.) Forage cell wall structure and digestibility. ASA-CSSA-SSSA, Madison, WI.

James, D. W., J. Preiss, and A. D. Elbein. 1985. Biosynthesis of polysaccharides. p. 107-207. *In* G. O. Aspinall (ed.) The polysaccharides, Vol. 3. Academic Press, New York.

Jung, H. G., and G. C. Fahey, Jr. 1983. Nutritional implications of phenolic monomers and lignin: A review. J. Anim. Sci. 57:206-219.

Jung, H. G. 1989. Forage lignins and their effects on fiber digestibility. Agron. J. 81:33-38.

Jung, H. G., and M. D. Casler. 1991. Relationship of lignin and esterified phenolics to fermentation of smooth bromegrass fibre. Anim. Feed Sci. Technol. 32:63-68.

Jung, H. G., and D. A. Deetz. 1993. Cell wall lignification and degradability. p. 315-346. *In* H. G. Jung, D. R. Buxton, R. D. Hatfield, and J. Ralph (ed.) Forage cell wall structure and digestibility. ASA-CSSA-SSSA, Madison, WI.

Jung, H. G., and M. P. Russelle. 1991. Light source and nutrient regime effects on fiber composition and digestibility of forages. Crop Sci. 31:1065-1070.

Jung, H. G., and T. Sahlu. 1986. Depression of cellulose digestion by esterified cinnamic acids. J. Sci. Food Agric. 37:659-665.

Jung, H. G., and K. P. Vogel. 1986. Influence of lignin on digestibility of forage cell wall material. J. Anim. Sci. 62:1703-1712.

Jung, H. G., F. R. Valdez, R. D. Hatfield, and R. A. Blanchette. 1992. Cell wall composition and degradability of forage stems following chemical and biological delignification. J. Sci. Food Agric. 58:347-355.

Kandler, O., and H. Hopf. 1980. Occurrence, metabolism, and function of oligosaccharides. p. 221-270. *In* J. Preiss (ed.) The Biochemistry of plants, Vol. 3. Academic Press, New York.

Kennedy, J. F., and C. A. White. 1983. Bioactive carbohydrates in chemistry, biochemistry and biology. Ellis Horwood Limited: Chichester, West Sussex, England.

Kerley, M. S., G. C. Fahey, Jr., J. M. Gould, and E. L. Iannotti. 1988. Effects of lignification, cellulose crystallinity and enzyme accessible space on the digestibility of plant cell wall carbohydrates by the ruminant. Food Microstructure 7:59-65.

Lam, T. B. T., K. Iiyama, and B. Stone. 1992a. Cinnamic acid bridges between cell wall polymers in wheat and *Phalaris* internodes. Phytochem. 31:1179-1183.

Lam, T. B. T., K. Iiyama, and B. Stone. 1992b. Changes in phenolic acids from internode walls of wheat and *Phalaris* during maturation. Phytochem. 31:2655-2658.

Lechtenberg, V. L., D. A. Holt, and H. W. Youngberg. 1971. Diurnal variation in nonstructural carbohydrates, in vitro digestibility, and leaf to stem ratio of alfalfa. Agron. J. 63:719-724.

Lechtenberg, V. L., D. A. Holt, and H. W. Youngberg. 1972. Diurnal variation in nonstructural carbohydrates of *Festuca arundinacea* (Schreb.) with and without N fertilizer. Agron. J. 64:302-305.

Lewis, N. G., and E. Yamamoto. 1990. Lignin: Occurrence, biogenesis and biodegradation. Annu. Rev. Plant Physiol. Plant Mol. Biol. 41:455-496.

Linden, J. C., A. R. Moreira, and D. H. Smith. 1980. Enzymatic hydrolysis of the lignocellulosic component from vegetative forage crops. Biotechnol. Bioeng. Symp. No. 10, p. 199-212.

Manners, D. J. 1985. Starch. p. 149-203. *In* P. M. Dey and R. A. Dixon (ed.) Biochemistry of storage carbohydrates in green plants. Academic Press, New York.

Masuda, Y., N. Sakurai, and K. Nishitani. 1982. Analytical studies on growing plant cell walls. p. 79-128. *In* S. P. Sen (ed.) Recent developments in plant science. Today and Tomorrow's Printers and Publishers, New Delhi, India.

McIlroy, R. J. 1967. Carbohydrates of grassland herbage. Herbage Abstr. 37:79-87.

McDonald, P. 1981. The biochemistry of silage. John Wiley and Sons, New York.

Meier, H., and J. S. G. Reid. 1982. Reserve polysaccharides other than starch in higher plants. p. 418-471. *In* F. A. Loewus and W. Tanner (ed.) Plant Carbohydrates I: Intracellular carbohydrates. Springer-Verlag, New York.

Mertens, D. R. 1987. Predicting intake and digestibility using mathematical models of ruminal function. J. Anim. Sci. 64:1548-1558.

Mertens, D. R. 1993. Kinetics of cell wall digestion and passage in ruminants. p. 535-570. *In* H. G. Jung, D. R. Buxton, R. D. Hatfield, and J. Ralph (ed.) Forage cell wall structure and digestibility. ASA-CSSA-SSSA, Madison, WI.

Monties, B. 1991. Plant cell walls as fibrous lignocellulosic composites: Relations with lignin structure and function. Anim. Feed Sci. Technol. 32:159-175.

Moore, K. J., J. O. Fritz, and E. H. Jaster. 1989. In situ degradation of cell wall neutral sugars of normal and brown midrib sorghum x sudangrass hybrids. p. 929-930. *In* Proc. XVI Int. Grassland Congr., Vol. III. Nice, France, 4-11 October 1989, Association Française pour la Production Fourragère, Versailles, France.

Morrison, I. M. 1979. Carbohydrate chemistry and rumen digestion. Proc. Nutr. Soc. 38:269-274.

Morrison, W. R., and J. Karkalas. 1990. Starch. p. 324-352. *In* P. M. Dey (ed.) Methods in plant biochemistry, Vol. 2. Academic Press, New York.

Moser, L. E. 1980. Quality of forage as affected by post-harvest storage and processing. p. 227-260. *In* C. S. Hoveland (ed.) Crop quality, storage, and utilization. ASA, Madison, WI.

Nishitani, K., and D. J. Nevins. 1988. Enzymic analysis of feruloylated arabinoxylans (Feraxan) derived from *Zea mays* cell walls. I. Purification of novel enzymes capable of dissociating Feraxan fragments from *Zea mays* coleoptile cell wall. Plant Physiol. 87:883-890.

Nishitani, K., and D. J. Nevins. 1989. Enzymic analysis of feruloylated arabinoxylans (feraxan) derived from Zea mays cell walls. II. Fractionation and partial characterization of feraxan fragments dissociated by a *Bacillus subtilis* enzyme (feraxanase). Plant Physiol. 91:242-248.

Nocek, J. E., and S. Tamminga. 1991. Site of digestion of starch in the gastrointestinal tract of dairy cows and its effect on milk yield and composition. J. Dairy Sci. 74:3598-3629.

O'Neill, M., P. Albersheim, and A. Darvill. 1990. The pectic polysaccharides of primary cell walls. p. 415-441. *In* P. M. Dey (ed.) Methods in plant biochemistry. Academic Press, London.

Owens, F. N., R. A. Zinn, and Y. K. Kim. 1986. Limits to starch digestion in the ruminant small intestine. J. Anim. Sci. 63:1634-1648.

Pell, A. N., and P. Schofield. 1993. Microbial adhesion and degradation of plant cell walls. p. 397-423. *In* H. G. Jung, D. R. Buxton, R. D. Hatfield, and J.

Ralph (ed.) Forage cell wall structure and digestibility. ASA-CSSA-SSSA, Madison, WI.

Perlin, A. S., and B. Casu, 1982. Spectroscopic methods. p. 133-193. *In* G. O. Aspinall (ed.) The polysaccharides, Vol. 1. Academic Press, New York.

Pidgen, W. J., and D. P. Heaney. 1969. Lignocellulose in ruminant nutrition. Adv. Chem. Series 95:245-260.

Pontis, H. G. 1990. Fructans. p. 353-369. *In* P. M. Dey (ed.) Methods in plant biochemistry, Vol. 2. Academic Press, New York.

Pontis, H. G., and E. Del Campillo. 1985. Fructans. p. 205-227. *In* P. M. Dey and R. A. Dixon (ed.) Biochemistry of storage carbohydrates in green plants. Academic Press, New York.

Preiss, J. 1988. Biosynthesis of starch and its regulation. p. 182-254. *In* J. Preiss (ed.) The biochemistry of plants, Vol. 14. Academic Press, New York.

Ralph, J., and R. F. Helm. 1993. Lignin/hydroxycinnamic acid/polysaccharide complexes: Synthetic models for regiochemical characterization. p. 201-246. *In* H. G. Jung, D. R. Buxton, R. D. Hatfield, and J. Ralph (ed.) Forage cell wall structure and digestibility. ASA-CSSA-SSSA, Madison, WI.

Reeves, J. B., III. 1987. Lignin and fiber compositional changes in forages over a growing season and their effects on in vitro digestibility. J. Dairy Sci. 70:1583-1594.

Reid, R. L., and T. J. Klopfenstein. 1983. Forages and crop residues: Quality evaluation and systems of utilization. J. Anim. Sci. 57(Suppl. 2):534-562.

Sarkanen, K. V., and C. H. Ludwig. 1971. Definition and nomenclature. p. 1-18. *In* K. V. Sarkanen and C. H. Ludwig (ed.) Lignins: Occurrence, formation, structure and reactions. John Wiley and Sons, New York.

Sawai, A., T. Kondô, and S. Ara. 1983. Inhibitory effects of phenolic acid esters on degradability of forage fibers. J. Japan. Grassl. Sci. 29:175-179.

Scott, R. W. 1979. Colorimetric determination of hexuronic acids in plant materials. Anal. Chem. 51:936-941.

Selvendran, R. R. 1983. The chemistry of plant cell walls. p. 95-147. *In* G. G. Birch and K. J. Parker (ed.) Dietry fibre. Applied Sci. Publishers, New York.

Selvendran, R. R., and M. S. DuPont. 1984. Problems associated with the analysis of dietary fibre and some recent developments. p. 1-68. *In* R. D.

King (ed.) Developments in food analysis techniques. Elsevier Applied Sci., London.

Selvendran, R. R., B. J. H. Stevens, and M. A. O'Neill. 1985. Developments in the isolation and analysis of cell walls from edible plants. p. 39-78. *In* C. T. Brett and J. R. Hillman (ed.) Biochemistry of plant cell walls. Cambridge University Press, Cambridge.

Smith, D. 1973. The nonstructural carbohydrates. p. 105-155. *In* G. W. Butler and R. W. Bailey (ed.) Chemistry and biochemistry of herbage. Academic Press, New York.

Smith, D. 1981. Removing and analyzing total nonstructural carbohydrates from plant tissue. Research Report R2107, Univ. of WI Agric. Exp. Sta., Madison, WI.

Streeter, C. L., and G. W. Horn. 1982. Effect of treatment of wheat straw with ammonia and peracetic acid on digestibility in vitro and cell wall composition. Anim. Feed Sci. Technol. 7:325-329.

Sturgeon, R. J. 1990. Monosaccharides p. 1-37. *In* P. M. Dey (ed.) Methods in plant biochemistry, Vol. 2. Academic Press, New York.

Sutton, J. D. 1979. Carbohydrate fermentation in the rumen - variations on a theme. Proc. Nutr. Soc. 38:275-281.

Taylor, R. L., and H. E. Conrad. 1972. Stoichiometric depolymerization of polyuronides and glycosaminoglycuronans to monosaccharides following reduction of their carbodiimide-activated carboxyl groups. Biochem. 11:1383-1388.

Theander, O. 1983. Advances in the chemical characterisation and analytical determination of dietary fibre components. p. 77-93. *In* G. G. Birch and K. J. Parker (ed.) Dietary fibre. Applied Sci. Publishers, London.

Van Soest, P. J. 1973. The uniformity and nutritive availability of cellulose. Fed. Proc. 32:1804-1808.

Van Soest, P. J. 1982. Nutritional ecology of the ruminant. O and B Books, Inc., Corvallis, OR.

Van Soest, P. J. 1993. Cell wall matrix interactions and degradation - session synopsis. p. 377-395. *In* H. G. Jung, D. R. Buxton, R. D. Hatfield, and J. Ralph (ed.) Forage cell wall structure and digestibility. ASA-CSSA-SSSA, Madison, WI.

Van Soest, P. J., and V. C. Mason. 1991. The influence of the Maillard reaction upon the nutritive value of fibrous feeds. Anim. Feed Sci. Technol. 32:45-53.

Van Soest, P. J., J. B. Robertson, and B. A. Lewis. 1991. Methods for dietary fiber, neutral detergent fiber, and nonstarch polysaccharides in relation to animal nutrition. J. Dairy Sci. 74:3583-3597.

Varel, V. H., and H. G. Jung. 1986. Influence of forage phenolics on ruminal fibrolytic bacteria and in vitro fiber degradation. Appl. Environ. Microbiol. 52:275-280.

Varner, J. E., and L. S. Lin. 1989. Plant cell wall architecture. Cell 56:231-239.

Vian, B. 1982. Organized microfibril assembly in higher plant cells. p. 23-43. *In* R. M. Brown, Jr. (ed.) Cellulose and other natural polymer systems. Plenum Publishing Corp, New York.

Volenec, J. J., and C. J. Nelson. 1984. Carbohydrate metabolism in leaf meristems of tall fescue: I. Relationship to genetically altered leaf elongation rates. Plant Physiol. 74:590-594.

Volenec, J. J. 1986. Nonstructural carbohydrates in stem base components of tall fescue during regrowth. Crop Sci. 26:122-127.

Wedig, C. L., E. H. Jaster, and K. J. Moore. 1987. Hemicellulose monosaccharide composition and in vitro disappearance of orchardgrass and alfalfa hay. J. Agric. Food Chem. 35:214-218.

Wedig, C. L., E. H. Jaster, and K. J. Moore. 1989a. Disappearance of hemicellulosic monosaccharides and alkali-soluble phenolic compounds of normal and brown midrib sorghum x sudangrasses fed to heifers and sheep. J. Dairy Sci. 72:104-111.

Wedig, C. L., E. H. Jaster, and K. J. Moore. 1989b. Disappearance of hemicellulosic monosaccharides and alkali-soluble phenolic compounds of normal and brown midrib sorghum x sudangrass silages fed to Holstein steers. J. Dairy Sci. 72:112-122.

Westhafer, M. A., J. T. Law, Jr., and D. T. Duff. 1982. Carbohydrate quantification and relationships with N nutrition in cool-season turfgrasses. Agron. J. 74:270-274.

White, B. A., R. I. Mackie, and K. C. Doerner. 1993. Enzymatic hydrolysis of forage cell walls. p. 455-484. *In* H. G. Jung, D. R. Buxton, R. D. Hatfield, and J. Ralph (ed.) Forage cell wall structure and digestibility. ASA-CSSA-SSSA, Madison, WI.

Wilkie, K. C. B. 1979. The hemicelluloses of grasses and cereals. Adv. Carbohydr. Chem. Biochem. 36:215-264.

Wilkie, K. C. B. 1985. New perspectives on non-cellulosic cell-wall polysaccharides (hemicelluloses and pectic substances) of land plants. p. 1-38. In C. T. Brett and J. R. Hillman (ed.) Biochemistry of plant cell walls. Cambridge University Press, Cambridge.

Windham, W. R., Barton, F. E., II, and D. S. Himmelsbach. 1983. High-pressure liquid chromatographic analysis of component sugars in neutral detergent fiber for representative warm- and cool-season grasses. J. Agric. Food Chem. 31:471-475.

Zeikus, J. G. 1980. Fate of lignin and related aromatic substrates in anaerobic environments. p. 101-109. In T. K. Kirk, T. Higuchi, and H. Chang (ed.) Lignin biodegradation: Microbiology, chemistry, and potential applications, Vol. 1. CRC Press, Boca Raton, LA.

CHAPTER 7

MINERALS IN FORAGES

Jerry W. Spears

INTRODUCTION

Minerals are required for virtually all vital processes in the body. A deficiency of each essential macro or micromineral in animals results in abnormalities that can only be corrected by supplementation of the deficient mineral. In addition to requirements for mammalian functions, ruminants fed forages are dependent on an adequate supply of a number of minerals to optimize rumen microbial activity and, thus, forage utilization.

In the past 25 years, considerable progress has been made in understanding the function of minerals, particularly trace minerals. However, in ruminants fed forage diets, our knowledge of requirements and factors affecting availability of minerals is extremely limited for a number of minerals. With the exceptions of copper and magnesium, little is known regarding dietary factors that affect mineral requirements of ruminants.

Forages provide an important source of minerals for ruminants. In some instances, forages may provide adequate quantities of all essential minerals required by ruminants. However, in other situations, forages are deficient in one or more mineral and supplementation is required for optimal animal performance and(or) health. Severe mineral deficiencies still occur to some degree, but marginal mineral deficiencies are probably much more widespread. Marginal mineral imbalances or deficiencies may result in no clinical deficiency signs and only small decreases in metabolic functions, but the overall impact on growth, reproduction, or health of ruminants can be substantial.

DISTRIBUTION AND CHEMICAL FORMS OF MINERALS IN FORAGES

Bioavailability of forage minerals to ruminants may be affected by the distribution of minerals within the forage and the chemical form of the elements present. Information regarding the chemical form of minerals in forages is sparse for many minerals. Table 1 shows the major forms of minerals believed to exist in plants. Information in Table 1 was compiled from reviews by Butler and Jones (1973) and Hazell (1985). The form and distribution of minerals in plant tissue also was reviewed by Little (1982).

J. W. Spears, Dep. of Animal Science, North Carolina State Univ., Box 7621, Raleigh, NC 27695.

Table 1. Chemical forms of minerals in plant materials.[a]

Mineral	Forms in Plants
Calcium	Calcium phosphate; calcium oxalate; possibly bound to pectin and lignin
Phosphorus	Inorganic phosphate; RNA; phospholipids; other phosphate esters; phytic acid
Sulfur	Sulfur amino acids; other organic sulfur-containing compounds; sulfate
Magnesium	Chlorophyll; bound to lignin
Sodium	Sodium ion
Potassium	Potassium ion
Chlorine	Chloride ion
Zinc	Anionic complexes
Copper	Neutral or anionic complexes
Selenium	Selenomethionine; selenate
Manganese	Organic chelates
Iron	Porphyrins; anionic complexes; ferric hydroxide
Iodine	Iodide ion
Silicon	Solid silica in plant cell wall; silicic acid
Molybdenum	Molybdenum-containing enzymes; molybdate ion

[a]From Butler and Jones (1973) and Hazell (1985).

A number of minerals have been found to be associated with the plant cell wall. Whitehead et al. (1985) isolated the cell wall fraction of perennial ryegrass (*Lolium perenne* L.), tall fescue (*Festuca arundinacea* Schreb.), white clover (*Trifolium repens* L.), and alfalfa (*Medicago sativa* L.) by mechanical disintegration of fresh or frozen forage. Table 2 presents the proportion of each mineral in tall fescue and white clover found in the cell wall. Only small amounts of phosphorus, sulfur, potassium, and magnesium were located in the cell wall while larger quantities of calcium, manganese, zinc, and copper were present. Differences among forages also were apparent, with a much higher percentage of the total calcium, manganese, zinc, and copper present in the cell wall of white clover compared to tall fescue.

A sizable amount of the total calcium, copper, zinc, and iron in alfalfa hay and grass silage was associated with the neutral detergent fiber (NDF) fraction (Kincaid and Cronrath, 1983). The association of certain minerals with fiber or other insoluble plant components could decrease the rate and extent of mineral release from forages in the gastrointestinal tract. Jones (1978) proposed that the carboxyl group of uronic acids and the carboxyl and hydroxyl groups of phenolic compounds are the principal groups in the plant cell wall involved in complexing metal cations.

Table 2. The proportion of minerals in tall fescue and white clover in the cell wall.[a]

Mineral	Tall Fescue	White Clover
	----- Percent of total mineral -----	
Phosphorus	4.9	5.0
Sulfur	3.6	5.2
Calcium	14.1	45.0
Magnesium	6.5	14.2
Potassium	1.7	0.9
Manganese	22.5	93.8
Zinc	29.0	45.4
Copper	26.5	40.6

[a]Whitehead et al. (1985).

AVAILABILITY OF MINERALS IN FORAGES

The ability of a forage to provide animals with an adequate supply of minerals is dependent on the mineral content and also the bioavailability of the mineral. Bioavailability of a mineral is defined as the proportion of the ingested element that is absorbed, transported to its site of action, and converted to a physiologically active form (O'Dell, 1984). Determining the concentration of most minerals in forage is relatively simple; however, it is difficult to accurately measure the bioavailability of forage minerals for specific functions in the animal. For one reason, the dietary concentration of a given mineral must be below the animal's requirement for a specific function if the mineral is to be absorbed and utilized with maximal efficiency. The body attempts to maintain mineral concentrations within the body in a fairly narrow range via homeostatic control mechanisms that reduce absorption or increase excretion (Miller, 1975).

Because of the difficulty in measuring utilization of various minerals by ruminants, researchers have frequently used such criteria as apparent absorption, mineral balance, tissue mineral concentrations, or solubilization of forage minerals in the rumen using in situ techniques to estimate bioavailability. Apparent absorption measurements are of questionable value in many situations. True absorption data in ruminants fed forages are limited because of the difficulty in measuring the endogenous fecal component (Thompson and Fowler, 1991).

A prerequisite for absorption is release of the mineral from the forage in a soluble form. Minerals may be absorbed in the ionic form or as soluble complexes or chelates depending on the particular mineral, but minerals are not absorbed when bound to insoluble substances. A number of studies (Playne et al., 1978; Rook et al., 1983; van Eys and Reid, 1987; Emanuele and

Staples, 1990; Ledoux and Martz, 1991; Puoli et al., 1991) reported the release of forage minerals in the rumen using in situ procedures. The proportion of the total mineral solubilized in the rumen has generally been high for most minerals.

Table 3 presents selective data from three studies (van Eys and Reid, 1987; Emanuele and Staples, 1990; Puoli et al., 1991) where ruminal release of minerals from forages was measured using the dacron bag technique. Van Eys and Reid (1987) measured ruminal release of macrominerals from tall fescue and tall fescue-red clover (*Trifolium pratense* L.) harvested at different maturities. Ruminal release of calcium decreased with increasing maturity; however, release of the other minerals studied was not greatly affected by stage of maturity. Values shown in Table 3 are means averaged across maturities. Emanuele and Staples (1990) measured ruminal release of calcium, magnesium, potassium, phosphorus, copper, and zinc from six different forages at incubation times ranging from 0 (water soluble) to 72 h. Although differences among forages in mineral release were observed, release of all minerals was high after 72 h of incubation. Calcium and zinc had the lowest percentage release of the minerals examined. Percentage release of calcium after 72 h ranged from 59 to 78% while zinc release ranged from 53

Table 3. Ruminal release of minerals from forages.

Mineral	Tall Fescue[a]	Switchgrass[b]	Bermudagrass[c] (*Cynodon dactylon* L.)	Alfalfa[c]
	------------------ Percent disappearance ------------------			
Calcium	71.0	30.3	78.1	59.3
Magnesium	91.9	83.9	96.5	95.2
Potassium	95.7	82.5	99.9	99.9
Phosphorus	88.3	60.8	65.3	85.1
Sulfur	81.9	47.2	NM[d]	NM[d]
Copper	NM[d]	NM[d]	75.8	92.9
Zinc	NM[d]	NM[d]	62.1	79.4

[a]van Eys and Reid (1987); 48-h incubation.
[b]Puoli et al. (1991); 24-h incubation.
[c]Emanuele and Staples (1990); 72-h incubation.
[d]Not measured.

to 81%. Puoli et al.(1991) also observed a low disappearance of calcium from switchgrass (*Panicum virgatum* L.) and big bluestem (*Andropogon gerardii* Vitm.) hays.

In situ studies have shown that minerals in forages consist of: 1) a fraction that is very soluble and released rapidly, 2) a fraction that is released slowly over a period of hours as the forage cell wall and (or) protein is degraded, and 3) at least for some elements, a fraction that is not released (Playne et al.,

1978; van Eys and Reid, 1987; Emanuele and Staples, 1990). The major portion of magnesium, potassium, phosphorus, and copper in forages appears to reside in the rapidly released fraction.

Emanuele et al. (1991) measured the solubility or release of macrominerals from forages within the rumen, abomasum, and intestines using a mobile dacron bag technique. After ruminal incubation for 24 h, bags were incubated for 1 h in an acid-pepsin solution that simulated the abomasal environment. Bags then were inserted into the duodenum of a cannulated cow and collected in the feces upon excretion. Most of the total release of calcium, magnesium, phosphorus, and potassium occurred in the rumen. Of the total forage calcium, 12.5 to 48.2% was released during incubation in the acid-pepsin solution. Total release of calcium varied from 47 to 91% depending on forage species. Total release of phosphorus, magnesium, and potassium was high for all six forages examined, ranging from 84 to 100%.

The in situ technique for estimating mineral availability from forages has limitations. Errors due to accumulation of bacterial residues in the bag and soluble minerals entering bags from the ruminal environment and binding to forage may occur. For example, Puoli et al. (1991) reported that approximately 80% of the phosphorus in switchgrass and big bluestem hay had disappeared if dacron bags were removed after 4 h of ruminal incubation; however, dacron bags recovered after a 60-h incubation indicated that only 16 to 32% of the phosphorus had disappeared. Because a mineral is released in a soluble form does not necessarily mean that the mineral will remain soluble until it is absorbed. Soluble minerals may interact with microbial cells, other minerals, or other forage components to form insoluble compounds.

Fiber in human diets is believed to reduce availability of certain minerals (Torre et al., 1991). The amount of fiber in the lower gut of humans is small compared to the quantity of undigested fiber constituents reaching the abomasum and small intestine of ruminants. Neutral detergent fiber and acid detergent fiber (ADF) fractions isolated from human foods have been shown to bind iron, copper, and zinc in vitro (Torre et al., 1991). Possible relationships between undigested fiber constituents and mineral availability in ruminants have received little attention. Kabaija and Smith (1988) studied the effect of dietary fiber on metabolism of trace minerals in sheep. The experimental diets contained 15.5, 33.5, or 60% NDF and fiber was varied by replacing corn (*Zea mays*) with ground guinea grass hay (*Panicum maximum* L.). Increasing dietary fiber greatly decreased apparent absorption of manganese, zinc, iron, and copper. Fiber level also affected the proportion of trace minerals soluble in different segments of the gastrointestinal tract following slaughter. Interactions between minerals that are believed to be of practical significance will be discussed later in this paper.

FACTORS AFFECTING MINERAL CONTENT OF FORAGES

The concentration of individual minerals in forages varies greatly depending on soil, plant, and management factors. Mineral analysis of forages is recommended regularly because of the many factors that affect forage mineral concentrations. Adams (1975) summarized mean mineral concentrations and standard deviations for forages analyzed at the Pennsylvania State Forage Laboratory between 1969 and 1973 (Table 4). These data clearly indicate that even within a rather small region of the United States, major variation exists in forage mineral levels. With some minerals, the standard deviation was as large as the mean value. Minson (1990) reviewed the world literature regarding mineral concentrations in forages.

Table 4. Mineral content of forages analyzed at the Pennsylvania State forage testing service (1969-1973).[a]

	Legume forage	Mixed, mainly legume	Grass forage	Mixed, mainly grass
Number of samples	992	4014	352	4119
Phosphorus[b]	3.0 ± 0.5	2.9 ± 0.6	2.2 ± 0.7	2.3 ± 0.6
Potassium[b]	25.5 ± 6.1	22.6 ± 6.0	16.8 ± 6.1	17.9 ± 5.5
Calcium[b]	11.8 ± 2.7	10.2 ± 2.8	4.9 ± 2.0	6.5 ± 2.8
Magnesium[b]	2.4 ± 0.7	2.2 ± 0.7	1.6 ± 0.6	1.8 ± 0.6
Sodium[b]	0.24 ± 0.17	0.18 ± 0.15	0.14 ± 0.19	0.13 ± 0.16
Manganese[c]	44.1 ± 49.2	48.1 ± 21.3	76.4 ± 64.1	57.3 ± 40.0
Iron[c]	221.7 ± 124.9	222.0 ± 142.7	184.4 ± 145.2	192.3 ± 138.9
Zinc[c]	18.1 ± 18.8	27.2 ± 12.7	27.6 ± 10.7	26.5 ± 12.8
Copper[c]	13.1 ± 8.2	13.1 ± 5.7	12.9 ± 8.4	12.0 ± 7.0

[a] From Adams (1975); Values are means ± standard deviations.
[b] g kg^{-1}.
[c] mg kg^{-1}.

Soil and plant factors that affect mineral content of forages have been reviewed (Fleming, 1973; Reid and Horvath, 1980). The mineral content of forages is greatly influenced by the quantity and the availability of minerals in the soil. Availability is very important because, for many minerals, only a small portion of the total mineral is available for uptake by the plant. The mineral composition of soil is affected by the origin of the parent rock, losses that occur through leaching and surface erosion, and applications of fertilizers and waste materials. Availability of minerals in the soil and uptake by plants is dependent on such factors as soil pH, and soil moisture and drainage conditions.

Soil pH, well within the physiological range observed under field conditions, has a major impact on the concentration of certain trace elements in forages (Fleming, 1973). Changes in the mineral content of herbage due to altering soil pH is dependent on soil type. Molybdenum and selenium concentrations in forage increase as soil pH increases from acid to neutral or alkaline conditions (Fleming, 1973; Underwood, 1981). Large decreases in forage manganese and cobalt concentrations have been observed due to relatively small increases in soil pH. Zinc and copper may decrease slightly with increasing soil pH. Macrominerals are less affected than microminerals by soil pH.

Legumes generally are higher in calcium, magnesium, potassium, copper, zinc, and cobalt than grasses, while grasses are higher in manganese and particularly silicon when grown under similar conditions (Fleming, 1973; Underwood, 1981; Minson, 1990). Differences also exist between grass and legume species in mineral uptake when grown under identical conditions on the same soil type. For example, tall fescue contained higher calcium and magnesium and lower potassium concentrations than orchardgrass (*Dactylis glomerata* L.) or canary grass (*Phalaris arundinacea*) when grown on each of five different soil types (Odom et al., 1980).

Stage of forage maturity affects the content of a number of minerals in forages. A rapid uptake of minerals by plants usually occurs during the early stages of growth. With advancing maturity, the dry matter (DM) content of the plant generally increases more rapidly than mineral uptake causing concentrations of many minerals to decrease (Fleming, 1973). Phosphorus and potassium decrease markedly with increasing maturity while calcium is not greatly altered by stage of maturity (Fleming, 1973; Underwood, 1981; Ammerman et al., 1982). Decreased forage concentrations of cobalt, copper, iron, molybdenum, and zinc also may occur with advancing maturity (Underwood, 1981). In addition to maturity, climatic and seasonal changes can influence forage minerals.

The effect of nitrogen (N) fertilization on forage mineral concentrations depends on the soil mineral supply (Fleming, 1973). If a mineral is low in soil, increased plant growth due to N fertilization may result in lower forage concentrations. If the soil mineral supply is adequate, uptake of certain minerals may increase in response to N fertilization due to an increase in the leaf to stem ratio. Phosphorus fertilization greatly increases herbage phosphorus when applied to soils deficient in phosphorus, but effects on other minerals are generally small. Potassium fertilization can decrease forage magnesium concentrations, especially when applied at high rates (Fleming, 1973).

MINERALS AND FORAGE QUALITY

Certain minerals can affect forage quality. A number of minerals are required by ruminal microorganisms for their normal growth and metabolism (Durand and Kawashima, 1980). Low concentrations of these minerals in forage may impair the ability of microorganisms to digest fiber and synthesize protein. Studies indicate that forage concentrations of sulfur and phosphorus can be low enough to impair fiber digestion (Durand and Komisarczuk, 1988). Recently, Lopez-Guisa and Satter (1992) suggested that divalent cations may serve as a link between the negatively charged surfaces of bacterial and plant cell walls and, thus, enhance fiber digestion.

A number of studies indicate that increasing concentrations of certain minerals via fertilization may affect forage quality differently than supplementing the mineral to unfertilized forage. Calcium fertilization of pangolagrass (*Digitaria decumbens* Stent.) grown on calcium-deficient soils reduced the ADF content and increased calcium from 2.2 to 3.8 g kg^{-1} (Rees and Minson, 1976). Dry matter digestibility was 4.6% higher and voluntary intake was 11.3% higher in sheep fed calcium-fertilized compared to those fed control pangolagrass. Calcium supplementation of the unfertilized or fertilized pangolagrass did not affect digestibility or intake.

Rees and Minson (1978) reported that sulfur fertilization of pangolagrass increased forage sulfur from 1.0 to 1.6 g kg^{-1}, but also increased the ADF and lignin contents. Sulfur fertilization increased cellulose and lignin digestibility, but reduced digestibility of neutral detergent solubles by sheep. Retention time of pangolagrass in the reticulo-rumen of sheep was reduced by 18.5% as a result of sulfur fertilization.

Sulfur fertilization of temperate forages grown on sulfur-deficient soils increased forage concentrations of sulfur, N, and sugars, and decreased lignin (Jones et al., 1970, 1982, 1990). Improvements in growth rate of lambs grazing subclover (*Trifolium subterraneum* L.)-grass pasture (Jones et al., 1990) and of lambs fed harvested subclover-grass or ryegrass (Jones et al., 1982) have been observed in response to increasing forage sulfur through fertilization.

In areas where forage yield responses to sulfur are not expected, forage quality may be improved via sulfur fertilization to increase plant sulfur. Digestibility of fiber was higher in steers fed sulfur-fertilized orchardgrass hay compared to control hay (Spears et al., 1985). Apparent digestibility of permanganate lignin was much higher (25.9 vs 43.3%) for sulfur-fertilized hay. The higher digestibility of permanganate lignin in sulfur-fertilized hay was associated with changes in the composition and apparent digestibility of phenolic compounds (Chestnut et al., 1986). Sulfur-fertilized orchardgrass contained lower concentrations of phenolic monomers after nitrobenzene oxidation of alkali-extracted NDF residues and less p-coumaric acid and ferulic acid than control orchardgrass. Apparent digestibilities of p-hydroxybenzaldehyde, vanillin, and syringaldehyde (obtained after nitrobenzene oxidation) were higher for sulfur-fertilized hay. Panditharatne et al. (1986) reported that sulfur fertilization of orchardgrass increased N, but slightly decreased the NDF and lignin content.

Fertilization of alfalfa with magnesium reduced cell wall constituents (Reid et al., 1979). The highest rate (448 kg ha^{-1}) of magnesium fertilization evaluated increased intake by lambs of hay harvested the first two years following fertilization. Increasing the magnesium content of timothy (*Phleum pratense* L.) from 0.8 to 1.6 g kg^{-1} via magnesium fertilization did not affect intake by sheep but reduced DM digestibility by 3.5% (Reid et al., 1984). Stepwise regression analysis of data obtained from digestibility and intake trials with 221 cool-season forages and 35 warm-season hays indicated that forage magnesium concentration had a negative effect on digestibility but a positive effect on intake (Reid and Jung, 1991).

MINERALS AND DISEASE RESISTANCE

In recent years, the nutritional effect on immunity and disease resistance has become increasingly clear. Selenium, copper, zinc, and cobalt deficiencies have been shown to alter various components of the immune system (Suttle and Jones, 1989). Reduced disease resistance has been observed in ruminants deficient in selenium, copper, and cobalt.

The effect of selenium on immune responsiveness and disease resistance has been reviewed recently (Stabel and Spears, 1993). In ruminants selenium deficiency has: 1) decreased the humoral immune response following viral challenge (Reffett et al., 1988), 2) decreased cell mediated immunity as measured by mitogen-stimulated lymphocyte proliferation (Larsen et al., 1988), 3) decreased the ability of neutrophils to kill ingested organisms (Boyne and Arthur, 1981), and 4) reduced random and chemotactic migration of neutrophils in vitro (Aziz et al., 1984).

Selenium supplementation or administration has decreased mortality rates in ruminants fed diets low in selenium when clinical deficiency signs, such as muscular dystrophy, were not apparent. It is evident from a number of studies that increased death losses due to selenium deficiency could easily go unnoticed under practical conditions. Injecting pregnant ewes at mid-gestation with selenium reduced lamb mortality rate from 31.3 to 13.8% (Hamdy et al., 1968). When lambs were slaughtered, a higher number of pneumonic lesions were observed in lambs from control ewes compared to those from selenium-injected ewes. Mortality of lambs from weaning, at 4 to 5 mo of age, until they were one year of age was reduced by oral supplementation of selenium in a 3-year study (Walker et al., 1979). Deaths in this study were largely attributed to pneumonia associated with *Pasteurella multocida* and mortality rate was highest in non-supplemented lambs during one year when severe cold weather occurred following shearing. Mortality rates, from birth to weaning, were reduced from 12.6 to 4.3% over a 2-year period in lambs born to ewes fed marginally selenium-deficient diets and given monthly injections of 4 mg of selenium (Knott et al., 1983). A 2-year study with beef cows and calves fed feedstuffs marginally deficient in selenium indicated that bi-monthly selenium-vitamin E injections reduced calf death losses (4.2 vs 15.3%) from birth to weaning (Spears et al., 1986). Most of the deaths in this study were attributed to diarrhea and subsequent unthriftiness. Selenium supplementation of low selenium diets did not improve parasite resistance in sheep experimentally infected with *Haemonchus contortus* (Jelinek et al., 1988) or *Ostertagia circumcincta* and *Trichostrongylus colubriformis* (McDonald et al., 1989).

Evidence suggests that selenium deficiency also may increase the susceptibility of dairy cows to intramammary infections. Intramuscular injection of cows fed a selenium-deficient diet with sodium selenite did not affect the incidence of clinical mastitis; however, the duration of clinical symptoms was reduced (Smith et al., 1984). Experimental mastitis, induced by intramammary challenge with *E. coli*, was more severe and of longer duration in selenium-deficient cows compared to selenium-adequate cows (Erskine et al., 1989). However, severity and duration of infection was not affected by selenium deficiency when mastitis was induced by intramammary challenge with *Staphylococcus aureus* (Erskine et al., 1990).

Neutrophils isolated from ewes and calves deficient in cobalt had a reduced ability to kill *Candida albicans* (MacPherson et al., 1989). Increased disease susceptibility also has been noted in cobalt-deficient lambs (Ferguson et al., 1988; MacPherson et al., 1989).

The ability of neutrophils to kill ingested *Candida albicans* was reduced in steers fed a copper-deficient diet (Boyne and Arthur, 1981). Peripheral blood granulocytes from ewes and calves fed copper-deficient diets had similar phagocytic capacity but reduced ability to kill ingested *Candida albicans* compared to controls adequate in copper (Jones and Suttle, 1981). Copper deficiency induced by feeding cattle either 5 mg molybdenum kg^{-1} or 500 mg iron kg^{-1} diet also impaired neutrophil function (Boyne and Arthur, 1986).

Copper deficiency in lambs grazing improved pastures resulted in increased susceptibility to bacterial infections and greater mortality (Woolliams et al., 1986). Over a 2-year period, death losses in lambs from birth to 24 weeks of age were much higher in a low copper line compared to a high copper line (28 vs 12%). Lamb survival was enhanced in the low copper line but not in the high copper line by administration of copper at six weeks of age. Many of the lamb deaths were associated with bacterial infections, with *Pasteurella hemolytica* and *E. coli* being the most commonly isolated organisms. The increased disease susceptibility in copper-deficient lambs may have been confounded with molybdenum as pastures contained from 1.2 to 3.1 mg molybdenum kg^{-1} DM (Woolliams et al., 1986). Copper depletion in calves produced by feeding a low copper diet under controlled conditions reduced serum IgM concentrations, but did not affect clinical signs (feed intake or rectal temperature) of respiratory disease following inoculation with IBR virus and *Pasteurella hemolytica* (Stabel and Spears, 1990).

Recent evidence suggests that dietary molybdenum may affect parasite resistance in lambs. In these studies (Suttle et al., 1992a, b), a basal diet containing 0.1 mg kg^{-1} was compared to the diet supplemented with 4.8 mg molybdenum kg^{-1}. Molybdenum supplementation reduced the number and length of adult worms recovered from the small intestine 11 days after lambs were exposed daily for 28 days to a trickle infection of the intestinal nematode, *Trichostrongylus vitrinus* (Suttle et al., 1992b). When the abomasal nematode, *Haemonchus contortus*, was given for 42 days, molybdenum addition reduced the number of worms recovered from the abomasum 14 days after infection from 4,167 to 907 (Suttle et al., 1992a). Furthermore, fecal egg counts were lower in lambs supplemented with molybdenum following infection with *Haemonchus contortus*. They suggested that molybdenum may have increased the inflammatory response resulting in increased parasite rejection.

Zinc deficiency is known to severely impair immune function in laboratory animals. A genetic disorder (lethal trait A46) of zinc metabolism has been reported in Holstein and Shorthorn calves that results in severe zinc deficiency due to impaired zinc absorption. Calves with lethal trait A46 exhibit thymus atrophy and impaired lymphocyte response to mitogen stimulation (Perryman et al., 1989). Lambs fed a semipurified diet severely deficient in zinc (3.7 mg kg^{-1}) had a lower percentage of lymphocytes and a higher percentage of neutrophils in peripheral blood as well as reduced in vitro lymphocyte blastogenesis (Droke and Spears, 1993). However, immune responses in lambs marginally deficient in zinc (8.7 mg kg^{-1} diet) did not differ from those observed in lambs fed adequate zinc (Droke and Spears, 1993).

MINERAL FUNCTIONS, REQUIREMENTS, AND DEFICIENCIES

Calcium

Approximately 99% of the total calcium in the body functions as a structural component of bones and teeth. The remaining 1% of body calcium is involved in such vital functions as blood clotting, membrane permeability, neuromuscular excitability, secretion of certain hormones, and enzyme activation. Calcium deficiency results in bones that are soft and weak, and reduced growth and milk production (McDowell, 1992).

Forages, with the possible exception of corn silage, usually contain adequate calcium for beef cattle and sheep, and few controlled studies have evaluated the effects of calcium supplementation on growth or bone strength in ruminants receiving forage diets. Long-term calcium supplementation did not affect gain or lambing percentages in ewes grazing hill pasture in Scotland containing 1.7 to 3.3 g calcium kg^{-1} (Gunn, 1969). Calcium supplementation seemed to improve the firmness and permanence of incisor teeth in ewes.

Although forages are generally high in calcium, the availability of calcium in some forages may be low because of the presence of calcium oxalate. In alfalfa, 20 to 33% of the calcium was present as insoluble calcium oxalate and apparently unavailable to the animal (Ward et al., 1979). True absorption of calcium in alfalfa hay was found to be only 24.6% in lactating dairy cows compared to 42.2% for an alfalfa-corn silage diet (Martz et al., 1990). Relative availability of calcium to cattle was about 20% lower in tropical grasses containing calcium oxalate than in grasses containing little calcium oxalate (Blaney et al., 1982).

Magnesium

Magnesium plays important roles in neuromuscular function, skeletal development, and activation of a number of enzymes involved in almost all aspects of metabolism (McDowell, 1992). A clinical deficiency of magnesium is unlikely in growing ruminants because they only require approximately 1 g magnesium kg^{-1} DM. However, magnesium supplementation has increased intake (Reid and Jung, 1991) and fiber digestion (Wilson, 1980) in ruminants fed forages containing more than 1 g magnesium kg^{-1}.

Lactating ruminants require approximately 2 g magnesium kg^{-1} diet early in lactation to prevent the onset of hypomagnesemic or grass tetany. Grass tetany occurs primarily in lactating beef cows or ewes and is characterized by low blood magnesium concentrations. Initial signs of grass tetany in cows are nervousness, reduced feed intake, and muscular twitching around the face and ears. Animals are uncoordinated and walk with a stiff gait. In the advanced stages, cows go down on their sides with their head back and go into convulsions. Death usually occurs unless the animal is treated intravenously or subcutaneously with a magnesium salt solution.

Grass tetany is most common in mature lactating ruminants grazing lush spring pastures or fed harvested forages low in magnesium. With early spring pastures, the problem is more one of low availability rather than low forage magnesium concentrations per se. Ruminants are dependent on a daily supply of magnesium from the gastrointestinal tract to maintain normal blood magnesium concentrations because homeostatic mechanisms are not efficient in regulating blood magnesium levels. Magnesium concentrations in bone are high, but mature animals lack the ability to mobilize large amounts of magnesium from bone.

Fertilization of pastures with high levels of N and potassium is associated with increased incidence of grass tetany. Forages conducive to causing grass tetany are generally high in N, potassium, and organic acids and low in soluble carbohydrates (Mayland et al., 1990). Factors affecting absorption of magnesium in ruminants were reviewed by Fontenot et al. (1989). A number of studies have shown that high dietary potassium reduces magnesium absorption (Greene et al., 1983; Wylie et al., 1985). The site of potassium inhibition of magnesium absorption appears to be the rumen, which is the major site of magnesium absorption. Wylie et al. (1985) reported that infusing potassium into the rumen reduced magnesium absorption, while potassium infusion into the abomasum or ileum had no effect on absorption. Evidence suggests that magnesium absorption from the rumen occurs by an active sodium-linked process (Martens and Rayssiguier, 1980) and sodium supplementation to a low sodium diet increased magnesium absorption in sheep (Martens et al., 1987).

Grass tetany can be prevented by providing magnesium in a free choice mineral if intake of magnesium from the supplement is adequate. Mineral supplements containing magnesium are unpalatable. The addition of palatable ingredients such as dried molasses, ground corn, or cottonseed meal will greatly increase consumption of magnesium supplements and, thus, reduce the likelihood of grass tetany (Robinson et al., 1989).

Phosphorus

Phosphorus is often discussed with calcium because the two minerals function together in bone formation. Approximately 80% of the body phosphorus is found in bones and teeth. Phosphorus also functions in: 1) cell growth and differentiation as a component of DNA and RNA, 2) energy utilization and transfer as a component of ATP, ADP, and AMP, 3) phospholipid formation, and 4) maintaining acid-base and osmotic balance. Signs of phosphorus deficiency include reduced growth and feed efficiency, depraved appetite, impaired reproduction, reduced milk production, and weak, fragile bones (Underwood, 1981).

Phosphorus-deficient soils are widespread and forages produced on these soils are low in phosphorus. Drought conditions and increased forage maturity also result in low forage phosphorus concentrations. In grazing livestock, phosphorus deficiency has been described as the most prevalent mineral deficiency throughout the world (McDowell, 1992). Early studies in South Africa and Texas with cattle grazing forages low in phosphorus showed large improvements in fertility and calf weaning weights due to phosphorus supplementation (Dunn and Moss, 1992).

The skeleton provides a large reserve of phosphorus that can be drawn on during periods of inadequate phosphorus intake in mature animals. Skeletal reserves can subsequently be replaced during periods when phosphorus intake is high relative to requirements. Controversy exists regarding phosphorus requirements of ruminants because of lack of data on true absorption coefficients for phosphorus and disagreements over endogenous fecal losses of phosphorus. Absorption values reported for phosphorus are variable (Suttle, 1987). True absorption values used in calculating dietary phosphorus requirements for beef cattle range from 64% (TCORN, 1991) to 85% (NRC, 1984). The endogenous loss of phosphorus constitutes the animal's maintenance requirement for phosphorus and is primarily of fecal origin in ruminants fed forage diets. Endogenous fecal losses of phosphorus consist largely of unabsorbed salivary phosphorus. Salivary phosphorus is affected by plasma phosphorus concentration, which is dependent on phosphorus intake, and factors that affect salivary flow such as DM intake and physical form of diet (TCORN, 1991). Therefore, fecal endogenous losses of phosphorus are not constant, but vary depending on dietary phosphorus intake and other factors that affect salivary phosphorus.

Some studies suggest that NRC (1984) recommendations for phosphorus may overestimate requirements for beef cattle. Call et al. (1978) fed Hereford heifers, beginning at approximately 7 months of age, a basal diet containing 1.4 g phosphorus kg^{-1} diet for 2 years. No differences between the two groups were detected in growth, rib bone morphology and phosphorus content, age at puberty, conception rate, or calving interval. The basal diet supplied approximately 66% of the NRC (1984) recommended phosphorus requirement. In a second study, Hereford heifers were fed low phosphorus diets from weaning through their fifth gestation and lactation (Call et al., 1986). The low phosphorus group received 6 to 12.1 g phosphorus d^{-1} while controls received 20.6 to 38.1 g phosphorus d^{-1} with phosphorus intake increasing as the cattle grew larger. Females fed the low phosphorus intake (approximately 68% of NRC requirements) remained healthy and growth and reproduction were similar to that observed in those fed supplemental phosphorus. When phosphorus intake of cows was reduced to 5.1 to 6.6 g d^{-1}, clinical signs of deficiency occurred within 6 months (Call et al., 1986). Deficiency signs observed were reduced feed intake, weight loss, spontaneous bone fractures, and impaired reproduction. Reproduction was not reduced until cows were fed the very low phosphorus diet for over a year. The phosphorus deficiency noted in cows fed 5.1 to 6.6 g phosphorus d^{-1} was corrected by increasing phosphorus intake to 11.7 to 12.6 g d^{-1}. They concluded that 12 g of phosphorus d^{-1} over a production year was adequate for

450 kg Hereford cows (Call et al., 1986). No measurements of milk production or calf weaning weights were given in these papers (Call et al., 1978, 1986). Diets used in these studies consisted of alfalfa, wheat straw, beet molasses, and dried beet pulp.

Under range conditions where low forage phosphorus concentrations occur, phosphorus supplementation only improved calving interval and calf weaning weights in one year of a 5-year study in New Mexico (Judkins et al., 1985). The one year where performance responses to phosphorus supplementation were noted was a drought year. Further studies are needed to better define phosphorus requirements of beef cattle and to determine if results obtained in Hereford cows can be extrapolated to other beef breeds when corrected for differences in mature size and level of milk production.

Sodium, Chlorine, and Potassium

Potassium, sodium, and chlorine will be discussed together because of their similar functions. All three of these macrominerals are involved in maintaining osmotic pressure, controlling water balance, and regulation of acid-base balance. Potassium and sodium also function in muscle contractions and nerve impulse transmission, while chlorine is necessary for the formation of HCl in gastric juice. Deficiency signs for sodium, chlorine, and potassium are rather nonspecific and include pica, and reduced feed intake, growth, and milk production (Underwood, 1981).

Because ruminants fed forage diets readily consume salt when provided free choice, it has traditionally been assumed that supplemental salt is needed. However, ruminants have a specific appetite for sodium and they will consume much more salt than they actually require (Morris, 1980). Chloride requirements in growing ruminants are low and a deficiency of chloride is not likely to occur under practical conditions (Neathery et al., 1981). Sodium deficiency is more likely to occur, but sodium supplementation to forage diets may not be necessary to prevent a sodium deficiency in some situations (Morris et al., 1980). In a review, Morris (1980) concluded that 1 g sodium kg^{-1} DM was adequate for lactating beef cows and that requirements for nonlactating beef cattle did not exceed 0.6 to 0.8 g sodium kg^{-1} DM.

Although supplemental salt may not always be necessary to prevent a sodium or chloride deficiency, salt is an excellent vehicle for getting other minerals or feed additives into grazing ruminants. Supplementation of both sodium and chloride is generally needed in lactating dairy cow diets because of the large quantities of these minerals secreted in milk.

Forages generally are excellent sources of potassium. In fact, high potassium in lush spring pastures appears to be a major factor associated with the occurrence of hypomagnesemia or grass tetany in beef cows (Mayland et al., 1990). As forages mature, the potassium content decreases and low concentrations of potassium have been observed in range forage and in accumulated tall fescue during the winter (Clanton, 1980). Potassium supplementation under winter range conditions has increased gain in stocker cattle and resulted in reduced weight loss in beef cows (Clanton, 1980).

Requirements for potassium in ruminants fed forage diets are not well established. Based on the limited data available, Clanton (1980) suggested that growing cattle under range conditions require 3 to 4 g potassium kg^{-1}, considering the low rates of gain observed under these conditions, while gestating beef cows require 5 to 7 g potassium kg^{-1} DM.

Sulfur

Rumen microorganisms are capable of synthesizing all organic sulfur-containing compounds (methionine, thiamine, and biotin) required by mammalian tissues from inorganic sulfur. Sulfur also is required by rumen microorganisms for their growth and normal cellular metabolism. A dietary deficiency of sulfur can dramatically depress microbial numbers as well as microbial digestion and protein synthesis. Supplementation to increase the sulfur content of spear grass hay from 0.40 to 0.75 g kg^{-1} significantly increased counts of rumen bacteria, protozoa, and sporangia of anaerobic fungi in sheep (Morrison et al., 1990). The greatest effect was observed in sporangial forms of rumen anaerobic fungi which increased from non- detectable levels up to 33 to 41 sporangia per mm^2 leaf surface with sulfur supplementation (Morrison et al., 1990).

Some rumen bacteria specifically require methionine, but most bacteria are able to synthesize the sulfur-containing amino acids from sulfide (Goodrich et al., 1978). Ruminal sulfide is derived from reduction of inorganic sulfur sources and from the degradation of sulfur-containing amino acids. Sulfide can be absorbed from the rumen and oxidized by tissues to sulfate, a less toxic form of sulfur. In situ studies have usually indicated a rapid release of sulfur from cool season grasses with some 60 to 70% of the total forage sulfur either water soluble or released within 3 h of rumen incubation (Glenn et al., 1985; van Eys and Reid, 1987). Muntifering et al. (1984) suggested that rapid solubilization of forage sulfur could result in considerable losses of sulfur from the rumen and, thus, perhaps a limitation of sulfur for microbial functions.

Sulfur supplementation of tall fescue hay containing 2 g sulfur kg^{-1} DM or greater has increased fiber digestibility in some studies (Spears et al., 1978; Guardiola et al., 1983), but not in others (Muntifering et al., 1984). In Australia, sulfur supplementation has increased gain in lambs (Wheeler et al., 1975) and steers (Archer and Wheeler, 1978) grazing sorghum x sudangrass (*Sorghum bicolor x S. sudanense*) containing 0.8 to 1.5 g sulfur kg^{-1}. The sulfur requirement of ruminants grazing sorghum x sudangrass may be increased because of the need for sulfur in the detoxification of cyanogenic glucosides found in most sorghum forages. Sorghum forages appear to be inherently low in sulfur relative to most forages and the sulfur content of sorghum x sudangrass did not increase in response to sulfur fertilization (Wheeler et al., 1980).

In some areas of the United States and other countries, sulfur-deficient soils exist and in these areas, sulfur fertilization has increased forage sulfur and lamb performance (Jones et al., 1982, 1990). As discussed in the section on minerals and forage quality, part of the animal responses to sulfur fertilization may be due to changes in forage quality rather than increased forage sulfur per se.

Cobalt

Cobalt functions as a component of vitamin B_{12} (cobalamin). The ruminant animal is not dependent on a dietary source of vitamin B_{12} because rumen microorganisms are capable of synthesizing B_{12} from dietary cobalt. Two vitamin B_{12}-dependent enzymes occur in mammalian tissues (Smith, 1987). Methylmalonyl CoA mutase is involved in the metabolism of propionate to succinate as it catalyzes the conversion of L-methylmalonyl CoA to succinyl CoA. The second vitamin B_{12}-requiring enzyme, 5-methyltetrahydrofolate homocysteine methyltransferase (methionine synthase), catalyzes the transfer of methyl groups from 5-methyltetrahydrofolate to homocysteine to form methionine and tetrahydrofolate. This reaction is important in the recycling of methionine following transfer of its methyl group.

Depressed appetite and failure to grow, or moderate weight loss, are early signs of cobalt deficiency. As the deficiency becomes more severe, ruminants exhibit: 1) severe unthriftiness, 2) rapid weight loss, 3) pale skin and mucous membranes due to pernicious anemia, and 4) fatty degeneration of the liver (Smith, 1987). The major metabolic defect in cobalt deficiency may be an inability to properly metabolize propionate due to reduced activity of methylmalonyl CoA mutase. This hypothesis is based on findings that propionate clearance from plasma is reduced and that plasma concentrations and urinary excretion of methylmalonate are greatly elevated in cobalt deficiency (Smith, 1987).

If propionate conversion to succinate is impaired in cobalt deficiency, one would expect a shortage of glucose since propionate is a major contributor to glucose supply via gluconeogenesis in ruminants. However, plasma glucose concentrations are not always affected by severe cobalt deficiency (Kennedy et al., 1991). Recent findings indicate that an inability by ruminal microorganisms to convert succinate to propionate is an early manifestation of cobalt deficiency (Kennedy et al., 1991). Ruminal and plasma succinate concentrations were greatly increased in lambs fed cobalt- deficient diets. Ruminal succinate increased 200-fold within 2 days after lambs were fed the deficient diet. Considerable amounts of succinate are produced by ruminal microorganisms, but succinate does not accumulate under normal conditions in the rumen because it is rapidly converted to propionate by propionate-producing bacteria. At least in some bacteria, the pathway involved in the conversion of succinate to L-methylmalonyl CoA and then to propionate is the reverse of that found in the liver of ruminants and involves vitamin B_{12}-dependent methylmalonyl CoA mutase (Barker, 1972). These findings suggest that ruminants fed diets deficient in cobalt exhibit altered ruminal fermentation due to lack of cobalt for vitamin B_{12} synthesis by the microorganisms.

Requirements for cobalt are approximately 0.1 mg kg^{-1} DM (Mills, 1981). Forage cobalt concentrations provide a fairly reliable indicator of cobalt adequacy. Cobalt concentrations in forages are greatly affected by soil cobalt level and also by soil pH. Increasing soil pH from 5.4 to 6.4 reduced the cobalt content of ryegrass from 0.35 to 0.12 mg kg^{-1} DM (Mills, 1981). Cobalt status of ruminants can be evaluated by measuring serum vitamin B_{12}, plasma methylmalonate, liver cobalt, or liver vitamin B_{12} concentrations (Mills, 1987). In cattle, measurement of serum B_{12} may be of limited value because of the presence of B_{12} analogues in bovine serum (Halpin et al., 1984).

Copper

Copper deficiency in grazing ruminants is widespread in many areas of the world. Signs that have been attributed to copper deficiency in ruminants include: 1) anemia, 2) depigmentation and impaired keratinization of hair and wool, 3) cardiac failure, 4) diarrhea (cattle), 5) low reproduction, 6) poor growth, and 7) neonatal ataxia in lambs (Underwood, 1981). Copper functions as an essential component of a number of enzymes including lysyl oxidase, ceruloplasmin, cytochrome oxidase, tyrosinase, and superoxide dismutase. Many of the deficiency signs described above can be explained biochemically by reduced activity of one or more of these enzymes during copper deficiency.

Copper requirements vary from less than 4 to well over 10 mg kg^{-1} diet, depending on other dietary components, especially molybdenum and sulfur. Thus, copper concentrations in forages are of limited value in assessing copper adequacy unless forage concentrations of copper antagonists such as molybdenum, sulfur, and iron are also considered. Liver copper concentrations less than 20 mg kg^{-1} DM or plasma concentrations less than 0.5 µg ml^{-1} are indicative of deficiency (Underwood, 1981). However, in the presence of high dietary molybdenum and sulfur, copper in liver and plasma may not accurately reflect copper status because the copper can exist in tightly bound forms unavailable for biochemical functions (Suttle, 1991).

For many years, important interactions that exist between copper, molybdenum, and sulfur in ruminants have been recognized. The antagonistic action of molybdenum on copper metabolism is exacerbated when sulfur is also high. In recent years, considerable progress has been made in understanding the nature of this three-way interaction (Gooneratne et al., 1989; Suttle, 1991). This interaction is believed to center around the formation of thiomolybdates in the ruminal environment. Thiomolybdates are formed by the following reactions:

$MoO_4^= + H^+ + HS^- \longrightarrow H_2O + MoO_3S^=$
(monothiomolybdate)
$MoO_3S^= + H^+ + HS^- \longrightarrow H_2O + MoO_2S_2^=$
(dithiomolybdate)
$MoO_2S_2^= + H^+ + HS^- \longrightarrow H_2O + MoOS_3^=$
(trithiomolybdate)
$MoOS_3^= + H^+ + HS^- \longrightarrow H_2O + MoS_4^=$
(tetrathiomolybdate)

Copper reacts with thiomolybdates in the rumen to form insoluble complexes that are poorly absorbed. Some thiomolybdates are absorbed and affect systemic metabolism of copper (Gooneratne et al., 1989). Systemically, thiomolybdates result in copper being tightly bound to plasma albumin and not available for biochemical functions. Certain copper-dependent enzymes may be inhibited directly by thiomolybdates (Mason, 1986). Price et al. (1987) found predominately tri- and tetrathiomolybdates in the solid phase of ruminal, duodenal, and ileal digesta. Small amounts of di- and trithiomolybdates were detected in the liquid phase of duodenal digesta. They concluded that tri- and tetrathiomolybdates were primarily responsible for reducing copper absorption while di- or trithiomolybdates were likely responsible for post-absorptive effects on copper metabolism (Price et al., 1987).

Independent from its role in the molybdenum-copper interaction, sulfur has been postulated to reduce copper absorption via formation of copper sulfide in the gut (Suttle, 1974). Increasing dietary sulfur in the inorganic or organic form from 1 to 3 or 4 g kg^{-1} diet reduced copper availability in sheep by 30 to 40% (Suttle, 1974). Reducing the sulfate content of drinking water high in sulfate from 500 to 42 mg L^{-1} by reverse osmosis increased the copper status of cattle (Smart et al., 1986). The presence of protozoa in the rumen reduced ruminal solubility of copper and decreased storage of copper in the liver (Ivan, 1988). This is thought to relate to the ability of protozoa to degrade protein resulting in the production of sulfide from breakdown of sulfur amino acids.

High iron concentrations in forages may contribute to development of a copper deficiency in the ruminant. A number of studies have shown that feeding levels of iron (250 to 1200 mg kg^{-1} as ferrous carbonate) typical of those often found in forages reduces copper status in cattle (Bremner et al., 1987; Phillippo et al., 1987a,b) and sheep (Prabowo et al., 1988). Interestingly, long-term (24 to 32 weeks) copper depletion induced by high dietary iron has not resulted in clinical signs of copper deficiency such as reduced growth, infertility in heifers, or changes in hair pigmentation or texture. In these studies, liver and plasma copper concentrations were reduced to levels indicative of severe deficiency. Iron-induced copper depletion has resulted in pancreatic damage (Fell et al., 1985) and impaired ability of neutrophils to ingest and kill *Candida albicans* in vitro (Boyne and Arthur, 1986).

Copper toxicity can be a problem in sheep receiving forages high in copper and low in molybdenum. The use of a mineral supplement high in copper also can cause toxicity problems. As little as 8 to 11 mg copper kg^{-1} diet may result in toxicity in sheep if dietary molybdenum is less than 1 mg kg^{-1} diet (NRC, 1985). Copper toxicity can be prevented or treated if detected early by increasing dietary molybdenum and sulfur concentrations. Tetrathiomolybdate administration also has been effective in treating and preventing copper toxicity in sheep (Gooneratne et al., 1981). Cattle are much less susceptible to copper toxicity and generally can tolerate over 100 mg copper kg^{-1} diet (NRC, 1984).

Iron

Iron is an essential component of a number of proteins involved in oxygen transport and (or) utilization. These proteins include hemoglobin, myoglobin, and a number of cytochromes and iron-sulfur proteins involved in the electron transport chain. Several enzymes also either contain iron or are activated by iron. A deficiency of iron results in anemia, anorexia, reduced growth, and listlessness (Miller et al., 1988).

Forages usually contain iron concentrations in excess of the requirements of ruminants. Iron deficiency in grazing ruminants is unlikely unless parasite infestations or diseases exist that cause chronic blood loss. In fact, Underwood (1981) states "Iron deficiency in grazing animals, in which iron is the primary limiting factor, has never been demonstrated unequivocally." Injecting nursing beef calves with iron dextran at 1 and 3 weeks of age increased hemoglobin concentrations and RBC numbers but calf gains were not affected by iron (Reece et al., 1985). Rice et al. (1967) reported higher hemoglobin concentrations in beef calves injected with iron dextran at a young age as well as increased weight gains. However, the gain response was transitory and was not reflected in heavier weaning weights.

Ruminants grazing pastures or being fed harvested silage or hay may be exposed to excessive levels of iron through forage, water, or soil ingestion. The possibility of high dietary iron contributing to copper depletion was discussed in the section on copper. The addition of between 100 and 1,000 mg iron L^{-1} of fermentation media reduced in vitro digestion of a tall fescue-based substrate by ruminal microorganisms (Harrison et al., 1992). Based on their in vitro findings, they calculated that iron concentrations of approximately 500 mg kg^{-1} may influence rumen microbial activity. Studies are needed to determine if the solubility of iron from forages and soil is sufficient to adversely affect ruminal fermentation. Research by Healy (1972) indicated that a significant amount of iron from various soil types was soluble in rumen fluid.

Manganese

Manganese functions as a component of the enzymes pyruvate carboxylase, arginase, and superoxide dismutase and as an activator for a number of enzymes (Hurley and Keen, 1987). Of the many enzymes that can be activated by manganese, only the glycosyltransferases are known to specifically require manganese. An inadequate intake of manganese in young ruminants can result in skeletal abnormalities (stiffness, twisted legs, enlarged joints, and reduced bone breaking strength) and ataxia (Hurley and Keen, 1987). In older animals, manganese deficiency causes low reproductive performance characterized by depressed or irregular estrus, low conception rate, abortion, stillbirths, and small birth weights. Skeletal abnormalities observed in manganese deficiency are believed to result from decreased activity of the glycosyltransferases which are involved in mucopolysaccharide synthesis (Hurley and Keen, 1987). How manganese affects reproduction is not clear.

Manganese requirements for growth are low (10 mg kg^{-1}) but requirements for normal reproduction are considerably higher (20 to 40 mg kg^{-1}). Absorption of manganese from ^{54}manganese chloride in lactating dairy cows was less than 1.0% (Van Bruwaene et al., 1984) and little is known concerning dietary factors that may influence manganese absorption. Some evidence suggests that high dietary calcium and phosphorus may increase manganese requirements (Hidiroglou, 1979).

Forages vary considerably in manganese content but concentrations are generally above 20 mg kg^{-1} (Minson, 1990). However, the availability of manganese in forages has not been studied to any extent. In some situations, manganese may limit reproduction in ruminants fed forage diets. Manganese supplementation improved pregnancy rate in ewes grazing pastures containing 26 to 48 mg kg^{-1} of manganese (Egan, 1972). Supplementing beef cows with 14 mg manganese kg^{-1} of diet from parturition until week 19 postpartum reduced services per conception from 1.6 to 1.1 but did not affect overall conception rate (DiCostanzo et al., 1986). In this study, cows were fed a corn silage-based diet containing 32 mg manganese kg^{-1} diet. Masters et al. (1988) reported that manganese supplementation did not improve gain, digestibility, or wool production in rams fed an oaten hay-based diet containing 13 mg manganese^{-1} diet. Testicular growth per unit of live weight gain was increased when dietary manganese was increased to 19 mg kg^{-1} diet (Masters et al., 1988).

Molybdenum

Molybdenum functions as a component of the enzymes xanthine oxidase, sulfite oxidase, and aldehyde oxidase. There is no evidence that molybdenum deficiency occurs in ruminants under practical conditions. However, molybdenum may enhance microbial activity in the rumen in some instances. The addition of 2 mg molybdenum kg^{-1} to a semipurified diet containing 0.36 mg molybdenum kg^{-1} diet increased cellulose digestion in lambs (Ellis et al., 1958). Recently, the addition of 10 mg molybdenum kg^{-1} diet to a high roughage diet containing 1.7 mg molybdenum kg^{-1} increased the rate of in situ DM disappearance from the rumen of cattle (Shariff et al., 1990). In situ DM disappearance was not improved by molybdenum supplementation when steers were fed a ground barley-based diet containing 1 mg molybdenum kg^{-1} diet (Shariff et al., 1990).

High forage concentrations of molybdenum (20 mg kg^{-1} or higher) can cause toxicity characterized by diarrhea, loss of weight, anorexia, stiffness, and changes in hair color in cattle (Ward, 1978). Sheep are less susceptible to molybdenum toxicity than cattle (Underwood, 1981). Molybdenosis can usually be overcome by providing large amounts of copper. Concentrations of molybdenum much lower than those needed to cause acute toxicity signs can result in copper deficiency as discussed in the previous section.

Recent studies suggest that some abnormalities normally attributed to copper deficiency may instead be due to molybdenum toxicity. The concentrations of molybdenum evaluated in these studies were much lower than those typically associated with toxicity. The addition of as little as 5 mg molybdenum kg^{-1} to a diet containing 0.1 mg molybdenum kg^{-1} caused copper depletion associated with reduced growth and feed efficiency, loss of hair pigmentation, changes in hair texture, and infertility in heifers (Bremner et al., 1987; Phillippo et al., 1987a, b). In these same experiments, cattle fed high dietary iron had similar copper status, based on liver copper, plasma copper, and superoxide dismutase and ceruloplasmin activities, to those fed molybdenum, but did not show clinical signs of copper deficiency. Feeding beef cows and their calves 5 mg molybdenum kg^{-1} diet reduced calf gains from birth until weaning by 28%, while calf gains were not affected by addition of 500 mg iron kg^{-1} diet (Gengelbach et al., 1992).

Molybdenum effects on reproduction have been particularly striking (Phillippo et al., 1987b). The addition of 5 mg molybdenum kg^{-1} diet starting at 13 to 19 weeks of age increased age and decreased live weight of heifers at puberty compared to control or iron-supplemented heifers. Peak LH concentrations after prostaglandin administration was much lower in molybdenum- supplemented heifers. Conception rate in heifers fed molybdenum was only 12 to 33% compared with 57 to 80% in control and iron-supplemented heifers (Phillippo et al., 1987b). Differences in conception rate were noted even though rate of gain was standardized across treatments at 0.6 kg day^{-1} during the breeding phase. When heifers that had received the iron-supplemented diet were switched to the molybdenum diet, conception rate was reduced within 12 weeks. In contrast, replacing the molybdenum diet with the iron- supplemented diet resulted in normal conception rates within 12 weeks. These changes in fertility occurred in the absence of detectable changes in copper status.

These results suggest that relatively low levels of molybdenum may exert direct effects on certain metabolic processes independent of alterations in copper metabolism. It is unclear how thiomolybdates fit into the overall picture. However, rats fed thiomolybdates develop diarrhea and skeletal abnormalities not observed in simple copper deficiency in the rat (Mills and Bremner, 1980). In many practical situations, supplemental copper may be more important in alleviating or preventing adverse effects of molybdenum than in providing copper for copper-requiring functions.

Selenium

Selenium was first recognized because of problems associated with severe selenium toxicity. In 1957, selenium was shown to be an essential trace mineral and in the past 25 years, it has become evident that naturally occurring selenium deficiency is a much greater problem than selenium toxicity. Forages grown in many areas of the United States and other countries are severely or at least marginally deficient in selenium.

White muscle disease in young ruminants is a common clinical sign of selenium deficiency that results in degeneration and necrosis in both skeletal and cardiac muscle (Underwood, 1981). Affected animals may show stiffness, lameness, or even cardiac failure. Other signs of selenium deficiency that have been observed include unthriftiness (often times with weight loss and diarrhea), anemia with presence of heinz bodies, and increased incidence of retained placenta in dairy cows (Underwood, 1981; Morris et al., 1984).

Reproductive performance has been inconsistently affected by selenium deficiency (Miller et al., 1988). Selenium deficiency can increase mortality and reduce live weight gains and wool growth in the absence of clinical deficiency signs (Spears et al., 1986; Langlands et al., 1991a, b, c).

In 1973, glutathione peroxidase (GSH-Px) was identified as the first known selenium metalloenzyme (Underwood, 1981). Glutathione peroxidase catalyzes the reduction of hydrogen peroxide and lipid hydroperoxides, thus preventing oxidative damage to body tissues. Decreases in GSH-Px activity associated with selenium deficiency can explain the occurrence of white muscle disease, heinz body anemia, and possibly other signs of selenium deficiency. A second selenometalloenzyme, iodothyronine 5'-deiodinase, was recently identified (Arthur et al., 1990). This enzyme catalyzes the deiodination of thyroxine (T_4) to the more metabolically active triiodothyronine (T_3) in tissues. Arthur et al. (1988) reported that selenium- deficient cattle had increased T_4 and decreased T_3 concentrations in plasma relative to selenium-supplemented cattle. Depressed activity of iodothyronine 5' deiodinase may explain the unthriftiness and poor growth often observed in selenium-deficient ruminants.

Selenium requirements under practical conditions are not well defined and little is known regarding dietary or physiological factors that may influence requirements. The selenium requirement of ruminants is generally considered to be between 0.1 and 0.3 mg kg^{-1} diet. Selenium concentrations in plasma, serum, and whole blood, and GSH-Px activities in plasma, whole blood, and erythrocytes have been used to assess selenium status. Glutathione peroxidase activities indicative of a selenium deficiency can vary from one laboratory to another depending on assay conditions. Furthermore, plasma GSH-Px may increase in response to disease (Stabel et al., 1989). Plasma or serum selenium concentrations between 0.07 and 0.10 µg ml^{-1} are considered adequate in dairy cows (Gerloff, 1992). In Australia, a positive relationship was observed between blood selenium concentrations and responsiveness to selenium supplementation in sheep (Paynter et al., 1979). However, Langlands et al. (1989) concluded from a large number of on farm studies with cattle in Australia that selenium concentrations in whole blood and plasma were poor indicators of responsiveness to selenium supplementation unless unthriftiness was apparent.

Organic selenium in forages may affect selenium status differently from supplemental selenium which is usually provided as selenite. Selenomethionine, the predominant form of selenium in forages, can be incorporated into nonspecific body proteins in place of methionine (Behne et al., 1991). Some organic selenium absorbed may end up in proteins other than enzymes that specifically require selenium, but selenomethionine appears to be absorbed more efficiently than selenite. Selenomethionine was approximately twice as available, based on erythrocyte GSH-Px, as cobalt selenite when fed to selenium-deficient heifers (Pehrson et al., 1989). In deficient areas, selenium can be provided in a free-choice mineral supplement. Alternate methods of supplementing selenium include injecting selenium every 3 to 4 months or at critical production stages and the use of boluses that are retained in the rumen and release selenium over a period of months.

Selenium is probably the most toxic mineral commonly supplemented to ruminant diets and excess supplementation, regardless of route of administration, can cause toxicosis and even death. The maximum tolerable level of selenium for ruminants has been estimated at 2 mg kg^{-1} diet (NRC, 1980), but lower levels could cause problems if provided for long periods.

Zinc

Over 200 enzymes from various sources have been shown to be zinc metalloenzymes (Hambidge et al., 1986). In addition, a number of enzymes are activated by zinc. Because of its involvement in so many enzymes, zinc is critical for the metabolism of many nutrients including proteins, nucleic acids, and carbohydrates. Zinc deficiency signs include reduced growth and feed intake, excessive salivation, and skin lesions (parakeratosis).

Severe zinc deficiency is rare, but has been observed in ruminants grazing forages (McDowell, 1992). The extent that marginal or subclinical zinc deficiency exists is unknown, but is likely more widespread. Based on zinc supplementation studies, subclinical zinc deficiency can result in impaired reproduction (Piper and Spears, 1982) and decreased weight gains (Mayland et al., 1980). No good indicator of zinc status is available for detecting marginal deficiencies.

Forages, especially low quality forages, often contain concentrations of zinc below NRC recommended levels. However, zinc requirements have been established largely in ruminants fed high quality forages or concentrate-based diets. Ruminants fed forages may have lower zinc requirements than animals fed concentrate-based diets because of lower rates of gain or perhaps higher requirements due to the close association of zinc with the plant cell wall. Not only are zinc requirements of ruminants fed forage diets poorly defined, but little is known regarding factors that affect zinc availability in forages. The inconsistent responses to zinc supplementation in grazing ruminants or in ruminants fed forage diets suggest that zinc requirements are affected by dietary or physiological factors.

Zinc supplementation increased gain in nursing calves grazing mature forage that contained 7 to 17 mg zinc kg^{-1} (Mayland et al., 1980). Increasing the zinc content of alfalfa hay above basal levels of 14 to 20 mg kg^{-1} by foliar application or fertilization with zinc did not affect performance or digestibility in growing lambs (Reid et al., 1987). Zinc addition to a hay-based diet containing 22 mg zinc kg^{-1} did not increase growth or spermatogenesis in ram lambs (Hatch et al., 1987). Zinc supplementation increased lambing percentage and lamb birth weights in ewes grazing pastures containing 13 to 34 mg zinc kg^{-1} (Masters and Fels, 1980). In later trials with ewes grazing pastures containing 10 to 30 mg zinc kg^{-1}, supplemental zinc only improved reproductive performance in one of six trials (Masters and Fels, 1985).

CONCLUSIONS

Mineral concentrations in forages vary greatly, and are affected by soil mineral level, soil pH, plant species, stage of forage maturity, and application of fertilizers or waste materials. Forage mineral concentrations are of limited value in assessing mineral status of ruminants, because little is known regarding availability and factors affecting availability of minerals in ruminants fed forage diets. Availability of minerals from forages is an area that should be emphasized in the future.

The relatively high solubilization of forage minerals in the rumen suggests that, in most instances, the release of minerals in a soluble form is not a major factor limiting absorption. Interactions of minerals with other minerals, other plant constituents, and(or) microbial cells following solubilization appear to be more important as regards mineral absorption.

Deficiency signs have been described for essential macro and microminerals. However, identifying minerals that limit performance or health of ruminants under practical conditions remains a major problem today. Future research is needed to better define criteria for diagnosing marginal deficiencies of certain minerals.

Considerable progress has been made in the past 25 years in defining requirements and elucidating the function(s) of minerals. A continuing need exists to enhance our understanding of how certain minerals function. There also is a need to more accurately determine requirements for many minerals by ruminants. Increased knowledge of mineral requirements and availability of minerals from forages in the future will allow for formulation of mineral supplements that maximize ruminant performance.

REFERENCES

Adams, R.S. 1975. Variability in mineral and trace mineral content of dairy cattle feeds. J. Dairy Sci. 58:1538-1548.

Ammerman, C.B., J.E. Moore, P.R. Henry, S.M. Miller, and F.G. Martin. 1982. Effect of age and sample preparation on mineral concentration of bermudagrass hay. J. Dairy Sci. 65:1329-1333.

Archer, K.A., and J.L. Wheeler. 1978. Response by cattle grazing sorghum to salt-sulfur supplements. Aust. J. Exp. Agric. Anim. Husb. 18:741-744.

Arthur, J.R., P.C. Morrice, and G.J. Becket. 1988. Thyroid hormone concentrations in selenium deficient and selenium sufficient cattle. Res. Vet. Sci. 45:122-123.

Arthur, J.R., F. Nicol, and G.J. Becket. 1990. Hepatic iodothyronine 5'-deiodinase. Biochem. J. 272:537-540.

Aziz, E.S., P.H. Klesius, and J.C. Frandsen. 1984. Effects of selenium on polymorphonuclear leukocyte function in goats. Am. J. Vet. Res. 45:1715-1718.

Barker, H.A. 1972. Coenzyme B_{12}-dependent mutases causing carbon chain rearrangements. p. 509-537. *In* P.D. Boyer (ed.) The enzymes. Vol. 6. Academic Press, New York.

Behne, D., A. Kyriakopoulos, S. Scheid, and H. Gessner. 1991. Effects of chemical form and dosage on the incorporation of selenium into tissue proteins in rats. J. Nutr. 121:806-814.

Blaney, B.J., R.J.W. Gartner, and T.A. Head. 1982. The effects of oxalate in tropical grasses on calcium, phosphorus and magnesium availability to cattle. J. Agric. Sci., Camb. 99:533-539.

Boyne, R., and J.R. Arthur. 1981. Effects of selenium and copper deficiency on neutrophil function in cattle. J. Comp. Path. 91:271-276.

Boyne, R., and J.R. Arthur. 1986. Effects of molybdenum or iron induced copper deficiency on the viability and function of neutrophils from cattle. Res. Vet. Sci. 41:417-419.

Bremner, I., W.R. Humphries, M. Phillippo, M.J. Walker, and P.C. Morrice. 1987. Iron-induced copper deficiency in calves: Dose-response relationships and interactions with molybdenum and sulfur. Anim. Prod. 45:403-414.

Butler, G.W., and D.I.H. Jones. 1973. Mineral biochemistry of herbage. p. 127-162. *In* G.W. Butler and R.W. Bailey (ed.) Chemistry and biochemistry of herbage. Vol. 2. Academic Press, New York.

Call, J.W., J.E. Butcher, J.T. Blake, R.A. Smart, and J.L. Shupe. 1978. Phosphorus influence on growth and reproduction of beef cattle. J. Anim. Sci. 47:216-225.

Call, J.W., J.E. Butcher, J.L. Shupe, J.T. Blake, and A.E. Olson. 1986. Dietary phosphorus for beef cows. Am. J. Vet. Res. 47:475-481.

Chestnut, A.B., G.C. Fahey, Jr., L.L. Berger, and J.W. Spears. 1986. Effects of sulfur fertilization on composition and digestion of phenolic compounds in tall fescue and orchardgrass. J. Anim. Sci. 63:1926-1934.

Clanton, D.C. 1980. Applied potassium nutrition in beef cattle. p. 17-32. *In* Proc. Third International Minerals Conf., Miami, FL. Jan. 17-18, 1980.

DiCostanzo, A., J.C. Meiske, S.D. Plegge, D.L. Haggard, and K.M. Chaloner. 1986. Influence of manganese, copper and zinc on reproductive performance of beef cows. Nutr. Rep. Int. 34:287-293.

Droke, E.A., and J.W. Spears. 1993. In vitro and in vivo immunological measurements in growing lambs fed diets deficient, marginal or adequate in zinc. J. Nutr. Immunol. 2:71-90.

Dunn, T.G., and G.E. Moss. 1992. Effects of nutrient deficiencies and excesses on reproductive efficiency of livestock. J. Anim. Sci. 70:1580-1593.

Durand, M., and R. Kawashima. 1980. Influence of minerals in rumen microbial digestion. p. 375-408. *In* Y. Ruckebusch and P. Thivend (ed.) Digestive physiology and metabolism in ruminants. MTP Press Limited, Lancaster, England.

Durand, M., and S. Komisarczuk. 1988. Influence of major minerals on rumen microbiota. J. Nutr. 118:249-260.

Egan, A.R. 1972. Reproductive responses to supplemental zinc and manganese in grazing Dorset Horn ewes. Aust. J. Exp. Agric. Anim. Husb. 12:131-135.

Ellis, W.C., W.H. Pfander, M.E. Muhrer, and E.E. Pickett. 1958. Molybdenum as a dietary essential for lambs. J. Anim. Sci. 17:180-188.

Emanuele, S.M., and C.R. Staples. 1990. Ruminal release of minerals from six forage species. J. Anim. Sci. 68:2052-2060.

Emanuele, S.M., C.R. Staples, and C.J. Wilcox. 1991. Extent and site of mineral release from six forage species incubated in mobile dacron bags. J. Anim. Sci. 69:801-810.

Erskine, R.J., R.J. Eberhart, P.J. Grasso, and R. Scholz. 1989. Induction of *Escherichia coli* mastitis in cows fed selenium-deficient or selenium supplemented diets. Am. J. Vet. Res. 50:2093-2100.

Erskine, R.J., R.J. Eberhart, and R. Scholz. 1990. Experimentally induced *Staphylococcus aureus* mastitis in selenium-deficient and selenium-supplemented dairy cows. Am. J. Vet. Res. 51:1107-1111.

Fell, B.F., L.J. Farmer, C. Farquharson, I. Bremner, and D.S. Graca. 1985. Observations on the pancreas of cattle deficient in copper. J. Comp. Path. 95:573-590.

Ferguson, E.G., G.B.B. Mitchell, and A. MacPherson. 1988. Cobalt deficiency and *Ostertagia circumcincta* infection in lambs. Vet. Rec. 124:20-21.

Fleming, G.A. 1973. Mineral composition of herbage. p. 529-566. *In* G.W. Butler and R.W. Bailey (ed.) Chemistry and biochemistry of herbage. Academic Press, New York.

Fontenot, J.P., V.G. Allen, G.E. Bunce, and J.P. Goff. 1989. Factors influencing magnesium absorption and metabolism in ruminants. J. Anim. Sci. 67:3445-3455.

Gengelbach, G.P., J.D. Ward, and J.W. Spears. 1992. Effect of maternal dietary iron, molybdenum and copper on copper status and cellular immunity of calves. J. Anim. Sci. 70(Suppl. 1):301 (abstr.).

Gerloff, B.J. 1992. Effect of selenium supplementation on dairy cattle. J. Anim. Sci. 70:3934-3940.

Glenn, B.P., D.G. Ely, S. Glenn, L.W. Douglass, L.S. Bull, and L.P. Bush. 1985. Effects of ammonium nitrate and potassium sulfate fertilization on rates of ruminal in situ disappearance of tall fescue and orchardgrass nitrogen and sulfur. Can. J. Anim. Sci. 65:631-645.

Goodrich, R.D., T.S. Kahlon, D.E. Pamp, and D.P. Cooper. 1978. Sulfur in ruminant nutrition. National Feed Ingredients Association, West Des Moines, IA.

Gooneratne, S.R., J. McC Howell, and J.M. Gawthorne. 1981. Intravenous administration of thiomolybdate for the treatment and prevention of chronic copper poisoning in sheep. Br. J. Nutr. 46:457-468.

Gooneratne, S.R., W.T. Buckley, and D.A. Christensen. 1989. Review of copper deficiency and metabolism in ruminants. Can. J. Anim. Sci. 69:819-845.

Greene, L.W., J.P. Fontenot, and K.E. Webb, Jr. 1983. Site of magnesium and other macromineral absorption in steers fed high levels of potassium. J. Anim. Sci. 57:503-510.

Guardiola, C.M., G.C. Fahey, Jr., J.W. Spears, U.S. Garrigus, O.A. Izquierdo, and C. Pedroza. 1983. The effects of sulfur supplementation on cellulose digestion in vitro and on nutrient digestion, nitrogen metabolism and rumen characteristics of lambs fed on good quality fescue and tropical star grass hays. Anim. Feed Sci. Technol. 8:129-138.

Gunn, R.G. 1969. The effects of calcium and phosphorus supplementation on the performance of Scottish Blackface hill ewes, with particular reference to the premature loss of permanent incisor teeth. J. Agric. Sci., Camb. 72:371-378.

Halpin, C.G., D.J. Harris, I.W. Caple, and D.S. Petterson. 1984. Contribution of cobalamin analogues to plasma vitamin B_{12} concentrations in cattle. Res. Vet. Sci. 37:249-251.

Hambidge, K. M., C.C. Casey, and N.F. Krebs. 1986. Zinc. p. 1-137. *In* W. Mertz (ed.) Trace elements in human and animal nutrition. Vol. 2. Academic Press, New York.

Hamdy, A.H., W.D. Pounder, A.L. Trapp, D.S. Bell, and A. Laqace. 1968. Effect on lambs of selenium administration to pregnant ewes. J. Vet. Med. Assoc. 143:749-751.

Harrison, G.A., K.A. Dawson, and R.W. Hemken. 1992. Effects of high iron and sulfate ion concentrations on dry matter digestion and volatile fatty acid production by ruminal microorganisms. J. Anim. Sci. 70:1188-1194.

Hatch, P.A., J.W. Spears, L. Goode, and B.H. Johnson. 1987. Influence of dietary zinc on growth and testicular development in ram lambs fed a high fiber diet. Nutr. Rep. Int. 35:1175-1183.

Hazell, T. 1985. Minerals in foods: Dietary sources, chemical forms, interactions, bioavailability. World Rev. Nutr. Diet. 46:1-123.

Healy, W.B. 1972. In vitro studies on the effects of soil on elements in ruminal, duodenal and ileal liquors from sheep. N. Z. J. Agric. Res. 15:289-305.

Hidiroglou, M. 1979. Manganese in ruminant nutrition. Can. J. Anim. Sci.

59:217-236.

Hurley, L.S., and C.L. Keen. 1987. Manganese. p. 185-223. *In* W. Mertz (ed.) Trace elements in human and animal nutrition. Vol. 1. Academic Press, New York.

Ivan, M. 1988. Effect of faunation on ruminal solubility and liver content of copper in sheep fed low or high copper diets. J. Anim. Sci. 66:1496-1501.

Jelinek, P.D., T. Ellis, R.H. Wroth, S.S. Sutherland, H.G. Masters, and D.S. Petterson. 1988. The effect of selenium supplementation on immunity and the establishment of an experimental *Haemonchus contortus* infection in weaner Merino sheep fed a low selenium diet. Aust. Vet. J. 65:214-217.

Jones, D.G., and N.F. Suttle. 1981. Some effects of copper deficiency on leukocyte function in sheep and cattle. Res. Vet. Sci. 31:151-156.

Jones, L.H.P. 1978. Mineral components of plant cell walls. Am. J. Clin. Nutr. 31:S94-S98.

Jones, M.B., J.H. Oh, and J.E. Ruckman. 1970. Effect of P and S fertilization on the nutritive value of subclover. Proc. N.Z. Grassl. Assoc. 32:69-75.

Jones, M.B., V.V. Rendig, D.T. Torell, and T.S. Inouye. 1982. Forage quality for sheep and chemical composition associated with sulfur fertilization on a sulfur deficient site. Agron. J. 74:775-780.

Jones, M.B., M.W. Demment, C.E. Vaughn, G.P. Deo, M.R. Dally, and D.M. Center. 1990. Effects of phosphorus and sulfur fertilization on subclover-grass pasture production as measured by lamb gain. J. Prod. Agric. 3:534-539.

Judkins, M.B., J.D. Wallace, E.E. Parker, and J.D. Wright. 1985. Performance and phosphorus status of range cows with and without phosphorus supplementation. J. Range Manage. 38:139-143.

Kabaija, E., and O.B. Smith. 1988. Trace element kinetics in the digestive tract of sheep fed diets with graded levels of dietary fiber. J. Anim. Physiol. Anim. Nutr. 59:218-224.

Kennedy, D.G., P.B. Young, W.J. McCaughey, S. Kennedy, and W.J. Blanchflower. 1991. Rumen succinate production may ameliorate the effects of cobalt-vitamin B_{12} deficiency on methylmalonyl CoA mutase in sheep. J. Nutr. 121:1236-1242.

Kincaid, R.L., and J.D. Cronrath. 1983. Amounts and distribution of minerals in Washington forages. J. Dairy Sci. 66:821-824.

Knott, R.W., J.L. Ruttle, and G.W. Southward. 1983. Effects of vitamin E and selenium injections on reproduction and preweaning lamb survival in ewes consuming diets marginally deficient in selenium. J. Anim. Sci. 57:553-558.

Langlands, J.P., G.E. Donald, J.E. Bowles, and A.J. Smith. 1989. Selenium concentrations in the blood of ruminants grazing in northern New South Wales. III. Relationship between blood concentration and the response in liveweight of grazing cattle given a selenium supplement. Aust. J. Agric. Res. 40:1075-1083.

Langlands, J.P., G.E. Donald, J.E. Bowles, and A.J. Smith. 1991a. Subclinical selenium insufficiency. I. Selenium status and the response in liveweight gains and wool production of grazing ewes supplemented with selenium. Aust. J. Exp.Agric. 31:25-31.

Langlands, J.P., G.E. Donald, J.E. Bowles, and A.J. Smith. 1991b. Subclinical selenium insufficiency. II. The response in reproductive performance of grazing ewes supplemented with selenium. Aust. J. Exp. Agric. 31:33-35.

Langlands, J.P., G.E. Donald, J.E. Bowles, and A.J. Smith. 1991c. Subclinical selenium deficiency. III. The selenium status and productivity of lambs born to ewes supplemented with selenium. Aust. J. Exp. Agric. 31:37-43.

Larsen, H.J., G. Qvernes, and K. Moksnes. 1988. Effect of selenium on sheep lymphocyte responses to mitogens. Res. Vet. Sci. 45:11-15.

Ledoux, D.R., and F.A. Martz. 1991. Ruminal solubilization of selected macrominerals from forages and diets. J. Dairy Sci. 74:1654-1661.

Little, D.A. 1982. Utilization of minerals. p. 259-283. In J.B. Hacker (ed.) Nutritional limits to animal production from pastures. Commonwealth Agricultural Bureaux, Slough, England.

Lopez-Guisa, J.M., and L.D. Satter. 1992. Effect of copper and cobalt addition on digestion and growth in heifers fed diets containing alfalfa silage or corn crop residues. J. Dairy Sci. 75:247-256.

MacPherson, A., G. Fisher, and J.E. Paterson. 1989. Effect of cobalt deficiency on the immune function of ruminants. p. 397-398. In L.S. Hurley, B. Lonnerdal, C.L. Keen and R.B. Rucker (ed.) Trace elements in man and animals - TEMA 6. Plenum Press, New York.

Martens, H., and Y. Rayssiguier. 1980. Magnesium metabolism and hypomagnesaemia. p. 447-466. *In* Y. Ruckebusch and P. Thivend (ed.) Digestive physiology and metabolism in ruminants. MTP Press Limited, Lancaster, England.

Martens, H., O.W. Kubel, G. Gabel, and H. Honig. 1987. Effects of low sodium intake on magnesium metabolism of sheep. J. Agric. Sci., Camb. 108:237-243.

Martz, F.A., A.T. Belo, M.F. Weiss, R.L. Belyea, and J.P. Goff. 1990. True absorption of calcium and phosphorus from alfalfa and corn silage when fed to lactating cows. J. Dairy Sci. 73:1288-1295.

Mason, J. 1986. Thiomolybdates: Mediators of molybdenum toxicity and enzyme inhibitors. Toxicology 42:99-109.

Masters, D.G., and. H.E. Fels. 1980. Effect of zinc supplementation on the productive performance of grazing Merino ewes. Biol. Trace Element Res. 2:281-290.

Masters, D.G., and H.E. Fels. 1985. Zinc supplements and reproduction in grazing ewes. Biol. Trace Element Res. 7:89-93.

Masters, D.G., D.I. Paynter, J. Briegel, S.K. Baker, and D.B. Purser. 1988. Influence of manganese intake on body, wool and testicular growth of young rams and on the concentration of manganese and the activity of manganese enzymes in tissues. Aust. J. Agric. Res. 39:517-524.

Mayland, H.F., R.C. Rosenau, and A.R. Florence. 1980. Grazing cow and calf responses to zinc supplementation. J. Anim. Sci. 51:966-974.

Mayland, H.F., L.W. Greene, D.L. Robinson, and S.R. Wilkinson. 1990. Grass tetany: A review of Mg in the soil-plant-animal continuum. p. 29-41. Proc. Pacific Northwest Anim. Nutr. Conf., Nov. 6-8, 1990, Vancouver, British Columbia.

McDonald, J.W., D.J. Overend, and D.I. Raynter. 1989. Influence of selenium status in Merino weaners on resistance to trichostrongylid infection. Res. Vet. Sci. 47:319-322.

McDowell, L.R. 1992. Minerals in animal and human nutrition. Academic Press, Inc., New York.

Miller, J.K., N. Ramsey, and F.C. Madsen. 1988. The trace elements. p. 342-400. *In* D.C. Church (ed.) The ruminant animal: Digestive physiology and nutrition. Prentice Hall, Englewood Cliffs, NY.

Miller, W.J. 1975. New concepts and developments in metabolism and homeostasis of inorganic elements in dairy cattle. A review. J. Dairy Sci. 58:1549-1560.

Mills, C.F. 1981. Cobalt deficiency and cobalt requirements of ruminants. p. 129-141. *In* W. Haresign (ed.) Recent advances in animal nutrition. Butterworths, Boston, MA.

Mills, C.F. 1987. Biochemical and physiological indicators of mineral status in animals: Copper, cobalt and zinc. J. Anim. Sci. 65:1702-1711.

Mills, C.F., and I. Bremner. 1980. Nutritional aspects of molybdenum in animals. p. 519-542. *In* M. Coughlan (ed.) Molybdenum and molybdenum containing enzymes. Pergamon Press, Oxford, England.

Minson, D.J. 1990. Forage in ruminant nutrition. Academic Press, New York.

Morris, J.G. 1980. Assessment of sodium requirements of grazing beef cattle: A review. J. Anim. Sci. 50:145-152.

Morris, J.G., R.E. Delmas, and J.L. Hull. 1980. Salt (sodium) supplementation of range beef cows in California. J. Anim. Sci. 51:722-731.

Morris, J.G., W.S. Cripe, H.L. Chapman, Jr., D.F. Walker, J.B. Armstrong, J.D. Alexander, Jr., R. Miranda, A. Sanchez, Jr., B. Sanchez, J.R. Blair-West, and D.A. Denton. 1984. Selenium deficiency in cattle associated with heinz bodies and anemia. Science 223:491-492.

Morrison, M., R.M. Murray, and A.N. Boniface. 1990. Nutrient metabolism and rumen micro-organisms in sheep fed a poor-quality tropical grass hay supplemented with sulfate. J. Agric. Sci., Camb. 115:269-275.

Muntifering, R.B., S.I. Smith, and J.A. Boling. 1984. Effect of elemental sulfur supplementation on digestibility and metabolism of early vegetative and fall-accumulated regrowth fescue hay by wethers. J. Anim. Sci. 59:1100-1105.

NRC. 1980. Mineral tolerance of domestic animals. National Academy Press, Washington, DC.

NRC. 1984. Nutrient requirements of beef cattle. (6th rev. ed.) National Academy Press, Washington, DC.

NRC. 1985. Nutrient requirements of sheep. (5th rev. ed.) National Academy Press, Washington, DC.

Neathery, M.W., D.M. Blackmon, W.J. Miller, S. Heinmiller, S. McGuire, J.M. Tarabula, R.P. Gentry, and J.C. Allen. 1981. Chloride deficiency in Holstein calves from a low chloride diet and removal of abomasal contents. J. Dairy Sci. 64:2220-2233.

O'Dell, B.L. 1984. Bioavailability of trace elements. Nutr. Rev. 42:301-308.

Odom, J.W., R.L. Haaland, C.S. Hoveland, and W.B. Anthony. 1980. Forage quality response of tall fescue, orchardgrass and *Phalaris* to soil fertility level. Agron. J. 72:401-402.

Panditharatne, S., V.G. Allen, J.P. Fontenot, and W.H. McClure. 1986. Yield, chemical composition and digestibility by sheep of orchardgrass fertilized with different rates of nitrogen and sulfur or associated red clover. J. Anim. Sci. 62:813-821.

Paynter, D.I., J.W. Anderson, and J.W. McDonald. 1979. Glutathione peroxidase and selenium in sheep. II. The relationship between glutathione peroxidase and selenium-responsive unthriftiness in Merino lambs. Aust. J. Agric. Res. 30:703-709.

Pehrson, B., M. Knutsson, and M. Gyllensward. 1989. Glutathione peroxidase activity in heifers fed diets supplemented with organic and inorganic selenium compounds. Swedish J. Agric. Res. 19:53-56.

Perryman, L.E., D.R. Leach, W.C. Davis, W.D. Mickelson, S.R. Heller, H.D. Ochs, J.A. Ellis, and E. Brummerstedt. 1989. Lymphocyte alterations in zinc-deficient calves with lethal trait A46. Vet. Immuno. Immunopath. 21:239-248.

Phillippo, M., W.R. Humphries, and P.H. Garthwaite. 1987a. The effect of dietary molybdenum and iron on copper status and growth in cattle. J. Agric. Sci., Camb. 109:315-320.

Phillippo, M., W.R. Humphries, T. Atkinson, G.D. Henderson, and P.H. Garthwaite. 1987b. The effect of dietary molybdenum and iron on copper status, puberty, fertility and oestrous cycles in cattle. J. Agric. Sci., Camb. 109:321-336.

Piper, E.L., and J.W. Spears. 1982. Influence of copper and zinc supplementation on mineral status, growth and reproductive performance of heifers. J. Anim. Sci. 55(Suppl. 1):319 (Abstr.).

Playne, M.J., M.G. Echevarria, and R.G. Megarrity. 1978. Release of nitrogen, sulfur, phosphorus, calcium, magnesium, potassium and sodium from four tropical hays during their digestion in nylon bags in the rumen. J. Sci. Food Agric. 29:520-526.

Prabowo, A., J.W. Spears, and L. Goode. 1988. Effects of dietary iron on performance and mineral utilization in lambs fed a forage-based diet. J. Anim. Sci. 66:2028-2035.

Price, J., A.M. Will, G. Paschaleris, and J.K. Chesters. 1987. Identification of thiomolybdates in digesta and plasma from sheep after administration of ^{99}Mo-labelled compounds into the rumen. Br. J. Nutr. 58:127-138.

Puoli, J.R., G.A. Jung, and R.L. Reid. 1991. Effects of nitrogen and sulfur on digestion and nutritive quality of warm-season grass hays for cattle and sheep. J. Anim. Sci. 69:843-852.

Reece, W.O., P.O. Brackelsberg, and D.K. Hotchkiss. 1985. Erythrocyte changes, serum iron concentration and performance following iron injection in neonatal beef calves. J. Anim. Sci. 61:1387-1394.

Rees, M.C., and D.J. Minson. 1976. Fertilizer calcium as a factor affecting the voluntary intake, digestibility and retention time of pangola grass (*Digitaria decumbens*) by sheep. Br. J. Nutr. 36:179-187.

Rees, M.C., and D.J. Minson. 1978. Fertilizer sulfur as a factor affecting voluntary intake, digestibility and retention time of pangola grass (*Digitaria decumbens*) by sheep. Br. J. Nutr. 39:5-11.

Reffett, J.K., J.W. Spears, and T.T.Brown, Jr. 1988. Effect of dietary selenium on the primary and secondary immune response in calves challenged with infectious bovine rhinotracheitis virus. J. Nutr. 118:229-235.

Reid, R.L., B.S. Baker, and L.C. Vona. 1984. Effects of magnesium sulfate supplementation and fertilization on quality and mineral utilization of timothy hays by sheep. J. Anim. Sci. 59:1403-1410.

Reid, R.L., and D.J. Horvath. 1980. Soil chemistry and mineral problems in farm livestock. A review. Anim. Feed Sci. Tech. 5:95-167.

Reid, R.L., and G.A. Jung. 1991. Plant/soil interactions in nutrition of the grazing animal. p. 48-63. *In* Proc. 2nd Grazing Livestock Nutrition Conf., Aug. 2-3, 1991, Steamboat Springs, CO.

Reid, R.L, G.A. Jung, C.H. Wolf, and R.E. Kocher. 1979. Effects of magnesium fertilization on mineral utilization and nutritional quality of alfalfa for lambs. J. Anim. Sci. 48:1191-1201.

Reid, R.L., G.A. Jung, W.L. Stout, and T.S. Ranney. 1987. Effects of varying zinc concentrations on quality of alfalfa for lambs. J. Anim. Sci. 64:1735-1742.

Rice, R.W., G.E. Nelms, and C.O. Schoonover. 1967. Effects of injectable iron on blood hematocrit and hemoglobin and weaning weight of beef calves. J. Anim. Sci. 26:613-617.

Robinson, D.L., L.C. Kappel, and J.A. Boling. 1989. Management practices to overcome the incidence of grass tetany. J. Anim. Sci. 67:3470-3484.

Rook, J.A., A.O. Akinsoyinu, and D.G. Armstrong. 1983. The release of mineral elements from grass silages incubated in sacco in the rumens of Jersey cattle. Grass Forage Sci. 38:311-316.

Shariff, M.A., R.J. Boila, and K.M. Wittenberg. 1990. Effects of dietary molybdenum on rumen dry matter disappearance in cattle. Can. J. Anim. Sci. 70:319-323.

Smart, M.E., R. Cohen, D.A. Christensen, and C.M. Williams. 1986. The effects of sulfate removal from the drinking water on the plasma and liver copper and zinc concentrations of beef cattle and their calves. Can. J. Anim. Sci. 66:669-680.

Smith, K.L., J.H. Harrison, D.D. Hancock, D.A. Todhunter, and H.R. Conrad. 1984. Effect of vitamin E and selenium supplementation on incidence of clinical mastitis and duration of clinical symptoms. J. Dairy Sci. 67:1293-1300.

Smith, R.M. 1987. Cobalt. p. 143-183. In W. Mertz (ed.) Trace elements in human and animal nutrition. Vol. 1. Academic Press, New York.

Spears, J.W., J.C. Burns, and P.A. Hatch. 1985. Sulfur fertilization of cool season grasses and effect on utilization of minerals, nitrogen and fiber by steers. J. Dairy Sci. 68:347-355.

Spears, J.W., D.G. Ely, and L.P. Bush. 1978. Influence of supplemental sulfur on in vitro and in vivo microbial fermentation of Kentucky 31 tall fescue. J. Anim. Sci. 47:552-559.

Spears, J.W., R.W. Harvey, and E.C. Segerson. 1986. Effects of marginal selenium deficiency and winter protein supplementation on growth. reproduction and selenium status of beef cattle. J. Anim. Sci. 63:586-594.

Stabel, J.R., and J.W. Spears. 1990. Effect of copper on immune function and disease resistance. p. 243-252. In C. Kies (ed.) Copper bioavailability and metabolism. Plenum Publishing Corp., New York.

Stabel, J.R., and J.W. Spears. 1993. Role of selenium in immune responsiveness and disease resistance. p. 333-356. In D.M. Klurfeld (ed.) Human nutrition-A comprehensive treatise. Vol. 8: Nutrition and immunology. Plenum Press, New York.

Stabel, J.R., J.W. Spears, T.T. Brown, Jr., and J. Brake. 1989. Selenium effects on glutathione peroxidase and the immune response of stressed calves challenged with *Pasteurella hemolytica*. J. Anim. Sci. 67:557-564.

Suttle, N.F. 1974. Effects of organic and inorganic sulfur on the availability of dietary copper to sheep. Br. J. Nutr. 32:559-568.

Suttle, N.F. 1987. The absorption, retention and function of minor nutrients. p. 333-361. *In* J.B. Hacker and J.H. Ternouth (ed.) The nutrition of herbivores. Academic Press, Australia.

Suttle, N.F. 1991. The interactions between copper, molybdenum and sulfur in ruminant nutrition. Annu. Rev. Nutr. 11:121-140.

Suttle, N.F., and D.G. Jones. 1989. Recent developments in trace element metabolism and function: Trace elements, disease resistance and immune responsiveness in ruminants. J. Nutr. 119:1055-1061.

Suttle, N.F., D.P. Knox, K.W. Angus, F. Jackson, and R.L. Coop. 1992a. Effects of dietary molybdenum on nematode and host during *Haemonchus contortus* infection in lambs. Res. Vet. Sci. 52:230-235.

Suttle, N.F., D.P. Knox, F. Jackson, R.L. Coop, and K.W. Angus. 1992b. Effects of dietary molybdenum on nematode and host during *Trichostrongylus vitrinus* infection in lambs. Res. Vet. Sci. 52:224-229.

TCORN. 1991. A reappraisal of the calcium and phosphorus requirements of sheep and cattle. Nutr. Abstr. Rev. (Series B) 61:573-612.

Thompson, J.K., and V.R. Fowler. 1991. The evaluation of minerals in the diets of farm animals. p. 235-259. *In* J. Wiseman and D.J.A. Cole (ed.) Feedstuff evaluation. Butterworths, Cambridge, England.

Torre, M., A.R. Rodriguez, and F. Saura-Calixto. 1991. Effects of dietary fiber and phytic acid on mineral availability. Crit. Rev. Food Sci. Nutr. 1:1-22.

Underwood, E.J. 1981. The mineral nutrition of livestock. Commonwealth Agricultural Bureaux, Slough, England.

Van Bruwaene, R., G.B. Gerber, R. Kirchmann, J. Colard, and J. Van Kerkom. 1984. Metabolism of ^{51}Cr, ^{54}Mn, ^{59}Fe and ^{60}Co in lactating dairy cows. Health Physics 46:1069-1082.

van Eys, J.E., and R.L. Reid. 1987. Ruminal solubility of nitrogen and minerals from fescue and fescue-red clover herbage. J. Anim. Sci. 65:1101-1112.

Walker, S.K., G.P. Hall, D.H. Smith, R.W. Ponzoni, and G.J. Judson. 1979. Effect of selenium supplementation on survival, liveweight and wool weight of young sheep on Kangaroo Island, South Australia. Aust. J. Exp. Agric. Anim. Husb. 19:689-706.

Ward, G.M. 1978. Molybdenum toxicity and hypocuprosis in ruminants: A review. J. Anim. Sci. 46:1078-1085.

Ward, G., L.H. Harbers, and J.J. Blaha. 1979. Calcium-containing crystals in alfalfa: Their fate in cattle. J. Dairy Sci. 62:715-722.

Wheeler, J.L., D.A. Hedges, K.A. Archer, and B.A. Hamilton. 1980. Effect of nitrogen, sulfur and phosphorus fertilizer on the production, mineral content and cyanide potential of forage sorghum. Aust. J. Exp. Agric. Anim. Husb. 20:330-338.

Wheeler, J.L., D.A. Hedges, and A.R. Till. 1975. A possible effect of cyanogenic glucoside in sorghum on animal requirements for sulfur. J. Agric. Sci., Camb. 84:377-379.

Whitehead, D.C., K.M. Goulden, and R.D. Hartley. 1985. The distribution of nutrient elements in cell wall and other fractions of the herbage of some grasses and legumes. J. Sci. Food Agric. 36:311-318.

Wilson, G.F. 1980. Effects of magnesium supplements on the digestion of forages and milk production of cows with hypomagnesaemia. Anim. Prod. 31:153-157.

Woolliams, C., N.F. Suttle, J.A. Woolliams, D.G. Jones, and G. Wiener. 1986. Studies on lambs from lines genetically selected for low and high copper status. Anim. Prod. 43:293-301.

Wylie, M.J., J.P. Fontenot, and L.W. Greene. 1985. Absorption of magnesium and other macrominerals in sheep infused with potassium in different parts of the digestive tract. J. Anim. Sci. 61:1219-1229.

CHAPTER 8

FUNGAL ENDOPHYTES, OTHER FUNGI, AND THEIR METABOLITES AS EXTRINSIC FACTORS OF GRASS QUALITY

C. W. Bacon

INTRODUCTION

Considerable improvements in forage quality of grasses have been made within the past two decades. These improvements are consequences of genetic manipulations of grass cultivars and improved management strategies which guarantee that there is more than an adequate amount of grasses for grazing. These improvements are integral factors in the concept of forage quality. Nevertheless, animals grazed on certain properly managed grasses may still have reduced performance. Reduced animal performance is characterized by low gains, reproduction difficulties, low acceptability, and toxicity syndromes. The factors in forage grasses responsible for reduced animal performance are called antiquality components, and consist of a large diverse group of chemical constituents which may be divided into constitutive and extrinsic or exogenous antiquality components. Constitutive antiquality components are those that are produced directly by and are, therefore, inherent factors of grass species. Extrinsic antiquality components are those that are produced on a grass species but by another biological entity and are exogenous to the normal metabolism of that grass species.

This review is concerned with extrinsic antiquality factors of grasses produced by fungi which are commonly referred to as mycotoxins. These antiquality factors are biological, originating from a variety of fungi, some of which have been associated not only with animal performance problems but also human toxicities since ancient times. In this regard, species of *Claviceps* and their toxins (ergot alkaloids) are historically perhaps the first recorded groups of extrinsic biological antiquality factors reported in forages.

The primary focus of this chapter is to critically evaluate an expanding body of literature which centers around toxins associated with fungi which live within forage grasses as endophytes, resulting in an antiquality component of an otherwise desirable forage species. Evidence will be presented which establishes that most of the cultivated forage grasses are descendants of

---

USDA/ARS, Toxicology and Mycotoxin Research Unit, Richard B. Russell Agricultural Research Center, Athens, GA 30613.

either wild or ancestral species that are naturally infected with endophytes. Further, a considerable discussion pertaining to positive interactions of endophytes with grasses will be presented which should prove helpful in understanding the symbiotic operation of these organisms in natural systems. This latter discussion is warranted since the association of endophytic fungi with most of the major forage grasses is natural and ecologically significant and their removal from grasses may, in certain instances, result in poor forage productivity which might indirectly affect forage quality.

In this chapter, information from basic and applied sources has been assembled to illustrate unique characteristics of the endophyte-grass-livestock interaction, which has an evolutionary basis and is fundamental to the disastrous effects on a trophic interaction, referred to here as grazing. Because of space limitations, this review emphasizes antiquality factors and toxins related to fungal endophytes. However, there is another large group of fungi that also affects grass quality. These fungi, discussed in part two, are referred to here as mycotoxic nonendophytic fungi. Mycotoxic nonendophytic fungi include nonmutualistic saprophytes living on dead matter, endophytic latent fungi, and localized systemic pathogens of grasses. Nonendophytic fungi will be discussed very briefly since it is this group that has been recently reviewed extensively (see Smith and Henderson, 1991), and historically the numerous early studies of this group formed the basis for our understanding of toxic fungi in general.

PART I: GRASS ENDOPHYTES

Biological Concepts

Endophytic Species of Fungi

The endophytic fungi in this review include fungi of the tribe Balansiae (family Clavicipitaceae, class Ascomycetes) and their related anamorphs (Figure 1). The species within this tribe were initially delineated only on the nature and degree of association with ovaries and vegetative parts of grasses (Diehl, 1950). However, this classification system fails to recognize phylogenetic relationships among species, and does not account for any relationships of species with only an anamorphic state.

A recent consideration of the phylogeny of the clavicipitaceous fungi as expressed in Figure 1 indicates two major features of this family, one group parasitic on insects (Cordycepitoideae typified by species of *Cordyceps*), the other two groups parasitic on grasses (Bacon and Hill, 1994). This phylogeny is based on a series of studies which, in addition to the host-fungus relationships of Diehl (1950), also include fungal morphology, conidiation, biochemistry, and molecular biology (Sampson, 1933; Diehl, 1950; Luttrell, 1979; Luttrell and Bacon, 1977; Latch et al., 1984; Rykard et al., 1984; Rykard et al., 1985; Bacon et al., 1986; White, 1987; White, 1988).

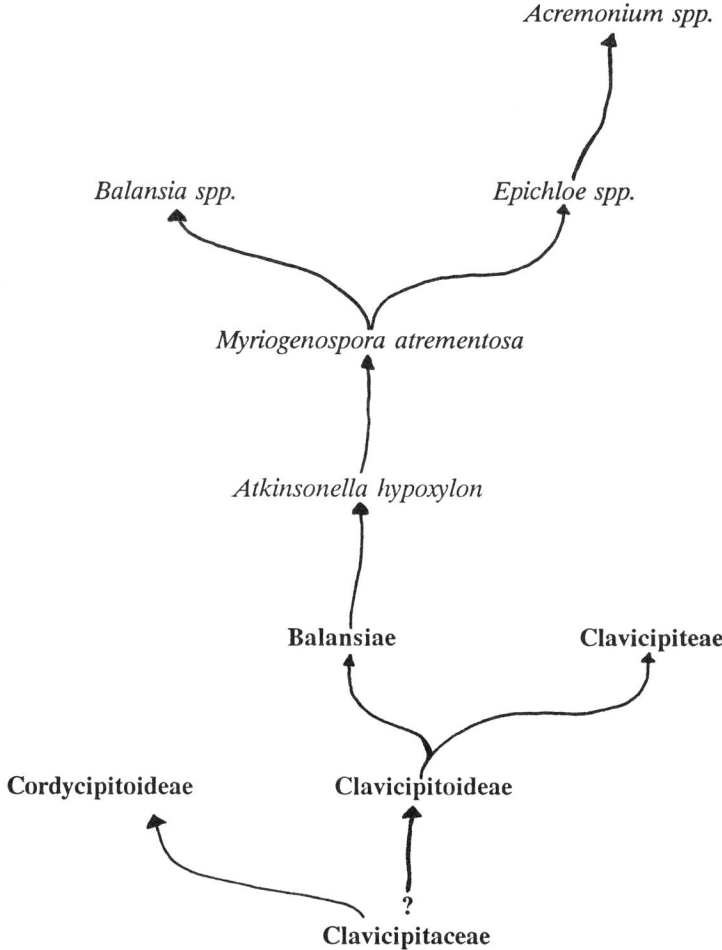

Figure. 1. Possible evolutionary lines of taxa among grass endophytes of the tribe Balansiae within the family Clavicipitaceae, subfamily Clavicipitoideae, beginning with an unknown (?) ancestor. The tribe Clavicipiteae and the subfamily Cordycipitoideae are illustrated for possible relationships within the family, although no cladistics is implied.

Considered from the above viewpoint, there are four major tendencies within the Clavicipitoideae: 1) a line indicating localized parasites of grass ovaries as shown by the *Claviceps* (Clavicipiteae) and also characterized by having a stage independent of the grass, the sclerotium; 2) the change from a strictly ovarian pathogen and location, to also include the foliage parasitic habit (systemic) with varying degrees of the endophytic habit, and the loss of an independent stage from the grass (presently shown by the monotypic species *(Atkinsonella hypoxylon)*; 3) development of an endophytic habit but slightly pathogenic *(Myriogenospora atrementosa)* which culminates with species of *Balansia* characterized by a completely endophytic habit; and 4) the loss of sexuality, they are nonpathogenic, and are strictly seedborne and this group culminates in the development of the completely endophytic mutualistic habit (the *Acremonium* species). Thus, the *Acremonium* endophytes which are fungi associated with important forage grasses, also represent the most advanced group of grass endophytes.

This phylogeny serves to unify the Clavicipitaceae into one group but yet separates the endophytic fungi into two very broad categories: 1) those that produce external stromata (a mass of fungus hyphae which produces spores) on their grass hosts, and 2) those that do not and which, for the most part, are imperfect fungi only upon isolation in culture.

The first group include the fungi *Atkinsonella, Balansia, Epichloe,* and *Myriogenospora* which form external primary anamorphic stromata within or around stems, leaves, or florets of grasses from which either effuse, pulvinate, or stalked perithecial stromata arise and are, therefore, fungi of the first category. Most of these endophytic species are not associated with major forage grasses, but are associated with invasive grass species, particularly rangeland species (Diehl, 1950; White and Cole, 1986; White et al., 1990; White et al., 1992) (Table 1). Fungi of these genera are taxonomically related and have been placed in the tribe Balansiae of the family Clavicipitaceae (Diehl, 1950), and referred to here as the balansioid endophytes. Other essential distinguishing characteristics of species of this tribe are included in early studies (Luttrell and Bacon, 1977; Rykard et al., 1984; Bacon and De Battista, 1991).

The species in the second category include the genera *Acremonium, Gliocladium, Phialophora,* and *Pseudocercosporella*. Also included in this latter category are those endophytes that have been observed in a variety of annual ryegrasses (Latch et al., 1988; Nelson and Read, 1990). Most of these endophytic species are associated with important forage species and therefore represent the greatest economic concern to forage quality (Table 1). Since these species reproduce asexually by conidia, they belong to the form-class Deuteromycetes, and are, for the sake of convenience, separated from the balansioid endophytes and referred to here as the acremonioid endophytes.

As presently defined, the genus *Acremonium* consists of *A. coenophialum* (Morgan-Jones and Gams, 1982) from two cultivars of tall fescue (*Festuca*

Table 1. Distribution of endophytes among the world's major forage grasses.*

Tribe/species	Common name	Endophyte genus
AGROSTIDEAE		
Agrostis alba, and other spp.	Bent grasses	*Acremonium*
Alopecurus pratensis	Meadow foxtail	*Acremonium***
Phelum pratense	Timothy	"
Sporobolus	Dropseed	*Balansia*
Stipa comata	Needle grass	*Acremonium*
AVENEAE		
Arrhenatherum elatium	Tall oatgrass	*Acremonium***
ANDROPOGONEAE		
Andropogon scoparium	Blue stem	*Balansia*
Sorghum vulgre, and other spp.	Forage sorghums	*Balansia*
CHLORIDEAE		
Bouteloua spp.	Grama grasses	*Balansia*
Chloris gayanam	Rhodesgrass	"
Cynodon dactylon	Bermudagrass	"
		"
FESTUCEAE		
Dactylis glomerata	Orchard grass	*Acremonium***
Elymus spp.	Wildryes	"
Eragrostis spp.	Lovegrasses	*Balansia*
Festuca spp.	Fescue grasses	*Acremonium*
Lolium spp.	Ryegrasses	"
Poa spp.	Bluegrasses	*Acremonium***
PANICEAE		
Axonopus offinis	Carpet grass	*Balansia* spp.
Panicum spp.	Panic grasses	*Balansia* spp.
Paspalum spp.	Dallisgrasses	"
Setaria italica	Foxtail millet	"

*The major forage grasses are compilations from Hoover et al., 1948; Bula et al., 1977; and Crowder, 1977. The distribution of endophytes is based on the work of Diehl, 1950; Kohlmeyer and Kohlmeyer, 1974; and White, 1987.

**Indicates that this endophytic species is *Epichloe typhina (A. typhinum)*.

arundinacea Schreb.) and from *Poa autumnalis* Muhl. ex Ell. (White and Bultman, 1987), *A. typhinum* from *F. rubra* L. (Morgan-Jones and Gams,

1982), *A. uncinatum* from *F. pratensis* L. (Gams et al., 1990), and *A. lolii* from *Lolium perenne* L. (Latch et al., 1988).

Since the *Acremonium* species are so agronomically important, much effort is being made to incorporate molecular data to clarify species relations. The essential premise is that *Epichloe typhina* which also produces an *Acremonium* state might be related to other *Acremonium* species. Ribosomal DNA was used to characterize nucleotide sequences among several species of endophytic fungi of grasses, including the *Acremonium* species (Schardl and Siegel, 1993). The results suggested a remarkable similarity of ribosomal DNA among all *Acremonium* organisms and that they were phylogenetically related to, and evolved from *E. typhina* (Schardl and Siegel, 1993). They also found that fungal mutualists did not necessarily coevolve with their grass hosts. In fact, as much dissimilarity occurred among *A. coenophialum* isolates from *F. arundinacea* as existed between *A. coenophialum* and *A. lolii* or *A. typhinum*. These findings raise doubts as to the true taxonomic derivation among species of *Acremonium* within sect. *Albo-lanosa*. Further, the ability of both the acremonioid and balansioid fungi to produce ergot alkaloids indicates at least a biochemical affinity among members of the Clavicipitaceae.

The Endophytic Habit

Symbiosis Endophytic fungi are those fungi which live their entire life cycle within the aerial portions of grasses and sedges by forming nonpathogenic and completely intercellular associations (Sampson, 1933; Diehl, 1950). The grass and fungus are symbiotically associated as a single ecological entity. Thus, in the discussions to follow, symbiotum (plural symbiota) or symbiotic will be used to refer to the traditional endophyte-infected grass terminology (Schardl et al., 1991) and nonsymbiotum to refer to noninfected grasses. The use of these terms will serve not only to emphasize the naturalness and importance of this association, but also to indicate potential problems which might occur if it is dismantled in attempts to remove the fungus as an antiquality factor.

The nature of the symbiosis varies among each of the major endophytic groups (Figure 1). Some symbioses are suggestive of incomplete, pathogenic relationships as shown by some of the *Balansia* species. The balansioid endophytes exist as free entities only very briefly during the spore stage and there is some question as to the significance of this spore as a dispersal unit. Unlike the balansioid endophytes, the acremonioid endophytes, as well as other forms of symbiotic associations (Ahmadjian and Paracer, 1986), do not live apart not even briefly, i.e., once a tall fescue or ryegrass symbiosis is established, the association is constant. The *Acremonium* symbioses are seed disseminated (Sampson, 1933), their association with the grass begins after germination of infected seed where upon the fungus infects developing seedlings, and symbiotic perennation is guaranteed by infecting the meristematic areas and all young vegetative organs except the roots.

Mutualism Regardless of the degree of the association, all endophytic fungi of grasses form biotrophic associations in which there is little or no destruction of either grass or fungus and as will be presented later, certain members exploit the biochemical properties of the other, indicating that in specific cases the relationship is mutualistic. Mutualisms are specific associations of two or more organisms that are characterized as interactions between individual organisms in which the genetic fitness of each participant is increased by the action of the other. This definition precisely describes the ecological nature of the symbiota and components of the mutualism are aptly referred to as plant or fungal mutualists.

Mutualisms are polymorphic in nature; thus, within a mutualistic population, there exists successful symbionts or true mutualists, unsuccessful symbionts whose fitness is not increased, and nonsymbiotic or noncohabitating individuals (Keeler, 1985). Most of the symbiotic grasses are of this first category and will be treated in this context. Examples of unsuccessful symbiotic grasses might reflect our inability to determine the survival value of specific traits under a precise environmental condition. It is rather doubtful that the third category of noncohabitating individuals exist, at least completely, since the fungi are only found in association with grasses. However, noninfected grasses of an infected species exist which may either occur naturally or they may be agricultural artifacts resulting from agronomic uses and practices, as for example, endophyte-free tall fescue (Bacon and Hill, 1994).

The tall fescue and perennial ryegrass symbiota are defensive mutualisms (Clay, 1988) because the overall competitive benefits (Hill et al., 1991a) are due to specific characteristics such as enhanced drought tolerance (Read and Camp, 1986; Arechavaleta et al., 1989), increased tillering and growth (Latch et al., 1985b; Hill et al., 1990), and increased resistance to herbivory from mammals and insects (Cheplick and Clay, 1988). The information available on *Balansia*-associated grasses suggest that they too are defensive mutualisms (Clay, 1984; Clay et al., 1985; Clay 1986).

The present data provide overwhelming evidence that interactions between the grass and fungus, as well as environment, increase the phenotypic variation among symbiotic plants (Clay, 1984; Kelley and Clay, 1987; Bradshaw, 1988; Hill et al., 1990). They also suggest that symbiotic plants do not have all mechanisms of fitness imparted upon them by the endophyte, but it is the sum of all the plants in the population which express the fitness characteristics (Bacon and Hill, 1994). Therefore, symbiotic populations of grasses are far more plastic and adaptable than nonsymbiotic populations. How fast symbiotic plants encroach upon nonsymbiotic plants in a mixed community will depend upon the types and severity of stresses imposed upon the plants in their environment. Undoubtedly, severe or repeated cattle grazing and other severe environmental stresses will eliminate nonsymbiotic grasses, while reduced grazing will have minimal effect on the competitiveness of a nonsymbiotic plant. Generally, the more ideal the growing conditions, the better the ability of the nonsymbiotic plant to compete with its symbiotic counterpart.

The endophytic habit is important not only from the standpoint that it is where the fungus resides but also because it is here that the fungus and (or) grass produce antiquality factors referred to as mycotoxins. Further, the nature of the endophytic habit is important because understanding will assist in formulating any possible control measures to remove the fungus. The endophytic habit was recognized as early as 1887 (De Bary, 1887) and described in the fungus *Epichloe typhina* (Pers. ex Fr.) Tul. The endophytic habit was subsequently established as a perennial feature of several grasses (Diehl, 1950), and this was followed by studies on the distribution of the endophyte within grasses (Sampson, 1933; Diehl, 1950). Most endophytes produce an external and transient sporulation structure (conidial stroma) upon which spores are produced which infect grasses.

Within an infected grass, the endophyte is intercellular and there is no evidence of any endophytic species producing mass destruction of host cells. However, this is not to imply that an endophyte does not 'create' a niche for itself within the grass. The intercellular spaces occupied by the fungus are not vacant spaces but spaces normally occupied by the middle lamellae of each cell. The endophyte apparently has the necessary enzymes to dissolve the middle lamellae as it grows between each layer of cells. There is no nutrient absorbing structure, typical of other parasitic fungi. Nutrient exchange takes place within this location, which consists of an external cellular plant matrix referred to as the apoplasm. The nutrients within the apoplasm may either be derived from the host due to normal cytoplasmic leakage and (or) from products derived from the digestion of the middle lamellae as the fungus grows between cells. This suggests that added nutrients will be available to the endophyte during periods of rapid plant growth, which would influence, and certainly correlates with, increased fungal growth and biochemical activities as evidenced by the periods of high ergot alkaloid content and herbivore toxicities (Belesky et al., 1987b; Belesky et al., 1988; Hill et al., 1990).

The extent of host colonization by an endophyte is endophyte-specific. Thus, species of *Balansia* and *A. hypoxylon* are found throughout all organs of the grass, except the roots. The species *Epichloe*, *Acremonium*, and *Myriogenospora* are found only within the sheath of leaves, meristematic areas, and inflorescence structures. Further, some endophytes appear to have a quantitative expression of hyphae within a host, e.g., the endophyte of the annual ryegrass while found in the sheath as discontinuous sparse fragments is found in much higher density within the seed (Latch et al., 1988; Nelson and Read, 1990). A single plant is usually not infected by more than one endophyte, but a grass species can serve as a host for several different endophytic species. However, double infection of a single grass plant by different genera has been reported, as in *Panicum anceps* infected by *M. atramentosa* and *B. henningsiana* (Luttrell and Bacon, 1977; Rykard, 1983). This may reflect occupation of different in situ locations. *Balansia* species are generalized and endophytic, whereas the infection by *M. atramentosa* is localized and systemic. Nevertheless, Latch et al., (1984) reported the co-occurrence of two endophytic fungi, *Gliocladium*-like and *Phialophora*-like, colonizing identical tissues along with *A. lolii* and *A. coenophialum* in plants of perennial ryegrass (*L. perenne*) and tall fescue, respectively.

Coevolution The reciprocal potential for genetic fitness derived from a mutualism serves as the driving force for the coevolution of endophytes and grasses. Plant-fungus associations appeared to have entered close relationships very early in their evolution. The Gramineae probably originated in the upper Cretaceous Period of the Mesozoic Era, much later than dicotyledonous plants. The fossil records of the first authentic species of grasses, fruits of *Stipa* (Cockerell, 1956) and *Phalaris* (Beetle, 1958), were obtained from the late Tertiary deposits, approximately 40 to 50 million years ago. The oldest fossil *Festuca* species was reported in the Miocene epoch of the Tertiary period (Thomasson, 1986). Thus, geologically speaking, we are dealing with a relatively young association which must have been initiated at least during the early Pliocene of the Cenozoic Era, approximately 25 million years ago.

To understand the need for a mutualistic relationship to develop between grasses and fungal endophytes, a complete understanding of the systematics and evolution of both the endophyte and species, their biochemical relatedness and requirements must be known. For more detailed discussions of this evolutionary aspect, as well as a model for fitness and other salient but theoretical evolutionary events within these mutualisms, the reader is referred to earlier reviews (White, 1988; Bacon and Hill, 1994). Briefly, the major impetus considered responsible for extending the adaptability of symbiotic grasses are the variety of secondary metabolites, most of which are produced by the fungus within the association.

Fungi are rated as highly adaptable because they have biochemical abilities to utilize a wide variety of substrates and produce precursors from intermediary metabolism for use in primary and secondary metabolism. On the other hand, grasses are one of the few groups of plants that lack the ability to produce excessive secondary metabolites (Zahner et al., 1983). In theory cohabitation and the establishment of a completely compatible association, although gradual, were based partially on the need for secondary metabolites that were lacking in the grass but were contributed by the fungus. The evolutionary events which resulted in the cohabitation of grasses with this group of fungi will probably remain unknown. It is clear, however, that the clavicipitaceous fungal mutualists meet the criterion that Law (1985) proposed for mutualism: 1) the inhabitants are genetically similar while their hosts are genetically diverse, 2) the inhabitants rarely or never undergo sexual reproduction, and 3) the inhabitants lack strong specificity to a particular host species.

Distribution of Endophytes Among Grasses

Endophytic fungi have been reported to infect grass species belonging to six subfamilies of the Gramineae as delineated by Gould and Shaw (1983) and several of these are considered important forage grasses throughout the temperate and tropical grazing zones of the world (Hoover et al., 1948; Bula et al., 1977; Crowder, 1977). Of the North American species of grasses, this represents a total of 232 associations (Bacon and De Battista, 1991). Thus, only 15% of all the New World's grass species are hosts for endophytes but whose distribution ranges from temperate to tropical zones. The distribution of symbiotic grasses appears to be limited to host habitat and climatic conditions. However, several grass species with widely different geographic distribution are infected with the same fungus at each end of their geographic extremes (Diehl, 1950).

Host-specific endophytes have not been demonstrated, although no comprehensive study of this phenomenon has been attempted. In natural grass communities, several host species may be found growing sympatrically but it is usually observed that only one species is infected by a particular fungal endophyte, suggesting that there is a compatibility factor. Experimental infections of grasses and sedges with *Atkinsonella hypoxylon* and *B. cyperi* established that these two fungi were broadly cross-compatible within a specific host population, but *B. cyperi* was less host-specific than *A. hypoxylon* (Leuchtmann and Clay, 1988).

As indicated above, the Balansioid endophytes are probably related to the *Acremonium* endophytes, so their distribution among the grasses will be discussed together (Table 1). Endophytes of this type are found associated with more grass species than any other endophyte (Latch et al., 1984; Williams et al., 1984; Halisky et al., 1985; White and Cole, 1985; White and Cole, 1986; Morgan-Jones et al., 1990; White et al., 1992). However, the distribution of this species within the subfamilies of the Gramineae is very narrow as 90% of the *E. typhina* is associated with the Festucoideae, mainly the Festuca species and other cool season grasses. This percentage includes the related *Acremonium* anamorphs. Both *E. typhina* and *Acremonium* sp. are the endophytes associated with major forage grasses. On the other hand, *Balansia* species are primarily associated with over 102 warm season grass and sedge species from all subfamilies except Bambusoideae. This distribution appears to favor two subfamilies since 97% of the *Balansia*-associated grasses include species of the Panicoideae (64%) and Eragrostoideae (33%) (Bacon and De Battista, 1991).

Endophytic fungi are found in approximately 15% of the North and South American species of grasses, many of which are used as forages (Bacon and De Battista, 1991). However, this percentage is probably much higher as it does not reflect a wide scale sampling of all the forage grass species for endophytes. The *Balansia*-associated grasses are not necessarily major forage species; they are, however, major North American rangeland species (Table 1). *Balansia epichloe* is the most cosmopolitan of the *Balansia* species, its host range varies from cool-season grasses of the Festucoideae to warm-season grasses of the Panicoideae, Eragrostoideae and Arundinoideae. Other *Balansia* species have been reported from only one or two grass tribes or species. Thus, *B. ambiens, B. oryzae,* and *B. pallida* are associated only with grasses of the tribe Oryzeae, while other species of *Balansia* are found infecting only panicoid or eragrostoid grass species. Two species, *B. aristidae* and *B. hemicrypta*, are restricted to the grass genus *Aristida*, while *Balansia oryzae* is one of two species found in association with rice, which is the only known association of an endophyte with a forage crop used for human food (Mohanty, 1964). Two minor genera, *Atkinsonella* and *Myriogenospora*, are even more restrictive in their associations with grasses (Diehl, 1950). Nevertheless, *M. atrementosa* is associated with the Paspalum species which are used as forage grasses. *Atkinsonella hypoxylon, B. pilulaeformis* and *B. cyperi* are associated with a variety of native grasses and nongraminaceous hosts that are found throughout the major grasslands of the temperate and cooler tropical regions of the world.

Positive Benefits Derived from Symbiotic Grass

Natural populations of symbiotic grasses persist longer and compete better than nonsymbiotic populations (Clay, 1986; Clay, 1987; Prestidge et al., 1982; West et al., 1989). However, as discussed earlier, the components of improved fitness of symbiotic populations are multifaceted and difficult to access. Most research on symbiotic grasses has emphasized variation of fitness at the population level, with relatively few studies partitioning that variation into its genetic and environmental components. Thus, it is relatively difficult or impossible to interpret which symbiotic component contributes to the phenotypic expression and any evolutionary significance of much of the fitness variation from such natural systems. The alternative approach of using symbiotic and nonsymbiotic clones (Arechavaleta et al., 1989; Hill et al., 1990; Hill et al., 1991a; Hill et al., 1991b; Agee and Hill, 1991) has greatly facilitated our approach and understanding of several aspects of the variation of mutualistic responses within a population. This approach should allow us to extrapolate to the effects of such variation on trophic interactions, one of which is measured as an antiquality factor. Examples of variation obtained from using clonal lines are indicated below.

In summary improved fitness includes increased growth rate and tiller density (Bradshaw, 1988; Hill et al., 1990; Hill et al., 1991a), changes in morphology (Diehl, 1950; Hill et al., 1990), resistance and toxicity to grazing animals (Byford, 1979; Wallner et al., 1983; Read and Camp, 1986), insect deterrence (Prestidge et al., 1982; Clay, 1988), nematode resistance (Kimmons et al., 1990; West et al., 1990a,b), disease resistance (White and Cole, 1985; Yoshihara et al., 1985), and drought tolerance (Read and Camp, 1986; Arechavaleta et al., 1989; Elmi and West, 1989). The chemical basis for each component of fitness is not completely understood, but data defining cause and effect relationships are eminent. There may be a chemical similarity between compounds which is responsible for each aspect of improved fitness although the compounds may be found in very distinct symbiota (Table 2). However, this information is far from complete as there are only a few experiments which document positive benefits from endophytes within other associations, particularly the balansioid symbiota.

All the positive benefits derived from the symbiota are expressed, or measured when the association is subjected to stresses. Characteristics which are associated with drought tolerance are not expressed unless the plant is exposed to prolonged drought conditions. For example, roots of symbiotic plants will grow faster and deeper into a soil profile under drought stress than its cloned nonsymbiotic ramet. However, there is no difference when both are grown at field capacity (Richardson et al., 1990). Further, total reserve carbohydrate content and structures are similar among plants, regardless of endophyte content, when soil water content is at field capacity (-0.1 bars), but sugar monomers increase and polymers decrease in symbiotic tall fescue once drought stress conditions are imposed (Richardson et al., 1992). Moreover, there is variation among symbiota, as for example, plant morphological changes are not constant among tall fescue genotypes (Hill et al., 1990). This variation is expressed specifically by tillering capacity, specific leaf weights, and crown weight which may increase, decrease, or remain the same depending upon the genotype of the infected plant. All of these factors may explain the basis of variation under stress.

Insect and Nematode Defense

The accumulation of antiquality components (Table 2) in symbiotic grasses is a clear example of a defensive mutualism derived from endophytic fungi. Detailed data expressing this are based on specific and generalized characteristics from insect feeding experiments, which suggest that at least antibiosis may be the functional defense mechanism. Since most insects are grazers, data obtained on insects may be appropriate for mammalian grazers. Thus, it is this category of mutualism which is directly related to the livestock toxicity syndromes discussed below. In the use of endophytes for forage improvements, we are dealing with a double edge sword: 1) a mutualism which is essential for the competitiveness of pasture and rangeland grasses, and 2) an antiquality component. The complex interaction described above must be considered if there are any attempts at removing the mammalian toxicity from the insect toxicity. This removal also may result in the total removal of all toxicity and related stress-sparing mechanisms.

There are numerous effects of symbiotic grasses on insect herbivory (Gaynor and Hunt, 1983; Barker, 1988; West et al., 1989; Siegel et al., 1991), and they suggest some specificity (Hardy et al., 1986; Siegel et al., 1991). The list of insects reported as being deterred or poisoned by symbiotic grasses includes several species of aphids (Latch et al., 1985a; Siegel et al., 1991), sod webworms (Funk et al., 1983), leafhoppers (Pottinger et al., 1985), chinch bugs (Funk et al., 1985), crickets (Asay et al., 1975; Ahmad et al.,

Table 2. Toxins and suspect toxins associated with symbiotic grasses.

Acremonioid symbiota	Balansioid symbiota
Ergovaline	Elymoclavine
Ergonovine	Ergonovine
Ergosine	Ergonine
Chanoclavine I	Chanoclavine I
Peramine	Isochanoclavine I
Indole acetic acid	Agroclavine
Ergosterol	Dihydroelymoclavine
Ergosinine	6,7-Secoagroclavine
Cyclopentanoid sesquiterpenoids*	Penniclavine
Ergonovine	Erytho 1-(3-indoly)propane-1,2,3-triol
Caffeic acid*	Threo 1-(3-indoly)propane-1,2,3-triol
p-Coumaric acid	3-Indole acetic acid
p-Hydroxybenzoic acid	3-Indole ethanol
Loline	3-Indole acetamide
N-acetylloline	Methyl-3-indolecarboxylate
N-formylloline	Ergobalansine
N-acetylnorloline	Ergobalansinine
Paxilline	
Lolitrem A	
Lolitrem B	
Lolitrem C	
Lolitrem D	

*These compounds were isolated from *Epichloe typhina (Acremonium typhinum)* and are grouped with the acremonioid symbiota. Hormane, norharmane, and halostachine were isolated from tall fescue (Yates, 1983), but the infection status of the grasses was not reported; therefore, they have been excluded from the list.

1985), corn flea beetle (Kirfman et al., 1986), black beetle (Siegel et al., 1987), bluegrass billbug (Ahmad et al., 1986), flour beetles (Clay, 1988), and Argentine stem weevil (Prestidge et al., 1982; Pottinger et al., 1985). Symbiotic grasses also are toxic to the lepidopteran larvae of fall armyworms (Hardy et al., 1985) and species of Crambis (Clay et al., 1985). Acremonioid grasses are reported as being nematicidal (Pedersen et al., 1988; Kimmons et al., 1990; West et al., 1990a), but this effect may be nulled if the symbiotic grasses also are associated with mycorrhizal infections (Barker, 1988).

Peramine is the chemical considered responsible for deterring the feeding activity of several insects, specifically the Argentine stem weevil (Rykard et al., 1985) and some species of aphids (Siegel et al., 1991). Peramine is a very simple alkaloid which has been reported in symbiotic perennial ryegrass and tall fescue (Rowan and Gaynor, 1986; Siegel et al., 1991). It is not known if the fungus makes this compound or if its synthesis requires a combination of fungus and grass. In addition to peramine, the loline alkaloids (N-formyl and N-acetyl loline) also have been implicated in the toxicity and deterring activity of one species of aphid, but affected another aphid species only if the infected grass contained other toxins (the ergopeptide alkaloids, peramine, and (or) the lolitrems) (Siegel et al., 1991). Several environmental factors also may be responsible for the final effect (McLean, 1970; Lyons et al., 1990a), and the co-occurrence of simple indoles, and terpinoids (Table 2) also may influence the toxicity.

Abiotic Stress Tolerances

Drought tolerance   Most of the research conducted on the effects of endophyte infection on abiotic stress tolerances has been done on tall fescue. Similar mechanisms might exist for other *Acremonium*-infected grass of the same tribe, i.e., the perennial ryegrass symbiotum. However, it might be inappropriate to use infected tall fescue as a model for the warm-season *Balansia*-infected grasses. Nevertheless, the occurrence of insect and mammalian herbivore deterrences and disease resistance mechanisms in both types of symbiotic grasses suggest that similar abiotic stress mechanisms also might exist in the balansioid symbiota. Again, it is important in this concept to define, separate, and recognize the important contributions made by the fungus from that of the grass. Only when this separation is made can the total genetic potential of forage quality or antiquality be assessed.

Experimental evidence for the occurrence of a drought stress mechanism in symbiotic grasses is an outgrowth of the initial observation of Read and Camp (1986) that two of three populations of symbiotic tall fescue were more drought resistant than tall fescue plants with a low level of infection. Turgor maintenance has been identified as a major mechanism through which tall fescue tolerates drought (West et al., 1989), but root growth (De Battista et al., 1990; Richardson et al., 1990), and stomatal responses (Belesky et al., 1987; Richardson et al., 1992) also are considered important. While control mechanisms for drought tolerance in the tall fescue symbiotum are unknown, it has been determined that only specific tissue types, young meristematic and elongating leaf, from the tall fescue symbiotum were capable of developing low osmotic potential in response to water stress (West et al., 1989). This suggests that the basic mechanism centers around the ability of grasses to develop low osmotic potential (West et al., 1990b). Further, since osmotic adjustment exists primarily in young and immature leaf blades, it may be the mechanism of persistence and tiller survival in the tall fescue symbiotum but only under intermittent drought (West et al., 1989; West et al., 1990b). The universal occurrence of this mechanism throughout populations of the acremonioid symbiota might vary. White et al. (1989) studied the expression of osmotic adjustment in two clones of the tall fescue symbiotum and concluded that cell wall elasticity appeared to explain the differences in turgor maintenance among the two clones.

The nature of the osmoticum responsible for this effect is unknown, but polyols, sugars, and amino acids are considered by many as likely candidates (Morgan, 1984). In addition to being non-metabolizable under stress, the essential substance or substances also must be osmotically active and compatible (nontoxic) with the normal plant physiological processes. Polyols were considered likely candidates since they are normal neutral metabolites of fungi, and have been reported as being osmotica in other plants (Lewis, 1967). In a study designed to examine the involvement of polyols in this mechanism, it was determined that in one genotype of symbiotic tall fescue, mannitol and arabitol were present (Richardson et al., 1992). However, the concentration of these polyols was not high enough to affect the overall osmotic pool in this genotype. It was further shown that in symbiotic grasses grown under drought stress, arabitol, glucose, and fructose were the only compounds which accumulated in sufficient quantities to affect the overall osmotic pool (Richardson et al., 1992).

Nitrogen Efficiency *Acremonium*-infected grasses appear to have some advantage over nonsymbiotic grasses in the area of nitrogen (N) utilization. The efficient utilization of low soil N by the tall fescue symbiota was reported by Arechavaleta et al. (1989). The amounts of dry matter produced by low N (11 mg/pot) were the same as the amount of dry matter produced by nonsymbiotic tall fescue at high (220 mg/pot), medium (73 mg/pot), and low (11 mg/pot) levels of N (Arechavaleta et al., 1989). Of the many enzymes responsible for N utilization, glutamine synthetase is primarily responsible for N efficiency. When the activity of this enzyme within the tall fescue symbiotum was compared to the activity in the uninfected plants grown under low N, it was discovered that glutamine synthetase was higher in the symbiotum. The high level of this enzyme in symbiota grown under low soil levels was considered an efficient means of utilizing N (Lyons et al., 1990b). The ability to efficiently utilize low N apparently is not present in seedlings of this symbiotum, but develops with grass maturity (Clay, 1987).

High levels of N not only increase the amount of dry matter produced in the tall fescue symbiotum (Arechavaleta et al., 1989; De Battista et al.,

1990; Hill et al., 1990), but also the amounts of the ergot alkaloid (Lyons et al., 1986; Arechavaleta et al., 1989; Hill et al., 1990). Thus, although there is an increase in growth from N, the corresponding increase in the amount of toxins in the foliage should reduce herbivory. The effects of soil N on the accumulation pattern of other herbivore toxins are unknown, but since they contain N, they may also be affected by high soil N, as is observed for the accumulation pattern of the ergot alkaloids produced by *Claviceps* sp of rye.

As in the case of high soil N, fungal toxins might also interact with low soil N to deter herbivory. This is based on the concept that the total N content of plants affects the degree of insect herbivory (Mattson, 1980). Therefore, herbivore feeding on grasses is increased when grown at low soil N as compared to high soil N. Lyons et al. (1990b) determined that the total free N concentration of leaves of tall fescue was decreased significantly by endophyte infection although the plants were fertilized with high rates of N (10 Mm total N/pot). Under high N, increased herbivory is prevented due to the combined action of increased amounts of toxins and low total N content of herbage. In a mixed population of grasses, herbivory of symbiota would be less than herbivory of uninfected grasses. Thus, under a wide range of soil N levels, toxins and N content would offset any tendency to overgraze infected grasses.

Effects of Endophytes on Forage Quality

Quality-Antiquality Evaluation of Endophyte-Infected Grasses

The concept of grazing implies a very large category of trophic interactions but from the standpoint of livestock herbivory, it describes a trophic interaction with plants that results in both low lethality and intimacy of a forage grass. Of course, this is complicated by the fact that we are dealing with a natural phenomenon within the context of a man made and managed system. Grasses comprise most of a ruminant's diet and the selection of a grass species is based on its nutritive effects on livestock production. Analyses of several symbiotic and nonsymbiotic pasture or potential pasture grasses of tall fescue cultivars for nutritive values for livestock, as well as actual grazing trials, indicate that there are decided differences in actual livestock performance (Table 3) (Schmidt et al., 1982; Hoveland et al., 1983; Aldrich et al., 1990; Chestnut et al., 1992; Porter et al., 1993). However, laboratory analyses of symbiotic and nonsymbiotic tall fescue indicate that the two are identical in terms of protein, in vitro dry matter disappearance (IVDMD), fiber and lignin (Table 3), as well as mineral contents (Chestnut et al., 1991b).

The greatest impact of symbiotic grasses is on the low productivity of livestock, particularly ruminant livestock (Table 3). However, most of the symbiotic grasses also are used as conservation/recreational species and are

Table 3. Forage characteristics and cattle performance on symbiotic and nonsymbiotic tall fescue.

Item	Symbiotic	Nonsymbiotic	Ref*
Grass:			
IVDMD, %	65.8	62.6	1
Crude protein, %	9.2	8.2	2
Neutral detergent fiber, %	74.5	72.0	2
Acid detergent fiber, %	40.7	40.0	2
Ash, %	5.6	6.72	2
Cattle:			
Grazing days	768.0[a]**	593.0[b]	3
Beef gain (ha^{-1})	384.0[a]	492.0[b]	3
Average daily gain (Kg)	0.50[a]	0.83[b]	3
Gains per tester steer (Kg)	84.0[a]	144.0[b]	3
Rectal temperature (C)	40.1[a]	39.3[b]	1
Average feed intake (Kg/d)	4.40	4.79	1
Respiration rate, breath/min	55.0[a]	53.0[a]	1
Skin vaporization, kcal/m^2/h	91.3	113.7	4
Prolactin, ng/ml	1.5	1.5	4
Triiodothyronine, ng/ml	57.2	60.6	4
Cortisol, ng/ml	57.2	60.6	4
Plasma melatonin, mean day-night difference, ng/ml	78.36[a]	48.10[b]	5

*References: [1]Schmidt et al., 1982; [2]Chestnut et al., 1991a; [3]Hoveland et al., 1983, 4 years means; [4]Aldrich et al., 1990; [5]Porter et al., 1993.

**Means in a row with the same letter are not significantly different at the $P< 0.05$ level.

grazed by wildlife. The effect of these grasses on wildlife resources is expected to mimic those observed in livestock production and small animal experiments. Thus, preliminary studies indicate that populations of small insectivorous and herbivorous mammals such as shrews, voles, and cotton rats were four to five-fold higher in pastures of nonsymbiotic tall fescue (Pelton et al., 1991). The effects of symbiotic grasses on deer, rabbits and other wildlife herbivores are sporadic and undocumented. As the emphasis on planting increased acreage of symbiotic grasses for turf and conservation purposes increases, the potential for increased toxicity and effects on wildlife also will increase.

Acremonium **Toxicity: Tall Fescue** Historically, tall fescue and perennial ryegrass symbiota were associated with poor animal performance problems and toxicities. The most common of these are fescue toxicosis and ryegrass staggers (Cunningham, 1943; Yates, 1962; Fletcher and Harvey, 1982). Plants with deterrents to ruminant herbivory should have a competitive edge over plants without such a mechanism, thus selection would favor symbiotic grasses with such mechanisms over uninfected plants or symbiotic grasses without this mechanism. The degree of toxicity to ruminant herbivory is expected to reflect the amount of toxin contained within grasses. The amount of toxin within a plant at a given location varies both qualitatively (Yates et al., 1985; Lyons et al., 1986; Arechavaleta et al., 1991), and quantitatively (Bacon et al.,1986; Rowan and Shaw, 1987; Belesky et al., 1987b; Hill et al., 1991b) but fluctuates seasonally (Fluckiger et al., 1976; Belesky et al., 1987b; Belesky et al., 1988). Both genotypes of the grass and fungus affect the final expression of ergot alkaloid content (Hill et al., 1990; Hill et al., 1991b; Kearney et al., 1991). The variation and fluctuation of endophyte-infected tall fescue (Thompson et al., 1989a) within a location is expected to affect animal performance similarly (Crawford et al., 1989; Thompson et al., 1989b; Chestnut et al., 1991b).

Cattle consuming *A. coenophialum*-infected tall fescue may show either severe or mild symptoms. Cattle showing severe tall fescue toxicosis resemble those showing classical ergotism and include gangrene of the extremities, and a slight nervousness or palsy in the flank region. These symptoms, commonly referred to as fescue foot, are usually observed under cool temperatures. During the early studies of toxicity on this grass, these were the only signs recognized (Yates, 1962). Fescue foot was observed to affect only a few animals within the herd, and toxicity from tall fescue was not considered a serious economic threat, especially since it was so sporadic.

When the endophyte and the recognition of mild or subclinical effects on cattle were revealed (Schmidt et al., 1982), it became apparent that symbiotic tall fescue affected all animals within a herd and imposed severe economic losses (Stuedemann and Hoveland, 1988). The syndrome resulting from the subclinical effects often is referred to as fescue summer toxicosis or the summer slump, since it is more evident in the warm periods of the year. Cattle with the mild or subclinical symptoms exhibited reduced weight gains and feed intake, reduced reproductive efficiency, reduced milk production, low heat tolerance or hyperthermia, increased respiration rates, and reduced circulating prolactin and serum cholesterol (Yates, 1983; Porter et al., 1990).

Cattle affected with fescue summer toxicosis are further characterized as having low plasma melatonin (Porter and Thompson, 1992), rough and long hair coats which last well into the spring-summer period (Yates, 1983), and decreased tolerance to light (Hemken et al., 1981; Bond et al., 1984). Cattle grazing symbiotic tall fescue seek shade or stand in water, and graze in the cooler periods of the day or at night. Another aspect of tall fescue summer toxicosis in cattle is the idiosyncratic development of fat necrosis which has been associated with high blood cholesterol of animals grazing symbiotic tall fescue (Stuedemann et al., 1985). Tall fescue summer toxicosis can be alleviated when the feed of cattle is changed, but average daily gains are still depressed if cattle are moved from symbiotic tall fescue to feedlots for finishing during warm weather.

Data showing that the greatest economic loss to the cattle industry is from the fescue summer toxicosis was estimated by Stuedemann and Hoveland (1988) to be $793 million annually in the USA which is primarily due to reduced reproduction rates and poor weight gains. According to Crawford et al. (1989), average daily gain of cattle was reduced by 45.5 g for each 10% increase in endophyte infestation level. The toxicity effects can be reduced by diluting the overall pasture infestation level with another forage such as clover (species of *Trifolium*) (Chestnut et al., 1991ab). In another study (Chestnut et al., 1991b), the average daily gains were depressed when the percentage of endophyte infestation levels was adjusted to 22% within by the addition of clover. This depression continued up to the 35% infection level, but no further reductions in average daily gain and beef production were observed when the infection was increased to 81%, the highest level used (Chestnut et al., 1991b).

Horses grazing on *A. coenophialum*-infected tall fescue have very specific symptoms, particularly reproductive and neuroendocrine effects. These symptoms are, as in cattle, related to toxicity induced by ergot alkaloids and can be prevented by the daily administration of perphenazine, a synthetic dopamine antagonist (Ireland et al., 1989). There has been no report of the gangrenous aspect, fescue foot, occurring in horses. The endophyte lowers circulating progesterone and prolactin levels in brood mares which also have extended gestation periods, retained, mineralized and thickened placentas, agalactia, and deliver foals that are dysmature, weak, or stillborn (Monroe et al., 1992; Putnam et al., 1991; McCann et al., 1992). These symptoms are, as in cattle, related to toxicity induced by ergot alkaloids and can be prevented by the daily administration of perphenazine, a synthetic dopamine antagonist (Porter and Thompson, 1992).

Sheep grazing on *Acremonium*-infected tall fescue have lower circulating prolactin, cholesterol, and as much as a 59% reduction in milk production (Henson et al., 1987; Bond et al., 1988). Ewes have lowered conception rates, but unlike sheep grazed on perennial ryegrass, the growth rate and feed intake are not reduced (Bond et al., 1984; Bond et al., 1988).

Acremonium **Toxicity: Perennial Ryegrass** Toxicity from perennial ryegrass infected with *A. lolii* affects mainly sheep (Keogh, 1973; Byford, 1979), but cattle, horses, and deer also are affected (Byford, 1979). This disease is a neuromuscular disorder and is referred to as ryegrass staggers (Keogh, 1973). The disorder occurs sporadically and primarily in New Zealand and southern Australia. It is distinct from the grass staggers (tetany) disease which is induced by a magnesium deficiency. It has been shown experimentally that sheep develop symptoms of ryegrass staggers within 7 to 14 d of being placed on toxic pastures (Fletcher, 1982). As is true of the symbiotic tall fescue, the syndrome in sheep is manifested in a variety of symptoms, including severe clinical symptoms of head nodding, trembling of the neck and shoulder muscles, swaying while standing, staggering, and a stilted gait with collapse if over-prodded (Keogh, 1973). Similar tetanic spasms are observed in cattle grazing perennial ryegrass; however, they may either collapse or assume a sitting posture if excited. These severe symptoms are associated with the level of endophyte (Hannah et al., 1990), increased amounts of dead basal dry matter in ryegrass stands, slow plant growth rate, or overgrazing (Keogh, 1973). There is a 2-10% loss of animals to ryegrass staggers which can account for at least half of the total profits. Subclinical effects include reduced average daily weight gains, depressed prolactin blood levels in ewes (Stilham et al., 1982; Fletcher and Barrel, 1984), and reduced testosterone levels in rams (Henson et al., 1987).

Symptoms of *Balansia* **toxicity** The *Balansia* species are mainly associated with invasive and rangeland grass species, e.g., species of *Sporobolus, Andropogon, Agrostis, Clamogrostis, Chloris, Eragrostis,* and *Panicum.*
Detailed studies of their effects on grass quality and cattle performance have not been conducted. Since the balansioid endophytes produce ergot alkaloids similar to those produced by the acremonioid endophytes (Porter et al., 1979a,b; Porter et al., 1981; Lyons et al., 1986; Rowan and Shaw, 1987), their effects and modes of action on grazing animals should be the same (Witters et al., 1975; Berde and Schild, 1978) . Indeed, ergotism in cattle has been reported in animals consuming *Balansia*-infected grasses (Hance, 1876; Nobindro, 1934; Bailey, 1903). Experimental feeding of the culture medium of one species, *B. epichloe*, reduced total prolactin concentrations in lactating Holstein cows (Wallner et al., 1983). Also, there was a decrease in the milk-induced rise in serum prolactin, although there was no effect on milk production (Wallner et al., 1983). This work was not only the first to establish that serum prolactin concentration was affected in cattle consuming a symbiotic grass but it also established that the ergot alkaloids produced by the fungus were the active toxins.

Toxins Associated with Symbiotic Grasses

The variety of defensive compounds (Table 2) found within symbiotic individuals of a population of endophyte-infected forage grasses will vary because of the process of natural selection under herbivory. Long-term herbivory, and specific types of herbivory (insect or mammalian), will lead to sub-populations of toxic grasses, each chemically defined and based on its specific mixture of deterring and toxic compounds. Since most endophytes are maternally transmitted, seed-sown pastures will reflect this diversity. This may serve to confound any potential cattle toxicities initially, but as cattle grazing within a location continues, the seed of toxic and deterring individual plants will be consumed the least, resulting in these individuals self-seeding, producing more of the toxic types.

The time required for the establishment of a population high in symbiotic individuals toxic to cattle may involve years. In areas where there are no pressures from herbivory, there should be a mixture of ecotypes, including individuals totally devoid of insect and mammalian toxins. Therefore, in experiments of short duration, it is important when studies of specific deterring or toxic mechanisms of symbiota are being conducted to consider this, and to define and use appropriate pastures. The major groups of toxins chemically identified with antiquality factors of symbiotic forage grasses are the ergot alkaloids and tremorgenic neurotoxins and these may be monitored to determine the percent distribution within a location.

The degree of chemical expression by the fungal mutualist depends not only on the biochemical competence of the fungus (Bacon et al., 1975; Bacon, 1988), but also upon the genotype of each plant (Hill et al., 1991b). The environmental factors of soil N (Gaynor and Hunt, 1983; Arechavaleta et al., 1989; Lyons et al., 1990b) and moisture (Arechavaleta et al., 1991) also interact to affect the accumulation of ergot alkaloids. Further, there is a fungus-grass genotype interaction since ergovaline content of one genotype of symbiotic tall fescue increases with increased leaf area while this relationship is not observed in other genotypes (Hill et al., 1990). This raises the issue as to whether the controlling mechanism is associated with the plant, the endophyte, or an interaction between the two. By inserting endophytes into a common tall fescue genotype and by conducting genetic studies between high- and low-ergovaline producing symbiotic plant genotypes (Johnson et al., 1986; Agee and Hill, 1991), it has been documented that at least the plant can regulate the expression of ergovaline production by the endophyte. In this regard, regulation may well reflect the individual variation of toxic precursors and primary metabolites released in the apoplasm from the plant. The apoplasm is the source of nutrients which are utilized by the fungi for growth and synthesis of secondary metabolites, many of which are toxic.

Ergot alkaloids   The ergot alkaloids (Table 2) have been shown to be produced directly by the fungus as in tall fescue endophyte (Yates et al.,

1985; Lyons et al., 1986), and the balansioid endophytes (Bacon et al., 1986), or indirectly in the case of perennial ryegrass since they are isolated only from symbiotic grasses (Rowan et al., 1986).

The ergot alkaloids found in symbiotic grasses consist of both the ergopeptine and clavine types (Porter et al., 1979b; Porter et al., 1981; Yates et al., 1985; Lyons et al., 1986). These two groups differ from each other in the presence (ergopeptide bond) or absence (clavine) of a peptide bond and attached amino acids. There is considerable controversy over the biological activity of each ergot alkaloid, but it is generally agreed that the ergopeptine alkaloids are more active than the clavine alkaloids.

The predominant ergot alkaloids found in the *Balansia*-infected grasses are of the clavine types, while the ergopeptide ergot alkaloids are found as the dominant ergot alkaloid in the *Acremonium*-infected grasses (Table 2). Both types of ergot alkaloids are biosynthetically derived from the same precursors, usually the simple clavine alkaloids (Floss, 1976), and there are biotypes of endophytes which reflect this variation (Bacon, 1988). Ergot alkaloids vary in concentration from 0.01 to 3.0 ug/g of plant (dry weight) (Gallager et al., 1984; Yates et al., 1985; Lyons et al., 1986) whose concentrations depend on the season, and the in planta physiological status (Bacon, 1988; Belesky et al., 1988), and grass-fungus interaction (Hill et al., 1991b; Agee and Hill, 1991). Livestock toxicity and possible mechanisms of action resulting from consuming toxic forages and specific ergot alkaloids have been reported in earlier reviews (Yates, 1983; Hemken et al., 1984; Porter and Thompson, 1992) and will not be repeated here.

Tremorgenic toxins The tremorgenic neurotoxins, commonly called the lolitrems, have been isolated only from the perennial ryegrass symbiotum. This group of toxins is apparently absent in symbiotic tall fescue and has not been examined for in the *Balansia*-infected grasses. This class of compounds is considered responsible for ryegrass staggers of sheep (Gallagher et al., 1982; Gallagher et al., 1984), and consists of four biologically active compounds, all containing a complex indole isoprenoid ring system (Gallagher et al., 1984). The major lolitrem is lolitrem B which ranges from 3 to 25 ug/g dry weight in perennial ryegrass herbage, and a smaller concentration also occurs in ryegrass seed (Gallagher et al., 1987). The lolitrems, unlike the ergot alkaloids, have not been isolated from cultures of the fungus *A. lolii*, but its indole isoprenoid precursor, paxilline, has, indicating that it is synthesized by the fungus in culture (Christopher and Mantle, 1987; Weedon and Mantle, 1987). In addition to finding paxilline in fungus cultures, it has been detected in ryegrass seed (Weedon and Mantle, 1987). Thus, it is unknown if the fungus only produces paxilline which is converted to the lolitrems by the plant, or if under culture or other conditions, there is an incomplete synthesis by the fungus. The latter might be the case, as other fungi can make tremorgenic neurotoxins, structurally related to the lolitrems, in culture (Lanigan et al., 1979) and otherwise (di Manna et al., 1976), which suggests that the *A. lolii* might be capable of synthesizing the lolitrem molecule in vivo but not in vitro.

Miscellaneous toxins There are a variety of chemically diverse compounds (Table 2) in symbiotic grasses which, as indicated above, were possibly one of the driving forces in evolution that led to the successful establishment of the associations. With the exception of the class of ergot alkaloids, there is presently no identified consistent and specific chemical which would suggest that there are other evolutionary biochemical relationships within the two broad types of symbiota.

The loline alkaloids are pyrrolizidine bases that are found in the tall fescue symbiotum in concentrations as high as 0.8% of the dry weight of tall fescue plants (Bush et al., 1982). These alkaloids have not been isolated from cultures of the fungus, and the production of similar pyrrolizidine bases by higher plants in general (McLean, 1970) imply that they may be products of the plant responding to infection from the endophyte. This implication is strengthened by the report of Belesky et al. (1987b), which indicates that the concentration of loline alkaloids reflects the extent of infection within the population. A limited number of studies suggests that the loline alkaloids are only mildly toxic (Bush et al., 1979; Strahan et al., 1988; Eichenseer et al., 1991) especially when that toxicity is compared with that of the usual pyrrolizidine alkaloids (McLean, 1970). Two of the forms present in symbiotic tall fescue, N-acetyl and N-formyl lolines, are considered more toxic than the other forms reported in this grass. However, these compounds may act synergistically or potentiate the activity of other toxins in symbiotic tall fescue, as suggested from their effects on insect herbivory (Siegel et al., 1991).

Other compounds reported in symbiotic grasses which may impose toxicity problems to livestock include 3-indole acetic acid, indole ethanol and related simple indoles (Porter et al., 1978), peramine (Rowan et al., 1986), the tetraenone steroid (Porter et al., 1975), the ergosterols (Davis et al., 1986), phenolic acid derivatives (Koshino et al., 1988), and sesquiterpenes (Yshihara et al., 1985). The effects of some of these on insects, laboratory animals, and fungi suggest toxicity (Yshihara et al., 1985; Davis et al., 1986; Dowd et al., 1988), while the toxicities of others have not been established.

PART II. MYCOTOXINS IN GRASSES FROM NONENDOPHYTIC FUNGI

Nonendophytic (nonmutualistic) fungi produce a wide diversity of mycotoxins (Cole and Cox, 1981) on many different substrates. However, the major focus of research on these toxins has been conducted on food commodities, although some emphasis has been placed on mixed animal feed. Thus, a complete listing of the presence or absence of mycotoxicoses, and (or) toxins occurring naturally on forage grasses is unavailable. The following discussion is based on information from the literature which indicates that the occurrence of a specific fungus taxon is found on grasses, although in certain

Table 4. Mycotoxins produced by nonendophytic fungi on or associated with growing or dried forage grasses.

Species	Mycotoxins	Animals affected	Forage type
Aspergillus: *A. fumigatus*	Fumigaclavines, fumigallin, fumigatin, fumitoxins, gliotoxin, fumitremorgens, verruculogen, ririditoxin, tryptoquivalines, and spinulosin	Cattle	Hay, grass litter
A. terrus	Citrovirdin, citrinin, gliotoxin, patulin, terreic acid, terretonin, territrems, cytochalasin E, and terrdionol	Cattle	Hay, leaves, stems
Claviceps: *C. purpurea, C. paspali, C. cynodontis*	Ergot alkaloids, and paspalinine	Cattle, pig, sheep, horses	Seed, seedheads,
Diplodia maydis	Diplodiatoxin	Cattle	
Fusarium: *F. sporotrichioides, F. poae, F. equisetti*	Butenolide, trichothecenes, zearalenone, and moniliforme	Cattle	Decaying leaves, leaves
Myrothecium: *M. roridum, M. leucotrichum, M. verrucaria*	Verucarins, and roridins	Sheep	Decaying leaves, rye stubble, hay
Penicillium: *P. crustosum, P. canescens, P. janczewskii*	Penitrem A, and penitrem B	Cattle	Litter
Phomopsis paspali	Cytochalasin H	Cattle	Litter
Pithomyces chartarum	Sporidesmins	Sheep	Ryegrass litter
Stachybotrys chartarum (=atra)	Epoxytrichothecene (Satratoxins)	Horses, pig, cattle, sheep	Hay, straw

cases the presence of its toxin has been inferred from either livestock behavior or demonstration of toxin production in culture by a fungus isolated from grasses (Table 4). However, there are several livestock mycotoxicoses associated with forage grasses many of which are of considerable economic importance. Included among these are ergotism, paspalum staggers, stachybotoxicosis, facial eczema and geeldikkop, myrotheciotoxicosis, fusariotoxicoses, and toxicoses induced by *Penicillium* and *Aspergillus* species.

Grasses serving as substrates for nonendophytic fungi include most of the major forage species throughout the world, and these grasses may become toxic either during their period of active growth (field fungi) or while in storage (storage fungi). Mycotoxin accumulation occurs during or after fungal growth; therefore, the factors which favor growth indirectly or directly are also those that favor mycotoxin synthesis. Mycotoxins are secondary metabolites whose synthesis appears to depend on qualitative and quantitative nutritional factors during growth, the genetics of an isolate, and an interaction with one or more environmental factors. Detailed discussions of these factors as well as other mycotoxicoses on forage plants other than grasses are contained in earlier reviews (Ciegler et al., 1971; Kadis et al., 1971; Kadis et al., 1972; Purchase, 1974; Wyllie and Morehouse, 1977; Smith and Henderson, 1991).

Grass Mycotoxicoses

Ergotism

The genus *Claviceps* Tulasne is almost exclusively found on grasses throughout most of the temperate and tropical areas of the world. Species of *Claviceps* are field fungi and are phylogenetically related to the endophytes of grasses but differ in the absence of the endophytic habit. This fungus is a localized ovarian replacement disease in which the individual ovaries of florets are usually destroyed by the fungus. The fungus then occupies this location within the florets where it intercepts nutrients normally translocated to the ovary. There is also an indication that the sclerotium also competes for nutrients with noninfected florets, reducing the final yield of seed directly by replacement and by nutrient intervention (Bacon and Luttrell, 1982). During this very brief parasitic phase, the initial infective hyphae enlarges to ten times its initial size, often outweighing (wet weight) normal seed of a noninfected grass three-fold. The resulting mass of mycelium produced is referred to as a sclerotium. The sclerotium is the major location of toxins referred to as ergot alkaloids. These ergot alkaloids are similar, if not identical to those produced and described for the *Acremonium* endophytes. However, since species of *Claviceps* are localized to the flowering stage toxicity from this group can occur only by ingesting sclerotia. Leaves and stems of *Claviceps*-infected grasses are not toxic.

The genus *Claviceps* includes 26 species according to Langdon (1954). At least four more species have been named since this compilation, increasing this number to 30 (Walker, 1957; Pantidou, 1959; Kulkarni et al., 1976; Frederickson and Mantle, 1991). Species of *Claviceps* show three morphological levels of host associations. The lowest level resembles the balansioid endophytes described above, while the most advanced type is the typical dark-colored elongated sclerotial parasite, and is typified by *C. purpurea* (Fr.) Tul. The intermediate is typified by *C. paspali* Stev. and Hall, and is characterized by straw-colored globose-shaped sclerotia. Most of these species are associated with specific grasses, some of which are major forage species and they include: several species of fescue, bermudagrass (*Cynodon dactylon* (L.) Pers.), several species of *Paspalum* L., and a varied assortment of valuable rangeland species of perennial and annual grasses. *Claviceps purpurea* is one the most cosmopolitan species, occurring on well over 250 hosts of primarily temperate annual and perennial grasses. Three species (*C. nigricans* Tul., *C. grohii* Groves, and *C. junci* Adams) are associated only with the closely related relatives of grasses, the Juncaceae and Cyperaceae. The species of some grasses have developed physiological races with a wide host range; other species have narrow host ranges and also are characterized by a variable sclerotial size and morphology.

The species of *Claviceps* show considerable variation relative to alkaloid content and there is no correlation between grass host species and the capacity to produce ergot alkaloids neither quantitatively nor qualitatively. Individual sclerotia may or may not contain ergot alkaloids which may reflect either a fungal genetic or host and environmental interaction. The ergot alkaloids include many of those produced by the endophytes (Table 2) as well as several others.

The basic animal disorder of ergotism is usually characterized as gangrene of the extremities resulting from an impairment of peripheral circulation by the ergot alkaloids. Other aspects of ergotism include reproductive disfunction, and a nervous syndrome characterized by muscle tremors and uncoordinated movements when excited.

Paspalum Staggers

The dallisgrasses (*Paspalum* spp.) are found primarily in the tropical sections of the world. These grasses are parasitized specifically by *C. paspali*. In addition to producing the ergot alkaloids, this species also produces the tremorgenic substances collectively referred to as the paspalitrems (A, B, and C), and paspalinine (Table 4). These substances are produced in the sclerotia; thus, livestock toxicity results from ingesting infected seed heads. Ingested seed heads produce a neurological disease called paspalum staggers in cattle and sheep with clinical signs including tremors, incoordination, and ataxia. This disease occurs in Australia, Italy, New Zealand, Portugal, South Africa, and the United States. The neurological disease is not caused by and is distinct from similar appearing conditions caused by ingesting ergot alkaloids (Mantle et al., 1977). Paspalum staggers is caused by one of several diterpene indole-type compounds referred to as paspalinine and paspalitrems, of which paspalitrem B is the most biologically active (Cole et al., 1977; Gallagher et al., 1980).

Stachybotryotoxicosis

Stachybotryotoxicosis is primarily a disease of horses, but sheep, poultry, cattle, pigs, and man also may become affected. Animals show clinical signs ranging from death to mouth necrosis, salivation, edema, and inflammation of the head. It is caused by the cellulose saprophyte *Stachybotrys chartarum* *(Ehrenb. and Link) Huges (=S. atra,* or *S. alternans)* which colonizes straw of various grasses, although it also occurs on hay, and silage. Thus, this species should be considered a storage fungus. Sporadic outbreaks of this disease has been reported in Hungary, France, South Africa, Romania, Russia, Finland, the United States, and Czechoslovakia and it occurs primarily in the fall of the year. The toxins were identified by Epply (1977) as macrocyclic trichothecenes, commonly referred to as satratoxins F, G, and H.

Facial Eczema and Geeldikkop

Both facial eczema and geeldikkop are types of a photodermatitis reaction aggravated by sunlight and are distinguished from other forms of photodermatitis diseases in that the liver is damaged by the mycotoxin which prevents the detoxification of one of the porphyrin pigments formed by decomposition of chlorophyll (Mortimer and Ronaldson, 1983). The two phosensitizations are caused by toxins from spores of the same fungi but on different grasses under field conditions. Facial eczema is caused by *Pithomyces chartarum* (Berk. and Curt.) Ellis that colonize accumulated debris from perennial ryegrass during the autumn with subsequent sporulation and production of toxins occurring under low temperatures from February to May (Smith and Crawley, 1964; Brooks, 1969). Geeldikkop is associated with *P. chartarum* growing on debris of species of *Panicum*, although this disease is classically associated with this fungus on *Tribulus terrestris* L., an annual herb. It is considered that the host plants contribute to the basic differences of the two diseases. The identity of the mycotoxin in both diseases has been established as sporidesmin, a diketopiperazine. Almost all the toxin is contained within the spores which is taken in by livestock consuming *Pithomyces*-infected pasture litter. The major distinction between the two diseases is the degree of liver pathology which in geeldikkop is characterized by crystalloid substances in bile ducts (Coetzer et al., 1985). Facial eczema occurs mainly in sheep, although cattle and deer also are affected. Geeldikkop has been reported only in sheep and occurs mainly in South Africa (Coetzer et al., 1985) and southwestern United States (Sperry et al., 1955; Taber et al., 1968) where it occurs on *Panicum* species. On this latter grass species the mycotoxicoses then is referred to as dikoor.

Myriotheciotoxicosis

Two field fungi of the genus *Myrothecium* Kunze ex Fries, *M. roridum* and *M. verrucaria,* are associated with grasses, primarily perennial ryegrass, where they produce the toxic trichothecenes roridins (roridin A, D, E, and H) and verrucarins (Verrucarin A, B, and J) (Vertinskii et al., 1967; di Menna et al., 1973). These toxins were considered antibiotics by those who first isolated and determined their chemical structures (Brian et al., 1948; Harri et al., 1962). Both the roridins and the verrucarins are acutely toxic to a wide range of animals when administered orally, and can cause severe dermatitis if applied topically. These toxic fungi occur world wide in soil and are commonly found on the leaves of pasture grasses (di Menna et al., 1973). There is considerable variation in the ability of isolates to produce these mycotoxins and there are considerable toxicological variations that each isolate produces in experimental animals. Myriotheciotoxicosis is very similar to poisoning of cattle by kikuyu grass (*Pennisetum clandestinm* L.) (Martinovich et al., 1972) which suggests that the etiologic agent of this toxin is also a species of *Myrothecium*. Information on the natural outbreaks of this toxicoses are limited to New Zealand (di Menna et al., 1973; Mortimer et al., 1971). However, the widespread occurrence of these fungi, the ease with which it can colonize dead grass residues, and the potency of their toxins suggest that this taxon can play an important role in deteriorating the quality of grasses.

Fusariotoxicoses

The genus *Fusarium* Link consists of several species that produce toxins on grasses. Of the five most important toxigenic species (Marasas et al., 1984), three are considered important producers of toxins in pastures. These species are *F. sporotrichioides* Sherb., *F. poae* (Peck) Wollenw., and *F. equisetti* (Corda) Gordon. All three are cosmopolitan saprophytes that occur in soils and in grass debris. They are distributed from temperate to tropical and subtropical areas of the world. Of the three, *F. sporotrichioides* appears to be restricted to temperate and cold areas (Marasas et al., 1984). The toxins from these species produce problems in cattle, pigs, and horses.

The compilation of the fusariotoxins (Table 4) by the three *Fusarium* species of grasses indicates considerable diversity in chemical structure, yet these toxins or their modifications all are produced by the three taxons. Thus, the reason for combining the three species into one livestock disease complex as fusariotoxicoses. However, this presents a problem of actually assigning the correct fungus to the toxin and livestock disease. Appropriate identification can be achieved by isolating and identifying the fungus, although this too might present even greater problems since the identification of these fungi is difficult. For a more definitive description of these fungi and their toxicology the work of Marasas et al. (1984) should be consulted. A fourth species of the five, *F. moniliforme*, might become important, especially in the tropical areas of the world, if the extent of a single report of finding one of its toxins, fumonisin B_1 in a grass sample is corroborated to include other pasture areas of the world (Mirocha et al., 1992). Fumonisin B_1 is carcinogenic and considered responsible for swine pulmonary edema, leucoencephalomalasia in horses (Bezuidenhout et al., 1988; Marasas et al., 1988; Colven and Harrison, 1992). This mycotoxin is usually produced by this species on corn and corn products. This fungus also produces several other toxins (Bacon and Williamson, 1992), as well as other unknown toxins (Marasas et al., 1984).

Penicillium Toxicoses

The genus *Penicillium* Link consists of a large assortment of species found growing on a wide variety of food, feed, and plant residues. Members of this species, like those of the *Fusarium*, also are difficult to identify and several species have been reduced to synonymy. Thus, it is very difficult to correctly associate a known animal toxicoses with a specific *Penicillium* species with certainty. This large genus produce a wide variety of toxins but on forage grasses the number is greatly reduced. Table 4 lists three species and their major toxins isolated from grasses. These metabolites are neurotoxins and cattle toxicities from ingesting grasses infected with these species or their toxin have been reported (Gallagher et al., 1980; Wilson et al., 1981; di Menna and Mantle, 1978). The toxicity signs of livestock ingesting these toxins on grasses resemble those of other tremorgenic and staggers syndromes mentioned above. Indeed, isolation of the toxin might be necessary before a correct etiologic agent can be identified. Other tremorgenic toxins produced by these species are paxilline, verruculogen, verucosidin, and the janthitrems (Ciegler and Pitt, 1970). Documentation that these toxins are related to field cases of livestock toxicity is sporadic or lacking. Since *P. crustosum* Thom is very cosmopolitan and most isolates of it produce penitrem A (Pitt, 1979a), it might be the major species of concern for livestock toxicity from *Penicillium*-infected grasses.

Aspergillus Toxicoses

The genus *Aspergillus* Micheli was erected to describe those fungi that produce spores in chains which originate from a spore-bearing structure called a head. Fungi in this genus are primarily saprophytic and have the ability to grow on many different substrates under a range of environmental conditions (Pitt, 1979b). Two species, *A. fumigatus* Fresenius and *A. terreus* Thom, have been associated with livestock performance problems on aspergilli-infected grasses. However, the significance and natural frequency of occurrence of problems in livestock from these two aspergilli on grasses have not been well documented. Nevertheless, among the substrates which favor the growth of these *Aspergillus* species is decaying plant material, e.g., grasses. Further, both fungi are soil-borne and frequent colonizers of silage and hay. The important toxic aspergilli are listed in Table 4, and these toxins are mainly tremorgenic toxins. The important toxins include the fumitremogens A, B, and C produced by *A. fumigatus* (Yamazaki et al., 1971; Cole et al., 1977), and the territrems A, B, and C produced by *A. terreus* (Styne and Vleggaar, 1985). Other toxins produced by *A. fumigatus* include the clavine alkaloids fumigaclavines A and C, and the ergot alkaloids agroclavine, erymoclavine, festuclavine, and chanoclavine (Yamazaki et al., 1971). Both fungi are cosmopolitan and are frequent colonizers of hay, silage, and grass litter, especially under damp conditions (Cole et al., 1977). Toxicoses in livestock caused by *A. fumigatus* on grasses include abortions, tremors, and neurotoxicosis (Wilson et al., 1981), and liver and kidney damage (Thornton et al., 1968). Deaths and a protein deficiency syndrome also were attributed to *A. fumigatus* (Cole et al., 1977).

In addition to these two genera of aspergilli, an osmiophilic toxic species *A. chevalieri* (Mangin) Thom & Church is also found on several types of hay (Forgacs and Carll, 1962). Isolates of this species have been shown to produce physcion and other toxic anthraquinones (Bachmann et al., 1979). However, this taxon is not considered a serious problem since its toxins are not effective when they are administered orally.

SUMMARY

Several forage and rangeland grass species either formed a completely compatible association with a group of closely related fungi which are referred to collectively as endophytes, or they served as hosts to pathogenic or saprophytic fungal species. In both types of fungi, mycotoxins accumulate in forage and rangeland grasses. Mycotoxins are extrinsic antiquality factors in grass species. Mycotoxin accumulation occurs after and during fungal growth which may take place either during the growth phase of grasses or when grasses have been processed into hay or undergoing natural decay. Since mycotoxins are secondary rather than primary metabolites, their synthesis depends on qualitative and quantitative nutritional factors during and after growth, the genetic potential of the isolate, and an interaction with one or more environmental factors. These factors affect the accumulation pattern of toxins from both endophytic and nonendophytic fungi on grasses.

Control of the nonendophyte should be directed towards the removal of the fungus or conditions responsible for mycotoxin accumulation. Control of endophytes is complex. Control measures designed to prevent the growth of either endophytic or nonendophytic fungi must consider the specific ecological niche of a fungus. Successful prevention of mycotoxin accumulation implies not only a safe grazing product but also that all the agronomic aspects of forage quality have not been affected. Since endophytes increase the agronomic performance of grasses, e.g., by preventing overgrazing and increased stress tolerance they must be controlled differently from saprophytic and pathogenic fungal species.

Symbiotic associations of fungal endophytes with several grasses are defensive mutualisms. The consequence of this type of symbiosis is the production of a variety of chemicals, randomly distributed among forage grasses which are defensive chemicals to grazing ungulates, lagomorphs, rodents, as well as insects and several other invertebrates. While not all antiquality components have been analyzed in symbiotic grasses, those that have appear to serve both as deterrents and toxins and are considered to be affected by several physical environmental factors. Current concepts suggest that symbiotic grasses are natural deterrences to destructive grazing of forage and rangeland grasses. Additionally, drought tolerance, N efficiency, increased rooting and herbage yield also are attributed to this symbiotic association. However, relative to livestock farming, symbiotic grasses are egregious, and pose serious economic threats to grazing livestock. These economically important concerns include reduced reproductive efficiency, reduced weight gains, and often death of grazing livestock. The removal of mutualistic fungi from grasses may relieve one specific class of antiquality factors, mycotoxins, but the agronomic properties of the grass may be considerably impaired. This enigma can be solved by a recognition of the evolutionary significance of the mutualistic association, then genetically defining and partitioning desirable traits into single components, along with a concerted grass breeding effort designed at incorporating desirable genetic components into nonsymbiotic individuals and, if necessary, molecular modification of the desired forage if only by, for example, using transformed endophytes (Murray et al., 1992; Tsai et al., 1992).

REFERENCES

Agee, C.S., and N.S. Hill. 1991. Variability in progeny from high-and low-ergovaline producing tall fescue parents. p. 185. *In* Agronomy abstracts. ASA, Madison, WI.

Ahmad, S., S. Govindarajan, C.R. Funk, and J.M. Johnson-Cicalese. 1985. Fatality of house crickets on perennial ryegrass infected with a fungal endophyte. Entomol. Exp. Appl. 39:183-190.

Ahmad, S., J.M. Johnson-Cicalese, W.K. Dickson, and C.R. Funk. 1986. Endophyte-enhanced resistance in perennial ryegrass to the bluegrass bill-bug *Sphenophorus parvalus*. Entomol. Exp. Appl. 41:3-10.

Ahmadjian, V., and S. Paracer. 1986. Symbiosis, an introduction of biological associations. University Press of New England, Hanover, NH.

Aldrich, C.G., M.T. Rhodes, J.L. Miner, J.A. Paterson, and M.S. Kerley. 1990. Effects of consumption of tall fescue infested with endophytic-fungus and supplementation with dopamine antagonist on heat dissipation, body temperatures and blood parameters in ruminants housed at elevated temperatures. p. 205-208. *In* S.S. Quisenberry and R.E. Joost (ed.) International Symposium on *Acremonium*/Grass Interactions. 5-6 Nov. 1990. Louisiana Agricultural Experiment Station, Baton Rouge.

Arechavaleta, M., C.W. Bacon, C.S. Hoveland, and D.E. Radcliffe. 1989. Effect of the tall fescue endophyte on plant response to environmental stress. Agron. J. 81:83-90.

Arechavaleta, M., C.W. Bacon, R.D. Plattner, C.S. Hoveland, and D.E. Radcliffe. 1991. Accumulation of ergopeptide alkaloids in endophyte-infected tall fescue grown under deficits of soil water and nitrogen fertilizer. Appl. Environ. Microbiol. 58:857-861.

Asay, K.H., T.R. Minnick, G.B. Garner, and B.W. Harmon. 1975. Use of crickets in a bioassay of forage quality in tall fescue. Crop Sci. 5:585-588.

Bachmann, M., J. Luthy, and C. Schlatter. 1979. Toxicity and mutagenicity of molds of the *Aspergillus glaucus* group. J. Agric. Food Chem. 27:1342-1347.

Bacon, C.W. 1988. Procedure for isolating the endophyte from tall fescue and screening isolates for ergot alkaloids. Appl. Environ. Microbiol. 54:2615-2618.

Bacon, C.W. 1990. Isolation, culture and maintenance of endophytic fungi of grasses. p. 259-282. *In* D.P. Labeda (ed.) The isolation and screening of microorganisms from nature. McGraw-Hill, New York.

Bacon, C.W., and J. De Battista. 1991. Endophytic fungi of grasses. p. 231-256. *In* D.K. Arora (ed.) Advances in applied mycology. Marcel Dekker, Inc., New York.

Bacon, C.W., and N.S. Hill. 1994. Symptomless grass endophytes: Products of coevolutionary symbioses and their role in the ecological adaptation of infected grasses. *In* S.C. Redlin (ed.) Systematics, ecology and evolution of endophytic fungi in grasses and woody plants. American Phytopathological Society, St. Paul, MN. (In press)

Bacon, C.W., and E.S. Luttrell. 1982. Competition between ergots of *Claviceps purpurea* and rye seed for photosynthates. Phytopathology 72: 1332-1336.

Bacon, C.W., P.C. Lyons, J.K. Porter, and J.D. Robbins. 1986. Ergot toxicity from endophyte-infected grasses: A review. Agron. J. 78:106-116.

Bacon, C.W., J.K. Porter, and J.D. Robbins. 1975. Toxicity and occurrence of *Balansia* on grasses from toxic fescue pastures. Appl. Microbiol. 29:553-556.

Bacon, C.W., and J.W. Williamson. 1992. Interactions of *Fusarium moniliforme*, its metabolites and bacteria with corn. Mycopathologia 117:79-82.

Bailey, V. 1903. Sleepy grass and its effect on horses. Science 17:392-393.

Barker, G.M. 1988. Mycorrhizal infection influences *Acremonium*-induced resistance to Argentine stem weevil in ryegrass. p. 199-203. In Proc. 40th N.Z. Weed and Pest Control Conf.

Beetle, A.A. 1958. *Piptochaetium* and *Philaris* in the fossil record. Bull. Torrey Bot. Club 85:179-181.

Belesky, D.P., O.J. Devine, J.E. Pallas, Jr., and W.C. Stringer. 1987a. Photosynthetic activity of tall fescue as influenced by a fungal endophyte. Photosynthetica 21:82-87.

Belesky, D.P., J.D. Robbins, J.A. Stuedemann, S.R. Wilkinson, and O.J. Devine. 1987b. Fungal endophyte infection-loline derivative alkaloid concentration of grazed tall fescue. Agron. J. 79:217-220.

Belesky, D.P., J.A. Stuedemann, and S.R. Wilkinson. 1988. Ergopeptine alkaloids in grazed tall fescue. Agron. J. 80:209-212.

Berde, B., and H.O. Schild (ed.). 1978. Ergot alkaloids and related compounds. Vol. 49. Handbook of experimental pharmacology. Springer-Verlag, New York.

Bezuidenhout, S.C., W.C.A. Gelderblom, C.P. Gorst-Allman, R.M. Horak, W.F.O. Marasas, G. Spiteller, and R. Vleggaar. 1988. Structure elucidation of the fumonisins, mycotoxins from *Fusarium moniliforme*. J. Chem. Soc. Chem. Commun. 1988:743-745.

Bond, J., G.P. Lynch, D.J. Bolt, H.W. Hawk, C. Jackson, and R.J. Wall. 1988. Reproductive performance and lamb weight gain for ewes grazing fungus-infected tall fescue. Nutr. Rep. Int. 37:1099-1102.

Bond, J., J.B. Powell, D.J. Undersander, P.W. Moe, H.F. Tyrell, and R.R. Oltjen. 1984. Forage composition and growth and physiological characteristics of cattle grazing several varieties of tall fescue during summer conditions. J. Anim. Sci. 59:584-593.

Bradshaw, A.D. 1988. Population differentiation in *Agrostis tenuis* Sibth. II. The incidence and significance of infection by *Epichloe typhina*. New Phytol. 58:310-315.

Brian, P.W., H.G. Hemming, and E.G. Jeffreys. 1948. Production of antibiotics by species of *Myrothecium* in New Zealand pastures and its relation to animal diseases. J. Gen. Microbiol. 79:81-87.

Brooks, P.J. 1969. *Pithomyces chartarum* in pastures and measures for prevention of facial eczema. J. Stored Prod. Res. 5:203-207.

Bula, R.J., V.L. Lechtenberg, and D.A. Holt. 1977. Potential of the world's forage for ruminant animal production. p. 7-28. *In* Potential of temperate zone cultivated forages. Winrock International Livestock Research and Training Center, Morrilton, AR.

Bush, L.P., J. Boling, and S.G. Yates. 1979. Animal disorders. p. 247-292. *In* R.C. Buckner and L.P. Bush (ed.) Tall fescue. ASA, Maidson, WI.

Bush, L.P., P.C. Cornelius, R.C. Buckner, D.R. Varney, R.A. Chapman, P.B. Burrus, C.W. Kennedy, T.A. Jones, and M.J. Saunders. 1982. Association of N-acetyl loline and N-formyl loline with Epichloe typhina in tall fescue. Crop Sci. 22:941-943.

Byford, M.J. 1979. Ryegrass staggers in sheep and cattle. N. Z. J. Agric. May 1979:65.

Cheplick, G.P., and K. Clay. 1988. Acquired chemical defenses in grasses: The role of fungal endophytes. Oikos 52:309-318.

Chestnut, A.B., P.D. Anderson, M.A. Cochran, H.A. Fribourg, and K.D. Gwinn. 1992. Effects of hydrated sodium calcium aluminosilicate on fescue toxicosis and mineral absorption. J. Anim. Sci. 70:2838-2846.

Chestnut, A.B., H.A. Fribourg, J.B. McLaren, D.G. Keltner, B.B. Reddick, R.J. Carlisle, and M.C. Smith. 1991a. Effects of *Acremonium coenophialum* infestation, bermudagrass, and nitrogen fertilizer or clover on steers grazing tall fescue pastures. J. Prod. Agric. 4:208-213.

Chestnut, A.B., H.A. Fribourg, J.B. McLaren, R.W. Thompson, R.J. Carlisle, K.D. Gwinn, M.C. Dixon, and M.C. Smith. 1991b. Effect of endophyte infestation level and endophyte-free tall fescue cultivar on steer productivity. Tenn. Farm Home Sci. 160:38-44.

Christopher, W.M., and P.G. Mantle. 1987. Paxilline biosynthesis by *Acremonium loliae*, a step toward defining the origin of lolitrem neurotoxins. Phytopathology 26:969-971.

Ciegler, A., S. Kadis, and S.J. Ajl (ed.). 1971. Microbial toxins. Vol. IV. Fungal toxins. Academic Press, New York.

Ciegler, A., and J.I. Pitt. 1970. Survey of the genus *Penicillium* for tremorgenic toxin production. Mycopathol. Mycol. Appl. 42:119-124.

Clay, K. 1984. The effect of the fungus *Atkinsonella hypoxylon* (Clavicipitaceae) on the reproductive system and demography of the grass *Danthonia spicata*. New Phytol. 98:165-175.

Clay, K. 1986. Grass endophytes. p. 188-204. *In* N.J. Fokkema and J. Van Den Heuvel (ed.) Microbiology of the phyllosphere. Cambridge University Press, Cambridge.

Clay, K. 1987. Effects of fungal endophytes on the seed and seedling biology of *Lolium perenne* and *Festuca arundinacea*. Oecologia 73:358-362.

Clay, K. 1988. Fungal endophytes of grasses: A defensive mutualism between plants and fungi. Ecology 69:10-16.

Clay, K., T.N. Hardy, and A.M. Hammond, Jr. 1985. Fungal endophytes of grasses and their effects on an insect herbivore. Oecologia 66:1-6.

Cockerell, T.D.A. 1956. The fossil flora of Florissant, Colorado. Amer. Mus. Natur. Histor. Bull. 24:71-110.

Coetzer, J.A., T.S. Kellerman, G.C.A. van der Weshuizen, W.F.O. Marasas, J.A. Minne, G.F. Bath, and P.A. Basson. 1985. The possible role of *Pithomyces chartarum* in the aetiology and pathogenesis of bovine hepatogenous photosensitivity disease in South Africa. p. 463-470. *In* J. Lacey (ed.) Trichothecenes and other mycotoxins. John Wiley and Sons, Chichester, England.

Cole, R.J., and R.H. Cox. 1981. Handbook of toxic fungal metabolites. Academic Press, New York.

Cole, R.J., J.W. Dorner, J.A. Lansden, R.H. Cox, C. Pape, B. Cunfer, S.S. Nicholson, and D.M. Bedell. 1977. Paspalum staggers: Isolation and identification of tremorgenic metabolites from sclerotia of *Claviceps paspali* stevens et Hall. Tet. Lett. 21:235-238.

Cole, R.J., J.W. Kirksey, J.W. Dorner, D.M. Wilson, J.C. Johnson, A.N.

Johnson, D.M. Bedell, J.P. Springer, K.K. Chetal, J.C. Clardy, and R.H. Cox. 1977. Mycotoxins produced by *Aspergillus fumigatus* isolated from moldy silage. J. Agric. Food Chem. 25:826-830.

Colvin, B.M., and L.R. Harrison. 1992. Fusarium-induced pulmonary and hydrothorax in swine. Mycopathologia 117:79-82.

Crawford, R.J., J.R. Forwood, R.L. Belyea, and G.B. Garner. 1989. Relationship between level of endophyte infection and cattle gains on tall fescue. J. Prod. Agric. 2:147-151.

Crowder, L.V. 1977. Potential of the world's forages for ruminant animal production. p. 49-78. *In* Potential of tropical zones cultivated forages. Winrock International Livestock Research and Training Center, Morrilton, AR.

Cunningham, I.J. 1943. A note on the cause of tall fescue lameness in cattle. N. Z. J. Sci. Technol. 24:167b-178b.

Davis, N.D., E.M. Clark, K.A. Schrey, and U.L. Diener. 1986. Steriod metabolites of *Acremonium coenophialum*, an endophyte of tall fescue. J. Agric. Food Chem.34:105-108.

De Bary, A. 1887. Comparative morphology and biology of fungi, mycetozoa and bacteria. Claredon Press, Oxford, UK.

De Battista, J.P., J.H. Bouton, C.W. Bacon, and M.R. Siegel. 1990. Rhizome and herbage production of endophyte-removed tall fescue clones and populations. Agron. J. 82:651-654.

Diehl, W.W. 1950. *Balansia* and Balansiae in America. p. 1-82. *In* USDA Agric. Monograph 4. U. S. Govt. Print. Office, Washington, DC.

di Menna, M.E., and P.G. Mantle. 1978. The role of *Penicillium* in ryegrass staggers. Res. Vet. Sci. 24:347.

di Menna, M.E., P.G. Mantle, and P.H. Mortimer. 1976. Experimental production of staggers syndrome in ruminants by a tremorgenic *Penicillium* from the soil. N.Z. Vet. J. 24:45-51.

di Menna, M.E., P.H. Mortimer, B.L. Smith, and M. Tulloch. 1973. The incidence of the genus *Myrothecium* in New Zealand pastures and its relation to animal diseases. J. Gen. Microbiol. 79:81-87.

Dowd, P.F., R.J. Cole, and R.F. Vesonder. 1988. Toxicity of selected tremorgenic mycotoxins and related compounds to *Spodotera frugiperda* and *Heliothis zea*. J. Antibiotics 61:1868-1872.

Eichenseer, H., D.L. Dahlman, and L.P. Bush. 1991. Influence of endophyte infection, plant age and harvest interval on *Rhopalosiphum padi* survival and its relation to quantity of N-formyl and N-acetyl loline in tall fescue. Entomol. Exp. Appl. 60:29-38.

Elmi, A.A., and C.P. West. 1989. Endophyte effect on leaf osmotic adjustments in tall fescue. p. 111. *In* Agronomy abstracts. ASA, Madison, WI.

Eppley, R.M. 1977. Chemistry of stachybotryotoxicosis, p. 285-293. *In* J.V. Rodrick, C.W. Hesseltine, and M.A. Mehlmann (ed). Mycotoxins in human and animal health. Pathotox Publishers, Park Forest South, IL.

Fletcher, L.R. 1982. Observation of ryegrass staggers in weaned lambs grazing different ryegrass pastures. N. Z. J. Exp. Agric. 10:203-207.

Fletcher, L.R., and G.K. Barrel. 1984. Reduced liveweight gains and serum prolactin levels in hoggets grazing ryegrasses containing *Lolium* endophyte. N. Z. Vet. J. 32:139-140.

Fletcher, L.R., and I.C. Harvey. 1982. An association of a *Lolium* endophyte with ryegrass staggers. N. Z. Vet. J. 29:185-186.

Floss, H. 1976. Biosynthesis of ergot alkaloids and related compounds. Tetrahedron 32:873-912.

Fluckiger, E., W. Doepfner, M. Marks, and W. Niederer. 1976. Effects of ergot alkaloids on the hypothalmic-pituitary axis. Postgrad. Med. J. 52:57-61.

Forgacs, J., and W.T. Carll. 1962. Mycotoxicoses. Adv. Vet. Sci. 7:273-283.

Frederickson, D.E., and P.G. Mantle. 1991. *Claviceps africana* sp. nov.; the distinctive ergot pathogen of sorghum in Africa. Mycol. Res. 95:1101-1107.

Funk, C.R., P.M. Halisky, M.C. Johnson, M.R. Siegel, and A.V. Stewart. 1983. An endophytic fungus and resistance to sod webworms. Bio/technol. 1:189-191.

Funk, C.R., P.M. Halisky, S. Ahmad, and R.H. Hurley. 1985. How endophytes modify turgrass performance and response to insect pests in turfgrass breeding and evaluation trials. p. 137-145. *In* F. Lemaire (ed.) Proc. 5th International Research Conference, Avignon. Versailles:

INRA.

Gallagher, R.T., J. Finer, J. Clardy, A. Leutwiler, F. Weibel, W. Acklin, and D. Agrigoni. 1980. Paspalinine, a tremorgenic metabolite from *Claviceps paspali*. Tet. Lett. 21:235-238.

Gallagher, R.T., A.D. Hawkes, P.S. Steyn, and R. Vleggaar. 1984. Tremorgenic neurotoxins from perennial ryegrass causing ryegrass staggers disorder of livestock: Structure and elucidation of lolitrem B. J. Chem. Soc., Chem Commun. 1984:614-616.

Gallagher, R.T., G.C.M. Latch, and R.G. Keogh. 1980. The janthitrems: fluorescent tremorgenic toxins produced by *Penicillium janthinellum* isolates from ryegrass pastures. Appl. Environ. Microbiol. 39:272-273.

Gallagher, R.T., G.S. Smith, M.E. di Menna, and P.W. Young. 1982. Some observations on neurotoxin production in perennial ryegrass. N. Z. Vet. J. 30:203-204.

Gallagher, R.T., G.S. Smith, and J.M. Sprosen. 1987. Distribution and accumulation of lolitrem B neurotoxin in perennial ryegrass plants. Proc. 4th Anim. Sci. Congress, Hamilton, New Zealand.

Gams, W., O. Petrini, and D. Schimdt. 1990. *Acremonium uncinatun*, a new endophyte of *Festuca pratensis*. Mycotaxon 37:67-71.

Gaynor, D.L., and W.F. Hunt. 1983. The relationship between nitrogen supply, endophytic fungus, and Argentine stem weevil resistance in ryegrass. p. 257-263. *In* Proc. N.Z. Grassland Assoc.

Gould, F.W., and R.B. Shaw. 1983. Grass systematics. Texas A&M University Press, College Station.

Halisky, P.M., D.C. Saha, and C.R. Funk. 1985. Prevalence of non-choke-inducing endophytes in turf and forage grasses. Phytopathology 75:1331 (Abstr.)

Hance, H.F. 1876. On a mongolian grass producing intoxication in cattle. J. Bot. 14:210-212.

Hannah, S.M., J.A. Paterson, J.E. Williams, M.S. Kerley, and J.L. Miner. 1990. Effects of increasing dietary levels of endophyte-infected tall fescue seed on diet digestibility and ruminal kinetics in sheep. J. Anim. Sci. 68:1693-1701.

Hardy, T.N., K. Clay, and A.M. Hammond, Jr. 1985. Fall armyworm

(Lepidoptera: Noctuidae): A laboratory bioassay and larval preference study for the fungal endophyte of perennial ryegrass. J. Econ. Entomol. 78:571-575.

Hardy, T.N., K. Clay, and A. M. Hammond, Jr. 1986. Leaf age and related factors affecting endophyte-mediated resistance to fall armyworm (Lepidoptera: Noctuidae) in tall fescue. Environ. Entomol. 15:1083-1089.

Harri, E., W. Loeffler, H.P. Sigg, H. Stahelin, C. Stoll, C. Tamm, and D. Wiesinger. 1962. Helv. Chim. Acta 45:839-853.

Hemken, R.W., J.A. Boling, L.S. Bull, and R.H. Hatton. 1981. Interaction of environmental temperature and anti-quality factors on the severity of summer fescue toxicosis. J. Anim. Sci. 58:710-714.

Hemken, R.W., J.A. Jackson, and L.B. Daniels. 1984. Toxic factors in tall fescue. J. Anim. Sci. 58:1011-1016.

Henson, M.C., E.L. Piper, and L.B. Daniels. 1987. Effects of induced fescue toxicosis on plasma and tissue catecholamine concentrations in sheep. Dom. Anim. Endocrin. 4:7-15.

Hill, N.S., D.P. Belesky, and W.C. Stringer. 1991a. Competitiveness of tall fescue as influenced by *Acremonium coenophialum*. Crop Sci. 31:185-190.

Hill, N.S., W.A. Parrot, and D.D. Pope. 1991b. Ergopeptide production by endophytes in a common tall fescue genotype. p. 1545-1547. *In* Agronomy abstracts. ASA, Madison, WI.

Hill, N.S., W.A. Parrot, and D.D. Pope. 1991c. Ergopeptine alkaloid production by endophytes in a common tall fescue genotype. Crop Sci. 31:1545-1547.

Hill, N.S., W.C. Stringer, G.E. Rottinghaus, D.P. Belesky, W.A. Parrot, and D.D. Pope. 1990. Growth, morphological, and chemical component responses of tall fescue to *Acremonium coenophialum*. Crop Sci. 30:156-161.

Hoover, M.M.,M.A. Hein, W.A. Dayton, and C.O. Erlanson. 1948. The main grasses for farm and home. p. 639-700. *In* Grass, the Yearbook of Agriculture 1948. USDA, U.S. Gov. Printing Office, Washington, DC.

Hoveland, C.S., S.P. Schmidt, C.C. King, Jr., J.W. Odom, E.M. Clark, J.A. McGuire, L.A. Smith, H.W. Grimes, and J.L. Holliman. 1983. Steer performance and association of *Acremonium coenophialum* on tall fescue pasture. Agron. J. 75:821-824.

Ireland, F.A., W.E. Loch, R.V. Anthony, and K. Worthy. 1989. The use of bromocriptine and perphenazine in the study of fescue toxicosis in pregnant pony mares. p. 201-206. *In* Eq. Nutr. Physiol. Symp. Proc., Stillwater, OK.

Johnson, M.C., L.P. Bush, and M.R. Siegel. 1986. Infection of tall fescue with *Acremonium coenophialum* by means of callus culture. Plant Dis. 70:380-382.

Kadis, S., A. Ciegler, and S.J. Ajl (ed.). 1971. Microbial toxins. Vol. VII. Algal and fungal toxins. Academic Press, New York.

Kadis, S., A. Ciegler, and S.J. Ajl (ed.). 1972. Microbial toxins. Vol. VIII. Fungal toxins. Academic Press, New York.

Kearney, J.F., W.A. Parrot, and N.S. Hill. 1991. Infection of somatic embryos of tall fescue with *Acremonium coenophialum*. Crop Sci. 31:979-984.

Keeler, K.H. 1985. Cost:benefit models of mutualism. p. 100-127. *In* The biology of mutualism:Ecology and evolution. Croom Helm, London.

Kelley, S.E., and K. Clay. 1987. Interspecific competitive interactions and the maintenance of genotypic variation within two perennial grasses. Evolution 41:92-103.

Keogh, R.G. 1973. Induction and prevention of ryegrass staggers in grazing sheep. N. Z. J. Exp. Agric. 1:55-57.

Kimmons, C.A., K.D. Gwinn, and E.C. Bernard. 1990. Nematode reproduction on endophyte-infected and endophyte-free tall fescue. Plant Dis. 74:757-761.

Kirfman, G.W., R.L. Brandenburg, and G.B. Garner. 1986. Relationship between insect abundance and endophyte infestation level in tall fescue in Missouri. J. Kans. Entomol. Soc. 59:552-554.

Kohlmeyer, J., and K. Kohlmeyer. 1974. Distribution of *Epichloe typhina* (Ascomycetes) and its parasitic fly. Mycologia 66:77-86.

Koshino, H., S. Terada, T. Yoshihara, S. Sakamura, T. Shimanuki, T. Sato, and A. Tajimi. 1988. Three phenolic acid derivatives from stromata of *Epichloe typhina* on *Phleum pratense*. Phytochemistry 27:1333-1338.

Kulkarni, B.G.P., V.S. Seshadri, and R.K. Hegde. 1976. The perfect stage of *Spacelia sorghi* McRae. Mysore J. Agric. Sci. 10:286-289.

Langdon, R.F.N. 1954. The origin and differentiation of *Claviceps* species. Univ. Queensland Papers, Dept. Botany 3:61-68.

Lanigan, G.W., A.L. Payne, and P.A. Cockrum. 1979. Production of tremorgenic toxins by *Penicillium janthinellum* Biourge: A possible etiological factor in ryegrass staggers. Aust. J. Exp. Biol. Med. Sci. 57:31-37.

Latch, G.C.M., M.J. Christensen, and G.J. Samuels. 1984. Five endophytes of *Lolium* and *Festuca* in New Zealand. Mycotaxon 20:535-550.

Latch, G.C.M., M.J. Christensen, and D.L. Gaynor. 1985a. Aphid detection of endophyte infection in tall fescue. N. Z. J. Agric. Res. 28:129-132.

Latch, G.C.M., W.F. Hunt, and D.R. Musgrave. 1985b. Endophytic fungi affect growth of perennial ryegrass. N. Z. J. Agric. Res. 28:165-168.

Latch, G.C.M., M.J. Christensen, and R.E. Hickson. 1988. Endophytes of annual and hybrid ryegrasses. N. Z. J. Agric. Res. 31:57-63.

Law, R. 1985. Evolution in a mutualistic environment. p. 145-170. *In* D.H. Bougher (ed.) The biology of mutualism: Ecology and evolution. Croom Helm, London.

Leuchtmann, A., and K. Clay. 1988. *Atkinsonella hypoxylon* and *Balansia cyperi*, epiphytic members of the Balansiae. Mycologia 80:192-199.

Lewis, D.H. 1967. Sugar alcohols (polyols) in fungi and green plants. New Phytol. 66:143-184.

Lewis, D.H. 1985. Symbiosis and mutualism: Crisp concepts and soggy semantics. p. 29-39. *In* D.H. Boucher (ed.) The Biology of mutualism. Oxford University Press, New York.

Luttrell, E.S. 1979. Host-parasite relationships and development of the ergot sclerotium in *Claviceps purpurea*. Can. J. Bot. 58:942-958.

Luttrell, E.S., and C.W. Bacon. 1977. Classification of *Myriogenospora* in the Clavicipitaceae. Can. J. Bot. 55:2090-2097.

Lyons, P.C., J.J. Evans, and C.W. Bacon. 1990a. Effects of the fungal endophyte *Acremonium coenophialum* on nitrogen accumulation and metabolism in tall fescue. Plant Physiol. 92:726-732.

Lyons, P.C., J.J. Evans, and C.W. Bacon. 1990b. Effects of the fungal endophyte *Acremonium coenophialum* on nitrogen accumulation and metabolism in tall fescue. Plant Physiol. 92:726-732.

Lyons, P.C., R.D. Plattner, and C.W. Bacon. 1986. Occurrence of peptide and clavine ergot alkaloids in tall fescue. Science 232:487-489.

Marasas, W.F.O., T.S. Kellerman, W.C.A. Glederblom, J.A.W. Coetzer, P.G. Thiel, and J.J. Van der Lugt. 1988. Leukoencephalomalacia in two horses induced by fumonisin B_1 isolated from *Fusarium moniliforme*. Onderstepoort J. Vet Res. 55:197-203.

Marasas, W.F.O., P.E. Nelson, and T.A. Tousson. 1984. Toxigenic *Fusarium* species. The Pennsylvania State University Press, University Park.

Mantle, P.G., P.H. Mortimer, and E.P. White. 1977. Mycotoxic tremogens of *Claviceps paspali* and *Penicillium cyclopium*: A comparative study of effects on sheep and cattle in relation to natural staggers syndrome. Res. Vet. Sci. 24:49-56.

Martinovich, D., P.H. Mortimer, and M.E. di Menna. 1972. Similarities between so-call kikuyu poisoning of cattle and two experimental mycotoxicoses. N.Z. Vet. J. 20:57-58.

Mattson, W.S. 1980. Herbivory in relation to plant nitrogen content. Ann. Rev. Ecol. Syst. 11:119-161.

McCann, J.S., A.B. Caudle, F.N. Thompson, J.A. Stuedemann, G.L. Heusner, and D.L. Thompson, Jr. 1992. Influence of endophyte-infected tall fescue on serum prolactin and progesterone in gravid mares. J. Anim. Sci. 70:217-223.

McLean, E.K. 1970. The toxic actions of pyrrolizidine (*Senecio*) alkaloids. Pharm. Rev. 22:429-483.

Mirocha, C.J., C.G. Mackintosh, U.A. Mirza, W. Xie, Y. Xu, and J. Chen. 1992. Occurrence of fumonisin in forage grass in New Zealand. Appl. Environ. Microbiol. 58:3196-3198.

Mohanty, N.N. 1964. Studies on ubatta disease of rice. Indian Phytopathology 17:308-316.

Monroe, J.L., D.L. Cross, L.W. Hudson, D.M. Hendricks, S.W. Kennedy, and W.C. Bridges. 1988. Effects of selenium and endophyte-contaminated fescue on the performance and reproduction in mares. Equine Vet. Sci. 8:148-153.

Morgan, J.M. 1984. Osmoregulation and water stress in higher plants. Annu.

Rev. Plant Physiol. 35:299-319.

Morgan-Jones, G., and W. Gams. 1982. Notes on Hyphomycetes. XLI. An endophyte of *Festuca arundinacea* and the anamorph of *Epichloe typhina*, new taxa in one of two new sections of *Acremonium*. Mycotaxon 15:311-318.

Morgan-Jones, G., J.F. White, Jr., and E.L. Piontelli. 1990. *Acremonium chilense*, an undescribed endophyte occurring in *Dactylis glomerata* in Chile. Mycotaxon 39:441-445.

Mortimer, P.H., J. Cambell, M.E. di Menna, and E.P. White. 1971. Experimental myrotheciotoxicosis and poisoning in ruminants by verrucarin A and roridin A. Res. Vet. Sci. 12:508-515.

Mortimer, P.H., and J.W. Ronaldson. 1983. Fungal-toxin-induced photosensitization. p. 361-370. *In* R.J. Keeler and A.T. Tu (ed.). Handbook of natural toxins. Vol. 1, Plant and fungal toxins. Marcel Dekker, New York.

Murray, F.R., G.C.M. Latch, and D.B. Scott. 1992. Surrogate transformation of perennial ryegrass, *Lolium perenne*, using genetically modified *Acremonium* endophyte. Mol. Gen. Genet. 233:1-9.

Nelson, L.R., and J.C. Read. 1990. Fungal endophyte infection in Italian Ryegrass. Plant Dis. 74:183.

Nobindro, U. 1934. Grass poisoning among cattle and goats in Assam. Indian Vet. J. 10:235-236.

Pantidou, M.E. 1959. *Claviceps* from Zizania. Can. J. Bot. 37:1233-1236.

Pedersen, J.F., R. Rodriguez-Kabana, and R.A. Shelby. 1988. Ryegrass cultivars and endophyte in tall fescue affect nematodes in grass and succeeding soybean. Agron. J. 80:811-814.

Pelton, M.R., H.A. Fribourg, J.W. Laudre, and T.D. Reynolds. 1991. Preliminary assessment of small wild mammal populations in tall fescue habitats. Tenn. Farm Home Sci. 160:68-71.

Pitt, J.I. 1979a. *Penicillium crustosum* and *P. simplicissimum*, the correct names for two common species producing tremorgenic mycotoxins. Mycologia 71:1166-1177.

Pitt, J.I. 1979b. The genus *Penicillium* and its teleomorphic states *Eupenicillium* and *Talaromyces*. Academic Press, New York.

Porter, J.K., C.W. Bacon, and J.D. Robbins. 1979a. Ergosine, ergosinine, and chanoclavine I from *Epichloe typhina*. J. Agric. Food Chem. 27:595-598.

Porter, J.K., C.W. Bacon, and J.D. Robbins. 1979b. Lysergic acid amide derivatives from *Balansia epichloe* and *Balansia claviceps* (Clavicipitaceae). J. Nat. Prod. 42:309-314.

Porter, J.K., C.W. Bacon, J.D. Robbins, and D. Betowski. 1981. Ergot alkaloid identification in Clavicipitaceae systemic fungi of pasture grasses. J. Agric. Food Chem. 29:653-657.

Porter, J.K., J.D. Robbins, C.W. Bacon, and D.S. Himmelsbach. 1978. Determination of epimeric 1-(3-indoyl)propane-1,2,3-triol isolated from *Balansia epichloe*. Lloydia 41:43-49.

Porter, J.K., J.A. Stuedemann, F.N. Thompson, and L.B. Lipham. 1990. Neuroendocrine measurements in steers grazed on endophyte-infected fescue. J. Anim. Sci. 68:3285-3292.

Porter, J.K., J.A. Stuedemann, F.N. Thompson, B.A. Buchanan, and H.A. Tucker. 1993. Melatonin and pineal neurochemicals in steers grazed on endophyte-infected tall fescue: Effects of metoclopramide. J. Anim. Sci. 71:1526-1531.

Porter, J.K., and F.N. Thompson, Jr. 1992. Effects of fescue toxicosis on reproduction in livestock. J. Anim. Sci. 70:1594-1603.

Porter. J.K., C.W. Bacon, J.D. Robbins, and H.C. Higman. 1975. A field indicator in plants associated with ergot-type toxicities in cattle. J. Agric. Food Chem. 23:771-775.

Pottinger, R.P., G.M. Barker, and R.A. Prestidge. 1985. A review of the relationships between endophytic fungi of grasses (*Acremonium* spp.) and Argentine stem weevil (*Listronotus bonarienses* Kuschel). p. 322-331. *In* Proc. 4th Australasian Conference Grassland Invertebrate Ecology. Lincoln College, Canterbury.

Prestidge, R.A., R.P. Pottinger, and G.M. Barker. 1982. An association of *Lolium* endophyte with ryegrass resistance to Argentine stem weevil. p. 119-122. In Proc. N. Z. Weed Pest Control Conf.

Purchase, I.H.F. (ed.). 1974. Mycotoxins. Elsevier, Amsterdam.

Putnam, M.R., D.I. Bransby, J. Schumacher, T.R. Boosinger, L. Bush, R.A. Shelby, J.T. Vaughan, D. Ball, and J.P. Brendemuehl. 1991. Effects of the fungal endophyte *Acremonium coenophialum* in fescue on pregnant mares and foal viability. Am. J. Vet. Res. 52:2071-2074.

Read, J.C., and B.J. Camp. 1986. The effect of the fungal endophyte *Acremonium coenophialum* in tall fescue on animal performance, toxicity, and stand maintenance. Agron. J. 78:848-850.

Richardson, M.D., G.W. Chapman, Jr., C.S. Hoveland, and C.W. Bacon. 1992. Sugar alcohols in endophyte-infected tall fescue under drought. Crop Sci. 32:1060-1061.

Richardson, M.D., N.S. Hill, and C.S. Hoveland. 1990. Rooting patterns of endophyte infected tall fescue grown under drought stress. p. 129. *In* Agronomy abstracts. ASA, Madison, WI.

Rowan, D.D., and D.L. Gaynor. 1986. Isolation of feeding deterrents against Argentine stem weevil from ryegrass infected with the endophyte. J. Chem. Ecol. 12:647-658.

Rowan, D.D., M.B. Hunt, and D.L. Gaynor. 1986. Peramine, a novel insect feeding deterrent from ryegrass infected with the endophyte *Acremonium loliae*. J. Chem. Soc., Chem.Commun. 1986:935-936.

Rowan, D.D., and G.J. Shaw. 1987. Detection of ergopeptine alkaloids in endophyte-infected perennial ryegrass by tandem mass spectrometry. N. Z. Vet. J. 35:197-198.

Rykard, D.M. 1983. Comparative morphology of the conidial states and host-parasite relationship in members of Balansiae (Clavicipitaceae). Ph.D. diss. University of Georgia, Athens.

Rykard, D.M., C.W. Bacon, and E.S. Luttrell. 1985. Host relations of *Myriogenospora atramentosa* and *Balansia epichloe* (Clavicipitaceae). Phytopathology 75:950-956.

Rykard, D.M., E.S. Luttrell, and C.W. Bacon. 1984. Conidiogenesis and conidiomata in the Clavicipitoideae. Mycologia 76:1095-1103.

Sampson, K. 1933. The systemic infection of grasses by *Epichloe typhina* (Pers.) Tul. Trans. Br. Mycol. Soc. 18:30-47.

Schardl, C.L., J.S. Liu, J.F. White, Jr., R.A. Finkel, and M.R. Siegel. 1991. Molecular phylogenetic relationships of nonpathogenic grass mycosymbionts and clavicipitaceous plant pathogens. Plant Systemat. Evol. 178:27-41.

Schardl, C.L., and M.R. Siegel. 1993. Genetics of *Epichloe typhina* and

Acremonium coenophialum. p. 169-185. *In* R. Joost and S. Quisenberry (ed.) *Acremonium*/grass interactions. Vol.44, Agric. Ecosystem Environ., Elsevier Science Publishers, Netherlands.

Schmidt, S.P., C.S. Hoveland, E.M. Clark, N.D. Davis, L.A. Smith, H.W. Grimes, and J.L. Hilliman. 1982. Association of an endophytic fungus with fescue toxicity in steers fed Kentucky 31 tall fescue seed or hay. J. Anim. Sci. 55:1259-1263.

Siegel, M.R., G.C.M. Latch, and M.C. Johnson. 1987. Fungal endophyte of grasses. Ann. Rev. Phytopath. 25:293-315.

Siegel, M.R., G.C.M. Latch, L.P. Bush, N.F. Fammin, D.D. Rowen, B.A. Tapper, C.W. Bacon, and M.C. Johnson. 1991. Alkaloids and insecticidal activity of grasses infected with fungal endophytes. J. Chem. Ecol. 16:3301-3315.

Smith, J.D., and W.E. Crawley. 1964. Disturbance of pasture herbage and spore dispersal of *Pithomyces chartarum* (Berk. and Curt.) M.B. Ellis. N.Z. J. Agric. Res. 7:281-287.

Smith, J.E., and R.S. Henderson. 1991. Mycotoxins and animal foods. CRC Press, Boca Raton, FL.

Sperry, O.E., R.D. Turk, G.O. Hoffman, and F.B. Stroud. 1955. Photosensitization of cattle in Texas. Texas Agric. Ext. Bull. 812.

Styne, P.S., and R. Vleggaar. 1985. The tryptoguivalines. Prog. Chem. Org. Nat. Prod. 48:62-65.

Stilham, W.D., C.J. Brown, L.B. Daniels, E.L. Piper, and H.E. Fetherstone. 1982. Toxic fescue linked to reduced milk output in ewes. Arkansas Farm Res. 31:9.

Strahan, S.R., R.W. Hemken, J.A. Jackson, Jr., R.C. Buckner, L.P. Bush, and M.R. Siegel. 1988. Performance of lactating dairy cows fed tall fescue forage. J. Dairy Sci. 70:1228-1231.

Stuedemann, J.A., and C.S. Hoveland. 1988. Fescue endophyte: History and impact on animal agriculture. J. Prod. Agric. 1:39-44.

Stuedemann, J.A., T.S. Rumsey, J. Bond, S.R. Wilkinson, L.P. Bush, D.J. Williams, and A.B. Caudle. 1985. Association of blood cholesterol with occurrence of fat necrosis in cows and tall fescue summer toxicosis in steers. Am. J. Vet. Res. 46:1990-1995.

Taber, R.A., R. Pittit, W.A. Taber, and J.W. Dollahite. 1968. Isolation of *Pithomyces chartarum* in Texas. Mycologia 60:727-730.

Thomasson, J.R. 1986. Fossil grasses: 1820-1986 and beyond. p. 159-167. *In* T.R. Soderstrom, K.W. Hilu, C.S. Campbell and M.E. Barkworth (ed.) Grass Systematics and Evolution. Smithsonian Institution Press, Washington, DC.

Thompson, R.W., H.A. Fribourg, and B.B. Reddick. 1989a. Sample intensity and timing for detecting *Acremonium coenophialum* in tall fescue pastures. Agron. J. 81:966-971.

Thompson, R.W., H.A. Fribourg, and B.B. Reddick. 1989b. Sampling intensity and timing for detecting *Acremonium coenophialum* in tall fescue. Agron. J. 81:966-971.

Thornton, R.H., G. Shirley, and R.M. Salisbury. 1968. A nephrotoxin from *Aspergillus fumigatus* and its possible relationship with New Zealand mucosal disease-like syndrome in cattle. N.Z. J. Agric. Res. 11:1-10.

Tsai, H.-F., M.R. Siegel, and C.L. Schardl. 1992. Transformation of *Acremonium coenophialum*, a protective fungal symbiont of the grass *Festuca arundinacea*. Curr. Genet. 22:399-406.

Vertinskii, K.I., K.A. Dzhilavyan, and V.P. Koroleva. 1967. A disease of sheep caused by the fungus *Myrothecium verrucaria* (In Russian). Byull. Inst. Eksp. Vet. 2:86-90.

Wallner, B.M., N.H. Booth, J.D. Robbins, C.W. Bacon, J.K. Porter, T.E. Kiser, R.W. Wilson, and B. Johnson. 1983. Effect of an endophytic fungus isolated from toxic pasture grass on serum prolactin concentrations in the lactating cow. Am. J. Vet. Res. 44:1317-1322.

Walker, J. 1957. A new species of *Claviceps* from Zizania. Can. J. Bot. 37:1233-1236.

Weedon, C.M., and P.G. Mantle. 1987. Paxilline biosynthesis by *Acremonium loliae*; a step towards defining the origin of lolitrem neurotoxins. Phytochemistry 26:969-971.

West, C.P., E. Izekor, A. Elmi, R.T. Robbins, and K.E. Turner. 1989. Endophyte effects on drought tolerance, nematode infestation and persistence of tall fescue. p. 23-27. *In* C.P. West (ed.) Proc. Arkansas Fescue Toxicosis Conference, No. 140. Arkansas Agric. Exp. Sta., University of Arkansas, Fayetteville.

West, C.P., E. Izekor, D.M. Oosterhuis, and R.T. Robbins. 1990a. The effect of *Acremonium coenophialum* on the growth and nematode infestation of tall fescue. Plant Soil 112:3-6.

West, C.P., D.M. Oosterhuis, and S.D. Wullschleger. 1990b. Osmotic adjustment in tissues of tall fescue in response to water deficit. Environ. Experimen. Bot. 30:1-8.

White, J.F., Jr. 1987. The widespread distribution of endophytes in the Poaceae. Plant Dis. 71:340-342.

White, J.F., Jr. 1988. Endophyte-host associations in forage grasses. XI. A proposal concerning origin and evolution. Mycologia 80:442-446.

White, J.F., Jr., and T.L. Bultman. 1987. Endophyte-host associations in forage grasses. VIII. Heterothallism in *Epichloe typhina*. Am. J. Bot. 74:1716-1722.

White, J.F., Jr., and G.T. Cole. 1985. Endophyte-host associations in forage grasses. I. Distribution of fungal endophytes in some species of Lolium and *Festuca*. Mycologia 77:323-327.

White, J.F., Jr., and G.T. Cole. 1986. Endophyte-host associations in forage grasses. V. Occurrence of fungal endophytes in certain species of *Bromus* and *Poa*. Mycologia 78:102-107.

White, J.F., Jr., P.M. Halisky, S. Sun, G. Morgan-Jones, and C.R. Funk, Jr. 1992. Endophyte-host associations in grasses. XVI. Patterns of endophyte distribution in species of the tribe agrostideae. Am. J. Bot. 79:472-477.

White, J.F., Jr., A.C. Morrow, and G. Morgan-Jones. 1990. Endophyte-host associations in forage grasses. XII. A fungal endophyte of *Trichacne insularis* belonging to *Pseudocercosporella*. Mycologia 82:218-226.

White, R.H. 1989. Water relations characteristics of tall fescue as influenced by *Acremonium coenophialum*. p. 167. *In* Agronomy abstracts, ASA. Madison, WI.

Williams, M.J., P.A. Backman, E.M. Clark, and J.F. White, Jr. 1984. Seed treatments for control of the tall fescue endophyte *Acremonium coenophialum*. Plant Dis. 68:49-52.

Wilson, B.J., C.S. Byerly, and L.T. Burka. 1981. Neurologic disease of fungal origin in three herds of cattle. J. Am. Vet. Med. Assoc. 179:480-485.

Witters, W.L., R.A. Wilms, and R.D. Hood. 1975. Prenatal effects of elymoclavine administration and temperature stress. J. Anim. Sci. 41:1700-1704.

Wyllie, T.D., and L.G. Morehouse (ed.). 1977. Mycotoxic fungi, mycotoxins, and mycotoxicoses: An encyclopedic handbook. Vol. 1, 2, and 3. Marcel Dekker, New York.

Yates, S.G. 1962. Toxicity of tall fescue forage: A review. Econ. Bot. 16:295-303.

Yates, S. G. 1983. Tall fescue toxins. p. 249-273. *In* M. Recheigel (ed.) Handbook of naturally occurring food toxicants. CRC Press, Boca Raton, FL.

Yates, S.G., R.D. Plattner, and G.B. Garner. 1985. Detection of ergopeptine alkaloids in endophyte infected, toxic Ky-31 tall fescue by mass spectrometry/mass spectrometry. J. Agric. Food Chem. 33:719-722.

Yamazaki, M., S. Suzuki, and K. Miyaki. 1971. Tremorgenic toxins from *A. fumigatus*. Chem. Pharm. Bull. 19:1739-1742.

Yoshihara, T., S. Togiya, H. Koshino, S. Sakamura, T. Shimanuki, T. Sato, and A. Tajimi. 1985. Three fungitoxic sesquiterpenes from stromata of *Epichloe typhina*. Tetra. Let. 26:5551-5554.

Zahner, H., H. Anke, and T. Anke. 1983. Evolution and secondary pathways. p. 153-171. *In* J.W. Bennett and A. Ciegler (ed.) Secondary metabolism and differentiation in fungi. Marcel Dekker, New York.

CHAPTER 9

INTRINSIC CHEMICAL FACTORS IN FORAGE QUALITY

Lowell Bush and Harold Burton

INTRODUCTION

This review is intended to examine some of the more widely recognized intrinsic chemical components of herbage that have impacted forage research during the past 25 years. Consequently, it is not a comprehensive review of chemicals produced and accumulated in herbage which are under genetic control of the forage plant. Lignin and polyphenols, phytoestrogens, rumen bloat, cyanogenic glucosides and alkaloids were major areas selected for inclusion in this presentation. Brown midrib studies were selected as the backdrop for lignin and polyphenol research. Phytoestrogens are not a major problem in the U.S. but molecular biology research to enhance phytoalexin production may impact phytoestrogen levels in the modified pathways and subsequent plants. Past rumen bloat research has suggested many different avenues of investigations. The elucidation of the etiology is not completed but the recent investigations seem well along toward a definitive conclusion. Since the description of stable foam in the rumen of bloated animals, the excitement of the science is evident in reading the literature and could well be used as an excellent model for a graduate education forum. The cyanogenic glucosides are represented by dhurrin in sorghum and related species. Plant development associated with toxicity and management of animals to minimize the negative impact on the animals have been understood for many years. More recently, investigations have had objectives to determine the biosynthesis and inheritance of dhurrin. Biosynthesis involves cytochrome P-450 and channeling of enzyme systems in the membrane. Both concepts were and are on the frontiers of plant science research and are included to illustrate the significance of fundamental research to production situations. The last group of compounds considered in this chapter is the alkaloids. The simple indole alkaloids of *Phalaris spp.* may be most associated with reduced palatability of *Phalaris* in the U.S., whereas sudden death syndrome of *Phalaris* ingestion apparently is not caused by the alkaloids. Lupine alkaloids in the U.S. are mainly a problem associated with weedy species in western rangeland, but lupine poisoning from herbage and seed is very important elsewhere in the world. Many different classes of alkaloids are present in tall fescue herbage. The diazaphenanthrene alkaloids and the loline group of pyrrolizidine alkaloids were considered intrinsic for this review. The

---

Lowell Bush and Harold Burton, Dep. of Agronomy, Univ. of Kentucky, Lexington, KY 40546-0091

diazaphenanthrene alkaloids probably have a minor role in animal health and performance. Interaction of both these alkaloid groups with extrinsic alkaloids of the fungal endophyte/grass association is yet to be determined and allows an opportunity for much creative research in the near future.

LIGNIN AND POLYPHENOLS

Polyphenols are produced in greatest amount of any group of plant allelochemicals formed from photosynthetically fixed carbon, and the structurally ill-defined lignins are derived from polyphenols. Phenolics are an ubiquitous heterogeneous group of plant metabolites (Fahey and Jung, 1989). Polyphenols in plants, including biosynthesis, distribution, nutritional ecology, and pharmacological properties have been presented in reviews by Butler (1989), Fahey and Jung (1989), and Mole (1989). In this section, the relationship between protein and dry matter (DM) digestion as altered by polyphenols and lignin in the forages and most specifically in the grasses associated with brown midrib genotypes will be discussed.

Brown Midrib and Chemical Composition

One of the earliest reports of brown pigment to develop in cells of the midrib and leaf sheaths of maize (*Zea mays* L.) was by Eyster (1926). He designated this character brown midrib (bm) and found it to be homozygous recessive. Jorgenson (1931) further described the *bm* character as a simple recessive trait, not identified with any chemical, but the pigmentation was found in leaf midrib, stem, tassel, cob, and roots and was linked with the purple aleurone character in maize. Subsequently, several *bm* genes have been identified in maize. Kuc's group (Kuc and Nelson, 1964; Gee et al., 1968) was the first to show that brown midrib mutants were associated with lower lignin and with less p-hydroxycinnamic acid bound to the lignin than normal genotypes. However, in the *bm* mutants, equal quantities of ferulic acid were present but about 50% less p-coumaric acid.

Muller et al. (1971) examined *bm1*, *bm3*, and the *bm1/bm3* double mutant for lignin and other structural components. On a whole plant basis, the *bm3* mutant had the lowest concentration of acid detergent lignin (ADL). The double mutant had less cellulose and acid detergent fiber (ADF) but the highest hemicellulose level compared to the single *bm* mutants. Normal genotypes and bm mutants had similar cell wall content (CWC) and crude protein content.

Porter et al. (1978) chemically induced brown midrib mutation (*bmr*) in sorghum (*Sorghum bicolor* L. Moench). These *bmr* mutants had brown pigmentation in the leaf midrib and stem of mature plants. Lignin in the stem was more than 50% lower in one line and as much as 25% lower in the leaves of another line. No consistent differences in other fiber components were observed. Fritz et al. (1981) evaluated three *bmr* genes from the Porter et al. (1978) study that had been incorporated in both grain sorghum and sudan grass and found the *bmr* condition expressed in both phenotypes. Cherney et al. (1986) evaluated the *bmr* gene in a sudangrass and two sorghum x sudangrass hybrids and reported a 23% reduction in lignin on a DM basis along with decreased p-coumaric acid and p-hydroxybenzaldehyde but increased levels of protocatechuic acid and vanillin. p-Coumaric acid is considered to be important in cross linkage among fiber components (Hartley, 1972) and the decreased p-coumaric acid content associated with *bmr* indicates differences in linkage between lignin and hemicellulose of normal and *bmr* genotypes. This difference may be causal to the increased in vitro dry matter disappearance (IVDMD) observed with *bmr* compared to normal genotypes.

Brown Midrib and Utilization

Stover silage made from *bm3* maize had 15% greater cell wall digestibility by sheep and the animals consumed 29% more DM compared to normal maize (Muller et al., 1972). Lechtenberg et al. (1974) used *bm3* maize to evaluate the effect of lignin on the in vitro rate of CWC and cellulose disappearance. The *bm3* maize stover was 35 to 40% lower in lignin than the normal genotype but they found that rate of disappearance was not affected by lignin concentration. The absolute amount of CWC disappearance per unit time was much greater for the *bm3* maize but the difference between genotypes was due to the higher concentration of digestible CWC in the *bm3*. Ad libitum intake by sheep was 29% greater in *bm3* than normal genotypes. Increased DM intake was not attributed to differences in the CWC digestion rate constant. They concluded that lignin prevents digestion of a portion of the CWC and cellulose without interfering with rate of digestion of the remaining CWC and cellulose. Porter et al. (1978) showed that in vitro CWC disappearance (IVCWCD) of stems of 13 different *bmr*-containing sorghums increased 17%, whereas the IVCWCD of leaves from these plants only increased 9%. Fritz et al. (1981) compared a normal genotype with the *bmr* genotype in sorghum x sudangrass forage and observed that a 20% reduction in lignin was associated with a 23% increase of IVDMD. These results are important as the IVDMD differences observed were not confounded with grain that may have been in silage or feed of earlier studies with whole plant and brown midrib genotypes.

During a 72 h digestion of sorghum x sudangrass hybrids, lignin in both normal and *bmr* genotypes converged to 138 g kg^{-1} NDF (Cherney et al., 1986). Therefore, apparent lignin disappearance was lower in *bmr* than the normal genotype. Rate constants for in vitro digestion for neutral detergent fiber (NDF), cellulose, and hemicelluloses were higher for *bmr* genotypes than normal genotypes. This observation is similar to that reported by Cherney et al. (1986) for sorghum, but is in disagreement with data from corn (Lechtenberg et al., 1974). Also, the ferulic acid/p-coumaric acid ratio was higher in the *bmr* than the normal genotype and supports the suggestion of a positive relationship between cell wall digestion and the ferulic acid/p-coumaric acid ratio (Hartley, 1972).

An induced brown midrib mutant in pearl millet [*Pennisetum americanum* (L.) Leeke] was associated with a 20% reduction in lignin on a whole plant basis and a 10% increase in IVDMD (Cherney et al., 1988). Concentration of the phenolic monomers varied; p-coumaric acid and p-hydroxybenzoic acid decreased but p-hydroxybenzaldehyde, syringic acid, vanillin, and vanillic acid increased compared to the normal genotype. On a DM basis, ferulic acid levels did not change. The p-coumaric acid/ferulic acid ratio in pearl millet, sorghum, and maize *bmr* genotypes were similar at 1.4, 1.5, and 1.5, respectively. These results strongly suggest that all *bmr* mutants have similar lignin and phenolic composition changes and that this particular trait should improve forage quality. Van Soest (1981) concluded that lignin prevented digestion of about 1.4 times its own weight of cell wall carbohydrate. Lignin did not reduce digestion by encrustation but by formation of chemical bonds between the lignin and carbohydrate.

Brown Midrib and Agronomic Characters

Despite the nutritional advantages of the brown midrib trait in maize, commercial development has been limited. Principal negative agronomic characters are decreased DM and increased lodging. Miller et al. (1983) reported a 23% reduction in grain yield, 10% reduction in stover yield, and a 16% reduction in DM yield in Minnesota for *bm3* maize lines compared to normal genotypes. Lee and Brewbaker (1984) evaluated 15 *bm3* hybrids and their isogenic normal counterparts and found an average 17% reduction in stover and 20% reduction in grain yield when grown in Hawaii. Both Miller et al. (1983) and Lee and Brewbaker (1984) concluded that the gain in stover digestibility because of the *bm3* trait did not offset yield losses to justify a separate *bm3* breeding program and that normal populations of maize offer more potential for silage breeding progress. Lee and Brewbaker (1984) also excluded any linkage of the *bm3* gene with yield-reducing genes and suggested that the pigmentation was directly responsible for reduced photosynthesis of the mutants. However, in the United Kingdom, Weller et al. (1985) found DM to be equal in *bm3* and normal genotypes of forage maize. The *bm3* genotype had reduced grain yield but this was accompanied by greater stover yield. Stem diameter, stem weight, and rind thickness were similar in both genotypes, but significantly more pressure was required to crush the stems of normal genotype plants compared to *bm3* plants. The authors concluded that background genotype was more important to lodging than the *bm3* gene and that *bm3* should not be incorporated into lines with a tendency to lodge.

Regulation of biosynthesis resulting in the reduced lignin content in brown midrib plants has not been elucidated. However, it may not be the same in all species. Bucholtz et al. (1980) found that sorghum with lower lignin content from *bmr-6* mutant contained more aldehyde groups and reduced cinnamyl alcohol dehydrogenase activity. Cinnamyl alcohol dehydrogenase catalyzes the last step in the series of reactions leading from phenylalanine to cinnamyl alcohols. The reduction of aldehyde lignin intermediates (cinnamyl aldehydes) to cinnamyl alcohols and the oxidation of the phenolic hydroxy group of the cinnamyl alcohol monomer is the first step in polymerization of these monomers to lignin. However, in maize, no difference was detected between *bm3* and a normal line for cinnamyl alcohol dehydrogenase but catechol O-methyltransferase activity was reduced 90% in the *bm3* line (Grand et al., 1985).

Tannins

Tannin classification generally is separated into two groups, condensed and hydrolyzable. Condensed tannins, also referred to as proanthocyanidins, are a polymeric flavan-3-ol in which the interflavan bonds are usually between ring B and ring A (C4 to C8) with a molecular weight from 500 to 3000. Hydrolyzable tannins are derived from gallic acid and polyols, usually glucose. These tannins may be hydrolyzed by acids or bases to yield a sugar and gallic acid. Ester, ether, and C-C bonds are formed in ellagitannins between the gallic acid units, and ellagic acid is formed upon acid hydrolysis (Haslam, 1981; Fahey and Jung, 1989).

The physiological and pharmacological properties of simple polyphenols have been reviewed by Fahey and Jung (1989) and Mole (1989). Only a brief summary will be presented here. The most observed effect of simple polyphenols in the diet has been reduced feed intake. Mechanism of the reduction is not known and may be either on nutrient digestion or disruption of metabolism. These mechanisms could include reduced palatability by binding to protein or taste receptors and reduced digestibility by binding to digestive enzymes (protein) or to substrate protein (Mole, 1989).

There are several possible explanations for observed adverse effects of tannins on ruminant nutrition as well as some potential beneficial effects. Most significant among the beneficial effects are prevention of bloat and protection of protein against rapid ruminal degradation. Price and Butler (1980) suggested that adverse effects of tannins may be mediated by 1) depressed feed intake, 2) binding with dietary protein, 3) complexing with digestive enzymes, 4) binding with endogenous protein, 5) interacting with the digestion tract, and 6) direct toxic effects. Fahey and Jung (1989) concluded that there was no direct evidence that tannins were harmful. In a series of reports, T.N. Barry concluded that high concentration of condensed tannins in *Lotus pedunculatus* Cav. depressed metabolizable energy intake by both depressing voluntary intake and organic matter digested (Barry and Duncan, 1984; Barry and Manley, 1984). However, condensed tannins may have a beneficial and detrimental nutritional effect. The beneficial effect comes from binding with protein and thus increasing the amino acids available for post-ruminal absorption. Tannins in excess of those bound to protein are "free" and probably responsible for decreased intake and ruminal carbohydrate digestion (Barry and Manley, 1986). In grazing sheep, reduced intake of high tannin *L. pedunculatus* resulted in reduced live weight gain and wool growth (Barry, 1985; Terrill et al., 1992).

More recently, Provenza et al. (1990) concluded that condensed tannins are feeding deterrents and that deterrence, at least in goats, is a learned response from postingestion malaise and not instinctive response recognition. This group also concludes that chemical structure of condensed tannins is significant to biological activity and that in deterring intake, the condensed tannins act as a toxin and not as a digestion inhibitor (Clausen et al., 1990).

Most recent studies of the influence of polyphenols on in vivo digestion have utilized brown midrib mutants. Wedig et al. (1988) compared sorghum x sudangrass hybrids with and without the *bmr* gene and found p-coumaric acid to be higher in the normal genotype and sheep ruminal digestibility of NDF, ADF, and cellulose to be lower than the *bmr* genotype. Only cellulose digestibility was lower for the total gastrointestinal tract comparison between the two forages. Fritz et al. (1988) and Wedig et al. (1989b) showed that rate of in situ cell wall digestion was the same for *bmr* and normal genotypes of sorghum x sudangrass hays and silage. The lack of difference in rate of digestion agrees with the earlier in vitro digestibility report by Lechtenberg et al. (1974). Again, *bmr* materials had a greater extent of digestion as had been demonstrated in vitro. In a subsequent paper to examine the hemicellulosic and alkali-soluble phenolic substances, Wedig et al. (1989a) found that arabinose, xylose, and uronic acids were more digestible in *bmr* genotypes but p-coumaric acid disappearance was higher in animals fed the normal genotypes. Grazing lambs showed a preference for *bmr* pearl millet genotypes compared to normal genotypes and in vivo digestibility and DM intakes also were greater for the *bmr* forage (Cherney et al., 1990).

PHYTOESTROGENS

Another important group of naturally occurring phenolic compounds in plants are the phytoestrogens. Estrogenic compounds isolated from plants have been known for less than 70 years. Phytoestrogens have been found in a broad variety of plants but most research is limited to a few economically important species. Amounts of phytoestrogens ingested may be high but the relative estrogenic activity is low and total inhibition of the reproduction process in consuming animals is not observed. Frequently, partial disruption of the reproduction process, often at the subclinical level, occurs and goes unnoticed (Lightfoot, 1974). Much of the research on phytoestrogens has come from Australia and New Zealand. The review by Adams (1989) covers much of the plant science as well as the pharmacological properties of the phytoestrogens, animal physiology and metabolism, and effects on animals including humans.

Most studied substances for phytoestrogen activity are in one of three groups: 1) coumestans, 2) isoflavones, and 3) isoflavans. These classes of compounds are formed from acetate-derived fragments and phenylpropanoids and this group contains the greatest number of plant phenolic compounds. Biosynthesis and properties of these compounds have been reviewed by Wong (1973) and Adams (1989).

Isoflavones

Isoflavonoid compounds are mainly found in *Fabaceae* and the best known isoflavone phytoestrogens are genistein, biochanin A, daidzein, and formononetin. These compounds are most important in white clover (*Trifolium repens* L.), red clover (*T. pratense* L.), and subterranean clover (*T. subterraneum* L.). Relative activities of these plant estrogens are low compared to diethylstilbestrol, but because they may occur in high concentrations in the forage and given the large amounts ingested, animal health problems are observed.

In subterranean clover, the presence of a glycosidase enzyme in the leaf is released upon maceration and the enzyme causes release of bound (glycosides of formononetin) phytoestrogen. As is the case with glycosidase for the cyanogenic glycosides of birdsfoot trefoil (*Lotus corniculatus* L.), this enzyme also is under single gene control (Francis and Millington, 1965). There are four isoflavone genotypes and they each seem to be a single recessive character. These genotypes are typified by 1) no isoflavones produced, 2) no hydrolyzing system produced, 3) no methylation of genistein or daidzein, and 4) no biosynthesis.

Formononetin concentration in red clover changed during the grazing season in New Zealand with highest concentration of 13.8 mg g^{-1} DM in March (Kelly et al., 1979) and at no time was the concentration below 3 mg g^{-1} DM, the level below which it is unlikely to have a detrimental effect on ewe fertility (Marshall, 1973). Hay making reduced formononetin by at least 50% but estrogenic activity apparently was reduced further. Formononetin concentration in red clover grown in Ireland decreased from 5.6 mg g^{-1} to 3.5 mg g^{-1} with increased time to first harvest during May and June (McMurray et al., 1986). Lamina tissue had higher concentrations of formononetin than petiole or stem tissue. Expanding lamina contained over 20 mg g^{-1} of formononetin. Stems had the lowest formononetin concentration followed in increasing amounts in petioles, expanded laminae, and expanding lamina. Ensiling red clover, with or without Na_2SO_4, resulted in increased biochanin A, formononetin, genistein, and daidzein during the first 15 d but only daidzein content remained above initial levels by 180 d of storage (Palfii et al., 1978).

Introduction of persistent subterranean clover to southern California was made more difficult by the observation that best persistence was associated with high formononetin content. However, two hardseeded Spanish cultivars with low formononetin were found that were as persistent as the most persistent Australian lines with high formononetin content (Graves et al., 1991). This observation of cultivar difference was not unexpected as Smith et al. (1986) found formononetin, biochanin A, and genistein concentrations to be dependent upon germplasm source, even though all cultivars tested had decreased phytoestrogen with later harvest dates. Gildersleeve et al. (1991) developed a seedling assay for formononetin, genistein, and biochanin A to rank germplasm for potential phytoestrogen activity and should make breeding programs for low isoflavone content more efficient.

Coumestans

This group of isoflavonoid phytoestrogens has been studied primarily in alfalfa (*Medicago sativa* L.) and white clover. Coumestans are represented by coumestrol, 4'-methoxycoumestrol and sativol in alfalfa, and repensol, trifoliol, and coumestrol in white clover. Many studies with alfalfa indicate that the concentration of coumestrol is dependent on the presence of foliar diseases (Loper et al., 1967). Coumestrol content was positively associated with lesion size and number of common leaf spot [*Pseudopeziza medicaginis* (Lib.)Sacc.] and rust (*Uromyces striatus* Schroet.). Selection for disease resistance resulted in lower coumestrol content. Wong and Latch (1971) observed a similar response with white clover. In healthy plants, levels of the phytoestrogens were very low or not detected but infection by four different pathogenic fungi resulted in readily detectable amounts of coumestrol, repensol, trifoliol, and formononetin. More recently, Barbetti and Nichols (1991) demonstrated that *Phoma medicaginis* Malbr. and Roum. and *Leptosphaeruline trifolii* (Rostr.) Petr. on annual *Medicago spp.* swards not only decreased forage yield but greatly increased coumestrol content in green forage and the end-of-season dry stem and fruits. Results of the above studies suggested that the fungus mycelium and certainly the urediospores of *Uromyces striatus* accumulated the phytoestrogens. Shore et al. (1992) reported that the exogenous esteroidal estrogens, estrone and 17β-estradiol, at very low concentrations (0.005 to 0.5 μg L^{-1}) increased shoot and root dry weight. However, the endogenous estrogen content of the plant (including phytoestrogens) did not increase until estrogen application was above 50 μg L^{-1} in the irrigation water. Estrogens in sewage water from both farm and municipal effluent ranged from 0.04 to 0.36 μg L^{-1}, within the range that modified plant growth.

Role of the phytoestrogens in plants is not readily apparent but some are intermediates in pathways for biosynthesis of phytoalexins. In alfalfa, formononetin and daidzein are metabolites in the pathway just prior to the phytoalexin medicarpin (reviewed by Dixon et al., 1992). Dixon's group has been investigating the molecular biology for modifying the phytoalexin content (disease resistance) of alfalfa. Much is being elucidated with respect to promoters, expression, and production of substances in the isoflavonoid pathway. Therefore, as these types of studies continue, not only must the end product (the phytoalexin and disease resistance) of genetic manipulation be considered but the intermediates in the pathway (phytoestrogens) as well. The cloning of genes for the isoflavonoid phytoalexin, medicarpin, may allow direct testing of the role of the phytoalexin in disease resistance. More important to the story of the phytoestrogens, daidzein and formononetin, will be the physiological role and their manipulation in the plant. As presented earlier in this section, some isoflavonoid phytoestrogen concentrations are apparently dependent upon the presence of foliar diseases. Perhaps the enhanced accumulation in the presence of the pathogens is a consequence of phytoalexin elicitation and the phytoestrogens are just intermediates in the plant-microbe interaction with the grazing animal a secondary organism and not directly in the ecological development of the plant.

Many studies have and continue to examine plants for estrogenic activity. The review by Farnsworth et al. (1975) remains one of the most comprehensive cataloging of phytoestrogens and antifertility agents found in plants.

BLOAT

Large quantities of gas, CO_2, and methane formed during fermentation of feed, and CO_2 released from acidification of bicarbonate, are produced during digestion of forage by the microflora in the rumen. Bloat is a distention of the ruminal cavity and occurs when the gases, 40 to over 200 L, are retained by development of a persistent foam within the rumen (Waghorn, 1991). Bloat most often occurs when animals graze legume pastures such as red clover, white clover, or alfalfa. Other legumes such as birdsfoot trefoil, sainfoin (*Onobrychis viciifolia* Scop.), arrowleaf clover (*Trifolium vesiculosum* Savi), and cicer milkvetch (*Astralagus cicer* L.) are considered to be bloat-safe legumes. Research progress on bloat during the last 30 to 40 years makes a very interesting story as different forage substances and conditions have been implicated in its etiology and as researchers have attempted to develop bloat-safe cultivars.

Bloat has been the object of several reviews (Clarke and Reid, 1974; Howarth, 1975; Czerkawski, 1986; Howarth et al., 1986; Hall and Majak, 1989). The review by Clarke and Reid (1974) is a very complete treatise on the early bloat research as it contains 413 citations. This chapter will only present findings related to legume bloat and with emphasis on the recent research. Significant ideas from earlier research have been included to provide continuity for the more recent investigations. Several investigators have contributed to the understanding of bloat but the groups in British Columbia and Saskatchewan, Canada and New Zealand have contributed and are continuing to contribute very substantially to elucidation of the etiology of bloat.

Stable Foam

An important development was the demonstration that stable foam was associated with the distention of the ruminal cavity and the observation that antifoaming agents reduced bloat in animals (Reid, 1960). Following these findings, considerable research was conducted to determine: 1) the foam volume potential of proteins from different forages, 2) the conditions in which stable foams form, and 3) the protein responsible for formation of these stable foams. Soluble protein content, especially ribulose 1,5-bisphosphate carboxylase (fraction I protein, an 18S protein), of bloat-inducing forages first was suspected as being the significant protein in stable foam formation (McArthur et al., 1964), but it was soon demonstrated that other soluble proteins could stabilize foams (Howarth et al., 1977). However, fraction I protein was not significantly correlated with bloat, whereas total N, soluble nonprotein N, and insoluble N in alfalfa herbage were positively associated with bloat. Also, Hall et al. (1988) sampled ruminal fluid prior to feeding and found no relationship between soluble protein N and occurrence of foam in the rumen. Yet a survey of fields in New Zealand indicated that soluble protein N of the forage was 35% higher on farms with severe animal bloat than on farms with no bloat (Ledgard et al., 1990).

Kendall (1966) demonstrated in vitro that tannins inhibited foam production and that addition of polyvinyl pyrrolidone (PVP) to the protein mixtures inhibited the effect of tannins. Also, PVP enhanced the foam production from the bloat-safe legumes, lespedeza, trefoil, and crownvetch. Jones et al. (1973) demonstrated that the qualitative presence of flavolans (polymeric flavanols, condensed tannins) was correlated with protein precipitants in bloat-safe legumes. Flavanols were found in two, *T. arvense* L. and *T. affine* L., of 44 *Trifolium* species examined. These findings stimulated studies to genetically produce bloat-safe cultivars in red clover, white clover, and alfalfa by increasing the condensed tannin content. However, as discussed in the polyphenol section earlier, tannins may decrease digestibility and be an undesirable constituent. Sarkar et al. (1976) did not find condensed tannins in cicer milkvetch, a bloat-safe legume. They also found differences among quantity and quality of the tannins present in legumes of low and high nutritive value and suggested that not only may the tannins be associated with bloat potential but with nutritive value of the forage. However, in a later report, Goplen et al. (1980) screened accessions, cultivars, breeding lines, and mutagenized populations of *Medicago falcata* L. and *M. sativa* and did not find any plant that contained detectable amounts of tannin. These results suggested a bloat-safe alfalfa could not be developed by simply adding tannins. This hypothesis should be revisited with the newer molecular biology techniques now available.

Digestion Rate and Foam Volume

Cooper et al. (1966) found a positive relationship between volume of foam formed and bloat potential in legume species and Pounden et al. (1959) determined there was at least 50% greater gas production from alfalfa and white clover than from orchardgrass (*Dactylis glomerata* L.). Another important avenue of investigation was initiated by Howarth et al. (1978, 1982) when they suggested bloat occurrence may be due to extent of cell disruption during grazing, mastication, and microbial digestion. They found that mesophyll cells in bloat-safe legumes were more resistant to mechanical rupture than mesophyll cells of bloat-inducing legumes. Fay et al. (1980) determined the rate and volume of gas and foam production of bloat-stimulating and bloat-safe legumes. They found that gas volume was greater when a mixture of solid and fluid ruminal contents was used compared to fluid only. Gas volume also was greater when whole and chewed leaves of bloat-inducing legumes were used compared to bloat-safe leaves. However, volumes of gas from homogenized leaves of both types were the same but the foam produced from the bloat-inducing legumes was greater than from bloat-safe legumes. They concluded that rate of disintegration and digestion of legumes by ruminal bacteria were important for development of pasture bloat. This stimulated much research into differences in mechanical disruption of leaf tissues in bloat etiology. Lees et al. (1981) concluded that bloat-safe legumes had strong cell walls and (or) a high degree of tissue strength, whereas bloat-inducing legumes had weak cell walls and low tissue strength. These results were obtained by crushing leaves by shaking with glass beads, sonication of leaf tissue and homogenization using a ground glass tissue grinder, and measuring chlorophyll in appropriate fractions to determine tissue damage and cell disruption. Their laboratory continued these investigations by measuring tissue disruption by ruminal microbes using a nylon bag technique. Bloat-inducing legumes were disrupted more rapidly than the bloat-safe legumes and these observations supported their earlier theory of cell rupture for bloat. However, they concluded that a 25 to 30% reduction in DM loss after 6 to 8 h would be required to develop a bloat-safe alfalfa. Digestion rate studies indicated that secondary and tertiary vein patterns and structure may be involved in bloat potential (Lees et al., 1982). Lees (1984) subsequently showed that the bloat-safe legumes had an adaxial cuticle, epidermal cell wall, and mesophyll cell wall that were thicker than bloat-inducing legumes. These characteristics were associated with resistance to cell disruption of the whole leaf and individual cells.

Rumen Prefeeding Condition

Majak et al. (1983) proposed that ruminal conditions prior to feeding predisposed an animal to bloat. Significant properties associated with bloat potential included chlorophyll and soluble protein content, ion quantity and type, and pH. Based upon these studies, alfalfa lines were developed having slow and fast initial rates of digestion and they were evaluated with sheep. Soluble protein, soluble carbohydrates, chlorophyll, and volatile fatty acids were lower in the rumen of sheep fed alfalfa having a slow initial digestion rate compared to the alfalfa having a fast digestion rate (Kudo et al., 1985) and prefeeding ruminal conditions from the slow initial digestion rate forage were not conducive for bloat formation. Differences between slow and fast initial digestion rate forages were apparent also when grown in different environments and this suggests that additional selection for slow initial rate of digestion may reduce bloat-causing potential of alfalfa.

Majak et al. (1986) found that prefeeding chlorophyll concentrations in the rumen were higher in animals that subsequently bloat than in those that did not bloat. This result suggests that more chloroplast and chloroplast fragments are suspended in ruminal fluid of animals that bloat and that the rate of chloroplast clearance in bloating animals is slower than in nonbloating animals (Hall et al., 1988). Because chloroplast membranes have a net negative charge at neutral pH (Muhlethaler, 1977), it is likely that ions in the ruminal fluid may be associated with protein release and colloidal behavior.

Prefeeding ruminal concentrations of sodium and potassium were negatively correlated and subsequent development of bloat was associated with a low ruminal level of sodium and a high level of potassium (Hall et al., 1988). Also, the concentration of divalent ions, magnesium and calcium, were positively associated with bloat. Divalent cations stabilize some of the proteins of the chloroplast membrane and may be the cause of the high chlorophyll content of ruminal fluid associated with predisposition for bloat by stabilization of the aggregate of chloroplast fragments by these "excess" divalent cations. Majak and Hall (1990) showed that sodium ruminal content postfeeding increased and potassium content decreased in both bloat-prone and bloat-resistant animals. However, initial (ca. 2 h) postfeeding sodium concentration decreased and potassium levels increased more in animals that were bloat-prone compared to bloat-resistant animals. By 8 h postfeeding, ruminal contents of animals had returned to prefeeding ion concentration levels and bloat-prone animals could be distinguished by their lower sodium and higher potassium concentrations. Calcium and magnesium are present in the rumen at lower concentrations than the monovalent ions but Hall et al. (1988) reported that higher concentrations in prefeeding ruminal contents were associated with a higher incidence of bloat. Over 30 years ago, Smith and Woods (1962) demonstrated that drenching lambs with magnesium increased the severity of bloat but that if a chelating agent was added with magnesium or calcium, bloat was reduced. Majak and Hall (1990) conclude that manipulation of cations through supplementation may provide a means of controlling bloat. Total electronegativity or redox potential of ruminal digesta was not clearly related to occurrence of bloat (Waghorn, 1991), further suggesting that if collodial charge is involved in bloat, quantity of individual cations may be important.

Bloat and Weather

Recent attempts to correlate forage composition or animal response for bloat development with weather factors has been informative but not greatly beneficial for elucidating the etiology of bloat (Hall and Majak, 1991). Bloat occurred frequently in animals fed fresh-cut alfalfa in autumn when daily temperatures, precipitation, and sunshine were much less than during incidences of summer bloat. Bloat occurred after a minimal temperature of -2.2°C (killing frost). Alfalfa composition on days bloat occurred compared to composition on no bloat days had the same kind of relationship as found in summer bloat studies. On days bloat occurred, the forage had higher moisture, chlorophyll, total N, and soluble N contents but lower ADF than on no bloat days. We may conclude from these studies that ambient weather conditions for the animal contribute little to occurrence of bloat but that forage composition as result of the weather is the significant contributor to occurrence of bloat.

CYANOGENIC GLUCOSIDES

General understanding of the genetic, physiological, biochemical, and cultural conditions that modify accumulation of cyanogenic compounds in plants has been increasing for nearly 200 years. However, it is only in the last few years that many details of inheritance, localization, and biosynthesis have been elucidated. Discussion of cyanogenesis in this article will be directed principally at dhurrin from sorghum and related plants with a few comments on *Trifolium spp*. For maximal utilization of a cyanogenic forage or for forage crop improvement, the localization, biosynthesis, and inheritance of these compounds must be understood.

Localization

Dhurrin (p-hydroxymandelonitrile-β-D-glucoside) content of young sorghum plants decreased 75% during the first 24 d of growth (Akazawa et al., 1960) and mature plants may not contain dhurrin. More recently, Haskins et al. (1987) concluded that young plants and regrowth tissue contain greater hydrogen cyanide potential (HCN-p) than mature sorghum plants and a plant height of sorghum x sudangrass hybrids or sudangrass should be at least 38 cm prior to grazing. Dhurrin levels in indiangrass [*Sorghastrum nutans* (L.) Nash] seedlings less than 20 cm in height was in the dangerous level (> 500 mg kg^{-1} DM), actually 750 mg kg^{-1}. However, if the plants were taller than 40 cm, dhurrin levels were in the safe zone (< 500 mg kg^{-1}), and if the plants were taller than 1 m, dhurrin levels were less than 200 mg kg^{-1}. Mulcahy and Stuart (1987) reported sweet sorghums had greater HCN-p than sorghum x sudangrass hybrids. Generally, the more DM accumulated, the less HCN-p was measured. Kojima et al. (1979) were the first to demonstrate cellular localization of the dhurrin and the enzymes responsible for cyanide production. They isolated epidermal and mesophyll protoplasts and bundle sheath strands and determined the dhurrin content and activities of catabolic enzymes in each tissue. Most of the dhurrin was located in vacuoles of the epidermal protoplasts and the enzymes responsible for release of the cyanide were found only in the mesophyll protoplasts. The bundle sheath strands contained neither dhurrin nor the catabolic enzymes necessary for cyanide production.

Biosynthesis

The biosynthesis of the cyanogenic compounds, dhurrin from sorghum and lotaustralin and linamarin of white clover, has been reviewed by several authors (Conn, 1980; Halkier et al., 1989; Poulton, 1991). A generalized biosynthetic pathway proceeds from the amino acid → N-hydroxyamino acid →

aldoxime → nitrile → α-hydroxynitriles and then to a glycoside or lipid derivative for in vivo storage. In the last few years, this system has been elucidated in sorghum. Dhurrin biosynthesis uses tyrosine as the amino acid precursor and proceeds in order via N-hydroxytyrosine, 2-nitro-3-(p-hydroxyphenyl)propionic acid, 1-*aci*-nitro-2-(p-hydroxyphenyl)ethane, (E)-p-hydroxyphenyl acetaldehyde oxime, (Z)-p-hydroxyphenylacetaldehyde oxime, p-hydroxyphenylacetonitrile, and p-hydroxymandelonitrile to the glycoside dhurrin (Halkier et al., 1991). McFarlane et al. (1975) demonstrated the in vitro formation of p-hydroxymandelonitrile from tyrosine by a microsomal fraction from sorghum seedlings. The reaction required O_2 and NADPH in addition to tyrosine. The enzymes were shown subsequently to be organized in the membrane to allow for efficient channeling of the flow of carbon atoms from tyrosine into p-hydroxymandelonitrile (Moller and Conn, 1980). Because of this channeling, exogenously-added radiolabelled compounds did not readily equilibrate with the endogenous membrane-bound compounds and isolation and identification of the intermediates in the pathway was difficult. Measurement of O_2 consumption, tyrosine utilization, and NADPH oxidation demonstrated that there were three oxygen-requiring hydroxylation steps in biosynthesis of p-hydroxymandelonitrile from tyrosine (Halkier et al., 1988, 1991). The N-hydroxylase which converts tyrosine to N-hydroxytyrosine and the C-hydroxylase converting p-hydroxyphenylacetonitrile to p-hydroxymandelonitrile are cytochrome P-450 monooxygenases (Halkier and Moller, 1991). These P-450 enzymes are inhibited by carbon monoxide and the inhibition is reversed by 450 nm light. These enzymes are also sensitive to selected P-450 inhibitors and to antibodies for NADPH-cytochrome P-450 reductase (Halkier and Moller, 1991). The other step for incorporation of an atom of oxygen is the N-hydroxylation and N-oxidation of N-hydroxytyrosine to 2-nitro-3-(p-hydroxyphenyl)propionate (Halkier et al., 1991), but this is not a P-450 dependent reaction. From these studies, it was determined that the hydroxylation of tyrosine by tyrosine N-monooxygenase was the rate-limiting step in dhurrin biosynthesis. In vivo glycosation of p-hydroxymandelonitrile to dhurrin is catalyzed by a soluble enzyme *B*-glycosyltransferase (Reay and Conn, 1974). This is the last step in dhurrin biosynthesis. Dhurrin is apparently stored in the vacuole of the cell within which it is synthesized (Halkier et al., 1988), mainly the epidermal cells (Kojima et al., 1979).

Cyanogenesis

Cyanogenesis is the actual release of HCN, the toxic entity to consuming animals. Upon cell disruption (wilting, chewing during ingestion, crushing), degradation of the cyanogenic glycoside is initiated. The first step is the cleavage of the sugar moiety by a β-glycosidase. Dhurrinase II is the glycosidase in green shoots of sorghum (Hosel et al., 1987). The enzyme is specific for dhurrin and its structural analog without the hydroxyl group. The product of this reaction, p-hydroxymandelonitrile, may decompose spontaneously or enzymatically in the presence of p-hydroxymandelonitrile lyase to HCN and p-hydroxybenzaldehyde. Nonenzymatic production of HCN proceeds rapidly at alkaline pH but the enzymatic release of HCN from plant macerates would be significant at the acidic pH of the macerate (pH 5.0 to 6.5). The β-glycosidase and the α-hydroxynitrile lyase are in the mesophyll cells (Kojima et al., 1979) and the maceration is required to bring the substrate and enzymes together. Enhanced enzymatic release of HCN in *Hevea brasiliensis* Muell. Arg. was greatest with a ratio between the hydroxynitrile lyase and the β-glycosidase of 4.8 (Selmar et al., 1989). HCN released within 5 min with the 4.8 enzyme ratio was equal to total decomposition (HCN release) at pH 13. Important in these considerations is not only the rate of release but the total amount of HCN produced.

Inheritance

Early reports of inheritance of cyanogenesis in sorghum, sudangrass, white clover, and *Lotus spp.* were reviewed by Nass (1972). Results of the early investigations with sorghum and sudangrass often were confusing if not conflicting, and the general conclusion was that several genes were involved. More recently, Gorz et al. (1986) reported that HCN-p of mature sorghum leaves of two breeding lines and their hybrids was controlled by a single gene pair. F_1 plants were intermediate in HCN-p between the two parents, suggesting that neither high nor low HCN-p was dominant. Subsequent investigations (Gorz et al., 1987) using reciprocal translocations identified the presence of one or more genes modifying dhurrin content on at least 5 of the 10 chromosome pairs of sorghum. The obvious conclusion of multigenic inheritance of dhurrin content is reasonable when considered with the many biosynthetic steps, each with an enzyme catalyst, involved in dhurrin formation.

The complex multigenic inheritance of HCN-p in sorghum is contrasted with simpler inheritance in birdsfoot trefoil and white clover. Cyanogenesis in birdsfoot trefoil is determined by a dominant gene; consequently, both cyanogenic and acyanogenic types of plants occur readily (Dawson, 1941). The presence of acyanogenic plants raises many ecological and physiological questions about the role of cyanogenesis in plants.

In white clover, inheritance of cyanogenesis is conditioned by two pairs of genes. Actually, plants with no HCN-p could occur because 1) they did not contain the cyanogenic glucoside, 2) they contained the cyanogenic glucoside but without the cyanogenic β-glucosidase, or 3) they contained neither the cyanogenic glucoside nor the β-glucosidase. One gene, *Ac*, was responsible for the formation of the cyanoglucosides, lotaustralin and linamarin. The *Ac* gene has incomplete dominance but action of the gene is not understood. Plants with *acac* have at least two steps blocked in conversion of the amino acids, isoleucine and valine, to the α-hydroxynitrile compounds and it is also suggested that the β-glucosyltransferase enzyme is missing (Hughes and Conn, 1976; Collinge and Hughes, 1982; Hughes et al., 1988).

The other important factor in white clover for cyanogenesis is the *Li* gene for the presence of linamarase, the *B*-glucosidase. *Li* seems to be dominant to *li* as *Lili* plants have about one-half the amount of linamarase as *LiLi* plants and *lili* plants have no linamerase activity (Kakes, 1985). The enzyme is in the cell wall of the leaf and accounts for up to 5% of the leaf soluble proteins. Much is known about the molecular biology of the *Li* locus which provides potential for additional modification of cyanogenesis in white clover (Hughes, 1991).

ALKALOIDS

Phalaris Alkaloids

Phalaris species and many other *Poaceae* contain biological amines that elicit an animal response similar to true alkaloids. These amines meet the generally accepted definition for protoalkaloids. Tryptophan is the main precursor of these alkaloids and the indole ring remains intact. The principal groups of alkaloids in *Phalaris* are gramine, tryptamine derivatives, and β-carboline derivatives (Corcuera, 1989). The stable precursors between tryptamine and gramine also are found in the herbage. In gramine biosynthesis, cleavage of the side chain between C_2 and C_3 occurs and 3-aminomethylindole is formed. The amino-N of tryptophan is the source of the amino-N of 3-aminomethylindole. The amino group is methylated sequentially to 3-methylaminomethylindole and then to gramine (Leete and Minich, 1977).

Biosynthesis

Tryptophan is decarboxylated via tryptophan decarboxylase to tryptamine. Tryptamine derivatives are formed by mono- or di- methylation of the amino-N or hydroxylation and methylation at C_5. Methylation is catalyzed by N-methyltransferase using S-adenosylmethionine as the methyl donor. The first N-methyltransferase or primary indolethylamine N-methyltransferase has lower Km's for both substrate than the secondary indolethylamine N-methyltransferase (Mack et al., 1988). The rate-limiting enzyme in the pathway is the pyridoxal phosphate-dependent tryptophan decarboxylase. The sequence of decarboxylation, methylation, and hydroxylation is not rigid and, consequently, many similar but distinct routes seem to lead to the tryptamine derivative alkaloids (Baxter and Slaytor, 1972). The simple β-carboline alkaloids also are formed from tryptophan via decarboxylation to tryptamine followed by N-alkylation, ring closure, and appropriate methylation and hydroxylation to the methyl and methoxy derivatives.

Alkaloid concentration in *Phalaris aquatica* L. changed rapidly in the first few days of seedling growth. No free indole compounds including tryptophan were detected in dry seed. Tryptamine and N-methyltryptamine were first detected at 3 d and reached maximal concentration at 5 d after initiation of germination (Mack et al., 1988). The precursor tryptophan reached maximal concentration 6 d after initiation of germination. N, N-dimethyltryptamine concentration was greatest 8 d after germination commenced.

Accumulation Sites

In reed canarygrass (*P. arundinacea* L.), leaf blades contained more than twice as much alkaloid as leaf sheath and pseudostems (Hagman et al., 1975). Alkaloid concentrations in leaf blade increased with time to first harvest of spring growth, whereas concentration decreased in leaf sheaths and pseudostems. Total alkaloid concentration was less than 3 mg g^{-1}. The upper one-third of the herbage contained higher concentrations of alkaloids than the middle and lower one-thirds and was highly correlated to total herbage alkaloid concentration. The lower one-third of the herbage always was lower in alkaloid content than the other two fractions. Regrowth herbage had higher alkaloid concentration than first-growth herbage, presumably because of the greater proportion of leaves in regrowth herbage (Woods et al., 1979). Added N fertilizer tended to increase alkaloid concentration, whereas addition of phosphorus or potassium to infertile soils reduced alkaloid concentrations (Moore et al., 1967; Marten et al., 1974). Increased day-night temperatures and light intensity increased alkaloid concentrations but increased day length did not increase alkaloid concentration (Moore et al., 1967).

Animal Response

Animal response to ingestion of *Phalaris spp.* containing the tryptamine alkaloids has included an acute cardiac disorder and sudden death; a chronic nervous disorder characterized by tremors, staggers or diarrhea; and low voluntary intake. Bourke et al. (1988) showed that 5-methoxy dimethyltryptamine and gramine induced the clinical signs of the nervous disorder but not acute sudden death syndrome. Clinical responses to the two alkaloids were similar but 5-methoxy dimethyltryptamine was 10 to 100 times more potent than gramine.

Low voluntary intake of reed canarygrass appeared to be associated with the presence of the tryptamine alkaloids (Roe and Mottershead, 1962). Relative palatability of reed canarygrass was negatively correlated with alkaloid concentration of herbage (Simons and Marten, 1971; Marten et al., 1973). More definitive studies by Marten et al. (1976, 1981) showed that palatability differences and the associated alkaloid concentration differences among reed canarygrass genotypes had substantial biological significance for grazing lambs and steers. They concluded that alkaloid concentrations at or above 2 mg g^{-1} of DM reduced gain of grazing lambs and that the diarrhea associated with grazing *Phalaris spp.* was likely due to the alkaloids.

Genetics

Plant genotypes have been identified that contain both quantitatively and qualitatively altered alkaloid content (Woods and Clark, 1971; Marten, 1973; Hovin and Marten, 1975; Marum et al., 1979; Woods et al., 1979). Woods and Clark (1971) concluded that the presence of tryptamine alkaloids was controlled by one dominant gene. Marum et al. (1979) grouped genotypes into 1) gramine only accumulators, 2) tryptamine and carboline derivative accumulators, and 3) methoxylated tryptamine and carboline derivative accumulators. Based on these groupings, they concluded that a two gene model best fit the inheritance data. A single dominant allele at one locus controlled synthesis for alkaloids of the tryptamine and carboline group. A second dominant allele at a second locus controlled synthesis of the methoxylated derivatives. The allele for the methoxylated derivatives was epistatic to the allele for the tryptamine and carboline derivatives. Gramine-accumulating genotypes occur only when both loci have homozygous recessive alleles. Coulman et al. (1977) and Woods et al. (1979) reported a high (0.72 to 0.97), narrow sense heritability estimate for gramine and this suggested that much progress could be made in development of improved quality lower alkaloid genotypes. Ostrem (1987) concluded that even with populations only containing gramine, heritability was sufficiently high to expect a response to selection for lowered alkaloid concentration. These genetic analyses fit the enzymatic data for biosynthesis modeled by Mack et al. (1988).

Lupine Alkaloids

Two separate problems are associated with animals grazing lupins. Lupinosis is a mycotoxicosis caused by the ingestion of toxins produced by the fungus *Phomopsis leptostromiformis* Kühn and will not be included in this chapter. Readers are referred to Allen (1986) for details. The second condition, referred to as alkaloid poisoning or lupine poisoning, is caused by ingestion of quinolizidine alkaloids produced by the plant. Alkaloids of *Lupinus spp.* are quinolizidine derivatives of usually bi-, tri-, and tetracyclic ring structure. These alkaloids occur in other genera and families but in *Lupinus* will occur in concentrations ranging up to 30 g kg^{-1}. Principal forage species are white lupine (*L. albus* L.), blue lupine (*L. angustifolius* L.), and yellow lupine (*L. luteus* L.) and the principal alkaloids are the bicyclic lupinine, the tricyclic cystinine, and the tetracyclic compounds lupanine and sparteine (Keeler, 1989). Aslanov et al. (1987) compiled a comprehensive list of over 100 lupine quinolizidine alkaloids. This large number of alkaloids in lupins is most likely a consequence of the many species in the genus Lupinus from two centers of diversity. Annual species from the Mediterranean area are the ancestors to our cultivated crop plant species and from western North America and South America come the vast majority of the species of the genus of which many are weedy rangeland plants. Ingestion of herbage from these wild species on rangeland often results in animal disorders. Chemistry and distribution of lupine alkaloids has been reviewed most recently by Aslanov et al. (1987).

Sites of Synthesis and Accumulation

Alkaloids are produced mainly in the leaflets and chlorophyllous petioles and not in roots of lupins (Wink, 1987). Biosynthesis is localized in chloroplasts of green mesophyll tissue and not epidermis (Wink et al., 1980; Wink and Hartmann, 1982; Wink and Mende, 1987). After synthesis, the alkaloids are translocated via the phloem throughout the plant where they accumulate principally in epidermal tissue (Wink and Witte, 1984; Wink et al., 1984; Wink and Mende, 1987). Accumulation in the epidermal cells requires metabolic energy and probably is catalyzed by transport protein (Wink and Mende, 1987). The alkaloids are readily translocated to the seed, and ripe seed will contain the vast majority of alkaloid present in the plant (Williams and Harrison, 1983). Alkaloid levels decrease during germination and early seedling development. Concentration of alkaloids increases in the vegetative tissues before flowering and is followed by a rapid accumulation in the seed pod and seed. Alkaloid synthesis does not occur in the seed (Williams and Harrison, 1983).

Environmental Effects

Environmental conditions and plant condition also influence alkaloid concentration. High-N status plants and N-fixing plants contained higher concentrations of alkaloids than low-N status plants (Johnson et al., 1987). Periodic defoliation had little effect on alkaloid concentration in high-N plants, whereas N-fixing plants had lowered alkaloid levels from defoliation treatments. Available N and defoliation treatments affected relative concentration of individual alkaloids as well as total alkaloid concentration. Total alkaloid concentration in regrowth leaves following defoliation of velvet lupine (*L. leucophyllus* Dougl.) was 15% lower than in the original leaves (Ralphs and Williams, 1988). Breakage of leaf cells stimulated alkaloid accumulation within 4 h and the stimulation occurred more quickly in N-fixing plants than in low-N status plants (Johnson et al., 1989). Leaf damage caused a 55% increase in alkaloid concentration within 24 h. In N-fixing plants, over half of this increase occurred within 4 h. Also, undamaged leaves on plants that had some leaves damaged had a 33% increase in alkaloid concentration. Because alkaloid biosynthesis occurs in the chloroplast, it is not unexpected that fluctuation of alkaloid content in the leaflet is influenced by light (Wink and Hartmann, 1982). Alkaloid content reached maximum levels 5 to 8 h after start of illumination and were at minimal levels during the night.

Biosynthesis

Biosynthesis of the lupine alkaloids is not well understood. The amino acid precursor is lysine which is decarboxylated by lysine decarboxylase to cadaverine and a free cadaverine is likely because C-2 of lysine becomes randomized at C-1 and C-5 in the diamine. Wink et al. (1980) proposed that three molecules of cadaverine are converted in the chloroplast to 17-oxosparteine by 17-oxosparteine synthase. This enzyme system is membrane bound and intermediates are not released, possibly channeled as has been shown for the biosynthesis of dhurrin. 17-Oxosparteine is the immediate precursor for lupanine. In this proposed pathway, the tetracyclic lupanine serves as the precursor for the tricyclic alkaloids plus being used for derivatization for the wide range of tetracyclic alkaloids. The bicyclic alkaloids may be formed separately or as premature release of a two cadaverine unit intermediate from the enzyme system imbedded in the membrane.

Animal Response

Lupine alkaloids induce a range of effects in animals including respiratory depression and failure, fibrillation, and general hypotension and inhibition of muscular transmission (Culvenor and Petterson, 1986). Death of livestock usually occurs in animals which consume, over a short period, large quantities of mature plants containing seed pods and seed. Chronic toxicity may be observed by reduced growth rate due to methionine deficiency (Culvenor and Petterson, 1986). Pigs fed diets containing 0.33 g kg^{-1} alkaloid reduced voluntary intake (Pearson and Carr, 1977) and Godfrey et al. (1985) concluded that growing pigs could tolerate 0.2 g kg^{-1} lupine alkaloids in their diet.

Velvet lupine is widespread on mountain rangeland of the western U.S. and has caused more sheep deaths than any other plant on the western mountain ranges (Kingsbury, 1964). The alkaloid, anagyrine, which is found in several lupine species in the western U.S., is the only known teratogenic lupine alkaloid. Ingestion of anagyrine between d 40 and 70 of gestation often results in congenital defects in calves. The condition is known as "crooked calf disease". Davis (1982) and Davis and Stout (1986) found 14 species from the western U.S. rangeland which had anagyrine concentration above 1.44 g kg^{-1}, the threshold level for induction of crooked calf disease.

Genetics

Improvement of quality of forage and seed of lupins was initiated near the beginning of the 20th century. Alkaloid-free or "sweet" mutants were discovered in the 1920's and have led to use of lupins as a cultivated crop. Genes for reduced alkaloid content are recessive (Waller and Nowacki, 1978; Harrison and Williams, 1982). The biosynthetic pathway from cadaverine to the many individual alkaloids is dependent upon many enzymes and several genes for lowered alkaloid content have been identified. Consequently, not all the identified genes have the same quantitative or qualitative effect on total alkaloid accumulation (Williams et al., 1984). Approaches and strategies for lowered alkaloid content, improved seed protein, and lack of pod shatter have been discussed by Keeler (1989). It is noteworthy that, both quantitatively and qualitatively, seed alkaloids are determined by the genotype of the female parent and that cell suspension cultures do not accumulate the same alkaloids as leaves (Wink et al., 1983). In cell suspension cultures, lupanine was the principal alkaloid but the cultures contained only 20% of the concentration of the leaves of these genotypes. Consequently, evaluation of alkaloid content for crop improvement cannot be made in cell suspension culture or seed, but must be made on the green herbage of the plant.

Tall Fescue Alkaloids

Animals grazing tall fescue (*Festuca arundinacea* Schreb.) in the U.S. often exhibit symptoms of fescue toxicosis. It has been established that grasses infected with fungal endophytes are responsible for the observed toxicosis symptoms (Stuedemann and Hoveland, 1988). At least five classes of anti-herbivore alkaloids, some extrinsic and some intrinsic, are known to be produced in tall fescue-fungal endophyte associations. Extrinsic chemicals include at least the ergots, indole diterpenoid, and azaindolizine alkaloids. The diazaphenanthrene alkaloids are known to be intrinsic. However, the pyrrolizidine alkaloids have been isolated only from the host-endophyte complex. These alkaloids have not been found in the host or in the fungus when each is cultured alone.

Diazaphenanthrene Alkaloids

The diazaphenanthrene alkaloids, perloline and perlolidine, are formed by the grass. Butler (1962) found genetic differences in perloline content of ryegrasses (*Lolium perenne* L. and *L. perenne* x *L. multiflorum* Lam.). Buckner et al. (1973) reported that *Lolium spp.* had less perloline than *Festuca spp.* and that progress could be made by conventional plant breeding to alter perloline content of herbage. Cornelius et al. (1974) concluded that perloline content was controlled primarily by a few genes with a high degree of dominance for low perloline content. Environmental factors affect perloline content more than most other alkaloids in forage crops and certainly more than the other alkaloids in tall fescue. Perloline content increased with increased N availability (Bush and Buckner, 1973; Robbins et al., 1973). Greater accumulation of perloline may also occur in hot dry weather. Many of the environmental responses were reviewed by Bush et al. (1979).

In vitro ruminal functions were inhibited by perloline (Bush et al., 1970, 1976). Lambs receiving diets supplemented with 5 g kg^{-1} perloline had lower cellulose and crude protein digestion than lambs on a control diet (Boling et al.,

1975). Perloline also increased animal body temperatures and increased N retention. These observations plus other bioassays suggest that perloline is only mildly toxic and probably is not a significant factor in fescue toxicosis most frequently observed. If the extrinsic alkaloids of the tall fescue/endophyte complex are altered to minimize tall fescue toxicosis, the apparent significance of perloline may increase. This is especially true if a mass action effect of alkaloids as cosubstrate for thiaminase occurs in the rumen as proposed by Dougherty et al. (1991).

The Loline Group of Pyrrolizidine Alkaloids

Pyrrolizidine alkaloids are found in many flowering plants (Smith and Culvenor, 1981), especially *Senecio spp.* of the western U.S. rangeland. Most of the pyrrolizidine alkaloids are esters of hydroxylated 1-methylpyrrolizidine and most of the hepatotoxic pyrrolizidine alkaloids are esters of an unsaturated pyrrolizidine base. The metabolism and toxicity to animals and humans of many of these alkaloids has been reviewed by Cheeke (1989). Interestingly, it is the metabolism of these alkaloids that produces the toxic entity, the dehydropyrrolizidine alkaloid. Dehydropyrrolizidine alkaloids are very reactive alkylating agents. The toxic effects of the pyrroles are thought to be by covalent binding to proteins and nucleic acids (Winter and Segall, 1989).

Saturated amino pyrrolizidines, lolines, and derivatives are found in greatest abundance of all the alkaloids in tall fescue. These alkaloids contain an oxygen bridge between carbons 2 and 7 of the pyrrolizidine base as well as being saturated. N-formylloline and N-acetylloline are found in greatest amounts with much lesser amounts of loline, N-methylloline, norloline, N-acetylnorloline, and N-formylnorloline present. These alkaloids are not known hepatoxins and their toxicity has not been elucidated fully. Proposed biosynthesis and toxicity has been reviewed earlier (Bush et al., 1993; Powell and Petroski, 1993).

Karimov and Kamilov (1961) first reported that loline compounds decreased blood pressure and decreased coronary blood flow in cats and dogs. More recently, loline and derivatives have been tested in insect bioassays and found to have feeding deterrent activity or be toxic (Johnson et al., 1985; Yates et al., 1989). Jackson et al. (1987) demonstrated that a partially purified fraction of pyrrolizidine alkaloids depressed feed intake, average daily gain, and serum prolactin levels in rats. Mika (1987) reported a negative correlation between voluntary intake and pyrrolizidine alkaloid content from tall fescue and meadow fescue (*Festuca pratensis* Huds.) haylage fed to steers. N-acetylloline has caused vasoconstriction in vitro (Oliver et al., 1990).

Dougherty et al. (1991) and Lauriault et al. (1990) showed that thiamine added to the diet of toxic tall fescue herbage increased dry matter intake. The large amounts of pyrrolizidine alkaloids in toxic tall fescue herbage could be the co-substrate for rumen thiaminase I (Brent and Bartley, 1984) and generation of thiamine deficiency in the animal. Other alkaloids in tall fescue could act as cosubstrate for thiaminase I to increase this proposed action and effect on the consuming animal.

Seasonal trends of N-acetylloline and N-formylloline accumulation are not influenced by growth conditions as dramatically as perloline (Bush et al., 1979; de Guglielmone et al., 1981; Stuedemann et al., 1985). An increased accumulation of lolines during late summer may have been associated with lesser DM accumulation due to water or temperature stress that did not alter alkaloid biosynthesis as much (Stuedemann et al., 1985). Putman et al. (1991) measured a similar response from April to August when accumulation was in excess of 3 g kg^{-1} DM. In severe water stress, Kennedy and Bush (1983) reported accumulation in excess of 11 g kg^{-1} but more importantly, Belesky et al. (1989) showed that moderate water stress increased the pyrrolizidine alkaloid concentration. Only as the relationship between the fungal endophytes and host is more fully understood will the site of pyrrolizidine alkaloid biosynthesis be known and only as fescue toxicosis is better elucidated will the role of the pyrrolizidine alkaloids in this animal malady be known.

CONCLUSION

Biology of a few intrinsic chemical constituents of herbage were used to illustrate the progress that has been made in the past 25 years in understanding plant/animal interactions. Continued improvement in forage utilization most likely will result from research advances made with molecular biology in both rumen microbes and the plant in conjunction with forage breeding and management. The specificity of genetic change that may be made with molecular biology may allow full utilization of a variable such as the brown midrib. However, as mentioned in the example for increased phytoalexin (medicarpin) production for disease tolerance and the potential for increased phytoestrogen production, it is recognized that some caution must be exercised in the application of molecular biology techniques.

Most intrinsic chemicals are not produced in isolated biochemical pathways and perturbation of one substance will often cause change in other chemicals that may have an impact on forage utilization. Forage researchers must continue to conduct and make use of sophisticated biochemical research techniques such as the enzymology and membrane science as described for dhurrin biosynthesis. Also, the interaction of grazing animals with individual biochemicals in forage must be better understood for improved forage utilization. Palatability, acceptability, and selectivity by grazing animals was an area considered for this review but insufficient data were available for critical evaluation. Animal responses to plant biochemicals appear to be both pre- and post-ingestive and this area offers tremendous opportunity for significant contribution to forage utilization in the future. As our understanding of the physiology of forage plants increases, some of the intrinsic chemicals may be found to be actually produced or symbiotically produced by endo- and ecto-plant microbes. Obviously, there is great potential for improved chemical quality of forages, but to realize this potential will require much cooperative effort of many research disciplines.

REFERENCES

Adams, N.R. 1989. Phytoestrogens. p. 24-51 *In* P.R. Cheeke (ed.) Toxicants of Plant Origin. Phenolics. Vol. 4. CRC Press, Boca Raton, FL.

Akazawa, T., P. Miljanich, and E.E. Conn. 1960. Studies on cyanogenic glycoside of *Sorghum vulgare*. Plant Physiol. 35:535-538.

Allen, J.G. 1986. Lupinosis, A review. p. 173-187. *In* Proc. 4th Int. Lupin Conf., Geraldton, Western Australia. 15-22 Aug. 1986. Australian Dept. of Agriculture, Geraldton, Western Australia.

Aslanov, K.A., Y.K. Kushmuradov, and A.S. Sadykov. 1987. Lupin alkaloids. p. 117-192. *In* A. Brossi (ed.) The alkaloids - Chemistry and pharmacology. Vol. 31. Academic Press, San Diego, CA.

Barbetti, M.J., and P.G.H. Nichols. 1991 Effect of *Phoma medicaginis* and *Leptosphaerulina trifolii* on herbage and seed yield and coumestrol content of annual *Medicago spp.* Phytophylactica 23:223-227.

Barry, T.N. 1985. The role of condensed tannins in the nutritional value of *Lotus pedunculatus* for sheep. 3. Rates of body and wool growth. Br. J. Nutr. 54:211-217.

Barry, T.N., and S.J. Duncan. 1984. The role of condensed tannins in the nutritional value of *Lotus pedunculatus* for sheep. 1. Voluntary intake. Br. J. Nutr. 51:485-491.

Barry, T.N., and T.R. Manley. 1984. The role of condensed tannins in the nutritional value of *Lotus pedunculatus* for sheep. 2. Quantitative digestion of carbohydrates and proteins. Br. J. Nutr. 51:493-504.

Barry, T.N., and T.R. Manley. 1986. Interrelationships between concentrations of total condensed tannin, free condensed tannin and lignin in *Lotus* sp. and their possible consequences in ruminant nutrition. J. Sci. Food Agric. 37:248-254.

Baxter, C., and M. Slaytor. 1972. Biosynthesis and turnover of N, N-dimethyltryptamine and 5-methoxy-N,N-dimethyltryptamine in *Phalaris tuberosa*. Phytochem. 11:2767-2773.

Belesky, D.P., W.C. Stringer, and R.D. Plattner. 1989. Influence of endophyte and water regime upon tall fescue accessions. II. Pyrrolizidine and ergopeptine alkaloids. Ann. Botany 64:343-349.

Boling, J.A., L.P. Bush, R.C. Buckner, L.C. Pendlum, P.B. Burrus, S.G. Yates, S.P. Rogovin, and H.L. Tooley. 1975. Nutrient digestibility and metabolism in lambs fed added perloline. J. Anim. Sci. 40:972-976.

Bourke, C.A., M.J. Carrigan, and R.J. Dixon. 1988. Experimental evidence that tryptamine alkaloids do not cause *Phalaris aquatica* sudden death syndrome in sheep. Aust. Vet. J. 65:218-220.

Brent, B.E., and E.E. Bartley. 1984. Thiamin and niacin in the rumen. J. Anim. Sci. 59:813-822.

Bucholtz, D.L., R.P. Cantrell, P.J. Axtell, and V.L. Lechtenberg. 1980. Lignin biochemistry of normal and brown midrib mutant sorghum. J. Agric. Food Chem. 28:1239-1241.

Buckner, R.C., L.P. Bush, and P.B. Burrus, II. 1973. Variability and heritability of perloline in *Festuca sp.*, *Lolium sp.*, and *Lolium-Festuca* hybrids. Crop Sci. 13:666-669.

Bush, L.P., J. Boling, and S. Yates. 1979. Animal disorders. p. 247-292. *In* R.C. Buckner and L.P. Bush (ed.) Tall fescue. Agron. Monogr. 20. ASA, Madison, WI.

Bush, L.P., and R.C. Buckner. 1973. Tall fescue toxicity. p. 99-112. *In* A.G. Matches (ed.) Antiquality components of forages. CSSA Spec. Publ. 4. Madison, WI.

Bush, L.P., H. Burton, and J.A. Boling 1976. Activity of tall fescue alkaloids and analogues in in vitro rumen fermentation. J. Agric. Food Chem. 24:869-872.

Bush, L.P., F.F. Fannin, M.R. Siegel, D.L. Dahlman, and H.R. Burton. 1993. Chemistry, occurrence and biological effects of saturated pyrrolizidine alkaloids associated with endophyte-grass interactions. Agric. Ecosystems Environ. 44:81-102.

Bush, L.P., C. Streeter, and R.C. Buckner. 1970. Perloline inhibition of in vitro ruminal cellulose digestion. Crop Sci. 10:108-109.

Butler, G.W. 1962. Genetic differences in the perloline content of ryegrass (*Lolium*) herbage. N.Z. J. Agric. Res. 5:158-162.

Butler, L.G. 1989. Sorghum polyphenols. p. 95-121. *In* P.R. Cheeke (ed.) Toxicants of plant origin. Phenolics. Vol. 4. CRC Press, Boca Raton, FL.

Cheeke, P.R. 1989. Pyrrolizidine alkaloid toxicity and metabolism in laboratory animals and livestock. p. 1-22. *In* P.R. Cheeke (ed.) Toxicants of plant origin. Alkaloids. Vol. 1. CRC Press, Boca Raton, FL.

Cherney, J.H., J.D. Axtell, M.M. Hassen, and K.S. Anliker. 1988. Forage quality characterization of a chemically induced brown-midrib mutant in pearl millet. Crop Sci. 28:783-787.

Cherney, J.H., K.J. Moore, J.J. Volenec, and J.D. Axtell. 1986. Rate and extent of digestion of cell wall components of brown-midrib sorghum species. Crop Sci. 26:1055-1059.

Cherney, D.J.R., J.A. Patterson, and K.D. Johnson. 1990. Digestibility and feeding value of pearl millet as influenced by the brown-midrib, low-lignin trait. J. Anim. Sci. 68:4345-4351.

Clarke, R.T., and C.S.W. Reid. 1974. Foamy bloat of cattle. A review. J. Dairy Sci. 57:753-785.

Clausen, T.P., F.D. Provenza, E.A. Burritt, P.B. Reichardt, and J.P. Bryant 1990. Ecological implications of condensed tannin structure: A case study. J. Chem. Ecol. 16:2381-2392.

Collinge, D.B., and M.A. Hughes. 1982. *In vitro* characterisation of the *Ac* locus in white clover (*Trifolium repens* L.). Arch. Biochem. Biophys. 218:38-45.

Conn, E.E. 1980. Cyanogenic compounds. p. 433-451. *In* W.S. Briggs (ed.) Ann. Rev. Plant Physiol. Vol. 31. Annual Reviews. Palo Alto, CA.

Cooper, C.S., R.F. Eslick, and P.W. McDonald. 1966. Foam formation from extracts of 27 legume species in vitro. Crop Sci. 6:215-216.

Corcuera, L.J. 1989. Indole alkaloids from *Phalaris* and other gramineae. *In* P.R. Cheeke (ed.) Toxicants of plant origin. Alkaloids. Vol. 1. CRC Press. Boca Raton, FL.

Cornelius, P.L., R.C. Buckner, L.P. Bush, P.B. Burrus, II, and J. Byars. 1974. Inheritance of perloline content in annual ryegrass x tall fescue hybrids. Crop Sci. 14:896-898.

Coulman, B.E., D.L. Woods, and K.W. Clark. 1977. Distribution within the plant, variation with maturity, and heritability of gramine and hordenine in reed canary grass. Can. J. Plant Sci. 57:771-777.

Culvenor, C.C.J., and D.S. Petterson. 1986. Lupin toxins - Alkaloids and phomopsins. p.188-198. *In* Proc. 4th Int. Lupin Conf., Geraldton, Western Australia. 15-22 Aug. 1986. Australian Dept. of Agriculture, Geraldton, Western Australia.

Czerkawski, J.W. 1986. An introduction to rumen studies. Pergamon Press, Oxford.

Davis, A.M. 1982. The occurrence of anagyrine in a collection of Western American lupines. J. Range Manage. 35:81-84.

Davis, A.M., and D.M. Stout. 1986. Anagyrine in western American lupines. J. Range Manage. 39:29-30.

Dawson, C.D.R. 1941. Tetrasomic inheritance in *Lotus corniculatus* L. J. Genet. 45:49-72.

de Guglielmone, A.E.R., A.M. Sanez, M.A. Carabelli, M.B. Gulielmoni, E.O. Basile, and J.O. Severuga. 1981. Festucas toxicas e inocuas: Diferencias en el contenido de alcaloides y su relacion con un ensayo preliminar a campo. Resista de los Asociacion Argentina Consorcios Regionales Experimentacion Agricola (CREA) 15:40-47.

Dixon, R.A., A.D. Choudhary, K. Dalkin, R. Edwards, T. Fahrendorf, G. Gowri, M.J. Harrison, C.J. Lamb, G.J. Loake, C.A. Maxwell, J. Orr, and N.L. Paiva. 1992. Molecular biology of stress-induced phenylpropanoid and isoflavonoid biosynthesis in alfalfa. p. 91-138. *In* H.A. Stafford and R.K. Ibrahim (ed.) Phenolic metabolism in plants. Recent Adv. Phytochemistry. Vol. 26. Plenum Press, New York.

Dougherty, C.T., L.M. Lauriault, N.W. Bradley, N. Gay, and P.L. Cornelius. 1991. Induction of tall fescue toxicosis in heat-stressed cattle and its alleviation with thiamin. J. Anim. Sci. 69:1008-1018.

Eyster, W.H. 1926. Chromosome VIII in maize. Science 64:22.

Fahey, Jr., G.C., and H.-J.G. Jung. 1989. Phenolic compounds in forages and fibrous feedstuffs. p. 123-190. *In* P.R. Cheeke (ed.) Toxicants of plant origin. Phenolics. Vol. 4. CRC Press, Boca Raton, FL.

Farnsworth, N.R., A.S. Bingel, G.A. Cordell, F.A. Crane, and H.H.S. Fong. 1975. Potential value of plants as sources of new antifertility agents II. J. Pharm. Sci. 64:717-754.

Fay, J.P., K.-J. Cheng, M.R. Hanna, R.E. Howarth, and J.W. Costerton. 1980. In vitro digestion of bloat-safe and bloat-causing legumes by rumen microorganisms: Gas and foam production. J. Dairy Sci. 63:1273-1281.

Francis, C.M., and A.J. Millington. 1965. Varietal variation in the isoflavone content of subterranean clover: Its estimation by a microtechnique. Aust. J. Agric. Res. 16:557-564.

Fritz, J.O., R.P. Cantrell, V.L. Lechtenberg, J.D. Axtell, and J.M. Hertel. 1981. Brown midrib mutants in sudangrass and grain sorghum. Crop Sci. 21:706-709.

Fritz, J.O., K.J. Moore, and E.H. Jaster. 1988. In situ digestion kinetics and ruminal turnover rates of normal and brown midrib mutant sorghum x sudangrass hays fed to nonlactating holstein cows. J. Dairy Sci. 71:3345-3351.

Gee, M.S., O.E. Nelson, and J. Kuc. 1968. Abnormal lignins produced by the brown midrib mutants of maize. II. Comparative studies on normal and brown midrib-1 dimethylformamide lignins. Arch. Biochem. Biophys. 123:403-408.

Gildersleeve, R.R., G.R. Smith, I.J. Pemberton, and C.L. Gilbert. 1991. Detection of isoflavones in seedling subterranean clover. Crop Sci. 31:889-892.

Godfrey, N.W., A.R. Mercy, Y. Emms, and H.G. Payne. 1985. Tolerance of growing pigs to lupin alkaloids. Aust. J. Exp. Agric. 25:791-795.

Goplen, B.P., R.E. Howarth, S.K. Sarkar, and K. Lesins. 1980. A search for condensed tannins in annual and perennial species of *Medicago*, *Trigonella*, and *Onobrychis*. Crop Sci. 20:801-804.

Gorz, H.J., F.A. Haskins, R. Morris, and B.E. Johnson. 1987. Identification of chromosomes that condition dhurrin content in sorghum seedlings. Crop Sci. 27:201-203.

Gorz, H.J., F.A. Haskins, and K.P. Vogel. 1986. Inheritance of dhurrin content in mature sorghum leaves. Crop Sci. 26:65-67.

Grand, C., P. Parmentier, A. Boudet, and A.M. Boudet. 1985. Comparison of lignins and of enzymes involved in lignification in normal and brown midrib (bm_3) mutant corn seedlings. Physiol. Veg. 23:905-911.

Graves, W.L., W.H. Weitkamp, M.R. George, G.R. Smith, B.L. Kay, and C.G. Pitera. 1991. Stand persistence and estrogenic activity of hardseeded subterranean clover cultivars for southern and central California. J. Prod. Agric. 4:111-114.

Hagman, J.L., G.C. Marten, and A.W. Hovin. 1975. Alkaloid concentration in plant parts of reed canarygrass of varying maturity. Crop Sci. 15:41-43.

Halkier, B.A., J. Lykkesfeldt, and B.L. Moller. 1991. 2-Nitro-3-(*p*-hydroxyphenyl)propionate and *aci*-1-nitro-2-(*p*-hydroxyphenyl)ethane, two intermediates in the biosynthesis of the cyanogenic glucoside dhurrin in *Sorghum bicolor* (L.) Moench. Proc. Nat. Acad. Sci. USA 88:487-491.

Halkier, B.A., and B.L. Moller. 1991. Involvement of cytochrome P-450 in the biosynthesis of dhurrin in *Sorghum bicolor* (L.) Moench. Plant Physiol. 96:10-17.

Halkier, B.A., C.E. Olsen, and B.L. Moller. 1989. The biosynthesis of cyanogenic glucosides in higher plants. J. Biol. Chem. 264:19487-19494.

Halkier, B.A., H. V.Scheller, and B.L. Moller. 1988. Cyanogenic glucosides: The biosynthetic pathway and the enzyme system involved. p. 49-61. *In* D. Evered and S. Harnett (ed.) Cyanide compounds in biology. Ciba Found. Symp. 140. John Wiley and Sons, New York, NY.

Hall, J.W., and W. Majak. 1989. Plant and animal factors in legume bloat. p. 93-106. *In* P.R. Cheeke (ed.) Toxicants of plant origin. Proteins and amino acids. Vol. 3. CRC Press, Boca Raton, FL.

Hall, J.W., and W. Majak. 1991. Relationship of weather and plant factors to alfalfa bloat in autumn. Can. J. Anim. Sci. 71:861-866.

Hall, J.W., W. Majak, A.L. Van Ryswyk, R.E. Howarth, and C.M. Kalnin. 1988. The relationship of rumen cations and soluble protein with predisposition of cattle to alfalfa bloat. Can. J. Anim. Sci. 68:431-437.

Harrison, J.E.M., and W. Williams. 1982. Genetical control of alkaloids in *Lupinus albus*. Euphytica 31:357-364.

Hartley, R.D. 1972. *p*-Coumaric and ferulic acid components of cell walls of ryegrass and their relationships with lignin and digestibility. J. Sci. Food Agric. 23:1347-1354.

Haskins, F.A., H.J. Gorz, and B.E. Johnson. 1987. Seasonal variation in leaf hydrocyanic acid potential of low- and high-dhurrin sorghums. Crop Sci. 27:903-906.

Haslam, E. 1981. Vegetable tannins. p. 527-556. *In* E.E. Conn (ed.) The biochemistry of plants. Vol. 7. Academic Press, New York.

Hosel, W., I. Tober, S.H. Eklund, and E.E. Conn. 1987. Characaterization of B-glucosidases with high specificity for the cyanogenic dhurrin in *Sorghum bicolor* (L.) Moench seedlings. Arch. Biochem. Biophys. 252:152-162.

Hovin, A.W., and G.C. Marten. 1975. Distribution of specific alkaloids in reed canarygrass cultivars. Crop Sci. 15:705-707.

Howarth, R.E. 1975. A review of bloat in cattle. Can. Vet. J. 16:281-294.

Howarth, R.E., K.-J. Cheng, W. Majak, and J.W. Costerton. 1986. Ruminant bloat. p. 516-527. *In* L.P. Milligan, W.L. Grovum, and A. Dobson (ed.) Control of digestion and metabolism in ruminants. Prentice-Hall, Englewood Cliffs, NJ.

Howarth, R.E., B.P. Goplen, S.A. Brandt, and K.-J. Cheng. 1982. Disruption of leaf tissues by rumen microorganisms: An approach to breeding bloat-safe forage legumes. Crop Sci. 22:564-568.

Howarth, R.E., B.P. Goplen, A.C. Fesser, and S.A. Brandt. 1978. A possible role for leaf cell rupture in legume pasture bloat. Crop Sci. 18:129-133.

Howarth, R.E., W. Majak, D.E. Waldern, S.A. Brandt, A.C. Fesser, B.P. Goplen, and D.T. Spurr. 1977. Relationships between ruminant bloat and the chemical composition of alfalfa herbage. 1. Nitrogen and protein fractions. Can. J. Anim. Sci. 57:345-357.

Hughes, M.A. 1991. The cyanogenic polymorphism in *Trifolium repens* L. (white clover). Heredity 66:105-115.

Hughes, M.A., and E.E. Conn. 1976. Cyanoglucoside biosynthesis in white clover (*Trifolium repens* L.). Phytochem. 15:697-701.

Hughes, M.A., A.L. Sharif, M.A. Dunn, and E. Oxtoby. 1988. The molecular biology of cyanogenesis. p. 111-124. *In* D. Evered and S. Harnett (ed.) Cyanide compounds in biology. Ciba Found. Symp. 140. John Wiley and Sons, New York.

Jackson, J.A., R.W. Hemken, L.P. Bush, J.A. Boling, M.R. Siegel, P.M. Zavos, and S.G. Yates. 1987. Physiological responses in rats fed extracts of endophyte infected tall fescue seed. Drug Chem. Toxicol. 10:369-379.

Johnson, M.C., D.L. Dahlman, M.R. Siegel, L.P. Bush, G.C.M. Latch, D.A. Potter, and D.R. Varney. 1985. Insect feeding deterrents in endophyte-infected tall fescue. Appl. Environ. Microbiol. 49:568-571.

Johnson, N.D., B. Liu, and B.L. Bentley. 1987. The effects of nitrogen fixation, soil nitrate, and defoliation on the growth, alkaloids, and nitrogen levels of *Lupinus succulentus* (Fabaceae). Oecologia 74:425-431.

Johnson, N.D., L.P. Rigney, and B.L. Bentley. 1989. Short-term induction of alkaloid production in lupins. Differences between N_2-fixing and nitrogen-limited plants. J. Chem. Ecol. 15:2425-2444.

Jones, W.T., L.B. Anderson, and M.D. Ross. 1973. Bloat in cattle. XXXIX. Detection of protein precipitants (flavolans) in legumes. N.Z. J. Agric. Res. 16:441-446.

Jorgenson, L.R. 1931. Brown midrib in maize and its linkage relations. Agron. J. 23:549-557.

Kakes, P. 1985. Linamarase and other *B*-glucosidases are present in the cell walls of *Trifolium repens* L. leaves. Planta 166:156-160.

Karimov, V.A., and I.K. Kamilov. 1961. Pharmacology of the new loline alkaloid and of its derivative. Dok. Akad. Nauk Uzb. SSR. 12:43-47.

Keeler, R. 1989. Quinolizidine alkaloids in range and grain lupins. p.133-167. *In* P.R. Cheeke (ed.) Toxicants of plant origin. Alkaloids. Vol. 1. CRC Press, Boca Raton, FL.

Kelly, R.W., R.J.M. Hay, and G.H. Shackell. 1979. Formononetin content of "Grasslands Pawera" red clover and its oestrogenic activity to sheep. N.Z. J. Exp. Agric. 7:131-134.

Kendall, W.A. 1966. Factors affecting foams with forage legumes. Crop Sci. 6:487-489.

Kennedy, C.W., and L.P. Bush. 1983. Effect of environmental and management factors on the accumulation of N-acetyl and N-formyl loline alkaloids in tall fescue. Crop Sci. 23:547-552.

Kingsbury, J.M. 1964. Poisonous plants of the United States and Canada. p. 333-341. Prentice Hall, Englewood Cliffs, NJ.

Kojima, M., J.E. Poulton, S.S. Thayer, and E.E. Conn. 1979. Tissue distributions of dhurrin and of enzymes involved in its metabolism in leaves of *Sorghum bicolor*. Plant Physiol. 63:1022-1028.

Kuc, J., and O.E. Nelson. 1964. The abnormal lignins produced by the brown-midrib mutants of maize. I. The brown-midrib-1 mutant. Arch. Biochem. Biophys. 105:103-113.

Kudo, H., K.-J. Cheng, M.R. Hanna, R.E. Howarth, B.P. Goplen, and J.W. Costerton. 1985. Ruminal digestion of alfalfa strains selected for slow and fast initial rates of digestion. Can. J. Anim. Sci. 65:157-161.

Lauriault, L.M., C.T. Dougherty, N.W. Bradley, and P.L. Cornelius. 1990. Thiamin supplementation and the ingestive behavior of beef cattle grazing endophyte-infected tall fescue. J. Anim. Sci. 68:1245-1253.

Lechtenberg, V.L., V.F. Colenbrander, L.F. Bauman, and C.L. Rhykerd. 1974. Effect of lignin on rate of in vitro cell wall and cellulose disappearance in corn. J. Anim. Sci. 39:1165-1169.

Ledgard, S.F., M.B. O'Connor, V.R. Carruthers, and G.J. Brier. 1990. Variability in pasture nitrogen fractions and the relationship to bloat. N. Z. J. Agric. Res. 33:237-242.

Lee, M.H., and J.L. Brewbaker. 1984. Effects of brown midrib-3 on yields and yield components of maize. Crop Sci. 24:105-108.

Lees, G.L. 1984. Cuticle and cell wall thickness: Relation to mechanical strength of whole leaves and isolated cells from some forage legumes. Crop Sci. 24:1077-1081.

Lees, G.L., R.E. Howarth, and B.P. Goplen. 1982. Morphological characteristics of leaves from some legume forages: Relation to digestion and mechanical strength. Can. J. Bot. 60:2126-2132.

Lees, G.L., R.E. Howarth, B.P. Goplen, and A.C. Fesser. 1981. Mechanical disruption of leaf tissues and cells in some bloat-causing and bloat-safe forage legumes. Crop Sci. 21:444-448.

Leete, E., and M.L. Minich. 1977. Biosynthesis of gramine in *Phalaris arundinacea*. Phytochem. 16:149-150.

Lightfoot, R.J. 1974. A look at recommendations for the control of infertility due to clover disease in sheep. Proc. Aust. Soc. Anim. Prod. 10:113-121.

Loper, G.M., C.H. Hanson, and J.H. Graham. 1967. Coumestrol content of alfalfa as affected by selection for resistance to foliar diseases. Crop Sci. 7:189-192.

Mack, J.P.G., D.P. Mulvena, and M. Slaytor. 1988. N,N-dimethyltryptamine production in *Phalaris aquatica* seedlings. Plant Physiol. 88:315-320.

Majak, W., and J.W. Hall. 1990. Sodium and potassium concentrations in ruminal contents after feeding bloat-inducing alfalfa to cattle. Can. J. Anim. Sci. 70:235-241.

Majak, W., J.W. Hall, and R.E. Howarth. 1986. The distribution of chlorophyll in rumen contents and the onset of bloat in cattle. Can. J. Anim. Sci. 66:97-102.

Majak, W., R.E. Howarth, K.-J. Cheng, and J.W. Hall. 1983. Rumen conditions that predispose cattle to pasture bloat. J. Dairy Sci. 66:1683-1688.

Marshall, T. 1973. Clover disease - What we know and what we can do. J. Agric. Western Aust. 14:198-206.

Marten, G.C. 1973. Alkaloids in reed canarygrass. p. 15-31. *In* A.G. Matches (ed.) Anti-quality components of forages. CSSA Spec. Publ. 4. CSSA, Madison, WI.

Marten, G.C., R.F. Barnes, A.B. Simons, and F.J. Wooding. 1973. Alkaloids and palatability of *Phalaris arundinacea* L. grown in diverse environments. Agron. J. 65:199-201.

Marten, G.C., R.M. Jordan, and A.W. Hovin. 1976. Biological significance of reed canarygrass alkaloids and associated palatability variation to grazing sheep and cattle. Agron. J. 68:909-914.

Marten, G.C., R.M. Jordan, and A.W. Hovin. 1981. Improved lamb performance associated with breeding for alkaloid reduction in reed canarygrass. Crop Sci. 21:295-298.

Marten, G.C., A.B. Simons, and J.R. Frelich. 1974. Alkaloids of reed canarygrass as influenced by nutrient supply. Agron. J. 66:363-368.

Marum, P., A.W. Hovin, and G.C. Marten. 1979. Inheritance of three groups of indole alkaloids in reed canarygrass. Crop Sci. 19:539-544.

McArthur, J.M., J.E. Miltimore, and M.J. Pratt. 1964. Bloat investigations - The foam stabilizing protein of alfalfa. Can. J. Anim. Sci. 44:200-206.

McFarlane, I.J., E.M. Lees, and E.E. Conn. 1975. The *in vitro* biosynthesis of dhurrin, the cyanogenic glycoside of *Sorghum bicolor*. J. Biol. Chem. 250:4708-4713.

McMurray, C.H., A.S. Laidlaw, and M. McElroy. 1986. The effect of plant development and environment on formononetin concentration in red clover (*Trifolium pratense* L.). J. Sci. Food Agric. 37:333-340.

Mika, V. 1987. Voluntary intake of low and high alkaloid tall and meadow fescues. Arch. Zootecnia. 36:151-156.

Miller, J.E., J.L. Geadelmann, and G.C. Marten. 1983. Effect of the brown midrib-allele on maize silage quality and yield. Crop Sci. 23:493-496.

Mole, S. 1989. Polyphenolics and the nutritional ecology of herbivores. p. 191-223. *In* P.R. Cheeke (ed.) Toxicants of Plant Origin. Phenolics. Vol. 4. CRC Press, Boca Raton, FL.

Moller, B.L., and E.E. Conn. 1980. The biosynthesis of cyanogenic glucosides in higher plants. Channeling of intermediates in dhurrin biosynthesis by a microsomal system from *Sorghum bicolor* (Linn) Moench. J. Biol. Chem. 255:3049-3056.

Moore, R.M., J.D. Williams, and J. Chia. 1967. Factors affecting concentrations of dimethylated indolealkylamines in *Phalaris tuberosa* L. Aust. J. Biol. Sci. 20:1131-1140.

Muhlethaler, K. 1977. Introduction to structure and function of the photosynthesis apparatus. p. 503-521. *In* A. Trebst and M. Avron (ed.) Encycl. Plant Physiol. N.S. Vol. 5. Springer-Verlag, Berlin.

Mulcahy, C., and P.N. Stuart. 1987. Chemical composition, *in vitro* digestibility, leaf:stem ratio, HCN potential and dry matter production of forage sorghums in south-east Queensland. Queensland J. Agric. Anim. Sci. 44:51-57.

Muller, L.D., R.F. Barnes, L.F. Bauman, and V.F. Colenbrander. 1971. Variations in lignin and other structural components of brown midrib mutants of maize. Crop Sci. 11:413-415.

Muller, L.D., V.L. Lechtenberg, L.F. Bauman, R.F. Barnes, and C.L. Rhykerd. 1972. In vivo evaluation of brown midrib mutant of *Zea mays* L. J. Anim. Sci. 35:883-889.

Nass, H.G. 1972. Cyanogenesis: Its inheritance in *Sorghum bicolor*, *Sorghum sudanense*, *Lotus*, and *Trifolium repens* - A review. Crop Sci. 12:503-506.

Oliver, J.W., R.G. Powell, L.K. Abney, R.D. Linnabary, and R.J. Petroski. 1990. N-Acetyl loline-induced vasoconstriction of the lateral saphenous vein (cranial branch) of cattle. p. 239-243. *In* S.S. Quisenberry and R.E. Joost (ed.) Proc. Int. Symp. *Acremonium*/Grass Interactions. Louisiana Agric. Exp. Stat., Baton Rouge, LA.

Ostrem, L. 1987. Studies on genetic variation in reed canarygrass, *Phalaris arundinacea* L. Hereditas. 107:235-248.

Palfii, Y., O.G. Malik, O.R. Dlyaboga, and M.I. Lun. 1978. Isoflavone-phytoestrogen content of red clover under the influence of different preservative techniques. Soviet Agric. Sci. 12:17-19.

Pearson, G., and J.R. Carr. 1977. A comparison between meals prepared from the seeds of different varieties of lupin as protein supplements to barley based diets for growing pigs. Anim. Feed Sci. Technol. 2:49-58.

Porter, K.S., J.D. Axtell, V.L. Lechtenberg, and V.F. Colenbrander. 1978. Phenotype, fiber composition and in vitro dry matter disappearance of chemically induced brown midrib (*bmr*) mutants of sorghum. Crop Sci. 18:205-208.

Poulton, J.E. 1991. Cyanogenesis in plants. Plant Physiol. 94:401-405.

Pounden, W.D., A.D. Pratt, N.A. Frank, H.R. Conrad, A. Hahn, A. Fetter, and R.R. Davis. 1959. Pasture bloat - Stable foam and gas production capacities of pasture plants in vitro. Vet. Med. 54:159-162.

Powell, R.G., and R.J. Petroski. 1993. The loline group of pyrrolizidine alkaloids. p. 320-338. *In* S.W. Pelletier (ed.) The alkaloids: Chemical and biological perspectives. Vol. 8. Springer-Verlag, Berlin.

Price, M.L., and L.G. Butler. 1980. Tannins and nutrition. p. 1-37. *In* Agric. Exp. Stat. Bull. 272. Purdue University, W. Lafayette, IN.

Provenza, F.D., E.A. Burritt, T.P. Clausen, J.P. Bryant, P.B. Reichardt, and R.A. Distel. 1990. Conditioned flavor aversion: A mechanism for goats to avoid condensed tannins in blackbrush. Am. Nat. 136:810-828.

Putman, M.R., D.I. Bransby, J. Schumacher, T.R. Boosinger, L.P. Bush, R.A. Shelby, J.T. Vaughan, and D. Ball. 1991. The effects of the fungal endophyte *Acremonium coenophialum* in fescue on pregnant mares and foal viability. Am. J. Vet. Res. 52:2071-2074.

Ralphs, M.H., and C. Williams. 1988. Alkaloid response to defoliation of velvet lupine (*Lupinus leucophyllus*). Weed Tech. 2:429-432.

Reay, P.F., and E.E. Conn. 1974. The purification and properties of a uridine diphosphate glucose: Aldehyde cyanohydrin β-glucosyltransferase from sorghum seedlings. J. Biol. Chem. 249:5826-5830.

Reid, C.S.W. 1960. Bloat: The foam hypothesis. p. 668-671. *In* C.L. Skidmore (ed.) Proc.8th Int. Grassl. Congr., Reading, England. 11-21 July 1960. Alden Press, Oxford.

Robbins, J.D., S.R. Wilinson, and D. Burdick. 1973. Loline alkaloids of tall fescue seed and forage. p. 98-107. *In* Prod. Fescue Toxicity Conf., Lexington, KY. 31 May - 1 June 1973. Univ. of Missouri, Columbia, MO.

Roe, R., and B. Mottershead. 1962. Palatability of *Phalaris arundinacea* L. Nature 193:255-256.

Sarkar, S.K., R.E. Howarth, and B.P. Goplen. 1976. Condensed tannins in herbaceous legumes. Crop Sci. 16:543-546.

Selmar, D., R. Lieberei, B. Biehl, and E.E. Conn. 1989. α-Hydroxynitrile lyase in *Hevea brasiliensis* and its significance for rapid cyanogenesis. Physiologia Plantarum 75:97-101.

Shore, L.S., Y. Kapulnik, B. Ben-Dor, Y. Fridman, S. Wininger, and M. Shemesh. 1992. Effects of estrone and 17β-estradiol on vegetative growth of *Medicago sativa*. Physiol. Plant. 84:217-222.

Simons, A.B., and G.C. Marten. 1971. Relationship of indole alkaloids to palatability of *Phalaris arundinacea* L. Agron. J. 63:915-919.

Smith, G.R., R.D. Randel., and C.Brandshaw. 1986. Influence of harvest date, cultivar, and sample storage method on concentration of isoflavones in subterranean clover. Crop Sci. 26:1013-1016.

Smith, K.J., and W. Woods. 1962. Relationship of calcium and magnesium to the occurrence of bloat in lambs. J. Anim. Sci. 21:798-802.

Smith, L.S., and C.C.J. Culvenor. 1981. Plant sources of hepatotoxic pyrrolizidine alkaloids. J. Natural Prod. 44:129-152.

Stuedemann, J.A., and C.S. Hoveland. 1988. Fescue endophyte: History and impact on animal agriculture. J. Prod. Agric. 1:39-44.

Stuedemann, J.A., T.S. Rumsey, J. Bond, S.R. Wilkinson, L.P. Bush, D.J. Williams, and A.B. Caudle. 1985. Association of blood cholesterol with occurrence of fat necrosis in cows and tall fescue summer toxicosis in steers. Am. J. Vet. Res. 46:1990-1995.

Terrill, T.H., G.B. Douglas, A.G. Foote, R.W. Purchas, G.F. Wilson, and T.N. Barry. 1992. Effect of condensed tannins upon body growth, wool growth and rumen metabolism in sheep grazing sulla (*Hedysarum coronarium*) and perennial pasture. J. Agric. Sci., Camb. 119:265-273.

Van Soest, P.J. 1981. Limiting factors in plant residues of low biodegradability. Agric. Environ. 6:135-143.

Waghorn, G.C. 1991. Electronegativity and redox potential of rumen digesta *in situ* in cows eating fresh lucerne. N. Z. J. Agric. Res. 34:359-361.

Waller, G.R., and E.K. Nowacki. 1978. Alkaloid biology and metabolism in plants. Plenum Press, New York.

Wedig, C.L., E.H. Jaster, and K.J. Moore. 1988. Effect of brown midrib and normal genotypes of sorghum x sudangrass on ruminal fluid and particulate rate of passage from the rumen and extent of digestion at various sites along the gastrointestinal tract in sheep. J. Anim. Sci. 66:559-565.

Wedig, C.L., E.H. Jaster, and K.J. Moore. 1989a. Disappearance of hemicellulosic monosaccharides and alkali-soluble phenolic compounds of normal and brown midrib sorghum x sudangrasses fed to heifers and sheep. J. Dairy Sci. 72:104-111.

Wedig, C.L., E.H. Jaster, and K.J. Moore. 1989b. Disappearance of hemicellulosic monosaccharides and alkali-soluble phenolic compounds of normal and brown midrib sorghum x sudangrass silages fed to Holstein steers. J. Dairy Sci. 72:112-122.

Weller, R.F., R.H. Phipps, and A. Cooper. 1985. The effect of the brown midrib-3 gene on the maturity and yield of forage maize. Grass Forage Sci. 40:335-339.

Williams, W., and J.E.M. Harrison. 1983. Alkaloid concentration during development in three *Lupinus* species and the expression of genes for alkaloid biosynthesis in seedlings. Phytochem. 22:85-90.

Williams, W., J.E.M. Harrison, and S. Jayasekera. 1984. Genetical control of alkaloid production in *Lupinus mutabilis* and the effect of a mutant allele *mutal* isolated following chemical mutagenesis. Euphytica 33:811-817.

Wink, M. 1987. Site of lupanine and sparteine biosynthesis in intact plants and in vitro organ cultures. Z. Naturforsch. 42:868-872.

Wink, M., and T. Hartmann. 1982. Diurnal fluctuation of quinolizidine alkaloid accumulation in legume plants and photomixotrophic cell suspension cultures. Z. Naturforsch. 37:369-375.

Wink, M., T. Hartmann, and L. Witte. 1980. Enzymatic synthesis of quinolizidine alkaloids in lupin chloroplasts. Z. Naturforsch. 35:93-97.

Wink, M., H.J. Heinen, G. Vogt, and H.M. Schiebel. 1984. Cellular localization of quinolizidine alkaloids by laser desorption mass spectrometry (LAMMA 1000). Plant Cell Rep. 3:230-233.

Wink, M., and P. Mende. 1987. Uptake of lupanine by alkaloid-storing epidermal cells of *Lupinus polyphyllus*. Planta Medica 53:465-469.

Wink, M., and L. Witte. 1984. Turnover and transport of quinolizidine alkaloids. Diurnal fluctuations of lupaine in the phloem sap, leaves and fruits of *Lupinus alba* L. Planta 161:519-524.

Wink, M., L. Witte, T. Hartmann, C. Theuring, and V. Volz. 1983. Accumulation of quinolizidine alkaloids in plants and cell suspension cultures: Genera Lupinus, Cytisus, Baptisa, Genista, Laburnum, and Sophora. Planta Medica 48:253-257.

Winter, C.K., and H.J. Segall. 1989. Metabolism of pyrrolizidine alkaloids. p. 23-40. *In* P.R. Cheeke (ed.) Toxicants of plant origin. Alkaloids. Vol. 1. CRC Press, Boca Raton, FL.

Wong, E. 1973. Plant phenolics. p. 265-322. *In* G.W. Butler and R.W. Bailey (ed.) Chemistry and biochemistry of herbage. Vol. 1. Academic Press, London.

Wong, E., and G.C.M. Latch. 1971. Effect of fungal diseases on phenolic contents of white clover. N.Z. J. Agric. Res. 14:633-638.

Woods, D.L., and K.W. Clark. 1971. Genetic control and seasonal variation of some alkaloids in reed canarygrass. Can. J. Plant Sci. 51:323-329.

Woods, D.L., A.W. Hovin, and G.C. Marten. 1979. Seasonal variation of hordenine and gramine concentrations and their heritability in reed canarygrass. Crop Sci. 19:853-857.

Yates, S.G., J.C. Fenster, and R.J. Bartelt. 1989. Assay of tall fescue seed extracts, fractions, and alkaloids using the large milkweed bug. J. Agric. Food Chem. 37:354-357.

CHAPTER 10

THE APPLICATION OF NEAR INFRARED REFLECTANCE SPECTROSCOPY (NIRS) TO FORAGE ANALYSIS

J. S. Shenk and M. O. Westerhaus

INTRODUCTION

The primary reason for the production of forage is to provide feed for livestock. Since the beginning of recorded history, forages have been fed and marketed on the basis of its appearance and weight. In the past 100 years, it became apparent that nutrient content as well as quantity measurements should be considered in deriving proper animal feeding programs. During this 100 year period, laboratory methods were developed and refined to provide nutrient information to the industry; however, nutritional evaluation of these products was expensive and time-consuming. Near-infrared spectroscopy (NIRS) analysis offered the promise of rapid, low-cost analysis of nutrient composition that could be applied to the increasing need for efficiency in the feeding of livestock.

BRIEF HISTORY

The first reports of NIRS appeared in the literature in 1939 (Gordy and Martin). The potential of NIRS for solving analytical problems was developed by Kaye in 1954. By 1950, it was believed by a number of researchers that hydrogenic (X-H) stretching vibrations were responsible for most of the absorption bands in the NIR region. The first commercial NIR moisture monitors were developed in the 1960s. Whetzel (1968) compiled a review of NIRS literature.

In 1968, Ben-Gera and Norris applied NIRS to the analysis of agricultural products. They recognized the potential of diffuse reflectance measurements in the NIR region for rapid analysis of grains and oilseeds. These agricultural materials were found to exhibit specific NIR absorption bands. Although their spectra were obtained from a Cary 14 scanning monochromator, Ben-Gera and Norris suggested that NIR discrete filter instruments could be used to analyze grains for protein and moisture, and soybeans for protein, oil, and moisture using diffuse reflectance measurements at 1680, 1940, 2100, 2180, 2230, and 2310 nm.

The paper by Norris et al. in 1976 demonstrated new uses for NIRS in the analysis of agricultural products. They showed that forages could be analyzed by NIRS for quality constituents, including sheep digestion and intake measurements. They proposed a new set of wavelengths at 1672, 1700, 1940, 2100, 2180, and 2336 nm for forage materials. Instrument manufacturers followed Norris's suggestions and developed NIR discrete filter instruments capable of forage analysis. Shenk and Hoover (1976), using a scanning monochromator, presented additional evidence that NIRS could provide rapid and accurate analysis of forage quality. In 1978, they developed an NIR instrument that was installed in a mobile van and taken to farms and hay markets. Barton and Burdick (1978) showed that an NIR instrument with tilting filters could analyze the major quality constituents

---

J. S. Shenk and M. O. Westerhaus, Dep. of Agronomy, Penn State Univ., University Park, PA 16802.

in forage products. In 1978, the USDA NIRS Forage Network was started to develop software, coordinate further research, and prepare the technology to be transferred to the private sector. This network of seven laboratories presented a summary of their research findings in USDA Agriculture Handbook 643 (1985). The project was later expanded to include grains, soybeans, and other agricultural products. A supplement to the Handbook was published in 1989.

In 1983, commercial companies began to market NIR instruments with sophisticated software for forage and feed analyses. Since forage and feed are more variable than grains and oilseeds, scanning monochromators, scanning filter instruments (tilting filters), or fixed filter instruments with more than ten filters were required to produce acceptable accuracy (Shenk and Barnes, 1977). The research before 1986 involved improvement of software and instrument design, calibrations for new applications and constituents, and feasibility studies by forage extension specialists and instrument manufacturers. Notable contributions were made by the state extension projects in Pennsylvania, Minnesota, Wisconsin, and Illinois. More recently, advances have been made in instrumentation and calibration techniques, standardization of instruments, monitoring of instrument performance, and the understanding of the spectra.

Research in the 1990's will attempt to improve calibration techniques, lower instrument cost without sacrificing accuracy or precision, and increase the sophistication of the software without requiring more expertise from the user.

NIR SPECTRA OF FORAGES

Theoretical Basis for Spectra

The spectra of forages are very similar to those of other materials. The NIR region (700 to 2500 nm) includes molecular absorptions of overtone (700 to 1800 nm) and combination (1800 to 2500 nm) bands. For each fundamental absorption band there exists a series of overtones with decreasing intensity. Two more molecular absorptions or overtones can combine to absorb at one combination band. NIR band intensities are weaker than their corresponding mid-infrared fundamentals by a factor of 10 to 100.

Molecular absorptions in the near-infrared region are primarily due to X-H bonds, where X is carbon, nitrogen (N), or oxygen. Absorptions by the X-H bonds are due to hydrogenic stretching, bending, or deformation vibrations. Other important molecular absorptions in the NIR region include the carbonyl carbon-to-oxygen double bond stretching vibrations, carbon-to-carbon stretching vibrations, and metal halides. The higher wavelengths also contain information on the structural bending and distortion within a molecule.

Stretching vibrations occur at higher frequencies (lower wavelengths) than bending vibrations. Stretching vibrations are either symmetric or asymmetric, and bending vibrations are either in-plane or out-of-plane. In-plane bending consists of scissoring and rocking, and out-of-plane bending consists of wagging and twisting. In general, stretching occurs at the highest frequency, followed by scissoring, wagging, and twisting and rocking. The major bands in the NIR region are second or third overtones of fundamental O-H, C-H, and N-H stretching vibrations in the mid-infrared region.

Monochromatic light produced by an NIR instrument interacts with finely ground material as absorption, diffraction, reflection, refraction, and transmission. Loss of energy from the sample can occur from absorption, internal refraction, internal scattering, specular reflection, and trapping losses due to wide angle solid ray reflection. Diffuse reflectance NIR instruments measure the energy that interacts with the sample before reaching the detectors.

An NIR absorption band occurs whenever the frequency of NIR radiation matches the frequency (or an overtone or combination) of bending and stretching vibrations of a molecular bond in the sample. The relationship between transmission of energy through a sample and the concentration of the absorbing molecular bonds is known as Beer's Law, which states that molecular bond concentration is linear with log 1/transmission. Although Beer's Law is valid only for transmission, it has proved useful for diffuse reflectance.

Spectral Bands

NIR reflectance measurements are usually transformed into log of inverse reflectance (log(1/R)). Although collected as individual data points, NIR data from scanning monochromators can be considered to be an estimate of a continuous NIR spectrum, composed of many overlapping absorption bands. The theoretical shape of an NIR absorption band is Lorentzian in the frequency scale. (NIR is usually displayed in the wavelength scale.) The slit shape of the instrument adds a smoothing function to the band that approaches a normal distribution. The final shape of a band is a convolution of Lorentzian and Gaussian distributions.

Bands are defined by three criteria - location, height, and width. The height of an absorption band is measured at the peak of the band. The band location is measured as the wavelength of its peak. Band width is measured as the width of the peak at half of the peak height. Many bands appear only as shoulders, which are noticeable deviations from the usual band shape that do not have a local maximum. A typical spectrum of a forage sample contains 7-10 peaks with many shoulders.

Because of the extensive overlapping of individual bands in forage spectra, band characteristics cannot be accurately estimated in log (1/R) form. In a composite band, consisting of several overlapping bands, a change in the height of one of the bands can cause a shift in the peak location of the composite band. Nevertheless, reasonable estimates of band locations can be obtained from derivatized spectra.

Adding to the difficulty of band criteria estimation is the fact that many factors contribute to a spectrum. These include a multiplicative response to changes in particle size or packing density, confounding of the 1100 to 1400 nm region with information from the visible region or soil particles, confounding of the 2300 to 2500 nm region with tails of fundamental absorption peaks in the mid-IR, and shifting of the baseline caused by differences in the ceramic reference material, sample holder glass thickness, or placement of the sample.

Spectral Decomposition

Peak characteristics can be estimated through a combination of spectral deconvolution and decomposition (Nadler, 1988; Nadler et al., 1989). First, spectral resolution is increased from 10 nm to 5 nm using spectral deconvolution. Peak locations then are estimated as the maxima of the deconvoluted spectrum.

Next, baseline and tailing effects of the visible and mid-IR spectrum are estimated. Finally, peak height and width are estimated by comparing a log(1/R) spectrum with its second derivative. The estimates are improved iteratively using the fourth derivative of the log(1/R) spectrum.

As an example, the spectra of low and high quality hay samples will be decomposed into individual bands. The major differences between the deconvoluted peaks will be related to the laboratory analyses of the samples.

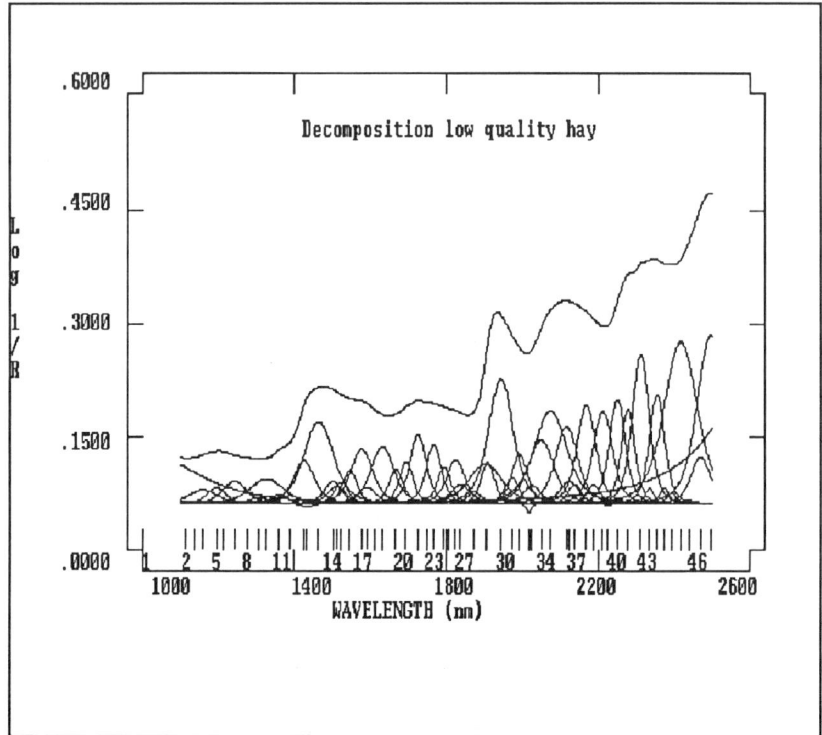

Figure 1 Decomposition of a low quality hay spectrum

Figure 1 is the spectrum and its decomposition of a low quality hay sample. The log(1/R) spectrum is smooth with very few peaks. Using spectral decomposition, the spectrum can be seen to have more than 45 mathematically identified absorption peaks.

Figure 2 is the spectrum and its decomposition of a high quality hay sample. The log(1/R) spectrum differs from the spectrum of low quality hay in the 2000 - 2200 nm region. This difference is affirmed by the deconvoluted peaks.

Figure 3 is an overlay plot of the absorption peak information at 1944 nm, 2140 nm, and 2314 nm. The low quality hay spectrum was from a sample with 910 g/kg dry matter (DM), 126 g/kg crude protein, and 406 g/kg acid detergent fiber (ADF). The high quality hay sample contained 900 g/kg DM, 231 g/kg

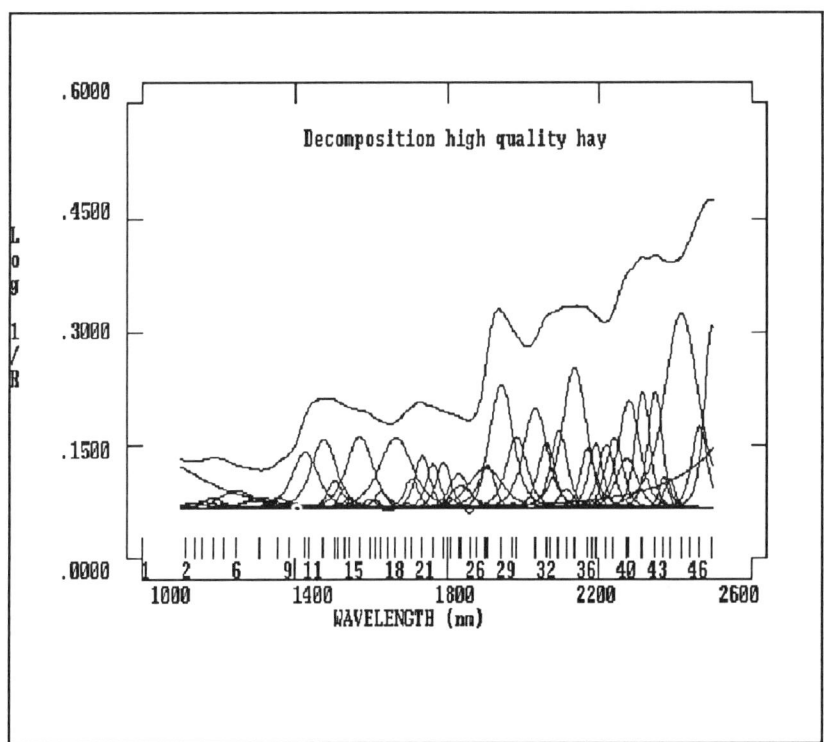

Figure 2 Decomposition of a high quality hay spectrum

crude protein, and 325 g/kg ADF. As expected, the low peak at 1944 nm, the low peak at 2140 nm, and the high peak at 2314 nm were from the low quality hay sample. These results coincide nicely with known band locations for water, protein, and fiber.

Special Considerations

Water and Hydrogen Bonding

The transmission spectrum of pure water appears simple in log 1/R form. When it is decomposed, however, many major and minor absorption peaks are evident. It is obvious that neither the first overtone band at 1450 nor the combination band at 1930 can be generated from a single Gaussian or Lorentzian absorption band. Each of the broad bands at 1450 and 1930 nm contains information on more than one hydrogen bonded subspecies. It is known that variations in hydrogen bonded molecular subspecies can cause band broadening and peak position shifts. Variations in hydrogen bonding (intra- and intermolecular) result in changes in the force constants of the X-H bonds. The largest change in the force constants of the fundamental vibrations occurs for the X-H stretching vibration. It is known that the band location of O-H vibrations is isolinear, and changes position linearly with changes in temperature.

This information has important implications in forage. The water bands at 1450 and 1930 nm consist of multiple overlapping bands. The apparent location

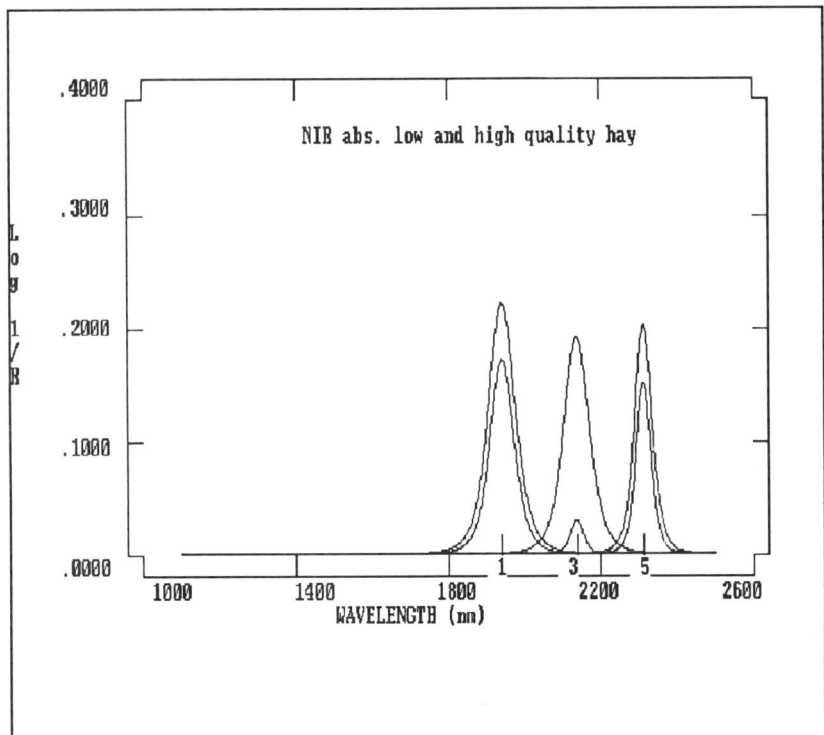

Figure 3 Absorption peaks of low and high quality hay at 1944, 2140, and 2314 nm.

of these composite bands changes as the spectra are measured from one sample type to another. This apparent shift in band position is due to the changes in the relative proportions of the individual bands making up the composite bands. In addition, the concentrations of the molecular subspecies in any particular sample are directly influenced by chemical interactions with other molecular species in the sample.

Particle Size Effects in Forages

The primary effect of particle size variation is to expand or contract by multiplying the spectrum by a scaling factor. Since this does not necessarily relate to sample chemistry, it is considered an interference that should be removed mathematically or ignored. Multiple linear regression can compensate for multiplicative scatter by "forcing" the regression intercept toward the mean reference value of the calibration set. Derivative transformations of the spectra also have been used to minimize the effects of particle size, although derivatives only remove base line shifts and cannot fully correct for multiplicative scatter effects.

Several mathematical procedures for removing particle size variation have been suggested. These include Fourier deconvolution (McClure and Williams, 1986), detrending and standard normal variate transformations (Barnes et al., 1989), multiplicative scatter corrections (Martens and Naes, 1989), principal components elimination (Osborne and Fearn, 1986), and simultaneous removal of particle size and water variation. These corrections for particle size may not always improve accuracy of NIRS analysis for two reasons. First, none of these procedures do a perfect job of removing particle size effects independent of absorption information, and second, particle size variation may provide useful information for the calibration. The mean particle size and distribution can be an indirect measure of forage chemistry, due to the interaction of the various plant structures, such as stems and leaves, with the drying and grinding procedure.

REFERENCE METHODS FOR FORAGE ANALYSIS

The NIR spectrum contains information about the major X-H chemical bonds in forage products. The spectrum is the summation of NIR absorption bands of functional groups from the major chemical and physical properties of a forage sample. In contrast, the currently accepted laboratory procedures used to calibrate NIR instruments are not well defined chemically and can be very difficult to relate to spectroscopic data. Unfortunately, the traditional laboratory procedures are the accepted means for measuring the quality of forage and have been in use by the agricultural industry for many decades.

The traditional chemical procedures for forages were developed to estimate the feeding value of a material for livestock production. These procedures fit into three general categories: 1) proximate analysis, developed more than 100 years ago to estimate general nutritive value; 2) the Van Soest detergent techniques, developed in the late 1960's as a broad analytical system with better nutritional and chemical definitions; and 3) specific analytical procedures designed to measure single chemical entities that could be used to better describe the nutritional profile of a feed substance.

When NIRS analysis for forages became available in the 1970's, researchers were faced with a dilemma. NIRS provided new information about the chemical composition of forages, based on molecular bonds. But workers in the forage industry were satisfied with current laboratory procedures because they had worked with them for many years and developed feed composition tables and ration balancing programs for them. The forage community required that any new analytical procedure be capable of duplicating the analyses of the traditional laboratory methods. It was difficult for NIRS researchers to relate spectroscopic information to traditional laboratory methods and balance livestock diets.

Two examples will be used to demonstrate this problem. The most widely used reference method for measuring moisture in forage products is oven drying overnight at 100°C or 135°C for 2 h. NIRS can be used to evaluate the validity of these reference methods. Careful examination of the NIR spectra of these apparently dry samples reveals that moisture is still present in a sample following the 100°C drying method. Both reference drying methods drive off all volatile compounds present in the sample. Studies by Windham et al. (1989) show that NIRS correlates better to the Karl Fisher method of moisture determination than to either oven drying method for several types of forage products. It has been suggested that oven drying methods should be used to estimate "crude" moisture content, and the Karl Fisher technique should be used for more accurate moisture determinations.

The second example of the difficulty in relating traditional methods to spectroscopic data is the estimation of protein content in forage products. The current reference method for estimating protein concentration of most forage products is to multiply N content by a constant to convert N to protein. Thus, crude protein is only a measure of total N content. NIRS, however, can measure N-H molecular bonds, which are part of the protein molecule. As was the case with moisture determinations, the reference method and NIRS measure different components of the sample. The strength of the relationship between NIRS and the reference method depends on the agreement between N-H bond information and total N content.

Chemical Reference Methods

Moisture

The moisture bands are among the most prominent features of forage spectra, due to the high extinction coefficient of the moisture O-H bands. This fact, plus the large range of moisture values and the unambiguity of the constituent, should result in highly accurate moisture calibrations from NIRS. However, basic studies revealed that moisture calibrations for forages unless calibrated with the Karl Fisher reference method were not as accurate as might be expected (Windham et al., 1989; Windham and Barton, 1991). As stated above, oven moisture methods are more appropriately termed oven "drying methods", since volatile materials are driven off with the moisture. The NIRS calibration error for moisture increases as the concentration of volatile compounds in the sample increases.

Protein

Kjeldahl N determination is the most well known and widely practiced method used to measure the total N in a sample of plant material. As stated above, a conversion factor is used to convert N content to protein content. Although the procedure has several disadvantages, it has become the reference procedure by which all other N determining methods are compared. The multiple step procedure is relatively costly, complicated, time-consuming, and hazardous.

In an attempt to improve upon the disadvantages of the Kjeldahl procedure, other methods for determining total N have been developed. Recently, the combustion method for N determination has gained popularity because it has fewer adverse environmental consequences.

The most important wavelengths for direct measurement of protein by NIRS include: the carbonyl stretch of the primary amide at 2060 nm (Maillard effect for bound protein); the 2168 to 2180 nm combination band consisting of N-H bend 2nd overtone, the C-H stretch/C=O stretch combination; and the C=O stretch/N-H in-plane bend/C-N stretch combination bands. The bands at 2050 to 2060 nm indicate N-H stretching vibrations of various types and are also useful. Aromatics containing amino acids/proteins also exhibit the aromatic C-H stretch 1st overtone in the 1640-1680 nm region. In certain cases, the N-H stretch 1st overtone regions are used for protein measurement in the 1500-1530 nm region.

Fiber

NIR instruments have been successfully calibrated to measure Van Soest detergent fiber components (Goering and Van Soest, 1970). Since cellulose is a major contributor of the fiber portion of forage, the absorption peaks at 2300, 2310, and 2340 nm are usually present in calibration equations. Since lignin consists of phenylpropanoid residues as building blocks, wavelengths associated with the regions from 1600-1800 nm and 2200-2300 would be expected to be present in the calibration equation. In 1987, Albrecht et al. predicted cellwall carbohydrates and starch with NIR. Because of the complex nature and mixtures of these substances in the cell wall due to species, maturities, and methods of storage, the accuracy of these equations are usually not as good as for the N fractions.

Minerals

Since NIRS measures absorption by molecular bonds, pure minerals, such as calcium and phosphorus, do not have NIR absorption bands. Organic complexes and chelates could be detected using NIRS, but ionic forms and salts do not absorb energy in the NIR region. Salts can indirectly be detected in high moisture samples due to changes in hydrogen bonding and resultant band shifts.

Calibrations for mineral content in forages suggest that NIRS absorption bands for organic acids might exist, although there is no direct evidence for this. Alternatively, it is possible that computer-directed calibrations make use of naturally occurring correlations between mineral concentrations and concentrations of constituents that NIR can measure, such as protein, fiber, and specular characteristics of the sample. The papers by Clark et al. (1987, 1989) indicate that more work is required to determine the relationship between the relatively high organic acid composition of forage crops and the individual mineral concentrations.

Estimating Animal Response

Many laboratory methods exist to predict the digestibility of energy and DM of forage. They range from chemical methods to in vitro techniques to enzymatic methods (Aufrere et al., 1988). These indirect methods have been shown to correlate well with absorption information in the NIR spectra (Coelho et al., 1988). Norris et al. (1976) showed that nutrient digestion by animals and DM intake could be predicted directly from the spectra of forage samples. Additional studies have been conducted by Ward et al. (1982) using esophageal-fistulated sheep, and Eckman et al. (1983) using sheep fed diets containing mixtures of forage and concentrates. Further response work has been documented by Holecheck et al. (1982) using esophageally fistulated cattle, Harpster et al. (1982) using sheep fed a hay diet, and Abrams et al. (1987) using dairy cattle. In nearly every case, the results showed that more than one reference method must be used for best prediction of animal response from a diet.

Many different wavelengths were used in these studies in the prediction of animal response. This is not surprising since animal response does not depend upon a single chemical entity, but depends on a balance of many ingredients, and each ingredient might absorb energy in several different bands. For this reason, scanning monochromators are recommended for predicting animal response.

Abrams et al. (1987) showed that NIRS could be more accurate in predicting animal response than any single current reference method or any combination of these methods. This is another example of the dilemma in relating traditional methods to NIRS. The NIR spectroscopic method can be more accurate in predicting animal response than current reference methods, but the forage industry is based on the numbers provided by the traditional reference methods.

Sample Presentation Methods

Sampling

Proper sampling is essential for any chemical technique in order to achieve an aliquot for analysis that properly represents the composition of the larger sample of interest. Increased accuracy and precision in the laboratory will not improve upon poor sampling technique, nor will it give more accurate numbers for estimating actual forage quality. Correct sampling procedures are discussed by Shenk et al. (1992) and Abrams (1989).

Drying

Two methods have been utilized for drying forage samples - oven and microwave. Since the two methods produce spectra with different characteristics, it is recommended that only one drying method be used for a calibration. The oven drying procedure producing the minimum chemical changes in the sample involves a forced-air convection oven at 65°C. Samples are left in the oven until they attain approximately 950 g/kg DM (8-12 h).

Grinding

Dried samples should be chopped in a commercial-grade kitchen type blender for 30 sec to 1 min before being ground through any grinders. The grinders most commonly used in forage work include the UDY, the Cyclotec, the Wiley, and the Brinkman, with 0.5 mm or 1.0 mm screens. The 1.0 mm screen is most commonly used because it combines reasonable sample throughput and acceptable particle size.

Samples which are to be analyzed in slurry form by transmission or diffuse reflectance may be pulverized or homogenized using cell disrupter devices or high speed blenders. The solid particulate material is suspended by vigorous mixing and shaking. Spectral measurements should be taken immediately after the sample is prepared.

Mixing and Packing

Dry samples require tumbling or thorough mixing to minimize the effects of repack error. Repack error is measured as the differences between sample spectral measurements of several aliquots of a sample. If this effect is not minimized, the correlation between the optical data and the reference chemical information is drastically reduced and the sensitivity of NIRS for any particular application is reduced. Repack averaging, or compression, and other methods of reducing these effects have been demonstrated in the literature (Shenk et al., 1992).

Method of Sample Presentation

A number of methods are available to present the sample to the instrument. The same device used to develop the calibration must be used on the host instrument and different instruments have different sample viewing devices. There are five basic methods of presenting the sample to the instrument. The small sample cup is used for fine ground uniform materials, large pack cups are used for heterogenous samples that are usually unground or high in moisture (Shenk, 1993), pour devices are used to scan samples as they pass by the optical sensors, liquid cups are used for true transmission or folded transmission, and fiber optic probes are used for remote measurements.

Natural Product Analysis

Traditional analysis of forage products by NIRS includes the drying and grinding of samples. The analysis of undried, ground silage by NIRS reflectance has been reported (Blosser, 1985; Blosser et al., 1986). Both of these publications indicated that NIRS has some potential to predict the composition of high moisture products. The advantage of analyzing samples in their undried, unground state is the major reduction in sample preparation time. These researchers reported that improvement in sample presentation and calibration methodology is needed to obtain optimal prediction accuracy. Since the writing of these papers, new instrumentation and software techniques have been developed for scanning undried, unground samples (Shenk and Westerhaus, 1992; Shenk, 1993).

New advances in sample presentation and software now provide the potential to analyze grains, oilseeds, and forages such as hay, silage, and fresh pasture in their natural form (Shenk, 1993). Forages are cut into 1.5 to 2.5 cm lengths and are packed by hand in large pack cups. Grains and oilseeds are analyzed by pouring the seeds into the instrument sample viewing device. Calibration of undried materials requires careful attention to DM determinations to obtain accurate prediction results. The accuracy of these natural calibrations is similar to the accuracy of calibrations obtained with dried and ground samples. The accuracy of the analysis of undried and unground hay core samples, haylage, corn grain, and soybeans in **Tables 1** and **2** are examples of this method. Other calibrations for fresh pasture, corn silage, and high moisture corn grain also are acceptable.

Table 1. Accuracy of NIRS calibrations for chopped hay and haylage selected by global and neighborhood Mahalanobis distance.

Constituent	Hay core samples			Wet haylage		
	n	SECV[a]	1-VR[b]	n	SECV	1-VR
		— g kg^{-1} —			— g kg^{-1} —	
Protein	150	8.0	0.90	205	9.0	0.90
Acid detergent fiber	150	16.0	0.87	205	21.0	0.85
Neutral detergent fiber	150	25.0	0.90	205	29.4	0.85
Dry matter	150	4.5	0.95	205	6.2	0.99

[a] Standard error of cross validation, an estimate of accuracy.
[b] Proportion of reference method variation explained by cross validation predicted values.

Studies have been conducted to make sure that natural product equation performance is acceptable across instruments. Single sample standardization appears to be satisfactory (Shenk and Westerhaus, 1991c).

Table 2. Accuracy of NIRS calibrations for corn grain and soybeans selected by global and neighborhood Mahalanobis distance.

Constituent	Corn grain			Soybeans		
	n	SECV[a]	1-VR[b]	n	SECV	1-VR
		— g kg^{-1} —			— g kg^{-1} —	
Protein	58	2.0	0.77	80	3.7	0.94
Fat	—	—	—	80	3.3	0.99
Dry matter	58	1.5	0.99	80	3.5	0.85

[a] Standard error of cross validation, an estimate of accuracy.
[b] Proportion of reference method variation explained by cross validation predicted values.

POPULATION STRUCTURING

Choosing Samples With Spectra

NIRS relies heavily on the definition of a target population, collection of representative samples with accurate laboratory reference values, and advanced statistical procedures to obtain the most accurate calibration. Populations must be broad enough to encompass most samples during routine analysis, yet not be so broad as to lower accuracy. Often, a large number of samples are available, but obtaining laboratory reference values for all samples would be very expensive.

Two procedures, CENTER and SELECT, have been developed to define a population and to select samples for calibration. Both procedures use principal components to condense the information in a full spectrum to a smaller number of scores.

Establishing Spectral Boundaries For a Product

The procedure CENTER (Shenk and Westerhaus, 1991a) was developed to rank spectra in a file according to their Mahalanobis distance from the average spectra of the file, using principal component scores. The Mahalanobis distance values, standardized by dividing them by their average value, are called "global H" values. The CENTER procedure first computes the principal components (eigenvectors) of the spectra and replaces the spectral data points with principal component scores. It then computes the average scores, and calculates the global "H" from each sample to that average. Spectra in the file are reordered from smallest global "H" to largest global "H". (This provides a representative split by taking every ith sample). Samples with extreme spectra have a large amount of influence in wavelength selection and coefficient size determination during calibration development. If samples with extreme spectra are valid extensions to the population, they should be retained. Including them in the calibration will broaden the range of samples that can be predicted by the calibration. If these samples are from another population, they should be removed from the calibration set. Including them in the calibration will lower the accuracy of the calibration for all samples.

Three natural product libraries of forage samples will be used to demonstrate this principle. The first file consists of 128 undried and unground pasture samples selected from a larger population of 250 samples. The second file consists of 200 undried and unground haylage samples selected from a larger population of over 500 samples. The third file consists of 150 hay samples selected from a larger population of over 250 samples obtained with a core sampler and scanned without grinding.

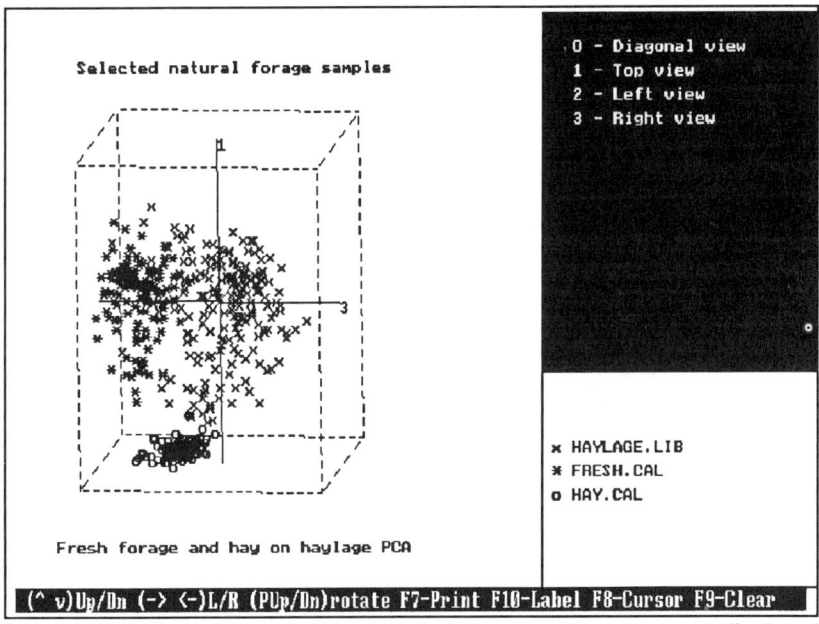

Figure 4 Natural forage samples (fresh forage, hay, and haylage) displayed against the haylage PCA file

The three libraries of samples were plotted on a 3-dimensional graph (i.e., hypersphere). The first three principal component scores were used to represent each spectra. Since haylage was found to have the most spectral variation, the first three principal components from the haylage spectra were used to obtain all the scores shown in **Figure 4**. The fresh forage and haylage spectra were separated along axis three, but were more similar to each other than to the hay file because the haylage and fresh samples were all high in moisture. The scores of the hay samples were clustered near the bottom of the plot and had much less variation than the other two groups of scores.

Defining Neighborhoods in the Calibration Population

The procedure SELECT (Shenk and Westerhaus, 1991b) was developed to identify samples needed for calibration from a larger set of samples. As in procedure CENTER, the first step was to compute the principal component and obtain the principal component scores. Next, a Mahalanobis distance is computed, not between the scores of a sample and the average scores, but between the scores of two samples. The Mahalanobis distance values are standardized as in procedure CENTER and called "neighborhood H" values. The neighborhood of a sample is defined as all sets of scores with neighborhood "H" distances to that sample less than a specified cutoff value. The procedure SELECT is based on the assumption that only one sample is required to represent all samples in a neighborhood. By changing the cutoff value, the number of neighborhoods (and samples selected for calibration) can be controlled.

The distribution of neighborhoods depends on two factors. First is the number of eigenvectors required to explain the variation in the population. The number of available neighborhoods increases as the number of dimensions (eigenvectors) increase. The CENTER and SELECT procedures base the number of dimensions on the number of samples and number of wavelengths. The second factor is the distribution of sample spectra. Neighborhoods near the average spectrum are more likely to contain samples than neighborhoods far from the average spectrum. The number of selected samples initially increases with "H" as the number of possible neighborhoods increases, then the number of selected samples decreases with "H" as fewer and fewer neighborhoods far from the average spectrum contain samples.

Developing a Product Library of Spectra

The accuracy of an NIRS analysis equation is directly related to the structure and distribution of the selected samples in the calibration file. It has been shown that in order for a calibration to be very useful, a group of samples for the product (product library) should be large and cover all reasonable samples of the product. In addition, a good calibration file of spectra should have a symmetric distribution of samples along the eigenvectors in the hypersphere.

A good product library is a group of samples that represent all reasonable samples of the product. Selecting a sample set for proper calibration is the single most important step to have a successful calibration. Samples analyzed by the reference method must be representative of the entire population to be analyzed. Calibration sets must not only uniformly cover both the spectral and constituent range, they must also be composed of a uniformly distributed set of sample types. Ideal calibration sets are composed of a proper number of samples with widely varying compositions and constituent values.

It was originally suspected, based on the wheat and oil seed data, that a minimum of fifty samples would be required to derive a calibration equation across forage samples of mixed species. These calibrations were successful for very small populations, but did not predict large randomly selected populations adequately. For example, it is now known that to measure adequately constituents across all types of wheat, screening of hundreds of samples is needed for a good robust calibration. The calibration samples used must contain all of the variance expected to be encountered in a "real world" analysis situation.

The next two figures show the trends across the three eigenvectors for protein and neutral detergent fiber (NDF). The constituent trends are displayed by the three symbols. The o, +, and sq were used to display the samples with low, medium, and high values or concentrations. **Figure 5** shows the distribution of protein across the hypersphere. The first observation is that although a protein trend existed, it was not as strong as for particle size and DM. In addition, the eigenvectors were different but had at least one eigenvector in common with both particle size and DM. The population seems to be rather symmetric but some empty neighborhoods can be seen.

Figure 6 was developed for fiber expressed as NDF. This relationship too is not as strong as for particle size and moisture but a good trend is present. The

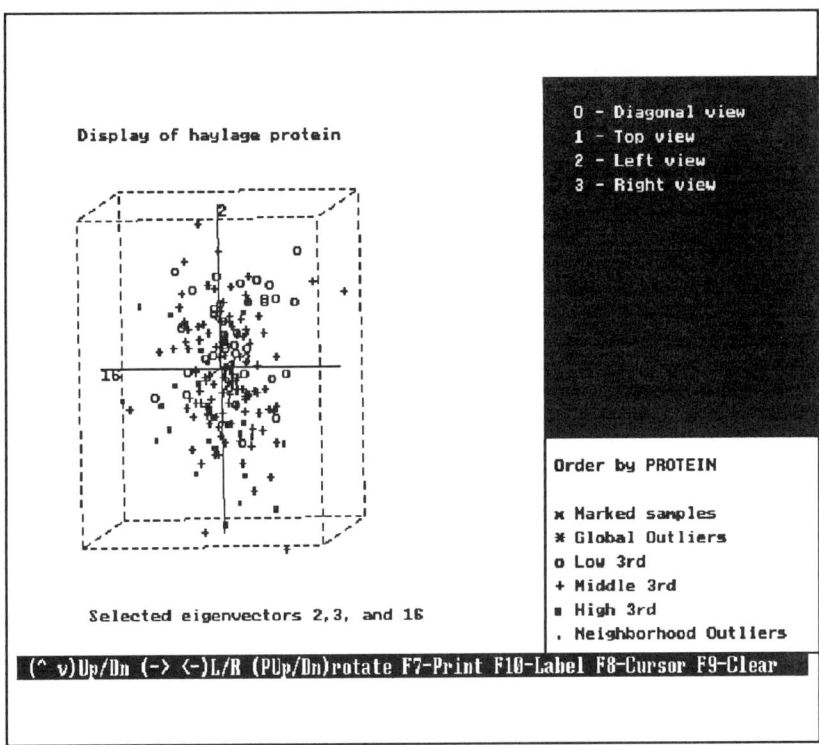

Figure 5 Display of haylage protein across the hypersphere

NDF variable has commonality with all three of the previous variables because it shares eigenvector 2 and 3. It, however, has its own new eigenvector 20 which makes it unique in this sequence. This figure provides a detailed look at this NDF distribution in the hypersphere.

The display of statistical concepts and evaluation of symmetry provides a better understanding of NIR spectra and the physical/chemical characteristics to be predicted. The three figures presented in this section show that these relationships can be easily displayed and understood in graphical form. In addition, both quantitative and qualitative information can be made available for study. From an educational perspective, the concepts of global "H" and neighborhood "H" can be shown even though the display is limited to three dimensions at a time. And last, the concept of population symmetry in the hypersphere can be displayed and evaluated as an aid in the development of product libraries with symmetric spectral and reference value distributions. Equations developed from using these concepts will be more accurate and robust for routine analysis.

CALIBRATION

New Mathematical and Statistical Tools for Forage Analysis

The spectral data produced by an NIR instrument represent the total chemical and physical properties of a forage sample. It can be obtained in seconds and is highly repeatable. However, it is only useful when interpreted in the form of

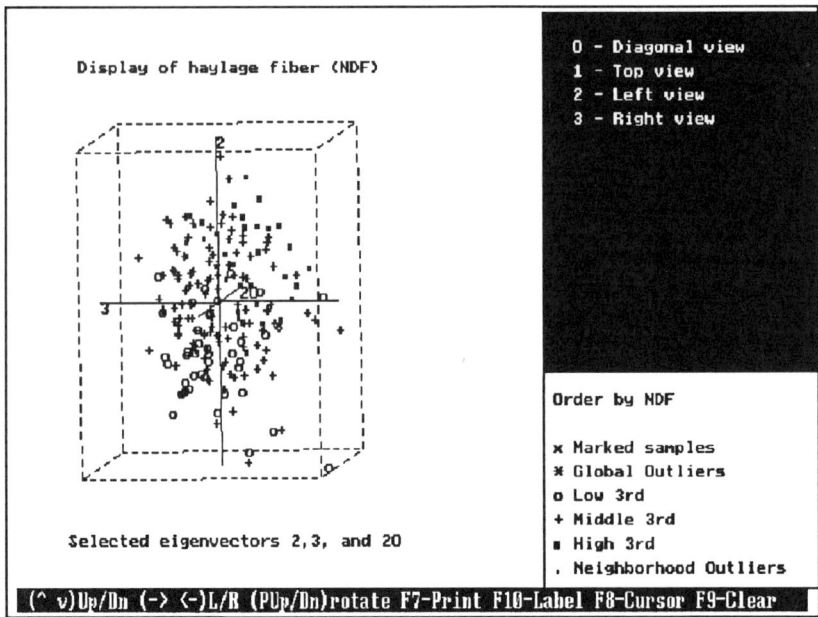

Figure 6 Display of haylage neutral detergent fiber (NDF) across the hypersphere

current reference methods. Calibration is the term used to describe the conversion of NIR absorption information into laboratory reference method (LRM) information. The accuracy of this conversion is measured as the standard error of calibration (SEC) and the standard error of prediction (SEP).

Predicting LRM values from NIR spectra is not simple. NIRS measurements are often made on whole samples without grinding, drying, or extracting to remove confounding constituents. The statistical model that relates NIR to LRM information can be highly complex. Also, spectra are obtained at 2 nm intervals with a 10 nm band pass, which introduces high intercorrelations into the data.

Chemical information in a spectrum is present at specific locations. However, physical properties of a sample, such as particle size, affect the whole spectrum. (Particle size variation results primarily in the whole spectrum being expanded or contracted by a multiplier.) This allows for whole-spectrum mathematical corrections for particle size that do not affect the localized chemical information. Two corrections for particle size are popular - standard normal variate (SNV) corrections and multiplicative scatter corrections (MSC). The SNV correction scales each spectrum to have a standard deviation of 1.0. This correction can be made on individual spectra. The MSC procedure requires a target spectrum, usually the average spectrum for a file of spectra. MSC then expands or contracts the spectrum and shifts it up or down to look most like the target spectrum. Both methods can be helpful for files of spectra with considerable particle size variation. However, both methods are affected by chemical information. They may not be helpful if the particle size variation is small compared to the variation due to chemical information.

NIRS calibrations with least squares regression are difficult because of the large number of highly intercorrelated spectral data points. Principal component analysis is ideally suited to reduce the spectral information into a smaller number of independent factors. These factors can be used to compute global and neighborhood "H" values. But they can also be used for principal component regression (PCR) calibrations.

Partial least squares (PLS) is a regression technique similar to PCR. The main difference is that PLS uses the chemical information to form factors useful for fitting the chemical information. In PCR, the factors are formed without the chemical information, and are designed just to explain the spectral information.

Neural networks have been applied to the NIRS calibration with some success (Westerhaus and Reeves, 1992). The main advantage of neural networks is their ability to fit nonlinear data. However, the iterative procedure takes days instead of minutes to compute and must be constantly tested with a validation set to determine the optimal predictive accuracy.

Old and New Regression Methods

Multiple regression

Multiple regression is the simplest statistical procedure used to predict one variable from one or more other variables. The mathematical model fit by multiple regression is $Y = B_0 + B_1X_1 + B_2X_2 + \ldots + B_pX_p$. Y is the variable to be predicted, B_0 is the equation intercept, B_1 through B_p are the scaling factors that relate changes in the X variables to Y, and p is the number of X variables. In our example, Y is a laboratory reference value and X_1 through X_p are NIR absorbance values.

Step-up regression

When not all the wavelengths are needed in the regression model, some form of wavelength selection is needed. The simplest form of wavelength selection is step-up regression. Step-up regression begins by selecting the X variable most highly correlated with Y. Next, the partial correlations of all remaining X variables to Y are computed, with the previously selected variable fixed in the model. The X variable with the highest partial correlation is added to the model. Additional variables are added one at a time until overfitting occurs.

Overfitting is a situation where too many wavelengths are added to the regression model. Adding wavelengths to the regression model always improves the fit of the calibration samples. When the number of wavelengths in the model equals the number of samples, the fit is perfect. Several rules exist to suggest how many wavelengths are enough. Experience shows that one term can be fit for every 10 samples. Another rule states that the partial F statistics for the regression coefficients must all be above 8.0. The recommended method for determining how many wavelengths is enough is to test each model on an independent set of samples. The SEP for the independent samples usually decreases with additional terms until the model is sufficiently complex, then increases as additional terms fit specific samples in the calibration set.

Stepwise regression

Stepwise regression is similar to step-up regression except wavelengths are not forced to remain in the model once selected. After a new wavelength is added to the model, each wavelength in the model is removed from the model, and is replaced with any other wavelength that improves the overall fit of the model. This is done one at a time for each wavelength in the model, cycling through the wavelengths several times, until no further changes can be found to improve the model fit. The stopping rule for adding wavelengths and the validation procedure is the same as for step-up regression. However, stepwise regression usually has a lower SEC and SEP than step-up regression because more combinations of wavelengths are evaluated.

Full Spectrum Calibration Methods

REG70

In 1985, we developed our first regression program using the full spectrum to develop calibrations, called REG70. The program used every 10th wavelength out of 700. The drawback to this procedure was the large number of samples required for calibration. With 70 wavelengths, a minimum of 200 calibration samples was recommended.

Why did the REG70 using 70 wavelengths work when step-up and stepwise regression seldom required more than 9 terms? One answer is that by spacing the used wavelengths 20 nm apart, the high intercorrelations caused by the 10 nm band pass was minimized. More importantly, REG70 did not select wavelengths based on their correlation to the lab data. It is wavelength selection, not estimating the B_i coefficients, that causes overfitting. In wavelength selection, hundreds of wavelengths and thousands of wavelength combinations are evaluated. These wavelengths can be viewed as having an ideal component and an error component. In wavelength selection, the maximum correlation also has an ideal component and an error component. After several terms are selected, the error component dominates the ideal component in the partial correlations, and overfitting occurs. In REG70, all wavelengths are preselected, and overfitting is less of a problem.

REG70 performed well, especially with large calibration sets where a complex regression model was needed. It became clear that including all the spectral information in the regression model resulted in accurate calibrations. The next step was to include the maximum amount of spectral information using principal components.

Principal component regression

The principal component procedure is basically a data reduction method. It is designed to reduce the intercorrelated spectral data points (wavelengths) to a smaller set of independent variables called principal components or eigenvectors. Singular-value decomposition is an excellent algorithm for obtaining the principal components. Each principal component is a set of loadings (weights), one for each wavelength. A score is computed by multiplying spectral absorbance values by these weights and summing the results. By using several principal components, spectra information can be condensed into a few scores. Principal components are constructed to give the maximum variability in the scores. The principal components also are constructed to be independent of each other. Note that this procedure does not involve the lab values in the calculations.

Principal component regression (PCR) is simply multiple regression on the scores. Since the scores are independent, the stepwise procedure is not needed.

The only question is how many principal component factors to use. Factors are usually added until the SEP for a validation set reaches a minimum. Since the scores are based on only the absorbance data and an extensive variable selection is not performed, overfitting is not a serious problem. When the PCR model is fit, the B_i coefficients can be converted into a set of coefficients for the original absorbance data, giving, for example, a 700 term equation based on 16 principal component factors.

Partial least squares regression

Partial least squares is a third calibration method that uses the full spectrum (Martens and Naes, 1989). It was first used for NIRS calibrations in the 1980's. PLS is similar to PCR, but differs in that the loading calculations include LRM values. PLS is usually better than PCR as a method for predicting LRM values because the LRM information helps form "relevant" factors, i.e., factors constructed to correlate highly with LRM values. Modified PLS is a procedure where the NIR residuals at each wavelength, obtained after each factor is calculated, are standardized (divided by the standard deviations of the residuals at a wavelength) before calculating the next factor. MPLS is often more stable and accurate than the standard PLS algorithm.

Neural networks

Calibrations for NIR instruments are often limited by use of linear mathematical models. Neural networks have been suggested as a calibration technique to resolve nonlinear relationships in spectral data. Westerhaus and Reeves (1992) compared the traditional calibration methods described above for forages with the neural networks procedure. The traditional calibration methods provided acceptable calibrations; however, in many cases, neural networks were more accurate. Even though they were shown to provide improvement in calibration accuracy, neural networks are not yet a popular method for deriving forage calibrations.

Calibration Procedures

Spectra data pretreatment

If NIR spectra were not affected by particle size, scatter coefficient, and path length variation, calibrations could be performed directly on the absorbance data. When these interferences are present, however, some form of spectral data pretreatment often helps the calibration. The full spectrum can be corrected by MSC or SNV, as discussed previously. Alternatively, local changes can be isolated from the rest of the spectrum by derivatizing the spectrum. Derivatives of a spectrum can be calculated several ways, but the most popular is the segment-gap method. The segment refers to the range of data points averaged together, and the gap refers to the distance between averages being subtracted. A first derivative is computed as the differences between two averages separated by the specified gap. A second derivative can be computed by applying the first derivative procedure to first derivative data.

The two arithmetic operations, averaging and subtracting, affect the precision of the spectral values in opposite ways. Averaging increases the precision of the numbers by averaging out random fluctuations. Subtraction, however, cancels the digits in common between the two numbers, leaving fewer significant digits. Increasing the gap results in fewer lost significant digits (because the spectral changes are greater), but loses the ability to isolate local changes. Since the averaging and subtraction can be performed in either order, the most precision is maintained by subtracting data points first, and averaging the differences second.

Mathematically, a first derivative spectrum is independent of the baseline. (A single number can be added to all spectral data points without affecting the derivatized spectra.) A second derivative spectrum is independent of an additive ramp (slanted line). (A straight line of any slope can be added to all spectral data points without affecting second derivative spectra.) A third derivative spectrum is independent of an additive quadratic curve, and so on. None of these derivative treatments are independent of the multiplicative changes resulting from scatter and path length changes, but derivatizing spectra minimizes the effect of scatter by emphasizing local changes.

There has been some discussion in the literature concerning what derivative is best. No single derivative gives the best predictions for all constituents and products. This is not surprising since for every calibration for segment-gap derivatized spectra, there exists an equivalent equation for the non-derivatized spectra. Each term in a first derivative calibration becomes two terms for the non-derivatized spectra, and each term in a second derivative calibration becomes three terms for the non-derivatized spectra. Thus, there is nothing special about the calibration itself derived from derivatized spectra. What is special is the sequence of steps followed deriving the calibration. For example, with a large amount of scatter or path length variability, chemical information is masked by the full spectrum scatter or path length effects. All correlations between single wavelength absorbance values and LRM values are small. The wavelength with the highest correlation is usually not one that would have been chosen without the scatter or path length interference. With derivatized data, however, absorbance values at a single wavelength can have very high correlation with LRM values. This provides a better start for step-up or stepwise regression. The same is true for the full spectrum calibration methods. The first factor of a derivatized calibration is usually much more strongly related to the LRM values than the first factor of a non-derivatized calibration. However, consistent with the fact that all calibrations for derivatized spectra can be transformed into an equivalent calibration for non-derivatized spectra, non-derivatized calibration can equal the accuracy of derivatized spectra, but usually at the expense of requiring additional terms or factors.

No single math treatment gives the best prediction for all variables and all products. In general, first and second derivative will perform the best with Detrend (Barnes et al., 1989) by improving prediction accuracy in many cases. At the moment, however, the only way to obtain the lowest prediction error is by trial and error.

Cross validation

Both PCR and PLS must use external validation or cross validation to prevent overfitting. Cross validation is a procedure for obtaining validation errors by partitioning the calibration set into several groups. A calibration is performed for each group, reserving that group for validation and calibrating on the remaining groups, until every sample has been predicted once. The validation errors are combined into a standard error of cross validation (SECV). All that is retained from the cross validation procedure is the number of PCR or PLS factors associated with the minimum SECV. Following cross validation, all samples are used for calibration using the number of factors determined by cross validation.

Cross validation is an efficient use of samples because all samples are used for both calibration and validation. Experience shows that validation with a small external set randomly selected from a calibration set can be quite variable. Changing the validation set changes the validation error and often the recommended number of factors. This also can be seen in cross validation by inspecting the validation errors of each group. In cross validation, however, the result is based on the average of several groups, not a single group.

Another advantage of cross validation is the availability of prediction residuals for outlier identification. High leverage samples (samples with extreme spectra) are fit very well by least squares methods, whether their LRM values are accurate or in error. With cross validation, however, true prediction residuals are obtained, often exhibiting large residuals for samples with extreme spectra. Samples with large cross validation residuals ("T" values above 2.5) are usually omitted, and cross validation performed again. The cycle of omitting samples with large residuals followed by another cross validation should be performed 2 or 3 times.

Cross validation works best for the full spectrum calibration methods, because full spectrum calibration models do not change drastically when a few samples are omitted. Calibration methods based on wavelength selection, however, often select different wavelengths when a few samples are omitted from the calibration set. It does not make sense to average prediction errors from models using different wavelengths.

The stepwise and step-up regression methods can utilize leverage correction. Leverage correction uses a mathematical short-cut that quickly computes what the prediction residual would have been if one sample had been omitted from the calibration. However, that sample is not removed from the wavelength selection process, where most of the overfitting takes place. Thus, leverage correction has limited usefulness.

External validation samples can be obtained in two ways. They can be randomly selected from the calibration samples before calibration is performed, or they can be a separate group of samples. If they are randomly split from the calibration samples, the validation samples can be used to detect overfitting and estimate prediction error. A better estimate of prediction error would be obtained from averaging the results of several random splits. And this is what cross validation does.

If the validation samples were from a separate group from the calibration samples, the validation samples can be used to test for equation robustness, and should not be used to detect overfitting or estimate prediction error. There is never a guarantee of prediction accuracy for samples not represented in the calibration set. For example, a researcher might collect hundreds of samples for calibration and use cross validation to determine the optimal number of factors and give an estimate of prediction error. Colleagues and reviewers, unfamiliar with cross validation, might request an estimate of prediction error on a traditional external validation set. The additional samples collected for this external validation set, however, might differ from the original set in some way, resulting in a higher estimated prediction error than originally found with cross validation. This is not a problem with cross validation. The problem is that the original calibration was not broad enough to accurately predict the new samples, and needs to be expanded.

Although cross validation can prevent overfitting within the calibration set and provide an estimate of prediction accuracy, it cannot assure the accuracy of samples not represented in the calibration set. However, the analyst can have confidence that a new sample is represented in the calibration if the global and neighborhood "H" values are small.

Population Diversity and Calibration Method

Prediction error tends to vary with the diversity of the product. This is because the chemical composition of samples in highly diverse populations differs both quantitatively and qualitatively. Calibrations for quantitative changes are straightforward. Calibrations for qualitative changes are not. Many reference methods measure a well defined chemical entity, such as N X 6.25. Often, these chemical entities do not absorb NIR energy directly. It is the bonds between these chemical entities and other elements that absorb NIR energy. When the chemical entity exists in several different compounds, qualitative differences can exist between the absorption patterns within a population of samples, lowering calibration accuracy. Other reference methods, such as fiber, are not well defined chemically.

The more spectrally diverse the population, the larger the standard deviation of the reference values, and the larger the number of eigenvectors required to describe the population. Much of the variability in a population is due to qualitative differences that are not totally resolved by the calibration process, resulting in higher calibration and prediction errors as population diversity increases.

Using these concepts and mathematical procedures, the papers by Shenk and Westerhaus (1991a, 1991b) showed that the lowest prediction errors were obtained when using the 0.6 standardized "H" limit to select samples for calibration. It is not known if size of neighborhoods, number of samples, or number of terms is most important for the development of a good population.

They also found that narrow populations require smaller neighborhoods than broad populations to ensure an adequate number of samples for calibration. The standardized "H" limits are relatively insensitive to population broadness and yield a more consistent degree of sample elimination across populations than the R^2 method proposed in the USDA Handbook 643.

When they compared the performance of MSR and MPLS regression on spectrally diverse populations, they found that the average improvement across populations and constituents for MPLS over MSR was 18%. This finding was also made by Frank (1987) using Monte-Carlo simulation.

Adding Samples with Special Features

Calibration sample sets must often contain samples with special features. These samples might be the result of different processing methods, an unusual growing season, differing harvesting techniques, contamination, or mixture with other products. Examples of these types of samples would include frost-damaged pasture, improperly fermented silage early in the harvest season, heat-damaged silage, forage samples contaminated with soil, and mixtures of weeds and forage. Samples with these features may not appear in the normal sample collection procedure.

Spectra also may acquire special characteristics during sample preparation. The best approach is to standardize sample processing and presentation for the product. However, small differences in grinding procedures, drying methods, and sample handling inevitably will occur. The calibration set must include samples with these special features. If it is desirable that a calibration support multiple processing methods, samples representing all methods must be added to the calibration set. Special consideration should be given to sample moisture and temperature as the sample is measured. Unless samples are stored in sealed moisture-proof pouches, water and other volatile compounds will equilibrate to the levels in the surrounding atmosphere. Samples stored for a long time will tend to equilibrate to a common DM value. Samples whose spectra exhibit low and high water content should deliberately be added to the calibration set.

Temperature has a major effect on the shape of water bands in forage products. The underlying cause behind the changes in band shape is explained by the fact that the apparent water band is actually a composite of several underlying bands, representing the different forms of water in the sample. As temperature changes, water in the sample changes from one form to another (i.e., there are different amounts of hydrogen bonding), resulting in a different composite band. A recommended temperature for samples during measurement should be established and an effort should be made to deliberately add temperature variability to the scans of calibration samples. This will allow accurate analyses of samples that deviate from the recommended temperature.

Using a Repeatability File in Calibration

A repeatability file is always helpful to minimize systematic differences among instruments and to minimize the effects of sample temperature on analysis variation in routine analysis (Westerhaus, 1990). Temperature is the major factor, resulting in unwanted variation in analysis repeatability across instruments after standardization. Changes in temperature affect the sample and the instrument. Other small but systematic differences in optics and electronics between instruments over time after standardization also may occur. The goal of the repeatability file is to make the calibration equation insensitive to changes in temperature and instruments, but not reduce the accuracy of the calibration.

There are two advantages to the repeatability file approach. First, although derived in a least squares setting, it also can be applied to PLS regression. Since PLS is based on regressions of the individual X's on Y, wavelengths with poor repeatability receive a lower PLS loading. The result is an emphasis on wavelengths with high repeatability. Principal component regression, however, does not benefit from the repeatability file approach. This is because the added noise is modeled along with the meaningful spectral information. In fact, repeatability on the test repeatability scans is actually worse for equations derived with a repeatability file for the first few terms, because more emphasis is placed on wavelengths with high variability.

Another advantage to the repeatability file approach is that many samples can be used and no lab data are required. It becomes easy to form large repeatability files that represent a variety of instruments and temperatures. The scaling by the degrees of freedom ratio keeps the repeatability file from overpowering the calibration samples. By combining spectra from several products, a comprehensive repeatability file can be formed that can be used for a variety of products.

In general, changing temperatures has more impact on the repeatability scans than changing instruments. Differences among instruments are made small by instrument standardization. Temperature effects in the spectra cannot be corrected; they must be ignored. The de-emphasis of spectral regions sensitive to temperature achieved with a repeatability file enables the derivation of NIRS prediction equations that easily transfer to new instruments and temperatures. By including scans from several instruments and temperatures in a repeatability file, it is possible to derive equations that predict accurately and consistently in a network of instruments where new instruments are being added over time.

Single and Multi-Product Calibrations

Traditionally, NIRS calibrations for forage have been limited to single products. Specific calibrations are developed for silage, sun-cured hay, and fresh forage. Silage is sometimes subdivided into grass/legume haylage and small grain silage of wheat (*Triticum aestivum* L.), barley (*Hordeum vulgare* L.), rye (*Secale cereale* L.), and oats (*Avena sativa* L.). These products require different processing prior to NIRS analysis. Silage and fresh forage is dried by microwave or forced-draft oven before grinding with a cyclone mill. Field-cured hay is ground with a cyclone mill. Hay too moist to be ground is first dried with a forced-draft or microwave oven. The classification of forage samples into these different products has been arbitrary, confusing, and often misleading.

It has been our experience that as a population was expanded to include new groups of samples, the calibration error increased. The primary reason for limiting a calibration to a single product was to improve the accuracy of prediction. Methods to establish the boundaries of single-product calibrations have been described previously (Shenk and Westerhaus, 1991a, 1991b). It has also been reported that broad-based equations for forage can be almost as accurate as single-product equations, and multi-product calibrations are common. Hay calibrations are almost always based on both grass and legume samples. The combination of grass and legume is necessitated by the mixture of species in fields used for hay and pasture. While grass calibrations are more accurate on grass samples and legume calibrations are more accurate on legume samples, the most accurate way to analyze the mixtures is to include both products in the calibration.

There are two advantages of developing and using multi-product calibrations. In a laboratory situation, the instrument operator is responsible for correctly classifying a sample and selecting the appropriate equation. Proper identification usually prevents an incorrect analysis. By combining related products into broad categories, selecting the correct calibration should be easier. Furthermore, the cost of obtaining reference laboratory values should be lower because fewer samples are required.

In the first study by Shenk and Westerhaus (1993a), a multi-product forage calibration based on 1746 calibration samples of hay and haylage was compared to single-product calibrations. When tested on 100 validation samples, the multi-product equation performed nearly as well as the single-product calibrations. A multi-product calibration usually requires more samples and more equation terms than a single-product calibration. Hundreds, even thousands, of samples are needed to support a large number of PLS terms without overfitting the data. In this study, the maximum of 16 PLS terms was selected by cross validation for almost every final calibration equation. The calibration error of the multi-product equation was generally smaller than the sampling error from a typical lot of hay or layer in a silo.

In a second study, a multi-product silage calibration was derived from 756 haylage and small grain silage samples and tested on 100 hay samples reserved for validation. The SEP was unacceptable for crude protein, ADF, and DM. Much of the error was bias. Correcting for bias helped, but the remaining error [SEP(C)] remained unacceptable for protein. These results were not unexpected, because there were no hay samples in the multi-product silage calibration.

In an attempt to improve upon these results, two additional series of calibrations were evaluated. Custom hay calibrations were developed from the 50, 100, 166, and 500 sample hay files. Expanded multi-product calibrations were developed from the 756 haylage and small grain silage samples combined with 50, 100, 166, or 500 hay samples. All evaluations were made on the 100 hay samples reserved for validation. Both the custom hay and expanded silage calibrations were more accurate as more hay samples were added to the calibration.

The expanded silage calibration was more accurate when only 50 hay samples were available, but the custom hay calibration was more accurate when 500 hay samples were available. This result is largely a function of the number of samples. Fewer samples are needed to update an existing calibration than to develop a new calibration. Thus, if only a few samples are available, they can best be used by combining them with samples of a similar product. If many samples are available, they should be used to develop a custom single-product calibration. The number of new samples required for a custom calibration to be more accurate than expanded multi-product equations depends on the spectral and chemical diversity of the new samples. For the highly diverse samples in this study, a group of 150 or fewer samples was best utilized by adding them to similar samples and deriving an expanded calibration; a group of more than 150 samples was best utilized by deriving a custom calibration with them.

Global, Local, and Expanded Calibration Equations

A global calibration equation is designed to analyze all reasonable samples of a given product. Global calibrations can be made up of single or multiple products. More precisely, a global calibration might have the capability of being used for 90-95% of all samples of a given product. The first step in creating a global calibration is to define the domain of samples to be covered by the calibration. Specific calibrations based on a small range of samples will typically perform better than broad-based calibrations as long as the samples to be analyzed are represented in the calibration set. However, only a small set of samples can be analyzed by a specific calibration. The goal in deriving a global calibration is to cover as broad a range of samples as possible while maintaining acceptable accuracy.

A local calibration equation is designed to analyze all reasonable samples in a small region of the product domain. These spectra are a subset of the samples within the global calibration library. If the local calibration is developed properly, it has the potential to be more accurate in this limited spectral region of the product domain than the global calibrations. For example, hay is a very complex product consisting of many types of species, maturities, and cuttings. A laboratory may only be analyzing hay samples for a small geographical region consisting of one major species such as alfalfa. By choosing a representative set of alfalfa samples from the region, and locating the spectra in the library with the standardized "H" for closest pairs, a calibration often can be developed that has greater accuracy than the global calibration.

An expanded calibration equation is a new concept. It is an equation derived from a global or local product library that can be updated with spectra and LRM values without having access to the spectra in the product library file. With this feature, equations are no longer closed but are open for expansion whenever it is needed. The PCR scores are stored in a file provided with the calibration. This file is used to reconstruct the product library spectra so that the new spectra can be added for a recalibration. This method of improving calibration equation accuracy is superior to slope and bias adjustments.

Expanding Calibrations for Laboratory Situations

Deriving a calibration for new samples is an expensive and time-consuming process. Often a laboratory analyzing samples with NIRS will receive new samples that are spectrally different from the library samples for that product. MATCH (Shenk and Westerhaus, 1993b) was designed to be useful in this situation. The spectra in the product library most like the new samples are identified and selected. Local calibrations are developed from these selected samples. New samples saved to test the local calibration can be used to expand the calibration for recalibration. Expanded calibrations have been shown to be as accurate as custom calibrations derived from a large number of samples. The MATCH program can successfully identify library samples that are similar to new samples. The monitoring procedure described in USDA Handbook 643 is satisfactory for evaluating calibrations developed from matched samples. If new samples are not well represented in the product library, the local calibration developed from the closest library samples can be expanded with new samples to achieve acceptable accuracy. In the study by Shenk and Westerhaus (1993b), a total of 30 out of 284 samples were needed to produce acceptable prediction accuracy, reducing potential laboratory costs by 89%.

Outlier Tests and Interpretation

Four different outlier tests can be made during routine analysis. The first is the neighborhood "H", second is the global "H", third is a "T" test, and fourth is a root mean square (RMS) test. The global "H" defines where the sample just scanned is relative to the spectral mean of the calibration population. The neighbor "H" determines if the sample has any neighbors in the calibration population. The "T" outlier indicates if the sample analyzed has extreme analytical values relative to the calibration population and the RMS test makes sure multiple subsamples have similar spectra before they are averaged. The outlier test is very useful to the instrument operator in providing confidence in the accuracy of the analysis provided by the NIR instrument.

Accuracy of NIRS Analysis Equations

In general the accuracy (SEP or SECV values) of NIRS equations falls into three categories. Accuracy is estimated as the agreement between the NIRS-predicted values and the laboratory reference method values of samples excluded from the calibration. The category with the highest accuracy includes procedures that measure single chemical entities with high precision. Examples would include crude protein and moisture in most forages. The category with slightly lower accuracy includes laboratory procedures that do not measure a single chemical entity, such as fiber or digestibility measurements for forages. For these constituents, samples with completely different chemical compositions can obtain the same laboratory analysis. The category with the lowest accuracy includes constituents that are present in small quantities or that do not directly absorb energy in the NIR region, such as the minerals in forages. Calibrations for these constituents usually rely on secondary correlations, that is, correlations between the constituent to be predicted and another component of the sample that can be measured by NIRS. Calibrations for the last two categories require that samples to be predicted are represented in the calibration set. NIRS analyses in the third group are no substitute for the laboratory analysis, but can be useful if the alternative is to assume an average value from a feed composition table.

The accuracy of NIRS equations is always of great concern. Although NIRS accuracy is often compared to laboratory reference method precision, the accuracy of NIRS analyses should be evaluated by the usefulness of the analysis. One consideration is the sampling error in obtaining a representative sample. An NIRS to laboratory procedure agreement of 0.5 is not necessary when the sample to sample variation from the same sampling population is 1.0. An NIRS analytical equation can be considered accurate enough when the SEP or SECV is less than the sampling error in the population from which the sample was drawn.

Partial least squares is recommended as the calibration method of choice because the SECV is obtained by predicting every sample in the calibration once. When using MSR, the SEP is obtained from a set of samples chosen randomly from the calibration set. This random set may or may not be appropriate. In fact, unpublished studies in our laboratory indicate that at least 10 random sets of samples must be chosen from a calibration set of 200 samples to get a valid view of the equation SEP.

COLOR ANALYSIS OF FORAGE

The analysis of color has been a long used measurement in the marketing of forages. The agriculture community has a good understanding of this criterion, especially in the marketing of hay. Often the color of hay is the most important criteria used in hay marketing to assess hay quality. The problem is that no quantitative standard or measurement exists to bring this measurement to the market place.

One instrument manufacturer has an instrument capable of making a color as well as a composition analysis. The VIS-NIR instrument is the NIRSystems 6500. This instrument has a silicon detector that can collect the visible as well as the NIR spectrum in the 30 sec of scanning time. This type of instrumentation can bring the color measurement to the marketplace of the future. At present, the instrument does not make the color measurement directly but it is calibrated with a second color instrument. A number of color instruments are available to make the color measurements.

Color is defined to cover the spectral region between 400 and 700 nm. This segment of the electromagnetic spectrum can be divided into 6 regions: region 1 from 400 to 425 is violet, region 2 from 425 to 492 is blue, region 3 from 492 to 575 is green, region 4 from 575 to 585 is yellow, region 5 from 585 to 645 is orange, and region 6 from 645 to 700 is red. These color characteristics are measured with three criteria known as L, A, and B. L is defined as brightness, A is the green-brown axis, and B is the yellow-blue axis. These measurements are made with high quality color instruments and their quantitative values can be used to calibrate the VIS-NIR instrument for these criteria.

One of the problems that will be encountered when the color measurement is provided to the market place will be the lack of standardization. The color values from a color instrument are usually expressed relative to a color standard. Farmers and agricultural advisors will need some simplified method of understanding these values and incorporating them into a marketing system. One simple method would be to devise an index of 1 to 10 for each of these three color criteria. These index values can be transformed into color chips on cards and distributed in the market place. With an instrument to make the quantitative measurement and a simplified visible color indexing system, this measurement can be provided and supported in the marketplace. Table 3 provides information on the accuracy of the color measurement for forage materials. Although not shown in Table 3, meaningful correlations have been obtained for the green/brown measurement even in the spectral range 1300 - 2400.

Table 3. Prediction of color with a VIS-NIR monochromator.

Product	Brightness (L)	Green/brown (A)	Yellow/blue (B)
	— 1-VR[a] —	— 1-VR —	— 1-VR —
Fresh forage	0.94	0.96	0.92
Hay	0.62	0.98	0.81
Haylage	0.90	0.87	0.85
Corn silage	0.70	0.94	0.75
High moisture corn	0.90	0.95	0.94

[a] Proportion of reference method variation explained by cross validation predicted values.

INSTRUMENT NETWORKS

Calibration development using NIR instrumentation for forage products is an involved and tedious undertaking. Therefore, it is advantageous to be able to transfer calibrations from the original or "master" instrument to a "host" instrument with minimal loss in performance. This is no minor problem due to the usual differences among instruments. General guidelines for calibration (or spectral) transfer procedures are described in the following text. As improvements are made in the performance specifications for spectrophotometer, the demands for increased uniformity among instruments increases. The end-user is becoming more aware of instrument design specifications and expects nearly identical performance from each instrument used.

One of the important goals of NIRS analysis has been uniformity of analysis. That is, a sample should obtain the same analysis value from all instruments. The first effort by instrument manufacturers was to accomplish this goal using a slope and bias adjustment applied to analysis values. In 1986 a second method was developed that could be used with instruments generating a spectra. This method adjusted the equation wavelengths and coefficients so that the modified equation could be used on different instruments without slope and bias adjustment to analytical values.

Managing instruments on a network represents a real challenge. First, there is the question of calibration accuracy over a much broader range of samples for a product. And second, there is the need to produce the same analysis on the same sample across instruments, laboratories, and operators. To meet these demands, system managers need additional training and understanding of the technology. They also need software designed to assist them in problem solving, monitoring performance, and management activities.

Standardized Instruments

Even if instruments meet the manufacturers' specifications, their spectra may not be alike enough to allow consistent predictions across instruments without adjustments to the spectra or equation. A standardization procedure has been developed to adjust the spectra of any monochromator, tilting filter, or discrete filter instrument to match that produced on a master instrument.

The first instruments were calibrated individually. To facilitate the sale of pre-calibrated instruments, manufacturers devised methods to improve the consistency of analyses when using a master calibration on multiple instruments. Most manufacturers choose to adjust the coefficients of calibration equations. The shortcoming of this method was that spectra were not the same among instruments. To produce consistent spectra, a method was developed to standardize the spectra of instruments. This allowed equations to be used on any instrument with the same spectral coverage without slope or bias adjustment. Spectral standardization also permitted the sharing of spectra among standardized instruments and across instrument repair.

The general goal of the standardization procedure is to make the repeatability of NIRS analyses across instruments acceptable (Shenk and Westerhaus, 1987; Shenk, 1990, 1991). Acceptability is obtained when the standard error of a difference (SED) across instruments is similar to the repeatability of NIRS analyses of subsamples drawn from the same container of material.

Characterization Standards

A set of standard samples must be selected to characterize the instruments. These samples are chosen to represent the products to be routinely analyzed by the network. For finely ground forage samples analyzed in small rotating sample holders, thirty different products ranging from hay to meat and bone meal are used to characterize the instruments. These samples are placed in sealed sample holders to prevent changes in sample moisture. The samples must be handled with care to prevent changes in particle distribution when standardizing an instrument. Samples and instruments are allowed to equilibrate to a room temperature of 24 ± 0.5 C for 1 h before scanning. Sealed samples are scanned, allowed to cool for 30 min, and scanned again. All duplicate scans are evaluated and must have root mean square corrected for bias [RMS(C)] values less than 150 micro-absorbance units before averaging for standardization.

Monochromator Standardization

All spectra are transformed by a first derivative math treatment. Individual wavelength corrections between master and host instrument are computed by: 1) finding the most highly correlated host instrument wavelength for each master instrument wavelength, 2) fitting a quadratic model to the highest correlation and the correlations of the two neighboring wavelengths, and 3) using the location of the maximum of the quadratic model as the host wavelength that most matched the master wavelength. Finally, a quadratic model is fit to predict the location of the master instrument's wavelengths on the host instrument. Interpolations are performed to shift the host instrument's data according to the quadratic model.

A photometric correction at each shifted wavelength is accomplished by predicting the photometric response of the master instrument from the photometric response of the host instrument using simple linear regression. The photometric response of the host instrument is then adjusted using the regression coefficients. All wavelength and photometric correction factors are stored in a file. This file is used to standardize each spectrum collected by the host instrument.

The intercept ($b0$) in the quadratic equation for wavelength adjustment improved upon the mechanical wavelength alignment performed at the factory. The wavelength slope ($b1$) term improved upon the optical alignment performed at the factory. A quadratic ($b2$) term was included to improve the wavelength alignment across monochromators produced by different manufacturers.

The standardized host instrument produces spectra very similar to those produced by the master instrument. RMS(C) values between spectra from a standardized instrument and the master instrument should be less than 700 micro OD when tested with an independent set of samples.

Simplified Monochromator Standardization

A simplified version of the general procedure has been published (Shenk and Westerhaus, 1991c) for the NIRSystems 4500, 5000, and 6500. First, wavelength alignment and linearization is checked using internal reference materials. Peaks in the visible region are compared with the nominal didymium peaks at 805.23 and 878.52 nm. Peaks in the NIR region are compared with the polystyrene peaks at 1143.63, 1681.27, 2166.40, and 2305.93 nm. Instruments should agree among themselves with an average wavelength error less than 0.20 nm, using new linearization constants (K and Phi values) if necessary.

Next, a representative sample is selected to correct for differences among each instrument's ceramic reference. These differences, if uncorrected, result in large prediction biases. Correction is accomplished by scanning the representative sample on two instruments. The instrument where the calibration equation was developed is designated the "master" instrument and the other instrument is designated the host instrument. The differences between the two scans are computed and stored in a file. Standardization is accomplished by adding these spectral differences to each scan on the host instrument.

This form of instrument standardization works best for spectra similar to the spectrum of the representative sample. Small differences in photometric linearity, detector speed, and wavelength alignment are treated as ceramic differences. These small differences are adequately corrected for samples similar to the representative sample, but remain uncorrected for samples different from the representative sample. The spectrum of the representative sample should therefore be similar to the average spectrum of all samples expected for routine analysis. Using this method, all instruments in the NIRSystems 6500 series will have RMS(C) values less than 500 and will meet the criteria of prediction repeatability outlined earlier.

Filter Instrument Standardization

Filter instruments (discrete or tilting) are standardized using the virtual master instrument (master monochromator spectra transformed by the VMR files) using 2 block PLS (Shenk, 1990). Each log(1/R) value of the virtual master filter instruments is estimated with a linear combination of many log(1/R) values of the host filter instruments. The generated PLS equations are stored in a VRT file for each host instrument. These files are used to standardize the spectra of host instrument(s). When filter instruments are adequately standardized to the master monochromator, RMS(C) values are less than 500 using an independent set of test samples.

Monitoring Network Performance

The monitoring of a network of instruments is similar to monitoring a single instrument. If all instruments are properly standardized in the network, then the monitoring represents a combined statistical evaluation of all instruments.

There are two criteria that can be monitored - consistency of spectra across instruments and consistency of equation performance across instruments. These tests should be routinely performed on all instruments in the network. The spectra consistency test is conducted with a small set of sealed samples that are scanned by all instruments in the network. If an instrument is clearly an outlier, an instrument diagnostic test should be performed. If the instrument fails diagnostics, contact the instrument manufacturer to have it repaired. If the instrument is working properly, restandardize it to bring it in line with the other instruments.

The second test is to evaluate equation performance. If the instruments pass the spectra test described above, they will usually pass the equation test if a repeatability file was used when the equations were developed. Limits are used to identify samples and instruments that are not in agreement.

The two most important aspects of an instrument network are accuracy of the analysis and repeatability of analysis across instruments. Accuracy, as measured by SEC or SEP, should be less than field sampling error. Repeatability across instruments should be less than subsampling error from the container holding the sample.

ROUTINE FORAGE ANALYSIS

Routine forage analysis requires a precalibrated instrument to perform fast and accurate analyses. The instrument can be configured to analyze up to 50 finely ground samples automatically or to analyze undried and unground samples with a sample transport attachment. Undried and unground samples should be analyzed in duplicate, checking the RMS agreement between subsamples and omitting extreme subsample scans. Instrument performance should be checked daily with a check cell. Instrument diagnostics should be performed at least once a month. Calibrations installed on an instrument can either be developed from spectra collected on that instrument or on another instrument. If the calibration spectra were collected on another instrument, the agreement between standardized instruments must be checked before the calibration is used.

The computer software should be set up so that samples with extreme analytical values (high "T" values) and unusual spectra (high global and neighborhood "H" values) are identified. Samples that have unusual spectra (large "H" values) should be saved for reference method analysis. After 10 or more sampleshave been identified in routine analysis, the calibration should be expanded to include these samples.

Many factors can cause NIRS analysis error to exceed the level computed when the calibration was developed. This increase in error can only be identified by routine monitoring of calibration performance. A simple monitoring procedure using confidence limits based on calibration errors has been presented in the 1989 USDA Handbook 643 (Shenk et al., 1989).

Control Limits

Two control limits are used. The first limit detects if a meaningful bias is occurring and the second limit detects if a meaningful increase in unexplained error is occurring. A brief example of the monitoring system is presented below.

Assuming 1) a difference between the NIRS analytical values and reference values in either bias or unexplained error is to be detected with 90% confidence, 2) the calibration set contains at least 100 samples, 3) a bias greater than the SEC and an unexplained error greater than two times the SEC are unacceptable, the following procedure is recommended:

1. Choose 9 samples at random from the group of samples to be analyzed.

2. Obtain analysis values by both the reference and NIRS methods.

3. Calculate bias control limits: $BCL = .6(SEC)$.

4. Calculate unexplained error control limits: $UCL = 1.3(SEC)$.

5. Compute the bias and unexplained error of the analytical values in the test set.

These tests should be made on a regular basis or whenever there is some reason to believe that a calibration may not be performing well. At a minimum, one sample out of every 100 should be randomly selected for reference method analysis after NIRS analysis. The test should be performed after nine monitoring samples have been accumulated.

Actions

If either of these control limits is exceeded, the cause must be identified before corrective action can be taken. Control limits can be exceeded because of changes in the reference method, changes in the instrument, changes in sample processing, changes in sample temperature, and changes in the sample population.

If the laboratory is suspect, the monitoring samples should be analyzed in blind duplicate or the analyses should be verified by another reputable laboratory. Many laboratory problems have been identified because of poor agreement with NIRS analyses.

If an instrument is not working properly or is not standardized properly, the problem should be referred to the manufacturer. Instrument standardization should always be checked after any instrument repair.

Changing sample processing from that used for the calibration samples often results in large prediction biases and small increases in unexplained error. While a bias correction can correct the bias component, only changing the sample processing or recalibration can correct the whole problem. Adjusting the bias only fixes the problem temporarily; another set of nine samples would probably suggest a different bias correction. The problem is in the calibration set and can only be corrected there.

Changes in sample temperature also can result in large prediction biases and small increases in unexplained error. If the sample temperature cannot be maintained at the temperature of the calibration samples when they were scanned, recalibration will be necessary. In this case, a few samples could be scanned at different temperatures and incorporated into the repeatability file without laboratory analysis.

If none of the above problems caused the control limits to be exceeded, the problem is that the calibration is not appropriate for the new samples. Even in this case, the bias will be the primary problem. Adjusting the bias only fixes the problem temporarily; another set of nine samples would probably suggest a different bias correction. The problem is in the calibration set and can only be corrected there.

Recalibration and Equation Expansion

Recalibration is necessary whenever: 1) the instrument changes, 2) samples are believed to be part of the original calibration population, but upon analysis, exhibit many large 'H' values, or 3) when monitoring NIRS analyses using the primary reference method, the test results continue to fall outside the control limits.

Samples for recalibration should be selected on the basis of their spectra. Neighborhood 'H' values should be used to omit samples already represented in the calibration set.

The selected samples should be analyzed by the laboratory reference methods and added to the calibration samples. Recalibration should be performed on the expanded calibration set. The new calibration must also be monitored to ensure that analytical accuracy is maintained. If the product being analyzed is highly variable, the calibration monitoring and recalibration procedure may need to be repeated a number of times. If the spectra of the calibration samples are not available but the calibration is expandable, the new spectra and reference values from the equation monitoring test can be used to expand the calibration to improve prediction accuracy.

When routine forage analysis is conducted with routine instrument diagnostics, computer outlier test for unusual spectra, and sample monitoring with recalibration and equation expansion, the accuracy of NIRS analysis can be maintained. This will be true for NIRS analysis provided by a research service laboratory, public forage testing laboratory, or private laboratory. Participating in a lab certification program, such as National Forage Testing Association program sponsored by The National Hay Association and The American Forage and Grassland Council, can provide additional assurance of calibration accuracy.

STATISTICS AND THEIR USE

Measurement of Centrality

The mean or average value is helpful in NIRS to establish a point of reference for the data. Bias is a special case of the mean, where bias is the difference between two means.

Formula to calculate the mean

$$\text{Mean} = \frac{\Sigma(\text{individuals})}{N}$$

Bias = mean of differences

Measure of Dispersion

The standard deviation (SD) is the primary statistic used in NIRS analysis to measure dispersion. The range also is useful but not quite as descriptive as the SD. The word standard refers to the formula used to calculate the statistic. The formula itself is the agreed-upon standard. Deviation is a measure of dispersion. Therefore, a SD is a measure of dispersion using an agreed upon formula.

Formula to calculate the standard deviation (SD, or s):

$$SD = \frac{\sqrt{[\Sigma(\text{individual} - \text{mean})^2]}}{\sqrt{[N - 1]}}$$

Formula to calculate an average standard deviation:

$$ASD = \frac{\sqrt{\Sigma[\Sigma(\text{replicate} - \text{mean})^2]}}{\sqrt{\Sigma[n - 1]}}$$

where the mean is over the n replicates, and the outer summation signs in the numerator and denominator are over samples.

Measurements of Uncertainty

Measurements of uncertainty are usually referred to in NIRS as error. The general formula used is the standard error of difference (SED). It can be applied to many sets of data: LAB vs LAB, predicted vs predicted, predicted vs LAB, spectra vs spectra. This has led to many different abbreviations and terms to represent the same statistic.

Formula to compute SED:

$$SED = \frac{\sqrt{[\Sigma(D)^2]}}{\sqrt{[N]}}$$

The standard error of differences corrected for bias [SED(C)] is computed as:

$$SED(C) = \frac{\sqrt{[\Sigma(D)^2 - (\Sigma D)^2/N]}}{\sqrt{[N-1]}}$$

Abbreviations used to describe uncertainty of NIRS analyses:

SEC = standard error of calibration

SECV = standard error of cross validation

SEV = standard error of validation

SEP = standard error of performance

RMS = root mean square

Measurements of Co-relationships Between Two Variables

A number of statistics are used to define this relationship in NIRS. The primary statistics are correlation (r) and regression. Regression methods will be presented later. Only the formula for simple linear regression is provided at this point.

Formula to compute correlation (r):

$$r = \frac{\sqrt{[(\Sigma(XY)^2 - \Sigma(X)*\Sigma(Y)/N)/(N-1)]}}{\sqrt{[(s_x)*(s_y)]}}$$

Fraction of explained variance is the square of r (r^2).

Slope (b):

$$b = \frac{(\Sigma(XY)^2 - \Sigma(X)*\Sigma(Y)/N)/(N-1)}{(s_x)*(s_x)}$$

Formula to compute simple linear regression

b = equation above

a = mean(Y) - mean(X)*b

Y = a + b(X) where Y is the NIRS analysis or predicted value and X is the LAB value or NIRS data point at a wavelength.

Outlier Identification

The detection of outliers is very important in NIRS analysis. In general the traditional test methods are used, but their interpretation and use is often from experience rather than hypothesis testing.

Regarding the formula to compute the "T" statistic for a difference between two values, here again these differences may come from LAB vs LAB, predicted vs predicted, predicted vs LAB, or spectra vs spectra.

$$t = \frac{\text{difference}}{\text{SED}} \quad \text{or} \quad \frac{\text{residual}}{\text{SEC}}$$

Statistics for Grouping Samples and Calibration

Statistical Formula for Grouping Samples Based on Spectra

It is well known that NIRS requires separate calibrations for different products to accurately predict reference method analyses. This may be due to the spectra or to the reference methods. In any event, NIRS analyses relate more closely to reference method analyses when confined to a group of similar samples. To be practical, however, each calibration must cover as many different samples as possible and maintain acceptable prediction accuracy. Grouping here means deciding which samples should be used to represent a product. The group must be broad enough to cover most of the examples anticipated for analysis, yet not so broad as to lower accuracy.

Grouping Benefits

1. Improved accuracy through product definition through 1) more accurate NIRS analyses (quantitative) and 2) more accurate product identification (qualitative).

2. Reduced cost of lab analyses.

Grouping Requirements

If NIRS measurements were made at only three wavelengths, the three numbers could be used as coordinates to plot the spectrum as a point in three dimensional space. A group of spectra could be viewed as a swarm of points in three dimensional space. The ideal swarm of samples representing a product would be compact (no outliers), not have any large gaps, and not have unnecessary duplication of samples. Spectra with more than three points can be viewed as points in multidimensional space, with each wavelength contributing a dimension. The two requirements for grouping spectra are: 1) identification and elimination of spectral extremes, and 2) identification and elimination of spectral twins.

Grouping statistics

Identification of spectral extremes and twins requires a measure of distance between two spectra, or between a spectrum and the average spectrum. Either R^2 or "H" can be used.

1. R^2 (spectral similarity)

$$R^2 = \frac{(x'y)(x'y)}{(x'x)(y'y)}$$

2. H (spectral distance)

$$H = x(X'X)^{-1}x'$$

When NIRS measurements are made at many wavelengths, as in scanning instruments, some form of data reduction is required before computing the "H" statistic. Several methods are available to reduce the number of NIR data points.

1. Use meaningful wavelengths only.

2. Use every ith wavelength.

3. Transform with principal components.

Grouping Algorithms

The CENTER algorithm ranks spectra according to distance from the average spectrum or specified spectrum. It uses either R^2 or "H" to compute distance. Two options are available for trimming extreme spectra.

1. Trim a specific number of spectra.

2. Trim spectra beyond a critical distance.

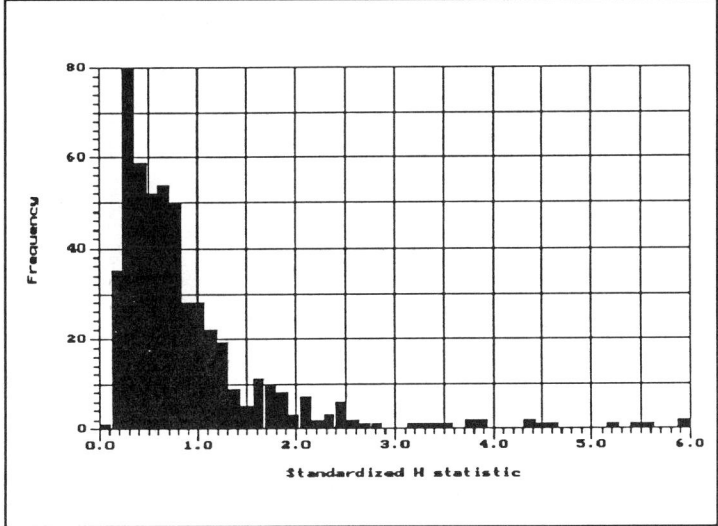

Figure 7 Histogram of standardized H values of 514 hay samples before CENTER.

The SELECT algorithm computes distances between all pairs of spectra. It uses either R^2 or "H" to compute these distances.

1. Identifies "neighbors" for each spectrum (distance less than a specified critical distance).

2. Finds the spectrum with the most "neighbors" and keeps that spectrum and eliminates its "neighbors".

3. Repeats the process with remaining spectra until no spectra are closer than the critical distance.

The SELECT algorithm can automatically vary the critical distance to keep a desired number of spectra.

The SELECT algorithm can be used to "fill in the gaps" in the library of existing spectra.

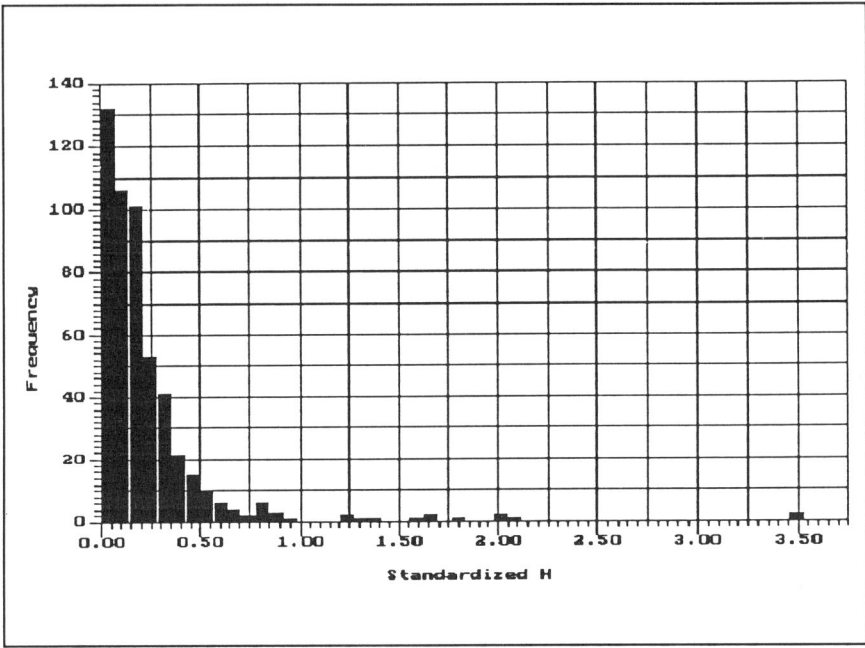

Figure 8 Histogram of standardized H values between closest pairs before SELECT.

Statistical Concepts and Formulae for Calibration

Removing Interferences from the NIR Data

Interferences are defined as spectral absorbencies or characteristics unrelated to the constituent of interest. Typical examples are moisture and particle size. Interferences can be partially removed by

1. Detrending (scanning instruments only). Detrend is a mathematical procedure that scales a spectrum to have a standard deviation of 1.0, then removes the linear and quadratic trends in the spectrum.

2. Derivatives (scanning instruments only). Derivatives (differences) are formed by subtracting nearby data points. A first derivative

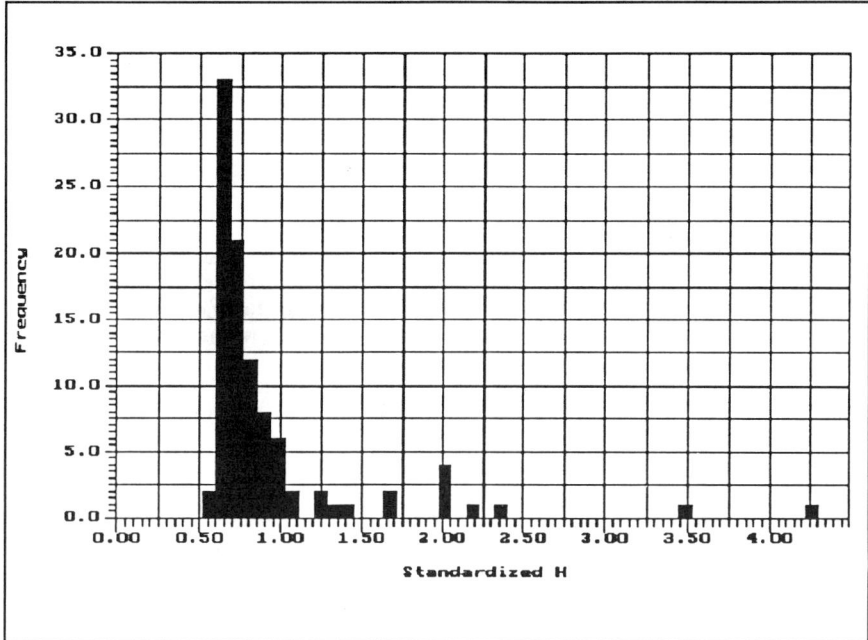

Figure 9 Histogram of standardized H values between closest pairs after SELECT.

with a five point gap is computed as point 1 minus point 6, point 2 minus point 7, and so on. This subtraction cancels the "signal" that is in common between the two points and doubles the "noise". A boxcar smooth is usually used to offset this decrease in the signal to noise ratio. A five point running average smooth is the average of five points.

3. Omitting portions of the spectrum. If certain regions of the spectrum are too noisy or swamped by particle size or another absorber (such as water), wavelengths in those regions can be ignored in the calibration process.

Linear calibration techniques - Linear equations have the form:

$$Y = b_0 + b_1X_1 + \ldots + b_nX_n$$

Three methods are available to generate calibration equations:

1. Stepwise regression. Stepwise regression adds wavelengths to the model one at a time, selecting the ones that most improve the fit of the model. Wavelengths in the model can be eliminated if they become unimportant.

2. PCR regression. Principal components are linear combinations of NIR data that maximize the differences between spectra. Each linear combination is uncorrelated with the previous ones.

Multiplying NIR data by these linear combinations form new NIR variables, which are used in principal component regression.

3. PLS regression. Partial least squares also forms linear combinations of the NIR data. However, PLS uses the reference method values to form new NIR variables that are highly correlated to the reference method.

Number of Terms (Knowing When to Quit)

One of the most difficult decisions in calibration is how many terms are enough. Too few and accuracy is lowered. Too many and overfitting occurs. The calibration is specific to the spectra in the calibration file and performs poorly on other spectra. A "good" number of terms can be determined in 4 ways:

1. A general guideline is to fit one term for every 10 samples but do not use an equation that has a coefficient with an 'F' statistic less than 8.0.

2. Cross validation estimates prediction error by splitting the calibration samples into groups. One group is reserved for validation and the remaining groups are used for calibration. The process is repeated until all groups have been used for validation once.

3. External validation measures the accuracy of the calibration equation on samples not used in the calibration.

4. Equation repeatability measures the consistency of repeated NIRS analyses on the same sample.

Handling Outliers

Outliers are samples (spectra and reference method values) that do not fit the calibration. Some samples are outliers because of mistakes, either in the spectra or in the reference method analysis. Other samples are outliers because they were misclassified and should be analyzed by another equation. Sometimes samples are outliers because the wrong mathematical model (equation) was used to fit the calibration samples. Unless outliers are verified to be free from errors, it is safer to eliminate them. If outliers are free from error, more samples like them should be included in the calibration set. If a new calibration model does not fit the outliers any better, the outliers should be split off to form a separate calibration. Two methods are available for dealing with outliers:

1. Downweighting. Downweighting starts with the usual calibration to obtain the calibration residuals. A second calibration is performed using fewer of the samples with large residuals.

2. Multi-pass outlier elimination of large 'T' and 'H' outliers.

THE POTENTIAL AND FUTURE OF NIRS ANALYSIS

Near-infrared radiation has been around since the beginning of the universe and it will be around long after our civilization has gone. Near-infrared radiation is only a small portion of the electromagnetic spectrum and the electromagnetic spectrum is part of one of the four primary forces of the universe. Given this impressive status, one does not need to wonder if NIR has a future. Its future is part of the ongoing existence of the universe. The potential and future of NIRS analysis is another question.

NIRS analysis only has a 25 year history. Karl Norris began to explore its properties in the mid-1960's. In this short 25 year period, the use of NIRS analysis has gone from the simple measurement of moisture in ground wheat samples using a single filter instrument to the analysis of fresh pasture grass using a highly sophisticated VIS-NIR monochromator with 10 Mb of software on high speed lap-top computers. Certainly we can say that NIRS has kept pace with the rapidly developing electronic industry, and the forage industry has benefitted greatly from its development.

The basic disadvantage with our current use of NIRS is that in the forage industry, it is a secondary analysis method and the calibrations rely entirely on a primary reference method. We still have not figured out how to use the NIRS measurement directly for feeding livestock. This then will be one of the major challenges for the next 25 years. For this reason, we added the first part of this chapter regarding the fundamentals of NIR spectroscopy. By more fully understanding the spectroscopic measurement itself, the knowledge to make use of the information directly in livestock feeding programs will be found. In this past 25 years we have learned much about the spectrum and its properties. We have developed algorithms to control and manage and sort the spectrum, but we still lack the ability to use the NIRS absorption information directly without referring to a reference method.

A second challenge will be to the industry manufacturing NIR instruments. Hopefully, the cost of the instrumentation and software will be reduced so that more people can take advantage of the technology. Secondly, we need the development of robust sensors to use in hay balers and forage choppers and other harvesting equipment to be able to access quality in the real-word environment. This will require miniaturization of the technology into small devices and even hand-held devices for the hay marketing industry. The beauty of NIRS is that these analytical devices are possible and will be able to replace our current forage analytical system based on wet chemistry analysis.

In our opinion, the potential and future of NIRS rest in the hands of the instrumentation industry, the academic research community, and the livestock industry made up of many people including the farmer. NIRS analysis is the electronic wave of the present and will be in the future. It is environmentally safe and its advantages far outweigh its disadvantages.

REFERENCES

Abrams, S.M. 1989. Sampling. p. 22. *In* G.C. Marten, J.S. Shenk, and F.E. Barton,II (ed.) Near infrared reflectance spectroscopy (NIRS): Analysis of forage quality. Agric. Handbook No. 643. USDA- ARS. U.S. Government Printing Office, Washington, DC.

Abrams, S.M., J.S. Shenk, M.O. Westerhaus, and F.E. Barton,II. 1987. Determination of forage quality by near infrared reflectance spectroscopy: Efficacy of broad based calibration equations. J. Dairy Sci. 70:806-813.

Albrecht, K.A., G.C. Marten, J.L. Halgerson, and W.F. Wedin. 1987. Analysis of cell-wall carbohydrates and starch in alfalfa by near infrared reflectance spectroscopy. Crop Sci. 27:586-588.

Aufrere, J., and B. Michalet-Doreau. 1988. Comparison of methods for predicting digestibility of feeds. Animal Feed Sci. Technol. 20:203-218.

Barnes, R.J., M.S. Dhanoa, and S.J. Lister. 1989. Standard normal variate transformation and de-trending of near-infrared diffuse reflectance spectra. Appl. Spectroscop. 43:772-777.

Barton, F.E.,II, and D. Burdick. 1978. Analysis of bermudagrass and other forages by near-infrared reflectance. p. 45-51. In Proc. 8th Res. and Ind. Conf., Coastal Bermudagrass Processors Assoc., Athens, GA.

Ben-Gera, I., and K.H. Norris. 1968. Direct spectrophotometric determination of fat and moisture in meat products. J. Food Sci. 33:64-67.

Blosser, T.H. 1985. High-moisture feedstuffs, including silage. p. 55-57. In G.C. Marten, J. S. Shenk, and F. E. Barton,II (ed.) Agricultural Handbook No. 643. USDA-ARS. U.S. Government Printing Office, Washington, DC.

Blosser, T.H., J.B. Reeves, III, and V.F. Colenbrander. 1986. Near infrared reflectance spectroscopy (NIRS) for predicting chemical composition of undried feedstuff. J. Dairy Sci. 69: 136 (Abstr.).

Clark, D.H., E.E. Cary, and H.F. Maryland. 1989. Analysis of trace elements in forages by near infrared reflectance spectroscopy. Agron. J. 81:91-95.

Clark, D.H., H.F. Maryland, and R.C. Lamb. 1987. Mineral analysis of forages with near infrared reflectance spectroscopy. Agron. J. 79:485-490.

Coelho, M., F.G. Hembry, F.E. Barton,II, and A.M. Saxton. 1988. A comparison of microbial, enzymatic, chemical and near-infrared reflectance spectroscopy. Methods in forage evaluation. Anim. Feed Sci. Technol. 20:219-231.

Eckman, D.D., J.S. Shenk, P.J. Wangsness, and M.O. Westerhaus. 1983. Prediction of sheep responses by near infrared reflectance spectroscopy. J. Dairy Sci. 66:1983-1987.

Frank, I.E. 1987. Intermediate least squares regression method. Chemometrics Intell. Lab. Syst. 1:233-242.

Goering, H.K., and P.J. Van Soest. 1970. Forage fiber analysis (Apparatus, reagents etc.). USDA Agricultural Handbook No. 379, U.S. Government Printing Office, Washington, DC.

Gordy, W., and P.C. Martin. 1939. The infrared absorption of HCl in solution. J. Chem. Phys. 7:99-102.

Harpster, H.W., E. Keck, P.J. Wangsness, S.M. Abrams, and J.S. Shenk. 1982. Predicting the feeding value of hay from infrared spectroscopy and chemical lab analysis. J. Anim. Sci. 55(Suppl. 1):310-311.

Holechek, J.L., J.S. Shenk, M. Vavra, and D. Arthur. 1982. Prediction of forage quality using near infrared reflectance spectroscopy on esophageal fistula samples from cattle on mountain range. J. Anim. Sci. 55:971-975.

Kaye, W. 1954. Near infrared spectroscopy. 1. Spectral identification and analytical application. Spectrochim. Acta 6:257-287.

Martens, H., and T. Naes. Multivariate Calibration. 1989. John Wiley and Co., New York.

McClure, W.F., and R.E. Williamson. 1986. Status of near infrared technology in the tobacco industry. p. 34-35. In Proc. 40th Tobacco Chemists Research Conf., Knoxville, TN.

Nadler, T.K. 1988. Identifying resonances in the near infrared absorption spectrum. M.S. Thesis, The Pennsylvania State University, State College, PA.

Nadler, T.K., S.T. McDaniel, M.O. Westerhaus, and J.S. Shenk. 1989. Deconvolution of near-infrared spectra. Appl. Spectroscop. 43:1354-1358.

Norris, K.H., R.F. Barnes, J.E. Moore, and J.S. Shenk. 1976. Predicting forage quaiity by infrared reflectance spectroscopy. J. Anim. Sci. 43:889-897.

Osborne, B.G., and T. Fearn. 1986. Near infrared spectroscopy in food analysis. John Wiley and Co., New York, p. 35-40.

Shenk, J.S. 1990. Standardizing NIRS instruments. p. 649-654. In R. Biston and N. Bartiaux-Thill (ed.) Proc. 3rd Int. Conf. NIR. Agriculture Research Center Publishing, Gembloux, Belgium.

Shenk, J.S. 1991. Networking and calibration transfer. p. 223-228. In Proc. 4th Int. Conf. NIR. Aberdeen, Scotland.

Shenk, J.S. 1993. Analysis of undried, unground forage with a visible-NIR monochromator. In Proc. XVII International grassland congress. New Zealand - Australia. In press.

Shenk, J.S., and R.F. Barnes. 1977. Current status of infrared reflectance. p. 57-62. In Proc. 34th South. Past. Forage Crop Improve. Conf., Auburn, AL.

Shenk, J.S., and M.R. Hoover. 1976. Infrared reflectance spectro-computer design and application. p. 122-124. In Proc. 7th Technicon Int. Cong., 2, Tarrytown, N.Y.

Shenk, J.S., and M.O. Westerhaus. 1987. Software standardization of NIR instrumentation. 14th Fed. Anal. Chem. Specros. Soc. No. 129.

Shenk, J.S., and M.O. Westerhaus. 1991a. Population definition, sample selection, and calibration procedures for near infrared reflectance spectroscopy. Crop Sci. 31:469-474.

Shenk, J.S., and M.O. Westerhaus. 1991b. Population structuring of near infrared spectra and modified partial least squares regression. Crop Sci. 31:1548-1555.

Shenk, J.S., and M.O. Westerhaus. 1991c. New standardization and calibration procedure for NIRS analytical systems. Crop Sci. 31:1694-1696.

Shenk, J.S., and M.O. Westerhaus. 1992. NIRS analysis of natural agriculture products. p. 235-240. *In* K.I. Hildrum, T. Isaksson, T. Naes, and A. Tandberg (ed.) Near infra-red spectroscopy. Bridging the gap between data analysis and NIR applications. Ellis Horwood, England.

Shenk, J.S., and M.O. Westerhaus. 1993a. Near infrared reflectance analysis with single- and multiproduct calibrations. Crop Sci.33:582-584.

Shenk, J.S., and M.O. Westerhaus. 1993b. Using near infrared reflectance product library files to improve prediction accuracy and reduce calibration costs. Crop Sci. 33:578-581.

Shenk, J.S., M.O. Westerhaus, and S.A. Abrams. 1989. Protocol for NIRS calibration: Monitoring analysis results and recalibration. p. 104-110. *In* G.C. Marten, J.S. Shenk, F.E. Barton,II (ed.) Near infrared reflectance spectroscopy (NIRS): Analysis of forage quality. Agric. Handbook No. 643. USDA-ARS. U.S. Government Printing Office, Washington, DC.

Shenk, J.S., J.J. Workman, Jr., and M.O. Westerhaus. 1992. Application of NIR spectroscopy to agriculture products. p. 383-481. *In* D.A. Burns and E.W. Ciurczak (ed.) Handbook of near-infrared analysis. Marcel Dekker, Inc.,NY.

Ward, R.G., G.S. Smith, J.D. Wallace, N.S. Urguhart, and J.S. Shenk. 1982. Estimates of intake and quality of grazed range forage by near infrared reflectance spectroscopy. J. Anim. Sci. 54:399-402.

Westerhaus, M.O. 1990. Improving repeatability of NIR calibrations across instruments. p. 671-674. *In* R. Biston and N. Bartiaux-Thill (ed.) Proc. 3rd Int. Conf. NIR. Agriculture Research Center Publishing, Gembloux, Belgium.

Westerhaus, M.O.,and J.B. Reeves,III. 1992. NIR calibrations of agricultural products with neural networks. p. 79-84. *In* K.I. Hildrum, T. Isaksson, T. Naes, and A. Tandberg (ed.) Near infra-red spectroscopy. Bridging the gap between data analysis and NIR applications. Ellis Horwood, England.

Whetzel, K.B. 1968. Near-infrared spectroscopy. Appl. Spectroscopy Res. 2:1-67.

Windham, W.R., and F.E. Barton, II. 1991. Moisture analysis in forage by near infrared reflectance spectra: Collaborative study of calibration methodology. J. Assoc. Off. Anal. Chem. 74:324-331.

Windham, W.R., D.R. Mertens, and F.E. Barton, II. 1989. Protocol for NIRS calibration: Sample selection and equation development and validation. p. 96-103. *In* G.C. Marten, J.S. Shenk, and F.E. Barton,II (ed.) Near infrared reflectance spectroscopy (NIRS): Analysis of forage quality. Agric. Handbook No. 643. USDA-ARS. U.S. Government Printing Office, Washington, DC.

Chapter 11

REGULATION OF FORAGE INTAKE

D. R. Mertens

INTRODUCTION

Forage quality can be assessed in terms of the animal performance that is elicited when the forage is offered to the animal. Animal performance is the product of supply, nutrient and energy concentration, intake, digestibility, and metabolism. Typically, forage quality is measured assuming that the forage is freely available and that deficient nutrients other than protein and energy are provided in a supplement. Of the remaining factors affecting forage quality, intake is the most important one affecting animal performance. Reid (1961) listed the amount of forage eaten as the first factor limiting the usefulness of forage-testing schemes. His list also included the effect of concentrates on the amount of forage eaten, degree of selective feeding, and effect of physical form as important information not provided by the forage testing systems used at that time. Unfortunately, intake, substitution rates, selection, and physical form still are not a part of routine forage evaluation systems in the 1990's.

Animal performance depends on the intake of digestible and metabolizable nutrients. Of the variation in digestible dry matter (DDM) or digestible energy (DE) intake among animals and feeds, 60 to 90% is related to differences in intake, whereas only 10 to 40% is related to differences in digestibility (Crampton et al., 1960; Reid, 1961). Intake generally accounts for twice as much variability in DDM intake as does digestibility (Milford, 1960; Ingalls et al., 1965). Milford and Minson (1966) observed that DDM intake is more closely correlated with dry matter intake (DMI) than any other feed or animal characteristic. Similarly, high correlations would occur between DMI and metabolizable (ME) or net energy (NE) intake, and between DMI and overall animal performance when forages are fed.

The primary importance of intake in assessing forage quality does not mean that digestibility or metabolizability of nutrients are invariant or unimportant. Metabolizability can vary depending on the amounts and ratios of absorbed nutrients, the individual and interacting biochemical pathways of nutrient conversion and the type of animal production (MacRae et al., 1985; Blaxter, 1989). However, differences in the efficiency of converting DE to ME are small compared to differences in intake and digestibility, and the metabolic conversion of DE to ME or NE is typically considered to be relatively constant within type of production, such as maintenance, growth, pregnancy, or lactation (Moe and Tyrrell, 1976; NRC, 1989).

Digestibility of forages is more variable than metabolizability, and feces typically are the greatest loss of ingested nutrients and energy. Extensive research has been devoted to measuring digestibility and relating it to feed characteristics because digestibility can be accurately measured with relative ease compared to DMI. Although intake is more important than digestibility in assessing forage quality, progress in understanding the basic factors that affect intake has been hampered by our inability to measure it accurately and to separate the influences of animal and diet on intake. Thus, we routinely use availability information (DE, ME, and NE)

USDA-Agricultural Research Service, US Dairy Forage Research Center, Madison, WI 53706 to economically evaluate feeds and formulate rations, but only the INRA (1989) system and NDF-Energy Intake System (Mertens, 1985; 1987) attempt to use variation in the intake potential of feeds when formulating rations for ruminants.

One of the greatest difficulties in predicting forage quality or animal responses to diets is related to intake. For evaluation, it is most desirable to have intake as a response to the forage; yet to predict animal performance, it is easiest to have intake as an input. Does intake determine animal performance (intake as an input) or does animal performance determine intake (intake as a response)? This dichotomy reflects part of the difficulty in measuring and tabulating the intake potential of forages. It also indicates the very real difference in predicting intake as a response to a diet of known ingredient and chemical composition, compared to estimating intake to formulate a ration to meet an optimal or target level of animal performance. In the first instance, the diet composition is the known input, and animal performance and intake are unknown responses. In the second instance, animal performance and intake are the known (assumed) inputs and diet composition is the unknown variable to be solved. When mixed rations are fed, animal nutritionists can, and probably should, formulate them to optimize performance or profit using intake as a known input. However, when forage quality is measured, intake is a response.

When intake is a response, a second dichotomy is raised. Is the intake response limited by the potential of the animal or by characteristics of the diet? Much of the confusion in the literature about the mechanisms by which forage characteristics influence intake can be related to lack of recognition that animal characteristics also can limit intake within or among experiments. To clarify the relationships between intrinsic forage characteristics and intake, it is important to develop mechanistic conceptual models of intake regulation in ruminants that are based on biological principles and defensible theoretical concepts. These models can serve as a basis for identifying feeding situations in which forage or animal characteristics limit intake. The objectives of this chapter are to: 1) review the mechanisms of intake regulation, 2) present integrated models of intake concepts that include both animal and dietary factors, and 3) discuss variation in intake and factors affecting the measurement of forage intake potential.

DEFINITIONS

In discussing concepts of intake regulation, several terms are used which sometimes are defined differently by various authors. The following definitions will be used in this chapter:

Appetite. The drive to consume feed. In this chapter, appetite will be defined as a response to meeting long term nutrient requirements. Forbes (1986) defined appetite as the drive to eat a specific nutrient.

Eating pattern. The amount and spacing of feed consumption during short-term intake regulation.

Eating rate. The amount of dry matter (DM) consumed per unit of time.

Hunger. The urge or desire to eat in response to short-term stimuli from feeding centers in the brain.

INRA (Institut National de la Recherche Agronomique) *ingestibility.* A feed characteristic that represents the maximum quantity that is eaten when fed to a reference animal under standardized *ad libitum* feeding conditions.

INRA intake capacity. An animal characteristic that represents the quantity of feed that the animal consumes voluntarily when fed a reference feed under standardized *ad libitum* feeding conditions.

INRA fill value. The ratio of the DMI of the reference forage divided by the

DMI of the test forage for each reference animal. Fill values of a forage differ for each reference animal; sheep fill units (SFU) = 75/(forage DMI for reference sheep), cattle fill units (CFU) = 95/(forage DMI for reference cattle), and lactating cow fill units (LFU) = 140/(forage DMI for reference lactating cows).

INRA reference forage. A pasture grass, cut at the grazing stage of first growth, containing 150 g crude protein (CP) and 250 g crude fiber per kg DM with an organic matter (OM) digestibility of 770 g kg^{-1}. The voluntary DMI of the reference forage is 75 g DM kg^{-1} BW$^{0.75}$ for reference sheep, 95 g DM kg^{-1} BW$^{0.75}$ for reference cattle, and 140 g DM kg^{-1} BW$^{0.75}$ for reference lactating cows.

INRA reference cattle. Friesian heifers, 16 to 18 months old, with an average body weight (BW) of 400 kg (BW range of 350 to 450 kg).

INRA reference lactating cow. An adult lactating dairy cow producing 25 kg of 4% fat-corrected milk per day in the fourth month of lactation, with a BW of 600 kg.

INRA reference sheep. A 1.5 to 4 year old Texel wether with an average weight of 60 kg (BW range of 40 to 75 kg).

INRA substitution ratio. Proportional reduction in forage intake (kg d^{-1}) per unit of concentrate intake (kg d^{-1}) when a forage diet is supplemented with concentrates. Typically, total DMI is increased when concentrates are added to forage diets, but forage intake is reduced.

Intake. The absolute amount of DM consumed per unit of time. Because it has the units of time, intake is a rate, not a pool. Traditionally, intake is measured over a period of 5 to 10 d and expressed as daily amounts consumed per unit of BW.

Intake, relative. Ratio of the intake of a test feed to that of a reference feed when fed to similar animals. To be most useful, relative intake should adjust for appetite differences among animals by comparing the intake of the test forage to the intake of a diet or high-quality forage that measures the long-term intake potential of the test animal.

Intake constraint. The daily amount of neutral detergent fiber (NDF) the animal consumes voluntarily when achieving maximal performance (maintenance + tissue balance + production). Intake constraint differs from INRA intake capacity in that the former is constrained to meet a variable target energy demand whereas the latter is defined for a specific energy demand of a particular reference animal.

Intake potential of the animal. Intake by the animal when energy demand of the animal and not the filling effects of the diet limit DMI. This differs from the *potential intake* of Forbes (1986) which he defines as the weight of food required to fulfill all of the animal's nutrient requirements.

Intake potential of the forage. Intake of a forage when the filling effects of the forage and not animal characteristics limit DMI.

Intake regulation, long-term. The mechanisms that regulate intake over periods of weeks or longer. If feed energy density and supply are not limiting, non-producing, mature animals usually balance energy intake with energy output and maintain a constant BW; whereas producing animals will balance energy intake with the energy demands corresponding to their productive potential.

Intake regulation, short-term. The events and stimuli that regulate meal eating patterns within days or a week.

Maximum DMI. The daily amount of feed eaten when the energy density of the forage or diet is the minimum allowed to meet the target production potential (maintenance + tissue balance + production) of the animal.

Palatability. The characteristic of a feed indicating its acceptability, usually associated with the gustatory, olfactory, or visual senses. Palatability affects the preference for a feed when several are available and the rate of eating and intake when a single feed is offered.

Preference. Relative acceptability of a feed when given the choice among two or more feeds that are available in a cafeteria-style feeding situation. Preference is a more specific indication of palatability that affects acceptability among feeds, but does not measure intake modification when no choice among feeds is allowed (a single feed is fed).

Refusal ratio. The fraction of the amount of feed offered in a stall-feeding situation that is not consumed.

Ruminal contents. Amount of wet material, DM, or volume in the rumen measured at any unspecified time when the animal is fed a diet not documented to limit intake by fill.

Ruminal fill. Volume of ruminal contents when measured immediately after the cessation of a meal when the animal is provided a diet that does not meet its energy demand and intake is limited by the bulk of the diet.

Satiety. The condition of having the urge to eat satisfied. The satiety signal from the brain stops a short-term eating interval or meal.

Selection. Specifically defined to indicate preferential consumption among feed subcomponents, such as leaves vs stems or immature plant tops vs mature plant bases.

THEORIES OF INTAKE REGULATION

To delineate those aspects of intake that are inherent characteristics of forage quality as opposed to those that are functions of animal characteristics, it is pertinent to review the mechanisms by which intake is regulated in animals. The focus of discussion in this chapter will be on the non-forage factors that influence or regulate intake for the purpose of providing background information that will be useful in interpreting intake as a component of forage quality. Intake is regulated in both the short and long term. Mechanisms for the short term stimulation of meal initiation and cessation within a day are quite different from the long term regulation of BW and performance over the lifetime of the animal. Failure to be clear about which type of intake regulation is being investigated or discussed often leads to unnecessary confusion and conflict.

Factors affecting intake, and the stimuli and mechanisms that regulate it, are incompletely known as indicated by the diversity of information in the reviews of Conrad (1966), Balch and Campling (1969), Baile and Meyer (1970), Baumgardt (1970), Campling (1970), Jones (1972), Baile and Forbes (1974), Journet and Remond (1976), Bines (1979), Waldo (1986), Grovum (1987), NRC (1987), and Owens et al. (1991). Although models that predict daily intake by summing meal events (Forbes, 1977a) may ultimately be most useful in studying intake mechanisms, concepts based on mechanisms of long term intake regulation can provide the fundamental basis for understanding biological principles affecting the intake of forages. Systems control theory can be used to study intake regulation and provide a conceptual model for assessing forage quality. In its simplest terms, control theory is used to analyze systems in which output controls input by means of a feedback mechanism.

Both neural (Bell, 1971) and hormonal (Bray, 1974; Booth et al., 1976; de Jong, 1986) control systems ensure that food intake equals energy output in mature animals so they maintain a relatively stable BW when food quality and supply are adequate. When energy density of the diet is altered, rats (Adolph, 1947) and ruminants (Dinius and Baumgardt, 1970) adjust intake to balance caloric input with output. Similarly, when animals are placed in cold environments (Stevenson, 1954) or are shorn (Minson and Ternouth, 1971), thereby increasing their energy demand, their intake increases in an attempt to match their increased energy expenditure. These observations suggest that animals monitor their energy expenditures or deficits and respond by consuming the energy that is needed.

Initially, there was scientific debate about which physiological need was regulated. The current consensus, based on numerous studies, is that energy intake is regulated rather than the intake of protein, fat, or carbohydrate. Evidence that the supply of readily metabolizable energy regulated intake was obtained by Booth (1972) who observed that the satiating effects of various nutrients were related to their energy yield and not their convertibility to glucose. The central role of energy also is consistent with the observation that excess energy cannot be easily dissipated; therefore, its intake must be regulated to balance requirements and obey the First Law of Thermodynamics (conservation of energy). Recent studies indicate that animals have some ability to dissipate excess energy intake by increasing basal metabolism via futile (substrate) cycles (Newsholme, 1985); however, the ability to excrete excess energy intake is limited and the major control of energy balance is through intake regulation. Conversely, excess protein can be catabolized easily and excreted in the urine, thereby negating the need to regulate its intake. The concept that protein intake is not regulated does not mean that protein concentration in feeds does not affect intake, only that the effect is indirect. This will be discussed in later sections.

Although animals regulate energy intake, the control system is not simple and direct. Animals have no mechanism for measuring energy per se, and caloric intake is not metered as it is ingested (Brobeck, 1960). Rather, daily intake is adjusted indirectly to maintain energy balance. If intake were regulated by a simple feedback mechanism, the delay between the intake of energy and its availability to tissues in the body would result in excess intake. Apparently, animals terminate meals before energy is digested and metabolized based on signals other than energy density. In the short term, intake may be greater or less than current energy expenditure, but in the long term, intake will balance expenditures and BW of mature animals is maintained when the quality and supply of the diet is adequate. Although some species may become fat at maturity when provided an adequate diet, they nevertheless maintain a stable, fatty body condition and do not continue to gain weight indefinitely.

Intake regulation is complicated because the body can store and recover energy in glycogen and fat reserves. Although meals are periodic, the need for energy by the body is continuous. Energy stores provide a buffer that can accommodate both short and long term perturbations in intake. Very poor correlations between energy intake and expenditure can occur using daily balances, whereas weekly or bi-weekly balances show good correlations between energy input and output (Edholm et al., 1955). Another factor indicating the complexity of short-term regulation of intake is related to psychogenic modulation. This will be discussed in more detail in a later section, but it is important to indicate at this point that meals can be initiated or postponed based on social interactions, management conditions, and feed characteristics.

The nature of the signal(s) that regulate intake is a major research issue that has not been resolved (Mogenson and Calaresu, 1978). Changes in glucose or other metabolites (chemostatic control), changes in body temperature (thermostatic control), and changes in body fat (lipostatic control) are the most commonly accepted mechanisms by which short term intake is regulated. In addition, changes in amino acids (aminostatic control), the modulating effects of taste (palatability or gustatory control), habit, experience, social and emotional factors, and gastric distension (fill limitation) have been proposed as mechanisms controlling intake. Some scientists accept the view that intake is under the control of multiple factors; others contend that a multifactor mechanism is unnecessarily complex and may simply reflect our lack of understanding of basic mechanisms.

Some of the confusion about intake regulation may be related to the differences in short and long term mechanisms. Each mechanism may have different signals and control systems. Short term intake may be controlled by glucostatic (or thermostatic, aminostatic, or psychogenic) control while long term intake may be regulated by lipostatic control (or fill limitation) (Mayer and Thomas, 1967). Multiple factors probably are involved in the satiety signals that terminate meals, but most researchers have concluded that relatively simple models can be used to describe long term intake regulation.

A set point for body fat has been proposed to explain the constant BW of mature animals and the normal growth rate of immature animals when fed a diet that is adequate in energy, protein, vitamins, and minerals (Baumgardt, 1970; Cabanac et al., 1971). No control center or stimuli has been conclusively associated with a set point, and the concept may be useful only in describing the outcome of intake regulation rather than in defining the mechanism of long term intake control. The concept of a set point is based on a static, rather than a dynamic, conceptual model of intake regulation.

Analysis of dynamic control systems indicates that equilibria, steady-states, or "set points" can occur from interactions among dynamic processes. These set points can be varied by changing any one of the many components in a control system. Variability is inconsistent with the classical definition of "set point" and Russek (1978) introduced the concept of "defended body weight" to indicate the variable BW animals will attempt to maintain in a specific situation. As Booth et al. (1976) indicated, a dynamic system can achieve stability of fat stores, which is mathematically equivalent to a set point, without the use of a set point control function that would require: 1) a stimulus from the fat stores, 2) a receptor for the stimulus, 3) a system for generating a precise reference fat store value, and 4) a comparator mechanism to compute the error between current state and the reference set point.

Although the concept of set point may not be useful or necessary for explaining the mechanisms of intake regulation, it can be useful in describing the steady state result of a dynamic intake control system. The steady state solution of a dynamic model is static, and static models offer advantages in terms of describing models algebraically and solving them analytically. Because we formulate rations or select forages to meet the long term needs of animals, factors associated with long term rather than short term intake regulation mechanisms are related more directly to forage quality. Although debate about intake regulation continues, apparently two or three mechanisms control long term intake in animals. These mechanisms will be discussed in detail.

Physiological Regulation

The intake potential of an animal depends on its species, sex, physiological state (maintenance, growth, pregnancy, and lactation), size, body shape, and health. In addition, certain environmental factors such as ambient temperature and photoperiod (Ingvartsen et al., 1992), as well as management treatments such as exogenous hormones or growth promoters, can directly influence an animal's energy demand and intake potential. Long term intake regulation involves mechanisms that control average daily intake over extended periods of time during which BW equilibrium is attained and maintained. Intake would be controlled during homeostasis to maintain long term physiological equilibrium. During homeorhesis, intake would be coordinated with the control of metabolism to support the changes in intake associated with physiological states, such as lactation or pregnancy (Bauman and Currie, 1980). Monteiro (1972) developed a closed-loop system model that assumes that intake is proportional to the energy demand associated with milk production. His model has a delay in response between intake and energy demand that could be interpreted as a homeorhetic mechanism associated with lactation.

When animals are fed high energy rations that are palatable, low in fill, and readily digested, intake is regulated to meet the energy demands of the animal, unless the diet is fermented too rapidly and digestive disorders occur. Physiological regulation of intake can be interpreted to mean that the DMI of the animal times the energy concentration in diet DM will equal the animal's energy demand. This mechanism of intake regulation can be described easily by a simple algebraic equation that can be rearranged to solve for the energy concentration or intake needed to meet a specific animal requirement or potential:

$I_e \times E = R$ Equation 1

$E = R/I_e$ Equation 2

$I_e = R/E$ Equation 3

where I_e is intake (kg d^{-1}) expected when energy demand is regulating intake, E is the energy concentration of the diet (Mcal kg^{-1}), and R is the animal's energy requirement or output (Mcal d^{-1}).

Equation 3 indicates that intake is a positive, linear function of the animal's energy requirement; therefore, intake will increase with increasing energy demand by the animal. Equation 3 also indicates that intake is a reciprocal function of the feed characteristic (available energy concentration). As energy concentration in the diet increases, intake will decline in a curvilinear manner. Conversely, if energy concentration of the diet decreases, intake must increase to meet energy demand. However, there is a maximum limit to intake by the animal. It is important to recognize that intake is not the only response in equation 1 that can be varied by the animal. If energy concentration is too low and intake cannot be adjusted to accommodate a target level of production potential, the animal has the ability to reduce energy output by reducing productivity or increasing the use of body reserves. Thus, the animal effectively changes its output (R) to match allowable energy input. This dichotomy, that the animal can vary either energy intake or energy output to achieve equilibrium, indicates the problem in using this theory of intake regulation alone to predict intake, evaluate forages, or formulate rations.

Equations 2 and 3 illustrate, mathematically, the dual nature of intake as it is used to formulate rations and (or) evaluate forage quality. In ration formulation, intake is an input that is assumed to be known and equation 2 is used to define the energy concentration needed in the ration. If intake is assumed to be lower than maximal, the energy concentration needed in the diet will be calculated to be greater than the minimum necessary to meet energy requirements. To no one's surprise, ration formulation systems that underestimate intake apparently work because they are self-fulfilling prophesies. Low estimates of intake result in high estimates of energy concentration needed in the diet which result in low intakes that match those used to formulate the ration. Caution should be exercised in the adoption of intake prediction equations to be used to estimate forage quality or formulate rations. If the database used to generate them contains diets with higher concentrations of energy than necessary to meet the animal's requirements, intakes predicted by these equations will be lower than the maximum that can be achieved by animals. These low estimates of intake can be used to formulate rations, but the rations may not be optimal and will underestimate the potential contribution that forages can make to mixed rations because they overestimate the levels of concentrates needed in rations.

Evaluation of forage quality is dependent on an estimate of intake potential of the forage because intake plays such a critical role in determining animal performance. Based on the concept that physiological energy demand regulates intake, equation 3 could be used to estimate the intake potential of rations that are high in energy in relation to animal requirements. Equation 3 illustrates two potential difficulties in quantifying the intake potential of high quality forages. First, for high quality diets, the animal's requirements and not characteristics of feed determine intake. Thus, intakes measured under these conditions do not measure forage or diet quality. Unless the energy demand of the animal is high, the true intake potential of the forage cannot be measured. Second, if the animal's demand exceeds the intake potential of the forage, something other than energy concentration of the feed is limiting intake.

When available energy concentration of the forage or ration is low relative to animal requirements, equation 3 will overestimate intake potential because it assumes that the animal's energy requirement (R) will be met. Under actual feeding conditions, the animal will reduce performance or lose weight to attain an equilibrium between energy intake and output when consuming low quality diets. Measuring intake when animals cannot meet their energy requirements will result in biased estimates of intake potential of feeds that are not useful in formulating rations to meet specific levels of animal production. Although animals will attempt to regulate intake to meet energy demand, intake must be limited by factors other than energy concentration of the diet when the animal cannot consume enough feed to meet its energy needs.

Physical Limitation

When animals are fed diets that are palatable, yet high in bulk (fill) and low in available energy concentration, intake is limited by some restriction of capacity in the digestive tract (Balch and Campling, 1962; Campling, 1970; Bines, 1971; Baile and Forbes, 1974). These diets result in intakes of energy that cannot meet the animal's potential demand and the animal reduces performance or loses weight to accommodate the limits of the diet. Lehman (1941) postulated that ballast associated with indigestible OM residues was a factor limiting intake. Conrad et al. (1964) and Owens et al. (1991) followed similar logic when suggesting that there is an upper limit on the amount of fecal OM that can be excreted per day. They hypothesized that intake was restricted by the capacity of the intestines to pass undigested residues. However, Mertens (1973) concluded that the capacity of the intestines probably does not limit intake based on the variability in fecal output and the increases in intake observed when pelleted diets were fed. When low quality, high fiber forages are pelleted, intake often increases, typically with decreases in digestibility. Thus, fecal output increases when pelleted feeds are fed which would be impossible if fecal output was limiting intake of unpelleted forage. Thus, capacity of the intestines probably is not the factor limiting the intake of low quality, high fiber forages. Grovum and Phillips (1978) confirmed the conclusion logically derived by Mertens (1973) by infusing methylcellulose in the intestines of sheep and observing that intake was not inhibited when fecal output was significantly increased.

Physical distension of the reticulorumen generally has been accepted as the major factor limiting intake of many forages and high fiber diets (Balch and Campling, 1962; Baile and Forbes, 1974). In long term regulation, stomach capacity is modified (within limits) by hypertrophy of organs or reduced constrictions associated with internal adipose deposits to achieve a balance between distension stimuli and animal performance. The level of distension required to elicit satiety probably varies with the physiological state (performance potential) of the animal. In short term regulation of intake, stretch receptors in the reticulorumen provide the stimulus to the brain satiety center that triggers the end of a meal. Some ambiguity in the role of distension in intake regulation can be expected because brain centers which control intake integrate a variety of inhibitory or stimulatory inputs (Forbes, 1986).

The physiological stimuli that result in physical control of intake provides a conceptual basis for defining "fill". Too often, fill is used to describe any or all measurements of the weight of ruminal contents. However, the term "fill" is useful in discussions of intake regulation only when it is used to indicate the occupied volume of the rumen when intake is controlled by distension. If stretch receptors are involved, it is contended that volume rather than weight of ruminal contents should be used to measure "fill". Fill can only be measured when the rumen is full, immediately after the end of a meal in which intake is limited by bulk. All other measurements of ruminal contents should be referred to as "ruminal contents" and not as "ruminal fill". Perhaps "ruminal capacity" would be a better term than "ruminal fill" because it implies an upper limit to occupied ruminal volume. However, "fill" has been used traditionally to describe ruminal contents when distension of bulky diets limits intake. Some confusion occurs in the research literature because fill has been used to describe ruminal contents when high energy, low fill diets were fed. It is inappropriate to call these ruminal contents "fill" because this implies that the limit to intake in these instances was due to the filling effect of the diet rather than the energy demand of the animal.

The relationship between ruminal contents and intake of feeds was shown by Blaxter et al. (1961) who observed that sheep offered poor, medium, or good hay had similar amounts of DM in the reticulorumen. Campling and Balch (1961) also provided evidence that mass in the rumen affects intake. Removing swallowed hay from the rumen increased the eating time of cows whereas adding digesta to the rumen during a meal decreased intake. Weston (1966) inserted coarsely ground roughage, sawdust, or finely ground polyvinyl chloride into rumens of sheep offered chopped forages. Intake decreased in relation to the amount of materials given intraruminally. The observation that ruminants eat to a constant ruminal fill (capacity) also was confirmed by the work of Ulyatt et al. (1967) and Thornton and Minson (1972).

The fill limitation concept of intake regulation indicates that when animals are provided an adequate supply of palatable rations that are high in fill, intake is limited by the intake capacity or constraint of the animal. Physical limitations to intake can be interpreted to mean that the intake of the animal times the diet's filling effect equals the animal's intake constraint. This mechanism of intake limitation can be easily described by a simple algebraic equation that can be rearranged to solve for the filling effect or intake allowed by a specific animal intake constraint:

$I_f \times F = C$ Equation 4
$F = C/I_f$ Equation 5
$I_f = C/F$ Equation 6

where I_f is intake (kg d^{-1}) expected when fill is limiting intake, F is the volume of the filling effect (L kg^{-1}) of the diet, and C is the animal's intake capacity or constraint (L d^{-1}).

Equation 6 indicates that intake is a linear function of the intake constraint of the animal and will increase as the animal's intake constraint increases. Equation 6 also indicates the potential difficulty facing an animal that has access to a diet that is high in fill. Because intake is a reciprocal function of the feed's filling effect, it will decrease in a curvilinear manner as the fill of the forage or diet increases.

However, intake can only go so low and allow the animal to obtain enough energy to survive or attain its performance potential. Equation 4 indicates that if appetite is large due to a high energy demand, the animal can accommodate a diet high in fill by increasing its intake constraint (C).

The possibility that the animal can vary either intake or capacity in response to a diet of a given filling effect illustrates the confusion associated with defining C. Although the animal can increase its constraint, it is obvious there is a maximum beyond which the gut cannot stretch and a passage rate it cannot exceed. This maximum intake capacity when animals are fed forages or rations that are so high in fill that performance or even maintenance of life cannot be sustained is of little importance in the practical evaluation of forage quality. It seems more logical to define C in relation to the intake that is needed to meet a target requirement. This capacity is more appropriately termed an intake constraint because the upper limit of intake capacity that is acceptable is constrained by energy requirements needed to meet the animal's performance potential. When energy requirements are not met, but intake capacity is not at its absolute maximum, an ingestive capacity of the animal is being measured that is between the intake constraint and the absolute maximum intake capacity of the animal.

Equations 5 and 6 indicate the dual nature of intake in formulating rations or estimating forage quality. In ration formulation, intake is assumed to be known and equation 5 can be used to calculate the filling effect of the ration that is allowed under the constraint of a given production target. If intake is estimated to be too low, the acceptable filling effect of the diet will be estimated to be too high. This will result in a low intake that does not meet the animal's energy needs. The safest strategy when formulating rations is to use equation 2 with a slightly low estimate of intake because it will at least guarantee that the energy concentration in the diet will be adequate to meet animal requirements. Conversely, using only equation 5 to formulate rations will not guarantee that the animal's energy needs are met but only indicates that its rumen will be full.

Equation 6 is probably the most important equation for the prediction of intake as a component of forage quality. When forages are offered as the sole diet and animal performance potential is high, it is often the filling effect of the forage that limits intake. When the filling effect of a forage or diet is high, equation 6 will adequately describe the intake potential of the feed regardless of whether the animal has a high or low energy demand. Although animals with higher energy demand may try to accommodate a high fill diet by increasing C, the filling effect of the diet will still be the factor limiting intake. However, when the filling effect of the forage is low, such as for high quality legumes, the energy demand of the animal used to evaluate forage quality can have a dramatic effect on the measurement of the intake potential of the forage. The intake potential (maximum intake) of a forage can be measured only when the animal's energy demand is high enough to ensure that the animal's energy demand is not limiting intake. This important criteria for measuring forage quality will be illustrated more thoroughly in the section on measuring forage intake potential.

Psychogenic Modulation

In humans and other animal species, taste, smell, texture, and visual appeal can affect both short and long term intake. In addition, emotional states, social interactions, and learning can modify the intake of foods. Mertens (1985) postulated that these same or similar factors affect the feed intake of animals and suggested that they be aggregated into a class of psychogenic modifiers or modulators of intake. The psychogenic regulation of food intake involves the animal's behavioral response to inhibitory or stimulatory factors in the feed or feeding environment that are not related to the feed's energy value or filling effect. Baumont et al. (1989) observed that the duration and rate of eating of forages were influenced by behavior and palatability.

The most commonly recognized feed characteristic that impacts psychogenic modulation of feed intake is palatability. Differences in forage palatability have been reported (Minson and Bray, 1986; Burns et al., 1988; Black et al., 1989). The definition of palatability used in this chapter is more broad than that used by Marten (1970) in his extensive review of the subject at the first National Conference on Forage Quality Evaluation and Utilization. He offered the following conceptual definition of Relative Forage Palatability: "A plant characteristic(s) eliciting a proportional choice among two or more forages conditioned by plant, animal, and environmental factors which stimulate a selective intake response by the animal; this characteristic(s) may also be described in terms of acceptability, preference, selective grazing, and relish conditioned by sensory impulse, and while it may influence voluntary intake, it is never synonymous with voluntary intake properly measured." His definition of palatability is more restrictive than that used in this chapter and is closer to my definition of preference. In this discussion, palatability will include any feed characteristic that stimulates or inhibits the intake of a feed whether it is fed alone or is a choice among alternative feeds.

The major reason that palatability has been defined in terms of preference among feeds is because intake differences among freely chosen feeds can be used to quantify preference (as opposed to palatability). When two feeds are provided to animals simultaneously, the amount of each feed consumed can be a quantitative measure of preference. Unfortunately, this measure of preference may not be related to any modulation of intake that occurs due to palatability when only one feed is offered. Sometimes we scientists define accurately what we can measure rather than measure what is needed for understanding. There is a need to define palatability and other psychogenic modulations of intake in a way that can be measured and investigated.

The psychogenic modulation of intake can be interpreted to mean that the expected or predicted intake potential based on the physiological or physical mechanisms is modified by some proportional factor. This mechanism of intake modulation can be described by a simple algebraic equation that can be rearranged to measure the psychogenic effect:

$I_a = I_p \times M$, Equation 7
$M = I_a/I_p$, Equation 8

where I_a is the actual or observed intake (kg d^{-1}) of the animal, I_p is the predicted intake potential (kg d^{-1}) of the animal and diet based on physiological or physical control mechanisms, and M is the proportional psychogenic modulation factor (dimensionless ratio).

It is proposed that psychogenic modulation is multiplicative rather than additive based on the logic that animals with greater intake potential will have a larger absolute change in intake. Equation 8 offers the potential for quantifying the modulating effects of management, disease, social interactions, and palatability on long term intake regulation. It is speculated that M will be less than 1.0 in most circumstances because most psychogenic effects inhibit intake. However, analyses of dynamic control systems of intake regulation indicate that diets with high palatability can result in increases in the dynamic set point or defended body weight (Russek, 1978) of the animal.

INTEGRATION OF INTAKE MECHANISMS

Static or Steady-State Models

The physiological, physical, and psychogenic mechanisms of intake regulation each establish independent controls on intake which need to be integrated into a common equation. The physical and physiological mechanisms of intake control provide limits at the opposite extremes of forage quality. When high energy, low fill rations or forages are fed, physiological energy demand regulates intake; whereas when high fill, low energy rations or forages are offered, physical fill limits intake. In any situation, the lesser of the two intake limits will predict the intake potential of a specified animal-feed combination. This system of controls can be described by an equation that can be combined with equation 7 to predict actual intake under conditions when feed availability is not limiting:

$I_p = \min(I_e, I_f)$, Equation 9

$I_a = \min(I_e, I_f) \times M$, Equation 10

where all terms are as previously defined.

Equation 9 implies that intake regulation is a discontinuous function of diet or forage characteristics because, in any specific situation, either energy demand or fill limits intake. In general, the energy value and filling effect of diets or forages are inversely related. This allows equations 3, 6, and 9 to be described by a graph with inversely related X-axes (Figure 1). The psychogenic modifier (M) in equation 10 would modify Figure 1 by a simple scalar adjustment to the Y-axis.

One of the characteristics of this simple static model of intake regulation based on dual control mechanisms (Figure 1) is the occurrence of a unique solution. Because fill and energy concentrations in feeds are inversely and curvilinearly

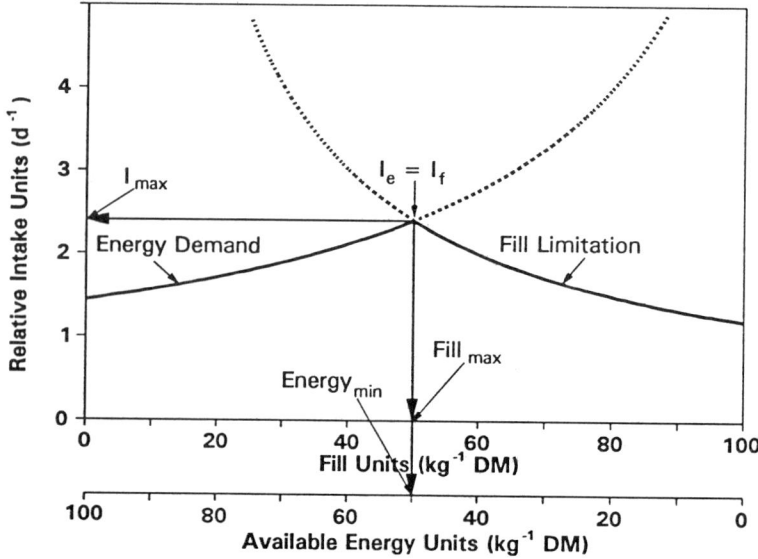

Figure 1. Illustration of the bi-phasic, discontinuous nature of intake regulation based on simple algebraic equations describing expected intakes when limited by physiological energy demand (I_e) or physical fill (I_f). Maximum intake (I_{max}) occurs at the intersection of the two theories of intake regulation and defines the diet having maximum fill ($Fill_{max}$) and minimum energy concentration ($Energy_{min}$) which meets the animal's energy requirement and maximizes ruminal fill.

related to intake potential, the intersection of equations 3 and 6, when $I_e = I_f$, is the maximum intake that can both meet the animal's energy demand and fill the rumen. Solving for the maximum constrained intake yields the following equations:

$I_{max} = I_p$, when $I_e = I_f$,   Equation 11
$R/E = C/F_{max}$,   Equation 12
$F_{max}/E = C/R$,   Equation 13
$F_{max} = (E \times C)/R$,   Equation 14

where I_{max} is the maximum intake (kg d^{-1}) that will provide the energy needed to meet the animal's requirements, F_{max} is the maximum filling effect (L d^{-1}) of a diet that can meet the animal's needs, and all other variables as previously defined.

If C and R are held constant and E (energy concentration) is increased, the filling effect of the diet (F_{max}) can be increased and still meet animal requirements. Increasing C has a similar impact on F_{max}. However, if the animal's energy requirement (R) is increased, the maximum filling effect of the diet (F_{max}) must be reduced to allow enough intake to meet requirements. If the goal is to formulate rations that maximize forage content or to select the lower limit of forage quality that meets the performance needs of the animal, equation 14 provides a quantitative approach that is based on sound biological principles. Remember that C in equation 14 is not the absolute maximum intake capacity of the animal because this fails to allow the animal to meet its needs; rather, C in equation 14 is the intake capacity that is constrained to meet the animal's requirements.

Equation 10 essentially summarizes the most commonly accepted theory of intake regulation in ruminants. This comprehensive framework of intake regulation can be summarized in the following statements:

1. The basic drive to eat (appetite) of an animal is determined by its genetic potential and physiological state which corresponds to the maximum rate of nutrient utilization by its tissues.
2. When the diet contains adequate concentrations of available energy, protein, vitamins, and minerals, the animal consumes feed at a level that matches its appetite, and animal potential is the limit for intake.
3. When diets of low nutritive value are offered, the animal consumes feed at a level that matches its gut capacity, and intake is restricted by the filling effect of the diet and the intake constraint of the animal. The tolerance of the animal for ruminal fill increases to accommodate increased nutrient demands, but animal performance also is compromised.
4. Modifying the dominant roles of physical limitation and physiological regulation on intake is the psychogenic stimuli associated with palatability, social interactions, disease, and feeding management.

This conceptual framework of intake regulation and the equations that can be used to describe it will be used in the remaining discussion to illustrate the difficulties in measuring intake as a component of forage quality.

NDF-Energy Intake System

The accepted static theories of intake regulation have practical utility in forage evaluation and ration formulation only when they can be related to specific feed and animal characteristics that are routinely measured. In addition, the theory can be tested only when the vague concepts of the feed's filling effect and available energy and the animal's intake constraint and energy requirements are quantitatively defined. Mertens (1985, 1987, 1992) developed and refined the concept that NDF and net energy of lactation (NE_L) can serve as proxies for the filling effect and available energy in the accepted theories of intake regulation. They can be used as starting points for relating mechanisms of intake regulation to feed and animal (dairy cow) characteristics simultaneously and can serve as reference points for relating intake mechanisms to a routinely measured characteristic of feeds (NDF). As more information is gained and digestion kinetics, specific gravity, volume, and rate of particle size reduction and passage can be routinely measured or estimated, NDF concentration can be refined to more accurately estimate the true filling effect of the forage or diet.

Using NDF to represent the filling effect of the diet requires that the intake capacity or constraint of the animal also be expressed in units of NDF. Based on experiments designed to detect the NDF concentration of diets that optimized intake and dairy cow performance, Mertens (1985) observed that DMI was maximized when NDF intake was 12.5 ± 1.0 g kg^{-1} BW d^{-1}. Using this value for C and predicting NE$_L$ from forage NDF concentration allows the accepted theory of intake regulation to be illustrated using a routinely measured characteristic of feeds (Figure 2). Preliminary evaluation suggests that the NDF intake constraint of mature beef cattle (Mertens, unpublished) and sheep (Mertens, 1973) is also approximately 12.5 g kg^{-1} BW d^{-1}.

Recent literature contains numerous reports (Briceno et al., 1987; Reid et al., 1988; Harlan et al., 1991; Llamas-lamas and Combs, 1991; Weiss and Shockey, 1991) that NDF is not related to intake for use in ration formulation or forage evaluation. These discrepancies can be explained by comparing their results to the accepted mechanism of intake regulation using the NDF-Energy Intake System. Assuming that psychogenic effects in most research experiments are minimized (M = 1), equation 10 indicates that intake is a function of either fill or available energy. Because the relationship between NDF and fill is the opposite of that between NDF and available energy, the relationship of NDF to intake will be different depending upon whether intake constraint or energy demand is limiting intake in any particular instance. When energy is limiting intake, intake will be positively correlated with NDF concentration in the feed whereas, when fill is

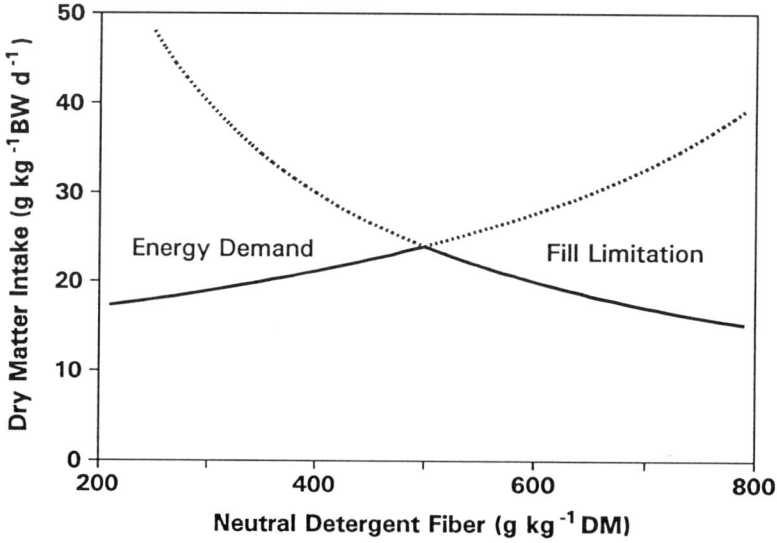

Figure 2. Expected relationship between the neutral detergent fiber (NDF) concentration of the diet and intake when regulated by energy demand or fill limitation. It is assumed that NDF is positively correlated with the filling effect and negatively correlated with the energy availability of the diet.

limiting intake, intake will be negatively correlated with NDF concentration (Figure 2). Indeed, when measured over a wide range of diet NDF concentrations by animals with low energy demands, expected intake (indicated by the solid discontinuous lines in Figure 2) will have essentially a zero linear correlation with NDF concentration in the forage or diet. This does not mean that intake is not related to NDF, only that the relationship is not a simple linear one.

The summarization by Jung and Linn (1988) provides an excellent data base for evaluating the NDF-Energy Intake System. They compiled experiments published in the Journal of Dairy Science from 1986 through 1988 and obtained about 100 cow-treatment combinations in which both NDF and intake were measured. Some of the NDF values published were high in comparison to those commonly observed (NRC, 1989; Mertens, 1992), probably due to starch contamination when amylase is not used in the NDF procedure. Even after the data base is edited by replacing atypical NDF values with values adjusted to correspond with the acid detergent fiber (ADF) values that were published, a poor overall correlation is observed (Figure 3). Jung and Linn observed a correlation of -0.12 and concluded that NDF was not useful in predicting intake. However, if the data are compared to the theoretical lines derived for the NDF-Energy Intake System, the opposite conclusion is reached and the explanation for the apparent discrepancy is evident.

As illustrated by Mertens (1992, Figure 25.2), any intake below and to the left of the fill limitation line is feasible. The relationship between intake and NDF would be proven to be invalid only if intake observations occurred above and to the right of the fill limitation line because all other observations will satisfy either equation 3 or 6. Only a few intakes occur in this region indicating that the accepted theory of intake using NDF is consistent with observations. The exceptions are related to diets containing finely ground, wet corn gluten feed which probably does not have the same filling effect per unit of NDF as long forages. The data base compiled by Jung and Linn (1988) indicates that most of the diets reported in the Journal of Dairy Science contain higher than needed levels of energy and that energy demand of the cows and not the filling effect of the diets limited intake.

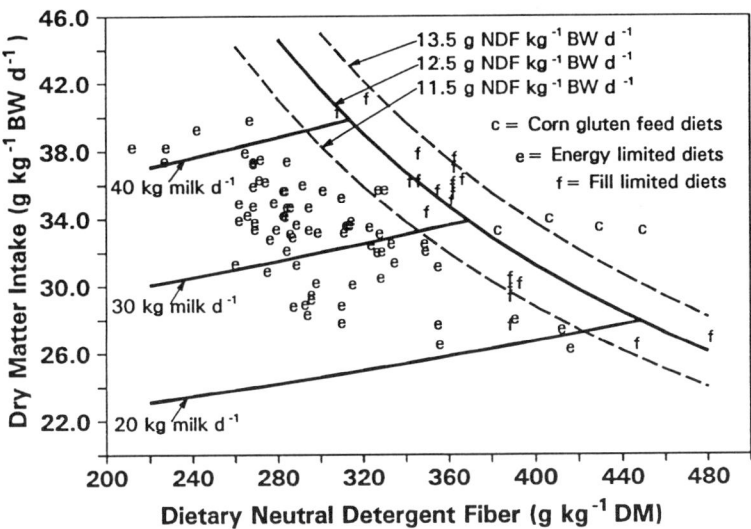

Figure 3. Comparison of the NDF-Energy Intake System predictions to the observed intakes of dairy cows reported in the Journal of Dairy Science as compiled by Jung and Linn (1988).

Diets were classified as energy (e) or fill (f) limited in Figure 3 by using equations 3 and 6 to calculate the expected intakes for each mechanism and labeling each diet by which intake was minimum as indicated by equation 9. The positive relationship between intake and NDF indicated by the energy demand lines for various levels of milk production (Figure 3) corresponded well with observed productions. The negative correlation between NDF and intake will be high only when the filling effect of the diet is high in relation to the cow's energy demand as indicated by the diets labeled with an "f" in Figure 3.

The prediction of intake and its use in forage evaluation is hampered by two important factors: 1) intake is influenced by characteristics of the animal, diet, and feeding situation and 2) prediction of intake is affected by the type of information that is known. Equations that attempt to predict intake solely as a function of animal characteristics (BW, production level, BW change, physiological state, etc.) or diet properties (fiber, bulk, energy density, chewing requirement, nutrient balance, etc.) will not have universal applicability and are doomed to failure. These equations fail because they implicitly assume that only one component of the intake regulation system (i.e., animal, diet, or feeding situation) limits intake in all instances. The NDF-Energy Intake System overcomes this deficiency by integrating animal characteristics, diet properties, and feeding situation attributes into a simple system of mathematical equations. The aggregation of information into concepts of intake constraint and energy demand of the animal, and filling effects and energy availability of the diet, does not prove they exist or suggest that they are not complex concepts that require additional evaluation and characterization. Mertens (1992) indicated the limitations of using a chemical entity (NDF) to describe a physio-biochemical entity (filling effect) and suggested that NDF must be adjusted for particle size differences to accurately serve as a proxy for the filling effect of a feed. Likewise, it can be speculated that kinetic characteristics, such as rates of digestion and passage, should be used to modify NDF to more correctly reflect the filling effect.

Intake prediction also is affected by the information that is known and the purpose for estimating intake. When a ration is formulated, an intake prediction is needed under the assumption that the animal's requirements are known and will be met. Diet properties are unknown, but it is implicitly assumed that they will be optimized for some function (profit, cost, production, etc.) under some set of known constraints. The NDF-Energy Intake System was designed to predict intake under these circumstances. Therefore, C in equations 6 and 14 was defined as a NDF intake constraint that was determined when 4% fat-corrected milk production was maximized (Figure 4). It is speculated that the system will predict intake most accurately when animals are near their maximal production. The NDF intake constraint is useful because it provides an upper limit to DMI and maximum forage in the ration that will allow the animal to maximize performance. Any intake less than this prediction is acceptable as long as the ration is formulated to have the increased energy density needed to meet the energy demands of the animal (Mertens, 1992).

When the diet is known and the performance of the animal is unknown, intake prediction is much more difficult because the animal can accommodate the diet by varying intake and (or) performance. Figure 4 indicates that NDF intake can exceed the NDF intake constraint for maximal performance when animals are fed suboptimal diets. This indicates that the NDF intake constraint is not a constant and suggests that additional research is needed to refine the NDF-Energy Intake System for predicting intake under suboptimal conditions. The adjustment of the intake constraint by the animal may even suggest to some that fill never limits intake.

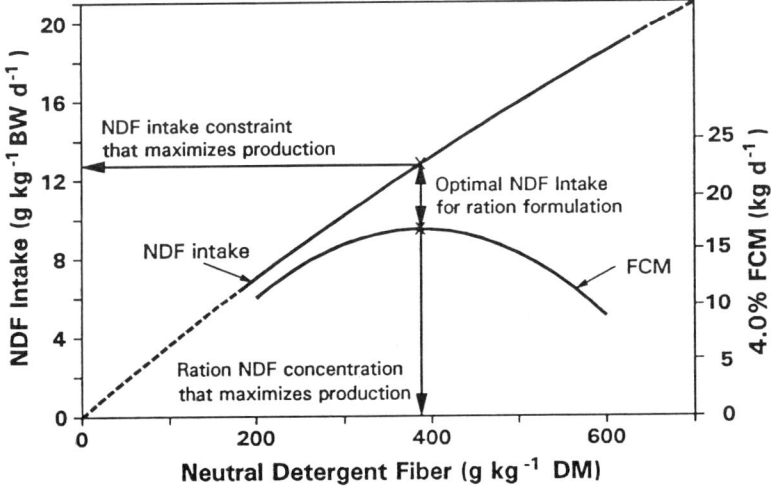

Figure 4. Relationship of neutral detergent fiber (NDF) concentration in the ration to NDF intake and 4% fat-corrected milk (FCM) production (Mertens, unpublished). The NDF intake constraint is determined experimentally as the NDF intake that maximizes FCM production.

However, if the animal cannot achieve its productive potential after it increases its intake constraint, the possibility that fill ultimately limited intake cannot be precluded.

The value of the NDF-Energy Intake System for predicting intake is that it not only provides a specific, quantitative system for estimating intake when formulating rations that maximize DMI and forage use, but also it serves as a framework for quantitatively understanding intake regulation. Although animals integrate information from a variety of sources and compromise between performance and intake for any specific situation, it seems logical that the simplest approach to developing a system for predicting intake would attempt to identify the most limiting factor and use a quantitative measure of it to estimate intake. The NDF-Energy Intake System assumes that energy demand by the animal or filling effect of the diet are first limiting factors. This simple model may not predict intake in all situations, but it provides a quantitative way of identifying when intake is limited by fill (when NDF intake is > 11.0 to 13.0 g BW^{-1} d^{-1}) or by energy demand (when NE intake equals the potential energy demand of the animal).

Optimization of Oxygen Utilization

The accepted static theories of intake regulation are based on the concept that it is beneficial for animals to increase the consumption of feed up to the level needed to match their maximum genetic and physiological potential. This assumes that animals seek to obtain a maximal growth and production rate, resulting in maximal feed intake unless limited by fill. Ketelaars and Tolkamp (1991) proposed an alternative theory of intake regulation that is based on the concept that animals optimize oxygen utilization. They speculate that feed intake is regulated to maximize the benefits (input of NE) relative to the costs (oxygen consumption) to the animal. They contend that oxygen free radicals, which are a byproduct of metabolic reactions, are toxins that damage cells and shorten the life span. They theorize that animals may regulate intake to optimize the amount of productive energy input per unit of oxygen consumed. Thus, the amounts and ratios of absorbed nutrients control intake according to their theory. Their thesis presents an extensive critique of the accepted theory of intake regulation, especially the concept that fill limits intake. Their proposal provides an alternative model of intake regulation and many of the objections they raise to the currently accepted theory of intake regulation will be discussed in the remaining sections.

Ketelaars and Tolkamp (1991) use the data from ARC (1980) as the primary validation set for their theory. They observed that the voluntary intake equation used by ARC to predict the intake of long forages agrees very closely with the intake that optimizes the ratio of net energy intake to oxygen consumption. Although their model is internally consistent with the data of ARC (1980), it is not evident from their presentation how ARC intake predictions or their model predictions compare with the accepted theory of intake regulation. The theory of Ketelaars and Tolkamp (1991) is based on the metabolizability (q-value) of feeds which is used to determine the NE output and oxygen input of the animal. Information about feeds grown in the UK (Ministry of Agriculture, Fisheries, and Food, 1990) were used to develop the relationship between NDF and q for grasses ($q = 1.023 - .0082*NDF$, $R^2 = .78$).

Intakes estimated by the equations of ARC (1980) for coarse diets and the theory of Ketelaars and Tolkamp (1991) were plotted on a common graph (Figure 5) with the fill limitation line proposed by Mertens (1987). The agreement between ARC and the fill limitation line (Mertens, 1987) is exceptionally high in the range of 500 to 800 g NDF kg^{-1} DM, considering the difference in animal species (sheep versus dairy cattle), diets (forages versus mixed rations), and feed characteristics used to predict intake (q versus NDF) between the two data sets. The difference between the ARC (1980) and Mertens' (1987) predictions when NDF is less than 500 g kg^{-1} DM probably is related to the few observations in the ARC data set in this range and the likelihood that the intake of sheep would be limited by energy demand in

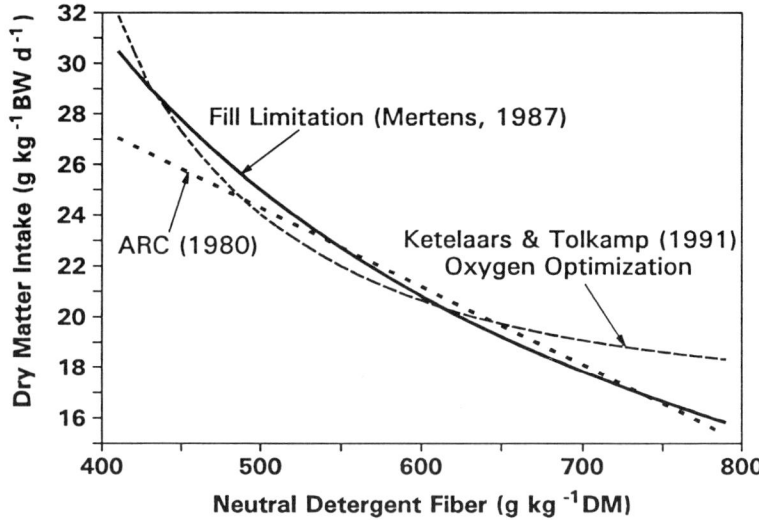

Figure 5. Differences in predicted intakes estimated by the empirical equations of ARC (1980), the NDF-Energy Intake System (Mertens, 1987), or the optimization of oxygen utilization proposed by Ketelaars and Tolkamp (1991).

that range of NDF (Figure 2). The theory of Ketelaars and Tolkamp predicts substantially higher intakes of high-fiber grass forages than that indicated by theories based on fill limitation (Mertens, 1987) or empirical equations based on diet metabolizability (ARC, 1980). This suggests it would over-predict the intake of low quality forages which often result in intakes that are lower than those predicted by fill limitation (see final section of this chapter). Their theory also over-predicts intake for high energy, low fiber diets because the fill limitation line (Mertens, 1987) in Figure 5 would normally be truncated by the energy demand line as shown in Figure 2. Figure 5 illustrates the value of modeling in providing the basis for designing experiments that would discriminate among the theories. Although additional research is needed, it appears that the concept of optimizing oxygen utilization may be more useful in describing the point of maximal intake for good to medium quality forages rather than be a substitute for the fill limitation of intake as Ketelaars and Tolkamp (1991) contend.

Dynamic Models of Intake Regulation

Dynamic models are well suited to the study of intake regulation because they offer the opportunity to evaluate concepts rigorously as they change over time. Steady-state solutions to dynamic models, although technically not dynamic, often provide quantitative insight into the operations of conceptual models. Waldo (1970) presented the key principle that resulted in our current understanding and use of digestion kinetics at the first National Conference on Forage Quality Evaluation and Utilization when he postulated that the curvilinear nature of digestion might be due to the presence of an indigestible fraction. With co-workers (Waldo et al., 1972), he developed the first-order model of digestion and combined it with rates of passage to calculate the filling effect of diets based on cellulose. Mertens (1973) used this model to relate digestion kinetics of NDF in forages to their intake. Ellis (1978) used the steady-state solution for digestibility proposed by Waldo et al. (1972) to estimate intake from fecal output as envisioned by Conrad et al. (1964) using rates of passage and digestion and the pool of undigested DM in the rumen. Pienaar et al. (1980) developed a more complex steady-state model for predicting intake from the flows of fermentable and unfermentable OM. Poppi et al. (1981b) investigated the relationship of intake to the ruminal retention time of DM using a steady-state model that described the flow of large and small particles through the rumen.

Baldwin et al. (1977) proposed a dynamic model of digestion (second-order) and passage (first-order) for evaluating nutritive value; however, intake was an input to this model that was not calculated as a model response. Mertens and Ely (1979, 1982) developed a dynamic model that included digestion as a first-order process as proposed by Waldo et al. (1972) and divided passage into first-order rates of particle size reduction and escape from the rumen. Intake was predicted by iteratively solving the model to estimate the intake that resulted in a ruminal fill of 15.8 g NDF kg^{-1} BW d^{-1}. Illius and Gordon (1991) modified the model of Mertens and Ely (1979) to scale rate of passage and particle size reduction to $BW^{0.27}$ and observed that model simulations agreed with intake observations for sheep and cattle.

Several dynamic models (Forbes, 1977b, 1983; Black et al., 1981; Bywater, 1984; Kahn and Spedding, 1984; Fisher et al., 1987) have attempted to integrate both physical and physiological mechanisms of intake regulation in a single model. Most of these models have attempted to use an either (or) approach for deciding whether intake is controlled by fill or energy demand, similar to that proposed for the static model (equation 9). Dynamic models, however, calculate the change in fill and energy supplied over time within a day to regulate intake and incorporate kinetic characteristics such as rates of digestion and passage into intake control mechanisms. Fisher et al. (1987) assumed that intake is restricted by ruminal distension and nutrient absorption, but incorporated a feedback function that adjusted intake for both mechanisms simultaneously. Intake rate was adjusted based on the ratio of maximal ruminal contents to current ruminal contents and optimal nutrient absorption flow to current nutrient absorption flow. Their function modulates the chemostatic effect based on the level of distention.

INTAKE VARIATION AND REGULATION MECHANISMS

Appreciation and knowledge of the mechanisms controlling intake in animals indicates that intake is not simply a feed attribute, but is a function of characteristics of the feed, animal, and feeding situation. To obtain valid estimates of forage quality, the extrinsic factors affecting intake must be controlled so the intrinsic characteristics of the forage that determine intake are measured. Raymond (1969) stressed the need for rigid standardization of the conditions under which intake is measured. This is necessary not only to reduce variation in intake measurements, but also to improve the comparability of measurements among experiment stations and to ensure that forage characteristics and not animal or feeding situation factors are being measured.

The multitude of factors that influence intake is indicated in Figure 6 which attempts to present a relatively complete listing of variables in a structured format. It is clear that many of these factors do not impact intake through a singular and distinct path. Numerous interactions occur that would make the figure extremely complicated if they were indicated. For example, rate of passage is described in Figure 6 as an animal characteristic, yet it is clear that rate of particle size degradation, rate of digestion, and plant morphology all affect rate of passage. Likewise, rate of digestion probably is affected by animal characteristics that influence the ruminal environment and its microbial population. Many of these factors have been discussed, but space does not allow the elucidation of all factors that influence intake regulation. Three factors will be discussed in this section because they have important impacts on the measurement, regulation, and description of intake as a component of forage quality.

Refusal Allowance and Selection

The intake of forages can be measured only if they are available *ad libitum*. This implies that animals are allowed to eat as much as they desire, i.e., the amount of feed offered is greater than that which the animal consumes. Unfortunately, *ad libitum* intake is complicated by the observation that animals eat forages selectively. This suggests that animals may have greater intake if more feed is offered (greater refusal allowed), thereby providing greater opportunity for selection. Greater opportunity for selection also could increase digestibility, assuming that animals select the less fibrous and more digestible components of the feed.

Blaxter et al. (1961) indicated that, to avoid bias in measuring intake, the same excess of feed should be offered to all animals and suggested that animals be fed at least 1.15 times their consumption on previous days. However, the amount of refusal allowed by investigators can vary from 0 to 30% of that consumed, with a typical range of 5 to 15% (Brown et al., 1968; Colovos et al., 1970; Jones et al., 1972; Dulphy and Demarquilly, 1983). Some researchers (Heaney et al., 1963; Minson and Milford, 1967) restricted intake to maintain a constant amount rather than proportion of refusals. When the amount of refusal is held constant, the

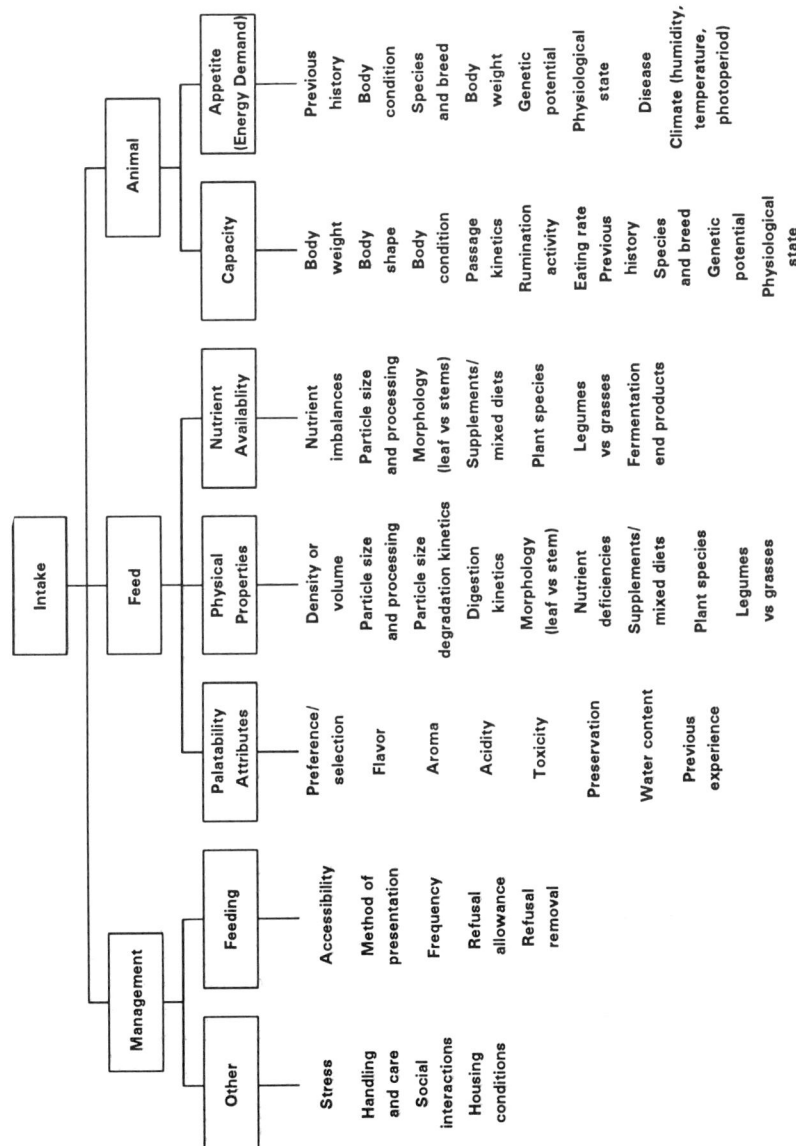

Figure 6. Classification of factors affecting the intake of feeds by ruminants.

proportion of refusal is higher for forages with low intakes which allows a greater degree of selection for low quality forages.

More recently, Zemmelink (1980) extensively reviewed the literature and concluded that variation in techniques among investigators resulted in measurements of intake that were not comparable. He observed that the ranking of intakes of tropical forages varied with the level of excess that was offered. This implies that selection of components within a forage provides the animal a wide latitude in varying intake and probably digestibility. It also indicates that the level of refusal that is observed in any feed intake studies should be reported with the results of the experiment to provide important information for the interpretation of intake and performance responses by the animal. When large refusal allowances are permitted, caution must be exercised in relating chemical composition or any other feed characteristic to intake measurements because the characteristics of the feed offered usually differs from that of the feed consumed. Differences between the feed consumed and that offered may explain some of the discrepancies that exist in the literature when feed characteristics are related to intake.

Nutrient Deficiencies or Imbalances

The effect of nutrient deficiencies on intake has been discussed by Forbes (1986). Intake often is depressed when the CP content of diets is below 60 to 80 g kg^{-1} DM (Blaxter and Wilson, 1962; Minson and Milford, 1967). This level of CP is near the requirement for protein and ammonia needed by bacteria in the rumen. If each kilogram of fermentable OM yields 150 to 250 g of microbial CP (NRC, 1985), this suggests that one gram of CP is needed for each 6 to 7 g of OM fermented in the rumen. If 85% of the total tract digestibility occurs in the rumen, and forage protein is 25% undegradable, the range in CP needed for effective digestion (ignoring recycling of N to the rumen and microbial inefficiency of N utilization) would range from 70 g CP kg^{-1} DM for forages with 400 g kg^{-1} DDM to 140 g CP kg^{-1} DM for forages with 800 g kg^{-1} DDM. The need for CP and other nutrients, such as minerals, vitamins, and microbial growth factors, suggests that some or all of the effects of nutrient deficiencies on intake could be mediated via physical limitation. If microbial fermentation is inhibited, feed will not disappear from the rumen as quickly, the apparent filling of the diet would be increased, and intake will decrease.

Egan (1965) infused casein into the intestines of sheep fed low N roughages and observed that intake and ruminal contents increased. Thornton and Minson (1973) reported that forages with low CP resulted in reduced intake and gut contents. These observations suggest that intake was not limited by fill; rather, some metabolic mechanism related to protein characteristics or the protein to energy ratio was limiting the intake of diets low in CP. It is difficult to envision a mechanism whereby increasing protein concentration would result in increased intake without including energy regulation in the theory. If protein is deficient in the diet relative to energy and animals were consuming feed in response to protein concentration, it would seem that they would increase intake of a low protein diet, not consume more when protein was supplemented. If it is energy intake that is actually regulated, however, the increase in intake with increasing protein supplementation can be resolved.

Because animals have limited ability to dissipate excess energy, diets that are inadequately balanced for nutrients in relation to energy pose a dilemma for the animal. If intake is increased to obtain more of the deficient nutrient, excess energy is absorbed as well and must be accumulated by the animal. Eventually, the excess energy accumulation results in reduced intake in an attempt to balance energy input and output. Thus, when diets low in protein relative to energy are fed to animals with high protein requirements (rapidly growing young animals, wool growth, or lactating females), intake may be effectively limited by protein deficiency via a mechanism which essentially operates to maintain energy equilibrium. Alternatively, Leng (1990) postulated that the metabolic efficiency of energy utilization of poor quality forages is reduced when protein is deficient which creates heat stress for animals in tropical environments that results in reduced intake and performance.

Body Weight and the Basis for Expressing Intake

In general, intake increases with increasing BW of the animal. Thus, some variation in intake can be removed by expressing intake as a ratio to animal weight. Traditionally, researchers in the US have expressed DMI as a percentage of BW which is equivalent to BW^{-1}, whereas European scientists typically express intake per unit of metabolic body size (MBS) which is equivalent to $BW^{0.73}$ (Brody, 1945) or $BW^{0.75}$ (Kleiber, 1961). At the first National Conference on Forage Quality Evaluation and Utilization, Waldo (1970) favored the use of MBS because it was accepted by scientists investigating energy metabolism as the base for expressing maintenance energy requirements. Thus, expressing intake on the same basis would allow the BW base to cancel, providing a simple measure of intake as a multiple of maintenance. However, a review of the physical and physiological mechanisms of intake regulation suggests that the BW base that is appropriate for each mechanism may not be the same.

In contrast to Kleiber (1961) who speculated that animals consume energy in proportion to MBS, Van Soest (1982) proposed that intake is proportional to BW^{-1}. He developed this concept from the relationship between gut contents and BW observed in herbivorous species. Parra (1978) and Demment and Van Soest (1985) summarized the data on gut contents of captured wild herbivores ranging in weight from 10^{-2} to nearly 10^4 kg and observed that wet gut contents were related to BW by the power 1.032 which was not significantly different from 1.00. Van Soest (1982) concluded that intake should be related to the same function of BW as is gut capacity. Implicit in this conclusion is the assumption that animals limit intake in relation to fill and when given the choice they will select diets that do not overburden the capacity of the gut to handle digesta. This differs from the assumption of Kleiber (1961) who assumed that animals had access to energy-rich feeds; therefore, intake would be related to MBS because energy requirements are related to MBS.

Animals have some ability, within limits, to adjust gut capacity to meet the short term demands imposed by a bulky diet that does not meet their energy needs. However, it seems inappropriate to base a conceptual model of intake regulation or a system of formulating rations on this exception. In general, small herbivores selectively consume diets high in available energy and low in fiber when allowed the opportunity suggesting that gut capacity is a limit that has affected the evolutionary development of herbivores (Demment and Van Soest, 1985). If gut contents were proportional to MBS or energy demand, small or lactating ruminants would have gut contents that are such a high proportion of BW (300 to 500 g kg^{-1} BW) that mobility and survival would be sacrificed. Nor does this relationship agree with change in body shape that occurs with age. In general, young animals have smaller body circumferences and depths in relation to their height or weight than do older animals (Brody, 1945).

Ketelaars and Tolkamp (1991) criticized the isometric relationship between gut contents and BW because all animals were not fed the same diet and in the wild, smaller species select more concentrated diets. They compared the gut contents of dwarf goats fed a grass hay to the empty BW equations developed by ARC (1980) and concluded that gut contents appear to be roughly proportional to basal metabolism (or MBS) in domestic species. This suggests that smaller species, and by implication all animals, adjust their gut capacity to meet their higher (relative to BW) energy requirements. However, indirect comparisons must be interpreted with caution. It is possible to select treatments (pelleted versus unpelleted forages) or forage qualities that bias comparisons of gut fill among animals with different energy requirements. As indicated by Figure 2, it is possible to select a forage the intake of which is limited by fill for a small animal with higher relative energy demand, but is limited by energy demand for a large animal. In this instance, the ruminal contents of the small animal would be greater than the large animal because the former was limited by fill and the latter was not. Direct comparisons between sheep and cattle when fed ten different roughages with DDM from 460 to 580 g kg^{-1} indicate that ruminal contents were proportional to BW (Hendricksen et al., 1981; Poppi et al., 1981a). Although it appears that gut contents are related to $BW^{1.0}$, the relationship with intake is not clear, possibly because diet and animal characteristics are confounded in many experiments. Colburn and Evans (1968) observed that the best reference base for Jersey steers fed orchardgrass and alfalfa hays was $BW^{0.54}$.

The relationship of intake to animal BW is important to anyone assessing forage quality. Some of the discrepancies and controversy in the literature about forage quality may be related to the use of different BW bases for expressing intake measurements. The conceptual model of intake regulation can be used to illustrate this point. Let us assume that two sheep, one weighing 40 kg and the other weighing 60 kg, consume forages to meet their energy requirements (proportional to MBS) when forage quality is high, but limit intake based on fill (proportional to BW) when forage quality is low. When the same intake predictions are expressed per unit of BW (Figure 7a) or MBS (Figure 7b), the relationship between NDF and intake appear to be different. Using MBS as the basis for expressing intake decreases animal variation for high quality forages because energy demand is regulating intake. However, the opposite effect is true when intake is expressed on a BW basis.

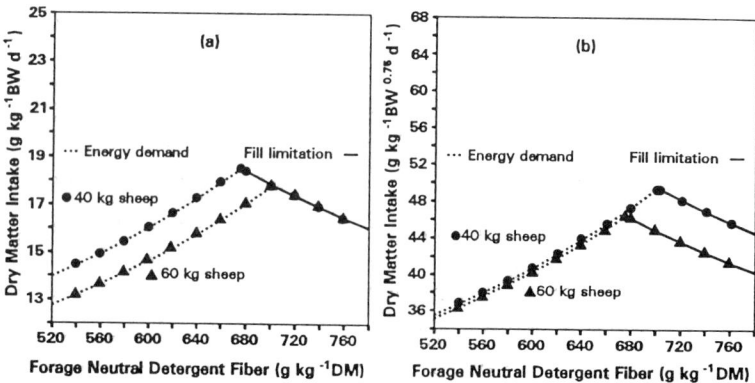

Figure 7. The same intake data are expressed on different body weight (BW) bases to illustrate the effects of BW on apparent variation in intake measurements. Expressing intake per unit of $BW^{1.0}$ reduces the variation in intake due to BW for high fiber forages as shown in figure (a). Conversely, expressing intake per unit of metabolic body size ($BW^{0.75}$) reduces variation in intake due to BW for low fiber forages as illustrated in figure (b).

This illustration indicates that no single BW basis can be used across a wide range in forage quality to remove animal variation unless both physical and physiological control systems are affected by the same BW relationship. For energy-rich forages, variation in intake among animals may be minimized by expressing intake per unit of MBS because intake of these forages probably is limited by energy demand and energy demand generally is related to MBS. Because the intrinsic intake potential of a forage should reflect its filling effect rather than the energy demand of the animal to which it is fed, it seems most appropriate to measure forage intake potential with animals that have high energy requirements to ensure that fill is limiting intake. Under these circumstances, it is suggested that forage intake potential should be expressed in terms of $BW^{1.0}$ to minimize the variation in intake associated with differences in animal size.

Measuring Forage Intake Potential

Intake is a function of animal characteristics, intrinsic feed properties, and attributes of the feeding situation. The number and complexity of factors affecting intake make the measurement of intake potential of a forage difficult to accomplish and to interpret. Standard research approaches would require that the problem be reduced to its simplest form for investigation. This suggests that all extraneous sources of variation be eliminated so only the effects of the forage on intake are measured and studied. It is easy to understand why conditions in the feeding situation that adversely affect voluntary intake should be minimized or eliminated. However, it is more difficult to control, or even document, the effect the animal has on intake because intake is the direct result of the interaction between the animal and the feed. If a forage with high digestibility is fed to a mature wether with relatively low energy requirements, is the intake potential of the forage measured? It can be argued that it is the low energy demand of the animal and not the quality of the forage that determines the intake that is measured. Alternatively, it could be argued that the high quality (digestibility) of the forage is limiting intake, although this is not the characteristic of a forage that most researchers or nutritionists would recognize as the primary factor limiting the intake potential of forages.

There is no doubt that animals contribute significant variation to the measurement of intake and may even be the limitation for intake when high quality forages are fed. Variation in intake attributable to animals is important within groups of animals in a trial, but it is even more important when comparing intakes of forages across trials when animal characteristics vary greatly. Osbourn et al. (1974) demonstrated that significant variation in intake among forages can be removed by using the intakes of animals on a reference forage as a covariate. Similarly, Abrams et al. (1987) observed that intake could be more accurately measured when animal effects on intake are removed.

A review of the mechanisms regulating intake suggests that different animal attributes can affect intake at the extremes of forage quality. For high quality forages, the animal's energy requirement (R in equation 3) determines intake whereas for low quality forages, the animal's intake constraint (C in equation 6) limits intake. Which attribute of the animal should be measured and used as a covariate? Will the appropriate attribute vary with the quality of the forages to be evaluated for intake potential? Although these questions need additional research, the observations of Osbourn et al. (1974) suggest that the energy requirement of animals is the greater source of variation in intake. This is indicated by the larger reduction in variation of intake for high quality compared to low quality forages when adjusted for intakes of a reference forage (Figure 8). Figure 3 also illustrates

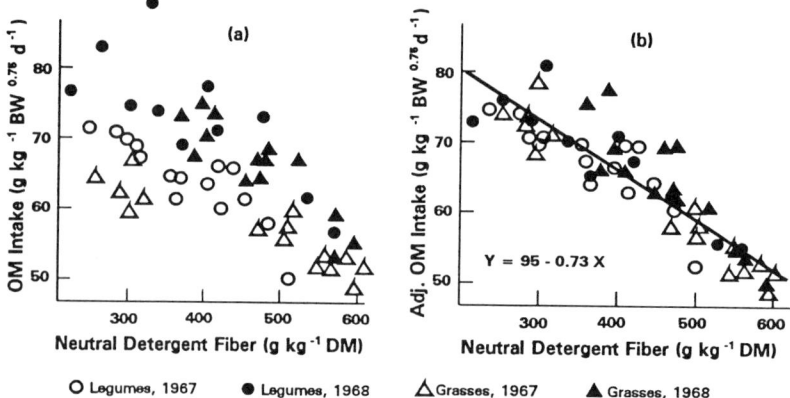

Figure 8. Reduction in variation associated by adjusting organic matter (OM) intake for variation among animals when fed a reference forage (from Osbourn et al., 1974).

the V-shaped distribution of intakes associated with low fiber, high quality diets suggesting that removing variation in intake associated with energy demand would improve the relationships between feed characteristics and intake potential.

French workers (INRA, 1989) concluded that voluntary intake as it is commonly measured has limited physiological meaning for the animal and is not adequate for the prediction of forage intake because it depends on both the intake constraint of the animal and the ingestibility of the feed. They developed a system in which the animal and feed components of intake regulation are separated but can be combined in a simple additive manner to formulate rations. Their system is similar to that of Hyppola and Hasunen (1970) in Finland and Kristensen (1983) in Denmark, but is more refined and is based on a larger set of animal data. The INRA system defines a fill unit (FU) based on a reference forage fed to reference animals. By feeding all test forages to similar reference animals, the difference in ingestibility of forages can be measured with minimal error associated with animal differences. Similarly, differences in intake constraint of animals were measured by feeding the reference forage, thereby minimizing the errors associated with forage differences. The resulting FU system is primarily of value for diets based on forages whose intake is regulated by physical limitations, but includes the effects of palatability and energy availability on intake as well.

Ingestibility of a forage is reported as fill values for reference sheep (SFU), reference cattle (CFU), and reference lactating cows (LFU). Most of the fill values for forages were determined using sheep, and regression equations were used to estimate CFU and LFU from SFU when forages were fed to two animal species. They reported that fill values are closely related to cell wall contents; however, NDF concentrations are not reported in their tables to allow direct comparisons (INRA, 1989). Regression equations between NDF and crude fiber or ADF (Mertens, unpublished) were used to estimate NDF and demonstrate the relationship between NDF and INRA fill units. The relationship between SFU and NDF is not as linear as expected (Figure 9) and this illustrates several characteristics of the FU system that are not readily apparent.

Although defined as fill units, the ingestibilities determined in the INRA (1989) system measure more than the filling effect of the feed. Because ingestibilities are determined by actual feeding trials (using reference animals), they include effects associated with the animal as well as those associated with forage characteristics

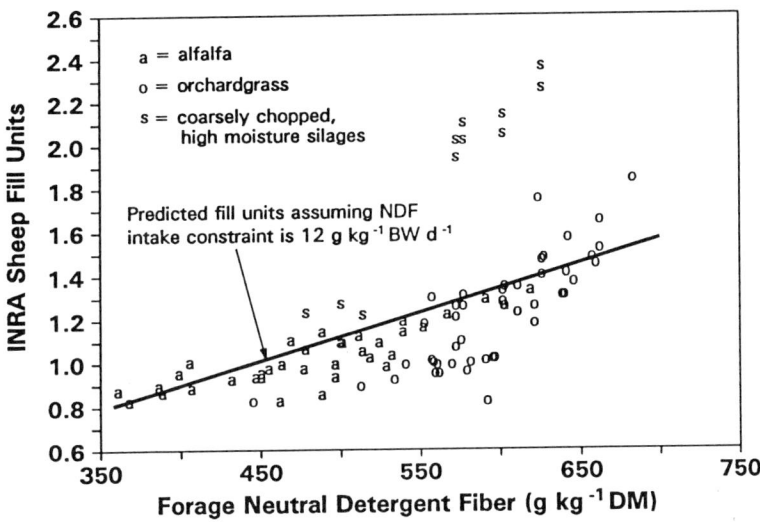

Figure 9. Relationship between neutral detergent fiber concentration of forages and sheep fill units reported by INRA (1989). Discrepancies associated with the high fill units for silages is probably related to differences in palatability rather than true differences in fill.

such as palatability, fill, and energy concentration. A linear relationship between NDF and filling effect assumes that all NDF is alike in producing fill and that other factors do not influence intake. The lower than expected SFU (higher intakes) of immature orchardgrass may reflect a more rapid rate of digestion and a smaller indigestible fraction for the fiber of these forages. The model of Waldo et al. (1972) demonstrates that fiber with slower rates of digestion and passage should result in greater filling effects in the rumen per unit of fiber. Thus, the higher than expected SFU for orchardgrass forages with high NDF could be due to a greater filling effect per unit of NDF, but it also could be related to lower palatability or tissue level nutrient imbalances for these mature or rain-damaged hays. The higher SFU of coarsely chopped, high moisture silages also may be related to poor palatability.

The INRA tables (1989) provide a unique data base for defining the psychogenic modulation factor (equation 7) associated with palatability. Most interfering variation is removed from this data set because reference animals were used in a standardized experimental design. Perhaps palatability can be quantitatively defined as the ratio of the ingestibility of one forage to another when both have the same energy concentration or filling effect. Assuming that NDF is related to the true filling effect of the forage (whether linearly or nonlinearly) and differences in the filling effects of NDF from various forages is negligible, any differences in intake among feeds with the same NDF concentration could be attributed to differences in palatability. Although these assumptions are speculative, they allow a method for estimating psychogenic modulation factors to be demonstrated.

This approach was used to characterize modulation factors in relation to fresh forage for the alfalfa and orchardgrass data shown in Figure 9. Barn-dried hay and direct-cut, finely chopped silage with additives resulted in modulation factors of 0.92 compared to fresh forage. Modulation factors for other treatments were: 0.88 for field-dried hay and direct-cut, finely chopped silages without additive, 0.86 for wilted, finely chopped silages, 0.84 for slightly rain damaged hays, 0.82 for extensively rain-damaged hays, 0.79 for direct-cut, medium chopped silages with additives, 0.57 for direct-cut, flail chopped silages with additives, and 0.55 for direct-cut, flail chopped silages without additives. These factors reflect a psychogenic modulation of intake above and beyond any differences in intake due to NDF or available energy concentration. Research is needed to validate this approach and determine if modulation factors can be used to describe palatability across forage species as well as across conservation or treatment effects within a forage type.

The work of Osbourn et al. (1974), Abrams et al. (1987), and the INRA (1989) indicate the importance of removing animal variation when attempting to measure the intake potential of forages. The concepts of intake regulation presented in this chapter indicate the complexity of intake measurement and the role that animals play in determining intake in any given situation. The simple model described by equations 3, 6, and 10 is useful in demonstrating the causes of variation in intake measurements among experiments and in explaining some of the conflicting conclusions that appear in the literature. For example, if a mature animal with moderate energy demand is fed high and low quality forages, it may have almost equal intakes of each (Figure 10). In this instance, equal intakes do not mean that these forages have equal filling effects or that both meet the animal's energy demand. It does, however, indicate that no single feed characteristic can be related to the intakes observed by a simple linear model because they were regulated by two different mechanisms resulting in a correlation of zero between NDF and intake in this example.

The curvilinear relationship between forage quality and intake observed by Mertens (1973), Reid et al. (1988), and Ketelaars and Tolkamp (1991) is accurately predicted by the simple model of intake regulation as demonstrated by the M's shown in Figure 10. The model also demonstrates how the variability in relationships between feed characteristics and intake can occur. Based on which sets of intakes that are used (indicated by M's in Figure 10), the correlation between

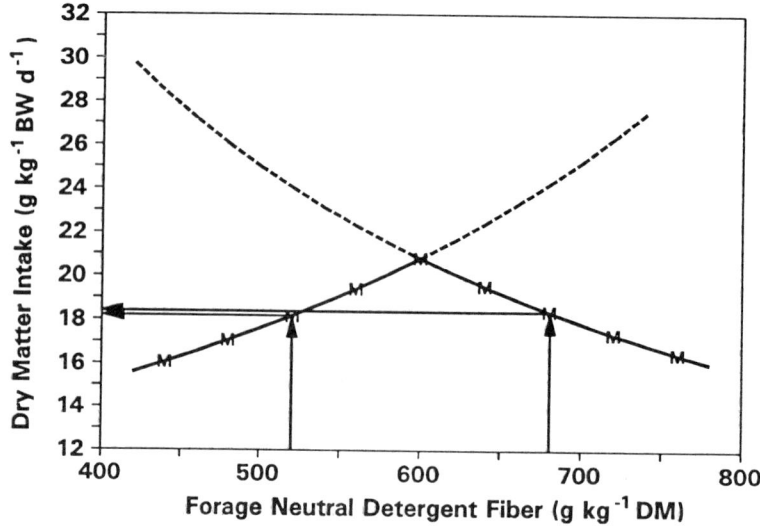

Figure 10. Predicted intake responses of a mature (M) wether fed forages differing in quality. Because two different mechanisms regulate intake, it is possible to observe similar intakes for forages of greatly differing neutral detergent fiber concentrations.

intake and NDF can vary from +1.0 (when NDF ranges from 450 to 600 g kg^{-1} DM) to zero (over the full range of NDF) to -1.0 (when NDF ranges from 600 to 750 g kg^{-1} DM).

The simple model also suggests that it is possible to obtain variable estimates of intake for the same forage due to differences in animal appetite or drive to eat. For example, if forages of high and low quality are fed to animals with high (H), medium (M), and low (L) energy requirements, the variation in intake expected is illustrated in Figure 11. The variation in intake of the high quality forage is quite dramatic and indicates the difficulty of assigning an intake potential to high quality forages. The use of reference animals, such as those in the INRA system, does not solve the problem of measuring the intake potential of high quality forages unless the energy demand of the reference animal is high. Most scientists would agree that the intake potential of the high quality forage should be defined by the highest intake observed (Figure 11). Differences in animal requirements can have significant, but much less, impact on the intake of low quality forages (Figure 11). The assignment of an intake potential for low quality forage is more controversial. Using logic similar to that used to measure the intake potential of the high quality forage would assign the highest intake as the intake potential of the low quality forage. However, I would argue that the lowest intake (associated with the animal with lowest requirements) represents the intake potential of the low quality forage because it comes the closest to meeting the animal's energy demand. The higher intake of the low quality forage by the animal with higher requirements results in an intake potential that over-evaluates this forage. Perhaps the definition of intake potential should be changed to be "the maximum intake of a forage which allows animals to meet their potential energy requirements."

When most scientists use the term "intake potential" they envision the intake of a forage when fill is limiting intake. For high quality forages, the intake potential

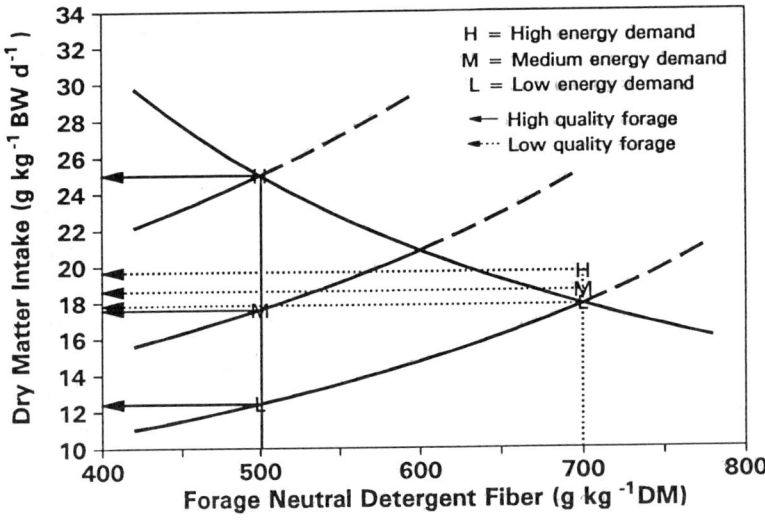

Figure 11. Expected intakes of sheep with low (L), medium (M), and high (H) energy demands when fed two grass forages with low (500 g kg^{-1} DM) and high (700 g kg^{-1} DM) neutral detergent fiber concentrations. Depending on the energy requirements of the animal, expected intake of the high quality forage may be greater than, equal to, or less than the intakes of the low quality forage. The true intake potential of a forage can be measured only when the energy demand of the animal matches forage quality.

will only be measured if the energy demand of the animal is extremely high so that energy demand does not limit intake (Figure 11). This conclusion helps answer the question of which animal characteristic (intake constraint or energy demand) should be measured during covariate periods to reduce animal effects on intake measurements. Although it could be argued that the intake constraint of animals should be determined because the filling effect of forages is the desired measurement, the most important concern is to assure that energy demand of the animal has not limited intake of the test forage during the experiment. It is clear that the V-shaped distribution intake (Figures 3 and 8) is related more to variation in energy demand which limits intake of high quality diets than to variation in intake constraint of animals when fill limits intake. The model of Mertens (1987) can be used to identify instances in which fill is not limiting intake. It is unlikely that fill limits intake when NDF intake is less than 11 g kg^{-1} BW d^{-1}, which is about one standard deviation below the average NDF intake that maximized DMI.

The fact that intake results from the interaction of the forage, animal, and feeding situation precludes the prediction of actual intakes based solely on feed characterization. Attempts to use more sophisticated chemical analyses or new technologies such as near infrared reflectance spectroscopy to improve the prediction of intake fail to appreciate that intake depends as much on the animal and feeding situation as it does on feed characteristics. The answer to the problem of predicting intake depends more on the measurement of the forage's intake potential when limited by fill and on mathematical models that integrate the effects of the animal, feed, and feeding situation than on more detailed analyses of the forage itself. Although elaborate dynamic models of digestion that include the kinetics of digestion, particle size reduction, volume or density changes, and passage will improve our understanding of the mechanisms of intake regulation, even the simple static model described by equations 1 through 10 provides a valuable framework on which to improve our estimation and use of the intake potential of forages.

UNCERTAINTIES IN INTAKE REGULATION THEORIES

Intake regulation is a complex function that is critical to the very survival of the animal. It is illogical to assume that a function so important would be controlled by a single mechanism because this would surely limit the survival of the individual and species when confronted with the myriad of environments that animals have faced over evolutionary time and individual animals encounter during their lifetimes. Because of the variation in animals, diets, feeding situations, and intake regulation mechanisms, it is simplistic to assume that any single experiment will eliminate a mechanism of intake from consideration, or prove that it is *the* mechanism of control.

The goal of this chapter is to describe the important concepts of intake regulation that are commonly accepted and to reduce them to a simple mathematical basis that can be used as a framework for evaluating intake as a component of forage quality assessment. The value of the NDF-Energy Intake System is that it integrates information about the animal, dietary, and feeding situation into a single system. Coefficients were developed using data from lactating dairy cow experiments in which rations were balanced to provide adequate protein, minerals, and vitamins. It is expected that this model may not function adequately at the extremes of high quality (<250 g NDF kg^{-1} DM) or of low quality (>700 g NDF kg^{-1} DM and <50 g CP kg^{-1} DM) rations. The simple model of intake regulation was presented to provide a framework for understanding the factors that influence intake and stimulate debate. To provide all the supporting and contradicting evidence for intake regulation based on the theories of physiological energy demand, physical fill limitation, and psychogenic modification would require the writing of a book rather than a chapter.

Some scientists are critical of the concept that only physical fill limits the intake of low energy, high fill diets and that only chemical factors or physiological energy demand regulates the intake of high energy, low fill diets for being too simplistic. (It should be noted that the model presented does not require that forage intake always be limited by fill or the concentrate intake always be limited by energy demand.) They suggest that we must go beyond NE and NDF when studying intake regulation and search for the physiological signals that control intake. However, there is little evidence that the pursuit of particular signals critical to the control of short term intake has provided information that is useful for predicting intakes or evaluating forages quantitatively. Animal physiological signals do not regulate intake in isolation. It is the interaction of the animal, diet, and feeding situation that provides the triggers for control signals. Thus, characteristics of all three factors must be measured and used to predict the intake potential of a specific forage, fed to a particular animal, under defined conditions.

Knowledge about intake regulation is critical to the productive use of forages. New information needs to be obtained, critically evaluated, and used to refine current theories or to develop new ones. The book of Forbes (1986) provides an excellent overview of the mechanism of intake control and the factors influencing those mechanisms. For alternative views about intake regulation, you are encouraged to read the papers of Egan (1965, 1970) and the reviews of Grovum (1987), Ketelaars and Tolkamp (1991), and Owens et al. (1991). These scientists question the concept that fill regulates intake and stress the importance of physiological signals, oxygen utilization, or post-ruminal (or tissue level) protein (or energy) status as the regulator of intake. These alternative hypotheses are welcomed and can be useful in forage quality evaluation when based on animal and dietary characteristics that can be measured and used quantitatively to predict intake over a wide range of forage qualities and animal productions.

Although the NDF-Energy Intake System is based only partially on the concept of fill limitation of intake (and does not require that the fill limitation be located in the rumen), data that apparently contradict the concept of fill limitation in the rumen must be addressed to avoid the appearance of a noncritical review of the literature. The importance of fill in limiting intake in ruminants has stimulated a long term debate among scientists. Controversy arises because investigators often attempt to determine if fill is *the only* mechanism of intake regulation rather than confirm that it is *a potential* mechanism of intake regulation. Physical limitation regulates intake only when the bulk of the diet is high in relation to the animal's energy demand. When researchers observe that forages or diets do not maximize ruminal fill, they have simply identified feeding situations in which the animal's energy demand, accessible feed supply, nutrient imbalances, or external factors limit intake. Because intake is not limited by fill in a particular circumstance does not mean that fill never limits intake.

Several researchers have observed that ruminants do not eat to a constant ruminal fill when offered some diets (Montgomery and Baumgardt, 1965; Ingalls et al., 1966; Egan, 1970; Doyle, 1984; Weston, 1989; Johnson and Combs, 1991; 1992). The simple model of intake regulation can explain many of the differences in ruminal contents with changes in the nutritive value of the feed or physiological status of the animal that have been reported. It is intuitive that, when a mature animal is fed a high quality, low fiber forage, ruminal contents will not constrain intake and do not represent "ruminal fill" (Figure 10). The negative relationship between ruminal contents and energy availability of the diet reported by Weston (1985) was often the result of animals consuming the high quality diets to meet their energy demand resulting in a concurrent reduction in ruminal contents. Doyle (1984) observed that sheep ate to a constant ruminal content for each particular forage, but the amount of ruminal contents was not constant among forages and did not always equal the maximum capacity of the rumen. The amount of ruminal contents often was least when the quality of the forage was highest, suggesting that energy demand limited intake in these situations. Ingalls et al. (1966) suggested that factors other than fill control the intake of high quality forages such as legume hay. His conclusion is consistent with the concept that energy demand limits the intake of highly digestible, low fiber, legume forages.

Many of the experiments designed to evaluate the fill limitation theory are not adequate tests because the intake of the control diet was not limited by ruminal fill. For example, the NDF-Energy Intake System suggests that the low fiber (<280 g NDF kg^{-1} DM) diets fed by Johnson and Combs (1991, 1992) did not limit intake due to fill; therefore, the addition of bladders to the rumen would have little effect under these circumstances, as they observed. It is not necessary to conclude that ruminal expansion occurred, only that the rumens of these cows contained unused volume that could be occupied by exogenous bladders without affecting intake. Likewise, in an experiment designed to show that neural receptors in the rumen did not sense distension comparable to the volume of the meal, sheep were fed pelleted alfalfa which was unlikely to provide the ruminal fill necessary to elicit a distension response (Grovum, 1979).

Owens et al. (1991) suggested that fill does not limit intake because no more than 60% of the ruminal water capacity of forage-fed or grazing ruminants was occupied by ruminal contents. The hypothesis that fill is not limiting in these situations is based on three assumptions: 1) water capacity of the rumen represents fill limitation volume, 2) ruminal liquid volume calculated by difference represents ruminal fill volume, and 3) diets used in the experiments were limited by fill. It is possible and even likely that water may distend the rumen beyond the physiological limit that functions to regulate intake due to fill. It is not uncommon to observe animals with rumens so tightly packed with contents after a meal that it is difficult to push the cannula plug into the rumen to remove it before the animal's rumen can be evacuated. It is difficult to believe that this level of distension would not limit intake even if this volume did not equal ruminal water capacity. In the experiments when ruminal water capacity was measured, it was not recorded whether rumens were full of contents when emptied to determine if full rumen volumes were compared to ruminal water capacity.

Rumen volume is often derived by subtracting ruminal DM contents from the total weight of ruminal contents. This assumes that DM in the rumen contributes nothing to ruminal volume and that the volume of ruminal contents is related only to the liquid volume in the rumen. This assumption leads to the conclusion that the density of ruminal contents (total ruminal contents divided by ruminal liquid volume) is greater than 1.10 g mL^{-1}. The buoyancy of gas trapped in digesta and the low density of large forage particles suggest that the density of ruminal contents should be less than that of water (1.0 g mL^{-1}). Thus, the liquid volume of ruminal contents determined by difference probably does not reflect the ruminal distension volume detected by the animal.

In some experiments cited when rejecting the fill limitation of intake, it is questionable whether intakes were limited by fill. Miner et al. (1990) fed their sheep a complete, pelleted commercial lamb-growing ration that would not be expected to have intake limited by fill. Likewise, Stanley et al. (1993) fed chopped alfalfa hay which may not have had the filling effect necessary to limit the intake of the pregnant beef cows. In their experiments, NDF intakes of nonlactating cows were <11.0 g kg^{-1} BW d^{-1}. Hannah et al. (1991) fed forage containing 790 g NDF kg^{-1} DM which should have limited intake by fill. However the NDF intake of the control steers was about 7.0 g kg^{-1} BW d^{-1} which is less than the 12.0 g kg^{-1} BW d^{-1} suggested as the average intake constraint. Interestingly, NDF intake increased to 12.0 g kg^{-1} BW d^{-1} when diets were supplemented. These observations indicate that there are conditions when the intake of low quality forages is not limited by fill. The value of the simple model is that it helps to identify situations in which fill is not limiting intake.

Other experiments indicate that fill limitation cannot be defined by a single constant NDF intake. Hightshoe et al. (1991) and Beck (1992) observed NDF intakes above 13.0 g kg^{-1} BW d^{-1} when low quality forages were supplemented. Their work confirms that the NDF intake constraint (Figure 4) can be exceeded when animals with relatively high energy demands are fed forages with high fiber concentrations. Intakes of NDF >13.0 g kg^{-1} BW d^{-1} in these trials also suggest that supplementation may have eliminated a post-ruminal factor limiting intake, but probably did not allow these animals to achieve maximal production. Similarly, when ground or pelleted high fiber feeds are fed (Vanzant et al., 1991), intakes of NDF can exceed 13.0 g kg^{-1} BW d^{-1}. As suggested by Mertens (1992), this probably occurs because ground feeds do not have the same volume per unit of NDF as does long forage. The simple model should be refined to adjust chemically measured NDF for differences in particle size that have an impact on the filling effect of ground forages and by-product feeds.

Research reported by Egan (1965, 1970), Doyle (1984), Leng (1990), and Owens et al. (1991) indicates that post-ruminal metabolism or physiology can impact the intake of low quality forages. In each situation, the estimated NDF intake of the low quality forage was <11.0 g kg^{-1} BW d^{-1}. This indicates that fill did not limit intake and suggests that the supply of amino acids to tissues, the balance between amino acids and energy at the tissue level, or the ratio of glucogenic to ketogenic end products of digestion was the factor limiting intake. It is also possible that intake of high fiber diets may be limited by the chewing time needed to process the fiber for passage through the digestive tract (Van Soest, 1982). Lower than expected intakes of low quality forages indicate that the simple model presented in this chapter will not predict intake for all forages and suggests that factors other than fill may need to be investigated to predict the intake of low quality forages. More research is needed to define the interrelationships among ruminal fill, digesta flow through the entire tract, and tissue metabolism that affect intake over a broad range of forage qualities.

Should the simple model of intake based on energy demand and fill limitation be condemned because it fails to predict the intake of extremely low quality forages or approved because it indicates that intakes of these forages are not limited by fill? Should the model be abandoned because it provides no information about the physiological signals controlling intake or accepted because it can be used to predict intake in many situations based on measurable dietary and animal characteristics? Should it be criticized because it is not a complex system that perfectly estimates intake in all situations or praised because it can serve as a framework for understanding intake regulation and predicting intake of many forages and optimal rations? Future research will decide which of these alternatives is correct and whether the simple model can or should be modified to explain an ever-widening range of intake situations. Until these questions are resolved, the aggregation of information into the concepts of intake constraint and energy requirements of the animal and filling effects and energy availability of the feed should not eliminate debate about factors influencing intake. Instead, the simple model presented in this chapter can serve as a framework for understanding general concepts of intake regulation.

Conclusion

The regulation of intake involves hunger and satiety signals that operate through various hormonal and neural mechanisms to control both short and long term voluntary intake. Homeostatic mechanisms serve to regulate intake so that BW and tissue stores are maintained during adult life. Homeorhetic mechanisms adjust intake to meet the specific needs of various physiological states such as growth, pregnancy, and lactation. It is commonly accepted that an animal's appetite, or drive to eat, is a function of its energy requirement which is determined by its genetic potential and physiological state. When high quality diets are fed, the animal eats to meet its energy demand, and intake is limited by the animal's genetic potential to utilize absorbed energy. However, when low quality diets are fed, the animal consumes feed to the level that matches the capacity of the gastrointestinal tract to handle digesta. The tolerance of the animal for ruminal distension can increase to accommodate high nutrient demands, but the animal also compromises performance to balance energy output with input. The dominant roles of physiological regulation and physical limitation on intake are modified by psychogenic stimuli related to palatability, social interactions, disease, and feeding management. Thus, intake is affected by characteristics of the animal, feed, and feeding situation.

Much of the confusion in the literature about intake measurement, regulation, and prediction can be related to lack of appreciation that the animal and feeding management can affect intake as much as forage characteristics. In addition, debates occur about whether fill is *the* mechanism rather than *a* mechanism of intake regulation without recognizing that characteristics of the diet and the animal determine whether fill limits intake in any specific instance. Just because ruminal contents can be increased by a treatment does not prove that intake is never limited by fill, only that intake was not limited by fill during the pre-treatment period. Simple static models can be useful for describing the biological concepts involved in regulating intake and clarifying the intrinsic forage characteristics that control intake in any particular situation.

Equations describing the physical and physiological mechanisms of intake regulation show that they each control intake at opposite ends of the forage quality scale. In addition, psychogenic stimuli can modulate intake over the full range of forage quality. Thus, intake is regulated by at least three different mechanisms and it is the responsibility of researchers to characterize their specific experimental situation carefully enough to identify which factor is dominant. Although complex dynamic models can be used, even simple static models can help identify the factors limiting intake in a given experiment. When measuring the intake potential (maximum intake) of forages, it becomes evident that intake should not be limited by the energy demand of the animal. For low quality forages, this is not a problem because even the energy demand of animals with low requirements, such as adults at maintenance, cannot be met and intake is limited by fill. However, when high quality forages are evaluated, they must be fed to animals with extremely high energy demands to ensure that the intrinsic filling effect of the forage and not the genetic potential and physiological status of the animal is limiting intake.

Although it is clear that intake potential is an intrinsic property that is crucial to forage evaluation and utilization, predicting the intake of a specific forage when fed to a specific animal may depend on more than feed characteristics. Thus, prediction of intake in general may not benefit from more sophisticated or detailed analysis of the feed because the feed alone does not determine intake. Measuring more properties of the feed is irrelevant if intake is limited by characteristics of the animal or feeding situation. Mathematical models are needed to integrate all components of intake regulation for the general prediction of intake. A simple model can be derived based on the currently accepted theories of intake regulation. Using NDF to represent the filling effect and NE to reflect the available energy of the forage, the simple static model can be a starting point for defining intake relationships. Although NDF may not completely describe the filling effect of the diet, it can serve as the basis for identifying intakes limited by fill. More rapid progress in forage evaluation for intake may be made if deviations from the general relationship between NDF and intake are identified, quantified, and used to refine the simple static model.

REFERENCES

Abrams, S.M., H.W. Harpster, P.J. Wangness, J.S. Shenk, E. Keck, and J. Rosenberger. 1987. Use of a standard forage to reduce effects of animal variation on estimates of mean voluntary intake. J. Dairy Sci. 70:1235-1240.

Adolph, E.F. 1947. Urges to eat and drink in rats. Am. J. Physiol. 151:110-125.

ARC (Agricultural Research Council). 1980. The nutrient requirements of ruminant livestock. Commonwealth Agric. Bureaux, Farnham Royal, England. 351 pp.

Baile, C.A., and J.M. Forbes. 1974. Control of feed intake and regulation of energy balance in ruminants. Physiol. Rev. 54:160-214.

Baile, C.A., and J. Mayer. 1970. Hypothalamic centres: Feedbacks and receptor sites in the short-term control of feed intake. p. 254-263. *In* A.T. Phillipson (ed.) Physiology of digestion and metabolism in the ruminant. Oriel Press Ltd. Newcastle upon Tyne, England.

Balch, C.C., and R.C. Campling. 1962. Regulation of voluntary intake in ruminants. Nutr. Abstr. Rev. 32:669-686.

Balch, C.C., and R. C. Campling. 1969. Voluntary intake of food. p. 554-579. *In* W. Lenkeit, K. Breirem, and E. Crasemann (ed.) Hanbuch der Tierernahrung Vol.1. Paul Parey, Hamburg.

Baldwin, R.L., L.J. Koong, and M.J. Ulyatt. 1977. A dynamic model of ruminant digestion for evaluation of factors affecting nutritive value. Agric. Systems 2:255-288.

Baumgardt, B.R. 1970. Control of feed intake in the regulation of energy balance. p. 235-253. *In* A.T. Phillipson (ed.) Physiology of digestion and metabolism in the ruminant. Oriel Press Ltd. Newcastle upon Tyne, England.

Bauman, D.E., and W.B. Currie. 1980. Partitioning of nutrients during pregnancy and lactation: A review of mechanisms involving homeostasis and homeorhesis. J. Dairy Sci. 63:1514-1529.

Baumont, R., J.P. Brun, and J.P. Dulphy. 1989. Influence of the nature of hay on its digestibility and the kinetics of intake during large meals in sheep and cows. p. 787-788. *In* Vol. II Proc. XVI Inter. Grassland Cong., 4-11 Oct., 1989, Nice, France. Assoc. francaise pour la prod. fourrage'ere, Versailles, France.

Beck, T.J., D.D. Simms, R.C. Cochran, R.T. Brandt, Jr., E.S. Vanzant, and G.L. Kuhl. 1992. Supplementation of ammoniated wheat straw: Performance and forage utilization characteristics in beef cattle receiving energy and protein supplements. J. Anim. Sci. 70:349-357.

Bell, F.R. 1971. Hypothalamic control of food intake. Proc. Nutr. Soc. 30:103-109.

Bines, J.A. 1971. Metabolic and physical control of food intake in ruminants. Proc. Nutr. Soc. 30:116-122.

Bines, J.A. 1979. Voluntary food intake. p. 23-48. In W.H. Broster and H. Swan (ed.) Feeding strategy for the high yielding dairy cow. Eur. Assoc. Anim. Prod. Publ. No. 25, Granada Publishing Ltd., London.

Black, J.L., D.E. Beever, G.J. Faichney, B.R. Howarth, and N. McC.Graham. 1981. Simulation of the effect of rumen function on the flow of nutrients from the stomach of sheep: Part 1 - Description of a computer program. Agric. Systems 6:195-219.

Black, J.L., W. F. Colebrook, S.G. Gherardi, and P.A. Kennedy. 1989. Diet selection and the effect of palatability on voluntary feed intake by sheep. p. 139-151. In Proc. 50th Minnesota Nutr. Conf., Minn. Agr. Ext. Ser., St. Paul.

Blaxter, K.L. 1989. Energy metabolism in animals and man. Cambridge Univ. Press, Cambridge, 336 pp.

Blaxter, K.L., F.W. Wainman, and R.S. Wilson. 1961. The regulation of food intake by sheep. Anim. Prod. 3:51-61.

Blaxter, K.L., and R.S. Wilson. 1962. The voluntary intake of roughages by steers. Anim. Prod. 4:351-358.

Booth, D.A. 1972. Postabsorptively induced suppression of appetite and the energostatic control of feeding. Physiol. Behav. 9:199-202.

Booth, D.A., F.M. Toates, and S.V. Platt. 1976. Control system for hunger and its implications in animals and man. p. 127-143. In D. Novin, W. Wyrwicka, and G.A. Bray (ed.) Hunger: Basic mechanisms and clinical implications. Raven Press, New York.

Bray, G.A. 1974. Endocrine factors in the control of food intake. Fed. Proc. 33:1140-1145.

Briceno, J.V., H.H. Van Horn, B. Harris, Jr., and C.J. Wilcox. 1987. Effects of neutral detergent fiber and roughage source on dry matter intake and milk yield and composition of dairy cows. J. Dairy Sci. 70:298-308.

Brobeck, J.R. 1960. Regulation of feeding and drinking. p. 1197-1206. In J. Field (ed.) Handbook of physiology. Vol. II, Section I, Neurophysiology. Am. Physiol. Soc., Washington, DC.

Brody, S. 1945. Bioenergetics and growth. Hafner Publishing Co., Inc., New York. 1023 pp.

Brown, R.H., R.E. Blaser, and J.P. Fontenot. 1968. Effect of spring harvest date on nutritive value of orchardgrass and timothy. J. Anim. Sci. 27:562-567.

Burns, J. C., D.H. Timothy, R.D. Mochrie, and D.S. Fisher. 1988. Relative grazing preference of *Panicum* germplasm from three taxa. Agron. J. 80:574-579.

Bywater, A.C. 1984. A generalised model of feed intake and digestion in lactating cows. Agric. Systems 13:167-186.

Cabanac, M., R. Duclaux, and N.H. Spector. 1971. Sensory feedback regulation of body weight: Is there a ponderstat? Nature 229:125-127.

Campling, R.C. 1970. Physical regulation of voluntary intake. p. 226-234. *In* A.T. Phillipson (ed.) Physiology of digestion and metabolism in the ruminant. Oriel Press Ltd., Newcastle upon Tyne, England.

Campling, R.C., and C.C. Balch. 1961. Factors affecting the voluntary intake of food by cows. 1. Preliminary observations of the effects, on the voluntary intake of hay, of changes in the amount of the reticulo-ruminal contents. Br. J. Nutr. 15:523-530.

Colburn, M.W., and J.L. Evans. 1968. Reference base, W^b, of growing steers determined by relating forage intake to body weight. J. Dairy Sci. 51:1073-1076.

Colovos, N.F., J.B. Holter, R.M. Koes, W.E. Urban, and H.A. Davis. 1970. Digestibility, nutritive value and intake of ensiled corn plant (*Zea mays*) in cattle and sheep. J. Anim. Sci. 30:819-824.

Conrad, H.R. 1966. Symposium on factors influencing the voluntary intake of herbage by ruminants: Physiological and physical factors limiting intake. J. Anim. Sci. 25:227-235.

Conrad, H.R., A.D. Pratt, and J.W. Hibbs. 1964. Regulation of feed intake in dairy cows. I. Change in importance of physical and physiological factors with increasing digestibility. J. Dairy Sci. 47:54-62.

Crampton, E.W., E. Donefer, and L.E. Lloyd. 1960. A nutritive value index for forages. J. Anim. Sci. 19:538-544.

de Jong, A. 1986. The role of metabolites and hormones as feedbacks in the control of food intake in ruminants. p. 459-478. *In* L.P. Milligan, W.L. Grovum, and A. Dobson (ed.) Control of digestion and metabolism in ruminants. Proc. Sixth Int. Symp. on Ruminant Physiology, Banff, Canada. Prentice-Hall, Englewood Cliffs, NJ.

Demment, M.W., and P.J. Van Soest. 1985. A nutritional explanation for body-size patterns of ruminant and non-ruminant herbivores. Am. Naturalist 125:641-672.

Dinius, D.A., and B.R. Baumgardt. 1970. Regulation of food intake in ruminants. 6. Influence of caloric density of pelleted rations. J. Dairy Sci. 53:311-316.

Doyle, P.T. 1984. Fibre intake, digestion and flow rates in sheep. p. 105-116. *In* S.K. Baker, J.M. Gawthorne, J.B. Mackintosh, and D.B. Purse (ed.) Ruminant physiology concepts and consequences. Univ. Western Australia, Perth, W.A.

Dulphy, J.P., and C. Demarquilly. 1983. Voluntary intake as an attribute of feeds. p. 135-156. *In* G.E. Robards and R.G. Packham (ed.) Feed information and animal production. Second symp. int. network of feed info. centres. Commonwealth Agric. Bureaux. Farnham Royal, UK.

Edholm, O.G., J.G. Fletcher, E.M. Widdowson, and R.A. McCance. 1955. The energy expenditure and food intake of individual men. Br. J. Nutr. 9:286-300.

Egan, A.R. 1965. Nutritional status and intake regulation in sheep. II. The influence of sustained duodenal infusion of casein or urea upon voluntary intake of low protein roughages by sheep. Aust. J. Agric. Res. 16:463-472.

Egan, A.R. 1970. Nutritional status and intake regulation in sheep. VI. Evidence for variation in setting of an intake regulation mechanism relating to the digesta content of the reticulo-rumen. Aust. J. Agric. Res. 21:735-746.

Ellis, W.C. 1978. Determinants of grazed forage intake and digestibility. J. Anim. Sci. 61:1828-1840.

Fisher, D.S., J.C. Burns, and K.R. Pond. 1987. Modelling ad libitum dry matter intake by ruminants as regulated by distension and chemostatic feedbacks. J. Theor. Biol. 126:407-418.

Forbes, J.M. 1977a. Interrelationships between physical and metabolic control of voluntary food intake in fattening, pregnant and lactating mature sheep: A model. Anim. Prod. 24:91-101.

Forbes, J.M. 1977b. Development of a model of voluntary food intake and energy balance in lactating cows. Anim. Prod. 24:203-214.

Forbes, J.M. 1983. Models for the prediction of food intake and energy balance in dairy cows. Livest. Prod. Sci. 10:149-157.

Forbes, J.M. 1986. The voluntary intake of farm animals. Butterworths, London. 206 pp.

Grovum, W.L., and G.D. Phillips. 1978. Factors affecting the voluntary intake of food by sheep. 1. The role of distension, flow-rate of digesta and propulsive motility in the intestines. Br. J. Nutr. 40:323-336.

Grovum, W.L. 1979. Factors affecting the voluntary intake of food by sheep. 2. The role of distension and tactile input from compartments of the stomach. Br. J. Nutr. 42:425-436.

Grovum, W.L. 1987. A new look at what is controlling food intake. p. 1-39. *In* F.N. Owens (ed.) Proc. Symp.: Feed intake by beef cattle. Oklahoma State Univ. Agric. Expt. Sta., Stillwater.

Hannah, S.M., R.C. Cochran, E.S. Vanzant, and D.L. Harmon. 1991. Influence of protein supplementation on site and extent of digestion, forage intake, and nutrient flow characteristics in steers consuming dormant bluestem-range forage. J. Anim. Sci. 69:2624-2633.

Harlan, D.W., J. B. Holter, and H.H. Hayes. 1991. Detergent fiber traits to predict productive energy of forages fed free choice to nonlactating dairy cattle. J. Dairy Sci. 74:1337-1353.

Heaney, D.P., W.J. Pidgen, D.J. Minson, and G.I. Pritchard. 1963. Effect of pelleting on energy intake of sheep from forages cut at three stages of maturity. J. Anim. Sci. 22:752-757.

Hendricksen, R.E., D.P. Poppi, and D.J. Minson. 1981. The voluntary intake, digestibility, and retention time by cattle and sheep of stem and leaf fractions of a tropical legume (*Lablab purpurens*). Aust. J. Agric. Res. 32:389-398.

Hightshoe, R.B., R.C. Cochran, L.R. Corah, D.L. Harmon, and E.S. Vanzant. 1991. Influence of source and level of ruminal-escape lipid in supplements on forage intake, digestibility, digesta flow, and fermentation characteristics in beef cattle. J. Anim. Sci. 69:4974-4982.

Hyppola, K., and O. Hasunen. 1970. Dry matter and energy standards for dairy cows. Acta Agralia Fennica 116:1-59.

Illius, A.W., and I.J. Gordon. 1991. Prediction of intake and digestion in ruminants by a model of rumen kinetics integrating animal size and plant characteristics. J. Agric. Sci., Camb. 116:145-157.

Ingalls, J.R., J.W. Thomas, E.J. Benne, and M. Tesar. 1965. Comparative response of wether lambs to several cuttings of alfalfa, birdsfoot trefoil, bromegrass and reed canarygrass. J. Anim. Sci. 24:1159-1164.

Ingalls, J.R., J.W. Thomas, M.B. Tesar, and D.L. Carpenter. 1966. Relations between ad libitum intake of several forages species and gut fill. J. Anim. Sci. 25:283-289.

Ingvartsen, K.L., H.R. Andersen, and J. Foldager. 1992. Random variation in voluntary dry matter intake and the effect of day length on feed intake capacity in growing cattle. Acta. Agric. Scand., Sect. A, Anim. Sci. 42:121-126.

INRA (Institut National de la Recherche Agronomique). 1989. Ruminant nutrition: Recommended allowances and feed tables, R. Jarrige (ed.) John Libbey Eurotext, Paris. 389 pp.

Johnson, T.R., and D.K. Combs. 1991. Effects of prepartum diet, inert bulk and dietary polyethylene glycol on dry matter intake of lactating dairy cows. J. Dairy Sci. 74:933-944.

Johnson, T.R., and D.K. Combs. 1992. Effects of inert rumen bulk on dry matter intake in early and midlactation cows fed diets differing in forage content. J. Dairy Sci. 75:508-519.

Jones, G.M. 1972. Chemical factors and their relation to feed intake regulation in ruminants: A review. Can. J. Anim. Sci. 52:207-239.

Jones, G.M., R.E. Larson, A.H. Javed, E. Donefer, and J.M. Gaudreau. 1972. Voluntary intake and nutrient digestibility of forages by goats and sheep. J. Anim. Sci. 34:830-838.

Journet, M., and B. Remond. 1976. Physiological factors affecting the voluntary intake of feed by cows: A review. Livestock Prod. Sci. 3:129-146.

Jung, H., and J. Linn. 1988. Forage NDF and intake. A critique. p. 39-48. *In* Proc. 48th Minnesota Nutr. Conf., Minn. Agr. Ext. Ser., St. Paul.

Kahn, H.E., and C.R.W. Spedding. 1984. A dynamic model for the simulation of cattle herd production systems: 2 - An investigation of various factors influencing the voluntary intake of dry matter and the use of the model in their validation. Agric. Systems 13:63-82.

Ketelaars, J.J.M.H., and B.J. Tolkamp. 1991. Toward a new theory of feed intake regulation in ruminants. Agric. Univ. Wageningen, The Netherlands. 254 pp.

Kleiber, M. 1961. The fire of life. An introduction to animal energetics. John Wiley and Sons, New York. 454 pp.

Kristensen, V.F. 1983. Styring af foderoptaglsen ved hjaelp af foderrationnens sammensaetning og valg af fodringsprincip. p. 7.1-7.35. *In* V. Ostergaard and A. Neimann-Sorensen (ed.) Optimale foderrationer til malkekoen. Fodervaerdi, foderoptagelse, omsaetning og produktion. Report 551, Statens Husdybrugsforsog. (Nat. Inst. Anim. Sci.), Copenhagen.

Lehman, F. 1941. Die lehre vom ballast. Z. Tierernahrung u. Futtermittelkunde. 5:155-173.

Leng, R.A. 1990. Factors affecting the utilization of 'poor-quality' forages by ruminants particularly under tropical conditions. Nutr. Res. Rev. 3:277-303.

Llamas-Lamas, G., and D.K. Combs. 1991. Effect of forage to concentrate ratio and intake level on utilization of early vegetative alfalfa silage by dairy cows. J. Dairy Sci. 74:526-536.

MacRae, J.C., J.S. Smith, P.J.S. Dewey, A.C. Brewer, D.S. Brown, and A. Walker. 1985. The efficiency of utilization of metabolizable energy and apparent absorption of amino acids in sheep given spring- and autumn-harvested dry grass. Br. J. Nutr. 54:197-209.

Marten, G.C. 1970. Measurement and significance of forage palatability. p. D1-D55. *In* R.F. Barnes, D.C. Clanton, C.H. Gordon, T.J. Klopfenstein, and D.R. Waldo (ed.) Proc. national conference on forage quality evaluation and utilization. 3-4 Sep. 1969. Nebraska Center for Continuing Education, Lincoln.

Mayer, J., and D.W. Thomas. 1967. Regulation of food intake and obesity. Science 156:328-337.

Mertens, D.R. 1973. Application of theoretical mathematical models to cell wall digestion and forage intake in ruminants. Ph.D. dissertation. Cornell Univ., Ithaca, NY (Diss. Abstr. 74-10).

Mertens, D.R. 1985. Factors influencing feed intake in lactating cows: From theory to application using neutral detergent fiber. p. 1-18. In Proc. Georgia Nutr. Conf., Univ. of Georgia, Athens.

Mertens, D.R. 1987. Predicting intake and digestibility using mathematical models of ruminal function. J. Anim. Sci. 64:1548-1558.

Mertens, D.R. 1992. Nonstructural and structural carbohydrates. p. 219-235. In H.H. Van Horn and C.J. Wilcox (ed.) Large dairy herd management. Am. Dairy Sci. Assoc., Champaign, IL.

Mertens, D.R., and L.O. Ely. 1979. A dynamic model of fiber digestion and passage in the ruminant for evaluating forage quality. J. Anim. Sci. 49:1085-1095.

Mertens, D.R., and L.O. Ely. 1982. Relationship of rate and extent of digestion to forage utilization. J. Anim. Sci. 54:895-905.

Milford, R. 1960. Criteria for expressing the nutritional values of subtropical grasses. Aust. J. Agric. Res. 11:121-137.

Milford, R., and D.J. Minson. 1966. Intake of tropical pasture species. p. 815-822. In Vol. I, Proc. IX Inter. Grassland Congr., 7-20 Jan., Sao Paulo, Da Agriculture de Estado de Sao Paulo, Brazil.

Miner, J.L., M.A. Della-Fera, and J.A. Paterson. 1990. Blockade of satiety factors by central injection of neuropeptide Y in sheep. J. Anim. Sci. 68:3805-3811.

Ministry of Agriculture, Fisheries, and Food. 1990. UK tables of nutritive value and chemical composition of feedingstuffs. D.I. Givens and A.R. Moss (ed.) Rowett Research Services, Ltd., Aberdeen.

Minson, D.J., and R.A. Bray. 1986. Voluntary intake and in vivo digestibility by sheep of five lines of *Cenchrus ciliaris* selected on the basis of preference rating. Grass Forage Sci. 42:47-52.

Minson, D.J., and R. Milford. 1967. The voluntary intake and digestibility of diets containing different proportions of legume and mature Pangola grass (*Digitaria decumbens*). Aust. J. Expt. Agric. Anim. Husb. 7:546-551.

Minson, D.J., and J.H. Ternouth. 1971. The expected and observed changes in the intake of three hays by sheep after shearing. Br. J. Nutr. 26:31-39.

Moe, P.W., and H.F. Tyrrell. 1976. Estimating metabolizable and net energy of feeds. p. 232-237. In P.V. Fonnesbeck, L.E. Harris, and L.C. Kearl (ed.) First int. symp. on feed composition, animal nutrient requirements and computerization of diets. Utah State Univ., Logan.

Mogenson, G.J., and F. R. Calaresu. 1978. Food intake considered from the viewpoint of systems analysis. p. 1-24. *In* D.A. Booth (ed.) Hunger models. Computable theory of feeding control. Academic Press, New York.

Monteiro, L.A. 1972. The control of appetite in lactating cows. Anim. Prod. 14:263-281.

Montgomery, J.J., and B.R. Baumgardt. 1965. Regulation of food intake in ruminants. II. Rations varying in energy concentrations. J. Dairy Sci. 48:1623-1628.

Newsholme, E.A. 1985. Substrate cycles and energy metabolism: Their biochemical, biological, physiological and pathological importance. p. 174-186. *In* P.W. Moe, H.F. Tyrrell, and P.J. Reynolds (ed.) Energy metabolism of farm animals. Eur. Assn. Anim. Prod. Publ. No. 32, Rowman and Littlefield, Totowa, NJ.

NRC (National Research Council). 1985. Ruminant nitrogen usage. National Academy Press, Washington, DC. 138 pp.

NRC (National Research Council). 1987. Predicting feed intake of food-producing animals. National Academy Press, Washington, DC. 85 pp.

NRC (National Research Council). 1989. Nutrient requirements of dairy cattle. National Academy Press, Washington, DC. 85 pp.

Osbourn, D.F., R.A. Terry, G.E. Outen, and S.B. Cammell. 1974. The significance of a determination of cell walls as the rational basis for the nutritive evaluation of forages. p. 374-380. *In* Vol. 3 Proc. XII Inter. Grassland Cong., 11-20 June 1974, Izdatelbstvo, Moscow.

Owens, F.N., J. Garza, and P. Dubeski. 1991. Advances in amino acid and N nutrition in grazing ruminants. p. 109-137. *In* Proc. 2nd Grazing Livestock Nutr. Conf., 2-3 Aug. 1991, Steamboat Springs, CO. Oklahoma State Univ. Agr. Exp. Sta. Misc. Pub. MP-133.

Parra, R. 1978. Comparison of foregut and hindgut fermentation in herbivores. p. 205-230. *In* G.G. Montgomery (ed.) The ecology of arboreal folivores. Smithsonian Institute, Washington, DC.

Pienaar, J.P., C.Z. Roux, P.J.K. Morgan, and L. Grattarola. 1980. Predicting voluntary intake of medium quality roughages. S. Afr. J. Anim. Sci. 10:215-225.

Poppi, D.P., D.J. Minson, and J.H. Ternouth. 1981a. Studies of cattle and sheep eating leaf and stem fractions of grasses. I. The voluntary intake, digestibility and retention time in the reticulo-rumen. Aust. J. Agric. Res. 32:99-108.

Poppi, D.P., D.J. Minson, and J.H. Ternouth. 1981b. Studies of cattle and sheep eating leaf and stem fractions of grasses. III. The retention time in the rumen of large feed particles. Aust. J. Agric. Res. 32:123-137.

Raymond, W.F. 1969. The nutritive value of forage crops. Adv. Agron. 21:1-108.

Reid, J.T. 1961. Problems of feed evaluation related to feeding dairy cows. J. Dairy Sci. 11:2122-2133.

Reid, R.L., G.A. Jung, and W.V. Thayne. 1988. Relationships between nutritive quality and fiber components of cool season and warm season forages: A retrospective study. J. Anim. Sci. 66:1275-1291.

Russek, M. 1978. Semi-quantitative simulation of food intake control and weight regulation. p. 195-226. In D.A. Booth (ed.) Hunger models. Computable theory of feeding control. Academic Press, New York.

Stanley, T.A., R.C. Cochran, E.S. Vanzant, D.L. Harmon, and L.R. Corah. 1993. Periparturient changes in intake, ruminal capacity, and digestive characteristics of beef cows consuming alfalfa hay. J. Anim. Sci. 71:788-795.

Stevenson, J.M. 1954. Diet and survival. p. 165-188. In M.I. Ferrer (ed.) Trans. of the Third Conference on Cold Injury. Josiah Macy Jr. Foundation, New York.

Thornton, R.F., and D.J. Minson. 1972. The relationship between voluntary intake and apparent retention time in the rumen. Aust. J. Agric. Res. 23:871-877.

Thornton, R.F., and D.J. Minson. 1973. The relationship between apparent retention time in the rumen, voluntary intake, and apparent digestibility of legume and grass diets in sheep. Aust. J. Agric. Res. 24:889-898.

Ulyatt, M.J., K.L. Blaxter, and I. McDonald. 1967. The relations between the apparent digestibility of roughages in the rumen and lower gut of sheep, the volume of fluid in the rumen and voluntary intake. Anim. Prod. 9:463-470.

Van Soest, P.J. 1982. Nutritional ecology of the ruminant. O & B Books, Inc. Corvalis, OR. 374 pp.

Vanzant, E.S., R.C. Cochran, and D.E. Johnson. 1991. Pregnancy and lactation in beef heifers grazing tallgrass prairie in the winter: Influence on intake, forage utilization, and grazing behavior. J. Anim. Sci. 69:3027-3038.

Waldo, D.R. 1970. Factors influencing the voluntary intake of forages. p. E1-E22. In R.F. Barnes, D.C. Clanton, C.H. Gordon, T.J. Klopfenstein, and D.R. Waldo (ed.) Proc. national conference on forage quality evaluation and utilization. 3-4 Sep. 1969, Nebraska Center for Continuing Education, Lincoln.

Waldo, D.R. 1986. Effect of forage quality on intake and forage-concentrate interactions. J. Dairy Sci. 69:617-631.

Waldo, D.R., L.W. Smith, and E.L. Cox. 1972. Model of cellulose disappearance from the rumen. J. Dairy Sci. 55:125-129.

Weiss, W.P., and W.L. Shockey. 1991. Value of orchardgrass and alfalfa silages fed with varying amounts of concentrates to dairy cows. J. Dairy Sci. 74:1933-1943.

Weston, R.H. 1966. Factors limiting intake by sheep. I. The significance of palatability, the capacity of the alimentary tract to handle digesta and the supply of glucogenic substrate. Aust. J. Agric. Res. 17:939-954.

Weston, R.H. 1985. The regulation of feed intake in herbage-fed ruminants. Proc. Nutr. Soc. Aust. 10:55-62.

Weston, R.H. 1989. Factors limiting the intake of feed by sheep. XV. Voluntary feed consumption and digestion in lambs fed chopped roughage diets varying in quality. Aust. J. Agric. Res. 40:643-661.

Zemmelink, G. 1980. Effect of selective consumption on voluntary intake and digestibility of tropical forages. Agric. Res. Rep. (Verl. landouwk. Onderz) 896, Wageningen.

CHAPTER 12

MEASUREMENT OF FORAGE INTAKE

J. C. Burns, K. R. Pond, and D. S. Fisher

INTRODUCTION

The daily quantity of dry matter (DM) that is consumed by an animal is a measurement critical to making nutritional inferences about feed and subsequent animal responses. Consequently, direct measurements or estimates of DM intake (DMI) have always been of major interest to nutritionists (Schneider and Flatt, 1975; NRC, 1984; Jarrige, 1989). Accurate measurement of DMI provides the basis for application of nutritional requirements to achieve a particular animal response. It undergirds the concept of diet formulation relative to determining animal daily responses. Precise, but often inaccurate, estimates of DMI allow relative comparisons of feeds, but are less useful in quantitative applications.

Once estimates of DMI are obtained, intake of specific nutrient entities can be determined by multiplying DMI by the nutrient concentration in the DM consumed. The coefficient of digestion can be calculated for the nutrient after accounting for physiological losses (Chapter 15).

The measurement of DMI integrates a number of factors such as plant chemical properties, plant and animal physical characteristics, and animal physiological processes (Chapter 14). Animal physiological processes are greatly dependent on animal species, breed, type, and class, as well as production status.

Feed intake is adjusted upward as energy requirements increase. Forages (grazed or stored) generally have sufficiently high fiber concentrations that physically limit intake through gut fill or distension before energy demands are met. High quality pastures can be similar to concentrate feeds and intake may be limited by nutrient balance (Fisher et al., 1990; Poppi et al., 1990). In physical limitation, the rate that forage exits the rumen becomes an important factor in daily intake regulation. The extent and rate of plant tissue comminution upon initial mastication and through subsequent rumination (chewing) and digestion by rumen microflora greatly influences digesta passage and the time interval between meals (Owens and Goetsch, 1986). Dry matter intake of forage is often limited by the ease (rate) with which a diet can be prehended and the residence time of particles in the digestive tract (Chapter 17). The diet selected is forage species-specific and altered by plant maturity and morphology (Blaser et al., 1986).

J. C. Burns, Dep. of Crop Science and Animal Science, North Carolina State Univ. and USDA, ARS; K. R. Pond, Dep. of Animal Science, North Carolina State Univ.; D. S. Fisher, Dep. of Crop Science, North Carolina State Univ. and USDA, ARS, Raleigh, NC 27695. The use of trade names in this publication does not imply endorsement by North Carolina ARS or USDA, ARS, of the products named, or criticism of similar ones not mentioned. All programs and services of the USDA are offered on a nondiscriminatory basis without regard to race, color, national origin, religion, sex, age, marital status, or handicap.

In addition to physical limitations on intake, there are metabolic controls termed chemostatic, lipostatic, and thermostatic (Chapter 11). The chemostatic effect limits intake of animals on diets high in digestible energy concentration, the lipostatic effect limits intake as excessive body fat accumulates, and the thermostatic effect adjusts intake upward as environmental temperatures decline and downward as temperatures increase. These physical and metabolic regulations function simultaneously in controlling daily DMI (Egan, 1970; Forbes, 1986; Fisher et al., 1990; Ketelaars and Tolkamp, 1992). In a production setting, these factors regulating DMI must be recognized and considered in diet selection (type and form) because high DMI is a requisite for high daily performance. The relative importance and interactive effects of these regulatory mechanisms must be understood to understand DMI over the full range of forage digestibilities.

THE IDEAL INTAKE MEASUREMENT

The Concept

An assay to determine the potential DMI of a forage should reflect the animal's physiological status and diet selection without restrictions due to feeding level. Estimates of DMI of a forage, other than at unrestricted levels of feeding, provide some integration of diet selectivity and lack of satiety.

The eating behavior of ruminants has been clearly characterized as selective, with a strong preference for green leaf and against dead and stem tissues (Minson, 1981b). This preference phenomena operates in both grazing (Allden and Whittaker, 1970; Sheath et al., 1987) and stall (Burns et al., 1985a) settings, but to different degrees. In both cases, diet selection is partially forage species-dependent. In a grazing setting, canopy morphology (including height) can be very influential in altering diet composition and quantity consumed, where as in a stall setting, the canopy morphology is largely destroyed (Demment et al., 1987; Demment and Laca, 1993). As a consequence, the influence of plant morphology on the diet selected is altered. In a stall setting, additional physical characteristics (length of chop, brittleness, dust, etc.) become important. Daily DMI is ultimately determined by animal adjustments in eating behavior and digestive processes in response to biological and physical differences in their diet.

The true assessment of the nutritive value of a forage resides with the diet that the animal can select from within the bounds of the quality and quantity of forage offered. The bounds for the grazing animal are established by the characteristics of the forage on offer and by the quantity (kg ha^{-1}) of herbage mass (HM) present at each meal, and for the animal in confinement, by the characteristics of the forage on offer, the quantity of forage fed at each feeding, and the number of feedings each day.

Greater selectivity by the animal for leafy tissue from within the bounds of forage on offer will generally result in a higher quality diet and a greater DMI. The high extreme would be represented by the consumption of only leaf tissue until satiated without limitation of foraging time.

In Confinement

The concepts presented above suggest that, ideally, DMI of each forage (treatment) be assessed over a range of forage offered. This approach has been suggested by Zemmelink (1980) but generally has not been applied in evaluating DMI of forages fed in confinement. It has, however, been used in pasture studies. In the latter case, a range of forage on offer has been generated by using a range of set stocking rates (animals per unit land area) (Wheeler, 1962), or controlled through varying animal numbers using the put and take method (Mott, 1960) to maintain a range of HM (kg ha^{-1}). In these trials, quality and quantity of the diet selected from within the bounds of the HM offered is integrated and expressed in terms of animal daily performance. Estimates of daily intake in grazing trials, however, are generally obtained through indirect methods (Chapter 13).

Conducting an intake assay in confinement that evaluates every forage (treatment) over a minimum of three levels of offer, such as 95, 115, and 135% *ad libitum* (Appendix Table 12-1), would evaluate the physical and chemical characteristics of the forage as well as the behavioral and physiological responses of the animal to the forage. This approach would provide the following valuable information:

(a) The daily DMI when fed to maximize consumption and minimize refusals of the fed forage (95% *ad libitum*) giving DMI of the forage offered vs. the potential diet that could be consumed.

(b) The daily DMI when selective consumption is maximized (135% *ad libitum*) giving the potential DMI of the diet selected from the forage offered.

(c) An assessment of the potential selectivity when integrating diet quality and quantity (the intake response vs. level of forage offered).

(d) Valid comparisons of maximal DMI of similar forages among laboratories through using a range of feeding levels.

(e) A response surface for intake, digestibility, and performance that would permit economic assessment of the feeding level that should be used in practice (number of feeding levels needed may vary based on the nature of the response).

Evaluating DMI over a range of feeding levels provides information that has applications to many different production situations. Utilizing a range of feeding levels is most important when quality is low or when there is potential for selection within the diet for higher quality portions (i.e., the selection of leaves over stems). It is less critical for high quality or homogeneous forages but forage species interactions may still be present. An alternative approach has been suggested (Minson, 1981a) of separating forages into leaf and stem and estimating the intake and digestibility of these two fractions. Good physical separation, however, is difficult to achieve.

It is not to be implied that the evaluation of DMI of a forage at one feeding level is incorrect or is of no value, only that the inference from such data is limited to the level evaluated. This point is emphasized here because it has also not been well understood in grazing trials. The notion exists that a valid comparison among grazed forages can be made only if each forage (treatment) is evaluated over a range of at least three levels of HM (Bransby, 1988). The use of only one level of HM does not yield invalid data, but as noted above, limits the inference that should be made to the HM at which each forage (treatment) was evaluated. Response data from such discrete treatments, whether from stall or grazing trials, have limited usefulness for subsequent economic analysis.

Source of Forage

In most stall intake trials, one source of forage will generally constitute a treatment or is a common base for several treatments with individual animals considered replicates. Ideally, the forage source should be replicated in space or time or both. This would require either a much larger experiment or repeating it a second year, but would provide the biological base for the broader inferences that are generally, but incorrectly, made from experiments using a single-source forage. A basis for broader inference generally occurs in grazing trials where pasture treatments are replicated (in space) and repeated several years (in time).

Impact of Maturity Changes

The maturation process in forage plants is the major contributor to the change in their nutritive value and, consequently, impacts DMI both within and among most forage species. Legumes, such as ladino white clover (*Trifolium repens* L.), are exceptions because they have little true stem available for consumption. The term "maturity" generally encompasses growth and maturation processes involving both physiological changes and morphological development stages. During maturation, physiological change occurs with the onset of cell development beginning with undifferentiated tissue and continuing through differentiation and secondary cell wall thickening (Harris, 1990).

Morphological changes occur with plant development, reflecting differing proportions of leaf, sheath, stem, and inflorescence, the latter tissue making a contribution following the boot stage. Maturation, although a continuing physiological process, is generally viewed and evaluated nutritionally as a discrete-stage phenomena with the harvest event based mainly on morphology. This has led to the oversimplified conclusion that maturity and nutritive value are negatively and linearly associated (Blaser et al., 1986).

Ideally, the nutritive value of forages would be assessed throughout the maturation process. This would reveal critical periods relative to potentially large changes in herbage yield, DMI, DM digestion, and digestible DMI. An example is noted for a forage sorghum [Sorghum bicolor (L.) Moench] hybrid (Figure 1). The results (Ademosum et al., 1968) from goats show that the relationship between maturation and nutritive value is negative, but not linear. Selection of three discrete maturity stages such as vegetative, boot, and fully headed, as frequently practiced in stall nutrition studies, can force a significant negative and linear relationship ($r = -0.99$) as noted by the dashed lines in Figure 1.

Obtaining animal response estimates at frequent intervals during the plant maturation process, as done by Ademosum et al. (1968), requires sacrifice of the number of days per animal on which each coefficient is based. It provides, however, a moving average of the nutritive value of forage over the total range of forage maturation. The use of conventional methods (See Appendix Table 12-1) would require an extremely large experiment that would exceed most facilities and budgets. However, specific intervals could be identified for which

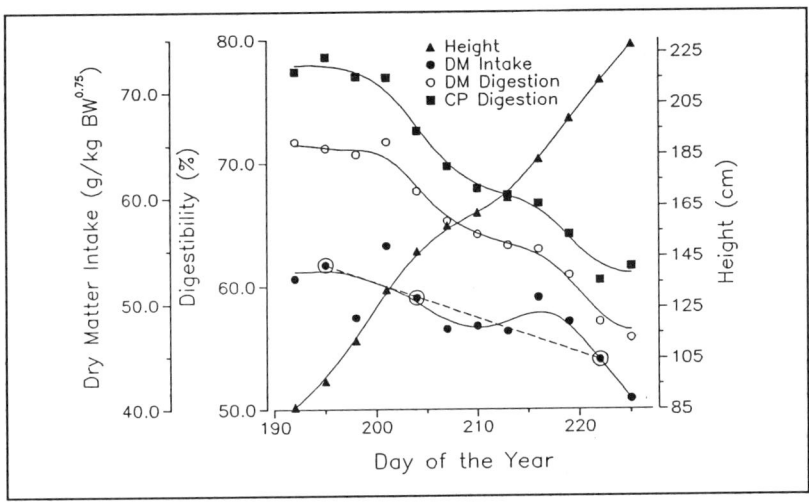

Figure. 1. Influence of maturity on height, DMI, DM digestion, and crude protein digestion of a sorghum hybrid [data from Ademosum et al. (1968) with goats using a continuous 3 d evaluation]. The dashed line shows the relationship for DM digestion when three discrete (circled points) maturities (vegetative, boot, and 40% headed) were selected.

a normal 21-d (7-d adjustment and 14-d measurement) intake trial (followed by a digestion phase) could be conducted. In such a trial, forage would either need to be stored dry or kept frozen and thawed before feeding. A 14-d harvest interval has been used by Burns et al. (1985b) to evaluate the maturity of switchgrass (*Panicum virgatum* L.) from late vegetative through heading. Longer intervals between evaluations results in less sharply identified changes in nutritive value with increasing maturity.

The practical importance of knowing how DMI varies with maturity (time) resides with economic considerations relative to harvesting the maximum forage yield (kg ha^{-1}) at a nutritional value that will permit acceptable daily animal responses. Forages compared in this manner give a proper assessment (including changes due to rainfall distribution, sunlight hours, location, etc. among years) of their potential nutritive value through maturation and provide especially valuable information to the producer inclined to fine-tune his profit margin.

On Pasture

A method that adequately estimates DMI of grazing animals remains essential to fully utilize the value of pasture research but continues to be elusive. Because the free grazing animal obtains its daily intake of a diet selected from within the bounds of the forage offered, the ideal method should incorporate both diet quality and the daily (24 h) quantity of the diet consumed. Such data would permit the application of nutrient requirements to expected animal daily performance as is done in confinement feeding of beef cattle (NRC, 1984). Further, it would provide a basis for comparison among forage species of the effects of canopy morphology on animal grazing behavior and hence the nutritive value of the daily diet selected. Also, assessment could be obtained on the likely HM by pasture species interaction for diet quality and for the daily quantity of the diet consumed. Accurate estimates of intake would have major implications on the recommendations provided to producers regarding the quantity of HM needed relative to the desired daily animal response and expected productivity per unit land area.

By definition, daily DMI of free-grazing animals can only be obtained directly by either weighing animals continuously or before and after each grazing event or through monitoring grazing behavior. The former represents an ideal method in theory. A direct method called the Animal Weight Telemetry System (Horn, 1981) based on the use of pressure transducers affixed under each hoof of the animal and held in place by a boot has been developed. This approach requires large investments in equipment and sophisticated computer software and has not been used in practice. Weighing animals to estimate intake may require appropriate adjustments for losses (respiration, defecation, urination, etc.) during grazing and for non-forage intake (supplemental minerals, water, soil, etc). Monitoring grazing behavior requires an estimate of grazing time, number of prehensile bites, ingested boli and bite weight, or ingested boli weight (Hodgson, 1982; Demment et al., 1987). This method has not been perfected but has attributes consistent with the ideal approach; i.e., the animals are relatively undisturbed, it is applicable to most pasture conditions (species, HM, etc), once the procedure is developed measurements are relatively easily taken, laboratory analyses are minimal, and the cost of the monitoring device is reasonable. Progress on this approach is being made and is encouraging (Luginbuhl et al., 1991b) but, at present, direct approaches are rarely used because of cost and equipment availability.

The assessment of pasture nutritive value in grazing trials is generally achieved by the comparison of average daily response (weight gain, milk or fiber production) per animal. This value provides a long-term and direct integration of the quantity (intake), quality (digestibility), and utilization (conversion) of the diet consumed for each treatment and is an expression of digestible DMI. Unfortunately, DMI, as such, is not generally determined.

Indirect methods to estimate DMI (Appendix Table 12-2) may seem adequate in theory, but in practice they all fall short of the ideal method because of a number of difficulties (Owens and Hanson, 1992). These difficulties take the form of high labor requirements, high frequency of disturbing experimental animals, sources of error that can introduce bias, large numbers of chemical analyses to be conducted, and, in some techniques, the complexity of modeling marker flow (Quiroz et al., 1988; Luginbuhl et al., 1993), calculating the parameters (Moore et al., 1992), and interpreting the data (Chapter 17). In spite of these limitations, and in the absence of viable alternatives, indirect methods serve a useful role in estimating intake of grazing animals.

INTAKE MEASUREMENT IN CONFINEMENT

Direct Measurements

Measuring the voluntary intake of animals is easiest to quantitate if the forage is brought to the animal and directly measured. Typically, animals are individually housed in separate pens, tie stalls, or digestion crates. Feed is brought to the animal one or more times per day and orts (refusals) are removed once (or more) per day. When planning to measure voluntary intake of animals in confinement, several questions should be considered and answered consistent with the objectives of the study. These are:

(a) When and how often should the animals be offered fresh feed?
(b) When and how often should refusals be removed and measured?
(c) How much feed should be offered or what level of refusal is the target?
(d) How long is necessary for adjustment of the animal to the feed?
(e) How many animals are needed?
(f) How should animals be housed to yield data applicable to production conditions?

Animals eat for discrete periods of time (meals) and may have 10 to 20 meals per day. In ruminants, most meals occur in daylight when the animals are most active (Forbes, 1986). In addition, intake can be stimulated by providing fresh feed, especially if the feed can spoil with time (i.e., silage). Some experiments involve special feeding devices that offer feed at similar time periods throughout 24 h such that 24 equal feedings (every h) or 12 equal feedings (every 2 h) are given to the animal (Croom et al., 1982). This may decrease the variability in intake with time but may create an artificial "steady-state" condition without discrete meals. In some studies, animals are fed two times per day and may be fed a higher portion of the total feed in the morning followed by a lesser portion at night. When given access to the feed for a minimum of 18 h, Blaxter et al. (1961) found no difference in intake whether fed at 6 h or at 12 h intervals. Feeding bulky conserved forage may require multiple feedings per day or special feeders equipped to contain the large volume of forage consumed by the animal each day.

Feed refusals are typically taken once per day just before the morning feeding (Chenost and Demarquilly, 1982). This cleans the feed trough for the new offer and provides an estimate of how much feed should be offered that day. High moisture feeds such as green chop or silage or feeds that spoil or where high selectivity occurs for plant parts or species may require more frequent removal of refusals. It is also very important to measure the DM of the refusals as they are removed because saliva, spilled drinking water, or weather can greatly influence water content and, therefore, the weight of the refusals. Drying and weighing the feed and refusals is required to obtain accurate measurements of DMI. These measurements can be taken daily or composited by animal and completed every week.

The level of feeding for each day should be based on the previous 2-d intake. Generally, 100% plus *ad libitum* feeding is practiced to make sure that intake is not limited. As previously discussed, level of refusals should be adjusted depending on the morphology and quality of the feed and depending on the species and production status of the animal to be fed. The larger the required intake, the higher the refusals will need to be, especially if the feed is of poor quality and selection is practiced by the animal.

Generally, intake should be determined for 2 wk after intake has stabilized. This evaluation period results in a measurement error of $\pm\ 2\%$ (Blaxter et al., 1961). Adjustment to a stable intake may take 10 to 15 d if the animal is not familiar with the feed or if the feed is a major change from previous diets (Blaxter et al., 1961; Schneider and Flatt, 1975; Chenost and Demarquilly, 1982). There can also be large differences among animals as to how long an adjustment period is needed before intake is stabilized. Adjustments of amount of feed offered during the initial period should not exceed 15 to 20% of the previous day's intake (Blaxter et al., 1961). Unfortunately, voluntary intake also fluctuates due to many factors such as season, weather, temperature, and photoperiod, as well as physiological state of the animal. A recommended protocol is detailed in Appendix Table 12-1.

The number of animals needed in such experiments can be estimated from the expected coefficient of variation (Chenost and Demarquilly, 1982). Without previous knowledge of expected variation among treatments, it is generally accepted that a latin square design or a minimum of four animals per treatment are needed. The required number of replicates can be calculated based on an estimated experimental error, the magnitude of the minimum treatment effect that the researcher wishes to detect, and the acceptable likelihood of a type II statistical error (Gomez and Gomez, 1984). It needs to be understood that the intake measurement is inherently variable. A real difference in apparent digestibility of four units can be detected with three to four sheep contrasted to eleven sheep required to detect an intake difference of 10 units. The number of sheep required to have an 80% chance of detecting a real difference ($P=0.05$) increases from 3 to 48 per treatment (forage) as expected real differences in intake units (g/weight$^{0.75}$) declines from 30 to 5 (Heaney et al., 1968). Although some workers have used up to 10 sheep per treatment, most have settled on a maximum of six because of the complexity involved. This number has permitted detection of a difference of 11 g kg^{-1} of body weight$^{0.75}$ (Chenost and Demarquilly, 1982).

Housing of the animals is also extremely important. Animals confined to crates or tie stalls are inactive and may develop leg and hoof problems. Also, it is extremely difficult to detect estrus in female animals when confined to individual stalls. Pressure from animal care committees and animal welfare advocates may limit the future use of crated or restrained animals. The development of electronic gates that allow an individual animal to eat from only one feed trough while socializing in groups that are housed together (Broadbent et al., 1970) is becoming the standard for intake measurements. Animals housed in such an arrangement should yield data that can be related to field conditions with much more confidence.

Empirical Estimates

A number of equations have been developed that employ regression techniques to estimate the quantity of forage consumed. Generally, terms in the regression equation for beef cattle include live weight (kg) and daily gain (kg d^{-1}) (Minson, 1990). The equations for dairy cattle are more complex including in addition a term for milk, or fat corrected milk (kg d^{-1}), and various other possible terms such as concentrate fed (kg d^{-1}), weeks after calving, month of lactation, DM digestibility, and ammonia N concentration depending on the experimental objectives and nature of treatments (Neal et al., 1984; Minson, 1990).

A method termed the "fill unit" system has been proposed by Jarrige et al. (1986) in which voluntary DMI (VDMI) is predicted based on a fill value (FV) of a forage and the feed intake capacity (IC) of an animal. This approach places VDMI in the form of a fill unit (FU) and makes it additive as are energy and protein. By definition, 1 kg of DM of a reference pasture grass has a FV of 1. Each category of animal has an IC expressed by one single value in FU. Equations have been developed to account for interactions between forages and concentrates.

These empirical approaches are useful in application to production enterprises, but generally provide little insight into understanding the basic biology of animal intake and generally do not make adjustments for environmental (weather) factors.

INTAKE MEASUREMENT OF GRAZING ANIMALS

Estimating the forage intake of free grazing animals is so difficult that all of the commonly used methods have limitations and consist of various compromises that may introduce error (Minson, 1990; Owens and Hanson, 1992). While none of the techniques are completely adequate, they each have value in specific situations and can yield valuable data if their shortcomings are recognized as discussed below.

Direct Measurements

Differences in Animal Mass

One of the most direct methods of determining intake is to weigh animals during eating (Horn, 1981) or before and after they eat. This has obvious limitations in most grazing settings but has been used successfully when ingestive behavior is of interest and grazing periods can be of limited duration (Le Du and Penning, 1982; Penning and Hooper, 1985; Penning et al., 1991). This technique has been most often used with sheep but can be used with larger ruminants as well. The chief limitation is the short-term nature of these measurements and the need to account for weight loss due to defecation and urination.

Differences in Herbage Mass

The difference in HM before and after grazing is sometimes used to estimate intake. The reduction in HM observed in a paddock due to grazing may be divided by the product of the number of animals and days grazed. This value may be assumed to be equal to the daily intake per animal. Problems with this technique can be minimized with management and sufficient labor (Meijs et al., 1982). The assumption that the decline in herbage mass is entirely due to consumption of forage is often an overestimation because trampling or consumption by nonexperimental animals (deer, rabbits, etc.) may remove forage from the harvested sward. The cutting height is seldom low enough to include all of the forage that may have been trampled. Using a low cutting height minimizes this problem but may result in the death of the sward in the sampled areas. Excessive soil contamination by either trampling or when harvesting, or both, may require adjustment for ash and results expressed on an organic matter basis.

Another consideration is the growth of the sward during grazing. This can result in an underestimation of intake. To minimize these problems, either the grazing period must be kept short (1 to 3 d) or cages should be placed in the pasture to prevent grazing in selected sample areas and used to estimate the growth rate of the pasture for the period in question. The assumption that the growth of the forage in the cage is equal to the growth in the rest of the paddock may not be true if the time period is too long (> 7 to 10 d). However, enough time must elapse for an adequate quantity of growth to occur or the sampling variance in the selected areas will prevent a sufficiently precise estimate of growth. Pasture cages also are applied under continuous stocking since the method of measuring HM before and after grazing is not available. In this case, the accumulation of herbage in the cage may be divided by the number of animal-days to estimate forage intake. It is also important under continuous stocking to keep the time frame between placing the cage and cutting the forage as brief as is practical.

The most serious limitation to the application of these difference methods is the intensive sampling required to provide an adequate estimate of the change in HM. Statistical methods may be used to estimate the number of samples for a particular experimental situation after a limited amount of experience in that environment (Gomez and Gomez, 1984). However, often it is not possible to obtain the number of samples needed to make the estimates as reliable as desired.

One area in which the difference in HM has been very useful as a method of estimating intake is in studies of ingestive behavior. For example, Dougherty et al. (1989) utilized the technique of tethered grazing and determined HM before and after grazing meals to estimate intake. The method fits very well into the application of tethered grazing because animals are closely managed and grazing time is short and regulated.

Another application of HM difference methods is in construction of artificial pastures utilizing perforated boards (Black and Kenney, 1984; Demment and Laca, 1993). These investigations of ingestive behavior allow precise sward description but are necessarily very short in duration (30 s). In pursuit of similar objectives, other researchers (Burlison et al., 1991) restrained sheep in modified digestion crates and gave them access to an area of sward 0.56 by 0.46 m. In both techniques, animal training is required and sheep were used.

Indirect Methods of Estimating Intake using Fecal Output and Diet Digestibility

Because of the difficulty in making direct determinations of DMI of grazing animals, a number of indirect methods have evolved. The most useful of these are described below and a recommended protocol is presented in Appendix Table 12-2.

Determining Fecal Output Directly

Total fecal collection, along with an estimate of diet DM digestibility, has been used to estimate DMI. Total fecal output can be measured directly by the use of a harness and a collecting bag. The advantages and disadvantages have been reviewed by Meijs (1981). In brief, the advantages of collection bags for determining fecal output are that they give rapid results and require only DM and ash determinations. Major disadvantages are significant reductions in animal performance, incomplete collection of feces due to losses (especially on diets low in DM), distortion of hind legs due to weight of feces in bag, high labor requirement, and influence on grazing behavior. In addition, an estimate of diet quality is required as needed for the marker methodology (see next section on emperical estimates). This technique is more practical with sheep because of the high DM of the feces and the smaller size of the animals.

Determining Fecal Output Indirectly

Fecal output (FO) can be estimated based on the ratio between the quantity of a marker dosed to an animal and its concentration in the feces [FO (g/d) = (ug of marker administered) / (ug of marker per g of feces)]. An estimate of the digestibility of the diet can be obtained and used to calculate DMI [DMI = FO/(1 - diet digestibility/100)]. The DMI equation requires an accurate estimate of diet digestibility and not just a correct ranking of the forages being tested. It is important to keep in mind that the use of inert markers is more appropriate for estimating fecal output than forage intake.

Daily Dosing of an Inert Marker Administering one or two doses of an inert marker each day is a commonly used method for determining fecal output. This is logistically a very labor intensive technique. Each time the animals are dosed with the marker, they must be either removed from the pasture and restrained or the animal trained to permit dosing through a rumen cannula. If the animals are restrained, a high labor input is required and the handling is stressful for the animals. The stress may alter grazing behavior, fecal output, and forage intake. Also, diurnal variation has been observed in the concentration of marker in feces with animals dosed once daily. This can be reduced but not eliminated by twice daily dosing; however, labor requirements and animal stress increase proportionally. Chromic sesquioxide impregnated paper has been administered to reduce diurnal variation (Telford, 1980) but is no longer commercially available. Various devices to infuse a marker continuously have been utilized to minimize diurnal variation but these devices are expensive and may require surgical modification of the animals (Corbett, 1978; Brandyberry et al., 1991). A timed-release bolus for continuous infusion of chromium sesquioxide (Cr_2O_3) will be discussed in a later section.

Various modifications of equipment and experimental design can reduce animal stress and labor requirements. Frequent sampling of feces and compositing may be useful in overcoming diurnal variation in marker concentrations. Two fecal samples daily are not always sufficient to estimate the mean marker concentration for the day when diurnal variation is present. Even though there are many difficulties with this procedure, an advantage is that the kinetic properties of the marker do not influence the estimate of fecal output.

Pulse Dosing of an Inert Marker In contrast to daily dosing, a marker may be administered once to an animal followed by frequent fecal collections in order to characterize the "pulse" in marker concentration found in the feces (Pond et al., 1986; Pond et al., 1987; Pond et al., 1989). The characteristics of the excretion curve make it possible to estimate rate of digesta passage, mean residence time, and digestive tract fill, as well as fecal output and, with an estimate of digestibility, forage intake. Rate of passage and mean residence time are just inverses in some analysis methods, but the additional explanatory power of this procedure over the daily dose procedure is substantial in some situations. However, there are more pitfalls with this procedure than with the daily dose. First, this procedure is labor intensive just as the daily dose procedure is. While the stress on the animals may not be as great as in daily dosing, the animals are disturbed frequently over the first 2 to 3 d for fecal collections that may require grab (rectal) sampling. The sampling schedule may alter grazing behavior (Fisher et al., 1986) and, consequently, the kinetic parameters estimated from the excretion curve must be interpreted with caution. A large number of chemical analyses are required to develop the marker excretion curve and the complexity of modeling marker flow to calculate kinetic parameters and proper interpretation of the resulting data adds a degree of difficulty (Quiroz et al., 1988; Moore et al., 1992; Luginbuhl et al., 1993). The problem of estimating diet digestibility is as serious a problem with this procedure as it is with daily dosing. An additional problem with this procedure is that the kinetic properties of the dosed material must be similar to those of the digesta. Consequently, rather than simply dosing inorganic chromium, the chromium is mordanted to the dietary component of interest. Typically, the dietary component of interest is forage fiber and the mordanted sample should have a particle size distribution similar to the ingested fiber. Collection of extrusa from an esophageal-cannulated animal followed by neutral detergent extraction and chromium mordanting are recommended. It is also important to mordant as little chromium as is practical (Pond et al., 1987) for detection in the feces. This is to prevent the mordanted chromium from seriously altering the weight of the forage particles and their kinetic properties and to minimize disassociation of the chromium from the forage particles (Quiroz et al.,

1988; Moore et al., 1992; Luginbuhl et al., 1993). This procedure has been successfully used in our laboratory. In addition, with proper labeling, rare earth elements (Pond et al., 1989) may be used as forage particle markers (Chapter 17).

Sampling Diet and Estimating its Digestibility

Associated with the use of inert markers in estimating DMI of the grazing animal is the requirement of knowing the digestibility of the animal's diet. Obtaining samples that represent the diet of the dosed animals can frequently be the major factor limiting accuracy of the marker technique.

The two basic approaches in obtaining a representative diet are a manual collection of forage by the experimenter or use of surgically altered (rumen or esophageal cannula) animals (Le Du and Penning, 1982). In the former case, the experimenter observes what the animals are grazing and then attempts to mimic their selection manually. This procedure is sometimes criticized for being subjective. The use of animals with either a rumen or esophageal cannula is intended to provide a diet similar to the one consumed by the tester animals. However, biased estimates of the diet of tester animals can result from the use of cannulated animals if adequate sampling is not achieved.

Use of the rumen cannulated animal requires the removal of sufficient rumen contents to expose the cardia. The animal is allowed to graze while the experimenter catches the ingested material as it drops from the cardia with entrance via the rumen cannula. The esophageal cannula has generally replaced the rumen cannula because it requires only removal of the cannula plug to allow collection of extrusa. In both cases, care should be used to avoid contamination of the sample with excess saliva or rumen contents. Saliva should not be drained from the sample because of DM losses (Little, 1972) and samples contaminated with rumen fluid should be discarded. The drying of extrusa for digestibility analyses is best achieved by freeze drying (Le Du and Penning, 1982). Estimates of digestibility can be obtained directly or using internal markers.

In Vitro or In Situ Dry Matter (Organic Matter) Disappearance After a representative sample is collected, the accuracy of the marker technique resides with the method used to estimate the absolute digestibility of the diet. The two-stage in vitro bioassay is the method of choice for estimating diet quality and has the broadest application (Le Du and Penning, 1982). Frequently 48 h fermentation values are used, but 72 h values may be more appropriate for C4 grasses when absolute values are important. In either case, in vitro values can also result in serious errors (Holechek et al., 1986; Dove et al., 1990). It is possible to develop regression equations for conversion of in vitro estimates to in vivo. If properly validated, this approach can be used to improve prediction of intake based on fecal output.

Internal Markers Internal markers are natural plant constituents that are neither digested nor absorbed by the animal. They can be used to estimate DM digestibility (DIG) of the diet by knowing the ratio of the marker in the diet (M_d) and feces (M_f).

$$\text{DIG}, \% = 100 - [100 * (M_d/M_f)]$$

Internal markers that have been evaluated are silica, lignin, fecal N, chromogen, indigestible neutral detergent fiber, and acid insoluble ash. Silica occurs naturally in forage and is indigestible and recovered in the feces. Silica concentrations in forage have been variable and inconsistent and subject to contamination by soil. These problems and a poor recovery of silica in feces have limited its use (Gallup et al., 1945). Lignin is a naturally occurring indigestible portion of the cell wall and reliable results as an internal marker have been noted (Ellis et al., 1946). Diurnal variation in lignin concentration in the feces can be high so many samples must be collected and analyzed. Fecal N was associated with forage digestibility (Holter and Reid, 1959) and an equation was developed (Schneider and Flatt, 1975) to determine DM digestibility using crude protein concentration in forage and feces. This method has not been evaluated for many forage species. Chromogens are plant pigments that are completely recovered in the feces. Chromogens, as markers (Reid et al., 1950, 1952) seem best for lush growing forage and least effective for drought or stressed (poorly pigmented) plant tissues. Indigestible neutral detergent fiber (after 144 to 196 h in vitro fermentation) has been successfully used as an internal marker (Lippke et al., 1986). Variable recovery and a recovery by particle size interaction are potential problems with this technique. Acid insoluble ash (ash after acid hydrolysis) has been used as an internal marker in feedlot diets and when grazing (Thoney et al., 1985). However, the concentration in most forages is low requiring analysis of larger samples of forage and feces than with the other markers.

Lignin has been the most widely used marker of the ones discussed above. However, an infallible and totally dependable internal marker is not available (Cochran et al., 1986). Such an internal marker would contribute greatly to improving the estimation of diet digestibility and DMI of the grazing animal. Further information on these methodologies, and discussion of other ratio techniques and the fecal index technique, can be found in Le Du and Penning (1982).

Empirical Estimates

Methods have been developed for grazing animals (steers, cows, calves, and sheep) that utilize animal responses in a back calculation to estimate daily requirements. For example, the effective feed unit (EFU) concept developed by Peterson and Lucas (1968) provides an estimate of the average daily EFU (which can be expressed as digestible DM, gross energy, digestible energy, metabolizable energy, net energy, total digestible nutrients (TDN), etc.) consumed per tester animal. This value can be converted into an estimate of daily DMI as proposed by Baker (1982) using energy values. In both cases, conversion factors for maintenance and weight gain or loss are required. The EFU is no better than the conversion factors employed. Regression equations have also been developed (Neal et al., 1984; Minson, 1990) that provide estimates of daily DMI as discussed under the previous major heading.

An innovative approach to estimating intake has been developed (Pienaar et al., 1980; Pienaar and Roux, 1989) utilizing observed in vitro rates of gas production to estimate mean retention time and digestibility. Fecal output and intake are then estimated. This technique is dependent on the availability of the appropriate equipment for measuring gas production. The use of gas production allows for more reliable and stable estimates of digestion rate than the traditional gravimetric procedures. Accuracy has been quite good (Pienaar et al., 1980) and the technique is rapid enough to allow use in some applied settings.

Subjective Assessments

The power of visual observation to assess short-term (<1 d) changes or differences that occur in plant-animal studies is generally not considered. Yet the mind possesses an appreciable capacity to integrate multiple factors that are difficult to quantitate with instrumentation. For example, in a study comparing short-term preference for six *Pennisetum* species, Burns et al. (1978) devised a relative preference score by assigning each entry a residual herbage score from 1 = not grazed to 10 = grazed to the stubble (10 cm). Because all forages had similar canopy heights at time of grazing, the residual forage provided an estimate of forage consumption (relative intake). This subjective score was compared to an occupancy score (objective measure) obtained by recording which forage entry animals were grazing every 2 min. The relative preference score gave results that were similar to the relative occupancy score but with much less effort. In a similar type study (Burns et al., 1988), but where HM of forage entries varied widely, a pregrazing canopy-height measurement was obtained as was a postgrazing stubble height measurement. Relative intake estimates were obtained by subtracting the postgrazing measurement from the pregrazing measurement (canopy removed) and was compared with a visual defoliation score ranging from 1 (none or a few bites taken) to 10 (leaves and stems removed to a 10 cm stubble). Correlations (r) between estimates of intake from the subjective defoliation score vs. canopy removed (objective) = 0.71 to 0.78 ($P \leq 0.01$). The subjective evaluation represented only a fraction of the cost and effort. This process does require experience and the resulting relationships would likely change among groups (locations) of researchers.

A major concern in using subjective assessments centers on the validity of the process compared with data obtained by objective methods. Since subjective data are generally not continuous, i.e., based on a low and high bound, special statistical procedures should be used to aid interpretation (Schiffman et al., 1981). Further, subjective data contain both treatment effects and observer bias which cannot be separated.

Observer bias arises from memory carry-over such that scores assigned to plots in one replicate, or at a previous evaluation, may unknowingly influence subsequent scores. Such bias results in artificially reduced variances causing small differences to become significant and may erroneously create differences. Observer bias can be partially prevented through a concerted effort on the part of the observer to score each plot independently and to use multiple (3) scorers.

In the studies cited above (Burns et al., 1978, 1988), subjective scores were obtained along with objective measurements. This point is important because it provides the researcher an assessment of how capable the observers were in assigning a useful score, if the score was of any value for the purpose intended, and provides the experience needed to decide in future studies if subjective scores can be used without objective measurements.

ADVANCES IN INTAKE MEASUREMENT IN CONFINEMENT

Several innovations have occurred that have aided the estimation of DMI or the interpretation of intake data that merit some mention.

Electronic Gates

An electronic gate (American Calan Inc., Northwood, NH) that can be keyed to allow access of a single animal to any designated manger has become widely used. The gate, originally called the "Broadbent Gate" (Broadbent et al., 1970), consists of a specially shaped, spring-loaded fiberglass door equipped with a circuit board and a solenoid locking bolt. When power is supplied to the circuit board and an animal with the right key (suspended around the neck with a plastic chain) faces the door, the solenoid is activated, withdrawing the locking bolt, freeing the door, and allowing the animal access. This occurs as the animal presses against the top of the door forcing the spring-loaded door open. Upon withdrawing its head, the door closes and the solenoid locking bolt latches. Forage is fed and refusals obtained as usual. The benefit of this innovation resides with freeing the animal from confinement and the restriction of conventional intake stalls, which may alter animal appetite and health. Animals remain socially active, can rest and ruminate unrestricted as a group, and obtain some exercise. Aside from the degree of convenience, it was proposed that the gates would avoid bias in estimating treatment effects and had potential for reducing experimental error (Broadbent et al., 1970).

Aside from the cost, several negative attributes warrant mentioning. Initially, animals must be trained. This is easily done but it takes time. First, animals are brought into the facility and given free access to all mangers (gates tied open) until they appear comfortable eating from them (3 to 5 d). Then all gates are closed. Equal success has been achieved by either placing tape over the latching bolt for 1 or 2 d so animals adjust to pushing the gates open or locking the gates immediately, forcing animals to seek their designated manger. Animals are then given about 10 d to become adjusted to their individual manger. This period provides time to give individual help to those few animals that become confused or are timid. Our success rate is near 100%. It is always a good idea to include several extra animals that are trained and can be used if the need arises.

It is essential that the gates be properly installed because dimensions are specific for certain ranges in animal size. The system should be checked several times a day to assure circuit boards are working and that the key has not become dislodged from the animals neck. In either event, the animal is denied access to its feed. A final point that can be serious is the ability of some animals to learn to steal feed from their neighbors. This is usually prompted initially by animals withdrawing their heads with feed in their mouth. While they continue chewing, another animal has the opportunity to sample the feed. The problem is aggravated if forage is fed long or if there is a strong preference among the treatments being evaluated. In the latter case, the animal on the least preferred treatment frequently remains unsatiated and begins to compete with animals on the better diets. Aggressive behavior may result, beginning with head butting, with the point of attack being the front of the shoulder of the feeding animal causing it to withdraw from its manger. The challenging animal will catch the gate with its muzzle before it can latch and consume the other animal's feed. This problem has been solved by adding short panels on both sides of the gate with the preferred feed. This reduces such aggressive behavior. If head butting continues, the panels assure that the gate closes and latches before it can be caught by the muzzle of the other animal.

Computerized Chew Meter

Devices to monitor jaw movement in intake trials with ruminants have been employed for some time (Johnstone-Wallace and Kennedy, 1944; Balch and Campling, 1962). More recently, interest has developed in recording detailed information about masticatory bites (Duranton and Bueno, 1982) and boli characteristics (Bae et al., 1981). However, manual analysis of chart recordings is extremely tedious, time consuming, and subject to many errors. The application of microcomputer technology to record input signals associated with jaw movement (Murphy and Jaster, 1984) was a beneficial advancement. The most recent innovation in this technology is the development of computer software to interpret the events and to assist in data reduction (Luginbuhl et al., 1987; Beauchemin et al., 1989). This brings the methodology to a point of practical application in intake studies. Events each hour are processed into the time devoted to the distinct activities of ruminating, eating, and resting. These data can be displayed in tabular or graphic form. Data processing provides hourly assessment of items such as total number of eating and ruminating chews, number of boli, chews per bolus, and number of eating and rumination chews per minute. Information also can be obtained for different segments of each meal as well as on consecutive meals.

The innovation resides with the software that has adequate data handling power. This technology provides a refined level for accessing the plant-animal interface. Aside from minimal computer needs, the major negative attributes of the technology are the custom fitting of the halter, which positions the sensing device on each animal, and the maintenance of the latex tube and soldered unions, both of which eventually fail. The exact relationship between jaw movements and boli characteristics remains to be established. Therefore, the use of this approach for estimating intake is currently limited.

Continuous-reading Mangers

The application of strain gages or pressure transducers to understanding aspects of DMI has occurred through the digitization of the output signal allowing continuous signal monitoring by microcomputer. Mangers placed on sensors fitted with strain gages (Baumont et al., 1989, 1990) or suspended from pressure transducers (Worley, 1987) and monitored continuously (every 10 s) during feeding have provided estimates of weight loss from the manger. This technology gives an estimate of both the total quantity of feed consumed and the rate of its consumption. The absolute weight loss for a short unit of time (min) during a meal, as opposed to an average weight loss per unit time calculated after completion of a meal, permits the assessment of intake kinetics. Comparisons can be made among animals fed the same feed or among feeds fed to the same animal. This technology must be computer assisted with appropriate software that damps extreme fluctuations to estimate the manger weight during the feeding period.

Application of the same technology has been used to develop an automated delivery system (PINPOINTER, UIS Corporation, Cookeville, TN) for the feeding of concentrates to individual animals. One PINPOINTER unit can be set to receive and provide feed individually to some 20 different animals. The quantity to be fed to each animal can be entered into the computer. When an animal enters the stall, it is identified and the designated quantity of feed is augured into a scale hopper, weighed, and delivered to the feeder for consumption. When the animal exits the stall, any residual feed is dumped to a residue scale hopper, weighed, and discarded. Weights are recorded for subsequent processing. The PINPOINTER system is not readily adapted to feeding unprocessed forages.

Ingestive Mastication

Although ingestive mastication is an area of study in its own right and will not be detailed here, the overall effect of the resistance of feed to comminution during ingestion is important in controlling DMI (Moseley and Manendez, 1989). Evidence exists of the diversity of particle sizes in masticated forage with variation occurring among forage species (Pond et al., 1984a; Weston and Kennedy, 1984). Methodologies of wet sieving (Grenet, 1984) and dry sieving (Smith and Waldo, 1969) have been developed and compared (Ehle, 1984). Dry sieving was noted to be less time consuming (Allen et al., 1984) and resulted in smaller mean particle sizes and less variation (Ehle, 1984). Wet sieving sorted particles according to length and dry sieving according to cross-sectional area (Ehle, 1984). Mathematical models to properly analyze weight distributions of sieved particles have been published (Gill et al., 1984; Pond et al., 1984b; Vaage et al., 1984; Fisher et al., 1988).

Recently, Luginbuhl et al. (1991a) reported on a microcomputer-assisted image analysis technique that measures particle length and width. This technique has been applied to sieved particles and provides the degree of particle size quantification that is important when accurate summary statistics for particle size and their distribution are needed. The reader interested further in particle size analysis is referred to "Techniques in particle size analysis of feed and digesta in ruminants" (Kennedy, 1984). As noted for computerized records of jaw movements, the exact relationship between particle size and DMI remains to be established and this limits the use of data describing ingestive mastication for estimating DMI.

RECENT ADVANCES IN INTAKE MEASUREMENT UNDER GRAZING

Alkanes as Internal and External Markers

The use of n-alkanes as markers is somewhat unique in that a combination of an internal and external marker can be used (Mayes et al., 1986). Plant alkanes are found in cuticular waxes and consist of predominantly odd-numbered chains of 25 to 35 carbons. These waxes are relatively indigestible and as chain length increases, the percent recovery in the feces increases (Barrowman et al., 1989; Ohajuruka and Palmquist, 1991). Analysis is relatively easy and precise, but requires a gas chromatograph. Since plant alkanes have odd numbered chain lengths, the alkanes with even numbered chain lengths may be fed as external markers. The use of C32 as the external marker and C33 as the internal compounds has been suggested because their percent recovery is high and similar and the errors in recovery cancel in the calculations of digestibility and fecal output as long as the recovery of the selected markers is similar (Mayes et al., 1986; Dove et al., 1989b).

The use of alkanes to estimate DM digestibility may improve upon the presently used IVDMD values which have been shown to cause errors in estimation of intake (Holechek et al., 1986; Dove et al., 1990). The use of alkanes can provide estimates of both digestibility and fecal output for the calculation of intake. The procedure can even allow for individual animal variation in digestibility of forages (Mayes et al., 1986; Dove et al., 1989a; Dove et al., 1990; Vulich et al., 1991; Bechet and Tulliez, 1992).

A problem has been pointed out (Laredo et al., 1991) that may require some modifications in the most common approach to using alkanes. Forage species contain variable quantities of alkanes and the concentrations of C33 may be too low in some tropical forages for use as the internal marker. This can require the use of a shorter chain length with a lower percent recovery in the feces and possibly result in errors in calculation of intake. This problem will be discussed further in the section entitled, "Future Considerations". However, the variation from species to species in alkane composition can be used to advantage by solving a set of simultaneous equations and estimating the species composition of the diet (Dove, 1991).

Controlled Release Devices

Over approximately the last decade, a controlled release device (CRD) has been developed and marketed (Captec, Nufarm, Auckland, New Zealand) for the delivery of chromium sesquioxide (Cr_2O_3). The CRD was developed to overcome the difficulties of once or twice daily dosing with Cr_2O_3 as well as the diurnal variation in output of Cr_2O_3. Some results have been promising (Laby et al., 1984; Parker et al., 1989) and resulted in the use of the CRD in some ambitious research projects (Barlow et al., 1988) in which there was no independent check of the validity of the CRD results. Unfortunately, the CRD was not tested in a wide enough range of environments with a broad enough range of forage species to uncover the problems that have recently surfaced with the device (Brandyberry et al., 1991; Buntinx et al., 1992). The problems may be primarily associated with the commercial version, but failure of the CRD has been so serious that it cannot be recommended for most experimental work at this time. Shortcomings of the CRD include large deviations between observed and manufacturer-specified release rates with both cattle and sheep (Brandyberry et al., 1991; Buntinx et al., 1992), variation in release rates (Buntinx et al., 1992), a feed by CRD interaction (Parker et al., 1989), and an animal by CRD interaction in release rate (Pond et al., 1990). Thus, release rate may be dependent on the particular animal and forage being tested.

This is, of course, disappointing because of the many advantages a CRD offers. However, the devices may be useful in the future if the current problems can be solved. As has been proposed previously (Brandyberry et al., 1991; Buntinx et al., 1992), the release rate may be calculated by means of total fecal collection from a subsample of the animals to be tested. Brandyberry et al. (1991) estimated that five steers in a 7-d total fecal collection trial would provide an estimate of the release rate with an accuracy of 5%. This should be done for each of the forages to be tested. The CRD is difficult and expensive to test adequately and the premature published conclusions that the device was useful will undoubtedly delay testing of improved devices by the relatively few research teams capable of collecting the appropriate data. One improved device that may fall into this category is a combination of the CRD with alkanes rather than Cr_2O_3 (Dove, 1991). Based on the history of the device with Cr_2O_3, it is not logical to assume that it will be useful until further testing has been done. Even after testing, the use of any CRD must proceed with caution in a research setting. In the first report of the alkane CRD, the conclusion that it was useful for accurate estimates of intake was based on only seven animals and one forage species (Dove, 1991). Further testing is essential and, eventually, retesting will be required on any commercially available devices and not just on the experimental prototypes.

Grazing Behavior

The activity of ingestive intake has been suggested as a method of estimating daily DMI (Coleman et al., 1989). Quantification occurs through the product of bite size x rate of biting x grazing time (Chapter 19; Hodgson, 1982; Demment et al., 1987). Mechanical methods for short-interval measurements have been reviewed (Hodgson, 1982), but recent advances have occurred through automation to aid quantification. A battery powered meter developed by Houessou et al. (1989) records total grazing time per day, time spent ruminating, and number of boli ruminated. The system operates based on changes in air pressure caused when the animal depresses a rubber pear while chewing. The meter stores the data on a chip and is readily down-loaded to a PC for further summary. Automation of data collection and summaries of grazing behavior is the major advancement over the frequently used mechanical vibracorder (Hodgson, 1982).

A modification of the stall chewing meter developed by Luginbuhl et al. (1987) has been field tested with the grazing animal and provides total grazing time, rumination time, resting time, number of boli ruminated, total jaw movements, and number of intake bites (Luginbuhl et al., 1991b). Eating chews can be calculated by difference between total jaw movements and eating bites plus ruminating chews.

The major factor limiting this technology for estimating DMI is obtaining an acceptable estimate of bite size or DMI per bite. Forwood et al. (1985) developed a conductivity transducing cannula that can provide an accurate reading of the number of boli swallowed. An estimate of boli weight (DM) from direct boli collection, coupled with intake bites per boli (Forbes, 1988) would provide an estimate of DMI per bite. If the estimate of bolus weight could be obtained within reasonable error, then a direct estimate of DMI is feasible from grazing behavior. Although limited, data indicate that boli weight may be dependent on animal weight, forage species, and canopy morphology (Forwood et al., 1991).

FUTURE CONSIDERATIONS

Plant Material with Low Alkane Concentrations

The longer chain length alkanes are not present in all plant material in high enough concentrations to be useful (Laredo et al., 1991). The use of the shorter chain length alkanes is complicated by a lack of recovery. It may be possible to administer a spectrum of even-numbered alkanes and adjust fecal concentrations of the shorter chain length alkanes to the longest chain alkane. This method would allow an estimation of percent recovery for each alkane as well as multiple calculations of digestibility, fecal output, and intake. This approach was suggested by Mayes et al. (1986) and is hinted at in the work of Dove et al.(1991) in which two even-numbered alkanes were dosed by a CRD and another by a once daily administration of impregnated paper. The adjustment for percent recovery may make the estimates of diet composition from fecal alkane concentrations with animals grazing polycultures more accurate (Dove, 1991). Adequate fecal sampling to compensate for diurnal variation is as important with this material as it is with the more common markers. Although twice daily dosing has been found to be sufficient in some cases (Dove et al., 1989a,b), in other cases it has seemed to be inadequate (Dove et al., 1991).

Pulse Dose of Alkanes

There is a possibility that a combination of alkane and pulse dose technologies may result in an increase in the information collected by alkane dose and an improvement in the reliability of the pulse dose technique. An alkane pulse dose has been done (Marais et al., 1992) by both a single dose of cellulose powder and by coating alkane on forage. However, the objectives of that study were not to assess rumen kinetics. Consequently, samples were not analyzed at the later times so the necessary curve fitting and mathematical analysis were not possible.

Animal Behavior and Intake

The effects of various aspects of animal behavior on grazing intake have been alluded to. For example, we may change the intake of an animal under study by our methods of determining intake. The *ad libitum* intake observed on pasture is an integration of many factors, for example, HM, plant morphology, forage composition, bite size, biting rate, grazing time, topography, animal class, animal physiological status, environment, animal preference, and the effects of our experimental procedures all influence *ad libitum* intake. It is especially difficult to determine the importance of animal preference on intake (Kenney and

Black, 1984; Minson and Bray, 1986; Burns et al., 1988). Although these experiments give some indication as to the relative magnitude of animal preference for one forage over another, the critical experiment will require two forages that differ only in preference. This will not be an easy experiment to conduct. Kenney and Black (1984) found large differences in preference due to physical form. Physical limitations to intake rate appeared to produce a large preference for the forage with the higher intake rate.

Further research is needed on the effects of preference on intake independent of forage composition and physical form. In order to conduct this research, two forages must be used that have only negligible differences in composition and physical form but substantial differences in preference. If preference for one forage over another is a significant factor affecting intake, perhaps it will be possible to breed for animal preference or plant characteristics that increase intake by altering preference alone.

CONCLUSION

Quantification of the DMI of animals is a measurement basic to the understanding of their nutritional status. Measurements made in confinement and subsequent interpretation generally are straightforward, providing appropriate methods are carefully followed. Estimates of DMI of the grazing animal are necessarily more complicated because there is no straightforward way to measure daily DMI. The DM available to the animal consists of the HM present over the total land area to which the animal has access. The animal determines the quality of its diet through selective grazing and quantity consumed of that diet at each meal. The daily DMI and associated intake of nutrients are the total of meals consumed during a given 24 h. In stall intake trials, the convention has been to evaluate discrete batches of forage providing animals 5 to 10% more forage DM than can be consumed. A more useful evaluation of forages would be to feed a range of maturities as well as a range in *ad libitum* intake (0 to 30%). This would characterize the maturation process of each forage species and would provide a means of challenging the genetic potential of the animal from an all forage diet.

In a grazing setting, indirect methods (mainly inert markers) have evolved to estimate DMI and permit relative comparisons of DMI among pasture treatments. The accuracy of these methods in estimating DMI resides with the marker methodology in estimating fecal output and in obtaining an accurate assessment of digestibility of the diet consumed. New technologies are emerging and must be continuously sought and applied to provide a quantitative estimate of daily intake of the grazing animal.

Suggested protocols have been developed for the estimation of DMI in confinement, as well as for the grazing animal. These protocols allow for research objectives that may vary, but the protocol will need to be modified as new innovations provide increased accuracy and ease of measuring forage intake.

REFERENCES

Ademosum, A. A., B. R. Baumgardt, and J. M. Scholl. 1968. Evaluation of a sorghum-sudangrass hybrid at varying stages of maturity on the basis of intake, digestibility and chemical composition. J. Anim. Sci. 27:818-832.

Allden, W. G., and I. A. M. Whittaker. 1970. The determination of herbage intake by grazing sheep: The interrelationship of factors influencing herbage intake and availability. Aust. J. Agric. Res 21:755-766.

Allen, M. S., J. B. Robertson, and P. J. Van Soest. 1984. A comparison of particle size methodologies and statistical treatments. p. 39-56. In D. M. Kennedy (ed.) Techniques in particle size analysis of feed and digesta in ruminants. Occasional publ. no. 1, Can. Soc. Anim. Sci., Edmonton, Alberta, Canada.

Barrowman, J. A., A. Rahman, M. B. Lindstrom, and B. Borgstrom. 1989. Intestinal absorption and metabolism of hydrocarbons. Progress Lipid Res. 28:189-203.

Bae, D. H., J. G. Welch, and A. M. Smith. 1981. Efficiency of mastication in relation to hay intake by cattle. J. Anim. Sci. 52:1371-1375.

Baker, R. D. 1982. Estimating herbage intake from animal performance. p. 77-93. In J. D. Leaver (ed.) Herbage intake handbook. The British Grassl. Society, Hurley, Maidenhead, Berkshire, UK.

Balch, C. C., and R. C. Campling. 1962. Regulation of voluntary food intake in ruminants. Nutr. Abstr. Rev. 32:669-686.

Barlow, R., K. J. Ellis, P. J. Williamson, P. Costigan, P. D. Stephenson, G. Rose, and P. T. Mears. 1988. Dry-matter intake of Hereford and first-cross cows measured by controlled release of chromic oxide on three pasture systems. J. Agric. Sci. (Camb.) 110:217-231.

Beauchemin, K. A., S. Zelin, D. Genner, and J. G. Buchanan-Smith. 1989. An automatic system for quantification of eating and ruminating activities of dairy cattle housed in stalls. J. Dairy Sci. 72:2746-2759.

Baumont, R., J. P. Brun, and J. P. Dulphy. 1989. Influence of the nature of hay on its ingestibility and the kinetics of intake during large meals in sheep and cows. p. 787-788. In R. Jarrige (ed.) Proc. XVI Inter. Grassl. Congr., Nice, France, 4-11 October 1989. Association Francaise pour la Production Fourragére, Versailles, France.

Baumont, R., N. Seguier, and J. P. Dulphy. 1990. Rumen fill, forage palatability and alimentary behaviour in sheep. J. Agric. Sci. (Camb.) 115:277-284.

Bechet, G., and J. Tulliez. 1992. Estimation of grass intake by lambs, with alkanes as markers. Ann. Zootech. 41:85.

Black, J. L., and P. A. Kenney. 1984. Factors affecting diet selection by sheep. II. Weight and density of pasture. Aust. J. Agric. Res. 35:565-578.

Blaser, R. E., R. C. Hammes, J. P. Fontenot, H. T. Bryant, C. E. Polan, D. D. Wolf, F. S. McClaugherty, R. G. Kline, and J. S. Moore. 1986. Growth stages of plants, forage quality and animal production. p. 9-14. *In* Forage-Animal Management Systems. Virginia Agric. Exp. Stn Bull.86-7, VPI and SU, Blacksburg.

Blaxter, K. L., F. W. Wainman, and R. S. Wilson. 1961. The regulation of food intake by sheep. Anim. Prod. 3:51-61.

Brandyberry, S. D., R. C. Cochran, E. S. Vanzant, and D. L. Harmon. 1991. Effectiveness of different methods of continuous marker administration for estimating fecal output. J. Anim. Sci. 69:4611-4616.

Bransby, D. I. 1988. Compromises in the design and conduct of grazing experiments. p. 53-67. *In* G. C. Marten (ed.) Grazing research: Design, methodology and analysis. CSSA Special Pub. 16. CSSA, Madison, WI.

Broadbent, P. J., J. A. R. McIntosh, and A. Spence. 1970. The evaluation of a device for feeding group-housed animals individually. Anim. Prod. 12:245-252.

Buntinx, S. E., K. R. Pond, D. S. Fisher, and J. C. Burns. 1992. Evaluation of the captec chrome controlled-release device for the estimation of fecal output by grazing sheep. J. Anim. Sci. 70:2243-2249.

Burlison, A. J., J. Hodgson, and A. W. Illius. 1991. Sward canopy structure and the bite dimensions and bite weight of grazing sheep. Grass Forage Sci. 46:29-38.

Burns, J. C., R. D. Mochrie, and D. H. Timothy. 1985a. Intake and digestibility of dry matter and fiber of flaccidgrass and switchgrass. Agron. J. 77:933-936.

Burns, J. C., K. R. Pond, R. D. Mochrie, and D. H. Timothy. 1985b. Mastication, intake and digestion of switchgrass hays. p. 123. *In* Agronomy Abstracts. ASA, Madison, WI.

Burns, J. C., D. H. Timothy, R. D. Mochrie, D. S. Chamblee, and L. A. Nelson. 1978. Animal preference, nutritive attributes, and yield of *Pennisetum flaccidum* and *P. Orientale*. Agron. J. 70:451-456.

Burns, J. C., D. H. Timothy, R. D. Mochrie, and D. S. Fisher. 1988. Relative grazing preference of *Panicum* Germplasm from three taxa. Agron. J. 80: 574-580.

Chenost, M., and C. Demarquilly. 1982. Measurement of herbage intake by housed animals. p. 95-112. *In* J. D. Leaver (ed.) Herbage intake handbook. The British Grassl. Society, Hurley, Maidenhead, Berkshire, UK.

Cochran, R. C., D. C. Adams, J. D. Wallace, and M. L. Galyean. 1986. Predicting digestibility of different diets with internal markers: Evaluation of four potential markers. J. Anim. Sci. 63:1476-1483.

Coleman, S. W., T. D. A. Forbes, and J. W. Stuth. 1989. Measurements of the plant-animal interface in grazing research. p. 37-51. *In* G. C. Marten (ed.) Grazing research: Design, methodology and analysis. CSSA spec. Publ. 16. CSSA, Madison WI.

Corbett, J. L. 1978. Measuring animal performance. p. 163-231. *In* L. 't Mannetje (ed.) Measurement of grassland vegetation and animal production. CAB, Farnham Royal, UK.

Croom, W. J., R. Gaines, and M. Froetschel. 1982. Simple, inexpensive device for feeding cattle and sheep at preset time intervals. J. Dairy Sci. 65:1047-1050.

Demment, M. W., and E. A. Laca. 1993. The grazing ruminant: Models and experimental techniques to relate sward structure and intake. p. 439-460. In Proc. VII World Conference on Anim. Prod. Vol. 1, Invited Papers. 28 June - 2 July 1993, Edmonton, Alberta. Canadian Soc. Anim. Prod., Edmonton, Alberta.

Demment, M. W., E. A. Laca, and G. B. Greenwood. 1987. Intake in grazing ruminants: A conceptual framework. p. 208-225. *In* Feed intake by beef cattle. OK Agric. Exp. Stn MP 121, Oklahoma State Univ., Stillwater.

Dougherty, C. T., N. W. Bradley, P. L. Cornelius, and L. M. Lauriault. 1989. Accessibility of herbage allowance and ingestive behavior of beef cattle. Appl. Anim. Behav. Sci. 23:87-97.

Dove, H. 1991. Using the n-alkanes of plant cuticular wax to identify plant species in the diet of herbivores. p. 58. *In* M. W. Zahari, Z. A. Tajuddin, N. Abdullah, and H. K. Wong (ed.) Proc. 3rd Int. Symp. Nutr. Herb. 25-30 August 1991. Penang, Malaysia. Malaysian Soc. Anim. Prod. Serdang, Selangor, Malaysia.

Dove, H., J. Z. Foot, and M. Freer. 1989a. Estimation of pasture intake in grazing ewes, using the alkanes of plant cuticular waxes. p. 1091-1092. *In* R. Jarrige (ed.) Proc. XVI Inter. Grassl. Congr., Nice, France, 4-11

October 1989. Association Francaise pour la Production Fourragére. Versailles, France.

Dove, H., R. W. Mayes, M. Freer, J. B. Coombe, and J. Z. Foot. 1989b. Faecal recoveries of the alkanes of plant cuticular waxes in penned and in grazing sheep. p. 1093-1094. *In* R. Jarrige (ed.) Proc.XVI Inter. Grassl. Congr., Nice, France, 4-11 October 1989. Association Francaise pour la Production Fourragére. Versailles, France.

Dove, H., R. W. Mayes, C. S. Lamb, and K. J. Ellis. 1991. Evaluation of an intra-ruminal controlled-release device for estimating herbage intake using synthetic and plant cuticular wax alkanes. p. 82. *In* M. W. Zahari, Z. A. Tajuddin, N. Abdullah, and H. K. Wong (ed.) Proc. 3rd Int. Symp. Nutr. Herb. 25-30 August 1991. Penang, Malaysia. Malaysian Soc. Anim. Prod. Serdang, Selangor, Malaysia.

Dove, H., J. A. Milne, and R. W. Mayes. 1990. Comparison of herbage intakes estimated from in vitro or alkane-based digestibilities. Proc. N. Z. Soc. Anim. Prod. 50:457-459.

Duranton, A., and L. Bueno. 1982. Un dispositif simple et fiable d'analyses des movement de machories chey les petits ruminants. Ann. Zootech. 31:489-492.

Egan, A. R. 1970. Nutritional status and intake regulation in sheep. VI. Evidence for variation in setting of an intake regulatory mechanism relating to the digesta content of the reticulorumen. Aust. J. Agric. Res. 21:735-746.

Ehle, F. R. 1984. Measurement of mean particle size of forages by wet and dry sieving techniques. p. 18-21. *In* P. M. Kennedy (ed.) Techniques in particle size analysis of feed and digesta in ruminants. Occasional publ. no. 1., Can. Soc. Anim. Sci., Edmonton, Alberta, Canada.

Ellis, G. H., G. Matrone, and L. A. Maynard. 1946. A 72 percent H_2SO_4 method for the determination of lignin and its use in animal nutrition studies. J. Anim. Sci. 5:285-297.

Fisher, D. S., J. C. Burns, and K. R. Pond. 1986. Sampling effects on grazing behavior in marked forage trials. p. 141. *In* Agronomy abstracts. ASA, Madison, WI.

Fisher, D. S., J. C. Burns, and K. R. Pond. 1988. Estimation of mean and median particle size of ruminant digesta. J. Dairy Sci. 71:518-524.

Fisher, D. S., J. C. Burns, and K. R. Pond. 1990. Modelling physical limitations to intake in ruminant digestion. p. 19-28. *In* A. B. Robson and D. P. Poppi (ed.) Proc. 3rd int. workshop modelling digestion and metabolism in farm animals. 4-6 September 1989. Canterbury, New Zealand. Lincoln College, Canterbury, New Zealand.

Forbes, J. M. 1986. The voluntary food intake of farm animals. Butterworths, London.

Forbes, T. D. A. 1988. Researching the plant-animal interface: The investigation of ingestive behavior in grazing animals. J. Anim. Sci. 66:2369-2379.

Forwood, J. R., A. M. B. da Silva, and J. A. Paterson. 1991. Forage and livestock affecting accuracy of electronic intake measurements of grazers. p. 187. In Agronomy Abstracts. ASA, Madison, WI.

Forwood, J. R., M. M. Hulse, and L. L. Ortbals. 1985. Electronic detection of bolus swallowing to measure forage intake of grazing livestock. Agron. J. 77:969-972.

Gill, M., J. France, R. C. Siddons, and M. S. Dhanoa. 1984. Compartmental models incorporating particle size for the estimation of digestion in the rumen. p. 142-153. In P. M. Kennedy (ed.) Techniques in particle size analysis of feed and digests in ruminants. Occasional publ. no. 1., Can. Soc. Anim. Sci., Edmonton, Alberta, Canada.

Gallup, W. D., C. S. Hobbs, and H. M. Briggs. 1945. The use of silica as a reference substance in digestion trials with ruminants. J. Anim. Sci. 4:68-71.

Gomez, K. A., and A. A. Gomez. 1984. Statistical procedures for agricultural research. John Wiley and Sons, New York.

Grenet, E. 1984. Wet sieving technique for estimating particle size in herbivore digesta. p. 167. In P. M. Kennedy (ed.) Techniques in particle size analysis of feed and digesta in ruminants. Occasional publ. no. 1., Can. Soc. Anim. Sci., Edmonton, Alberta, Canada.

Harris, P. J. 1990. Plant cell wall structure and development. p. 71-90. D. E. Akin, L. G. Ljungdahl, J. R. Wilson, and P. J. Harris. (ed.) Microbial and plant opportunities to improve lignocellulose utilization by ruminants. Elsevier Sci. Publ. Co., New York.

Heaney, D. P., G. I. Pritchard, and W. J. Pigden. 1968. Variability in ad libitum forage intakes by sheep. J. Anim. Sci. 27:159-164.

Hodgson, J. 1982. Ingestive behaviour. p. 113-138. In J. D. Leaver (ed.) Herbage intake handbook. The British Grassl. Society, Hurley, Maidenhead, Berkshire, UK.

Holechek, J. L., H. Woffard, D. Arthun, M. L. Galyean, and J. D. Wallace. 1986. Evaluation of total fecal collection for measuring cattle forage intake. J. Range Manage. 39:2-4.

Holter, J. A., and J. T. Reid. 1959. Relationship between the concentration of crude protein and apparently digestible protein in forages. J. Anim. Sci. 18:1339-1349.

Horn, F. P. 1981. Direct measurement of voluntary intake of grazing livestock by telemetry. p. 367-372. In J. W. Wheeler and R. D. Mochrie (ed.) Forage evaluation: Concepts and techniques. Armidale, Australia, 27-31 October 1980, Griffin Press Limited, Netley, South Australia.

Houessou, S., M. Prudhon, P. Bosc, G. Bechet, and G. Molenat. 1989. Effets de la duree de presence au paturage sur le comportement alimentaire d'ovins en crau. p. 1261-1262. In R. Jarrige (ed.) Proc. XVI Inter. Grassl. Congr., Nice, France, 4-11 October, 1989. Association Francaise pour la Production Fourragére. Versailles, France.

Jarrige, R. 1989. Ruminant Nutrition: Recommended allowances and feed tables. INRA, John Libbey, Paris.

Jarrige, R., C. Demarquilly, J. P. Dulphy, A. Hoden, J. Robelin, C. Beranger, Y. Geay, M. Journet, C. Malterre, D. Micol, and M. Petit. 1986. The INRA "fill unit" system for predicting the voluntary intake of forage-based diets in ruminants: A review. J. Anim. Sci. 63:1737-1758.

Johnstone-Wallace, D. B., and K. Kennedy. 1944. Grazing management practices and their relationship to the behavior and grazing habits of cattle. J. Agric. Sci. (Camb.) 34:190-197.

Kennedy, P. M. 1984. Techniques in particle size analysis of feed and digesta in ruminants. Occasional publ. no. 1., Can. Soc. Anim. Sci., Edmonton, Alberta, Can.

Kenney, P. A., and J. L. Black. 1984. Factors affecting diet selection by sheep. I. Potential intake rate and acceptability of feed. Aust. J. Agric. Res. 35:551-563.

Ketelaars, J. J. M. H., and B. J. Tolkamp. 1992. Toward a new theory of feed intake regulation in ruminants. 1. Causes of differences in voluntary feed intake: Critique of current views. Livestock Prod. Sci. 30:269-296.

Laby, R. H., C. A. Graham, S. R. Edwards, and B. Kautzner. 1984. A controlled release intraruminal device for the administration of fecal dry-matter markers to the grazing ruminant. Can. J. Anim. Sci. 64:337-338.

Laredo, M. A., G. D. Simpson, D. J. Minson, and C. G. Orpin. 1991. The potential for using n-alkanes in tropical forages as a marker for the determinations of dry matter intake by grazing ruminants. J. Agric. Sci. (Camb.) 117:355-361.

Le Du, Y. L. P., and P. D. Penning. 1982. Animal-based techniques for estimating herbage intake. p. 37-75. *In* J. D. Leaver (ed.) Herbage intake handbook. The British Grassl. Soc., Hurley, Maidenhead, Berkshire, UK.

Lippke, H., W. C. Ellis, and B. F. Jacobs. 1986. Recovery of indigestible fiber from feces of sheep and cattle on forage diets. J. Dairy Sci. 69:403-412.

Little, D. A. 1972. Studies on cattle with oesophageal fistulae: The relation of the chemical composition of feed to that of the extruded bolus. Aust. J. Exp. Agric. Anim. Husb. 12:126-130.

Luginbuhl, J-M., D. S. Fisher, K. R. Pond, and J. C. Burns. 1991a. Image analysis and nonlinear modeling to determine dimensions of wet-sieved, masticated forage particles. J. Anim. Sci. 69:3807-3816.

Luginbuhl, J-M., K. R. Pond, and J. C. Burns. 1993. Whole tract digesta kinetics and comparison of techniques for the estimation of fecal output in steers fed Coastal bermudagrass hay at four levels of intake J. Anim. Sci. 71:(In Press).

Luginbuhl, J-M., K. R. Pond, J. C. Burns, D. S. Fisher, and J. C. Russ. 1991b. A computer interface system to monitor the ingestive and ruminating behavior of grazing ruminants. p. 177. *In* Proc. 2nd Grazing Livestock Nutrition Conference. 2-3 August 1991, Steamboat Springs, CO. Oklahoma Agric. Exp. Sta. MP-133, Oklahoma State Univ., Stillwater.

Luginbuhl, J-M., K. R. Pond, J. C. Russ, and J. C. Burns. 1987. A simple electronic device and computer interface system for monitoring chewing behavior of stall-fed ruminant animals. J. Dairy Sci. 70:1307-1312.

Marais, J. P., M. de Figueiredo, and D. L. Figenschou. 1992. The behaviour of insoluble n-alkane markers in the rumen. p. X24. *In* Programme and Proceedings. 31st Annual Cong. of the South African Soc. of Anim. Prod. 13-16 April 1992. Zithabiseni, Kwandebele. South African Society of Animal Production, Germiston, Republic of South Africa.

Mayes, R. W., C. S. Lamb, and P. M. Colgrove. 1986. The use of dosed and herbage n-alkanes as markers for the determination of herbage intake. J. Agric. Sci. (Camb.) 107:161-170.

Meijs, J. A. C. 1981. Herbage intake by grazing dairy cows. Agric. Res. Rep. 909. Centre Agric. Publ. and Documents, Wageningen, Netherlands.

Meijs, J. A. C., R. J. K. Walters, and A. Keen. 1982. Sward methods. p. 11-36. *In* J. D. Leaver (ed.) Herbage intake handbook. The British Grassl. Society, Hurley, Maidenhead, Berkshire, UK.

Minson, D. J. 1981a. The measurement of digestibility and voluntary intake of forages with confined animals. p. 159-176. *In* J. W. Wheeler and R. D. Mochrie (ed.) Forage evaluation: Concepts and techniques, Armidale, Australia, 27-31 October 1980, Griffin Press Limited, Netley, South Australia.

Minson, D. J. 1981b. Nutritional differences between tropical and temperate pastures. p. 143-157. *In* F. H. W. Morley (ed.) World Animal Science B1. Grazing Animals. Elsevier Sci. Publ. Co., New York.

Minson, D. J. 1990. Forage in ruminant nutrition. Academic Press, San Diego.

Minson, D. J., and R. A. Bray. 1986. Voluntary intake and in vivo digestibility by sheep of five lines of *Cenchrus ciliaris* selected on the basis of preference rating. Grass Forage Sci. 41:47-52.

Moore, J. A., K. R. Pond, M. H. Poore, and T. G. Goodwin. 1992. Influence of model and marker on digesta kinetic estimates for sheep. J. Anim. Sci. 70:3528-3540.

Moseley, G., and A. A. Manendez. 1989. Factor affecting the eating rate of forage feeds. p. 789-790. *In* R. Jarrige (ed.) Proc. XVI Inter. Grassl. Congr., Nice, France, 4-11 October 1989. Association Francaise pour la Production Fourragére, Versailles, France.

Mott, G. O. 1960. Grazing pressure and the measurement of pasture production. p. 606-611. *In* Proc. 8th Inter. Grassl. Cong. Reading, UK. 11-21 July 1960. Alden Press, Oxford.

Murphy, M. R., and E. H. Jaster, 1984. A computerized system for recording and analyzing chewing behavior. J. Dairy Sci. 67 (Suppl. 1):204 (abstr.).

Neal, H. D. S. C., C. Thomas, and J. M. Cobby. 1984. Comparison of equations for predicting voluntary intake by dairy cows. J. Agric. Sci. (Camb.) 103:1-11.

NRC. 1984. Nutrient requirements of beef cattle. National Academy Press, Washington, DC.

Ohajuruka, O. A., and D. L. Palmquist. 1991. Evaluation of n-alkanes as digesta markers in dairy cows. J. Anim. Sci. 69:1726-1732.

Owens, F. N., and A. L. Goetsch. 1986. Digesta passage and microbial protein synthesis. p. 196-223. *In* L. P. Milligan, W. L. Grovum, and A.

Dobson (ed.) Control of digestion and metabolism in ruminants. Prentice-Hall, Englewood Cliffs, NJ.

Owens, F. N., and C. F. Hanson. 1992. External and internal markers for appraising site and extent of digestion in ruminants. J. Dairy Sci. 75:2605-2617.

Parker, W. J., S. N. McCutcheon, and D. H. Carr. 1989. Effect of herbage type and level of intake on the release of chromic oxide from intraruminal controlled release capsules in sheep. N. Z. J. Agric. Res. 32:537-546.

Penning, P. D., and G. E. Hooper. 1985. An evaluation of the use of short-term weight changes in grazing sheep for estimating herbage intake. Grass Forage Sci. 40:79-84.

Penning, P. D., A. J. Parsons, R. J. Orr, and T. T. Treacher. 1991. Intake and behaviour responses by sheep to changes in sward characteristics under continuous stocking. Grass Forage Sci. 46:15-28.

Peterson, R. G., and H. L. Lucas, Jr. 1968. Computing methods for the evaluation of pastures by means of animal responses. Agron. J. 60:682-687.

Pienaar, J. P., and C. Z. Roux. 1989. Use of the gamma function in equations which describe ruminal fermentation and outflow rates for the prediction of voluntary intake and protein degradation. S. Afr. J. Anim. Sci. 19:99-106.

Pienaar, J. P., C. Z. Roux, P. J. K. Morgan, and L. Grattarola. 1980. Predicting voluntary intake on medium quality roughages. S. Afr. J. Anim. Sci. 10:215-225.

Pond, K. R., J. C. Burns, and D. S. Fisher. 1987. External markers - Use and methodology in grazing studies. p. 49-53. *In* Proc. Grazing Livestock Nutrition Conf., 23-24 July 1987. Jackson, WY. Univ. of Wyoming, Laramie.

Pond, K. R., J. C. Burns, D. S. Fisher, and R. A. Quiroz. 1986. Appropriate markers and methodology for grazing studies. p. 62-66. *In* Proc. 42nd Southern Pasture and Forage Crop Improvement Conf. Athens, GA. 15-17 April 1986. National Tech. Info. Ser., Springfield, VA.

Pond, K. R., W. C. Ellis, and D. E. Akin. 1984a. Ingestive mastication and fragmentation of forages. J. Anim. Sci. 58:1567-1574.

Pond, K. R., W. C. Ellis, J. H. Matis, and A. G. Deswysen. 1989. Passage of chromium-mordanted and rare earth-labeled fiber: Time of dosing kinetics. J. Anim. Sci. 67:1020-1028.

Pond, K. R., J-M. Luginbuhl, J. C. Burns, D. S. Fisher, and S. E. Buntinx. 1990. Estimating intake using rare earth markers and controlled release devices. p. 73-81. *In* Proc. 45th Southern Pasture and Forage Crop Improvement Conf. Little Rock, AR, 12-14 June 1989. National Tech. Info. Ser., Springfield, VA.

Pond, K. R., E. A. Tolley, W. C. Ellis, and J. H. Matis. 1984b. A method for describing the weight distribution of particles from sieved forage. p. 123-133. *In* P. M. Kennedy (ed.) Techniques in particle size analysis of feed and digesta in ruminants. Occasional publ. no. 1, Can. Soc. Anim. Sci., Edmonton, Alberta, Canada.

Poppi, D. P., M. Gill, J. France, and R. A. Dynes. 1990. Additivity in intake models. p. 29-46. *In* Proc. 3rd int. workshop modelling digestion and metabolism in farm animals. 4-6 September 1989. Canterbury, New Zealand. Lincoln College, Canterbury, New Zealand.

Quiroz, R. A., K. R. Pond, E. A. Tolley, and W. L. Johnson. 1988. Selection among nonlinear models for rate of passage studies in ruminants. J. Anim. Sci. 66:2977-2986.

Reid, J. T., P. G. Woolfolk, W. A. Hardison, C. M. Martin, A. L. Brundage, and R. W. Kaufmann. 1952. A procedure for measuring the digestibility of pasture forage under grazing. J. Nutr. 46:255-269.

Reid, J. T., P. G. Woolfolk, C. R. Richards, R. W. Kaufmann, J. K. Loosli, L. K. Turk, J. I. Miller, and R. E. Blazer. 1950. A new indicator method for the determination of digestibility and consumption of forages by ruminants. J. Dairy Sci. 33:60-71.

Schiffman, S. S., M. L. Reynolds, and F. W. Young. 1981. Introduction to multi-dimensional scaling: Theory, methods, and application. Academic Press, New York.

Schneider, B. H., and W. P. Flatt. 1975. Pasture consumption and digestibility. p. 107-121 and 178-188. *In* The evaluation of feeds through digestibility experiments. The University of Georgia Press. Athens, GA.

Sheath, G. W., P. V. Rattray, and D. C. Smeaton. 1987. Influence of pasture quantity and quality on intake and production of sheep. p. 33-43. *In* F. P. Horn, J. Hodgson, J. J. Mott, and R. W. Brougham (ed.) Grazing-lands research at the plant-animal interface. Winrock Inter. Inst. Agric. Develop., Morrilton, AK.

Smith, L. W., and D. R. Waldo. 1969. Method for sizing forage cell wall particles. J. Dairy Sci. 52:2051-2053.

Telford, J. P. 1980. Factors affecting intake and digestibility of grazed forage.

Ph.D. diss. Texas A&M Univ., College Station (Diss. Abstr. 80-08054).

Thoney, M. L., B. A. Palhof, M. R. Carlo, D. A. Ross, N. L. Firth, R. L. Quass, D. J. Perosio, D. J. Duhaime, S. R. Rollins, and A. Y. M. Nour. 1985. Sources of variation of dry matter digestibility measured by the acid insoluble ash marker. J. Dairy Sci. 68:661-668.

Vaage, A. S., J. A. Shelford, and G. Moseley. 1984. Theoretical basis for the measurement of particle length when sieving elongated feed particles. p. 76-83. In P. M. Kennedy (ed.) Techniques in particle size analysis of feed and digesta in ruminants. Occasional Publ. no. 1., Can. Soc. Anim. Sci., Edmonton, Alberta, Canada.

Vulich, S. A., E. G. O'Riordan, and J. P. Hanrahan. 1991. Use of n-alkanes for the estimation of herbage intake in sheep: Accuracy and precision of the estimates. J. Agric. Sci. (Camb.) 116:319-323.

Weston, R. H., and P. M. Kennedy. 1984. Various aspects of reticulorumen digestive function in relation to diet and digesta particle size. p. 1-17. In P. M. Kennedy (ed.) Techniques in particle size analysis of feed and digesta in ruminants. Occasional Publ. no. 1., Can. Soc. Anim. Sci., Edmonton, Alberta, Canada.

Wheeler, J. L. 1962. Experimentation in grazing management. Herb. Abstr. 32:1-7.

Worley, R. R. 1987. Some physical aspects of digestion of forages by steers and binding of ytterbium and hafnium as digesta particle markers. Ph.D. diss. Texas A&M Univ., College Station (Diss. Abstr. 88-02156).

Zemmelink, G. 1980. Effect of selective consumption on voluntary intake and digestibility of tropical forages. Agric. Res. Rept. 896. Centre Agric. Publ. and Documents, Wageningen, Netherlands.

Appendix Table 12-1. Protocol for Determining Forage Dry Matter Intake in Confinement

Item	Comment
A. Background Information Required (for Methods Section of publications)	
1. Forage	
(a) Species identification	Common name, latin name, and cultivar
(b) Growing conditions	Specify fertility/stress conditions/soil type
(c) Growth stage	Give physiological maturity (veg., boot, etc)
(d) Proportion of mixture	Hand separate species, give as % DM
(e) Leaf:Stem:Dead	Hand separate; give as % DM
(f) Physical Form	Chopped (range of lengths); ground (screen size); long (range of lengths)
2. Animal	
(a) Species and sex	Use exp. animals similar to application
(b) Weight and condition	At initiation and weigh weekly at same hour
(c) Immediate history	If history has bearing on daily response
(d) Standardize	Feed common forage one to two weeks before trial, deworm if needed
(e) Provide water	Continuous or twice per day
(f) Provide mineral	Separate boxes, trace mineralized salt and dicalcium phosphate
B. Purpose of Intake Study is to measure:	
(a) Potential of diet	Feed $\geq$ 135%
(b) Forage offered (performance)	Feed 110 to 120%
(c) Forage offered (maintenance)	Feed 95 to 100%
(d) Response from c to a	Trt ranges: 95 to >135%
C. Designs often used	Completely randomized, randomized complete block, latin square, and reversals
D. Daily event	

Day	Activity
1	Weigh animals, score condition, feed experimental diet
2 to 7 (preliminary)	Adjustment period for normal forage (Adj. feeding to *ad libitum*; if short 1 d, add; if excess 3 d, reduce)
(2 to 14, preliminary)	Adjustment period longer for silage or if anti-quality factor present
7 and 14	Weigh animals
7 to 21	14 d intake measurement (Sample daily constant % of forage offered; save in air-tight container)
8 to 22	Collect daily total refusal from each animal, weigh and composite in individual-animal container
22	Weigh animals

E. Sample collection
1) Forage fed — Sample daily (to give 2 to 3 kg for period); collect as composite in sealed container
2) Refusal — At end of collection period, dry (55°C), weigh, mix, and either subsample (500 g) or store total wt. in air tight container
3) Waste (lost from manger but recovered) — Collect total during period that refusals are collected (if minimal, collect at end of trial, if excessive, collect more often), place in cloth bag and dry (55°C for 2 d)

F. Sample preparation (As fed forage, refusal, waste)
1) If air equilibrated — Weigh before grinding
2) Reduce large samples — Use chopper or Wiley mill (4-mm screen)
3) Subsample — Mix thoroughly, randomly remove 400 g, pass through a 1-mm screen, store in air tight container
4) Excess material — Discard excess refusal and waste. Grind excess 'as fed' forage and store in air tight container, place in cold storage for future reference material

G. Analyses (As fed forage, refusal, waste)
1) Dry matter (DM) — Determine for 'as fed' forage as soon as possible (avoid moisture changes after feeding), and for refusal and waste. Adjust silage DM for volatile losses
2) Samples < 88% DM — Re-dry all samples to prevent molding. Use original DM to compute intake and second DM to compute nutrient composition on a DM basis

H. Calculations
1) Dry Matter Intake (DMI)

$$\text{Daily DM Fed (g)} = \frac{(\% \text{ DM Feed})}{100} * \frac{(\text{Total Feed Fed})}{\text{Days in Period}}$$

$$\text{Daily DM Refusal (g)} = \frac{(\% \text{ DM Refusal})}{100} * \frac{(\text{Total Refusal})}{\text{Days in Period}}$$

$$\text{Daily DM Wasted (g)} = \frac{(\% \text{ DM Waste})}{100} * \frac{(\text{Total Waste})}{\text{Days in Period}}$$

Daily DMI (g) = (DM Fed) - (DM Refusal) - (DM Wasted)

2) Intake of other (X) constituents (including organic matter)

$$\text{Daily Intake of X} = \frac{(\text{DM Fed} * \%X)}{100} - \frac{(\text{DM Refusal} * \%X)}{100} - \frac{(\text{DM Waste} * \%X)}{100}$$

3) Expressing DMI

In most experiments where comparisons are within species/breeds and animal live weights are similar, express as percent of body weight (i.e., kg 100 kg^{-1}). In experiments with comparisons among animal species differing widely in live weight (goats vs. cattle vs. elephants), the use of metabolic body weight is appropriate.

Appendix Table 12-2. Protocol for Determining Forage Dry Matter Intake (Indirect Estimates) for Grazing Animals

Item	Comment
A. Background Information Required (include in Methods Section of publications)	
1. Forage	
(a) Species identification	Common name, Latin name, and cultivar
(b) Growing condition	Specify fertility/stress conditions/soil type
(c) Pasture available	Estimate herbage mass (kg ha^{-1})
(d) Proportion of mixture	Hand separate species: give as % DM or visually estimate
(e) Leaf:Stem:Dead	Hand separate; give as % DM
2. Animals	
(a) Species, sex, state	Use exp. animals similar to application
(b) Weight and condition	At start and end of sampling; est. body condition
(c) Immediate history	If history has bearing on daily response
(d) Provide water	Continuous or twice per day
(e) Provide mineral	Trace mineralized salt and others as needed
B. Purpose of Intake Study	
1) Comparing forages	Species characteristics dominate; herbage mass less important
2) Comparing herbage mass	Herbage mass dominate; canopy characteristics less important
3) Comparing both 1 and 2	Herbage mass and canopy structure important
C. Designs frequently used	Completely random and randomized complete block
D. Fiber for Marking	
1) Source	Diet from each treatment; obtain as close to experimental period as possible
2) Diet	Obtain using esophageal cannulated animals; collect bulk sample retaining saliva
3) Preparation	Extract diet with neutral detergent solution, wash fiber, attach marker, dry and pack into gelatin capsules

E. Daily Events

Day	Activity
-2, -1	Sample for herbage mass (kg ha^{-1})
0	Pulse dose; Admin. orally (gelatin capsules via balling gun, liquid marker via syringe) beginning 0600 to two testers per paddock; obtain fecal samples [0-h grab (or rectal) sample (100 g fresh) and samples at 1100, 1500, 1900 and 2300 h] and place in freezer

4) Expressing DMI

In most experiments where comparisons are within species/breeds and animal live weights are similar, express as percent of body weight (i.e., kg 100 kg^{-1}). In experiments with comparisons among animal species differing widely in live weight (goats vs. cattle vs. elephants), the use of metabolic body weight is appropriate.

1	Obtain a) diet from each paddock (0700-0900 h), retain saliva, quick freeze masticate in liquid N, and store in freezer and b) fecal grab samples at 0600, 1000, 1300, 1700, 2000, and 2300 h
2,3	Obtain a) diet from each paddock and b) fecal samples at 0600, 1300, and 1900 h
4	Obtain a) diet from each paddock and b) fecal samples at 0600 and 1600 h
5, 6	Sample for herbage mass (kg ha^{-1})

F. Sample handling
  1) Forage
   a) Herbage mass — Oven-dry (75°C), weigh, and discard
   b) Separation — Do fresh or store in freezer, and separate later, freeze, freeze dry, weigh, and grind through 1-mm screen
  2) Diet (Masticate) — Freeze dry, subsample, and grind in a udy mill to pass a 0.5-mm screen, return to freezer
  3) Feces — Freeze dry, grind in a coffee mill for marker analysis or Wiley mill to pass a 1-mm screen for chemical analysis

G. Analyses
  1) Diet Digestibility — Estimate from freeze-dried masticate using in vitro DM (or organic matter) disappearance (IVDMD) fermented for 48 or 72 h as appropriate (accuracy is of major concern). Internal markers or in vitro-in vivo regression are other options if properly validated
  2) Marker — Estimate marker concentration in each fecal sample

H. Calculations
1) Marker concentration fit to appropriate model for each animal to estimate variables below. Use mean of testers in each paddock for analysis of variance.
   a) Fecal DM output (g d^{-1})
   b) Rate of passage (h^{-1})
   c) Mean retention time (h)
   d) Digestive tract fill (g)
2) Dry Matter Intake (DMI) Calculation

$$\text{DMI (g d}^{-1}) = \frac{\text{Fecal DM output (g d}^{-1})}{1 - (\text{Diet IVDMD}/100)}$$

3) Intake of other (X) constituents (including organic matter)
   Daily intake of X = DMI * Proportion of X in Masticate

CHAPTER 13

PREDICTION OF INTAKE AS AN ELEMENT OF FORAGE QUALITY

Dennis J. Minson and John R. Wilson

INTRODUCTION

The primary nutritional factor controlling animal production is the quantity of feed (dry matter; DM) that an animal eats each day when excess feed is offered (voluntary feed intake; VI or *ad libitum* intake). Despite recognition of this by farmers and their advisors, VI has been completely ignored in most textbooks on nutrition until the last decade. When VI is considered, it is usually relegated to a chapter near the end of the book (e.g., Standing Committee on Agriculture, 1990). The failure of most scientists to recognize the importance of voluntary intake has led to an unnecessary and undesirable gulf between the science and practice of feeding grazing animals.

The principle aim of nutrition studies, conducted at the beginning of the 19th century, was to determine the value of different feeds (relative to hay) and to relate these differences to the chemical composition of the feed (Tyler, 1959). Allowing animals to eat different quantities of feed only confused the interpretation of the results of these experiments. In 1916, Osborne and Mendel recognized that *ad libitum* feeding gave rise to variable results and asked the fundamental question "does one animal grow because it eats more or the other fail because it eats less?" Unfortunately, their question did not lead to the measurement of the VI of different feeds but to the paired feeding design for nutrition experiments. This lack of interest in measuring VI led to the belief that all animals ate about 3% of body weight and that there were no differences in VI among feeds. This view prevailed well into the second half of the 20th century.

HISTORICAL

There has never been a concerted attempt to determine the different chemical and physical components that limit the VI of forages. There are no reports in the scientific literature of the discovery of yet another factor related to VI, only research papers containing comments on VI being affected by certain factors. Very rarely do the words "voluntary intake" appear in the abstracts, let alone the title of these papers, so it is difficult to ensure that the true "discoverer" for each chemical component or physical attribute has been

---

D. J. Minson and J. R. Wilson, Division of Tropical Crops and Pastures, CSIRO, Cunningham Laboratory, 306 Carmody Road, St. Lucia, Brisbane, Queensland 4067, Australia.

correctly identified. The factors related to VI will now be considered according to their date of discovery.

1924-Phosphorus (P)

Sir Arnold Theiler and his team (1924) can without doubt claim to have discovered that low P levels in forage limit VI. In their classic study, two groups of six cattle (initial weight, 205 kg) were used in a double change-over experiment lasting 16 mo. All the animals were fed hay (0.7 g/kg DM P) *ad libitum* plus a fixed quantity of an energy and protein supplement low in P. Half the cattle received 84 g/d bonemeal. Records were kept of the VI of the hay and bone chewing. The cattle receiving bonemeal ate more hay, gained more weight, and stopped bone-chewing within 2 mos (Figure 1). After 10 mos, the P-supplemented cattle were eating 70% more hay than the unsupplemented animals, a convincing demonstration that P deficiency reduces VI.

1935-Cobalt (Co)

The key role of Co on VI of forage was convincingly demonstrated for the first time by Lines (1935) working in South Australia. He put six ewes, in an advanced stage of "coast disease" (salt sickness), into pens and fed them weighed quantities of cereal hay which had been grown on land where "coast disease" was prevalent. Over the next 18 wk, the ewes became progressively weaker and more lethargic, exhibited a steady decline in appetite, and three died. "When the survivors were *in extremis*, and their daily intake of food had declined to 400-600 gm per day, one animal was given 1 mg per day of cobaltous nitrate. Within three days, this animal had improved in health and its appetite increased materially, steadily rising to 900-1000 gm per day". "Seven weeks later, the same amount of cobalt was given to a second animal". "The response of this animal was similar to the one previously mentioned".

1951-Nitrogen (N)

The VI of mature grass hay is low and one of the factors limiting intake can be the low level of crude protein. This was first demonstrated by Clark and Quin (1951) working in South Africa who found that the VI of a poor quality grass hay could be increased by up to 75% if sprayed with a mixture of urea and molasses but VI was depressed if only molasses was applied. In subsequent work with oat (*Avena sativa*) straw, adding urea alone increased VI by 39% (Campling and Freer, 1961). This increase in VI was associated with a 39% reduction in the time the cattle spent ruminating the urea-supplemented straws (Campling and Freer, 1961), possibly because the ruminal microflora had softened the fiber to an extent that less chewing was required to reduce the straw particles to a size where they could pass out of the rumen. Digesting forages for 48 h *in vitro* has been found to reduce their resistance to mechanical fracture by 55% (Minson, 1986).

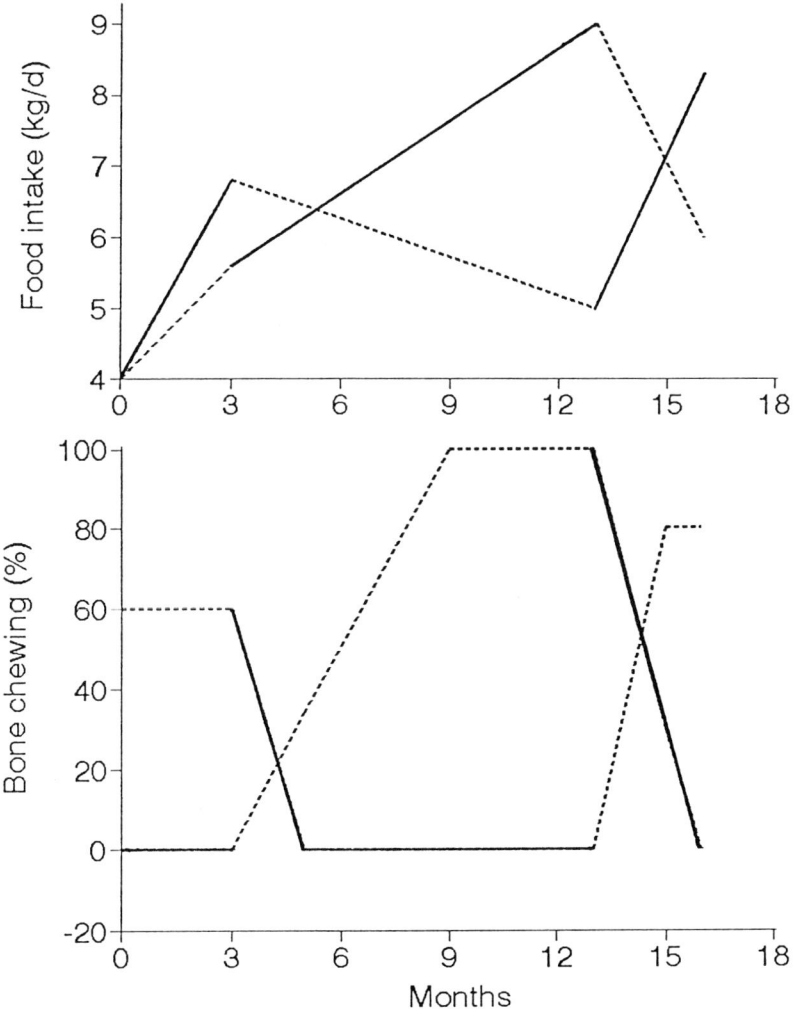

Figure 1. Influence of phosphorus on feed consumption and bone chewing, P supplemented ———, no additional P --- (derived from Theiler et al., 1924).

1957-Particle Size of Forage

Kick et al. (1937) reported that the time cattle spent ruminating was considerably reduced when ground hay was fed. However, it was another 20 yr before it was recognized that this treatment had any effect on VI. Using groups of between 10 and 15 steer calves, Webb et al. (1957) found that grinding and pelleting increased the VI of hays made from timothy (*Phleum pratense*), alfalfa (*Medicago sativa*) and *Sericea lespedeza* (Table 1). This effect of reduced particle size on VI has been confirmed by many subsequent studies covering a wide range of temperate and tropical forages (see Minson, 1990). The increase in VI is not constant but appears to be related to the resistance of forage to physical breakdown. Thus, grinding and pelleting has a larger effect on the VI of mature forage than when young forage is processed (Heaney et al., 1963).

Table 1. Voluntary intake (g/kg $BW^{0.75}$) by steer calves of forages with different particle size (Webb et al., 1957).

Species	Form		Increase in VI, %
	Chopped or long	Ground and pelleted	
Timothy-alfalfa	84	124	48
Alfalfa	85	122	44
Sericea lespedeza	89	102	15

1960-Silage versus Hay

The VI of silage is lower than that of hay made from the same forage. For silage made from fresh forage, the VI is about two thirds that of hay (Moore et al., 1960). By wilting the forage before it is ensiled, VI can be increased (Moore et al., 1960; Murdoch, 1960), an effect confirmed in many subsequent studies reviewed by Minson (1990).

It was initially thought that the lower VI of silage made from fresh forage was caused by the products of the fermentation process and (or) the extensive deamination of the protein. However, it became increasingly apparent that the lower VI was associated with the longer time that fresh cut silage was retained in the rumen. This high retention time appears to be caused by the difficulty animals have in regurgitating the partially digested silage which is probably swallowed before it is thoroughly chewed. This hypothesis is supported by many studies which show that if the forage is finely chopped before (or after) ensiling, VI is increased, efficiency of regurgitation is improved, and less time is spent ruminating (see Minson, 1990).

1960-Rate of Digestion

In 1960, Donefer et al. showed that the cellulose in legumes was digested more rapidly *in vitro* than the cellulose in grasses and that VI was correlated ($r = 0.83$) with the rate of digestion. Similar relations have subsequently been published for a range of feeds with rate of digestion being measured by the *in vitro* or *in sacco* techniques (Orskov et al., 1988). In many of these studies, a large proportion of the difference in VI can be accounted for by the potential indigestibility of the forage (Orskov et al., 1988), suggesting that rate of digestion *per se* may not be the primary factor controlling VI. This possibility is supported by the observation that the rate of digestion *in vitro* of leaf and stem fractions of tropical grasses is similar although the VI of the leaf fraction is 46% higher than that of the stem fraction of similar digestibility (Laredo and Minson, 1973).

1961-Digestibility

Blaxter et al. (1961) was the first group to demonstrate that the VI of forage was positively correlated with digestibility. Using three contrasting hays fed to sheep *ad libitum*, they concluded "that the range of quality of the foods as judged by digestibility was considerable and that appetite was correlated with food quality". They presented a graph relating *ad libitum* DM intake to the apparent digestibility of the dietary energy. Similar positive regressions have subsequently been published for many temperate and tropical forages (see Minson, 1990). It should be noted that Blaxter et al. (1961) were very careful to avoid saying that VI of forage was controlled by digestibility because the relation between VI and digestibility is no more than a simple correlation caused by both VI and digestibility being related to the fiber content of the forage.

The relationship between VI and digestibility has proved very useful in plant breeding and selection programs where digestibility can be used for selecting forages of higher VI, but breaks down when used to compare grasses with legumes or chopped with pelleted forages (Heaney et al., 1963).

1961-Magnesium (Mg)

McLeese et al. (1961) observed that the VI of a semipurified diet by sheep was reduced when the diet was deficient in Mg. The body contains very little labile Mg and VI can fall by 67% within 3 to 4 days of feeding a diet deficient in Mg and recover rapidly once Mg is fed. This loss of appetite appears to be caused by reduced cellulolytic activity of the ruminal bacteria, and a reduction in cellulose digestibility of 24 to 30% (see Minson, 1990).

1962-Selenium (Se)

Low levels of Se in the diet were shown by Muth et al. (1958) to be the cause of nutritional muscular dystrophy or white-muscle disease in young ruminants. Four years later, McLean et al. (1962) working in New Zealand, reported that with some but not all forages, Se increased the growth rate of lambs and, in the case of white clover (*Trifolium repens*), intake was increased by 38%. Subsequent work showed that white clover had a lower level of Se than many other forages (see Minson, 1990).

1962-Water Content

Graziers have a strong feeling that very wet forage has a low feeding value but it was not until 1962 that this was linked to a low VI. Davies (1962) showed that once the water content of forage exceeded about 80%, any further increase in water content led to a decrease in VI of DM, an effect confirmed in many subsequent studies (see Minson, 1990). The decrease in VI was not caused by water *per se* because putting large quantities of water into the rumen via a ruminal fistula had no effect on VI (Davies, 1962). It has recently been suggested that the lower intake is caused by the wet forage being swallowed before it has been properly chewed and, hence, before maximal particle breakdown has occurred. Forage that is prematurely swallowed would require more breakdown during rumination, be retained longer in the rumen and, hence, be eaten in smaller quantities (Minson, 1990).

1964-Lignin

Lignin is an almost completely indigestible component of forages that increases in concentration as the plant matures. Hawkins et al. (1964) reported that the VI of bermuda grass (*Cynodon dactylon*) was correlated with the lignin concentration of the forage ($r = -0.68$ and -0.61). Subsequent work with both temperate and tropical grasses confirmed this effect (see Minson, 1990). However, when both grasses and legumes were included in the same study, the correlation was only 0.13 (Van Soest, 1965).

1964-Fiber

In 1964, Hawkins et al. reported that the VI of bermuda grass was negatively correlated with the crude fiber concentration in the forage ($r = -0.87$), a result confirmed by Wilson et al. (1966), using 58 mixed temperate forages, ($r = -0.79$). Using samples of grasses and legumes of known VI, Van Soest (1965) showed that VI was significantly correlated with the concentration of neutral detergent fiber (NDF) ($r = -0.65$) and acid detergent fiber (ADF) ($r = -0.53$). In the following year, Wilson et al. (1966), using 58 mixed forages, reported that VI also was related to the level of modified ADF ($r = -0.84$). The fiber content of forage can be measured in many other ways and the relations between VI and fiber content have recently been reviewed (Minson, 1990).

1969-Sulfur (S)

Sulfur is an essential element of all proteins but it was not until 1969 that S was shown to affect VI (Playne, 1969). Sheep fed a mixture of spear grass (*Heteropogon contortus*) and Townsville stylo (*Stylosanthes humilis*) containing 0.6 g/kg DM S ate 359 g/d. However, when supplemented with sodium sulfate, VI was 534 g/d, a 49% increase due to the supplemental sulfur. Similar increases in intake have subsequently been reported in many other studies (see Minson, 1981).

1973-Leaf versus Stem

Farmers have for a long time recognized that the leaves of forages are of greater quality than the stems. This difference was generally believed to be caused by lower fiber and higher digestibility of the leaf fraction. However, in 1973, a method was developed for separating chopped dried forage into leaf and stem fractions (Laredo and Minson, 1973). These fractions had similar digestibilities but the VI of the leaf fraction was 46% higher than that of the stem fraction. The higher VI was caused by a lower resistance of leaf particles to breakdown during eating and the shorter retention time of the leaf fraction in the rumen (Minson, 1990).

1975-Sodium (Na)

The essential nature of Na has been known for hundreds of years but only relatively recently has it been shown that Na affected VI (Morris and Murphy, 1972). When a basal diet containing 0.065 g/kg DM Na was fed to calves, the daily intake was 2.15 kg/d but increased to 2.99 kg/d when a Na supplement was provided.

CHEMICAL AND PHYSICAL CHARACTERISTICS IMPORTANT TO REGULATION OF VOLUNTARY INTAKE

The VI of forage by ruminants is controlled by many factors as shown in the previous section of this review. These factors may be divided into two groups: 1) nutrients essential for maximal microbial activity in the rumen and the maintenance of metabolic functions of the animal, and 2) components of the diet which slow down the passage of forage through the rumen. In addition to these two groups of factors, VI will be decreased by the presence of toxic compounds which are reviewed in Chapters 8 and 9. Voluntary intake of forage is also controlled by the demand of the animal for nutrients; VI is increased by lactation, shearing, cold stress and depressed by heat stress and internal parasites (Minson, 1990). The effect on VI of essential nutrients and components that slow the passage of forage through the rumen will be considered in the following sections.

Essential Nutrients

In the previous section, VI was shown to be decreased by low levels of N, S, P, Mg, Na, Co, and Se in the forage. The concentrations of copper (Cu) and iodine (I) also can affect the production of grazing animals but there is, at present, no evidence that these elements affect VI. In the case of calcium (Ca), iron (Fe), zinc (Zn), and manganese (Mn), there is little or no evidence that these elements are ever deficient in grazed forage and limit VI (Minson, 1990).

Although it has been clearly established that at least seven elements can limit the VI of forage, no equations or graphs have been published for predicting VI from the concentration of these elements in forage. Until equations are developed, it will not be possible to take all factors into account when predicting VI from laboratory analyses.

The development of equations relating VI to the concentration of essential elements will often be very difficult. In the case of Mg, the quantity of labile Mg in the body is very low and VI rapidly adjusts to a change in level of dietary Mg (see previous section). However, with other elements (e.g., P and Na), the body contains large reserves which can act as a source of "internal supplements" when animals eat forage deficient in these elements. The size of the reserves is unknown and there is no information on the rate at which these nutrients can be released into the blood. The present lack of information on mineral reserves in ruminants and their rate of mobilization is associated with the traditional role of most ruminant nutritionists who are only interested in estimating the recommended allowances of the various elements for use in diet formulation.

Rate of Passage Through the Rumen

Many equations have been published relating VI of forage to the fiber content (Table 2) and digestibility of the diet (Table 3). Other equations have been developed which relate VI to the concentration of the readily available fractions of the plant (Table 4). Voluntary intake also has been predicted by near infrared reflectance spectroscopy (Norris et al., 1976) but the predicted values may be seriously biased when the forages studied are different from those used to calibrate the instrument (Minson et al., 1983).

One may well ask how is it possible for VI to be controlled by so many different fractions in the plant? The answer is that VI is not controlled by any of these chemical components but by the time the forage is retained in the rumen. Most particles of forage are broken down to a size sufficiently small to pass through a 1 mm screen before they leave the rumen (Poppi et al., 1985) and chewing during eating and ruminating is responsible for most of the breakdown of particles in chopped forages (McLeod and Minson, 1988). It is known that the time forages are retained in the rumen is correlated with the lignin content (Thornton and Minson, 1972) so that it is probable that the correlations between VI and chemical composition are actually caused by the resistance of forages to breakdown being correlated with chemical composition.

Table 2. Equations for predicting the voluntary intake of forage by sheep (g/kg $BW^{0.75}$/d) from the concentration of fibrous fraction in the forage.

Component	Equation	r	RSD	Reference
Neutral detergent fiber (g/kg)				
temperate forage	VI = 110.4 - (1716/100 - x)	-0.65	-	Van Soest (1965)
	VI = 95 - 0.073 x	-0.88	-	Osbourn et al. (1974)
	VI = 165.7 - 0.15 x	-0.90	± 5.9	Seoane et al. (1982)
temperate silage	VI = 132.7 - 0.105 x	-0.87	± 6.8	Laforest et al. (1986)
Modified acid detergent fiber (g/kg)				
temperate forage	VI = 118.6 - 0.181 x	-0.82	± 7.0	Clancy and Wilson (1966)
Crude fiber (g/kg)				
temperate forage	VI = 98.3 - 0.11 x	-0.62	±11.8	Chenost (1966)
cut in 1965	VI = 110.0 - 0.17 x	-0.89	± 6.2	Wilson and McCarrick (1967)
cut in 1966	VI = 109.8 - 0.18 x	-0.62	±10.3	Wilson and McCarrick (1967)
Lignin (%)				
tropical grass	VI = 2761 - 835L + 82.2L^2	0.97	-	Thornton and Minson (1972)
Cellulose solubility in cupriethylenediamine (g/kg)				
temperate forage	VI = 201.1 - 0.372 x	0.74	-	Johnson et al. (1965)

Table 3. Prediction of voluntary intake by sheep (g/kg $BW^{0.75}$/d) from the *in vivo* digestibility coefficient of the dry matter (DMD), organic matter (OMD), or energy (ED) in forage.

Species	Equation		r	RSD	Reference
Orchard grass *(Dactylis glomerata)*	Log	VI = 0.45DMD + 1.512	0.50	-	Demarquilly (1965)
Meadow fescue *(Festuca pratensis)*	Log	VI = 2.23DMD + 0.364	0.72	-	Demarquilly (1965)
	Log	VI = 1.03DMD + 1.090	0.62	-	Demarquilly (1965)
Italian ryegrass *(Lolium multiflorum)* Tetraploid		VI = 18·2OMD - 20.5	-	± 1.6	Osbourn et al. (1966)
Diploid		VI = 14·2OMD - 53.6	-	± 1.8	Osbourn et al. (1966)
Perennial ryegrass *(Lolium perenne)*	Log	VI = 0.77DMD + 1.246	0.44	-	Demarquilly (1965)
Alfalfa *(Medicago sativa)*	Log	VI = 1.01DMD + 1.699	0.82	± 7.3	Demarquilly (1965)
		VI = 358DMD + 170DMD^2 - 85.3	-	± 7.3	Troelsen and Campbell (1969)
Temperate forage		VI = 136OMD - 25.6	-	-	Minson et al. (1964)
	Log	VI = 0.21DMD + 1.426	0.16	-	Demarquilly (1965)
		VI = 158ED - 20.5	-	± 6.3	Blaxter et al. (1966)
		VI = 151.1OMD - 36.16	0.86	± 7.8	Chenost (1966)
		VI = 105.7DMD - 10.5	0.77	± 7.8	Clancy and Wilson (1966)
		VI = 130DMD - 11.2	0.79	± 7.9	Hovell et al. (1986)
		VI = 34.8DMD + 100DMD^2 + 2.37	-	± 8.5	Troelsen and Campbell (1969)

Table 4. Equations for predicting the voluntary intake of forage by sheep (g/kg $BW^{0.75}$/d) from the concentration of readily digestible components in the forage.

Component/Species	Regression	r	RSD	Reference
Solubility in water (g/kg), temperate forage	VI = 13.25 + 0.358 x	0.94	± 4.9	Seoane et al. (1982)
Pepsin-soluble dry matter (g/kg), *Panicum* spp.	VI = 24.0 + 0.194 x	0.67	± 7.7	Minson and Haydock (1971)
Crude protein (g/kg), temperate forage	VI = 33.5 + 0.22 x	0.86	± 7.2	Wilson and McCarrick (1967)
	VI = 31.7 + 0.21 x	0.67	± 9.7	Wilson and McCarrick (1967)
	VI = 36.6 + 0.31 x	0.90	± 6.2	Seoane et al. (1982)
temperate silage	VI = 17.7 + 0.237 x	0.57	±14.6	Wilkins et al. (1971)
	VI = 35.8 + 0.239 x	0.92	± 5.4	Laforest et al. (1986)
Dry matter soluble in 1 N sulfuric acid (g/kg), temperate forage	VI = 9.4 + 0.339 x	0.76	-	Johnson et al. (1965)

In practical terms, equations relating VI to chemical composition or digestibility should be used with great care and any predicted values treated with great suspicion. They are probably most reliable when used to rank forages in plant breeding programs and are least satisfactory when applied to a wide range of forage species grown under different conditions.

Resistance to Chewing

Laboratory methods have been developed that simulate the physical resistance of forage to breakdown by chewing. These are based on 1) the energy required to grind dry forage through a 1 mm screen, or 2) the extent that particles in fresh forage are reduced in size by a mechanical masticator.

Grinding Energy

Grinding energy (also called fibrousness index) may be determined by measuring the electrical energy required to pulverize 5 g of oven-dried forage through a 1 mm screen in a laboratory mill (Chenost, 1966). The relation between VI and grinding energy (x) for 25 temperate legume and grass hays was:

$$VI\ (g/kg\ BW^{0.75}) = 91.74 - 55 \log x\ ;\ r = -0.90;\ RSD \pm 6.6$$

For 30 leaf and stem samples of 5 tropical grasses, there was a negative correlation ($r = -0.81$) between VI and grinding energy (Laredo and Minson, 1973).

Mechanical Mastication

Troelsen and Bigsby (1964) developed the first mechanical masticator. When the forage was cycled as a slurry through a gear water pump for 10 min and particle distribution determined by wet sieving, the final particle size index (x) was correlated with VI

$$VI\ (g/kg\ BW^{0.75}) = 353\ x - 14;\ r = 0.94$$

This result was encouraging but the technique was not used in routine work because it was so laborious. Simpler and more rapid methods have now been developed by CSIRO in which wet forage is masticated between a pair of spring-loaded fluted rollers for a fixed length of time. The proportion of the forage particles retained on a 1 mm screen following wet sieving then is measured. When applied to samples of mature subterranean clover (*Trifolium subterraneum*) of known VI, there was a higher correlation ($r = -0.93$) than with either the Troelsen and Bigsby ($r = -0.89$) or dry-grinding technique ($r = -0.87$) (Baker *et al.*, 1992). This is most encouraging and further work on this physical approach to predicting VI should be encouraged.

PLANT ANATOMICAL FACTORS IMPACTING ON FORAGE INTAKE

Significance of Anatomy

Plant anatomy influences feed intake partly through its effect on forage cell wall content and digestibility, thereby contributing to the predictive relationships already described (Tables 2 and 3). The differences in quantity of cell wall and digestibility arise from changes in the proportion of thin to thick-walled cell types and in the degree of lignification.

Another important effect of anatomy on intake arises from the variations in structural organization of organs and tissues. These influence the size of fiber structures and their resistance to breakdown to particles sufficiently small (< 1 mm screen size) to pass rapidly from the rumen. This structural effect is supported by differences in the relationships between VI and digestibility for comparisons of species and plant components (Minson, 1987) and by the loss of such differences when feeds are ground and pelletted. It is further emphasized by the scatter of VI among forages at comparable levels of cell wall content (Van Soest, 1982) and by the good correlations between VI and predictive measures based on the effort required to cause physical destruction of feed, such as artificial wet mastication and dry grinding.

Despite their possible importance in controlling VI, no attempt has been made to use anatomical factors, such as the cross-sectional area of vascular bundles, to predict intake. There are several reasons for this: 1) the measurements are time consuming, 2) a combination of characteristics contribute to particle size and ease of breakdown, and 3) intake experiments require large quantities of feed containing different proportions of leaf, sheath, and stem, each with a different anatomy.

The main thrust of anatomical studies in the past has been to examine cell wall limitations to digestion to explain family, genera, species, and cultivar variation (Akin, 1989; Wilson, 1993). These studies have defined the limits to high nutritive quality imposed by the structural/functional relationships of organs (e.g., leaf, stem) and tissues (e.g., epidermis, sclerenchyma, vascular bundle) (Niklas, 1989; Wilson, 1990), and provide breeding criteria for improving digestibility (Schank et al., 1973; Ehlke and Casler, 1985).

Fewer studies have examined how resistance of plants to comminution during digestion and rumination is related to structural characteristics. Of particular interest is the volume of large fiber particles and their ease of breakdown. Clearly, high cell wall content, often described in chemical terms as fiber (viz. crude fiber, NDF) *per se* would not limit intake if all this cell wall came from thin-walled mesophyll cells (c. 20 to 30 μm diameter) or stem parenchyma cells (c. 100 μm diameter). These cells easily break into particles ("fines") that pass the smallest sieve (0.15 mm) used in sieving experiments, and could readily leave the rumen.

It is the plant fiber in the botanical sense, the vascular and sclerenchyma fiber strands, that are the main contributors to the larger particulate fractions (> 1.0 mm screen size) in the rumen. Together with the epidermis, they maintain leaf and stem integrity, and require chewing during eating and rumination for disruption. When leaves and stems are disrupted, these individual fiber structures, or composites of them, become the particles that must be reduced in size to escape from the rumen. Thus, in selecting for improvement of intake using indices based on resistance to grinding or mastication, the normal structural integrity of the plant will be modified.

Anatomical studies help to indicate the modifications needed and help predict where changes can or cannot be made. The following discussion focuses on 1) the organization of tissues within leaf and stem that limits their ease of comminution and the fiber structures released that give rise to the large particle fraction in the rumen, and 2) the arrangement of cells within the fiber structures that make digestion an inefficient mechanism for size reduction of particles.

Leaf and Stem Structures Affecting Particle Breakdown

Epidermal Features

The epidermis is designed to resist normal physical stress (flexure, wilting, expansion, compression) and invasion by microorganisms. It is the first structure in leaf and stem which must be disrupted to get reduction in width of the material eaten. This may occur through splitting it apart or by its shedding from these organs. These processes may be aided by digestion and weakening of the underlying cells.

Splitting of leaves appears to be influenced by the wall profile of adjoining epidermal cells. Seen in paradermal view, these can be lobed (Figure 2a), straight-sided (Figure 2b) or sinuous as in a dovetail joint (Figure 2c) (Wilson, 1993). Most tropical C4 grasses have the structurally strong sinuous form (Watson and Dallwitz, 1980) making the leaf epidermis more resistant to splitting. This is demonstrated by easier epidermal splitting and loss for temperate Italian ryegrass (*Lolium multiflorum*) (straight-sided) by comparison with tropical grasses, green panic (*Panicum maximum* var. *trichoglume*) and bermuda (*Cynodon dactylon*) (sinuous) (Pond et al., 1984; Wilson et al., 1989a). Also, Sakurai (1963) observed epidermal disruption of a number of grasses when fed to animals, and divided them into groups in which it was easy or difficult. With only one exception, these groups corresponded to straight-sided and sinuous types, respectively (Wilson et al., 1989a). Watson and Dallwitz (1980) mention variation in sinuosity within tropical species which perhaps could be exploited to decrease leaf resistance to breakdown. The lobed form of epidermis (Figure 2a) is characteristic of legume leaves and appears to fragment easily and slough off within a short time of being eaten (Brazle and Harbers, 1977; Kelly and Sinclair, 1989).

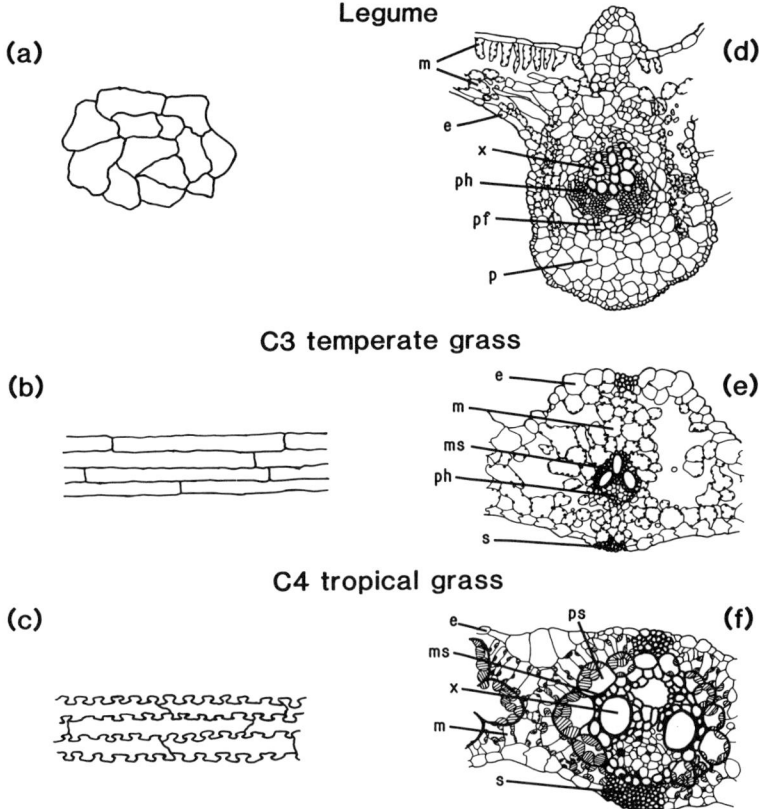

Figure 2. Paradermal view of epidermal cells (a-c) and cross-sections of a main vein (d-f) for a legume (*Macroptilium atropurpureum* [DC.] Urban), temperate grass (*Lolium multiflorum* L.), and a tropical grass (*Panicum maximum* var. *trichoglume* - PCK type). e - epidermis, m - mesophyll, ph - phloem, s - sclerenchyma, p - parenchyma, x - xylem, ps - parenchyma bundle sheath, ms - mestome sheath [from Wilson, 1993].

In stems, the inherent weakness of the epidermis with straight-sided walls seems to be overcome by the greater thickness of the epidermis walls and by strong lignification which produces an indigestible middle lamella binding adjoining walls together. For example, in stems of green panic and ryegrass the epidermal forms (sinuous and straight-sided, respectively) are the same as for the leaf but the radial walls joining together the epidermal cells in the stems are 1.8 to 2.4 μm thick compared to the leaves at only 0.3 to 0.5 μm (J. R. Wilson, unpublished data). When chopped (50 mm) stem pieces of these species were digested for 3 wk in nylon bags in the rumen, their stems did not differ in breakdown and still had 82 to 94% by weight of their particulate fraction as whole, unsplit stem pieces; the whole stem fraction was even as high as 38% by weight when the stems were chewed and then digested for 3 wk. With ryegrass leaves, complete shredding of the epidermis occurred after only 6 h of digestion (Wilson et al., 1989a).

Shedding of the leaf epidermis allows quicker breakdown of particle width (Wilson et al., 1989a), and enables mesophyll cells to be lost to the fine particle fraction. Shedding occurs easily by chewing and (or) short times of digestion when the epidermis is attached to the leaf vascular structures only by thin-walled mesophyll cells as in most legumes (Figure 2d) and many sown temperate pasture grasses (e.g., ryegrass, Figure 2e; Wilson, 1993). When both adaxial and abaxial epidermis is shed, as in many temperate grasses, then the width of leaf pieces is quickly reduced to the small diameter (<0.12 mm) of isolated vascular bundles. In legumes, the vascular skeleton is left unsupported, allowing breaking away of the minor veins (Wilson, 1991). Variation in ease of shedding is apparent among species of legumes associated with the differences in cuticle, epidermal, and mesophyll cell wall thickness (Lees, 1984).

Shedding does not occur easily in most tropical grasses which, because of their C4 anatomy, have the abaxial and adaxial epidermis linked to the vascular bundles by thick-walled, lignified sclerenchyma and parenchyma bundle sheath cells (Figure 2f; Wilson et al., 1989a), an "I" girder structure (Wilson, 1990). There are variations among species or genera in the occurrence of this structure; some grasses may have only a single flange "T" girder holding just the abaxial epidermis to the leaf. Other grasses may have the girder disrupted by a single layer of thin-walled cells between the sclerenchyma and bundle sheath (Wilson, 1990) which we predict could facilitate breakage of the girder and, hence, epidermal shedding. Generally, these leaf epidermal characteristics are well known and are sufficiently stable to be used as keys in taxonomic descriptions (Watson and Dallwitz, 1980).

Shedding of the epidermis in stems of both tropical and temperate grasses is possible when they are young, and the epidermal cells and underlying sclerenchyma are weakly lignified (Figure 3a). However, in old grass stems, the epidermis, sclerenchyma ring, and peripheral vascular bundles form a solid outer rind of thick-walled, lignified cells (Figure 3b; Wilson, 1993). This multi-girder (pipe) structure (Wilson, 1990) means that stems need more chewing and rumination to reduce particle size (McLeod et al., 1990).

The epidermis of legume stems, young or old, is easily disrupted and shed from the lignified stele because it is underlain by rows of thin-walled, digestible chlorenchymatous cells (Figure 3c, d; Wilson, 1993).

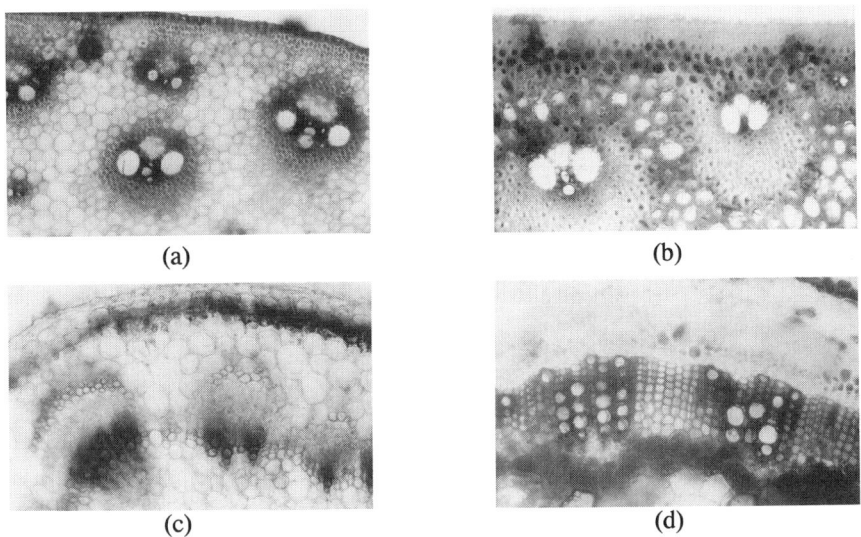

Figure 3. Cross-sections of the outer cortical region of grass and legume stem: young (a) and old (b) stem of *Sorghum bicolor* and young (c) and old (d) stem of *Stylothanthes humilis*.

Vascular and Sclerenchyma Fiber Strands

Once the epidermis is fully breached, leaves can break into isolated vascular strands or sometimes groups of bundles as occurs in grasses with large midribs. The initial size of these strands affects how much chewing is needed before they are reduced below the critical 1 mm mesh size. Kelly and Sinclair (1989) measured the average size of particles retained on a 1 mm sieve as 3.9 mm long x 0.34 mm wide for grass, and 2.3 mm long x 1.24 mm wide for legume. Comparison of these sizes with the width of vascular strands in grass and legume leaves (Table 5) indicates that, except for the midrib of grasses, the normal width of the isolated strands would allow them, on a width basis, to pass a 1 mm sieve without chewing. However, where the epidermis is not easily disrupted, as in many tropical grasses, leaf particles will comprise a number of bundles and require more chewing to reduce the width. In relation to length, the initial shorter length of most classes of veins in legume leaves (Table 5) would reduce the quantity of chewing needed before they reach the critical length. Another advantage of the legumes is their reticulate venation with many angular junctions between the veins which provide natural breakage points (Wilson, 1991). Hence, we would predict from their anatomical structure that legume leaves will naturally break into small, chunky particles as often observed in the rumen (e.g., Moseley and Jones, 1984) and require less rumination (Kelly and Sinclair, 1989).

Table 5. Sizes of vascular structures in leaves of grasses and legumes and total vascular bundles as a proportion of leaf cross-sectional area (CSA).

Leaf Attribute	Tropical grass[1] (*Panicum maximum*)	Temperate grass[1] (*Lolium multiflorum*)	Tropical legume[2]		Temperate legume[2]	
			Neonotonia wightii	*Leucaena leucocephela*	*Trifolium pratense*	*Medica sativa*
Midrib						
length (mm)	100–600	100–400	61±2.6	12±0.5	28±1.9	21±0.8
width (mm)	0.8±0.05[6]	0.7±0.05	0.38±0.01	0.26±0.01	0.38±0.03	0.30±0.02
Main veins[3]						
length (mm)	100–600	100–400	25±1.7	1.7±0.02	8.5±0.07	9.0±0.39
width (mm)	0.07–0.11	0.06–0.11	0.21±0.004	0.08±0.003	0.05±0.004	0.06±0.006
Minor veins[4]						
length (mm)	100–600	100–400	6.0±0.4	1.6±0.06	1.7±0.2	1.2±0.03
width (mm)	0.016–0.02	0.03–0.05	<0.13	<0.08	<0.05	<0.06
Vascular bundles[5] (% CSA)[7]	33.3±0.24	9.3±0.12	2.6±1.2	3.2±0.5	5.7±0.1	3.3±0.3

[1] Wilson et al., 1989a.
[2] J.R.Wilson, unpublished data for legumes grown in the same environment (Brisbane, Australia).
[3] Grasses - major vascular bundles; legumes - main laterals.
[4] Grasses - minor vascular bundles; legumes - sublaterals interconnecting laterals.
[5] Grasses - includes sclerenchyma and parenchyma bundle sheath; legumes - includes phloem fibers.
[6] Standard error of mean.
[7] Cross-sectional area.

Length reduction by chewing is of greater importance for grass leaves. Most veins, however minor, run the full length of the blade or sheath (100 to 600 mm) and, thus, many chews are needed to reduce length to < c. 3.9 mm required for them to enter the small particulate fraction. The veins of C4 tropical grasses have an added disadvantage being c. 3x the cross-sectional area of those of C3 temperate grasses (Table 5) because of their thick-walled parenchyma bundle sheath (Figure 2f) which is not easily broken down (Harbers, 1985) and remains as an additional sheath around the vascular tissue.

In stems of both legumes and grasses, the vascular structure can only split widthwise into separate vascular strands when the stem is very young and the vascular strands are present as isolated bundles (Figure 3a, c). More mature stems usually have a continuous ring of heavily lignified vascular tissue (Figure 3b, d), so there is no natural separation into narrow fibers as in the leaf. The circumference of this continuous vascular ring for a stem 5 mm in diameter is equivalent to a planar particle 30 mm wide, and it is as long as the length of each stem internode. Thus, extensive chewing would be required to reduce both width and length to pass a 1 mm screen, and this is what is observed (McLeod et al., 1990).

Significance of Cell Organization within Fiber Tissues

It is the composite structure of many thick-walled and lignified cells in these vascular and sclerenchyma strands in both leaf and stem that makes these fiber particles physically strong and difficult to reduce in size. A very small particle (only 0.25 mm diameter) of stem sclerenchyma can have as many as 1200 tightly packed cells in cross-section. Longitudinally, the strands in grasses are like laminated beams (Wilson, 1990), a structure which confers great compressive and longitudinal strength with no natural points of breakage, except perhaps at stem nodes where vascular structures divide and branch off to leaves. The cells in these strands laterally and vertically are joined together without intercellular spaces by a heavily lignified middle lamella with strong chemical bonding which gives no point of weakness. This middle lamella appears entirely resistant to microbial digestion (Engels, 1989) and confines digestion to the lumen surface of cells. Because of it, the fiber strands cannot separate into individual cells. Wilson (1993) calculated that these structural factors make the wall material of sclerenchyma strands up to 180 x less accessible to ruminal microbes than the walls of mesophyll cells. Hence, it is not surprising that digestion is ineffective as a means of reducing the length of vascular strands (Wilson et al., 1989c).

Again, legume leaves and stems seem to have a natural advantage over those of grasses in that this lignified cell structure is confined to the xylary tissues only, whereas in grasses, it also extends to sclerenchyma vascular caps, sclerenchyma ring or strands, and in stems to the parenchyma cells between bundles (Wilson, 1993).

MODIFICATIONS TO ANATOMY TO FACILITATE PARTICLE BREAKDOWN

Leaves of legumes have an anatomical structure that is advantageous to breakdown to small particles. Further modifications are not needed, except perhaps for some species in the opposite direction to slow breakdown and decrease the risk of bloat (Lees, 1984). For grass leaves, some structural modifications would increase rate of leaf breakdown. These modifications would include thinner cuticle, thinner-walled epidermal cells, less sinuous or a straight-sided epidermal wall profile, and an absence or natural disruption of the girder structure. These changes would make epidermal disruption easier and help reduce width of leaf particles and their particulate volume in the rumen.

Thinner cuticle and thinner walls of sclerenchyma and parenchyma cells at the stem periphery may increase the effectiveness of chewing for stem breakdown. Variation in these wall thicknesses is clearly seen in genotypes of grasses (Wilson et al., 1989b). Control of plant development to keep stems for a longer period in the stage when vascular bundles are discrete and not a solid ring would have a large effect, allowing stems to fracture easily into small slivers like leaves. Perhaps there is genotype variation in the completeness of the ring structure within species. Perhaps the functional need for strong stem structures could be avoided by reducing culm elongation and keeping stems encircled by leaf sheaths to provide strength (Niklas, 1990).

Few structural changes can be suggested to facilitate breakage in length of the fiber strands. Shorter stem internodes and leaves would reduce the original length of such structures, perhaps reducing the need for chewing. This should not be a diasadvantage since very short (3 to 6 cm) swards of ryegrass are regarded as optimal for animal production (Parsons et al., 1983). Selection for lower proportions of vascular and sclerenchyma tissues in leaf and stem will increase digestibility and indirectly reduce resistant particulate matter and improve intake. However, although these structures contribute most of the content of cell wall in leaves (e.g., cocksfoot, 61%, Wilson, 1990), and stems (e.g., sorghum, 79%, Wilson et al., 1994), they occupy only a small cross-sectional area (Akin, 1989). Thus, area measures of tissue proportions need to be accurate and well replicated to determine small but highly important variations.

Possibly the most important structural change, if it could be achieved, would be a significant weakening of the middle lamella region of adjoining cells in fiber strands, through less lignin or modified chemical bonding. An alternative would be to find or engineer microbes to digest this region. Both of these would facilitate cell separation and minimize all problems of plant structure and their effects on particle size.

Finally, it should be stressed that in breeding plants for reduced resistance of leaves and stems, and their component tissues, to comminution, anatomical structures whose inherent functional role is to provide mechanical strength will be modified. If our current forage species are structurally over-engineered for their domesticated use in sown pastures, then perhaps this can be achieved without performance penalty.

INTAKE UNDER GRAZING CONDITIONS

The intake of forage by grazing animals is affected by all the factors described for cut forage but there are additional factors that only occur in the field. These factors relate to the water content of the forage, forage yield, allowance, and sward heterogeneity.

Water Content of Forage

The water content of forage varies with stage of growth, level of N, and weather conditions. No attempt has been made to relate the VI of grazing animals to the moisture content of selectively grazed forage but several studies with cut forage have shown that VI is decreased when moisture content exceeds about 800 g/kg (Davies, 1962; John and Ulyatt, 1987). In both of these studies, there were large day-to-day differences in water content of the forage and the sheep tended to eat the same quantity of wet forage each day. It is not known how long this effect lasts, nor whether animals can adjust their intake to allow for the high water content of some forage, and how rapidly this adjustment occurs. More research is required on this subject.

Forage Mass

When cut forage is fed *ad libitum* to ruminants, the quantity eaten is controlled by the chemical and physical attributes of the forage. However, when forages are grazed, an additional dimension must be taken into account when predicting VI, *viz.* forage availability to the grazing animal. The volume of pasture that can be encompassed by the tongue in a single bite is limited, as is the number of bites that an animal will take each day (Hendricksen and Minson, 1980). The quantity of forage eaten by grazing animals will be lower than the level set by the chemical and physical structure of the forage, if the density of the forage in the canopy falls below that required by animals to achieve the critical bite size. For dairy cows, the critical bite size is about 0.3 g organic matter (Stobbs, 1973). Although the critical bite size has not been determined for other ruminants, the concept of a critical bite size provides a useful basis for understanding the way ruminants react to different types of swards.

Studies with artificial swards prepared by threading tillers through boards with holes at different spacings have shown that bite size is positively related to both sward height and density but is best described by the mass per unit area effectively covered by one bite, provided forage mass exceeds about 1,000 kg/ha (Black and Kenney, 1984). Field observations have shown that where the yield of young desired forage exceeds 1,500 to 2,000 kg/ha, sheep have no difficulty in satisfying their appetite (Allden and Whittaker, 1970); this is equal to a sward height of 6 cm or more in perennial ryegrass (Penning et al., 1984). Pastures 10 cm high have been recommended for maximal VI with dairy cattle (Johnstone-Wallace and Kennedy, 1944) but there is a need for more critical studies with swards of various initial heights grazed back to different final heights.

Forage Allowance

For strip or paddock grazed pastures there is usually a curvilinear relation between the quantity of forage eaten and the quantity offered. For maximal intake, the allowance of desired forage must be about twice the maximal intake or 60 g OM/kg liveweight/d (Figure 4). However, in some studies, the intake of forage continues to increase as more forage is offered (Penning et al., 1986).

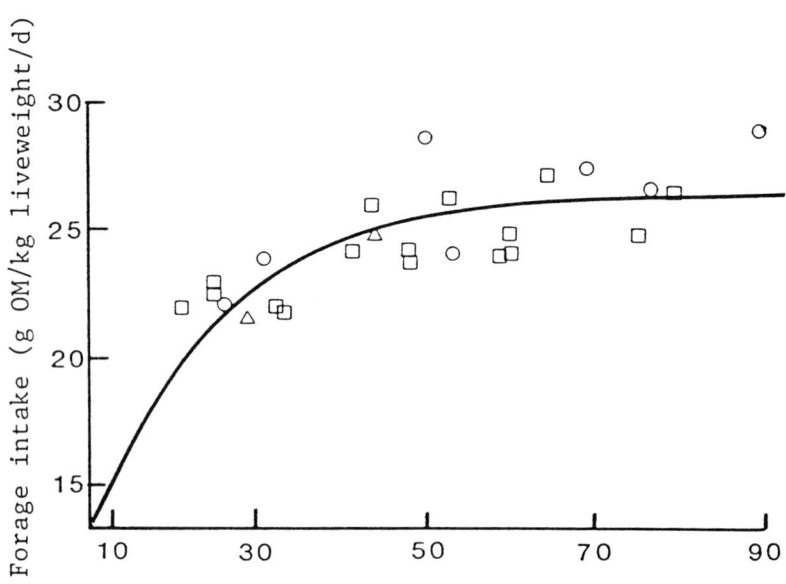

Figure 4. The effect of forage allowance on intake by calves (O), beef steers (△), and dairy cows (□). Data from Ernst et al. (1980).

Sward Heterogeneity

Swards are rarely uniform and where these pastures are lightly grazed, ruminants have an opportunity of selecting the preferred parts of the sward. Some parts of the sward may be completely rejected by grazing animals and in these cases, intake is related to the quantity of the preferred part of the diet and not to the total DM on offer (Minson, 1990).

SUMMARY AND CONCLUSIONS

The prediction of the VI of forage by ruminants would be simple if intake was controlled by only one chemical or physical attribute of forage. Unfortunately, the problem is not that simple and VI is <u>correlated</u> with at least 24 different characteristics of the forage. Although these correlations are statistically significant in one or more of the published studies, a high correlation is no <u>proof</u> that the factor involved actually controls VI as opposed to being correlated with a second attribute which actually determines VI. It is fortunate that this distinction exists between a correlation and a cause because it is obviously highly improbable that VI would be controlled by every major chemical component of the diet (Tables 2 and 4).

A high correlation between VI and a chemical or physical attribute of the forage is a useful method of identifying <u>potential</u> factor(s) controlling VI. However, proof that VI is controlled by a particular attribute can only be achieved by designing experiments in which the VI is measured for diets containing different levels of just one attribute. This method has proved effective with the essential elements P, Co, N, S, Mg, Se, and Na. Another approach is to take forages with different VI's and see whether they differ in the attribute(s) thought to affect VI. This approach was used to show that VI was not controlled by level of pepsin-soluble DM in the forage (Table 6).

There is a very real problem in obtaining valid evidence about the factors that control the VI of forage. Reliance on correlation is unwise in critical work and in the future, conclusions based on correlations must be treated with great care unless the hypothesis is supported by definitive experiments. These experiments will often require large animal feeding facilities and be expensive to conduct but are essential if false conclusions are to be avoided. There is no quick and easy solution to the question of what controls the VI of forages.

Eventually, it may be possible to discard many factors as having no direct effect on intake, but we will never be in a position where intake is related to just one factor. There is sound evidence that intake can be decreased by a deficiency of a number of different elements and the rate of reduction in particle size. All these factors will have to be included in an equation aimed at predicting intake in all situations. However, such an equation would be so complex and expensive to use that it would lack the speed and ease that most workers require of any method of predicting VI.

Table 6. Pepsin soluble dry matter of six varieties of Panicum with contrasting voluntary intakes (Minson and Haydock, 1971).

Variety	Species	Voluntary intake g/kg $BW^{0.75}$/d	Pepsin soluble DM g/kg DM
Kabulabula	*P. coloratum*	49.1	182
CPI 13372	*P. coloratum* var. *Makarikarienses*	55.1	175
Green panic	*P. maximum* var. *trichoglume*	55.2	168
Burnett	*P. coloratum* var. *Makarikarienses*	58.1	185
Coloniao	*P. maximum*	64.3	188
Hamil	*P. maximum*	67.3	171

The predictive equation might be simplified if only required for limited situations. There is a need to predict VI in the following limited situations: 1) diagnosing the cause of poor production by animals when feed supply is judged to be adequate; 2) when selecting forage of higher VI in plant breeding programs; and 3) estimating the quantity of forage consumed by animals in grazing experiments. These three situations will be briefly considered.

When attempting to diagnose the cause of poor animal production, the forage would be analyzed for the essential elements and the level of these compared with published values for "recommended levels". If the concentrations are above those levels and not in the toxic range, then these elements are not limiting intake. However, if the concentration of an element in the forage is below the recommended level, intake might be limited by a deficiency if the labile reserves of the element in the animal are insufficient to maintain adequate levels in the blood. The second group wishing to predict intake are plant breeders. They can simplify the prediction if they assume that the plants they produce will be adequately fertilized. Mineral analysis can be omitted in most selection programs, except those specifically aimed at increasing the level of a mineral marginally low in some forages (e.g., Mg or Na). Selection for intake in plant breeding programs is usually restricted to reducing the total quantity of fiber measured as NDF, the lignin associated with this fiber or a measure of the combined effect of these two chemical attributes, *in vitro* digestibility. Although digestibility is only correlated and does not control VI, it is a criterion that has been used successfully in many plant breeding programs in the past. The third reason for predicting VI is a desire to know how much forage is removed by grazing animals from an experimental area or farm. Where this is the aim, it is probably simpler and more accurate to estimate VI from one of the published equations relating intake of DM to the animal liveweight and level of production (Minson, 1990) than to use a prediction equation based on feed composition.

REFERENCES

Akin, D.E. 1989. Histological and physical factors affecting digestibility of forages. Agron. J. 81:17-25.

Allden, W.G., and I.A. McD. Whittaker. 1970. The determinants of herbage intake by grazing sheep: The interrelationship of factors influencing herbage intake and availability. Aust. J. Agric. Res. 21:755-766.

Baker, S.K., L. Klein, D.J. Minson, and D.B. Purser. 1992. Voluntary intake and the energy required to shear or comminute dry, mature subterranean clover. Proc. Nutr. Soc. 17:74, Abstr.

Black, J.L., and P.A. Kenney. 1984. Factors affecting diet selection by sheep. II. Height and density of pastures. Aust. J. Agric. Res. 35:365-378.

Blaxter, K.L., F.W. Wainman, and R.S. Wilson. 1961. The regulation of food intake by sheep. Anim. Prod. 3:51-61.

Blaxter, K.L., F.W. Wainman, and J.L. Davidson. 1966. The voluntary intake of food by sheep and cattle in relation to their energy requirement for maintenance. Anim. Prod. 8:75-83.

Brazle, F.K., and L.H. Harbers. 1977. Digestion of alfalfa hay observed by scanning electron microscopy. J. Anim. Sci. 45:506-512.

Campling, R.C., and M. Freer. 1961. The effect of urea on the voluntary intake of oat straw by cattle. Proc. Nutr. Soc. 20: xvi, Abstr.

Chenost, M. 1966. Fibrousness of forage: Its determination and its relation to feeding value. p.406-411. *In* Proc. X Inter. Grassl. Congr., Helsinki, Finland, 7-16 July 1966.

Clancy, M.J., and R.K. Wilson. 1966. Development and application of a new chemical method for predicting the digestibility and intake of herbage samples. p.137-140. *In* Proc. X Inter. Grassl. Congr., Helsinki, Finland, 7-16 July 1966.

Clark, R., and J.I. Quin. 1951. Studies on the alimentary tract of the merino sheep in South Africa. xxii - The effect of supplementing poor quality grass hay with molasses and nitrogenous salts. Onderstepoort J. Vet. Res. 25:93-103.

Davies, H.L. 1962. Intake studies in sheep involving high fluid intake. Proc. Aust. Soc. Anim. Prod. 4:167-171.

Demarquilly, C. 1965. Factors affecting the voluntary intake of green forage by sheep. p.877-885. *In* Proc. IX Inter. Grassl. Congr., Sao Paulo, Brazil, 7-20 January 1965.

Donefer, E., E.W. Crampton, and L.E. Lloyd. 1960. Prediction of the nutritive value index of a forage from *in vitro* rumen fermentation data. J. Anim. Sci. 19:545-552.

Ehlke, N.J., and M.D. Casler. 1985. Anatomical characteristics of smooth bromegrass clones selected for *in vitro* dry matter digestibility. Crop Sci. 25:513-517.

Engels, F.M. 1989. Some properties of cell wall layers determining ruminant digestion. p. 80-87. *In* A. Chesson and E.R. Orskov (ed.) Physico-chemical characteristics of plant research for industrial and feed use. Elsevier Applied Sci., London.

Ernst, P., Y.L.P. Le Du, and L. Carlier. 1980. Animal and sward production under rotational and continuous grazing management - A critical appraisal. p.119-127. *In* W.H. Prins and G.H. Arnold (ed.) The role of nitrogen in intensive grassland production. Purdoc, Wageningen, The Netherlands.

Harbers, L.H. 1985. Ultrastructural utilization of plants by herbivores. Food Microstruct. 4:357-364.

Hawkins, G.E., G.E. Parr, and J.A. Little. 1964. Composition, intake, digestibility and prediction of digestibility of coastal Bermudagrass hay. J. Dairy Sci. 47: 865-870.

Heaney, D.P., W.J. Pigden, D.J. Minson, and G.I. Pritchard. 1963. Effect of pelleting on energy intake of sheep from forages cut at three stages of maturity. J. Anim. Sci. 22:752-757.

Hendricksen, R., and D. J. Minson. 1980. The feed intake and grazing behaviour of cattle grazing a crop of *Lablab purpureus* cv. Rongai. J. Agric. Sci. (Camb.) 95:547-554.

Hovell, F.D.D., J.W.W. Ngambi, W.P. Barber, and D.J. Kylie. 1986. The voluntary intake of hay by sheep in relation to its degradability in the rumen as measured by nylon bags. Anim. Prod. 42:111-118.

John, A., and M.J. Ulyatt. 1987. Importance of dry matter content to voluntary intake of fresh grass forages. Proc. N.Z. Soc. Anim. Prod. 47:13-16.

Johnson, R.R., B.A. Dehority, and J.L. Parsons. 1965. Relationships between *in vitro* measurements on forages and their nutritive value. p.773-778. *In* Proc. IX Inter. Grassl. Congr., Sao Paulo, Brazil, 7-20 January 1965.

Johnstone-Wallace, D.B., and K. Kennedy. 1944. Grazing management practices and their relationship to the behaviour and grazing habits of cattle. J. Agric. Sci. (Camb.) 34:190-197.

Kelly, K.E., and B.R. Sinclair. 1989. Size and structure of leaf and stalk components of digesta regurgitated for rumination in sheep offered five forage diets. N.Z. J. Agric. Res. 32:365-374.

Kick, C.H., P. Gerlaugh, A.F. Schalk, and E.A. Silver. 1937. The effect of mechanical processing of feeds on the mastication and rumination of steers. J. Agric. Res. 55:587-597.

Laforest, J.P., J.R. Seoane, G. Dupuis, L. Phillip, and P.M. Filpot. 1986. Estimation of the nutritive value of silages. Can. J. Anim. Sci. 66:117-127.

Laredo, M.A., and D.J. Minson. 1973. The voluntary intake, digestibility and retention time by sheep of leaf and stem fractions of five grasses. Aust. J. Agric. Res. 24:875-888.

Lees, G.L. 1984. Cuticle and cell wall thickness: Relation to mechanical strength of whole leaves and isolated cells from some forage legumes. Crop Sci. 24:1077-1081.

Lines, E.W. 1935. The effect of the ingestion of minute quantities of cobalt by sheep affected with "Coast Disease": A preliminary note. J. Counc. Sci. Ind. Res. (Aust.) 8:117-119.

McLean, J.W., G.G. Thomson, C.E. Iversen, K.T. Jagusch, and B.M. Lawson. 1962. Sheep production and health on pure-species pastures. Proc. N.Z. Grassl. Assoc. 24:57-70.

McLeese, D.M., M.C. Bell, and R.M. Forbes. 1961. Magnesium - 28 studies in lambs. J. Nutr. 74:105-114.

McLeod, M.N., and D.J. Minson. 1988. Large particle breakdown by cattle eating alfalfa and ryegrass. J. Anim. Sci. 66:992-999.

McLeod, M.N., P.M. Kennedy, and D.J. Minson. 1990. Resistance of leaf and stem fraction of tropical forage to chewing and passage in cattle. Br. J. Nutr. 63:105-119.

Minson, D.J. 1981. Effect of chemical and physical composition of herbage eaten upon intake. p. 167-181. *In* J.B. Hacker (ed.) Nutritional limits to animal production from pasture. CAB, Farnham Royal, England.

Minson, D. J. 1986. The resistance to fracture in forage - effect of digestion. Annual report 1985-86. CSIRO Division of Tropical Crops and Pastures, Brisbane, Australia.

Minson, D.J. 1987. Plant factors affecting intake. p. 137-144. *In* R.W. Snaydon (ed.) Managed grasslands, B. Analytical studies. Elsevier Sci. Publ. B.V., Amsterdam.

Minson, D.J. 1990. Forage in ruminant nutrition. Academic Press, San Diego, CA.

Minson, D.J., and K.P. Haydock. 1971. The value of pepsin dry matter solubility for estimating the voluntary intake and digestibility of six Panicum varieties. Aust. J. Exp. Agric. Anim. Husb. 11:181-185.

Minson, D.J., C.E. Harris, W.F. Raymond, and R. Milford. 1964. The digestibility and voluntary intake of S22 and HI ryegrass, S170 tall fescue, S48 timothy, S215 meadow fescues and Germinal Cocksfoot. J. Br. Grassl. Soc. 19:298-305.

Minson, D.J., K.L. Butler, N. Grummitt, and D.P. Law. 1983. Bias when predicting crude protein, dry matter digestibility and voluntary intake of tropical grasses by near-infrared reflectance. Anim. Feed Sci. Technol. 9: 221-237.

Moore, L.A., J.W. Thomas, and J.F. Sykes. 1960. The acceptability of grass/legume silage by dairy cattle. p.701-704. In Proc. VIII Int. Grassl. Congr., Reading, England, 11-21 July 1960.

Morris, J. G., and G. W. Murphy. 1972. The sodium requirements of beef calves for growth. J. Agric. Sci. (Camb.) 78:105-108.

Moseley, G., and J.R. Jones. 1984. The physical digestion of perennial ryegrass (*Lolium perenne*) and white clover (*Trifolium repens*) in the foregut of sheep. Br. J. Nutr. 52:381-390.

Murdoch, J.C. 1960. The effect of prewilting herbage on the composition of silage and its intake by cows. J. Br. Grassl. Soc. 15: 70-73.

Muth, O. H., J. E. Oldfield, L. F. Remmert, and J. R. Schubert. 1958. Effects of selenium and vitamin E in white muscle disease. Science 128:1090.

Niklas, K.J. 1989. The cellular mechanics of plants. Am. Scient. 77:344-349.

Niklas, K.J. 1990. The mechanical significance of clasping leaf sheaths in grasses: Evidence from two cultivars of *Avena sativa*. Ann. Bot. 65:505-512.

Norris, K.H., R.F. Barnes, J.E. Moore, and J.S. Shenk. 1976. Predicting forage quality by infrared reflectance spectroscopy. J. Anim. Sci. 43:889-897.

Orskov, E. R., G. W. Reid, and M. Kay. 1988. Prediction of intake by cattle from degradation characteristics of roughages. Anim. Prod. 46:29-34.

Osborne, T.B., and L.B. Mendel. 1916. A quantitative comparison of casein, lactalbumin, and edestin for growth or maintenance. J. Biol. Chem. 26:1-23.

Osbourn, D.F., D.J. Thomson, and R.A. Terry. 1966. The relationship between voluntary intake and digestibility of forage crops, using sheep. p. 363-366. *In* Proc. X Int. Grassl. Congr., Helsinki, Finland, 7-16 July 1966.

Osbourn, D.F., R.A. Terry, G.E. Outen, and S.B. Cammell. 1974. The significance of a determination of cell walls as the rational basis for the nutritive evaluation of forage. p.374-380. *In* V. G. Iglovikov and A. P. Movsisyants (ed.) Proc. XII Inter. Grassl. Congr., Moscow, USSR, 10-20 June 1974.

Parsons, A.J., E.L. Leafe, B. Collett, P.D. Penning, and J. Lewis. 1983. The physiology of grass production under grazing. II. Photosynthesis, crop growth and animal intake of continuously-grazed swards. J. Appl. Ecol. 20:127-139.

Penning, P.D., G.L. Steel, and R.H. Johnson. 1984. Further development and use of an automatic recording system in sheep grazing studies. Grass Forage Sci. 40:79-84.

Penning, P.D., G.E. Hooper, and T.T. Treacher. 1986. The effect of herbage allowance on intake and performance of ewes suckling twin lambs. Grass Forage Sci. 41:199-208.

Playne, M.J. 1969. Effects of sodium sulphate and gluten supplements on the intake and digestibility of a mixture of spear grass and Townsville lucerne hay by sheep. Aust. J. Exp. Agric. Anim. Husb. 9:393-399.

Pond, K.R., W.C. Ellis, and D.E. Akin. 1984. Ingestive mastication and fragmentation of forages. J. Anim. Sci. 58:1567-1574.

Poppi, D.P., R.E. Hendricksen, and D.J. Minson. 1985. The relative resistance to escape of leaf and stem particles from the rumen of cattle and sheep. J. Agric. Sci. (Camb.) 105:9-14.

Sakurai, M. 1963. Histological study on the decomposition of pasture grass tissue by livestock digestion. Grassl. Div. Nat. Inst. Anim. Ind., Kaoto-Tosan Agric. Expt. Stn. Res. Rep. No. 15.

Schank, S.C., M.A. Klock, and J.E. Moore. 1973. Laboratory evaluation of quality in subtropical grasses: II. Genetic variation among *Hemarthrias* in *in vitro* digestion and stem morphology. Agron. J. 65:256-258.

Seoane, J.R., M. Cote, and S.A. Vissar. 1982. The relationship between voluntary intake and the physical properties of forages. Can. J. Anim. Sci. 62:473-480.

Standing Committee on Agriculture. 1990. Feeding standards for Australian livestock - Ruminants. CSIRO, Melbourne, Australia.

Stobbs, T. H. 1973. The effect of plant structure on the intake of tropical pastures. I. Variation in the bite size of grazing cattle. Aust. J. Agric. Res. 24:809-819.

Theiler, A., H.H. Green, and P.J. Du Toit. 1924. Phosphorus in the livestock industry. South Africa Dept. Agric. J. 8:460-504.

Thornton, R.F., and D.J. Minson. 1972. The relationship between voluntary intake and mean apparent retention time in the rumen. Aust. J. Agric. Res. 23:871-877.

Troelsen, J.E., and F.W. Bigsby. 1964. Artificial mastication - A new approach for predicting voluntary forage consumption by ruminants. J. Anim. Sci. 23:1139-1142.

Troelsen, J.E., and J.B. Campbell. 1969. The effect of maturity and leafiness on the intake and digestibility of alfalfas and grasses fed to sheep. J. Agric. Sci. (Camb.) 77:145-154.

Tyler, C. 1959. The historical development of feeding standards. p. 8-18. In Scientific principles of feeding farm live stock. Farmer and Stock-Breeder Publications Ltd., London.

Van Soest, P.J. 1965. Symposium on factors influencing the voluntary intake of herbage by ruminants: Voluntary intake in relation to chemical composition and digestibility. J. Anim. Sci. 24:834-843.

Van Soest, P.J. 1982. Nutritional ecology of the ruminant. O & B Books, Inc., Corvallis, OR.

Watson, L., and M.J. Dallwitz. 1980. Australian grass genera. Anatomy, morphology and keys. Aust. Nat. Univ. Res. Sch. Biol. Sci., Canberra, Australia.

Webb, R.J., G.F. Cmarik, and H.A. Cate. 1957. Comparison of feeding three forages as baled hay, chopped hay, hay pellets and silage to steer calves. J. Anim. Sci. 16:1057-1058.

Wilkins, R.J., K.J. Hutchinson, R.F. Wilson, and C.E. Harris. 1971. The voluntary intake of silage by sheep. J. Agric. Sci. (Camb.) 77: 531-537.

Wilson, J.R. 1990. Influence of plant anatomy on digestion and fibre breakdown. p. 99-117. In D.E. Akin, L.G. Ljungdahl, J.R. Wilson, and P.J. Harris (ed.) Microbial and plant opportunities to improve the utilization of lignocellulose by ruminants. Elsevier Sci. Publ. Co., New York.

Wilson, J.R. 1991. Plant structures: Their digestive and physical breakdown. p. 207-216. *In* Y.W. Ho, H.K. Wong, N. Abdullah, and Z.A. Tajuddin (ed.) *In* Recent advances on the nutrition of herbivores. Malay. Soc. Anim. Prod., Kuala Lumpur.

Wilson, J.R. 1993. Organization of forage plant tissues. p. 1-32. *In* H.G. Jung, D.R. Buxton, R.D. Hatfield, and J. Ralph (ed.) Forage cell wall structure and digestibility. ASA, CSSA, and SSSA, Madison, WI.

Wilson, J.R., D.E. Akin, M.N. McLeod, and D.J. Minson. 1989a. Particle size reduction of the leaves of a tropical and a temperate grass by cattle. II. Relation of anatomical structure to the process of leaf breakdown through chewing and digestion. Grass Forage Sci. 44:65-75.

Wilson, J.R., K.L. Anderson, and J.B. Hacker. 1989b. Dry matter digestibility *in vitro* of leaf and stem of buffel grass (*Cenchrus ciliaris*) and related species and its relation to plant morphology and anatomy. Aust. J. Agric. Res. 40:281-291.

Wilson, J.R., M.N. McLeod, and D.J. Minson. 1989c. Particle size reduction of the leaves of a tropical and temperate grass by cattle. I. Effect of chewing during eating and varying times of digestion. Grass Forage Sci. 44:55-63.

Wilson, J.R., D.R. Mertens, and R.D. Hatfield. 1994. Isolates of cell types from sorghum. Digestion, cell and anatomical characteristics. J. Sci. Food Agric. [in press]

Wilson, R.K., and R.B. McCarrick. 1967. A nutritional study of grass swards at progressive stages of maturity. Irish J. Agric. Res. 6:267-279.

Wilson, R.K., T.A. Spillane, and M.J. Clancy. 1966. The influence of fibre content on herbage intakes by ruminants. Irish J. Agric. Res. 5:142-143.

CHAPTER 14

PROCESSES OF DIGESTION AND FACTORS INFLUENCING DIGESTION OF FORAGE-BASED DIETS BY RUMINANTS

Neal R. Merchen and Leslie D. Bourquin

INTRODUCTION

Digestion is the process of degradation of macromolecules in food to simple compounds that can be absorbed from the gastrointestinal tract. Digestion in mammals is accomplished via two strategies. In one case, digestion is mediated by acid hydrolysis in the gastric stomach and by digestive enzymes in the small intestine. In the second approach, digestion occurs by fermentative metabolism of dietary components by microbes occupying certain compartments of the gut.
Virtually all mammals practice both types of digestion but the degree of emphasis on one strategy or the other varies considerably. Herbivorous species are characterized by the presence of a site of extensive fermentation in the alimentary tract. In some animals (e.g., the horse), the primary site of fermentation is in the hindgut (cecum/colon). In others, fermentation occurs largely in a modified compartment of the forestomach. Ruminants, of course, are animals that have achieved maximum capability for forestomach fermentation. Although ruminants are characterized overwhelmingly by the role of fermentation in the reticulo-rumen, digestion in the abomasum and small intestine are also vital processes.

SITES OF DIGESTION AND PROCESSES INVOLVED

Ruminal Microbial Digestion and Fermentation

Microbial digestion and synthesis of microbial components in the rumen requires certain conditions provided by the host. These include retention of digesta and ruminal microbes for prolonged periods of time, anaerobiosis, constant temperature (39°C), neutral to slightly acidic pH (5.5 to 7.0), and removal of end-products (Stevens et al., 1980). In most circumstances, this environment is closely controlled by mechanisms such as the type and quantity of food consumed, saliva secretion during eating and rumination, mixing via ruminal contractions, diffusion/secretion of materials (urea, bicarbonate) into the rumen, absorption of end-products (volatile fatty acids [VFA], ammonia),

---

N. R. Merchen, Dep. of Animal Sciences, Univ. of Illinois, 1207 W. Gregory, Urbana, IL 61801; L. D. Bourquin, Dep. of Food Science and Human Nutrition, Michigan State Univ., East Lansing, MI 48824.

and passage of undigested residues and microbial cells out of the rumen (Leng, 1973).

Ruminal microbes are well-adapted to ferment a variety of carbohydrates including sugars, starches, and the complex polysaccharides found in plant cell walls (cellulose, hemicelluloses, pectins). The end-products of fermentation by a mixed population of ruminal microorganisms are the three primary VFA (acetic, propionic, and butyric acids), methane, and carbon dioxide. The details of these pathways are well established and have been reviewed (Leng, 1973; Baldwin and Allison, 1983). A portion of the dietary protein also is degraded and fermented by ruminal microbes with the resultant production of additional VFA and of smaller amounts of branched-chain fatty acids (isovalerate, isobutyrate, 2-methylbutyrate) and ammonia. The VFA produced in the rumen are absorbed from that site and provide 50 to 80% of the total metabolizable energy available to the host. Ruminal microorganisms also synthesize substantial amounts of protein that can be digested postruminally.

Losses of energy and possibly protein occur in conjunction with fermentative microbial metabolism. In general, about 75% of the energy in the fermented substrate (carbohydrate) is recovered in VFA that the host can utilize (Ulyatt, 1973). The remaining energy contained in the original substrate is lost during fermentation by three principal routes: 1) heat of combustion of end-products not utilized by the host (largely methane), 2) heat produced due to substrate fermentation, and 3) heat produced by assimilation of substrate components into bacterial cells (Leng, 1973). These losses are tolerable when the fermented substrate consists of materials that are otherwise unavailable to the host (e.g., cell wall polysaccharides) but greatly compromise the efficiency with which ruminants utilize other carbohydrates and high quality proteins that could be readily digested by hydrolytic/enzymatic processes.

Small Intestinal Digestion

The supply of nutrients reaching the small intestine of the ruminant is greatly altered relative to the diet because of extensive fermentation in the rumen. In animals fed forage diets, only 5 to 8% of the readily digested carbohydrate appears to escape ruminal digestion (Ulyatt and MacRae, 1974). Other nutrients of dietary origin that reach the postruminal gut include most plant lipids and variable proportions of cell wall carbohydrates and protein that are undegraded by ruminal microbes. In addition to these dietary contributions, substantial quantities of microbial cells of ruminal origin enter the intestinal tract.

Digestion in these sites is mediated by the occurrence of acid hydrolysis in the abomasum and by the activity of approximately 40 digestive enzymes (Langer and Snipes, 1991). Enzymatic digestion is initiated by secretions of the abomasal mucosa but occurs largely in the small intestine where it is mediated by both pancreatic and intestinal enzymes. Details of these processes are well-documented, have been reviewed (Argenzio, 1984; Merchen, 1988), and will not be re-examined here. There are probably few limitations to digestion of dietary and microbial nutrients that reach the small intestine when high forage diets are fed (Ørskov and Kay, 1987) other than that of unfermented cell wall residues. Extensive digestion of these nutrients may be assisted by the more or less continuous nature of digestion in the ruminant. Grazing animals and those fed ad libitum invest a large proportion of their time in eating or ruminating. These activities, coupled with the large pool of digesta in the reticulo-rumen, result in flow of digesta into the postruminal tract on a fairly continuous basis. This, in turn, stimulates continuous release of digestive secretions into the abomasum and small intestine.

Microbial Digestion and Fermentation in the Hindgut

Ruminants gain some benefit from the occurrence of secondary fermentation in the cecum and proximal colon. These compartments, collectively referred to here as the hindgut, are modestly enlarged and are characterized by the presence of an active microbial population. When forage diets are fed, the principal substrates entering the hindgut are undigested cell wall materials and undigested components of bacterial cells. These substrates are more refractory than those entering other segments of the gut, although susceptibility of cell wall residues to microbial attack may be improved following exposure to abomasal and intestinal digestion and by reductions in particle size relative to the feed. Digestion in the hindgut is mediated strictly by microbial activity; no digestive enzymes have been associated with cecal or colonic mucosa (Ulyatt et al., 1975).

Digestion in the hindgut is typically less extensive than in other sites because of the more refractory substrate arriving at this site and because retention time is much shorter than in the rumen (7 to 18 h; Grovum and Hecker, 1973). Viable bacterial counts of cecal contents are considerably lower than for ruminal contents and range from 10^7 to 10^9/g of digesta (Kern et al., 1974; Ulyatt et al., 1975). Important bacterial species in the hindgut are identical to those of the reticulo-rumen (Mann and Ørskov, 1973; Sharpe et al., 1975) although relative numbers vary. Protozoa appear to be absent from the cecum of ruminants (Kern et al., 1974). The physico-chemical environment of the hindgut is likewise similar to that of the forestomach; pH is neutral to slightly acidic (range 6.0 to 7.7; Faichney, 1968) and total VFA concentrations are in the range of 60 to 90% of those typically observed in the rumen (Faichney, 1968; Sharpe et al., 1975).

Experiments with sheep demonstrated that from 17 to 60% of dietary cellulose and hemicelluloses enter the hindgut (Beever et al., 1972; Thomson et al., 1972; Ulyatt and MacRae, 1974) and that 19 to 65% of those components entering disappear there. Ulyatt et al. (1975) summarized data indicating that from 5 to 30% of the digestible cellulose consumed disappears in the hindgut. Hindgut fermentation is a quantitatively less important means of digestion than are either ruminal fermentation or small intestinal digestion but it clearly makes significant contributions to digestion and to the supply of energy to the host.

IMPLICATIONS OF SITE OF DIGESTION

Site of digestion of many nutrients affects the nature of absorbed end-products and the extent of nutrient losses occurring during digestion. Fermentation is a relatively costly strategy for digestion because microbial metabolism of carbohydrates to produce VFA is accompanied by significant losses of energy as methane and heat. Methane production results in losses of 5 to 12% of the gross energy of the diet (Reid et al., 1980) with greater proportions lost when high forage diets are fed. Likewise, energy losses as heat of fermentation contribute to inefficiencies in utilization of dietary energy. Fermentation losses are acceptable when the primary substrates are cell wall components that are otherwise unavailable to the host, but postruminal digestion of nonstructural carbohydrates would be likely to result in recovery of absorbed end-products (e.g., glucose) that retain a greater proportion of the energy in the digestible substrate that the host animal can utilize.

Efficiency of utilization of dietary protein also can be impacted by fermentative vs hydrolytic/enzymatic digestion. Ruminal microbes may degrade high quality dietary proteins so extensively that net losses of protein occur in the rumen. The profile of amino acids (AA) reaching the lower gut is often radically altered when compared to the diet because of degradation of feed protein and synthesis of microbial protein in the rumen (Merchen and Titgemeyer, 1992). Qualitative and quantitative differences in end-product absorption as a result of differences in site of digestion of certain nutrients (especially non-structural carbohydrates and protein) are probably quite important in predicting animal responses to some diets and feeding practices.

Mackie (1987) described research in ruminant nutrition as largely the study of interactions between the host, the microorganisms present in the reticulo-rumen, substrates available, and end-products of digestion. The ruminant-microbial relationship is regarded by microbial ecologists as a "cooperation" model (Mackie, 1987). The relationship is exemplified by the fact that the microbial partner endows the host with the capability of obtaining energy from compounds in plant cell walls that would otherwise be unavailable. The disadvantage to the host is that all ingested nutrients are exposed to fermentation and its attendant losses.

The advantages of fermentative digestion are maximized when forage diets are fed. Consumption of diets containing large proportions of cell wall materials minimizes fermentation losses of energy from nutrients that could be digested otherwise. In experiments in which forages have been fed and disappearance of energy measured in different segments of the gastrointestinal tract, it is reported that an average of 56% of apparently digestible energy (DE) disappears in the stomach (range 46 to 64% across 13 diets; Beever et al., 1971, 1972; Ulyatt and MacRae, 1974). Corresponding values (% of DE) for disappearance in the small and large intestines were 32% (range 24 to 39%) and 12% (range 4 to 16%), respectively.

DIGESTION OF FORAGE POLYSACCHARIDES

The past twenty-five years have witnessed a tremendous research effort to characterize the process of plant polysaccharide digestion by ruminants and to describe factors influencing the extent of digestion of plant cell wall polysaccharides. During this time, significant advances in knowledge of both plant cell wall chemistry and gastrointestinal microbiology have increased our basic understanding of the process of fiber digestion. Improved methods for determining cell wall composition and structure as well as improved animal experimentation techniques also have been instrumental in furthering our understanding of fiber utilization. While it is obvious that improvements in our comprehension of microbial aspects of fiber utilization have been central to furthering our knowledge on fiber digestion, numerous reviews (Chesson and Forsberg, 1988; Hespell, 1988; Weimer, 1992; White et al., 1993; and references in Akin et al., 1990) have given more justice to this topic than would be possible in this chapter.

Digestibility of Polysaccharide Components

Storage Polysaccharides

Forages generally contain only modest quantities of storage polysaccharides although immature temperate forages may contain significant concentrations. Fructosans are the primary storage polysaccharides present in leaves and stems of temperate grasses, whereas tropical grasses and all legumes utilize starch as the major storage polysaccharide (Van Soest, 1982). Both fructosans and starch present in forages are rapidly and extensively fermented by ruminal microorganisms (Morrison, 1979; Chesson and Forsberg, 1988). Given the ease of digestion of storage polysaccharides, the remainder of this section will focus on utilization of forage structural polysaccharides.

Structural Polysaccharides

The major structural polysaccharides of forage cell walls include pectic substances, hemicelluloses, and cellulose. Pectic substances are a minor constituent in grasses, but are found in greater quantities in legumes, often comprising greater than 100 g/kg of dry matter (DM). Pectic substances are present in highest concentrations in the middle lamella, especially in dicotyledonous plants. Lower concentrations of pectic substances are found in the primary cell wall. No deposition of pectic polysaccharides occurs during secondary thickening of cell walls (Lam et al., 1990). Legume pectic substances are rapidly and extensively (>850 g/kg) digested in the rumen (Chesson and Monro, 1982; Titgemeyer et al., 1992).

The hemicelluloses represent a diverse group of polysaccharides that include (glucurono)arabinoxylans, xyloglucans, glucomannans, and mixed-linked glucans. Arabinoxylans are the predominant hemicellulosic polysaccharides present in primary cell walls of grasses, whereas xyloglucans are the most abundant hemicelluloses in legume primary cell walls. Glucuronoarabinoxylans are the major hemicellulosic polysaccharides deposited during secondary thickening of cell walls in both grasses and legumes (Chesson and Forsberg, 1988). Hemicelluloses usually contribute a greater proportion of total polysaccharides in secondary cell walls than in primary cell walls (Chesson, 1990). Forage hemicelluloses are not digested by ruminants as extensively as pectic substances. The extent of digestion of hemicelluloses varies considerably and is dependent upon factors such as plant genotype, stage of maturity, and feeding practices. Different hemicellulose fractions present in a single forage source likely are digested to differing extents. For example, xyloglucans appear to be more extensively fermented than glucuronoarabinoxylans in alfalfa (*Medicago sativa;* Titgemeyer et al., 1992). Such differences point to variations in digestibilities of primary and secondary cell walls, which will be discussed in later sections.

Cellulose is the polysaccharide component present in greatest concentrations in both primary and secondary cell walls of all forages. Cellulose synthesized during secondary thickening of cell walls is considerably more crystalline than cellulose in primary cell walls (Lam et al., 1990), although crystallinity alone is of questionable importance as an effector of cellulose digestibility in intact plant cell walls (Richards, 1976). As for hemicelluloses, forage cellulose is incompletely digested by ruminants. The extent to which cellulose is digested is dependent upon a host of plant and feeding practice variables that will be discussed later.

The comparative digestibilities of cellulose and hemicelluloses in forages has been the subject of considerable debate. Several studies have found hemicelluloses to have a somewhat lower extent of digestion than cellulose (Beever et al., 1971; Thomson et al., 1972) whereas other studies report the opposite (Beever et al., 1981). These discrepancies may be related to the methods used for quantifying cellulose and hemicellulose digestibilities. Experiments that base cellulose and hemicellulose contents on detergent fractionation generally find hemicelluloses to be more extensively digested than cellulose (Beever et al., 1981), whereas the opposite trend usually is found when other methods are used to determine hemicellulose and cellulose composition (e.g., Beever et al., 1971; Thomson et al., 1972; Ben-Ghedalia and Miron, 1984). This observation has been confirmed in studies where cellulose and hemicellulose digestibilities have been estimated by both methods (Ben-Ghedalia and Miron, 1984; Bourquin et al., 1990) and is probably a consequence of hemicellulosic contamination of acid detergent residues (Morrison, 1980). Ben-Ghedalia and Miron (1984) concluded that the hemicellulose fraction estimated as NDF minus ADF accounted for the more available portion of the hemicellulosic polysaccharides; hence, the digestion of hemicelluloses estimated by monosaccharide analysis more closely represented the true digestibility of hemicelluloses. Regardless of the methods used for determining digestibilities, the magnitude of difference in cellulose and hemicellulose digestibilities is minor compared to differences in digestibilities of these fractions due to factors such as plant species or stage of maturity.

In recent years, numerous experiments have been conducted to measure digestibilities of individual cell wall monosaccharides. Results of these studies generally rank the digestibilities of major cell wall monosaccharides in the order: total uronic acids = arabinose = galactose > glucose > xylose. Although the digestibility of total uronic acids is usually high, indirect evidence suggests that glucuronate is digested to a considerably lesser extent than galacturonate (Gaillard, 1962; Ben-Ghedalia and Miron, 1984). This is expected because galacturonate is restricted to pectic substances in forages and glucuronate is found only in hemicellulosic glucuronoarabinoxylans. Tremendous disparities often are found in digestibilities of individual monosaccharide components of legumes and extensively digested grasses (Gordon et al., 1983; Titgemeyer et al., 1992), whereas digestibilities of different cell wall monosaccharides are quite similar in poorly degraded substrates such as cereal straws (Gordon et al., 1983). The smaller differences in digestion of monosaccharides in cereal straw may occur because straws are a chemically more homogeneous material than are more extensively digested forages (Gordon et al., 1983). Differential extents of digestion of cell wall monosaccharides often are interpreted to indicate the selective removal of cellulose and the more extensively substituted hemicelluloses from cell walls during the course of digestion. However, Chesson and Forsberg (1988) point out that observed differences in monosaccharide digestibilities can be explained by compositional differences between primary cell walls, which are extensively digested, and lignified secondary-thickened walls, which are poorly digested.

Ruminal versus Postruminal Digestion

In fresh or dried forages fed in either the long or chopped form, ruminal digestion generally accounts for at least 90% of total tract cellulose digestion, but often a somewhat smaller (sometimes as low as 60%) proportion of total tract hemicellulose digestion (Armstrong and Smithard, 1979). The remainder of total cellulose and hemicellulose digestion occurs in the hindgut, as no significant digestion of structural carbohydrates occurs in the small intestine of ruminants (Hoover, 1978). Several factors have been demonstrated to shift the site of digestion of cell wall polysaccharides such that a greater proportion is digested in the hindgut. These factors include increased forage maturity, physical processing, increased level of intake, and dietary supplementation of rapidly fermented carbohydrates (Hoover, 1978; Armstrong and Smithard, 1979).

The greater contribution to digestion of hemicelluloses compared to cellulose by the hindgut has been hypothesized to result from the greater susceptibility of hemicelluloses to mild acid hydrolysis in the abomasum and duodenum. Acetyl groups and hydroxycinnamic acids esterified to hemicelluloses may be especially susceptible to acidic conditions. However, it is more plausible that the greater hemicellulose digestion in the lower tract is simply a consequence of the comparatively high hemicellulose content of slowly degraded plant tissues.

Plant Anatomical Considerations

Plant morphological fractions (e.g., leaf and stem) and different plant cell types (e.g., mesophyll, phloem, xylem) vary considerably both in cell wall content and composition. Consequently, different plant fractions and cell types also demonstrate heterogeneity in digestibility by ruminants. This section will briefly review the digestion of different plant anatomical fractions by ruminants.

Leaf versus Stem

Leaf and stem fractions of forages differ considerably in chemical composition and in digestibility. Leaves generally contain greater concentrations of protein and lower concentrations of cell wall polysaccharides and lignin than do stems of all forages. A large difference in composition of leaf and stem fractions usually is observed for legumes (Albrecht et al., 1987; Titgemeyer et al., 1992), whereas leaf and stem fractions of grasses and cereal straws are more similar in composition (Laredo and Minson, 1973; Ramanzin et al., 1986; Jung and Vogel, 1992). The *in vitro* or *in situ* digestibility of leaves usually is considerably greater than that of stems. The greater extent of digestion observed for leaves is due to both their lower cell wall content and a greater digestibility of the cell wall polysaccharides of leaves (Jung and Vogel, 1992; Titgemeyer et al., 1992). The magnitude of difference in digestibilities of leaf and stem fractions is positively correlated with increasing level of forage maturity (Mowat et al., 1965; Albrecht et al., 1987). In fact, at immature stages of growth, *in vitro* DM digestibility of grass stems can be slightly greater than that of leaves (Mowat et al., 1965).

Composition and digestibilities of leaf and stem fractions also vary by position on the plant. Leaves from the lower portion of corn (*Zea mays* L.) plants had lower cell wall contents and higher digestibilities than leaves higher on the plant (Deinum, 1976). Wilson (1990) related the higher digestibility of leaves at the base of *Panicum maximum* var. *Trichoglume* to the lower cross-sectional areas of vascular tissue and sclerenchyma and a thinner cuticular layer compared to the upper leaves. Lower stems of alfalfa contained greater concentrations of total cell wall material and lower concentrations of pectic substances than apical stems (Hatfield, 1992), suggesting that lower stems would be less digestible than upper stems. Pritchard et al. (1963) found that upper stems of mature timothy (*Phleum pratense* L.), orchardgrass (*Dactylis glomerata* L.), and smooth bromegrass (*Bromus inermis* Leyss) had lower *in vitro* DM digestibilities than lower stems. The greater digestibilities of the lower stems probably resulted from a higher content of extensively fermentable pith, as the bottom internodes of mature grasses have a solid central core while the top internodes are hollow (Wilson, 1990).

When offered to ruminants, the voluntary intake of isolated leaves is considerably greater than that of stems for both grasses (Laredo and Minson, 1973; Poppi et al., 1980) and legumes (Hendricksen et al., 1981). This is illustrated by selected results from Poppi et al. (1980; Table 1) who measured the intake and digestibility of leaf and stem fractions prepared from two C_4

Table 1. Average voluntary intake, apparent digestibilities of dry matter (DM) and neutral detergent fiber (NDF), and ruminal retention time of NDF by cattle of leaf and stem fractions of Pangola grass (*Digitaria decumbens*) and Rhodes grass (*Chloris gayana*) harvested at two stages of maturity (adapted from Poppi et al., 1980).

	Fraction		
Item	Leaf	Stem	Significance Level
Voluntary intake, g DM/kg body weight$^{0.9}$	28.4	21.0	$P<0.01$
Apparent digestibility, g/kg DM	536	541	NS
NDF	618	553	$P<0.01$
Ruminal retention time of NDF, h	32.4	45.7	$P<0.01$

grasses, Pangola grass and Rhodes grass, at two stages of physiological maturity (6 and 12 wk regrowths). Averaged across both forage source and stage of maturity, cattle consumed 35% more DM when offered leaf fractions vs stem fractions. Contrary to the greater DM digestibilities generally found when leaf fractions are fermented *in vitro* or *in situ*, the apparent total tract digestibility of DM was similar when cattle consumed leaf and stem fractions. However, NDF apparent digestibility was greater for leaves than for stems. The lower than expected digestibility of leaves was the result of the considerably shorter ruminal retention time of leaf fractions relative to stem fractions. Similar *in vivo* digestion and rate of passage results were reported for other tropical grasses (Laredo and Minson, 1973) and a tropical legume (Hendricksen et al., 1981).

Plant Cell Type Composition and Degradability

Plant tissues are comprised of a variety of specialized cell types which differ in function, composition, and potential degradability by ruminal microorganisms. Owing to the difficulties in separating specific cell types from plant tissues, few quantitative studies on the composition of different cell types have been published. The detailed composition of cell types isolated from the leaves of perennial ryegrass (*Lolium perenne*) and Italian ryegrass (*Lolium multiflorum*) have been reported (Chesson et al., 1985; Gordon et al., 1985). The yield and chemical composition of mesophyll, epidermis, and fiber (primarily composed of sclerenchyma and vascular tissue) cell walls isolated from leaf blades of early-cut perennial ryegrass are summarized in Table 2. Although the total

Table 2. Yield and chemical composition of mesophyll, epidermis and fiber cell walls isolated from leaves of perennial ryegrass[a].

Item	Cell Fraction		
	Mesophyll	Epidermis	Fiber[b]
	...........g/kg of leaf DM...........		
Cell fraction yield	664	153	182
	...........g/kg of cell fraction DM...........		
Monosaccharide residues			
Arabinose	64	58	38
Fucose	1	2	1
Galactose	25	20	8
Glucose	594	566	509
Mannose	3	2	1
Rhamnose	8	5	2
Zylose	110	167	274
Uronic acid	47	76	38
Total carbohydrate	852	896	871
Total phenolics	27	27	50
Nitrogen x 6.25	58	22	17
Acetyl groups	9	11	19
Ash	43	19	22
Total non-carbohydrate	138	78	107
Total recovery	990	974	978

[a]Adapted from Chesson et al. (1985) and Gordon et al. (1985).
[b]Fraction composed primarily of sclerenchyma and vascular tissues (Chesson et al., 1985).

carbohydrate content of the cell type fractions was similar, mesophyll cell walls contained greater quantities of arabinose, galactose, and glucose than fiber cell walls. However, xylose content was considerably higher in fiber cell walls than in mesophyll cell walls, reflecting the high hemicellulose content of secondary cell walls. Fiber cell walls contained approximately double the concentration of total phenolics as mesophyll cell walls. Concentrations of esterified phenolic acids and acetyl groups, compounds known to be covalently linked to hemicellulosic polysaccharides, also were considerably greater in fiber cell walls relative to mesophyll. Thus, secondary cell walls (fiber cell walls) were enriched in phenolic material and xylans (which have a low degree of substitution with other saccharides) relative to primary cell walls. Epidermis cell walls generally were intermediate in composition to mesophyll and fiber cell walls.

The ruminal degradation of these cell wall fractions was reported by Chesson et al. (1986). Both mesophyll and epidermis cell walls were rapidly and fully degraded within 12 h. Conversely, fiber cell walls were more slowly degraded in the rumen. Only 400 g/kg of DM was lost from fiber cell walls in 12 h (Chesson et al., 1986). Fiber cell walls were 850 g/kg digested after 48 h of ruminal incubation. No evidence was found for the preferential degradation of any cell wall polysaccharide component during fermentation, as all monosaccharide constituents were lost at a rate similar to the rate of DM loss (Chesson et al., 1986). These results support the hypothesis that different extents of digestion of cell wall monosaccharides during ruminal fermentation of intact forages is a consequence of the differential extent of digestion of the various cell types in the forage.

Most studies on the distribution and digestibility of different plant cell types have relied on microscopic techniques, which yield only qualitative information. Using such methods, the general ranking of digestibilities of the various cell types present in forages follows the order: phloem = mesophyll = undifferentiated parenchyma > epidermis > parenchyma bundle sheath > sclerenchyma > lignified vascular tissue (Wilson, 1990). The non-degradable component of cell wall material is primarily contributed by sclerenchyma and lignified vascular tissue and occasionally stem epidermis and parenchyma (Wilson, 1990). Histochemical techniques reveal that these tissues generally contain the highest concentrations of phenolic material of the various plant cell types (Akin, 1989).

Knowledge of the digestibility of different cell types is of little value without knowledge of the contribution of the various cell types to the total weight of the cell wall material. Because of their difficulty in execution, few direct fractionation studies of the type reported by Gordon et al. (1985) have been reported. Most estimates of the relative contribution of cell types to total cell wall material have been based on the cross-sectional areas occupied by specific cell types as a proportion of the entire tissue. The average distribution of cell types present in leaf blades of cool- and warm-season grasses reported by Akin (1989) is presented in Table 3. Mesophyll cells are the most abundant cell type

Table 3. Mean proportions (g/kg) of cell types in cross-sections of cool- and warm-season grass leaf blades (from Akin, 1989).

Cell type	Cool-season (n=6)	Warm-season (n=9)
Total vascular tissue	150	220
Lignified vascular tissue	70	40
Parenchyma bundle sheath	60	150
Phloem	20	20
Epidermis	230	350
Sclerenchyma	60	60
Mesophyll	570	380

present, accounting for 57 and 38% of the cross-sectional area of leaf blades from cool- and warm-season grasses, respectively. Epidermis cells also account for a large proportion of the cross-sectional area of leaf blades, averaging 23 and 35% for cool- and warm-season grasses. Vascular tissues and sclerenchyma do not account for a great percentage of leaf blade cross-sectional area. However, these measures do not account for the different thicknesses of cell wall types. Lignified vascular tissue plus sclerenchyma accounted for only 4 and 15% of the cross-sectional areas of leaf blades from Italian ryegrass and orchardgrass, but these modest area yields translated into 430 and 610 g/kg of the total cell wall weight of these leaf tissues (Wilson, 1990). Vascular tissue plus sclerenchyma cells also accounted for 890 g/kg of the indigestible cell wall material in orchardgrass leaf blades, whereas mesophyll cells contributed 62% of the cross-sectional area and only 40 g/kg of the indigestible cell wall content of the same substrate (Wilson, 1990).

The reticulate venation pattern of legume leaves makes estimates of cell type proportions from a single leaf cross-section inaccurate (Wilson, 1990). Thus, few estimates of cell type composition of legume leaves are available. In his review, Wilson (1990) reported cross-sectional areas of vascular tissue plus sclerenchyma of 5% for alfalfa and 15% for lespedeza (*Lespedeza juncea*). Anatomy of the leaf sheath has been studied less than leaf blade anatomy, but Wilson (1990) concluded that the sheath has a higher percent sclerenchyma than the blade but a similar percentage of vascular tissue. Also, the proportion of sclerenchyma and vascular tissue in the sheath increases considerably between the vegetative and flowering stages of growth (Wilson, 1990). Forage stems contain greater cross-sectional proportions of vascular tissue and sclerenchyma than either the leaf blade or sheath (Wilson, 1990), which would help explain the low digestibility of stems relative to leaves.

Constraints to Cell Wall Polysaccharide Degradation

A primary objective of research conducted on the topic of cell wall digestibility by ruminants has been to identify plant structural factors that limit extent of cell wall degradation by ruminal microorganisms. Several factors have been hypothesized to qualify as constraints to cell wall digestion. Chesson and Forsberg (1988) classified such factors as being either mural (internal) or extramural (external) to the plant cell wall, an excellent distinction that also will serve as a basis of classification for this discussion.

Extramural Factors

The most readily identified external factor limiting the digestion of cell wall material is the cuticular layer present on the plant surface. The cuticle is indigestible by ruminal microorganisms (Akin, 1979) and thus serves as a barrier preventing the access of microorganisms to the outer surface of plant material. The presence of the cuticular layer necessitates that ruminal microorganisms attack plant tissue at the inner surfaces of broken cells. The quantity of cuticle is increased under conditions of high temperature, light, and aridity, and it is found in greater concentrations on the abaxial surface of leaves (Wilson, 1990). A considerable increase in cuticle development on the leaf sheath of grasses occurs within 5 d after the sheath is exposed to the environment during elongation, which has been associated with a dramatic decline in sheath DM digestibility (Wilson, 1990). The epidermal cell wall layer adjacent to the cuticle often is undegraded by ruminal microorganisms (Wilson, 1990).

Another extramural factor that has been confirmed to restrict the access of microbial cells to plant cell walls is the warty layer lining the inner surface of lignified cell walls (Engels and Brice, 1985). The chemical nature of this layer is unknown. However, cell wall material of barley straw underlying the warty layer could only be degraded at positions where the structure of the warty layer was disrupted (Engels and Brice, 1985).

The suberized lamella towards the outside of the radial and tangential walls of parenchyma bundle sheath cells in certain types of C_4 plants (Wilson, 1990) also would qualify as an external limitation to cell wall digestion. This layer can restrict microbial access to these cells for up to 24 to 48 h, thereby contributing to the slow digestion of parenchyma bundle sheath cells in some C_4 plants (Wilson, 1990).

A final factor that can be considered an extramural limitation to cell wall digestion is the poorly digested middle lamella and primary cell wall of secondary-thickened plant cells. Engels and Schuurmans (1992) reported that newly synthesized secondary walls of corn internodes were completely digestible whereas the middle lamella and primary cell walls of the same cells were indigestible. This phenomenon probably is related to the initiation of lignification in the middle lamella and primary cell wall (which are known to contain the highest lignin concentrations in secondary thickened cell walls) during synthesis of the secondary cell wall. Engels and Schuurmans (1992) proposed that, unless mechanically damaged, these poorly degraded primary cell walls would serve as a perfect barrier preventing access to the underlying secondary walls.

Mural Factors

The structure of cell wall polysaccharides often has been hypothesized to limit their extent of degradation. However, no solid evidence confirms this to be a significant factor affecting the digestion of intact plant cell walls. Richards (1976) concluded that cellulose crystallinity played no significant role in limiting the extent of cellulose digestion by ruminants. Likewise, no convincing evidence has been found to confirm that certain aspects of hemicellulose structure have any effect on its extent of degradation. As discussed earlier, the more extensive degradability of cell wall arabinose and galactose residues relative to xylose probably can be attributed to differences in composition of hemicelluloses in extensively vs poorly degraded plant tissues. Esterified acetyl groups and *p*-coumaric acid often are found in greater concentrations in fermented residues than in original cell walls (Morris and Bacon, 1977; Titgemeyer et al., 1991). However, like xylose, these components also have been shown to be present in greatest concentrations in the least degradable plant tissues (Harris et al., 1980;

Gordon et al., 1985). Thus, a direct negative effect of esterified acetyl groups and phenolic acids on cell wall digestibility has not been established. However, because of their relationship with the least degradable cell wall components, these compounds may be useful as markers for predicting the potential digestibility of different forages.

Fermented residues of lignified cell wall material such as cereal straws generally have glycosyl linkage patterns similar to the unfermented substrates, suggesting that no preferential degradation of any cell wall polysaccharide fraction occurs (Gordon et al., 1983). The distribution of alkali-labile substituents on xylose residues also does not change appreciably during fermentation of forage cell walls, indicating that the distribution of ester-linked substituents on xylose residues (e.g., acetyl groups) does not have a significant impact on cell wall polysaccharide digestion (Chesson et al., 1983).

The only internal chemical factor that has conclusively been demonstrated to negatively affect the extent of cell wall polysaccharide digestion is lignin (Akin and Chesson, 1989). As previously stated, plant cell types having the lowest extents of digestion (e.g., lignified vascular tissue and sclerenchyma) have been found to contain the highest phenolic concentrations by histochemical methods. Chesson (1988) hypothesized that the modest accumulation of total phenolic material observed during the digestion of poorly-fermented forages could be much more significant if the total accumulation of phenolic material was localized to the outer surface of the cell wall. Accumulation of phenolic material on plant cell surfaces would hypothetically impart a change in surface chemistry sufficient to inhibit further cell wall digestion by ruminal microorganisms, thereby protecting the underlying cell wall from further attack. This hypothesis has yet to be substantiated by careful analysis of the surface chemistry of plant cell walls undergoing ruminal degradation.

It is well documented that forage digestibility declines with increasing lignin concentrations (Van Soest, 1982). Jung and Vogel (1986) found that a curvilinear model best described the relationship between digestibility and lignin concentration, indicating that the inhibitory effect of lignin on digestion declined at higher lignin concentrations. The relationship between lignin content and digestibility is weaker when compared across plant classes. For example, the lignin content of legumes is approximately twice that of grasses at the same DM digestibility (Van Soest, 1982). This discrepancy can, in large measure, be attributed to the lower cell wall content of legumes which results in relatively high digestibility in spite of highly lignified cell walls.

Factors Affecting Extent and Site of Cell Wall Digestion

Plant Maturity

The predominant feature of increasing physiological maturity of most forages is a tremendous reduction in the leaf to stem ratio (Mowat et al., 1965; Troelsen and Campbell, 1969; Albrecht et al., 1987). Among grasses, both leaf and stem increase in cell wall and lignin contents with advancing maturity (Jung and Vogel, 1992). Consequently, the digestibilities of both leaves and stems of grasses decline with increasing forage maturity, although the rate of decrease is greater for stems than leaves (Mowat et al., 1965). Albrecht et al. (1987) found that alfalfa stems increased in cell wall and lignin contents with increasing maturity, but alfalfa leaves maintained relatively constant composition across maturity levels. Thus, alfalfa stem digestibility decreases considerably with increasing maturity, whereas digestibility of alfalfa leaves is little affected (Mowat et al., 1965; Albrecht et al., 1987).

Ruminants fed mature forages have lower voluntary intakes and digestibilities than when fed immature forages (Troelsen and Campbell, 1969; Ventura et al., 1975). The results of Hogan et al. (1969) illustrate these effects (Table 4). Hogan et al. (1969) examined the composition and utilization by sheep of *Phalaris tuberosa* harvested at three stages of maturity. As expected, concentrations of cell wall constituents and lignin increased and soluble carbohydrate content decreased during forage maturation. Voluntary intake of *Phalaris* decreased 25% and total tract digestibility of cell wall constituents decreased 37% between the lowest and highest stages of maturity (Hogan et al., 1969). The lower digestibility of the more mature grass also was associated with a shift in site of cell wall digestion toward the hindgut, which accounted for 25% of total tract cell wall digestion for the most mature grass (Hogan et al., 1969).

Table 4. Composition, voluntary intake by sheep, and cell wall digestibility of *Phalaris tuberosa* harvested at three stages of maturity (from Hogan et al., 1969).

Item	Stage of Maturity		
	I	II	III
	..........g/kg of organic matter..........		
Concentration of			
Cell wall constituents	436	628	749
Lignin	30	41	74
Soluble carbohydrates	209	151	113
Voluntary intake[a]	61.4	53.7	46.3
Digestion of cell wall constituents			
Total tract digestion, g/kg intake	822	758	515
Ruminal digestion, g/kg digested	953	858	753

[a] g organic matter/kg body weight$^{0.75}$/ day. From Ulyatt (1973).

Grasses versus Legumes

Legumes usually contain lower cell wall concentrations but greater lignin concentrations than grasses at similar maturities. Also, hemicelluloses comprise a smaller percentage of total cell wall in legumes than in grasses (Van Soest, 1982). Temperate, but not tropical, legumes generally are consumed in greater quantities than grasses by ruminants (Minson, 1990). The greater intake potential of temperate legumes is related to their lower cell wall contents and their lower resistance to breakdown during eating and rumination. Thornton and Minson (1973) found that the voluntary intake of legumes was 28% higher than that of grasses of equal digestibility, an increase which was associated with a 17% shorter ruminal retention time of digesta for legumes. White clover (*Trifolium repens*) has a faster rate of DM and NDF digestion (Aitchison et al., 1986) and a faster rate of particle size reduction in the rumen (Moseley and Jones, 1984) than perennial ryegrass. Rumination activity (chews/g cell wall constituents) was significantly lower for sheep consuming red clover (*Trifolium pratense* L.) compared to reed canarygrass (*Phalaris arundinacea* L.) or smooth bromegrass (*Bromus inermis* L.; Chai et al., 1985).

Digestibility of legume cell walls is usually lower than that of grass cell walls, an effect that is related to the higher lignin concentration in cell walls of legumes (Van Soest, 1982). Smith et al. (1972) found a similar lignin:cellulose ratio in indigestible residues of grasses (0.94) and legumes (1.09), suggesting that lignin protects similar quantities of cell wall polysaccharides from digestion in both grasses and legumes. Comparisons of site of digestion of legume vs grass cell walls are hampered by the inherently small number of substrates that can be tested in such experiments. Thomson and Beever (1980) concluded that no significant differences in site of cell wall or OM digestion were apparent among grasses and legumes.

Temperate versus Tropical Forages

Tropical forages generally contain greater cell wall and lignin concentrations than temperate forages (Van Soest, 1982). Minson (1990) found that voluntary intake by sheep of temperate forages was usually greater than that of tropical forages, a difference that applied to both grasses and legumes. However, Reid et al. (1988) did not find differences in voluntary intake of C_3 and C_4 grasses by cattle. Dry matter digestibility of temperate forages is usually greater than that of tropical forages (Reid et al., 1988; Minson, 1990). Minson and McLeod (1970) estimated that temperate grasses were an average of 13 percentage units higher in DM digestibility than tropical grasses. The difference was estimated to be only 4 percentage units greater for temperate vs tropical legumes (Minson and Wilson, 1980). Lagasse et al. (1990) found greater voluntary intakes and total tract NDF digestibilities when cattle consumed vegetative orchardgrass compared to vegetative bermudagrass (*Cynodon dactylon*). However, both extent and site of NDF digestion were similar when cattle consumed mature orchardgrass or mature bermudagrass at 1.5% of body weight (Jones et al., 1988).

The greater digestibility usually observed for temperate grasses has been attributed both to differences in anatomical structure of C_3 vs C_4 grasses and to the higher temperature at which tropical grasses are normally grown. Temperate grass leaves usually contain smaller proportions of slowly digested parenchyma bundle sheath and epidermis cells and larger proportions of rapidly digested mesophyll cells than do tropical grass leaves (Akin, 1989; Table 3). The significance of these anatomical variations has been confirmed by the greater digestibility of C_3 vs C_4 species of the genus *Panicum* grown under the same environmental conditions (Wilson et al., 1983). Higher environmental temperature during growth increased cell wall and lignin contents and decreased digestibility of a variety of forages without altering anatomical composition of leaves or stems (Wilson et al., 1991).

Processing Effects

Method of conservation (dehydration, ensiling) usually has little effect on structural carbohydrate composition of forages, although the soluble carbohydrate fraction of forages is reduced considerably by ensiling (McDonald and Edwards, 1976). Alterations in structural carbohydrate composition of hays are primarily a consequence of leaf loss during harvest. Drying has no apparent effect on the ruminal or total tract digestibilities of soluble or structural carbohydrates (Thomson and Beever, 1980). Extent and site of digestion of structural carbohydrates does not differ significantly between hays and wilted silages (Beever et al., 1971; Merchen and Satter, 1983a).

Grinding and pelleting of forages results in large alterations in extent and site of digestion of cell wall carbohydrates. Voluntary intake of ground forages is consistently higher than forages fed in the long or chopped form. The intake response to forage processing increases with advancing forage maturity (Thomson and Beever, 1980). Increased forage intake coupled with the reduced particle size of ground forages results in shorter ruminal residence times and reduced extents of digestion of ground forages (Thomson and Beever, 1980). Digestibility depressions consequent to grinding are greater for grasses (up to 15 percentage units) than for legumes (3 to 6 percentage units; Thomson and Beever, 1980). The hindgut makes a proportionately larger contribution to total tract digestibility of cell wall carbohydrates when ruminants consume ground forages (Beever et al., 1972; Thomson et al., 1972).

Chemical treatments of forages and crop residues include hydrolytic (e.g., sodium hydroxide, ammonia) and oxidative (e.g., ozone, sulfur dioxide) treatments. Both treatment types result in considerable alterations in cell wall structure and composition. Hydrolytic treatments lead to partial solubilization of hemicelluloses and lignin and hydrolysis of acetic, phenolic, and uronic acid esters, whereas oxidative treatments attack and degrade lignin and can result in extensive solubilization of cell wall polysaccharides (Fahey et al., 1993). Chemical treatments increase extent of digestion of cell wall polysaccharides by ruminants but generally do not affect site of cell wall digestion (Fahey et al., 1993). Monocotyledonous plants respond to both hydrolytic and oxidative treatments, whereas dicotyledons usually respond better to oxidative treatment.

More thorough discussions of physical and chemical treatments of forages are presented in Chapter 23.

Level of Intake

Increasing level of feed intake is associated with reductions in the extent of digestion of DM and cell wall polysaccharides by ruminants (Robertson and Van Soest, 1975; Tyrrell and Moe, 1975). At high intakes, depressions in digestibility of structural carbohydrates are 2 to 3 times greater than those of non-structural carbohydrates (Tyrrell and Moe, 1975). Digestibility depression results from reduced extent of ruminal digestion resulting from decreased ruminal residence time of digesta (Grovum and Williams, 1977; Staples et al., 1984). Reductions in total tract digestibilities often are less pronounced than reductions in extent of ruminal digestion due to an increased contribution of the lower tract to total digestion (Staples et al., 1984; Bourquin et al., 1990). The effects of increasing level of intake are illustrated by the results of Staples et al. (1984). Increasing intake by steers of mixed diets from 55 to 100% of ad libitum intake reduced apparent total tract digestibilities of DM and NDF by 8 and 11 percentage units, respectively. The fraction of total DM digestion occurring post-ruminally increased slightly with increasing levels of intake. These effects were associated with reductions in residence time of DM in the rumen and in the total digestive tract (Staples et al., 1984).

Associative Effects

Associative effects refer to non-additive differences in digestibilities of feedstuff components of mixed diets compared to digestibilities determined for the same components when fed alone (Moe, 1981). Digestibilities of components in mixed diets may be additive at maintenance levels of intake, but negative associative effects often occur when mixed forage-concentrate diets are fed at high intakes (Joanning et al., 1981; Moe, 1981). When mixed corn silage-corn grain diets were fed to steers at 2.4 to 3.1 times maintenance, observed NDF digestibilities were 7.8 and 13.7 percentage units lower than expected for diets containing immature and mature silage, respectively (Joanning et al., 1981). Site of digestion of structural carbohydrates often is shifted toward the lower tract when ruminants consume increasing proportions of grain in mixed forage-concentrate diets (DeGregorio et al., 1982; Brink and Steele, 1985; Bourquin et al., 1990).

Theories that have been advanced to explain the negative effects of dietary non-structural carbohydrates on fiber digestion are: 1) a preference by rumen microorganisms for non-structural carbohydrates rather than structural carbohydrates; 2) a decrease in ruminal pH caused by rapid non-structural carbohydrate digestion; and 3) competition for essential nutrients resulting in preferential proliferation of non-structural carbohydrate-digesting microorganisms (Hoover, 1986). Mould and Ørskov (1983/84) concluded that ruminal digestion of cell wall carbohydrates was inhibited both by low ruminal pH and by the availability of a readily fermentable energy source. Structural carbohydrate digestion generally is markedly depressed when ruminal pH is depressed below 6.0 (Mould et al., 1983/84; Hoover, 1986).

Positive associative effects have been noted when different forage sources, such as grasses and legumes, are fed in combination (Minson and Milford, 1967; Hunt et al., 1985). These effects usually are only observed when one forage source supplies a nutrient, most often protein, that is deficient in the other forage source. Hunt et al. (1985) observed positive associative effects for DM and NDF digestibilities of alfalfa-tall fescue (*Festuca arundinacea*) combinations fed to sheep. Soofi et al. (1982) found positive associative effects for DM intake and NDF digestibility when sheep consumed combinations of alfalfa hay and soybean stover (*Glycine max* Merrill). Positive responses in intake and digestibility also are commonly noted when protein supplements are provided to ruminants consuming low quality forages (Weston, 1967; Ventura et al., 1975).

RUMINAL DEGRADATION OF FORAGE PROTEIN

Protein (nitrogen [N]) nutrition of ruminants is a complex process that is greatly impacted by events occurring in the forestomach and small intestine of the animal. Dietary proteins and nonprotein N are broken down in the rumen to ammonia, which can be used by ruminal microorganisms as a precursor for biosynthesis of AA and microbial protein. Microbial protein subsequently is digested in the small intestine by the host and serves as an important source of AA that can be used for synthesis of protein in animal products. In addition to microbial protein, the supply of AA to the host originates from dietary protein that escapes ruminal degradation and some endogenous protein. Under most circumstances, microbial protein provides the majority of the AA to the host with most of the remainder accounted for by undegraded dietary protein. When microbial protein production is limited or when AA requirements are high, ruminal microbial protein is insufficient to meet the AA needs of the host. Likewise, if excessive degradation of protein occurs in the rumen, more ammonia may be produced than the microbial population can utilize. In such a case, excess ammonia will be lost via absorption from the rumen, conversion to urea in the liver, and excretion in urine. Hence, conversion of dietary protein to microbial protein is sometimes a wasteful process. The extent of degradation of protein in a given feedstuff or diet may become a limiting factor in the process of supplying AA to the host and, consequently, on productivity of the animal.

Data on *in vivo* estimates of the extent of degradation of protein in various forages is presented in Table 5. All estimates were obtained in cattle or sheep fitted with abomasal or duodenal cannulas and fed the forages as the sole dietary ingredient. Substantial variation in degradability values can be observed both within and among forages and *in vivo* data for degradation of protein in many forages is limited or nonexistent. Some variation is certainly associated with problems encountered in measuring digesta flows and in separation of protein of microbial vs dietary origin. In addition, differences among the forages themselves and differences imposed by the feeding situation contribute to the range of values observed here.

Table 5. *In vivo* measurements of ruminal undegraded protein (UDP) in various forages.

Species	Form	Animal	UDP, g/kg CP intake	Reference
Dried Forages				
Legumes				
Alfalfa	Hay, chopped	Sheep	280	Mathers and Miller, 1981
Alfalfa	Hay, chopped	Sheep	260	Kennedy et al., 1982
Alfalfa	Hay, chopped	Sheep	252	Merchen and Satter, 1983a
Alfalfa	Hay, chopped	Sheep		Kawas et al., 1990
	Pre-bloom		64	
	Mid-bloom		121	
	Late-bloom		189	
	Full-bloom		225	
Alfalfa	Hay, chopped	Sheep	260	Atwell et al., 1991
Subterranean clover (*Trifolium subterranean*)	Hay, chopped different maturities	Sheep		Hume and Purser, 1974
	I (immature)		270	
	II		550	
	III		470	
	IV (mature)		570	

Table 5. *In vivo* measurements of ruminal undegraded protein (UDP) in various forages (continued).

Species	Form	Animal	UDP, g/kg CP intake	Reference
Persian clover (*Trifolium resupinatum*)	Hay, chopped	Sheep	530	Hogan et al., 1989
Grasses				
Italian ryegrass	Hay (artificially dried)	Sheep		Beever et al., 1981
	Chopped		300	
	Ground/pelleted		539	
Timothy	Hay (artificially dried)	Sheep		Beever et al., 1981
	Chopped		324	
	Ground/pelleted		530	
Brome (*Bromus inermis*)	Hay, chopped	Sheep	400	Kennedy et al., 1982
Perennial ryegrass	Hay, chopped	Sheep	368	Hogan et al., 1989
Pangola grass	Hay, chopped	Sheep	400	Hogan et al., 1989

Table 5. *In vivo* measurements of ruminal undegraded protein (UDP) in various forages (continued).

Species	Form	Animal	UDP, g/kg CP intake	Reference
Silages				
Legumes				
Alfalfa	Wilted (470 g DM/kg as is)	Sheep	229	Merchen and Satter, 1983a
Alfalfa	Direct-cut (210 g DM/kg as is) + formic acid/ formaldehyde	Steers		Glenn et al., 1989
	Low intake		264	
	High intake		298	
Alfalfa	Pre-bloom Wilted (320 g DM/kg as is)	Sheep	132	Charmley and Veira, 1990a
	Wilted + heat		217	
Alfalfa	Mid-bloom Direct-cut (170 g DM/kg as is)		216	Charmley and Veira, 1990a
	Direct-cut + heat		304	
Alfalfa	Wilted (280 g DM/kg as is)	Sheep	189	Charmley and Veira, 1990b
	Wilted + heat		246	

Table 5. *In vivo* measurements of ruminal undegraded protein (UDP) in various forages (continued).

Species	Form	Animal	UDP, g/kg CP intake	Reference
Grasses				
Perennial ryegrass	Wilted (260 g DM/kg as is) Wilted + formic acid/ formaldehyde	Sheep	169 669	Siddons et al., 1979
Perennial ryegrass	Wilted (210 g DM/kg as is) Wilted + formic acid Wilted + formic acid/ formaldehyde	Heifers	120 180 200	Rooke et al., 1983
Perennial ryegrass	Wilted (220 g DM/kg as is) + formic acid/ formaldehyde	Cows	240	Rooke et al., 1987
Perennial ryegrass	Direct-cut (220 g DM/kg as is) + formic acid	Steers	220	Dawson et al., 1988
Perennial ryegrass	Direct-cut (170 g DM/kg as is) Wilted (380 g DM/kg as is)	Cows	245 237	Teller et al., 1992

Table 5. *In vivo* measurements of ruminal undegraded protein (UDP) in various forages (continued).

Species	Form	Animal	UDP, g/kg CP intake	Reference
Orchardgrass	Direct-cut (18% DM) + formic acid/ formaldehyde	Steers		Glenn et al., 1989
	Low intake		358	
	High intake		298	
Mixed grass	Direct-cut	Cows		Narasimhalu et al., 1989
	(200 g DM/kg as is)		327	
	Wilted		297	
	(400 g DM/kg as is)			
Fresh Forages				
Legumes				
White clover	Fresh, harvested	Steers		Beever et al., 1986
	Primary growth		160	
	First regrowth		231	
	Second regrowth		251	
White clover	Fresh, harvested, primary growth	Steers		Beever et al., 1987
	Control		166	
	Formaldehyde-treated		189	
White clover	Fresh, harvested, regrowth	Steers		Beever et al., 1987
	Control		183	
	Formaldehyde-treated		263	

Table 5. *In vivo* measurements of ruminal undegraded protein (UDP) in various forages (continued).

Species	Form	Animal	UDP, g/kg CP intake	Reference
Grasses				
Perennial ryegrass	Fresh, harvested,	Steers		Beever et al., 1986
	Primary growth		275	
	First regrowth		266	
	Second regrowth		220	

In spite of these caveats, data in Table 5 and data obtained from similar studies will be used to develop generalizations about the influence of factors associated with the plant (maturity, type of plant) and post-harvest treatment (method of conservation, physical processing, heat damage) on degradability of forage protein. One generalization that may be made is that, under most circumstances, forage protein is quite extensively degraded by ruminal microbes. Across 52 observations on individual forages reported in Table 5, the average value for ruminally undegraded protein (UDP) is 284 g/kg crude protein (CP) intake. This average includes values for several forages in which extent of degradation has been reduced by chemical or physical treatments that are imposed for that purpose. If one considers only legumes and grasses fed fresh or conserved as hays or silages that have not been treated (formaldehyde, heat, grinding and pelleting) in fashions intended to reduce ruminal protein degradation, the average UDP value is reduced to 267 g/kg CP intake. This means that utilization of about 730 g/kg CP in these forages is dependent upon conversion of products of ruminal protein degradation into microbial protein.

Factors Influencing Ruminal Degradation of Forage Protein

Plant Maturity

Because of compositional changes that occur in conjunction with advancing maturity, feeding of more mature forages usually results in decreased DM intakes and reduced postruminal supplies of many nutrients. Advanced forage maturity is often associated with a decreased extent of degradation of forage protein in the rumen, although some exceptions have been noted. Hume and Purser (1974) partitioned digestion of N compounds in the gastrointestinal tract of sheep fed subterranean clover (*Trifolium subterranean*) hays harvested at four stages of maturity. Maturities were defined as pre-wilted (maturity I), wilted (II), post-wilted (III), and mature (IV) and these stages corresponded to progressively greater senescence of the plant leaf. Extent of ruminal degradation was greater for the maturity I clover (730 g/kg CP intake) than for any of the more mature clovers (average = 470 g/kg CP intake). It is important to emphasize that decreased degradability associated with increased maturity may not be expressed as an increased net supply of N or AA to the small intestine. Decreased CP content of plants at later maturities and decreased production of microbial protein resulting from lower intakes of digestible OM when mature forages are fed often result in reduced net flows of N into the intestine despite reduced ruminal degradability of forage protein (Hume and Purser, 1974; Kawas et al., 1990). In addition, these studies have revealed lower small intestinal digestibilities of N from more mature forages. It is likely that some of the plant N, while protected from ruminal degradation, is also more refractory to postruminal digestion.

Kawas et al. (1990) fed alfalfa hays harvested at pre-, early-, mid-, and late-bloom to sheep and reported extensive ruminal degradation at all maturities; however, UDP increased from 60 to 230 g/kg of dietary CP with advancing maturity (Table 5). As previously mentioned, quantity of N disappearing in the small intestine was decreased due to advanced maturity because of decreases in both total duodenal N flows and in digestibility of the N reaching the duodenum.

Much less pronounced effects of stage of harvest were reported by Beever et al. (1986) when cattle were fed perennial ryegrass or white clover harvested daily and fed as fresh forage. In this work, both forages were fed as primary growth (May) or as mid- (July) or late- (September) season regrowths. Ruminal degradation of forage CP was extensive for all forages (range 730 to 840 g/kg CP); however, all forages could be regarded as very high quality. Dry matter digestibilities ranged from 770 to 830 g/kg for the forages and CP concentrations were high in all cases. It is probable that stage of harvest would have less impact on digestion characteristics if a high quality is maintained. Even under these circumstances, there was a modest increase in UDP (from 160 to 250 g/kg CP; Table 5) as maturity progressed for white clover. Extensive ruminal degradation of protein in fresh forages and the high CP level of such forages is frequently associated with a net loss of N during ruminal digestion (Beever et al., 1986; Beever and Siddons, 1986; Ulyatt et al., 1988).

Method of Conservation

The method by which harvested forages are conserved prior to feeding might influence ruminal protein degradation due to characteristic changes in the composition of the CP fraction of the forage during harvest and storage. Ohshima and McDonald (1978) reported that 75 to 90% of the total N in fresh herbage is present as protein N. Following cutting and during drying in the field, transformations occur in N compounds due to activity of plant proteases. The resultant proteolysis results in increases in soluble and non-protein N as well as an altered distribution of AA in the non-protein N fraction in comparison to the parent protein (Sullivan, 1973). Ensiling and the attendant fermentation results in further proteolysis and metabolism of AA. In silages that have undergone a "desirable" fermentation, the N fraction is highly soluble and non-protein in nature, with low concentrations of ammonia N. Much of the non-protein N in silages is present as free AA and low molecular weight peptides (McDonald and Whittenbury, 1973; McDonald and Edwards, 1976; Ohshima and McDonald, 1978). It is sometimes inferred that utilization of forage CP may be inversely related to changes in N solubility occurring during harvest and storage.

As mentioned previously, consumption of high quality forages in the fresh form frequently results in a net loss of N from the reticulo-rumen because of extensive ruminal degradation (Beever et al., 1986; Beever and Siddons, 1986; Ulyatt et al., 1988). Few direct comparisons of fresh forage to conserved forages have been made with regard to site of N digestion. Beever et al. (1976) fed sheep perennial ryegrass harvested fresh and stored frozen or artificially dried by one of several regimens that resulted in treatments with a wide range of N solubilities. Drying resulted in increased flows of N to the small intestine and increased quantities of N disappearing from the small intestine. Flores et al. (1986) demonstrated that sheep fed direct-cut alfalfa silage (210 g DM/kg as is) had decreased flows of total AA to the duodenum compared to fresh (frozen) alfalfa.

Wilting hay-crop silages to 350 to 400 g DM/kg as is before ensiling does not seem to increase the fraction of protein escaping ruminal degradation in comparison to conserving the same forages as direct-cut silages, even though direct-cut silages often undergo greater proteolysis during ensiling. In three studies summarized in Table 5 (Narasimhalu et al., 1989; Charmley and Veira, 1990a; Teller et al., 1992), the average UDP value of wilted silages was 222 g/kg CP vs 262 g/kg CP for direct-cut silages. Charmley et al. (1990) showed that harvesting perennial ryegrass as wilted (350 to 370 g DM/kg as is) silage did not increase flow of total N to the duodenum of steers when compared to good quality untreated direct-cut (230 g DM/kg as is) silage. Other experiments in which wilted silages (ranging from 280 to 470 g DM/kg as is) have been fed have likewise resulted in reports of extensive (730 to 830 g degraded/kg CP) ruminal degradation (Siddons et al., 1979; Merchen and Satter, 1983a; Charmley and Veira, 1990b).

Harvesting alfalfa as good-quality hay (ca. 190 g CP/kg DM) did not result in a substantial improvement in UDP value when compared to the same crop conserved as wilted silage (252 vs 229 g/kg CP; Merchen and Satter, 1983a). It is likely that forage proteins that are most susceptible to proteolysis during ensiling are also most susceptible to microbial degradation in the rumen. Hence, changes in N solubility due to method of forage conservation might be anticipated to have only modest effects on extent of degradation of forage protein.

Physical Processing

A significant amount of research has been conducted to investigate the effects of physical processing (e.g., chopping, fine grinding/pelleting) on digestion of forage components. In general, processing hay by coarse chopping before feeding has little effect on site or extent of digestion (Galyean and Owens, 1991). However, reducing particle size of forage by fine grinding and pelleting can increase ruminal dilution rate of particulate digesta (Thomson and Beever, 1980; Rode et al., 1985). This may be associated with increased escape of feed protein from ruminal degradation. In addition, heat generated during pelleting may provide some protection of feed protein in the rumen. Generally, grinding and pelleting enhances flow of N and AA to the small intestine (Thomson and Beever, 1980).

An example of these effects is presented in Table 6. Beever et al. (1981) fed Italian ryegrass or timothy in either coarsely chopped or ground and pelleted forms to sheep and fractionated duodenal AA flows into microbial, endogenous, and feed components. In this experiment, grinding and pelleting resulted in substantial increases in flow of feed AA to the small intestine. Supply of UDP was estimated to increase by an average of about 72% for the two forages. This observation is consistent with other reports (Coelho da Silva et al., 1972; Beever and Thomson, 1981; Rode et al., 1985). However, this increase may not result in an appreciable increase in supply of intestinally

Table 6. Nitrogen intakes and duodenal flows of N and AA in sheep fed Italian ryegrass or timothy in chopped or ground and pelleted forms.

Item	Italian ryegrass		Timothy	
	Chopped	Pelleted	Chopped	Pelleted
		g/d		
N intake	22.3	22.2	25.5	22.3
Duodenal flows				
Total N	27.8	31.4	29.0	30.0
Total AA	139	161	148	159
Microbial AA	88	81	87	75
Feed AA[a]	37	63	44	67

[a]Corrected for estimated endogenous AA flows

absorbed AA to the animal. First, net synthesis of microbial protein may decrease when ground and pelleted forages are fed (Coelho da Silva et al., 1972; Beever et al., 1981). Secondly, net disappearance of the N or AA entering the small intestine is sometimes decreased, possibly due to heat damage to protein that might occur during pelleting (Coelho da Silva et al., 1972; Beever and Thomson, 1981).

Legumes versus Grasses

Direct comparisons of grasses and legumes are difficult to make. However, in general, it appears that protein in legumes is somewhat more extensively degraded in the rumen than that in grasses. Across observations for forages in Table 5 that had not been treated (formaldehyde or heat) to allow for increased ruminal escape, the average UDP value for legume hays was 311 g/kg CP intake (n=13) while that for grass hays was 358 g/kg CP intake (n=5). This difference also existed for legume (192 g/kg CP intake; n=4) versus grass (224 g/kg CP intake; n=8) silages. While recognizing that there are limitations in making these comparisons across experiments, these differences are consistent with those observed in individual experiments.

Net losses of N often occur during ruminal digestion of good-quality legumes (particularly alfalfa) such that duodenal N flows in animals fed these forages are lower than intake (Kennedy et al., 1982, 1986; Beever et al., 1986; Ulyatt et al., 1988). Such losses also may occur when grasses are fed but are often diminished because grasses contain somewhat less N than legumes (at similar maturities) and the protein in grasses is usually less extensively degraded in the rumen. Kennedy et al. (1982) reported that UDP of chopped alfalfa hay (260 g/kg CP) was lower than that of chopped bromegrass (400 g/kg CP). Likewise, ruminal degradation of protein in white clover was more extensive than that in perennial ryegrass when fed as fresh harvested forage (Beever et al., 1986) or when consumed by grazing animals (Ulyatt et al., 1988). Glenn et al. (1989) reported lower average UDP values for alfalfa silage (281 g/kg CP) than for orchardgrass silage (328 g/kg CP). It can generally be anticipated that efficiency of utilization of legume protein is compromised more greatly than that of grasses due to attendant losses of N that accompany extensive degradation of protein in the rumen.

HEAT DAMAGE IN CONSERVED FORAGES

Decreased digestibilities of N in some stored forages (hays, hay crop silages) were first noted by investigators more than 40 years ago. Van Soest (1965) then demonstrated that heating of forages in the laboratory resulted in increased concentrations of ADF, lignin, and of N in the ADF (acid detergent insoluble N; ADIN) in those forages. Subsequently, samples of an array of forages (fresh materials, hays, and hay crop silages) were collected from various sites in the US. Digestibilities of all forages and their components had been measured in *in vivo* studies. These forages were subjected to a battery of laboratory analyses and relationships between components of these analyses and apparent digestibilities of dietary components were established (Goering et al., 1972). Apparent digestibility of N was most highly related to the ADIN content of the forage as described by the following equation:

$$Y = 72.96 - 1.02\ X\ (r^2 = 0.86)$$
where X = ADIN, % of total N
and Y = apparent N digestibility, %

Subsequent studies by these workers and by scientists at Michigan State (see review of Thomas et al., 1982) clearly established that extent of ADIN formation was related to extent of heating that occurred during storage or processing.

These highly significant relationships between extent of heating, ADIN formation, and apparent N digestibility strongly imply that ADIN content of a feedstuff is reflective of the formation of indigestible N-containing complexes (Maillard products) resulting from heat and that some decrease in nutritive (especially protein) value accompanies ADIN formation. Further research has provided evidence of reduced performance in animals fed excessively heated forages (Pierson et al., 1971; Yu et al, 1977; Thomas et al., 1982). Because of these observations, ADIN has frequently been used as an assessor of heat damage in forages (particularly low-moisture hay crop silages) and has been suggested for use in defining the unavailable protein fraction of feedstuffs in systems designed to model ruminant N metabolism (Van Soest et al., 1982; NRC, 1985).

There is no question that excessive heating and consequent ADIN formation should be avoided and that excessively heated forages should be discounted in value. However, because there may be little relationship between apparent N digestibility and N supply in the small intestine of the animal (Merchen, 1990), established relationships between ADIN and apparent N digestibility may not directly reflect the true value of protein in a heated forage. Also, given that moderate heating often enhances protein value by decreasing ruminal degradability, development of ADIN at modest levels might not signal any undesirable effects and may, in fact, reflect desirable effects of heating.

There is some indication that modest heating of low-moisture alfalfa silage, while increasing ADIN, may result in an increase in supply of N absorbed from the small intestine (Merchen and Satter, 1983b). In this study, alfalfa silages harvested at DM contents of 290, 400, and 660 g DM/kg as is were utilized. Temperature of the forages was monitored during ensiling and it was noted that the material ensiled at 660 g DM/kg as is remained at a higher temperature (>40°C) for an extended period. This was reflected by the higher ADIN content of this forage relative to the two higher moisture silages (129 g ADIN/kg N vs 75 g ADIN/kg N). As might have been predicted, apparent N digestibility decreased about 5 percentage units when lactating cows were fed the 660 g DM/kg material. However, the supply of N disappearing from the small intestine (as % of N intake) was actually increased in comparison to the higher moisture silages. This was the result of less extensive degradation of dietary CP in the rumen and reduced losses of N due to ammonia absorption.

Several investigations have been conducted into the use of modest heat application to direct-cut or wilted alfalfa before ensiling as a means of reducing proteolysis during ensiling and (or) providing protection to the plant proteins from ruminal degradation. Such approaches have been shown to be effective in reducing extent of proteolysis and formation of soluble N during ensiling (Mandell et al., 1989; Charmley and Veira, 1990a). Increases in the proportion of dietary CP that escapes ruminal degradation have been observed when heat-treated forages were fed to sheep (Charmley and Veira, 1990a, 1990b) and quantities of total N disappearing in the small intestine were likewise increased. This work may offer intriguing possibilities for improvement in utilization of CP in high quality ensiled forages although further work is needed to define optimal heat treatment regimens and to establish the economic viability of such an approach.

MICROBIAL PROTEIN PRODUCTION ON FORAGE-BASED DIETS

In most situations, the most important source of protein reaching the lower gut of ruminants is the microbial protein produced in the rumen. Microbial protein is of high quality with a well balanced AA profile (Merchen and Titgemeyer, 1992). About 80% of the AA-N is utilized, as measured in N retention studies with intragastrically-nourished lambs receiving microbial cells as the sole source of protein (Ørskov and Kay, 1987). As discussed previously, protein in many forages is extensively degraded and this increases the dependency of the host on microbial protein as a source of AA for production. Table 7 contains data summarized from several studies in which the proportion of microbial N in total N reaching the postruminal gut (MN:TN) and the efficiency of net microbial protein synthesis (E_{MCP}, g N/kg OM apparently digested in the stomach [OMD_A]) have been reported in animals consuming forage diets. In these experiments, the average value of MN:TN is 0.64, indicating that nearly two thirds of the CP reaching the small intestine is of microbial origin and emphasizing the importance of microbial protein when such diets are fed. Values for MN:TN in these experiments range from 0.47 to

Table 7. *In vivo* measurements of the proportion of microbial N in total abomasal or duodenal N (MN:TN) and of efficiency of net microbial protein synthesis (E_{MCP}) in animals fed forage-based diets.

Diet	Animal	MN:TN[a]	E_{MCP}, g MN/kg OMD_A [b]	Reference
Italian ryegrass	Sheep			Beever et al., 1981
Chopped		0.63	17.2	
Ground/pelleted		0.50	19.6	
Timothy	Sheep			Beever et al., 1981
Chopped		0.59	20.3	
Ground/pelleted		0.47	18.0	
Alfalfa	Sheep			Merchen and Satter, 1983a
Hay, chopped		0.66	41.9	
Silage (470 g DM/kg as is)		0.63	35.2	
Perennial ryegrass	Steers			Rooke et al., 1983
Silage (210 g DM/kg as is)				
Control		0.81	27.2	
+ Formic acid		0.75	27.2	
+ Formic acid/formaldehyde		0.72	26.5	
White clover, fresh	Steers			Beever et al., 1987
Primary growth				
Control		0.70	45.4	
+ Formaldehyde		0.72	62.5	

Table 7. *In vivo* measurements of the proportion of microbial N in total abomasal or duodenal N (MN:TN) and of efficiency of net microbial protein synthesis (E_{MCP}) in animals fed forage-based diets (continued).

Diet	Animal	MN:TN[a]	E_{MCP}, g MN/kg OMD_A[b]	Reference
First regrowth	Sheep			Kawas et al., 1990
Control		0.65	42.4	
+ Formaldehyde		0.65	71.1	
Alfalfa hay				
Pre-bloom		0.75	34.0	
Early bloom		0.73	27.0	
Mid-bloom		0.54	24.0	
Full bloom		0.49	25.0	
Perennial ryegrass	Dairy cows			Teller et al., 1992
Silage				
Direct-cut (170 g DM/kg as is)		0.54	28.6	
Wilted (380 g DM/kg as is)		0.58	32.0	

[a]Proportion of microbial N in total abomasal or duodenal N.
[b]Efficiency of net ruminal microbial protein synthesis, expressed as g microbial N per kg OM apparently digested (OMD_A) in the stomach.

0.81. Although some of this variation undoubtedly results from methodological problems associated with measurement of digesta flow and with markers of microbial protein, some dietary factors appear to affect MN:TN values.

The average value for E_{MCP} from studies summarized in Table 7 is 32.9 g N/kg OMD_A. Again, a large range of values (17.2 to 71.1) is noted although most values are in the range of 20 to 35. Some distinction should be made between net microbial protein synthesis (e.g., the quantity of microbial protein entering the lower gut) and true microbial protein synthesis. The difference between the two quantities consists of the appreciable intraruminal recycling of microbial N that occurs due to cell death and lysis, protozoal predation, and other factors (van Nevel and Demeyer, 1977; Leng and Nolan, 1984). Net microbial protein synthesis typically represents only 30 to 50% of the microbial protein truly synthesized in the rumen (Demeyer and van Nevel, 1979; Leng and Nolan, 1984) and pursuit of strategies that might diminish this discrepancy should be considered as a priority in attempts to improve efficiency of protein utilization by ruminants. Dietary factors that may influence E_{MCP} are discussed in the following section.

Factors Influencing Microbial Protein Synthesis on Forage Diets

Stage of Maturity

Feeding forages at advanced stages of maturity results in decreases in the quantity of microbial protein entering the postruminal gut of sheep (Hume and Purser, 1974; Kawas et al., 1990). Net microbial protein production is a function of E_{MCP} and of the quantity of OM digested in the rumen. Reductions in net synthesis associated with more mature forages result largely from reductions in ruminal OM digestion although E_{MCP} is also sometimes at least modestly reduced as well. Hume and Purser (1974) reported that duodenal microbial CP supply decreased from 16 to 8.4 g/d when immature subterranean clover hay vs mature hay was fed. This was associated with a decrease of 43% in the amount of OMD_A. Values for E_{MCP} for the immature and mature hays were 29.5 and 27.3 g N/kg OMD_A, respectively. Likewise, Kawas et al. (1990) observed that net microbial protein synthesis decreased by 48% when full-bloom alfalfa hay was fed compared to pre-bloom hay. In that experiment, the quantity of OMD_A decreased by 30% while E_{MCP} decreased from 34.0 to 25.0 g N/kg OMD_A for pre- vs full-bloom hays, respectively.

Method of Conservation

Effects of method of conservation on E_{MCP} and net microbial protein synthesis have been inconsistent. Frequently, increased microbial CP flows to the small intestine have been observed when fresh forages have been fed compared to ensiled forages or when dehydrated forages or hays are fed vs fresh or ensiled forages (Beever et al., 1974; Thomson and Beever, 1980). Negative effects on net microbial protein production seem to be greatest for direct-cut unwilted silages and little difference in net production or E_{MCP} is observed when good quality wilted silages are compared to hays (Merchen and Satter, 1983a).

Several hypotheses have been proposed to explain these effects (Thomson and Beever, 1980). Fresh forages contain higher proportions of water-soluble carbohydrates than ensiled forages and, consequently, increased quantities of total carbohydrate are fermented in the rumen when fresh vs ensiled forages are fed. This is likely to account for improvements in net synthesis and in E_{MCP} observed for fresh vs ensiled materials. Feeding dehydrated forages or hays often results in decreased rate of ruminal degradation of dietary CP in the rumen and this may have beneficial effects on the rate of supply of N-containing substrates to the microbes relative to energy availability from carbohydrate fermentation. Some consideration also might be given to differences in the nature of the OM disappearing in the rumen, particularly when unwilted silages are fed. Organic acids produced by extensive fermentation during ensiling of unwilted silages often comprise 10 to 15% of the total OM in such forages. Absorption of these compounds from the rumen contributes substantially to total OM disappearing at that site but does not provide substrate for microbial fermentation. Thus, actual production of energy that may be used by the microbes for protein synthesis may be lower for these forages compared to fresh or dried forages even though quantity of OM disappearing from the stomach may be similar.

Physical processing

Fine grinding and pelleting usually decreases the proportion of microbial N in total N reaching the lower gut (Table 7); however, this effect is due largely to an increase in the quantity of nonmicrobial (dietary) N escaping the rumen (Thomson and Beever, 1980; Beever et al., 1981; Table 7). In most cases, net microbial protein production is unaffected or slightly decreased when ground and pelleted forages are fed although it has been increased in some studies (Rode et al., 1985). Likewise, E_{MCP} may be unchanged (Beever et al., 1981) or increased (Rode et al., 1985) due to grinding. The quantity of OM disappearing in the stomach is usually decreased by grinding and pelleting due to an increased fractional turnover rate of particulate digesta (Thomson and Beever, 1980; Rode et al., 1985). These responses probably contribute to observed increases in E_{MCP}.

CONCLUSIONS

The past twenty-five years have seen a dedicated effort from researchers towards the topic of digestion and utilization of nutrients in forages. Much progress has occurred in characterization of the physico-chemical nature of plant polysaccharides and the process of digestion of plant cell wall polysaccharides. A great deal of information has been compiled on factors that influence extent of digestion of plant cell walls and ruminal degradation of forage protein. Significant advances in knowledge of cell wall chemistry, gastrointestinal microbiology, and ruminal N metabolism have created major changes in our concepts regarding the process of digestion and in our interpretation of digestibility data.

Further work on fiber digestion should continue to seek to identify and elucidate mechanisms that create primary limitations to microbial degradation of fiber. A major objective of the study of variation among plants in cell wall composition should relate to effects of such variation on digestion characteristics. Further documentation of structural constraints to microbial digestion of cell wall materials will be necessary. These approaches need to be coupled with strategies such as genetic selection of plant varieties, genetic modification of ruminal microorganisms, and animal feeding and management practices to fully capitalize on the potential of forages as a feedstock.

Extensive ruminal degradation of protein in high quality forages clearly creates inefficiencies in utilization of protein. Future research needs to emphasize development of approaches for utilization of ruminally degraded protein in animals fed harvested forages (e.g., through control of rate of ruminal protein degradation or by implementation of feeding strategies that will lead to more efficient microbial capture of protein degradation products). Our knowledge of the effectiveness with which protein in high quality forages is utilized by the grazing animal is still rudimentary; further evaluation is required before strategies for optimizing its use can be fully developed.

REFERENCES

Aitchison, E. M., M. Gill, M. S. Dhanoa, and D. F. Osbourn. 1986. The effect of digestibility and forage species on the removal of digesta from the rumen and the voluntary intake of hay by sheep. Br. J. Nutr. 56:463-476.

Akin, D. E. 1979. Microscopic evaluation of forage digestion by rumen microorganisms - A review. J. Anim. Sci. 48:701-710.

Akin, D. E. 1989. Histological and physical factors affecting digestibility of forages. Agron. J. 81:17-25.

Akin, D. E., and A. Chesson. 1989. Lignification as the major factor limiting forage feeding value especially in warm conditions. p. 1753-1760. In R. Desroches (ed.) Proc. XVI International Grassland Congress, Nice. Association Francaise pour la Production Fourragere, INRA, Versalles.

Akin, D. E., L. G. Ljungdahl, J. R. Wilson, and P. J. Harris. 1990. Microbial and plant opportunities to improve lignocellulose utilization by ruminants. Elsevier Science Publishing Co., Inc., New York.

Albrecht, K. A., W. F. Wedin, and D. R. Buxton. 1987. Cell-wall composition and digestibility of alfalfa stems and leaves. Crop Sci. 27:735-741.

Armstrong, D. G., and R. R. Smithard. 1979. The fate of carbohydrates in the small and large intestines of the ruminant. Proc. Nutr. Soc. 38:283-294.

Argenzio, R. A. 1984. Digestion and absorption of carbohydrate, fat, and protein. p. 301-310. In M. J. Swenson (ed.) Dukes' physiology of domestic animals. Cornell University Press, Ithaca, NY.

Atwell, D. G., N. R. Merchen, E. H. Jaster, G. C. Fahey, Jr., and L. L. Berger. 1991. Site and extent of nutrient digestion by sheep fed alkaline hydrogen peroxide-treated wheat straw-alfalfa hay combinations at restricted intakes. J. Anim. Sci. 69:1697-1706.

Baldwin, R. L., and M. J. Allison. 1983. Rumen metabolism. J. Anim. Sci. 57 (Suppl. 2):461-477.

Beever, D. E., S. B. Cammell, and A. S. Wallace. 1974. The digestion of fresh, frozen and dried perennial ryegrass. Proc. Nutr. Soc. 33:73A-74A.

Beever, D. E., J. F. Coelho da Silva, J. H. D. Prescott, and D. G. Armstrong. 1972. The effect in sheep of physical form and stage of growth on the sites of digestion of a dried grass. 1. Sites of digestion of organic matter, energy and carbohydrate. Br. J. Nutr. 28:347-356.

Beever, D. E., M. S. Dhanoa, H. R. Losada, R. T. Evans, S. B. Cammell, and J. France. 1986. The effect of forage species and stage of harvest on the processes of digestion occurring in the rumen of cattle. Br. J. Nutr. 56:439-454.

Beever, D. E., H. R. Losada, D. L. Gale, M. C. Spooner, and M. S. Dhanoa. 1987. The use of monensin or formaldehyde to control the digestion of the nitrogenous constituents of perennial ryegrass (*Lolium perenne* cv. Melle) and white clover (*Trifolium repens* cv. Blanca) in the rumen of cattle. Br. J. Nutr. 57:57-67.

Beever, D. E., D. F. Osbourn, S. B. Cammell, and R. A. Terry. 1981. The effect of grinding and pelleting on the digestion of Italian ryegrass and timothy by sheep. Br. J. Nutr. 46:357-370.

Beever, D. E., and R. C. Siddons. 1986. Digestion and metabolism in the grazing ruminant. p. 479-497. *In* L. Milligan, W. L. Grovum, and A. Dobson (ed.) Control of digestion and metabolism in ruminants. Prentice-Hall, Englewood Cliffs, NJ.

Beever, D. E., and D. J. Thomson. 1981. The effect of drying and processing red clover on the digestion of the energy and nitrogen moieties in the alimentary tract of sheep. Grass Forage Sci. 36:211-219.

Beever, D. E., D. J. Thomson, and S. B. Cammell. 1976. The digestion of frozen and dried grass by sheep. J. Agric. Sci. (Camb). 86:443-452.

Beever, D. E., D. J. Thomson, E. Pfeffer, and D. G. Armstrong. 1971. The effect of drying and ensiling grass on its digestion by sheep. Br. J. Nutr. 26:123-134.

Ben-Ghedalia, D., and J. Miron. 1984. The digestion of total and cell wall monosaccharides of alfalfa by sheep. J. Nutr. 114:880-887.

Bourquin, L. D., K. A. Garleb, N. R. Merchen, and G. C. Fahey, Jr. 1990. Effects of intake and forage level on site and extent of digestion of plant cell wall monomeric components by sheep. J. Anim. Sci. 68:2479-2495.

Brink, D. R., and R. T. Steele. 1985. Site and extent of starch and neutral detergent fiber digestion as affected by source of calcium and level of corn. J. Anim. Sci. 60:1330-1337.

Chai, K., P. M. Kennedy, L. P. Milligan, and G. W. Mathison. 1985. Effects of cold exposure and plant species on forage intake, chewing behavior and digesta particle size in sheep. Can. J. Anim. Sci. 65:69-76.

Charmley, E., M. Gill, and C. Thomas. 1990. The effect of formic acid treatment and the duration of the wilting period on the digestion of silage by young steers. Anim. Prod. 51:497-504.

Charmley, E., and D. M. Veira. 1990a. Inhibition of proteolysis at harvest using heat in alfalfa silages: Effects on silage composition and digestion by sheep. J. Anim. Sci. 68:758-766.

Charmley, E., and D. M. Veira. 1990b. Inhibition of proteolysis in alfalfa silages using heat at harvest: Effects on digestion in the rumen, voluntary intake and animal performance. J. Anim. Sci. 68:2042-2051.

Chesson, A. 1988. Lignin-polysaccharide complexes of the plant cell wall and their effect on microbial degradation in the rumen. Anim. Feed Sci. Technol. 21:219-228.

Chesson, A. 1990. Nutritional significance and nutritive value of plant polysaccharides. p. 179-195. In J. Wiseman and D. J. A. Cole (ed.) Feedstuff evaluation. Butterworths, Inc., London.

Chesson, A., and C. W. Forsberg. 1988. Polysaccharide degradation by rumen microorganisms. p. 251-284. In Hobson, P. N. (ed.) The rumen microbial ecosystem. Elsevier Applied Science, London.

Chesson, A., A. H. Gordon, and J. A. Lomax. 1983. Substituent groups linked by alkali-labile bonds to arabinose and xylose residues of legume, grass and cereal straw cell walls and their fate during digestion by rumen microorganisms. J. Sci. Food Agric. 34:1330-1340.

Chesson, A., A. H. Gordon, and J. A. Lomax. 1985. Methylation analysis of mesophyll, epidermis, and fibre cell-walls isolated from the leaves of perennial and Italian ryegrass. Carbohydr. Res. 141:137-147.

Chesson, A., and J. A. Monro. 1982. Legume pectic substances and their degradation in the ovine rumen. J. Sci. Food Agric. 33:852-859.

Chesson, A., C. S. Stewart, K. Dalgarno, and T. P. King. 1986. Degradation of isolated grass mesophyll, epidermis and fibre cell walls in the rumen and by cellulolytic rumen bacteria in axenic culture. J. Appl. Bacteriol. 60:327-336.

Coelho da Silva, J. F., R. C. Seeley, D. J. Thomson, D. E. Beever, and D. G. Armstrong. 1972. The effect in sheep of physical form on the sites of digestion of a dried lucerne diet. 2. Sites of nitrogen digestion. Br. J. Nutr. 28:43-61.

Dawson, J. M., C. I. Bruce, P. J. Buttery, M. Gill, and D. E. Beever. 1988. Protein metabolism in the rumen of silage-fed steers: Effect of fishmeal supplementation. Br. J. Nutr. 60:339-353.

DeGregorio, R. M., R. E. Tucker, G. E. Mitchell, Jr., and W. W. Gill. 1982. Carbohydrate fermentation in the large intestine of lambs. J. Anim. Sci. 54:855-862.

Deinum, B. 1976. Effect of age, leaf number and temperature on cell wall and digestibility of maize. p. 29-41. *In* H. Veenman and B. V. Zonen (ed.) Carbohydrate research in plants and animals. Misc. Papers 12. Landbouwhogeschool Wageningen, The Netherlands.

Demeyer, D. I., and C. J. van Nevel. 1979. Effect of defaunation on the metabolism of rumen micro-organisms. Br. J. Nutr. 42:515-524.

Engels, F. M., and R. E. Brice. 1985. A barrier covering lignified cell walls of barley straw that restricts access by rumen microorganisms. Curr. Microbiol. 12:217-224.

Engels, F. M., and J. L. L. Schuurmans. 1992. Relationship between structural development of cell walls and degradation of tissues in maize stems. J. Sci. Food Agric. 59:45-51.

Fahey, Jr., G. C., L. D. Bourquin, E. C. Titgemeyer, and D. G. Atwell. 1993. Postharvest treatment of fibrous feedstuffs to improve their nutritive value. p. 715-766. *In* H. G. Jung, D. R. Buxton, R. D. Hatfield, and J. Ralph (ed.) Forage cell wall structure and digestibility. ASA-CSSA-SSSA, Madison, WI.

Faichney, G. J. 1968. Volatile fatty acids in the caecum of the sheep. Aust. J. Biol. Sci. 21:177-180.

Flores, D. A., L. E. Phillip, D. M. Veira, and M. Ivan. 1986. Digestion in the rumen and amino acid supply to the duodenum of sheep fed ensiled and fresh alfalfa. Can. J. Anim. Sci. 66:1019-1027.

Gaillard, B. D. E. 1962. The relationship between the cell-wall constituents of roughages and the digestibility of the organic matter. J. Agric. Sci. (Camb.) 59:369-373.

Galyean, M. L., and F. N. Owens. 1991. Effects of diet composition and level of feed intake on site and extent of digestion in ruminants. p. 483-514. *In* T. Tsuda, Y. Sasaki, and R. Kawashima (ed.) Physiological aspects of digestion and metabolism in ruminants. Academic Press, Inc., San Diego, CA.

Glenn, B. P., G. A. Varga, G. B. Huntington, and D. R. Waldo. 1989. Duodenal nutrient flow and digestibility in Holstein steers fed formaldehyde- and formic acid-treated alfalfa or orchardgrass silage at two intakes. J. Anim. Sci. 67:513-528.

Goering, H. K., C. H. Gordon, R. W. Hemken, D. R. Waldo, P. J. Van Soest, and L. W. Smith. 1972. Analytical estimates of nitrogen digestibility in heat damaged forages. J. Dairy Sci. 55:1275-1280.

Gordon, A. H., J. A. Lomax, and A. Chesson. 1983. Glycosidic linkages of legume, grass and cereal straw cell walls before and after extensive degradation by rumen microorganisms. J. Sci. Food Agric. 34:1341-1350.

Gordon, A. H., J. A. Lomax, K. Dalgarno, and A. Chesson. 1985. Preparation and composition of mesophyll, epidermis and fibre cell walls from leaves of perennial ryegrass (*Lolium perenne*) and Italian ryegrass (*Lolium multiflorum*). J. Sci. Food Agric. 36:509-519.

Grovum, W. L., and J. F. Hecker. 1973. Rate of passage of digesta in sheep. 2. The effect of level of food intake on digesta retention times and on water and electrolyte absorption in the large intestine. Br. J. Nutr. 30:221-230.

Grovum, W. L., and V. J. Williams. 1977. Rate of passage of digesta in sheep. 6. The effect of level of food intake on mathematical predictions of the kinetics of digesta in the reticulorumen and intestines. Br. J. Nutr. 38:425-436.

Harris, P. J., R. D. Hartley, and K. H. Lowry. 1980. Phenolic constituents of mesophyll and non-mesophyll cell walls from leaf laminae of *Lolium perenne*. J. Sci. Food Agric. 31:959-962.

Hatfield, R. D. 1992. Carbohydrate composition of alfalfa cell walls isolated from stem sections differing in maturity. J. Agric. Food Chem. 40:424-430.

Hendricksen, R. E., D. P. Poppi, and D. J. Minson. 1981. The voluntary intake, digestibility and retention time by cattle and sheep of stem and leaf fractions of a tropical legume (*Lablab purpureus*). Aust. J. Agric. Res. 32:389-398.

Hespell, R. B. 1988. Microbial digestion of hemicelluloses in the rumen. Microbiol. Sci. 5:362-365.

Hogan, J. P., P. M. Kennedy, C. S. McSweeney, and A. C. Schlink. 1989. Quantitative studies of the digestion of tropical and temperate forages by sheep. Aust. J. Exp. Agric. 29:333-337.

Hogan, J. P., R. H. Weston, and J. R. Lindsay. 1969. The digestion of pasture plants by sheep. IV. The digestion of *Phalaris tuberosa* at different stages of maturity. Aust. J. Agric. Res. 20:925-940.

Hoover, W. H. 1978. Digestion and absorption in the hindgut of ruminants. J. Anim. Sci. 46:1789-1799.

Hoover, W. H. 1986. Chemical factors involved in ruminal fiber digestion. J. Dairy Sci. 69:2755-2766.

Hume, I. D., and D. B. Purser. 1974. Ruminal and post-ruminal protein digestion in sheep fed on subterranean clover harvested at four stages of maturity. Aust. J. Agric. Res. 26:199-208.

Hunt, C. W., J. A. Paterson, and J. E. Williams. 1985. Intake and digestibility of alfalfa-tall fescue combination diets fed to lambs. J. Anim. Sci. 60:301-306.

Joanning, S. W., D. E. Johnson, and B. P. Barry. 1981. Nutrient digestibility depressions in corn silage-corn grain mixtures fed to steers. J. Anim. Sci. 53:1095-1103.

Jones, A. L., A. L. Goetsch, S. R. Stokes, and M. Colberg. 1988. Intake and digestion in cattle fed warm- or cool-season grass hay with or without supplemental grain. J. Anim. Sci. 66:194-203.

Jung, H. G., and K. P. Vogel. 1986. Influence of lignin on digestibility of forage cell wall material. J. Anim. Sci. 62:1703-1712.

Jung, H.-J. G., and K. P. Vogel. 1992. Lignification of switchgrass (*Panicum virgatum*) and big bluestem (*Andropogon gerardii*) plant parts during maturation and its effect on fibre degradability. J. Sci. Food Agric. 59:169-176.

Kawas, J. R., N. A. Jorgensen, and C. D. Lu. 1990. Influence of alfalfa maturity on feed intake and site of nutrient digestion in sheep. J. Anim. Sci. 68:4376-4386.

Kennedy, P. M., R. J. Christopherson, and L. P. Milligan. 1982. Effects of cold exposure on feed protein degradation, microbial protein synthesis and transfer of plasma urea to the rumen of sheep. Br. J. Nutr. 47:521-535.

Kennedy, P. M., R. J. Early, R. J. Christopherson, and L. P. Milligan. 1986. Nitrogen transformations and duodenal amino acid content in sheep given four forage diets and exposed to warm and cold ambient temperatures. Can. J. Anim. Sci. 66:951-957.

Kern, D. L., L. L. Slyter, E. C. Leffel, J. M. Weaver, and R. R. Oltjen. 1974. Ponies vs steers: Microbial and chemical characteristics of intestinal ingesta. J. Anim. Sci. 38:559-564.

Lagasse, M. P., A. L. Goetsch, K. M. Landis, and L. A. Forster, Jr. 1990. Effects of supplemental alfalfa hay on feed intake and digestion by Holstein steers consuming high-quality bermudagrass or orchardgrass hay. J. Anim. Sci. 68:2839-2847.

Lam, T. B.-T., K. Iiyama, and B. A. Stone. 1990. Primary and secondary walls of grasses and other forage plants: Taxonomic and structural considerations. p. 43-69. In D. E. Akin, L. G. Ljungdahl, J. R. Wilson, and P. J. Harris (ed.) Microbial and plant opportunities to improve lignocellulose utilization by ruminants. Elsevier Science Publishing Co., Inc., New York.

Langer, P., and R. L. Snipes. 1991. Adaptations of gut structure to function in herbivores. p. 349-384. In T. Tsuda, Y. Sasaki, and R. Kawashima (ed.) Physiological aspects of digestion and metabolism in ruminants. Academic Press, Inc. San Diego, CA.

Laredo, M. A., and D. J. Minson. 1973. The voluntary intake, digestibility, and retention time by sheep of leaf and stem fractions of five grasses. Aust. J. Agric. Res. 24:875-888.

Leng, R. A. 1973. Salient features of the digestion of pastures by ruminants and other herbivores. p. 82-129. In G. W. Butler and R. W. Bailey (ed.) Chemistry and biochemistry of herbage. Vol. 3. Academic Press, Inc. London.

Leng, R. A., and J. V. Nolan. 1984. Nitrogen metabolism in the rumen. J. Dairy Sci. 67:1072-1089.

Mackie, R. I. 1987. Microbial digestion of forages in herbivores. p. 233-265. In J. B. Hacker and J. H. Ternouth (ed.) The nutrition of herbivores. Academic Press, Inc. Orlando, FL.

Mandell, I. B., D. N. Mowat, W. K. Bilanski, and S. N. Rai. 1989. Effect of heat treatment of alfalfa prior to ensiling on nitrogen solubility and in vitro ammonia production. J. Dairy Sci. 72:2046-2054.

Mann, S. D., and E. R. Ørskov. 1973. The effect of rumen and post-rumen feeding on the caecal microflora of sheep. J. Appl. Bact. 36:475-484.

Mathers, J. C., and E. L. Miller. 1981. Quantitative studies of food protein degradation and the energetic efficiency of microbial protein synthesis in the rumen of sheep given chopped lucerne and rolled barley. Br. J. Nutr. 45:587-604.

McDonald, P., and R. A. Edwards. 1976. The influence of conservation methods on digestion and utilization of forages by ruminants. Proc. Nutr. Soc. 35:201-211.

McDonald, P., and R. Whittenbury. 1973. The ensilage process. p. 33-60. *In* G. W. Butler and R. W. Bailey (ed.) Chemistry and biochemistry of herbage. Vol. 3. Academic Press, Inc. London.

Merchen, N. R. 1988. Digestion, absorption and excretion in ruminants. p. 172-201. *In* D. C. Church (ed.) The ruminant animal: Digestive physiology and nutrition. Prentice-Hall, Englewood Cliffs, NJ.

Merchen, N. R. 1990. Effects of heat damage on protein digestion by ruminants: Alternative interpretations. Proc. Distillers Feed Res. Conf. 45:57-65.

Merchen, N. R., and L. D. Satter. 1983a. Digestion of nitrogen by lambs fed alfalfa conserved as baled hay or as low moisture silage. J. Anim. Sci. 56:943-951.

Merchen, N. R., and L. D. Satter. 1983b. Changes in nitrogenous compounds and sites of digestion of alfalfa harvested at different moisture contents. J. Dairy Sci. 66:789-801.

Merchen, N. R., and E. C. Titgemeyer. 1992. Manipulation of amino acid supply to the growing ruminant. J. Anim. Sci. 70:3238-3247.

Minson, D. J. 1990. Forage in ruminant nutrition. Academic Press, Inc., San Diego, CA.

Minson, D. J, and M. N. McLeod. 1970. The digestibility of temperate and tropical grasses. p. 719-722. *In* M. J. T. Norman (ed.) Proc. XI International Grassland Congress, Surfers Paradise, Queensland, Australia. University of Queensland Press, St. Lucia.

Minson, D. J., and R. Milford. 1967. The voluntary intake and digestibility of diets containing different proportions of legume and mature Pangola grass (*Digitaria decumbens*). Aust. J. Exp. Anim. Husb. 7:546-551.

Minson, D. J., and J. R. Wilson. 1980. Comparative digestibility of tropical and temperate forage - A contrast between grasses and legumes. J. Aust. Inst. Agric. Sci. 46:247-249.

Moe, P. W. 1981. Energy metabolism of dairy cattle. J. Dairy Sci. 64:1120-1139.

Morris, E. J., and J. S. D. Bacon. 1977. The fate of acetyl groups and sugar components during the digestion of grass cell walls in sheep. J. Agric. Sci. (Camb.) 89:327-340.

Morrison, I. M. 1979. Carbohydrate chemistry and rumen digestion. Proc. Nutr. Soc. 38:269-274.

Morrison, I. M. 1980. Hemicellulosic contamination of acid detergent residues and their replacement by cellulose residues in cell wall analysis. J. Sci. Food Agric. 31:639-645.

Moseley, G., and J. R. Jones. 1984. The physical digestion of perennial ryegrass (*Lolium perenne*) and white clover (*Trifolium repens*) in the foregut of sheep. Br. J. Nutr. 52:381-390.

Mould, F. L., and E. R. Ørskov. 1983/84. Manipulation of rumen fluid pH and its influence on cellulolysis *in sacco*, dry matter degradation and the rumen microflora of sheep offered either hay or concentrate. Anim. Feed Sci. Technol. 10:1-14.

Mould, F. L., E. R. Ørskov, and S. O. Mann. 1983/84. Associative effects of mixed feeds. I. Effects of type and level of supplementation and the influence of the rumen fluid pH on cellulolysis *in vivo* and dry matter digestion of various roughages. Anim. Feed Sci. Technol. 10:15-30.

Mowat, D. N., R. S. Fulkerson, W. E. Tossell, and J. E. Winch. 1965. The *in vitro* digestibility and protein content of leaf and stem portions of forages. Can. J. Plant Sci. 45:321-331.

Narasimhalu, P., E. Teller, M. Vanbelle, M. Foulon, and F. Dasnoy. 1989. Apparent digestibility of nitrogen in rumen and whole tract of Friesian cattle fed direct-cut and wilted grass silages. J. Dairy Sci. 72:2055-2061.

NRC. 1985. Ruminant nitrogen usage. National Academy Press, Washington, DC.

Ohshima, M., and P. McDonald. 1978. A review of the changes in nitrogenous compounds of herbage during ensilage. J. Sci. Food Agric. 29:497-505.

Ørskov, E. R., and R. N. B. Kay. 1987. Non-microbial digestion of forages by herbivores. p. 267-280. *In* J. B. Hacker and J. H. Ternouth (ed.) The nutrition of herbivores. Academic Press, Inc. Orlando, FL.

Pierson, D. C., R. D. Goodrich, J. C. Meiske, and J. G. Linn. 1971. Influence of heat damage on haylage quality. J. Anim. Sci. 33:296 (Abstr.).

Poppi, D. P., D. J. Minson, and J. H. Ternouth. 1980. Studies of cattle and sheep eating leaf and stem fractions of grasses. I. The voluntary intake, digestibility and retention time in the reticulo-rumen. Aust. J. Agric. Res. 32:99-108.

Pritchard, G. I., L. P. Folkins, and W. J. Pigden. 1963. The *in vitro* digestibility of whole grasses and their parts at progressive stages of maturity. Can. J. Plant Sci. 43:79-87.

Ramanzin, M., E. R. Ørskov, and A. K. Tuah. 1986. Rumen degradation of straw. 2. Botanical fractions of straw from two barley cultivars. Anim. Prod. 43:271-278.

Reid, J. T., O. D. White, R. Anrique, and A. Fortin. 1980. Nutritional energetics of livestock: Some present boundaries of knowledge and future research needs. J. Anim. Sci. 51:1393-1415.

Reid, R. L., G. A. Jung, and W. V. Thayne. 1988. Relationships between nutritive quality and fiber components of cool season and warm season forages: A retrospective study. J. Anim. Sci. 66:1275-1291.

Richards, G. N. 1976. Search for factors other than 'lignin-shielding' in protection of cell-wall polysaccharides from digestion in the rumen. p. 129-135. *In* H. Veenman and B. V. Zonen (ed.) Carbohydrate research in plants and animals. Misc. Papers 12. Landbouwhogeschool Wageningen, The Netherlands.

Robertson, J. B., and P. J. Van Soest. 1975. A note on digestibility in sheep as influenced by level of intake. Anim. Prod. 21:89-92.

Rode, L. M., D. C. Weakley, and L. D. Satter. 1985. Effect of forage amount and particle size in diets of lactating dairy cows on site of digestion and microbial protein synthesis. Can. J. Anim. Sci. 65:101-111.

Rooke, J. A., H. A. Greife, and D. G. Armstrong. 1983. The digestion by cattle of grass silages made with no additive or with the application of formic acid or formic acid and formaldehyde. Grass Forage Sci. 38:301-310.

Rooke, J. A., N. H. Lee, and D. G. Armstrong. 1987. The effects of intraruminal infusions of urea, casein, glucose syrup and a mixture of casein and glucose syrup on nitrogen digestion in the rumen of cattle receiving grass-silage diets. Br. J. Nutr. 57:89-98.

Sharpe, M. E., M. J. Latham, and B. Reiter. 1975. The immune response of the host animal to bacteria in the rumen and caecum. p. 193-204. *In* I. W. McDonald and A. C. I. Warner (ed.) Digestion and metabolism in the ruminant. University of New England Publishing Unit, Armidale, Australia.

Siddons, R. C., R. T. Evans, and D. E. Beever. 1979. The effect of formaldehyde treatment before ensiling on the digestion of wilted grass silage by sheep. Br. J. Nutr. 42:535-545.

Smith, L. W., H. K. Goering, and C. H. Gordon. 1972. Relationships of forage compositions with rates of cell wall digestion and indigestibility of cell walls. J. Dairy Sci. 55:1140-1147.

Soofi, R., G. C. Fahey, Jr., L. L. Berger, and F. C. Hinds. 1982. Digestibilities and nutrient intakes by sheep fed mixtures of soybean stover and alfalfa. J. Anim. Sci. 54:841-848.

Staples, C. R., R. L. Fernando, G. C. Fahey, Jr., L. L. Berger, and E. H. Jaster. 1984. Effects of intake of a mixed diet by dairy steers on digestion events. J. Dairy Sci. 67:995-1006.

Stevens, C. E., R. A. Argenzio, and E. T. Clemens. 1980. Microbial digestion: Rumen versus large intestine. p. 685-706. *In* Y. Ruckebusch and P. Thivend (ed.) Digestive physiology and metabolism in ruminants. AVI Publishing Co., Westport, CT.

Sullivan, J. T. 1973. Drying and storing herbage as hay. p. 1-31. *In* G. W. Butler and R. W. Bailey (ed.) Chemistry and biochemistry of herbage. Vol. 3. Academic Press, Inc. London.

Teller, E., M. Vanbelle, M. Foulon, G. Collignon, and B. Matatu. 1992. Nitrogen metabolism in rumen and whole digestive tract of lactating dairy cows fed grass silage. J. Dairy Sci. 75:1296-1304.

Thomas, J. W., Y. Yu, T. Middleton, and C. Stallings. 1982. Estimations of protein damage. p. 81-98. *In* F. N. Owens (ed.) Protein requirements of cattle: Symposium. Oklahoma State University Press, Stillwater.

Thomson, D. J., and D. E. Beever. 1980. The effect of conservation and processing on the digestion of forages by ruminants. p. 291-308. *In* Y. Ruckebusch and P. Thivend (ed.) Digestive physiology and metabolism in ruminants. AVI Publishing Co., Westport, CT.

Thomson, D. J., D. E. Beever, J. F. Coelho da Silva, and D. G. Armstrong. 1972. The effect in sheep of physical form on the sites of digestion of a dried lucerne diet. 1. Sites of organic matter, energy and carbohydrate digestion. Br. J. Nutr. 28:31-41.

Thornton, R. F., and D. J. Minson. 1973. The relationship between apparent retention time in the rumen, voluntary intake, and apparent digestibility of legume and grass diets in sheep. Aust. J. Agric. Res. 24:889-898.

Titgemeyer, E. C., L. D. Bourquin, and G. C. Fahey, Jr. 1992. Disappearance of cell wall monomeric components from fractions chemically isolated from alfalfa leaves and stems following *in situ* ruminal digestion. J. Sci. Food Agric. 58:451-463.

Titgemeyer, E. C., M. G. Cameron, L. D. Bourquin, and G. C. Fahey, Jr. 1991. Digestion of cell wall components by dairy heifers fed diets based on alfalfa and chemically treated oat hulls. J. Dairy Sci. 74:1026-1037.

Troelsen, J. E., and J. B. Campbell. 1969. The effect of maturity and leafiness on the intake and digestibility of alfalfas and grasses fed to sheep. J. Agric. Sci. (Camb.) 73:145-154.

Tyrrell, H. F., and P. W. Moe. 1975. Effect of intake on digestive efficiency. J. Dairy Sci. 58:1151-1163.

Ulyatt, M. J. 1973. The feeding value of herbage. p. 131-178. *In* G. W. Butler and R. W. Bailey (ed.) Chemistry and biochemistry of herbage. Vol. 3. Academic Press, London.

Ulyatt, M. J., D. W. Dellow, C. S. W. Reid, and T. Bauchop. 1975. Structure and function of the large intestine of ruminants. p. 119-133. *In* I. W. McDonald and A. C. I. Warner (ed.) Digestion and metabolism in the ruminant. University of New England Publishing Unit, Armidale, Australia.

Ulyatt, M. J., and J. C. MacRae. 1974. Quantitative digestion of fresh herbage by sheep. 1. The sites of digestion of organic matter, energy, readily fermentable carbohydrate, structural carbohydrate, and lipid. J. Agric. Sci. (Camb.) 82:295-307.

Ulyatt, M. J., D. J. Thomson, D. E. Beever, R. T. Evans, and M. J. Haines. 1988. The digestion of perennial ryegrass (*Lolium perenne* cv. Melle) and white clover (*Trifolium repens* cv. Blanca) by grazing cattle. Br. J. Nutr. 60:137-149.

van Nevel, C. J., and D. I. Demeyer. 1977. Determination of rumen microbial growth in vitro from ^{32}P-labelled phosphate incorporation. Br. J. Nutr. 38:101-114.

Van Soest, P. J. 1965. Use of detergents in analysis of fibrous feeds. III. Study of effects of heating and drying on yield of fiber and lignin in forages. J. Assoc. Off. Agric. Chem. 48:785-790.

Van Soest, P. J. 1982. Nutritional ecology of the ruminant. O & B Books, Inc., Corvallis, OR.

Van Soest, P. J., C. J. Sniffen, D. R. Mertens, D. G. Fox, P. H. Robinson, and U. Krishnamoorthy. 1982. A net protein system for cattle: The rumen submodel for nitrogen. p. 265-279. In F. N. Owens (ed.) Protein requirements for cattle: Symposium. Oklahoma State University Press, Stillwater.

Ventura, M., J. E. Moore, O. C. Ruelke, and D. E. Franke. 1975. Effect of maturity and protein supplementation on voluntary intake and nutrient digestibility of Pangola digitgrass hays. J. Anim. Sci. 40:769-774.

Weimer, P. J. 1992. Cellulose degradation by ruminal microorganisms. Crit. Rev. Biotechnol. 12:189-223.

Weston, R. H. 1967. Factors limiting the intake of feed by sheep. II. Studies with wheaten hay. Aust. J. Agric. Res. 18:983-1002.

White, B. A., R. I. Mackie, and K. C. Doerner. 1993. Enzymatic hydrolysis of forage cell walls. p. 455-484. In H. G. Jung, D. R. Buxton, R. D. Hatfield, and J. Ralph (ed.) Forage cell wall structure and digestibility. ASA-CSSA-SSSA, Madison, WI.

Wilson, J. R. 1990. Influence of plant anatomy on digestion and fibre breakdown. p. 99-117. In D. E. Akin, L. G. Ljungdahl, J. R. Wilson, and P. J. Harris (ed.) Microbial and plant opportunities to improve lignocellulose utilization by ruminants. Elsevier Science Publishing Co., Inc., New York.

Wilson, J. R., R. H. Brown, and W. R. Windham. 1983. Influence of leaf anatomy on the dry matter digestibility of C_3, C_4, and C_3/C_4 intermediate types of *Panicum* species. Crop Sci. 23:141-146.

Wilson, J. R., B. Deinum, and F. M. Engels. 1991. Temperature effects on anatomy and digestibility of leaf and stem of tropical and temperate forage species. Neth. J. Agric. Sci. 39:31-48.

Yu, Y., G. K. Macleod, J. B. Stone, and D. G. Grieve. 1977. Influence of heating on nutritive value of alfalfa-bromegrass for sheep. J. Dairy Sci. 60:1436-1439.

CHAPTER 15

MEASUREMENT OF IN VIVO FORAGE DIGESTION BY RUMINANTS

R. C. Cochran and M. L. Galyean

INTRODUCTION

Digestibility can be defined simply as the fraction of a feedstuff or dietary constituent that is lost on passage through the digestive tract. In its most basic form, digestibility is determined by measuring the quantity of feed consumed and the quantity of feces voided after an animal has had sufficient time to become accustomed to a diet. Conventional measurements of digestibility have contributed significantly to the development of systems intended to describe the nutritive value of feedstuffs (Van Soest, 1982). Currently, the net energy system represents the most useful approach to describing nutrient requirements and the nutritive value of feedstuffs (National Research Council, 1981b, 1984; Agricultural Research Council, 1990). According to the classical energy partitioning scheme that provides the basis for the net energy system, the single greatest energy loss during feedstuff utilization by ruminants consuming forage-based diets is that voided by the animal as feces (Garrett and Johnson, 1983). Although measurement of dry matter (DM) or organic matter (OM) digestibility is not strictly equivalent to accounting for fecal energy loss (i.e., determination of digestible energy), these items are closely related (Moir, 1961). Indeed, the conventional digestion trial is an integral component of the determination of digestible energy. Therefore, because of the magnitude of fecal energy loss and its importance in defining nutritive value, the measurement of digestibility in vivo is a fundamental technique that should be understood thoroughly by those studying the nutrition of ruminants consuming forage-based diets.

Previous research that has examined sources of error in digestion experiments (Schneider and Lucas, 1950; Donefer, 1966; Barnes, 1968) noted considerable variation among laboratories in digestion coefficients determined on the same hay. In some cases, variation among laboratories was three to six times greater than within-laboratory variation (Donefer, 1966). Although several factors likely contributed to the variation in these studies, procedural variability in the conduct of the digestion trials was undoubtedly an important factor. Such observations highlight the importance of using accepted, standardized methodology as much as possible in the measurement of digestibility. The aim of this chapter

---

R. C. Cochran, Dep. of Animal Sciences and Industry, Kansas State Univ., Manhattan, KS 66506; M. L. Galyean, New Mexico State Univ., Clayton Livestock Research Center, Clayton, NM 88415.

will be to review information regarding procedures used to conduct digestion trials and to describe a general experimental protocol that could be used as a guide for determining the digestibility of forages by ruminants.

PREPARATION FOR CONDUCTING A DIGESTION TRIAL

Housing, Restraint, and Collection Materials

Housing

When planning a digestion experiment, initial consideration should be given to the facilities and collection materials. Very few authors have undertaken the complete description of facilities dedicated to conducting digestion experiments (Balch et al., 1962; Schneider and Flatt, 1975). From the limited descriptions available, and from knowledge of factors that can influence intake and(or) digestion, one can generally define optimal housing conditions. Environmental factors (e.g., temperature, photoperiod, muddy pens) are known to influence intake and, in many cases, digestibility (National Research Council, 1981a). As such, the most suitable housing for conducting digestion trials would consist of a fully enclosed building in which environmental variables can be controlled. Temperature is possibly the most critical environmental component to control. Intake and passage rate increase and digestibility decreases as temperature falls below the thermoneutral zone (National Research Council, 1981a; Kennedy et al., 1986). Therefore, temperature in an environmentally controlled building should be held relatively constant during a digestion experiment and should fall within the thermoneutral zone of the species under evaluation (approximately 0 to 16 C for a mature cow and -3 to 20 C for a mature ewe; National Research Council, 1981a). Adequate air flow should be considered, particularly as it impacts both temperature and humidity within the facility.

Because photoperiod can affect intake (Forbes, 1986), a consistent lighting pattern should be maintained throughout the course of a digestion experiment. Intake is generally greater with longer photoperiod (Forbes, 1982). Furthermore, Peters et al. (1980) noted that heifers consumed more feed with a 16-h light:8-h dark pattern than they did with continuous lighting. A lighting pattern that involves both light and dark cycles with proportionately more light than dark during each 24-h cycle is most desirable. Except for studies designed expressly to evaluate digestion under ad libitum feeding conditions, maintaining adequate intake and preventing feed refusal may be items of concern when conducting digestion trials; thus, conditions that are conducive to maximal intake may be advantageous. One possible exception to this recommendation is when the experimental protocol calls for sampling during the night (e.g., "spot" or "grab" sampling) because short-term exposure to light during the night can affect intake patterns (Schanbacher and Crouse, 1981). Depending on the experimental objectives, one may want to leave animals under continuous lighting to minimize disturbance during night sampling periods.

Many researchers are faced with the prospect of conducting digestion experiments under less-than-optimal conditions (e.g., grazing livestock, enclosed buildings without full environmental control, etc.). Under such conditions, the researcher should be cognizant of the effect that environment can have on intake and digestion and employ experimental designs that minimize potential confounding with environment. In particular, latin square or crossover experiments that assume no treatment x experimental period interaction should be avoided. In addition, when digestion experiments are conducted without environmental control, some environmental information should be presented in associated reports (e.g., mean temperature and its variability, season, light/dark cycles, etc.).

Animal Restraint

After housing requirements are satisfied, the method used to restrain animals during experimental periods should be considered. Most digestion trials are conducted with the individual animal as the experimental unit and, therefore, require quantitative collection of each animal's excreta and measurement of daily intake. As a result, approaches to animal "restraint" may include individual stalls in which animals are permitted some freedom of movement, tie stalls, or "metabolism" stalls. The method of choice will likely depend on the manner in which excreta is collected and perceived concerns regarding the impact of method of restraint on intake. A concern expressed regarding the use of metabolism stalls, the most restrictive approach to restraint, is that they tend to decrease feed intake and may be uncomfortable for the animal (Schneider and Flatt, 1975). Exercise, particularly between collection periods, will help minimize behavioral problems and the development of sore feet and legs in animals housed in metabolism stalls. Similarly, selection of animals with a docile temperament and exposure of animals to the metabolism stalls before the experiment will help minimize intake problems (Schneider and Flatt, 1975). Some workers have used special floor material (such as heavy rubber strips) in metabolism stalls in an attempt to improve animal comfort (Horn et al., 1954). The obvious advantage of using a metabolism stall is that it provides a straightforward, efficacious method for collecting excreta and monitoring intake.

Early researchers frequently used tie stalls and hired laborers to manually collect feces during digestion trials (Balch et al., 1951). Because of rising labor costs and the tedious nature of this responsibility, metabolism stalls have been designed that obviated the need for manual collection of excreta. Numerous designs have been proposed for metabolism stalls (Briggs and Gallup, 1949; Hobbs et al., 1950; Bratzler, 1951; Horn et al., 1954). The most desirable designs allow easy adjustment of length and width to restrict movement and appropriately position animals to ensure quantitative, uncontaminated collection of excreta (Nelson et al., 1954). The stalls described by Nelson et al. (1954) were constructed of materials that clean easily and were modified to permit movement of the stall when not in use (Erwin et al., 1956). With loose stalls or tie stalls, separate collection devices are required to permit collection of feces and urine.

Although urine collections are not required to conduct a digestion trial per se, some experimental objectives necessitate urine collection (e.g., measurement of N balance, determination of metabolizable energy).

Collection Materials

Fecal collection bags were developed originally when large tubes of coated material, intended to direct feces into a pan behind a metabolism stall, were completely sealed at the bottom. Subsequently, fecal bags were found to be well suited to total fecal collection under grazing conditions (Cook et al., 1952; Border et al., 1963; Schneider and Flatt, 1975). Special designs are required when fecal collection bags are used with females to avoid contamination of the feces by urine (Hobbs et al., 1950; Van Es and Vogt, 1959). This may involve simple deflection of the urine (Ballinger and Dunlop, 1946; Owen and Ingleton, 1961; Kartchner and Rittenhouse, 1979) or more elaborate collection of urine via receptacles designed for use in conjunction with a fecal collection bag (Balch et al., 1951; Gorski et al., 1957; Stillwell et al., 1983). Use of urinary catheters also is an effective means of either diverting or collecting urine from females. Urine collection devices also have been designed for use with males (Hobbs et al., 1950; Cook et al., 1952; Bredon et al., 1961b; Border et al., 1963). As noted for fecal collection bags, urine collection devices can be used effectively under grazing conditions (Cook et al., 1952; Border et al., 1963). If urine collections are to be conducted using males in tie stalls, application of suction will likely be necessary unless the urinal is constructed to hold significant quantities of urine (Border et al., 1963). One advantage of metabolism stalls is that they are usually elevated, which allows urine collection via gravity flow.

Animal Considerations

Animal Selection and Training

A critical element of ensuring success when conducting digestion experiments is to select animals with a disposition that is amenable to such work. In addition, sufficient training of selected animals should occur before an experiment begins in order to minimize animal disturbance during the experiment. A selection/training procedure that has worked well at several research institutions begins with selecting young animals based on perceived disposition while they are undergoing routine handling (at weaning for example). At initial sorting, it is advisable to retain 25 to 30% more animals than will be used in the experiment. The entire group of animals then undergoes an initial training phase that consists of being "broke" to lead and being exposed to considerable handling by humans (e.g., grooming, hand-feeding, etc.). During this initial training phase, those animals that are excessively nervous or are not adapted well to handling are culled. If possible, one or two acceptable animals should be held in reserve as spares. The remaining animals (including spare animals) are then exposed to conditions similar to those of the digestion experiment. For example, animals may be placed in metabolism stalls or fitted with fecal collection bags for sufficient time to permit them to become comfortable with the equipment and(or) surroundings. Such training will likely require a few days to several weeks. The longer training periods may be necessary when range livestock are first brought into confinement and must learn activities such as drinking from mechanical watering cups.

A policy of selecting animals that are as similar as possible to the population in question is strongly encouraged when considering choice of species, age, sex, and physiological status of animals to be used in digestion experiments. In some situations, financial or logistical constraints may make it impossible to select animals like the population of interest. Under such conditions, it is important to remember that the experimental animals serve as a model, and attempts should be made to carefully define the limitations of applying information derived with the model animals to the population of interest. Regardless of the similarity between the experimental animals and the population of interest, it is imperative that the animals used in the digestion experiment be as similar to each other as possible. Ideally, experimental animals will have similar genetic background, physiological status, age, weight, sex, and previous nutritional history.

Number of Animals

A final item for consideration in animal selection is the number of animals needed to conduct a digestion trial. Previous estimates of the number of animals required for determining total tract digestion with sheep and cattle have usually fallen in the range of three to six per feed (Carberry and Chatterjee, 1936; Hodgson and Knott, 1937; Forbes et al., 1946; Bredon et al., 1961a; Heaney et al., 1965; Hattan and Owen, 1970). Whereas Schneider and Flatt (1975) recommended using as many animals as possible, they implied that four, five, or six animals per treatment were adequate, and that one should avoid using less than three per treatment. For digestion trials conducted under grazing conditions, measuring diet selection with esophageally or ruminally fistulated livestock is widely practiced. Holechek et al. (1982) and Holechek and Vavra (1983) noted that a minimum of four fistulated animals per treatment should be used to measure chemical composition of selected forage. Although the estimates noted above can be helpful as general guidelines, the best approach to determine the number of replicates is to base such decisions on: 1) variability of the trait under evaluation (assuming such data are available); 2) size of the treatment difference that one desires to detect; and 3) how willing one is to accept (i.e., the probability) that a significant treatment effect may go undetected. Although most statistics textbooks provide the simple formula needed to determine number of replicates, convenient tabular listings also are available (Berndtson, 1991). For example, using the data of Galyean et al. (1976) as an example (mean = 80.8, SD = 1.71, CV = 2.1%), four animals were needed to detect a 5% difference from a control treatment in total tract OM digestibility with 80% certainty. Conversely, with four replicates, only a difference 30% from control could be detected for ruminal OM digestion (CV = 12.3). Data from previous experiments can provide an estimate of the coefficient of variation that is used in the tables provided by Berndtson (1991). Attention by investigators to the number of replicates needed to detect differences could increase the power of digestion experiments. Without such attention, an experiment may provide little, if any, useful information.

Schedules and Records

Although often considered as both mundane and trite, appropriate record-keeping and establishment of a routine schedule when conducting a digestion trial are critical to the success of the endeavor. A data collection and feeding schedule should be established before beginning a digestion trial and all individuals involved with the experiment should be aware of and follow the schedule. One of the principal sources of error in the determination of digestion coefficients arises from irregular expulsion of feces ("end-period" error; Blaxter et al., 1956). Blaxter et al. (1956) suggested that irregular feeding intervals give rise to increased variability in the rate of feces production. Thus, irregularity in feeding patterns and irregularity in time of feces collection could exacerbate "end-period" errors. Harris (1970) and Schneider and Flatt (1975) provide a detailed schedule of daily activities that can be adopted for digestion trials. Aside from establishing a schedule, it is quite helpful to prepare a format for record-keeping before an experiment begins. This process helps ensure that all necessary data will be collected and that sufficiently detailed observations will be recorded. Given the low cost of personal computers and the robustness of the current generation of spreadsheets, use of these tools affords an effective means of tracking and eventually summarizing data collected during a digestion experiment.

METHODS FOR CONDUCTING A DIGESTION TRIAL

Period Structure and Length

After animals have undergone sufficient training to be acquainted with their general surroundings, feeding, watering, and collection devices, they are ready to begin the actual experiment. Digestion experiments are generally considered to consist of two phases, the preliminary period and the collection period. The purpose of the preliminary period is to establish the intake level, ensure that all the residues of the previous diet are voided from the gastrointestinal tract (Van Soest, 1982; Merchen, 1988), and to allow the ruminal microbial population time to adapt to the diet under evaluation. In many situations, particularly with significant changes in diet, more time will be required for the microbial population to adapt than to clear the digestive tract of residues from the previous diet (Burroughs et al., 1950; Lloyd et al., 1956; Kaufmann et al., 1980). Numerous reports exist regarding optimal length of preliminary and collection periods. Although preliminary periods as long as 24 to 30 d have been suggested in some cases (Swift, 1925; Nicholson et al., 1956), most recommendations fall in the 10 to 14 d range (Staples and Dinusson, 1951; Blaxter et al., 1956; Lloyd et al., 1956; Davis et al., 1958; Harris, 1970; Van Soest, 1982; Merchen, 1988).

Generally, use of preliminary periods of less than 10 d is not advised. Indeed, Kaufmann et al. (1980) suggested that at least 10 to 14 d are required for adaptation to a new diet.

The collection period is the period when feed, orts, and excreta are collected quantitatively. Generally, "end-period" errors have been observed to decrease significantly with increased length of collection period (Blaxter et al., 1956). In fact, Blaxter (1967) noted that the analytical and biological error in measuring fecal energy excretion was nearly cut in half by increasing the length of collection from 2 to 4 d; a collection period of 8 d decreased errors by approximately seven-fold from that of a 2-d collection period. Accordingly, Clanton (1961) noted that many early ruminant nutritionists suggested that collection periods should be 10 to 14 d in length. Subsequent research suggested that there is little advantage to extending collection periods beyond 6 to 7 d compared with a 10-d collection period (Staples and Dinusson, 1951; King et al., 1960; Clanton, 1961). Most recent texts that address aspects of conducting digestion trials suggest that collection periods should be between 5 and 10 d in length (Harris, 1970; Van Soest, 1982; Merchen, 1988). The mode for length of collection period is probably 7 d. Despite pressures on time and facilities, use of fecal collection periods less than 4 d in length would not be recommended, and longer periods are desirable. One notable exception to this situation is when markers are being continuously or frequently administered and spot samples of feces or digesta (e.g., fecal grab samples or samples of intestinal digesta) are collected at different times of day. In such cases, total number of samples collected is as important as total length of the collection period. Theurer et al. (1981) reported similar means and standard deviations for nutrient digestion coefficients when samples of abomasal digesta were collected at 12-h intervals for 6 d vs 4-h intervals for 2 d (total of 12 samples per method).

Feeding Practices and Measurement of Nutrients Consumed

As noted previously, one purpose of the preliminary period is to establish the intake level. Level of intake can significantly influence rate of passage (Warner, 1981) and, subsequently, digestibility (Maynard et al., 1979). Therefore, to avoid confounding digestion measurements with intake effects, level of intake should be the same across treatments when the aim of an experiment is to compare digestion coefficients per se. Given that ruminal digesta passage is a function of both ruminal mass and outflow rate (Grovum, 1983), and that ruminal capacity is correlated positively with body weight (Purser and Moir, 1966), attempts to equalize intake level should be done relative to animal weight. That is, if equal intakes are used, treatments should be fed such that all animals receive the same amount relative to body weight. Because there is potential for variability in the relationship between body weight or size and ruminal capacity, the importance of selecting animals with similar characteristics should not be overlooked.

To ensure that the level of intake established will actually be consumed and to try and make the data as physiologically valid as possible, nutritionists frequently feed animals at a level slightly less than ad libitum intake. This approach also helps avoid the need to quantify and analyze refused feed (i.e., orts). To determine this level of intake, Schneider and Flatt (1975) recommended establishing the ad libitum intake in a "transition" or "standardizing" period that precedes the preliminary period. This period would likely require at least 10 to 14 d. After ad libitum intake was established, the scientist could then determine that some percentage of ad libitum intake would be used *across all treatments*. Allowing each treatment to be fed as a percentage of its own ad libitum intake during a standardization period simply "locks in" the potentially confounding intake effect. Typically, average intake across all treatments should be based on the treatment with the least intake. If a standardization period is used, the length of the preliminary period could be shortened (an additional 6 to 7 d at restricted intake would likely be adequate). Alternatively, if one has a solid base of prior knowledge regarding potential intake of the feedstuffs to be evaluated, an intake level below ad libitum could be set arbitrarily and used throughout the preliminary and collection periods. Regardless of the approach used, once the general intake level was set, it could be expressed relative to average body weight of the experimental animals and fed accordingly. In long experiments (e.g., Latin square designs), particularly where there is potential for weight gain or loss, one may want to calculate intake levels based on initial body weight; one possible exception would be where animals are re-equilibrated to a standard diet after each collection period.

Frequency of feeding is an additional item that should be considered when conducting a digestion trial. Although ad libitum feeding allows consumption of many meals throughout the day, few animals are offered feed more than two or three times daily under production conditions. For this reason, and to save labor (assuming automated feeders are not available), many nutritionists offer feed either once or twice daily during a digestion trial. Although the effect of frequency of feeding on diet digestibility has been reported to be small in some instances (Satter and Baumgardt, 1962; Bunting et al., 1987), researchers using techniques which assume the existence of steady-state conditions (e.g., digesta flow estimations) often prefer use of frequent feeding intervals. Generally, frequent feeding is accomplished via an automated feeding system. Such effort is likely helpful, although it should be recognized that given the meal-feeding nature of ruminants, steady-state conditions may never fully exist (Owens and Hanson, 1992). Regardless of feeding interval chosen, it should be the same across treatments and throughout the experiment. In addition, Blaxter et al. (1956) suggested that when animals are fed more than once daily, feeding intervals should be evenly spaced.

In some situations it is not feasible to control intake (e.g., grazing animals). Additionally, digestibility is sometimes secondary to some primary characteristic under evaluation (e.g., supplementation trials where changes in intake may be of principal concern). Under these conditions one must carefully characterize the amount of nutrients in the *diet consumed* as opposed to the *diet fed* or *diet available*. For animals fed in confinement, this involves weighing and sampling refused feed. Orts can be sampled in a manner that collects either a fixed percentage of the refusal each day or a fixed amount each day. Collecting a percentage of the refused feed may vary from collecting all the orts (i.e., 100% if absolute amounts are small) to collecting a small percentage of the orts if the absolute amount of feed refused is large. When using the fixed percentage approach, collected material is weighed, dried (for preservation and DM determination), and composited with other ort samples from that animal. The entire composite would subsequently be processed and a subsample retained for chemical analysis. If the absolute amount of refused feed is quite large, another alternative is to representatively sample the orts, but retain a fixed amount of the material. This can be done by placing the material, after weighing, in a container that is large enough to allow mixing and sampling. A representative sample is then collected, dried, and ground. Subsamples from each day's orts could then be composited in the same proportion as the amount of feed refused for that day (relative to total feed refused during the collection period). Because of the considerable lag time in ruminants between offering a feed and its appearance in the feces, collection of feed and orts usually begins before excreta collection. A common practice with forage-fed ruminants is to stagger the sampling of feed offered and orts remaining by 2 d and 1 d, respectively, before the first fecal collection. Although some experimental objectives may call for measurement of digestibility under ad libitum feeding, it is still important to remember that level of intake must be well established (e.g., consistent) before initiating fecal collection.

Several approaches have been used to measure diet consumed under grazing conditions; however, the procedure generally considered to yield the most realistic values of diet chemical composition is sampling with fistulated animals (Van Dyne and Torell, 1964; Holechek et al., 1982). Holechek et al. (1982) suggested that a minimum of a 4-d collection period (one sample daily) using four esophageally fistulated animals per treatment was needed to estimate diet chemical composition with reasonable accuracy. Lesperance et al. (1960) and Olson (1991) reported that diet sampling with either ruminally fistulated or esophageally fistulated animals yielded acceptable estimates of diet selection. Presumably, similar numbers of animals and collection events are required for sampling with ruminally fistulated animals. One advantage of collecting diet samples with ruminally fistulated livestock is that the length of the collection period can be somewhat extended and total sample size can be considerably larger than when using esophageally fistulated livestock. When sampling with esophageally fistulated livestock, length of collection events must be carefully monitored to avoid excessive constriction of the fistula. Generally, collection events of 20 to 30 min are used with esophageally fistulated livestock. Regardless of method of sampling, care must be taken to avoid contamination of diet samples with ruminal contents. When sampling with ruminally fistulated animals, the rumen must be evacuated carefully before allowing animals to graze (Lesperance et al., 1960; Olson, 1991). Olson (1991) also wiped down the rumen wall with a wet sponge before sampling. Given the degree of particle attachment to the reticuloruminal wall after evacuation, it may be desirable to use a small amount of water to wash the reticulorumen and then to suction the rinse water out before releasing the animals to graze. When sampling with esophageally fistulated animals, care must be taken to prevent animals from ruminating during sampling events. Fasting the animals overnight before sampling is believed to aid in preventing rumination and subsequent contamination of diet samples; however, some researchers have avoided this practice because of concern that fasting might influence diet selection. In contrast, Langlands (1967) and Sidahmed et al. (1977) reported that moderate fasting (up to 22 h) exerted little effect on chemical composition of esophageal extrusa. Thus, it seems that short-term fasting may be a tool that could be applied safely when collecting diet samples with esophageally fistulated animals. Choice of method of sampling may be influenced by logistical and(or) maintenance considerations. Generally, animals with esophageal fistulas are considered to require more intensive care than ruminally fistulated livestock. For example, many types of esophageal cannulas require removal and rotation on a regular basis to ensure that the fistula remains in good condition. In contrast, collecting diet samples with many esophageally fistulated livestock can be successfully handled by one or two individuals. To collect diet samples with large numbers of ruminally fistulated livestock would require a significant increase in the labor force or the development of an expanded sampling scheme which would increase the time required to collect samples.

The final component that should be considered with regard to measuring nutrients consumed is the method of sampling harvested feedstuffs fed in confinement trials. Generally, two options exist for sampling such feeds. One approach is to collect daily samples as each feedstuff is weighed for feeding and to form a "running" composite of each feedstuff. A second approach is to weigh and bag individual meals before the digestion experiment begins and to retain a representative number of the samples for chemical analysis. The latter approach has been suggested as appropriate when a fixed level of intake is set in advance (Goering and Van Soest, 1970; Schneider and Flatt, 1975). In addition, such an approach likely requires that daily intake be small enough and storage space sufficient enough to permit storage before and during the digestion experiment. This approach could be problematic when large ruminants are used to evaluate digestibility of forages with low bulk density (i.e., large samples and storage space required) or fermented feeds (refrigeration/freezing required if weighed in advance). When forage digestibility is evaluated and the forage has been processed (for example, coarsely chopped) to minimize selectivity, a running composite can be easily formed by simply retaining material collected in the same fashion that it is weighed for feeding. However, if baled forages are fed in "long-stem" form, sampling should be done with a core sampling device (Schneider and Flatt, 1975; Jurgens, 1982). As a minimum, each bale used in the digestion trial should be sampled once, although multiple samples from each bale are desirable (Schneider and Flatt, 1975). When sampling baled hay, core samples should be collected from the butt-end of rectangular bales and from the outside to the center of round bales. Regardless of approach used to collect feed samples, it is critical to remember that such samples must be representative of the material that is fed.

Measuring Excreta Output

Total Fecal Collection

To determine total tract DM or OM "apparent" digestion, one need only measure fecal DM or OM output. Some experimental objectives necessitate measurement of urine output (e.g., metabolizable energy determinations, N balance) or intestinal DM or OM flow (to determine site of digestion). As noted above, urine collections are easily accomplished either in metabolism stalls or via collection devices strapped to the animal. Measurement of site of digestion involves use of intestinally fistulated animals and various flow and microbial markers. Readers interested in determination of site of digestion are referred to the discussions presented by Merchen (1988) and Galyean and Owens (1991). To measure fecal output, one can either physically collect and weigh the total fecal excretion, or output can be estimated via the use of inert markers. Alternatively, total tract digestion can be determined directly via internal marker ratios in situations where diet intake is not measured directly (e.g., with grazing animals; Galyean et al., 1987). The relative accuracy of total collection compared with marker procedures is discussed in a subsequent section (see section regarding sources of error and design considerations).

As noted previously, total fecal collection can be successfully accomplished with metabolism stalls or with fecal collection bags. When using metabolism stalls, feces may be weighed and sampled one or more times daily. When sampling only once daily, it is common to empty fecal pans from each animal more than once daily and place the contents in separate, sealable containers (e.g., garbage cans with plastic liners and lids). Weighing and sampling is then conducted at the end of each 24-h cycle. Juko et al. (1961) and Fuller and Cadenhead (1969) did not observe significant changes in chemical composition or heat of combustion of feces allowed to stand at room temperature for either 7 or 24 h. Thus, the practice of allowing fecal material to stand at room temperature for up to one day does not seem to compromise results of digestion experiments. When using fecal collection bags, the practice of emptying the bags once daily is also quite common, particularly with sheep or younger cattle. As noted by Balch et al. (1962), however, it may be desirable to empty fecal collection bags several times daily with large cattle; use of at least two bags per animal facilitates this procedure. One disadvantage of using fecal bags for total collection, particularly with large cattle, is that extended collection periods (> 7 d) may result in significant soreness under the tail and over the withers. Considerable caution should be used to ensure optimal adjustment of harnesses and fecal collection bags and to ensure that the devices fit the animal appropriately.

After weighing, feces must be subsampled for subsequent chemical analysis. Two approaches may be used for subsampling feces. First, a fixed percentage of each animal's daily fecal output may be saved and used to form a composite for that animal. The percentage retained will depend on the amount of feces excreted and the size of the sample needed for subsequent analyses. Amount of sample retained has varied from complete collection with some small ruminants like sheep, to collecting only 4 to 20% of daily fecal excretion (Schneider and Flatt, 1975) of cattle. Schneider and Flatt (1975) suggested that the most commonly used approach for proportional sampling is to collect 5% of daily excretion. The second approach to subsampling feces is to collect a fixed amount of feces from each animal daily. Following drying and grinding, each animal's ground subsamples can be composited in proportion to each day's fecal output. Hence, each day's fecal output is expressed as a percentage of total output during the collection period. Then, the ground subsample from each day is added to each animal's final composite in the same proportion as that day's fecal output relative to the entire collection period. Juko et al. (1961) collected daily fecal samples of 100, 200, 400, and 600 g by collecting numerous 10-g subsamples from the total fecal excretion on a given day after it had been mixed thoroughly. They observed no difference in composition of feces as a result of sample size and ultimately recommended that a 300-g sample was most appropriate based on convenience for weighing/handling and adequacy of the sample size for subsequent analyses.

Although total collection of feces can be tedious and, in some cases, labor-intensive, it should usually be the method of choice whenever possible (Galyean et al., 1987). However, in some experiments, total fecal collections are either not feasible or are logistically difficult. Under such conditions, the use of external or internal markers affords an opportunity to estimate digestibility without conducting total fecal collections. Furthermore, in grazing trials where intake is not known, internal markers are used in conjunction with either external markers or total fecal collection to determine digestibility and intake (Cordova et al., 1978).

External Markers

When using external markers to estimate fecal output, markers may be administered continuously, frequently, or in a single pulse-dose (Galyean et al., 1987; Pond et al., 1987; Owens and Hanson, 1992). When delivered in a single pulse-dose, fecal marker excretion over time can be described mathematically and used to predict fecal output (Pond et al., 1987). Galyean et al. (1987) reported that considerable variability has been observed in the accuracy of fecal output predictions based on pulse-dosing procedures. Furthermore, both Galyean et al. (1987) and Pond et al. (1987) noted that an additional limitation of pulse-dosing procedures was the extensive sampling required following dosing (multiple samples at different times of the day over several days). An alternative to the pulse-dosing procedure is to continuously or frequently (at least daily) administer an external marker and then collect fecal "grab" samples after marker concentration has reached equilibrium in the feces. Average marker concentration in the grab samples is divided into the dose of marker to estimate fecal output. Several days are required for marker concentrations to achieve equilibrium in the feces. Brandyberry et al. (1991) noted that with continuous infusion of either CoEDTA or $YbCl_3$, approximately 100 to 120 h were required to achieve relative equilibrium. This concurs with observations of Owens and Hanson (1992), who indicated that 5 to 7 d are required for a marker to reach a plateau. After the marker has reached equilibrium, "grab" samples of feces can be collected to determine fecal output. Brandyberry et al. (1991) did not observe differences in the amount of fecal output determined by total collection or continuous infusion (7-d grab sampling period following equilibration). Similarly, Siddons et al. (1985) reported accurate estimations of fecal output via continuous infusion of Yb acetate although CrEDTA underestimated output. A common problem encountered when frequent marker administration is used to estimate fecal output is diurnal variation in fecal marker concentration (Galyean et al., 1987; Owens and Hanson, 1992); this is particularly true when markers are administered only once daily (Raleigh et al., 1980; Prigge et al., 1981). Continuous or frequent administration of marker (> twice daily) has reduced diurnal variation (Raleigh et al., 1980; Galyean et al., 1987; Brandyberry et al., 1991). Owens and Hanson (1992) reported that diurnal variation may exist, at times, even with frequent administration of marker. However, they also suggested that regularity in the pattern of diurnal variation, such as that reported by Hopper et al. (1978), may permit prediction of an adjusted mean marker concentration via mathematical description of the marker curves.

Numerous reviews exist in the literature regarding the suitability of different external markers for use in monitoring digestibility or digesta kinetics (Ellis et al., 1982; Galyean et al., 1987; Pond et al., 1987; Owens and Hanson, 1992). In reviewing marker usage, many authors have described the characteristics of an ideal marker. Owens and Hanson (1992) summarized such characteristics as follows: "...an ideal marker 1) must not be absorbed, 2) must not affect or be affected by the digestive tract or its microbial population, 3) must flow parallel with or be physically similar to or intimately associated with the material it is to mark, and 4) must have a specific method of estimation." Implicit in their description of the first desirable characteristic is the assumption that a marker must be fully recoverable at sites distant from the site of administration (e.g., small intestine, feces). Therefore, regardless of marker chosen, either the percentage recovery must be known or marker recovery must be complete in order to generate accurate digestibility data. Obviously, the most desirable approach is to work with a marker that is recovered quantitatively in the feces. However, Lucas (1952) presented a simple formula that allows one to quickly assess the degree of error that will exist if one attempts to estimate digestion via a marker that is not fully recoverable. This approach would be helpful for a marker that was not quantitatively recoverable but where actual recovery was determined. However, it should be recognized that the commitment of time and animals to accurately determine fecal marker concentration and recovery will be the same as for physically determining total fecal output. Indeed, the preferred method of determining marker recovery would be via total fecal collection. Requirements regarding length of preliminary or collection periods and animal numbers are not lessened simply because one is interested in determining recovery of a marker. Thus, care should be taken to avoid the use of recovery estimates from one or two animals to adjust fecal recoveries of the animals in an experiment. Adjustments based on such an approach are as likely to bias digestion coefficients as they are to improve their accuracy. If one desires to estimate fecal output using markers for a large number of animals and has concern regarding the recovery and(or) release rate of a marker, a full-scale total-collection trial should be conducted for determination of marker recovery (with adequate number of animals). In such cases, the conditions of the digestion trial must match the conditions of the larger trial as much as possible in order to ensure applicability of marker recovery data to the population of interest.

Internal Markers

Internal markers are inherent dietary constituents that, because of their resistance to digestion, may be used to monitor digestibility of other constituents in the diet (Cochran et al., 1987). If intake is known, an internal marker can be used in the same manner as described previously for external markers. However, under conditions when diet intake is not known, an internal marker may be used to estimate digestibility directly, which can be used in subsequent estimations of intake. When an internal marker is used in this fashion, concentration of the marker and any other nutrient of interest (DM, OM, N, etc.) are determined in representative samples of diet consumed and feces excreted. By evaluating the changes in nutrient concentration in feed and feces relative to the internal marker, one can calculate diet digestibility (Schneider and Flatt, 1975). Several comprehensive reviews of internal markers used in digestion experiments are available (Streeter, 1969; Kotb and Luckey, 1972; Cochran et al., 1987). Although new internal markers have been proposed during the past 25 to 30 yr (Van Soest et al., 1966; Berger et al., 1979; Waller et al., 1980; Penning and Johnson, 1983a,b; Cochran et al., 1988), the dominant theme of most recent internal marker research is that no internal marker proposed to date demonstrates quantitative recovery across a wide variety of diets (Cochran et al., 1986; Galyean et al., 1987; Cochran et al., 1987; Judkins et al., 1990; Sunvold and Cochran, 1991). As a result, it seems imperative that researchers validate or define recovery of internal markers for the diet(s) with which they are working before application in the calculation of digestibility (Cochran et al., 1987; Judkins et al, 1990; Sunvold and Cochran, 1991).

The relative impact of total fecal collection versus intermittent grab sampling (used in marker-based techniques) on animal well-being and behavior is occasionally debated. Hatfield et al. (1993) compared animal response in wethers subjected to both approaches for determining fecal output. Based on the results from their experiments, it appears that when appropriate techniques of animal training and management are used, neither technique imposes stress levels that would be considered to be "detrimental to the animal's well being."

Processing Fecal Samples

Once fecal material has been collected (whether by total collection or grab sampling), it must be appropriately processed to permit long-term storage and subsequent analysis for different nutrient constituents. Probably the most common method of processing feces after sampling is drying in a forced-air oven followed by grinding in a sample mill. Nutrient losses upon drying have been evaluated by a number of researchers. Although documentation of carbon (Kleiber et al., 1936) and DM (Raymond and Harris, 1954) losses in feces during oven-drying are available, such losses appear to be small. For both feces and forages, DM losses were greater when drying occurred at lower temperatures (< 50 C; Raymond and Harris, 1954). Conversely, fresh forages are subject to artifact formation when dried at temperatures close to 100 C (Raymond and Harris, 1954; Danley and Vetter, 1971). Indeed, Van Soest (1982) noted that drying temperatures above 60 C have been shown to promote artifact production although he also suggested that feces is less subject to artifact production than forage. As such, drying temperatures close to 50 C are commonly employed in digestion experiments. For forages collected with esophageally or ruminally fistulated animals, freeze-drying has been suggested as preferable to oven-drying (Burritt et al., 1988); this is particularly true for material collected during the growing season (Burritt et al., 1988). Special precautions also should be taken in processing fermented feeds to avoid loss of volatile components. Although DM loss upon drying feces is not large, the amount of N loss during drying typically runs between 5 to 10% (Kleiber et al., 1936; Gallup and Hobbs, 1944; Raymond and Harris, 1954; Juko et al., 1961) and can range as high as 30% (French, 1930). Thus, for studies with the aim of monitoring N balance, determining N in fresh fecal samples is recommended (Schneider and Flatt, 1975). Following desiccation, samples are typically ground before conducting laboratory analyses. Although some experimental objectives may require use of non-standard grind sizes, a screen size of 1 mm (2 mm in some laboratories) is commonly used when grinding samples for laboratory analyses. Given the potential for grind size to impact results of laboratory analyses, standardization of grind size should be encouraged.

CALCULATION OF DIGESTION COEFFICIENTS

Mathematical calculations used to determine digestion coefficients are relatively direct. The following represents a summary of some key formulas that would be used to calculate digestibility either by total fecal collection or marker-based procedures. The fundamental calculation used in a digestion experiment solves for digestion of a nutrient when refused feed does not have to be accounted for:

$$\% \text{ nutrient digestion} = \frac{\text{nutrient consumed (kg)} - \text{nutrient in feces (kg)}}{\text{nutrient consumed (kg)}} \times 100 \quad (1)$$

In those trials where significant amounts of feed have been refused and the orts differ in chemical composition from the feed offered, the following formula can be used to determine the digestion coefficient for a particular nutrient:

$$\% \text{ nutrient digestion} = \frac{(\text{nutrient fed (kg)} - \text{nutrient refused (kg)}) - \text{nutrient in feces (kg)}}{(\text{nutrient fed (kg)} - \text{nutrient refused (kg)})} \times 100 \quad (2)$$

When intake is known and an external marker is continuously or frequently fed, fecal output can be calculated as follows:

$$\frac{\text{fecal DM output (g/d)}}{} = \frac{\text{marker dose (g/d)}}{\text{concentration of marker in feces (g/g of DM)}} \quad (3)$$

Once fecal output has been determined using an external marker, the calculation of digestion can proceed as described in equations 1 or 2. If one uses an internal marker when intake is known, the internal marker can be used to calculate fecal output as described in equation 3 by simply determining the amount of marker consumed by each animal (concentration of marker in diet x amount of diet consumed). If intake is not known, then digestion coefficients for different nutrients can be calculated from the following:

$$\% \text{ nutrient digestion} = 100 - 100 \times \frac{\% \text{ marker in feed} \times \% \text{ nutrient in feces}}{\% \text{ marker in feces} \times \% \text{ nutrient in feed}} \quad (4)$$

In equation 4, if the "nutrient" of interest was DM, both the % nutrient in the feed and feces would be 100%. As noted earlier, calculations presented that are based on the use of markers assume that the marker is recovered completely in the feces. If this assumption is incorrect, adjustments for incomplete recovery must be made for digestion or output calculations to be correct.

SOURCES OF ERROR AND DESIGN CONSIDERATIONS

Consideration of potential sources of error is critical to the design of any experiment. In the case of digestibility measurements, variation attributable to animals and analytical and technical errors are of major concern. Generally, measurements that require several steps to yield the final answer are subject to greater random variation than more direct measurements. Errors associated with each step of a measurement process are cumulative. For example, to measure fecal OM output by the ratio of fecal marker concentration to dose given, a constant dose of marker is administered, fecal samples are collected after some theoretical equilibrium is reached, and fecal samples are analyzed for marker concentration. Errors associated with the dose should be small (weighing the daily dose of marker), unless ruminal marker delivery devices are used. Conversely, errors associated with the fecal samples could be considerable. First, diurnal variation in marker excretion introduces sampling error, the magnitude of which depends on the length of collection and frequency of marker dosing. Second, recovery of marker in the feces may not be complete, thereby violating a basic assumption of the technique. Third, analytical errors can occur with analysis of marker concentration and moisture/ash. In contrast, direct measurement of fecal output involves only errors associated with weighing feces, and sampling and analysis of fecal moisture/ash. Although errors may cancel among observations in multi-step techniques, yielding an ultimate estimate not different from that obtained with more direct methods, variation will be greater for the multi-step technique. As a result of greater variation with our example of the marker-based technique to measure fecal output, greater numbers of observations would be required to detect treatment differences than for direct measurement of fecal output.

Analytical and Technical Errors

Errors associated with sample collection and laboratory analysis can be reduced by implementing routine quality control programs. The principal objective of such efforts is to avoid constant sources of error. For example, gravimetric measurements are central to most digestion studies, so analytical balances should be checked for accuracy before and during collection periods. Sampling procedures for diets, digesta, feces, etc. should ensure that representative aliquots are obtained for analyses. Hence, feces and digesta samples should be mixed thoroughly before subsampling, and composited samples should reflect weighting for the mass/volume of each collection that comprises the composite. As noted earlier, all personnel involved in a study should be cognizant of established routines for feeding, sample collection, and sample analysis. Processing (e.g., drying, grinding, mixing) should be applied consistently across samples. Analytical instruments used for subsequent analysis of samples should be checked routinely for accuracy and precision. The use of known standards to test equipment and procedures can reduce the chance of random and constant errors. As noted above, multi-step procedures typically result in greater variation among replicates than methods that have fewer steps; for some methods, this may necessitate increased replication and greater care in evaluation of laboratory data. If more than one procedure is available to obtain a desired measurement, the procedure that minimizes variation is the logical choice.

Variation attributable to length of the collection period is often overlooked as a source of error. Fecal excretion, urine flow, and other biological functions are inconstant (discontinuous) by nature (Blaxter, 1967). Hence, estimates obtained from short periods of total collection can deviate substantially from mean values derived from longer periods. However, because of the discontinuous nature of digesta flow, spot sampling (e.g., in marker-based studies) may be more effective if done frequently within a day for fewer days vs infrequently within a day for more days.

Animal Variation

Animal-to-animal variation is typically the single largest source of variation in digestion experiments. For example, in a 4 x 4 Latin square experiment with high-concentrate diets (Galyean et al., 1976), the mean square for total tract OM digestion associated with animal was 4.8 times greater than the mean square associated with collection period (time). Similarly, the mean square for animal associated with measurements of ruminal OM digestion (lignin as a marker) was eight times greater than the mean square associated with collection period. Hence, in environmentally controlled settings, the most important variable to control is likely to be animal variation. Careful selection of animals for genetic and physiological uniformity is critical. Certain types of experimental designs may be more appropriate than others for handling animal-to-animal variation.

Experimental Design Considerations

As noted above, designs that allow for control of animal variation seem best suited for digestion experiments. Hence, use of Latin square, crossover, and block (by animal) designs in such experiments is commonplace and justifiable. Nonetheless, investigators should be aware of the assumptions inherent to such designs. In multi-blocking designs like Latin squares and crossovers, rows and columns are assumed to be independent, and treatment is assumed not to interact with either rows or columns. In environmentally controlled settings, collection period (e.g., rows) is unlikely to interact with treatment. When collection period is subject to environmental effects, this assumption may be less tenable.

Repeated measures approaches may be beneficial in some digestion studies, but most such designs do not consider animal as a direct source of variation. Further, if time periods in which measurements are taken interact with treatments, small numbers of observations per time x treatment subclass result. Hence, such designs are most logical in controlled settings where time is unlikely to interact with treatment. Many investigators use time as a block and rerandomize the same animals to treatments in multiple blocks conducted over time. The usual approach is to restrict allotment such that animals do not repeat on the same treatment in subsequent blocks. This approach ignores correlated responses that arise from using the same animal over time (blocks) that are considered by the repeated measures design, and it also has the potential for block (time) x treatment interactions.

Thoughtful consideration of potential sources of error, coupled with advance planning and statistical consultation, should increase the quality of data obtained from digestion experiments. Given the costs in labor, time, and money needed for such experiments, planning efforts should be viewed as a prerequisite.

CONCLUSION

The measurement of in vivo digestibility in ruminants has added significantly to our understanding of the nutritive value of forages. Furthermore, considerable literature is available which can guide one in the selection of appropriate methods for conducting reliable digestion trials. Indeed, techniques are currently available that can yield accurate measurements of in vivo digestion under some conditions (e.g., total collection confinement trials). However, additional developmental research would be beneficial for those situations where digestion must be estimated using markers (e.g., grazing trials, site of digestion trials). Because of the time and labor required to accurately measure in vivo digestion, alternative approaches for predicting digestibility have been developed (see associated chapters) and will continue to be pursued. As future technologies are developed for evaluating forage quality, the measurement of in vivo digestion will continue to play an important role as a "control" methodology for validating the accuracy of alternative technologies. Because in vivo digestion measurements provide valuable insight regarding nutritive value and serve an important validation function, students of ruminant nutrition should strive to be well versed in the procedures used for measuring in vivo digestibility and in the subsequent interpretation of digestibility data.

REFERENCES

Agricultural Research Council. 1990. The Nutrient Requirements of Ruminant Livestock. C.A.B. International, Oxon, U.K.

Balch, C.C., S. Bartlett, and V.W. Johnson. 1951. Apparatus for the separate collection of faeces and urine from cows. J. Agric. Sci. 41:98-101.

Balch, C.C., V.W. Johnson, and C. Machin. 1962. Housing and equipment for balance studies with cows. J. Agric. Sci. 59:355-358.

Ballinger, C.E., and A.A. Dunlop. 1946. An apparatus for the collection of faeces from the cow. New Zealand J. Sci. Tech. A. 27:509-520.

Barnes, R.F. 1968. Variability within and among experiment stations in the determination of in vivo digestibility and intake of alfalfa. J. Anim. Sci. 27:519-524.

Berger, L., T. Klopfenstein, and R. Britton. 1979. Effect of sodium hydroxide on efficiency of rumen digestion. J. Anim. Sci. 49:1317-1323.

Berndtson, W. W. 1991. A simple, rapid and reliable method for selecting or assessing the number of replicates for animal experiments. J. Anim. Sci. 69:67-76.

Blaxter, K.L., N. McC. Graham, and F.W. Wainman. 1956. Some observations on the digestibility of food by sheep, and on related problems. Br. J. Nutr. 10:69-91.

Blaxter, K.L. 1967. Nutrition balance techniques and their limitations. Nutr. Soc. Proc. 26:86-96.

Brandyberry, S.D., R.C. Cochran, E.S. Vanzant, and D.L. Harmon. 1991. Technical note: Effectiveness of different methods of continuous marker administration for estimating fecal output. J. Anim. Sci. 69:4611-4616.

Border, J.R., L.E. Harris, and J.E. Butcher. 1963. Apparatus for obtaining sustained quantitative collections of urine from male cattle grazing pasture or range. J. Anim. Sci. 22:521-525.

Bratzler, J.W. 1951. A metabolism crate for use with sheep. J. Anim. Sci. 10:592-601.

Bredon, R.M., C.D. Juko, and B. Marshall. 1961a. The nutrition of zebu cattle. Part III. Digestibility techniques: Investigation of the effects of combination of dry faeces, length of digestibility trials and number of animals required. J. Agric. Sci. 56:99-103.

Bredon, R.M., B. Marshall, and C.D. Juko. 1961b. The nutrition of Zebu cattle. Part I. Equipment for the separate collection of urine and faeces from steers. J. Agric. Sci. 56:91-92.

Briggs, H.M., and W.D. Gallup. 1949. Metabolism stalls for wethers and steers. J. Anim. Sci. 8:479-482.

Bunting, L.D., M.D. Howard, R.B. Muntifering, K.A. Dawson, and J.A. Boling. 1987. Effect of feeding frequency on forage fiber and nitrogen utilization in sheep. J. Anim. Sci. 64:1170-1177.

Burritt, E.A., J.A. Pfister, and J.C. Malechek. 1988. Effect of drying method on the nutritive composition of esophageal fistula forage samples: Influence of maturity. J. Range Manage. 41:346-349.

Burroughs, W., P. Gerlaugh, and R.M. Bethke. 1950. The influence of alfalfa hay and fractions of alfalfa hay upon the digestion of ground corncobs. J. Anim. Sci. 9:207-213.

Carbery, M., and I.B. Chatterjee. 1936. Studies on the determination of digestibility coefficients: II. The estimation and computation of digestibility coefficients from individual tests and their order of precision. Indian J. Vet. Anim. Husb. 6:87-99.

Clanton, D.C. 1961. Comparison of 7- and 10-collection periods in digestion and metabolism trials with beef heifers. J. Anim. Sci. 20:640-643.

Cochran, R.C., D.C. Adams, J.D. Wallace, and M.L. Galyean. 1986. Predicting digestibility of different diets with internal markers: Evaluation of four potential markers. J. Anim. Sci. 63:1476-1483.

Cochran, R.C., E.S. Vanzant, K.A. Jacques, M.L. Galyean, D.C. Adams, and J.D. Wallace. 1987. Internal markers. p. 39-48. In M.B. Judkins, D.C. Clanton, M.K. Petersen, and J.D. Wallace (ed.) Proc. Grazing Livestock Nutrition Conf., Univ. of Wyoming, Laramie.

Cochran, R.C., E.S. Vanzant, and T. Delcurto. 1988. Evaluation of internal markers isolated by alkaline hydrogen peroxide incubation and acid detergent lignin extraction. J. Anim. Sci. 66:3245-3251.

Cook, C.W., L.A. Stoddart, and L.E. Harris. 1952. Determining the digestibility and metabolizable energy of winter range plants by sheep. J. Anim. Sci. 11:578-590.

Cordova, F.J., J.D. Wallace, and R.D. Pieper. 1978. Forage intake by grazing livestock: A review. J. Range Manage. 31:430-438.

Danley, M.M., and R.L. Vetter. 1971. Changes in carbohydrate and nitrogen fractions and digestibility of forages: Method of sample processing. J. Anim. Sci. 33:1072-1077.

Davis, C.L., J.H. Byers, and L.E. Luber. 1958. An evaluation of the chromic oxide method for determining digestibility. J. Dairy Sci. 41:152-159.

Donefer, E. 1966. Collaborative in vivo studies on alfalfa hay. J. Anim. Sci. 25:1227-1231.

Ellis, W.C., C. Lascano, R. Teeter, and F.N. Owens. 1982. Solute and particulate flow markers. p. 37-56. In F.N. Owens (ed.) Protein requirements for beef cattle: symposium. Oklahoma Agric. Exp. Sta. Misc. Pub. MP-109.

Erwin, E.S., I.A. Dyer, M.E. Ensminger, and W. Moore. 1956. A portable metabolism stall for beef steers. J. Anim. Sci. 15:435-510.

Forbes, E.B., R.F. Elliott, R.W. Swift, W.H. James, and V.F. Smith. 1946. Variation in determinations of digestive capacity of sheep. J. Anim. Sci. 5:298-305.

Forbes, J.M. 1982. Effects of lighting pattern on growth, lactation and food intake of sheep, cattle and deer. Livestock Prod. Sci. 9:361-374.

Forbes, J.M. 1986. The voluntary food intake of farm animals. Butterworths and Co. Ltd., London.

French, R.B. 1930. The use of preservatives to prevent loss of nitrogen from cow excreta during the day of collection. J. Agric. Res. 41:503-506.

Fuller, M.F., and A. Cadenhead. 1969. The preservation of faeces and urine to prevent losses of energy and nitrogen during metabolism experiments. p. 455-460. *In* K.L. Blaxter, D. Kielanwoski, and G. Thorbes (ed.) Energy metabolism of farm animals. Oriel Press Limited, Newcastle upon Tyne.

Gallup, W.D., and C.S. Hobbs. 1944. The desiccation and analysis of feces in digestion experiments with steers. J. Anim. Sci. 3:326-332.

Galyean, M.L., D.G. Wagner, and R.R. Johnson. 1976. Site and extent of starch digestion in steers fed processed corn rations. J. Anim. Sci. 43:1088-1094.

Galyean, M.L., and F.N. Owens. 1991. Effects of diet composition and level of feed intake on site and extent of digestion in ruminants. p. 483 - 514. *In* T. Tsuda, Y. Sasaki, and R. Kawashima (ed.) Physiological aspects of digestion and metabolism in ruminants. Academic Press, Inc., San Diego, CA.

Galyean, M.L., L.J. Krysl, and R.E. Estell. 1987. Marker-based approaches for estimation of fecal output and digestibility in ruminants. p. 96 - 113. *In* F.N. Owens (ed.) Feed intake by beef cattle. Oklahoma State Univ., MP 121.

Garrett, W.N., and D.E. Johnson. 1983. Nutritional energetics of ruminants. J. Anim. Sci. 57:478-497.

Goering, H.K., and P.J. Van Soest. 1970. Forage fiber analyses (apparatus, reagents, procedures and some applications). ARS Agric. Handbook 379. Washington, DC.

Gorski, J., T.H. Blosser, F.R. Murdock, A.S. Hodgson, B.K. Soni, and R.E. Erb. 1957. A urine and feces collecting apparatus for heifers and cows. J. Anim. Sci. 16:100-109.

Grovum, W.L. 1983. Integration of digestion and digesta kinetics with the control of feed intake - a physiological framework for a model of rumen function. p. 244-268. *In* F.M.C. Gilchrist and R.I. Mackie (ed.) Herbivore nutrition in the subtropics and tropics. Science Press, South Africa.

Harris, L.E. 1970. Nutrition research techniques for domestic and wild animals. Volume 1. Lorin E. Harris, Logan, UT.

Hatfield, P.G., J.W. Walker, J.A. Fitzgerald, H.A. Glimp, and K.J. Hemenway. 1993. The effects of different methods of estimating fecal output on plasma cortisol, fecal output, forage intake, and weight change in free-ranging and confined wethers. J. Anim. Sci. 71:618-624.

Hattan, G.L., and F.G. Owen. 1970. Efficiency of total collection and chromic oxide techniques in short-term digestion trials. J. Dairy Sci. 53:325-329.

Heaney, D.P., G.I. Pritchard, and W.J. Pigden. 1965. Between-animal variability in ad libitum forage intakes by sheep. J. Anim. Sci. 24:909.

Hobbs, C.S., S.L. Hansard, and E.R. Barrick. 1950. Simplified methods and equipment used in separation of urine from feces eliminated by heifers and by steers. J. Anim. Sci. 9:565-570.

Hodgson, R.E., and J.C. Knott. 1937. Variations between animals used in digestion experiments. Am. Dairy Sci. Assn. Western Div. Ann. Mtg. 23:31-36.

Holechek, J.L., and M. Vavra. 1983. Fistula sample numbers required to determine cattle diets on forest and grassland ranges. J. Range Manage. 36:323-326.

Holechek, J.L., M. Vavra, and R.D. Pieper. 1982. Methods for determining the nutritive quality of range ruminant diets: A review. J. Anim. Sci. 54:363-376.

Hopper, J.T., J.W. Holloway, and W.T. Butts, Jr. 1978. Animal variation in chromium sesquioxide excretion patterns of grazing cows. J. Anim. Sci. 46:1096-1102.

Horn, L.H., Jr., M.L. Ray, and A.L. Neumann. 1954. Digestion and nutrient-balance stalls for steers. J. Anim. Sci. 13:20-24.

Judkins, M.B, L.J. Krysl, and R.K. Barton. 1990. Estimating diet digestibility: A comparison of 11 techniques across six different diets fed to rams. J. Anim. Sci. 68:1405-1415.

Juko, C.D., R.M. Bredon, and B. Marshall. 1961. The nutrition of zebu cattle. Part II. The techniques of digestibility trials with special reference to sampling, preservation and drying of faeces. J. Agric. Sci. 56:93-97.

Jurgens, M.H. 1982. Animal feeding and nutrition. 5th ed. Kendall/Hunt Publishing Company, Dubuque, IA.

Kartchner, R.J., and L.R. Rittenhouse. 1979. A feces-urine separator for making total fecal collections from the female bovine. J. Range Manage. 32:404-405.

Kaufmann, W., H. Hagemeister, and G. Dirksen. 1980. Adaptation to changes in dietary composition, level and frequency of feeding. p. 587-602. *In* Y. Ruckebusch and P. Thivend (ed.) Digestive physiology and metabolism in ruminants. AVI Publishing Co., Inc., Westport, CT.

Kennedy, P.M., R.J. Christopherson, and L.P. Milligan. 1986. Digestive responses to cold. p. 285-306. *In* L.P. Milligan, W.L. Grovum, and A. Dobson (ed.) Control of digestion and metabolism in ruminants. Prentice-Hall, Englewood Cliffs, NJ.

King, W.A., J. Lee, III, H.J. Webb, and D.B. Roderick. 1960. Comparison of 6- and 10-day collection periods for digestion trials with dairy heifers. J. Dairy Sci. 43:388-392.

Kleiber, M., R.W. Caldwell, and H. Johnson. 1936. Losses of N and C in drying feces of cattle. Soc. Exp. Biol. Med. Proc. 31:128-130.

Kotb, A.R., and T.D. Luckey. 1972. Markers in nutrition. Nutr. Abst. Rev. 42:813-845.

Langlands, J.P. 1967. Studies on the nutritive value of the diet selected by grazing sheep. Anim. Prod. 9:167-175.

Lesperance, A.L., V.R. Bohman, and D.W. Marble. 1960. Development of techniques for evaluating grazed forage. J. Dairy Sci. 43:682-689.

Lloyd, L.E., H.E. Peckman, and E.W. Crampton. 1956. The effect of change of ration on the required length of preliminary feeding period in digestion trials with sheep. J. Anim. Sci. 15:846-853.

Lucas, H.L. 1952. Algebraic relationships between digestion coefficients determined by the conventional method and by indicator methods. Science 116:301-302.

Maynard, L.A., J.K. Loosli, H.F. Hintz, and R.G. Warner. 1979. Animal nutrition. 7th ed. McGraw-Hill Book Company, New York.

Merchen, N.R. 1988. Digestion, absorption and excretion in ruminants. p. 172-201. *In* D.C. Church (ed.) The ruminant animal. Prentice Hall, Englewood Cliffs, NJ.

Moir, R.J. 1961. A note on the relationship between the digestible dry matter and the digestible energy content of ruminant diets. Austr. J. Exp. Agric. Anim. Husb. 1:24-26.

National Research Council. 1981a. Effect of environment on nutrient requirements of domestic animals. National Acad. Press, Washington, DC.

National Research Council. 1981b. Nutritional energetics of domestic animals. National Acad. Press, Washington, DC.

National Research Council. 1984. Nutritional requirements of beef cattle. National Acad. Press, Washington, DC.

Nelson, A.B., A.D. Tillman, W.D. Gallup, and R. MacVicar. 1954. A modified metabolism stall for steers. J. Anim. Sci. 13:504-510.

Nicholson, J.W.G., E.H. Haynes, R.G. Warner, and J.K. Loosli. 1956. Digestibility of various rations by steers as influenced by the length of preliminary feeding period. J. Anim. Sci. 15:1172-1179.

Olson, K.C. 1991. Diet sample collection by esophageal fistula and rumen evacuation techniques. J. Range Manage. 44:515-519.

Owen, J.B., and J.W. Ingleton. 1961. A method of collecting faeces from ewes. Anim. Prod. 3:63-64.

Owens, F.N., and C.F. Hanson. 1992. Symposium: External and internal markers. J. Dairy Sci. 75:2605-2617.

Penning, P.D., and R.H. Johnson. 1983a. The use of internal markers to estimate herbage digestibility and intake. 1. Potentially indigestible cellulose and acid insoluble ash. J. Agric. Sci. (Camb.) 100:127-131.

Penning, P.D., and R.H. Johnson. 1983b. The use of internal markers to estimate herbage digestibility and intake. 2. Indigestible acid detergent fibre. J. Agric. Sci. (Camb.) 100:133-138.

Peters, R.R., L.T. Chapin, R.S. Emery, and H.A. Tucker. 1980. Growth and hormonal response of heifers to various photoperiods. J. Anim. Sci. 51:1148-1153.

Pond, K.R., J.C. Burns, and D.S. Fisher. 1987. External markers: Use and methodology in grazing studies. p. 49-53 *In* M.B. Judkins, D.C. Clanton, M.K. Petersen, and J.D. Wallace (ed.) Proc. Grazing Livestock Nutr. Conf., Univ. of Wyoming, Laramie.

Prigge, E.C., G.A. Varga, J.L. Vicini, and R.L. Reid. 1981. Comparison of ytterbium chloride and chromium sesquioxide as fecal indicators. J. Anim. Sci. 53:1629-1633.

Purser, D.B, and R.J. Moir. 1966. Rumen volume as a factor involved in individual sheep differences. J. Anim. Sci. 25:509-515.

Raleigh, R.J., R.J. Kartchner, and L.R. Rittenhouse. 1980. Chromic oxide in range nutrition studies. Sta. Bull. 641. Oregon State Univ., Corvallis.

Raymond, W.F., and C.E. Harris. 1954. The laboratory drying of herbage and faeces, and dry matter losses possible during drying. J. Br. Grassl. Soc. 9:119-130.

Satter, L.D., and B.R. Baumgardt. 1962. Changes in digestive physiology of the bovine associated with various feeding frequencies. J. Anim. Sci. 21:897-900.

Schanbacher, B.D., and J.D. Crouse. 1981. Photoperiodic regulation of growth: A photosensitive phase during light-dark cycle. Am. J. Physiol. 241:E1-E5.

Schneider, B.H., and H.L. Lucas. 1950. The magnitude of certain sources of variability in digestibility data. J. Anim. Sci. 9:504-512.

Schneider, B.H., and W.P. Flatt. 1975. The evaluation of feeds through digestibility experiments. The Univ. of Georgia Press, Athens.

Sidahmed, A.E., J.G. Morris, W.C. Weir, and D.T. Torell. 1977. Effect of the length of fasting on intake, in vitro digestibility and chemical composition of forage samples collected by esophageal fistulated sheep. J. Anim. Sci. 46:885-890.

Siddons, R.C., J. Paradine, D.E. Beever, and P.R. Cornell. 1985. Ytterbium acetate as a particulate-phase digesta-flow marker. Br. J. Nutr. 54:509 - 519.

Staples, G.E., and W.E. Dinusson. 1951. A comparison of the relative accuracy between seven-day and ten-day collection periods in digestion trials. J. Anim. Sci. 10:244-250.

Stillwell, M.A., R. Senft, and L.R. Rittenhouse. 1983. Total urine collection from free-grazing heifers. J. Range Manage. 36:798-799.

Streeter, C.L. 1969. A review of techniques used to estimate the in vivo digestibility of grazed forage. J. Anim. Sci. 29:757-768.

Sunvold, G.D., and R.C. Cochran. 1991. Technical note: Evaluation of acid detergent lignin, alkaline peroxide lignin, acid insoluble ash, and indigestible acid detergent fiber as internal markers for prediction of alfalfa, bromegrass, and prairie hay digestibility by beef steers. J. Anim. Sci. 69:4951-4955.

Swift, R.W. 1925. A biometric study of length of metabolism experiments with cattle. J. Dairy Sci. 8:270-281.

Theurer, B., S. Rahnema, J.A. Garcia, and M.C. Young. 1981. Effect of 2- versus 6-day collections for the determination of ruminal and postruminal digestion in steers. J. Anim. Sci. 52:134-137.

Van Dyne, G.M., and D.T. Torell. 1964. Development and use of the esophageal fistula: A review. J. Range Manage. 17:7-19.

Van Es, A.J.H., and J.E. Vogt. 1959. Separate collection of feces and urine of cows. J. Anim. Sci. 18:1220-1223.

Van Soest, P.J. 1982. Nutritional ecology of the ruminant. O & B Books, Inc., Corvallis, OR.

Van Soest, P.J., R.H. Wine, and L.A. Moore. 1966. Estimation of the true digestibility of forages by the in vitro digestion of cell walls. p. 438 - 441. Proc. 10th Int. Grassl. Congr., Helsinki, Finland.

Waller, J., N. Merchen, T. Hanson, and T. Klopfenstein. 1980. Effect of sampling intervals and digesta markers on abomasal flow determinations. J. Anim. Sci. 50:1122-1126.

Warner, A.C.I. 1981. Rate of passage of digesta through the gut of mammals and birds. Nutr. Abstr. Rev. 51:789-820.

APPENDIX

General Protocol for Digestion Experiments Conducted in Confinement

I. General Preparation

1. Select animals of the same age, weight, genotype, physiological history, and nutritional background. Ensure that animals are appropriately trained and previously exposed to the collection environment and collection apparatus. Address potential health concerns (e.g., parasites) well in advance of beginning the experiment. Maintain animals on a similar diet.

2. Determine the environmental conditions that will exist in the available housing. Note the temperature and light/dark cycles and adjust if necessary. If complete environmental control is not available, be prepared to collect some environmentally related data.

3. Ensure that all equipment is both available and in proper working order. This might include evaluation of metabolism stalls, watering devices, fecal bags (if used), and feed, sample, and animal scales.

4. Prepare sample collection receptacles (bags, pans, vials, etc.) and label in advance. This would include receptacles for feed samples, orts, fecal samples and, if collected, urine.

5. Establish collection and feeding schedules. Develop data sheets that allow easy tracking of amounts fed, refused, and excreted by each animal daily and the associated collection/feeding times. Also include information regarding the experiment identification, weather data, person(s) collecting data, and notes for any unusual events. Data sheets also should be prepared for DM determinations.

6. Weigh animals and randomize to treatments. This can be done on the morning of the first day of the experiment, although the first day of the experiment likely will be less hectic if this is done the day before the experiment begins.

II. Experimental Schedule (adapted from Harris, 1970)

Day	Activity
1	Weigh and randomize animals if not done previously. Place animals in metabolism stalls and begin feeding experimental diets.
2-14	Feed at fixed level of intake or establish voluntary intake if digestion is to be determined under ad libitum feeding conditions. Weigh and record amount of feed refused.
15	Collect forage samples
16	" Collect orts
17	" " Collect feces
18	" " "
19	" " "
20	" " "
21	" " "
22	" "
23	"
24	Weigh animals in the early morning.

III. Daily Activities

1. Ensure that all animals are in good health and equipment is functioning properly.

2. Collect and weigh orts. Sample orts and prepare for drying at 50 C. Ensure that ort collection occurs at the same time each day and that animals are fed shortly thereafter.

3. If not pre-weighed, weigh each animal's daily feed allotment, and deliver the feed to all animals at the same time. If feed is weighed

on a daily basis, take representative samples of the feedstuff(s) during the weighing process and save as part of the "running" composite. Dry daily samples. Ensure that feeding occurs at the same time each day.

4. Collect feces at the same time each day. If using fecal bags, remove the bag, weigh the bag plus fecal contents, empty the bag and scrape (or use a clean bag if more than one bag is available per animal), tare the bag, and then place the bag back on the animal. If using fecal pans, empty the pan into a tared receptacle, and weigh entire contents. After total fecal output has been weighed, mix thoroughly (flat-blade hoe and a large metal container for holding feces works well) and subsample feces. Prepare feces for freezing or drying at 50 C. If urine is collected and N balance determined, fresh samples of feces should be preserved (e.g., frozen) for N determination.

GENERAL PROTOCOL FOR DIGESTION EXPERIMENTS CONDUCTED UNDER GRAZING CONDITIONS

I. General Preparation

Generally, preparation for a trial conducted under grazing conditions is similar to that described for confinement trials with the following exceptions:

1. Animals must be selected and trained both for diet sampling and fecal collections. With carefully designed collection schedules, ruminally fistulated animals can be used for both diet and fecal collections. However, use of separate sets of animals for diet sampling (e.g., esophageally or ruminally fistulated animals) and fecal collections avoids the need to impose excessive stagger between periods when diet and fecal samples are collected.

2. It is imperative that grazing animals be adequately trained to carry fecal bags before they are used in an experiment.

3. Prepare portable facilities to be used in gathering and handling animals each day while on pasture. Alternatively, if fecal bags are not used and if animals have undergone considerable training, sampling can be accomplished in the pasture.

II. Experimental Schedule

Day	Activity		
1	Weigh and randomize animals if not done previously. If feed other than pasture is to be consumed, restrain animals individually and begin supplementation treatments.		
2-14	Unrestricted grazing and, if applicable, supplementation (individual basis).		
15	Collect forage samples via fistula	Collect samples of supplements fed (if applicable)	
16	"	"	
17	"	"	Collect feces
18	"	"	"
19	"	"	"
20	"	"	"
21	"	"	"
22			"
23			"
24	Weigh animals in the early morning.		

III. Daily Activities

Daily activities for a digestion trial conducted under grazing conditions are similar to those for a confinement trial with the following exceptions. Diet sampling via fistulated animals needs to be done in a manner that prevents contamination with ruminal contents (see earlier discussion). Sampling should occur at the same time of day for all treatments. In addition, to have samples represent average diet quality, it may be valuable to collect samples at different times of day (Langlands, 1967). Samples collected using fistulated animals should be frozen immediately after collection and subsequently lyophilized. Following lyophilization and grinding, samples should be composited on an equal weight basis within animal.

CHAPTER 16

ESTIMATION OF DIGESTIBILITY OF FORAGES
BY LABORATORY METHODS

William P. Weiss

INTRODUCTION

Accurate data on digestibility of forages would greatly assist diet formulation and economic valuation of different forages. Although the value of accurate digestibility data is unequivocal, obtaining actual data is time consuming, expensive, and requires large amounts of the test forage. Digestibility is not constant among or within forages; therefore, reference data on digestibility (Schneider, 1947; National Academy of Sciences, 1971) are of limited worth for diet formulation. These problems have led to the development of several biological and chemical methods that can be used to estimate digestibility of feeds. The validity of methods used to estimate digestibility depends upon how the data are to be used. Relative differences among samples within a species can be used as selection criteria for plant breeders. Differences among and within forage types also can be used to determine the relative economic and nutritional value of forages. When relative differences are the primary objective, accuracy (how close the estimate is to the actual value) is not a major concern; however, precision (a measure of within laboratory variation) is important so that small differences can be detected. When formulating diets and predicting animal response, accurate estimates of digestibility are required. Precision also should be high so that the number of replicate analyses needed is small. When the same laboratory is used for all analyses, reproducibility (a measure of variation among laboratories) may not be important. For collaborative studies or when samples may be sent to different labs, reproducibility is important.

The objectives of this paper are to: 1) discuss different methods of estimating digestibility of forages; 2) outline factors that affect the accuracy, precision, and reproducibility of estimated values and how these factors can be controlled; and 3) compare results from various methods of estimating digestibility. Several excellent reviews have been published on estimating digestibility of feeds (Johnson, 1966; Barnes, 1973; Morrison, 1976; Barber et al., 1984; van Es, 1986). This review will focus on recent publications.

Dep. of Dairy Science, Ohio Agricultural Research and Development Center, The Ohio State Univ., Wooster 44691.

IN VITRO BIOLOGICAL PROCEDURES

In Vitro Disappearance

In vitro dry matter disappearance (IVDMD) is determined by incubating a feed in the presence of ruminal contents and a buffer solution. The first reported use of this technique was in 1919 by Waentig and Gierisch (Hungate, 1966). Their IVDMD values were about 50% less than expected in vivo values. Since that time, in vitro techniques have been researched and modified extensively with much of the pioneering research being conducted in the 1940's and 1950's (Johnson, 1963). A major problem with early in vitro systems was the inability to maintain a desirable pH which limited incubation times to about 8 h. An important advancement in in vitro technology occurred when McDougall (1948) published a paper describing the mineral composition of sheep saliva. With the advent of McDougall's buffer (and its modifications), long term in vitro incubations became possible.

By the late 1950's, several different in vitro systems had been developed (Johnson, 1963). Warner (1956) established three criteria for the evaluation of different in vitro systems:

1) The maintenance of a normal microbiological population.
2) The maintenance of normal rates of digestion.
3) The ability to predict in vivo results.

For forage evaluation, the relationship between in vivo and in vitro digestibility is of the utmost importance. Baumgardt et al. (1958) were first to develop in vivo-in vitro relationships for predicting dry matter (DM) digestibility of forages. The in vitro method used by Baumgardt et al. (1958) consisted of incubating the fermentation mixture in dialysis tubing immersed in buffer. Walker (1959) reported relationships between disappearance in an all glass in vitro system and in vivo digestibility for forages. The dialysis method used by Baumgardt et al. (1958) was not highly precise and was difficult to maintain for long periods of time (48 to 72 h). The all glass system used by Walker (1959) produced relatively high precision, and IVDMD values and in vivo DM digestibilities were correlated ($r=0.93$), but IVDMD values were consistently lower than in vivo estimates. The major advance in in vitro techniques came with the development of the two-stage procedure by Tilley and Terry (1963). This procedure with minor modifications is still in widespread use.

Two-Stage In Vitro Method

Basically, the original Tilley and Terry (1963) method consisted of incubating a small amount of feed (ca. 0.5 g) with 10 ml of ruminal fluid and 40 ml of a buffer solution in small flasks under anaerobic conditions for 48 h. After 48 h, fermentation was stopped by the addition of mercuric chloride. After centrifugation, the residue was digested with acid-pepsin.

After a 48 h digestion period, the sample was centrifuged and the residue dried to determine IVDMD. Blanks were run concurrently to correct for DM provided by the inoculum. Tilley and Terry (1963) reported that the standard error between replicated samples within a run was 6.6 g/kg of digestible DM and across different runs was about 12 g/kg of digestible DM. They also reported that in vivo digestibilities of several different grasses and legumes could be predicted from IVDMD values with a high degree of accuracy. Since the original procedure was published, modifications have been introduced to increase laboratory through-put, precision, and accuracy.

Laboratory methods that are to be used for routine forage analysis or for screening a large number of samples must be relatively rapid, easy to perform, and have acceptable precision and accuracy. Alexander and McGowan (1966) and Minson and McLeod (1972) present detailed descriptions on modifications of the original two-stage method to increase laboratory through-put without sacrificing accuracy or precision. Using either the Alexander and McGowan or the Minson and McLeod modifications, laboratories should be able to analyze between 200 and 300 samples per week. The main modification of Alexander and McGowan (1966) was the omission of the centrifugation step between the fermentation and pepsin stages. Minson and McLeod (1972) incubated samples in centrifuge tubes which eliminated transfer steps. Both papers describe equipment necessary to increase the productivity of technicians.

High precision (intralaboratory variation) is important so that analyses do not have to be replicated excessively. For high precision, samples should be ground through a 1 mm screen (McLeod and Minson, 1969; Lentz and Buxton, 1992), and sample size (0.5 g) should be kept constant (McLeod and Minson, 1969). Variation caused by analytical technique can be controlled relatively easily; however, the major cause of poor precision is variation in the activity of the inoculum (Barnes, 1967).

Donor Animal Effects

Source of inoculum has a major influence on the precision and accuracy of IVDMD values. Factors affecting the impact of donor animal on IVDMD values include: animal-to-animal variation, species of donor animal, feeding management, and probably most importantly, diet fed to the donor animal.

Intrinsic variation among donor animals may decrease precision. Ayres (1991) reported that donor animal (four sheep all fed the same diet) significantly affected in vitro organic matter (OM) disappearance; however, Engels and van der Merwe (1967) reported that donor animal (four sheep fed the same diet) had no effect on in vitro disappearance. Nelson et al. (1972) reported that in some situations, donor animal was a significant source of variation. In general, when both the diet fed to the donor animal and the sample had low concentrations of crude protein (CP), variation among donor animals was greater than when diets fed to donor animals and samples contained higher concentrations of CP.

Species of donor animal can affect accuracy of IVDMD values, but it does not appear to have a large effect on precision. Horton et al. (1980) reported that ruminal fluid collected from sheep produced different IVDMD values than did ruminal fluid collected from cattle. The animal species effect was not consistent. When donor animals were fed wheat (*Triticum aestivum* L.) straw or a 1:1 mixture of alfalfa (*Medicago sativa* L.) and barley (*Hordeum vulgare* L.) pellets, ruminal fluid from sheep produced IVDMD values that were higher (ca. 80 g/kg) than those obtained when cattle were the source of ruminal fluid. When a medium quality smooth bromegrass (*Bromus inermis* Leyss.)-alfalfa hay was fed, ruminal fluid from cattle produced higher values (ca. 40 g/kg) than did fluid collected from sheep. Relative ranking for a variety of forage substrates (legumes, warm and cool season grasses, and straws) were similar, but not identical among donor animal species. Grant et al. (1974) reported that IVDMD values were similar when ruminal fluid was collected from water buffaloes, European cattle (*Bos taurus*), and Zebu cattle (*Bos indicus*). In general, large quantities of ruminal fluid are easier to obtain from cattle than from sheep. This difference may make cannulated cattle the donor animal of choice for commercial laboratories. For research purposes, the donor animal should be the same species as the target animal.

The time ruminal fluid is collected, relative to feeding, should be kept constant. Generally, IVDMD values decrease when inoculum is collected more than 16 to 18 h postfeeding (Ayres, 1991). When donor animals are fed once daily, ruminal fluid should be collected at a specific time relative to feeding. The activities of most fibrolytic enzymes peak 8 to 12 h postfeeding (Williams, 1988), which suggests that this would be the optimal time to collect inoculum. Another means of reducing this variation is to feed donor animals several (at least three) times each day which essentially removes this effect (Alexander and McGowan, 1966).

The diet fed to ruminants affects the microbiological population and chemical environment within the rumen. Diet of the donor animal also influences IVDMD. Feeding donor animals a relatively high concentrate diet (barley) reduced IVDMD of alfalfa hay and of the basal diet (approximately 60:40 barley: alfalfa hay) as compared to donor animals fed alfalfa hay (Calder, 1970). Horton et al. (1980) reported that feeding a barley:alfalfa diet (1:1) to donor animals produced lower IVDMD values than when donor animals were fed medium quality grass-legume hay. Conversely, Nik-Khah and Tribe (1977) reported no difference in IVDMD values when donor animals were fed diets ranging from all hay to 90:10 concentrate:hay (ingredient composition of the concentrate was not given, but the concentrate was relatively high in fiber). Although the amount of concentrate fed to donor animals can affect the accuracy of in vitro values, precision usually is not affected greatly.

Type of forage fed to donor animals when the diet is composed predominantly of forage does not influence precision greatly. The exception to this generality is low protein forages. Generally, when the forage fed to donor animals has less than 100 g/kg of CP (DM basis), precision is low. Nelson et al. (1972) reported that when corn (*Zea mays* L.) silage (86 g/kg of CP) was the donor diet, the coefficient of variation (CV) of replicated analyses for a variety of substrates averaged 6.4, but when perennial ryegrass (*Lolium perenne* L.) that contained 110 g/kg of CP was the donor diet, the CV averaged 2.8. The amount of CP necessary in donor diets for high precision depends on the type of substrate. When donor animals were fed a grass containing 70 g/kg of CP, precision decreased linearly as the amount of CP in the substrate decreased (Bezeau, 1965). Bezeau (1965) found that precision was independent of substrate protein concentration when the basal diet contained at least 110 g/kg of CP. Many researchers have reported improved precision when nitrogen (N) was added to incubation fluid (Alexander and McGowan, 1966; Engels and van der Merwe, 1967; Nelson et al., 1972; Dhanoa and Deriaz, 1984). Generally about 10 mg of N (from urea and (or) ammonium sulfate) was added to the incubation mixture. Addition of N to the incubation mixture reduces variation among donor animals (Nelson et al., 1972) and reduces the variation caused by feeding donor animals different diets (Engels and van der Merwe, 1967; Nelson et al., 1972). When high precision is the goal, augmentation of McDougall's buffer with about 10 mg of N per flask or using a nutrient-buffer solution such as that developed by Goering and Van Soest (1970) appears to be warranted; however, addition of N and other nutrients may affect accuracy.

Species and quality of forage fed to donor animals influence the accuracy of in vitro values more than they influence precision. Quality of alfalfa (CP ranged from 140 to 180 g/kg) fed to donor animals had no effect on IVDMD values or precision (Bezeau, 1965). Several studies, however, have reported that when ruminal fluid collected from animals fed alfalfa was used, in vitro DM and fiber digestibilities were higher than when ruminal fluid from animals fed grasses or low quality roughages was used (Bezeau, 1965; Engels and van der Merwe, 1967; Nelson et al., 1972; Jung and Varel, 1988; Ayres, 1991). No consistent effect of grass type (warm vs. cool season) has been reported. Nelson et al. (1972) reported that when bermudagrass (*Cynodon dactylon* L.), a warm season grass, was fed to donor animals, IVDMD values for a variety of forages were higher than when perennial ryegrass, a cool season grass, was fed, but when another warm season grass (bahiagrass, *Paspalum notatum* Flugge) was fed, IVDMD values were less than when perennial ryegrass was fed. Jung and Varel (1988) reported no differences between fiber digestibilities (in vitro) when smooth bromegrass or switchgrass (*Panicum vigatum* L.), a warm season grass, was fed to the donor animal. The primary conclusion reached from these studies is that no single donor diet will produce accurate results for all possible test forages. Substrate by donor diet interactions exist which means that relative ranking of different forages is affected by the diet fed to donor animals (Knipfel and Troelsen, 1966; Engels and van der Merwe, 1967; Nelson et al., 1972; Horton et al., 1980; Ayres, 1991). Because forage type fed to donor animals affects IVDMD values, samples should be analyzed using inoculum from donor animals that have been fed a common diet. The forage fed to donor animals should be common to the area so that similar feed can be obtained over time.

Statistical Considerations

Intralaboratory variation can be kept acceptably low by following the above standard procedures. Typically, the average deviation between duplicated analyses within a run (no variation caused by ruminal fluid) is between 4 and 10 g/kg of digestible DM (Tilley and Terry, 1963; Alexander and McGowan, 1966; Barnes, 1967; Nelson et al., 1972; Ayres, 1991). Ayres (1991) established statistical criteria for rejecting values when the difference between duplicates within a run exceeded about 30 g/kg of digestible OM. When single samples are replicated over runs (donor animal conditions kept constant), the deviation is about 1.5 to 2 times higher than when samples are replicated within a run (Tilley and Terry, 1963; Alexander and McGowan, 1966). Standard error between runs when samples are replicated within a run has been reported to be between 6 and 8 g/kg of digestible DM or OM (Ayres, 1991). The degree of precision between runs is comparable to the degree of variation in in vivo digestibilities found among animals. Some researchers have recommended that standards be included with each in vitro run to account for among run variation; however, this approach may actually add bias. Genizi et al. (1990) and Ayres (1991) recommend that standards not be used to reduce among run variation. Ayres (1991) states that each laboratory must first determine the normal amount of run-to-run variation (95% confidence intervals) and when the IVDMD of the standards is outside the 95% confidence interval, the run can be corrected based on the deviation of the standards (assuming all standards deviate similarly). When standards deviate greatly or in different directions, the entire run should be discarded.

Accuracy

High precision is important when attempting to find relative differences among samples and for reducing the amount of replication needed; however, high precision does not necessarily imply high accuracy. Some analytical methods can very precisely produce inaccurate values. Accurate estimates of digestibility are important when balancing diets, for determining the true economic value of feedstuffs, and for predicting animal performance. Determining accuracy is extremely difficult because many factors such as DM intake (Tyrrell and Moe, 1975) and total diet composition (Hoover, 1986) can affect in vivo digestibility of forages. Different situations have different requirements for accuracy.

Many studies have shown strong statistical correlations (r > 0.9) between in vivo and in vitro digestibility data (e.g., Tilley and Terry, 1963; Alexander and McGowan, 1966; Troelsen, 1970; McLeod and Minson, 1974; Aerts et al., 1977; Valdes and Jones, 1987; Givens et al., 1989; Genizi et al., 1990; Navaratne et al., 1990; Aufrère et al., 1992). A strong correlation does not necessarily mean that IVDMD equals in vivo digestibility; usually an equation must be derived to convert in vitro data to in vivo digestibility. Since a regression equation must be used to obtain estimates of in vivo digestibility, regression error is introduced. Regression error can be expressed as standard error of the estimate, also known as residual standard deviation (RSD), or as standard error of prediction (SE_p) (Neter and Wasserman, 1974). The RSD is a measure of how precisely the population used to generate the equation fits the equation. The SE_p is a measure of the error associated with predicting a value for a new observation. Standard error of prediction is always larger than RSD. When large populations of feeds are used to develop regression equations, RSD range from about 20 to 40 g/kg of digestible DM or OM (Barnes, 1973). With smaller populations, RSD have been as high as 60 g/kg of digestible DM (Ayres, 1991). Because of the variation in analytical techniques and variation caused by donor animals (including diet), no universal equation can be developed. Each lab must generate their own equations. A new equation may be needed if the species of donor animal changes or the diet fed to the donor animal changes appreciably (Horton et al., 1980).

Calibration equations can be developed by following one of three different methods. One way is for each laboratory to determine both in vitro and in vivo digestibility coefficients for a diverse population of feeds. This method is extremely time-consuming, expensive, and not practical for commercial laboratories. The database probably would be quite limited and may be appropriate only for feeds grown under a few different conditions within a relatively small geographic area. Furthermore, in vivo digestibility data from a single source does not include many important sources of variation, thereby limiting the inference space of the equation. A more practical and better method is to obtain a large set of diverse feeds that have known in vivo digestibilities (a calibration set) and measure IVDMD on those samples (Goldman et al., 1987). The in vivo data are then regressed on the in vitro data. Goldman et al. (1987) reported a SE_p of about 30 g/kg of digestible DM or OM for an equation derived from a large calibration set of diverse samples (n = 422). This method would require relatively large samples of many different feeds so that different laboratories could generate calibration equations. Indirect calibration also can be used to estimate in vivo digestibility from IVDMD values (Genizi et al., 1990). For this method, samples of known IVDMD from a central laboratory are analyzed at another laboratory. An equation then is derived to convert in vitro data generated from the independent laboratory to estimates of in vitro data from the central lab. The central laboratory must have an accurate in vitro-in vivo equation that is then used to convert the estimated in vitro data to in vivo estimates. Indirect calibration appears to be a reliable method of generating estimated in vivo values (Genizi et al., 1990).

Prediction error can be reduced by using different regression equations (Table 1) for different forage species (McLeod and Minson, 1969; McLeod and Minson, 1976; Omed et al., 1989) or by including forage species in the regression model (Goldman et al., 1987). The reduction in SE_p is usually quite small when separate equations are used for legumes and grasses as compared to a single equation (Tilley and Terry, 1963; Valdes and Jones, 1987; Omed et al., 1989). A single equation for both concentrate feeds and forages should not be derived because of unacceptably high SE_p (Omed et al., 1989). Corn silage samples produce an in vitro-in vivo equation very different from equations derived from grass or legume data (Valdes and Jones, 1987) so a corn silage equation is needed. Including low quality roughages, such as straw, in data sets with grasses and legumes increases the SE_p by 10 to 20 g/kg of digestible DM or OM (Aufrère and Michalet-Doreau, 1988; Navaratne et al., 1990). Based on the above information, separate equations are needed for: 1) legumes and grasses (both warm and cool season); 2) corn silage; 3) concentrate feeds; and 4) low quality roughages. The equations in Table 1 should not be adopted directly by laboratories that are conducting in vitro analyses. Because of the many confounding variables, each laboratory should generate its own equations.

Table 1. Sample equations for converting IVDMD values (two-stage method) to in vivo OM digestibility (all values expressed as g/kg, DM basis).[1]

Feed	Intercept	Slope	SE_p^2	Reference
C-3 grasses	124	0.82	22.7	Aerts et al., 1977
C-3 grasses	5.2	1.01	14.6	Terry et al., 1978
C-3 grasses	-136	1.20	18.5	Omed et al., 1989
C-3 grasses	172	0.71	24.0	Moss and Givens, 1990
C-4 grasses	115	0.83	24.0	McLeod and Minson, 1969
C-4 grasses	-125	1.27	37.8	Navaratne et al., 1990
Legumes	-4.1	1.02	16.0	Terry et al., 1978
Legumes	-9.8	1.03	19.4	Omed et al., 1989
C-3 grasses and legumes[3]	-10.1	0.99	23.1	Tilley and Terry, 1963
C-3 grasses and legumes	-48.2	1.08	19.3	Omed et al., 1989
Corn silage	29.3	0.58	21.1	Aufrère et al., 1992
Concentrates	-26.6	1.10	50.1	Omed et al., 1989

[1] In vivo = a + b * IVDMD.
[2] Standard error of prediction.
[3] Equation for predicting digestible DM, not OM.

For research purposes, accuracy is generally more important than analytical cost and time. Because of the potential for interactions between diet of the donor animal and substrate, the diet fed to the donor animal should mimic experimental treatments as closely as possible. For example, if the digestibility of different alfalfa varieties is of interest, the donor animal should be fed alfalfa. If the effect of protein supplementation on digestibility is of interest, donor animals should be fed diets containing the different dietary treatments and the buffer should not be supplemented with N. Associative effects between forage and grain influence forage digestibility; therefore, if concentrate supplementation is of interest, the donor animals should be fed diets with appropriate forage:concentrate ratios. For commercial applications, a compromise between accuracy, cost, and turn-around time must be reached. A commercial laboratory cannot maintain donor animals that are fed an almost infinite number of different diets. Therefore, these types of laboratories should maintain donor animals on a constant diet over time. The diet should be predominantly forage, but be supplemented with adequate CP and minerals so that they are not limiting. Low quality forages should be avoided because poor precision can result. High quality legumes can be relatively expensive and may reduce differences in digestibility among samples. The best compromise is probably good quality grass-legume hay containing about 140 g/kg of CP. A calibration equation then would be generated using in vitro data obtained from donor animals fed that particular diet. If the diet fed to the donor animal changes appreciably (e.g., grass to legume), then a new calibration equation should be generated.

Accuracy of the two-stage method is its salient feature; however, the method has limitations. One limitation of the method is the need for ruminally cannulated animals. To overcome this requirement, El Shaer et al. (1987) used sheep feces as an inoculum source. Subsequent testing found that inoculum from feces and ruminal fluid produced similar IVDMD values (Omed et al., 1989). The two-stage method has a relatively long assay time. The original method of Tilley and Terry (1963) required 48 h fermentation followed by 48 h pepsin digestion. The pepsin digestion step can be deleted when the residue after in vitro fermentation is extracted with neutral detergent (ND) solution (Goering and Van Soest, 1970). This procedure results in an estimate of true digestibility, but because of the constant digestibility of neutral detergent soluble material, equations can be used to convert this value to IVDMD. The accuracy of the original Tilley and Terry method suggests that a 48 h fermentation period is the overall optimal incubation period when accurate estimates of in vivo digestibility are desired. The length of the in vitro fermentation, however, can be altered depending upon what data are desired. No single time point will be best for all forages and applications because residence time in the rumen is not constant. For research purposes, the fermentation end point should be based on estimated or actual ruminal turnover data. For example, high producing dairy cows have relatively rapid turnover times and a 48 h in vitro incubation may overestimate in vivo digestibility. A beef cow fed at maintenance will have a relatively slow turnover rate and a 48 h in vitro fermentation may underestimate actual digestibility. For most forages, a 48 h incubation will not produce maximal digestibilities. When estimates of maximal digestibility are desired, incubation periods exceeding 96 h usually are required. Longer incubation times are associated with reduced variation between duplicate samples (Barnes, 1967). Long incubation times (>72 h) produce less accurate values, however, because of the disparity between in vivo turnover time and the in vitro incubation time (McLeod and Minson, 1969).

In Vitro Digestion Kinetics

Digestibility of fiber is dependent on duration of fermentation. Effect of time can be used to determine rates of fiber digestion by stopping fermentation at various times (e.g., McLeod and Minson, 1969; Smith et al., 1971). Accurate data concerning the digestion kinetics of feed are useful in the development of dynamic models for predicting digestibility. Very little work has been conducted examining factors affecting the kinetic data generated in vitro. Certain factors could have a large effect on lag time and rate, but only a minimal effect on 48 or 72 h disappearance values. Grant and Mertens (1992b) conducted a series of experiments to determine sources of variation in the in vitro assay when digestion kinetic measures were being estimated. They recommended that in vitro systems used to measure digestion kinetics should be gassed continuously with carbon dioxide, and the media should include a reducing agent and nutritional additives (minerals and N) so that the system does not limit microbial action. Grant and Mertens (1992a) developed and tested different buffering systems for use in measuring in vitro disappearance kinetics. These buffer systems will be useful in monitoring the effects of different ruminal pH on digestion.

Fermentation Endproducts

Several different techniques have been proposed to estimate digestibility based on production of specific fermentation endproducts such as volatile fatty acids (VFA) and total gases. Menke et al. (1979) reported a high correlation between digestible OM and gas production in vitro (r = 0.96, RSD = 19 g/kg of digestible OM). The method used in that study was essentially a standard in vitro incubation, but the incubation vessel was a syringe. Gas production (change in volume) was measured after 24 h of incubation. The method was not highly precise and required several replicates (triplicates) and gas production within a run had to be corrected based on standards. Regression equations were needed to convert gas production to digestible OM. Because of the empirical nature of the regression equations, different laboratories would have to generate their own equations based on feedstuffs with known digestibilities. The advantage of this method over the two-stage method is that measuring gas production is quicker (24 h vs. more than 48 h) and requires less technician time (fewer weighing steps and no transfer or centrifugation steps). This method, however, has not undergone the rigorous testing of the Tilley and Terry method. Additional research is needed to quantify various sources of error and to independently verify the method.

Volatile fatty acid production in vitro is correlated highly to digestibility (Dennison and Phillips, 1983). The relationship between VFA production and digestibility was not constant among forages; therefore, different equations would be needed for different forages. This method appears to have no advantage over the in vitro disappearance assay and would require more sophisticated equipment (gas chromatograph). Specific information on VFA production, however, could be useful in generating data on the metabolizable energy value of forages.

Heat generated during fermentation has been used to estimate digestibility (Arieli and Werner, 1989). Feeds were incubated using the standard in vitro method and samples were removed from the flasks at various times. Samples were placed in a calorimeter and fermentation heat measured. Heat of fermentation was correlated highly ($r=0.98$) to OM digestibility (Arieli and Werner, 1989). This method has the potential of providing estimates of available energy (not just digestibility), but because a calorimeter is required, commercial application of the method is limited. Additional research is needed on this method.

ENZYMATIC METHODS

Biological in vitro assays have the problems of uncontrollable variation (e.g., variation in the activity of rumen fluid) and they require access to animals (in most cases, ruminally cannulated animals). To overcome these problems, enzymatic digestion techniques have been developed. These assays can be standardized and much of the run-to-run variation and inter- and intra-laboratory variation removed. The first reported use of enzymatic (cellulase and (or) pepsin) digestion to estimate digestibility of forage did not produce encouraging results (Donefer et al., 1963). The hydrolytic activity of the cellulase (cellulase-36, source not identified) was very low, and produced DM disappearance values similar to those found when forages were extracted with water. Since that study, different methods have been developed that have resulted in high correlations between enzymatic digestion and in vivo or in vitro digestibility for a variety of forages (Guggolz et al., 1971; Jones and Hayward, 1973; Jones and Hayward, 1975; McQueen and Van Soest, 1975; Terry et al., 1978; Dowman and Collins, 1982; DeBoever et al., 1988; Dickerson et al., 1988; Steg et al., 1990).

Several different methods have been published describing the use of enzymes to estimate digestibility of forages. One-stage methods consist of incubating a feed sample (generally ground through a 1 mm screen) for 48 to 72 h in a buffer solution containing cellulase. The solution is filtered and the residue is weighed. Disappearance of DM is generally substantially less than either in vitro or in vivo digestibility (Jones and Hayward, 1973). The relatively low disappearance values obtained from one-stage methods led to the development of two-stage procedures. Treatment of samples with pepsin in HCl prior to cellulase treatment increased disappearance of DM as compared to cellulase treatment alone (Jones and Hayward, 1975; Bughrara and Sleper, 1986). Cellulase digestion followed by pepsin digestion did not increase DM disappearance as much as did pretreatment with pepsin. Generally, pretreatment with HCl-pepsin involves incubating the sample in a mixture of HCl (approximately 0.1 N) and pepsin (0.2% wt/vol) for 24 to 48 h at approximately 40 to 50°C. The pepsin solution is removed and cellulase solution added. Extracting samples with ND solution prior to cellulase treatment also increased DM disappearance as compared to no pretreatment (Roughan and Holland, 1977; Bughrara and Sleper, 1986). Pretreatment involves incubating the sample in hot, not boiling, ND solution for about 1 h at 100 to 110°C. The ND solution is removed and cellulase solution added. For both pepsin and ND pretreatment, cellulase incubation usually lasts 24 to 48 h. Pretreatment of forages prior to cellulase treatment probably increases the access of the enzyme to cell wall components, perhaps by solubilizing some hemicelluloses (pepsin-HCl treatment), removing interfering compounds such as proteins and starch, or by disrupting the structure of the cell wall. Pretreatment with ND solution requires less time than pretreatment with pepsin-HCl, but slightly higher precision with the pepsin-HCl method than with the ND method has been reported (DeBoever et al., 1988; Davis et al., 1990). Both methods produce DM disappearance values that are correlated to in vivo or in vitro digestibility estimates (Bughrara and Sleper, 1986). For forages containing appreciable quantities of starch (e.g., corn silage), pretreatment with amylase is recommended (Dowman and Collins, 1982). DeBoever et al. (1988) reported that starch also could be removed from high starch samples by altering the standard pepsin-HCl pretreatment to include a short, high temperature pepsin-HCl treatment (45 min at 80°C).

Cellulases from different sources vary greatly in their ability to digest forages (Jones and Hayward, 1975; McQueen and Van Soest, 1975; Clark and Beard, 1977; Dowman and Collins, 1982; Gabrielsen, 1986; Coelho et al., 1988; DeBoever et al., 1988; Bughrara et al., 1992). Cellulases from *Trichoderma* sp., particularly *T. reesei* (formerly called *T. viride*), typically digest the greatest amount of DM from a variety of forages as compared to other sources of cellulase (Jones and Hayward, 1975; McQueen and Van Soest, 1975; Clark and Beard, 1977; Dowman and Collins, 1982; Coelho et al., 1988; DeBoever et al., 1988). Gabrielson (1986) reported, however, that cellulase from *Pennicillium funicillium* exhibited greater hydrolytic activity than did cellulases from *T. reesei* when smooth bromegrass was the substrate. Differences among hydrolytic activity of various cellulases exist even when excess enzyme is present (Gabrielson, 1986). This finding suggests that the hydrolytic activity of certain cellulases is limited by assay conditions or the chemistry and (or) structure of forage cell walls and not by enzyme concentrations. Although different sources of cellulase result in different DM disappearance values (in some cases extremely large differences), many sources of cellulase have resulted in disappearance values that are correlated highly to in vivo or in vitro digestibility data. Because digestibilities obtained using enzymes are often much lower than those obtained from in vitro or in vivo assays, regression equations are needed to predict digestibility from enzymatic disappearance. The regression coefficients are affected by forage species (Terry et al., 1978; Dowman and Collins, 1982; Gabrielsen, 1986; DeBoever et al., 1988; Dickerson et al., 1988; Omed et al., 1989), method of pretreatment (Bughrara et al., 1989; Davis et al., 1990), and source of enzyme (Clark and Beard, 1977; Dowman and Collins, 1982; Gabrielsen, 1986). Because of the many factors affecting the regression equation, individual laboratories probably would need to generate their own equations when accurate estimates of digestibility are required. As with in vitro digestibility, a sample set with known in vivo digestibilities would need to be analyzed for enzymatic disappearance and various regression equations generated. Separate regression equations probably would be needed for grasses, legumes, and high starch forages. Prediction errors when enzymatic disappearance was used to estimate in vivo digestibility have been reported to be higher (Clark and Beard, 1977; Terry et al., 1978; DeBoever et al., 1988) and lower (Jones and Hayward, 1975; Dowman and Collins, 1982; Davis et al., 1990) than when in vitro disappearance is used. Two of the studies that reported lower SE_p for enzymatic disappearance than for in vitro used samples from a single feedstuff class (e.g., cool season grasses). When all studies are examined, SE_p for disappearance using cellulase when preceded by ND or pepsin-HCl treatment ranged from about 20 to 40 g/kg of digestible DM.

When a simple relative ranking of forage digestibility is the objective, enzymatic digestion shows great promise. Within a forage species, enzymatic digestion ranked samples consistent with in vitro data for alfalfa (Bughrara et al., 1989), birdsfoot trefoil (*Lotus corniculatus* L.) (Casler and Sleper, 1991), corn silage (Aufrère et al., 1992), and grasses (Davis et al., 1990). In most of these studies, samples were treated with pepsin or ND solution prior to cellulase treatment; however, Davis et al. (1990) concluded that when an estimate of actual digestibility is not needed, no pretreatment was necessary to obtain accurate rank data for birdsfoot trefoil. The method proposed by Davis et al. (1990) would require significantly less time and labor, thereby increasing sample through-put. Before cellulase incubation without pretreatment with pepsin or ND solution can be recommended as a routine analysis to rank forages, data are needed from other forage species.

IN SITU DISAPPEARANCE

In situ or in sacco disappearance is measured by placing feedstuffs in a fabric bag and then incubating the bag in the rumen of an animal. The first reported use of this technique was by Quin et al. (1938) who used silk pouches. Since that time, the method has gained widespread application and is used frequently to measure rate and extent of digestion. In situ data have been used for relative comparisons among feeds and as an intrinsic characteristic of feeds for use in diet formulation. Excellent reviews on the procedure have been published (Lindberg, 1983; Nocek, 1988).

As with any analytical method, in situ disappearance procedures must be standardized to increase precision. Furthermore, in situ results obtained using different analytical techniques must be compared to in vivo results to determine which method is most accurate. Many studies have been conducted to determine sources of variation for the in situ method (Van Keuren and Heinemann, 1962; Figroid et al., 1972; Mehrez and Ørskov, 1977; Van Hellen and Ellis, 1977; Playne et al., 1978; Ehle et al., 1982; Lindberg, 1983; Weakley et al., 1983; Nocek, 1985; van der Koelen et al., 1992). Considerably fewer studies have been conducted to determine how to make in situ data more accurate.

Bag Characteristics

Fabric type, size and uniformity of pores, and bag size can affect in situ results. The initial experiments used bags made of silk, but this material was soon replaced by synthetic fabrics. Nylon has a high concentration of N as compared to dacron (Weakley et al., 1983) which could be a disadvantage when using in situ techniques to measure protein digestion. When using woven fabrics, the type of fabric probably has little effect on in situ results if the fabrics have similar pore size.

Pore size had a significant effect on in situ results. Theoretically, pores must be large enough to allow free exchange of fluid and microorganisms between the bag and the ruminal milieux, but small enough to prevent the loss of indigestible particles or the entry of feed particles. In practice, obtaining the perfect compromise for pore size is extremely difficult. When forages were incubated in woven fabric bags with pore sizes 10 μm or less, total and cellulolytic bacterial counts, protozoal counts, and pH within the bags were depressed (Lindberg et al., 1984; Meyer and Mackie, 1986). In situ DM disappearance generally increases as pore size increases above 10 μm (Lindberg and Varvikko, 1982; Uden and Van Soest, 1984; Nocek, 1985). On the other hand, as pore size increases above about 10 μm, loss of particles caused by washout increases (Lindberg and Knutsson, 1981). When bags with large pore sizes are used (>100 μm), influx of material into the bag can reduce apparent disappearance greatly, even resulting in negative digestibilities (Van Hellen and Ellis, 1977). The general conclusion from most studies is that for high precision, woven material bags should have a pore size 40 to 60 μm (Lindberg and Varvikko, 1982; Nocek, 1988). This recommendation does not hold true for other types of fabric. Marinucci et al. (1992) showed that disappearance from bags made of a polyester with a uniform pore size (50 μm) was lower than that from bags made of dacron (a woven fabric) with a maximal pore size of 50 μm. They showed that the physical actions of ruminal incubation changed the size of the pores in dacron bags but not in the polyester bags. Van Hellen and Ellis (1977) reported bags made of a controlled porosity (0.2 to 5 μm) membrane resulted in in situ DM disappearances similar to bags made of nylon with a pore size of approximately 75 μm. Insufficient data are available at this time to make general recommendations for these types of fabrics.

The size of the bag probably has minimal independent effect on in situ results, but the ratio of sample size to bag surface area has a large impact on in situ data. As sample size increases relative to surface area of the bag, rate and extent of disappearance decrease (Mehrez and Ørskov, 1977). For woven fabric bags with pore sizes between 40 and 60 μm, sample mass to bag surface area should be no greater than 10 mg of DM/cm^2 (Lindberg, 1983; Nocek, 1988). The sample mass to bag surface area ratio for bags with larger or smaller pore sizes has not been studied adequately, but Lindberg (1983) suggests that for bags with pore sizes of about 10 μm, the ratio should be approximately 5 mg of DM/cm^2. The size of the bag should be determined by the amount of sample needed after incubation to conduct the desired assays and maintain the correct mass to surface area ratio.

Sample Characteristics

Because of the physical effects of mastication and rumination, particle size of samples within a bag cannot mimic what is occurring in the rumen. Homogeneity of the sample, initial particle size of the forage, and the desired application of the in situ data should be considered when determining the particle size of the sample. For forages, many studies have shown that grinding samples through screens ranging from about 0.3 to 5 mm had minimal effect on in situ disappearance measures (Van Keuren and Heinemann, 1962; Playne et al., 1978; Nocek and Kohn, 1988), especially when extent of digestion (48 h) was the desired measure. Increasing particle size may increase lag time and decrease the extent of disappearance for short-term incubations (Nocek and Kohn, 1988). Since few data are available on the accuracy of in situ disappearance, precision should be used as the determinant of the proper analytical technique. Wide variability in particle size decreases the precision of in situ measures (Nocek and Kohn, 1988). Until data are available on the effect of particle size on the accuracy of in situ analysis, samples should be milled through a relatively small screen (< 5 mm screen) to increase precision.

The method of processing samples prior to grinding on in situ incubation has not been studied extensively. Playne et al. (1978) reported that oven drying (60°C), freeze drying, or no drying had little effect on in situ disappearance of esophageal samples. Drying at temperatures above 60°C has been shown to alter the chemical composition of samples and should be avoided.

Dietary Effects

The diet fed to the recipient animal affects in situ disappearance measures. The single most important dietary effect is caused by the forage to concentrate ratio (more specifically the amount of starch) fed to the recipient animal. Several studies have shown that increasing the amount of grain fed to the recipient animal decreases extent and (or) rate of in situ DM disappearance of forages (Miller and Muntifering, 1985; Flachowsky and Schneider, 1992; Marinucci et al., 1992). On the other hand, de Faria and Huber (1984) reported no effect on in situ DM disappearance of corn silage, alfalfa hay, and orchardgrass (*Dactylis glomerata* L.) hay when incubated in cows fed varying ratios of alfalfa silage and corn grain. In general, the effect of increasing grain fed to the recipient animal on in situ measures is greater for lower quality forages and roughages. However, interactions between the type of feed being assayed and the diet fed to the recipient animal are prevalent (Lindberg, 1981). Because of interactions between sample type and the forage to concentrate ratio of the diet, ranking of different feedstuffs based on in situ disappearance probably will depend on the amount of grain being fed to the recipient animal. Absolute values for extent and rate of in situ DM disappearance also can be affected by the amount of grain being fed to the recipient animal.

The type of forage fed to the recipient animal can affect in situ disappearance of forages, but results have been inconsistent. Van Keuren and Heinemann (1962) reported that in situ disappearance of orchardgrass DM was higher when recipient animals were fed alfalfa hay as compared to alfalfa:orchardgrass (65:35) pasture, but reported opposite results for a second experiment. In those studies, ranking of forages based on in situ digestibility (alfalfa, orchardgrass, and sudangrass) was not affected by diet fed to the recipient animal. Hopson et al. (1963) reported that in situ disappearance of cellulose from different forages [alfalfa, smooth bromegrass, and timothy (*Phleum pratense* L.)] was not affected by the species of grass (smooth bromegrass or timothy) fed to the recipient cow. Cellulose disappearance was greater, however, when forages were incubated in rumens of animals fed alfalfa. Neathery (1969) reported in situ disappearance was greater for a variety of feedstuffs (alfalfa, bermudagrass, cottonseed hulls, corn cobs, soybean (*Glycine max* L.) straw, and corn stalks) when incubated in animals fed bermudagrass than when incubated in animals fed alfalfa. Ranking of feeds based on disappearance was not affected by type of forage fed to the recipient animal in any of these studies; however, absolute values were influenced by diet (Van Keuren and Heinemann, 1962; Hopson et al., 1963; Neathery, 1969). Based on these data, when a relative ranking is the desired outcome, the type of forage fed to the recipient animals probably is not important, but when actual values are required for diet formulation or other purposes, the type of forage fed should be similar to the type of forage being evaluated.

Analytical Techniques

Many technical aspects of the in situ assay have not yet been standardized. This lack of standardization is a major source of variation in the assay. The method of inserting bags into the rumen, the location of the bags within the rumen, and rinsing technique can affect in situ results. For kinetic studies, several bags are placed in the rumen and allowed to incubate for various times. The standard method was to place all bags in the rumen at the same time, and then remove bags at various times. Alternatively, bags can be placed into the rumen at various times and all bags removed at the same time (complete exchange method). Paine et al. (1982) found no difference between methods for rate or extent of in situ DM disappearance of forages, but outlined several benefits of the complete exchange method. Benefits include more efficient use of labor and potentially reduced variation in bag rinsing technique. Nocek (1985) reported, however, that the use of the standard method as compared to the complete exchange method resulted in slower rates of DM disappearance for soybean meal. Nocek (1985) postulated that when bags were removed, the digestive processes occurring in the remaining bags were disrupted. In general, because of the more efficient use of labor and the potential for less variation in rinsing technique, the complete exchange method is superior to the standard method. The standard method may be more accurate when in situ results are being extrapolated to once a day feeding situations. When animals are fed once daily, conditions in the rumen show a definite diurnal pattern which may affect rate of digestion (e.g., pH is lower shortly after feeding than at other times). When all bags are placed in the rumen at the time of feeding, the diurnal pattern exerts its effects on all bags. When bags are placed into the rumen at different times, the diurnal pattern exerts effects differently on different bags. Feeding animals more frequently and feeding a total mixed diet will greatly diminish diurnal patterns.

Generally, bags are placed in the ventral sac of the rumen, but few data are available comparing in situ disappearance when bags are placed in different areas of the rumen. Regardless of where bags are placed in the rumen, they must be in contact with ruminal contents. When this contact is reduced, disappearance is reduced (Udén and Van Soest, 1984; Marinucci et al., 1992). In those studies, contact between the bag and ruminal contents was limited by placing bags in a rigid container prior to incubation, but the results suggest that overloading the rumen with bags in such a manner to limit contact between individual bags and ruminal contents may affect in situ disappearance results. The value of weighting the bags (attaching the bags to chains or placing the in situ bags in a large mesh bag containing a weight) to prevent floating in the rumen is uncertain. Rodriquez (1968) reported that the use of weights reduced bag-to-bag variation whereas Mehrez and Ørskov (1977) reported no effect.

Probably the single most variable component of the in situ assay (other than biological variation) is the method used to wash the bags after incubation. A common method of rinsing bags is to soak or agitate bags in warm water until the rinsing solution is clear. This method can be quite subjective. Cherney et al. (1990) compared the effects of rinsing bags by hand to using a washing machine and found little difference in disappearance values, but machine rinsing for 2 min produced the lowest SE. In that study, only one technician was used to rinse the bags and all bags were rinsed at the same time, thereby removing those sources of variation.

Correction for Microbial Contamination

Bacterial attachment to particles within bags can inflate the amount of DM present, thereby reducing apparent extent and rate of DM digestion. Olubobokun et al. (1990) reported that microbial matter accounted for 100 to 200 g/kg of the DM found in in situ bags. The rate of in situ DM digestion was 29 to 42% higher when residue weights were corrected for bacterial attachment (using diaminopimelic acid as the marker) than when uncorrected weights were used (Olubobokun et al., 1990). Correcting for bacterial attachment also essentially eliminated lag time. On the other hand, Varvikko and Lindberg (1985) reported that correcting for bacterial contamination (using ^{15}N as a marker) had only a minimal effect on DM disappearance of perennial ryegrass. The effect was more pronounced when straw was the substrate. Nocek (1988) suggests that microbial correction is especially important for lower quality forages, but the data of Olubobokun et al. (1990) suggest that the correction also may be important for higher quality forages. The degree of bacterial contamination is not constant among different substrates. Forage species (Varvikko and Lindberg, 1985; Olubobokun et al., 1990) and N fertilization of forages (Messman et al., 1991) have been shown to affect the degree of bacterial contamination. This inconsistency can lead to misinterpretation of treatment differences for uncorrected DM disappearance. Correcting for microbial attachment, however, adds an additional variable to the assay. A discussion of appropriate microbial markers is beyond the scope of this paper, but prior to correcting for microbial attachment, sufficient consideration must be given to the choice of an appropriate marker (Broderick and Merchen, 1992). An inappropriate marker may add more bias to in situ results than not correcting for microbial attachment.

Statistical Considerations

After the factors outlined above are standardized, there are three additional sources of variation: variation among identical samples incubated in the same animal at the same time; variation among identical samples incubated in the rumen of different animals at the same time; and variation among identical samples incubated in the same animal at different times. Mehrez and Ørskov (1977) using barley as the substrate reported that the variation was greatest among animals, followed by among-day variation, with among-bag variation being the least important source of variation. Michalet-Doreau and Ould-Bah (1992) reported similar results for dehydrated alfalfa meal. The variances for barley (Mehrez and Ørskov, 1977) were four to five times greater than those for alfalfa meal (Michalet-Doreau and Ould-Bah, 1992). Presumably, starch from the barley made washing and filtering difficult which would add to the analytical error. van der Koelen et al. (1992), using total mixed diets as the substrate, reported that among-bag variation was most important and among-animal variation least important. All three studies reported that variation among bags, animals and time were significant and must be considered when determining the amount and type of replication needed. For widest application of results, incubations should be replicated over animals, time, and within animals and time. Based on the variances determined by van der Koelen et al. (1992), duplicate bags should be replicated in two or three animals at two separate time points (8 to 12 total bags per treatment) to obtain the necessary statistical power to detect differences of about 10% in DM disappearance. Playne et al. (1978) reached a similar conclusion.

Typically, the CV for in situ DM disappearance for bags replicated across animals and time range from 2 to 10 with 5 being average. This degree of error is quite similar to in vitro disappearance. As with the in vitro disappearance method, the use of standards has been suggested as a means of reducing variation among animals and time periods. Ayres (1991) presents guidelines for the use of standards for in vitro disappearance; the same guidelines should be applied to in situ disappearance data. That is, standards could be used to reject an entire run, but until population statistics are derived for the standards, they should not be used to adjust in situ disappearance values.

The in situ method often is used to generate rates and extents of digestion. Sampling scheme (Fadel, 1992), choice of kinetic model (Nocek and English, 1986), and method of deriving kinetic estimates (Mertens and Loften, 1980; Nocek and English, 1986; Messman et al., 1991) affect the estimates of pool size, rate, extent, and lag time. Fadel (1992) reported that sampling schedule did not greatly affect goodness-of-fit, but affected the estimates for lag time and rate. He reported that a schedule that includes many samples during the early phase of incubation followed by fewer sampling points was the optimal design. Nocek and English (1986) concluded that residue remaining in bags must be corrected for indigestible matter, but no single kinetic model worked best for all feeds. Estimates of kinetic data can be obtained by regressing the natural logarithm of residue mass on time or by using iterative nonlinear fitting programs. Mertens and Loften (1980) reported that iterative programs produced equations with a better fit of the data than did the logarithm regression method; however, Messman et al. (1991) reported that neither method consistently produced better fitting equations. Regardless of the method chosen to derive kinetic estimates, the actual data should be plotted and compared visually to the derived model to assess the adequacy of the model.

Accuracy of In Situ Disappearance

Very few papers have been published comparing in situ data to actual in vivo data. One reason for the paucity of data is that deciding what in vivo measure to compare to in situ data is difficult. In situ disappearance measures only events occurring in the rumen, whereas in vivo digestion experiments usually measure total tract digestibility. Ruminal digestibility can be measured in vivo using duodenally cannulated animals, but information on rate of passage and microbial contamination are needed. Obtaining these measures introduce additional errors. Lindberg (1983) reported that in situ disappearance of OM was 10 to 50% higher than in vivo ruminal disappearance. After they adjusted for microbial contribution, the estimates were much more similar. The accuracy of the estimates were dependent upon the pore size of the bags.

The duration of the incubation period can be varied depending on the desired use of the data. When estimates of digestion rate and extent are desired, a series of bags are needed. These kinetic parameters then can be incorporated into models so that effective degradability can be estimated (Mir et al., 1991). Effective degradability is an estimate of the extent of ruminal digestion that occurs with specific rates of passage. When a single time point will be used to estimate digestibility, the length of the incubation should be chosen based on estimated passage rate. The proper passage rate will depend on the DM intake of the animal being fed, type of feedstuff, and composition of the total diet. No single end point will be correct for all circumstances. In situ DM disappearance values (72 h) for forages were higher than digestibilities determined in vivo using sheep and cattle fed at or above maintenance requirements (Judkins et al., 1990; Messman et al., 1991; Flachowsky and Schneider, 1992). When in situ DM disappearance was measured at 48 h, Lindberg (1983) and Flachowsky and Schneider (1992) reported values fairly similar to in vivo DM digestibility values (in vivo digestibility determined by feeding cattle and sheep at or above their maintenance requirements). Judkins et al. (1990), however, reported that 48 h in situ DM disappearance was substantially higher than in vivo digestibility (above maintenance). In most studies, in situ DM disappearance (at either 48 or 72 h) was correlated to DM digestibility in vivo and ranked different forages fairly consistently with in vivo data (Chenost et al., 1970; Judkins et al., 1990; Messman et al., 1991; Flachowsky and Schneider, 1992); however, because of large numerical differences, in situ disappearance estimates should not be used as an estimate of in vivo digestibility. Regression equations can be developed to predict in vivo total tract digestibility from in situ data.

Total tract digestibility can be simulated by using the mobile bag technique. In this technique, bags are first incubated in the rumen and then inserted into the duodenum via a canula. The bags are collected at the ileum or rectum (Erasmus et al., 1990; Varvikko and Vanhatalo, 1990). Several experiments have been conducted using the mobile bag technique to evaluate the protein fraction of feeds. Few data are available on ability of this technique to estimate total tract DM digestibility.

CHEMICAL COMPOSITION AND STATISTICAL MODELS

Determining the chemical composition of feeds is usually much easier, quicker, and cheaper than conducting in vitro, enzymatic, or in situ digestibility. Literally hundreds of experiments have been conducted to determine the relationships between various feed fractions and digestibility. The objective of this section is not to cite or list all the individual equations that have been derived, but instead to discuss various types of equations and the feed fractions best suited for estimating digestibility. Minson (1982) and Fonnesbeck et al. (1984) produced extensive compilations of equations that can be used to estimate digestibility from chemical composition data.

Forage Constituents

For purposes of predicting digestibility, feed constituents or fractions should be categorized as uniform or nonuniform based on the Lucas test. The Lucas test is conducted by regressing the concentrations of a digestible nutrient in a feed on the concentrations of the total nutrient in the feed. The slope of the line represents the true digestibility of the nutrient and the intercept corresponds to metabolic fecal output (Van Soest, 1982). Uniform fractions have a constant digestibility and metabolic fraction (within an animal) regardless of feedstuff. The digestibility of nonuniform fractions varies among and within feeds. Theoretically, if forages could be described completely in terms of uniform fractions, digestibility could be estimated accurately by simply determining the composition of the forage.

Feed analysis systems that can be used to describe feeds in terms of uniform fractions are needed. One reason the proximate analysis system of describing feed in terms of crude fiber and nitrogen-free extract has been largely abandoned is because the nutritional value of those two fractions is so variable. The detergent system of fiber analysis (Van Soest et al., 1991) separates feeds into a combination of uniform and nonuniform fractions. Lignin and neutral detergent solubles (NDS) are uniform fractions. Neutral detergent fiber (NDF) and acid detergent fiber (ADF) are nonuniform fractions. Van Soest (1982) summarized several experiments designed to determine the true digestibility of NDS and found that it averaged about 98% (at maintenance feeding). Lignin, when measured using the ADF-sulfuric acid method, is a uniform fraction because it is essentially indigestible.

Empirical Equations

The use of empirical equations is probably the most common way commercial feed testing laboratories estimate digestibility. Empirical equations are derived by regressing digestibility on the concentration of a particular chemical fraction (usually ADF, NDF, crude fiber, lignin, cellulose, or CP). Fiber components are correlated negatively to digestibility and CP is correlated positively to digestibility (Minson, 1982; Fonnesbeck et al., 1984). A high correlation between a feed constituent and digestibility does not necessarily imply a direct cause and effect relationship, nor does a high correlation imply that the equation is accurate (Weiss, 1993). The correlation coefficient (r) or coefficient of determination (r^2) are indices of how well a population of data fits an equation; they give no information on how well a single observation fits the model. Empirical equations represent statistical relationships; hence, they are population-dependent. The regression coefficients are based on the database used in the regression; a different dataset will produce different coefficients. Even when datasets contain similar types of samples, regression coefficients vary among experiments. For example, three equations derived to estimate OM digestibility (g/kg) of grasses from their concentrations of ADF (g/kg) had slopes ranging from -1.1 to -1.4 and intercepts ranging from 1060 to 1180 (Sullivan, 1964; Bosman, 1970). Because of the dependency of the empirical coefficients on the population used to derive the equation, new samples should be very similar to that population. However, the nutritional value of feedstuffs, especially forages, changes continuously because of growing conditions and postharvest factors. No forage grown in the present will be truly represented by a population of samples gathered in the past.

More accurate estimates are obtained when empirical equations are generated from well defined populations (Minson, 1979). Different equations should be used for major forage classifications (grasses, legumes, and corn silage). Forages, therefore, must be classified which can be difficult for forage mixtures. For some forages (e.g., warm season grasses, small grain forages, and annual legumes), very limited data are available on relationships between composition and digestibility and equations are difficult to find. Within forage classifications, geographical and climatic conditions affect composition and digestibility. Equations generated for specific regions will be more accurate than equations generated from data gathered over a diverse area.

Acid detergent fiber is used most frequently by U.S. feed testing laboratories to estimate digestibility (Nichols and Dixon, 1984). Reported correlations between concentrations of ADF and digestibility of OM or DM vary between -0.5 and -0.95 (Minson, 1982). Generally, higher correlations are found when grasses and legumes are separated than when several types of forage are included in the population (Clancy and Wilson, 1966). Even though the correlation coefficients are relatively large, prediction error can be high. The average SE_p for ADF-based equations range from 20 to 60 g/kg of digestible DM (Minson, 1982). Abrams (1988) partitioned the error in predicting digestible DM from ADF and found that more than half of the total error was caused by lack-of-fit. In other words, more than half the error in predicting digestibility from ADF was caused by the equation. Abrams (1988) also placed an economic cost to U.S. dairy producers of about 100 million dollars (based on 1987 U.S. prices) on that error. Equations based on ADF probably have gained widespread use, not because of their accuracy, but because the ADF assay is rapid and inexpensive. Acid detergent solution does not partition feed into uniform fractions. The digestibilities of ADF and the non-ADF fractions are not constant among feeds which makes ADF mechanistically incorrect for predicting digestibility (Van Soest et al., 1991). However, if proper precautions are followed, ADF can be used to predict digestibility. The unknown sample must be similar to the samples used to generate the equation. The ADF concentration of the unknown should not differ greatly from the mean ADF concentration of the population used to derive the equation. Finally, the accuracy of the equation should be checked using an independent data set of similar samples (Weiss, 1993).

Empirical equations based on NDF (Bosman, 1970; Harlan et al., 1991) and crude fiber (Sullivan, 1964) are no better than ADF equations with respect to accuracy. Empirical equations based on lignin tend to have lower prediction error than ADF equations (Sullivan, 1964; McLeod and Minson, 1971; Joshi, 1972), but the reduced error probably does not justify the expense and hazards of lignin assays. Crude protein has been used to estimate digestibility, but error is unacceptably high (Minson and Kemp, 1961; Fonnesbeck et al., 1984).

Theoretically-Based Equations

A robust method of estimating digestibility requires the use of theoretically correct coefficients instead of coefficients derived empirically (Goering and Van Soest, 1970; Minson, 1976; Osbourn, 1978; Weiss, 1993). Theoretically-based equations will be population independent; the same equation should work equally well for all types of feeds. Goering and Van Soest (1970) presented a summative model based on the uniformity of the NDS fraction [1].

$$DDM = (0.98 \times NDS) + (k_{d\text{-}NDF} \times NDF) - M \qquad [1]$$

where DDM = digestible DM, $k_{d\text{-}NDF}$ = the digestibility coefficient for NDF, and M = metabolic fecal excretion (approximately 130 g/kg). Theoretically, model [1] is sound, but $k_{d\text{-}NDF}$ is variable and usually unknown. Goering and Van Soest (1970) derived an empirical equation based on the lignin:ADF ratio to estimate NDF digestibility which then was used in equation [1].

Conrad et al. (1984) developed a theoretical model that separates feeds into NDS and potentially digestible NDF. The surface area of NDF that is covered by lignin is calculated using the surface area law (mass raised to the two-thirds power), and this proportion is multiplied by lignin-free NDF to obtain an estimate of potentially digestible NDF. The potentially digestible NDF then is added to the digestible NDS fraction. An expanded model that accounted for variations in protein digestibility caused by heat damage (Thomas et al., 1982) also was derived by Conrad et al. (1984). Based on data published by Girard and Dupuis (1988), Weiss et al. (1992) revised the Conrad et al. (1984) model. The model has been tested on a wide array of feeds (several different types of forages and concentrates) and was accurate and not biased. The SE_p when all feeds were included in the model was 25 g/kg of TDN (Weiss et al., 1992).

Summative models have the advantage of being able to be expanded to include additional sources of variation. The concentration of ash in forages can be extremely variable (Mertens, 1973). Ash can be subtracted from the NDS component so that digestible OM can be determined. Summative models also can include a lipid fraction so that digestible energy can be estimated. The lipid content of most forages is usually quite low (<30 g/kg of DM), but for universal equations (forages and concentrates), variation in the lipid concentrations can significantly impact digestible energy values (Weiss, 1993).

Currently, even the best theoretically-sound models do not account for all factors that influence digestibility. Particle size (Osbourn et al., 1976), density, grinding resistance, and hydration capacity (Mir et al., 1990) are related to digestibility. Current models do not include terms that account for physical characteristics of forages. The rates of digestion and passage vary among feeds. Both these rates influence the amount of feed that is actually digested (Allen and Mertens, 1988). The effects of variable rates of passage and digestion should be included in summative models. Associative effects among feedstuffs can affect digestibility (Hoover, 1986) and these need to be incorporated into models. Weiss et al. (1992) and Weiss (1993) suggested possible mechanisms of incorporating these factors into models, but these suggestions have not been tested because of the paucity of accurate kinetic data.

COMPARISONS AMONG METHODS

Several studies have been conducted comparing different methods of predicting digestibility (in vitro, in situ, enzymatic, and chemical) to actual in vivo digestibility. Only studies where all analyses, including in vivo digestibility, were conducted at a single location will be discussed. The largest number of comparison have been conducted by Givens and coworkers in Great Britain (Givens et al., 1989; Givens et al., 1990a; Givens et al., 1990b; Moss and Givens, 1990). Grasses were the predominant forage used in those studies. Based on prediction error, in vitro and enzymatic digestion (pepsin-cellulase or NDF-cellulase) were more accurate predictors of digestibility than was the use of chemical constituents and empirical equations. They found little difference between the enzyme methods and IVDMD. Aerts et al. (1977) reported that in situ (48 h) and IVDMD were substantially better at predicting digestibility of several different types of forages than was the use of chemical measures. Navaratne et al. (1990) found that in vitro and in situ methods were better than enzymatic methods for predicting the digestibility of lower quality tropical forages. Coelho et al. (1988) reported that IVDMD was the best method of estimating digestibility, followed by enzymatic methods, and lastly the use of chemical components. For corn silage, there was little difference among in vitro, enzymatic, or chemical methods in their ability to estimate digestibility, but none of the methods were very good (Aufrère et al., 1992).

In summary, most studies reported that methods utilizing rumen fluid (in vitro or in situ) were better than chemical methods for estimating digestibility (Table 2). Enzyme methods were consistently better than chemical methods and often as good as in vitro and in situ methods. Perhaps the best advice when choosing a method was offered by Judkins et al. (1990). They state that caution should be exercised when choosing a method because no single one probably will work for all situations.

Table 2. Error (SE_p) associated with predicting in vivo DM digestibility by using various methods (values expressed as g/kg, DM basis).

Feed	IVDMD	In situ	Enzymatic	Chem. comp.[1]	Reference
C-3 grass hays	20	15	--	42 (NDF) 38 (ADF)	Aerts et al., 1977
C-3 grasses	35	--	31	45 (NDF) 44 (ADF) 53 (Lig)	Givens et al., 1990a
C-3 grass silages	32	36	42	51 (ADF) 44 (Lig)	Barber et al., 1990
C-4 grasses	38	41	51	--	Navarante et al., 1990
Legumes	27	--	--	22 (ADF) 37 (Lig)	McLeod and Minson, 1976
Corn silages	21	--	20	23 (NDF) 21 (ADF)	Aufrère et al., 1992
Corn silages and C-3 grass silages	38	34	--	44 (NDF) 38 (ADF)	Aerts et al., 1977
Concentrates, forages, and roughages	54	--	32	92 (NDF) 132 (ADF)	Aufrère and Michalet-Doreau, 1988

[1] Chemical composition; the component in parenthesis was the independent variable; Lig = lignin.

SUMMARY AND RECOMMENDATIONS

The best currently available method for estimating digestibility is IVDMD. Many sources of variation have been identified and reduced so that analytical precision is adequate. Methods to convert results to actual digestibility are available so that accurate results can be obtained. The problems with the in vitro method are its long analysis time, cost, and the need for a cannulated animal. Enzymatic digestion holds great promise as a means of estimating digestibility on a commercial basis. The method is relatively inexpensive and can be conducted with commercially available enzymes. Before this method can be recommended widely, however, it needs to undergo the same degree of study as the IVDMD method. Currently, the enzymatic method is best suited for relative ranking of feeds. Equations will be needed to convert enzymatic digestion values to estimates of in vivo digestibility. These equations will be difficult to derive until a standard enzyme source is agreed upon. In situ disappearance appears similar to in vitro disappearance in terms of accuracy, and is somewhat easier to perform than the in vitro method (no incubation media needs to be prepared). Much progress has been made in standardizing the in situ method, but equations are still needed to convert in situ data to estimated in vivo digestibility. Estimating digestibility using composition data is the least accurate method. Empirical equations based on composition data should be used with a great deal of caution. Prior to using empirical equations, samples should be matched to an equation that was derived from very similar samples. Theoretically-based equations are more robust than empirical equations and should have widespread application. The disadvantage of these models is that more composition data are required than for empirical equations. Equations based on chemical composition only account for chemical factors that affect digestion. These models must be expanded to include physical characteristics of feeds, animal factors, and associative effects. A great need exists for accurate data on digestion kinetics for incorporation into dynamic models. The accuracy of in vitro and in situ methods of determining digestion kinetic variables must be validated. Quantitative data also are needed on factors affecting digestion kinetics.

REFERENCES

Abrams, S. M. 1988. Sources of error in predicting digestible dry matter from the acid-detergent fiber content of forages. Anim. Feed Sci. Technol. 21:205-208.

Aerts, J. V., D. L. De Brabander, B. G. Cottyn, and F. X. Buysse. 1977. Comparison of laboratory methods for predicting the organic matter digestibility of forages. Anim. Feed Sci. Technol. 2:337-349.

Alexander, R. H., and M. McGowan. 1966. The routine determination of in vitro digestibility of organic matter in forages - an investigation of the problems associated with continuous large-scale operation. J. Br. Grassl. Soc. 21:140-147.

Allen, M. S., and D. R. Mertens. 1988. Evaluating constraints on fiber digestion by rumen microbes. J. Nutr. 118:261-270.

Arieli, A., and D. Werner. 1989. A comparison between fermentation heat of forages and organic matter digestibility determined by in vitro incubation with rumen fluid. Anim. Feed Sci. Technol. 23:333-341.

Aufrère, D. G., C. Demarquilly, J. Andrieu, J. C. Emile, R. Giovanni, and P. Maupetit. 1992. Estimation of organic matter digestibility of whole maize plants by laboratory methods. Anim. Feed Sci. Technol. 36:187-204.

Aufrère, J., and B. Michalet-Doreau. 1988. Comparison of methods for predicting digestibility of feeds. Anim. Feed Sci. Technol. 20:203-218.

Ayres, J. F. 1991. Sources of error with in vitro digestibility assay of pasture feeds. Grass Forage Sci. 46:89-97.

Barber, G. D., D. I. Givens, M. S. Gridis, N. W. Offer, and I. Murray. 1990. Prediction of the organic matter digestibility of grass silage. Anim. Feed Sci. Tech. 28:115-128.

Barber, W. P., A. H. Adamson, and J.F.B. Altman. 1984. New methods of forage evaluation. p. 161-176. In W. Haresign and D.J.A. Cole (ed.) Recent advances in animal nutrition. Butterworths, London.

Barnes, R. F. 1973. Laboratory methods of evaluating feeding value of herbage. p. 179-214. In G. W. Butler and R. W. Bailey (ed.) Chemistry and biochemistry of herbage. Academic Press, London.

Barnes, R. F. 1967. Collaborative in vitro rumen fermentation studies on forage substrates. J. Anim. Sci. 26:1120-1130.

Baumgardt, B. R., J. L. Cason, and R. A. Markley. 1958. Comparison of several laboratory methods as used in estimating the nutritive value of forages. J. Anim. Sci. 17:1205 (Abstr.)

Bezeau, L.M. 1965. Effect of source of inoculum on digestibility of substrate in in vitro digestion trials. J. Anim. Sci. 24:823-825.

Bosman, M.S.M. 1970. Methods of predicting herbage digestibility. 2. Mededelingen, Instituut voor Biologische en Scheikundige Onderzoek Landbouwgewassen, Wageningen, No. 413.

Broderick, G.A., and N.R. Merchen. 1992. Markers for quantifying microbial protein synthesis in the rumen. J. Dairy Sci. 75:2618-2632.

Bughrara, S. S., and D. A. Sleper. 1986. Digestion of several temperate forage species by a prepared cellulase solution. Agron. J. 78:94-98.

Bughrara, S. S., D. A. Sleper, R. L. Belyea, and G. C. Marten. 1989. Quality of alfalfa herbage estimated by a prepared cellulase solution and near infrared reflectance spectroscopy. Can. J. Plant Sci. 69:833-839.

Bughrara, S. S., D. A. Sleper, and P. R. Beuselinck. 1992. Comparison of cellulase solutions for use in digesting forage samples. Agron. J. 84:631-636.

Calder, F.W. 1970. Effect of barley supplement to the ration of donor animals used in in vitro digestibility determination. Can. J. Anim. Sci. 50:265-267.

Casler, M. D., and D. A. Sleper. 1991. Smooth bromegrass digestibility tests: Fungal cellulase vs. in vitro rumen fermentation. Crop Sci. 31:1335-1338.

Chenost, M., E. Grenet, C. Demarquilly, and R. Jarrige. 1970. The use of the nylon bag technique for the study of forage digestion in the rumen and for predicting feed value. p. 697-701. In Proc. XI Int. Grassland Congr., Brisbane, Australia.

Cherney, D.J.R., J. A. Patterson, and R. P. Lemenager. 1990. Influence of in situ bag rinsing technique on determination of dry matter disappearance. J. Dairy Sci. 73:391-397.

Clancy, M. J., and R. K. Wilson. 1966. Development and application of a new chemical method for predicting the digestibility and intake of herbage samples. p. 445-453. In Proc. X Int. Grassland Congr., Helsinki, Finland.

Clark, J., and J. Beard. 1977. Prediction of the digestibility of ruminant feeds from their solubility in enzyme solutions. Anim. Feed Sci. Technol. 2:153-159.

Coelho, M., F. G. Hembry, F. E. Barton, and A. M. Saxton. 1988. A comparison of microbial, enzymatic, chemical and near-infrared reflectance spectroscopy methods in forage evaluation. Anim. Feed Sci. Technol. 20:219-231.

Conrad, H. R., W. P. Weiss, W. O. Odwongo, and W. L. Shockey. 1984. Estimating net energy lactation from components of cell solubles and cell walls. J. Dairy Sci. 67:427-436.

Davis, D. K., R. L. McGraw, D. A. Sleper, and P. R. Beuselinck. 1990. Using a cellulolytic complex to estimate in vitro digestibility of birdsfoot trefoil. Can. J. Plant Sci. 70:487-493.

De Boever, J. L., B. G. Cottyn, J. I. Andries, F. X. Buysse, and J. M. Vanacker. 1988. The use of a cellulase technique to predict digestibility, metabolizable and net energy of forages. Anim. Feed Sci. Technol. 19:247-260.

de Faria, V. P., and J. T. Huber. 1984. Influence of dietary protein and energy on disappearance of dry matter from different forage types from dacron bags suspended in the rumen. J. Anim. Sci. 59:246-252.

Dennison, C., and A. M. Phillips. 1983. Forage evaluation by analysis after fermentation in vitro. S. Afr. J. Anim. Sci. 13:222-224.

Dhanoa, M. S., and R. E. Deriaz. 1984. Variability of the in vitro digestibility of standard herbage samples. Grass Forage Sci. 39:17-25.

Dickerson, Jr., R. L., B. E. Dahl, and G. Scott. 1988. Cellulase vs rumen fluid for in vitro digestibility of mixed diets. J. Range Manage. 41:337-339.

Donefer, E., P. J. Niemann, E. W. Crampton, and L. E. Lloyd. 1963. Dry matter disappearance by enzyme and aqueous solutions to predict the nutritive value of forages. J. Dairy Sci. 46:965-970.

Dowman, M. G., and F. C. Collins. 1982. The use of enzymes to predict the digestibility of animal feeds. J. Sci. Food Agric. 33:689-696.

Ehle, F. R., M. R. Murphy, and J. H. Clark. 1982. In situ particle size reduction and the effect of particle size on degradation of crude protein and dry matter in the rumen of dairy steers. J. Dairy Sci. 65:963-971.

El Shaer, H. M., H. M. Omed, A. G. Chamberlain, and R.F.E. Axford. 1987. The use of faecal organisms from sheep for the in vitro determination of digestibility. J. Agric. Sci. (Camb.) 109:257-259.

Engels, E.A.N., and F. J. van der Merwe. 1967. Application of an in vitro technique to South African forages with special reference to the effect to certain factors on the results. S. Afr. J. Agric. Sci. 10:983-995.

Erasmus, L. J., P. M. Botha, P. Lebzien, and H. H. Meissner. 1990. Composition and intestinal digestibility of rumen fermented feeds and duodenal digesta in cows using bag techniques. J. Dairy Sci. 73:3494-3501.

Fadel, J. G. 1992. Application of theoretically optimal sampling schedule designs for fiber digestion estimation in sacco. J. Dairy Sci. 75:2184-2189.

Figroid, W., W. H. Hale, and B. Theurer. 1972. An evaluation of the nylon bag technique for estimating rumen utilization of grains. J. Anim. Sci. 35:113-120.

Flachowsky, G., and M. Schneider. 1992. Influence of various straw-to-concentrate ratios on in sacco dry matter degradability, feed intake, and apparent digestibility in ruminants. Anim. Feed Sci. Technol. 38:199-217.

Fonnesbeck, P. V., M. F. Wardeh, and L. E. Harris. 1984. Mathematical models for estimating energy and protein utilization of feedstuffs. Bulletin 508, Utah Agricultural Experiment Station, Utah State Univ., Logan.

Gabrielsen, B. C. 1986. Evaluation of marketed cellulases for activity and capacity to degrade forage. Agron. J. 78:838-842.

Genizi, A., A. Goldman, A. Yulzari, and N. G. Seligman. 1990. Evaluation of methods for calibrating in vitro digestibility estimates of ruminant feeds. Anim. Feed Sci. Technol. 29:265-278.

Girard, V., and G. Dupuis. 1988. Effect of structural and chemical factors of forages on potentially digestible fiber, intake and true digestibility by ruminants. Can. J. Anim. Sci. 68:787-799.

Givens, D. I., J. M. Everington, and A. H. Adamson. 1989. The digestibility and metabolisable energy content of grass silage and their prediction from laboratory measurements. Anim. Feed Sci. Technol. 24:27-43.

Givens, D. I., J. M. Everington, and A. H. Adamson. 1990a. The nutritive value of spring-grown herbage produced on farms throughout England and Wales over 4 years. II. The prediction of apparent digestibility in vivo from various laboratory measurements. Anim. Feed Sci. Technol. 27:173-184.

Givens, D. I., J. M. Everington, and A. H. Adamson. 1990b. The nutritive value of spring-grown herbage produced on farms throughout England and Wales over 4 years. III. The prediction of energy values from various laboratory measurements. Anim. Feed Sci. Technol. 27:185-196.

Goering, H. K., and P. J. Van Soest. 1970. Forage fiber analyses (apparatus, reagents, procedures, and some applications). USDA/ARS Agricultural Handbook No. 379, Washington, DC.

Goldman, A., A. Genizi, A. Yulzari, and N. G. Seligman. 1987. Improving the reliability of the two-stage in vitro assay for ruminant feed digestibility by calibration against in vivo data from a wide range of sources. Anim. Feed Sci. Technol. 18:233-245.

Grant, R. J., and D. R. Mertens. 1992a. Development of buffer systems for pH control and evaluation of pH effects on fiber digestion in vitro. J. Dairy Sci. 75:1581-1587.

Grant, R. J., and D. R. Mertens. 1992b. Impact of in vitro fermentation techniques upon kinetics of fiber digestion. J. Dairy Sci. 75:1263-1272.

Grant, R.J., P.J. Van Soest, and R.E. McDowell. 1974. Influence of rumen fluid source and fermentation time on in vitro true dry matter digestibility. J. Dairy Sci. 57:1201-1205.

Guggolz, J., R. M. Saunders, G. O. Kohler, and T. J. Klopfenstein. 1971. Enzymatic evaluation of processes for improving agricultural wastes for ruminant feeds. J. Anim. Sci. 33:167-170.

Harlan, D.W., J.B. Holter, and H.H. Hayes. 1991. Detergent fiber traits to predict productive energy of forages fed free choice to nonlactating dairy cattle. J. Dairy Sci. 74:1337-1353.

Hoover, W. H. 1986. Chemical factors involved in ruminal fiber digestion. J. Dairy Sci. 69:2755-2766.

Hopson, J. D., R. R. Johnson, and B. A. Dehority. 1963. Evaluation of the dacron bag technique as a method for measuring cellulose digestibility and rate of forage digestion. J. Anim. Sci. 22:448-453.

Horton, G.M., D.A. Christensen, and G.M. Steacy. 1980. In vitro fermentation of forages with inoculum from cattle and sheep fed different diets. Agron. J. 72:601-605.

Hungate, R. E. 1966. The rumen and its microbes. Academic Press, New York, NY.

Johnson, R. R. 1963. Symposium on microbial digestion in ruminants: In vitro rumen fermentation techniques. J. Anim. Sci. 22:792-800.

Johnson, R. R. 1966. Techniques and procedures for in vitro and in vivo rumen studies. J. Anim. Sci. 25:855-875.

Jones, D.I.H., and M. V. Hayward. 1973. A cellulase digestion technique for predicting the dry matter digestibility of grasses. J. Sci. Food Agric. 24:1419-1426.

Jones, D. I. H., and M. V. Hayward. 1975. The effect of pepsin pretreatment of herbage on the prediction of dry matter digestibility from solubility in fungal cellulase solutions. J. Sci. Food Agric. 26:711-718.

Joshi, D. C. 1972. Different measures in the prediction of the nutritive value of forages. Acta Agric. Scand. 22:243-247.

Judkins, M. B., L. J. Krysl, and R. K. Barton. 1990. Estimating diet digestibility: A comparison of 11 techniques across six different diets fed to rams. J. Anim. Sci. 68:1405-1415.

Jung, H.G., and V.H. Varel. 1988. Influence of forage type on ruminal bacterial populations and subsequent in vitro fiber digestion. J. Dairy Sci. 71:1526-1535.

Knipfel, J.E., and J.E. Troelsen. 1966. Interaction between inoculum donor diet and substrate in in vitro ruminant digestion studies. Can. J. Anim. Sci. 46:91-95.

Lentz, E. M., and D. R. Buxton. 1992. Digestion kinetics of orchardgrass as influenced by leaf morphology, fineness of grind, and maturity group. Crop Sci. 32:482-486.

Lindberg, J.E. 1981. The effect of basal diet on the ruminal degradation of dry matter, nitrogenous compounds and cell walls in nylon bags. Swedish J. Agric. Res. 11:159-169.

Lindberg, J. E. 1983. Factors affecting predictions of rumen degradability using the nylon bag (in sacco) technique and a comparison between in vivo and in sacco degradability measurements. Ph.D. diss., Swedish University of Agric. Sci., Uppsala.

Lindberg, J. E., A. Kaspersson, and P. Ciszuk. 1984. Studies on pH, number of protozoa and microbial ATP concentrations in rumen-incubated nylon bags with different pore sizes. J. Agric. Sci. (Camb.) 102:501-504.

Lindberg, J. E., and P. G. Knutsson. 1981. Effect of bag pore size on the loss of particulate matter and on the degradation of cell wall fibre. Agric. Environ. 6:171-182.

Lindberg, J. E., and T. Varvikko. 1982. The effect of bag pore size on the ruminal degradation of dry matter, nitrogenous compounds and cell walls in nylon bags. Swedish J. Agric. Res. 12:163-171.

Marinucci, M. T., B. A. Dehority, and S. C. Loerch. 1992. In vitro and in vivo studies of factors affecting digestion of feeds in synthetic fiber bags. J. Anim. Sci. 70:296-307.

McDougall, E. I. 1948. Studies on ruminant saliva. 1. The composition and output of sheep's saliva. Biochem. J. 43:99-109.

McLeod, M. N., and D. J. Minson. 1969. Sources of variation in the in vitro digestibility of tropical grasses. J. Br. Grassl. Soc. 24:244-249.

McLeod, M. D., and D. J. Minson. 1971. The error in predicting pasture dry matter digestibility from four different methods of analysis for lignin. J. Br. Grassl. Soc. 26:251-256.

McLeod, M. N., and D. J. Minson. 1974. Predicting organic-matter digestibility from in vivo and in vitro determinations of dry-matter digestibility. J. Br. Grassl. Soc. 29:17-21.

McLeod, M. N., and D. J. Minson. 1976. The analytical and biological accuracy of estimating the dry matter digestibility of different legume species. Anim. Feed Sci. Technol. 1:61-72.

McQueen, R., and P. J. Van Soest. 1975. Fungal cellulase and hemicellulase prediction of forage digestibility. J. Dairy Sci. 58:1482-1491.

Mehrez, A. Z., and E. R. Ørskov. 1977. A study of the artificial fibre bag technique for determining the digestibility of feeds in the rumen. J. Agric. Sci. (Camb.) 88:645-650.

Menke, K. H., L. Raab, A. Salewski, H. Steingass, D. Fritz, and W. Schneider. 1979. The estimation of the digestibility and metabolizable energy content of ruminant feedingstuffs from the gas production when they are incubated with rumen liquor in vitro. J. Agric. Sci. (Camb.) 93:217-222.

Mertens, D. R. 1973. Application of theoretical mathematical models to cell wall digestion and forage intake in ruminants. Ph.D. diss., Cornell Univ., Ithaca, NY (Diss. Abstr. 74:10882).

Mertens, D. R., and J. R. Loften. 1980. The effect of starch on forage fiber digestion kinetics in vitro. J. Dairy Sci. 63:1437-1446.

Michalet-Doreau, B., and M.Y. Ould-Bah. 1992. In vitro and in sacco methods for the estimation of dietary nitrogen degradability in the rumen: A review. Anim. Feed Sci. Technol. 40:57-86.

Messman, M. A., W. P. Weiss, and D. O. Erickson. 1991. Effects of nitrogen fertilization and maturity of bromegrass on in situ ruminal digestion kinetics of fiber. J. Anim. Sci. 69:1151-1161.

Meyer, J.H.F., and R. I. Mackie. 1986. Microbiological evaluation of the intraruminal in sacculus digestion technique. App. Environ. Microbiol. 51:622-629.

Miller, B. G., and R. B. Muntifering. 1985. Effect of forage: concentrate on kinetics of forage fiber digestion in vivo. J. Dairy Sci. 68:40-44.

Minson, D. J. 1976. Relation between digestibility and composition of feed. Miscellaneous Papers 12, Landouwhogeschoul Wageningen, The Netherlands, p. 101-114.

Minson, D. J. 1979. Relationships of conventional and preferred fractions to determined energy values. p. 72-78. *In* W. J. Pigden, C. C. Balch, and Michael Graham (ed.) Proc. Standardization of Anal. Method. for Feeds, Ottawa, Canada.

Minson, D. J. 1982. Effect of chemical composition on feed digestibility and metabolizable energy. Nutr. Abstr. Rev. (B) 52:592-614.

Minson, D. J., and C. D. Kemp. 1961. Studies in the digestibility of herbage. 9. Herbage and faecal nitrogen as indicators of herbage organic matter digestibility. J. Br. Grassl. Soc. 16:76-79.

Minson, D. J., and M. N. McLeod. 1972. The in vitro technique: Its modification for estimating digestibility of large numbers of tropical pasture samples. Tech. Paper No. 8, Div. Tropical Pastures, Commonwealth Sci. Indust. Res. Org., Australia.

Mir, P. S., Z. Mir, and J. W. Hall. 1990. Physical characteristics of feeds and their relation to nutrient components and dry matter disappearance in sacco. Anim. Feed Sci. Technol. 31:17-27.

Mir, D.J., Z. Mir, and J.W. Hall. 1991. Comparison of effective degradability with dry matter degradability measured at mean rumen retention time for several forages and forage:concentrate diets. Anim. Feed Sci. Technol. 32:287-296.

Morrison, I. M. 1976. New laboratory methods for predicting the nutritive value of forage crops. World Rev. Anim. Prod. 12:75-82.

Moss, A. R., and D. I. Givens. 1990. Chemical composition and in vitro digestion to predict digestibility of field-cured and barn-cured grass hays. Anim. Feed Sci. Technol. 31:125-138.

National Academy of Sciences. 1971. Atlas of nutritional data on United States and Canadian feeds. NAS-NRC, Washington, DC.

Navaratne, H.V., M.N. Ibrahim, and J. B. Schiere. 1990. Comparison of four techniques for predicting digestibility of tropical feeds. Anim. Feed Sci. Technol. 29:209-221.

Neathery, M. W. 1969. Dry matter disappearance of roughages in nylon bags suspended in the rumen. J. Dairy Sci. 52:74-78.

Nelson, B. D., H. D. Ellzey, C. Montgomery, and E. B. Morgan. 1972. Factors affecting the variability of an in vitro rumen fermentation technique for estimating forage quality. J. Dairy Sci. 55:358-366.

Neter, J., and W. Wasserman. 1974. Applied linear statistical models. Richard D. Irwin, Inc., Homewood, IL.

Nichols, E. L., and R. C. Dixon. 1984. Forage energy prediction equations used by private laboratories. p. 32-35. *In* Proc. 44th Ann. Mtg. Amer. Feed Mfg. Assoc. Nutr. Council, Arlington, VA.

Nik-Khah, A., and D. E. Tribe. 1977. A note on the effect of diet on the inoculum used in digestibility determination in vitro. Anim. Prod. 25:103-106.

Nocek, J. E. 1985. Evaluation of specific variables affecting in situ estimates of ruminal dry matter and protein digestion. J. Anim. Sci. 60:1347-1358.

Nocek, J. E. 1988. In situ and other methods to estimate ruminal protein and energy digestibility: A review. J. Dairy Sci. 71:2051-2069.

Nocek, J. E., and J. E. English. 1986. In situ degradation kinetics: Evaluation of rate determination procedure. J. Dairy Sci. 69:77-87.

Nocek, J. E., and R. A. Kohn. 1988. In situ particle size reduction of alfalfa and timothy hay as influenced by form and particle size. J. Dairy Sci. 71:932-945.

Olubobokun, J. A., W. M. Craig, and K. R. Pond. 1990. Effects of mastication and microbial contamination on ruminal in situ forage disappearance. J. Anim. Sci. 68:3371-3381.

Omed, H. M., R.F.E. Axford, A. G. Chamberlain, and D. I. Givens. 1989. A comparison of three laboratory techniques for the estimation of the digestibility of feedstuffs for ruminants. J. Agric. Sci. (Camb.) 113:35-39.

Osbourn, D. F. 1978. Principles governing the use of chemical methods for assessing the nutritive value of forages: A review. Anim. Feed Sci. Technol. 3:265-275.

Osbourn, D. F., D. E. Beever, and D. J. Thomson. 1976. The influence of physical processing on the intake, digestion and utilization of dried herbage. Proc. Nutr. Soc. 35:191-200.

Paine, C. A., R. Crawshaw, and W.P. Barber. 1982. A complete exchange method for the in sacco estimation of rumen degradability on a routine basis. p. 177-178. In D. J. Thomson, D. E. Beever, and R. G. Gunn (ed.) Forage protein in ruminant animal production. Occasional Publ. No. 6., Br. Soc. Anim. Prod., Surrey, England.

Playne, M. J., W. Khumnualthong, and M. G. Echevarria. 1978. Factors affecting the digestion of oesophageal fistula samples and hay samples in nylon bags in the rumen of cattle. J. Agric. Sci. (Camb.) 90:193-204.

Quin, J. I., J. G. van der Wath, and S. Myburgh. 1938. Studies on alimentary tract of Merino sheep in South Africa. 4. Description of experimental technique. Onderstepoort J. Vet. Sci. Anim. Ind. 11:341-360.

Rodriguez, H. 1968. The in vivo bag technique in digestibility studies. Rev. Cuban Cienc. Agric. 2:77-81.

Roughan, P. G., and R. Holland. 1977. Predicting in vivo digestibilities of herbages by exhaustive enzymic hydrolysis of cell walls. J. Sci. Food. Agric. 28:1057-1064.

Schneider, B. H. 1947. Feeds of the world: Their digestibility and composition. Agricultural Experiment Station, West Virginia University, Morgantown, WV.

Smith, L. W., H. K. Goering, D. R. Waldo, and C. H. Gordon. 1971. In vitro digestion rate of forage cell wall components. J. Dairy Sci. 54:71-76.

Steg, A., S.F. Spoelstra, J.M. van der Meer, and V.A. Hindle. 1990. Digestibility of grass silage. Netherlands J. Agric. Sci. 38:407-422.

Sullivan, J. T. 1964. The chemical composition of forages in relation to digestibility by ruminants. USDA/ARS Pub. 34-62, Washington, DC.

Terry, R. A., D. C. Mundell, and D. F. Osbourn. 1978. Comparison of two in vitro procedures using rumen liquor-pepsin or pepsin-cellulase for prediction of forage digestibility. J. Br. Grassl. Soc. 33:13-18.

Thomas, J. W., Y. Yu, T. Middleton, and C. Stallings. 1982. Estimations of protein damage. p. 81-98. In Protein requirements for cattle: Symposium. MP-109. Division of Agric., Oklahoma State Univ., Stillwater.

Tilley, J.M.A., and R. A. Terry. 1963. A two-stage technique for the in vitro digestion of forage crops. J. Br. Grassl. Soc. 18:104-111.

Troelsen, J. E. 1970. Digestible energy in forage by in vivo and in vitro assays. Can. J. Anim. Sci. 50:557-562.

Tyrrell, H. F., and P. W. Moe. 1975. Effect of intake on digestive efficiency. J. Dairy Sci. 58:1151-1163.

Udèn, P., and P. J. Van Soest. 1984. Investigations of the in situ bag technique and a comparison of the fermentation in heifers, sheep, ponies, and rabbits. J. Anim. Sci. 58:213-221.

Valdes, E. V., and G. E. Jones. 1987. A comparison of in vitro and in vivo dry matter digestibility techniques for the evaluation of forage quality. Can. J. Anim. Sci. 67:573-576.

van der Koelen, C. J., P. W. Goedhart, A. M. van Vuuren, and G. Savoini. 1992. Sources of variation of the in situ nylon bag technique. Anim. Feed Sci. Technol. 38:35-42.

van Es, A.J.H. 1986. Energy values of feeds for livestock and their prediction. Netherlands J. Agric. Sci. 34:405-412.

Van Hellen, R. W., and W. C. Ellis. 1977. Sample container porosities for rumen in situ studies. J. Anim. Sci. 44:141-146.

Van Keuren, R. W., and W. W. Heinemann. 1962. Study of a nylon bag technique for in vivo estimation of forage digestibility. J. Anim. Sci. 21:340-345.

Van Soest, P. J. 1982. Nutritional ecology of the ruminant. O & B Books, Inc., Corvallis, OR.

Van Soest, P. J., J. B. Robertson, and B. A. Lewis. 1991. Methods for dietary fiber, neutral detergent fiber, and nonstarch polysaccharides in relation to animal nutrition. J. Dairy Sci. 74:3583-3597.

Varvikko, T., and J. E. Lindberg. 1985. Estimation of microbial nitrogen in nylon-bag residues by feed ^{15}N dilution. Br. J. Nutr. 54:473-481.

Varvikko, T., and A. Vanhatalo. 1990. The effect of differing types of cloth and of contamination by non-feed nitrogen on intestinal digestion estimates using porous synthetic-fibre bags in a cow. Br. J. Nutr. 63:221-229.

Walker, D. M. 1959. The in vitro digestion of roughage dry matter. J. Agric. Sci. (Camb.) 53:192-197.

Warner, A.C.I. 1956. Criteria for establishing the validity of in vitro studies with rumen micro-organisms in so-called artificial rumen systems. J. Gen. Microbiol. 14:733-748.

Weakley, D. C., M. D. Stern, and L. D. Satter. 1983. Factors affecting disappearance of feedstuffs from bags suspended in the rumen. J. Anim. Sci. 56:493-507.

Weiss, W. P. 1993. Predicting energy values of feeds. J. Dairy Sci. 76:1802-1811.

Weiss, W. P., H. R. Conrad, and N. R. St. Pierre. 1992. A theoretically-based model for predicting total digestible nutrient values of forages and concentrates. Anim. Feed Sci. Technol. 39:95-110.

Williams, A. G. 1988. Factors affecting the formation of polysaccharide-degrading enzymes by rumen micro-organisms. Anim. Feed Sci. Technol. 21:191-195.

CHAPTER 17

METHODOLOGY FOR ESTIMATING DIGESTION AND PASSAGE KINETICS OF FORAGES

W. C. Ellis, J. H. Matis, T. M. Hill, and M. R. Murphy

INTRODUCTION

The nutritive value of a feedstuff is determined by its content of chemical entities and their transformations to nutrients required by the animal. Digestive transformations are determined by intrinsic attributes of the forage and by their interactions with the kinetic processes of digestion. Quantitative expressions of the kinetics of digestion and passage are needed to more precisely estimate the quantity and composition of nutrients digested from forages and their subsequent efficiency of utilization by the animal.

The objective of this chapter is to review methodology for estimating parameters describing the kinetics of physical and chemical digestion in ruminants. The kinetic processes of primary concern are: 1) flow of digesta through segments of the ruminant's digestive system, several of which involve fermentative digestion; and 2) digestion of specific entities in the forage. Recent reviews of this subject by Allen and Mertens (1987), Sauvant and Ramangasovina (1991), and Mertens (1993a,b) have emphasized deterministic, age-independent kinetics of this flow and digestion. This review will additionally emphasize age-dependent models of digestion and passage of fragments derived from mastication during ingestion and rumination. The need to simultaneously consider the fragmentation process has been emphasized by Sauvant and Ramangasovina (1991).

PASSAGE AND DIGESTION OF FORAGE FRAGMENTS

The prehended forage is fragmented by mastication during its ingestion (Pond et al., 1984) and subsequent rumination (Pond et al., 1987) to yield the array of forage fragments which undergo digestion in the rumen. These fragments are highly varied with respect to their mass, size, shape, origin with respect to plant anatomy, plant physiology and chemical composition. Chemical entities within the individual tissues of the masticated fragments of forage are highly varied with respect to their chemical and physical nature, solubility, buoyancy, accessibility to colonizing microbes, and potential for digestion.

Certain chemical entities, i.e., some proteins, may be rapidly solubilized by saliva and ruminal fluids, and are rapidly digested external to the fragment. Other

---

W. C. Ellis, 241 Kleberg Center, Dep. of Animal Science, Texas A&M Univ., College Station, TX 77843; J. H. Matis, Blocker Building, Dep. of Statistics, Texas A&M Univ., College Station, TX 77843; T. M Hill, Dept. of Animal Sciences, Univ. of Maine, Orono, ME 04469; M. R. Murphy, 1207 West Gregory Drive, Dep. of Animal Sciences, Univ. of Illinois, Urbana, IL 61801.

entities are only solubilized by digestion and, hence, are digested within the labyrinthine architecture of the fragment. The architecture of the fragment undergoing digestion is important in determining physical access and colonization by microbes of the varied component tissues of the plant fragment. The labyrinthine structure of the fragment may limit diffusion of acidic fermentation end products and thereby dynamically alter the micro-environment of the "nitch" occupied by the microbe (Hungate, 1966). Interactions between the structure of the fragment and the microbes colonizing its internal tissues are also important in effecting entrapment of gases of fermentation. The resulting fermentation gas-based buoyancy appears to be a major force which positions fragments within digesta flow-paths leading to further comminution via rumination (Sutherland, 1989) and in flowpaths leading to escape from the reticulorumen.

Some tissues may include entities (i.e, lignin and the highly lignified carbohydrates) incapable of being digested while other entities within the same forage fragment may be potentially digestible. Entities, E, which are indigestible, E_i, disappear from the digestive site only by escape, i.e., as the chemical form consumed. Entities capable of being chemically modified from their dietary forms, primarily by hydrolysis, are referred to as potentially digestible entities, E_d. A schematic representation of the dynamics of digestion and flow of potentially digestible and indigestible entities is illustrated in Figure 1.

The fraction of E_i escaping per unit time expresses its rate of escape, k_e. Because E_i and E_d are components of the same forage fragment, it may be assumed that k_e of E_i and E_d are equal for insoluble entities. The E_d digested within segments of the digestive tract where influx is mixed with resident fragments (a mixing compartment) is determined by the extent of disappearance of E_d by digestion (D) relative to its total disappearance by digestion plus escape. If first order rates (to be defined later) are assumed for both, k_d, digestion and escape, then the fractional digestibility, D, of E_d, DE_d, is determined by rates of digestion and escape of E_d.

$$DE_d = k_d \text{ of } E_d / (k_d \text{ of } E_d + k_e \text{ of } E_d) \qquad \text{Equation 1}$$

Conversely, the fraction of the potentially digestible entity which is undigested, UE_d, due to passage is:

$$UE_d = k_e \text{ of } E_d / (k_e \text{ of } E_d + k_d \text{ of } E_d) \text{ or } 1-DE_d \qquad \text{Equation 2}$$

Due to the chemical and physical nature of structural carbohydrates (cell wall constituents), their hydrolytic digestion is slow relative to that of most nonstructural carbohydrates and intracellular entities. Digestion of structural carbohydrates is accomplished only by enzymes of microbial origin. To sustain a population of structural carbohydrate-digesting microbes, the nutrient flux from fermentation must be sufficient to support microbial growth rates in excess of the rate that microbes disappear from the site by lysis and escape. Thus, rate of passage of fragments must be slowed sufficiently that fermentation of their slowly digesting carbohydrates will yield the flux of nutrients required for growth of a significant microbial population. The k_e of the forage fragment must be slowed

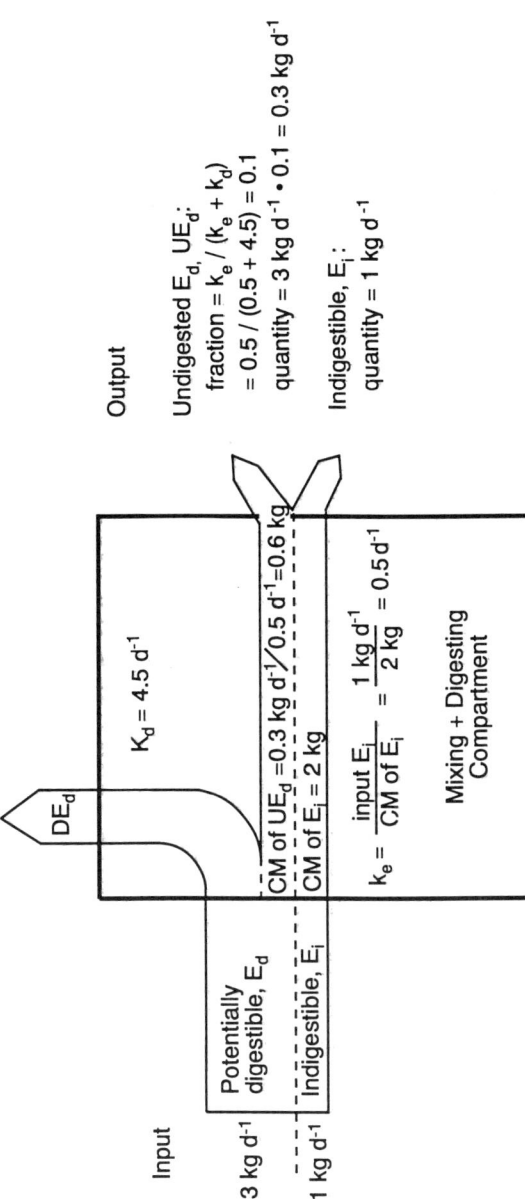

Figure 1. Effect of a mixing-digesting compartment on the dynamics of competing rates of digestion, k_d, and rates of escape, k_e, in determining the extent of digestion of potentially digestible feed entities (DE_e), compartmental mass (CM) of undigested, potentially digestible entities (UD_d) and indigestible entities (E_i), and compartmental output.

sufficiently that k_d considerably exceeds k_e for extensive digestion of structural carbohydrates.

The primary mechanism whereby herbivores slow the passage of forage fragments within their digestive system is via dilution of forage intake with residues of previous meals. Such dilution is achieved by mixing compartments. A mixing compartment is characterized by: 1) an enlarged tract segment which allows the physical accumulation of a larger mass of resident fragments than exists in adjacent segments, 2) mixing (or dilution) of newly ingested fragments with the total mass of resident fragments in the compartment, resulting in 3) competition (a form of dilution) for escape among all fragments within the mixing compartment. Assuming equal opportunity for escape by resident fragments of all ages, and assuming steady state conditions (influx of E_i equaling outflux of E_i and a constant compartmental mass of E_i), the expected rate of escape, in units of t^{-1}, of E_i is determined by the dilution of influx, mass•t^{-1}, divided by the compartmental mass of E_i. This dilution effect is referred to as fractional turnover because the flux of E_i continually displaces or "turns over" a fraction of the constant compartmental mass of E_i, a fraction which equals the intake rate of E_i if the compartmental mass is to remain constant.

k_e of E_i, d^{-1} = intake rate of E_i, k_g•d^{-1}/compartmental mass of E_i, kg   Equation 3

Thus, in Figure 1, a mixing compartment having a mass of 2 kg of E_i, and an intake rate of 1 kg d^{-1} of E_i, results in a k_e of 0.5 d^{-1}. This method for estimating the fractional rate of escape can be referred to as the flux/compartmental pool method. In order to estimate disappearance specifically via escape from the compartment, calculations must be based on indigestible entities because only these disappear totally via escape. It is essential to define the entity involved in such calculations (indigestible) with respect to the specific process sought (escape by passage) in order to interpret the rates in terms of a specific route of disappearance (escape).

Because of uncertainties in partitioning digestibility of potentially digestible entities in a compartment, the flux/compartmental pool method is not readily applicable for estimating rates of digestion. Partitioning of the total compartmental mass of E_d into DE_d and UE_d requires a kinetic model to estimate k_d or k_e. In Figure 1, a k_d of 4.5 d^{-1} is assumed in order to illustrate the calculation of UE_d and DE_d. Alternatively, compartmental mass (kg) of UE_d can be estimated from determined output of UE_d(kg d^{-1}) divided by determined k_e (g d^{-1}).

Microbial growth occurs throughout the gastrointestinal tract of animals, except for sites of extremely low pH, < 2. Conditions most favorable for sufficient growth of microbial populations to impact digestion of structural carbohydrates occur primarily in the non-HCl-secreting gastric regions and in the cecum-colon. The ruminant, with its large ruminal capacity for forage fragments and its coordinated motility sequences for mixing fragments, provides a most efficient mixing compartment for increasing residence time to allow extensive digestion of relatively slowly digesting entities such as the structural carbohydrates.

DIGESTA FLOW MODELS

Compartmental Flow Models

The mathematics describing the effects of mixing compartments have been extensively developed for easily mixed fluids and for inert and otherwise identical particles (Shipley and Clark, 1972). The flow of digesta through the digestive tract of animals has been extensively reviewed (Warner, 1981). Compartmental models of digesta flow were initially based on fluid models (Blaxter et al., 1956; Brandt and Thacker, 1958; Hungate; 1966, Grovum and Williams, 1973). These early models all assumed: 1) instantaneous and complete mixing of influxing fragments, 2) equal opportunity for escape of all resident fragments regardless of their age in the compartment, and 3) constant input, outflow, and compartmental mass (steady state). These assumptions yield an exponential distribution of residence times for the population of particles in the compartment, i.e., simple dilution of intake by the compartmental mass determining the competition for escape. Initial models of digesta flow assumed mass action dilution turnover as has been described for the flux/compartmental pool method (equation 3). Other models have been proposed: 1) by Matis (1972) and Matis et al.(1989) which utilized non-exponential distributions of resident times, gamma functions, to account for age-dependent processes required for escape of fragment residues, 2) by Dhanoa et al. (1985) to account for multiple sub-compartments within a mixing compartment, and 3) by France et al. (1985) who proposed distributed time lags and, more recently, diffusion (France et al., 1993) as a deterministic basis for age-dependent flow.

An understanding of these compartmental models and their applicability to digestive systems requires a clear understanding of the assumptions of each model and the implications these assumptions have relative to current concepts of the biology of the digestive process. Two recent reviews have discussed these considerations with emphasis on age-independent processes (Mertens, 1993a,b). Age-independent processes will be further illustrated here as the basis for subsequently illustrating age-dependent processes.

Age-Independent Mixing Compartments

The term "age-independent" is used to denote turnover rate constants caused by mass action dilution. An idealized mixing compartment is illustrated by the physical model in Figure 2. It is "ideal" because of the following assumptions concerning the mixing of inert and uniform particles: 1) Instantaneous, continuous, and complete mixing of influxing with all resident particles in the compartment (an age-independent mixing compartment may be defined as the mass that is instantaneously mixed with the influx): 2) Steady state conditions, i.e., constant influx equaling constant efflux and thus a constant volume of the mixing compartment; and 3) Equal opportunity for escape of particles of all resident ages or, alternatively, the probability for escape of each particle being determined only by its ratio to the mass of other particles in the compartment (see equation 3).

An ideal mixing compartment containing 2,000 particles, P_c, and an influx rate of 1,000 particles per day, $P_i\ d^{-1}$, would have a turnover rate of 0.5 d^{-1} (1,000

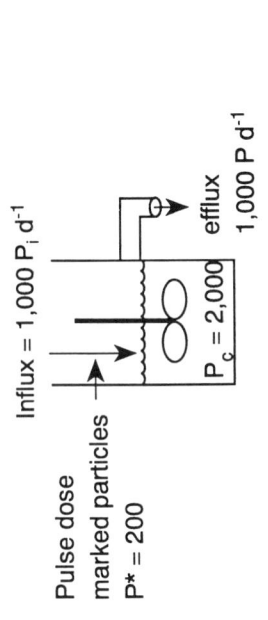

Figure. 2. A. Physical illustration of the dilution of a pulse dose of marked particles, P*, due to initial dilution by resident particles in the mixing compartment, P_c, and by dilution turnover, k_e, due to continued influx of feed particles, P, in an idealized mixing compartment in which influx is instantaneously mixed with P_c and influx = efflux. B. Quantification of initial dilution, CØ, of dose by compartmental mass and dilution via continued flux (turnover or escape), k_e. C. The exponential distribution of residence times associated with mass action dilution turnover of particles within and particles which have escaped from the compartment. D. Expression of age-independent dilution of P^*/P_c expected due to turnover rate and residence time, t.

P_i d^{-1} / 2,000 P_c) as illustrated in Figure 2A. Calculations of the dilutions associated with initial mixing and with continued influx are illustrated in Figure 2B. Because the inert particles disappear from the compartment only by escape, the rate terms of "turnover", "dilution", and "escape" are equally appropriate. If 200 marked particles, P^*, are pulse-dosed into the system in such a manner that steady state conditions are not perturbed, the 200 particles of P^* are instantaneously mixed with the 2,000 resident particles in the compartment, P_c, and uniquely displace (or turnover) a like number of 200 particles of P_c. The proportion (or specific activity, P^*/P_c) of marked particles in the compartment immediately after dosing (initial concentration) equals the dose of marked particles divided by the number of particles in the compartment at the time of dosing, or 200 P^*/2,000 P_c equals 0.1 (Figure 2C). Subsequent to the pulse dose of P^*, continued influx of P, is assumed to continuously and indiscriminately dilute P^* and dilute all resident particles initially ingested with the meal containing P^*. Thus, assuming steady state conditions, influx = efflux, the dilution rate due to turnover can be computed as the dilution of flux rate (influx or efflux) divided by the number of resident particles in the compartment, P_i/P_c, as illustrated in Figure 2B and, in this example, is 0.5 d^{-1}. This is the influx/compartmental pool method illustrated in Figure 1 and in equation 3.

Alternatively, k_e of the marked fragment, P^*, can be estimated from the rate of dilution of P^* in the compartment due to continued flux of P_i through the compartment. In Figure 2B, the initial dilution of P^* that would have resulted from its instantaneous mixing with P_c, CØ, can be estimated by extrapolating the observed exponentially distributed residence times to the time of pulse dose. The number of particles in the compartment (P_c) may be computed as P^*/CØ. Alternatively, the CØ and k_e could be estimated simultaneously by models having appropriate assumptions concerning the dilution and turnover of P^*.

The expected values for the dilution of P^* in P_c subsequent to the dose of P^* are reflected by the quantity of P^* remaining in the compartment, or, under steady state conditions, the proportion, P^*/P_c, of marked particles remaining in the compartment. This distribution is the expected distribution of residence times. The distribution of residence times for a perfect mixing, mass action diluting, mixing compartment (Figure 2) is graphically illustrated in Figure 2C. The illustrated residence time distribution is exponentially distributed as the result of influx of P_i continuously diluting a diminishing quantity of P^* in the compartment so that the quantity of P^* escaping decreases with time while the fraction of P^* escaping with time remains constant, i.e., age-independent. Thus, the expected proportion, P^*/P_c, remaining in the compartment after a specified residence time following pulse dose, t, is given by the exponential expression:

$$(P^*/P_c)_{(t)} = CØ \cdot e^{-k_e \cdot t} \qquad \text{Equation 4}$$

As illustrated in Figure 2C, an exponential distribution of residence times yields a curvilinear plot on an arithmetic scale or a linear plot on a logarithmic scale (either natural logarithms, ln, or base 10 logarithms). Because expressing P^*/P_c as its logarithmic value results in a linearization of the residence time distribution, the escape rate, k_e, can be estimated from discrete data obtained at two sampling times as:

$$k_e = [(\ln (P^*/P)_{t2}) - (\ln (P^*/P)_{t1})] / (t2 - t1), \text{ where } t2 > t1 \qquad \text{Equation 5}$$

More accurate estimates of CØ and k_e are obtained by fitting the exponentially distributed residence time model (equation 4) to multiple values obtained over a series of times after dosing. The resident mass of particles in the compartment and their escape rate, k_e, may be estimated by the flux/compartmental pool method (equation 3) when compartmental pool (mass, quantity, or volume of an indigestible entity) and flux (influx or efflux) of an indigestible entity are known. Alternatively, compartmental mass and escape rate can be estimated by fitting equation 4 to marker concentration data obtained following a pulse dose of marked particles. Both methods will yield similar estimates of compartmental mass and k_e if the mixing compartment system conforms to the three assumptions needed to generate exponentially-distributed resident times.

Other expressions of kinetics involving age-independent or exponentially distributed residence times are mean compartmental residence time, MCRT, and half-life, $t_{1/2}$.

$$\text{MCRT} = 1/k_e \qquad \text{Equation 6}$$

The MCRT is an expression of the mean time that the dose of P^* resides in the compartment. The MCRT frequently is referred to as the turnover time or the clearance time. It should be noted that the MCRT calculated as the inverse of k_e represents the MCRT only for exponentially distributed residence times and does not apply to non-exponentially distributed residence times. The MCRT should not be confused with the half-life of P^* which is computed as the ln of a 1/2 reduction in marker concentration, i.e., ln 0.5 ≈ -0.693, divided by the marker's turnover rate, k_e.

$$t_{1/2} \approx 0.693/k_e \qquad \text{Equation 7}$$

A marker having a k_e of 0.05 h^{-1} would have a MCRT of 20 h and a half-life of about 13.86 h.

Non-Mixing Flow

In contrast to mixing segments of the gastrointestinal digesta, other segments are contained in a relatively narrow bore, are propulsed by non-mixing types of motility, or contain relatively dry digesta which is resistant to mixing. These properties are not conducive to mixing of digesta as described for mixing compartments. Digesta flow through such segments is more polar, with some back-flow due to imperfect laminar polar flow. Three possible types of generally polar flow are illustrated in Figure 3. Perfect laminar, non-turbulent, polar flow as illustrated in Figure 3A yields a "spike" output due to all particles flowing in a polar manner with no mixing or differential flow rates inducing a distribution

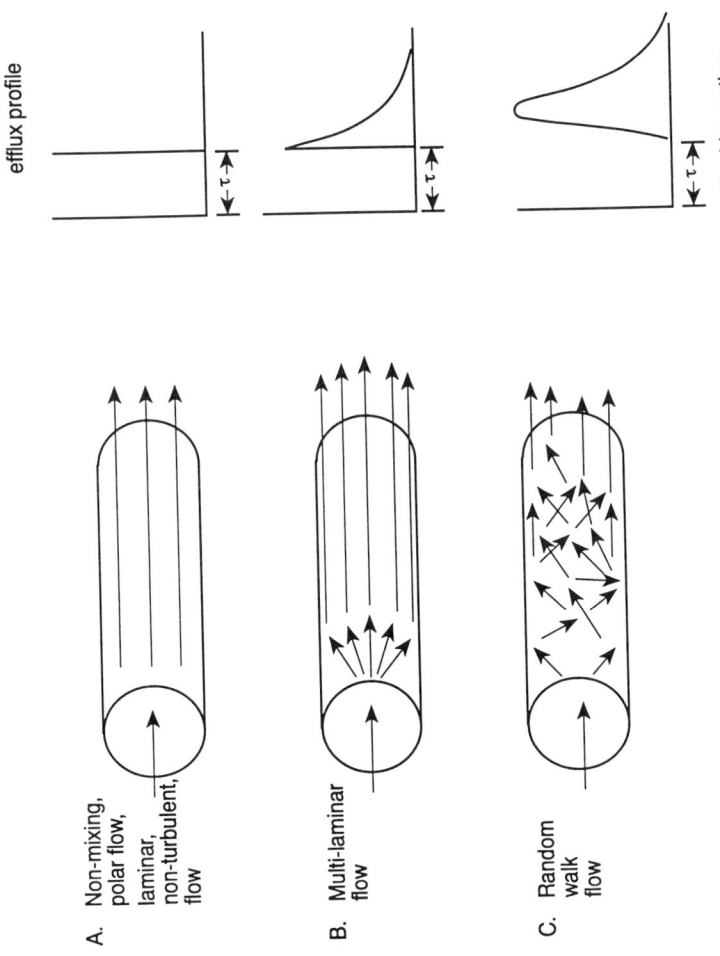

Figure. 3. Illustration of three possible types (A, B, and C) of non-compartmental, mixing flow and their expected efflux profiles.

of residence times. Imperfect, multi-laminar flow with some back-flow will yield a "spiked" initial emergence with trailing due to back-flow as illustrated in Figure 3B. Random obstacles to laminar flow, such as haustration within the digestive tract, will result in random diversions from laminar flow by digesta and yield a distribution of flow rates as illustrated in Figure 3C. When non-mixing and mixing compartments are in sequence, non-mixing flow results in a time delay between dosing and first detection of marked fragments emerging from the digestive system. Blaxter et al. (1956) used the term τ to denote time delay between dosing and first detection of the marked fragment in the feces and assumed τ represented residence time associated with flow through the intestines. In contrast, France et al. (1985) has used the term "time delay lags" with the symbol τ to apply to processes involving residence time within mixing compartments without corresponding flow. The term "time delay" or "τ" should not be confused with the term "transit time" which is more commonly used analogously to MCRT, i.e., mean residence time in the system of mixing compartments. Grovum and Williams (1973) used the term "transit time" to represent non-mixing flow in a manner analogous to the τ used by Blaxter et al. (1956). Dhanoa et al.(1985) applied the term "transit time" to MCRT due to sequential flow through multiple mixing compartments.

Age-Dependent Mixing Compartments

Matis (1972) noted that feedstuff fragments in the rumen are subjected to many digestive processes which require time before such fragments acquire properties required to increase the probability of their escape. He proposed non-exponentially distributed residence times to model the effects of age of fragments in the compartment on their distribution of residence time, i.e., an age-dependent distribution of residence times.

Matis (1972) choose the family of integer gamma functions to model effects of such an age-dependent process. This family is denoted as Gn with n corresponding to the integer of the gamma function. A G1 represents an exponential distribution of residence times and is age-independent. Gamma functions of G2 or greater represent age-dependent distributions of residence times. As the integer value of the gamma function increases beyond 2, the modelled effects of age slow the rate that the age-dependent escape rate approaches the asymptotic age-independent rate parameter (Figure 4). Thus, the order of integer gamma function provides a shape parameter for the distribution of residence times while the asymptotic rate, λ, characterizes the age-independent escape rate ultimately achieved. Matis (1972) postulated nil or extremely slow escape rates for masticated fragments initially entering the rumen digesta and, with aging, the escape rates would progressively increase to achieve some maximum rate of escape unaffected by further aging in the compartment.

Physical models of an age-dependent compartment are illustrated in Figure 5 and assumptions concerning their corresponding residence time distributions are illustrated in Figure 6. An additional conceptual representation of these same physical models is also given by Lalles et al.(1991). Note that in an age-independent compartment the distributions of residence time for fragments remaining in and emerging are identical (model 1, Figure 6). In contrast, the

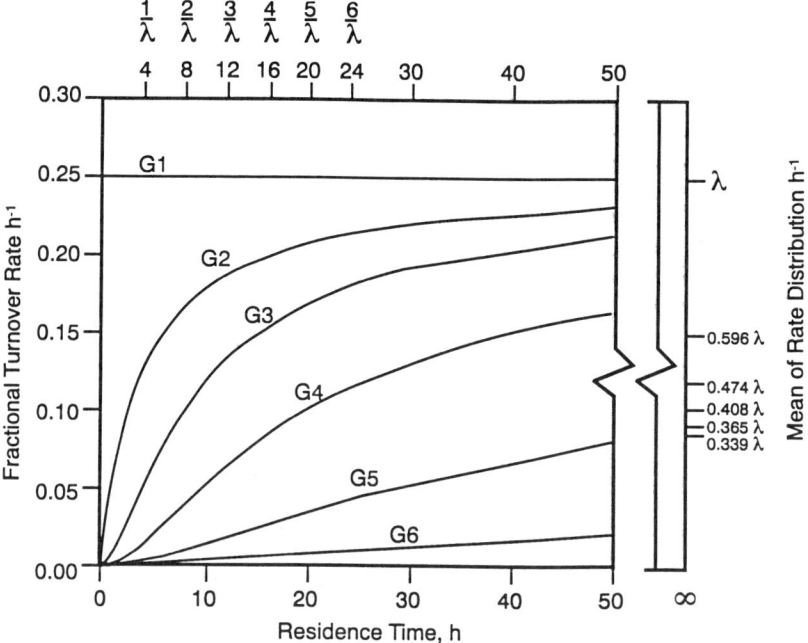

Figure. 4. Distribution of fractional turnover rates for various orders of gamma functions (G1, G2,...G6) versus residence time of a fragment in a mixing compartment. Note that G1 is an age-independent, rate constant and that progressively higher orders of age-dependency are modelled by increasing the order of discrete gamma functions, from G2 through G6 being illustrated. The asymptotic rate for gamma functions greater than G2, λ, equals k for the age-independent G1 function. The mean compartmental residence time for gamma distributions equals the numeric order of gamma function divided by λ. The mean (mean rate) for functions greater than G1 equals the constants indicated on the right vertical axis multiplied by λ.

distribution of residence times for particles remaining in and particles emerging from an age-dependent compartment differ (model 3 of Figure 6). This difference is the result of the negligible initial escape and progressively more rapid escape of fragments with advancing residence time. Roux and Pienaar (1984) proposed and used a non-integer gamma function to describe both passage and digestion of forage residues in sheep as occurring from a single age-dependent mixing compartment. Although this is similar to the integer gamma functions proposed by Matis (1972), the non-integer gamma function only can be used in single compartment models and poses problems for more complex systems.

France et al.(1985) used the term "distributed time delay lags" to describe the distribution of lag times which would result from conceptually discrete subpopulations of ingested fragments before they undergo age-independent turnover as illustrated by model 5 in Figures 5 and 6. Although conceptual and philosophical arguments may be advanced for this as a different modelling approach, this model is mathematically similar to the two compartment, age-dependent model earlier proposed by Matis (1972). The age-dependent compartment in the model of Matis (1972) assumes that the probability for escape by the population of forage fragments increases with time while that of France et al. (1985) assumes an age-dependent distribution of sub-populations of particles, with different intrinsic properties, that determines their subsequent lag time before becoming eligible to escape and undergo age-independent turnover.

Multi-Compartment Models

The earliest compartmental models of digesta flow through the ruminant were applied to marker excretion data from sheep (Blaxter et al., 1956; Grovum and Williams, 1973) and cattle (Brandt and Thacker, 1958). These models (model 1 in Figures 5 and 6) contained two, age-independent mixing compartments in sequence with a discrete time delay, τ. The G1→G1→τ→O model was selected by Blaxter et al. (1956) solely on its ability to describe the excretion pattern of marked particles in feces.

The residence time distributions within each of the age-independent compartments of the G1→G1→τ→O model (model 1, Figure 6) are identical due to assumed instantaneous mixing of the marker within each compartment. Marker is instantaneously mixed with resident digesta (compartmental mass) in the first compartment, undergoes turnover related to the compartmental mass of digesta in the first compartment, and immediately flows into and undergoes turnover related to the compartmental mass of the second compartment. The output profile of marker concentration from the second sequential compartment exhibits an initial ascending pattern due to initial dilution of marker by the compartmental mass of the second compartment. The subsequent descending pattern of marker concentrations is due to turnover within the two sequential compartments exceeding the initial dilution in the second sequential compartment. If the two compartments differ appreciably in compartmental mass, two age-independent turnover rates k can be resolved which are inversely related to their compartmental mass (Figure 1).

It is not possible to order the sequence of turnover compartments, nor the time delay segment, based on the output profile from the system of mixing compartments and time delay segments. Blaxter et al.(1956) tentatively assigned the faster turnover compartment to digesta in the rumen and designated its turnover rate k_1, with k_2 being assigned to the slower turnover compartment. In contrast, Grovum and Williams (1973) assigned the slower turnover rate to the rumen and designated its turnover rate k_1.

Other investigators have proposed that three sequential, exponentially distributed mixing compartments provide an improved statistical fit to observed data. Generally, one would expect to improve fit as the result of adding more parameters to a model, provided that the quality and quantity of data are adequate. However, the addition of more than two sequential exponentially distributed compartments commonly results in near equal estimates for two of the three age-independent rate parameters, suggesting that: 1) the data are inadequate, 2) less than three distinguishable compartments actually exist, or 3) that the two similar rates represent a single age-dependent compartment (Pond et al., 1988; Mertens 1989).

Matis (1972) proposed a two compartment model having sequential age-dependent and age-independent mixing compartments. This model (model 2, Gn→G1→τ→O) is physically illustrated in Figure 5 with residence time distributions illustrated in Figure 6. Age-dependent processes such as hydration, microbial colonization, and fragmentation by rumination are fast relative to mass action dilution turnover and such processes should occur in the initial entry compartment. Therefore, Matis (1972) assigned age-dependent distributions of residence time to the faster turnover compartment and assumed this to be the initial entry compartment. The asymptotic, age-dependent rate parameter for escape from the age-dependent compartment was designated λ_1 to distinguish this age-dependent rate parameter from the age-independent turnover rate from the second or terminal compartment k_2. Because the λ_1 is an estimate of age-independent turnover rate after infinite time, the rate parameter, λ_1, is not comparable to k and k_2, and compartmental mean residence time and mean distribution of turnover rates must be computed differently than for age-independent compartments.

The use of age-dependent residence time distributions have a number of advantages, as discussed by Pond et al. (1988) and Matis (1987). The age-dependent process: 1) is consistent with biological evidence for digesta flow paths in ruminal digesta (Sutherland, 1989), 2) being stochastic, compensates for deviations from more rigorous assumptions of age-independent mixing compartments such as instantaneous and complete mixing, 3) yields estimates of τ (and, conversely, a MCRT of n/λ_1) that are more consistent with the actual data (Figure 6 of Pond et al., 1988), and 4) resolves statistically valid estimates of turnover from two compartments having different distributions of residence times which are not resolvable by the G1→G1→τ→O model (rate λ_1 vs k_2 corresponding to Gn vs G1 respectively; Table 7 of Pond et al., 1988).

The models of Dhanoa et al. (1985) and France et al. (1985) also model age-dependent processes but use different mathematical approaches. Dhanoa et al. (1985) utilized a distribution of multiple, sequential age-independent compartments with each sequential compartment having a relatively small fixed difference in the magnitude of its age-independent fractional turnover rates. Flow through such multiple compartments yields a distribution of age-independent rates (model 4, G1→G1→(G1→G1)$_{n-1}$, Figures 5 and 6).

France et al. (1985) proposed the concept of distributed time delay lags (τ_1, τ_2, τ_3,..., τ_n) preceding an age-independent turnover compartment (model 5, (Fτ→G1→τ→O), in Figures 5 and 6). It appears that France et al. (1985) recommend the use of a gamma distribution of time delay lags. Use of a gamma distribution of time delay lags results in estimates of a mean time delay lag equal to the mean compartmental residence time for the initial entry compartment (Gn) of the Gn→G1→O model. However, no specific residence time distributions are recommended by France et al.(1985) and this model has not been subsequently used by that research group (France et al., 1988). Differences between the Gn→G1→τ→O model of Matis (1972) and the ((Fτ→G1→τ→O),→G1) model of France et al. (1988)

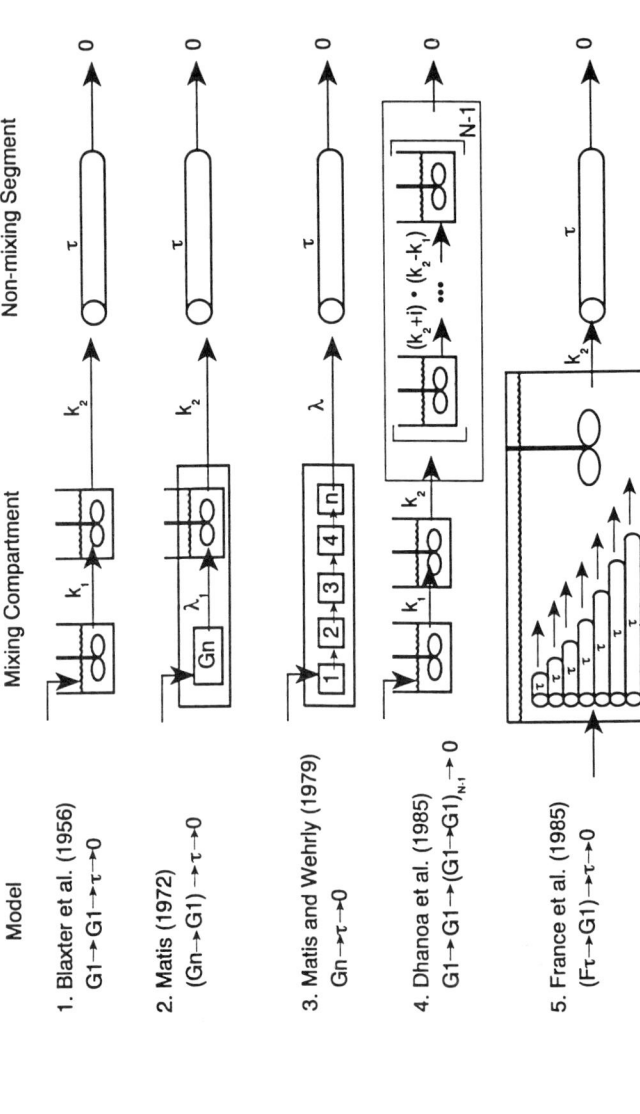

Figure 5. Five types of proposed models for digesta flow in the ruminant and their representation by physical models of mixing compartment and non-mixing flow. G1 = exponentially distributed residence times, k_1 and k_2 = exponential turnover rates for faster and slower compartments respectively, Gn = age-dependent distributed residence times, λ and λ_2 = age dependent turnover rate parameter for one compartment and faster turnover compartment, respectively, τ = time delay for escape, Fτ = distribution of τ and O = output.

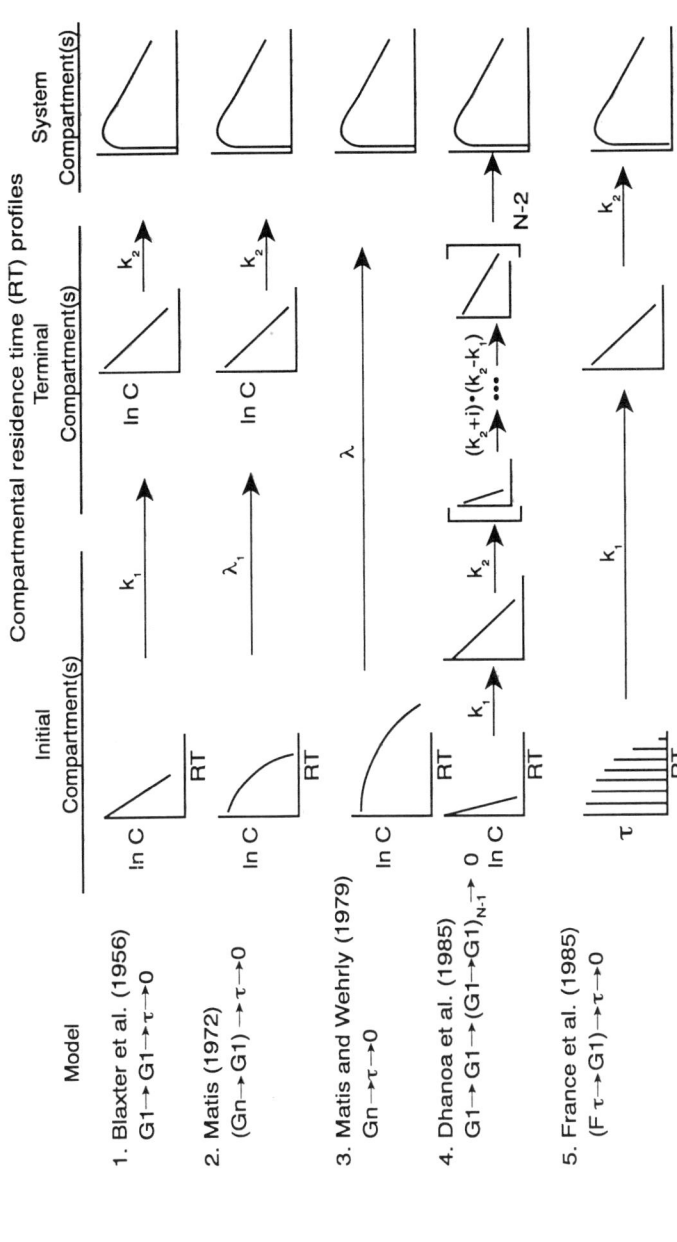

Figure 6. Five types of proposed models for digesta flow in the ruminant and the distribution of residence times in each compartment and the compartmental system. G1 = exponentially distributed residence times, k = exponential turnover rate, Gn = age-dependent distributed residence times, λ = age dependent turnover rate, τ = time delay for escape, Fτ = distribution of τ, and O = output.

are in modelling philosophy (Matis, 1987), not goodness-of-fit or interpretation of results. France et al. (1993) have subsequently proposed a model based on diffusion mechanisms to describe initial non-exponential turnover of fragments from rumen digesta; however, this model cannot be fitted to data.

The output profiles from all the mixing compartment systems illustrated in Figure 6 are similar and, when fitted to data of adequate quality with values distributed in ranges critical to parameters estimated by each model (Fadel, 1992), will yield similar statistical fits. Because different assumptions underlie each individual mathematical model, some models have different requirements for critically distributed values than do other models. Therefore, quantity, quality and distribution of data within an experiment can adversely affect estimation of individual parameters of different models and account for the fitting problems observed by Dhanoa et al. (1985) and Mertens (1989).

DIGESTA FLOW MARKERS

Particulate flow markers are required to identify the residence time distribution of residues from a discrete meal, i.e., a "dosed" meal. In order to estimate the flux of feed entities, such markers must be unique to the "dosed" meal and are, therefore, most commonly external markers. The use of markers for apprising site and extent of digestion in ruminants has recently been reviewed by Owens and Hanson (1992).

External

Ideally, the external flow marker should remain associated with the specific undigested nutrient in question, a virtually impossible requirement for nutrients other than in purified diets. Insoluble entities within the tissues of forage fragments are contained within and digested from the fragment. Therefore, the dilution of an external marker which remains associated with the undigested forage fragment will reflect the turnover of insoluble E_d or E_i constituents of the fragment. Another essential criterion for flow markers is that the flow and digestion of the marked forage fragments not differ from similar unmarked fragments.

When properly used, rare earth elements can be appropriate flow markers (Owens and Hanson, 1992). The rare earths are metabolically inert, indigestible, and (within their binding affinities for the fragments to be marked) are resistant to displacement from feedstuff residues within the normal pH range of ruminal digesta. Rare earth elements will be displaced from their feedstuff binding sites by protons at pH's comparable to the more acidic abomasal and duodenal digesta (Ellis and Beever, 1984). However, such displacement is of little consequence, at least for ruminants, because the ruminal digesta is the primary if not sole source of variation in flow of particulate matter and solutes. Components of the digesta also have comparable flow rates subsequent to the abomasum (Siciliano-Jones and Murphy, 1986; Wylie, 1987).

Recommended procedures for the use of rare earths are based on the principle of saturating all binding sites for rare earths on the feedstuff and then removing unbound and loosely-bound rare earths. Loosely-bound rare earths may be removed with a pH 4 acetic acid solution to simulate a competitive ruminal environment for displacement of the rare earth. Binding capacities are positively related to levels of fiber, lignin, and pectic substances in the feedstuff and typically range from 5 to 40 mg of rare earth element per gram of feedstuff DM. Thus, a level of 50 mg of rare earth element as a soluble sulfate, chloride, or acetate is recommended per gram of feedstuff to be marked. After soaking for a minimum of 12 h, unbound or loosely bound rare earth is then removed by soaking for 1 to 3 h in an acetic acid solution of pH 4 (approximately 0.01 M, simulating the ruminal environment) to remove unbound and loosely-bound rare earth which may otherwise be displaced by protons of ruminal acetic acid.

The validity of markers for estimating flow of particulate matter has been evaluated by a variety of indirect methods and conflicting conclusions have been reported (Owens and Hanson, 1992). A more direct test of the validity of external particulate flow markers would be to compare estimates of gastrointestinal mean residence time obtained for the flow marker with these estimated for an internal E_i. Results of one such comparison is indicated in Figure 7 (Wylie et al., 1986; Wylie, 1987). The gastrointestinal mean compartmental residence time, GMCRT, was estimated in sheep both by the dilution of rare earths (using an age-dependent, one-compartment model; model G2→τ→O, Figure 11) and by dilution of daily influx of indigestible neutral detergent fiber, INDF, by the gastrointestinal pool of INDF (equation 3) estimated at slaughter. Estimates of GMCRT did not differ by method of determination and both methods yielded similar responses to the correction of a dietary deficiency of crude protein. However, the response by the solute marker was only 0.58 that estimated for INDF. Faichney et al. (1989) utilized a similar approach in sheep and reached a similar conclusion with regard to the validity of specifically bound rare earth as flow markers for INDF. Agreement between these two independent methods supports the validity of specifically bound rare earth as particulate flow markers in ruminants. Owens and Hanson (1992) conclude that, "Despite imprecision in marker procedures, inherent variation may be small relative to other sources of variation (e.g., gut physiology, diet, environment, and feed intake). Even though absolute values may be imprecise and inaccurate, marker-based estimates usually provide reliable information about the direction and extent of kinetic changes induced by treatments."

Worley (1987) reported that hafnium (Hf), a period four element, was strongly bound to forages and concentrates and resistant to displacement by pH's as low as two. When bound to feeds, the Hf was completely recovered on neutral detergent fiber, NDF, suggesting that binding of Hf renders insoluble some E_d that otherwise would be solubilized in the procedure for determining NDF. The Hf was bound at all levels tested (up to 5% of the DM) and, hence, the addition of Hf must be limited in order to not adversely affect the fragment's density and overly affect its digestion. Hafnium may be a desirable marker for concentrates because of its large binding affinity. Because of its resistance to displacement by protons, Hf also may be a useful marker for measuring the flow of fragments through acidic segments of the digestive tract, such as in monogastrics.

Although there are relatively large differences among rare earths in their binding affinities, no significant difference in k_e has been observed for the rare earth markers Europium, Ytterbium, Terbium, Samarium, Lutetium, Lanthanum,

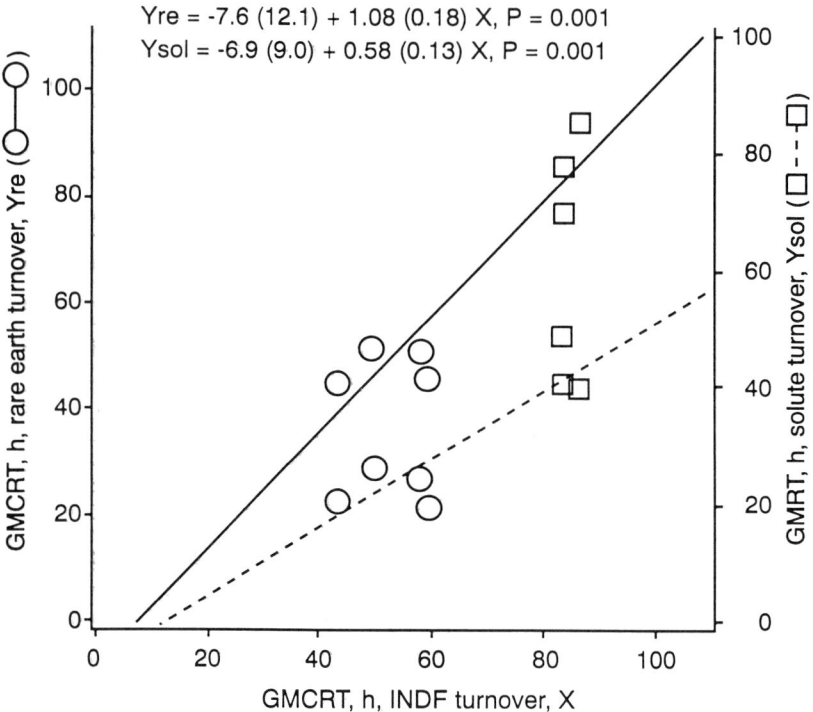

Figure 7. Relationships between gastrointestinal mean residence time, GMCRT, estimated by the rare earth compartmental turnover method, Yre, and the solute compartmental turnover method, Ysol, or estimated by the indigestible neutral detergent fiber flux-compartmental pool method, INDF, in sheep fed diets containing 60 g kg^{-1} crude protein, O, or 100 g kg^{-1} crude protein, □. Note that GMCRT estimated for the rare earth applied marker closely approximates that estimated by INDF turnover and that responses in CMCRT to supplementary protein are similar whether estimated by applied rare earth markers or the intrinsic indigestible fiber turnover method, i.e., a regression coefficient of 1.08 ± 0.18 for Yre vs X.

Samarium, Neodymium, and Terbium when applied to different samples of the same forage or concentrate (Cabello et al., 1990; Hill, 1991; Moore et al., 1992; see Table 12 later).

The number of acid resistant binding sites for rare earths on feedstuffs is relatively low (2 to 30 mg rare earth/g DM). Thus, relatively large doses of feedstuff must be marked and dosed in order to yield analytically reliable levels after several days of dilution turnover. Neutron activation analysis (Pond et al., 1985), plasma emission spectroscopy (Combs and Satter, 1992) and flameless atomic absorption spectrophotometry are preferred methods of analysis because of their sensitivity. In order to maximize sensitivity, digesta samples can be dried and the assay conducted on the ash (neutron activation analysis) or its leachate (emission and absorption spectrophotometry).

Internal Markers

Because internal markers unquestionably flow with undigested feed residues, they should be ideal flow markers. Because they are not unique to a given meal, indigestible entities of forages can only be used as flow markers where flux and compartmental mass is independently determined (equation 3). Although internal markers such as ^{14}C-labelled substrate can be used as a flow marker, ^{14}C is incorporated into potentially digestible as well as indigestible entities. Chemical or digestion procedures should be used to remove any potentially digestible ^{14}C before dosing (Smith, 1989) and some digestion model must be used to resolve disappearance of potentially digestible from disappearance of indigestible entities in digesta samples.

Dosing Procedures

Pulse-dosing of marked feedstuffs into the ruminal digesta via a ruminal cannula is frequently used. However, because of the functional implications of "un-mixing" (Sutherland, 1989) and differential flow from the rumen related to meal eating (Reid et al., 1979; Pond et al., 1989; Hill, 1991), dosing as a component of a normal meal is recommended. Dosing during the initial or final phases of a meal may affect estimates of flow (Pond et al., 1989; Hill, personal communication).

FITTING COMPARTMENTAL MODELS TO DATA AND INTERPRETING MODEL PARAMETERS

When fitted to data of good quality, the age-dependent or multi-compartment models all can be expected to yield statistically similar fits and estimates of total gastrointestinal tract residence time, i.e., compartmental residence time (as some inverse of the turnover rates) plus a discrete time delay, τ. Indeed, simple algebraic methods provide estimates of mean compartmental residence times similar to compartmental methods (Lalles et al., 1991). From a nutritional viewpoint, the objective of compartmental models is to resolve total gastrointestinal tract residence time into residence times within particular segments having unique nutritional significance, for example, sites of 1) gastric fermentation, 2) post-gastric hydrolytic digestion, 3) post-gastric fermentations, and 4) post-gastric time delay and sites which constrain digesta flow. Currently, inferences concerning the flow process within sub-segments of the total gastrointestinal digesta can only be made by measuring marker flow through such segments. The applicability of models to data and interpretations of estimated parameters will be illustrated with a sample data set.

Model Expressions

The mathematical expressions of the models are given by Pond et al.(1988) and reproduced in Tables 1 through 4. Expressions for deriving other parameters from rate related parameters estimated by these models are summarized in Table 5.

Sample Data

The sample data were collected from a 550 kg steer having a 10 cm ruminal fistula and a 2.5 cm ID "T" type duodenal cannulae. The steer was fed bermudagrass hay [*Cynodon dactylion* (L.) pers.] ad libitum for 10 d prior to, and during, digesta collection. The bermudagrass hay to be dosed was soaked in a pH neutral solution of Nd_3SO_4 providing 50 mg Nd/g forage DM for 12 h and then rinsed with water to remove unbound and loosely-bound Nd. The dried, marked hay was mixed with a small portion of a meal offered and consumed by the steer within 10 min. The steer then was offered fresh hay twice daily, 120% in excess of that consumed during the preceding 12 h interval. The bermudagrass hay contained 740 g kg^{-1} NDF and 115 g kg^{-1} crude protein on a DM basis. Daily intake during the 9 d experimental period averaged 7.3 kg DM and daily fecal output averaged 3.7 kg DM. The steer consumed 305 g of hay marked with 22.1 mg Nd kg^{-1} DM for a total dose of 6,748 mg of Nd. Samples of dorsal rumen raft digesta, ventral rumen digesta, duodenal digesta, and feces were collected at intervals and the concentrations of Nd in the samples were determined by neutron activation analysis (Pond et al., 1985). It was assumed that the flow of Nd represented the flow of undigested particles. Models described in Tables 1 through 4 were fitted to data using the NLIN procedure of SAS (1984).

Marker Flow Patterns

Patterns of marker concentration in digesta from various segments of the gastrointestinal tract subsequent to consumption of the marker containing meal are illustrated in Figure 8. The concentration values are depicted on an arithmetic scale to illustrate their true scale of variation. Successive values are joined by solid lines to aid in discerning periodicity. The dashed line represents the expected concentrations derived from fitting the $(G2 \rightarrow G1) \rightarrow \tau \rightarrow O$ model to the data as will be described later.

In the raft digesta, large, cyclic variations in marker concentration occurred. The mean intra-peak period during the first 72 h was approximately 12 h and suggested that such periodicity may be related to consumption of main meals by such animals (Deswysen and Ellis, 1988). Selective and non-steady state flow of fragments from the ruminal digesta (Reid et al., 1979) have been related to consumption of mean meals.

Considering the inertia of ruminal digesta and the slow mixing of influx (Deswysen and Erhlein, 1981), it is apparent that many age-independent compartments could exist if defined as the mass as the mass with which the flux instantaneously mixes. Thus, many conceptual compartments could exist as the

Table 1. Total dose (D) remaining in one compartment models with various gamma residence time distributions.

Model	Residence time distribution	Model expression[a,b,c,d,e,f]
G1→	Gamma 1	De^{-kt}
G2→	Gamma 2	$De^{-\lambda t}(1 + \lambda t)$
G3→	Gamma 3	$De^{-\lambda t}[1 + \lambda t + (\lambda t)^2/2]$
Gn→	Gamma n	$De^{-\lambda t} \sum_{i=0}^{n-1} (\lambda t)^i/i!$

[a] Parameter n, n=1, 2,..., is integer value of gamma distribution.
[b] Exponential is equivalent to gamma 1.
[c] k = rate parameter for exponentially distributed residence times.
[d] λ = rate parameter for gamma distributed residence times.
[e] t = time after dose of marker.
[f] To incorporate time delay, τ, substitute t-τ for t.

Table 2. Fractional concentration of marker in material emerging (O) from one compartment models with various gamma residence time distributions.

Model	Residence time distribution[a,b]	Model expressions[c,d,e,f,g]
G1→O	Gamma 1	Ce^{-kt}
G2→O	Gamma 2	$C\lambda te^{-\lambda t}/0.59635$
G3→O	Gamma 3	$C\lambda^2 t^2 e^{-\lambda t}/(2 \times 0.47454)$
G4→O	Gamma 4	$C\lambda^3 t^3 e^{-\lambda t}/(6 \times 0.40857)$
G5→O	Gamma 5	$C\lambda^4 t^4 e^{-\lambda t}/(24 \times 0.36528)$
G6→O	Gamma 6	$C\lambda^5 t^5 e^{-\lambda t}/(120 \times 0.33929)$

[a] Parameter, n, n=1, 2,..., is called order of age-dependency.
[b] Exponential is equivalent to gamma 1.
[c] C = Initial concentration in compartment mass (CM) assuming instantaneous mixing of dose (D), i.e., C = D/CM.
[d] k = rate parameter for exponentially distributed residence times.
[e] t = time after dose of marker.
[f] To incorporate time delay, τ, substitute t-τ for t.
[g] λ = rate parameter for gamma distributed residence times.

Table 3. Total dose (D) remaining in two compartment models (C1 and C2) with various gamma residence time distributions (G1,G2,G3,Gn) in C1 and C2.

Model	Residence time distribution C1	C2	Model expression[a,b,c,d,e,f]
G1→G1→	G1	G1	$D(k_1 e^{-k_2 t} - k_2 e^{-k_1 t})/(k_1 - k_2)$
G2→G1→	G2	G1	$D\{\delta^2 e^{-k_2 t} + e^{-\lambda_1 t}[1-\delta^2 + (1-\delta)\lambda_1 t]\}$
G3→G1→	G3	G1	$D\{\delta^3 e^{-k_2 t} + e^{-\lambda_1 t}[1-\delta_3 + (1-\delta^2)\lambda_1 t + (1-\delta)(\lambda_1 t)^2/2]\}$
Gn→G1→	Gn	G1	$D\left[\delta^n e^{-k_2 t} + e^{-\lambda_1 t} \sum_{i=0}^{n-1} (1-\delta^{n-1})(\lambda_1 t)^i/i!\right]$

[a] D = Initial dose.
[b] k_1 = Slower rate parameter for exponentially distributed residence times.
[c] k_2 = Faster rate parameter for exponentially distributed residence times.
[d] t = Time after dose of marker. To incorporate time delay (τ), substitute (t-τ) for t.
[e] λ_1 = rate parameter for gamma distributed residence times for faster turnover compartment.
[f] $\delta = \lambda_1/(\lambda_1 - k_2)$.

Table 4. Fractional concentration of marker in material emerging from terminal compartment (C2) of two compartment (C1 and C2) models with various gamma residence time distribution in the initial compartment (C1).

Model	Residence time distribution C1	C2	Model expression[a,b,c,d,e,f]
G1→G1→O	G4	G1	$C_2 k_1 (e^{-k_2 t} - e^{-k_1 t})/(k_1 - k_2)$
G2→G1→O	G2	G1	$C_2[\delta^2 e^{-k_2 t} - e^{-\lambda_1 t}(\delta^2 + \delta\lambda_1 t)]$
G3→G1→O	G3	G1	$C_2[\delta^3 e^{-k_2 t} - e^{-\lambda_1 t}(\delta^3 + \delta^2 \lambda_1 t + \delta\lambda_1^2 t^2/2)]$
Gn→G1→O	Gn	G1	$C_2\left[\delta^n e^{-k_2 t} - e^{-\lambda_1 t} \sum_{i=1}^{n} \delta^i (\lambda_1 t)^{n-i}/(n-i)!\right]$

[a] C_2 = initial concentration in second compartment if dose (D) had been introduced into volume (V) of second compartment and instantaneously mixed, i.e., $C_2 = D/V$.
[b] k_1 = rate parameter for exponentially distributed residence times.
[c] k_2 = Slower rate parameter for exponentially distributed residence time.
[d] t = time after dose of marker. To incorporate time delay, substitute (t-τ) for t.
[e] λ_1 = rate parameter for gamma distributed residence times of G2 or greater.
[f] $\delta = \lambda_1/(\lambda_1 - k_2)$.

Table 5. The calculation of secondary parameters from rate related parameters.

Item	Factor or expression
A. Rate, k, or mean rate, $\bar{k}$, of residence time distribution:	
1. G1 distributed residence times	k
2. G2 distributed residence times	$\bar{k}=\lambda_1 \cdot F$; F=.59635
3. G3 distributed residence times	$\bar{k}=\lambda_1 \cdot F$; F=.47454
4. G4 distributed residence times	$\bar{k}=\lambda_1 \cdot F$; F=.4085686
5. G5 distributed residence times	$\bar{k}=\lambda_1 \cdot F$; F=.365276
6. G6 distributed residence times	$\bar{k}=\lambda_1 \cdot F$; F=.339291
B. Compartmental mass (CM) of undigested DM at sampling site:	
1. One-compartment:	
a. Age-independent (CM)	M dose/D
b. Age-dependent (CMS)	M dose/D
2. Two-compartment:	
a. Age-independent, turnover (CM2)	M dose/C_2
b. Age-dependent (CM1). From CM2 assuming:	
CM1 $\xrightarrow{\lambda_1}$ CM2 $\xrightarrow{k_2}$	$CM2 \cdot (k_2/(\lambda_1 \cdot F))$
c. Two-compartment system (CMS), assuming:	
CM1 $\xrightarrow{\lambda_1}$ CM2 $\xrightarrow{k_2}$	$CM_2 \cdot k_2$
C. Compartmental flux of DM undigested at sampling site, mass·time^{-1}:	
1. Age-independent, one-compartment	CM·k
2. Age-dependent, one-compartment	$CM \cdot (\lambda_1 \cdot F)$
3. Two-compartment	$CM2 \cdot k_2$
D. Residence times:	
1. Compartmental (MCRT):	
a. G1, age-independent (MCRT1)	1/k
b. Gn[a], age-dependent (MCRT2)	n/λ_1
c. System (MCRTS):	
1. One compartment age-dependent	n/λ_1
2. Two-compartment: assuming CM1→CM2→	
a. Age-independent, age-independent	$1/k_1 + 1/k_2$
b. Age-dependent, age-independent	$n/\lambda_1 + 1/k_2$
2. Residence time, time delay	τ
3. Residence time, gastrointestinal	MCRT + τ

[a] n = order of integer gamma distribution.

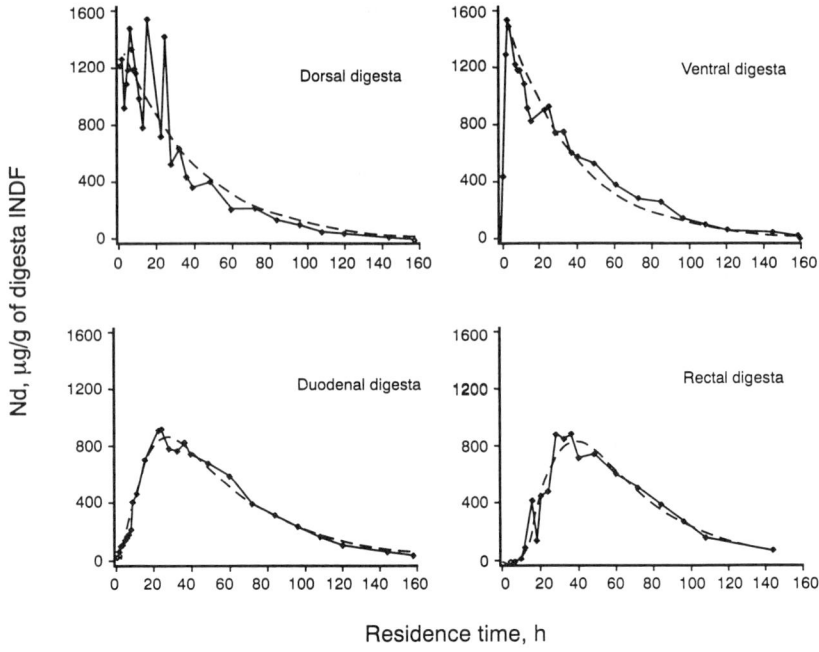

Figure 8. Observed values (diamonds) and expected values (dashed line) from fitting the G2→G1→τ→O model to marker concentrations in digesta sampled from the dorsal most, mid-dorsal rumen digesta (raft), the ventral most of the ventral rumen digesta (ventral), the ascending duodenum (duodenum), and the feces (rectum). The marker was bound to hay which was consumed as a 10 min initial portion of a larger meal.

result of such slow and imperfect mixing. Indeed, the sequential multiple, age-independent compartmental model of Dhanoa et al.(1985) identified as many as 20 exponentially distributed residence time compartments. Incomplete mixing occurs not only for ingested marked forage mixing with resident digesta but also for unmarked forage fragments of subsequent meals mixing with resident marked forage fragments. The use of constant infusion of marker would not circumvent this slow mixing problem as claimed by Cruickshank et al.(1989).

In digesta obtained from the ventral most portion of the ventral sac, the pattern of marker concentration initially increased until approximately 3 h post-dose and generally declined thereafter. This pattern indicated that the ventral sac digesta was the recipient compartment of marker flowing from an initial entry compartment, presumably the dorsal raft digesta as observed by Ehrlein (1980) and as postulated (Ellis et al., 1991) from fermentation based buoyancy mechanisms proposed by Sutherland (1989).

The pattern of marker concentration in digesta sampled from the ascending duodenum indicated that negligible time delay was associated with marker flow through the mass of digesta of the omasum and abomasum. As compared to ruminal digesta, less periodicity was apparent in duodenal digesta, especially during the first 30 h post-dose. The pattern of marker concentration appearing in the feces demonstrated a time delay associated with digesta flow from the ascending duodenum.

Deviations in observed marker concentrations versus those expected from fitting the (G2→G1)→τ→O model to data were large, especially in ruminal raft digesta. Much of the deviation is suggested to be due to processes occurring in the raft digesta which do not conform to the model assumptions of: 1) instantaneous mixing, 2) steady-state conditions (flux and mass of digesta), and 3) processes which differentially affect flow within ill-defined physical segments of the digesta. In view of obvious short-term deviations from model assumptions of steady-state conditions and instantaneous mixing, interpretation of model parameters must be considered average parameters expressing the cumulative effects of the 4 to 6 d of data to which the model was fitted.

The marker concentration patterns in duodenal digesta clearly appear to result from "multi-compartment"-like flow processes in the ruminal digesta. However, quantification of the dynamics of such flow-paths by sampling ruminal digesta is difficult because such "compartments" appear to be diffuse and non-static (Sutherland, 1989). Alternatively, sampling duodenal digesta and fitting models to such data appear to offer more promise in estimating the flow of ruminal digesta and intraluminal processes.

Postgastric Flow: Two Compartment Models

Fit of G1→G1→τ→O and G2→G1→τ→O models to the pattern of marker concentration appearing in the feces and duodenal digesta is illustrated in Figure 9. The statistics for parameters estimated by models of varied orders of age-dependency are summarized in Table 6. As compared to the two-compartment, age-independent model, G1→G1→τ→O, incorporating age-dependency into the faster turnover compartment of the two-compartment model, G2→G1→τ→O, clearly improved fit to segments of the data representing escape of marker during its shorter residence times (< 20 h). Compared to the two compartment models with age-independent distributions of residence time in each compartment, G1→G1→τ→O, the inclusion of the lowest order of age-dependency in the faster turnover compartment, G2→G1→τ→O: 1) improved fit to the data (reduced model EMS), primarily due to improved fit to the pattern of residence times of less than 20 h. As for other data (Matis, 1972; Pond et al., 1988), an order of age dependency of 2 or 3 appeared adequate; 2) incorporation of age-dependency, and increasing order of age-dependency, apportioned less residence time to non-mixing flow, τ, with corresponding increases in the mean compartmental residence time (n/λ_1, Table 6) in the faster turnover compartment, λ_1 (see also Figure 6 of Pond et al., 1988). For example, in the duodenal data, incorporating G2 age dependency in the faster turnover compartment increased MCRT from 13.37 h ($1/k_1$) to 14.74 h ($2/\lambda_1$) or

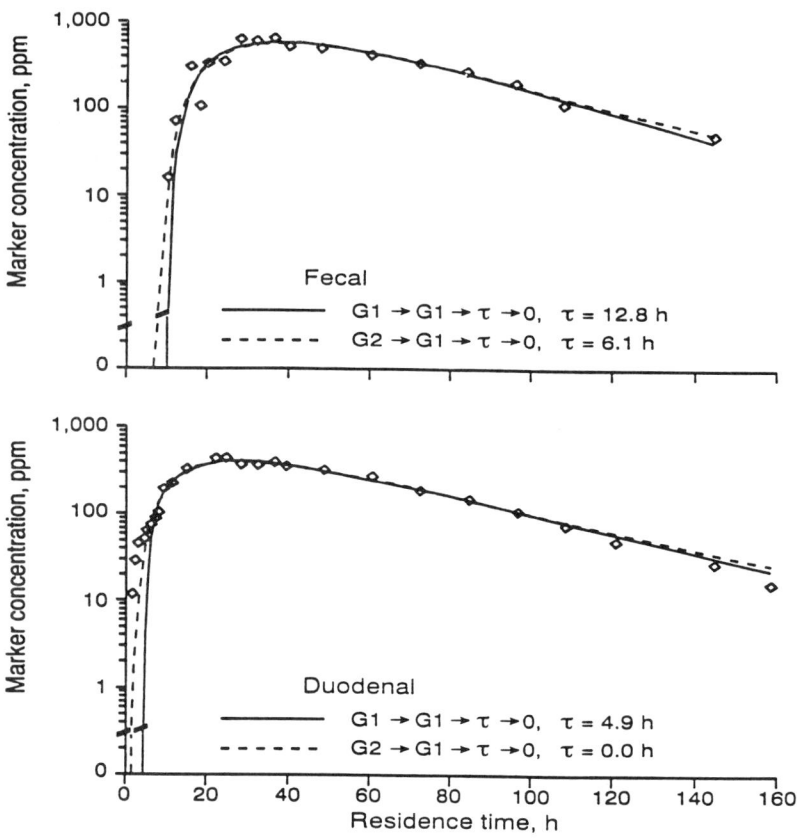

Figure 9. Observed values (diamonds) and expected values from fitting the G1→G1→τ→0 and G2→G1→τ→0 models to marker concentrations in digesta sampled from the duodenum and feces. The marker was bound to hay which was consumed as a 10 min initial portion of a larger meal.

by 1.37 h as compared to a decrease in τ from 4.9 to 1.2 or 3.7 h; 3) in the case of fecal data, the $k_1 \simeq k_2$ problem was avoided. The condition of $k_1 \simeq k_2$ may arise due to: a) fitting a model specifying two sequential age-independent compartments to data resulting from other than two age-independent, sequential compartments, b) the existence of two statistically indistinguishable compartment having approximately equal mass and, hence, turnover rates, ($k_1 \simeq k_2$); or c) the existence of one age-dependent compartment rather than two distinct age-independent compartments. The occurrence of two essentially equal rate parameters is inconsistent with the model assumptions of two sequential compartments. Thus,

Table 6. Parameters and their asymptotic standard errors estimated by fitting two-compartment models to excretion pattern of marker in duodenal and fecal digesta.

Sampling site and model	Parameter[a,b,c,d,e]				Model EMS
	C_2	k_1 or λ_1	k_2	τ	
	g Nd/kg DM	------------------h^{-1}------------------		h	
Duodenal:					
G1→G1→τ→O	737±96.4	0.0748±0.0173	0.0252±0.0042	4.9±0.47	898
G2→G1→τ→O	679±45.1	0.1357±0.0154	0.0229±0.0022	1.2±0.63	611
G3→G1→τ→O	628±31.1	0.2121±0.0215	0.0208±0.0017	0	587
G4→G1→τ→O	594±26.9	0.3086±0.0341	0.0193±0.0016	0	704
G5→G1→τ→O	574±22.6	0.3991±0.0431	0.0182±0.0013	0	639
G6→G1→τ→O	629±26.5	0.4234±0.0456	0.0211±0.0016	0	694
G1→G1→(G1→G1)→O	-	0.1618	0.0207	7.3±2.2	498
Feces:					
G1→G1→τ→O	1558±17·10[6]	0.0397±0.4437	0.0397±443	12.8±1.6	5,262
G2→G1→τ→O	1050±247	0.1062±0.0315	0.0261±0.0082	8.5±0.008	4,507
G3→G1→τ→O	967±167	0.1552±0.0411	0.0240±0.0061	6.1±2.5	4,341
G4→G1→τ→O	935±143	0.1942±0.0443	0.0226±0.0053	4.1±2.9	4,270
G5→G1→τ→O	917±131	0.2276±0.0501	0.0221±0.0056	2.4±3.2	4,229
G6→G1→τ→O	904±124	0.2573±0.0553	0.0220±0.0044	0.7±3.6	4,203
G1→G1→(G1→G1)→O	-	0.1314	0.0225	16.73	4,185

Footnotes for Table 6 appearing on preceding page.

[a] C_2 = initial concentration in the slower turnover (larger), age-independent compartment of two compartment models
[b] k_1 is the age-independent rate parameter for the faster turnover (smaller) compartment of two compartment models (Gn→G1→τ→O).
[c] λ_1 is the age-dependent rate parameter for the faster turnover (smaller) compartment of two compartment models (G2→G1→τ→O, G3→G1→τ→O, G4→G1→τ→O, G5→G1→τ→O, and G6→G1→τ→O).
[d] k_2 is the age-independent rate parameter for the age-independent, slower turnover (larger) age-independent compartment (G1) of two-compartment models.
[e] τ = time delay for escape.

when two sequential compartments are approximately equal, estimation of the faster rate parameter, and the related initial concentration parameter C_2, will be subject to considerable error of estimation as illustrated for fecal data in Table 6. The ratio of the two rate parameters should exceed a factor of 1.5 or more for reliable estimation of the faster rate (Pond et al., 1988; Cruickshank et al., 1989).

The effects of increasing the order of age-dependency of the two compartment model in partitioning residence time between time delay, τ, and age-dependent turnover, λ, is illustrated graphically in Figure 10.

Postgastric Flow: One Compartment Models

Models of marker emergence from an age-dependent mixing compartment with τ, Gn→τ→O, or without τ, Gn→O (Table 2) are appropriate models for post-gastric flow of digesta. Fit of Gn→τ→O models to duodenal and fecal data are illustrated in Figure 11 and estimated parameters are summarized in Table 7. The one compartment model, G2→τ→O, yielded similar statistical fit to duodenal and fecal data (Table 7) as did models (G1-G6)→G1→τ→O (Table 6). However, incorporation of age-dependence greater than G2 into the one-compartment models fitted to post-gastric flow resulted in increased error mean squares, apparently because of the dominance of age-independent distribution of residence times during the last one-half or more of the excretion curve for postgastric data (Figure 9).

The two-compartment models are more flexible because they merge the dominant age-independent distributions of residences associated with faster turnover processes and the age-dependent processes associated with the slower turnover processes of ruminal digesta. Because of their singular distributions of residence times, the one-compartment age-dependent models are very robust and, consequently, are recommended where data values are lacking during the early residence times as illustrated in Figure 11. Due to a deficiency of values at residence times for first emergence of the marker, the apportionment of residence times between τ and age-dependent turnover could not be resolved in this data set by multiple compartment models involving a time delay.

Due to slow and imperfect mixing of fragments of forage between their

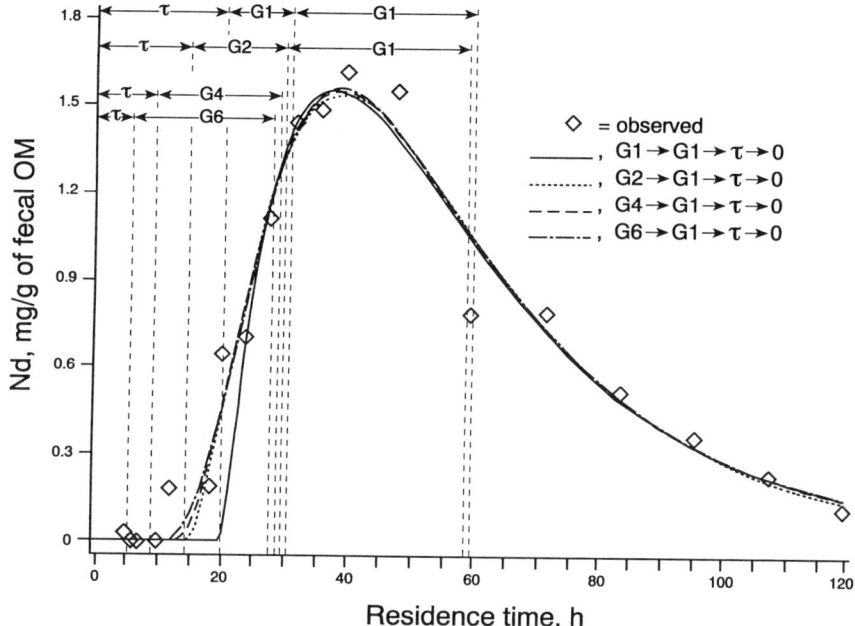

Figure 10. Effect of fitting two compartment models of progressively greater orders of age-dependency, G1 . . .G6 in the faster turnover compartment upon partition of the expected residence time between time delay and λ_1. Note that increasing orders of age-dependency progressively partitions greater mean compartmental residence time into the faster turnover compartment primarily by reducing the estimated time delay.

entry and sampling sites within the ruminal digesta, one could assume the Gn→τ→0 model would be applicable to ruminal digesta where marker entry site does not equal sample site (Ellis et al., 1979; 1984). Such an assumption is applicable to the present data since the marker was applied to the forage on offer and entered ruminal digesta via the esophageal orifice. The digesta was sampled at a distal site. However, variations in marker concentration in the earliest subsequent samples of ruminal digesta are generally too large to detect such dose and sampling site differences (Dixon et al., 1983).

Turnover Within Ruminal Digest

One compartment models expressing marker remaining (Gn→, Table 1) are appropriate if sampling from the dose site and if instantaneous mixing within a single mixing compartment is assumed. The incorporation of age-dependency into one-compartment models did not appreciably improve fit to the present ruminal

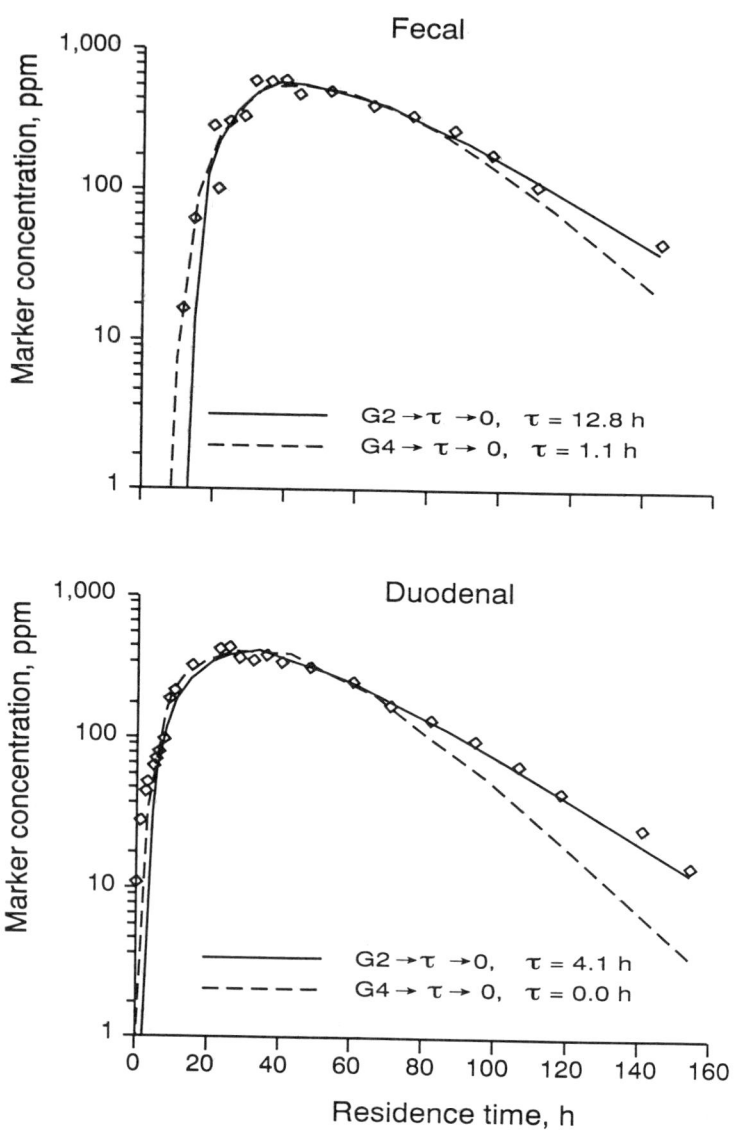

Figure 11. Observed values (diamonds) and expected values from fitting various one-compartment, age-dependent models to marker concentrations in digesta sampled from the duodenum and feces. The marker was bound to hay which was consumed as a 10 min initial portion of a larger meal.

Table 7. Parameters and their asymptotic standard errors estimated by fitting one-compartment models to excretion pattern of marker (Nd) in duodenal and fecal digesta.

Sampling site and model	Parameter[a,b,c,d]				Model EMS
	C	λ_1	$\bar{k}$	τ	
	g Nd/kg DM		h^{-1}	h^{-1}	h
Duodenum:					
G2→τ→O	682±17	0.0406±0.0016	0.0242	4.1±0.4	817
G3→τ→O	773±23	0.0627±0.0026	0.0297	0	1,131
G4→τ→O	806±31	0.0932±0.0051	0.0381	0	1,786
G5→τ→O	898±94	0.1333±0.0081	0.0487	0	1,238
G6→τ→O	892±40	0.1571±0.0101	0.0533	0	7,238
Feces:					
G2→τ→O	929±49	0.0397±0.0033	0.0237	12.8±1.2	4,933
G3→τ→O	039±54	0.0586±0.0041	0.0278	6.4±1.6	4,713
G4→τ→O	098±62	0.0737±0.0053	0.0301	1.1±2.2	5,458
G5→τ→O	210±72	0.0925±0.0066	0.0338	0	5,928
G6→τ→O	238±82	0.1225±0.0101	0.0416	0	6,866

[a] C = initial concentration in age-dependent compartment.
[b] λ = age-dependent rate parameter.
[c] τ = time delay.
[d] $\bar{k}$ = mean rate for age-dependent distributed rate parameter, λ.

digesta data (Table 8), probably because of large variation in marker concentration in the earliest samples of dorsal digesta and the dominance of age-independent turnover subsequent to 10 to 20 h post-dose in both dorsal and ventral digesta.

Marker concentrations during the earliest residence times, especially in the raft data, suggest an ascending pattern of marker concentration. It has been common practice to disregard such early values (that do not conform to an expected exponential distribution of residence times) and fit an age-independent one-compartment model only to those values which conform to an exponential distribution of residence times (Grovum and Williams, 1973; Dixon et al., 1983; Cruickshank et al., 1989).

Alternatively, effects of slow mixing, dose site and sampling site can be accounted for by fitting two-compartment models which assume the sampled digesta is emerging from the second compartment (Table 4, Gn→G1→τ→O). In this case, it is imperative that the fitting procedure utilize a grid option (see section on fitting models to data) that considers a large range of initial values for the faster turnover rate so that the fitting procedure can choose one of these fast turnover rates to account for the value in the ascending pattern. This approach, illustrated in Figure 12 and Table 9, also avoids the subjective decision as to which values

Table 8. Parameters and their asymptotic standard errors estimated by fitting one-compartment models to dilution pattern of marker (Nd) in dorsal and ventral ruminal digesta when the initial 0.8 h value for marker is deleted.

Model and sample site	Parameter[a,b,c,d] C_2	λ or k	k or $\bar{k}$	Model EMS
	g Nd/kg DM	h^{-1}	h^{-1}	
Ruminal Raft				
G1→	659±51	0.0244±0.0045	0.0244	11,362
G2→	585±37	0.0491±0.0067	0.0393	10,058
G3→	565±34	0.0765±0.0091	0.0363	9,802
G4→	557±32	0.1051±0.0112	0.0429	9,827
G5→	552±32	0.1345±0.0133	0.0491	9,932
G6→	550±32	0.1644±0.0153	0.0558	10,058
Ruminal Ventral				
G1→	656±30	0.0274±0.0028	0.0273	3,445
G2→	566±27	0.0523±0.0053	0.0312	5,116
G3→	538±28	0.0782±0.0078	0.0371	6,530
G4→	524±28	0.1047±0.0104	0.0428	7,592
G5→	516±29	0.1315±0.0128	0.0480	8,393
G6→	511±30	0.1587±0.0152	0.0538	9,017

[a] C_2 = initial concentration in the slower turnover (larger), age-independent compartment.
[b] λ = rate parameter for age-dependent (G2 or greater) residence time distributed turnover compartment.
[c] k = age-independent turnover rate.
[d] $\bar{k}$ = mean age-dependent turnover rate.

conform to an exponential distribution of residence times.
    Deleting initial values can lead to errors in estimating initial dilution, CØ, if the turnover rate is extrapolated to dosing time (Ellis et al., 1979; Cruickshank et al., 1989). In the current data set, mixing was relatively rapid between the dosing site (esophageal orifice) and the sampling site (mid-dorsal and ventral most digesta of the ventral sac), so similar estimates of the initial concentration in the second age-independent compartment was obtained for the Gn→G1→τ→O and for one-compartment (G1→) models. The problem of slow mixing between dose and sample site was more apparent in the raft of the ruminal digesta than in the ventral segments of ruminal digesta.

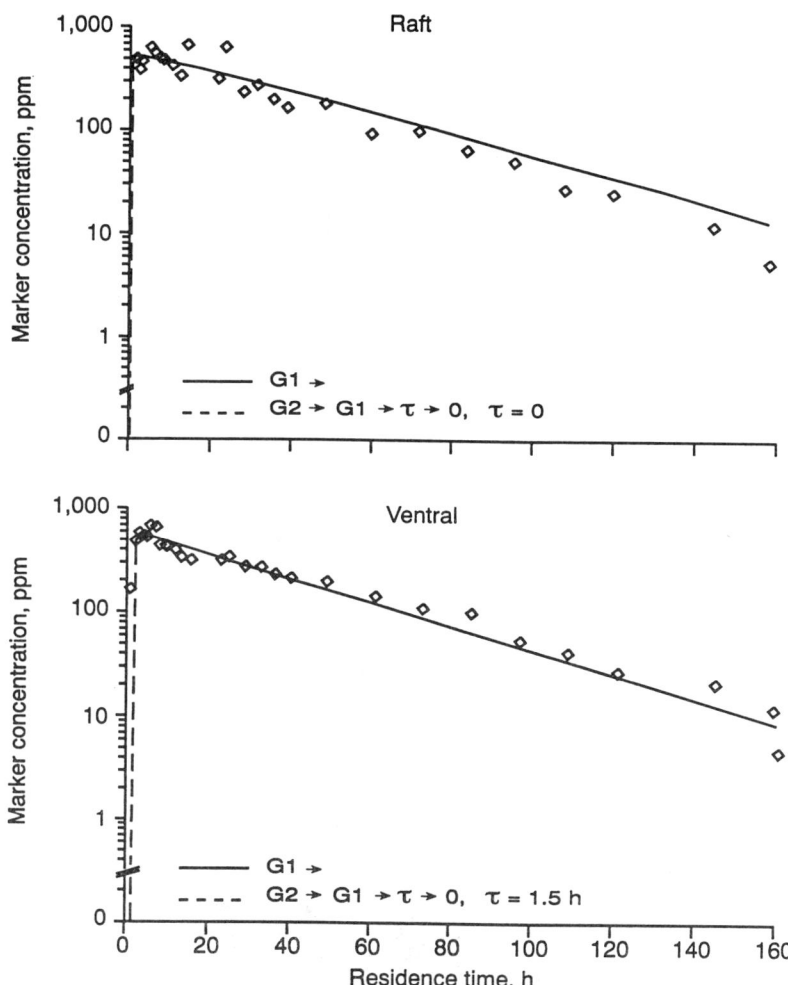

Figure 12. Observed values (diamonds) and expected values from fitting one (G1→) and two-compartment models (G2→G1→τ→O) to ruminal raft and ventral digesta. The marker was bound to hay which was consumed as a 10 min initial portion of a larger meal.

Table 9. Parameters and their asymptotic standard errors estimated by fitting two-compartment models to dilution pattern of marker in dorsal and ventral ruminal digesta including the initial 0.8 Hr value for marker.

Sample site and model	C_2	Parameters[a,b,c,d,e] k_1 or λ_1	k_2	τ	Model EMS
	g Nd/kg DM	----------h^{-1}----------		h	
Ruminal, Raft:					
G1→G1→τ→O	649±87	2.6±21.7	0.0242±0.0045	0	11,398
G2→G1→τ→O	646±157	4.7±105	0.0241±0.0045	0	11,411
G3→G1→τ→O	646±292	6.5±332	0.0241±0.0045	0	11,413
G1→G1→(G1→G1)→O	-	0.1169	0.0445	0	9,190
Ruminal, Ventral:					
G1→G1→τ→O	625±29	15.2±XL	0.0274±0.0033	1.7±0.04	4,501
G2→G1→τ→O	627±3,652	10.6±19,851	0.0274±0.0034	1.5±568	4,689
G3→G1→τ→O	636±27	2.8±.98	0.0274±0.0029	0.1±0.2	3,459
G1→G1→(G1→G1)→O	-	0.0651	0.0368	0	10,378

[a] C_2 = initial concentration in the slower turnover (larger), age-independent compartment.
[b] k_1 is the age-independent rate parameter for the faster turnover (smaller) compartment of age-independent, two-compartment model (G1→G1→τ→O).
[c] λ_1 is the age-dependent rate parameter for the faster turnover (smaller) compartment of the age-dependent, two-compartment models (G2→G1→τ→O, G3→G1→τ→O, G4→G1→τ→O, G5→G1→τ→O, and G6→G1→τ→O).
[d] k_2 is the age-independent rate parameter for the age-independent, slower turnover (larger) compartment (G1).
[e] τ = time delay for escape.
[f] XL = extremely large SE due to $\lambda_1 \approx k_2$.

Apportionment of Mean Compartmental Residence Time

The apportionment of mean compartmental residence (MCRT) based on sampling the various sites is summarized in Table 10. The two compartment models G2→G1→τ→O fitted to marker concentration remaining in ruminal digesta (sites R and V) essentially apportioned all the MCRT to age-independent turnover from the raft (42.3 h) and ventral digesta sites (36.5 h). Estimates of MCRT within the compartmental mixing systems (MCRTS) based on marker departure from ruminal raft (42.7 h) and ventral digesta (36.7 h) were only slightly smaller than MCRT estimated to be in the slower turnover compartment (MCRT2) based on marker flow through the duodenal digesta (43.7 h) and marker flow to the feces (38.3 h) when estimated by model G2→G1→τ→O.

Given the uncertainties associated with slow mixing and non-steady state conditions in ruminal digesta data, the MCRT2 estimated by the G2→G1→τ→O and G3→G1→τ→O models were all of similar magnitude throughout the tract (41.5, 36.5, 43.7, and 41.7 h, Table 9). The MCRT1 for the faster turnover compartment was first clearly detected in duodenal digesta (14.7 h) and increased to approximately 19 h when estimated from fecal data. Thus, post-abomasal flow appears primarily associated with a multi-laminar type flow as will be illustrated later for both solutes and different sized forage fragments (Figure 13 later). The MCRT for post-abomasal flow does not differ among solutes or different sized forage fragments and averages 1.1 h. (see Figure 13, later). In contrast, Faichney and Boston (1983) and Faichney (1984) obtained data with sheep which indicated considerable post-gastric MCRT in pregnant sheep. Thus, sheep, and especially pregnant sheep, may differ appreciably from cattle in this regard. Such possible differences are consistent with observations suggesting that selective grazers such as sheep have a greater relative mass of post-gastric digesta than do non-selective grazers such as cattle (Hoffmann, 1988).

Apportionment of Compartmental Mass

Models of digesta flow have largely been compared on the basis of how well they fit the excretion pattern of marker emerging from the total gastrointestinal tract (Dhanoa et al., 1985; Mertens, 1989). A more sensitive anatomical and physiological criterion would be to compare the compartmental mass estimated by models to the mass estimated by independent methods. Because extent of digestion differs within different segments of the digesta, compartmental mass of different segments of digesta is comparable only in terms of an indigestible entity such as INDF. The compartmental masses of indigestible fiber, as estimated by one and two compartment models fitted to data from different digesta segments are summarized in Table 11. These estimates of compartmental mass of INDF were calculated using model estimates of initial marker concentration in earlier Tables 6 and 7 (g Nd/kg DM) converted in Table 11 to g Nd/kg INDF by division by the fraction INDF/DM determined in digesta sampled from ruminal dorsal digesta, 0.47; ruminal ventral digesta, 0.4; duodenal digesta, 0.5; and feces, 0.7.

The compartmental mass, CM, of INDF comprising the slower turnover compartment, CM2, was computed as dose of marker, μg, divided by the initial

Table 10. Compartmental residence times estimated by fitting various models to marker concentrations at different sampling sites.

Sample site and model	Compartmental mean residence time[a,b,c]			Time delay τ
	C1, n/λ_1	C2, $1/k_2$	Mixing system	
	------------------------------h------------------------------			
Raft:				
G1→G1→τ→O	0.4	42.3	42.7	0
G2→G1→τ→O	0.4	41.5	42.7	0
G1→τ	-	-	40.9	0
G2→τ	-	-	40.7	0
G1→G1→(G1→G1)→O	8.6	22.5	31.1	0
Ventral:				
G1→G1→τ→O	0.07	36.5	36.6	1.7
G2→G1→τ→O	0.2	36.5	36.7	1.5
G1→	-	40.9	40.9	0
G2→τ	-	-	40.7	0
G1→G1→(G1→G1)→O	15.4	27.2	42.6	0
Duodenal:				
G1→G1→τ→O	13.4	39.7	58.4	4.9
G2→G1→τ→O	14.7	43.7	58.3	1.2
G2→τ→O	-	-	49.3	0
G1→G1→(G1→G1)→O	6.2	37.0	43.2	7.3
Fecal:				
G1→G1→τ→O	25.2	25.2	50.4	12.8
G2→G1→τ→O	18.8	38.3	57.1	8.5
G2→τ→O	-	-	50.4	12.8
G1→G1→(G1→G1)→O	7.6	44.4	52.0	16.7

[a] C1 = Age-dependent initial compartment of two-compartment model (MCRT1).
[b] C2 = Age-independent, terminal compartment of two-compartment model (MCRT2).
[c] C1 + C2 for two-compartment models or n/λ for one-compartment models (MCRTS).

dilution (C_2, µg marker/g INDF, see Table 4). Calculation of the CM of the faster turnover compartment differed according to assumptions of the model. Assumptions of sequential flow through two compartments implies (CM1 • k_1) = (CM2 • k_2) so CM1 = (CM2 • k_2) / k_1. The CM2 estimated by fitting the G1→G1→τ→O model to duodenal data was thus 4,578 g of INDF so CM1 equals 4,578 • 0.0252 / 0.0748 (Table 6) or 1,542 g and the compartmental mass for the compartmental system, CMS, equaled 4,578 + 1,542 or 6,120 g of INDF (Table 11).

Model G2→G1→τ→O also assumes flow through two sequential compartments, an age-dependent compartment involved with fragment modifications required to increase the probability of their turnover escape to the second compartment, and a second compartment whose age-independent turnover provides for escape from ruminal digesta (see Figure 5 and Figure 22 later). Although the two compartments of the G2→G1→τ→O model are modelled as functionally discrete compartments, the two compartments are analytically distinguished only on the basis of residence time distributions and may not be physically distinct compartments. Alternatively, the two functionally distinct compartments probably share the same space and may be indistinguishable in terms of mass. If it is assumed that the functional CM1 = CM2, then the CM of the system of mixing compartment, CMS, also equals CM2 or 4,969 g INDF as estimated by dose of marker divided by C_2 estimated by fitting the G2→G1→τ→O model to duodenal digesta. Such an assumption is in accord with estimates of CM2 that are similar whether estimated from data obtained from ruminal or post-ruminal segments of digesta (Table 11).

When the G2→G1→τ→O model was fitted, the compartmental mass of the slower turnover compartment, CM2, was similar whether estimated from marker disappearance from raft digesta or marker appearance in duodenal digesta, or the feces (4,911 vs 4,969 and 4,500 g INDF, respectively). The difference in CMS estimated from sampling duodenal (6,375 g INDF) versus the mean of the ruminal data (4,609 g INDF) is 1,766 g INDF, a value which approximates the MCRT1 estimated from sampling duodenal digesta (1,406 g INDF) and the feces (1,854 g of INDF). The CMS of the intervening omasal and abomasal digesta may contribute some of the difference in estimated mass based on ruminal as opposed to duodenal sampling. However, the major portion of the MCRT1 appears to reflect residence time within ruminal digesta that is not associated with turnover escape from the anatomical rumen (see physical representation of G2→G1→τ→O model in Figure 5).

Although referred to (and modelled) as two sequential compartments, the "compartments" of the G2→G1→τ→O model may not be compartments in a traditional sense but appear to represent two sequential phases of digestion. Functionally, the two physical compartments appear to be the result of "un-mixing" effects of differences in fermentation-based buoyancy of fragments between the raft rumination pools and the escape pools as described by Sutherland (1989) and further developed by Ellis et al.(1991). The physical separation of digesta into time delay raft and turnover pools may differ due to diet and the dynamics related to meal interval. For example, the raft may be a relatively small proportion of the ruminal digesta and of relatively short duration in sheep fed highly digestible temperate grasses (Sutherland, 1989), larger in cattle fed forages of intermediate

Table 11. Compartmental mass and output of indigestible fiber (INDF) estimated by fitting various models to the pattern of marker concentration in dorsal, ventral, duodenal digesta, and feces.

Sampling site and model	C_2 or C	Compartmental mass of INDF at or preceding sample site			Rate[d]	Output	
		CM1[a]	CM2[b]	CMS[c]	k, $\bar{k}$, or k_2	CM1·k, CMS·$\bar{k}$, or CM2·k_2	
	g Nd/ kg INDF	---------g INDF---------			h^{-1}	g INDF/hr	
Raft:							
G2→G1→τ→O	1,374[d]	-	4,911	4,911	0.0241	118[e]	
G1→τ	1,402	4,813	-	4,812	0.0244	117[e]	
Ventral:							
G2→G1→τ→O	1,567	-	4,306	4,306	0.0274	118[e]	
G1→τ	1,630	4,115	-	4,115	0.0273	113[e]	
Duodenal:							
G1→G1→τ→O	1,474	1,542	4,578	6,120	0.0252	115[e]	
G2→G1→τ→O	1,358	1,406	4,969	6,375	0.0229	114[e]	
G2→τ→O	1,364	-	-	4,947	0.0242	120[f]	
Feces:							
G1→G1→τ→O	2,226	3,031[g]	3,031[g]	6,063	0.0397	120[e]	
G2→G1→τ→O	1,500	1,854	4,500	6,354	0.0261	117[e]	
G3→G1→τ→O	1,381	1,589	4,886	6,475	0.0240	117[e]	
G4→G1→τ→O	1,336	1,439	5,051	6,490	0.0226	114[e]	
G5→G1→τ→O	1,310	1,369	5,151	6,520	0.0221	114[e]	
G2→τ→O	1,327	-	-	5,085	0.0237	120[f]	

Footnotes for Table 11 appearing on preceding page.

[a] CM1 is compartmental mass of age dependent compartment of two compartment model.
[b] CM2 is compartmental mass of age independent compartment of two compartment model.
[c] CMS is mass of compartmental system, CM1 + CM2, of two compartment models or mass of one compartment, age dependent model.
[d] k= age-independent rate of one-compartment age independent model, k=mean age dependent rate of one compartment, age dependent model, k_2=age independent rate of age independent compartment of two compartment model.
[e] CM2 • k
[f] CMS • $\bar{k}$

digestibility (Dixon et al., 1983), and even larger in cattle fed less digestible semi-tropical forages such as in this sample data set. Observation and evacuation of ruminal digesta from this steer indicated that the physical "raft" constituted essentially all (~ 90%) of the digesta mass in the rumen.

The distributed time delay lag model of France et al.(1985) (model (Fτ→G1)→τ→O, in Figures 5 and 6) is mathematically equivalent to the model of Matis (1972). Conceptually, France et al. (1985) postulates sub-population of individual fragments which posses unspecified intrinsic properties leading to an age-dependent distribution of time delay lags. In contrast, the model of Matis (1972) assumes an age-dependent process whereby the probability of escape for all fragments increases with residence time. Philosophical differences between such "mechanistic" and "stochastic" models have been presented (Matis, 1987).

Similarities in CM2 estimated from data obtained throughout the tract are consistent with the suggestion that CM2 of the ruminal digesta is the dominant, if not sole, compartment contributing to digesta passage in cattle (Ellis et al., 1979; 1984; France et al., 1988; Krysl et al., 1988). Undoubtedly, some post-gastric compartments add to the residence time of digesta in the tract. The results illustrated subsequently in Figure 13 for Holstein steers fed an all hay diet correspond to a MCRT for multi-laminar flow in the order of 0.5 to 1.5 h between the abomasum and the duodenum and 1.5 to 5.0 h between the duodenum and rectum for a total range of 2 to 6.5 h in post-gastric MCRT. The large variation in post-gastric residence time could be due to variations in compartmental mass of the "hind-gut" mixing compartments among individual ruminants (Deswysen et al., 1989b), similar to that described by Hoffmann (1988) for different classes of ruminants (i.e., concentrate selectors, intermediate feeders, and bulk grazers).

Fecal Output

Although assumptions of the different models result in some variation in estimates of compartmental mass, all models yield similar estimates of output from the compartmental system. All two-compartment models are formulated to estimate

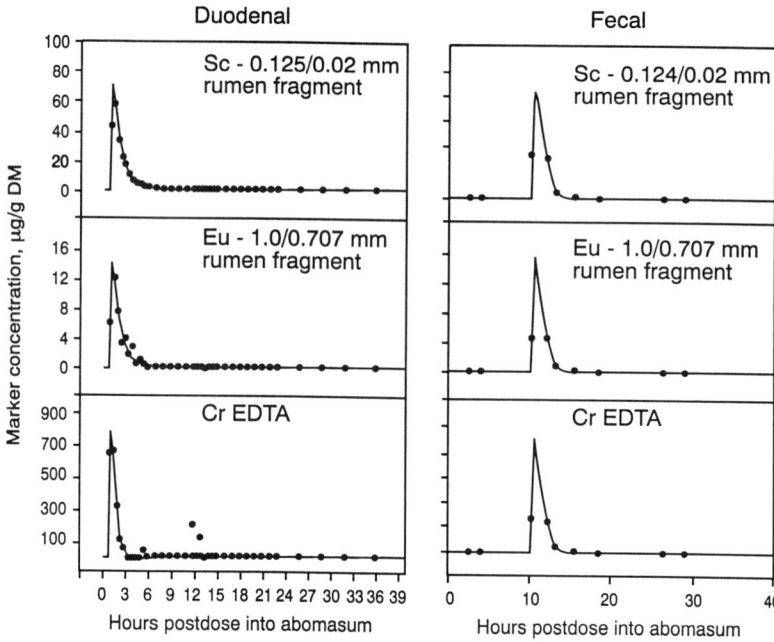

Figure 13. Observed values (dots) and expected values from fitting the G2→τ→O model to appearance of various markers dosed into the abomasum and sampled from duodenal digesta or feces. Rare earths, Sc or Eu, were applied to masticated fragments of bermudagrass separated from ruminal digesta via sieving with passage and retaining sieves of 1.0/0.7 mm (1.0/0.7 mm) or 0.125/0.02 mm (0.125/0.02 mm) respectively. Cr-ethylenediaminetetraacetic acid, Cr EDTA, was dosed as a solute marker.

the expected dilution of the marker dose had it been introduced into and instantaneously diluted by the compartmental mass of the terminal or larger compartment (the slower turnover, age-independent compartment). The validity of the one and two-compartment, age-dependent models for estimating fecal output has been consistently demonstrated (Ellis et al., 1980; Coffey et al., 1988; France et al., 1988; Krysl et al., 1988; Moore et al., 1992). The one compartment model has the advantage of fewer parameters, is more robust, and will estimate comparable yield of feces (Krysl et al., 1988). However, due to the dominance of exponentially distributed residence times within ruminal digesta, the G2→G1→τ→O model is preferred. France et al.(1989) and Moore et al.(1992) compared the accuracy of various models in estimating fecal output and concluded that the multi-compartment models reviewed here achieved similar accuracy.

Partitioning Gastrointestinal Residence Time

The above sample data set illustrates that ruminal digesta was the major source of compartmental mass and mean compartmental residence time in a mature steer fed an all forage diet, and that complete accounting of compartmental mass and residence time cannot be accomplished by sampling marker departure from the ruminal digesta. Alternatively, sampling of digesta which has escaped from the rumen is recommended for partitioning ruminal residence time into multiple compartments. Sampling from the abomasum or duodenum are options, with sampling from the ascending duodenum preferred due to its more constricted bore which reduces the opportunity for digesta evading sampling. The results in Figure 13 indicate that flow through the omasum and abomasum contributes relatively little τ (0.5 - 0.7 h) or MCRT (1.5 - 2.5 h).

The marker flow profile in Figure 10 illustrates the value of an age-dependent (or multiple, sequential compartment) flow process to model the more rapid turnover flow and the need for an appropriate number and distribution of sampling times to resolve this compartmental mixing flow from τ. The following distribution of samples are indicated: 1) 2 to 3 samples to establish a baseline before the anticipated appearance of analytically detectable (significant increases) in marker concentration, 2) 2 to 3 samples within the time span of anticipated initial appearance of analytically detectable concentrations of the marker to delineate τ from λ_1, 3) an additional 2 to 4 samples during the ascending phase of the marker profile to provide for estimating and partitioning λ_1 from k_2, and 4) more widely spaced (6 to 12 h) sampling during the declining phase of the marker profile, with the last sample taken beyond when marker concentration is anticipated to approach or reach analytically detectable limits. Thus, some 10 to 15 strategic samples are required to partition residence time due to τ, λ_1 and k_2. A deficiency of samples in the initial ascending profile will result in difficulties in achieving unique convergence via the fitting procedure and(or) will over estimate λ_1 and C_2.

It is frequently considered desirable to estimate ruminal turnover from fecal data. Figure 14 compares the patterns of marker appearance in duodenal and fecal digesta which is associated with compartmental flow for the current data set. Although the MCRTS ($2/\lambda_1 + 1/k_2$) was similar for either sampling site (58 vs 59 h), the MCRT1 ($2/\lambda_1$) was considerably shorter when estimated via sampling duodenal as compared to fecal digesta (14.7 vs 18.8, Table 9). In mature steers, differences in partitioning of MCRT estimated by duodenal versus fecal sampling appears due to the contribution of relatively small but rapid turnover pools in post-duodenal digesta. The ability to more accurately resolve such differences is severely limited by large deviations from model-assumed steady state conditions. Variations in ruminal outflow associated with meal eating and with accumulation of digesta in the rectum are suggested to be major sources of deviation from steady state assumptions. Lalles et al.(1991) reported a comparison of parameters estimated by various models fitted to a large number of data sets (n=29) obtained with 9 to 20 wk old calves using ^{141}Ce and ^{169}Yb specifically bound to concentrate or hay (200 g kg^{-1} diet) respectively, and for digesta sampled from the duodenum and feces. The mean MCRT2 for the slower turnover compartment estimated by the G2→G1→τ→O model (MRT1 in Lalles et al., 1991) fitted to duodenal data was

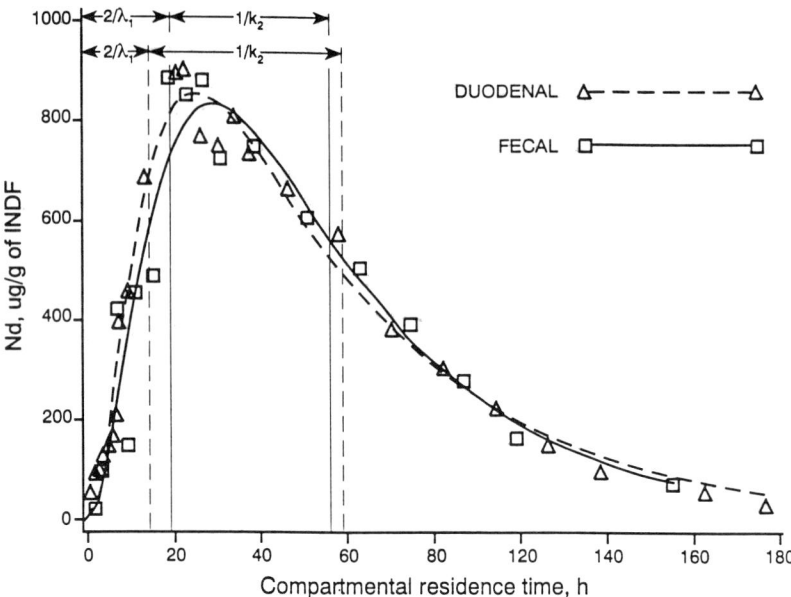

Figure 14. Observed and modelled compartmental excretion pattern of Nd-marked hay at the duodenum and in the feces of a steer fed an all hay diet. The distribution of residence times due to compartmental mixing are those estimated after first appearance of the marker by the G2→G1→τ→O model when fitted to observed marker concentrations at the respective digesta sampling sites. Note that the mean compartmental residence time of the mixing system is essentially equal (57 versus 58 h) whether sampling from the duodenum or feces. However, compared to sampling from the duodenum, sampling from the feces resulted in apportioning more mean compartmental residence time to the faster turnover compartment, $2/\lambda_1$.

poorly correlated with MCRT2 estimated from fecal data, averaging 21.5% smaller for ^{141}Ce (forage marker) and 5.0% larger for ^{169}Yb (concentrate marker), respectively. A number of data sets with mature cattle (Poore et al., 1990, 1991) suggest that ruminal turnover might be estimated from sampling fecal digesta.

Differential Flow of Individual Feed Residues

Results summarized in Table 12 indicate differential flow for fragments derived from grazed forage and from protein supplements. The results suggest that turnover of the residues from the supplement is positively related to turnover of the grazed forage residues. Such results imply that the fractional degradation of the protein supplement will not be constant but will be materially altered via the turnover rate of the associated forage residue. The results in Table 12 also illustrate

Table 12. Mean rates of escape[a] of rare earth markers specifically applied to masticate and protein supplements for grazing cattle[b].

Pasture: Marker-fragment	Pasture masticate	Supplement, ruminally dosed	SEM[c]	Cr/Hf or Tb/Eu
		----------k_e--------------		
Bermuda pasture:				
Tb-masticate	0.027	-	0.001	-
Eu-masticate	0.027	-	0.001	1.004
Hf-supplement	-	0.031	0.001	-
Cr-mordanted supplement	-	0.029	0.002	0.930
Ryegrass pasture (1988)				
Tb-masticate	0.056	-	0.003	-
Hf-supplement	-	0.110	0.040	-
Cr-mordanted supplement	-	0.079	0.019	0.720
Ryegrass pasture (1989)				
Tb-masticate	0.052	-	0.004	-
Hf-supplement	-	0.119	0.005	-

[a] Estimated by fitting G2→τ→O model to marker appearance in duodenal digesta.
[b] Adapted from Hill, 1991.
[c] n = 21

the repeatability of estimates of turnover for the same fragment by use of multiple doses of fragment bound markers and the slower turnover for Cr-mordanted fragments as compared to rare earth marked fragments.

Compartmental Models: Recommendations

Model assumptions of instantaneous mixing and steady state conditions are major limitations in fitting compartmental models to digesta flow data and in interpreting the estimated model parameters. The assumption of instantaneous mixing is in stark contrast to "un-mixing" processes which appear to regulate turnover from the ruminal digesta (Sutherland, 1989). Current methods for estimating the kinetics of digesta flow involve least-squares fitting of dilution data obtained over periods of days. Therefore, the kinetic parameters estimated must be interpreted as averages over the period of data collection. Such "averages" of daily kinetic rates are useful for the nutritionist who expresses as daily averages the nutrient requirements resulting from non-steady state metabolic processes in the animal.

Models of the form $Gn \to G1 \to \tau \to O$ are recommended because, at least in cattle, they appear to identify biologically functional pools of forage particles having flow paths successively through age-dependent and age-independent ruminal processes. The age-dependent flow-paths appear to correspond to the "large particle rumination" pools as conceived by Hungate (1966) and later mechanistically described by Sutherland (1989). Residence time in such a rumination pool appears to be of a similar magnitude as that resolved for MCRT1 of the $Gn \to G1 \to \tau \to O$ model (Ellis et al., 1991). The age-dependent flow paths from the Gn pool appear descriptive of the flow processes involving interactions between rumination and fermentation-based buoyancy. The age-independent flow process involving slower turnover of the mass of less buoyant, smaller particles appear to be well described by mass action competition for escape. This residence time appears to be resolved by the MCRT2 of the $Gn \to G1 \to \tau \to O$ model.

MICROBIAL DIGESTION

Extent and Rate of Digestion

The extent of digestion of an entity is determined by: 1) the proportion of the entity which is digestible, E_d, or conversely indigestible, E_i, after exposure to agents of digestion for a defined period of time, 2) the rate or rates of digestion, k_d, of E_d, 3) the rate of escape, k_e, of E_d, and 3) the time lag before detectable digestion of E_d is initiated, τ_d. These attributes are usually estimated by fitting appropriate models to data resulting from the measurement of the undigested entity, UE_d, remaining after exposure to digestion. It is conceivable that the digestion of intrinsically labelled material could be followed directly by appearance for digestion end-products, DE_{de}. This digestion process is conducted under conditions in which disappearance via escape is prevented. The route of disappearance is usually assumed to reflect hydrolysis of polymeric E_d or initial metabolism of monomeric forms of E_d; however, disappearance may also be by various chemical or physical routes depending on the methods employed.

Digestion Procedures

Two basic procedures have been used for measuring the disappearance of E_d. In vitro procedures involve the exposure of known amounts of forage fragments for varied residence times to an inoculum of microbes removed from the gastrointestinal tract, or to preparations of enzymes capable of hydrolyzing the E_d of interest, i.e; the agent(s) of digestion. If the E_d of interest is unique to the forage (e.g., cellulose), then the quantity of undigested, potentially digestible entity, UE_d, is simply determined in the mixture of undigested fragments and residual agents of digestion. If the E_d of interest is common to both the forage and the agent of digestion (e.g., protein is common to both forages and microbes), then a second stage must be employed to separate UE_d of forage from E_d associated with the agents of digestion. Alternatively, some marker of E_d which is unique to either the feed or the agent of digestion may be employed.

A second method, the in situ method, involves exposure of known initial quantities of forage to varied, successive residence times in ruminal digesta. The sample is enclosed in containers having porosities which: 1) allow normal rates of influx and colonization by normal rumen microbes, 2) allow efflux of end products of digestion, and 3) prevent influx of UE_d other than that supplied by the initial sample of forage. If the UE_d of interest is common to both forage and microbes, then a second stage or a unique marker must be employed to resolve the UE_d of undigested feed from the UE_d of microbial origin.

Rate of digestion of E_d and the concentration of E_i are measured to estimate intrinsic characteristics of the forage. In order that the derived estimates reflect only attributes of the substrate, it is imperative that the rate of digestion be limited solely by attributes of the substrate and not by other physical or chemical attributes of the digesting environment. The relative merits and applicability of in vitro and in situ methods with respect to obtaining meaningful kinetic data are discussed in greater detail by Mertens (1993a,b).

In vitro methods are especially applicable where the objective is to estimate extent of digestion following some fixed residence time and the results are to be interpreted in relative terms, for example, evaluation of cultivars or specified alternatives. However, questions arise concerning the effects of the in vitro environment on rates of digestion and the degree to which the microbial ecosystem in these systems corresponds to that in vivo.

If the objective is to describe the rate of digestion of relatively slowly digesting E_d such as fiber, then the rumen in situ method is preferable because aspects of the ruminal environment, except escape, are more faithfully simulated. A major problem with the rumen in situ method is defining a container porosity which will: 1) allow normal microbial colonization of all forage fragments within the sample, 2) allow normal diffusion of fermentation end products from the container, 3) prevent physical escape from the container of small fragments of the initial sample, and 4) prevent physical influx from ruminal digesta of small forage fragments not derived from the initial sample, and 5) avoid occlusion of pores of the container after longer periods of incubation (Van Milgen et al., 1992). Container porosities as small as 5 µm have been demonstrated to allow normal rates of fiber digestion (Van Hellen and Ellis, 1977). Porosities greater than 20 µm allowed detectable loss of sample when it had been ground to pass a 1-mm screen (Van Hellen and Ellis, 1977). Similarly, porosities greater than 20 µm (Van Hellen and Ellis, 1977) or 35 µm (Uden et al., 1974) allow detectable influx of small particles from ruminal digesta. Although materials of less than 35 µm appear preferable, porosities of 40-50 µm are commonly used without any correction for physical influx or efflux as a compromise to availability and cost.

A sample weight of 5 to 10 mg/cm^2 of the in situ container has been found to maximize fiber digestion (Van Hellen and Ellis, 1977; Lindberg, 1981; Nocek, 1988). Variation in the extent and rate of digestion estimated by rumen in situ methods occurs due to diet, individual animal, and site within the ruminal digesta (Van Hellen and Ellis, 1977).

Models of Hydrolytic Digestion

Models for describing hydrolytic digestion differ with respect to assumptions concerning: 1) the number of entities or groups of entities having homogeneous rates of digestion, 2) the expected distribution of residence times (or lifetimes) for the entity, 3) the factors affecting access by the microbe to the entity throughout the entire forage fragment, 4) the partition of potentially digestible versus indigestible entities that are accessible to agents of digestion, and 5) the partitioning of residence time between hydrolysis itself and that associated with a lag time prior to hydrolysis. Following the terminology used for passage models, models of hydrolytic digestion are summarized and illustrated in Figure 15. The model expressions are summarized in Table 13.

The G1→ digestion model expresses the expected digestion of an entity by simple mass action where the rate of digestion is a constant fraction (age-independent rate constant, k_d) of a constantly diminishing quantity of undigested entity, UE_d, with advancing residence time, t. The expected quantity of UE_d remaining at t=x, $UE_{d(x)}$, then equals the initial quantity of E_d, $E_{d(0)}$, multiplied by the fraction of E_d hydrolyzed during residence time t, $e^{-k_d \cdot t}$, here abbreviated E1. Digestion time delay, τ_d, can be incorporated into this and other models by substituting $t-\tau_d$ for t. The G1, E_i→ model partitions the total entity remaining after a specified residence time, $UE_{(x)}$, into the expected UE_d, $E_{d(0)} \cdot E1$, and the indigestible entity, E_i. The G1,G1→ model partitions disappearance of E_d as digestion of two potentially digestible entities distinguished by different exponentially distributed rates of digestion, k_1 and k_2. To avoid over-parameterization of the model, the proportion, P and (1-P), of E_d disappearing at each rate is estimated using a single parameter, P.

Non-exponential distributions of residence times such as the gamma 2 distribution (G2) as described for rates of escape, can be used to model age-dependent distributions of life times. Thus, the G2→ model describes the expected age-dependent disappearance of a potentially digestible entity by an integer gamma distribution of order 2. Discrete proportions of Ed may be modelled as being simultaneously digested with different distributions of life times, hence the Gn, G1→ models. By providing different distributions for the residence times for each entity, these models have been useful in resolving the disappearance of ^{14}C labelled forage fiber from ruminal digesta into two rates, inferred to be k_d and k_e (Smith, 1989). Inclusion of age-dependency into one compartments of the two may allows statistical resolution of two rates in data in which two the age-independent rates are indistinguishable via the G1,G1→ model.

In view of the complexity of chemical and physical attributes of E_d in masticated forage tissues, one could postulate numerous and complex biological processes of digestion having age-dependent distributions of residence time. Incorporation of age-dependent residence time distributions for digestion is not only biologically justified but provides flexibility in describing more complex patterns of digestion without over-parameterizing models with too many conceptual entities. Alternatively, one could postulate a heterogenous distribution of entities, each having characteristic rates of age-independent digestion (Ellis et al., 1984; Mahloogi, 1985; Van Milgen et al., 1993b). The simultaneous disappearance of

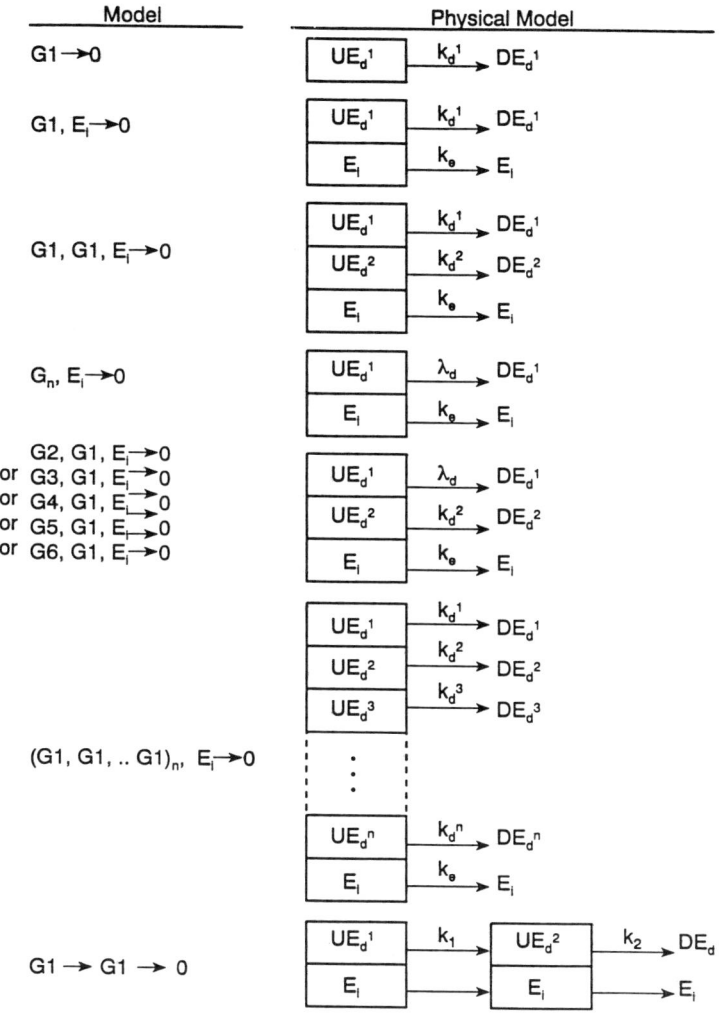

Figure 15. Schematic representation of some models for digestion of potentially digestible entities, E_d. Models differ according to the distribution of lifetimes of undigested E_d, UE_d, in one compartment models. Lifetime distributions corresponding to age-independent rates of digestion, k_d, of potentially digestible entities are identified by G1. Lifetime distributions of UE_d which disappear by age-dependent rate parameters of digestion, λ_d, are identified by G2, G3, G4, G5 and G6 and represent orders of integer gamma distributed lifetimes of UE_d. E_i represents indigestible entities, and O represents output via digestion. The two compartment model assumes two sequential forms of E_d.

Table 13. Digestion models[a]

Model[a]	Expression[b,c,d,e,f,g,h,i,j]

Exponential distribution of lifetimes for UE from one E_d:

G1→O $\quad UE_{d(x)} = E_{d(o)} \cdot E_i$

G1,E$_i$→O $\quad UE_{d(x)} = (E_{d(o)} \cdot E1) + E_i$

Age-dependent distribution of lifetimes for UE from one E_d:

G2,E$_i$→O $\quad UE_{d(x)} = E_{d(o)} \cdot E\lambda \cdot (1+\lambda t)$

Exponential distribution of lifetimes for UE from two E_d occurring at P and 1-P proportions:

G1,G1,E$_i$→O $\quad UE_{d(x)} = E_{d(o)} \cdot ((P \cdot E1) + ((1-P) \cdot E2) + E_i$

Age-dependent plus age-independent distribution of lifetimes for UE from two E_d occurring at P and 1-P proportions:

G2,G1,E$_i$→O $\quad UE_{d(x)} = E_{d(o)} \cdot ((P \cdot E\lambda \cdot (1+\lambda t)) + ((1-P) \cdot E2) + E_i$

G3,G1,E$_i$→O $\quad UE_{d(x)} = E_{d(o)} \cdot ((P \cdot E\lambda \cdot (1+\lambda t+\lambda t^2/2)) + ((1-P) \cdot E2) + E_i$

G4,G1,E$_i$→O $\quad UE_{d(x)} = E_{d(o)} \cdot ((P \cdot E\lambda \cdot (1+\lambda t+\lambda t^2+\lambda t^3/3)) + ((1-P) \cdot E2) + E_i$

G5,G1,E$_i$→O $\quad UE_{d(x)} = E_{d(o)} \cdot ((P \cdot E\lambda \cdot (1+\lambda t+\lambda t^2+\lambda t^3+\lambda t^4/24)) + ((1-P) \cdot E2) + E_i$

G6,G1,E$_i$→O $\quad UE_{d(x)} = E_{d(o)} \cdot ((P \cdot E\lambda \cdot (1+\lambda t+\lambda t^2+\lambda t^3+\lambda t^4+\lambda t^5/120)) + ((1-P) \cdot E2) + E_i$

Heterogeneous distribution of rates for exponentially distributed lifetimes of UE:

(G1,G1,...G1)n1→O $\quad UE_{d(x)} = E_{d(o)} \cdot (1+\beta \cdot t)^{-\alpha} + E_i$

Footnotes for Table 13 appearing on preceding page.

a   Time delay can be added to all models by substituting $(t-\tau)$ for t.
b $UE_{d(x)}$ = Undigested, potentially digestible entity remaining at $t = x$.
c $E_{d(0)}$ = Potentially digestible entity present at $t = 0$.
d E_i = Indigestible entity.
e $E1 = e^{-k_1 t}$, where E1 = entity 1 having an age-independent rate of digestion of k_1.
f $E2 = e^{-k_2 t}$, where E2 = entity 2 having an age-independent rate of digestion of k_2.
g $E\lambda = e^{-\lambda t}$, where $E\lambda$ = an entity having an age-dependent rate of digestion of λ.
h   P = proportion of E_d digested via age-dependent rate.
i β = Scaling parameter for distribution of age-independent rates of digestion.
j α = Shape parameter for distribution of age-independent rates of digestion.

the population of such entities could be modelled by assuming some heterogeneous distribution of digestion rates. Like the age-dependent, gamma-distributed residence times associated with passage, a model based on gamma-distributed simultaneous rates of digestion provides flexibility and robustness (due to minimal parameters) in fitting data. A physical model of the heterogeneous model with a gamma distribution of rates is graphically illustrated in Figure 15.

Indigestible Entities

Conceptually, indigestible entities could be estimated directly as the quantity of entity remaining after long periods of fermentation (Wilkens, 1969; Waldo et al., 1972). However, analytical definition of conceptual indigestible entities is difficult. Lippke et al. (1984) observed erratic recoveries of INDF in the feces of sheep (82.9 to 138%). The INDF was determined as the residual NDF after 8 d of fermentation of the ground forage or resulting feces. It was suggested that differences in size of the particle undergoing fermentation might influence accessibility of the rumen microbes to E_d within the fragment's interior. However, more finely grinding the samples confounds physical losses of NDF via the filtration procedures used.

Digestion of NDF is detectable after long periods of exposure to ruminal microbes (Robinson et al., 1986) and indicates the need for strategic and prolonged periods of fermentation in order to estimate INDF via fitting models to data to estimate E_i (Van Milgen et al., 1991; Fadel, 1992). Prolonged fermentation may present other problems, such as mineral precipitation occluding bag pores (Van Milgen et al., 1992).

Kinetics of Fiber Digestion and Interpretation of Results

Mahloogi (1985) compared various models for describing the digestion of fiber components from different size fragments derived from the mastication of bermudagrass and corn silage by cattle. A gamma distribution of exponential distributed rates proved superior to rectangular or triangular distributions. The heterogeneous rate model with a gamma distribution of expected exponential digestion rates provided a superior fit to all data sets, even though some individual

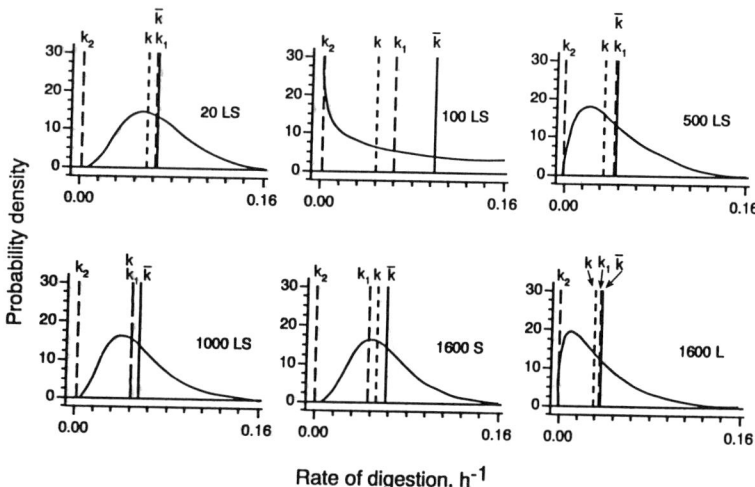

Figure 16. Distribution of digestion rates as estimated by the heterogeneous rate models fitted to ruminal in situ digestion of potentially digestible fiber from various sized masticated fragments of bermudagrass (20LS, 100LS, 500LS, 1000LS, 1600S, and 1600L, where the numerical value represents the fragment's retaining sieve size in microns, L=leaf origin, S=stem origin, and LS=mixed leaf and stem fragments). The rate of digestion, k, was estimated from fitting the single digestion rate model, $G1 \to \tau \to O$; the two age-independent digestion rates, k_1 and k_2, were estimated by fitting the $G1,G1 \to \tau \to O$ model and the mean rate, $\bar{k}$, was estimated by the $(G1, G1, \ldots G1)_n, \to \tau \to O$ model fitted to the data. Note the wide variation in the distribution of rates as estimated by the $(G1, G1, \ldots G1)_n, \to \tau \to O$ model and differences in estimates of rates by models.

data sets were described equally well by the exponentially distributed models, $G1,\tau \to$ and $G1,G1,\tau \to$ (Ellis et al., 1984). The distribution of rates estimated by fitting the heterogeneous rate model to data revealed a wide variation in the distribution of such rates. This variation in distribution of rates, expressed as their probability density, is illustrated in Figure 16 for some of the masticated forage fragments. The mean rate for the heterogeneous rate model ($\bar{k}_d = \alpha \cdot \beta$) was computed to allow comparisons to the k of the G1→ and k_1 and k_2 of G1→,G1→ exponential models in Figure 16. Note the large variations in distribution of heterogenous rates for digestion of different size masticated fragments and, in some cases, the lack of agreement between the exponential rates (k, k_1, and k_2) and the mean of the heterogeneous distribution of rates ($\bar{k}_d$).

The heterogeneous rate model contains an added mechanism, namely a distribution of rates. The heterogeneous rates model provided superior or equal, but in no case inferior, fit as compared to that of the deterministic mass action models. Van Milgen et al. (1993b) reported that the G1→ model and heterogeneous rate model provided equal fit to a data set of organic matter digestion. Van Milgen et al.(1993b) recommended the G1→ model because of its easily interpretable rates in contrast to the parameters of the heterogeneous rate models ($\alpha \cdot \beta$). However, the mean rate for the heterogeneous rate models ($\bar{k}_d = \alpha \cdot \beta$) will have the same mechanistic interpretation as k_d of the mass action model.

When fitted to data for digestion of NDF in masticated fragments, the heterogeneous rate model yielded considerably different estimates of τ_d and E_i than did the exponentially distributed models (Ellis et al., 1984). Differences among models in fit and parameter estimates are further illustrated for one data set in Figure 17. The superior fit of the heterogeneous rate model to the initial and terminal values in Figure 17 illustrates the flexibility of this model as compared to the more rigid assumptions associated with the exponential models (Matis, 1987). As a consequence, the heterogeneous model estimated a longer τ and an intermediate E_i as compared to the mass action-based models. That τ_d was, indeed, related to the size of masticated fragment as indicated in Figure 18. Mertens (1989) reported that,the heterogeneous rate model was equal to or superior to other models in fitting a large number of data sets but reported problems in fitting the model to some data sets. Van Milgen et al.(1993b) also reported difficulty in fitting the heterogeneous rate model to data. This problem will be addressed in a later section addressing the fitting of data.

Use of the heterogeneous rate model appears appropriate in view of the heterogeneous nature of the chemical entities and their physical distribution in the diverse tissues which comprise ingested fragments of forage. Its use in practice is justified by its flexibility in fitting data with widely divergent distributions of residence time and its, yielding of a correct estimate of $\bar{k}_d$. Thus, even when fitted to exponentially distributed data, $\bar{k}_d$ estimated by the heterogeneous rate model equalled the k estimated by the exponentially distributed model.

Effects of Fragment Size and Ruminative Mastication

Various observations suggest that the size of masticated fragments entering ruminal digesta could materially affect microbial access to, and digestion of, E_d distributed throughout the fragment. These effects could be appreciable if mean residence times in the "rumination pool" are on the order of 15% of the total ruminal residence time (Table 10, MCRT1 and MCRTS estimated by the G2→G1→τ→O fitted to duodenal data). The major effect of size of forage fragment digestion appears to be positively related to the digestion time delay for fiber.

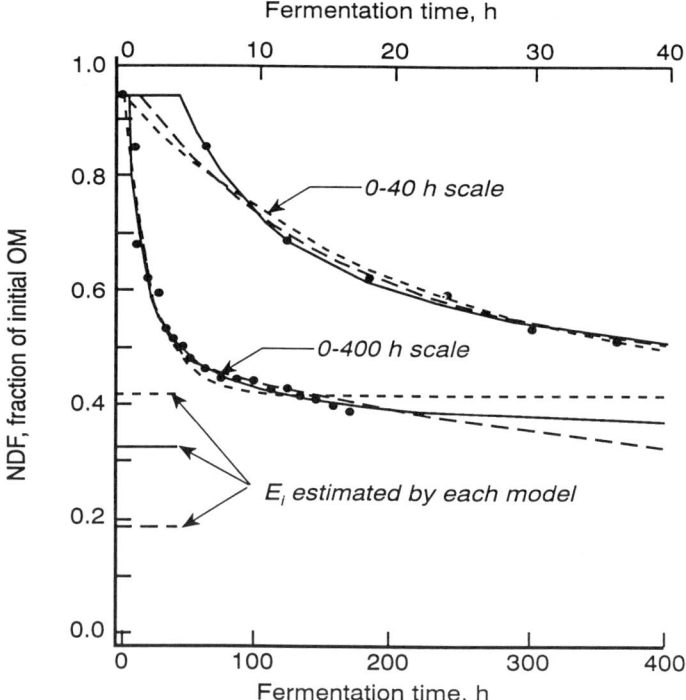

Figure 17. Observed values (dots) and expected values (lines) from fitting the G1→τ→O, the G1, G1,→τ→O and the (G1, G1,... G1)$_n$, →τ→O models to a data set. Model estimates, parameters, and fit of models is also indicated. The data set is that for masticated fragments of bermudagrass which passed a 0.1 mm sieve and were retained on a 0.02 mm sieve.

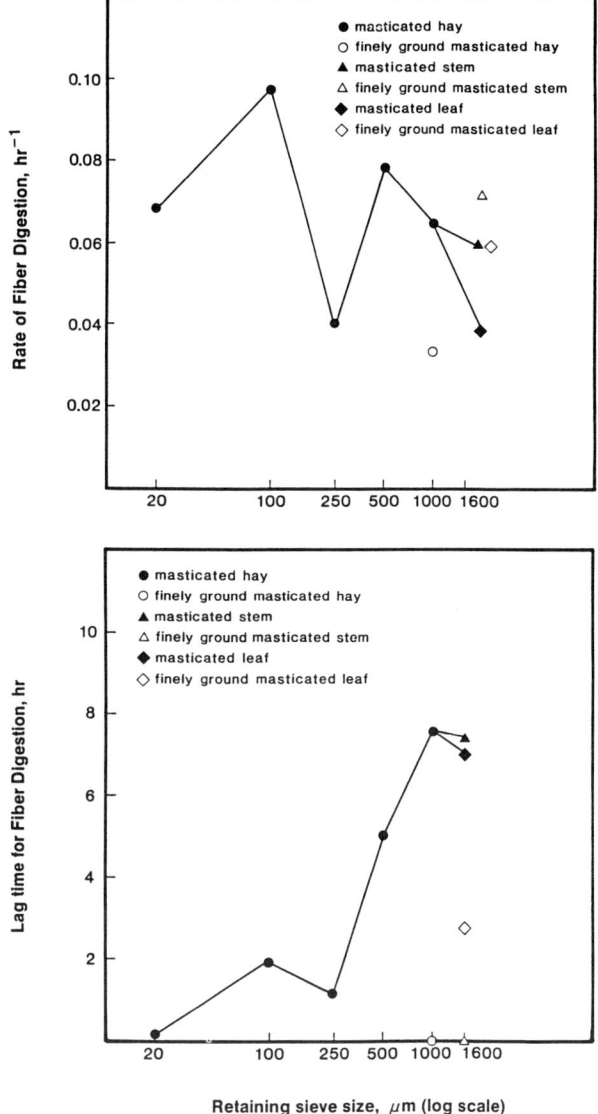

Figure 18. Relationships between size of masticated fragments of grazed bermudagrass and the rate of digestion and digestion time delay for digestion of NDF by fitting the heterogeneous rate models. Note that size of fermented fragment had no effect on rate of digestion of NDF but was positively related to digestion time delay. Digestion time delay was eliminated or materially reduced when the fragment was ground through a laboratory mill.

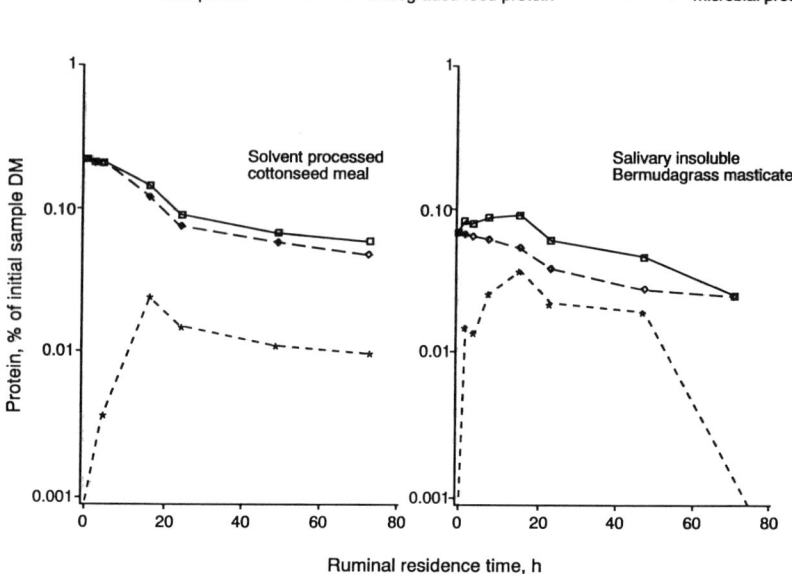

Figure 19. Concentrations of total protein, undegraded feed protein, and microbial protein (estimated from diaminopimelic acid) in two feeds during the time course of rumen in situ fermentation digestion (Figure 18).

Kinetics of Forage Cell Contents Digestion

The cell contents of the plant tissues (e.g., proteins, nucleic acids, and lipids) are common to both plants and microbes and their origin must be analytically distinguished in order to specifically determine undigested entities of the forage in the presence of the ruminal microorganisms. This necessity is illustrated in Figure 19 for two feeds, a relatively low protein forage and a high protein concentrate. Samples of both cottonseed meal and masticated bermudagrass were rapidly colonized by microbes in situ, achieving maximal concentrations of microbial protein of 1 to 5% of the samples' original DM before declining after approximately 20 h. This rapid colonization of the bermudagrass masticate actually resulted in an increase in total protein due to the rapid rate of accumulation of microbial protein within the in situ container exceeding the rate of degradation of the salivary insoluble protein of bermudagrass masticate. Thus, had accumulation of microbial protein not been corrected for, the rate of degradation of this feed protein could not have been determined. Insufficient correction for microbial colonization could result in erroneous estimates of time delay, especially in forages containing relatively low levels of protein (< 15%). Methodology especially related to these problems has been reviewed by Nocek (1988).

The diverse nature of colonization patterns illustrated in Figure 19 indicate a need to model the colonization-migration pattern of ruminal microbes in situ systems. The age-dependent process appears to be an appropriate model for the immigration, colonization, and migration processes occurring within the in situ container. Expectedly, immigration into the container would initially be slow, and then increase to some maximal extent of colonization followed by a slow migration from the sample. An age-dependent model should express the microbial flux at the substrate site for such an immigration-colonization-migration process. The age-dependent model formulated to express age-dependent flux from a mixing compartment, G2→O, was fitted to data and its estimated parameters were converted to secondary parameters appropriate to the process. The use of such a model to express maximal concentration of microbial protein, C_{max}, and the time required to achieve maximal concentration of microbial protein, MT, is illustrated in Figure 20.

Fitting the G2→ model to data for the six feeds indicated that the time to achieve maximal concentration of microbial protein, varied from 0.87 to 6.5 h and the time required to achieve maximal concentration of microbial protein varied from 3.7 to 46 h. In general, the maximal concentration of microbial protein was related to the fiber content of the feed. Animal products were slowly and sparsely populated. In view of the dynamics of the immigration-colonization-migration, it is difficult to incriminate the in situ procedure on the basis of comparing intraluminal bacteria concentrations to those in sacculus (Meyer and Mackie, 1986).

The digestion profiles of feed protein in six different feeds are illustrated in Figure 21. In this data set, diaminopimeleic acid was used as the microbial marker with the ratio of diaminopimelic acid in rumen to microbial protein being determined in the rumen fluid in which the in situ containers were incubated (Ellis et al., 1990 and 1992b). The masticate of forage was further extracted with saliva to remove salivary-soluble protein. The undegraded, salivary insoluble protein in the masticate was calculated by subtracting estimated microbial protein from total protein in the 50 μm porosity container. The patterns (two host rumens and two incubations per feed sample) of ruminally undegraded protein of in situ fermentation over time were poorly described by the single exponential entity model (G1→τ→O), with all sets being well described by the heterogeneous rate model (G1,G1,... G1)$_n$,→τ→O. Four data sets were described equally well by either the (G1,G1,... G1)$_n$,→τ→O or the G2,G1→τ→O model. Incorporation of age-dependency in the two entity models (G2,G1,→τ→O) provided a slightly improved fit as compared to the bi-exponential model (G1,G1,→τ→O). Estimates of digestion time delay, τ_d, for salivary insoluble feed proteins were not significantly different from zero (see Figure 21).

Digestion Time Lag

The usual biological rationale for a digestion time lag, τ_d, is that it represents the time for microbial access, colonization, and occurrence of analytically detectable digestion. A discrete digestion time delay, τ_d, has primarily been used as a means

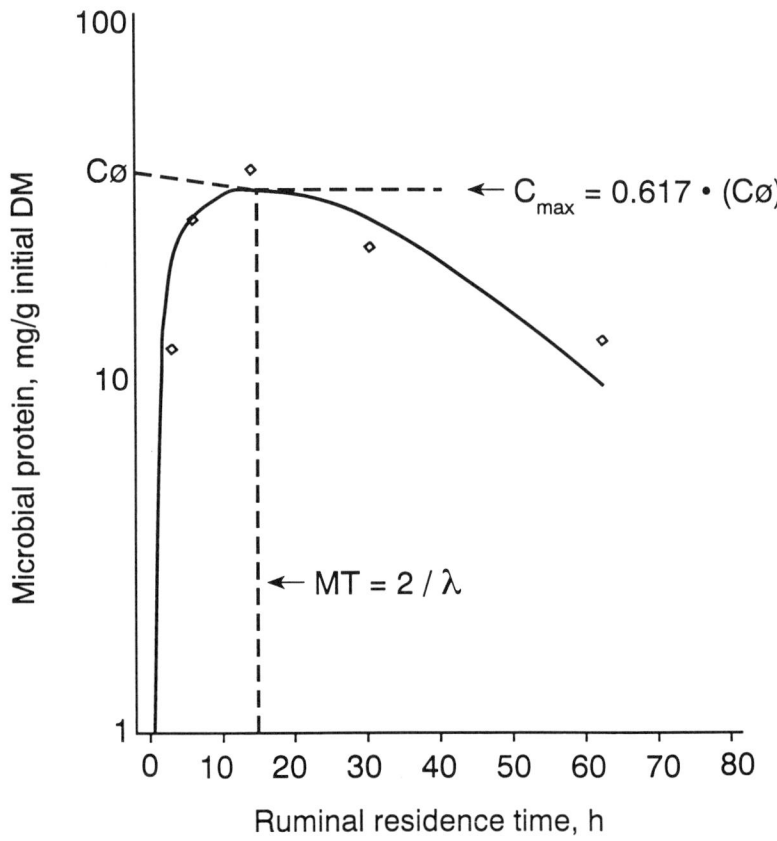

Fig. 20. Illustration of fit of G2→O model to the profile of microbial protein in situ and a method for estimating maximal concentration of microbial protein, Cø, the time required to achieve maximal levels of microbial protein, C_{max}, and the mean time to achieve C_{max}, MT.

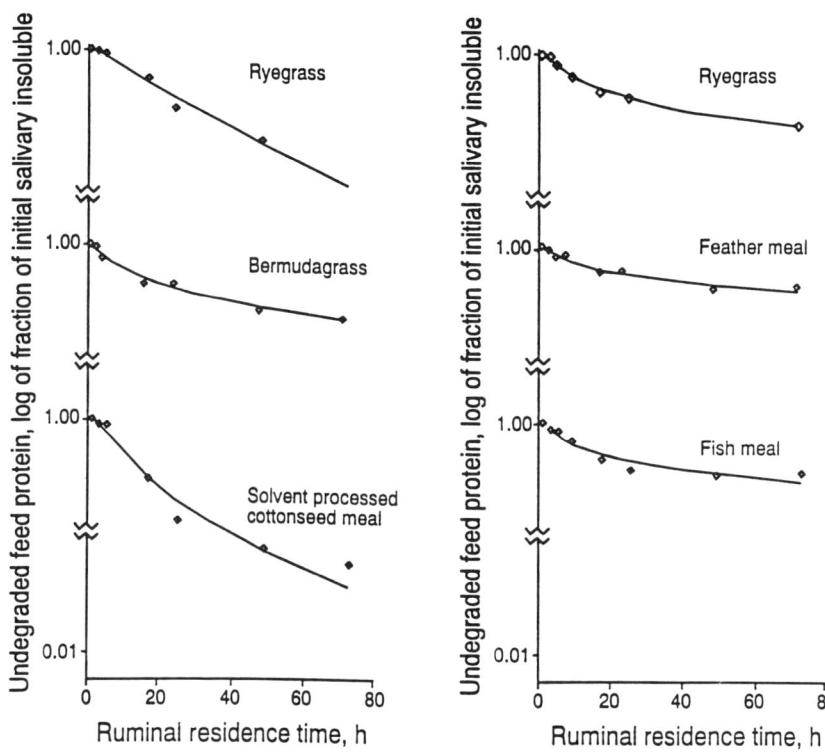

Figure 21. Ruminal in situ digestion of feed protein from six feeds. Note the non-exponential disappearance of feed protein from most feeds.

of equating initial values of E_d to values expected from some subsequent exponential distribution of undigested E_d, UE_d, (Mertens, 1993b). It is not biologically rational that such an abrupt initiation of mass action related rates should occur following a discrete digestion time lag. Van Milgen et al. (1991) questioned the biological rationale for a discrete time delay for onset of digestion and proposed a two-compartment sequential model to describe the hydration/digestion postulated to be involved in the digestion of dry matter. Later results (Van Milgen et al., 1993a) suggested that hydration did not appear to significantly delay the digestion process. However, the rationale for including a discrete time delay in digestion as an intrinsic property of the chemical entity is not evident and the two compartment model affords a method for describing a less abrupt initiation of digestion.

The need for a τ_d appears most evident in digestion of relatively insoluble entities such as fiber. In contrast, soluble and relatively soluble entities, i.e., protein and starch, appear to be digested without a significant or with small τ_d (Figure 21). The relatively long τ_d (>1 h) associated with masticated fragments, but not with the same fragments after fine grinding (see Figure 18), suggests that rapid digestion of readily fermentable tissues within large fragments may result in internal microenvironments too acidic for fiber digesting microbes and hence cause a time lag colonization of the interior of larger fragments by acid-intolerant, fiber digesting microbes.

Recommended Rate of Digestion Methodology.

Because of the complex and poorly defined interactions among feeds and the ruminal microbial ecosystem, the in situ method is recommended for determining the kinetics of digestion. Where rates of digestion are to be determined for entities common to feed and microbes, it is essential to correct for contributions due to microbial colonization during the fermentation. Potentially fermentable entities of feeds are extremely complex in their physical and chemical attributes and may be fermented with or without a time delay, with single or multiple age-constants, age-dependent distributions of lifetimes, or with a heterogeneous distribution of simultaneously occurring age-independent rates. Sufficient observations and replications should be planned to investigate the appropriate kinetic model and to estimate its parameters. Sufficient replication should be conducted to evaluate repeatability and to test for animal and diet interaction. Increased emphasis should be given to estimating digestion rates of feed particles as they actually enter the rumen, especially masticated fragments of forages, in contrast to use of samples ground through a laboratory mill.

FITTING MODELS TO DATA

All of the model expressions, except for the simple G1→0 model, require the use of nonlinear regression. This implies that parameters such as λ_1, k_1, or k_2 do not enter into the models as simple linear functions. Before the current widespread availability of computer software, nonlinear models often were analyzed using various "curve-peeling" procedures, which combine logarithmic transformations of the models with simple linear regression analyses. As a rule, procedures which transform the dependent variables, such as the observed concentrations, in order to simplify the analysis are out-of-date and should be avoided because they distort the natural weighting of the data. Instead, nonlinear regression procedures should be used (see Seber and Wild, 1989).

There are two main classes of nonlinear regression procedures. One class requires analytical formulas for the models, such as the expressions in Tables 1 through 4. A well known software package called SAS (1984) was used for fitting models to data reported in this paper. The nonlinear procedure of SAS is called NLIN. It has a grid search option which often simplifies the required specification of initial values. The NLIN also has an option called DUD, the acronym for "doesn't use derivatives", which, as the name implies, eliminates the need for user definition of the partial derivations of the model. A version of SAS called PCSAS also is available for IBM-compatible personal computers.

A new class of nonlinear procedures eliminates the need for analytical expressions for these special compartmental models. One example of such a procedure is KINETICA (Allen, 1990; Allen and Matis, 1992) which runs on personal computers with or without a math coprocessor. In KINETICA, the model is specified symbolically by naming the compartments and indicating the nature of their interconnectedness. This class of procedures estimates all of the parameters, as in SAS, by utilizing various numerical analysis techniques for all required internal computations. Other software packages for personal computers which have such nonlinear procedures are PCNONLIN (1985) and SAAM/CONSAM. This class of procedures also has the advantage of facilitating the fitting of models simultaneously from various parts of the digestive tract for combined estimation.

Mechanistic regression models based on digestion and passage kinetics are inherently nonlinear and, in general, nonlinear models require the use of nonlinear regression procedures. Therefore, experimenters who use deterministic models have learned to accept nonlinear regression procedures, even though they are much more difficult to implement than linear regression procedures. Sometimes experimenters are faced with a choice between various nonlinear, deterministic models, and often model selection is determined more by the simplicity of the corresponding statistical analyses than by the intrinsic merits of the various models. For example, the heterogeneous rate models are occasionally rejected, not for underlying theoretical reasons but because they are frequently sensitive to initial starting parameters. Ideally, one should specify first the most appropriate mechanistic model and then structure a suitable statistical approach for the analysis. For example, Bates and Watts (1988) give numerous suggestions for determining good starting values and for transforming parameters to facilitate the nonlinear analysis. Therefore, if the analysis of the desired model presents problems (e.g., with convergence), one could try alternative starting values and(or) reparameterizations to overcome the problem. In the case of the heterogeneous rate model with a gamma distribution of rates, α is the scale parameter and β is the shape parameter of the distribution whose mean age-independent rate is the product of α and β. To avoid solutions unique to the initial parameters selected, one should use a grid selection containing a wide array of initial starting values for α and β whose products equal one of several possible mean rates. Subsequent fittings should then be conducted using the most appropriate arrays of initial parameters indicated. Because the distributions may be so varied (see Figure 14, for example), the magnitudes of α and β may be quite disparate. One should not reject a mechanistic model solely because of difficulties with an initial statistical analysis until further remedial measures are attempted. New software, including that previously mentioned, also is available to enhance the statistical analysis and to simplify its implementation.

All of these sotware packages contain an option to weight the data. This is a useful option in cases where the weighting function is determined by mechanistic theory. However, weighing is questionable in cases where the weights must be determined empirically. The examples of this paper fall into the later category, and hence an un-weighted analysis was used, as in Matis et al. (1989).

INTEGRATING DIGESTION AND PASSAGE KINETICS

Digestion in ruminants is achieved via three sequential phases of digestion: 1) microbial fermentation in the forestomachs, 2) hydrolytic digestion in the abomasum and the small intestine, and 3) microbial digestion in the hindgut. Estimates of residence times (Table 10) and mass (Table 11) of E_i from marker flow to the duodenum suggests that forage fragment flow through at least two compartments within the forestomach. Because these functional compartments appear to exist almost entirely within some of the same mass of ruminal digesta, such compartments lack readily definable boundaries (boundaries being required to define discrete compartments). Interactions among pools, such as reversible flow, are unclear. Such ill-defined mixing-digesting compartments should, therefore, be referred to as mixing-digesting pools.

Pools of Ruminal Digesta

The functional bases for two sequential pools of ruminal digesta has been suggested (Ellis et al., 1991) to be caused by the factors determining the fragments buoyancy, the principles illustrated by Sutherland (1989). On initial ingestion, fragments of forage tissues are positioned in the rumination pool due to their initial buoyancy; primarily due to gases within the vascular tissues of the fragment. With subsequent colonization and fermentation of the more readily digestible non-vascular tissues, production and entrapment of fermentation gases within the fragment's architecture results in an acquired fermentation-based buoyancy. Larger fragments have more of readily digestible tissues which are internal to the labyrinthine passages (Pond et al., 1989) and, therefore, acquire greater fermentation-based buoyancy which maintains their initial position in the rumination pool. With increased residence time in the rumination pool, readily fermentable tissues are expended and ruminative mastication reduces the size, complexity and associated gas entrapping properties of the fragment so that the fragment becomes less buoyant. These less buoyant fragments are subtended to more ventral regions of the rumination pool by more buoyant fragments of more recently ingested fragments. With further rumination and continued loss of readily fermentable tissues, the fragment continue to lose buoyancy and sediment into the ruminal turnover pool.

Sutherland (1989) describes the positioning of fragments within the rumination pool as due to a balance between the forces of fermentation based buoyancy acting to "un-mix" digesta in opposition to the mixing forces of the coordinated motility of the rumen and reticulum. Because of the interacting forces of buoyancy, rumination and mixing motility, the rumination and ruminal turnover pools are dynamic in their size. These dynamics are exacerbated by the meal eating habits of the ruminant and differences in forages composition. Within the intermeal interval, such functional pools do not conform to steady state conditions and lack discreet boundaries within the digesta. A conceptualization of such pools is illustrated schematically in Figure 22.

The rumination and ruminal turnover pools may appear similar to the

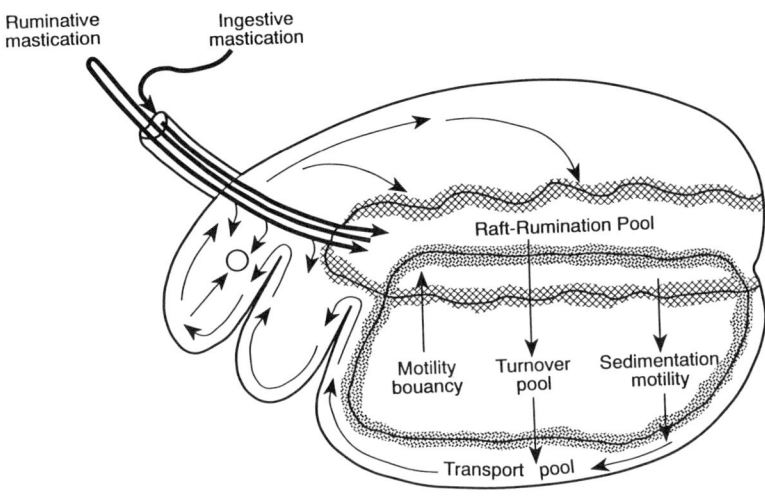

Figure 22. A conceptualization of the functional nature the raft-rumination pool and the ruminal turnover pool in cattle and the forces causal of such separate pools.

"rumination-large particle pool" and "small particle-liquid pool" pool concepts of Hungate (1966). However, pool separation based on the forces of fermentation-related buoyancy may be independent of fragment size if such "large" fragments are devoid of rapidly fermentable substrates or lack gas-entrapping internal structures. Consequently, there may be little difference in size of fragment in the dorsal and ventral ruminal digesta for cattle receiving corn silage as illustrated in Figure 23. The data in Figure 23 also illustrate the dynamic nature of the distribution of fragments by size within, and escaping from the ruminal digesta and the effects of meal eating patterns.

Multiple, Age-Independent Mixing-Digesting Pools

Assumptions concerning the kinetics of flow and digestion in a system of mixing-digesting pools will materially affect residence time of the fragment at the site of fermentation and, consequently, estimates of extent of digestion of E_d. In view of the dynamic nature of the force shaping the raft-rumination and turnover pools of ruminal digesta, a number of different systems of pools could be conceived involving number, sequence, and interchange among pools. The simplest system is two sequential pools with irreversible flow and a time delay as illustrated by the $Gn \rightarrow G1 \rightarrow \tau \rightarrow O$ model; a model with fits the data well (see Figure 9 and Table 6).

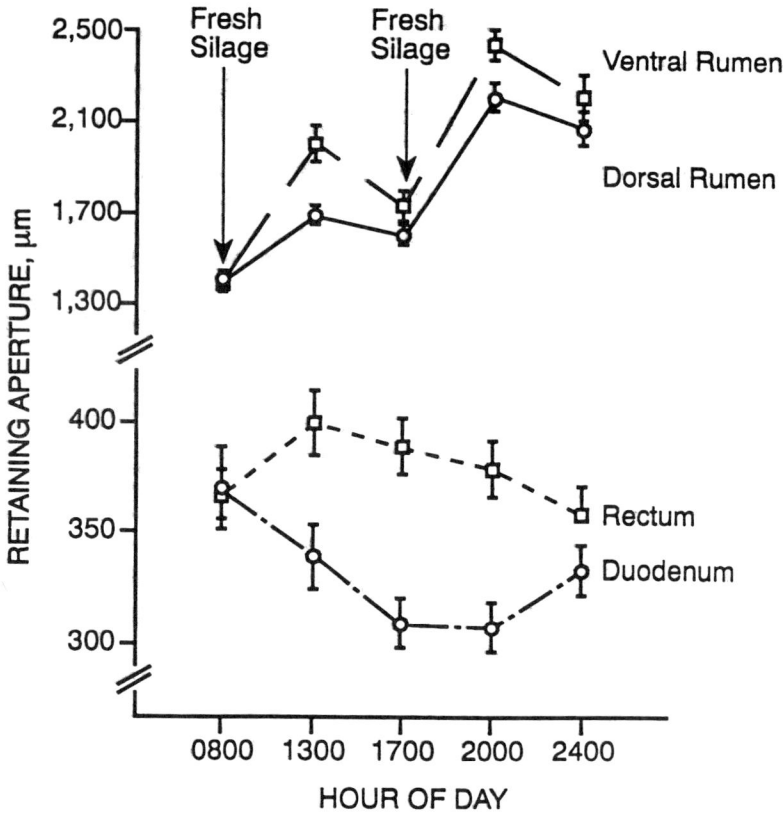

Figure 23. Variation in size, mean retaining aperture, of fragments of corn silage within and escaping from, the ruminal digesta of cattle offered fresh corn silage ad libitum twice daily (adapted from Deswysen et al. 1989a).

Digestion in and flow of E_d through two sequential and irreversible mixing-digesting pools is graphically illustrated in Figure 24.

The extent of digestion of potentially digestible fiber, DE_d, within each pool, ^1P and ^2P, equals the proportion of total disappearance (digestion plus escape) which disappears via digestion, a fraction equalling $k_d / (k_d + k_e)$ for age-independent processes. For simplicity, example calculations of DE_d in Figure 24 utilize age-independent expressions of rates of digestion and passage. The rates are those previously discussed for the sample data set involving bermudagrass hay. Escape of fragments from the rumination pool, ^1P, was assumed to be an age-dependent process expressed by the mean age-independent rate parameter, $^1\overline{k}_e$, which, in this example, equals the $\lambda_1 \cdot 0.59635$ (for gamma 2 distribution of residence times, see Table 5). The value for λ_1 of 0.1357 represents a typical

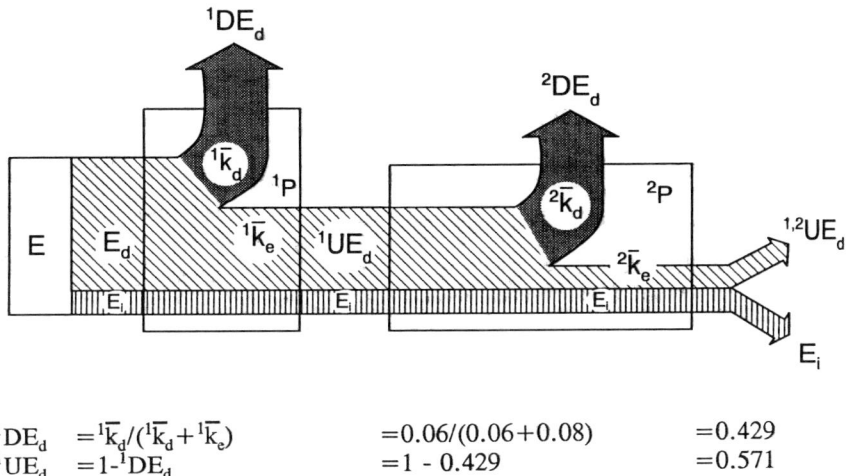

$$^1DE_d = {}^1\overline{k}_d/({}^1\overline{k}_d + {}^1\overline{k}_e) \quad = 0.06/(0.06+0.08) \quad = 0.429$$
$$^1UE_d = 1 - {}^1DE_d \quad = 1 - 0.429 \quad = 0.571$$

$$^2DE_d = {}^2\overline{k}_d/({}^2\overline{k}_d + {}^2\overline{k}_e) \quad = 0.06/(0.06+0.02) \quad = 0.75$$
$$^2UE_d = {}^1UE_d(1 - {}^2DE_d) \quad = 0.571(1-0.75) \quad = 0.193$$

$$^{1,2}DE_d = {}^1DE_d + ({}^1UE_d \cdot {}^2DE_d) \quad = 0.429 + (0.571 \cdot 0.75) \quad = 0.857$$
$$^{1,2}DE_d = 1 - {}^1UE_d \cdot {}^2UE_d \quad = 1 - (0.571 \cdot 0.25) \quad = 0.857$$

Digestibility of total fiber, E $\quad = DE_d \cdot {}^{1,2}DE_d/Ed + E_i = 0.857/E_d + E_i$

Figure 24. Diagrammatic representation of ruminal digestion of potentially digestible fiber, E_d, as the result of competitive rates of digestion, ${}^1\overline{k}_d$ and ${}^2\overline{k}_d$, and escape, ${}^1\overline{k}_e$ and ${}^2k_e$, in two sequential mixing-digesting pools, ${}^1P$ and ${}^2P$. E_i = indigestible fiber; ${}^1DE_d$ and ${}^2DE_d$ = fraction of E_d digested from pools 1 and 2 respectively and ${}^1UE_d$ = fraction of E_d escaping from ${}^1P$ to ${}^2P$.

value for λ_1 in cattle consuming warm season forages. For the current data (Table 6), the ${}^1\overline{k}_e$ is 0.08 h^{-1}. Rate of digestion of E_d in the rumination pool was assumed the result of a heterogeneous distribution of rates, the mean of which, ${}^1\overline{k}_d$, in this example equals 0.06 h^{-1} (see Figure 17). The fractional extent of digestion of E_d in the rumination pool, ${}^1DE_d$, equals ${}^1\overline{k}_d/({}^1\overline{k}_d + {}^1\overline{k}_e)$ or 0.426 if a digestion time lag, τ_d, is not considered.

The fraction of E_d escaping the rumination pool, ${}^1UE_d$ or 0.574 of E_d in this example, undergoes digestion in the second mixing-digesting pool. The E_d disappears from the second ruminal mixing-digesting pool via an age-independent rate of escape, ${}^2k_e$, of 0.02 h^{-1} (see Table 6, G2→G1→τ→O model) and a mean rate of a heterogeneous distribution of rates, ${}^2\overline{k}_d$, of 0.06 h^{-1}. The fractional extent of digestion of ${}^1UE_d$ in the second ruminal pool, ${}^2P$, equals 0.75. As a consequence of irreversible flow through two sequential mixing-digestion pools, the fraction of E_d digested is the sum of the fractional extent of digestion of ${}^1DE_d$ in ${}^1P$, $1 \cdot {}^1DE_d$, and the fractional extent of digestion of ^1UD_d in ^2P, $^1UE_d \cdot {}^2DE_d$, which, in this example, equals 0.429 in ^1P plus 0.428 (0.571 • 0.75) in ^2P or 0.857 of the E_d.

The digestibility of potentially digestible NDF, PDNDF, of SEM all forage diets suggests that DPDNFD is relatively constant and relatively complete, in the order of 0.90. The DPDNFD in the total tract averaged 91.4%±5.7(SEM) for 8 different forages (Ellis et al., 1991). In another all forage data set, the DPDNFD prior to the duodenum averaged 0.90 (Ellis et al. 1992a) with no detectable post-duodenal digestion of PDNDF. Poore et al. (1991) also reported that DPDNFD averaged 0.9 for digestion prior to the duodenum in dairy cows fed diets of 300 g kg^{-1} concentrate or less.

A DPDNFD of 0.9 is considerably greater than could be expected from flow through only one mixing-digestion pool. In a single mixing-digestion pool, to achieve a DPDNFD of 0.9 would require a $\bar{k}_e$ of 0.0066 h^{-1} if a $\bar{k}_d$ of 0.06 h^{-1} is assumed ($\bar{k}_e = (\bar{k}_d \cdot (1-DEd)/DE_d)$. Alternatively, a $\bar{k}_d$ of 0.18 h^{-1} would be required if a $\bar{k}_e$ of 0.02 is assumed ($\bar{k}_d = \bar{k}_e \cdot DE_d/(1-DE_d)$). The $\bar{k}_d$ and $\bar{k}_e$ required to yield the observed DPDNFD in a single mixing-digesting pool are more than two-fold different than those commonly observed. Obviously, some mechanism involving multiple and sequential mixing-digestion pools is required to increase residence time necessary for extensive digestion of slowly digesting entities such as potentially digestible fiber. The assumption of two irreversible and sequential mixing-digesting pools yield estimates of DPDNDF (Figure 24) consistent with more direct estimates of DPDNDF.

The assumption of irreversible flow may appear difficult to justify for digesta pools which are the product of such dynamic forces; however, models involving reversible flow between the pools indicate little back flow to the rumination pool occurs (Poppi personal communication). It appears that forces which so effectively constrain escape from the rumination pool are equally effective in constraining re-entry of fragments which have escaped.

Age-Dependent Passage and Digestion

For simplicity in illustration, the above example assumes age-independent kinetics. The flow of fragments from the ruminal digesta pools can be conceptualized and described equally well as flow through multiple pools or through a single age-dependent pool (compare model G2→G1→τ→O, Table 6 with model G2→τ→O, Table 7). A heterogeneous distribution of rates model represents an age-dependent process because of changes in the distribution of digestion rates with residence time (see Figure 17). Models to integrate age-dependent digestion and passage are difficult and have not been accomplished to date. However, the effects of different assumptions concerning age-dependent expressions of digestion and passage can be estimated for discrete residence times and are illustrated in Figure 25.

Age-independent digestion and escape of E_d results in a first order decline in UE_d and a constant fraction of DE_d disappearing via digestion (Figure 25A). In contrast, age-dependent processes of digestion and flow result in age-related changes in UEd and in the proportion of total disappearance via digestion (Figure 25B). Due to slow initial passage and rapid initial digestion, the initial fractional

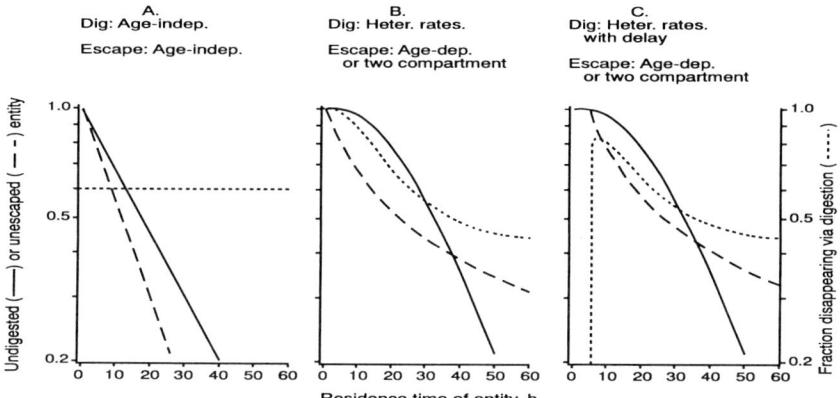

Figure 25. Effects of model assumptions concerning residence time distributions for digestion and escape of potentially digestible entities. A. Age-independent, exponentially distributed residence times for digestion, $k_d = 0.06$ h^{-1}, and escape, $k_e = 0.04$ h^{-1}. B. Heterogeneous distribution of rates of digestion, $\alpha = 0.6$, $\beta = 0.1$ and $\overline{k}_d = 0.06$ h^{-1}, and age-dependent rate of escape, $\overline{k}_e = 0.084$ h^{-1} for G3 distribution. C. Heterogeneous distribution of rates of digestion, $\alpha = 0.6$, $\beta = 0.1$ and $\overline{k}_d = 0.06$ h^{-1}, with time delay, τ_d, = 5 h and age-dependent rate of escape, $\overline{k}_e = 0.084$ h^{-1} for G3 distribution.

digestibility of E_d approaches 1 during the earliest residence times of the fragment. Consequently, a digestion time lag will have less impact when it occurs in combination with age-dependent escape than with age-independent escape (Figure 25C). With progression of residence time, the rate of escape increases and the rate of digestion decreases so that, as a consequence, fractional digestibility decreases.

Validity of Integrated Digestion and Flow Models

The validity of methodology for estimating passage and digestion parameters should be assessed by how well the integrated results agree with estimates of the same processes obtained by independent methodology. A number of studies provide some comparative results. Moore et al. (1989) obtained good agreement between estimates of apparent ruminal digestibility for NDF and starch by kinetic versus in vivo procedures (33 and 34, and 63 and 67, respectively). They utilized

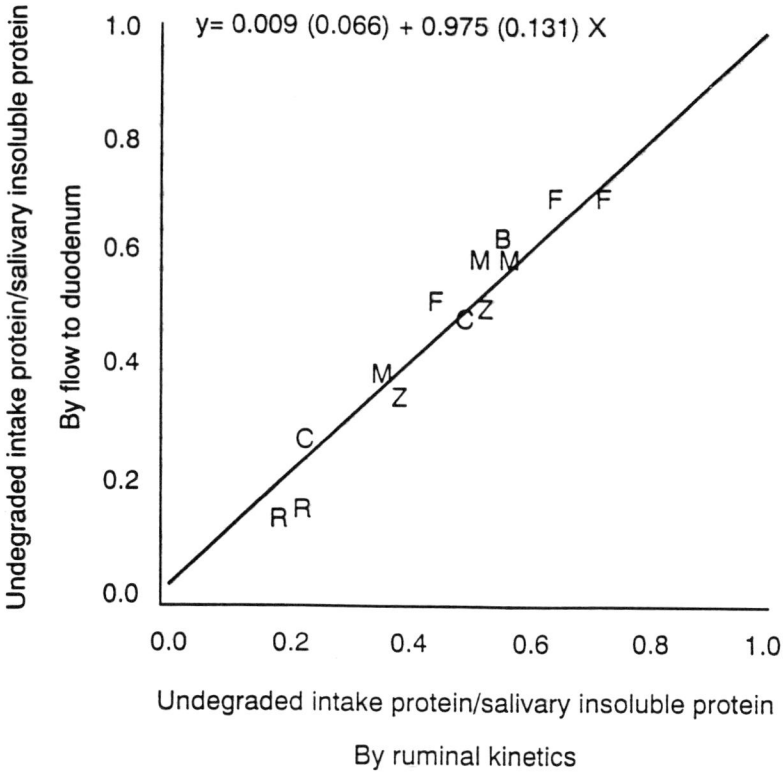

Figure 26. Relationship between fractional ruminal undegraded feed proteins estimated by escape to the duodenum and that estimated by ruminal kinetics. The mean rate of ruminally digested protein, $\bar{k}_d$, was estimated by fitting rumen in situ data to the heterogeneous rates model the mean rate of ruminal escape, $\bar{k}_e$, was estimated by fitting the G2→τ→O model to rare earth marker appearance at the duodenum. R = ryegrass masticate, C = solvent processed cottonseed meal, M = menhaden fish meal, Z = corn, F = feather meal, and B = ring dried blood meal.

ruminal in situ digestion methods, specifically bound rare earths as flow markers, and Cr_2O_3 in the total diet as the digestion marker. Ruminal digestion was estimated from duodenal digesta collected via a cannula.

Figure 26 illustrates good agreement between estimates of the fraction of ruminally undegraded protein, RUP, when estimated by appearance in duodenal digesta (in vivo method) as compared to a ruminal kinetic method (Hill et al., 1991; Ellis et al., 1992b). The in vivo method involved continuous infusion into the rumen of rare earth markers and ^{35}S for estimating digesta flow and labelling microbial protein respectively. The kinetic method utilized a ruminal in situ method for estimating rate of degradation of feed protein in grazed forage and protein supplements, with determination of diaminopimelic acid in donor rumen fluid to correct for microbial colonization. In the kinetic method, rates of ruminal escape of forage and supplement residues were estimated individually by using different rare earths initially applied to masticated samples of the grazed forage and supplement, the rare earth method samples being dosed via ruminal cannula. Ruminal escape rate was estimated by fitting the G2→τ→O model to marker apperance in the duodenal digesta. The rate of escape to the duodenum of rare earths initially bound to forage and supplement differed significantly by type of forage grazed and forage type versus supplement (see Table 12).

The results in Figure 26 indicate close agreement between RUP estimated by the kinetic methods and by independent in vivo methods. Each method utilized independent methodology and provides evidence that the methodology discussed here is valid.

SUMMARY

Potentially digestible entities, E_d, continue to be digested after long periods of time (weeks) while indigestible entities, E_i, remain undigested after similarly long periods of time. The digestibility of a chemical entity, E, is then determined by its content of E_i and the extent of digestion of E_d. The extent of digestion of E_d within a mixing-digesting site is determined by the proportion of E_d which disappears from that site by digestion relative to its disappearance via digestion plus passage. Kinetic methods are required to estimate rates of digestion, k_d, and escape, k_e, of E_d, and to partition E into E_d and E_i. Estimation of k_d, and partitioning of E into E_d and E_i requires vitro methods to measure the effects of controlled residence times on defined amounts of E by the agents (rumen microbial enzymes) of digestion. Estimation of k_e requires in vivo methods to desirable flow of masticated forage fragments through the ruminal digesta, the initial and dominant site of digestion in ruminants.

Markers which remain bound to the fragment are required to reflect flow of fragments from specified meals through segments of the digesta involved in their digestion. Specifically bound rare earths are the most appropriate markers for describing the flow of digesta through the less acidic segments of the tract, e.g. the rumen. Compartmental models are used to estimate the mean residence times of markers and to partition such into residence times for functional turnover pools of digesta. A data set was used to illustrate the effects of assumptions concerning age-independent and age-dependent distribution of residence times. Results of fitting various models to these data illustrate that specifically bound rare earths flow through at least two functional pools of ruminal digesta. Due to a lack of physical separation, these functional pools cannot be resolved by analyzing marker disappearance from ruminal digesta but require estimation from marker appearance at the most accessible post-ruminal site which can be representatively sampled, usually the duodenum. The k_e of fragments of forages and supplements differ and must be simultaneously and individually measured.

Models to estimate k_d, E_d, and E_i are summarized with respect to assumptions concerning the digestion process. A model assuming a heterogeneous distribution of digestion rates is emphasized because of its flexibility in fitting data, its use of fewer parameters, and its apparent biological applicability. This heterogeneous rates model also has the advantage of providing rate estimates for poorly understood processes that may be erronously estimated by inappropriate deterministic models.

The integration of digestion and flow kinetics indicates that some form of mixing-digesting flow, such as sequential pools, must occur in rumen digesta to provide the mean compartmental residence time necessary to achieve the observed extents of digestion for slowly digesting E_d. Data are summarized which illustrate excellent agreement between ruminally undegraded protein estimated by flow to the duodenum and that estimated more indirectly by the kinetic procedures described here. Good agreement between independent methodologies demonstrated the validity of the kinetic approach and illustrated the individual methods which must be utilized to avoid confounding estimates by kinetic methods.

Previous methods for integrating the kinetics of digestion and passage have assumed these processes to be age-independent, i.e., related to mass action and expressed as exponential distributions of residence times or lifetimes. In contrast, considerable evidence suggests that both digestion and passage, and their competitive determination of extent of digestion, may be age-dependent. Future research should emphasize an improved undestanding of the physical and chemical digestion of ingested forage fragments and the interaction involving subsequent physical fragmentation, flow and chemical digestion of the succession of derived fragments within the ruminal digesta. An improved understanding of the mechanisms regulating within day variation of fragment flow from ruminal digesta is also considered of high priority.

REFERENCES

Allen, M. S., and D. R. Mertens. 1987. Evaluating constraints on fiber digestion by rumen microbes. J. Nutr. 116:261-270.

Allen, D. M. 1990. Fitting compartmental models to data. p 113-126. In A. B. Robson, and D. P. Poppi (ed.) Modelling digestion and metabolism in farm animals. Lincoln University. Canterbury, New Zealand.

Allen, D. M., and J. H. Matis. 1992. KINETICA, a program for kinetic modeling in biological sciences. Dept. of Statistics Tech. Report #317, Univ. of Kentucky, Lexington.

Bates, D. M., and D. G. Watts. 1988. Nonlinear regression analysis and its applications. Wiley, New York.

Blaxter, K. L., N. M. McGraham, and F. W. Wainman. 1956. Some observations on the digestibility of food by sheep and on related problems. Br. J. Nutr. 10:69-91.

Brandt, C. S., and E. J. Thacker. 1958. A concept of rate of food passage through the gastrointestinal tract. J. Anim. Sci. 17:218-223.

Cabello, L., T. M. Hill, S. D. Martin, and W. C. Ellis. 1990. Ruminal escape turnover of unfermented residue of grazed bermudagrass and supplements in calves. J. Anim. Sci. 68(Suppl. 1):593 (Abstr.)

Combs, D. K., and L. D. Satter. 1992. Determination of markers in digesta and feces by direct current plasma emission spectroscopy. J. Dairy Sci. 75:2176-2183.

Coffey, K. P., E. E. Pickett, J. A. Paterson, C. W. Hunt, and S. J. Miller. 1988. Methods of ytterbium analysis for predicting fecal output and flow rate constants in cattle. J. Range Manag. 41:426-429.

Cruickshank, G. J., D. P. Poppi, and A. R. Sykes. 1989. Theoretical considerations in the estimation of rumen outflow rate from various sampling sites in the digestive tract. Br. J. Nutr. 62:229-239.

Deswysen, A. G., and H. J. Ehrlein. 1981. Silage intake, rumination and pseudo-rumination activity in sheep studies by radiography and jaw movement recordings. Br. J. Nutr. 46:327-332.

Deswysen, A. G., and W. C. Ellis. 1988. Site and extent of neutral detergent fiber digestion, efficiency of ruminal digesta flux and fecal output as related to variations in voluntary intake and chewing behavior in heifers. J. Anim. Sci. 66:2678-2686.

Deswysen, A. G., K. R. Pond, E. Rivera-Villarreal, and W. C. Ellis. 1989a. Effects of time of day and monensin on the size distribution of particles in digestive tract sites of heifers fed corn silage. J. Anim. Sci. 67:1773-1783.

Deswysen, A. G., P. A. Dutilleul, and W. C. Ellis. 1989b. Quantitative analysis of nycterohemeral eating and ruminating patterns in heifers with different voluntary intake and effects of monensin. J. Anim. Sci. 67:2751-2761.

Dhanoa, M. S., R. C. Siddons, J. France, and D. L. Gale. 1985. A multi-compartment model to describe marker excretion patterns in ruminant faeces. Br. J. Nutr. 53:663-671.

Dixon, R. M., J. J. Kennelly, and L. P. Milligan. 1983. Kinetics of [^{103}Ru]phenanthroline and dysprosium particulate markers in the rumen of steers. Br. J. Nutr. 49:463-473.

Ellis, W. C., J. H. Matis, and C. Lascano. 1979. Quantitating ruminal turnover. Fed. Proc. 38:2702-2706.

Ellis, W. C., J. H. Matis, B. Rector, and L. Rittenhouse. 1980. Models for estimating fecal output by a single dose marker technique. J. Anim. Sci. 51 (Suppl. 1):235 (Abstr.)

Ellis, W. C., and D. E. Beever. 1984. Methods for binding rare earths to specific feed particles. p 154-165. *In* P.M. Kennedy (ed.) Techniques in particle size analysis of feed analysis of feed and digesta in ruminants. Occasional Publ. no. 1. Canadian Society of Animal Science, Edmonton.

Ellis, W. C., J. H. Matis, K. R. Pond, and M. Mahloogi. 1984. Physical and chemical digestion of forage fragments with emphasis on stochastic, heterogenous rate models. p 34-42. *In* Proc. Second Int. Symp. on Modeling Ruminant Digestion and Metabolism. Davis, CA.

Ellis, W. C., T. M. Hill, S. D. Martin, F. M. Rouquette, Jr., and J. H. Matis. 1990. Feed-rumen microbial protein transactions in situ. J. Anim. Sci. 68(Suppl. 1):593(Abstr.)

Ellis, W. C., P. Kennedy, and J. H. Matis. 1991. Passage and digestion of plant tissues in herbivores. p 227-236. *In* Y. W. Ho, H. K. Wong, N. Abdullah, and A. Z. Tajuddin (ed.) Recent advances on the nutrition of herbivores. Third Int. Symp. on the Nutrition of Herbivores. 25-30 August, 1991. Malaysian Society Animal Production. Peneng, Malaysia.

Ellis, W. C., T. M. Hill, S. Martin, and J. H. Matis. 1992a. Digestion of potentially digestible fiber in grazing calves. J. Anim. Sci. 70(Suppl. 1):190(Abstr.)

Ellis, W. C., T. M. Hill, and F. M Rouquette. 1992b. The determination of undergraded intake protein in supplements and effects of grazed forage. J. Anim. Sci. 70(Suppl. 1):188(Abstr.)

Erhlein, H.J., and Inst. Wiss. Film C. 1980. Forestomach motility in ruminants. Film C 1328 des IWF, Goettingen, W. Germany. Publikation von H.J. Erhlein, Publikationen zu Wissenschagtlichen Filmen, Sektion Medizin, Serie 5, No. 9/C 1328. p 1-29.

Fadel, J. G. 1992. Application of theoretically optimal sampling schedule designs for fiber digestion estimation in sacco. J. Dairy Sci. 75:2184-2189.

Faichney, G. J., and R. C. Boston. 1983. Interpretation of the faecal excretion patterns of solute and particle markers introduced into the rumen of sheep. J. Agric. Sci. (Camb.) 101:575-581.

Faichney, G. J. 1984. The kinetics of particulate matter in the rumen. p 173-175. *In* L.P. Milligan, W. L. Grovum, and A. Dobson (ed.) Control of digestion and metabolism in ruminants. Prentice-Hall, Englewood Cliffs, NJ.

Faichney, G. J., C. Poncet, and R. C. Boston. 1989. Passage of internal and external markers of particulate matter though the rumen of sheep. Repro. Nutr. Devel. 29:325-338.

France, J., J.H.M. Thornley, M.S. Dhanoa, and R. C. Siddons. 1985. On the mathematics of digesta flow kinetics. J. Theor. Biol. 113:743-758.

France, J., M. S. Dhanoa, R. C. Siddons, J. H. M. Thornley, and D. P. Poppi. 1988. Estimating the production of faeces by ruminants from faecal marker concentration curves. J. Theor. Biol. 135:383-391.

France, J., J. H. M. Thornley, R. C. Siddons, and M. S. Dhanoa. 1993. On incorporating diffusion and viscosity concepts into compartmental models for analysing faecal marker excretion patterns in ruminants. Br. J. Nutr. 70:369-378.

Grovum, W. L., and V. J. Williams. 1973. Rate of passage of digesta in sheep. 4. Passage of marker through the alimentary tract and the biological relevance of rate-constants in concentration of marker in faeces. Br. J. Nutr. 30:377-389.

Hill, T.M. 1991. Effects of source of supplemental nutrients on forage intake, digestive kinetics and protein supply to the small intestine of grazing calves. Ph.D. diss. Texas A&M University, College Station.

Hoffmann, R. R. 1988. Anatomy of the gastro-intestinal tract. p 14-43. *In* D. C. Church (ed.) The ruminant animal digestive physiology and nutrition. Prentice Hall, Englewood Cliffs, NJ.

Hungate, R. E. 1966. The Rumen and Its Microbes. Academic Press, New York.

Krysl, L. J., M. L Galyean, R. E. Estell, and B. F. Sowell. 1988. Estimating digestibility and faecal output in lambs using internal and external markers. J. Agric. Sci. (Camb.) 111:19-25.

Lalles, J. P., E. Delval, and C. Poncet. 1991. Mean retention time of dietary residues within the gastrointestinal tract of the young ruminant: a comparison of non-compartmental (algebraic) and compartmental (modelling) estimation methods. Anim. Feed Sci. Technol. 35:139-159.

Lindberg, J. E., and P. G. Knutsson. 1981. Effect of bag pore size on the loss of particulate matter and on the degradation of cell wall fiber. Agric. Environ. 6:171-180.

Lippke, H., W. C. Ellis, and B. F. Jacobs. 1986. Recovery of indigestible fiber from feces of sheep and cattle on forage diets. J. Dairy Sci. 69:403-412.

Mahloogi, M. 1985. Fiber digestion from different size particles produced by mastication. Ph. D. diss. Texas A&M University, College Station.

Matis, J. H. 1972. Gamma time-dependency in Blaxter's compartment model. Biometrics 28:597-602.

Matis, J. H. 1987. The case of stochastic models of digesta flow. J. Theor Biol. 124:371-602.

Matis, J. H., and T. E. Wehrly. 1979. Stochastic models of compartmental systems. Biometrics 35:199-220.

Matis, J. H. and T. E. Wehrly, and W. C. Ellis. 1989. Some generalized stochastic compartmental models for digesta flow. Biometrics 45:703-720.

Mertens, D. R. 1989. Evaluating alternative models of passage and digestion kinetics. p 79-97. *In* A. B. Robson, and D. P. Poppi (ed). Modelling digestion and metabolism in farm animals. 3rd Int Workshop. 4-6 Sept., 1989. Lincoln University, Canterbury, New Zealand.

Mertens, D. R. 1993a. Kinetics of cell wall digestion in ruminants. p 535-570. *In* H. G. Jung, D. R. Buxton, R. D. Hatfield, and J. Ralph (ed.) Forage cell wall structrue and digestibility. American Society of Agronomy. Madison, WI.

Mertens, D. R. 1993b. Rate and extent of digestion. p 13-51. *In* J.M. Forbes, and J. France (ed.) Quantitative aspects of ruminant digestion and metabolism. C.A.B. International. Oxon, United Kingdom.

Meyer, J. H. F., and R. I. Mackie. 1986. Microbiological evaluation of the intraluminal in sacculus digestion technique. Applied & Enviro. Mbio. 51:622-629.

Moore, J. A., M. H. Poore, and R. S. Swingle. 1989. Calculated ruminal extent of digestion as influenced by in situ particle size, forage dosing time, and sampling site in Holstein cows. J. Anim. Sci. 67(Suppl. 1):554(Abstr.)

Moore, J. A., K. R. Pond, M. H. Poore, and T. G. Goodwin. 1992. Influence of model and marker on digesta kinetic estimates for sheep. J. Anim. Sci. 70:3528-3540.

Nocek, J. E. 1988. In situ and other methods to estimate ruminal protein and energy digestibility: a review. J. Dairy Sci. 71:2051-2069.

Owens, F. N., and C. F. Hanson. 1992. External and internal markers for appraising site and extent of digestion in ruminants. J. Dairy Sci. 75:2605-2617.

PCNONLIN. 1985. PCNONLIN User's Guide, Statistical Consultants Inc.,

Lexington Ky.

Pond, K. R., W. C. Ellis, and D. E. Akin. 1984. Ingestive mastication and framentation of forages. J. Anim. Sci. 58:1567-1574.

Pond, K. R., W. C. Ellis, W. D. James, and A. G. Deswysen. 1985. Analysis of multiple markers in nutrition research. J. Dairy Sci. 68:745-750.

Pond, K. R., W. C. Ellis, C. E. Lascano, and D. E. Akin. 1987. Fragmentation and flow of grazed coastal bermudagrass through the digestive tract of cattle. J. Anim. Sci. 65:609-618.

Pond, K. R., W. C. Ellis, J. H. Matis, H. M. Ferreiro, and J. D. Sutton. 1988. Compartmental models for estimating attributes of digesta flow in cattle. Br. J. Nutr. 60:571-595.

Pond, K. R., W. C. Ellis, J. H. Matis, and A. G. Deswysen. 1989. Passage of chromium mordanted and rare earth labeled fiber-time of dosing kinetics. Accepted for publication, J. Ani. Sci. 67:1020-1028.

Poore, M. H., J. A. Moore, and R. S. Wingle. 1990. Differential passage rates and digestion of neutral detergent fiber from grain and forages in 30, 70, and 90% concentrate diets fed to steers. J. Anim. Sci. 68:2965-2973.

Poore, M. H., J. A. Moore, T. P. Eck, and R. S. Swingle. 1991. Influence of passage model, sampling site, and marker dosing time on passage of rare earth-labeled grain through holstein cows. J. Anim. Sci. 69:2646-2654

Reid, C. S. W., A. John, M. J. Ulyatt, G. C. Waghorn, and L. P. Milligan. 1979. Chewing and the physical breakdown of feed in sheep. Ann. Rech. Vet. 10:205-207.

Robinson, P.H., J. G. Fadel, and S. Tamminga. 1986. Evaluation of mathematical models to describe neutral detergent residue in terms of its susceptibility to degradation in the rumen. Ani. Feed. Sci. & Technol. 15:249-271.

Roux, C. Z, and J. P. Pienaar. 1984. p. 176-177. In P. M. Kennedy (ed). Techniques in particle size analysis of feed and digesta in ruminants. Canadian Society of Animal Science. Edmonton.

SAS. 1984. SAS User's Guide: Statistics. Statistics Analysis System Institute, Inc., Cary NC.

Sauvant, D., and B. Ramangasoavina. 1991. Rumen modelling. p 283-296. In J. P. Jouany (ed.) Rumen microbial metabolism and ruminant digestion. Institut National De La Recherche Agronomique, Paris, France.

Seber, G. A. F., and C. J. Wild. 1989. Nonlinear regression. Wiley. New York.

Shipley, R. A., and R. C. Clark. 1972. Tracer methods for in vivo kinetics theory and applications. Academic Pres, Inc. New York, NY.

Siciliano-Jones, J., and M. R. Murphy. 1986. Passage of inert particles varying in length and specific gravity through the postruminal digestive tract of steers. J. Dairy Sci. 69:2304-2311.

Smith, L. W. 1989. A review of the use of intrinsically ^{14}C and rare earth-labeled neutral detergent fiber to estimate particle digestion and passage. J. Anim. Sci. 67:2123-2128.

Sutherland, T.M. 1989. Particle separation in the forestomach of sheep. p. 43-73. *In* A. Dobson, and M. H. Dobson (ed.) Aspects of Digestive Physiology in Ruminants. Cornell Univ. Press, Ithaca, NY.

Uden, P., R. Parra, and P. J. Van Soest. 1974. Factors influencing reliability of the nylon bag technique. J. Dairy Sci. 57(Suppl. 1):358(Abstr.)

Van Hellen, R. W., and W. C. Ellis. 1977. Sample container porosities for rumen in situ studies. J. Anim. Sci. 44:141-146.

Van Milgen, J., M. R. Murphy, and L. L. Berger. 1991. A compartmental model to analyze ruminal digestion. J. Dairy Sci. 74:2515-2529.

Van Milgen, J., M. L. Roach, L. L. Berger, M. R. Murphy, and D. M. Moore. 1992. Technical note: mineral deposits on dacron bags during ruminal incubation. J. Anim. Sci. 70:2551-2555.

Van Milgen, J., Larry L. Berger, and M. R. Murphy. 1993a. An integrated, dynamic model of feed hydration and digestion, and subsequent bacterial mass accumulation in the rumen. Br. J. Nutr. 70:471-483.

Van Milgen, J., L. L. Berger, and M. R. Murphy. 1993b. Digestion kinetics of alfalfa straw assuming heterogeneity of the potentially digestible fraction. J. Anim. Sci. 71:1917-1923.

Waldo, D. R., L. W. Smith, and E. L. Cox. 1972. Model of cellulose disapperance from the rumen. J. Dairy Sci. 55:125-129.

Warner, A.C.I. 1981. Rate of passage of digesta through the gut of mammals and birds. Nutr. Abstr. Rev. 51:789-820.

Wilkins, R. J. 1969. The potential digestible cellulose in forage and faeces. J. Agric. Sci. (Camb.) 83:57-64.

Worley, R. R. 1987. Some physical aspects of digestion of forages by steers and binding of ytterbium and hafnium as digesta particles markers. Ph. D. diss. Texas A&M University, College Station.

Wylie, M. J., W. C. Ellis, and J. H. Matis. 1986. Validity of rare earths as flow markers for undigested feed residues. J. Anim. Sci. 63(Suppl. 1):5 (Abstr.)

Wylie, M.J., 1987. The flow of feed residues through the gastrointestinal tract of ruminants. Ph. D. diss. Texas A&M University. Texas A&M University, College Station.

CHAPTER 18

MODELING FORAGE QUALITY CHANGES IN THE GROWING CROP

Gary W. Fick, Paul W. Wilkens, and Jerome H. Cherney

INTRODUCTION

One of the characteristic attributes of a mature science is the capacity to predict. In the past 25 years, forage scientists have made great progress in understanding forage quality regulation by genes and environment and in measuring forage chemistry and animal performance. Still, much remains to be learned, and we are really only in the initial stages of the process of predicting and controlling nutritive quality of forages growing in the field. Further progress promises important economic and ecological benefits through improved forage utilization.

Forages, especially perennial forages, are prime protectors of soil and water, and they also reduce costs by allowing less frequent crop establishment and by having lower requirements for machinery and storage with grazing. The utilization of forages, however, is limited by the wide variability and unpredictability of forage quality. For the manager of high producing livestock, it is easier to plan production systems with concentrate feeds that are much less variable than forages, even though they are more costly in both environmental and economic terms. Forages need to have more predictable and uniform nutritive characteristics to increase their attractiveness and utilization as dietary components.

The starting point for most work on forage quality prediction is the growing crop at the time of harvest. To better control forage quality in the growing crop, we must improve our predictive capabilities. Predictive capability is demonstrated by the construction and application of quantitative models, and thus modeling is the central theme of this chapter. This modeling approach is examined in three ways: a) the search for predictive bases of estimating the quality of growing forages is reviewed; b) the present hypotheses and capabilities of the models are examined, especially their strengths and weaknesses; and c) the future is considered, including probable developments in mechanistic modeling and field applications.

By focusing on the growing crop, two other dimensions of this chapter are defined. First of all, attention is drawn to the dynamic nature of forage quality, i.e., how it is changing with time. Secondly, to predict forage quality change, primary consideration is given to dynamic integrators of environmental,

---

G. W. Fick and J. H. Cherney, Dep. of Soil, Crop and Atmospheric Sciences, Cornell Univ., Ithaca, NY 14853; P. W. Wilkens, Int. Fert. Develop. Center, Muscle Shoals, AL 35662.

physiological, and morphological attributes of the forage system that can be used as predictors. Thus, we give considerable attention to the prediction of the chemical characteristics of forage that are usually used to estimate forage quality and animal performance. We are not looking for an alternative to forage testing but for procedures that can be applied when forage testing is not possible. We also give some attention to how modeling and forage testing can be used in concert to increase the control of forage quality.

THE SEARCH FOR PREDICTORS OF CHANGE IN FORAGE QUALITY

Age

The most frequently cited paper from the early work on predicting forage quality in the growing crop comes from Cornell University (Reid et al., 1959). At the time of publication, feed composition tables based on proximate analysis provided the data to balance diets. Results were imprecise, and there is no doubt that those authors were looking for an alternative to use in ration formulation. For first-cut temperate forages, both perennial grasses and legumes, they found very high correlations between the chronological age of the forage and its digestibility (Table 1, Eq. 1), and their recommendation to use regressions based on age at harvest instead of forage testing were of some influence for at least 15 years.

Reid et al. (1959) cited age-based regressions to predict forage quality from Scandinavia (Jarl and Helleday, 1951; Homb, 1952). The earlier paper reported both chemical fractions and digestion coefficients for those fractions in first growth and aftermath of "clover-rich forage." The mean effect of age on digestible organic matter (Table 1, Eq. 2) was similar to that reported by Reid et al. (1959), but variability among individual experiments was quite high (Figure 1). This early paper also reported data for crude protein (CP) and ether extract and is still a useful reference for current modelers.

Reid et al. (1959) also cited a study from Norway (Homb, 1952) that is likewise of interest to those modeling forage quality. Using forage of predominantly timothy (*Phleum pratense* L.) and red clover (*Trifolium pratense* L.), Homb (1952) reported regression equations based on age at harvest for CP, ether extract, ash, Ca, P, carotene, sugars, lignin, and several other fractions. In most cases, separate analyses were done for clover and grass. First growth and aftermath forage were also distinguished. In some cases, curvilinear relationships were reported (Table 1, Eq. 3). It should be noted that the independent variable X was set to 23 d of age on the day of head emergence in timothy. This probably reduced the year-to-year variability for the first harvest, but not for aftermath, which was regressed on days since the previous harvest. Digestion coefficients were also reported (Table 1, Eq. 4), and again, they showed the robustness of the pattern used by Reid et al. (1959).

The equation of Reid et al. (1959) became known as the "Cornell formula" (Kane and Moore, 1959) and it prompted similar analyses all over the world.

Table 1. Early equations[†] to relate the quality of the first growth of temperate forage species and the chronological age of the herbage.

Source:	(Reid et al., 1959)
Eq. 1:	$Y = 850 - 4.8 \cdot X$; where Y is **digestible dry matter** in g kg^{-1} DM and X is days after 30 April
Accuracy:	"Standard error of the estimate"[‡] = 16.5 g kg^{-1} DM
Region:	Central New York state, before 12 July
Source:	(Jarl and Helleday, 1951)
Eq. 2:	$Y = 828 - 4.4 \cdot X$; where Y is **digestible organic matter** in g kg^{-1} DM and X is days after 31 May
Accuracy:	Not reported, but see Figure 1
Region:	Sweden, between 5 June and 5 July
Source:	(Homb, 1952)
Eq. 3:	$Y = 195.7 - 3.43 \cdot X + 0.0175 \cdot X \cdot X$; where Y is the **crude protein** in timothy in g kg^{-1} DM and X is the age of herbage relative to 23 d of age at timothy head emergence
Accuracy:	Root mean square error (RMSE) = 65.7 g kg^{-1} DM
Region:	Norway, "from early growth to late maturity"
Source:	(Homb, 1952)
Eq. 4:	$Y = 854 - 4.68 \cdot X$; where Y is the **digestible organic matter** in g kg^{-1} DM and X is the same as for Eq. 3
Accuracy:	Root mean square error (RMSE) = 26.6 g kg^{-1} DM
Region:	Same as for Eq. 3
Source:	(Kane and Moore, 1959)
Eq. 5:	$Y = 740.2 - 3.93 \cdot X$; where Y and X are **digestible dry matter** and age, as in Eq. 1
Accuracy:	Root mean square error = 21 g kg^{-1} DM
Region:	Maryland
Source:	(Reid, 1973)
Eq. 6:	$Y = 3.22 - 0.017 \cdot X$; where Y is **daily dry matter intake** in kg (100 kg body weight)$^{-1}$ and X is the same as for Eq. 1
Accuracy:	Not reported
Region:	Same as for Eq. 1

[†] The equations have been converted to SI units. Accuracy is in terms of calibration. True prediction on independent data sets was not tested.
[‡] It is not clear if this is the root mean square error (RMSE), sometimes incorrectly called a standard error, or the actual standard error which is RMSE/$\sqrt{n}$; here n = 94.

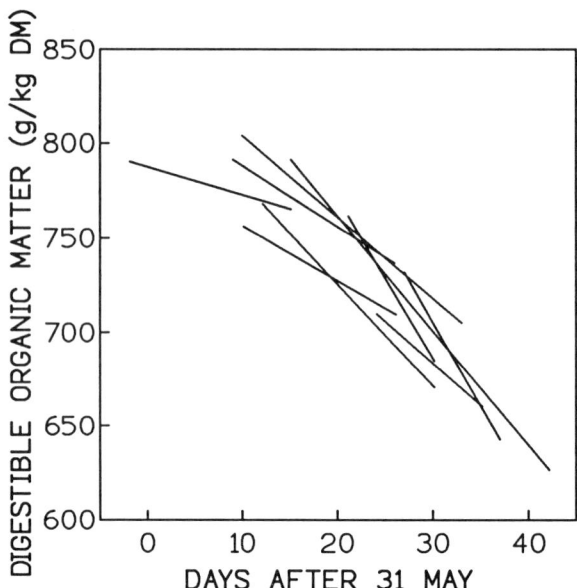

Figure 1. Digestible organic matter as a function of age of the herbage in the spring as presented in the original study of these factors (Jarl and Helleday, 1951). Each line represents a separate experiment.

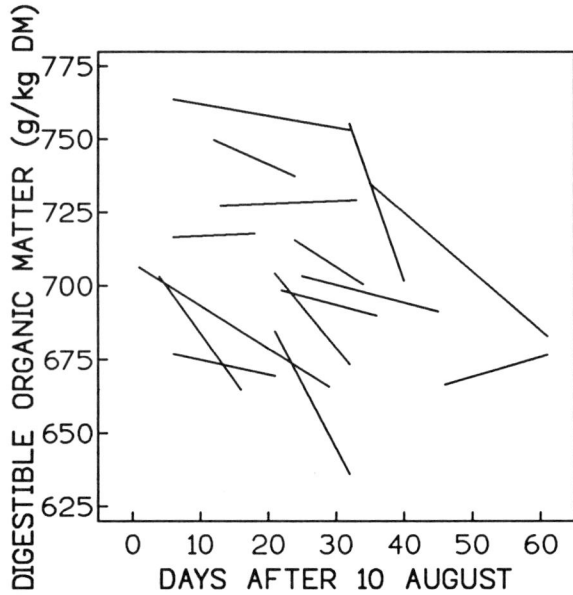

Figure 2. The association of herbage age and digestible organic matter was found to be weak or non-existent after the first harvest (Jarl and Helleday, 1951). Each line represents a separate experiment.

We will not attempt to review that work, but a nearly immediate evaluation from Beltsville (Kane and Moore, 1959) showed that the parameters were not constant (Table 1, Eq. 5). Reid himself summarized the application of the "Cornell formula" (Reid, 1973). He concluded that the slope of the equation (the daily decline in digestibility) became less steep and the intercept lower as the growing season became earlier or warmer. (A lower intercept also indicates that the starting date for counting time could be earlier.) Thus, the relationship was not general but required calibration for each location. Reid (1973) also reported an equation to predict intake (Table 1, Eq. 6), but it was not validated.

Besides reproductive growth, most early studies also considered later growths and found only weak relationships (Figure 2). Thus, relationships based on chronological age can be expected to vary in both time and space, and their attractiveness is their simplicity. They demonstrate that factors regulating the change of quality when the crop is reproductive are correlated with time, but they provide very little information about what the controlling factors might be.

Stage of Maturity

Stage of maturity, or morphological development, is another obvious factor associated with forage quality dynamics. In fact, Homb (1952) named his independent variable "stage of maturity," even though it was based solely on time adjusted to a common value of 23 d at heading of timothy. For red clover, the corresponding value was 35 d at first bud. The distinction between age and stage of maturity was often ignored because of their usual confounding. To be used as a basis of prediction, stage must be quantified. Deinum et al. (1968) were among the first to assign a value 1 to vegetative and 2 to reproductive perennial ryegrass (*Lolium perenne* L.). They reported that an increase in stage reduced digestible organic matter (DOM) by 78 and CP by 69 g kg^{-1} of dry matter (DM). Winch (1971) formulated a five-stage numerical scale for stage of development in forage legumes, and one year later, Gengenbach and Miller (1972) described a quantitative system for alfalfa (*Medicago sativa* L.) staging based on the weighted average of a random sample of stems in vegetative (stage 1), bud (stage 2), flower (stage 3), or seed pod (stage 4) conditions. The approach was refined by Kalu and Fick (1981) to include ten stages. This system has become the basis for most of the current procedures for predicting forage quality from stage of development, and it will be described in more detail later.

Leafiness

In addition to age and stage, leaf fraction of herbage has long been known to be associated with forage quality. Almost 100 years ago, Widtsoe (1897) collected data on leaf fraction and stage of development of alfalfa harvested at approximately weekly intervals. Examination of the data clearly shows the negative correlation of age and stage with CP and their positive correlation with crude fiber. Strong positive correlations are obvious for leaf fraction and forage quality. Widtsoe (1897) concluded that leaves are high in nutritional value and thus important contributors to overall herbage quality.

Along with Widtsoe (1897), a few papers should be considered milestones. Salmon et al. (1925) verified the findings of Widtsoe (1897), but it was not until the publication of Woodman and Evans (1935) that leaf content was stressed as the primary factor controlling quality of alfalfa. These authors also observed that alfalfa leaves are relatively uniform in quality throughout their life. Terry and Tilley (1964) reported essentially constant in vitro digestibility (IVD) for alfalfa leaves (810 to 830 g kg^{-1} DM) while IVD of alfalfa stems dropped from 850 to 560 from 2 April to 25 June. Terry and Tilley (1964) also reported IVD for orchardgrass (*Dactylis glomerata* L.), perennial ryegrass, timothy, and tall fescue (*Festuca arundinacea* Schreb.). Leaf blades of these grasses declined in quality with advancing age and maturity but at a slower rate than leaf sheaths and stems. These relationships appeared to vary somewhat across species. Mowat et al. (1965) confirmed the species interactions the following year. At about the same time, Smith (1964) determined that white clover (*Trifolium repens* L.), a legume whose herbage consists mainly of leaves, had very high digestibility (> 760 g kg^{-1} DM). Brink and Fairbrother (1992) recently confirmed this. Faix (1974) found that birdsfoot trefoil (*Lotus corniculatus* L.) and crownvetch (*Coronilla varia* L.) had nearly constant leaf quality throughout their life. Wilson and Minson (1983) showed that this was also true of the tropical legume siratro [*Macroptilium atropurpureum* (DC.) Urb.].

Leaf content was not used as a predictor of forage quality in any of these studies. However, the data are available to calibrate regression equations, and we have done so with IVD of total herbage as reported by Terry and Tilley (1964). For individual species with limited numbers of data pairs (data not shown), calibrations were sometimes very good. Alfalfa (n=6) had a calibration error (RMSE) for the second order polynomial of 16.8 g kg^{-1} DM, and perennial ryegrass (n=5) had a calibration error of 16.2 g kg^{-1} DM. When data for all five species were pooled (Figure 3), the RMSE increased to 43.5 with a linear fit (42.9 with quadratic), and perennial ryegrass stood out for its high IVD relative to leafiness.

Although leaf fraction is obviously associated with forage quality, its direct measurement in herbage is so tedious that it is impractical except in research. Our experience (Fick and Holthausen, 1975; Kalu et al., 1988, 1990) indicates that standard forage testing is much faster. Our attempt to substitute air-flow segregation of ground forage samples for direct separation of leaf fraction was initially promising (Rao et al., 1987), but subsequent efforts to calibrate the procedure required so many additional hand separations that little was gained (unpublished). However, forage leafiness can be estimated by simulation models, and leaf content may be fundamental in a mechanistic approach to the prediction of forage chemistry and quality.

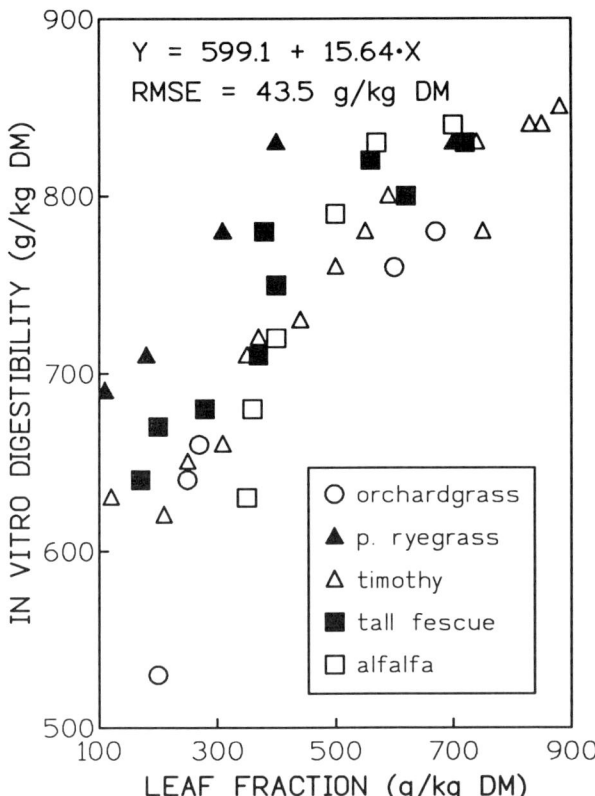

Figure 3. Relationship between in vitro digestibility of temperate forages and the leaf fraction of the herbage (Terry and Tilley, 1964).

Environmental factors

As early as 1959, Minson and Harris speculated that photoperiod, or possibly temperature, might explain the higher digestibility of orchardgrass observed in northern Great Britain when harvests were made on the same day across the whole country. The systematic analysis of environmental effects on forage quality started with Deinum (1966). In this and the more detailed subsequent paper (Deinum et al., 1968), regression analyses were used to calculate the contribution of several environmental factors to forage quality change. In 1974, Deinum and Dirven summarized the whole series of studies and offered a conceptual model of how environment regulates forage quality. The factors they analyzed were temperature, light supply, light duration (photoperiod), fertilization, moisture supply, and harvest management (age of herbage) (Table 2). Both temperate and tropical grasses were considered. With stage of development listed as the only interacting factor in Table 2, the paucity of information about interactions is quite striking.

Table 2. The average effect of a unit increase in several environmental factors on leafiness and digestibility of grasses (Deinum and Dirven, 1974).

Factor	Stage of development	Leafiness	Digestibility		
			Leaf	Stem	Total
		- - - - - g kg^{-1} DM - - - - -			
Temperature	Vegetative	-2.0	-6.5	-10.0	-8.5
(°C)	Reproductive	-7.0	-8.0	-15.0	-13.0
Light supply (MJ m^{-2} d^{-1})		0	+0.5	+0.7	+0.6
Light duration (h d^{-1})		No direct effect found then			
N fertilization (100 kg N ha^{-1})		+20.0	None	None	None
Other nutrients		Generally no effect			
Moisture supply		Generally small increases			
Age of herbage	Vegetative	-2.0	-1.4	-1.8	-1.5
(d)	Reproductive C$_3$ spp.	-9.0	-2.5	-4.5	-4.0
	Reproductive C$_4$ spp.	-8.0	-4.0	-5.0	-5.0

Their results were quite informative regarding the relative impact of various factors. For example, they showed that nitrogen (N) fertilization had the greatest influence on leaf fraction and that temperature had the greatest effect on digestibility (Table 2). Increased temperature decreased digestibility while increased light supply increased digestibility. Wilson (1982) and Wilson and Minson (1983) also reported that an increase of 1°C over the normal range for growth decreased digestibility of temperate legumes by 2.0 g kg^{-1} DM and digestibility of tropical legumes by 2.3 to 2.8 g kg^{-1} DM. Their estimate for the effect of temperature on digestibility of tropical grasses [-6.0 g kg^{-1} DM (Wilson, 1982)] showed a slightly faster rate of decline than that of Deinum and Dirven (1974) (Table 2).

Deinum et al. (1981) later investigated the effects of photoperiod. They harvested first growth timothy at 1- to 2-wk intervals within a latitude range of 51 to 69°N. The maximum June photoperiod ranged from 17.2 to 24.0 h. As photoperiod increased, growth rate increased and digestibility declined more rapidly. However, at the same stages of development, digestibility was 80 to 170 g kg^{-1} DM higher at the longest as compared to the shortest photoperiods. Thus, for reproductive timothy, the effect of 1 h d^{-1} more light could be as much as 20 or 25 g kg^{-1} DM of increased digestibility.

Table 2 reflects the state of knowledge that existed when various workers started to develop more complex models of forage growth and quality: a) the age factor was most thoroughly studied, b) stage needed better quantification, c) leafiness was promising but not practically measurable, and d) key environmental factors had been identified individually, but their interaction with each other and with the physiology of crop developmental stage was still poorly understood.

Early models

In the 1960's and especially in the 1970's, computer models of forage crops drew attention to the work outlined above as modelers searched for a basis for simulating forage quality. It is beyond the scope of this chapter to consider the methods of crop growth modeling. There are many good discussions on that subject (Penning de Vries and Laar, 1982; Wisiol and Hesketh, 1987; Hanks and Ritchie, 1991). We should note, however, that they simulate the passage of time by computing the value of state variables for specified time steps, usually 1 h or 1 d. For example, one early model kept track of daily changes in pasture plant density (plants m^{-2}), herbage yield, and herbage height (Smith and Williams, 1973). The SIMED alfalfa model estimated hourly yields of leaves, stems, roots, and nonstructural carbohydrate found in each plant part (Holt et al., 1975). Plant growth calculations were typically based on estimates of photosynthesis, which in turn depended on leaf area index estimated from leaf yield. Thus, in the modeling context, variables that were difficult to get from the field were routinely computed and readily available for hypothetical estimates of forage quality. The challenge was to find in the literature, or collect from the field, appropriate data to formulate and validate models.

One of the first of the more complex models to calculate forage quality dealt with hay harvest strategies. Increasing yield and decreasing digestibility were considered under conditions of uncertain rainfall patterns and varying machinery complements (Millier and Rehkugler, 1972). The "Cornell formula" was used to calculate the digestible DM of the first cutting. The second and third cuttings were given constant digestibilities of 600 and 685 g kg^{-1} DM, respectively. The approach was simple and reflected the predominance of studies of age or calendar date as the controller of quality. Edelsten and Newton (1975) also related digestibility of British pastures to day of the year.

Smith and Williams (1973) used a slightly more complicated approach where digestibility fraction (D) of Mediterranean-zone annual range was computed in the following manner:

$D = 0.802 - 0.17 \cdot PSIO - LEV$
$PSIO = EXP((0.0005 - 0.00096 \cdot H) \cdot HW)$
$LEV = 0.02 \cdot I$

where PSIO is silica fraction of the herbage, LEV is reduction in digestibility due to intake, H is height of the pasture (cm), HW is herbage on offer (kg ha^{-1}), and I is daily herbage intake (kg ha^{-1}). Higher silica indicates closer grazing and lower quality. The authors derived these relationships from their own work in Australia (Smith and Williams, 1973). In essence, they described an n-dimensional response surface, here with the dimensions of D, PSIO, H, HW, and I.

Many early pasture models calculated digestibility of live and dead material separately with functions based on age and herbage mass. The model of Smith and Williams (1973) was soon refined by Rice et al. (1974). They classified herbage into 90 living and 250 dead categories, one per day of potential grazing season. The weighted average of herbage mass in each category times its quality (digestibility or N concentration) gave the quality of the feed on offer. From the onset of germination, digestibility of age categories declined linearly from 800 to 200 g kg^{-1} of DM with this model. Although this concept was simple, the model was computationally very sophisticated because the mass in each age category could be varied by the patterns of germination, forage growth, and selective grazing.

There were many forage production models published in that era, all with quality components that were relatively simple in concept if not in computational mass (Seligman, 1975). Our first attempts at modeling forage quality were aimed at incorporating a more realistic and dynamic description without amassing a large number of state variables [there were 340 in Rice et al. (1974)]. From the large body of data for perennial ryegrass/white clover pastures in New Zealand, we constructed a model (CANPAS) with a function showing the potential daily decrease in digestibility of living herbage (Fick, 1978). Values ranged from about 8 g kg^{-1} DM d^{-1} in summer to almost zero in midwinter. The relationship was specific for the Southern Hemisphere, but it can be generalized for north or south latitudes as a function of time from the summer solstice. The actual computed digestibility was constrained to lie between maximum and minimum possible values [as with Rice et al. (1974)]. The model produced reasonable simulations of stockpiled (autumn-saved) pastures showing the known trade-off between higher yields from earlier accumulation dates and higher quality from later accumulation dates (Figure 4). Selective grazing and decomposition of dead herbage were also incorporated into this model (Fick, 1978; Fick, 1980).

Working at the same time, Edelsten and Corrall (1979) produced a model almost identical to CANPAS in concept. They fit their function for digestible organic matter (DOM) to a prediction equation:

DOM = 736 + 25·sine(OMEGA) + 16·cosine(OMEGA) - 1.08·J - 7.15·Y
OMEGA = 2·PI·I/365; R^2 = 0.656 (for calibration)

where J is the days since last harvest, Y is the amount harvested (Mg ha^{-1}), PI = 3.1416, and I is the day of the year. The variable OMEGA positions their function relative to the summer solstice.

There are many other examples, including more recent ones, of calculating forage quality in the context of a forage/livestock simulation model, but to our knowledge they have all been as descriptive and non-mechanistic as the examples presented above. It is also fair to say that most of them have been time dependent, although the better ones generalize the relationship to the summer solstice. It was in this context that we began a program to expand the bases for forage quality prediction.

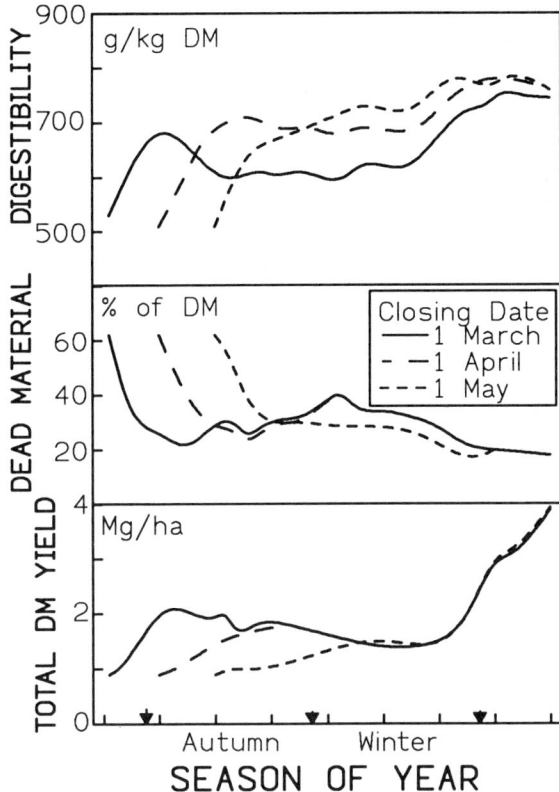

Figure 4. The CANPAS pasture model (Fick, 1978) simulated the yield, percent of dead material in the herbage, and herbage digestibility of New Zealand (Southern Hemisphere) pastures stockpiled in the autumn. "Closing date" refers to the date stockpiling was initiated, and the model predictions demonstrate that starting too early leads to high dead material and low quality while starting too late leads to higher quality but yields that are too low.

CURRENT APPROACHES AND THEIR EVALUATION

Age- and Weather-Based Predictions

Our work on prediction of quality of growing forage crops started in the 1970's when the senior author became a member of a national research team charged with the development of an integrated pest management program for the alfalfa weevil (*Hypera postica* Gyll.). Computer modeling of the crop, the pest, and its natural enemies, hay crop harvesting, and the economic components of the system were emphasized. Because of this beginning, alfalfa is the forage crop that has received most of our attention. From this beginning also came an immediate interest in the economic value of forage that had undergone varying amounts of leaf loss caused by insects and, consequently, that was variable in quality. It soon became obvious that early harvesting was an effective control for the insect pest, and we were assigned the additional responsibility for describing the harvest-date effect on forage quality and value.

Because this project was national in scope, our review of the literature gave us concepts to pursue, but the regression models we found were too site-specific to satisfy our needs. We chose to work with three bases for prediction: weather/environmental factors, stage of development, and leafiness. The environmental approach seemed expedient. Weather data were already required to drive the alfalfa growth, insect development, and forage harvesting models. Weather input data also gave our models generality, allowing the same models to be applied across the whole country. By the early 1980's, we had enough data to develop regression equations for use in alfalfa systems models (Onstad and Fick, 1983; Onstad and Shoemaker, 1984).

In the first of two studies, Onstad and Fick (1983) used accumulated air temperature heat units (or growing degree days, GDD) above a 5°C base, accumulated hours of light, and the chronological age of herbage as the independent variables. All variables were accumulated from the last spring occurrence of a mean daily temperature of 2°C, or from the previous harvest in the case of regrowth. Six regression models were considered that included the natural logarithm and the square and cube of up to two variables. Calibration data came from our work with alfalfa in central New York state in the 1970's, and CP, in vitro true digestibility (IVTD), and leaf fraction in the herbage were predicted. As far as we know, this was the first study where weather measurements were actually used as predictors of forage quality. Of 16 equations selected for validation with an independent data set, ten had only temperature as the independent variable, two had only light (leaf quality equations), and four had both. To our surprise, the multiple regression procedure that we used never selected age as a predictor of quality.

We also compared first, second, and third cuts of alfalfa and found the same equations applying to all. There was one exception. Leaf fraction of the first growth was considerably lower than regrowth at similar values of accumulated heat or light (Fick et al., 1988a). Previously, we had used simple models which calculated total herbage yield without separate treatment of leaves and stems (Fick and Onstad, 1983a, 1983b). We could more accurately predict concentrations of CP and IVTD from temperature data if we first calculated leaf fraction with the temperature-based equations and then quality of leaves and stems separately. Tests of those models on independent quality data showed RMSE (true prediction error) of 45 and 32 g kg^{-1} DM for IVTD and CP, respectively.

Our second analysis of environmental factors as predictors of forage quality (Fick and Onstad, 1988) was based on a national data set collected specifically for this purpose with the cooperation of colleagues in six states (California, Georgia, Kentucky, New Mexico, New York, and Wisconsin). Again, only alfalfa samples were analyzed, but the list of predicted quality measures was expanded to CP, IVTD, neutral detergent fiber (NDF), acid detergent fiber

Table 3. The best fitting equations developed from a national data set to predict several forage quality measurements from either chronological age of the herbage in days or the accumulated average daily air temperatures above a 5°C base as reported by Fick and Onstad (1988).

Quality measurement $g\ kg^{-1}\ DM$	Equation			RMSE of calibration $g\ kg^{-1}\ DM$	R^2
Age-based equations					
CP	Y = 376.	$- 6.88 \cdot X$	$+ 0.0595 \cdot X \cdot X$	30.2	0.56
IVTD	Y = 979.	$- 9.36 \cdot X$	$+ 0.0884 \cdot X \cdot X$	41.8	0.48
NDF	Y = 171.	$+ 8.68 \cdot X$	$- 0.0700 \cdot X \cdot X$	48.4	0.50
ADF	Y = 125.	$+ 8.00 \cdot X$	$- 0.0728 \cdot X \cdot X$	39.9	0.45
ADL	Y = 16.5	$+ 2.02 \cdot X$	$- 0.0173 \cdot X \cdot X$	10.4	0.48
Temperature-based equations					
CP	Y = 850.	$- 102. \cdot \ln(X + 1.)$		31.1	0.52
IVTD	Y = 1163.	$- 138. \cdot \ln(X + 1.)$		37.1	0.58
NDF	Y = 205.	$+ 0.335 \cdot X$		44.0	0.57
ADF	Y = -510.	$+ 131. \cdot \ln(X + 1.)$		33.2	0.61
ADL	Y = 0.597	$+ 0.175 \cdot X$	$- 0.000096 \cdot X \cdot X$	8.3	0.67

(ADF), and acid detergent lignin (ADL). In addition to weather variables for accumulated average daily temperature and hours of light used in the previous study, we also considered latitude, herbage age, and stage of development (to be discussed later) as the independent variables. Alfalfa samples were separated into leaf and stem fractions so that equations could be calibrated for leaves and stems as well as total herbage. Equations were separately developed based on age, heat sums, and stage of development. Then, the best fitting equations were selected by forward stepwise multiple regression using a pool of independent variables made up of those listed above plus their squares, cubes, and natural logarithms. We compared equations for first growth and aftermath and found no significant differences, so the results reported (Fick and Onstad, 1988) apply to all alfalfa harvests during the year.

One of our objectives was to compare equations based on chronological age of herbage with those based on temperature (Table 3). We were interested because of the widespread use of age as a predictor and the predominance of temperature over age in our first study (Onstad and Fick, 1983). Based on calibration errors (RMSE) for our best fitting equations, heat sums were slightly superior to age as a predictor of all forage characteristics tested except CP. More details, including leaf and stem equations not repeated here, are given in the original paper (Fick and Onstad, 1988).

Some additional interesting points emerge from the equations of Table 3. Probably of most importance for subsequent discussion are the relatively large values for RMSE, which reflect the large amount of variability in the original data. This variability apparently masked differences due to season of the year (first growth vs. aftermath) and cultivar (there were six cultivars represented in the forage samples). Also of interest is the fact that all age-based equations are curvilinear. That contrasts to the predominantly linear equations in Table 1; however, many early equations from Norway (Homb, 1952) were also curvilinear (Eq. 3, Table 1), and we used a greater range in age to develop our equations (Fick and Onstad, 1988) than did many earlier studies. Even though the range of input data for calibration was large, extrapolation is a problem with these equations. Polynomial forms are inaccurate at high values of the independent variable because they predict unrealistic quality improvement. Logarithmic forms were selected because they appear to be relatively robust at high values of X, but the intercepts of those equations become inaccurate as X approaches zero. When these equations are used to predict forage quality, they must be appropriately constrained (Wilkens and Fick, 1988).

In the Atlantic region of Canada, Bootsma (1984) had also considered temperature as a basis for forage management. On average, alfalfa reached early bud, late bud, and early bloom at 350, 400, and 450 GDD ($5°C$ base), respectively. Red clover was at early bloom at about 450 GDD. Timothy showed variability according to cultivars: 'Clair' reached 50% head at 400 GDD while 'Champ' and 'Climax' took about 450 and 500, respectively. Although simple rules based on weather data could be derived from these relationships, Bootsma (1984) noted that "Accumulated GDD are only an approximate method of predicting forage crop maturity. Variation in GDD requirements of forage crops among seasons and locations is evidence that an improved system needs to be developed."

Temperature-based quality prediction equations are attractive for weather-driven simulation models, and they have been used in that way (Rotz et al., 1989). The relatively high prediction error is a limitation for such use, but it may be possible to correct this problem by more accurate determination of the base temperature. The $5°C$ value was chosen by Selirio and Brown (1979) and repeated by us (Onstad and Fick, 1983; Fick and Onstad, 1988) and Bootsma (1984) primarily because of the availability of temperature data sets collected by weather services. Moline and Wedin (1969) used $7.8°C$ soil temperature as a base. Jeney (1972) and Fagerberg (1988a) selected $0°C$ as the base temperature. Sharratt et al. (1989) more recently showed that base temperature is variable depending on growth environment. Determination of a "least-squares" base temperature (Warrington and Kanemasu, 1983) for each harvest might improve accuracy of prediction equations, but they might not be general from one region to another.

Light was the other environmental factor identified by Deinum and Dirven (1974) as important in forage quality regulation. Light duration sometimes appeared as a factor in our regression analyses (Onstad and Fick, 1983), but our results were in general agreement with those of Deinum and Dirven (1974), showing light had a much weaker effect than temperature. However, two recent models have successfully used light duration as a predictor variable. Thompson et al. (1989) used multiple regression with weather measurements as independent variables to predict the digestibility of first-cut grasses in England. Heat units (or GDD, accumulated average daily air temperatures above a 0°C base from 1 January), photoperiod change (average change in hours of light in the previous week), and sunshine duration (hours of bright sunshine in the previous week) were generally significant predictors, but precipitation (drought stress) was not, presumably because there was little drought stress during the first growth of spring. They showed that regionally calibrated models were somewhat more accurate than equations for all of England. Overall RMSE values (for true prediction) were 21 and 19 g kg^{-1} DM for national and regional equations, respectively.

Sunshine duration was also used to predict a forage quality characteristic by McGechan (1990). The fraction of water soluble carbohydrates (WSC) in herbage was needed for a model of ensiling. Based on data for perennial ryegrass, WSC concentration was modeled to increase linearly from 120 to 240 g kg^{-1} DM from 4 wk before head emergence to 4 wk after head emergence, becoming constant at the higher value thereafter. To this baseline value was added an amount of WSC dependent on sunshine:

Total WSC = baseline WSC + 2.7·(H_1 + 0.16·H_2)

where H_1 and H_2 are hours of bright sunshine for yesterday and the day-before-yesterday, respectively. At continuous light, this would boost WSC by about 75 g kg^{-1} DM, and this much carbohydrate would dilute the apparent concentration of other chemical fractions in the forage. Supporting physiological evidence comes from Rickman et al. (1985) who reported that wheat (*Triticum aestivum* L.) appeared to have a carbohydrate pool dependent on solar radiation amount and buffered to change in about 5 to 7 d when light levels were permanently changed. Using temperature instead of light, Greenfield and Smith (1973) showed that nonstructural carbohydrates in alfalfa herbage also reached equilibrium in a new environment in about 6 d. Nonstructural carbohydrate levels in forage are not yet routinely incorporated into models, but they appear to be important.

Age- and weather-based prediction models of forage quality seem to be practical and reasonably accurate where a body of data is available for calibration. It also seems that greater accuracy is achieved with more localized calibrations. Temperature or heat-sum predictions have been more robust than age-based predictions, and they might be even better if more accurate base temperatures could be specified. Use of hours of bright sunshine in recent days as a predictor of forage quality lends mechanistic realism to even simple quality prediction models, but this refinement is in need of wider testing and requires weather records for cloudiness.

Mean Stage

As indicated above, the negative correlation of developmental stages with forage quality was widely recognized, but old qualitative staging systems did not facilitate prediction of forage quality. Numbers were associated with stages of crop development at least as far back as 1941 for cereals (Feekes, 1941), and a similar system for forage legumes was published by 1971 (Winch, 1971). These systems were based on the classic developmental stages of vegetative, bud (for legumes) or stem elongation (for grasses), flowering (for legumes) or head emergence (for grasses), and seed development, each divided into decimal substages representing a sequence of observable morphological events. Depending on how closely observations were to be made, the number of stages could proliferate. For example, one system for grasses described at least 51 stages and substages (Simon and Park, 1983). Other systems were very simple, based on the length of emerged grass inflorescence (Hides et al., 1983) or tiller length (Warndorff and Dovrat, 1987). Another shortcoming was a typical discontinuity in the numerical scale as the crop passed from one main stage to another. For example, sequential substages in the early legume system (Winch, 1971) were stage 2.2 (visible buds), 2.3 (visible color in buds), and 3.1 (first flower to 10% bloom). Incremental changes were 0.1, 0.1, and 0.8, respectively.

In addition, traditional descriptions of stage of development for forages applied only to the most mature shoots in a stand rather than the average stage of all the harvestable shoots.

Gengenbach and Miller (1972) corrected this last problem by calculating a mean stage of alfalfa with a random sample of all shoots in the stand, but we found their four-stage system was inadequate at early stages of regrowth (Fick and Liu, 1976). The breakthrough for linking stage to quality came when Bernard Kalu used a unique approach to define the stages of alfalfa development (Kalu, 1976). Knowing that forage quality was declining in approximately linear fashion in the spring (Reid et al., 1959), Kalu wanted to linearize the relationship between stage and time. He thus defined the sequential stages (Table 4) as the most developed shoot in ten harvests of the first growth made at weekly intervals from 21 May to 23 July 1975 at Ithaca, NY. [The vegetative stages were defined for convenience based on lengths of 15 and 30 cm (6 and 12 inches)]. Kalu was also careful to calculate a mean stage representative of all harvestable shoots in the herbage. The resulting relationship between mean stage and time was not perfectly linear (Kalu and Fick, 1981), but the mean stage slowed its increase at about the same time that forage quality showed its decline with resulting high correlations between quality and stage (R^2 = 0.98 and 0.99 for IVTD and CP concentrations, respectively).

Table 4. The ten stages of alfalfa development used for determining mean stage of development. See Fick and Mueller (1989) for details.

Stage name	Stage number	Stage description
Early vegetative	0	No reproduction, shoots ≤ 15 cm long
Mid-vegetative	1	" " shoots 16-30 cm long
Late vegetative	2	" " shoots ≥ 31 cm long
Early bud	3	1 to 2 nodes with visible buds
Late bud	4	≥ 3 nodes with visible buds
Early flower	5	One node with an open flower
Late flower	6	≥ 2 nodes with open flowers
Early seed pod	7	1 to 3 nodes with green seed pods
Late seed pod	8	≥ 4 nodes with green seed pods
Ripe seed pod	9	Nodes with mostly brown seed pods

This remarkable finding prompted additional research and the eventual publication of a manual on alfalfa staging (Fick and Mueller, 1989). Other papers extended the approach to additional quality characteristics (Kalu and Fick, 1983) and to a national data base (Fick and Onstad, 1988). One misinterpretation of the new approach relating to descriptive terminology is common. The new system is not a numerical replacement of descriptive terms. For example, one should not say that an alfalfa stand at "early flower" is at stage 5. The reason for using numbers is to calculate a mean stage that represents all shoots in the canopy. Two methods of calculating mean stage have been used, one with average stage weighted for number of shoots in each stage (Mean Stage by Count or MSC), and the other with average stage weighted for dry weight of shoots in each stage (Mean Stage by Weight or MSW). The only important difference between the two methods in time and labor requirement is for the oven drying and weighing used only with MSW.

Our early work indicated that MSC would fail to show progress in development when basal buds begin to elongate in maturing stands, but MSW could be applied to all stands (Kalu and Fick, 1981). Thus, we generally measured MSW in our own studies. Data collected by Mueller (1990) showed the frequency distributions of stems in stages at about "first flower" (MSW of approximately 3.0) and at "full bloom" (MSW of approximately 5.0). Diversity of individual shoots, seasonal variability, and non-normal distributions are clear (Figure 5) and emphasize the increased precision gained by averaging for the canopy instead of only noting the most developed stage.

Our own work in the formulation of prediction equations led to the national study mentioned above (Fick and Onstad, 1988) in which MSW was included as one of the predictors (Table 5). It was our conclusion that MSW was the most accurate of any single basis for prediction (compare Table 3).

The strong correlation between mean stage and alfalfa forage quality was soon confirmed by others (Buxton et al., 1985; Buxton and Hornstein, 1986; Sanderson and Wedin, 1988), including good values for MSC (Buxton et al., 1985; Buxton and Hornstein, 1986). Partly in response to this, we developed an equation to convert from MSC to MSW so the national equations (Table 5) could be used even with MSC data (Mueller and Fick, 1989):

$MSW = 0.456 + 1.153 \cdot MSC$, $r^2 = 0.982$, $n = 569$, $RMSE = 0.311$

Table 5. The best fitting equations developed from a national data set to predict several forage quality measurements from mean stage by weight (MSW) as reported by Fick and Onstad (1988).

Quality measurement $g\ kg^{-1}$ DM	Equation	RMSE of calibration $g\ kg^{-1}$ DM	R^2
CP	$Y = 371. - 75.8 \cdot X + 7.60 \cdot X \cdot X$	31.1	0.64
IVTD	$Y = 1000. - 163.0 \cdot \ln(X + 1.)$	31.5	0.70
NDF	$Y = 134. + 177.0 \cdot \ln(X + 1.)$	36.6	0.70
ADF	$Y = 125. + 131.0 \cdot \ln(X + 1.)$	29.1	0.70
ADL	$Y = 19.1 + 32.3 \cdot \ln(X + 1.)$	9.1	0.59

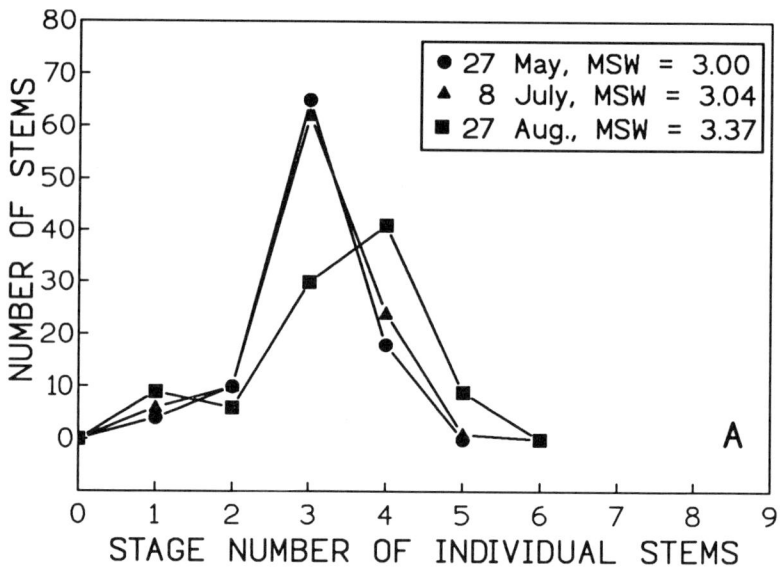
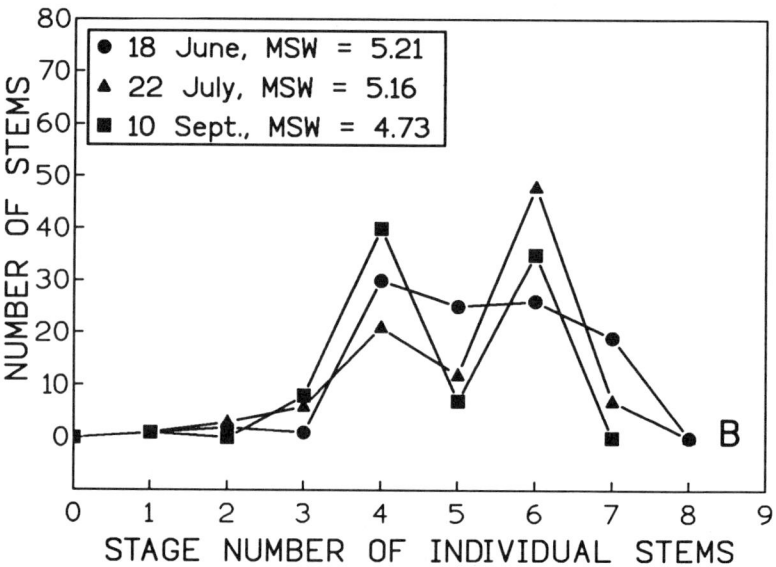

Figure 5. The number of stems in each alfalfa stage category at different times of the year for stands that were at about "first flower" (A) or "full bloom" (B). Especially note the heterogeneity at about the same apparent stage with the older canopies in (B).

Our data were from New York state. Sanderson (1992b) validated the equation for Iowa and Texas and found a small bias but low prediction error (RMSE = 0.288). His conclusion was that the equation may apply over a wide geographic and environmental range.

Our original intention was to develop a research-oriented procedure that would provide insight into quality regulation and concepts for the computer simulation of forage quality in alfalfa. The correlations between mean stage and alfalfa forage quality were for us unexpectedly high, and we redirected our efforts to see if mean stage could be used to predict forage quality while scouting alfalfa fields. Because it is faster and simpler, MSC was chosen for these studies. The information obtained would be useful both for timing the first harvest of the season to optimize forage quality and in supplemental economic calculations about the use of early harvest as a control tactic in integrated pest management of alfalfa.

Our first attempt to do this involved a simplified staging system with only five stage categories (Fick and Sniffen, 1985). With half the stages, prediction errors doubled, making the approach too inaccurate to be useful. One possible problem was that the variable amount of grass in the forage mixtures we encountered on farmers' fields in our field test might be altering the stage-quality relationship for alfalfa. Mueller (1990) tested this possibility with timothy and smooth bromegrass (*Bromus inermis* Leyss.) in mixtures with alfalfa and concluded that mixtures did not affect the alfalfa-stage to alfalfa-quality relationship. (Of course, grass in the mixture reduced the quality of the total herbage.) Our second attempt at field use involved the standard MSC system with all ten stages (Allen and Fick, 1990), to be discussed below.

One effect of the definition of alfalfa mean stage has been the extension of the concept to other forage species. Buxton and Marten (1989) showed that the correlations of several forage quality characteristics in four temperate grass species were better with age (r^2 = 0.88 to 0.95) than with GDD (r^2 = 0.77 to 0.92) or stage (r^2 = 0.74 to 0.89). Stage was based on the method of Simon and Park (1983), and they concluded that it would be necessary to consider the proportion of stems at each stage (i.e., calculate a mean stage) to improve the quality-to-stage relationship. Nordkvist and Åman (1986) also found that a discontinuous staging scale without mean stage showed no better correlation to alfalfa IVD than herbage age. In Sweden, Fagerberg (1988a,b) used a 24-stage system of Torssell (1984), modified to give a continuous scale and allow calculation of MSW for alfalfa, red clover, and timothy. Although the stages were unique, the calculation of MSW used the computation method of Kalu and Fick (1981). The best relationship between MSW and metabolizable energy or CP was for alfalfa (R^2 = 0.84 to 0.89); corresponding values for red clover and timothy were 0.51 to 0.91 and 0.35 to 0.92, respectively (Fagerberg, 1988a, b).

It seems that stage of development may be a better predictor of quality in C_3 grasses than in C_4 grasses. Using a grass staging system based on the extent of head emergence (Hides et al., 1983), a promising predictive relationship between developmental stage and forage quality was found for Italian ryegrass (*Lolium multiflorum* Lam.). Likewise, staging systems used for C_3 cereals showed good correlations with CP and IVDMD for wheat in the spring in Arkansas (West et al., 1991), but no relationship was found for a pennisetum hybrid (*Pennisetum americanum* Schum. x *P. purpureum* Schum.), a C_4 grass species, with stage based on tiller length (Warndorff and Dovrat, 1987).

Ohlsson and Wedin (1989) directly modified the alfalfa stage descriptions of Kalu to develop a new system for red clover. Moore et al. (1991) also formulated a new staging system for perennial forage grasses. Like traditional schemes for grasses, it includes substages for up to 50 categories of individual shoots. The numerical increment between grass substages is constant, and thus the procedure should describe the smooth development of many different species. At this writing, we do not know that MSC or MSW values based on the system of Moore et al. (1991) have been applied to the prediction of grass forage quality. It is probable that stage-to-quality associations will be more successful than age-to-quality relationships only in those species where stage and quality respond to the environment in the same way. When plant phenology is strongly programmed by photoperiod, which is typical of many species, but plant chemistry is more responsive to temperature, the association of stage and quality will be broken by unusual weather. Alfalfa may be relatively unique in the degree to which the same environmental factors drive maturation and the decline of quality, but even here we have evidence that unusual weather alters the situation (see below).

Leafiness

Despite the tediousness of measuring leaf fraction in forage samples noted above, we analyzed leafiness as a predictor of forage quality in both the work with Kalu et al. (1988, 1990) and for the national data set previously described (Fick and Onstad, 1988). This work confirmed, at least for alfalfa, the high correlation between leaf fraction and several measures of forage quality.

Kalu et al. (1988, 1990) returned to the original New York data set used to develop the MSW equations and regressed the various forage quality characteristics on leaf fraction of the alfalfa herbage. A linear fit was found for all five characteristics with the minimum r^2 value in excess of 0.82. Except for IVTD, the calibration errors measured by RMSE were on the order of those for MSW (Table 6). In fact, ADF and ADL showed somewhat better fits to leafiness than to MSW.

With the national data set, a multiple regression procedure was used to select the best fitting equation with up to three predictors from a pool of 15. The procedure stopped with two predictors because additional variables never improved the regression. For all five quality characteristics (Table 6), leaf fraction of the herbage was selected. For all except ADL, MSW was also selected. For ADL, leaf fraction and the heat sum (GDD with a 5°C base) were selected as the most important predictors. Comparison of the calibration errors (Table 6) showed that including leaf fraction as a predictor always improved the regressions in the highly variable national data set (Fick and Onstad, 1988).

Table 6. Regression equations for several alfalfa quality measurements on alfalfa leaf fraction (LF) (taken from Kalu et al., 1988, 1990) and comparison of the associated calibration errors (RMSE) for the same New York data set fit to mean stage by weight (MSW) or a national data set fit to MSW or LF+MSW (Fick and Onstad, 1988).

Equation	LF RMSE	Comparative RMSE		
		New York MSW	National data MSW	LF+MSW
------------------ g kg^{-1} DM -------------------				
CP = 8.3 + 380·LF	29.6	24.5	27.4	22.6
IVTD = 498.4 + 500·LF	35.0	19.0	31.5	23.4
NDF = 700.7 - 540·LF	27.1	22.0	36.6	27.9
ADF = 565.3 - 460·LF	19.6	25.0	29.1	20.6
ADL = 129.8 - 120·LF	8.2	8.4	9.1	†

† The best equation for ADL involved LF and growing degree days (GDD) with an RMSE of 6.2 g kg^{-1} DM.

In summarizing our work and related studies done by others in which attempts have been made to develop regression models to predict forage quality, heat summation and mean stage have received sufficiently widespread attention to now be regarded as sound bases for prediction. It also appears, at least for legumes, that leafiness is an important predictor if it can be determined. Unfortunately, that remains a serious limitation. The quantification of these factors has been an important step forward in the attempt to predict forage quality while the crop is growing in the field.

THE FUTURE OF QUALITY PREDICTION MODELING

Attempts to predict nutritional quality of forages will probably continue to develop along two lines. Both will depend on forage chemical analyses for validation, and both will be applied in circumstances where forage testing is not available or feasible. One of these circumstances is the practical world of field scouting and everyday decision making regarding cutting or grazing management. More and more sophisticated management aids are becoming available in the field, and they will increasingly require estimates of forage quality. The technology of NIRS has a place here, but it will probably remain impractical for most short term, tactical decisions in crop management. Thus, it is likely that simple quality prediction models based on available environmental data and (or) crop characteristics will have more utility. The other situation requiring forage quality models is forage systems analysis, which often involves prediction into the future and hypothetical systems. Models will be needed to compute reasonable quality values for simulations applied to forages. Ideally, these models should be mechanistic explanations of the dynamics of tissue chemistry in the growing forage crop.

Practical Applications of Forage Quality Models

One forage management decision where quality prediction may be used is the timing of harvests. Before forage quality models can be used in this way, it is necessary to define a goal, a plant ideotype. Going back to Mitchell (1960), most ideotypes for forage species were based on characteristics associated with yield and persistence. More recently, ideotypes have stressed forage quality, and indicated potential to improve quality without sacrificing yield in alfalfa (Coors et al., 1986; Lenssen et al., 1991), timothy (Ames-Gottfred et al., 1990), and smooth bromegrass (Casler, 1986). In general, ideotypes should have high digestibility through a reduction in fiber content or a reduction in the cellulose and (or) lignin components of fiber. While these are excellent goals for plant breeders now and in the future, field applications for timing harvests require that the ideotype descriptions be a) discrete and b) currently attainable.

The desirable ideotype for a particular forage species depends on its intended use. We will limit the following discussion by only considering ideotypes for high-producing dairy cattle. In a diet including corn silage, the alfalfa ideotype has the following composition: NDF = 400, ADF = 300, and CP = 200 g kg^{-1} DM (Lacefield, 1989). This ideotype is diet-dependent. For example, a diet with alfalfa haylage as the sole forage source (no corn silage) has an optimal alfalfa NDF of 450 g kg^{-1} DM (L.E. Chase, 1993, personal communication). Annual and perennial grass ideotypes have the following composition: NDF = 500, ADF = 280, and CP = 200 g kg^{-1} DM (J.H. Cherney, unpublished). Since these forage ideotypes are currently attainable and dependent on crop management, what we need is a prediction procedure to estimate the timing of these nutrient concentrations.

These ideotypes are based on the value of the forage to the ruminant. For low to moderate milk production per animal, an alfalfa NDF value considerably higher than 400 g kg^{-1} DM is acceptable. Relative forage value of alfalfa for high producing dairy cattle declines rapidly as it moves away from the ideotype (Wilks et al., 1993). Lower fiber values risk animal health problems (Fick and Sniffen, 1985). Forages with greater than 550 g kg^{-1} DM as NDF make it difficult to balance a diet with commonly available ingredients and result in more off-farm inputs, thus compounding any nutrient management problems. As animal scientists improve animal production potential, the importance of attaining the forage ideotype increases. This indicates that predicting forage quality in the future will only become more critical to forage producers.

Crop consultants and (or) producers will want predictive schemes that are fast and easy to use. Thus, models based on leafiness are not practical, and MSW models are also seen by many as too time consuming for on-farm use. This led to on-farm testing of MSC as a forage quality predictor for alfalfa in New York (Allen and Fick, 1990). Predictions were found to be biased, requiring recalibration for each harvest (G.W. Fick, unpublished). Sanderson (1992b) also tested the national equations for predicting alfalfa forage quality (Fick and Onstad, 1988) using data sets from Iowa and Texas. Prediction equations were again found to be biased, indicating that calibration for specific environments and (or) periodic recalibration was necessary. Prediction of quality also is complicated by any deviation from optimal alfalfa growing conditions, such as drought or phosphorus deficiency, where changes in stage of development only partially account for changes in fiber levels (Halim et al., 1989; Sanderson, 1992a).

Mean stage count in combination with other predictors has been suggested as a practical method of predicting alfalfa quality in Nevada (Lewis and Balliette, 1991). An equation for CP used GDD (with a 4.4°C base temperature), plant height, day of the year, and MSC as variables ($r^2 = 0.72$), and an equation for ADF used GDD, plant height, and day of year ($r^2 = 0.74$). These equations were not validated. Accumulated temperatures appeared to be the most important variable in the Nevada study, but there was only one year of data. Sanderson (1992b) showed that alfalfa stem ADF and NDF were more closely correlated to accumulated temperatures for his data from Iowa than were MSW or MSC, but addition of the Texas data showed temperature-based models to be inaccurate and imprecise.

Hintz and Albrecht (1991) developed a variety of prediction equations for alfalfa quality using mean stage and plant morphology. They described a fast and simple prediction method using plant height and maturity stage of the most mature stem in a sample (Table 7), and found that they provided better prediction equations than those based on MSW or MSC. Validation data collected on first-cutting alfalfa in Wisconsin in 1992 indicated that predicted values using the Hintz-Albrecht equation for NDF fit actual values reasonably well (K.A. Albrecht, 1992, personal communication).

Another fast and simple method for predicting quality uses an alfalfa sample collected for alfalfa weevil damage assessment (J.H. Cherney, unpublished). Fifty stems are randomly selected throughout a field (Waldron, 1990), and while this sample inevitably contains larger stems than the average stem in a field, it is a very precise sample (due to many sampling points) that correlates well to the actual quality in the field ($r^2 = 0.90$, n = 159 for 1992 samples). Promising preliminary data in 1991 relating proportion of stems with visible buds to NDF led to periodic sampling of 13 sites prior to first harvest in 1992.

Table 7. The best fitting equations developed from a Wisconsin data set to predict several forage quality measurements from height of the tallest stem (X_1) and stage of the most mature stem (X_2) according to Hintz and Albrecht (1991).

Quality measurement	Equation	RMSE of calibration	R^2
g kg-1 DM		g kg-1 DM	
CP	Y = 307.1 - 0.9·X_1 - 8.9·X_2	21.7	0.74
NDF	Y = 168.9 + 2.7·X_1 + 8.1·X_2	26.2	0.89
ADF	Y = 115.7 + 2.1·X_1 + 7.9·X_2	22.0	0.88
ADL	Y = 15.8 + 0.5·X_1 + 2.5·X_2	6.5	0.84

A total of 155 non-replicated "weevil" samples were collected along with an additional 155 samples for forage quality analysis. Each quality sample came from four areas in a field. Data from the 13 sites were summarized into five general environments in New York state (Figure 6). Results from 1992, which was unusually cool and late at the Ithaca and Mt. Pleasant locations, showed that indicators based on stage of maturity can be very poor predictors of forage quality. Alfalfa varied widely in the time of transition from a vegetative to a reproductive stage (May 20 to June 8). When NDF reached 400 g kg^{-1} DM, the proportion of "weevil sample" stems with visible buds ranged from 0.00 (three sites) to 1.00 (one site). Initial results from the same region in 1993 were similar to those of 1992. Clearly, caution is necessary when predicting quality from stage of development under unusual conditions.

Plant height (HT, in cm), maturity stage of the most mature stem, and GDD with a 5°C base were also determined for the same 1992 New York samples. (Because the New York quality samples were taken at a different stubble height, validation of the Wisconsin equations was not possible.) The best-fitting regression equations for NDF (g kg^{-1} DM) with all locations combined were the following:

NDF = -14.63 + 6.83·HT - 0.0239·HT·HT; R^2 = 0.79,
 RMSE = 34.1 g kg^{-1} DM
NDF = 105.8 + 0.7742·GDD; R^2 = 0.85, RMSE = 28.7 g kg^{-1} DM

Temperature clearly gave the better fit, but there were differences between locations (Figure 7).

A long term solution to the problems of accurate forage quality prediction involves defining model parameters from a thorough understanding of their relationships to animal performance. Short term solutions for on-farm use will probably involve a combination of forage analysis and short range forage quality prediction. Most attention may be given to the spring growth of alfalfa or grass, because the first harvest is typically 50% or more of the total seasonal yield. The first harvest also is where the largest mistakes are made in harvest date relative to forage quality. In grasses, this is in part due to a very rapid increase in NDF over a short time span (Cherney et al., 1993). A small prediction error in grass equations could lead to a very large error in predicted NDF. Relatively little work has been done on predicting the quality of grasses, and this certainly needs future attention.

The protocol that we propose here for dealing with the problems of predicting the forage quality of spring growth for on-farm use should apply to grasses, legumes, and their mixtures. It involves a combination of forage testing and prediction modeling. Much of the difference in the spring growth of forages between any two environments is associated with differences in the date of spring growth initiation. All variation in forage quality between sites due to initiation of spring growth and subsequent weather can be measured 2 or 3 wk before the start of the harvest season by simply analyzing forage samples to determine quality. This analysis could be relatively rapid, using NIRS technology. From this measured starting point, an equation based on GDD and historic weather data would allow prediction of the date a forage stand reaches 400 g kg^{-1} NDF (or whatever goal would be considered appropriate).

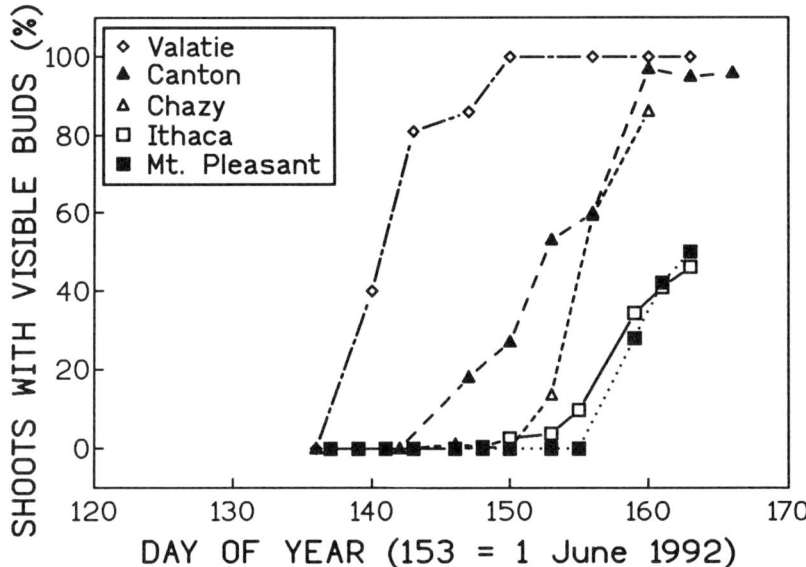

Figure 6. The percentages of shoots with visible buds in a 50-stem sample for alfalfa weevil scouting as a function of the day of the year. Data are from New York state with locations of increasing latitude falling in the sequence Valatie, Mt. Pleasant ≈ Ithaca, Chazy, Canton. Mt. Pleasant and Ithaca were unusually late.

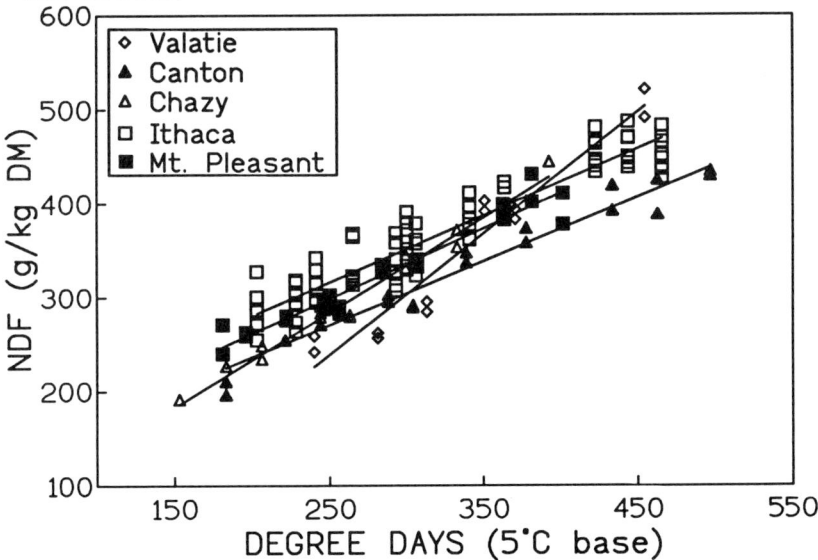

Figure 7. Neutral detergent fiber (NDF) as a function of growing degree days for five locations in New York in the spring of 1992. The equation for the Valatie location had a significantly higher slope than other locations except Chazy.

The error involved in predicting the change in NDF for a 2-to-3-wk period is much smaller than the error involved in predicting from the onset of spring growth. The period between testing forage quality and the date of anticipated harvest provides enough time to plan adequately for the spring harvest. If this combination of forage testing and modeling is successful, it would allow reasonable control of forage quality, at least until other predictive models are further refined.

Toward More Mechanistic Forage Quality Models

One feature of all of the regression models discussed above is that they are not general. The literature has numerous examples of site-specific quality and yield studies and the agronomic factors affecting them. Simulation modeling seeks to systematically analyze these data for mechanistic understanding that can predict forage crop growth, development, and quality dynamics over a wide range of spatial and temporal variation. In theory, mechanistic simulation models are the most general and robust, and hence, of the greatest utility.

Forage crop models are relatively complex, even without the components of forage quality. Most forage simulation models use empirical equations to derive quality estimates from easily measured characteristics, and as already shown, no single set of empirical equations are best at predicting forage quality in every case. All approaches are most valid for the region where they were developed, and this site specificity requires calibration and recalibration as location and weather patterns change (Sanderson, 1992b). This clearly points out the need for more research into fundamental relationships from which we can derive a more mechanistic approach to quality prediction.

The most fundamental elements of forage quality dynamics are the chemical and physical changes occurring in plant cells and tissues. Many of these changes are microscopically visible at the level of plant anatomy, and considerable progress has been made in our understanding of the relationship of plant anatomy and quality (Akin, 1989; Kühbauch and Bestajovsky, 1989). Enzymatic processes must govern these changes, and forage scientists are only starting to work with plants at this level (Kidambi et al., 1990). Even if we understood environmental regulation of plant chemistry, there might be too many orders of magnitude of difference between the cause and the effect for it to be useful in whole-plant models (Penning de Vries and Laar, 1982). Nevertheless, a literature review of how environment, enzymes, and plant chemistry are interrelated would be of use to modelers of forage quality.

Another approach to increasing the mechanistic basis of models is more thorough analysis of predictors now in use. Is a predictor the fundamental cause of quality change, or is it only a proxy for something else? Perhaps the factor that appears most like a proxy is harvest date or age. The easiest way to create spurious correlation is to associate factors that both change in time. Age has a perfect correlation with time. Many of the factors used in quality prediction models, as well as the various measures of quality themselves, are also correlated with time and with each other. Inter-correlation of predictor variables in multivariate analysis can result in unstable statistical models

(Johnson, 1972), and attempts to improve quality prediction by simply increasing the number of independent variables will probably fail for that reason.

Nevertheless, we should take a very careful look at age as a fundamental factor in forage quality change. In our 1992 data discussed above, all morphological measures of alfalfa were stable in a very cool spring while alfalfa quality continued to decline. Morphologically-based predictions completely failed. However, advancing time was associated with decreases in forage quality.

Liu (1977) reported similar results when short photoperiods prevented bud formation in alfalfa. Quality still declined, although branching as well as age were correlated with the change. Kalu (1982) found that the leaf fraction, uninfluenced by branching, still showed quality changes associated with age even when leaves from shoots of common stage number were compared.

In another field trial from New York, quality predictions based on MSW showed significant bias when alfalfa development was delayed by herbicides (Fick et al., 1988b). Forage quality had declined faster than predicted from the rate of stage development. We reasoned that aging, perhaps acting through attrition of enzymes or cell contents, or through progressive lignification of the cell wall, might be the cause. Age and stage are usually confounded, but the study by Liu (1977) with short photoperiods allowed us to determine the effect of age independent of MSW. We then recalibrated the equations of Kalu and Fick (1983), forcing inclusion of the age effect from Liu (1977). Only CP equations failed to show at least a 20% reduction in the amount of bias. Because of the small data set (n = 6), only a linear effect for age was calculated, but several of the relationships appeared to be curvilinear. A more accurate calibration will require a larger data set where age and stage are not confounded. Such confounding is the rule in the large body of data that now exists relating age to forage quality. Age may be a proxy for unidentified variables causing quality change, or it may actually be a primary factor. We need a better understanding of aging and its impact on enzymes and plant chemistry to know for sure.

Temperature is a factor that may be fundamental in the regulation of forage quality, and it has received some attention at more mechanistic levels. Temperature adaptation is known to produce different enzymes (isozymes), which may account for species adaptations to different temperature regimes as well (Kidambi et al., 1990). The apparent adjustment of the base temperature of alfalfa as the season progresses (Sharratt et al., 1989) is probably due to enzymatic adaptation. The enzymatic effects of temperature will be difficult to model, but it may be possible to improve temperature-based models simply through better estimates of base temperatures and GDD (Warrington and Kanemasu, 1983).

Light is another environmental factor that must have fundamental effects on forage quality. It directly influences levels of nonstructural carbohydrates in leaves via photosynthesis, and light levels indirectly influence several characteristics of leaves (e.g., mass, specific leaf area, thickness) that probably are related to quality. Although forage crop modelers have tended to ignore total nonstructural carbohydrate (TNC) or water-soluble carbohydrate (WSC) as measures of forage quality (presumably with the hope that it will be reduced to some constant level by post-harvest respiration or silage fermentation), this information is required by both ensiling (Pitt et al., 1985; McGechan, 1990) and rumen (Allen and Mertens, 1988; Hyer et al., 1991; Illius and Gordon, 1991) models. Nonstructural carbohydrate is computed in some forage crop growth models (Holt et al., 1975), and can be linked to forage quality (Thompson et al., 1989; McGechan, 1990). This needs to be done routinely and more mechanistically in the future.

It is also worth noting that light supply and nonstructural carbohydrates are related to the metabolism of N in plants. So far we are not aware of models to predict N fractions from these relationships. However, there is a substantial literature on modeling the role of N in plant growth (Keulen et al., 1989; Wolf et al., 1989; Greenwood et al., 1990), including pastures (McCaskill and Blair, 1990) and forage legumes (Lemaire et al., 1985). Protein fractions for forage quality analysis can be added to crop growth simulators in much the same manner as carbohydrate fractions.

Stage of development is another factor that is probably a proxy for real causes of forage quality change. It has been successful as a predictor of forage quality, possibly because it integrates the effects of so many factors, at least for alfalfa. Alfalfa phenology is therefore difficult to predict from weather measurements alone. Stress also influences alfalfa phenology. Stress from drought (Brown and Tanner, 1983; Wilson, 1983; Halim et al., 1989), fertility (Sanderson, 1992a), insects (Fick and Liu, 1976; Mueller, 1990), and disease (Wilkens, 1989) affect MSW or MSC in such a way that the impact of stress on forage quality is partially accounted for. Thus, temperature and (or) age will not be adequate for predicting alfalfa phenology under the wide range of common field conditions. We expect the same is true for forage grasses (Buxton and Marten, 1989).

Factors correlated with stage of development (and probably more basically related to forage quality than stage itself) include the proportion of stem and leaf in the herbage. The chemical differences between leaves and stems are well documented in forage quality research (Woodman and Evans, 1935; Kalu and Fick, 1983; Twidwell et al., 1988; Buxton and Marten, 1989), but the relationships are complicated. The switch to stem production at the start of reproduction causes major changes in grass physiology and chemistry (Kemp et al., 1989), and shoot height or length for individual species is directly related to increased proportion of stems. However, tiller length was a poor predictor of quality for a C_4 grass, possibly because interactions between tiller length and leaf area shading of basal stems cancelled any relationship (Warndorff and Dovrat, 1987). Likewise, studies with planting density or multileaf characteristics in alfalfa showed that plant morphology has a complex linkage to forage quality (Volenec et al., 1987; Volenec and Cherney, 1990), and more leafiness can be linked to lower yield (Kephart et al., 1989). In the future, characteristics like leaf fraction or stem length must be given mechanistic meaning in general models.

In the meantime, mean stage of development as a proxy has the advantage of integrating the many factors that influence leafiness and (or) steminess into a single useful measurement. This concept needs to be extended and tested on more species, and this should be possible with improved staging systems for grasses (Moore et al., 1991). It may also be possible to use the mean stage of a species like alfalfa to predict the quality of other forages growing in the same environment. This would be especially helpful in cases where photoperiod controls stage of development but forage quality shows more complex patterns.

The complexity of the mechanisms that regulate forage quality should be amenable to simulation modeling, and recent advances will facilitate such modeling. Already mentioned was the opportunity to link soluble carbohydrate and N fractions currently simulated in some models to forage quality characteristics based on those values. To do this with mechanistic detail, linkages to photosynthesis, leaf development, and canopy structure will be necessary. Thornley (1991) recently produced a theoretical model of the leaf life cycle that should be helpful, and applications are being clarified in models of grazing systems [(Parsons et al., 1988); D.R. Buckmaster, 1992, personal communication]. There is also a recent monograph on modeling plant phenology (Hodges, 1990), which emphasizes annuals, but useful general concepts from it can be applied to perennial forages as well. Another example is a herbage digestibility model (Illius, 1985) that uses "relative maturity" to link yield and quality. The all-to-common practice has been to formulate statistical models of quality without reference to yield, but yield and quality must both be known for most practical applications. Better linkages of forage growth and quality are present in the models developed from GROWIT (Smith and Loewer, 1983; Brown et al., 1986; Loewer, 1987), including consideration of soil fertility and mineral levels in herbage. Although it will certainly be challenging (Petit et al., 1992), such soil factors as pH and drainage may soon be simulated as well.

At present, functional or empirical relationships are most often used when predicting herbage quality because of a lack of basic understanding of genetic and environmental influences on cellular composition. As our understanding increases, increasingly mechanistic methods of quality prediction and the dynamics of quality change will be used to better predict the nutritive value of standing forage.

SUMMARY

Age of herbage, growing degree days, recent light levels, stage of development, and leafiness have been used to develop quantitative relationships that can predict forage quality for situations where forage testing is not possible. All of these relationships can give good predictions, but they are not general and require local calibration and recalibration. Computer simulation models available today likewise compute forage quality from empirical relationships that would be improved by substituting mechanistic concepts. To do this, more research into the basic relationships of plant growth and chemistry will be needed. Better simulation models based on improved understanding are already being developed for application in forage systems analysis. Of the simpler approaches to quality prediction, mean stage of development has proven to be fairly robust with substantial testing. Its measurement may still be too tedious for routine field use, and methods based on early forage testing plus measurements of stem lengths or computations from weather records and forecasts may allow projections into the future for the purpose of timing forage management to optimize forage quality.

REFERENCES

Akin, D.E. 1989. Histological and physical factors affecting digestibility of forages. Agron. J. 81:17-25.

Allen, M.S., and D.R. Mertens. 1988. Evaluating constraints on fibre digestion by rumen microbes. J. Nutr. 118:261-270.

Allen, S.J., and G.W. Fick. 1990. On-farm testing of mean stage by count as a predictor of alfalfa forage quality. p. 185. *In* Agronomy abstracts. ASA, Madison, WI.

Ames-Gottfred, N.P., D.L. Smith, and A.R. McElroy. 1990. Growth and quality characteristics of timothy genotypes grown in simulated swards. Can. J. Plant Sci. 70:330 (abstr.).

Bootsma, A. 1984. Forage crop maturity zonation in the Atlantic region using growing degree-days. Can. J. Plant Sci. 64:329-338.

Brink, G.E., and T.E. Fairbrother. 1992. Forage quality and morphological components of diverse clovers during primary spring growth. Crop Sci. 32:1043-1048.

Brown, J.W., and C.B. Tanner. 1983. Alfalfa stem and leaf growth during water stress. Agron. J. 75:799-805.

Brown, W.F., L.E. Moser, and T.J. Klopfenstein. 1986. Development and validation of a dynamic model of growth and quality of cool season grasses. Agric. Syst. 20:37-52.

Buxton, D.R., and J.S. Hornstein. 1986. Cell-wall concentration and components in stratified canopies of alfalfa, birdsfoot trefoil, and red clover. Crop Sci. 26:180-184.

Buxton, D.R., J.S. Hornstein, W. F. Wedin, and G. C. Marten. 1985. Forage quality in stratified canopies of alfalfa, birdsfoot trefoil, and red clover. Crop Sci. 26:273-279.

Buxton, D.R., and G.C. Marten. 1989. Forage quality of plant parts of perennial grasses and relationship to phenology. Crop Sci. 29:429-435.

Casler, M.D. 1986. Causal effects among forage yield and quality measures of smooth bromegrass. Can. J. Plant Sci. 66:591-600.

Cherney, D.J.R., J.H. Cherney, and R.F. Lucey. 1993. In vitro digestion kinetics and quality of perennial grasses as influenced by forage maturity. J. Dairy Sci. 76:790-797.

Coors, J.G., C.C. Lowe, and R.P. Murphy. 1986. Selection for improved nutritional quality of alfalfa forage. Crop Sci. 26:843-848.

Deinum, B. 1966. Climate, nitrogen and grass. I. Research into the influence of light intensity, temperature, water supply and nitrogen on the production and chemical composition of grass. Meded. Landb. Wageningen 66-11:1-91.

Deinum, B., A.J.H. van Es, and P.J. Van Soest. 1968. Climate, nitrogen and grass. II. The influence of light intensity, temperature and nitrogen on in vivo digestibility of grass and the prediction of these effects from some chemical procedures. Neth. J. Agric. Sci. 16:217-223.

Deinum, B., J. Beyer, P.H. Nordfeldt, A. Kornher, O. Østgård, and G. Bogaert. 1981. Quality of herbage at different latitudes. Neth. J. Agric. Sci. 29:141-150.

Deinum, B., and J.G.P. Dirven. 1974. A model for the description of the effects of different environmental factors on the nutritive value of forages. p. 1:338-346. *In* V.G. Iglovikov and A.P. Mosisyants (ed.) Proc. 12th Int. Grassl. Congr., Moscow, U.S.S.R., 11-20 June 1974. Izd-vo MIR, Moscow, U.S.S.R.

Edelsten, P.R., and A.J. Corrall. 1979. Regression models to predict herbage production and digestibility in a non-regular sequence of cuts. J. Agric. Sci., Camb. 92:575-585.

Edelsten, P.R., and J.E. Newton. 1975. A simulation model of intensive lamb production from grass. Grassland Research Institute Tech. Rep. No. 17, Hurley, U.K.

Fagerberg, B. 1988a. Phenological development in timothy, red clover and lucerne. Acta Agric. Scand. 38:159-170.

Fagerberg, B. 1988b. The change in the nutritive value in timothy, red clover and lucerne in relation to phenological stage, cutting time and weather conditions. Acta Agric. Scand. 38:347-362.

Faix, J.J. 1974. The effect of temperature and daylength on the quality and morphological components of three legumes. Ph.D. thesis, Cornell Univ., Ithaca, NY (Diss. Abstr. 74-24223).

Feekes, W. 1941. De tarwe en haar milieu. Versl. XVII Tech. Tarwe Comm. 12:560-561.

Fick, G.W. 1978. Computer simulation models of pasture production in Canterbury: Description and user's manual. Agricultural Economics Research Unit Res. Rep. No. 89, Lincoln College, NZ.

Fick, G.W. 1980. A pasture production model for use in a whole farm simulator. Agric. Syst. 5:137-161.

Fick, G.W., D.A. Holt, and D.G. Lugg. 1988a. Environmental physiology and crop growth. p. 163-194. In A.A. Hanson, D.K. Barnes, and R.R. Hill, Jr. (ed.) Alfalfa and alfalfa improvement. Agronomy Monogr. 29. ASA, CSSA, and SSSA, Madison, WI.

Fick, G.W., and R.S. Holthausen. 1975. Significance of parts other than blades and stems in leaf-stem separations of alfalfa herbage. Crop Sci. 15:259-262.

Fick, G.W., and B.W.Y. Liu. 1976. Alfalfa weevil effects on root reserves, development rate, and canopy structure of alfalfa. Agron. J. 68:595-599.

Fick, G.W., B.W.Y. Liu, and S.C. Mueller. 1988b. Age as a factor for predicting alfalfa forage quality. p. 160. In Agronomy abstracts. ASA, Madison, WI.

Fick, G.W., and S.C. Mueller. 1989. Alfalfa: Quality, maturity, and mean stage of development. Information Bull. 217, College of Agric. and Life Sci., Cornell Univ., Ithaca, NY.

Fick, G.W., and D.W. Onstad. 1983a. Simple computer simulation models for forage-management applications. p. 483-485. In J.A. Smith and V.W. Hays (ed.) Proc. 14th Int. Grassl. Congr., Lexington, KY, 15-24 June 1981. Westview Press, Boulder, CO.

Fick, G.W., and D.W. Onstad. 1983b. ALSIM 1 (LEVEL ZERO): Description, performance, and user instructions for a base-line model of alfalfa yield and quality. Agron. Mimeo 83-26, Dept. of Agronomy, Cornell Univ., Ithaca, NY.

Fick, G.W., and D.W. Onstad. 1988. Statistical models for predicting alfalfa herbage quality from morphological or weather data. J. Prod. Agric. 1:160-166.

Fick, G.W., and C.J. Sniffen. 1985. Estimating neutral detergent fiber in alfalfa herbage. p. 311-315. In Forage and Grassland Conference, Hershey, PA, 3-6 March. American Forage and Grassland Council, Lexington, KY.

Gengenbach, B.C., and D.A. Miller. 1972. Variation and heritability of protein concentration in various alfalfa plant parts. Crop Sci. 12:767-769.

Greenfield, P.L., and D. Smith. 1973. Influence of temperature change at bud on composition of alfalfa at first flower. Agron. J. 65:871-874.

Greenwood, D.J., G. Lemaire, G. Gosse, P. Cruz, A. Draycott, and J.J. Neeteson. 1990. Decline in the percentage N of C3 and C4 crops with increasing plant mass. Ann. Bot. 66:425-436.

Halim, R.A., D.R. Buxton, M.J. Hattendorf, and R.E. Carlson. 1989. Water-stress effects on alfalfa forage quality after adjustment for maturity differences. Agron. J. 81:189-194.

Hanks, J., and J.T. Ritchie (ed.). 1991. Modeling plant and soil systems. Agronomy Monogr. 31. ASA, CSSA, and SSSA, Madison, WI.

Hides, D.H., J.A. Lovatt, and M.V. Hayward. 1983. Influence of stage of maturity on the nutritive value of Italian ryegrass. Grass Forage Sci. 38:33-38.

Hintz, R.W., and K.A. Albrecht. 1991. Prediction of alfalfa chemical composition from maturity and plant morphology. Crop Sci. 31:1561-1565.

Hodges, T. 1990. Predicting crop phenology. CRC Press, Boca Raton, FL.

Holt, D.A., R.J. Bula, G.E. Miles, M.M. Schreiber, and R.M. Peart. 1975. Environmental physiology, modeling and simulation of alfalfa growth. I. Conceptual development of SIMED. Purdue Univ. Agric. Exp. Stn. Res. Bull. 907, West Lafayette, IN.

Homb, T. 1952. Chemical composition and digestibility of grassland crops (in Norwegian). Norges Landbrukshøgskoles Fôringsforsøk Beretning 71, Mariendals Boktrykkeri, Gjøvik, Norway.

Hyer, J.C., J.W. Oltjen, and M.L. Galyean. 1991. Development of a model to predict forage intake by grazing cattle. J. Anim. Sci. 69:827-835.

Illius, A.W. 1985. A simulation model of seasonal pattern of herbage digestibility and its interaction with intensity of defoliation. p. 982-984. *In* Proc. 15th Int. Grassl. Congr., Kyoto, Japan, 24-31 August 1985. Jap. Soc. Grassl. Sci., c/o Nat. Grassl. Res. Inst., Tochigi-ken, Japan.

Illius, A.W., and I.J. Gordon. 1991. Prediction of intake and digestion in ruminants by a model of rumen kinetics integrating animal size and plant characteristics. J. Agric. Sci., Camb. 116:145-158.

Jarl, F., and T. Helleday. 1951. Studies of the changes in the chemical composition, digestibility and nutritive value of herbage at different growth stages (in Swedish). K. Lantbr. Tidskr. 90:315-335.

Jeney, C. 1972. The influence of air temperature on the growth and development of lucerne. Takarmany-bazis 12(2):19-36.

Johnson, J. 1972. Econometric methods. 2nd ed. McGraw Hill, New York.

Kalu, B.A. 1976. Age and time of year effects on alfalfa (*Medicago sativa* L.) quality and morphological development. M.S. thesis, Cornell Univ., Ithaca, NY.

Kalu, B.A. 1982. Morphological stage of development and forage quality of field grown alfalfa (*Medicago sativa* L.). Ph.D. thesis, Cornell University, Ithaca, NY. (Diss. Abstr. 82-28396).

Kalu, B.A., and G.W. Fick. 1981. Quantifying morphological stage of development of alfalfa for studies of herbage quality. Crop Sci. 21:267-271.

Kalu, B.A., and G.W. Fick. 1983. Morphological stage of development as a predictor of alfalfa herbage quality. Crop Sci. 23:1167-1172.

Kalu, B.A., G.W. Fick, and P.J. Van Soest. 1988. Agronomic factors in evaluating forage crops. I. Predicting quality measures of crude protein and digestibility from crop leafiness. J. Agron. Crop Sci. 161:135-142.

Kalu, B.A., G.W. Fick, and P.J. Van Soest. 1990. Agronomic factors in evaluating forage crops. II. Predicting fiber components (NDF, ADF, ADL) from crop leafiness. J. Agron. Crop Sci. 164:26-33.

Kane, E.A., and L.A. Moore. 1959. Digestibility of Beltsville first-cut forages as affected by date of harvest. J. Dairy Sci. 42:936 (abstr.).

Kemp, D.R., C.F. Eagles, and M.O. Humphreys. 1989. Leaf growth and apex development of perennial ryegrass during winter and spring. Ann. Bot. 63:349-355.

Kephart, K.D., D.R. Buxton, and R.R. Hill, Jr. 1989. Morphology of alfalfa divergently selected for herbage lignin concentration. Crop Sci. 29:778-782.

Keulen, H. van, J. Goudriaan, and N.G. Seligman. 1989. Modelling the effects of nitrogen on canopy development and crop growth. Soc. Exp. Biol. Seminar Ser. 31:83-104.

Kidambi, S.P., J.R. Mahan, and A.G. Matches. 1990. Purification and thermal dependence of glutathione reductase from two forage legume species. Plant Physiol. 92:363-367.

Kühbauch, W., and J. Bestajovsky. 1989. Distribution of lignin in stems of alfalfa and red clover. Wirtschaftseigene Futter 35:19-28.

Lacefield, G.D. 1989. Cutting and grazing management of alfalfa. p. 13-22. *In* Managing alfalfa for profit. Proc. 19th Nat. Alfalfa Symp., 14 March 1989, Hershey, PA. Certified Alfalfa Seed Council, Davis, CA.

Lemaire, G., P. Cruz, G. Gosse, and M. Chartier. 1985. Relationship between dynamics of nitrogen uptake and dry matter growth for lucerne (*Medicago sativa* L.) (in French). Agronomie 5:685-692.

Lenssen, A.W., E.L. Sorensen, G.L. Posler, and L.H. Harbers. 1991. Basic alfalfa germplasms differ in nutritive content of forage. Crop Sci. 31:293-296.

Lewis, S.R., and J.F. Balliette. 1991. Alfalfa quality prediction for Nevada. Publ. 91-10, Nevada Coop. Exten., College of Agric., Univ. of Nevada, Reno.

Liu, B.W.Y. 1977. Statistical models for prediction of alfalfa quality. Ph.D. thesis, Cornell Univ., Ithaca, NY. (Diss. Abstr. 78-07789).

Loewer, O.J. 1987. GRAZE: A model of selective grazing by beef animals. Agric. Syst. 25:297-309.

McCaskill, M.R., and G.J. Blair. 1990. A model of S, P and N uptake by a perennial pasture. 1. Model construction. Fert. Res. 22:161-172.

McGechan, M.B. 1990. Operational research study of forage conservation systems for cool, humid upland climates. Part 1: Description of model. J. Agric. Eng. Res. 45:117-136.

Millier, W.F., and G.E. Rehkugler. 1972. A simulation - The effect of harvest starting date, harvesting rate and weather on the value of forage for dairy cows. Trans. ASAE 15:409-413.

Minson, D.J., and C.E. Harris. 1959. The digestibility of pure herbage species. p. 59-60. *In* Ann. Rep. Grassl. Res. Inst. 1957/8. Grassl. Res. Inst., Hurley, U.K.

Mitchell, K.J. 1960. The structure of pasture in relation to production potential. Proc. N. Z. Soc. Anim. Prod. 20:82-92.

Moline, W.J., and W.F. Wedin. 1969. Yield and quality evaluations of first-cutting Vernal alfalfa. Iowa State J. Sci. 43:261-272.

Moore, K.J., L.E. Moser, K.P. Vogel, S.S. Waller, B.E. Johnson, and J.F. Pedersen. 1991. Describing and quantifying growth stages of perennial forage grasses. Agron. J. 83:1073-1077.

Mowat, D.N., R.S. Fulkerson, W.E. Tossell, and J.E. Winch. 1965. The *in vitro* digestibility and protein content of leaf and stem portions of forages. Can. J. Plant Sci. 45:321-331.

Mueller, S.C. 1990. The effects of botanical composition on prediction of alfalfa (*Medicago sativa* L.) quality from mean stage of development. Ph.D. thesis, Cornell Univ., Ithaca, NY. (Diss. Abstr. 90-18160).

Mueller, S.C., and G.W. Fick. 1989. Converting alfalfa development measurements from mean stage by count to mean stage by weight. Crop Sci. 29:821-823.

Nordkvist, E., and P. Åman. 1986. Changes during growth in anatomical and chemical composition and in-vitro degradability of lucerne. J. Sci. Food Agric. 37:1-7.

Ohlsson, C., and W.F. Wedin. 1989. Phenological staging schemes for predicting red clover quality. Crop Sci. 29:416-420.

Onstad, D.W., and G.W. Fick. 1983. Predicting crude protein, in vitro true digestibility, and leaf proportion in alfalfa herbage. Crop Sci. 23:961-964.

Onstad, D.W., and C.A. Shoemaker. 1984. Management of alfalfa and the alfalfa weevil (*Hypera postica*): An example of systems analysis in forage production. Agric. Syst. 14:1-30.

Parsons, A.J., I.R. Johnson, and A. Harvey. 1988. Use of a model to optimize the interaction between frequency and severity of intermittent defoliation and to provide a fundamental comparison of the continuous and intermittent defoliation of grass. Grass Forage Sci. 43:49-59.

Penning de Vries, F.W.T., and H.H. van Laar (ed.). 1982. Simulation of plant growth and production. Pudoc, Wageningen, the Netherlands.

Petit, H.V., A.R. Pesant, G.M. Barnett, W.N. Mason, and J.L. Dionne. 1992. Quality and morphological characteristics of alfalfa as affected by soil moisture, pH and phosphorus fertilization. Can. J. Plant Sci. 72:147-162.

Pitt, R.E., R.E. Muck, and R.Y. Leibensperger. 1985. A quantitative model of the ensilage process in lactate silage. Grass Forage Sci. 40:279-303.

Rao, K.V., P.J. Van Soest, and G.W. Fick. 1987. Prediction of leaf fraction, neutral detergent fiber, and intake of ground forage samples by airflow segregation. Crop Sci. 27:601-603.

Reid, J.T. 1973. Quality hay. p. 532-548. *In* M.E. Heath, D.S. Metcalfe, and R.E. Barnes (ed.) Forages, the science of grassland agriculture. 3rd ed. Iowa State Univ. Press, Ames.

Reid, J.T., W.K. Kennedy, K.L. Turk, S.T. Slack, G.W. Trimberger, and R.P. Murphy. 1959. Effect of growth stage, chemical composition, and physical properties upon the nutritive value of forages. J. Dairy Sci. 42:567-571.

Rice, R.W., J.G. Morris, B.T. Maeda, and R.L. Baldwin. 1974. Simulation of animal functions in models of production systems: Ruminants on the range. Fed. Proc. 33:188-195.

Rickman, R.W., B.L. Klepper, and C.M. Peterson. 1985. Wheat seedling growth and development response to incident photosynthetically active radiation. Agron. J. 77:283-287.

Rotz, C.A., J.R. Black, D.R. Mertens, and D.R. Buckmaster. 1989. DAFOSYM: A model of the dairy forage system. J. Prod. Agric. 2:83-91.

Salmon, S.C., C.O. Swanson, and C.W. McCampbell. 1925. Experiments relating to the time of cutting alfalfa. Kansas Agric. Exp. Stn. Tech. Bull. 15, Manhattan.

Sanderson, M.A. 1992a. Mean stage and quality of alfalfa grown under different phosphorus levels. p. 185. *In* Agronomy abstracts. ASA, Madison, WI.

Sanderson, M.A. 1992b. Predictors of alfalfa forage quality: Validation with field data. Crop Sci. 32:245-250.

Sanderson, M.A., and W.F. Wedin. 1988. Cell wall composition of alfalfa stems at similar morphological stages and chronological age during spring growth and summer regrowth. Crop Sci. 28:342-346.

Seligman, N.G. 1975. A critical appraisal of some grassland models. p. 60-97. *In* G.W. Arnold and C.T. de Wit (ed.) Critical evaluation of systems analysis in ecosystems research and management. 2nd ed. Pudoc, Wageningen.

Selirio, I.S., and D.M. Brown. 1979. Soil moisture-based simulation of forage yield. Agric. Meteorol. 20:99-114.

Sharratt, B.S., C.C. Sheaffer, and D.G. Baker. 1989. Base temperature for the application of a growing-degree-day model to field grown alfalfa. Field Crops Res. 21:95-102.

Simon, U., and B.H. Park. 1983. A descriptive scheme for stages of development in perennial forage grasses. p. 416-418. *In* J.A. Smith and V.W.

Hays (ed.) Proc. 14th Int. Grassl. Congr., Lexington, KY, 15-24 June 1981. Westview Press, Boulder, CO.

Smith, D. 1964. Chemical composition of herbage with advance in maturity of alfalfa, medium red clover, ladino clover, and birdsfoot trefoil. Wisconsin Agric. Exp. Stn. Res. Report 16, Madison.

Smith, E.M., and O.J. Loewer, Jr. 1983. Mathematical-logic to simulate growth of two perennial grasses. Trans. ASAE 26:878-883.

Smith, R.C.G., and W.A. Williams. 1973. Model development for a deferred-grazing system. J. Range Manage. 26:454-460.

Terry, R.A., and J.M.A. Tilley. 1964. The digestibility of leaves and stems of perennial ryegrass, cocksfoot, timothy, tall fescue, lucerne and sainfoin, as measured by an in vitro procedure. J. Brit. Grassl. Soc. 19:363-372.

Thompson, N., P.R. Keiller, and C.W. Yates. 1989. Predicting the digestibility of grass grown for first-cut silage. Grass Forage Sci. 44:195-204.

Thornley, J.H.M. 1991. A model of leaf tissue growth, acclimation and senescence. Ann. Bot. 67:219-228.

Torssell, B.W.R. 1984. Conditions for plant production (in Swedish). p. 156-157. *In* Report 136. Dep. Plant Husbandry, Swedish Univ. Agric. Sci., Uppsala, Sweden.

Twidwell, E.K., K.D. Johnson, J.H. Cherney, and J.J. Volenec. 1988. Forage quality and digestion kinetics of switchgrass herbage and morphological components. Crop Sci. 28:778-782.

Volenec, J.J., and J.H. Cherney. 1990. Yield components, morphology, and forage quality of multifoliolate alfalfa phenotypes. Crop Sci. 30:1234-1238.

Volenec, J.J., J.H. Cherney, and K.D. Johnson. 1987. Yield components, plant morphology, and forage quality of alfalfa as influenced by plant population. Crop Sci. 27:321-326.

Waldron, J.K. 1990. New York State integrated pest management program -- 1990 alfalfa IPM scouting procedures. IPM Bull. 301A, New York State IPM Program, Cornell Univ., Ithaca, NY.

Warndorff, M., and A. Dovrat. 1987. The effect of tiller length and age on herbage quality of hybrid pennisetum canopies. Neth. J. Agric. Sci. 35:21-28.

Warrington, I.J., and E.T. Kanemasu. 1983. Corn growth response to temperature and photoperiod. I. Seedling emergence, tassel initiation, and anthesis. Agron. J. 75:749-754.

West, C.P., D.W. Walker, R.K. Bacon, D.E. Longer, and K.E. Turner. 1991. Phenological analysis of forage yield and quality in winter wheat. Agron. J. 83:217-224.

Widtsoe, J.A. 1897. Alfalfa or lucern: Its chemical life history. Utah Agric. Exp. Stn. Bull. 48, Logan.

Wilkens, P.W. 1989. The physiological and quantitative response of alfalfa to *Phoma medicaginis*. Ph.D. thesis, Cornell University, Ithaca, NY. (Diss. Abstr. 89-24557).

Wilkens, P.W., and G.W. Fick. 1988. FORVAL: A computer program using chemical analyses and market data to price hay. J. Agron. Educ. 17:122-127.

Wilks, D.S., R.E. Pitt, and G.W. Fick. 1993. Modeling optimal alfalfa harvest scheduling using short-range weather forecasts. Agric. Syst. 42:277-305.

Wilson, J.R. 1982. Environmental and nutritional factors affecting herbage quality. p. 111-131. *In* J.B. Hacker (ed.) Nutritional limits to animal production from pastures. CAB, Farnham Royal, U.K.

Wilson, J.R. 1983. Effects of water stress on herbage quality. p. 470-472. *In* J.A. Smith and V.W. Hays (ed.) Proc. 14th Int. Grassl. Congr., Lexington, KY. 15-24 June 1981. Westview Press, Boulder, CO.

Wilson, J.R., and D.J. Minson. 1983. Influence of temperature on the digestibility of the tropical legume *Macroptilium atropurpureum*. Grass Forage Sci. 38:39-44.

Winch, J.E. 1971. Growth stage key for legumes. p. (unnumbered). *In* C. James (ed.) A manual of assessment keys for plant diseases. Canada Dept. Agric. Publ. No. 1458 and American Phytopathological Society, St. Paul, MN.

Wisiol, K., and J.D. Hesketh (ed.). 1987. Plant growth modeling for resource management. Vol. I: Current models and methods. CRC Press, Boca Raton, FL.

Wolf, J., C.T. de Wit, and H. van Keulen. 1989. Modeling long-term crop response to fertilizer and soil nitrogen. I. Model description and application. Plant Soil 120:11-22.

Woodman, M.A., and R.E. Evans. 1935. Nutritive value of lucerne. IV. The leaf-stem ratio. J. Agric. Sci., Camb. 25:578-597.

CHAPTER 19

FORAGING BEHAVIOR IN GRAZING ANIMALS AND ITS IMPACT ON PLANT COMMUNITIES

J. Hodgson, D.A. Clark, and R.J. Mitchell

INTRODUCTION

In this paper, information from a range of sources is used to show the developing understanding of the factors influencing forage intake and diet composition in grazing animals, and the impact of these animals on the plant communities on which they depend. This is not intended to be an exhaustive review but, rather, a selective evaluation of progress over the last twenty-five years and the current status of knowledge and theory in a fascinating and complex area of research. Attention will be concentrated primarily upon the ways in which the behavioral characteristics of grazing animals influence their responses to vegetation. This is a particularly important feature of the ecology of pastoral systems which cannot easily be investigated in conventional nutrition studies. The review will draw on information from studies on the ecology of both natural and managed pastoral systems; the interplay between the two is proving to be particularly productive in the development of both field data and hypotheses.

INGESTIVE BEHAVIOR AND HERBAGE INTAKE

Early arguments about the importance of behavioral limits to herbage intake in grazing animals rested on the demanding nature of the food-gathering process (Johnstone-Wallace and Kennedy, 1944). Much of the early evidence was circumstantial, and the critical evidence for the importance of behavioral limits to herbage intake had to await the studies of Chacon and Stobbs (1976) which involved the removal of digesta from the rumens of grazing cattle. Allden and Whittaker (1970), following Allden (1962), first defined herbage intake in terms of the components of ingestive behavior (equation 1), and this simple concept provided the basis for active research on the interrelationships between behavioral variables and the sward and animal characteristics influencing them.

J. Hodgson, Massey University, Palmerston North, New Zealand; D.A. Clark, Dairying Research Corporation, Hamilton, New Zealand; R.J. Mitchell, Ag Research, Palmerston North, New Zealand.

Daily Herbage Intake (HI) = Grazing Time (GT)
x Rate of Biting (RB)
x Intake per Bite (IB).....................(1)

The field studies of Allden and Whittaker (1970) and Stobbs (1973a,b) led to an understanding of the influence of the structure of the sward canopy upon IB, in addition to the effects of quantitative variables like herbage mass or herbage allowance (Holmes, 1987; Nicol and Nicoll, 1987; Poppi et al., 1987; Rattray et al., 1987). They also demonstrated, but did not fully explain, the essentially reciprocal relationship between IB and RB, and showed that declining HI with declining herbage mass or height is the consequence of imperfect compensation in terms of GT in response to declining rate of intake (RI = IB x RB).

Studies of this kind provided information on the effects of animal species, genotype, and physiological state upon ingestive behavior and upon behavioral responses to varying sward conditions (Arnold and Dudzinski, 1967; Jamieson and Hodgson, 1979; Forbes and Hodgson, 1985). They also provided the basis for early theories on the functional relationships between the effects of the physical and physiological characteristics of herbivores upon the range and effectiveness of their adaptation to a range of forage opportunities (Allden and Whittaker, 1970; Jarman, 1974; Demment and Van Soest, 1985).

Studies to identify causation rather than simply to investigate association between sward characteristics and behavioral responses have been of relatively recent origin, and have relied for their success largely on the development of techniques for the control and manipulation of both swards and animals under close constraint. Examples are the hand-constructed "sward board" used originally by Black and Kenney (1984) and later by Laca et al. (1992a) to construct artificial swards from harvested plants, the grazing cage used by Burlison et al. (1991) to control access to small areas of pasture, and the "sward box" technique of Mitchell et al. (1991) and Illius et al. (1992) to grow sets of seedling swards. All of these procedures have allowed a substantial degree of control of intra-sward variability, a major source of difficulty in many behavior studies (Burlison et al., 1991), at the expense of severe limitations in sward area and in close confinement of the animals concerned. It has therefore been necessary to concentrate on very short-term estimates of the rate of herbage intake in these studies. However, they have made it possible to investigate ingestive responses in some detail, with particular reference to the dimensions of individual mouthfuls of herbage and the factors influencing these dimensions.

In addition, some experimenters have developed useful compromises between uniform artificial "swards" and the variation associated with grazed pastures. Dougherty et al. (1987) used tethered cattle to graze prepared swards for up to 3 h. Their use of trained animals and balanced change-over designs proved to be a powerful technique to evaluate the effect of variation in the herbage canopy during grazing on components of ingestive behavior.

Further development of the mechanistic view of ingestive behavior now becomes possible, for example equations 2 and 3 from Burlison et al. (1991):

Intake per Bite (IB) = Bite Volume (BV)
x Bulk Density of Herbage in Grazed Strata ..... (2)
Bite Volume (BV) = Bite Depth (BD) x Bite Area (BA) .... (3)
where BD is defined as the vertical distance between the sward surface and the cut ends of defoliated leaves and stems, and BA as the vertical projection of the area of vegetation encompassed by a single bite.

These simple equations provide a much stronger conceptual base for understanding the influence of sward characteristics on ingestive behavior and their interaction with animal variables (Laca et al., 1993; Mitchell et al., 1993a). Studies have universally shown that BD is much more responsive than BA to variations in sward conditions and that, in most circumstances, it is the major determinant of both BV and IB. This holds true for a range of herbivore species and a range of plant species and sward conditions (Burlison et al., 1991; Mitchell et al., 1991, 1993a; Laca et al., 1992b; Gong et al., 1993).

Bite Dimensions and Intake per Bite

Early work on temperate swards indicated that BD increased rectilinearly with increasing sward height up to surprisingly high levels (Milne et al., 1982; Burlison et al., 1991). More recent critical work has largely substantiated this evidence though, in some circumstances, the relationship may be asymptotic (Mitchell et al., 1991; Laca et al., 1992b), as in Figure 1. The reasons for these differences are not clear, but may reflect the better control of intra-sward variability and plant morphology in more recent studies.

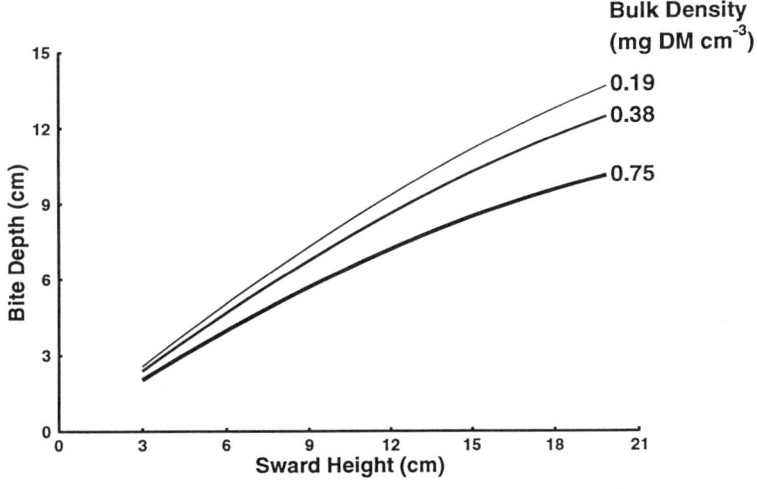

Figure 1. Bite depth of young deer and sheep (combined data) in relation to the surface height and bulk density of leafy seedling swards of Sorghum (*Sorghum bicolor* (L.) Moench). (from Mitchell et al., 1991)

There has been less success in identifying the morphological characteristics of plants which might control the relationship between sward height and BD. Early field studies by Barthram and Grant (1984) with swards dominated by perennial ryegrass *(Lolium perenne* L) indicated that the interface between leaf lamina and pseudostem may act as a deterrent to deep penetration within the sward canopy (in this context, the pseudostem of grasses is defined as the concentric layers of leaf sheath found on vegetative tillers, below the level of the ligule of the youngest fully expanded leaf). Similarly, cattle grazing tall fescue (*Festuca arundinacea* Schreb.) swards appeared to restrict BD to the upper surface of the pseudostem layer (Dougherty et al., 1992). Results of this kind led to the tentative suggestion that BD may be constrained by the effort required to detach plant material of relatively high structural strength in the basal layers of the sward (Hodgson, 1985). However, the accumulating evidence lacks consistency. There are several examples of studies in which BD did not penetrate to the base of the leafy upper layer of grass and grass/legume swards (Milne et al., 1982; Burlison et al., 1991), and others in which animals appeared to penetrate readily into the stem and pseudostem layers of both lucerne (*Medicago sativa* L.) (Laca et al., 1992b) and grass swards (Mitchell et al., 1993b). The potential influence of the structural strength of plant components on foraging strategy will be considered in more detail later.

Illius and Gordon (1987) describe variations in BD in geometric terms, relating this to the degree to which leaves of different length can be displaced horizontally before they slip out of the grip of the teeth (= "bite": see also Laca et al., 1993). Their theoretical treatment of the allometry of jaw size and body size provides an elegant explanation of the greater ability of small than of large animals to tolerate grazing on short swards, and *vice versa* on tall swards. In this respect, the observation by Wade (1991) that dairy cows under both continuous stocking and rotational grazing conditions tended to remove a constant one-third of the height of a sward at any one grazing is intriguing. The concept of a constant proportionality of herbage removal is supported by the results of Mursan et al. (1989) and Laca et al. (1992b) with cattle and, with less consistency, the results of Milne et al. (1982) and Burlison et al. (1991) with sheep, although the actual proportion of height removed appears to be greater for trimmed experimental swards (Demment et al., 1993) than for field conditions (Wade, 1991). Similar results have been obtained with horses (Hughes and Gallagher, 1992). All of the above examples relate to studies with temperate pasture species.

In assessing the implications of the above results, it is important to bear in mind the relatively limited variations in herbage bulk density and structural strength within the swards used in the more critical studies. Mitchell et al. (1993b), working with a series of sward boxes, managed to create variations in pseudostem length and structural strength, leaf to stem proportions, and quantity of dead material in the basal layers, and demonstrated significant effects upon BD, grazing height, and the depth of penetration within both leaf and pseudostem zones. It seems inconceivable that BD would not be influenced by vertical changes in plant morphology, even in relatively simple vegetative swards. Within limits, the depth of the leafy stratum may still be a better description of sward conditions than overall sward height when considering likely effects upon BD. Burlison et al. (1991) did not observe any marked difference in the relationship between BD and sward height for vegetative and reproductive swards of several gramineous species. These results may have been confounded by simultaneous variation in a number of plant characteristics, but are supported by recent evidence from our laboratories (Y. Gong, unpublished data).

Bite area has proved to be a more difficult response variable to work with than BD. Generally speaking, it shows less sensitivity to sward change, and a proportionately greater degree of apparently random variation (Burlison et al., 1991; Hughes et al., 1991; Mitchell et al., 1991, 1993a). It was hypothesized early that animals might adjust BA, in the face of varying leaf and stem population density or structural strength, to maintain some consistency in the effort required to tear off a mouthful of herbage (Hodgson, 1985; Hughes et al., 1991). However, although there is some evidence to suggest that BA is reduced as biting forces increase (Chambers et al., 1981; Hughes et al., 1991; Laca et al., 1993), the results of the few reports of direct or indirect measurements of biting forces have been equivocal. Also, there is no clear evidence of inverse proportionality in the response of BA to variations in tiller population density or herbage bulk density (Laca et al., 1992b; Mitchell et al., 1993b) which would be expected to minimize changes in biting effort.

On short swards, BA is likely to be limited directly by the difficulty of clamping plants between incisors and dental pad (Illius and Gordon, 1987; Hughes et al., 1991; Laca et al., 1993), thus setting a constraint independent of the biting effort required. On taller swards, an inverse relationship between population density and BA may be apparent (Burlison et al., 1991; Hughes et al., 1991; Laca et al., 1992b). There is generally also a positive relationship between sward height and BA, the effect being greater on sparse than on dense swards (Mitchell et al., 1991).

All of these results provide circumstantial evidence for interaction between the effects of sward height and structural strength on BA. A better understanding of the implications of this interaction will require further investigation of the biting forces involved in grazing swards of differing structure and maturity (Hughes et al., 1991). However the technical difficulties involved should not be underestimated (Vincent, 1982).

There have also been difficulties in investigating the influence of herbage bulk density on intake, partly because it is not easy to isolate its effects from those of concomitant variations in sward height and plant morphology, and partly because it may influence intake per bite in two distinct ways: indirectly through its influence on bite dimensions, as above, and directly through its multiplier effect on the volume of herbage ingested.

Initial evidence suggested that, in tropical swards, intake per bite and the short-term rate of herbage intake were more sensitive to variations in herbage bulk density than to variations in sward height (Stobbs, 1973a). The reverse appeared to be the case in temperate swards (Hodgson, 1981), reflecting characteristic differences in the structure and morphology of plants in the two environments (Stobbs, 1973b). However, the results of more recent controlled studies, covering alternative animal species and contrasting swards, demonstrate a continuous pattern of response in intake per bite over wide ranges of variation in both sward height and bulk density (Burlison et al., 1991; Mitchell et al., 1991, 1993a). Although the results of Burlison et al. (1991) and Laca et al. (1992b) suggest that the effects of variations in sward height and bulk density upon intake per bite are independent and additive, the results of Mitchell et al. (1991) (Figure 2) and of more recent studies in our laboratories indicate strong evidence of an interaction between the two.

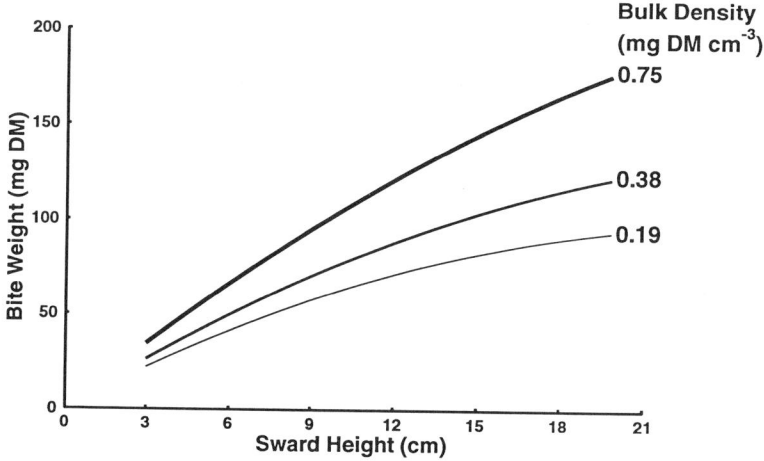

Figure 2. Bite weight of young deer and sheep (combined data) in relation to the surface height and bulk density of leafy seedling swards of Sorghum (from Mitchell et al., 1991).

These results should provide an objective basis for modelling intake responses to variations in sward characteristics, although there is need for further work on the effects of variations in the size, structure, structural strength, and geometry of individual sward components. The question of sward heterogeneity in both horizontal and vertical dimensions now assumes greater importance in research terms; some of its effects are considered later. The product of sward height and herbage bulk density is, of course, herbage mass, the sward variable most exhaustively investigated in field studies. However, the nature of the responses outlined above means that this variable is not adequate as a predictor of intake responses (Laca et al., 1992b).

Biting Rate and Grazing Time

Biting rate and grazing time are often regarded as the primary compensating responses of the animal to limitations in intake per bite (Hodgson, 1981). However, it is now clear from the results of several studies that the general rate of jaw movement (prehension, biting, and chewing) in grazing animals is remarkably constant (Penning, 1986). Variations in biting rate (bites min^{-1}) reflect variations in the relative proportions of the three jaw activities, and are therefore largely influenced by the manipulation necessary to graze effectively in swards of different structure (Penning, 1986; Laca et al., 1993). On taller swards, this can result in substantial reductions in bite rate, and the compensating changes which occur in intake per bite and bite rate result in relatively small variations in the short-term rate of herbage intake (Penning, 1986). However, on shorter swards, any increase in bite rate is unlikely to balance the reduction in intake per bite (Penning, 1986).

Extensions of grazing time in response to limitations in intake rate are well documented (Hodgson, 1981; Penning, 1986). In general terms, grazing time may increase as soon as sward conditions start to limit the short-term rate of herbage intake (Penning, 1986) and may reach double the base level of about 8 h d^{-1} in extreme conditions for both sheep and cattle (Hodgson, 1982). However, there is currently no very good basis for quantifying this relationship or for relating it directly to sward conditions except in a superficial sense. Increases in grazing time are seldom great enough to compensate for reductions in intake rate (Hodgson, 1981; Penning, 1986).

Grazing time is usually broken up into several well-defined periods of activity during the day and night. Bite rate varies both between and within periods, normally being greater in the morning and evening than during the day, and declining over the course of a grazing period (Hodgson, 1982). Procedures designed to investigate bite dimensions have largely precluded the opportunity to study responses in grazing time and periodicity. Although intake per bite may be the primary animal response to variations in sward conditions, the value of work in this area will ultimately depend upon the effectiveness with which functional responses can be used to predict daily herbage intake through associated information on adaptive changes in grazing time (Gordon and Lascano, 1993). This is a subject which requires increased attention in the future, not least because it provides the link between interests in behavioral and nutritional controls to forage consumption (Forbes, 1980).

DIET SELECTION

The preceding discussion treated swards as essentially homogeneous collections of plant material. This is, of course, far from reality under most practical conditions, where sward heterogeneity in both vertical and horizontal dimensions, and over time, is the norm (Gordon and Lascano, 1993). Thus, although the degree of sward control and uniformity achieved in recent critical studies has greatly aided the establishment of principles in defining behavioral responses (and to that extent has been directly analogous to the earlier development of controlled studies on aspects of digestive function), there is now a need to consider the application of these principles to a heterogeneous nutritional environment. This involves consideration of the causes and consequences of choice of diet by grazing animals.

At the 1969 Nebraska Conference, Marten (1970) presented a detailed review of information on the measurement and significance of forage palatability, defining the term as "a plant characteristic(s) eliciting a proportional choice among two or more forages conditioned by plant, animal and environmental factors which stimulate a selective intake response by the animal." This definition makes the point that "palatability" is a relative rather than an absolute term, and that its expression is dependent upon the balance within a set of interacting factors influencing any specific choice by an animal. As a consequence, the term can be difficult to deal with in a conceptual as well as a practical sense. This has led to caution in its use in a grazing context, and to a shift in focus from the concept of palatability itself to the practical manifestation of discriminatory behavior. In this discussion, the definitions of feeding preference ("... demonstrated under free choice and equal opportunity ...") and diet selection (" ... preference modified by accessibility within the vegetation canopy ...") as suggested by Hodgson (1979) are used to distinguish the components of discrimination. The distinction is important, particularly when considering the effects of alternative sward and animal variables on ingestive behavior, but it is not always made with clarity. It leads logically to an acknowledgement of the influence of canopy structure upon the intensity and balance of selection effects in particular cases and, hence, the need for objective information on canopy structure in grazing studies.

Van Dyne et al. (1980) summarized the voluminous literature on diet selection in grazing and browsing ungulates, and defined dietary composition in terms of the proportions of grasses, herbs, and shrubs using triangular coordinates. This review provided a useful summary of existing information and clearly identified differences in the selective behavior of different ungulate species in general terms. However, it did not provide insight into the vegetation characteristics influencing diet selection and foraging strategy, primarily because of the absence of information on the distribution of specific plant species or components within sward canopies, factors which are now acknowledged to exert a major impact on selection strategy and success. Nor did it define the grazing strategy of alternative animal species in terms of patch or stratum choice as a preliminary determinant of opportunity for selection and, therefore, failed to differentiate adequately between concepts of preference and selection. These are the issues which are discussed in more detail in this section.

Selection Strategies

Concepts of selection strategy now focus on the importance of visual and olfactory/gustatory cues, mediated by the effects of physical and structural characteristics of vegetation influencing ease and rate of intake. These cues are likely to be important at all levels of scale in the spectrum of choice available to the animal, but will be considered here primarily in the context of choice within plant communities. They may involve appraisal by the animal not only of the immediate effects of vegetation characteristics upon feeding success, but also the potential delayed effects upon health or nutritional status. Provenza and Balph (1990) review these factors in relation to alternative conceptual models of diet selection in grazing ruminants.

Arnold (1966) first investigated the influence of the senses of sight and olfaction on the selective behavior of grazing animals. The degree of surgical or behavioral interference necessary for these studies can make the results difficult to interpret (Arnold, 1966), and attention is now concentrated on the importance of specific vegetation characteristics in animal appraisal processes. Information about visual cues is still at an anecdotal and essentially qualitative stage (Bazeley, 1990), and is largely related to considerations of plant size, color and brightness.

The use of learning models to explain the diet selection processes of ruminants in response to variations in the biochemistry and nutritional characteristics of plants is quite recent (Provenza and Balph, 1987, 1988). Their value in explaining flexibility of diet choice has been well argued by these authors, whose models are based on the assumption that diets are modified in response to the positive and negative consequences of foraging. Conditioned aversions to foods may result from ingesting toxic compounds, but ruminants appear to have difficulty learning to avoid toxic foods unless they affect the emetic system (Provenza and Balph, 1990). There is some evidence for conditioned preferences in other species, but little for ruminants.

Provenza and Balph (1987, 1988) argue that in ruminants, the byproducts of microbial fermentation such as volatile fatty acids (VFA) and ammonia are candidates for producing conditioned food preferences. These compounds increase rapidly after a meal and their positive consequences may produce strong preferences. They suggest that this mechanism may be the primary reason why ruminants prefer leaf to stem, or green to dead material. However, simpler explanations for these choices are possible. Leaf is usually much easier to harvest than stem; and green material is often present in much greater proportions in the upper strata of swards which must be removed before the strata high in dead matter are available.

In view of the potential complexity of the influence of plant biochemistry upon animal preference (Arnold and Hill, 1972), it is not surprising that there have been few attempts to investigate the relative importance of biochemical and physical characteristics in specific circumstances. It may be questioned whether the normal criteria of nutritional adequacy in forage plants impact upon preference (Arnold and Hill, 1972) except in so far as they may correlate with biochemical or physical characteristics which are amenable to immediate appraisal by the animal. In these terms, the fibrous structure of the plant and its impact upon structural strength is likely to be of substantial importance and to correlate inversely with a nutritional criterion like digestibility.

Research on aspects of the biochemistry of dietary preference is particularly active, and covers herbivores ranging from insects to ungulates (e.g.,

Crawley, 1983). It is providing valuable insights into the importance of secondary compounds in plant defenses and in the selection strategies of animals, with particular reference to constituents like the condensed tannins, the alkaloids, and fungal toxins (Barry and Blaney, 1987), which have potential aversive effects at the gustatory level. The importance of these effects varies in different plant taxa, being greater in the legumes and shrub families than in the grasses and sedges (Malachek and Balph, 1987). Although it is possible to define the effects of these secondary compounds on dietary preferences in quantitative terms (e.g., Marten et al., 1973), it is not easy to incorporate them into more general models of selective behavior because of the difficulty of balancing their effects against those of variations in the physical characteristics of plants.

Kenney and Black (1984) and Black and Kenney (1984) have shown clearly that choice between alternative forage sources is strongly influenced by potential intake rate, whether this is modified by varying the height and bulk density of sward canopies or by varying the physical treatment of dried forages. These results were based on small numbers of observations with sheep, but they have subsequently been confirmed in larger experiments with both sheep and cattle and appear to hold good over substantial ranges of variation in both canopy height and bulk density (Clark, unpublished, quoted in Illius and Gordon, 1990; Illius et al., 1992; Demment et al., 1993).

Plant Morphology and Sward Canopy Structure

The influence of early nutritional experience on subsequent dietary choice is well established (Malachek and Balph, 1987), and there is some evidence that sheep and cattle can learn foraging skills. Flores et al. (1989) showed that lambs with experience of browsing on shrubs have a higher ingestion rate on shrubs than lambs previously foraging on grass. Newman et al. (1992) reported a particularly extreme example of the influence of immediate grazing experience on dietary preference, in which sheep which had been confined to grazing on white clover (*Trifolium repens* L.) showed an initial preference for grass when offered the choice, whereas sheep confined to grass showed the opposite preference.

Finally, recent evidence shows clearly that grazing ruminants demonstrate a regular pattern of sampling behavior (Bazeley, 1990; Illius et al., 1992), thus continuously reinforcing choice but limiting the impact of the process of discrimination.

The influence of sward canopy structure on the assessment of selective grazing has been a matter of contention for some time, with particular reference to the vertical and horizontal distribution of plant components and the interpretation of differences in the botanical composition of sward and diet.

Where observations have been made of the vertical distribution of sward canopy components, it has often been shown that, at least for temperate pastures, there is little difference between the composition of the diet and that of the upper strata of the canopy within which animals were known to be grazing (Hodgson, 1981; Milne et al., 1982; Illius et al., 1992). This has led to the concept of "passive selection", implying that differences between the composition of diet and sward are the consequence of an essentially non-discriminatory grazing at the sward surface superimposed on a non-random distribution of sward components (Figure 3). This may be particularly the case for apparent selection between leaf, pseudostem, and dead material or between legume and grass in well-managed temperate swards. It would be ridiculous to argue that non-discriminatory grazing is the general case (e.g., Hendrickson and Minson, 1980; L'Huillier et al., 1984; Grant et al., 1985), but the above evidence points to the need for objectivity in interpreting field data.

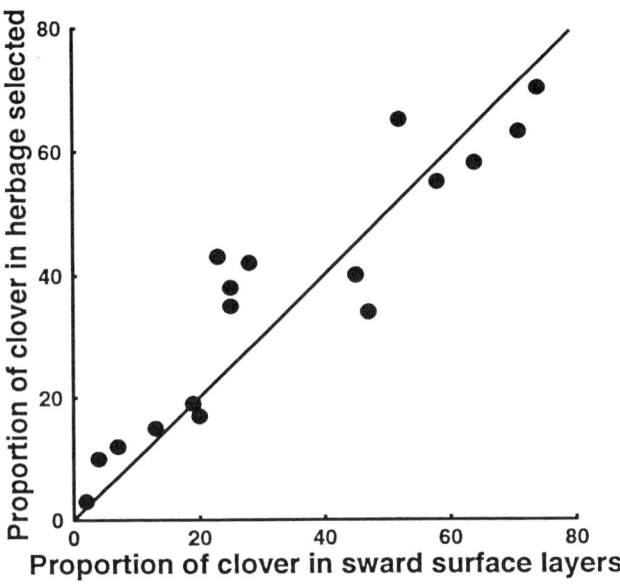

Figure 3. The relationship between the white clover content of the surface horizons of mixed white clover/perennial ryegrass swards, and that of the diet of grazing sheep. (from Milne et al., 1982)

Variations in the degree of selectivity may be due to differences in the maturity of vegetation and its constituent species (L'Huillier et al., 1984), to the characteristics of particular plant species (Grant et al., 1985), and to differences in grazing strategy between animal species (Clark et al., 1982; Grant et al., 1985; Nicol et al., 1987; Hodgson et al., 1991). These aspects will be considered in the following sections.

Apparent discrimination between the leaf, stem, and dead components of plants may be difficult to interpret in the absence of information on the spatial distribution of these components. However, the ability of animals to discriminate strongly between leaf and mature stem or between live leaf and standing dead material, if necessary by penetrating through the upper strata of the sward canopy to graze preferred material at the base of the canopy, is well established (Gardener, 1980; L'Huillier et al., 1984; Grant et al., 1985). The avoidance of mature seed heads and stems of grasses in preference for leaves seems likely to reflect differences in structural strength and shear strength and, therefore, in resistance to grazing, rather than any direct perception by the animal of nutritional differences.

Evidence from field studies demonstrates the influence of both spatial distribution and maturity of plant components on selective grazing. Thus, Gammon and Roberts (1978), working in African veld conditions, showed that the chances of defoliation of tillers of several grass species increased with increasing height until culm development, at which stage there was a sharp decline. In the USA, Heitschmidt et al. (1990) found that inter-specific variations in tiller height influenced defoliation intensity in mixed-grass prairie.

There may also be differences among plant species in the influence of culm development on the mechanics of the grazing process. For example, Forbes (1988) quotes examples of a depression in intake per bite in cattle following culm development in a tall-flowering species like bromegrass (*Bromus inermis* Leyss), but not in a species like bermudagrass (*Cynodon dactylon* (L.) Pers.) with no tall flower canopy. On the other hand, Burlison et al. (1991) and Gong (unpublished data), both working with several plant species, showed little change in the general response relationship between intake per bite and surface height as swards passed from the vegetative to the flowering stage.

Selective Grazing

Marten (1970) counselled caution in any attempt to identify plant species or individual plant characteristics as "palatable" or "unpalatable", preferring to emphasize the complexity of the matrix of variables within which specific dietary preferences might be determined. This philosophy is followed here, although this is not in any sense to deny the potential value of general concepts of relative preference in vegetation management. Evidence on selective grazing behavior has been reviewed recently by Stuth (1991), Gordon and Lascano (1993), and Owen-Smith (1993) for both intensive and extensive grassland systems. Here, attention will be focused upon an assessment of evidence on selective grazing in the specific context of sown grass/legume pastures, in order to illustrate the complexity of understanding required in a relatively simple situation.

It is generally accepted that, in temperate pastures, white clover is selectively grazed in preference to companion grasses, usually perennial ryegrass (Curll and Wilkins, 1982). In these conditions, the degree of selection observed appears to be influenced by the relative maturities of the foliage of legume and grass (Grant et al., 1985), and grazing animals, particularly sheep, are capable of picking out legumes from the lower strata of mixed swards (Laidlaw, 1983; L'Huillier et al., 1984; Grant et al., 1985). Selection is also greater where grass and legume are distributed in discrete patches than when the two are intimately intermixed (Clark and Harris, 1985). Indeed, in the latter case there may be very little evidence of the deliberate exercise of choice by the grazing animal (Milne et al., 1982).

The influence of stage of maturity on selection pressure between grass and legume raises questions about the species-specific nature of the criteria influencing preference. In the above examples, preference may have had much more to do with the relative maturity and the potential rate of intake of individual species than any specific biochemical characteristics. An example of differences in the potential rate of intake of herbage from grass and legume swards is shown in Table 1 (from Black and Kenney, 1984; Kenney and Black, 1986). Differences observed elsewhere have been less clear-cut, and largely reflect differences in canopy height between grasses and legumes (Illius et al., 1992; Gong et al.,1993).

Table 1. Potential intake rates of sheep grazing prepared swards of Wimmera ryegrass (*Lolium rigidum* Gaudi) (Black and Kenny, 1984) and Subterranean clover (*Trifolium subterraneum* L.) (Kenney and Black, 1986)

	Wimmera ryegrass	Subterranean clover
Herbage mass (t DM ha^{-1})	1-8	3-7
Maximum intake rate (g DM min^{-1})	5-6	20-30

In studies reported recently by Illius et al. (1992), it was shown clearly that, other things being equal, sheep demonstrated a preference for "patches" of perennial ryegrass/white clover sward containing 40 to 50% clover compared with patches of either higher or lower legume content. The authors linked this preference pattern to variations in the rate of intake of vegetation across the range of proportions of grass and clover. The evidence may also be explained in terms of a trade-off between preference and anti-preference characteristics in the legume, where preference factors (possibly structural, and related to potential rate of intake) predominated up to intermediate levels of legume content but were superseded by anti-preference factors (possibly secondary compounds) at higher levels. However, at this stage, it becomes difficult to distinguish cause and effect in the links between preference and rate of intake. There is supportive evidence from earlier field studies of a swing from white clover selection at low sward concentrations to relative avoidance at high concentrations (Milne et al., 1982; Clark and Harris, 1985); however, difficulties of vegetation control make these results less clear-cut.

In one series of field trials, Hodgson and Clark (1989) showed that the chance of defoliation of individual plants of white clover growing as isolates in a perennial ryegrass sward were determined primarily by the size of individual leaves and their position within the sward canopy, and were not influenced by biochemical (degree of cyanogenic activity) or visual (leaf mark) cues. This study provided no support for the view of Cahn and Harper (1976) that the selective behavior of sheep may be influenced by leaf mark patterns in different morphs of white clover, but in the latter case there was no examination of possible links between leaf mark and canopy distribution of alternative morphs.

The above results, if they can be shown to be of general application, indicate that preferential grazing behavior will tend to push patches of vegetation in mixed grass/legume pastures towards compositional extremes, although effects on overall mean composition may be small. Much of the current evidence from controlled studies relates to plant communities or ranges of variation in plant characteristics which are representative of temperate conditions. It is important now to obtain more extensive information on tropical and sub-tropical forages, where contrasts in plant morphology and maturity are typically more extreme than in temperate communities, and to extend the range of studies to incorporate browse species. The extension of these arguments to browsing animals is not straightforward, and concepts of ease of access and potential rate of intake must be tempered by information on the biochemistry and structural strength of the components of browse (Malachek and Balph, 1987).

ANIMAL FACTORS

It has been amply demonstrated that ingestive behavior and grazing selectivity can be substantially influenced by interactions between the effects of variation in body size, mouth dimensions and productive state within and between animal species, and by species - specific differences in behavior, some of which may relate to variations in body shape and jaw structure (Jarman and Sinclair, 1979; Illius and Gordon, 1987).

The work of Illius and Gordon (1987) provides an overview of the allometry of teeth dimensions and body size in grazing ungulates. The authors demonstrated in theoretical terms the greater sensitivity of large vs small animals (and of animals with large vs small mouths) to reductions in sward height, supporting and generalizing the field observations of Allden and Whittaker (1970), Bell (1970), and Gordon (1989).

These studies provide a basis for predicting the effects of intra- and inter-specific variations in body size and mouth conformation on responses to variations in the structural characteristics of pastures, but there have been few critical tests of these effects. Mitchell et al. (1991, 1993a) observed only small differences between sheep and deer in ingestive behavior responses to controlled variations in sward height and herbage bulk density. The results of Gong et al. (1993) indicated that goats increased intake per bite to a greater extent than sheep when offered turves of several grass species at a reproductive rather than a vegetative stage of growth.

In contrast to the generalizations of Illius and Gordon (1987), Collins and Nicol (1986) showed that the herbage intakes of sheep and cattle declined at approximately the same relative rate in response to the progressive depletion of the sward on which they grazed in common. Both species, however, maintained their intakes better on short pastures than did goats (Collins and Nicol, 1986). In the studies of Jamieson and Hodgson (1979), on the other hand, calves appeared to be less sensitive than lambs to reductions in herbage mass or sward height. In these studies, responses were measured in terms of daily herbage intake rather than intake per bite, and it is possible that the results were also influenced by species differences in the sensitivity of response in bite rate or grazing time.

No other direct comparisons of sheep and cattle have been reported, but the results of Laca et al. (1992b) with cattle are qualitatively very similar to those of Mitchell et al. (1991, 1993a) with sheep and deer under similar experimental conditions. In combination, these studies provide a firm basis for modelling ingestive behavior objectively.

The structural limitations of large jaws on short swards may be overcome to some extent by the articulation of two sets of incisors, as in the equine. However, responses in the components of ingestive behavior to variations in sward height have been qualitatively similar in horses (Hughes and Gallagher, 1993) to those in ruminants.

In general, bite dimensions and intake per bite increase with increasing body size and incisor arcade breadth (Illius and Gordon, 1987). Incisor breadth scales to a substantially smaller exponent of body weight than do maintenance energy requirements (Illius and Gordon, 1987), thus increasing the demands for higher nutrient concentration in the diet of smaller animals (Jarman and Sinclair, 1979). However, the close correlation between incisor breadth and body size in both inter- and intra-specific comparisons makes it difficult to determine the relative impacts of the two variables on intake per bite. Intuitively, teeth and jaw dimensions would be expected to be more important than body size in determining bite dimensions and intake per bite, although in the study of Penning et al. (1991), the reverse was the case. The definitive study, involving deliberate reduction or augmentation of the incisor arcade, has not to the authors' knowledge been carried out.

Increased nutrient demand (for example in comparisons between lactating and non-lactating animals, shorn and unshorn sheep, or between animals of different nutritional history) will usually increase forage intake (Hodgson, 1985), but the effects on the individual components of intake appear to be variable and difficult to predict. Dougherty et al. (1989) showed that lengthening the interval between grazings for cows from 1 to 3 h resulted in an immediate increase in biting rate and rate of herbage intake on lucerne pastures, but not on tall fescue.

Inter-specific differences in selective grazing are well documented (see Van Dyne et al., 1980), but it is not always clear whether these contrasts reflect differences in site selection or in bite selection within sites. Grant et al. (1985) demonstrated over a series of indigenous swards that cattle were relatively indiscriminate surface grazers compared to sheep, which tended to be more selective and to penetrate to a greater depth within the vegetation canopy. The contrasts between sheep and cattle were confirmed by Collins (1988) working under more closely controlled conditions on sown swards. In this study, goats concentrated their attention on the vegetation at intermediate levels in the canopy, a result somewhat at variance with the view of goats as shallower grazers than sheep (Gong et al., 1993).

Sheep generally avoid grazing mature grass seed heads and stems, but this pattern of behavior is less consistent in cattle and goats (Grant et al., 1985; Gong et al., 1993). This species difference is linked to the greater degree of penetration of the sward canopy by sheep than by the other two species, although it is not easy to ascribe cause and effect. Observations by Y. Gong (unpublished data) in this laboratory indicate that sheep are also able to avoid severing seed stalks grasped within individual mouthfuls of herbage.

There is little conclusive evidence of differences in the grazing choices of animals differing in age or productive state. Hodgson and Jamieson (1981) found no difference in the digestibility of the herbage selected from simple perennial ryegrass swards by lactating and non-lactating cattle grazing together, although there were indications in the same study that the selective activity of calves may have changed with experience of grazing conditions.

EFFECTS ON PLANT COMMUNITIES

The immediate consequences of grazing in most circumstances will be the depletion of the leafy strata of the sward canopy, and a consequent reduction in both the mass and the nutritive value of the remaining herbage. These effects are most obvious in pastoral systems involving high stocking rates and sub-division. Longer term consequences to the botanical composition and nutritive value of plant communities depend on the degree of differential grazing pressure imposed on different plant species, and the grazing tolerance and competitive ability of these species (Briske, 1990; Briske and Silvertown, 1993).

Plant Responses to Defoliation

The factors influencing selective grazing activity were discussed in general terms earlier, and will not be reconsidered here. It is worth re-emphasizing the point, however, that selection operates primarily at the level of the vegetation patch rather than at the more detailed level of the individual plant or plant unit. This is true both of relatively short, dense temperate pastures (Clark and Harris, 1985) and of taller, more open communities (Gammon and Roberts, 1978; Briske, 1986).

In all of these communities, the evidence indicates that variations in the frequency or severity of defoliation are likely to be directly related to the size of individual plants and their proximity to the surface of the vegetation canopy (Gammon and Roberts, 1978, Figure 4; Bircham and Hodgson, 1983; Briske, 1986), in accord with the evidence on the behavior patterns of grazing animals. There are, of course, exceptions to this generalization, usually associated with plant species with particularly effective grazing defenses, or mature vegetation with strongly developed contrasts (Gammon and Roberts, 1978; Gardener, 1980; L'Huillier el al., 1984). The generalization also breaks down in underutilized pasture as a consequence of the tendency of animals to concentrate grazing on the shorter and more juvenile vegetation on previously grazed areas. In these circumstances, the likelihood of defoliation will be greater for short than for tall plants (Hodgson and Ollerenshaw, 1969; Gibb and Ridout, 1988), increasing the chances of reinforcement of changes in species balance on a patch basis. Differential defoliation may also be more subtle than this, as for example in the report by Bootsma et al. (1990) in which the perennial ryegrass and white clover components of a mixed pasture grazed by deer were defoliated at similar frequency, but the clover was consistently defoliated to a lower level within the sward canopy.

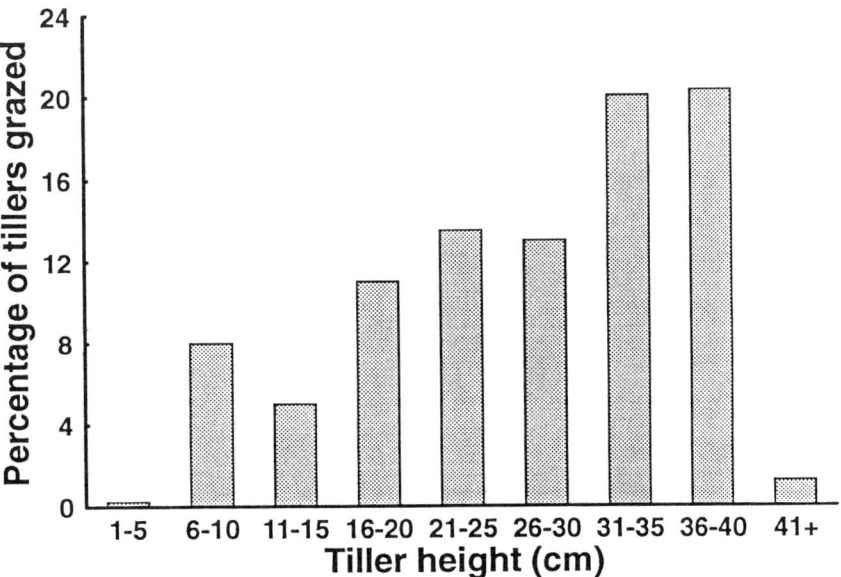

Figure 4. The influence of tiller height on chance of defoliation in *Hyperthelia dissoluta* (Steud.) growing in *Hyperthelia* veld (from Gammon and Roberts, 1978).

In principle, continued selective grazing pressure would be expected to drive the botanical composition of a specific plant community to extreme. In practice, however, species variation in grazing avoidance or grazing tolerance may exert substantial counter-pressure to maintain effective community stability.
 Though strictly outside the terms of this review, mention should be made of the plant characteristics influencing survival and competitive ability in grazed communities.
 Retention of residual leaf area and meristematic zones - associated with grazing avoidance - is one such mechanism. Others are the ability to rapidly mobilize energy reserves and deploy new leaf area, characteristics of successful species in disturbed environments (Grime, 1988).

Briske (1986) emphasized the importance of grazing avoidance as a plant survival strategy in studies on bunch-grass communities in Texas. A more extreme example comes from the work of Clements (1989) in Queensland on the influence of legume morphology on the survival of growing points in mixed grass/legume communities following grazing by cattle (Figure 5). Work in both the USA and Australia (Caldwell et al., 1981; Hodgkinson et al., 1985) has demonstrated that differences in regrowth capability can be more important than differential defoliation in determining species survival in mixed grass communities. Similar arguments may apply in the case of alternative grass species in temperate swards (Tavakoli et al., 1993).

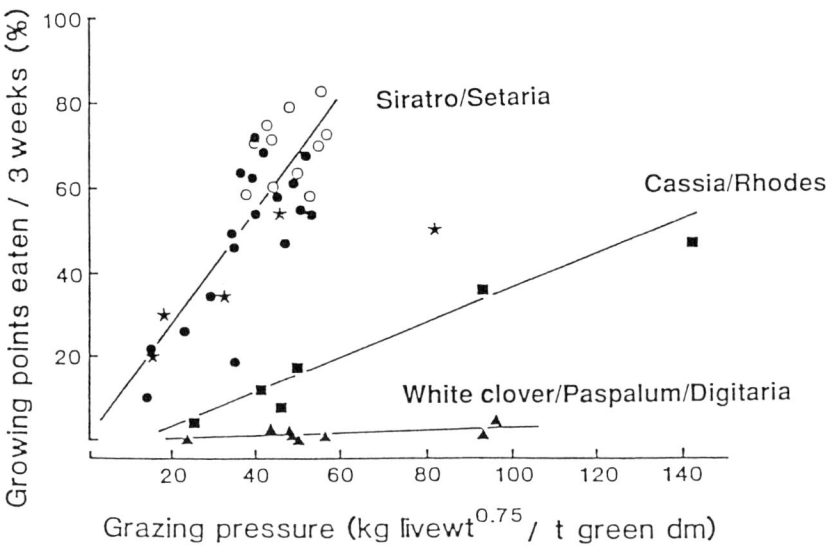

Figure 5. Percentage of growing points grazed for three legumes growing in legume/grass associations: ● Siratro *(Macroptilium atropurpureum)* with setaria *(Setaria sphacelata* cv Nandi), rotationally grazed; o Siratro with setaria, continuously stocked; ★ *Centrosema virginianum* with setaria (cv Narok and Nandi), rotationally grazed; ■ round-leaved cassia (*Cassia rotundifolia* cv Wynn) with Rhodes grass (Chloris gayana cv Callide) rotationally grazed; ▲ white clover with *Paspalum notatum* and *Digitaria* spp. rotationally grazed (from Clements, 1989).

Investigations of dynamic ecology in plant communities have been based largely on observations on marked plants or plant components to define patterns of defoliation frequency and severity, and to establish patterns of regrowth and development (e.g., Davies, 1981; Briske, 1990). These procedures draw heavily on the techniques of classical population ecology (Harper, 1977). They are laborious, but they can generate enormous amounts of information on many facets of the plant-animal interface. The need now is to seek means of developing generalized plant responses from the data sets specific to particular communities.

Consequences to Grazing Animals

In general terms, moderate grazing pressures may enhance plant diversity in comparison with ungrazed or lightly grazed communities, but high grazing pressures can reduce both plant diversity and primary production (Hodgkinson and Mott, 1987; McIvor, 1993). However, there are substantial differences in stability and resilience, and hence in resistance to the adverse effects of high grazing pressures, in both native and introduced plant communities in the major climatic zones of the world (Hodgson and Grant, 1981; Archer and Smeins, 1991; Briske and Silvertown, 1993; Walker, 1993).

Even where forage supply and demand are more or less in balance on an annual basis, the tendency of selective grazing to maintain or reinforce vegetation contrasts is clear in, for example, the bimodal distribution of grazed and ungrazed areas associated with excreta (Gibb and Ridout, 1988), the patchy distribution of clover in mixed swards (Clark and Harris, 1985), and the extreme patterns of plant maturity which develop in circumstances where herbage supply exceeds demand on a seasonal basis (Gammon and Roberts, 1978; Gardener, 1980; Grant et al., 1985). In all of these circumstances, the influence of vegetation changes upon the nutritive value of the diet will depend upon the selective grazing strategy of the animals concerned, and upon grazing management.

As already indicated, in most situations the ability of sheep to graze selectively and concentrate nutrients in the diet will be greater than that of cattle, or goats, with deer probably intermediate in ability (Van Dyne et al., 1980; Hodgson et al., 1991; Gong et al., 1993). However, although selective grazing may confer an advantage to the animal in terms of nutrient concentration in the diet, a concomitant reduction in the rate of forage intake as a consequence of smaller bites or lower bite rate may serve to minimize or even reverse the advantage in terms of nutrient intake. Furthermore, the additional searching activities may be counter-productive in energy terms (Murray, 1992). Two examples illustrate the point. In the first, a study on a series of native hill plant communities in Scotland, the grazing strategy of sheep ("nutrient concentrators") was more successful than that of cattle ("rate maintainers") in limiting fluctuations in daily nutrient intake on senescent autumn pastures with marked morphological contrasts, but was less successful on mature summer pastures of high mass but smaller contrasts (Grant et al., 1985; Hodgson et al., 1991). In the second, a study on the Island of Mull reported by Gordon (1989), large herbivores (cattle and male deer) were observed to concentrate their grazing activity on relatively tall, mature vegetation, leaving the close-grazed *Agrostis/Festuca* "greens" of high nutritive value but low potential intake rate to smaller herbivores (sheep and female deer).

Evidence from other natural ecosystems (Owen-Smith, 1993) suggests that animal species contrasts in discriminatory ability may be overlaid by the influence of vegetation structure on the foraging ability of ungulates of different size, and has given rise to theories about the mutual advantages of niche differentiation to individual species within guilds of herbivores (Bell, 1970; Jarman and Sinclair, 1979). Confirmatory evidence is limited, but Gordon (1988) has demonstrated the facilitatory effect of cattle grazing on subsequent vegetation use by deer. There is, however, good evidence from controlled grazing studies on the contrasting effects of different animal species on the same plant community. One example is the much more effective control by cattle than by sheep of *Nardus stricta*, a particularly aggressive invader species in Scottish hill plant communities because of its low preference rating (Hodgson and Grant, 1981). A second is the use of goats to enhance the clover content of mixed white clover/perennial ryegrass swards (Clark et al., 1984), a consequence of the relatively shallow surface-grazing habits of goats in contrast to sheep (Gong et al., 1993).

Management decisions are particularly important to the maintenance of species balance, maturity, and nutritive value in plant communities. Heady (1964) suggested that range vegetational trends were related more to variations in grazing intensity than to contrasts in plant palatability, and the same arguments are used in more intensive situations (Bryant and Sheath, 1987). The major management decisions relate to the timing and severity of defoliation in relation to patterns of plant growth and maturity and, in the more intensive conditions, the use of fertilizer. However, Hodgkinson and Mott (1987) and Walker (1993) argue that, once regressive changes are established on rangeland, hysteresis effects mean that vegetation recovery may be slow even after de-stocking.

There is increasing awareness of the factors influencing plant demography and population ecology in mixed plant communities, and this has been a particularly fruitful area of interchange between ecologists with agricultural and conservation interests. Nevertheless, although some generalizations are now possible, the understanding needed to provide effective management control is probably confined to a relatively few specific communities.

CONCLUSIONS: MODELLING THE PLANT-ANIMAL INTERFACE

The developing understanding of the interactions between populations of forage plants and grazing animals has encouraged rapid progress in modelling these interactions in both directions across the plant-animal interface, but there is still some way to go in the development of effective dynamic models which simulate sequential effects between system components. Detailed appraisal of grazing models is not appropriate here, but the following general comments follow from discussion in earlier sections of this review and are relevant to further modelling activity.

The allometry of mouth and body size provides a logical basis for the choice of specific plant communities or pasture conditions by different animal species or size classes (Illius and Gordon, 1987). However, although the patterns of response in ingestive behavior by different animal species to variations in sward canopy conditions are superficially similar, there are enough indications of species contrasts in specific components of behavior to justify caution in attempting generalization.

The accumulated evidence from controlled studies confirms the dominant influence of variation in bite depth on intake per bite, and of variation in sward canopy height on bite depth, but emphasizes the need for further study on "real" swards to define causation in the control of bite dimensions. There is now a clearer basis for modelling bite area in relation to variations in sward structure.

It seems entirely appropriate to consider bite rate as a function of sward conditions rather than simply as a compensating response to changes in intake per bite. The implications of the proportionate distribution of jaw activity between prehension, biting, and mastication are now becoming clear.

Variation in grazing time must be seen as the primary compensating response when intake per bite and rate of forage intake are limited by sward conditions. However, recent developments in research methodology have done little to augment understanding of the underlying mechanisms influencing grazing time and periodicity, their relationships to other animal activities, and their sensitivity to circadian rhythms.

Concepts of potential intake rate define one dimension of patch choice within plant communities. The influence of plant secondary compounds on expression of choice is also clear, although it is not at present easy to quantify these effects under grazing conditions, or relate them in any quantitative sense to the effects of sward structural characteristics.

The expression of choice must be constantly reinforced by extensive sampling behavior, thus limiting the sharpness of discrimination in grazing activity. Discriminatory grazing may also be modified by the spatial distribution of plant components in the sward canopy, and by contrasts between components. This is clearly an area which requires further investigation.

Although searching and selective grazing activity may confer an advantage to the animal in terms of nutrient concentration in the diet, a reduction in rate of forage intake and a concomitant increase in energy expenditure may serve to minimize or even reverse the advantage in terms of rate of nutrient intake. These issues are clearly of relevance to any attempts to fit conceptual models of foraging strategy (Krebs and Davies, 1984) to grazing ruminants.

Models of tissue dynamics (Johnson and Parsons, 1985) provide the basis for projecting the impact of grazing on pasture production and, if they can accommodate species differences in the dynamics of response to defoliation, may be appropriate for modelling aspects of community change. However, success in this respect would seem to be dependent upon three prerequisites:

a) incorporation of a facility to simulate the dynamics of natality and mortality in appropriate units of tissue turnover.

b) recognition of the importance of units of production in the simulation of sward canopy structure as a basis for interactive defoliation activity.

c) development of procedures for defining and simulating preference ranking of individual production units which, in combination with the definition of spatial distribution in (b) above, will provide an interactive basis for simulating selective defoliation.

Finally, it is clear that, to be effective, models of intake control need to make allowance for the trade-off between facilitatory and inhibitory factors ascribable to the physical, structural and nutritional characteristics of herbage and their combined effects on physical, chemostatic, and behavioral controls (Hodgson, 1985). This concept of mutual balance was first postulated by McClymont (1967) but has not been effectively incorporated into theoretical models of intake control, although Forbes (1980) provided a framework within which it might operate. One of the difficulties in dealing adequately with this area of interest is that nutritionists and behavioral ecologists have not so far pooled resources in order to provide comparative information on the relative importance of and ranges of sensitivity to the groups of plant characteristics involved in influencing behavioral and nutritional constraints. We suggest that one major outcome of the current Conference might be a resolve that, over the next 25 years, nutritionists and ecologists should develop common programs to investigate one of the most interesting interface areas in animal science today. Given the predominance of pastorally-based livestock systems on a World scale, its potential importance cannot be overstated.

REFERENCES

Allden, W.R. 1962. Rate of herbage intake and grazing time in relation to herbage availability. Proc. Aust. Soc. Anim. Prod. 4:163-166.

Allden, W.G., and I.A. McD Whittaker. 1970. The determinants of herbage intake by grazing sheep: The interrelationship of factors influencing herbage intake and availability. Aust. J. Agric. Res. 21:755-766.

Archer, S., and F.E. Smeins. 1991. Ecosystem-level processes. p. 109-139. *In* R.K. Heitschmidt and J.W. Stuth (ed.) Grazing management. An ecological perspective. Timber Press, Portland, OR.

Arnold, G.W. 1966. The special senses in grazing animals. II. Smell, taste, touch and dietary habits in sheep. Aust. J. Agric. Res. 17:531-542.

Arnold, G.W., and M.L. Dudzinksi. 1967. Studies on the diet of the grazing animal. II. The effect of physiological status in ewes and pasture availability on herbage intake. Aust. J. Agric. Res. 18:349-359.

Arnold, G.W., and J.L. Hill. 1972. Chemical factors affecting selection of food plants by ruminants. p. 71-101. *In* J.B. Harborne (ed.) Phytochemical ecology. Proc. Phytochemical Soc. Symposium No 8, Academic Press, London.

Barry, T.N., and B.J. Blaney. 1987. Secondary compounds of forages. p. 91-119. *In* J.B. Hacker and J.H. Ternouth (ed.) The nutrition of herbivores. Proc. Second International Symposium on the Nutrition of Herbivores, Brisbane, Australia. 6-10 July 1987 Academic Press, Sydney.

Barthram, G.T., and S.A. Grant. 1984. Defoliation of ryegrass-dominated swards by sheep. Grass Forage Sci. 39:211-219.

Bazeley, D.R. 1990. Rules and cues used by sheep foraging in monocultures. p. 43-367. *In* R.M. Hughes (ed.) Behavioural mechanisms of food selection. Proc. NATO Advanced Research Workshop on Behavioural Mechanisms of Food Selection. Greynog, Wales. 17-21 July 1989. Springer-Verlag, Berlin.

Bell, R.H.V. 1970. The use of the herb layer by grazing ungulates in the Serengati. p. 111-123. *In* A. Watson (ed.) Animal populations in relation to their food resources. Blackwell, Oxford.

Black, J.L., and P.A. Kenney. 1984. Factors affecting diet selection by sheep. II. Height and density of pastures. Aust. J. Agric. Res. 35:565-578.

Bircham, J.S., and J. Hodgson. 1983. The influence of sward condition on rates of herbage growth and senescence in mixed swards under continuous stocking management. Grass Forage Sci. 38:323-331.

Bootsma, A., A.M. Ataja, and J. Hodgson. 1990. Diet selection by young deer grazing mixed ryegrass/white clover pastures. Proc. N.Z. Grassl. Assoc. 51:187-190.

Briske, D.D. 1986. Plant responses to defoliation: Morphological considerations and allocation priorities. p. 425-427. *In* P.J. Joss, P.W. Lynch, and O.B. Williams (ed.) Proc. Second Int. Rangelands Congress, Adelaide. Australian Academy of Sciences, Canberra.

Briske, D.D. 1990. Vegetation dynamics in grazed systems: A hierarchical perspective. p. 1829-1833. *In* Proc. XVI Int. Grassld. Congr., Nice, France, 4-11 Oct. 1989. Association Francaise pour la Production Fourragere, INRA, Versailles.

Briske, D.D., and J.W. Silvertown. 1993. Plant demography and grassland community balance: The contribution of population regulation mechanisms. Proc. XVII Int. Grassl. Congr. New Zealand and Queensland 8-21 Feb. 1993 (in press).

Burlison, A.J., J. Hodgson, and A.W. Illius. 1991. Sward canopy structure and the bite dimensions and bite weight of grazing sheep. Grass Forage Sci. 46:29-38.

Bryant, A.M., and G.W. Sheath. 1987. The importance of grazing management to animal production in New Zealand. Proc. 4th Animal Sci. Congr. of the Asian-Australasian Assoc. of Anim. Prodn. Soc., p. 13-17.

Cahn, M.G., and J.L. Harper. 1976. The biology of the leaf mark polymorphism in *Trifolium repens* L. 2. Evidence for the selection of leaf marks by rumen fistulated sheep. Heredity 37:327-333.

Caldwell, M.M., J.H. Richards, D.A. Johnson, R.S. Novak, and R.S. Dzurec. 1981. Coping with herbivory: Photosynthetic capacity and resource allocation in two semi arid *Agropyron* bunch grasses. Oecologia 50:14-24.

Chacon, E., and T.H. Stobbs. 1976. Influence of progressive defoliation of a grass sward in the eating behaviour of cattle. Aust. J. Agric. Res. 27:709-727.

Chambers, A.R.M., J. Hodgson, and J.A. Milne. 1981. The development and use of equipment for the automatic recording of ingestive behaviour in sheep and cattle. Grass Forage Sci. 36:97-105.

Clark, D.A., and P.S. Harris. 1985. Composition of the diet of sheep grazing swards of differing white clover content and spatial distribution. N.Z. J. Agric. Res. 28:233-240.

Clark, D.A., M.G. Lambert, and M.P. Rolston. 1982. Diet selection by goats and sheep on hill country. Proc. N.Z. Soc. Anim. Prod. 42:155-157.

Clark, D.A., M.P. Rolston, M.G. Lambert, and P.J. Budding. 1984. Pasture composition under mixed sheep and goat grazing on hill country. Proc. N.Z. Grassl. Assoc. 45:160-166.

Clements, R.J. 1989. Rates of destruction of growing points of pasture legumes by grazing cattle. p. 1027-1028. *In* Proc. XVI Int. Grassl. Congr. Nice, France. 4-11 Oct 1989. Association Francaise pour la Produccion Fouragere, INRA, Versailles.

Collins, H.A. 1989. Single and mixed grazing of cattle, sheep and goats. Ph.D. Thesis, Lincoln College.

Collins, H.A., and A.M. Nicol. 1986. The consequences for feed dry matter intake of grazing sheep, cattle and goats to the same residual herbage mass. Proc. N.Z. Soc. Anim. Prod. 46:125-128.

Crawley, M.J. 1983. Herbivory. The dynamics of animal plant interactions. Blackwell, Oxford.

Curll, M.L., and R.J. Wilkins 1982. Frequency and severity of defoliation of grass and clover by sheep at different stocking rates. Grass Forage Sci. 37:291-297.

Davies, A. 1981. Tissue turnover in the sward. p. 179-227. *In* J. Hodgson, R.D. Baker, A. Davies, A.S. Laidlaw, and J.D. Leaver (ed.) Sward measurement handbook. British Grassland Society, Hurley, Berks.

Demment, M.W., and P.J. Van Soest. 1985. A nutritional explanation for body-size patterns of ruminant and nonruminant herbivores. Am. Nat. 125:641-672.

Demment, M.W., R.A. Distel, T.C. Griggs, E.A. Laca, and G.P. Deo. 1993. Selective behaviour of cattle grazing ryegrass swards with horizontal heterogeneity in patch height and bulk density. Proc. XVII Int. Grassl. Congr., New Zealand and Queensland (in press).

Doughery, C.T., N.W. Bradley, P.L. Cornelius, and L.M. Lauriault. 1987. Herbage intake rates of beef cattle grazing alfalfa. Agron. J. 79:1003-1008.

Dougherty, C.T., N.W. Bradley, P.L. Cornelius, and L.M. Lauriault. 1989. Ingestive behaviour of beef cattle offered different forms of lucerne (*Medicago sativa* L.). Grass Forage Sci. 44:335-342.

Dougherty, C.T., M.W. Bradley, L.M. Lauriault, J.E. Arias, and P.L. Cornelius. 1992. Allowance-intake relations of cattle grazing vegetative tall fescue. Grass Forage Sci. 47:211-219.

Flores, E.R., F.D. Provenza, and D.F. Balph. 1989. Relationship between plant maturity and foraging experience of lambs grazing hycrest crested wheatgrass. Appl. Anim. Behav. Sci. 23:279-284.

Forbes, J.M. 1980. A model of the short-term control of feeding in the ruminant: Effects of changing animal and feed characteristics. Appetite 1:21-41.

Forbes, T.D.A. 1988. Researching the plant-animal interface: The investigation of ingestive behaviour in grazing animals. J. Anim. Sci. 66:2369-2379.

Forbes, T.D.A., and J. Hodgson. 1985. Comparative studies of the influence of sward conditions on the ingestive behaviour of cows and sheep. Grass Forage Sci. 40:69-77.

Gammon, D.M., and B.R. Roberts. 1978. Patterns of defoliation during continuous and rotational grazing of the Matopos Sandveld of Rhodesia. 3. Frequency of defoliation. Rhodesia J. Agric. Res. 16:147-164.

Gardener, C.J. 1980. Diet selection and liveweight performance of steers on *Stylosanthes hamata* - native grass pastures. Aust. J. Agric. Res. 31:379-392.

Gibb, M.J., and M.S. Ridout. 1988. Application of double normal frequency distribution fitted to measurements of sward height. Grass Forage Sci. 43:131-136.

Gong, Y., J. Hodgson, M.G. Lambert, A.C.P. Chu, and I. Gordon. 1993. Comparison of response patterns of bite weight and bite dimensions between sheep and goats grazing a range of contrasting herbage. Proc. XVII Int. Grassl. Congr. New Zealand and Queensland 8-12 Feb. 1993 (in press).

Gordon, I.J. 1988. Facilitation of red deer grazing by cattle and its impact on red deer performance. J. Appl. Ecol. 25:1-10.

Gordon, I.J. 1989. Vegetation community selection by ungulates on the Isle of Rhum. III. Determinants of vegetation community selection. J. Appl. Ecol. 26:65-79.

Gordon, I.J.,and C. Lascano. 1993. Foraging strategies of ruminant livestock on intensively managed grasslands: Potential and constraints. Proc. XVII Int. Grassl. Congr., New Zealand and Queensland, 8-21 Feb. 1993 (in press).

Grant, S.A., D.F. Suckling, H.K. Smith, L. Torvell, T.D.A. Forbes, and J. Hodgson. 1985. Comparative studies of diet selection by sheep and cattle: The hill grasslands. J. Ecol. 73:987-1004.

Grant, S.A., L. Torvell, R.H. Armstrong, and M.M. Beattie. 1987. The manipulation of mat-grass pasture by grazing management. p. 623-664. *In* M. Bell and R.G.M. Bunce (ed.) Agriculture and Conservation in the Hills and Uplands. Institute of Terrestrial Ecology, Grange-over-Sands.

Grime, J.P. 1979. Plant strategies and vegetation processes. John Wiley and Sons, Chichester.

Harper, J.L. 1977. The population biology of plants. Academic Press, London.

Heady, H.F. 1964. Palatability of herbage and animal preferences. J. Range Manage. 17:76-82.

Hendricksen, R., and D. J. Minson. 1980. The feed intake and grazing behaviour of cattle grazing a crop of *Lablab purpureus* cv Rongai. J. Agric. Sci., Camb. 95:547-554.

Heitschmidt, R.K., D.D. Briske, and D.L. Price. 1990. Pattern of interspecific tiller defoliation in a mixed grass prairie grazed by cattle. Grass Forage Sci. 45:215-222.

Hodgkinson, K.C., and J.J. Mott. 1987. On coping with grazing. p. 171-192. *In* F.P. Horn, J. Hodgson, J.J. Mott, and R.W. Brougham (ed.) Grazinglands research at the plant-animal interface. A compendium of papers originally presented in abbreviated form at the XV Int. Grassl. Congr., Kyoto, Japan, 30 Aug. 1985. Winrock International.

Hodgkinson, K.C., J.J. Mott, and M.M. Ludlow. 1985. Coping with grazing: A comparison of two savanna grasses differing in tolerance to defoliation. p. 1089-1091. Proc. XV Int. Grassl. Congr., Kyoto, Japan, 24-31 Aug. 1985. Science Council of Japan and Japanese Society of Grassland Science.

Hodgson, J. 1981. Variations in the surface characteristics of the sward and the short-term rate of herbage intake by calves and lambs. Grass Forage Sci. 36:49-57.

Hodgson, J. 1982. Ingestive behaviour. p. 113-118. *In* J.D. Leaver (ed.) Herbage intake handbook. British Grassl. Soc., Hurley, Berks.

Hodgson, J. 1985. The control of herbage intake in the grazing ruminant. Proc. Nutr. Soc. 44:339-346.

Hodgson, J., and D.A. Clark. 1989. The influence of physical and biochemical characteristics upon the selection of white clover by grazing sheep. p. 1049-1050. *In* Proc. XVI Int. Grassld. Congr. Nice, France, 4-11 Oct 1989. Association Francaise pour la Produccion Fourragere, INRA, Versailles.

Hodgson, J., and S.A. Grant. 1981. Grazing animals and forage resources in the hills and uplands. p. 41-57. *In* J. Frame (ed.) The effective use of forage and animal resources in the hills and uplands. Occ. Symp. No. 5, Br. Grassl. Soc. Edinburgh, 2-4 Sept. 1980. British Grassland Society.

Hodgson, J., and W.S. Jamieson. 1981. Variations in herbage mass and digestibility, and the grazing behaviour and herbage intake of adult cattle and weaned calves. Grass Forage Sci. 36:39-48.

Hodgson, J., T.D.A. Forbes, R.H. Armstrong, M.M. Beattie, and E.A. Hunter. 1991. Comparative studies of the ingestive behaviour and herbage intake of sheep and cattle grazing indigenous hill plant communities. J. Appl. Ecol. 28:205-227.

Hodgson, J., and J.H. Ollerenshaw. 1969. The frequency and severity of defoliation of individual tillers in set-stocked swards. J. Br. Grassl. Soc. 24:226-234.

Holmes, C.W. 1987. Pastures for dairy cows. p. 133-143. In A.M. Nicol (ed.) Livestock feeding at pasture. N.Z. Soc. Anim. Prod. Occ. Pub. No 10. N.Z. Soc. Anim. Prod., Ruakura, Hamilton.

Hughes, T.P., and J.R. Gallagher. 1993. The influence of sward height on the mechanics of grazing and intake rate by race horses. Proc. XVII Int. Grassld. Congr., New Zealand and Queensland 8-12 Feb 1993 (in press).

Hughes, T.P., A.R. Sykes, D.P. Poppi, and J. Hodgson. 1991. The influence of sward structure on peak bite force and bite weight in sheep. Proc. N.Z. Soc. Anim. Prod. 51:153-158.

Illius, I.A., D.A. Clark, and J. Hodgson. 1992. Discrimination and patch choice by sheep grazing grass-clover swards. J. Anim. Ecol. 61:183-194.

Illius, I.A., and I.J. Gordon. 1987. The allometry of food intake in grazing ruminants. J. Anim. Ecol. 56:989-999.

Illius, A.W., and I.J. Gordon. 1990. Constraints on diet selection and foraging behaviour in mammalian herbivores. p. 369-393. In R.M. Hughes (ed.) Behavioural mechanisms of food selection. Proc. NATO Advanced Research Workshop on Behavioral Mechanisms of Food Selection. Greynog, Wales. 17-21 July 1989. Springer-Verlag, Berlin.

Jamieson, W.S., and J. Hodgson. 1979. The effects of variation in sward characteristics upon the ingestive behaviour and herbage intake of calves and lambs under a continuous stocking management. Grass Forage Sci. 34:273-282.

Jarman, P.J. 1974. The social organisation of antelope in relation to their ecology. Behaviour 48:215-266.

Jarman, P.J., and A.R.E. Sinclair. 1979. Feeding strategy and the pattern of resource partitioning in ungulates. p. 130-163. In A.R.E. Sinclair and M. Norton-Griffiths, (ed.) Serengeti. Dynamics of an ecosystem. Univ. of Chicago Press, Chicago.

Johnson, I.R., and A.J. Parsons, 1985. A theoretical analysis of grass growth under grazing. J. Theoretical Biol. 112:345-367.

Johnstone-Wallace, D.B., and K. Kennedy. 1944. Grazing management practices and their relationship to the behaviour and grazing habits of cattle. J. Agric. Sci., Camb. 34:190-197.

Kenney, P.A., and J.L. Black. 1984. Factors affecting diet selection by sheep. I. Potential intake rate and acceptability of feed. Aust. J. Agric. Res. 35:551-563.

Kenney, P.A., and J.L. Black 1986. Effect of simulated sward structure on the rate of intake of subterranean clover by sheep. Proc. Aust. Soc. Anim. Prod. 16:251-254.

Krebs, J.R., and N.M. Davies. 1984. Behavioural ecology: An evolutionary approach (2nd ed.) Blackwell Scientific Publications, Oxford.

Laca, E.A., E.D. Ungar, N.G. Seligman, and M.W. Demment. 1992b. Effects of sward height and bulk density on bite dimensions of cattle. Grass Forage Sci. 47:91-102.

Laca, E.A., E.D. Ungar, N.G. Seligman, M.R. Ramey, and M.W. Demment. 1992a. An integrated methodology to study short-term grazing behavior of cattle. Grass Forage Sci. 47:81-90.

Laca, E.A., M.W. Demment, R.A. Distel, and T.C. Griggs. 1993. A conceptual model to explain variation in ingestive behaviour within a feeding patch. Proc. XVII Int. Grassl. Congr., New Zealand and Queensland 8-21 Feb 1993 (in press).

Laidlaw, A.S. 1983. Grazing by sheep and the distribution of species through the canopy of a red clover-perennial ryegrass sward. Grass Forage Sci. 38:317-321.

L'Huillier, P.J., D.P. Poppi, and T.J. Fraser. 1984. Influence of green leaf distribution on diet selection by sheep and the implications for animal performance. Proc. N.Z. Soc. Anim. Prod. 44:105-107.

Malachek, J.C., and D.F. Balph. 1987. Diet selection by grazing and browsing livestock. p. 121-132. *In* J.B. Hacker and J.H. Ternouth (ed.) The nutrition of herbivores. Second Int. Symp. Nutr. Herbivores, Brisbane, Academic Press, Sydney.

Marten, G.C. 1970. Measurement and significance of forage palatability. p. D1-D55. *In* R.F. Barnes, D.C. Clanton, C.H. Gordon, T.J. Klopfenstein, and D.R. Waldo. Proc. Nat. Conf. Forage Quality Eval. and Util. 3-4 Sept. 1969. Nebraska Center for Continuing Education, Lincoln, Nebraska.

Marten, G.C., R.F. Barnes, A.B. Simons, and F.J. Wooding. 1973. Alkaloids and palatability of *Phalaris arundinacea* L. grown in diverse environments. Agron. J. 65:199-201.

McClymont, G.L. 1967. Selectivity and intake in the grazing ruminant. p. 129-137. *In* C.F. Code and H. Werner (ed.) Handbook of physiology, Section 6: Alimentary canal, Vol. 1. Control of food and water intake. American Physiological Soc., Washington, DC.

McIvor, J.G. 1993. Distribution and abundance of plant species in pastures and rangelands. Proc. XVII Int. Grassl. Congr. New Zealand and Queensland. 8-21 Feb. 1993 (in press).

Milne, J.A., J. Hodgson, R. Thompson, W.G. Souter, and G.T. Barthram. 1982. The diet ingested by sheep grazing swards differing in white clover and perennial ryegrass content. Grass Forage Sci. 37:209-218.

Mitchell, R.J., J. Hodgson, and D.A. Clark. 1991. The effect of varying leafy sward height and bulk density on the ingestive behaviour of young deer and sheep. Proc. N.Z. Soc. Anim. Prod. 51:159-165.

Mitchell, R.J., J. Hodgson, and D.A. Clark. 1993a. The independent effects of sward height and bulk density on the bite parameters of Romney ewes and red deer hinds. Proc. XVII Int. Grassl. Congr. New Zealand and Queensland 8-21 Feb 1993 (in press).

Mitchell, R.J., J. Hodgson, D.A. Clark, and C.B. Anderson. 1993b. The bite dimensions of sheep on swards differing in leaf:pseudostem ratio, dead matter content and/or tiller strength. N.Z. J. Agric. Res. (in press).

Murray, M.J. 1991. Maximising energy retention in grazing ruminants. J. Appl. Ecol. 60:1029-1045.

Mursan, A., T.P. Hughes, A.M. Nicol, and T. Suguira. 1989. The influence of sward height on the mechanics of grazing in steers and bulls. Proc. N.Z. Soc. Anim. Prod. 41:233-236.

Newman, J.A., A.J. Parsons, and A. Harvey. 1992. Not all sheep prefer clover: Diet selection revisited. J. Agric. Sci., Camb. 119:275-283.

Nicol, A.M., and G.B. Nicoll 1987. Pastures for beef cattle. p. 119-132. *In* A.M. Nicol (ed.) Livestock feeding on pasture. N.Z. Soc. Anim. Prod. Occ. Pub. No 10. N.Z. Soc. Anim. Prod., Hamilton.

Nicol, A.M., D.P. Poppi, M.R. Alam, and H.A. Collins. 1987. Dietary differences between goats and sheep. Proc. N.Z. Grassl. Assoc. 48:199-205.

Owen-Smith, N. 1993. Comparative foraging strategies of grazing ungulates in African savanna grasslands. Proc. XVII Int. Grassl. Congr., New Zealand and Queensland, 8-21 Feb. 1993 (in press).

Penning, P.D. 1986. Some effects of sward conditions on grazing behaviour and intake by sheep. p 219-226. In O. Gudmundsson (ed.) Grazing research at northern latitudes. Proc. NATO Adv. Workshop on Grazing Research at Northern Latitudes, 5-10 Aug 1985, Iceland, Plenum Press, New York.

Penning, P.D., A.J. Rook, and R.J. Orr. 1991. Patterns of ingestive behaviour of sheep continuously stocked on monocultures of ryegrass or white clover. Appl. Anim. Behav. Sci. 31:237-250.

Poppi, D.P., T.P. Hughes, and P.J. L'Huillier. 1987. Intake of pasture by grazing ruminants. p. 55-64. In A.M. Nicol (ed.) Livestock feeding at pasture. N.Z. Soc. Anim. Prod. Occ. Pub. No 10. N.Z. Soc. Anim. Prod., Hamilton.

Provenza, F.D., and D.F. Balph. 1987. Diet learning by domestic ruminants: Theory, evidence and practical implications. Appl. Anim. Behav. Sci. 18:211-232.

Provenza, F.D., and D.F. Balph. 1988. The development of dietary choice in livestock on rangelands and its implications for management. J. Anim. Sci. 66:2356-2368.

Provenza, F.D., and D.F. Balph. 1990. Applicability of five diet selection models to various foraging challenges ruminants encounter. p. 423-460. In R.M. Hughes (ed.) Behavioural mechanisms of food selection. Proc. NATO Advanced Research Workshop on Behavioural Mechanisms of Food Selection. Greynog, Wales. 17-21 July 1989. Springer-Verlag, Berlin.

Rattray, P.V., K.F. Thompson, H. Hawker, and R.M.W. Sumner. 1987. Pastures for sheep production. p. 89-103. In A.M. Nicol (ed.) Livestock feeding on pasture. N.Z. Soc. Anim. Prod. Occ. Pub. No 10. N.Z. Soc. Anim. Prod., Hamilton.

Stobbs, J.H. 1973a. The effects of plant structure on the intake of tropical pastures. I. Variation in the bite size of grazing cattle. Aust. J. Agric. Res. 24:809-819.

Stobbs, J.H. 1973b. The effect of plant structure on the intake of tropical pastures. II. Differences in sward structure, nutritive value, and bite size of animals grazing *Setaria anceps* and *Chloris gayana* at various stages of growth. Aust. J. Agric. Res. 24:821-829.

Stuth, J.W. 1991. Foraging behaviour. p. 65-83. *In* R. K. Heitschmidt and J.W. Stuth (ed.) Grazing Management. An Ecological Perspective. Timber Press, Portland, OR.

Tavakoli, H., J. Hodgson, and P.D. Kemp. 1993. Response to defoliation of tall fescue (*Festuca arundinacea* Schreb.). Proc. XVII Int. Grassl. Congr. New Zealand and Queensland 8-21 Feb. 1993 (in press).

Van Dyne, G.M., M.R. Brockington, Z. Szozs, J. Daek, and C.A. Ribic. 1980. Large herbivore sub-system. p. 269-537. *In* A.I. Bremeyer and G.M. Van Dyne (ed.) Grasslands, ecosystems and man. Cambridge University Press, Cambridge.

Vincent, J.F.V. 1982. The mechanical design of grass. J. Mater. Sci. 17:856-860.

Wade, M.H. 1991. Factors affecting the availability of vegetative *Lolium perenne* to grazing dairy cows with special reference to sward characteristics, stocking rate and grazing method. Ph.D. Thesis, University of Rennes.

Walker, B.H. 1993. Stability in rangelands: Ecology and economics. Proc. XVII Int. Grassl. Cong. New Zealand and Queensland 8-21 Feb. 1993. (in press).

CHAPTER 20

CHANGES IN FORAGE QUALITY DURING HARVEST AND STORAGE

C. Alan Rotz and Richard E. Muck

INTRODUCTION

In many regions of the world, the climate is not suitable for forage growth during much of the year. In these regions, forage must be conserved through harvest and storage to feed animals during the months when fresh forage is not available. A primary goal in forage conservation is to maintain the crop dry matter (DM) and nutrients with minimal loss. The amount of loss that occurs is influenced by the type and size of equipment and storage facilities used, management decisions, and weather. The major constraint for reducing losses is the cost of the required technology. Costs of production must be balanced with system performance (including losses) to select the best forage systems.

Many harvest and storage systems are used, but the major options are dry hay and silage production. Hay can be harvested in bales of various sizes, shapes, and densities. To produce dry hay, forage is normally dried in the field to a moisture content of less than 200 g/kg. Hay may be harvested at moisture contents up to 280 g/kg, but further processing or treatment is required for proper storage. Field curing of dry hay in thin, wide swaths requires 3 to 5 d of drying. In heavier windrows, field drying may require 6 to 7 d. Rain during the curing process can prolong the drying many more days. Hay that lays in the field more than two weeks is often not suitable for animal feed.

Losses during hay harvest and storage range from 15 to 100% of the initial forage DM. For hay dried under relatively good drying conditions, losses are 15 to 18% (Rees, 1982; Rotz and Abrams, 1988). Rain damage increases the loss by up to 30% (Rotz and Abrams, 1988) and with extended poor drying conditions, the whole crop can be lost. Average losses in hay making are estimated between 24 and 28% (Hodgson et al., 1947; Hoglund, 1964; Waldo and Jorgensen, 1981; Wilkinson, 1981; Buckmaster et al., 1990). Most of this loss occurs during harvest with about 5% loss during the storage of dry hay.

To make silage, forage is normally wilted in the field to a moisture content between 500 and 650 g/kg which reduces field curing time to between 1 and 4 d. Silage is stored in tower and bunker silos, stacks, bags and wrapped large round bales. The handling of wetter material and the reduced field curing time lead to lower harvest losses than in hay systems, but storage losses are greater. Average losses in silage production are 14 to 24% with about half of this loss occurring during storage (Hodgson et al., 1947; Hoglund, 1964; Waldo and Jorgensen, 1981; Wilkinson, 1981; Buckmaster et al., 1990).

C.A. Rotz, Agricultural Engineer, U.S. Dairy Forage Research Center, USDA/Agricultural Research Service, 206 Farrall Hall, Michigan State Univ., East Lansing, MI 48824; R.E. Muck, Agricultural Engineer, U.S. Dairy Forage Research Center, USDA/ARS, 1925 Linden Drive West, Univ. of Wisconsin, Madison, WI 53706.

Losses in hay and silage production include: 1) the physical detachment of forage material and 2) the internal depletion or degradation of plant nutrients. Physical separation is normally caused by harvest equipment, but rain also may disassociate forage material. This loss is primarily leaves. Because leaves have a higher concentration of nutrients, leaf loss causes a reduction of the nutrient concentration in the remaining forage. Nutrients depleted from forage are primarily soluble carbohydrates, but other soluble nutrients may be lost. Soluble carbohydrates are used by plant and microbial respiration whereas all soluble nutrients are susceptible to leaching by rainfall. These losses cause a substantial reduction in highly digestible DM, feed nutrients, and energy. Overall, the sum of these losses (physical detachment and nutrient depletion) typically cause a gain in neutral detergent fiber (NDF) concentration of 30 to 120 g/kg DM and either a small reduction or a small gain in crude protein (CP) concentration (Hodgson et al., 1947; Johnson et al., 1984; Rotz and Abrams, 1988).

The economic value of forage losses can be substantial. This value is a function of the amount of material lost, the effect of the loss on the nutrient concentration of the remaining forage, and the end use of the forage. As such, a comprehensive analysis of forage systems is required to determine the value of losses. On Michigan dairy farms with an annual milk production of 8000 kg/cow-yr, the average value lost ranges from $14/t DM of alfalfa (*Medicago sativa* L.) produced in silage systems to $35/t DM in dry hay systems (Buckmaster et al., 1990). The lost value is even greater for higher producing dairy herds or commercial hay systems where hay is priced according to weight and forage quality. Based upon recent forage production statistics, the value of harvest and storage losses in the U.S. exceeds $2 billion/yr.

Forage losses can be divided into five major categories. These are: 1) plant respiration loss; 2) rain damage; 3) machine-induced loss; 4) hay storage loss, and 5) silo storage loss. Losses and quality changes occurring in each of these categories will be discussed. The discussion includes a brief description of the processes involved, the factors influencing loss, and typical losses and quality changes occurring in each category.

RESPIRATION LOSS

Physical, biological, and chemical changes occur in forage as it dries in the field. The primary biological or metabolic loss is due to plant respiration, but other minor changes also occur. This section discusses all DM losses and nutrient changes during field curing except those caused by rainfall or machine treatment.

Factors Which Affect Respiration

Respiration performs a vital role in living plants. The hydrolytic and respiratory enzymes present in living cells continue to function after the crop is mowed until some lethal condition intervenes (Sullivan, 1973). As the crop dries, respiration decreases because diffusion processes supplying oxygen within the plant are inhibited and the enzyme-catalyzed reactions which require the presence of water are reduced (Parkes and Greig, 1974). Respiration decreases nearly in proportion to the decrease in moisture content expressed on a wet basis (Wood and Parker, 1971) approaching zero between 260 and 400 g/kg moisture (Greenhill, 1959; Wolf and Carson, 1973). Within this moisture range, plant enzymatic activity ceases or at least drops below a detectable level. Rewetting of the crop by dew or rain reactivates enzyme activity and thus prolongs respiration. Pizarro and James (1972) found that rewetted grass respired at a similar rate as non-rewetted grass at a similar moisture content.

Prolonged field curing also may allow the development of bacteria, yeasts, and fungi on forage (Pizarro and Warboys, 1972). Respiration of these microbial organisms increases the overall respiration rate (Honig, 1979). Since respiration by plant tissue and microbial flora use the same substrate, they have similar effects on loss and nutrient change in the crop (Pizarro and James, 1972). Microbial respiration effects are small during field curing except over extended poor drying conditions.

Plant respiration increases with temperature from no respiration at the freezing point of the forage plant. This temperature is about -4°C for alfalfa, but it may be slightly lower for some grass species (Wilkinson and Hall, 1966). Respiration rate increases exponentially to a temperature of 27°C. At higher temperatures, the rate levels off reaching a maximum at about 45°C. At 55°C, enzymatic activity and respiration cease in alfalfa (Wolf and Carson, 1973).

Plant maturity may influence respiration rate during wilting, but consistent effects are not reported. Pizarro and James (1972) found a marked decline in the respiration rate of perennial ryegrass (*Lolium perenne* L.) with increasing maturity. Respiration rate in grass at the emergence of inflorescence was nearly five times that in grass 30 d after anthesis was complete. In other studies, no variation in respiration rate was found across maturities of ryegrass (Wood and Parker, 1971) and alfalfa (Wilkinson and Hall, 1966). The differing results may be due to more rapid drying of the younger crop. With faster drying, respiration rate declines more quickly, offsetting some of the effect of the higher initial respiration rate (Melvin and Simpson, 1963).

Respiration rate also may vary across species, but little work has directly compared respiration among species as they dry. Morris (1972) found similar respiration-induced losses among five grass species dried slowly in the laboratory. Wilkinson and Hall (1966) reported that wheatgrass (*Agropyron*) appeared to have a higher respiration rate than that of alfalfa.

Conditioning treatments used to speed forage drying may alter respiration rate. Mechanical crushing can stimulate a small increase in respiration, but this effect is not consistently reported (Simpson, 1961; Wood and Parker, 1971). The faster drying obtained by conditioning causes a more rapid drop in respiration rate which likely offsets any initial increase.

Under suitable conditions, photosynthesis may continue for a short time after the crop is mowed (Greenhill, 1959). If photosynthesis occurs, carbohydrate DM is added to the plant offsetting some of the respiration loss. Published data suggest that very small DM increases occur when climatic conditions promote rapid drying, but under cloudy conditions which extend plant life, the gain may be as much as 5% (Rees, 1982).

Dry Matter and Nutrient Losses

Initial plant respiration can be considered as the complete oxidation of hexose sugar to carbon dioxide and water (Parkes and Greig, 1974). The exothermic relationship is:

$$C_6H_{12}O_6 + 6\ O_2 \rightarrow 6\ H_2O + 6\ CO_2 + 2870\ kJ$$

Soluble carbohydrates (primarily sugars) in the plant tissue provide the principal substrates for respiration. The water, carbon dioxide, and heat produced by respiration leave the plant, causing a DM loss. Respiration rate rapidly declines when the readily available carbohydrates are depleted. Substrates other than carbohydrates (fat and protein) are then used, but a more complex conversion is required (Greenhill, 1959). Depletion of available carbohydrates normally does not occur during field wilting unless the crop is heavily damaged by rain.

Most studies show good agreement between theoretical and measured values of DM loss, CO_2 production, and heat production (Greenhill, 1959; Wilkinson and Hall, 1966; Wood and Parker, 1971). In the one exception, Parkes and Greig (1974) measured DM loss that was six times that determined based upon measured CO_2 production. They suggested that a fermentation process occurred that converted sugar to lactic acid, ethanol, and CO_2.

Dry matter losses due to plant respiration are difficult to measure during field curing and measured values vary widely. Losses measured in field-cured alfalfa vary between -8 and 19% of the initial crop DM (Rotz and Abrams, 1988). The loss is influenced by drying conditions with the greatest loss in material dried in a warm, humid environment. For alfalfa dried under good weather conditions with less than 4 d of field curing, average DM loss due to respiration appears to be 3 to 4% (Rotz et al., 1984; Rotz and Sprott, 1984; Rotz et al., 1987; Rotz and Abrams, 1988). Dry matter loss of ryegrass and white clover (*Trifolium repens* L.) dried in the laboratory averages 5 to 7%. In warm humid conditions, the loss in grass species can exceed 10% (Morris, 1972).

Carbohydrates are the principal compounds used in respiration with little loss of total nitrogen (N) and fiber. In ryegrass, fructosans and total soluble fructose residues decrease throughout drying (Melvin and Simpson, 1963). Sucrose content may decrease early in the drying process but increase appreciably during the latter part of drying. Sucrose may be synthesized from glucose or fructose in leaf tissue to provide the net increase in sucrose. Glucose content of ryegrass does not vary consistently during drying. In alfalfa, the greatest losses are glucose and fructose with some decrease in sucrose (Sullivan, 1973).

Little loss of total N normally occurs during field drying, but some protein breaks down to simpler, nonprotein nitrogen (NPN)(Brady, 1960; Melvin and Simpson, 1963; Carpintero et al., 1979). Protein breakdown in the presence of sufficient moisture is very rapid and the extent of degradation is influenced by the time required to dry the crop. With very rapid wilting (less than 6 h), protein breakdown can be negligible (Carpintero et al., 1979). Based upon the work of Macpherson (1952), NPN increases 5 to 6 g/kg of total N each hour during field wilting until the crop dries to 600 g/kg moisture. At moisture contents below this level, further degradation of protein is negligible. A moderate increase in temperature during drying increases the rate of protein breakdown. However, warmer temperatures normally provide faster drying so the net effect on degraded protein is small.

Protein degraded during wilting forms a rapid accumulation of amides, particularly asparagine (Macpherson, 1952; Melvin and Simpson, 1963). Some of the nitrogenous constituents are decomposed to give CO_2 and ammonia with the latter then converted to amide. Other NPN forms include free amino acids, volatile bases, and peptides. Protein changes during wilting result in decreased true protein solubility (Brady, 1960; Rotz and Abrams, 1988).

In summary, plant respiration during field drying results primarily in the loss of soluble carbohydrates. Consequently, the concentration of CP, crude fiber, acid detergent fiber (ADF), NDF, lignin, and many other forage plant constituents not affected by respiration increase in proportion to the loss of other DM. Within the CP fraction, some true protein is converted to soluble NPN.

RAIN DAMAGE

When rain occurs during field curing, the loss of yield and quality of forage can be high. Yield or DM losses of up to 30% are commonly reported (Shepherd et al., 1954; Schukking and Overvest, 1979; Van Bockstaele et al., 1979; Wilman and Owen, 1982; Collins, 1985; Fonnesbeck et al., 1986; Rotz and Abrams, 1988) with a loss of over 50% reported with heavy rain damage (Collins, 1983). Most of the DM lost consists of highly soluble and digestible plant nutrients so that forage quality is disproportionately reduced. With heavy DM losses (greater than 25%), the crop is normally not acceptable for animal feed, resulting in a complete loss as feed.

Factors Which Affect Loss

The amount of loss incurred by rain damage varies widely, dependent upon many crop and environmental factors. The major factors which affect loss are the characteristics of the rain including amount, intensity, and duration. Crop factors such as moisture content at the time of rain, maturity, leaf-to-stem ratio, swath density, plant species, and the crop conditioning treatment also influence loss. Rain can indirectly cause other losses. Additional mechanical treatments may be needed for timely drying of the crop. Also, delayed harvest can retard crop regrowth, further reducing DM yield over the harvest season.

Environmental Factors

Leaf loss is one form of damage caused by rain. The impact of raindrops on the crop causes some leaves to sever from the stem and fall to the soil surface. This loss is important in legume species but less so in grass species. Reported values for leaf loss in legumes vary widely. In a field study, Shepherd et al. (1954) measured increased leaf loss of 8.8% in alfalfa subjected to two showers and 36% with three showers. Under artificial rainfall, leaf losses were less than 4% of the crop DM (Rotz et al., 1991b). Using a combination of artificial and natural rainfall, Collins (1983) found that leaf loss in alfalfa and red clover (*Trifolium pratense* L.) increased with the amount of rainfall, averaging 5.7, 8.0, and 10.0% for 25, 41, and 62 mm of rain, respectively.

The leaching of soluble plant nutrients is the predominant loss from rain damage. The influence of rain and crop characteristics on leaching loss is difficult to measure in the field under natural rainfall. Field measurements often include leaf loss as well as leaching loss and in some cases increased mechanical losses following rain. Published data on leaching loss also include plant and microbial respiration losses following the rain. When the plant is rewetted, respiration of plant tissue and microflora on the plant continue until the crop is again below about 300 g/kg moisture. Since both leaching and respiration remove many of the same plant constituents, it is impossible to separate the two losses. Field studies show a trend for loss to increase in proportion to the amount of rain. In general, the increased loss is about 0.7%/mm of rain, but it may be as high as 1%/mm (Shepherd et al., 1954; Van Bockstaele et al., 1979; Collins, 1983; Rotz and Abrams, 1988).

More controlled studies using artificial rainfall indicate that leaching loss increases with the amount and duration of rainfall and tends to decrease with increased rain intensity. Rotz et al. (1991b) found that alfalfa exposed to rain at a rate of 18 mm/h lost about 0.1% DM/mm of rain. With the rain amount held constant at 18 mm, the loss was about 3% DM for the first hour of rainfall and 1.1%/h for longer durations. With an increase in rain intensity, duration for a given amount of rain decreased reducing the loss per unit of rain.

Experiments which use artificial rainfall tend to show lower losses than those which occur under natural rain. Lower losses are due to differences in rainfall duration and the kinetic energy of the raindrops. Natural rainfall normally occurs over several hours and may persist for several days. Artificial rain often is applied in a relatively short period of 5 min to 1 h, and the crop begins to redry soon after the water is applied (Collins, 1982; Collins, 1983; Rotz et al., 1991b). With a shorter duration, losses are expected to be less.

The kinetic energy absorbed by the crop during rain is related to the size and velocity of raindrops. Artificial raindrops are normally smaller in size and/or lower in velocity than those of natural rain. A five-fold increase in drop diameter can increase the kinetic energy of the drop by 600 times (Rotz et al., 1991b). Although not proven, the energy absorbed on impact likely has a large effect on leaf shatter and the leaching of soluble nutrients. Since droplet size varies widely among natural rain occurrences, this also may explain some of the difference in loss incurred among showers and storms of different intensities.

Crop Factors

Many crop characteristics affect the amount of loss that occurs with rain, but few show consistent effects. Perhaps the greatest effect is from crop moisture content at the time of rainfall. Most experimental data indicate that rain early in the drying process causes little loss (Murdoch and Bare, 1963; Van Bockstaele et al., 1979). McGechan (1989b) noted an increase in loss with a decrease in crop moisture. Artificial rainfall on alfalfa of 400 g/kg or less moisture caused slightly greater leaching of plant nutrients and 15% greater leaf loss compared to alfalfa with 650 g/kg moisture (Rotz et al., 1991b).

Among legumes, plant species does not appear to affect leaf loss except that loss appears to be higher in plants with a greater leaf to stem ratio. Collins (1983) measured similar losses for both alfalfa and red clover with a trend toward greater loss in the more leafy clover. Stage of maturity did not affect leaf loss from either species. In another study, less rain-induced leaf loss occurred in legume-grass mixtures than occurred in pure stands of alfalfa, red clover, and birdsfoot trefoil (*Lotus corniculatus* L.) (Collins, 1985). Losses that occurred in the mixtures were primarily from the legumes.

Leaching losses likely vary among forage species, but few studies have compared species under the same conditions. Extensive comparisons of leaching losses among alfalfa, red clover, and birdsfoot trefoil show inconsistent results (Collins, 1982, 1983, 1985). In the 1982 study, leaching of nonstructural carbohydrates and cell contents was highest in red clover and lowest in alfalfa. In the 1983 study, leaching losses were the same in alfalfa and red clover, and in the 1985 study, leaching losses were highest in alfalfa, moderate in red clover, and relatively low in birdsfoot trefoil. In spite of inconsistencies among the studies, one trend was evident: forage plants with the lowest NDF content (highest concentration of cell contents) consistently had the greatest loss.

Leaching losses have not been directly compared between grass and legume species. Collins (1985) reported that DM losses from legume-smooth bromegrass (*Bromus inermis* Leyss.) mixtures were about 20% greater than those in respective pure legume stands. This suggests that grass was more susceptible to leaching. In the same experiment, however, fiber concentration increases in the forages indicated a greater leaching of nutrients from the pure legume crops. Field losses measured in Coastal bermudagrass (*Cynodon dactylon* (L.) Pers) were very low per unit of rainfall (0.13%/mm) compared to those commonly measured for legumes and other grasses (Hart and Burton, 1967). In a series of field trials, rain-induced losses appeared to be greater in Italian ryegrass (*Lolium multiflorum* Lam.) than in the perennial ryegrass species (Van Bockstaele et al., 1979). Based upon the limited available data, species effects appear most closely related to the initial concentration of cell contents in the forage.

Crop maturity also influences the concentration of cell contents in the forage and thus may influence loss. Collins (1983) found slightly less loss in both alfalfa and red clover at a late bloom stage of maturity compared to that at bud stage. In an extensive field study with a grass-white clover mixture, stage of maturity did not affect the extent of reduction in digestible DM caused by rain (Wilman and Owen, 1982). Under artificial rain, Rotz et al. (1991b) found no correlation between alfalfa maturity and loss. These results indicate that crop maturity effects are often small and relatively unimportant.

Conditioning treatments applied to hay to speed drying may increase the crop's susceptibility to rain loss. Crushing or crimping has a small effect, perhaps increasing rain loss by 20% (Murdoch and Bare, 1963; Rotz et al., 1987). Forage cut with a flail mower or conditioner can be more susceptible to damage. Rain-induced losses were reported to double with this type of treatment of grass crops, but little difference was found in alfalfa (Svensson, 1978; Rotz and Sprott, 1984).

Swath density may affect loss because forage spread in a thin swath is more exposed to the rain. Rotz and Sprott (1984) found slightly greater rain-induced loss in alfalfa left in a wider swath. In a grass crop, Wilman and Owen (1982) found swath thickness to have no effect on the extent of reduction in digestibility caused by rain. Under better controlled conditions with artificial rainfall, leaching loss was not greatly affected by swath thickness of alfalfa, but leaf loss in heavy swaths was about half that in light swaths (Rotz et al., 1991b).

Other Factors

The time of day when rain occurs appears to have little effect on loss when all other conditions are equal. Rain-induced losses were similar when artificial rain was applied early in the day with rapid redrying following the rain or in the evening with little drying overnight (Rotz et al., 1991b).

Delayed regrowth of the forage crop is an indirect loss from rain damage. Regrowth is retarded by about the same number of days as the length of the wilting period (Schukking and Overvest, 1979). This delay is caused by shading of the stubble by the cut forage and damage to the initial regrowth by delayed machine operations. A delay in the application of N fertilizer also can be a factor on grass crops. Delayed regrowth may affect the quality of the following harvest, but this effect is difficult, if not impossible, to predict.

Extra machinery operations required to dry the rewetted crop cause loss indirectly related to rain damage. Machine losses and their effect on forage quality are addressed later in this chapter. Losses caused by extra operations used following rain may be attributed to rain damage.

In summary, major factors that affect rain-induced losses are the crop species, crop moisture content, and the amount of rain. Other minor influences come from crop maturity and the form of conditioning applied to the crop. Trends in the effects of various factors on rain damage can be established considering all available information just discussed. Typical losses for a mechanically conditioned alfalfa crop are illustrated in Figure 1. This typical loss includes leaf shatter and leaching caused by rain and the plant and microbial respiration caused by the delay in drying.

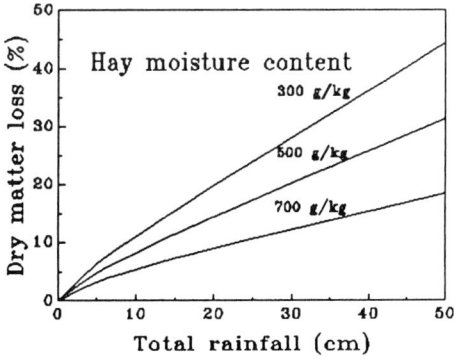

Figure 1. Typical rain-induced losses for mechanically conditioned alfalfa.

Effects on Forage Quality

Dry matter losses caused by rain damage can greatly alter the quality of the remaining forage. Leaf loss has a minor effect relative to other losses. Since leaves contain a higher concentration of important nutrients for the animal, any loss of leaves results in an overall reduction in nutrient concentration. For example, alfalfa consisting of 50% leaves, a leaf CP content of 280 g/kg, and a stem CP content of 120 g/kg has an overall CP content of 200 g/kg. A 10% loss of leaves reduces the CP concentration in the remaining forage to 196 g/kg. Although a substantial loss of DM and nutrients occurs, the nutrient concentration in the remaining forage is reduced only slightly.

Leaching loss and the microbial respiration resulting from rain damage have a more marked effect on the concentration of many plant nutrients. Rain leaches the more soluble nutrients from forage. These include readily available carbohydrates, soluble N, soluble minerals, and lipids (Fonnesbeck et al., 1986). These nutrients come primarily from the cell contents of the forage plant, and they are highly digestible nutrients for the animal. As a result, rain damage causes a substantial decrease in digestibility and an increase in fiber concentration (Collins, 1983; Rotz and Abrams, 1988).

Experimental data indicate that the portion of nutrients lost varies among the major soluble nutrients. Nutrient loss is likely related to the solubility and concentration of plant constituents. Crop species, crop maturity, the growing environment, and other factors affect nutrient concentration and solubility. Available data are sufficient to draw general conclusions on nutrient losses, but the effect of various factors on specific nutrient changes cannot be predicted. These effects are expected to be relatively small, inconsistent, and relatively unimportant in the analysis of rain damage.

Dry matter leached by rain is primarily nonstructural carbohydrate. Data collected by Fonnesbeck et al. (1986) indicate that about half of the DM lost is available carbohydrate. Data from Collins (1982) generally support this. Collins' data also suggest that loss from red clover is higher in nonstructural carbohydrates than the loss from alfalfa and birdsfoot trefoil. The lost carbohydrates are primarily sugars with a little loss of starch (Collins, 1982).

Measurements of N or CP loss from leaching vary widely. Reported data indicate that the loss from alfalfa consists of between 100 and 500 g CP/kg DM (Collins, 1982, 1983, 1985; Fonnesbeck et al., 1986; Rotz and Abrams, 1988; Rotz et al., 1991b). In two studies, Collins (1983, 1985) found a higher portion of the loss from alfalfa to be N compared to the loss from red clover or birdsfoot trefoil. In another study, the opposite was found (Collins, 1982). Limited data on loss from grass indicate that CP loss is similar to that for legumes (Murdoch and Bare, 1963). Data reported by Rotz et al. (1991b) did not indicate consistent effects of rain and crop characteristics on the portion of the loss which was CP. As an average or typical value, about 300 g/kg of the leached DM can be assumed to be CP.

Nitrogen leached from the plant tissue is highly soluble, so the concentration of water-insoluble protein is increased (Rotz and Abrams, 1988). The concentration of acid detergent insoluble protein (ADIP) also is increased. Rotz et al. (1991b) found that the increase in ADIP was often a little greater than that explained by elution of soluble protein. Rain damage, therefore, also may reduce the solubility of some of the remaining protein.

The remainder of the leaching loss consists of soluble minerals and lipids. Data of Fonnesbeck et al. (1986) indicate that for alfalfa, soluble minerals and total lipids each make up about 100 g/kg of the leached DM. In legume and legume-grass mixtures, Collins (1985) measured a small (1.5 g/kg) decrease in calcium concentration with little change in phosphorus, potassium, and magnesium.

MECHANICAL LOSSES

Several machine operations are normally required to harvest forage, and each operation causes additional loss. Mechanical losses affect forage quality in two ways. First, the DM lost contains nutrients that would have benefitted the animal consuming the forage. Second, the loss affects the nutrient concentration in the remaining forage. Mechanical losses often include more leaf material than stem material. Since leaves have a greater concentration of many important nutrients, a change in the portion of leaves causes a change in the nutrient concentration or quality of the remaining forage.

Operations used in forage harvest include mowing and conditioning, swath manipulation, and harvest by baling or chopping. Forage losses incurred vary widely among the operations. The amount of loss is influenced by the type of machines used and the adjustment of those machines. Many crop factors also influence the loss. These include crop species, maturity, moisture content, portion of crop that is leaves, and swath structure. The following discussion of the major operations includes the factors that influence loss and typical DM losses and quality changes.

Mowing and Conditioning

Most harvested forage is mowed with a mower-conditioner, a machine which combines mowing and conditioning in one operation. Mowing is most often done with either a cutterbar or rotary disk device. Other devices such as rotary drum and flail mowers are sometimes used, but high power requirements and greater losses deter their use. Many types of mechanical conditioning devices are employed to enhance forage drying. The most common device used to condition alfalfa is intermeshing rubber rolls, but other forms of intermeshing and non-intermeshing rubber and steel rolls are used. Flail type conditioning devices often are preferred for mowing grass forage crops. Most conditioning devices provide similar improvements in drying when properly adjusted (Straub and Bruhn, 1975; Rotz and Sprott, 1984; Shinners et al., 1991).

Losses reported for mowing and conditioning generally vary between 1 and 5% of the DM yield (Savoie et al., 1982; Koegel et al., 1985; Savoie, 1988; Shinners et al., 1991). This is a typical range in loss with well adjusted cutterbar and rotary disk mowers. Rotary mowers may at times cause a little more loss than that from cutterbar mowers, but under most conditions, losses are similar (Rotz and Sprott, 1984; Koegel et al., 1985; Shinners et al., 1991). Losses are much higher with flail mowing machines, averaging between 6 and 11% of yield (Barrington and Bruhn, 1970; Honig, 1979; Rotz and Sprott, 1984). Flail conditioning devices adjusted for very aggressive conditioning also may double the loss during alfalfa mowing (Koegel et al., 1985).

Among roll-type conditioning machines, roll design has little effect on loss (Shinners et al., 1991). Adjustment of roll pressure and clearance has more effect than the configuration or material used in the rolls. Crimping rolls, which are rarely used today, may cause a little more loss than the more common crushing-type rolls (Boyd, 1959; Kepner et al., 1960). The difference in loss between a well adjusted mower-conditioner and a similar mowing device without a conditioner appears to be about 1% (Kepner et al., 1960; Dobie et al., 1963; Goss et al., 1964; Rotz et al., 1987). When the machine is adjusted for more severe conditioning, loss may increase an additional 1 to 2% of yield.

Crop maturity influences mowing and conditioning loss in both legume and grass crops (Savoie, 1988; Shinners et al., 1991). The loss can double as forage matures from a late vegetative to a full bloom stage of development. The moisture content of the standing crop decreases with maturity. Perhaps the lower moisture and/or a weakened attachment of leaves cause greater loss.

Crop characteristics such as species and the amount of leaves likely affect loss, but these effects are not well documented. Savoie (1988) found mowing and conditioning loss in timothy (*Phleum pratense* L.) to be about 40% less than that in alfalfa with all other conditions similar. Loss in a mixed stand fell between those of the pure stands. When comparing full harvest systems, Ciotti and Cavallero (1979) found similar loss differences among alfalfa, orchardgrass (*Dactylis glomerata* L.), and mixed stands. In legume crops, greater loss is expected in crops with more leaves. Koegel et al. (1985) observed greater loss in more leafy second and third cuttings than occurred in first cutting alfalfa.

Since mowing and conditioning loss is normally small, the loss has only a small effect on the quality of the remaining forage. Buckmaster et al. (1990) estimated that 75% of this loss is leaves in alfalfa. Because leaves are higher in CP and lower in fiber than stem material, mowing and conditioning loss causes a small decrease in CP concentration and a small increase in fiber concentration in the remaining forage.

Swath Manipulation

A variety of machine treatments are sometimes used to manipulate the forage swath in an attempt to speed the field curing process. These treatments are generally categorized as tedding, swath inversion, and raking. Tedding machines use rotating tines to spread and fluff the swath to improve drying. Swath inverters lift and flip the swath over to expose the more wet forage on the bottom. The swath structure and dimensions are not affected much by inversion, but a small amount of widening and fluffing may occur. Raking operations are used to roll the forage off the soil surface into a more narrow windrow for pickup by a baler or forage harvester.

Tedding and Inversion

Loss caused by tedding is reported between 1 and 3% of the crop yield, but much greater loss can occur (Murdock and Bare, 1960; Ciotti and Cavallero, 1979; Savoie et al., 1982). The beating action of the tedder is most damaging to legume crops due to a more delicate attachment of leaves and stems. Tedding loss can be three to four times greater in alfalfa than in grass crops (Ciotti and Cavallero, 1979; Savoie, 1988). Within legume crops, a greater loss is expected as the leaf portion increases. Field measurements indicate less loss in first cutting alfalfa which normally has less leaves than later cuttings (Savoie et al., 1982; Davis et al., 1989). Laboratory measurements show less potential for loss in birdsfoot trefoil than alfalfa and less loss as the crops mature (MacAulay and Bilanski, 1968; Raghavan and Bilanski, 1973). These effects also may be due to a difference in the portion of leaves on the plant.

Leaf loss is greatly influenced by crop moisture content, particularly in legume crops. As the crop dries, leaf shatter increases exponentially with a very high potential loss in forage containing less than 300 g/kg moisture (Savoie, 1988). Laboratory measurements indicate that this loss may exceed 20% of the yield if tedding is applied to very dry legumes (MacAulay and Bilanski, 1968; Raghavan and Bilanski, 1973; Savoie, 1988).

Tedding loss affects the quality of the remaining forage by decreasing the portion of leaves remaining with the crop. Savoie (1988) found the portion of leaves in the lost material to increase from 60 to 80% as the moisture content of alfalfa at the time of tedding decreased from 800 to 300 g/kg. With a high tedding loss in alfalfa, up to a 10 g/kg DM decrease in CP and a 20 g/kg DM increase in NDF concentrations can occur as a result of the operation.

Swath inverting machines provide a more gentle method for manipulating field curing swaths. Although many designs are used, these machines normally include a pickup device that lifts the swath onto a platform or moving belts.

The swath is turned and dropped back to the soil surface inverted from its original position (Savoie and Beauregard, 1990). Measured increases in field loss due to swath inversion are small, varying between 0 and 1.5% of yield in predominantly alfalfa crops (Davis et al., 1989; Savoie and Beauregard, 1990). When a swath is inverted onto the same soil surface where a swath was removed, succeeding operations may pick up some of the material missed by the swath inverter. Thus, the actual loss may be very small. The limited available data show little relationship between the amount of loss and crop characteristics, including moisture content.

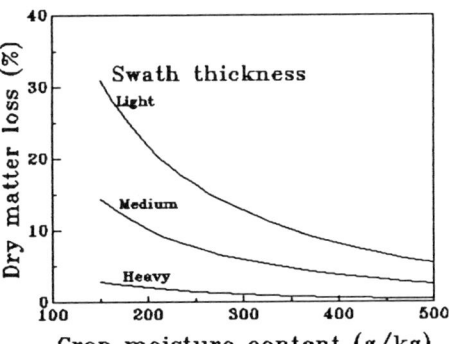

Figure 2. Typical dry matter losses from raking forage with a side-delivery rake.

Raking

Raking loss varies widely with reported losses ranging from 1 to 20% of crop yield (Dobie et al., 1963; Gordon et al., 1969; Savoie et al., 1982; Buckmaster, 1993). Raking loss is influenced by crop moisture content (Buckmaster, 1993) and the density of the forage laying in the swath (Klinner and Shepperson, 1975; Rotz and Abrams, 1988). Loss increases as the crop moisture content decreases, particularly below 300 g/kg (Figure 2). When the crop is spread over much of the field surface, it is more difficult to gather with the rake, and loss is increased. Although comparisons of raking loss in grass and legume crops have not been made, available data indicate similar losses among species at similar moisture contents.

The amount of raking loss varies among machine designs, but this influence is not adequately investigated. Rice (1966) compared six rake designs on grass and found little consistent difference in loss among machines. There was a trend toward less loss with a wheel rake compared to the side-delivery rake. This result is not consistent with other reports (Friesen, 1978; Buckmaster, 1993), so the difference was likely due to machine adjustment. Rice (1966) found a trend for greater loss with a traverse chain design, and Savoie et al. (1982) found greater loss with a rotary windrower. These two designs use tines to sweep hay into a windrow. The sweeping action likely allows more crop to become entangled with the stubble and lost. Side-delivery rake designs provide a rolling and wrapping action that reduces entanglement with the stubble and increases entanglement among plants.

Raking loss normally has a small effect on the quality of the remaining forage. Measured quality of forage lost during raking indicates that just a little more leaf DM is lost than stem DM (Rotz and Abrams, 1988; Buckmaster, 1993). With similar loss of leaves and stems, the quality of the remaining forage is affected little. The portion of leaves lost and the resulting effect on quality may be greater when the raked crop contains less than 300 g/kg moisture (Buckmaster, 1993).

Baling and Chopping

Loss during hay baling typically varies between 2 and 5% of yield with a little greater loss from large round balers than from small rectangular balers (Whitney, 1966; Friesen, 1978; Koegel et al., 1985; Rotz and Abrams, 1988;

Shinners et al., 1992). Baler loss includes pickup and chamber losses. Pickup loss is material dropped as the windrow is lifted into the machine. Chamber loss is that disassociated and dropped during the formation of the bale in the baling chamber.

Pickup loss varies between 1 and 3% of crop yield (Whitney, 1966; Friesen, 1978; Koegel et al., 1985; Rotz and Abrams, 1988; Shinners et al., 1992). This loss is primarily influenced by the structure of the swath or windrow. Pickup loss expressed as a portion of yield decreases as the heaviness or thickness of the swath increases. When forage is mowed and laid in a swath by a mower-conditioner, the forage plants are generally oriented in the same direction with little entanglement. Raked hay is rolled into a tighter windrow where plants become more entangled. Loss is likely greater during the pickup of less entangled swaths (McGechan, 1989b). Pickup loss can be very high when the machine is pulled at a faster speed than the rotating speed of the pickup device. The machine tends to overrun the swath causing excessive loss, as high as 5% (Friesen, 1978).

Crop species and moisture content may affect pickup loss, but these effects appear to be small. Although differences among grass and legume species have not been documented, a comparison of Friesen's (1978) data to other sources indicates that pickup loss may be slightly less in grass crops. Data of Rotz and Abrams (1988) indicate that the quality of material lost is similar to that picked up, i.e., the loss has no effect on the leaf fraction or nutrient concentration in remaining forage.

Chamber loss from a small rectangular baler varies between 1 and 3% of yield for most conditions (Friesen, 1978; Koegel et al., 1985; Rotz and Abrams, 1988; Shinners et al., 1992). This loss is primarily shattered leaves. The amount of chamber loss is largely influenced by crop moisture content with greater loss in drier material. Shattered leaves are very dry, containing less than 100 g/kg of moisture. When hay is baled at night, leaf moisture is higher and similar to stem moisture. With a more uniform moisture content near 180 g/kg, chamber loss can be reduced by 50% (Friesen, 1978).

Chamber loss is higher in large round balers, and it varies considerably among baler designs. For a variable chamber baler, the loss is about 40% greater than that in a small rectangular baler (Koegel et al., 1985). For a fixed chamber baler, the loss can be three times that of a small rectangular baler. Chamber loss is very sensitive to the feed rate of hay. Excessive loss occurs at low feed rates because the hay must be rolled in the chamber many more revolutions to form a bale and the rolling dislodges small particles.

Chamber loss consists of about 80% leaf material and 20% stem particles (Buckmaster et al., 1990). Since the loss is mostly high quality leaf material, excessive chamber loss has more effect on the quality of remaining forage than most other machine losses. By maintaining the chamber loss below 3%, the effect on forage quality is relatively small.

To make silage, a forage harvester is used to chop the crop for easier handling and better packing in a silo. Losses from a forage harvester include pickup and drift losses. Pickup loss is similar to that for balers (Whitney, 1966). Drift loss occurs as the chopped material exits the spout of the harvester and travels toward a trailing wagon or truck. As the machine moves through the field, the spout can become momentarily misdirected causing the flow of material to miss the trailing wagon or truck. A cross wind increases this loss by diverting a portion of the fine particles away from the wagon. Machine adjustment and operator error also contribute to more loss.

Drift loss typically varies between 1 and 3% of the harvested yield (Whitney, 1966; Straub et al., 1986). Dry forage is more susceptible to drift loss than wet forage. The effect of crop moisture on drift has not been measured, but Buckmaster et al. (1990) estimate an exponential increase from

0.5 to 5% as moisture content decreases from 750 to 500 g/kg. The quality of the lost material is expected to be similar to that harvested, so the loss has little effect on the quality of the remaining forage.

HAY STORAGE LOSSES

Respiration causes a low level of DM and nutrient loss to continue during hay storage (Wilkenson and Hall, 1966; Wood and Parker, 1971). Since hay moisture is below 400 g/kg, plant enzymatic activity is very low (Honig, 1979); respiration is primarily from microorganisms (bacteria, fungi, and yeasts) on the hay. Respiration transforms DM to heat and gases which leave the hay. In dry hay stored under cover, little heating occurs and DM loss over 6 mo of storage is about 5% (Collins et al., 1987; Rotz and Abrams, 1988; Rotz et al., 1991c). Similar loss occurs with either large round bales or small bales stored in a shed (Collins et al., 1987). Respiration reduces forage quality by removing some of the most digestible nutrients for the animal.

Hay stored outside and unprotected experiences the same loss as hay stored inside plus an additional loss from weathering of hay on the exposed surface. The additional loss is often 10 to 15% DM for large round bales. Factors which affect losses are different for the two types of storage.

Inside Storage

Dry Matter Loss

For protected hay, the major factor influencing loss during hay storage is the moisture content of hay as it enters storage. Hay with less than 150 g/kg moisture is relatively stable and little respiration occurs. In hay containing more moisture, microbial respiration causes the hay to heat during the first 3 to 5 weeks of storage. The amount of heating and the associated loss increase with moisture content. Dry matter losses during the first month of storage vary from 1% in hay of 150 g/kg moisture to 8% in 300 g/kg moisture hay (Nelson, 1966, 1968; Rotz et al., 1991c). For hay containing more than 300 g/kg moisture, excessive loss and even spontaneous combustion can occur.

Greater heating occurs as the hay density in bales is increased (Nelson, 1966; Buckmaster et al., 1989a). The heat developed per unit of hay DM, however, is nearly independent of density. The increased heating primarily comes from packing more DM into the bale. The loss per unit DM tends to increase with density, but this effect is small.

Heating during the first month of storage helps dry the hay. After the first month, hay normally contains less than 180 g/kg moisture and thus is relatively stable during the remaining storage period. Although a major portion of the loss may occur in the first month, a small loss of about 0.5% DM per month continues throughout storage (Rotz et al., 1991c). Consequently, storage loss also is affected by the length of the storage period.

Crop species appears to have little influence on DM and nutrient losses during inside storage. Studies conducted by Nelson (1966, 1972) show similar moisture content and density effects on changes in alfalfa and native hays during storage. The native hay was a mixture of grasses and clover. Greenhill et al. (1961) found similar temperature and moisture effects on DM loss and quality changes in white clover, alfalfa, and ryegrass hays.

Crop maturity may affect respiration and the resulting storage losses; however, consistent effects have not been reported. Nelson (1968) reported about 25% more heating and DM loss in bud-stage alfalfa compared to more mature alfalfa with no difference between half-bloom and full-bloom. Other studies found no difference in respiration rate across maturities of alfalfa and ryegrass (Wilkinson and Hall, 1966; Wood and Parker, 1971).

Quality Changes

Dry matter loss and heating during hay storage affect the concentration of most nutrient constituents. Much of the lost DM is nonstructural carbohydrate respired to carbon dioxide and water. The primary carbohydrates lost are sugars, particularly sucrose (Greenhill et al., 1961). As carbohydrate is depleted, proteins and fats also are used but at a slower rate. All of this loss is highly digestible DM.

Reported changes in CP concentration during hay storage range from a moderate loss (Davies and Warboys, 1978; Collins et al., 1987) to no change (Nelson, 1966, 1968, 1972) and even a small gain (Rotz and Abrams, 1988). An important factor affecting CP loss is the length of storage. Most protein data are for hay monitored over short storage periods of less than 60 d. During the first month when carbohydrate loss is greatest, protein loss is relatively small. The change in CP concentration is often insignificant and a small increase can occur with high carbohydrate loss in high-moisture hay (Rotz and Abrams, 1988; Rotz et al., 1991c). As storage proceeds, the rate of carbohydrate loss declines but protein loss continues. The differential rate of loss leads to a decrease in CP concentration over 6 to 9 mo of storage (Davies and Warboys, 1978; Collins et al., 1987).

The loss of CP presumably occurs due to slow volatilization of ammonia contained in the hay or produced through microbial respiration. The loss is nearly independent of hay moisture and DM loss. Data for untreated small hay bales indicate a loss of about 2.5 g CP/kg DM during each month of storage (Davies and Warboys, 1978; Collins et al., 1987; Rotz and Abrams, 1988).

The loss of more soluble N components causes small increases in the water insoluble N and acid detergent insoluble N (ADIN) concentrations (Rotz and Abrams, 1988). In addition, the heating of high-moisture hay causes the formation of further ADIN through Maillard reactions. Maillard or browning reactions are the chemical polymerization of sugars and other carbohydrates with amino acids. The polymers produced are measured as increased ADF, and ADIN increases as a fraction of total N. The formation of ADIN is proportional to the number of degree-days the hay is above 35°C (Thomas et al., 1982). Therefore, ADIN increases with the loss of more soluble protein and the heat damage of remaining protein (Buckmaster et al., 1989a).

Fiber concentrations consistently show an increase during hay storage (Greenhill et al., 1961; Collins et al., 1987; Rotz and Abrams, 1988; Buckmaster et al., 1989a). The increase primarily is due to the loss of non-fiber constituents with little or no loss of fiber. Components not lost during storage include ADF, NDF, crude fiber, lignin, and ash.

The digestibility of forage DM and many nutrient constituents decreases during storage (Davies and Warboys, 1978; Collins et al., 1987; Rotz and Abrams, 1988). The decrease occurs because the most digestible constituents are used in respiration. Heating also reduces the digestibility of CP through the formation of ADIN (Thomas et al., 1982). The digestibility of fiber changes less than that of many other plant components (Nelson, 1968). With the loss of readily available carbohydrates and a decrease in DM digestibility, energy content of the forage also is reduced.

Quality changes are relatively small in hay of less than 200 g/kg moisture. Typical changes in alfalfa hay during 6 mo of storage are a 10 to 20 g/kg decrease in digestible DM, no change in CP and ADIN concentrations, and a 10 g/kg increase in NDF concentration (Rotz and Abrams, 1988). Only in higher moisture hay do substantial changes occur that affect the diet and performance of animals consuming the forage. For example, in hay of 250 g/kg moisture, a DM loss of 8% causes about a 30 g/kg decrease in digestible DM content with a 35 g/kg increase in NDF content.

Outside Storage

Losses and quality changes in large round bales stored outside vary widely. Reported DM losses range from 3 to 40% (Rider et al., 1979; Verma and Nelson, 1983; Belyea et al., 1985; Collins et al., 1987; Brasche and Russell, 1988; Huhnke, 1988, 1990; Harrigan and Rotz, 1992). Many factors affect the loss, and most of these have not been examined in a common experiment.

Of the factors affecting loss during outside storage, weather, length of storage, and storage method have the greatest impact. An interaction occurs among these factors in that the storage method has less effect during short storage periods and/or in dry climates. The loss is again primarily caused by microorganisms in the hay, and the biological activity is greatest when the hay is moist and warm. Loss is less in hay stored over winter periods in northern climates or in hay stored in more arid climates where the hay remains relatively dry. For a given set of conditions, the loss appears to be nearly proportional to the length of storage. Reported losses in alfalfa hay range from 0.5 to 1.5% DM/mo of storage in drier climates (Rider et al., 1979; Huhnke, 1988) and 1.0 to 3.0% DM/mo in wetter climates (Verma and Nelson, 1983; Belyea et al., 1985; Collins et al., 1987; Brasche and Russell, 1988; Harrigan and Rotz, 1992).

Storage method affects loss by providing different levels of protection from the environment. Major options for outside storage of round bales include the placement of bales on the soil without covers, elevated off the soil without covers, and elevated with covers. When stored outside, the center of bales is preserved similar to hay stored in a shed (Harrigan and Rotz, 1992). Much greater loss occurs in the outer 10 to 20 cm of the bale where hay is exposed to the environment. When set on damp soil, high moisture levels develop in the bottom of the bale causing considerable deterioration and loss in that portion of the bale. Compared to elevated bales, setting bales on soil causes an average additional loss of about 3% DM (Verma and Nelson, 1983; Collins et al., 1987; Huhnke, 1988, 1990; Russell et al., 1990).

When rain occurs on an exposed bale, some of the rain is absorbed in the outer layer of the bale. The moisture content in this portion of the bale is increased to between 250 and 400 g/kg (Russell and Buxton, 1985; Huhnke, 1988; Harrigan and Rotz, 1992). The moisture content inside the bale also may increase a small amount. The increased moisture leads to greater microbial activity and loss. A plastic wrap around the bale circumference greatly reduces the moisture accumulation, particularly in bales elevated from the soil. A plastic wrap on an elevated bale reduces storage loss to about 35% of the loss in an uncovered bale (Rider et al., 1979; Verma and Nelson, 1983; Belyea et al., 1985; Collins et al., 1987; Brasche and Russell, 1988; Huhnke, 1988, 1990; Harrigan and Rotz, 1992).

Forage species may have some effect on loss. When stored under the same conditions, bermudagrass and smooth bromegrass hay in unprotected bales experienced about 30% less loss than that in alfalfa bales (Rider et al., 1979; Brasche and Russell, 1988). The apparent reason for the difference is that bales of grass hay shed rain more easily. This is supported by lower moisture contents in grass hay following storage. Since grass normally contains greater amounts of fiber, grass hay also may have less soluble matter to lose.

When round bales contain more than 180 g/kg of moisture, storage loss is increased (Verma et al., 1986). The outer bale surface dries to the moisture conditions of the environment with little additional loss, but the interior of the bale goes through a heating process with the associated loss described above for bales stored inside a shed. Other factors such as bale density, bale orientation, bale binding material, and the design of the baler may at times affect loss (Huhnke, 1990; Russell et al., 1990). With the limited available data, general conclusions across conditions cannot be drawn. These effects appear to be small with good harvest and storage practices.

Quality changes in large round bales are variable, but reported data indicate some relatively consistent trends. The DM lost is once again the most digestible portion of hay resulting in a decrease in digestible DM content (Russell and Buxton, 1985; Brasche and Russell, 1988; Huhnke, 1990; Russell et al., 1990). Fiber concentrations generally increase. In some studies, the increased fiber content is small, indicating that some of the lost DM is fiber (Verma and Nelson, 1983; Harrigan and Rotz, 1992). In other studies, the increased fiber content, particularly NDF, is very high, indicating no loss of fiber (Collins et al., 1987; Russell et al., 1990). Changes in lignin content show similar variations (Russell and Buxton, 1985; Brasche and Russell, 1988; Russell et al., 1990).

Crude protein changes are even more variable than those for fiber. Reported values range from a substantial increase to a substantial decrease in concentration with an average of little change in concentration (Rider et al., 1979; Verma and Nelson, 1983; Russell and Buxton, 1985; Collins et al., 1987; Huhnke, 1990; Russell et al., 1990; Harrigan and Rotz, 1992). ADIP concentration generally increases with the heating of hay and the loss of more soluble protein. More must be learned about the factors that affect forage quality in order to develop more quantitative understanding of how these factors affect hay quality in bales stored outside.

SILO STORAGE LOSSES

In contrast to hay, a moist forage is preserved by ensiling in an anaerobic environment at a low pH. Creating a silo environment without oxygen is essential for stopping plant respiration, preventing aerobic microbial growth, and stimulating the growth of lactic acid bacteria. The lactic acid bacteria ferment sugars producing principally lactic and acetic acids which lower silage pH. A low pH is essential for inhibiting plant enzyme activity and preventing the growth of undesirable anaerobic microorganisms.

Ensiling Phases and Processes

The effect of a particular plant, microbial, or chemical process on silage quality depends on the nature of the process, ensiling time, and environmental conditions. The silo storage period can roughly be divided into four major phases: pre-seal, active fermentation, stable phase, and feedout. Each phase is dominated by several major processes (Figure 3).

The pre-seal phase represents the time between the start of filling and the sealing of the silo. Oxygen is present, and the dominant process affecting forage quality is plant respiration. In the living plant, a portion of the energy liberated by respiration is captured in adenosine triphosphate. Biosynthesis is assumed to be limited in the harvested forage so most of the energy is released as heat (McDonald, 1981). In the silo, this heat is not readily dissipated so forage temperature is raised. Thus, respiration in the pre-seal phase not only causes silage DM loss, but the rates of many other ensiling processes are affected by the temperature rise. If heating from plant respiration is sufficient to raise temperatures above 35°C, Maillard or browning reactions may become significant. In these reactions, principally amino acids and sugars are polymerized, increasing ADIN content, and releasing heat.

Other plant enzymes also are active in this period. Enzymes released by cell rupture during chopping break down protein to peptides and free amino acids, and carbohydrates to sugars. Ruptured plant cells provide substrate for

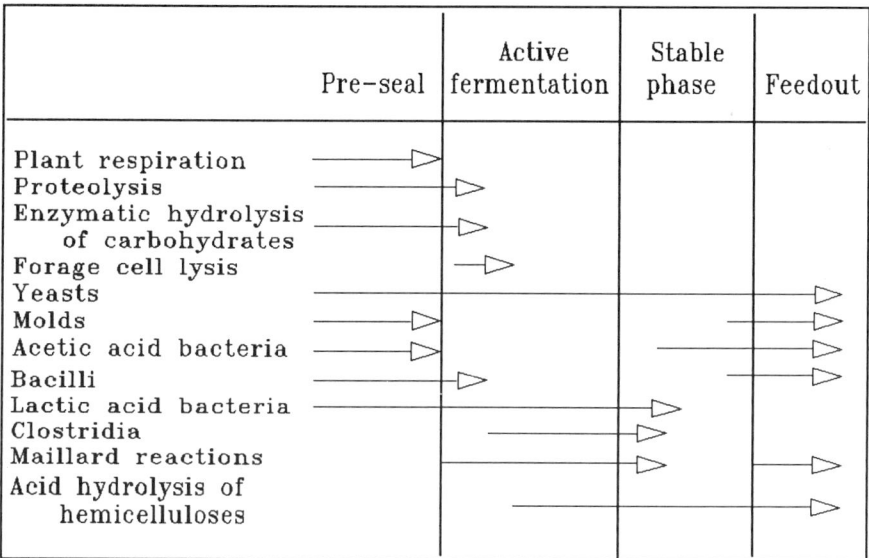

Figure 3. Major phases during ensiling when various plant, microbial, and chemical processes are most active.

aerobic microorganisms and populations of yeasts, molds, and aerobic bacteria increase. Their activity usually causes little change in forage quality at this stage of ensiling unless populations of approximately 10^8 colony forming units (CFU) of yeasts or bacteria/g forage and/or 10^6 CFU of mold/g forage are reached (Pitt et al., 1991). Similarly, facultative anaerobes such as lactic acid bacteria and enterobacteria will grow.

Once the silo is sealed, plant respiration removes the remaining oxygen in a matter of hours (Pitt et al., 1985). After anaerobic conditions prevail, the active fermentation phase occurs. Within several hours, plant cells begin to lyse releasing more cell contents (Pitt et al., 1985). In wet crops, this signals the beginning of effluent loss (including highly digestible soluble carbohydrates and N fractions contained therein) from the silo. With cell lysis, proteolysis and enzymatic breakdown of carbohydrates occur at their highest rates. These rates decline with time such that little activity occurs after one or two weeks.

Plant cell lysis also provides substrate for anaerobic microbial growth. Enterobacteria and lactic acid bacteria normally dominate all of the other microorganisms within the first 1 to 3 d after sealing. However, once pH has dropped to 5, the enterobacteria decline rapidly, leaving the lactic acid bacteria the principal microorganisms in the silage (Muck, 1991). Both groups ferment sugars primarily. The enterobacteria largely produce acetic acid whereas the lactic acid bacteria mainly produce lactic acid. Other products from these two groups include ethanol, 2,3-butanediol, succinic acid, formic acid, and mannitol (McDonald, 1981). The lactic acid bacteria grow actively for 1 to 4 wk lowering crop pH usually to between 3.8 and 5.0, dependent upon crop moisture, buffering capacity, and sugar content. Once pH drops sufficiently to inhibit microbial growth or the substrate is exhausted, lactic acid bacteria become inactive and their population slowly declines (Muck, 1991).

The third principal anaerobic bacteria, clostridia, have a much more detrimental effect on silage quality if silage pH is not sufficiently low. These strict anaerobes ferment sugars, lactic acid, and amino acids producing butyric acid and amines. These fermentations result in significant DM loss, and the fermentation products reduce the palatability of the silage (McDonald, 1981).

Many of the yeasts present when the silo is sealed are solely aerobic microorganisms; their populations decline rapidly once oxygen is not present. However, fermentative yeasts are normally present and may increase in numbers sufficient to affect silage quality (Muck et al., 1992). These yeasts typically ferment sugars to ethanol, causing DM loss (McDonald, 1981). Other aerobic microorganisms rarely have a significant effect during this period. Acetic acid bacteria and mold populations decline rapidly in the absence of oxygen (Muck et al., 1992). Some bacilli may ferment sugars, forming products similar to the enterobacteria (McDonald, 1981).

Maillard reactions may continue during the active fermentation phase if the temperature attained in pre-seal was sufficiently high. Once pH is low, chemical or acid hydrolysis of hemicelluloses may occur. Rates of acid hydrolysis are low (Dewar et al., 1963) and will continue throughout storage.

After anaerobic microbial activity has essentially ceased because of low pH or lack of substrate, the stable phase of storage begins. Even if silos are well sealed, slow diffusion of air occurs through the plastic or concrete. During active fermentation, the oxygen may be utilized by plant respiration. As the plant enzymes are inactivated, oxygen is principally used by aerobic and facultative anaerobic microorganisms. This microbial growth results in the loss of the most digestible components of the crop. Growth of certain molds can produce mycotoxins harmful to ruminants (Woolford, 1990). As acids are utilized by aerobic microorganisms, silage pH increases which may allow the development of listeria, an animal and human pathogen (Woolford, 1990). Growth of lactic acid bacteria in the presence of oxygen can result in the conversion of lactic acid to acetic acid (Condon, 1987).

When the silo is opened, the final phase of feedout occurs. Much more oxygen is present at the open face, and this oxygen penetrates into the silage mass. With greater oxygen availability, aerobic microbial growth can increase substantially, causing heating as well as the DM loss and health hazards described previously. The initial heating is generally caused by yeasts or acetic acid bacteria (Courtin and Spoelstra, 1990; Woolford, 1990). Bacilli and molds develop later if the forage is not readily consumed (Muck and Pitt, 1992).

Factors Affecting the Ensiling Processes

Plant Respiration

A major cause of silage DM loss is plant and microbial respiration. Plant respiration effects on forage quality in the field were discussed earlier. Rates and factors affecting respiration in the silo are similar at least during the pre-seal phase (Pitt et al., 1985). Potential activity probably declines with time in the silo but this has not been measured. Both proteolysis and starch hydrolysis activity last for approximately a week under anaerobic conditions (McKersie and Buchanan-Smith, 1982; Muck, 1987, 1990). Proteolytic activity has been measured after 21 d of ensiling, even though increases in soluble NPN in the silage ceased after approximately 3 d (McKersie and Buchanan-Smith, 1982). Consequently, reduced rates of plant respiration may occur even after active fermentation has ceased.

Proteolysis

Proteolysis in grass and legume silages may convert most of the true protein to soluble NPN (primarily peptides, free amino acids, and ammonia) via the activity of plant proteases. The N in wilted forages entering the silo normally consists of 100 to 350 g/kg soluble NPN (Kemble and MacPherson, 1954; Muck, 1987). At feedout, soluble NPN is 250 to 850 g/kg of total N

(Kemble, 1956; Muck, 1987; Albrecht and Muck, 1991). The level of proteolytic activity depends on forage species, pH, time, temperature, and moisture content.

Proteolysis varies among different legume silages. Among alfalfa, red clover, and birdsfoot trefoil, activity is highest in alfalfa (McKersie, 1985). Albrecht and Muck (1991) reported that proteolysis is inversely correlated with tannic acid content. Soluble NPN averaged 600 g/kg of total N in non-tannin-containing legume silages compared to 300 g/kg of total N in legumes with 30 g tannic acid equivalents/kg DM. Of the three non-tannin-containing legumes studied, soluble NPN was 100 to 200 g/kg higher in alfalfa than in red clover or cicer milkvetch (*Astragalus cicer* L.).

Overall proteolytic activity in legumes has an optimum at approximately pH 6.0 (Brady, 1961; McKersie, 1985), although the pH optima of specific proteolytic enzymes vary considerably (McKersie, 1981). Activity at pH 4.0 is reported to be 15 to 35% of optimum. Therefore, reduction of silage pH reduces but does not eliminate proteolytic activity. As indicated earlier, proteolytic activity decreases with time with little activity observed after 1 wk. This is reported in both alfalfa (McKersie and Buchanan-Smith, 1982; Muck, 1987) and ryegrass (Kemble, 1956). As a result, a rapid drop in pH (preferably the first day) is necessary to substantially alter the level of soluble NPN in a silage.

Increasing temperatures between 10 and 40°C increases proteolysis rates exponentially (McKersie, 1981). Temperature has less effect on the amount of proteolysis during ensiling, however, because a temperature increase also increases bacterial fermentation and the subsequent rate of pH decline. Increasing temperature from 15 to 35°C increases soluble NPN in alfalfa silage by only 100 g/kg of total N (Muck and Dickerson, 1988).

The moisture content of the crop affects proteolysis. Initial proteolysis rates in alfalfa are correlated linearly with moisture content with a rate of 0.9 mg N/g DM/h at 800 g/kg moisture and declining to 0 at 240 g/kg moisture content (Muck, 1987). The amount of proteolysis in alfalfa silage is not as strongly correlated with moisture content because variation in the pH time-course in wetter silages (>500 g/kg moisture) often offsets moisture effects on proteolysis.

Enzymatic and Acid Hydrolysis

Plant enzymes convert nonstructural carbohydrates to simple sugars. These sugars then are available for fermentation by lactic acid bacteria. Fructosans, the main storage carbohydrate in grasses, are largely hydrolyzed to fructose during ensiling (Henderson et al., 1972). Starch, the storage carbohydrate in legumes, is hydrolyzed presumably by α-amylases to maltose (Doehlert and Duke, 1983) during ensiling so that starch content is normally only 10 to 20 g/kg DM at feedout (Muck, 1990).

The activity of the amylases is affected by pH, starch concentration, glucose, time, and temperature. Optimum activity of alfalfa leaf amylase is reported to be at pH 7.25 with the activity at pH 4.0 approximately 10% of optimal activity (Muck, 1990). In alfalfa ensiling experiments, starch hydrolysis rates were found to be linearly correlated with starch concentration (little activity below 12 mg/g DM; 1.6 mg/g DM/h at 70 mg/g DM) and inhibited by the addition of glucose at 50 to 100 mg/g DM (Muck, 1990). Hydrolysis rates declined linearly with time, dropping by approximately 50% over 4 d (Muck, 1990). Moisture content had no effect on rates except below 400 g/kg moisture (Muck, 1990).

Both enzymatic and acid hydrolysis of structural carbohydrates, primarily the hemicellulosic fraction, probably occurs in both grasses and legumes. In both forage types, NDF reductions during storage of 10 to 30 g/kg DM were observed (Jones et al., 1992). Hemicellulases have been isolated from Italian ryegrass, perennial ryegrass, and orchardgrass (Dewar et al., 1963). Their activity declined with time such that little or no activity was observed after 7 d of incubation, but activity was enhanced at neutral pH and higher temperatures (Dewar et al., 1963). Acid hydrolysis also was measured by Dewar et al. (1963). Rates were low (10^{-5} g sugar/g hemicellulose/h), proportional to the hydrogen ion concentration, and they increased linearly with temperature. This study, and that of Morrison (1979), suggest that enzymatic hydrolysis of hemicelluloses is more important in grasses than acid hydrolysis.

In alfalfa, the opposite may be true. Morrison (1989) found a small reduction in the hemicellulosic fraction during ensiling and a greater loss of cellulose. Jones et al. (1992) found little evidence of cellulose degradation, but loss of hemicellulosic sugars was closely related to pH decline. Pitt et al. (1985) found that hemicellulase activity in alfalfa needed to be 10% of that in grasses for accuracy of predictions. Therefore, acid hydrolysis of hemicellulose appears to be more significant in alfalfa than enzymatic hydrolysis. Because of the discrepancy between Morrison (1989) and Jones et al. (1992), losses of cellulose during ensiling are uncertain.

Effluent

As the plant cells lyse during silo storage, effluent is produced. The amount of effluent that leaves the silo is dependent primarily on the moisture content of the crop and the type of silo. Most studies of effluent have been on unwilted or lightly wilted grasses in bunker or clamp silos. The moisture content at which no effluent leaves these silos ranges between 670 and 710 g/kg (McDonald, 1981). Based on a literature review, Waldo (1977) indicated that no effluent should be released from corn silage of 700 g/kg moisture in tower silos of less than 12 m. The moisture content should be reduced by 10 g/kg for every 3 m of additional height to prevent effluent loss. Thus, higher pressures and/or densities increase effluent production at a given moisture content. Effluent volumes tend to increase quadratically with increasing moisture content. At 800 g/kg moisture, effluent production of 70 to 220 L/Mg from grass may be expected (McDonald, 1981).

Effluent production is highest during the first week of ensiling and declines thereafter (Mayne and Gordon, 1986). The DM content of effluent varies inversely with flow rate. Effluent collected from Italian ryegrass silage with 810 g/kg moisture after three days of ensiling contained 50 g DM/kg and increased to 90 g DM/kg after 63 d (McDonald, 1981). Mayne and Gordon (1986) found 80 to 100 g DM/kg in effluent from grass silage with 780 g/kg moisture. These results suggest that a silage containing 800 g/kg moisture would lose approximately 40 g DM/kg of silage DM from effluent, based on 100 L of effluent/Mg of silage containing 80 g DM/kg of effluent.

Effluent contains many soluble compounds: sugars, fermentation products, soluble protein and NPN fractions, and ash. In a study reported by McDonald (1981), the DM of ryegrass effluent after 3 d of ensiling consisted of 220 g/kg CP, 250 g/kg sugar, and 300 g/kg ash. By 63 d, the CP content increased to 310 g/kg DM whereas sugar and ash declined to 40 and 210 g/kg DM, respectively. Much of the change over the 60 d was accounted for by increases in fermentation products. Effluent is high in minerals. Potassium, calcium, phosphorus, and magnesium averaged 86, 31, 8, and 6 g/kg DM, respectively, in grass silage effluent (Purves and McDonald, 1963).

Aerobic Microorganisms

The most significant effect of aerobic microorganisms on silage quality is respiration. The substrate for microbial respiration is dependent on the organism. Yeasts and acetic acid bacteria typically consume only soluble compounds such as sugars and fermentation products (Courtin and Spoelstra, 1990). Bacilli use proteins and some polysaccharides. Molds degrade the widest range of substrates including the structural carbohydrates and lignin (McDonald, 1981).

Aerobic microbial respiration, although not desirable, produces fewer immediate losses than plant respiration in the harvested crop. Aerobic microbes respire approximately half of the sugars consumed. The remainder is converted into cell mass which may be lost after the microorganisms die (Muck et al., 1991). Of the respired sugars, only 58% of the released energy ends up as heat in contrast to plant respiration in the harvested forage where virtually all energy is released as heat (McDonald, 1981).

Growth rates of these microorganisms are dependent on species, pH, temperature, substrate, and fermentation products. In general, yeasts and molds are less sensitive to low pH (i.e., 4.0) than bacteria (Muck et al., 1991). Molds are much slower growing than yeasts (Muck et al., 1991) and have much lower populations during silo storage due to their greater sensitivity to the absence of oxygen (Muck et al., 1992). Bacilli are apparently more sensitive to either low pH and/or fermentation products (Muck and Pitt, 1992) so that they seldom initiate aerobic deterioration.

Fermentation products are known to inhibit various aerobic microorganisms. Fungi are sensitive to the undissociated concentrations of volatile fatty acids (VFA), particularly the longer chain VFA (Woolford, 1975). In well-fermented silages, yeast and mold growth rates are reduced in proportion to the undissociated concentration of acetic acid (Muck et al., 1991). Clostridial silages contain significant levels of butyric acid and thus are known to be very stable aerobically (Woolford, 1990). Lactic acid is inhibitory to yeasts and even more so to acetic acid bacteria (Courtin and Spoelstra, 1990).

The apparent succession of aerobic microorganisms in silage exposed to air appears to be as follows (Muck and Pitt, 1992). Yeasts and/or acetic acid bacteria use sugars and fermentation products as substrates, potentially raise silage temperatures to 40 to 45°C, and cease to be a factor when those substrates are used up. Use of the fermentation products raises silage pH. Bacilli then become the dominant microorganisms, consuming more complex substrates and maintaining temperatures in the 40's. Finally, molds complete the deterioration process and they may raise temperatures above 50°C.

Aerobic microorganisms may cause other changes in forage quality. Molds such as *Aspergillus* and other species can produce mycotoxins during storage (Woolford, 1990). In higher pH silages (above 5.0 generally), *Listeria monocytogenes*, a human and animal pathogen, may multiply (Woolford, 1990).

Anaerobic Bacteria

Utilization of sugars by lactic acid bacteria produces the least adverse change in forage quality of any of the processes. In contrast to the aerobic organisms, only 10 to 20% of the sugars assimilated go into cell mass with the rest being fermented for energy (Pitt et al., 1985). Table 1 shows various fermentation pathways and the subsequent DM and energy losses. Lactic acid bacteria cause little loss of gross energy. Homofermentative pathways (i.e., those producing only lactic acid) result in no DM loss whereas heterofermentative pathways produce up to 24% DM loss. In most natural lactic acid bacterial fermentations with a mix of hetero- and homofermentative pathways, energy density per unit silage DM increases slightly.

Dry matter loss depends on the strain of lactic acid bacteria and the environmental conditions. Most natural lactic acid bacteria in silage are either heterofermentative or facultative homofermenters (Sneath et al., 1986). The facultative strains normally are homofermentative but become heterofermenters

Table 1. Selected silage fermentation pathways and accompanying DM and energy losses (McDonald, 1981).

Pathway	Dry matter loss (%)	Energy loss (%)
Homofermentative lactic acid bacteria		
1 glucose → 2 lactate	0.0	0.7
1 fructose → 2 lactate	0.0	0.7
Heterofermentative lactic acid bacteria		
1 glucose → 1 lactate + 1 ethanol + 1 CO_2	24.0	1.7
3 fructose → 1 lactate + 3 acetate + 2 mannitol + 1 CO_2	4.8	1.0
Clostridia		
1 glucose → 1 butyrate + 2 CO_2 + 2 H_2	51.1	20.9
2 lactate → 1 butyrate + 2 CO_2 + 2 H_2	51.1	18.4
Yeasts		
1 glucose → 2 ethanol + 2 CO_2	48.9	0.2

when sugar content and growth rate are low (Christensen et al., 1958). However, homofermentation in *Streptococcus bovis* at slow growth rates has been observed when pH is low (Russell and Hino, 1985) and, similarly, fermentation in silage has been observed to become more homofermentative with decreasing pH (Pitt et al., 1985).

The amount of sugars converted by lactic acid bacteria to fermentation products is dependent on the sugar content, moisture level, and buffering capacity of the crop. Lactic acid bacteria stop growing from either the lack of substrate or low pH. This is illustrated by alfalfa silage made at several moisture levels with and without glucose addition (Figure 4). At low moisture levels, added sugar had no effect on final pH, indicating that low pH stopped bacterial growth. Under wetter conditions, the glucose addition lowered silage pH, suggesting that fermentation in the untreated silages was limited by the sugar content. The pH at which bacterial growth ceased also decreased with increasing moisture content (Muck, 1990).

The buffering capacity of the crop (equivalents of acid needed to drop pH from 6.0 to 4.0) determines approximately the amount of fermentation acids necessary to reach a given pH. This has been demonstrated in alfalfa where silage pH was negatively correlated with the sugar-to-buffering capacity ratio (Melvin, 1965). In general, the buffering capacities are lowest for corn (150 to 250 meq/kg DM), intermediate for grasses (250 to 500), and highest for legumes (400 to 600) (McDonald, 1981). This suggests that the highest levels of fermentation products would be seen in alfalfa and other legumes and the lowest in corn. However, low sugar contents in alfalfa relative to other species means that grass silages often have the highest levels of fermentation products, occasionally reaching 200 g/kg DM (e.g., Morgan et al., 1980).

Most lactic acid bacteria ferment only sugars and some organic acids. However, some strains ferment serine and arginine, producing ammonia (Ohshima and McDonald, 1978). More importantly, many lactic acid bacteria multiply in the presence of oxygen. They do not respire oxygen as aerobic bacteria, but various silage strains can use oxygen, often converting lactic acid to acetic (Condon, 1987). This is one cause of high acetic acid content silages.

Poor fermentation is normally associated with the development of clostridia. Clostridia can be divided into three groups: those that produce butyric acid by fermenting sugars and lactic acid; those that ferment amino acids producing various acids, ammonia, and amines; and those that ferment both sugars and amino acids (McDonald, 1981). As shown in Table 1, clostridial fermentation of glucose and lactic acid results in butyric acid and substantial DM and energy losses. Amino acids are fermented via several mechanisms: deamination, decarboxylation, and oxidation/reduction. Deamination increases ammonia content but results in no DM loss. Decarboxylation causes DM loss from carbon dioxide release and produces amines which are detrimental to forage intake by animals. Oxidation/reduction reactions produce organic acids, ammonia, and carbon dioxide (McDonald, 1981).

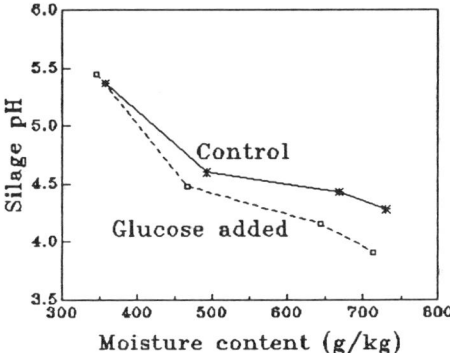

Figure 4. Alfalfa silage pH with or without the addition of glucose at the rate of 70 to 120 g/kg of DM (Muck, 1990).

Clostridia prefer warmer silage temperatures with most silage species having optima near 37°C (McDonald, 1981). The principal means of inhibiting clostridial growth is through rapid lowering of silage pH. The pH at which clostridial growth is completely inhibited is dependent on the water activity of the silage, which is a function of the crop species and its moisture content. The wetter the crop, the lower the pH necessary to prevent clostridial growth. The water activity of legumes at a given moisture level is lower than that of grasses so a lower pH is required to prevent clostridial growth in grasses. From a practical standpoint, clostridial growth is normally prevented by ensiling the crop with less than 700 g/kg of moisture. Exceptions to this general observation can occur if basic compounds such as ammonia or urea are added at ensiling, raising silage pH.

Enterobacteria normally do not have a large effect on silage quality. These facultative anaerobes compete with the lactic acid bacteria for sugars early in ensiling, producing primarily acetic acid. Their fermentation pathways are similar to the heterofermentative lactic acid bacteria, producing DM loss but little loss of gross energy. Unlike the lactic acid bacteria, enterobacteria have pH optima around 7.0, and their populations generally decline rapidly below pH 5.0 (McDonald, 1981). Spoelstra (1987) found that some strains reduce nitrate to nitrite and nitric oxide, a form of silo gas. Consequently, some loss of CP could result from their activity.

Maillard Reactions

As in the storage of high-moisture hay, Maillard reactions can increase silage ADF and ADIN. Hemicelluloses can be complexed in these reactions (Figure 5) so that the difference between NDF and ADF declines. As indicated in Figure 5, Maillard reaction rates increase exponentially with temperature. Little activity is expected at temperatures below 35°C. Since the Maillard reaction releases heat, the process can further increase silage temperature and cause more extensive rates of browning.

Moisture content is another factor affecting browning. For normal ensiling moisture contents of 400 to 700 g/kg, moisture does not affect the

Maillard reaction significantly (Goering et al., 1973). However, moisture affects the heat capacity of the silage. Based on a herbage heat capacity of 1.89 kJ/kg DM/°C (McDonald, 1981), the heat capacities of a silage at 700, 500, and 300 g/kg moisture are 11.67, 6.08, and 3.69 kJ/kg DM/°C, respectively. Thus a given amount of browning raises temperature more than three times as much at 300 g/kg moisture content than at 700 g/kg. As a result, Maillard products are most common in drier silages. Temperatures high enough to cause spontaneous combustion can be reached. Silo fires are a potential hazard in crops ensiled with less than 450 g/kg moisture.

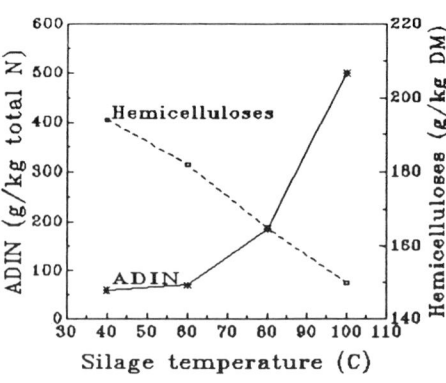

Figure 5. Effect of temperature on the extent of browning assayed as ADIN and the associated decline in hemicelluloses (Goering et al., 1973).

Susceptibility to browning varies among forages. The cause of this variation does not appear to be related to total N content, species, or initial ADIN (Goering et al., 1973). Analyses of forage silages by various researchers indicate that 18 to 40% of the silages had signs of overheating (Goering, 1976). With the move toward ensiling wetter silages over the last 20 yr, fewer silages with Maillard products are expected.

Overall Losses and Quality Changes in the Silo

The importance of the various processes in the silo relative to losses and silage quality is difficult to assess due to the number of processes occurring simultaneously, and their interactions. One means of integrating all factors and assessing their importance is through computer modeling.

Most computer programs have dealt with only one or two phases of ensiling. The earliest models principally focused on fermentation (Neal and Thornley, 1983; Pitt et al., 1985; Leibensperger and Pitt, 1987; Meiering et al., 1988). These models looked primarily at the growth of lactic acid bacteria and clostridia and at plant enzyme activity during active fermentation. Recently, modeling activity has turned to aerobic losses during storage and feedout. Pitt (1986) and McGechan (1989a) studied losses due to air infiltration during storage. Current efforts are on simulation of aerobic losses at feedout (Parsons, 1991; Muck and Pitt, 1992; Pitt and Muck, 1993). These modeling efforts have indicated that individual processes during silo storage can be reasonably simulated.

A comprehensive model of silo storage was developed to simulate losses from filling through feeding (Buckmaster et al., 1989b). This silo model was integrated in a simulation model of the dairy forage system (DAFOSYM) to study losses as affected by farm operations and environmental conditions (Rotz et al., 1989). Losses during pre-seal, the stable phase, and feedout were assumed to be solely by respiration. Oxygen movement into the silo was by diffusion. Changes in DM, fiber content, and N fractions during active fermentation were based on simulations of the Pitt et al. (1985) fermentation model. The comprehensive silo model reasonably predicted whole silo DM losses compared to actual studies. The effects of silo size and type on DM losses for silos emptied over 360 d are shown in Figure 6. A bottom-unloaded, oxygen-limited silo provided the lowest losses whereas the top-unloaded, concrete stave silo had approximately two percentage units greater losses over most capacities. Except at small silo capacities, the bunker silo had the greatest DM losses.

Recently, effluent production was added to the model to compare direct-cut ensiling of alfalfa with normal practices of ensiling alfalfa wilted to 650 g/kg moisture (Rotz et al., 1993). Harvesting over 26 yr of weather conditions for East Lansing, Michigan was simulated. Both harvest systems placed the alfalfa in bunker silos and direct-cut alfalfa was treated with formic acid to control fermentation. Distribution of the average silo losses are shown in Table 2. Normal wilted silage production had no effluent loss, and most losses were split between that caused by oxygen infiltration during storage and aerobic losses at feedout. Ensiling of direct-cut material increased overall losses and shifted losses more to events in pre-seal and active fermentation.

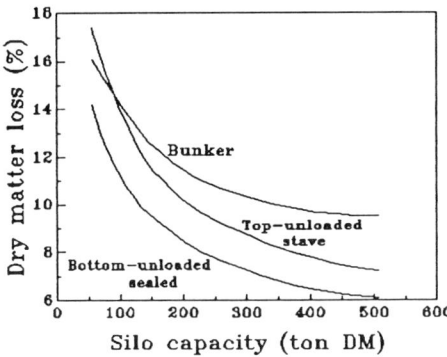

Figure 6. Effect of silo size and type on predicted DM loss for silos emptied over 360 d (Buckmaster et al., 1989).

Currently, there are no data to confirm the relative losses in each of the phases of ensiling listed in Table 2. However, data show that failure to cover bunker silos increased organic matter loss in the top 50 cm from 270 to 410 g/kg in a farm silo survey (Dickerson et al., 1990). Losses 2 m below the surface were substantially less under both covered and uncovered silos. Fermentation losses predicted by the model also appear reasonable. Bacterial inoculants, which provide a homofermentative silage fermentation, have been reported to improve DM recovery from silos by approximately 2.5 percentage units (Muck and Bolsen, 1991). Because some reduction in aerobic losses also may result from inoculant use, the fermentation losses in Table 2 are reasonable. Overall, these results tend to corroborate the importance of aerobic microbial respiration on silage DM losses suggested by the model of Buckmaster et al. (1989b).

Table 2. Simulated average DM losses during storage of wilted (650 g/kg moisture) and direct-cut silage predicted by the comprehensive silo model of DAFOSYM (Rotz et al., 1993).

	DM lost (%)	
	Wilted	Direct-cut
Effluent	0.0	4.7
Aerobic respiration during filling	0.8	1.3
Fermentation	0.7	1.5
Aerobic respiration during storage	5.0	4.7
Aerobic respiration during emptying	5.2	3.8
Total	12.1	17.7

Respiration not only has a dominating effect on DM losses in the silo but also on forage quality. Digestible carbohydrates are largely lost through respiration, increasing the concentration of other forage components. For example, little loss of N occurs in the silo unless effluent losses and/or nitrate in the incoming crop are significant. Consequently, a 10 to 20 g/kg DM increase in CP should occur dependent on the CP of the crop entering the silo and the DM lost during silo storage. Changes in cell wall or NDF content, however, also are dependent on the amount of enzymatic and acid hydrolysis of structural carbohydrates during storage. In laboratory studies where DM losses are minimized, NDF typically declines 10 to 60 g/kg DM in grasses and legumes (Spoelstra, 1990; Jones et al., 1992). Thus, under real farm conditions, NDF levels in silages may range from a 10 g/kg DM decline to a 40 g/kg DM increase, dependent on the respiration loss relative to the amount of cell wall hydrolysis. With good silo management, ADF changes little, so the concentration of ADF increases 20 to 50 g/kg DM with the loss of other carbohydrates. Carbohydrates lost are highly digestible, so DM digestibility and total digestible nutrient (TDN) concentration decline 20 to 70 g/kg DM.

METHODS TO REDUCE LOSSES AND IMPROVE QUALITY

Many options are available to reduce the DM and nutrient losses that occur during forage harvest and storage. These include equipment modifications to reduce loss, new equipment systems to speed field curing, and chemical treatments to enhance field drying and preservation in storage. The primary restraint to implementing new technologies that reduce losses or enhance quality is economic feasibility. Most technologies increase the costs of forage production. The improvement in yield and quality must have a value greater than the added cost to encourage adoption by growers. Other constraints include labor availability, environmental impact, and public perception.

Equipment Systems

Reduction of losses and preservation of forage quality always have been among the major considerations in the development and design of forage equipment. Other considerations include cost of production, labor requirements, power or fuel requirements, and operator safety. A design to reduce or eliminate loss may be compromised at times to meet these other equally or more important considerations. Progress has been made over the past fifty years in designing machines to reduce losses. An example is the development of the mower-conditioner. Much effort was devoted to balancing the effectiveness in improving drying with the loss caused by the operation. Further improvements are expected in the future.

Small improvements can be made to reduce the loss caused by various machine treatments. Such improvements may be done by designing machines for more gentle handling of forage or the capture and recycling of shattered forage material. One of the greatest opportunities for improvement is the reduction of chamber losses in large round balers. For most machine operations, the associated forage loss is low so the opportunity for reducing loss is small.

The largest harvest losses and quality changes occur during field curing, primarily from rain damage. Thus, new systems which speed field curing have the greatest potential for benefit. One technique under development is maceration and mat drying of forage. Maceration is the most severe form of mechanical conditioning. Plant stems are shredded, fully exposing the internal moisture. Drying restraints imposed by the internal cell structure, the epidermis, and the cuticle are removed. Macerated alfalfa can dry to a moisture content suitable for baling in 4 to 6 h of favorable drying conditions (Rotz et al., 1990).

Macerated forage contains many fine particles that are very susceptible to loss during field curing and harvest. To reduce the potential loss, the shredded material can be pressed into a mat that is laid on the field surface for rapid curing (Koegel et al., 1988). The dried mat then can be picked up with minimal loss. Commercial equipment for maceration and mat drying is not yet available, but research and development of the process is continuing. In order to implement the process on the farm, new equipment and procedures are required for forage harvesting, handling, and storage.

Fast drying is a major advantage of the mat process. The drying rate of alfalfa mats is two to three times greater than the rate of conventional swaths so that hay can be made with one day of field curing. Maceration of forage appears to greatly reduce respiration through very rapid drying and severe rupturing of plant tissue (Rotz et al., 1991b). Macerated forage is very susceptible to rain damage with three times the loss of alfalfa conditioned with intermeshing rubber rolls (Rotz et al., 1991b). However, with rapid drying, rain damage should not occur often.

The mat process can potentially provide a substantial improvement in forage quality. A long-term analysis indicates a 10% improvement in harvested yield through reduced field losses (Rotz et al., 1990). The reduction of loss provides a 20 to 30 g/kg DM average decrease in NDF concentration and a small improvement in harvested protein. In addition, shredding the forage increases fiber digestibility up to 15%. The improvement in digestion allows the animal to obtain more energy from the forage and perhaps to increase its intake of forage. With this quality enhancement, macerated forage produces a higher quality forage for high producing dairy cows than obtained through conventional forage systems.

Disadvantages of the mat system include high power requirements and equipment costs. The long-term benefits of faster drying, reduced loss, and improved forage quality appear to outweigh the added fuel and equipment costs (Rotz et al., 1990). In hay production, the system can return up to $4 for each dollar spent on increased equipment and fuel costs. The system appears less economical in silage production, but additional benefit may be obtained through reduced silo losses acquired by faster and greater packing in the silo.

Chemical Drying Aids

Chemicals can be used to speed the field curing of both legume and grass forage crops. Chemical treatments improve drying by opening stomata, desiccating the plant prior to mowing or by modifying the epicuticular waxes. Several chemical treatments are known to enhance forage drying, but most are not practical or economical for field use (Harris and Tulberg, 1980).

The first practical method for using chemicals to speed the field curing of forage came with the use of aqueous potassium carbonate. The exact mechanism by which this treatment improves drying is not known. The solution is thought to form a continuous film over the plant surface which extends down the cavities between the wax platelets. The film joins the liquid phase moisture within the parenchyma, pectin, and cuticular membrane, and thus enables moisture transfer in the liquid phase through the wax layer by capillary forces (Harris and Tulberg, 1980). When sprayed on the crop as it is mowed, the solution can increase drying rate. This process of chemical conditioning is effective on legume forage species but not on grass species.

Potassium carbonate is the most common active ingredient in chemical drying agents. A 28 g/L solution of potassium carbonate in water applied at an appropriate rate speeds drying under good drying conditions. Replacing half of the potassium carbonate with sodium carbonate is less expensive, but equally effective to potassium carbonate alone. The treatment is more effective as more solution is applied, but the economically optimal amount is about 300 L/ha for yields under 3.5 Mg/ha and 470 L/ha for greater yields. Mixing more chemical per unit of water does not improve drying performance nor compensate for a decrease in solution application rate (Rotz and Davis, 1986).

Chemical conditioning can improve the drying of all cuttings of alfalfa, but it is most effective on summer harvests (Rotz et al., 1987). In the Midwest, chemical conditioning provides about a 40% increase in the drying rate of first cutting alfalfa with up to a 120% increase on second and third cuttings. Poor drying conditions in the fall limit the increase to about 20%. The typical reduction in field curing time for chemical conditioning compared to mechanical conditioning is 5 daytime hours on first cutting and 7 to 8 h on later cuttings.

Chemical conditioning of alfalfa can provide an average increase in harvested yield of over 10% in hay production (Rotz et al., 1989). The process appears to have little effect on respiration rate and the resultant losses (Rotz et al., 1984; Rotz et al., 1987). When rain occurs during field curing, the treatment has little or no influence on rain-induced losses (Rotz et al., 1987; Rotz et al., 1991b). The primary benefit comes through the occasional avoidance of rain damage through faster drying. The improvement in yield and quality is of marginal economic benefit in hay production with little benefit in silage production.

Hay Preservatives

In hay making systems, field losses can be reduced by baling hay at a moisture content near 250 g/kg. Baling moist hay reduces baler chamber losses providing a small (up to 2%) improvement in harvested yield and a small improvement in harvested quality. Raking and pickup losses also may be reduced a small amount. Field curing time on the average is reduced about 1 d which reduces the potential for rain damage. With all of these factors combined, harvested yield is increased an average of 7%. However, the moist hay deteriorates rapidly in storage, offsetting the benefit of reduced field losses unless treated to enhance preservation.

Materials used for the preservation of high-moisture hay include propionic acid, organic acid mixtures (may include propionic, acetic, fumaric, citric, benzoic, lactic, and formic acids), buffered acid mixtures, anhydrous ammonia, and bacterial inoculants. Propionic acid (or an effective organic acid mixture) normally reduces mold growth (Rotz et al., 1991c). With application rates of 10 to 20 g/kg of hay weight, acid treatments can reduce the heating of high-moisture hay (Knapp et al., 1976; Davies and Warboys, 1978; Rotz et al., 1991c). Some report similar heating in treated, damp hay and dry (< 180 g/kg moisture) hay while others report a little more heating (Rotz et al., 1991c).

Propionic acid treatment reduces storage loss in damp hay during the first few months of storage, but the loss is higher than that in dry hay (Knapp et al., 1976; Davies and Warboys, 1978; Rotz et al., 1991c). Over 6 mo of storage, reported losses are similar in treated and untreated hays at similar moisture levels (Rotz et al., 1991c). Acid-treated hay maintains a higher moisture content throughout storage. Apparently the more moist environment in the hay maintains a little higher level of microbial activity even with the acid treatment. Over a 6 mo storage period, the loss in acid-treated hay catches up, providing little difference in losses and nutrient changes between treated and untreated high-moisture hays.

Dry matter lost during storage is very digestible (Buckmaster et al., 1989a). When compared to hay at similar moisture levels, propionic acid treatment provides smaller decreases in *in vitro* DM and cell wall digestibilities and lower levels of ADF, NDF, and ADIN after short storage periods (Knapp et al., 1976; Rotz et al., 1991c). Consistent improvement in hay quality is not reported with propionic acid treatment of hay stored for more than 4 mo (Davies and Warboys, 1978; Rotz et al., 1991c). When compared to dry hay, acid-treated damp hay is often higher in fiber content and less green in color (Rotz et al., 1991c).

Propionic and similar acids promote corrosion in balers and bale handling equipment. To reduce corrosion, buffered acid products have largely replaced the use of straight acids. The acid is blended with ammonia or other compatible chemical to increase the pH of the treatment. The less volatile buffered products may be as effective as propionic acid when equivalent amounts of propionate are retained in the hay. One buffered product was ineffective at application rates below 5 g/kg of hay (Rotz et al., 1991c).

Routine use of acid treatment on high moisture hay appears uneconomical. A thorough analysis of various harvest strategies indicate that the limited benefit received does not justify the cost of the treatment (Rotz et al., 1992). The treatment can only be economical when it is used to avoid heavy rain damage of the crop. The economic benefit can be improved by greatly increasing the effectiveness of the treatment over long storage periods and/or reducing the treatment costs.

Anhydrous ammonia is perhaps the most effective hay preservative (Rotz et al., 1986; Rotz et al., 1992). Storage DM loss is reduced or eliminated in hay of up to 350 g/kg moisture when wrapped in plastic and treated with ammonia at 10 g/kg of hay weight or more. Ammonia treatment prevents heating, and it may eliminate mold development while the hay is covered. Ammonia treatment increases CP concentration by adding NPN. With less storage loss, the increase in ADF and NDF which normally occurs during storage is reduced. Increases in *in vitro* DM, hemicellulose, and cellulose digestion and energy content are reported.

Although anhydrous ammonia provides the most effective and economical preservation of forage, animal and human safety concerns deter its use. Ammonia treatment of forage has caused toxicity to animals (Rotz et al., 1986). Toxicity most often occurs when ammonia is used on high quality hay at higher than recommended application rates (greater than 30 g/kg of DM). Direct exposure to anhydrous ammonia can cause severe burns, blindness, and death.

Bacterial inoculants are sometimes applied to hay. Inoculation with a few strains of *Lactobacillus* had no effect on mold, color, heating, DM loss, and quality change in high-moisture hay (Rotz et al., 1988). In another study, both *Lactobacillus* and *Bacillus* inoculants improved hay appearance with little effect on DM loss and quality compared to untreated hay of similar moisture (Tomes et al., 1990). Until a more tangible benefit is shown, the economic value of these products cannot be addressed.

Silage Additives

A wide variety of additives is used in silage making. The principal additives include bacterial inoculants, enzymes, and NPN. Each class of products has different effects on silage quality.

The most common silage additives in the U.S. are bacterial inoculants which supplement the natural lactic acid bacteria of the crop. Inoculant bacteria have been selected from forages for fast growth rate and homofermentativeness. When the inoculant bacteria dominate fermentation, the resulting silage has less acetic acid and ethanol, more lactic acid, and a lower pH than expected from the unaided natural fermentation. This shift in fermentation should improve DM recovery as indicated in Table 1. A survey of inoculant studies found an average 2.5 percentage unit improvement in DM recovery at feedout when the inoculant was successful (Muck and Bolsen, 1991). This should be due principally to the shift in fermentation products; however, some reduction in aerobic respiration also may be represented. An additional benefit from inoculant use is a small reduction in proteolysis, particularly in the ammonia fraction (Muck and Bolsen, 1991).

Inoculant success is primarily related to the size of the natural lactic acid bacterial population. Inoculants were successful in approximately two-thirds of the reported studies on grasses and legumes since 1985; these products have been successful less than half the time in corn and sorghum silages (Muck and Bolsen, 1991). This corresponds well with natural lactic acid bacterial numbers, which are typically higher on corn than on alfalfa or grass.

Most enzyme products consist of a mixture of cellulases, hemicellulases, pectinases, and amylases and are targeted for reducing the fiber content of a silage. A survey of published studies since 1985 found that these products reduced ADF and NDF in 80% of the grass experiments but in only 40% of the alfalfa trials (Muck and Bolsen, 1991). The failure on alfalfa was presumably from differences in cell wall structure. Typical reductions in fiber contents range from 10 to 50 g/kg DM. These products also appear to improve DM recovery; six of nine studies reported improvements, averaging 6.6 percentage units (Muck and Bolsen, 1991). The cause of reduced aerobic loss during silo storage is uncertain. When effective, enzyme-treated silages consolidate more than untreated silages, which should reduce porosity and subsequent oxygen movement into the silage mass. On the negative side, increased consolidation could cause additional effluent loss in wetter silages.

Both ammonia and urea are common additives to corn silage. These additives boost the CP content and make silages more aerobically stable by killing aerobic microorganisms. With lower volatile losses, urea is more efficient in increasing CP, whereas ammonia is more effective in improving aerobic stability (Muck and Bolsen, 1991). These additives increase crop pH at ensiling and thereby cause more fermentation and fermentation products, particularly acetic acid. Both compounds, but especially ammonia, improve DM and fiber digestibility and reduce proteolysis. In spite of these benefits, animal performance has been enhanced infrequently (Muck and Bolsen, 1991). Also, DM recoveries with NPN additives have been reduced in 11 of 16 recent trials, presumably due to increased fermentation losses by either lactic acid bacteria or clostridia (Muck and Bolsen, 1991).

SUMMARY

Forage harvest and storage as hay or silage are an important part of animal agriculture in many regions of the world. Hay and silage are used to feed animals during portions of the year when forage cannot be grown or throughout the year to avoid the need for pasture systems. Substantial losses of forage DM and nutrients occur during the harvest and storage processes. An important aspect in the selection of forage systems is their impact on the resulting yield and quality.

Many forms of loss occur. These can be categorized as field losses (respiration, rain damage, and mechanical damage) and storage losses. Plant and microbial respiration remove carbohydrates from the forage causing an increase in protein and fiber concentrations. Rain damage removes a portion of the leaves, leaches soluble nutrients, and induces further plant and microbial respiration. Mechanical losses normally cause greater loss of leaves than stems.

Because leaves contain higher concentrations of most nutrients important to the animal, a decrease in the leaf to stem ratio reduces the overall quality of forages. Microbial respiration continues to remove the most digestible forage nutrients during storage. Storage loss is primarily nonstructural carbohydrates but some loss of protein and change in protein solubility occur.

Table 3. Typical DM losses and quality changes during hay and silage production.

Type of forage, Type of loss	Dry matter loss (% DM) Range	Normal	Change in nutrient concentration (g/kg DM) CP	NDF	TDN
Legume crops					
Respiration[a]	1 - 7	4	9	17	-17
Rain damage[a], 5 mm	3 - 7	5	- 4	14	-15
25 mm	7 - 27	17	-17	60	-70
50 mm	12 - 50	31	-35	140	-142
Mowing/conditioning	1 - 4	2	- 7	12	-14
Tedding	2 - 8	3	- 5	9	-12
Swath inversion	1 - 3	1	0	0	0
Raking	1 - 20	5	- 5	10	-12
Baling, small bale	2 - 6	4	- 9	15	-19
round bale	3 - 9	6	-17	31	-38
Chopping	1 - 8	3	0	0	0
Hay storage, inside	3 - 9	5	- 7	21	-21
outside	6 - 30	15	0	50	-70
Silo storage, sealed	6 - 14	8	14	7	-37
stave	7 - 17	10	18	17	-47
bunker	10 - 16	12	23	27	-56
Grass crops					
Respiration[a]	2 - 8	5	8	32	-18
Rain damage[a], 5 mm	1 - 3	2	- 2	9	- 5
25 mm	4 - 14	8	-13	53	-30
50 mm	8 - 27	15	-27	110	-60
Mowing/conditioning	1 - 2	1	0	0	0
Tedding	1 - 3	1	- 2	4	- 4
Swath inversion	1 - 3	1	0	0	0
Raking	1 - 20	5	- 3	5	- 6
Baling, small bale	2 - 6	4	- 5	9	-10
round bale	3 - 9	6	-10	18	-20
Chopping	1 - 8	3	0	0	0
Hay storage, inside	3 - 9	5	-13	32	-18
outside	5 - 22	12	0	80	-48
Silo storage, sealed	6 - 14	8	8	9	-37
stave	7 - 17	10	12	22	-47
bunker	10 - 16	12	15	36	-56

[a]Respiration loss includes plant and microbial respiration for crop cured without rain damage. Rain damage includes leaf loss, nutrient leaching, and microbial respiration resulting from rain damage.

Typical losses and quality changes in forage harvest and storage are listed in Table 3. Typical ranges and the normal DM losses are estimated based on a comprehensive review of reported losses. Changes in CP, NDF, and TDN are estimated from the expected change in leaf-to-stem ratio and the relative rates of removal of nutrients by respiration and rain damage.

Because interactions occur among various losses, component losses should not be simply added to obtain total system loss. A more comprehensive study of weather and machine impacts on losses and the interaction among losses is performed with DAFOSYM, the dairy forage system model (Rotz et al., 1989; Buckmaster et al., 1990). Long-term simulations with DAFOSYM provide typical losses for a wide variety of alfalfa harvest systems over a full range of crop moisture conditions (Figure 7). In addition, the model integrates DM and nutrient losses with animal performance to predict the economic return for the farm. The change in economic return can be used to represent the value of forage loss expressed as a portion of the initial value (Rotz et al., 1991a). The value loss across systems also is shown in Figure 7 for a typical dairy farm with a herd milk production of 8172 kg/cow-yr. For most dry hay and silage systems, about 30% of the initial crop value is lost by the time it is fed.

Dry matter and nutrient losses are only one of several considerations when evaluating and selecting forage harvest systems. The overall goal for most producers is to harvest forage with minimal nutrient loss at the lowest cost. Other factors like timeliness of harvest and labor availability also must be considered. Speeding the drying process, reducing the number of machine operations, and modification and adjustment of machines can all be used to reduce losses. Only by balancing the losses and costs with added benefits can the best system for forage harvest be selected. Best options vary with climate, crops grown, farm infrastructure, and farm management styles.

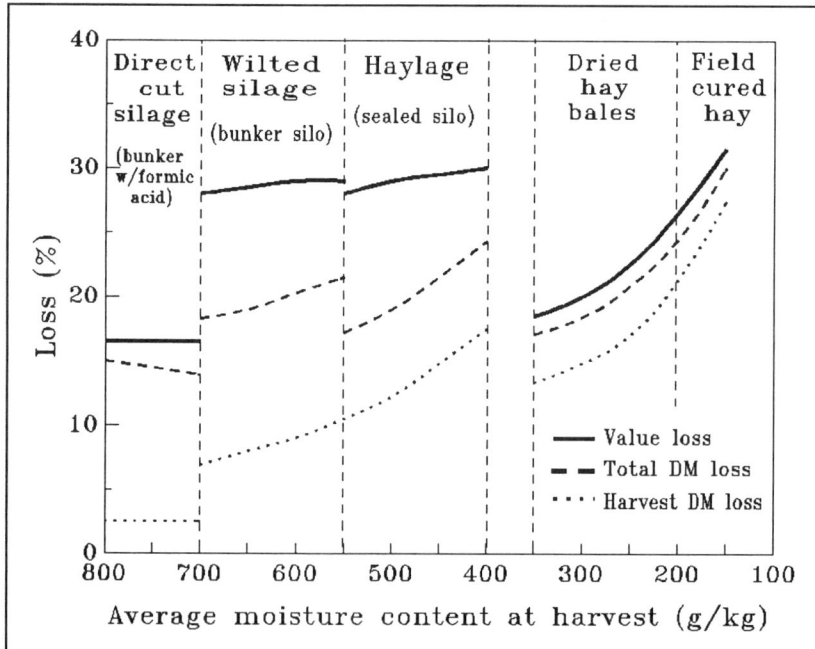

Figure 7. Typical harvest and storage DM losses and total value loss for various alfalfa production systems used in Michigan (Rotz et al., 1991a).

REFERENCES

Albrecht, K.A., and R.E. Muck. 1991. Proteolysis in ensiled forage legumes that vary in tannin concentration. Crop Sci. 31:464-469.

Barrington, G.P., and H.D. Bruhn. 1970. Effect of mechanical forage-harvesting devices on field-curing rates and relative harvesting yields. Trans. ASAE 13:874-878.

Belyea, R.L., F.A. Martz, and S. Bell. 1985. Storage and feeding losses of large round bales. J. Dairy Sci. 68:3371-3375.

Boyd, M.M. 1959. Hay conditioning methods compared. Agric. Eng. 40:664-667.

Brady, C.J. 1960. Redistribution of nitrogen in grass and leguminous fodder plants during wilting and ensilage. J. Sci. Food Agric. 11:276-284.

Brady, C.J. 1961. The leaf protease of *Trifolium repens*. Biochem. J. 78:631-640.

Brasche, M.R., and J.R. Russell. 1988. Influence of storage methods on the utilization of large round hay bales by beef cows. J. Anim. Sci. 66:3218-3226.

Buckmaster, D.R. 1993. Alfalfa raking losses as measured on artificial stubble. Trans. ASAE 36(3):645-651.

Buckmaster, D.R., C.A. Rotz, and J.R. Black. 1990. Value of alfalfa losses on dairy farms. Trans. ASAE 33:351-360.

Buckmaster, D.R., C.A. Rotz, and D.R. Mertens. 1989a. A model of alfalfa hay storage. Trans. ASAE 32:30-36.

Buckmaster, D.R., C.A. Rotz, and R.E. Muck. 1989b. A comprehensive model of forage changes in the silo. Trans. ASAE 32:1143-1152.

Carpintero, C.M., A.R. Henderson, and P. McDonald. 1979. The effect of some pre-treatments on proteolysis during the ensiling of herbage. Grass Forage Sci. 34:311-315.

Christensen, M.D., M.N. Albury, and C.S. Pederson. 1958. Variation in the acetic acid-lactic acid ratio among the lactic acid bacteria. Appl. Microbiol. 6:316-318.

Ciotti, A., and A. Cavallero. 1979. Haymaking losses in cocksfoot, lucerne and a cocksfoot-lucerne mixture in relation to conditioning and degree of drying at harvest. p. 214-220. *In* C. Thomas (ed.) Forage conservation in the 80's. Proc. Occasional Symposium No. 11, Brighton, UK., 27-30 Nov., Br. Grassl. Soc., Hurley, UK.

Collins, M. 1982. The influence of wetting on the composition of alfalfa, red clover, and birdsfoot trefoil hay. Agron. J. 74:1041-1044.

Collins, M. 1983. Wetting and maturity effects on the yield and quality of legume hay. Agron. J. 75:523-527.

Collins, M. 1985. Wetting effects on the yield and quality of legume and legume-grass hays. Agron. J. 77:936-941.

Collins, M., W.H. Paulson, M.F. Finner, N.A. Jorgensen, and C.R. Keuler. 1987. Moisture and storage effects on dry matter and quality losses of alfalfa in round bales. Trans. ASAE 30:913-917.

Condon, S. 1987. Responses of lactic acid bacteria to oxygen. FEMS Microbiol. Rev. 46:269-280.

Courtin, M.G., and S.F. Spoelstra. 1990. A simulation model of the microbiological and chemical changes accompanying the initial stage of aerobic deterioration of silage. Grass Forage Sci. 45:153-165.

Davies, M.H., and I.B. Warboys. 1978. The effect of propionic acid on the storage losses of hay. J. Br. Grassl. Soc. 33:75-82.

Davis, R.J., C.A. Rotz, and D.R. Buckmaster. 1989. Effect of swath manipulation on losses and drying of alfalfa. p. 329-332. In Proc. 1989 Forage and Grassland Conference, Guelph, Ontario, Canada. 22-25 May, Am. Forage Grassl. Council, Georgetown, TX.

Dewar, W.A., P. McDonald, and R. Whittenbury. 1963. The hydrolysis of grass hemicelluloses during ensilage. J. Sci. Food Agric. 14:411-417.

Dickerson, J.T., K.K. Bolsen, C. Lin, and G. Ashbell. 1990. Rate and extent of top spoilage losses in horizontal silos. p. 107-108. In Proc. Ninth Silage Conf., Univ. Newcastle upon Tyne, Newcastle upon Tyne, UK.

Dobie, J.B., J.R. Goss, R.A. Kepner, J.H. Meyer, and L.G. Jones. 1963. Effect of harvesting procedures on hay quality. Trans. ASAE 6:301-303.

Doehlert, D.C., and S.H. Duke. 1983. Specific determination of α-amylase activity in crude plant extracts containing ß-amylase. Plant Physiol. 71:229-234.

Fonnesbeck, P.V., M.M. Garcia de Hernandez, J.M. Kaykay, and M.Y. Saiady. 1986. Estimating yield and nutrient losses due to rainfall on field drying alfalfa hay. Anim. Feed Sci. Technol. 16:7-15.

Friesen, O. 1978. Evaluation of hay and forage harvesting methods. In G.R. Quick (ed.) Grain and forage harvesting. Am. Soc. Agric. Eng., St. Joseph, MI.

Goering, H.K. 1976. A laboratory assessment on the frequency of over-heating in commercial dehydrated alfalfa samples. J. Anim. Sci. 43:869-872.

Goering, H.K., P.J. Van Soest, and R.W. Hemken. 1973. Relative susceptibility of forages to heat damage as affected by moisture, temperature and pH. J. Dairy Sci. 56:137-143.

Gordon, C.H., R.D. Holdren, and J.C. Derbyshire. 1969. Field losses in harvesting wilted forage. Agron. J. 61:924-927.

Goss, J.R., R.A. Kepner, and L.G. Jones. 1964. Hay harvesting with self-propelled windrower compared with mowing and raking. Trans. ASAE 7:357-361.

Greenhill, W.L. 1959. The respiration drift of harvested pasture plants during drying. J. Sci. Food Agric. 10:495-501.

Greenhill, W.L., J.F. Couchman, and J. De Freitas. 1961. Storage of hay III - Effect of temperature and moisture on loss of dry matter and changes in composition. J. Sci. Food Agric. 12:293-297.

Harris, C.E., and J.N. Tulberg. 1980. Pathways of water loss from legumes and grasses cut for conservation. Grass Forage Sci. 35:1-11.

Hart, R.H., and G.W. Burton. 1967. Curing Coastal bermuda grass hay: Effects of weather, yield, and quality of fresh herbage on drying rate, yield, and quality of cured hay. Agron. J. 59:367-371.

Harrigan, T.M., and C.A. Rotz. 1992. Net, plastic and twine wrapped large round bale storage loss. Paper 921572. Am. Soc. Agric. Eng., St. Joseph, MI.

Henderson, A.R., P. McDonald, and M.K. Woolford. 1972. Chemical changes and losses during the ensilage of wilted grass treated with formic acid. J. Sci. Food Agric. 23:1079-1087.

Hodgson, R.E., J.B. Shepherd, W.H. Hosterman, L.G. Schoenleber, H.M. Tysdal, and R.E. Wagner. 1947. Comparative efficiency of ensiling, barn curing, and field curing forage crops. Agric. Eng. 28:154-156.

Hoglund, C.R. 1964. Comparative storage losses and feeding values of alfalfa and corn silage crops. Agric. Econ. Report 947, Agric. Econ. Dept., Michigan State Univ., East Lansing, MI.

Honig, H. 1979. Mechanical and respiration losses during pre-wilting of grass. p. 201-204. *In* C. Thomas (ed.) Forage conservation in the 80's. Proc. Occasional Symposium No. 11, Brighton, UK., 27-30 Nov., Br. Grassl. Soc., Hurley, UK.

Huhnke, R.L. 1988. Large round bale alfalfa hay storage. Appl. Eng. Agric. 4:316-318.

Huhnke, R.L. 1990. Round bale bermudagrass hay storage losses. Appl. Eng. Agric. 6:396-400.

Johnson, D.G., D.E. Otterby, R.G. Lundquist, J.A. True, F.A. Benson, R.E. Smith, L.K. Lindor, and R.C. Stommes. 1984. Yield and quality of alfalfa as affected by harvesting and storage. J. Dairy Sci. 67:2475-2480.

Jones, B.A., R.D. Hatfield, and R.E. Muck. 1992. Effect of fermentation and bacterial inoculation on lucerne cell walls. J. Sci. Food Agric. 60:147-153.

Knapp, W.R., D.A. Holt, and V.L. Lechtenberg. 1976. Propionic acid as a hay preservative. Agron. J. 68:120-123.

Kemble, A.R. 1956. Studies on the nitrogen metabolism of the ensilage process. J. Sci. Food Agric. 7:125-130.

Kemble, A.R., and H.T. Macpherson. 1954. Liberation of amino acids in perennial rye grass during wilting. Biochem. J. 58:46-49.

Kepner, R.A., J.R. Goss, and J.H. Meyer. 1960. Evaluation of hay conditioning effects. Agric. Eng. 41:299-304.

Klinner, W.E., and G. Shepperson. 1975. The state of haymaking technology - A review. J. Br. Grassl. Soc. 30:259-266.

Koegel, R.G., K.J. Shinners, F.J. Fronczak, and R.J. Straub. 1988. Prototype for production of fast-drying forage mats. Appl. Eng. Agric. 4:126-129.

Koegel, R.G., R.J. Straub, and R.P Walgenbach. 1985. Quantification of mechanical losses in forage harvesting. Trans. ASAE 28:1047-1051.

Leibensperger, R.Y., and R.E. Pitt. 1987. A model of clostridial dominance in ensilage. Grass Forage Sci. 42:297-317.

MacAulay, J.D., and W.K. Bilanski. 1968. Mechanical properties affecting leaf loss in birdsfoot trefoil. Trans. ASAE 11:568-571.

Macpherson, H.T. 1952. Changes in nitrogen distribution in crop conservation. II. Protein breakdown during wilting. J. Sci. Food Agric. 3:365-367.

Mayne, C.S., and F.J. Gordon. 1986. Effect of harvesting system on nutrient losses during silage making. 2. In-silo losses. Grass Forage Sci. 41:341-351.

McDonald, P. 1981. The biochemistry of silage. John Wiley and Sons, Chichester, UK.

McGechan, M.B. 1989a. Alternative models of air infiltration into silage clamps. Dept. Note 24, Scottish Centre Agric. Eng., Penicuik, UK.

McGechan, M.B. 1989b. A review of losses arising during conservation of grass forage: Part 1, Field losses. J. Agric. Eng. Res. 44:1-21.

McKersie, B.D. 1981. Proteinases and peptidases of alfalfa herbage. Can. J. Plant Sci. 61:53-59.

McKersie, B.D. 1985. Effect of pH on proteolysis in ensiled legume forage. Agron. J. 77:81-86.

McKersie, B.D., and J. Buchanan-Smith. 1982. Changes in the levels of proteolytic enzymes in ensiled alfalfa forage. Can. J. Plant Sci. 62:111-116.

Meiering, A.G., M.G. Courtin, S.F. Spoelstra, G. Pahlow, H. Honig, R.E. Subden, and E. Zimmer. 1988. Fermentation kinetics and toxic gas production of silages. Trans. ASAE 31:613-621.

Melvin, J.F. 1965. Variations in the carbohydrate content of lucerne and the effect on ensilage. Aust. J. Agric. Res. 16:951-959.

Melvin, J.F., and B. Simpson. 1963. Chemical changes and respiratory drift during the drying of ryegrass. J. Sci. Food Agric. 14:228-234.

Morgan, C.A., R.A. Edwards, and P. McDonald. 1980. Intake and metabolism studies with fresh and wilted silages. J. Agric. Sci., Camb. 94:287-298.

Morris, R.M. 1972. The rate of water loss from grass samples during hay-type conservation. J. Br. Grassl. Soc. 27:99-105.

Morrison, I.M. 1979. Changes in the cell wall of laboratory silages and the effect of various additives on these changes. J. Agric. Sci., Camb. 93:581-586.

Morrison, I.M. 1989. Influence of some chemical and biological additives on the fibre fraction of lucerne on ensilage in laboratory silos. J. Agric. Sci., Camb. 111:35-39.

Muck, R.E. 1987. Dry matter level effects on alfalfa silage quality. I. Nitrogen transformations. Trans. ASAE 30:7-14.

Muck, R.E. 1990. Dry matter level effects on alfalfa silage quality. II. Fermentation products and starch hydrolysis. Trans. ASAE 33:373-381.

Muck, R.E. 1991. Silage fermentation. p. 171-204. *In* G. Zeikus and E.A. Johnson (ed.) Mixed cultures in biotechnology. McGraw-Hill, New York.

Muck, R.E., and K.K. Bolsen. 1991. Silage preservation and silage additive products. p. 105-126. *In* K.K. Bolsen (ed.) Field guide for hay and silage management in North America, NFIA, Des Moines, IA.

Muck, R.E., and J.T. Dickerson. 1988. Storage temperature effects on proteolysis in alfalfa silage. Trans. ASAE 31:1005-1009.

Muck, R.E., and R.E. Pitt. 1992. Aerobic losses at the silo face. Paper 921003, Am. Soc. Agric. Eng., St. Joseph, MI.

Muck, R.E., R.E. Pitt, and R.Y. Leibensperger. 1991. A model of aerobic fungal growth in silage. 1. Microbial characteristics. Grass Forage Sci. 46:283-299.

Muck, R.E., S.F. Spoelstra, and P.G. van Wikselaar. 1992. Effects of carbon dioxide on fermentation and aerobic stability of maize silage. J. Sci. Food Agric. 59:405-412.

Murdoch, J.C., and D.I. Bare. 1960. The effect of mechanical treatment on the rate of drying and loss of nutrients in hay. J. Br. Grassl. Soc. 15:94-99.

Murdoch, J.C., and D.I. Bare. 1963. The effect of conditioning on the rate of drying and loss of nutrients in hay. J. Br. Grassl. Soc. 18:334-338.

Neal, H.D. St C., and J.H.M. Thornley. 1983. A model of the anaerobic phase of ensiling. Grass Forage Sci. 38:121-134.

Nelson, L.F. 1966. Spontaneous heating and nutrient retention of baled alfalfa hay during storage. Trans. ASAE 9:509-512.

Nelson, L.F. 1968. Spontaneous heating, gross energy retention and nutrient retention of high-density alfalfa hay bales. Trans. ASAE 11:595-600.

Nelson, L.F. 1972. Storage characteristics and nutritive value of high-density native hay bales. Trans. ASAE 15:201-205.

Ohshima, M., and P. McDonald. 1978. A review of the changes in nitrogenous compounds of herbage during ensilage. J. Sci. Food Agric. 29:497-505.

Parkes, M.E., and D.J. Greig. 1974. The rate of respiration of wilted rye grass. J. Agric. Eng. Res. 19:259-263.

Parsons, D. 1991. Modelling gas flows in a silage clamp after opening. J. Agric. Eng. Res. 50:208-218.

Pitt, R.E. 1986. Dry matter losses due to oxygen infiltration in silos. J. Agric. Eng. Res. 35:193-205.

Pitt, R.E., and R.E. Muck. 1993. A diffusion model of aerobic deterioration at the exposed face of bunker silos. J. Agric. Eng. Res. 55:11-26.

Pitt, R.E., R.E. Muck, and R.Y. Leibensperger. 1985. A quantitative model of the ensilage process in lactate silages. Grass Forage Sci. 40:279-303.

Pitt, R.E., R.E. Muck, and N.B. Pickering. 1991. A model of aerobic fungal growth in silage. 2. Aerobic stability. Grass Forage Sci. 46:301-312.

Pizarro, E.A., and D.B. James. 1972. Estimates of respiratory rates and losses in cut swards of *Lolium perenne* (S321) under simulated haymaking conditions. J. Br. Grassl. Soc. 27:17-21.

Pizarro, E.A., and I.B. Warboys. 1979. The effect of the wilting period on the microflora of harvested pasture plants. p. 53-60. *In* C. Thomas (ed.) Forage conservation in the 80's. Proc. Occasional Symposium No. 11, Brighton, UK., 27-30 Nov., Br. Grassl. Soc., Hurley, UK.

Purves, D., and P. McDonald. 1963. The potential value of silage effluent as a fertiliser. J. Br. Grassl. Soc. 18:220-222.

Raghavan, G.S., and W.K. Bilanski. 1973. Mechanical properties affecting leaf loss in alfalfa. Can. Agric. Eng. 15:20-23.

Rees, D.V.H. 1982. A discussion of sources of dry matter loss during the process of haymaking. J. Agric. Eng. Res. 27:469-479.

Rice, B. 1966. A comparison of six types of hay making machines. Irish J. Agric. Res. 5:43-53.

Rider, A.R., D. Batchelder, and W. McMurphy. 1979. Effects of long term outside storage on round bales. Paper 791538. Am. Soc. Agric. Eng., St. Joseph, MI.

Rotz, C.A., and S.M. Abrams. 1988. Losses and quality changes during alfalfa hay harvest and storage. Trans. ASAE 31:350-355.

Rotz, C.A., S.M. Abrams, and R.J. Davis. 1987. Alfalfa drying, loss and quality as influenced by mechanical and chemical conditioning. Trans. ASAE 30:630-635.

Rotz, C.A., J.R. Black, D.R. Mertens, and D.R. Buckmaster. 1989. DAFOSYM: A model of the dairy forage system. J. Prod. Agric. 2:83-91.

Rotz, C.A., L.R. Borton, and J.R. Black. 1991a. Harvest and storage losses with alternative forage harvesting methods. p. 210-213. *In* Forages: A versatile resource. Proc. 1991 Forage and Grassland Conf., Columbia, MO. 1-4 April, Am. Forage Grassl. Council, Georgetown, TX.

Rotz, C.A., D.R. Buckmaster, and L.R. Borton. 1992. Economic potential of preserving high-moisture hay. Appl. Eng. Agric. 8:315-323.

Rotz, C.A., and R.J. Davis. 1986. Sprayer design for chemical conditioning of alfalfa. Trans. ASAE 29:26-30.

Rotz, C.A., R.J. Davis, and S.M. Abrams. 1991b. Influence of rain and crop characteristics on alfalfa damage. Trans. ASAE 34:1583-1591.

Rotz, C.A., R.J. Davis, D.R. Buckmaster, and M.S. Allen. 1991c. Preservation of alfalfa hay with propionic acid. Appl. Eng. Agric. 7:33-40.

Rotz, C.A., R.J. Davis, D.R. Buckmaster, and J.W. Thomas. 1988. Bacterial inoculants for preservation of alfalfa hay. J. Prod. Agric. 1:362-367.

Rotz, C.A., R.G. Koegel, K.J. Shinners, and R.J. Straub. 1990. Economics of maceration and mat drying of alfalfa on dairy farms. Appl. Eng. Agric. 6:248-256.

Rotz, C.A., R.E. Pitt, R.E. Muck, M.S. Allen, and D.R. Buckmaster. 1993. Direct-cut harvest and storage of alfalfa on the dairy farm. Trans. ASAE 36(3):621-628.

Rotz, C.A., and D.J. Sprott. 1984. Drying rates, losses and fuel requirements for mowing and conditioning alfalfa. Trans. ASAE 27:715-720.

Rotz, C.A., D.J. Sprott, and J.W. Thomas. 1984. Interaction of mechanical and chemical conditioning of alfalfa. Trans. ASAE 27:1009-1014.

Rotz, C.A., D.J. Sprott, and J.W. Thomas. 1986. Anhydrous ammonia injection into baled forage. Appl. Eng. Agric. 2:64-69.

Russell, J.B., and T. Hino. 1985. Regulation of lactate production in *Streptococcus bovis*: A spiraling effect that contributes to rumen acidosis. J. Dairy Sci. 68:1712-1721.

Russell, J.R., and D.R. Buxton. 1985. Storage of large round bales of hay harvested at different moisture concentrations and treated with sodium diacetate and/or covered with plastic. Anim. Feed Sci. Technol. 13:69-81.

Russell, J.R., S.J. Yoder, and S.J. Marley. 1990. The effects of bale density, type of binding and storage surface on the chemical composition, nutrient recovery and digestibility of large round hay bales. Anim. Feed Sci. Technol. 29:131-145.

Savoie, P. 1988. Hay tedding losses. Can. Agric. Eng. 30:39-42.

Savoie, P., and S. Beauregard. 1990. Hay windrow inversion. Appl. Eng. Agric. 6:138-142.

Savoie, P., C.A. Rotz, H.F. Bucholtz, and R.C. Brook. 1982. Hay harvesting system losses and drying rates. Trans. ASAE 25:581-585.

Schukking, S., and J. Overvest. 1979. Direct and indirect losses caused by wilting. p. 210-223. *In* C. Thomas (ed.) Forage conservation in the 80's. Proc. Occasional Symposium No. 11, Brighton, UK., 27-30 Nov., Br. Grassl. Soc., Hurley, UK.

Simpson, B. 1961. Effect of crushing on the respiration drift of pasture plants during drying. J. Sci. Food Agric. 12:706-712.

Shinners, K.J., R.G. Koegel, and R.J. Straub. 1991. Leaf loss and drying rate of alfalfa as affected by conditioning roll type. Appl. Eng. Agric. 7:46-49.

Shinners, K.J., R.J. Straub, and R.G. Koegel. 1992. Performance of two small rectangular baler configurations. Appl. Eng. Agric. 8:309-313.

Shepherd, J.B., H.G. Wiseman, R.E. Ely, C.G. Melin, W.J. Sweetman, C.H. Gordon, L.G. Schoenleber, R.E. Wagner, L.E. Campbell, G.D. Roane, and W.H. Hosterman. 1954. Experiments in harvesting and preserving alfalfa for dairy cattle feed. Tech. Bull. No. 1079, USDA, Washington, DC.

Sneath, P.H.A., N.S. Mair, and M.E. Sharpe. 1986. Bergey's manual of systematic bacteriology, Vol. 2, Williams and Wilkins, Baltimore, MD.

Spoelstra, S.F. 1987. Degradation of nitrate by enterobacteria during silage fermentation of grass. Neth. J. Agric. Sci. 35:43-54.

Spoelstra, S.F. 1990. Effects of cell wall degrading enzymes on silage composition at different ensiling conditions. p. 27-28. *In* Proc. Ninth Silage Conference, Faculty Agric., Univ. Newcastle upon Tyne.

Straub, R.J., and H.D. Bruhn. 1975. Evaluation of roll design in hay conditioning. Trans. ASAE 18:217-220.

Straub, R.J., R.G. Koegel, and K.J. Shinners. 1986. Mechanical losses from alfalfa harvesting strategies. Paper 861035. Am. Soc. Agric. Eng., St. Joseph, MI.

Sullivan, J.T. 1973. Drying and storing herbage as hay. p. 1-31. *In* G.W. Butler and R.W. Bailey (ed.) Chemistry and biochemistry of herbage. Vol. 3. Academic Press, London, England.

Svensson, K. 1978. Yield and quality losses in forage harvesting - some results of Swedish experiments. p. 310-314. *In* G.R. Quick (ed.) Grain and forage harvesting. Am. Soc. Agric. Eng., St. Joseph, MI.

Thomas, J.W., Y. Yu, T. Middleton, and C. Stallings. 1982. Estimations of protein damage. p. 81-98. *In* F.N. Owens (ed.) Protein requirements for cattle: Symposium. 19-21 Nov. 1980, MP109, Division of Agric., Oklahoma State Univ., Stillwater.

Tomes, N.J., S. Soderlund, J. Lamptey, S. Croak-Brossman, and G. Dana. 1990. Preservation of alfalfa hay by microbial inoculation at baling. J. Prod. Agric. 3:491-497.

Van Bockstaele, E.J., T.J. Behaeghe, and A.E. De Baets. 1979. Studies on the field losses of wilting grass. p. 205-207. *In* C. Thomas (ed.) Forage conservation in the 80's. Proc. Occasional Symposium No. 11, Brighton, UK., 27-30 Nov., Br. Grassl. Soc., Hurley, UK.

Verma, L.R., and B.D. Nelson. 1983. Changes in round bales during storage. Trans. ASAE 26:328-332.

Verma, L.R., B.D. Nelson, and C.R. Montgomery. 1986. Preservation of high moisture ryegrass hay with ammonia. Appl. Eng. Agric. 2:76-81.

Whitney, L.F. 1966. Hay losses from baler and chopper components. Trans. ASAE 9:277-278.

Wilkinson, J.M. 1981. Losses in the conservation and utilization of grass and forage crops. Ann. Appl. Biol. 98:365-375.

Wilkinson, R.H., and C.W. Hall. 1966. Respiration heat of harvested forage. Trans. ASAE 9:424-427.

Wilman, D., and I.G. Owen. 1982. Effects of stage of maturity, nitrogen application and swath thickness on the field drying of herbage to the hay stage. J. Agric. Sci., Camb. 99:577-586.

Waldo, D.R. 1977. Suggestions for improving silage. How far with forages for meat and milk production? p. 69-91. *In* Proc. 10th Research-Industry Conf., Columbia, MO. Am. Forage Grassl. Council, Georgetown, TX.

Waldo, D.R., and N.A. Jorgensen. 1981. Forages for high animal production: Nutritional factors and effects of conservation. J. Dairy Sci. 64:1207-1229.

Wolf, D.D., and E.W. Carson. 1973. Respiration during drying of alfalfa herbage. Crop Sci. 13:660-662.

Wood, J.G.M., and J. Parker. 1971. Respiration during the drying of hay. J. Agric. Eng. Res. 16:179-191.

Woolford, M.K. 1975. Microbiological screening of the straight chain fatty acids (C_1-C_{12}) as potential silage additives. J. Sci. Food Agric. 26:219-228.

Woolford, M.K. 1990. The detrimental effects of air on ensilage. J. Appl. Bact. 68:101-116.

CHAPTER 21

ASSESSING FORAGE QUALITY USING INTEGRATED MODELS OF INTAKE AND DIGESTION BY RUMINANTS

A.W. Illius and M.S. Allen

INTRODUCTION

Forage is the food of animals, and it is on their ability to utilize forage that the definition of forage quality must depend. The academic question of what constitutes and determines forage quality attracts the attention of a wide range of disciplines and a corresponding diversity of viewpoint. Yet we are all faced with recognizing that the roles played by plant and animal characteristics in determining forage quality are inseparable. It is axiomatic that the concept of forage quality arises from the *interaction* of plant physico-chemical properties with the animal's faculties for ingestion, digestion, nutrient absorption, and utilization. In other words, *realized* forage quality is a shared property of plant material and of the consumer. An integrated definition of the determinants of realized forage quality could thus be proposed as 'the set of animal and plant variables that determine the rate at which nutrients can be assimilated'.

The many interactions of these variables do indeed create a complex system and, accordingly, the appropriate way to understand forage quality is at the systems level. This explicitly treats the system's behavior as the integrated response of the components. Likewise, it is recognized that essential properties of the system will be lost by reducing the level of study to that which separates, for example, plant chemistry from animal responses. Of course, systems analysis cannot make any progress without detailed work on the components and mechanisms which constitutes the system. But equally, deeper and deeper knowledge of mechanisms becomes progressively more useless as it becomes more distanced from the original context. We have set out this integrated viewpoint as a declaration of the whole problem, to define the appropriate methodology for tackling the problem, and to introduce an examination of the contribution which can be made by integrated modeling. This chapter is about achieving a proper integration of plant and animal properties, and some examples of these are reviewed briefly below.

Plant variables used as measures of quality range from the simplistic, such as nitrogen content, to the complex, such as a complete specification of chemical composition. They may be aggregate variables, combining plant and animal properties, such as the digestibility of the dry matter (DM) of the whole plant; or delineated into components such as the indigestible fraction of the cell wall and the fermentable substrates of the cell cytoplasm and of the cell wall. Consideration may be given to the physical attributes

---

A.W. Illius, Inst. of Cell, Animal and Population Biology, University of Edinburgh, West Mains Rd., Edinburgh, EH9 3JT, UK; M.S. Allen, Dep. of Animal Science, Michigan State University, East Lansing, MI 48824.

of the cell wall and of whole particles of plant material which are thought to determine the accessibility of fermentable substrates to microorganisms. The presence of other dietary ingredients, such as concentrated feeds, may modify digestion of forages.

Animal variables may affect intake directly, as would, for example, gut capacity or rate of utilization of end-products. There may be consequential effects on digestion caused by modification of the kinetics of digestion and passage. For example, animals with different intakes, perhaps arising from differing metabolic requirements for a given nutrient, are likely to experience altered digestibility (Moe et al., 1965) due to the effect of intake on rumen retention time (Riewe and Lippke, 1970; Van Soest et al., 1984); animal body size and, separately, the degree of maturity affect the rates and duration of most physiological processes (Taylor and Murray, 1987); animals habituated to diets promoting amylolytic activity over cellulolytic activity show reduced cell wall digestion rates (e.g., Linberg, 1981; Prins et al., 1984). Further effects of animal origin arise from modification of intake rate by external interference, such as rationing or meal patterning, or by sward variables affecting intake rate (Hodgson, Chapter 19).

Integrated modeling aims to incorporate such plant and animal variables in a single framework to examine how their interaction determines the system's behavior. The ultimate objective of modeling is likely to be prediction of forage quality and animal performance, i.e., intake, digestibility, nutrient supply, and utilization. Modeling also has at least two important subsidiary objectives. First, modeling is a way of figuring out the consequences of one's assumptions. Second, it helps the experimenter to identify areas where knowledge is lacking or estimates are too imprecise. In these ways, modeling is an aid to disciplined thinking, providing a framework within which assumptions are formalized and weaknesses in knowledge are exposed.

Given enough resources, there is virtually no limit to the amount of relevant and irrelevant detail which can be built into a model, and so some thought needs to be put into deciding which variables and interactions actually need to be included in the model, and which may be excluded. Here, the criterion of relevance is applied: include those which significantly affect the attainment of the model's objectives. This in itself has proved to be a significant challenge, and the history of digestion modeling reveals the accumulation, testing, and occasional deletion of detail.

HISTORICAL PERSPECTIVE

In an attempt to adjust digestibility for the residence time of feeds in the gastrointestinal tract, Blaxter et al. (1956) created the first integrated model of digestion and passage. This model mathematically described both passage of digesta through the gastrointestinal tract using a first order compartmental model and digestion of feed using first order kinetics and potentially digestible and indigestible DM. Digestibility was calculated by integrating the product of the cumulative digestion curve and the fecal excretion curve of a pulse-dosed marker, recognizing that digestion is a competition between rate of passage and rate of digestion in the gastrointestinal tract. The authors calculated rate of potentially digestible DM digestion given the indigestible DM fraction, intake, and rate of passage characteristics of the feed and digestibility of the feed. The model is highly aggregated and many plant and animal factors that are known to affect intake and digestibility cannot be accommodated, decreasing the accuracy of prediction.

In the early 1970's, two models were published that were highly influential for subsequent generations of integrated models: Baldwin et al. (1970) and Waldo et al. (1972). These models are very different in complexity and represent divergent schools of thought for integrated models. Baldwin et al. (1970) modeled the rumen with chemically defined substrates and emphasized the stoichiometry of fermentation and the prediction of fermentation end-products. Digestion was a second order process affected by microbial mass. This model influenced many subsequent "metabolic" or "research" models which described the process of digestion as the appearance of fermentation end-products, or are purely concerned with intermediary metabolism (e.g., Gill et al., 1984). The simple model of Waldo et al. (1972) partitioned cellulose into that which is potentially digestible by rumen microbes and that which is indigestible due to plant factors. The substrate was thus defined biologically rather than chemically. Digestion of the potentially digestible fraction and passage was assumed to occur at a first-order rate. This model of cellulose digestion was concerned with cellulose disappearance rather than appearance of fermentation end-products. Many physical models of ruminal processes have incorporated multiple particle pools and particle size reduction in the rumen with this model as a repeating unit for each particle size pool included.

Although many integrated models have been developed, few have had the objectives of predicting DM intake and digestibility. Metabolic models generally have focused on providing absorbed nutrient fluxes to the animal that appeared to be reasonable when compared to literature values and have required intake as a model input. Physical models have been primarily concerned with prediction of intake and digestibility but are less flexible for controlling intake by metabolic mechanisms. As research progresses and integrated models continue to evolve, differences between physical and metabolic models will become less distinct. As such, it is useful to compare both types of integrated models to examine the variation in structure.

MODEL STRUCTURE

The overall structure of integrated models varies due to combinations of relatively few main components and sub-components. Levels of aggregation vary greatly among the models. The most aggregated dynamic model is that of Blaxter et al. (1956) which fractionated feed only into that which is potentially digestible and indigestible. One of the least aggregated models is that of Baldwin et al. (1970) which described the feed in chemical terms. Factors affecting level of aggregation and, thus, structure can be grouped into those associated with substrate fractionation, digestion kinetics, passage kinetics, and intake.

Substrate Fractionation

Integrated models vary greatly regarding fractionation of substrate. Some have considered total DM and others cell wall only (Table 1). Metabolic models must include all DM to accurately predict fermentation end-products. Most models distinguish between potentially digestible cell wall and indigestible cell wall. The concept of an indigestible fraction of DM was considered in the digestion model of Blaxter et al. (1956). McAnally (1942) observed that the digestion of cellulose in feces suspended in silk bags in the rumen of sheep ceased after 5 d. Subsequent work by Wilkins (1969) found that cellulose digestibility of a range of grasses reached a maximum value at 6 d incubation *in vitro* or when suspended in the rumen. The author concluded that plant factors appear to limit further digestion as the residue consisted only of lignified and cutinized

Table 1. Comparative structure of integrated models for feed fractionation.

Structure	Reference*
Feed fractions	
All DM:	1, 2, 4, 8, 9, 10, 11, 12, 13, 15, 16, 17, 19, 20, 21
Organic matter only:	7
Fiber only:	6, 14, 18
Cellulose only:	3, 5 (model 2)
Fractionation	
No further fractionation:	14, 17
Fermentable organic matter	
No further fractionation:	1, 10
Soluble or cell contents:	
No further fractionation:	7, 8, 15, 19
Multiple fractions:	2, 4, 9, 11, 12, 13, 16, 20, 21
Insoluble or fiber:	
No further fractionation:	2, 3, 5 (model 2), 7, 8, 9, 11, 12, 16, 18, 19, 20, 21
Cellulose and hemicellulose:	4, 13
Fast- and slow-fermenting NDF:	6, 15
Indigestible fraction	
DM:	1, 8, 10
Organic matter:	7
Fiber:	3, 5 (model 2), 6, 9, 11, 12, 18, 19, 20, 21
Lignin and insoluble ash:	4, 13, 16

* 1- Blaxter et al. (1956), 2- Baldwin et al. (1970), 3- Waldo et al. (1972), 4- Baldwin et al. (1977a), 5- Baldwin et al. (1977b), 6-Mertens and Ely (1979), 7- Pienaar et al. (1980), 8- Forbes (1980), 9- Black et al. (1980-81), 10- Poppi et al. (1981), 11- France et al. (1982), 12- Bywater (1984), 13- Murphy et al. (1986), 14- Mertens (1987), 15- Fisher et al. (1987), 16- Baldwin et al. (1987), 17-Kennedy and Murphy (1988), 18- Allen and Mertens (1988), 19- Illius and Gordon (1991), 20- Hyer et al. (1991a,b), 21- Sniffen et al. (1992).

tissue. Baldwin et al. (1970) used the concept of potentially fermentable holocellulose but did not define it or account for an indigestible holocellulose fraction in the model. Direct descendants of this model (Baldwin et al., 1977a,b; Murphy et al., 1986; Baldwin et al., 1987) included a lignin pool but all hemicellulose and cellulose were treated as potentially fermentable. The concept of potentially fermentable substrate is well agreed upon. However, implementation of the concept may be difficult as measurement of the fraction is problematic. Potentially digestible cellulose of forages increased with ball milling (Dehority et al., 1961) and potentially digestible neutral detergent fiber (NDF) was affected by grinding (Cherney et al., 1988). An accurate assessment of potentially digestible and indigestible fractions is necessary whether the objectives of the model are to predict fermentation products or DM clearance from the rumen.

Mertens and Ely (1979) partitioned the potentially digestible cell wall into fast- and slow-fermenting fractions. This followed the suggestion of Akin et al. (1974) and Akin and Amos (1975) that plants have three groups of morphological tissue types based upon rates of disappearance: nearly indigestible vascular bundles and sclerenchyma tissues, slow-digesting bundle sheaths and epidermal cells, and fast-digesting mesophyll and phloem.

Digestion Kinetics

Digestion kinetics have been described as first order processes in which a constant fraction of potentially fermentable substrate digests per unit time, or second order in which the fraction of potentially fermentable substrate which digests per unit time is dependent upon a constant fraction of the microbial mass (Table 2). Waldo et al. (1972) proposed that digestion of potentially digestible cellulose follows first order kinetics citing linear semilog plots of potentially digestible fiber disappearance in the literature (Gill et al., 1969; Smith et al., 1971, 1972). However, the work cited was done *in vitro* and presumably under conditions of enzyme excess. In order for first order kinetics to apply, microbial activity must not be changing or must be in excess. It is doubtful that microbial enzymatic activity is always in excess in the rumen: differences in substrate digestibility have been shown with bags suspended in the rumen of animals receiving different diets (Lindberg, 1981). However, the task of accurately modeling ruminal microbial activity is difficult as data are not available from the literature. Second order models assume that microbial activity is a constant fraction of microbial mass. This assumption is tenuous, particularly when all microbial activity is lumped into one pool as in the models of Baldwin et al. (1977a,b), Black et al. (1980-81), France et al. (1982), and Murphy et al. (1986). With this structure, changes in diet composition such as increasing starch and decreasing fiber will increase the percent fiber digested per hour due to increased microbial mass, even though the pool size of fiber digesting microbes decreases. Baldwin et al. (1987) attempted to correct this flaw by partitioning the microbial mass into alpha-hexose- and beta-hexose-utilizing microbes.

There is a lag period before fiber digestion begins because microbial enzymes do not have instantaneous access to potentially digestible fiber when feeds enter the rumen. Feed particles must be wetted with rumen fluid before microbes can gain physical access to fermentation sites. This lag phenomenon has been described mathematically as a discrete process in which all potentially fermentable fiber endures a delay, after which fermentation begins at a first order rate (Mertens, 1977) and as a continuous process in which fiber enters the rumen initially unavailable for attachment and passage and becomes available at a first order rate after which digestion can occur (Allen and Mertens, 1988). Most models do not consider this lag phenomenon for digestion, with the exception of Forbes (1980), which included a discrete lag for digestion of DM in the gastrointestinal tract, and Illius and Gordon (1991), which included a discrete lag for fermentation of potentially fermentable fiber in the rumen.

Table 2. Comparative structure of integrated models for digestion kinetics.

Structure	Reference*
Reaction order	
1st order:	1, 3, 5, 6, 7, 8, 10, 12, 15, 17, 18, 19, 21
2nd order:	2, 4, 9, 11, 13, 16, 20
Microbial pools	
not applicable:	1, 3, 5 (model 2), 6, 7, 8, 10, 14, 15, 17, 18
one:	4, 9, 11, 12, 13, 19**, 20
more than one:	2, 16, 21
Digestion lag	
none:	1, 2, 3, 4, 5, 6, 7, 9, 10, 11, 12, 13, 15, 16, 17, 20, 21
discrete:	8, 19
continuous:	18 (model 3)
Fractional rate constants	
do not vary:	8, 9, 11, 16, 17, 20, 21
vary with feed:	1, 3, 6, 7, 12***, 19
vary with particle size:	2, 4, 5, 10, 13, 15
vary with intake:	1

* Same as Table 1.
** Used only for the partial determination of rumen fill.
***Function of feed cell wall and crude protein content.

Models vary in application of digestion rate constants to fermentable substrate. Mertens and Ely (1979) used different digestion rate constants for the fast and slow potentially digestible fiber fractions which varied by forage source. Illius and Gordon (1991) included one potentially digestible fiber fraction but varied rates of digestion by forage. In models using second order digestion kinetics, the fraction of substrate fermented per hour is a function of the microbial pool size which, in turn, is affected by substrate intake and passage. However, fractional rate of fiber fermentation varies greatly with forage maturity (Smith et al., 1972). In addition, particle size may greatly affect rate of starch fermentation (see Nocek, 1988). The use of constant fractional rates of fermentation of fiber and starch for all feeds can cause large errors in the prediction of microbial yield, fermentation products, intake, and digestibility.

Mertens and Ely (1979) and Illius and Gordon (1991) used the same digestion rate constants for all particle size pools. Although evidence exists to justify separate rates (McLeod and Minson, 1969; Cherney et al., 1988; Fisher et al., 1989), sufficient data were not available for the feeds used by these models. Fisher et al. (1987) adapted the model developed by Mertens and Ely (1979) to include different digestion rate constants for the three particle size fractions using simulated data. Baldwin et al. (1987) included two particle size pools but allowed digestion from the small pool only. As only small particles were allowed to pass from the rumen, the large particle pool effectively functions as a continuous lag for digestion and passage for that fraction of the diet which is specified as large particles. Thus, increasing the proportion of small particles in the diet should not affect ruminal digestibility of organic matter.

Passage Kinetics

Structural representations of the flow of feeds through the gastrointestinal tract vary greatly among models (Table 3). Particulate flow is controlled by the number of particle size pools, the fractional rate of particle size reduction from one pool to smaller pools, the fractional rate of passage of particulate pools from the rumen, and lag phase for particulate passage.

Particle size of digesta in the rumen is much coarser than feces, indicating a mechanism for selective retention (Hungate, 1966). Several models partition feed into two or three particle size pools and constrain large particle passage to more accurately represent passage kinetics. The number of particle size pools required to accurately represent passage is not known as there have been no comparisons of models with identical structure other than the number of particulate pools. Models which vary digestion rate with particle size (Fisher et al., 1987) or which include particulate breakdown as a function of the animal (Illius and Gordon, 1991) require multiple particle size pools. However, there may be no advantage to multiple particle size pools in less complex models as the passage rate of the aggregate pool can be adjusted to accomplish the same purpose.

Most models with multiple particle size pools have applied the same fractional rate of size reduction for all feeds and animals. Illius and Gordon (1991) scaled rate of particle size reduction from the large to the small particle size pool by body size, based on literature data for sheep and cattle (Poppi et al., 1981) and mule deer (*Odocoiles hemionus* R.) and elk (*Cervus elaphus* L.) (Spalinger et al., 1986) for across-species comparisons. The authors also scaled the rate of particle size reduction with the indigestible fiber fraction of the feed to account for slower particle size reduction rates with increasing plant maturity.

Fractional particulate passage rates have been held constant across animals for some models (Mertens and Ely, 1979), and varied as a function of animal size in others (Hyer et al., 1991a; Illius and Gordon, 1991). No integrated models vary fractional rate of passage within a particle size for feeds.

Passage lag is a constraint on intake when physical limitations exist as it increases the size of the total fiber pool in the rumen (Allen and Mertens, 1988). Integrated models have not considered a lag effect on passage with the exception of Forbes (1980) and Illius and Gordon (1991). The latter found that a 1% increase in discrete lag time decreased intake by nearly 1% but had no effect on digestibility, as the same lag term was used for digestion as well.

Forages intrinsically vary in rate of clearance from the rumen due to passage. This was demonstrated by Aitchison et al. (1986) by examining indigestible fiber turnover in sheep fed different forages. Variable clearance from the rumen by passage for different forages can be accommodated in models by altering the initial particle size

Table 3. Comparative structure of integrated models for passage kinetics.

Structure	Reference*
Organ compartments	
gastrointestinal tract:	1, 8
rumen only:	2, 3, 5, 6, 7, 9, 10, 11, 14, 16, 17, 18, 20
rumen and intestines:	4, 12, 13, 15, 19, 21
Number of rumen particle size pools	
not applicable:	1, 8, 14
one pool:	3, 7, 9, 11, 18, 20, 21
large non-escape and small escape:	2, 4, 5, 10, 12, 13, 16, 17
large non-escape and medium and small escape:	6, 15
large and medium escape pools:	19
large, medium, and small escape:	17
Rate of particle size reduction	
not applicable:	1, 3, 7, 8, 9, 11, 14, 18, 20, 21
do not vary:	2, 4, 5, 6, 10, 13, 15, 16, 17
vary with feed:	12
scaled with body size:	19
scaled with maturity:	19
Fractional rates of particle passage	
do not vary:	2, 3, 4, 5, 6, 7, 8, 10, 11, 13, 15, 16, 17, 18
vary with feed:	1, 9**, 12
scaled with body weight:	19
scaled with fraction of mature body weight:	20
related to buoyancy:	17, 18
Lag phase of passage from rumen	
none:	2, 3, 4, 5, 6, 7, 8, 9, 10, 11, 12, 13, 15, 16, 20, 21
discrete:	19
continuous:	17, 18

* Same as Table 1.
** Function of modulus of fineness and acid detergent lignin content of diet.

profile of feeds, the number of particle size fractions included, particle size breakdown rates, fractional passage rates, pools on which passage rates act upon, and by including a lag phase for passage.

Although intake and rumen retention time are negatively related (Riewe and Lippke, 1970; Van Soest et al., 1984), none of the models reviewed adjust rate of particle size breakdown or passage for level of intake. This neglects one of the useful benefits of integrated models - predicting digestibility of feeds at different intakes.

Intake

The models that have included intake as an output have predicted intake under conditions where either only physical fill is assumed to be limiting or where physical limits are combined with metabolic control. A summary of comparative structure of integrated models for intake appears in Table 4. Although Blaxter et al. (1956) calculated gastrointestinal tract DM fill, most models since have used rumen fill of DM or fiber, assuming that the rumen is the primary control point for the physical limitation of intake. Although most models predict rumen fill as DM, Mertens and Ely (1979) included cell wall only, assuming that cell wall best represents the space-filling properties of the forage and that animals will eat to maximum fiber fill.

Several attempts have been made to integrate physical and metabolic controls of intake. Forbes (1980) simulated ruminant feeding behavior in which feeding begins when the flux of energy-yielding substrates drops below energy requirements and terminates when there is an excess of energy or when the amount of digesta reached the available space within the abdomen. Bywater (1984) also assumed that feed intake is regulated either by the difference between energy intake and energy expenditure or by gut fill. The models of Bywater and Forbes are similar in their description of metabolic regulation of intake, but Bywater treats gut fill as the capacity of the rumen to contain digesta cell wall rather than the capacity of the

Table 4. Comparative structure of integrated models for intake.

Structure	Reference*
Intake	
not considered:	1, 3, 5, 17
model input:	2, 4, 9, 10, 11, 13, 16, 18, 21
model output:	6, 7, 8, 12, 14, 15, 19, 20
Regulation	
physical only:	6, 7, 19, 20
physical and metabolic:	8, 12, 14, 15
Physical fill factor	
ruminal DM:	8, 15, 19, 20
ruminal organic matter:	7
ruminal fiber:	6, 12
diet fiber content:	14

* Same as Table 1.

gastrointestinal tract to contain DM. Fisher et al. (1987) incorporated metabolic control of intake into the basic model of Mertens and Ely (1979). Intake control was achieved by combining the effects of rumen distension with those of absorbed nutrient flux in a double exponential function.

Mertens (1987) presented a simple integrated model of intake and digestion in which the metabolic control of intake was related to energy requirements divided by the energy density of the diet. The physical control of intake is described as daily rumen fill capacity divided by the fill effect of the diet. Daily rumen fill capacity was assumed to be a constant fraction of body weight and NDF was used to predict the energy content of the diet as well as the fill effect. A major limitation of this model is that it does not account for major differences among forages and animals affecting digestion and passage kinetics and, therefore, digestibility and intake (Allen, 1990).

The successful integration of the many possible mechanisms of intake control is a difficult task, and insights gained from previous work may have been limited by the simplistic representation of metabolic events. A recent investigation by Poppi et al. (1990) employed a sophisticated model of intermediary metabolism, based on that of Gill et al. (1984). Six mechanisms were identified as being important in control of intake: rate of intake, rumen fill, rate of fecal output, heat production, substrate cycling, and genetic potential for protein deposition. The authors developed a model to incorporate all mechanisms by limiting intake to the lowest predicted by each individual mechanism. They found that the pathways regulating intake varied with the type of diet. As expected, intakes of low digestibility forages were limited by pathways with a physical basis and high digestibility forages were limited to pathways with a metabolic basis. The degradation of excess ATP (taken as an indicator of nutrient imbalance) limited intake in four out of the five diets tested. The authors make the point that two or more pathways simultaneously constrained intake in three of the five diets tested.

Table 5. Sensitivity elasticity* of predicted intake and digestibility, or percentage change, of model parameter values (adapted from Illius and Gordon, 1991).

Parameter	Sensitivity elasticity	
	Intake	Digestibility
Lag time	-0.095	0
Digestion rate of:		
cell contents	+0.025	+0.023
cell wall, rumen	+0.086	+0.055
cell wall, cecum	0	+0.037
Passage rate, small	+0.632	-0.055
Particle breakdown rate	+0.010	0
Large particle escape rate	+0.014	0
Liquid passage rate	+0.089	0
Digestible cell wall[+]	+0.489	+0.606
Proportion of large particles	-0.129	+0.037
Microbial growth rate	-0.129	0
Proportion of microbes passing with solid phase	-0.01	
Gut DM contents	+1	0
	Percentage change	
Allow passage during lag	+2.4%	-5.1%
Reduce large particle digestion rate to half that of small	-2.87%	-1.3%
Change proportion of MRT in rumen and hindgut to 0.65 and 0.3 (cf 0.75 and 0.2)	+10%	+0.9%

* Percentage change in output from a 1% increase in parameter value
[+] Increase in the amount of cell wall constituents which are digestible; corresponding decrease ... indigestible cell wall.

SENSITIVITY ANALYSIS

One way to assess the relative importance of variables in a model is to determine the sensitivity of predictions to changes in each input in turn, the others being held constant. Mertens (1993) compared sensitivities in a number of models, and this may give some indication of the effect of model structure on sensitivity. Certainly, sensitivity is likely to be influenced by the structure of the model, and so it would not be sensible to combine estimates of sensitivity from a range of models. Instead, to examine the sensitivities of intake and digestibility both to a wide range of variables and to aspects of model structure, we present the results from a single model (Illius and Gordon, 1991) in Table 5.

The model shows appreciable sensitivity to the digesta DM load (one-to-one effect on intake, none on digestibility), the content of digestible cell wall, and passage rate. Digesta load and passage rate had been estimated from data in the literature, and the residual standard deviation (SD) of the equation used to predict passage rate digestible; corresponding decrease in indigestible cell wall.
was 18.25%. Therefore, a change of 1 SD in passage rate would change intake by 11.5% and digestibility by -1%. Likewise, an increase of 1 SD in rumen DM load would increase intake by 24.3% without changing digestibility.

PERFORMANCE OF INTEGRATED MODELS

Comparisons with Reality and Validation

The common objective of all integrated models is to describe the relationships among animal and plant factors to obtain realistic absorbed nutrient fluxes, digestibility of feeds or feed fractions, and feed intakes. Various factors other than intake and digestibility have been examined to validate models and few direct comparisons can be made among them. In this section, we discuss the ability of models to predict digestibility and intake of forages.

Digestibility

The first integrated model to compare output to independent experimentally measured digestibility data was Model 2 of Baldwin et al. (1977b). The comparison was for cellulose digestion as affected by intake only, using 17 observations of ruminal cellulose digestion. The authors reported an average prediction error of 20%, defined as the difference between predicted and observed values divided by the observed values. A systematic error was found in which rumen cellulose digestibility was over-predicted when forage digestibility was low and under-predicted when forage digestibility was high. The authors concluded that a more complex model was required which included microbial interactions with feed components. It is important to note that in this comparison, the validation data contained forages of different type and maturity as well as intake, yet kinetic parameters such as rate of digestion, rate of passage, and rate of particle size breakdown were kept constant for all observations. This followed from the view that fermentation is primarily a chemical, not physical, process, and is governed by the quantities of microbes and of plant chemical constituents, and by the rates of microbial utilization of substrates. In this study, the model structure was rejected even though it may well have been accepted had the values of the rate constants reflected differences among feeds. The authors did not include a validation of a more complex model including microbial growth (Model 3).

The model of Mertens and Ely (1979) was the first to predict digestibility of forages based on their individual kinetic characteristics. The model validation (Mertens and Ely, 1982) included data from various studies with sheep and cattle and included 166 forages varying in DM digestibility from 41 to 80%. The digestibility values predicted by the model were not highly related to observed (r^2 = 0.52). In addition, there was a bias in the prediction in which forages of low digestibility were under-predicted and forages of high digestibility were over-predicted. Although the authors attributed the variation in digestibility explained by the model to differences in rates of digestion and potentially digestible fractions, NDF of the forage undoubtedly accounted for a great deal of variation. In fact, with nearly the same data set (Mertens, 1973), NDF alone accounted for more variation (r^2 = 0.6) than digestibility predicted by the model. This is probably due to the introduction of variation by using the same passage rates and particle breakdown rates across forages and body weights. This effectively demonstrates the importance of including animal factors and the interaction among plant and animal factors in integrated models. Beever et al. (1980-81) evaluated the model of Black et al. (1980-81) and found that the apparent digestibility of organic matter in the rumen was under-predicted by 9.9% for chopped ryegrass (*Lolium perenne* L.) and over-predicted by 15.6% for pelleted ryegrass. The authors concluded that 'detailed studies of the effect of physical form of the diet on rumen function are obviously needed'.

Murphy et al. (1986) found relatively close agreement between observed and predicted values for organic matter, hemicellulose, and cellulose ruminal digestion coefficients for four data sets with sheep fed alfalfa (*Medicago sativa* L.) at two levels of intake, subterranean clover (*Trifolium subterraneum* L.), and forage oats. The average prediction error was calculated to be 7.5%, 5.6%, and 14.0% for organic matter, hemicellulose, and cellulose digestibility, respectively. However, the model generally over-predicted organic matter digestibility and under-predicted hemicellulose and cellulose digestion.

The model of Baldwin et al. (1987) was found to overestimate digestibility by less than 3% for 600 g/kg grain (corn and barley) diets but by 12 to 19% for 900 g/kg grain diets. The authors suggested that the disagreement between observed and predicted values may be due to a depression in microbial activity at reduced pH on high grain diets.

Illius and Gordon (1991) used 25 observations of forage digestibility from the literature for sheep and cattle with published rate and extent of digestion values to validate their model. They found that the model explained 69.8% of the variation in the data, but a significant bias underestimated low-quality forages and the error of prediction for DM digestibility was over 16%.

Intake

Although Blaxter et al. (1956) and Waldo et al. (1972) used integrated models to predict gut fill, the first integrated model to predict intake and compare the results to experimental data was that of Mertens and Ely (1979). This model used *in vitro* rate of fermentation data from 166 forages to predict forage intakes ranging from 1.2 to 4.4% of body weight (Mertens and Ely, 1982). The model under-predicted low DM intakes, over-predicted high DM intakes, and explained only 26% of the variation in the data. The authors suggested that the over-prediction at high DM intakes was probably due to intakes that were limited by the energy demand of the animal. Under-prediction at low DM intakes was attributed to rate of passage which was too low, increased ruminal capacity, or low *in vitro* rates of digestion. However, an evaluation of the scatter plot of predicted versus observed DM intake suggests that a correction of the bias will not significantly increase the coefficient of determination; large variation exists for observed intakes at all predicted intakes. The poor relationship between predicted and observed intakes independent of bias is probably due to the use of constants for fractional rates of passage and particle size reduction for all forages and animals as well as constant ruminal capacities for all sheep and all cows. As the amount of variation in intake explained by the model is considerably less than that observed by using dietary concentrations of fiber (NDF, ADF) alone (Mertens, 1973), the use of constant rates of passage and particle size breakdown limits the ability of the model to improve the accuracy of prediction.

Hyer et al. (1991b) used a reference data set from the literature containing 42 data points to evaluate their model. They found that intake was systematically under-predicted for low digestibility forages and over-predicted for high digestibility forages. The authors suggested that the hypothesis that ruminal fill limits intake is inadequate for highly digestible forages.

Illius and Gordon (1991) found that predicted intake agreed well with a data set from the literature containing 25 points explaining 61% of the variance in observed values. Bias in the predicted values was not present, since the regression of best fit was: observed intake = 0.98 predicted intake, with intercept not significantly different from zero. The variation in intake explained by the model is greater than that predicted by NDF alone ($r^2=0.10$, not significant) or indigestible NDF ($r^2=0.18$).

CURRENT LIMITATIONS TO THE DEVELOPMENT OF MODELS

The improvement of model performance will depend on extending and improving the depiction of the components and interactions to which the system is sensitive. This is true where the current level of aggregation in modeled variables fails to capture the essence of the important interactions. For example, models assuming that intake is limited by the physical constraints of gut capacity and digesta flow generally ignore the effect which voluntary limitation of intake will play when nutrient intake or heat load exceeds some physiological threshold. Combining these processes in a way which accurately reflects reality is not a trivial task. This section examines current weaknesses and deficiencies and suggests developments which might overcome them.

Specification of Animal Variables

The first step in the description of animal effects is to scale physiological functions allometrically with body weight (W). For example, the expression of intake or digesta load as a percentage of

W employs this principle, the allometric exponent simply being 1. While many models make some reference to the effects of animal size (e.g., by specifying maintenance energy needs as a function of $W^{0.73}$), few models carry this very far. Hyer et al (1991a) applied Taylor's (1980) size scaling rules to passage and fill parameters in order to relate a set of reference data obtained from sheep to the equivalent values in cattle. Allometric exponents to be used in such scaling were derived by Illius and Gordon (1991, 1992). They showed that the time taken to comminute large fiber particles, and the small particulate retention time, both scale with $W^{0.27}$, similar to other temporal variables such as the time between successive heart beats or intestinal contractions (Clark, 1927; Taylor and Murray, 1987). The duration of physiological events is predictably longer in large animals. It was assumed that reticular contraction rate and duration are determinants of the rate of passage from the rumen. The analysis of data from 39 species of ruminant and non-ruminant herbivores unequivocally supported the expected scaling rule. The time taken to comminute large particles obeys the same rule, as demonstrated by the relationship of large particle retention time and $W^{0.27}$ (Illius and Gordon, 1991). The scaling approach works well for interspecific differences, but has a less clear theoretical basis and is less likely to be successful with intraspecific variation, such as that within a group of dairy cows. The latter sort of variation is potentially a major source of error in model predictions, and is of considerable economic importance. Attention needs to be directed to isolating these sources of among-animal variation and modeling the causal variables.

Modeling Passage Rate

Current descriptions of the passage rate are poorly related to either plant or animal factors, and thus do not account for effects of forage characteristics on passage, diurnal variation in fractional outflow rate, or effects of intake on passage. The lack of connection between the physico-chemical properties of the plant and the processes involved in passage may partly be resolved by considering the interaction of digestion, gas production, and flotation of particles. Following consumption, fiber particles are initially buoyant, as a result of trapped air. As the air dissolves during the lag phase, when hydration and microbial attachment precede digestion, the particles becomes heavier than rumen fluid. When digestion commences, carbon dioxide and methane gases from fermentation can form gas bubbles in the particle, reducing the functional specific gravity sufficiently to cause the particle to float to the top of the rumen. There it is likely to become trapped in the fibrous mat of the upper strata and is unlikely to be available for passage (Hungate, 1966, Hooper and Welch, 1985). As the digestible fraction of the particle declines, bouyancy declines and only then do particles descend to the ventral rumen and become likely to be passed out (Sutherland, 1987). Thus, the probability of passage depends on the time-course of digestion, flotation, and entrapment in the dorsal rumen. Functional specific gravity is determined by factors such as fermentation rate, concentration of gases in the rumen fluid, the effects of rumen motility in redistributing particles and dislodging gas bubbles, and comminution of particles where that exposes undigested tissues. These interactions have been modeled at an aggregated level (Allen and Mertens, 1988; Kennedy and Murphy, 1988), but a more detailed implementation to achieve a plant-based specification of passage rate has not yet been attempted.

Gill (1987) presented data showing diurnal variation in passage in animals fed a single daily meal. Thus, although rumen fill is greatest after the meal, fractional outflow is lowest, increasing continuously until the time of the next meal. While the time-course of particle flotation could partly explain this pattern, it is unlikely to account fully for the phenomenon. The time-course of flotation is probably governed by attributes of the forage and, therefore, need not take the 24 h required to generate a diurnal pattern of passage. Temporal patterns of rumen motility have been recorded, particularly in response to eating and rumination, and there may be an interaction between VFA production rates and motility (see Shinozaki, 1959; also Grovum, 1986). Passage rate also may be modified by level of intake, since the efficiency of digesta flux per contraction is increased at higher intakes (Deswysen and Ellis, 1988; Okine and Mathison, 1991). The combined effects of intake, meal patterning and digestion, motility and contraction frequency, duration and efficiency are not yet sufficiently clear to allow reliable modeling of diurnal variation in passage rate.

Modeling Digesta Load

Model predictions of intake under *ad libitum* feeding are highly sensitive to assumptions about mean and maximal digesta load (Table 5). To assume a standard load for the purpose of comparing foods is a considerable simplification at other than the physical maximal intake. In particular, it will not reflect animal differences in digesta load arising from differences in appetite, rationing, past nutritional experience, or homeorhetic changes due to pregnancy or stage of lactation. What is needed is a way of modeling digesta load as a dynamic variable, able to respond to temporal changes in nutrient requirements and voluntary or externally-limited intake. The rate and bounds of this response require definition.

Variation in bulk density of digesta during comminution and digestion implies that the physical filling effect of forage is dynamic. It also differs from the normal depiction of digesta load, which is expressed in terms of its mass and, therefore, with equal importance given to each component of the digesta. This can be justified if the mass of digesta is regarded as the subject of homeostatic control, but not if 'fill' effects act via the stretching effect of bulk on the rumen wall. This distinction underlies a philosophical problem with the very use of digesta load as a regulator. Since it is hard to tell whether or when digesta load actually limits intake, and how it interacts with other variables such as nutrient demand, then the use of digesta load as a regulator is a convenient solution to a lack of knowledge. Clearly, incorporation of metabolic and other influences on intake and digesta load is required for further progress.

Specification of Plant Variables

Modeling Digestion Rate and Potential Digestibility of Cell Wall

The assumptions used in most physical models are that first-order kinetics apply to a uniform digestible cell wall fraction, and that the remainder is undigestible. This is a convenient assumption, because these characteristics can easily be estimated using *in situ* (nylon bag) techniques. However, there are a number of problems with this. Digestion rates and potential digestibility measured in nylon bags may underestimate *in vivo* values (Aitchison et al., 1986), and this has been identified as a likely source of bias in predictions of digestibility of lignified forages (Illius and Gordon, 1991). That said, the discrepancies estimated by Aitchison et al. (1986) may have been exaggerated by the omission of a lag term from estimation of digestion rate and by using chromium-mordanted fiber to estimate passage rate. Estimated digestion rate and extent may deviate from *in vivo* kinetics by being sensitive to the fineness of grind of the sample (Dehority et al., 1961), while *in vivo*, digestion kinetics may possibly differ between particle sizes. In principle, there may be a range of digestion rates associated with the size spectrum of particles (see Mahlooji et al., 1984). In practice, consistent differences have been harder to identify *in situ*, provided disruption of epidermis is adequate for microbial access (Ford et al., 1987; Kyle and Hovell, 1987).

The omission from physical models of explicit links between a fuller catalogue of plant characteristics and the processes of microbial attachment and utilization highlights the need for greater understanding of these essential processes. This would overcome current weaknesses in modeling the interactive effects of nutrient balance on microbial growth and forage digestion, breakdown, and passage rates. The depiction of digestion rate in metabolic models by second-order kinetics combining microbial poulation size with substrate utilization rate is hardly credible, since no reference is made to the physical properties of cell wall which control microbial access and growth.

Availability of Data

Despite the ease of estimating digestion kinetics of forages *in situ*, there are surprisingly few comparative data available. There are few studies on grasses and legumes at different growth stages, and almost none on tropical grasses and legumes at any growth stage. This is a severe limitation on the evaluation of realized forage quality in models. There are too few estimates of particle breakdown rate and passage rate to investigate relationships with plant characteristics. Descriptions of cell wall morphology and composition need to be quantitative if they are to be useful for modeling purposes.

WHAT HAVE INTEGRATED MODELS TAUGHT US ABOUT REALIZED FORAGE QUALITY

Physical models of digestion kinetics tend to ignore the detailed chemistry of a forage and, instead, use the bioassay technique of *in situ* or *in vitro* degradation. Thereafter, model output should allow some evaluation of realized forage quality, and the sensitivity analysis (Table 5) gives an estimate of the most important plant factors. Foremost among these is cell wall composition: a 1% increase in the proportion of the cell wall which is digestible increases digestibility by about 0.6%, and intake by 0.5%. The digestion rate of the cell wall is of much smaller importance: a doubling of the rate constant would only increase DM digestibility by 5%. Plant or animal factors affecting passage rate have a sizeable effect on intake, easily outweighing the negative effect on digestibility.

What do animal factors really contribute, and is our emphasis of *realized* forage quality justified? To test this, we re-examined the data set collated by Illius and Gordon (1991) to compare model predictions with observed intake and *in vivo* digestibility. New predictions were obtained for a single animal size (240 kg) to compare with the range of animal types (40 kg sheep to 450 kg cattle) used to produce the original observations. Omitting modeled relationships between size-related phenomena such as passage and breakdown rates tests the contribution which animal specifications currently make to the accuracy of prediction. Prediction of intake was less well correlated with observed (r^2=0.254) than predictions employing animal specifications (r^2=0.61). Observed intake was negatively correlated almost as closely with the indigestible NDF content (r^2=0.18) as with model predictions omitting animal effects. Animal specification, therefore, appears to contribute substantially to the prediction of intake. Conversely, prediction of digestibility was only slightly affected by ignoring animal size (r^2=0.621 cf. r^2=0.698 without and with animal size effects). For comparison, the negative correlation of indigestible NDF content with observed digestibility gave r^2=0.515. One way of interpreting these results is to admit that, while depiction of the animal contribution accounts for some of the variation in realized forage quality, there is some way to go before the description of both animal and plant factors gives entirely satisfactory results. Moreover, accurate prediction is much harder over a narrow range of animal and forage types, by comparison with the rather broad range so far considered.

PROSPECTS FOR THE FUTURE

An important use of integrated models in research is to evaluate the system response to mechanisms elucidated in component research. Both will need to address the two major sources of unexplained variation in intake and digestibility: the relationships between plant physico-chemical characteristics, microbial colonization, and substrate utilization; and the animal factors determining the interplay of nutrient requirements, appetite, digesta load, and the consequential effects on passage and digestion. More attention must be payed to the interaction of substrates, particularly energy and protein. This interaction occurs both in the rumen, influencing fermentation products, and in the host animal, where effects on intermediary metabolism alter appetite and food utilization. This will require more effort on the integration of rumen and metabolic models. Indeed, the integration of physical and metabolic factors influencing intake and digestion remains a major challenge, for it must be remembered that intake control is ultimately a *psychological* process in which neural mechanisms integrate a wide range of positive and negative stimuli.

There is also a case for testing experimentally our assumptions about model structure before further significant progress may be made with modeling. After all, new models appear in the literature with much the same basic structure and degree of aggregation as the "Baldwin" and "Waldo" models, yet there has been insufficient experimental work to test the underlying assumptions of these models. An example is the buoyancy theory of particle passage: does buoyancy matter or not, and can it lead us to more accurate models of passage? Another example is the effect on digestion rate of microbial biomass:

some models ignore it, assuming that enzyme activity is not limiting in the rumen, while other models let it dominate digestion, ignoring plant factors. Modelers would recognize that their activity is a relatively straightforward one, compared with the difficulties of carrying out and interpreting the sort of nutritional experiments on which their models depend.

REFERENCES

Aitchison, E., M. Gill, J. France, and M. S. Dhanoa. 1986. Comparison of methods to describe the kinetics of digestion and passage of fibre in sheep. J. Sci. Food Agric. 37:1065-1072.

Akin, D. E., D. Burdick, and G. E. Michaels. 1974. Rumen bacterial interactions with plant tissue during degradation revealed by transmission electron microscopy. Appl. Microbiol. 27:1149-1156.

Akin, D. E., and H. E. Amos. 1975. Rumen bacterial degradation of forage cell walls investigated by electron microscopy. Appl. Microbiol. 29:692-701.

Allen, M.S. 1990. Feeding the super-producing dairy cow. p. AL1-AL18. Proc. 29th Ann. Congress of S. Afr. Soc. of Anim. Prod. 27-29 Mar., Stellenbosh, South Africa.

Allen, M. S., and D. R. Mertens. 1988. Evaluating constraints on fiber digestion by rumen microbes. J. Nutr. 118:261-270.

Baldwin, R. L., H. L. Lucas, and R. Cabrera. 1970. Energetic relationships in the formation and utilization of fermentation end-products. p. 319-334. In A. T. Phillipson (ed.) Physiology of digestion and metabolism in the ruminant. Oriel Press, Ltd., Newcastle upon Tyne, England.

Baldwin, R. L., L. J. Koong, and M. J. Ulyatt. 1977a. A dynamic model of ruminant digestion for evaluation of factors affecting nutritive value. Agric. Systems 2:255-288.

Baldwin, R. L., L. J. Koong, and M. J. Ulyatt. 1977b. The formation and utilization of fermentation end-products: Mathematical models. p. 348-391. In R.T.J. Clark and T. Bauchop (ed.) Microbial ecology of the gut. Academic Press, New York.

Baldwin, R. L., J. H. M. Thornley, and D. E. Beever. 1987. Metabolism of the lactating cow. II. Digestive elements of a mechanistic model. J. Dairy Res. 54:107-131.

Beever, D. E., J. L. Black, and G. J. Faichney. 1980-81. Simulation of the effects of rumen function on the flow of nutrients from the stomach of sheep: Part 2 - Assessment of computer predictions. Agric. Systems 6:221-241.

Black, J. L., D. E. Beever, G. J. Faichney, B. R. Howarth, and N. McC. Graham. 1980-81. Simulation of the effects of rumen function on the flow of nutrients from the stomach of sheep: Part 1 - Description of a computer program. Agric. Systems 6:195-219.

Blaxter, K. L., N. McC. Graham, and F. W. Wainman. 1956. Some observations on the digestibility of food by sheep, and on related problems. Br. J. Nutr. 10:69-91.

Bywater, A. C. 1984. A generalised model of feed intake and digestion in lactating cows. Agric. Systems 13:167-186.

Cherney, J. H., D. J. R. Cherney, and D. R. Mertens. 1988. Fiber composition and digestion kinetics in grass stem internodes as influenced by particle size. J. Dairy Sci. 71:2112-2122.

Clark, A. J. 1927. Comparative physiology of the heart. Cambridge University Press.

Dehority, B. A., and R. R. Johnson. 1961. Effect of particle size upon the *in vitro*-cellulose digestibility of forages by rumen bacteria. J. Dairy Sci. 44:2242-2249.

Deswysen, A. G., and W. C. Ellis. 1988. Site and extent of neutral detergent fiber digestion, efficiency of ruminal digesta flux and fecal output as related to variations in voluntary intake and chewing behavior in heifers. J. Anim. Sci. 66:2678-2686.

Fisher, D. S., J. C. Burns, and K. R. Pond. 1987. Modeling *ad libitum* DM intake by ruminants as regulated by distension and chemostatic feedbacks. J. Theor. Biol. 126:407-418.

Fisher, D.S., J.C. Burns, and K.R. Pond. 1989. Kinetics of *in vitro* cell-wall disappearance and *in vivo* digestion. Agron J. 81:25-33.

Forbes, J. M. 1980. A model of the short-term control of feeding in the ruminant: Effects of changing animal or feed characteristics. Appetite 1:21-41.

Ford, C. W., R. Elliott, and P. J. Maynard. 1987. The effect of chlorite delignification on digestibility of some grass forages and on intake and rumen microbial activity in sheep fed barley straw. J. Agric. Sci., Camb. 108:129-136.

France, J., J. H. M. Thornley, and D. E. Beever. 1982. A mathematical model of the rumen. J. Agric. Sci., Camb. 99:343-353.

Gill, S. S., H. R. Conrad, and J. W. Hibbs. 1969. Relative rate of *in vitro* cellulose disappearance as a possible estimator of digestible DM intake. J. Dairy Sci. 52:1687-1690.

Gill, M., J.H.M. Thornley, J.M. Black, J.D. Oldham, and D.E. Beever. 1984. Simulation of the metabolism of absorbed energy-yielding nutrients by the young sheep. Br. J. Nutr. 52:621-649.

Gill, M. 1990. Future for whole-animal metabolism models. p. 333-344. *In* A.B. Robson and D.P. Poppi (ed.) Proc. 3rd Int. Workshop on Modeling Digestion and Metabolism in Farm Animals. 4-6 Sept. 1989. Lincoln University, Canterbury, New Zealand.

Grovum, W.L. 1986. The control of motility of the rumenoreticulum. p.18-40 *In* L.P. Milligan and A. Dobson (ed.) Control of digestion and metabolism in ruminants. Prentice-Hall, Englewood Cliffs, NJ.

Hooper, A. P., and J. G. Welch. 1985. Effects of particle size and forage composition on functional specific gravity. J. Dairy Sci. 68:1181-1188.

Hungate, R.E. 1966. The rumen and its microbes. Academic Press, NY.

Hyer, J. C., J. W. Oltjen, and M. L. Galyean. 1991a. Development of a model to predict forage intake by grazing cattle. J. Anim. Sci. 69:827-835.

Hyer, J. C., J. W. Oltjen, and M. L. Galyean. 1991b. Evaluation of a feed intake model for the grazing beef steer. J. Anim. Sci. 69:836-842.

Illius, A. W., and I. J. Gordon. 1991. Prediction of intake and digestion in ruminants by a model of rumen kinetics integrating animal size and plant characteristics. J. Agric. Sci., Camb. 116:145-157.

Illius, A. W., and I. J. Gordon. 1992. Modeling the nutritional ecology of ungulate herbivores: Evolution of body size and competitive interactions. Oecologia, Berlin 89:428-434.

Kennedy, P. M., and M. R. Murphy. 1988. The nutritional implications of differential passage of particles through the ruminant alimentary tract. Nutr. Res. Rev. 1:189-208.

Kyle, D. J., and F. D. Hovell. 1987. The effect of grinding of a sample on the loss of material from nylon bags incubated in the rumen of sheep. Anim. Prod. 44:496 (Abstr.).

Lindberg, J.E. 1981. The effect of basal diet on the ruminal degradation of DM, nitrogenous componds and cell walls in nylon bags. Swedish J. Agric. Res. 11:159-169.

Mahlooji, M., W.C. Ellis, J.H. Matis, and K.R Pond. 1984. Rumen microbial digestion of fiber as a stochastic process. Can. J. Anim. Sci. (Supp.) 64:114-115.

McAnally, R. A. 1942. Digestion of straw by the ruminant. Biochem. J. 36:392-399.

McLeod M.N., and D.J. Minson. 1969. Sources of variation in the *in vitro* digestibility of tropical grass. J. Br. Grass. Soc. 24:244-249.

Mertens, D.R. 1973. Applications of theoretical mathematical models to cell wall digestion and forage intake in ruminants. Ph.D. diss, Cornell Univ., Ithaca, NY. (Dissertation abstr. #DCJ74-10882).

Mertens, D.R. 1977. Dietary fiber components: Relationship to the rate and extent of ruminal digestion. Fed. Proc. 36:187-192.

Mertens, D.R. 1987. Predicting intake and digestibility using mathematical models of ruminal function. J. Anim. Sci. 64:1548-1558.

Mertens, D.R. 1993. Kinetics of cell-wall digestion and passage in ruminants. p. 535-570. In H.G. Jung, D.R. Buxton, R.D. Hatfield, and J. Ralph (ed.) Forage cell wall structure and digestibility. ASA-CSSA-SSSA, Madison, WI.

Mertens, D.R., and L.O. Ely. 1979. A dynamic model of fiber digestion and passage in the ruminant for evaluating forage quality. J. Anim. Sci. 49:1085-1095.

Mertens, D.R., and L.O. Ely. 1982. Relationship of rate and extent of digestion to forage utilization-A dynamic model evaluation. J. Anim. Sci. 54:895-906.

Moe, P.W., J.T. Reid, and H.F. Tyrrell. 1965. Effect of level of intake on digestibility of dietary energy by high-producing cows. J. Dairy Sci. 48:1053-1061.

Murphy, M. R., R. L. Baldwin, and M. J. Ulyatt. 1986. An update of dynamic model of ruminant digestion. J. Anim. Sci. 62:1412-1422.

Nocek, J.E. 1988. In situ and other methods to estimate ruminal protein and energy digestibilty: A review. J. Dairy Sci. 71:2051-2069.

Okine, E. K., and G. W. Mathison. 1991. Reticular contraction attributes and passage of digesta from the rumenoreticulum in cattle fed roughage diets. J. Anim. Sci. 69:2177-2186.

Pienaar, J. P., C. Z Roux, P. J. K. Morgan, and L. Grattarola. 1980. Predicting voluntary intake on medium quality roughages. S. Afr. J. Anim. Sci. 10:215-225.

Poppi, D. P., D. J. Minson, and J. H. Ternouth. 1981. Studies of cattle and sheep eating leaf and stem fractions of grasses. III. The retention time in the rumen of large feed particles. Aust. J. Agric. Res. 32:123-137.

Poppi, D.P, M. Gill, J. France, and R. Dynes. 1990. Additivity in intake models. p. 29-46. In A.B. Robson and D.P. Poppi (ed.) Proc. 3rd Int. Workshop on Modeling Digestion and Metabolism in Farm Animals. 4-6 Sept. 1989. Lincoln University, Canterbury, New Zealand.

Prins, R.A., A. Lankhurst, and W. van Hoven. 1984. Gastointestinal fermentation in herbivores and the extent of plant cell-wall digestion. p. 408-434. *In* F.M.C. Gilchrist and R.I. Mackie (ed.) Herbivore nutrition in the subtropics and tropics. Science Press, Johannesburg, South Africa.

Riewe, M. E., and H. Lippke. 1970. Considerations in determining the digestibility of harvested forages. p. F1-F17. Proc. Nat. Conf. on Forage Quality Evaluation and Utilization. 3-4 Sep. 1969. Nebraska Center for Continuing Education, Lincoln, NE.

Shinozaki, K. 1959. Studies on experimental bloat in ruminants. Effects of various volatile fatty acids introduced into the rumen on the rumen motility. Tohoku J. Agric. Res. 9:237-250.

Smith, L. W., H. K. Goering, D. R. Waldo and C. H. Gordon. 1971. *In vitro* digestion rate of forage cell wall components. J. Anim. Sci. 54:71-76.

Smith, L. W., H. K. Goering, and C. H. Gordon. 1972. Relationships of forage composition with rates of cell wall digestion and indigestibility of cell walls. J. Dairy Sci. 55:1140-1147.

Sniffen, C.J., J.D. O'Connor, P.J. Van Soest, D.G. Fox, and J.B. Russell. 1992. A net carbohydrate and protein system for evaluating cattle diets: II. Carbohydrate and protein availability. J. Anim. Sci. 70:3562-3577.

Spalinger, D.E., C.T. Robbins, and T.A. Hanley. 1986. The assessment of handling time in ruminants: The effect of plant chemical and physical structure on the rate of breakdown of plant particles in the rumen of mule deer and elk. Can. J. Zool. 64:312-321.

Sutherland, T. M. 1987. Particle separation in the forestomachs of sheep. p. 43-73. *In* A. Dobson and M.H. Dobson (ed.) Aspects of digestive physiology in ruminants. Cornell Univ. Press, Ithaca, NY.

Taylor, St C. S. 1980. Genetic size-scaling rules in animal growth. Anim. Prod. 30:161-165.

Taylor, St C. S., and J. I. Murray. 1987. Genetic aspects of mammalian survival and growth in relation to body size. p. 487-533. *In* J.B. Hacker and J.H. Ternouth (ed.) The nutrition of herbivores. Academic Press, Sydney, Australia.

Van Soest, P. J., D. G. Fox, D. R. Mertens, and C. J. Sniffen. 1984. Discounts for net energy and protein - Fourth revision. p. 121-136. Proc. Cornell Nutr. Conf., Cornell Univ. Press, Ithaca, NY.

Waldo, D. R., L. W. Smith, and E. L. Cox. 1972. Model of cellulose disappearance from the rumen. J. Dairy Sci. 55:125-129.

Wilkins, R. J. 1969. The potential digestibility of cellulose in forage and faeces. J. Agric. Sci., Camb. 73:57-64.

CHAPTER 22

ALTERATION OF PLANTS VIA GENETICS AND PLANT BREEDING

K. P. Vogel and D. A. Sleper

FORAGE QUALITY, EVOLUTION, AND PLANT BREEDING

Plant breeding is man-directed evolution. Plant breeders manipulate the genetic resources of a species, i.e., its germplasm, to produce plants that are of increased value to humanity. The same analogy applies to animal improvement programs. All of our major food crops and all of our domestic animals and their respective breeds, strains, or cultivars were developed by this process. Although humans have successfully manipulated the genetic resources of plants and animals for several thousand years, the science of genetics and breeding was not developed until this century.

Breeding work on most forage crops did not began until the 1930's and initial work was focused on developing strains that had good establishment capability, persistence, high forage yields, and had good insect and disease resistance. These are essential attributes of forages (Burton, 1986). This initial breeding work resulted in the development of grasses such as 'Coastal' bermudagrass (*Cynodon dactylon* L.), 'Lincoln' bromegrass (*Bromus inermis* Leyss.), and 'Kentucky 31' tall fescue (*Festuca arundinacea* Schreb.). Limited animal evaluation was involved in the development of these cultivars.

Breeding for improved forage quality was not an important research objective of most forage programs until the last 25 yr. Kneebone (1960) published a review of grass breeding and did not discuss breeding for improved forage quality. However, some earlier reviews on grass breeding including the extensive review of Hanson and Carnahan (1956) included minor sections on breeding for improved forage quality. It was not until the pioneering research of Dr. Glenn Burton and his associates at Tifton, GA demonstrated the economic value of improved digestibility in bermudagrass that breeding for forage quality became a major research objective of some grass breeding programs (Burton, 1972a; Chapman et al., 1972).

K. P. Vogel, USDA-ARS, 344 Keim Hall, Univ. of Nebraska, P. O. Box 830937, Lincoln, NE 68583-0937; D. A. Sleper, Dep. of Agronomy, 16A Waters Hall, Univ. of Missouri, Columbia, MO 65211.

Genes that are available for plant breeders to manipulate using conventional breeding methods are those that a species has accumulated during its evolutionary history. In contrast to traits such as seed production and rhizomatous spread, positive forage quality traits such as high digestibility may not be advantageous to a species. Any trait that discourages excessive utilization of their herbage could be advantageous to forage plants. Even breeding for high digestibility is breeding to remove the effects of factors that inhibit digestibility. Anti-quality factors such as alkaloid content, hydrocyanic acid precursors, and the presence of endophytes and their associated deleterious effects appear to enhance the evolutionary fitness of a herbaceous species. It is likely that herbivory by insects was more important than consumption by ruminants in this evolutionary process (Jones, 1981; Molyneux and Ralphs, 1992). Breeding for improved forage quality can be viewed as changing plants to reduce their fitness to compete in the "wild" but increasing their fitness for use in agriculture.

There are many facets to breeding for improved forage quality. A recent study utilized the Dephi survey technique to attempt to rank forage quality breeding objectives in terms of their importance for grasses and legumes utilized by livestock for meat, milk, and wool production (Wheeler and Corbett, 1989). In this procedure, repeated surveys are used to arrive at a consensus. Improved digestibility was the most important criterion on each of the four lists while high comminution was ranked second in all lists except legumes for wool where it was ranked third. High comminution relates to physico-chemical characteristics conducive to high rate of passage or high outflow rate from the rumen, promoting higher intake. For improving livestock gains, high non-structural carbohydrate content also was an important breeding objective. For improving wool production, high S-amino acid content of forage protein was ranked two or three. Protein content was in the middle of the list of 11 factors ranked. High relative palatability, high lipid content, and erect growth were ranked as least important. It was generally considered that mineral content and anti-quality constituents should be monitored rather than making them specific breeding objectives.

Although genetic and breeding research on forage quality has been conducted for over 50 yr and numerous papers have been written on the topic, only a limited number of cultivars with improved forage quality have been released for use in commercial agriculture. Cultivars with improved forage quality have not been developed for most forage species. The objectives of this report are to review the progress that has been achieved by breeding for forage quality, assess the potential for future genetic gains in forage quality, and identify breeding strategies and procedures, both conventional and molecular, that should be the most effective in achieving breeding objectives.

PLANT BREEDING AND TRANSFORMATION

Until very recently, the only genes that were available to a breeder for improving a species in a conventional breeding program were the genes that were in the plants of a species or its close relatives. Genes can be moved between plants of closely related species with varying degrees of difficulty.

Moving genes between unrelated species is not possible using conventional breeding methods. Molecular genetic approaches have and are making it possible to clone genes from virtually any living organism and insert the cloned gene into another organism including forage plants. The transformed plants express the cloned genes and produce the gene products of the inserted gene. Molecular genetics and transformation procedures give plant breeders the potential to use genes from any organism to improve a plant species.

Conventional Plant Breeding Procedures

Conventional plant breeding involves manipulating the genes of a species so that desired genes are packaged together in the same plant and as many deleterious genes as possible are excluded. The two main components of the plant breeding process are selection and hybridization. Selection of the plants to be mated is the critical component of the breeding process. The other component, hybridization or mating, can usually be done in a routine manner although for some species, the procedures are tedious and require a high degree of skill (Barnes, 1980; Burson, 1980; Hovin, 1980; Taylor, 1980). Breeding systems have been developed and continue to be developed that can be used to improve virtually all forage species. Recent reviews have described the relative theoretical and practical efficiencies of these systems (Hanna and Bashaw, 1987; Vogel and Pedersen, 1993). In brief, forage breeders have an array of breeding procedures that they can use to improve forage species. The critical problem involved in improving forage quality is having an effective and consistent selection procedure. A breeder must be able to differentiate and rank plants before breeding progress can be achieved.

Genetic studies are usually conducted before plant breeders initiate long-term breeding projects to obtain information on the inheritance of the specific traits that the breeder wants to improve. It is necessary to determine if the trait(s) is controlled by only a few genes that are expressed in a qualitative manner or if they are controlled by many genes and are expressed in a quantitative manner. Eye color in humans is a trait that is controlled by a few genes and is inherited in a qualitative manner while adult height and weight are controlled by many genes and are inherited in a quantitative manner. Qualitative traits such as eye color usually are not affected by environmental factors while quantitatively inherited traits are influenced by environmental factors. Breeders also often attempt to determine if genetic variation for quantitatively inherited traits is due to additive or non-additive genetic effects. Additive genetic effects are due to the accumulative effects of genes that are expressed in an additive manner while non-additive effects are those in which gene action results in heterosis. After breeders have determined the existence of genetic variation for a trait, they need to know the stability of the expression of those differences over environments.

A few basic equations can be used to express many of the concepts involved in plant breeding (Allard, 1964; Falconer, 1981). Assuming selection is conducted on an individual plant basis, the heritability estimate (h_x^2) for a trait "x" (Equation 1) is the ratio of the additive genetic variation (σ_{ax}^2) for that trait divided by the phenotypic variance (σ_P^2) (Falconer, 1981). Except for a few forage species for which it is possible to produce commercial F_1 hybrids, forage breeders have to utilize additive genetic variation. Additive genetic variation is used as the numerator in Equation 1 to provide an estimate of heritability in the narrow sense. Plant breeders and geneticists use various mating and evaluation strategies to obtain estimates of the additive genetic and phenotypic variance (Hallauer and Miranda, 1981). Narrow sense heritability estimates are used to predict gain from selection and also provide an estimate of the proportion of the total variation for a trait that can be attributable to genetic differences among individuals or families. Heritability estimates can range from 1.0 for a trait such as eye color that is not affected by environment to less than 0.10 for traits that are highly influenced by environment variables.

$$h_x^2 = \sigma_{ax}^2 / \sigma_P^2 \qquad [1]$$

$$G_x = i\ h_x\ \sigma_{ax} \qquad [2]$$

$$CG_x = i\ h_y\ r_{xy}\ \sigma_{ax} \qquad [3]$$

Gain from selection for a trait is the gain that is achieved by selecting individuals for that trait and intermating the selected plants to produce their progeny. The mean difference between the progeny of selected and unselected plants is the realized gain from selection. The expected or predicted gain from selection (G_x) for trait "x" is the product of the standardized selection differential (i), the square root of the heritability of the trait (h_x), and the square root of the additive genetic variation (σ_{ax}) (Equation 2). The standardized selection differential is simply the proportion of selected plants expressed in units of standard deviations from the mean. The genetic gain that can be achieved in a single breeding cycle is dependent upon the relative magnitude of the factors in Equation 2.

Selection for one trait can have an effect on another trait if the traits are genetically correlated. The expected correlated response (CG_x) for trait "x" if selection is practiced for trait "y" is given in Equation 3. The genetic correlation is (r_{xy}). If the genetic correlation is large and the heritability for trait y is also large, then substantial gains from selection can be achieved for a particular trait by indirect selection for the correlated trait. Indirect selection can be as effective as direct selection if h_y is 25% larger than h_x and the genetic correlation is 0.8 or larger. Correlated responses to selection can be important in breeding for improved digestibility in terms of selection criteria.

The stability of a trait over environments is important because it will influence the area of adaptation of an improved cultivar. Breeders can obtain estimates of the genotype x environment interactions by growing cultivars or experimental strains in an array of environments. Variance component analyses and regression procedures are used to determine the relative magnitude of genetic, environment, and genotype x environment interaction

(GxE) effects. If genetic effects are significant and GxE effects are non-significant for a specific trait even though environment effects are large, then the trait is stable over environments. If GxE effects are larger than genotypic effects, it may be necessary for breeders to develop cultivars for specific environments.

Molecular Genetics and Plant Transformation

Improving plants via molecular genetics involves isolating and cloning a gene from any source that regulates a specific metabolic activity and inserting that gene along with necessary promoter sequences into the DNA of the target organism. Many of the genes that have been cloned for herbicide, insect, and disease resistance are from microorganisms and produce enzymes that degrade herbicides, metabolites that are toxic to insects, or inhibit virus replication (Goodman et al., 1987; Gasser and Fraley, 1989). Genes that can block the expression of specific metabolic pathways also can be developed and cloned (Verma et al., 1987). Since many of the factors that reduce forage quality are substances that inhibit digestibility or are toxic, development of anti-sense genes that block the expression of specific metabolites would be a highly feasible approach (Iiyama et al., 1993). Anti-sense RNA for the messenger RNA for polygalacturonase in tomato fruits has been successfully utilized (Gasser and Fraley, 1989). Genes or anti-sense genes that regulate the production of plant metabolites such as lignin (Bailey, 1991) or plant development (Poethig, 1990) appear to be logical targets for developing improved forage plants via genetic transformation.

Methods of incorporating cloned, foreign DNA into the DNA of plant cells currently can be accomplished by using *Agrobacterium tumefaciens* (currently only with legumes and other dicots), micro-injection, polyethylene glycol, electroporation, particle guns or other mechanical means (Gasser and Fraley, 1989; Kaeppler et al., 1992; Hodges et al., 1993). These DNA transformation procedures must be used in cell or protoplast culture systems that are capable of being regenerated into intact plants. Prior to initiation of a molecular transformation program, it is first necessary to develop an appropriate cell or protoplast culture system for each specific species. These systems are often the most variable part of the entire process (Hodges et al., 1993). In a recent review of genetic transformation of forage crops, Hodges et al. (1993) indicated that, to date, alfalfa (*Medicago sativa L.*), white clover (*Trifolium repens* L.), birdsfoot trefoil (*Lotus corniculatus* L.), orchardgrass (*Dactylis glomerata* L.), and tall fescue (*Festuca arundinaceae* Screb.) have been transformed with foreign DNA. There undoubtedly will be other forage species transformed in the near future. The critical problems in genetic transformation of forage plants are determining what traits to alter, developing the appropriate cloned genes, and the development of improved transformation procedures for monocots. Methods for determining if transformed forage plants are safe when fed to livestock and are ecologically safe also need to be developed and utilized. Forage breeders do not need to develop plants that have the potential of becoming super-weeds.

GENETIC MODIFICATION OF QUALITY AND ANTI-QUALITY TRAITS

Forage Digestibility

Plant breeders must be able to screen large numbers of plants for the traits undergoing selection. Breeding for improved digestibility first became feasible when reliable, repeatable, *in vitro* dry matter (DM) digestibility methods were developed. The nylon bag *in situ* procedure enabled Burton et al. (1967) to select Coastcross-1 bermuda grass from an array of bermudagrass plants. However, the development of the two-stage *in vitro* procedure by Tilley and Terry (1963) was the critical development that allowed breeders to screen large numbers of plants for differences in digestibility (see Chapters 15 and 16). Subsequent developments including the use of cellulase enzymes to replace rumen fluid (Gabrielsen et al., 1988; Casler and Sleper, 1991) and the development of near infrared reflectance spectroscopy (NIRS) (see Chapter 10) have given breeders the capability to rapidly screen large numbers of samples. For plant breeders, the two-stage Tilley and Terry procedure (1963) remains the standard on which other *in vitro* DM digestibility methods are calibrated.

Direct Selection for Digestibility

Genetic variability for forage digestibility has been found for virtually every forage species for which a well designed and conducted trial has been completed. Some of the genetic differences are due to single gene effects while in other instances, differences in digestibility are due to the action of numerous genes, i.e., differences in digestibility are inherited in a quantitative manner. Many forage species are complex polyploids and can have up to seven or more sets of basic genomes (Hanson and Carnahan, 1956). A trait that may be simply inherited in a diploid may appear to be inherited in a quantitative manner in a hexaploid or octaploid species simply because of the number of genes segregating.

Single genes that can significantly change forage digestibility are the brown midrib genes in maize, sorghum, and sudangrass (Barnes et al., 1971; Fritz et al., 1981). These genes are recessive, are expressed only in the homozygous state, and produce altered lignins (Kuc and Nelson, 1964; Gee et al., 1968). The brown midrib mutants of corn, sorghum, and sudangrass are almost 10 percentage units higher in *in vitro* DM disappearance (IVDMD) than their normal counterparts at similar stages of maturity. To date, no brown midrib strains or hybrids of these crops have been released for commercial use even though the nutritional advantage of these lines has been documented (Colenbrander et al., 1973; Lush et al., 1984), primarily because forage yield is lower (7 to 29%) than for normal lines (Lee and Brewbaker, 1984). From our perspective, increasing the forage yield of brown midrib maize and sorghum lines to improve economic returns ha^{-1} would be the most effective and efficient use of breeding resources to improve these plants as forage crops. Brown midrib mutants have not been found in polyploid forage species, probably because of the improbability of the recessive mutation being found in the homozygous state in each of the genomes comprising a polyploid species.

Genes that control leaf surfaces and other anatomical features can affect forage digestibility. The recessive gene, *tr*, in pearl millet (*Pennisetum americanum* (L.) Leeke) removes tricomes from the leaves, leaving the leaves with a smooth waxy surface (Burton et al., 1977). This gene affects forage utilization by livestock by increasing palatability but causes a reduction in digestibility of intact leaves. It also increases the plants susceptibility to leaf rusts but reduces transpiration which aids in drought tolerance (Burton et al., 1977). In sorghum (*Sorghum bicolor* (L.) Moench), the bloom (a whitish, waxy covering) on the leaves can be removed by a single recessive gene. Green intact leaf sections of three bloomless sorghums were 22% more digestible than bloom covered green leaves of their normal isogenic lines but when dried and ground, bloom-covered leaves were slightly higher (14 g kg^{-1}) in IVDMD (Cummins and Dobson, 1972). Based on these results, breeding for altered leaf surface of forage species needs to be justified by improved animal performance.

Single loci can affect forage digestibility by altering leaf to stem ratios. The dwarf (d_2) gene of pearl millet shortens stem internodes (Burton et al., 1969). When evaluated in comparative feeding and grazing trials with isogenic tall counterparts, the dwarf gene reduced internode length by 25 to 43% which increased percentage of leaf by 20%, increased forage digestibility by more than 10%, and decreased forage yield by 30%. When fed to animals as hay, the hay from dwarf lines increased average daily gains by 20%. However, when grazed, pastures seeded to the dwarf plants had reduced carrying capacity (15%) but produced equivalent gains per hectare (467 vs 480 kg ha^{-1}) as the tall counterpart (Burton et al., 1969). It is likely that internodes could be shortened in other forage species with similar effects. Results on the quantitative inheritance of IVDMD indicate that it should be possible to improve IVDMD without decreasing forage yield (see following section) so breeding for shortened internodes may not be the most effective method of increasing animal productivity per unit of land.

Genetic differences among plants of a forage species appear to be inherited in a quantitative manner for most forage species with the exception of the few major genes described in the above sections. Genetic variation for *in vitro* digestibility has been reported for perennial ryegrass (*Lolium perenne* L.), orchardgrass (*Dactylis glomerata* L.), bermudagrass (*Cynodon dactylon* (L.) Pers.), smooth bromegrass (*Bromus inermis* Leyss.), crested wheatgrass (*Agropyron* spp.), intermediate wheatgrass (*Thinopyrum intermedium* (Host) Barkw. and D.R. Dewey), switchgrass (*Panicum virgatumi* L.), indiangrass (*Sorghastrum nutans* (L.) Nash), maize (*Zea mays* L.), forage sorghums, alfalfa (*Medicago sativa* L.), and other species (Cooper, 1962; Gill et al., 1967; Roth et al., 1970; Burton and Monson, 1972; Coulman and Knowles, 1974; Vogel et al., 1981a,b; Pedersen et al., 1982; Lamb et al., 1984; Vogel et al., 1986, 1993b; Deinum and Struik, 1989; Marten, 1989; Hopkins et al., 1993).

In most of the forages that have been evaluated for genetic variation in IVDMD, the phenotypic range in digestibility has been as large as 100 g kg^{-1} and the heritabilities are usually 0.3 or larger. These results clearly indicate that it should be feasible to improve the digestibility of most forage crops if adequate sampling and laboratory protocols are followed and efficient breeding systems are utilized. Although genetic studies indicate that it should be feasible to develop cultivars with improved digestibility for almost every forage species, to date, only a limited number of forage cultivars with improved digestibility as validated in animal trials have been released.

Dr. Glenn Burton and his colleagues at Tifton, GA have been extremely successful in developing bermudagrasses with improved forage digestibility and increased yield. These improvements have been obtained by developing F_1 hybrid bermudagrasses that, although sterile, can be vegetatively propagated on a large scale. It is only necessary to identify a single superior plant to have a new cultivar. Coastcross-1 bermudagrass, a sterile F1 interspecific hybrid, was the first pasture grass bred for improved forage quality. It does not yield any more than Coastal bermudagrass, but in grazing trials and in feeding trials with pelleted hay it produced up to 30% higher average daily gains and up to 50% more live weight gain ha^{-1} (Burton et al., 1967; Burton, 1972a; Chapman et al., 1972). Averaged over grazing seasons, it was 60 g kg^{-1} higher in digestibility than Coastal (Table 1).

Table 1. Improvements in bermudagrass digestibility, yield, and beef production per hectare.[+]

Cultivar	IVDMD	Available herbage	ADG	Gain ha^{-1}
	g kg^{-1}	kg ha^{-1}	kg	kg ha^{-1}
Coastal	542	2900	0.48	498
Coastcross-1[1]	606	2280	0.72	746
Coastal	565	930	0.59	713
Tifton 78[2]	574	1190	0.68	985
Tifton 78	571	2450	0.65	789
Tifton 85[3]	594	2750	0.67	1156

+ADG = average daily gain; available herbage = mean herbage available for grazing in pastures during trials.
[1] Chapman et al. (1972). Animal gains averaged over stocking rates. Grazing season was 168 d.
[2] Hill et al. (1993). Grazing season was 168 d for 3 yr.
[3] Hill et al. (1994). Grazing season was 168 d. IVDMD values were based on esophageal samples. Trial was conducted from 1989-1991.

Breeding work on improving both IVDMD and forage yield of bermudagrasses has continued at Tifton over the past 20 yr since the release of Coastcross-1. The development of 'Tifton 78' demonstrated that it is feasible to improve both yield and IVDMD in bermudagrass (Burton and Monson, 1988; Hill et al., 1993). Tifton 78 is a 'Tifton 44' x 'Callie' F_1 hybrid and is taller and spreads more rapidly than Coastal. It also is higher yielding and was 10 g kg^{-1} higher in IVDMD than Coastal averaged over 3 yr in a grazing trial (Table 1). Tifton 78 can be stocked heavier than Coastal. Steers grazing Tifton 78 had significantly higher average daily gains than steers grazing Coastal. The combination of higher stocking rates and higher average daily gains results in significant improvements in total beef production per hectare (Table 1). 'Tifton 85' is the latest bermudagrass cultivar that has been released by the USDA-ARS program at Tifton (Burton et al., 1993) and it represents a remarkable breeding achievement. In a small plot trial conducted during 1989 to 1991, the IVDMD (g kg^{-1}) and annual DM yield (Mg ha^{-1}), respectively, of bermudagrass cultivars were as follows: Coastal (502, 11), Tifton 44 (513, 10.4), Tifton 78 (557, 11.3), Tifton 85 (573, 14.7) (Hill et al., 1994). In a grazing trial conducted during the same period at Tifton, Tifton 85 produced over 1000 kg ha^{-1} of gain when grazed by beef yearlings because of its high yield and high IVDMD (Table 1).

The development of Tifton 78 and particularly Tifton 85 are conclusive proof that it is possible to significantly improve both DM yields and forage digestibility by breeding. It should be noted that the improvements in both forage yield and digestibility were achieved by direct selection for both traits. The biochemical basis for the improved forage quality of the high IVDMD bermudagrasses has not been determined. It also needs to be noted that it took 20 yr to develop, test, and release Tifton 85 from the time Coastcross-1 was released.

Most of the forage species used in the temperate areas of the world are sexual, cross-pollinated perennial species. The first cultivar of a sexual, perennial species that was bred for improved digestibility was 'Trailblazer' switchgrass (*Panicum virgatum* L.) (Vogel et al., 1991). A series of experiments was conducted during the development and evaluation of Trailblazer. These experiments demonstrated that the IVDMD of switchgrass was improved by breeding by using a form of modified mass selection with selected plants polycrossed in isolation (Vogel et al., 1981b) and that the IVDMD of switchgrass plants sampled at panicle emergence was an excellent predictor of the IVDMD of their progeny in pastures throughout the grazing season (Tables 2 and 3) (Gabrielsen et al., 1990). A small improvement in IVDMD (3 to 4 percentage units) significantly improved animal performance (17 to 24%) as measured by both average daily gains and beef production per hectare in a replicated grazing study conducted for 3 yr (Table 3) (Anderson et al., 1988). Differences in animal performance were not due to differences in selectivity but to intrinsic differences in quality (Ward et al., 1989). Additional research demonstrated that changes in forage digestibility were achieved by increasing the extent of cell wall digestibility and not by changing the cell wall to cell solubles ratio or the rate of cell wall digestibility (Moore et al., 1993). The improvement in IVDMD was accompanied by correlated changes in the molar ratio of ferulic acid and p-coumaric acid, demonstrating that these components of plant cell walls were heritable and genetically correlated with IVDMD (Gabrielsen et al., 1990).

Table 2. Forage yields and IVDMD of switchgrass and intermediate wheatgrass strains in small plot trials in eastern Nebraska.

Species/strain	Forage yield	IVDMD
	kg ha^{-1}	g kg^{-1}
Switchgrass (1978-1980)[1]		
Pathfinder	9184	542
Trailblazer	9632	582
SE	NS	8
Intermediate wheatgrass (1986-1987)[2]		
Oahe	6720	604
Slate	6720	597
Manska	7168	616
SE	224	6

[1] From Vogel et al., 1984.
[2] From Moore et al., 1994.
[3] NS = not significantly different.

Table 3. Performance of beef yearlings grazing switchgrass and intermediate wheatgrass strains in eastern Nebraska and gross returns per acre.

Species/strain	ADG[3]	Gain ha^{-1}	$/ha[4]
	kg	kg	
Switchgrass (1982-1983, 1985)[1]			
Pathfinder	0.58	283	437
Trailblazer	0.73	350	540
SE	0.07	13	
Intermediate wheatgrass (1989-1990)[2]			
Oahe	1.05	256	395
Slate	1.05	260	400
Manska	1.22	298	459
SE	0.05	20	

[1] From Anderson et al., 1988.
[2] From Moore et al., 1994.
[3] ADG = average daily gain.
[4] Based on a price of $1.54 kg^{-1}

Two cool-season grasses with improved digestibility have been released recently. 'Badger' bromegrass was released in 1992 (Casler, 1992) by the Wisconsin Agricultural Experiment Station. It averages 10 to 30 g kg^1 higher in IVDMD than 'Rebound' smooth bromegrass and when grazed by ewes and lambs (*Ovis aries*) in a replicated trial in 1985, produced 11% higher daily gains (Casler, 1992). 'Manska' intermediate wheatgrass was developed by Dr. John Berdahl, USDA-ARS geneticist at Mandan, ND, by reselection in several sources of the unreleased strain 'Mandan 759' for agronomic traits (Berdahl et al., 1993). In a series of trials over several environments, it consistently had high IVDMD (Berdahl et al., 1993; Vogel et al., 1993b). It was included in a replicated grazing trial at Mead, NE along with the two cultivars, and another experimental strain (NE TI 1) that had been bred for high yield and high IVDMD in the USDA-ARS program at Nebraska (Moore et al, 1994). In the grazing trial, beef yearlings grazing Manska produced over 14% more gain per hectare than yearlings grazing the two most widely used cultivars, Oahe and Slate (Tables 2 and 3) (Vogel and Moore, 1993; Moore et al., 1994). Although NE TI 1 had similar yields and IVDMD values as Manska, gains of cattle grazing it were only slightly higher than for cattle grazing Slate and Oahe.

An alfalfa cultivar with improved IVDMD also has been developed whose improvement in digestibility has been validated in animal trials (Emile et al., 1993). The experimental cultivar, '632P', was compared to the cultivar, 'Europe', in a feeding trial with lactating dairy cows in which first or second cut alfalfa was fed. The intake was larger for 632P than for 'Europe' and the cows fed 632P produced more milk than cows fed 'Europe'.

Table 4. Dry matter digestibility and milk production from Holstein cows fed hays of two alfalfa strains differing in *in vitro* organic matter digestibility (from Emile et al., 1993).

Trait	Alfalfa strain	
	6328P	Europe
In vitro digestibility, g kg^1	677	627
Dry matter intake, kg d^1	18	16.1
Milk production, kg d^1	22.5	21.1

The limited numbers of cultivars specifically bred and released to date for improved forage digestibility demonstrate that forage digestibility of warm- and cool-season grasses and legumes can be developed by direct selection for IVDMD. The predictions of Howarth and Goplen (1983) "that the in vitro digestion technique does not have much potential for improving the digestibility of cool season forages" has been proven to be false.

In addition, the results of the animal evaluation trials in which the improved cultivars were compared with standard cultivars demonstrate that small changes in IVDMD can have significant economic impact on the profitability of livestock production systems. Since management costs will be identical among similarly yielding cultivars of a species except for possible differences in seed costs, genetic improvements in digestibility that lead to improved gains can be considered to be 100% profit. The economic value of a percentage unit (1% or 10 g kg^{-1}) improvement in digestibility will vary with the productive potential of a unit of land and the market value of livestock products. Trailblazer and Manska both were evaluated in eastern Nebraska and they were similar in forage yield to the check cultivars to which they were compared. If the market value of beef yearlings is assumed to be $1.54 kg^{-1}, then in regions that have the forage production potential of eastern Nebraska, a percentage unit improvement in IVDMD as determined in small plot trials has an economic value of $25 ha^{-1} (Tables 2 and 3) (Vogel and Moore, 1993).

Indirect Selection for Digestibility

Forage digestibility also can be improved by indirect selection for other traits. Improvements in forage digestibility have been achieved by selecting for changes in maturity (lateness), palatability, and cell wall composition or fractions.

Late maturing cultivars may not differ in digestibility from early maturing strains of the same species if sampled at the same physiological stage of development, but will be higher in digestibility on specific dates during the pre-flowering growing season because of their slower rate of development. In pearl millet, late maturing, near-isogenic populations had higher forage yields, higher digestibility, better seasonal distribution of forage yield, and were more persistent than their earlier maturing counterparts (Burton et al., 1968). 'Tiflate' pearl millet is a late maturing cultivar that was released because of its improved performance that is largely attributable to its late maturity (Burton, 1972b). It is photoperiod-sensitive and requires a short day for flowering. It will not set seed in the continental USA, but will produce abundant seed in the tropics.

'Morpa' weeping lovegrass (*Eragrosits curvula* (Schrad.) Nees) was developed from plants of a plant introduction that survived the winter at Woodward, OK (Voight, 1971). The surviving plants were probably all the same genotype because of their uniformity and because Morpa is an obligate apomixix (produces seed by asexual reproduction). Morpa was selected for improved palatability in palatability trials. In a subsequent grazing trial, cattle grazing Morpa had 12% higher gains than cattle grazing common weeping lovegrass (Voight et al., 1970). Later research indicated that Morpa had higher forage digestibility than common lovegrass (Voight, 1975) and was also 6 to 8 d later in maturity (Voight, 1971). It is likely that the improved palatability and digestibility of Morpa was due to its later maturity.

In the USDA-ARS switchgrass breeding program at Nebraska, a single cycle of selection for high IVDMD and high forage yield in 'Cave-in-rock' switchgrass improved forage digestibility about a percentage unit. In subsequent evaluations, the high IVDMD Cycle 1 selection population was about 2 d later in maturity than the base population which was the original cultivar (Hopkins, 1993; K. P. Vogel, personal communication). Gabrielsen et al. (1990) reported that in switchgrass, IVDMD declined from 2.7 to 3.5 g kg^{-1} d^{-1} prior to flowering in a 3 yr study (Figure 1). In the evaluation trials of Cave-in-rock lines, the plants in plots were staged for maturity using the system of Moore et al. (1991) prior to harvest. After IVDMD was adjusted to a common maturity stage by regression procedures, there were no differences between the Cycle 1 stain selected for high IVDMD and the original population. The differences in IVDMD were apparently achieved by indirect selection for maturity. In the development of Trailblazer switchgrass, all plants were sampled at a common maturity stage so IVDMD per se was improved.

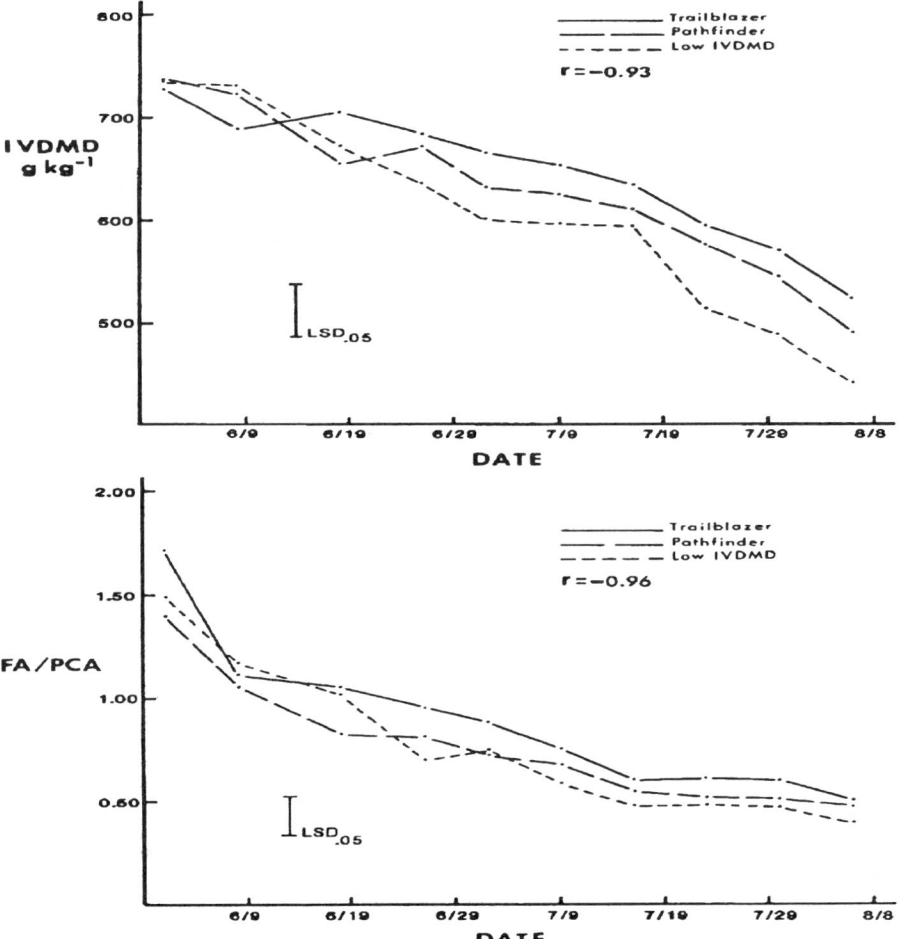

Figure 1. In vitro dry matter digestibility (IVDMD) and ferulic/p-coumaric acid (FA/PCA) ratio in grazed switchgrass forage of three different stains during 1984. Correlation coefficients (r) are for linear (IVDMD) and quadratic (FA/PCA) regressions with maturity, respectively, analyzed across strains. The LSD is for comparisons of strains. (From Gabrielsen et al., 1990).

Since heritability for heading date is usually large (0.80 or larger) for most species and heritability for IVDMD is low (0.3 or less), a rapid way to improve digestibility during the growing season is to simply select for late maturity. Potential gain will be limited by winter survival problems for plants that mature too late.

Plants can vary in digestibility due to differences in maturity, leaf to stem ratios, fertile to sterile tiller ratio, digestibility of plant parts, cell wall percentage of each component, and digestibility of cell walls (Casler, 1986; Buxton and Marten, 1989; Carpenter and Casler, 1990; and see Chapters 3, 4, and 18). These factors all can be considered to be components of whole plant digestibility and successful selection for any single factor will improve whole plant digestibility. In some instances, selection for a single component trait can result in greater initial gains than direct selection for IVDMD as demonstrated by Carpenter and Casler (1990) in smooth bromegrass. However, IVDMD is the only criterion that integrates all components of digestibility. If there is genetic variation for any component of digestibility in a species, the digestibility of its forage can likely be improved by breeding for improved IVDMD. If a breeder attempts to improve digestibility by selecting for one of its components such as acid detergent lignin, the potential gain is limited by the genetic variation for that trait alone.

Stability of Digestibility

Environmental factors including heat, drought, and light intensity can affect the digestibility of forage (Buxton and Casler, 1993). Genotype x environment interaction studies have been completed on crested wheatgrass (Lamb et al., 1984), intermediate wheatgrass (Vogel et al., 1986, 1993b), switchgrass (Hopkins, 1993; Hopkins et al., 1994), bromegrass (Casler et al., 1994), and timothy (McElroy and Christie, 1986). In each of these studies, IVDMD was more stable over environments than forage yield. In the crested wheatgrass and intermediate wheatgrass studies, GxE interaction effects for IVDMD were not significant. In the switchgrass study, 20 elite switchgrass strains were evaluated in Nebraska, Iowa, and Indiana. Although there were significant differences among strains for forage yield at individual locations, over locations the strains did not differ because of large GxE interaction effects. In contrast, GxE interaction effects were either small or not significant for IVDMD and there were significant differences among strains over environments. In the timothy GxE trials that were conducted in Canada, the GxE effects for IVDMD were significant but the variance component for the GxE effect was only 0.10 the size of the genetic variance for IVDMD. Based on a series of experiments in Wisconsin, South Dakota, and Canada, Casler et al. (1994) indicated that GxE effects for digestibility are relatively unimportant in that region for smooth bromegrass. In brief, forage digestibility is relatively stable over environments which means that strains should rank the same in digestibility over years and locations in which they are adapted.

Altering Plant Digestibility Via Molecular Genetics

Research is in progress to modify plant cell walls via molecular genetics including isolating and cloning genes regulating lignin synthesis (McIntyre et al., 1993a, b). The cloned genes are being utilized in anti-sense RNA and ribozyme technology to alter the lignification process. Transformed plants with altered lignin composition or structural changes will undoubtedly be developed. These transformed plants will require extensive evaluation by both animal scientists and agronomists to document any superiority and also to document their safety for use by livestock in specific environments.

Digestibility Summary

Although an effective method of evaluating forages for improved IVDMD has been available for over 25 yr and numerous studies have documented the existence of genetic variation for IVDMD, only a few forage cultivars have been developed and released with improved digestibility. Why the lack of breeding progress for improved digestibility? There are several possible reasons. Some breeding programs have placed emphasis on traits other than digestibility because they did not recognize the economic value of breeding for improved digestibility or believed other traits were more important. In some instances, breeders have used inappropriate selection criteria or methods or have simply misinterpreted their results and have come to the erroneous conclusion that it is not feasible to improve forage digestibility in specific species. In addition, there has been a lack of commitment to animal evaluation trials to validate the results of the laboratory and small plot research. Even when there has been animal evaluation, the results have sometimes been misinterpreted.

As an example, Kamstra et al. (1973) reported the results of an evaluation trial in which they compared the progeny of smooth bromegrass clones differing in digestibility. In a genetic study, the progeny of bromegrass clones were evaluated for differences in IVDMD. Plants whose progeny differed significantly in IVDMD were identified. These plants were used to produce 2-clone synthetics which then were tested in sward trials along with the cultivar 'Sac'. The hay from the trial was fed to wether lambs. In the spaced-planted breeders progeny test, the high IVDMD synthetic, SD1, was 5 percentage units higher in IVDMD than the low IVDMD synthetic, SD2 (Table 5). Plots in the sward trial were 0.135 ha with 4 replicates. There were no differences among the high and low IVDMD synthetics for IVDMD or *in vivo* digestibility for hay from the sward plots although both were higher in IVDMD and in *in vivo* digestibility than Sac. The *in vivo* digestibility and IVDMD values were similar except the *in vivo* values were 30 to 40 g kg^{-1} larger. Kamstra et al. (1973) expressed concern about the failure to obtain expected differences in IVDMD and *in vivo* digestibility in the sward trial and questioned the applicability of results from breeder's nurseries to field situations. They noted that the average daily gains ranked in the order predicted by the progeny test (Table 5) but failed to grasp the significance of these data. The average daily gains were the best estimate of forage quality that they had available and these data clearly showed that selection for IVDMD resulted in improved animal performance. To get enough hay for the feeding trial, they had to have large plots and it is likely that their failure to detect differences in IVDMD were due to sampling errors. Since the sheep ate most of the hay produced per plot, the sampling errors were likely negated. The point is that animal performance is the best measure of forage quality.

Table 5. Digestibility and daily gains of wether lambs fed hays of two experimental smooth bromegrass strains and the cultivar 'Sac' (from Kamstra et al., 1973).

Test and trait	Sac	SD1	SD2
Breeder progeny test			
IVDMD, g kg^{-1}		642^a	587^b
Hay from sward trial			
IVDMD, g kg^{-1}	628^b	656^a	668^a
In vivo lamb trial			
Hay digestibility, g kg^{-1}	590	625	627
Hay intake, kg/d^{-1}	0.77	0.82	0.88
Daily gain, kg	0.025^b	0.062^a	0.047^b

[a,b] Means with a different superscript letter differ significantly at the 0.05 level of probability.

Attempting to breed forages for improved forage quality without conducting animal evaluation trials can lead to erroneous conclusions. In the USDA-ARS grass breeding program at the University of Nebraska, three grazing trials have been conducted to evaluate strains differing in IVDMD. In a switchgrass trial (Anderson et al., 1988), results were as expected, i.e., the strain with the highest IVDMD produced the highest average daily gains. In an intermediate wheatgrass trial (Berdahl et al., 1993; Moore et al., 1994), one of the high IVDMD experimental strains, 'Manska', produced the animal performance expected while another one, 'NE TI 1', did not. In a crested wheatgrass trial, the cultivar that had the highest IVDMD in small plots, 'Ruff', produced lower daily gains than the cultivar 'Nordan' in some years (Vogel et al., 1993a). Again, the point is that genetic gains in improving digestibility need to be validated in animal trials.

Improved Rate of Passage

Improving the rate of passage of forage or the ease of comminution has been suggested as a means of improving forage quality (Wheeler and Corbett, 1989; Mertens, 1993). However, it is clear from Mertens (1993) recent review that factors which affect rate of passage are not completely understood and that laboratory tests that could be used on large numbers of breeder samples are not available. Wheeler and Corbett (1989) suggested that it would be desirable to develop techniques that could be used to measure ease of comminution such as energy required to pulverize forage samples, tensile strength, and shear strength. It will be necessary to establish the relationships between such measurements and rate of passage before they should be used in breeding programs. In addition, the relationship between rate of passage traits and other nutritional characteristics need to be determined. Grass breeders have finite resources. At the present time, investing those resources to improve digestibility would likely produce greater economic gains than breeding for improved rate of passage.

Protein Concentration and Quality

The proteins in forages primarily are enzymes that are involved in metabolic processes such as photosynthesis and respiration. Genetic variation for protein content has been found in forage plants but the genetic variation is less than the genetic variation for IVDMD and the GxE interaction variances are larger (Vogel et al., 1981a, 1986, 1993b; Lamb et al., 1984). Since the proteins in forages are involved in metabolic activities and are not storage proteins, it is probably not feasible to make major changes in protein quality or protein composition of forage plants by breeding and still have viable, vigorous plants. Small changes in protein concentration may be feasible. It is unlikely that changes in protein quality are feasible using conventional breeding approaches. Attempts are in progress to insert genes in forage plants that would result in improved protein quality. These include inserting chicken ovalbumin genes into alfalfa (Schroeder et al., 1991). Expression of these and other genes in forages may result in production of proteins for which there is no sink or storage site. This may affect the normal metabolism and vigor of forage plants.

Minerals

Mineral imbalances in forages can lead to nutritional diseases in livestock utilizing this forage. The major problem involving minerals is grass tetany caused by Mg and Ca deficiencies in C3 forage species. Most of the genetics and breeding research on minerals in forages has been on Ca, Mg, and K concentrations, which are the principal minerals involved in the grass tetany syndrome. Significant genetic differences in concentrations of these minerals have been found in all cool-season species evaluated to date including tall fescue, ryegrass, crested wheatgrass, reed canarygrass (*Phlaris arundinaceae* L.), and orchardgrass (Sleper et al., 1989). In general, the heritabilities for Mg and Ca are higher than those for K and their concentration is more stable over environments than is K concentration. Heritability estimates and genetic variances indicate that it should be possible to alter the mineral concentration of forages by breeding (Sleper et al., 1989).

Mass selection was successfully used by Hides and Thomas (1981) to increase the Mg concentration of annual ryegrass in three cycles of divergent selection. After the three cycles of selection, the high Mg population was

35% higher in Mg concentration than the control population while the low Mg population was 24% lower in Mg concentration than the control population. A tall fescue cultivar, 'Martin', has been released with increased Mg concentration (D. A. Sleper, personal communication). Crested wheatgrass cultivars that differ significantly in Mg concentration and the grass tetany potential ratio (K/(Ca + Mg) meq basis) also have been reported (Vogel et al., 1989b, 1993a). The incidence of grass tetany is usually not a production problem when this ratio is 2.2 or less (Kemp and t'Hart, 1957; Butler, 1963).

Animal evaluation trials have been conducted on ryegrass and crested wheatgrass cultivars differing in mineral concentrations. Mosely and Baker (1991) reported that when grazed by lactating ewes, a cultivar of Italian ryegrass bred for high Mg content, Bb 2067, resulted in significantly lower incidences of grass tetany than a control cultivar, 'RvP'. In the control pastures, the incidence of grass tetany was 21% in the first 10 d but was only 2 to 5% in the Bb 2067 pastures. Ewes grazing Bb 2067 also maintained Mg levels in their blood serum while Mg levels in the ewes grazing the control pastures dropped 35%. In a crested wheatgrass grazing trial, beef yearlings grazing the cultivar, 'Nordan', had 5% higher serum Mg levels than steers grazing the cultivar, 'Ruff' (Vogel et al., 1993a). These differences were associated with a less desirable K/(Ca + Mg) ratio for Ruff than for Nordan (2.6 vs 2.3) (Vogel et al., 1993a). In the crested wheatgrass trial, both cultivars had similar Mg concentrations in their forage but Ruff had higher K concentrations.

Based on previous genetic and breeding studies, it should be possible to alter mineral concentrations of most C3 forages by breeding. Since Mg and Ca concentrations are more heritable and more stable over environments than K concentrations, emphasis should be placed on breeding to increase Mg and Ca concentration in the forage. It may not be feasible to develop forages that will totally eliminate the risk of losses due to grass tetany under all environmental conditions, but it should be possible to develop forages that will reduce the incidence of grass tetany by breeding for increased Mg and Ca concentrations.

Diseases and Endophytes

Foliar Diseases

Burton (1981) reported that *Colletotrichum graminicola* (Ces.) G.W. Wils., which killed an estimated 10% of sudangrass leaves at Tifton, GA in the summer of 1952, increased lignin content of the leaves by 20%. Rust caused by *Puccinia sugstriata* Wll. & Barth var. *indica* on pearl millet made the forage highly unpalatable to livestock (Burton, 1981). Gross et al. (1975) reported that IVDMD of smooth bromegrass plants inoculated with either *Drechslera bromi* (Died.) Shoem. or *Rhynchosporium secalis* (Oud.) Davis was lower compared to uninoculated plants. Linear regression analyses showed that a 1% increase in disease-induced lesions resulted in a 0.04% decrease in IVDMD. Karn and Krupinsky (1983) reported that intermediate wheatgrass plants affected with leaf diseases had lower in vitro organic matter digestibility

(7.5 percentage units) and higher NDF contents than healthy plants in field and greenhouse studies. These and other reports clearly document that foliar diseases of forage plants that kill or damage plant tissue reduce the digestibility and palatability of forage plants.

Almost every improved forage cultivar on the market today is superior in disease resistance to common strains or earlier cultivars. Additional genetic gains in disease resistance can be made for virtually every forage crop. Genetic sources of resistance have been reported for almost every disease of important cool-season forage grasses (Braverman, 1986). A similar situation probably exists for warm-season grasses and legumes. In addition to the genetic gains for disease resistance that can be achieved by conventional breeding methods, it is likely that plants with improved resistance to specific diseases will be developed by transforming plants with specific genes. Improved disease resistance may be among the first uses of molecular genetics in forage plant improvement. In general, improvements in disease resistance will likely result in improved forage digestibility.

Endophytes

Animal disorders caused by livestock feeding on endophyte (*Acremonium coenophialum* Morgan Jones and Gams)-infected fescue and ryegrass results in major economic losses in many temperate areas of the world. In the USA alone, over 95% of the 12 million plus hectares of tall fescue is endophyte-infected (Shelby and Dalrymple, 1987). Several reviews on the endophyte and the consequences of its association with tall fescue have been published recently (Studemann and Hoveland, 1988; Pedersen et al., 1990; Bacon, 1993; Clay, 1993; Hoveland, 1993; Schmidt and Osborn, 1993) and, consequently, we will address the problem only from the viewpoint of breeding tall fescue with improved forage quality.

Alkaloids produced in endophyte-infected plants produce three major types of animal disorders, fescue foot, bovine fat necrosis, and fescue toxicosis, when consumed by livestock. Endophyte-free tall fescue can be developed since the disease is only transmitted through the seed and breeding stock free of the endophyte can be obtained by using old seed or with seed treatments. However, endophyte-infected tall fescue has several important agronomic advantages over uninfected tall fescue (Pedersen et al., 1990). Endophyte infection improves seedling performance and survival, is associated with insect and nematode resistance, improves nitrogen assimilation, and improves seed set. Survival of infected fescue is probably higher than that of endophyte free fescue, especially in drought or heat-stressed environments.

Breeders would like to keep the desirable aspects of endophyte infection but eliminate the undesirable aspects. Evidence indicates that different associations of endophytes in different fescue genotypes results in different levels of alkaloids being produced (Siegel et al., 1987; Hill et al., 1991). The alkaloids that have been extracted from infected tall fescue differ in their toxicity to ruminants and insects (Siegel et al., 1987). It may be possible to develop endophyte-plant combinations that produce alkaloids that provide resistance to insects and nematodes but do not produce alkaloids that are toxic to ruminants. This will require research on the genetics of the endophyte and its interaction with the host plant. Integrated research teams of biochemists and fungal and plant geneticists using molecular genetic approaches will be necessary to develop the desired endophyte and fescue or ryegrass combinations. At the present time, an inadequate knowledge base of the interactions and the genetics of the interacting organisms prevents breeders from solving the endophyte problem other than eliminating it from ryegrasses and fescues intended for use as forages.

Although endophyte-related problems of fescue and ryegrass have received the most attention, endophytes are known to occur on other species. White (1987) examined 100 genera of grasses using herbarium collections and found typical endophyte mycelium in 22 species of 9 genera demonstrating that endophytes are widespread in the Poaceae. Some of the species in which the mycellium were found included *Elymus canadensis* L., *E. virginicus* L., *Stipa robusta* (Vasey) Scrib., *Bromus anomalus* Rupr., and *Digitaria insularis* (L.) Mez. The widespread distribution of endophytes and the economic value or cost of their association with grasses will make research on improving the desirable aspects of this association increasingly important.

Anti-quality Compounds

Plant toxins provide a competitive advantage to plants that is usually directed at protecting the plant from predation by insects or diseases, and usually poisoning of ruminants or other livestock is a secondary result of this defensive mechanism (Jones, 1981; Molyneux and Ralphs, 1992). Plant toxins usually seem to be directed toward insects (Molyneux and Ralphs, 1992). As an example, dhurrin [(S)-p-hydroxymandelonitrile β-D-glucopyranoside] is a secondary metabolic product in *Sorghum* and *Sorghastrum* spp. and yields hydrocyanic acid (HCN) when hydrolyzed. It is produced at a metabolic "cost" to the plant. Woodhead and Bernays (1978) reported a positive correlation between the HCN released when sorghum plants were bitten by *Locusta migratoria* L and the unpalatability of individual plants to the locust. Phenolic acids liberated from phenolic esters by enzymes during feeding also had a contributing deterrent effect (Woodhead and Bernays, 1978).

The toxic and antinutrient compounds synthesized by legumes and other forages can be categorized as alkaloids, amino acids, cyanogens, isoflavone and coumestran estrogenic principles, nitro compounds, protease inhibitors, phytohemagglutenins, saponins, selenium compounds, and tannins (Smolenski et al., 1981). Genetic variation has been found in forage species for many of the anti-quality compounds and breeders have been successful in developing cultivars with low levels of the anti-quality compounds. The forage of these improved cultivars is generally safe for livestock utilization.

Success stories include the development of reed canarygrass cultivars with reduced levels of deleterious alkaloids (Marten et al., 1981; Kalton et al., 1989a,b) and the development of 'Blanco' blue lupine (Forbes et al., 1964) which lacks the poisonous alkaloid found in blue lupine (*Lupinus angustifolius* L.). Blanco also has white flowers which distinguishes it from the toxic, blue-flowered lupine. Low coumarin cultivars and germplasm of sweetclover have been released (Goplen, 1981; Gorz et al., 1992). Sudangrasses with decreased potential for prussic acid poisoning have been developed and released (Haskins et al., 1990). In each of these species, the genetic changes were achieved by using genes from within the species or related species.

Breeding to reduce the bloat potential of forages such as alfalfa or white clover to date has been unsuccessful. The substances in these plants that cause bloat in ruminants have not been clearly determined although several substances including saponins and proteins and structural components such as chloroplast membranes have been implicated (Howarth et al., 1986). Breeders lack definitive selection criteria for bloat. Until the causes of bloat are clearly delineated, forage breeders will not be able to successfully address this problem.

The anti-quality compounds that are found in forages are the products of specific pathways that have been characterized in many species. Using molecular genetic approaches to blocking those pathways using anti-sense genes or other "blocking" genes appears to be a highly effective means of developing forages with reduced or zero levels of anti-quality compounds. Since most of these compounds are not necessary for growth or development, blocking their synthesis may not affect forage yield or other agronomic traits. Molecular genetic approaches may eliminate the need to screen extensive germplasm arrays using conventional breeding methods for genes that reduce anti-quality compounds in forages.

CONCLUSIONS

Breeding improved perennial forages with improved forage quality and forage yield requires long-term investments in sustained research. It has been clearly demonstrated in both cool and warm-season species that forage quality can be significantly improved by breeding if the breeder is committed to improving forage quality and works with cooperative teams of agronomists and animal scientists. However, to date, only a limited number of cultivars have been developed and released with improved forage quality. These cultivars have demonstrated that improvements in forage quality usually result in greater improvement in forage profitability than similar improvements in forage yield. Breeders have the selection methods and breeding tools to make significant improvements in the quality of all forage plants.

REFERENCES

Allard, R. W. 1964. Principles of plant breeding. John Wiley & Sons, New York.

Anderson, Bruce, J. K. Ward, K. P. Vogel, M. G. Ward, H. J. Gorz, and F. A. Haskins. 1988. Forage quality and performance of yearlings grazing switchgrass strains selected for differing digestibility. J. Anim. Sci. 66:2239-2244.

Bacon, C. W. 1993. Abiotic stress tolerances (moisture, nutrients) and photosynthesis in endophyte-infected tall fescue. Agric. Ecosystems Environ. 44:123-141.

Bailey, James E. 1991. Toward a science of metabolic engineering. Science 252:1668-1674.

Barnes, D. K. 1980. Alfalfa. p. 177-187. *In* Walter R. Fehr and Henry H. Hadley (ed.) Hybridization of crop plants. ASA-CSSA. Madison, WI.

Barnes, R. F., L. D. Muller, L. F. Bauman, and V. F. Colenbrander. 1971. In vitro dry matter disappearance of brown mid-rib mutants of maize (*Zea mays* L.). J. Anim. Sci. 33:881-884.

Berdahl, J. D., R. E. Barker, J. F. Karn, J. M. Krupinsky, I. M. Ray, K. P. Vogel, K. J. Moore, T. J. Klopfenstein, B. E. Anderson, R. J. Haas, and D. A. Tober. 1993. Registration of 'Manska' pubescent intermediate wheatgrass. Crop Sci. 33:881.

Braverman, S. W. 1986. Disease resistance in cool-season forage, range, and turf grass II. Bot. Rev. 52:1-112.

Burson, Byron L. 1980. Warm-season grasses. p. 695-708. *In* Walter R. Fehr and Henry H. Hadley (ed.) Hybridization of crop plants. ASA-CSSA. Madison, WI.

Burton, G. W., R. W. Hart, and R. S. Lowrey. 1967. Improving forage quality in bermudagrass by breeding. Crop Sci. 7:329-332.

Burton, G. W., W. G. Monson, J. C. Johnson, Jr., R. S. Lowrey, H. E. Chapman, and W. H. Marchant. 1969. Effect of the D2 dwarf gene on the forage yield and quality of pearl millet. Agron. J. 61: 607-612.

Burton, G. W. 1972a. Registration of Coastcross-1 bermudagrass. Crop Sci. 12:125.

Burton, Glenn W. 1972b. Registration of Tiflate pearl millet. Crop Sci. 12:128.

Burton, G. W. 1981. Nutrient composition of forage crops. Effects of genetic factors. p. 1-9. *In* USDA-ARS Agric. Review and Manuals. ARM-S-21. U.S. Gov. Print. Office, Washington, DC.

Burton, Glenn W. 1986. Developing better forages for the south. J. Anim. Sci. 63:955-961.

Burton, Glenn W., R. N. Gates, and G. M. Hill. 1993. Registration of Tifton 85 bermudagrass. Crop Sci. 33:644-645.

Burton, Glenn W., Joel B. Gunnells, and R. S. Lowrey. 1968. Yield and quality of early and late-maturing, near-isogenic populations of pearl millet. Crop Sci. 8:431-434.

Burton, G. W., W. W. Hanna, J. C. Johnson, Jr., D. B. Leuck, W. G. Monson, J. G. Powell, H. D. Wells, and N. W. Widstrom. 1977. Pleiotropic effects of the *tr* trichomeless gene in pearl millet on transpiration, forage quality, and pest resistance. Crop Sci. 17: 613-616.

Burton, Glenn W., and Warren G. Monson. 1972. Inheritance of dry matter digestibility in bermudagrass, *Cynodon dactylon* (L.) Pers. Crop Sci. 12:375-378.

Burton, Glenn W., and W. G. Monson. 1988. Registration of 'Tifton 78' bermudagrass. Crop Sci. 28:187-188.

Butler, E. J. 1963. The mineral element content of spring pasture in relation to the occurrence of grass tetany and hypomagnesaemia in dairy cows. J. Agric. Sci. 60:329-340.

Buxton, D. R., and M. D. Casler. 1993. Environmental and genetic effects on cell wall composition and digestibility. p. 685-714. *In* H. G. Jung, D. R. Buxton, R. D. Hatfield, and J. Ralph (ed.) Forage cell wall structure and digestibility. ASA-CSSA-SSSA. Madison, WI.

Buxton, D. R., and G. C. Marten. 1989. Forage quality of plant parts of perennial grasses and relationship to phenology. Crop Sci. 29:429-434.

Carpenter, J. A., and M. D. Casler. 1990. Divergent phenotypic selection response in smooth bromegrass for forage yield and nutritive value. Crop Sci. 30:17-22.

Casler, M. D. 1986. Causal effects among forage yield and quality measures of smooth bromegrass. Can. J. Plant Sci. 66:591-600.

Casler, M. D. 1992. Registration of Badger smooth bromegrass. Crop Sci. 27:1073-1074.

Casler, M. D., J. F. Pedersen, G. C. Eizenga, and S. D. Stratton. 1994. Germplasm and cultivar development. Chapter 15. *In* L. E. Moser, D. R. Buxton, and M. D. Casler (ed.) Cool-season forage grasses. ASA Monograph. (In press).

Casler, M. D., and D. A. Sleper. 1991. Fungal cellulase vs. in vitro rumen fermentation for estimating digestibility in smooth bromegrass breeding. Crop Sci. 31:1335-1338.

Chapman, H. P., W. H. Marchant, P. R. Utley, R. E. Hellwig, and W. G. Monson. 1972. Performance of steers on Pensacola bahiagrass, Coastal bermuda grass, and Coast-cross-1 bermudagrass and pellets. J. Anim. Sci. 34:373-378.

Clay, K. 1993. Evolution and ecology of endophyte-grass symbioses. Agric. Ecosystems Environ. 44:39-64.

Colenbrander, V. F., V. L. Lechtenberg, and L. F. Bauman. 1973. Digestibility and feeding value of brown midrib corn stover silage. J. Anim. Sci. 37:294-297.

Cooper, J. P. 1962. Selection for digestibility in herbage grasses. Nature 195:1276-1277.

Coulman, B. E., and R. P. Knowles. 1974. Variability for in vitro digestibility of crested wheatgrass. Can. J. Plant Sci. 54:651-657.

Cummins, D. G., and J. W. Dobson, Jr. 1972. Digestibility of bloom and bloomless sorghum as determined by a modified in vitro technique. Agron. J. 64:682-683.

Deinum, B., and P. C. Struik. 1989. Genetic variation in digestibility of forage maize, *Zea-Mays* L., and its estimation by near infrared reflectance spectroscopy, NIRS. An analysis. Euphytica 42:89-98.

Emile, Jean-Claude, Gérard Genier, and Pierre Guy. 1993. Alfalfa energetic value improvement for dairy cows. *In* Proc. XVII International Grassland Congr., Palmerston North, New Zealand and Rockhampton. Queensland, Australia. 8-28 Feb. 1993. (In press).

Falconer, D. S. 1981. Introduction to quantitative genetics. 2nd. ed. Longman, New York.

Forbes, Ian, Jr., Glenn W. Burton, and Homer D. Wells. 1964. Registration of Blanco blue lupine. Crop Sci. 4:448.

Fritz, J. O., R. P. Cantrell, V. L. Lechtenberg, J. D. Axtell, and J. M. Herte. 1981. Brown midrib mutants in sudangrass and grain sorghum. Crop Sci. 21:706-709.

Gabrielsen, B. C., K. P. Vogel, B. E. Anderson, and J. K. Ward. 1990. Alkali-labile lignin phenolics and forage quality in three switchgrass strains selected for differing digestibility. Crop Sci. 30:1313-1320.

Gabrielsen, B. C., K. P. Vogel, and D. Knudsen. 1988. Comparison of in vitro dry matter digestibility and cellulase digestion for deriving Near Infrared Reflectance Spectroscopy calibration equations using cool-season grasses. Crop Sci. 28:44-47.

Gasser, C. S., and R. T. Fraley. 1989. Genetically engineered plants for crop improvement. Science 244:1293-1299.

Gee, M. S., O. E. Nelson, and J. Kuc. 1968. Abnormal lignins produced by the brown midrib mutants of maize. II. Comparative studies on normal and brown-midrib-1 dimethylformamide lignins. Arch. Biochem. Biophys. 123:403-408.

Gill, Herman Chaverra, R. L. Davis, and R. F. Barnes. 1967. Inheritance of *in vitro* digestibility and associated characteristics in Medicago sativa L. Crop Sci. 7:19-21.

Goodman, R. M., H. Hauptl, A. Crossway, and V. Knauff. 1987. Gene transfer in crop improvement. Science 236:48-54.

Goplen, B. P. 1981. Norgold - a low coumarin yellow blossom sweetclover. Can. J. Plant Sci. 61:1019-1021.

Gorz, H. J., F. A. Haskins, G. R. Manglitz, R. R. Smith, and K. P. Vogel. 1992. Registration of N28 and N29 germplasms. Crop Sci. 32:510.

Gross, D. F., G. J. Mankin, and J. G. Ross. 1975. Effect of diseases on the in vitro digestibility of smooth bromegrass. Crop Sci. 15:273-275.

Haskins, F. A., H. J. Gorz, and K. P. Vogel. 1990. Registration of NP31 and NP32, two populations of sudangrass selected for low dhurrin content. Crop Sci. 30:759-760.

Hallauer, Arnel R., and J. B. Miranda. 1981. Quantitative genetics in maize breeding. Iowa State University Press. Ames, IA.

Hanna, W. W., and E. C. Bashaw. 1987. Apomixis: Its identification and use in plant breeding. Crop Sci. 27:1136-1139.

Hanson, A. A., and H. L. Carnahan. 1956. Breeding perennial forage grasses. Agric. Research Service, U.S. Dept. Agric. Tech. Bul. 1145. Washington, DC.

Hides, David H., and Thomas A. Thomas. 1981. Variation in the magnesium content of grasses and its improvement by selection. J. Sci. Food Agric. 32:990-991.

Hill, G. M., G. W. Burton, and P. R. Utley. 1993. Forage quality and steer performance on Tifton 78 and Coastal bermudagrass pastures. *In* Proc. XVII International Grassland Congr., Palmerston North, New Zealand and Rockhampton, Queensland, Australia. 8-28 Feb. 1993. (In press).

Hill, G. M., R. N. Gates, and G. W. Burton. 1994. Forage quality and performance of steers grazing Tifton 85 and Tifton 78 bermudagrass pastures. J. Anim. Sci. (In press).

Hill, N. S., W. A. Parrott, and D. D. Pope. 1991. Ergopeptide alkaloid production by endophytes in a common tall fescue genotype. Crop Sci. 31:1545-1547.

Hodges, Thomas K., Keerti S. Rathore, and Jianying Peng. 1993. Advances in genetic transformation of plants. *In* Proc. XVII International Grassland Congr., Palmerston North, New Zealand and Rockhampton, Queensland, Australia. 8-28 Feb. 1993. (In press).

Hopkins, Andrew A. 1993. Genetic variation among switchgrasses for agronomic, forage quality, and biofuel traits. Ph.D. Dissertation. Univ. of Nebraska-Lincoln, NE. (Diss. Abstr. 1307530).

Hopkins, A. A., K. P. Vogel, and K. J. Moore. 1993. Predicted and realized gains from selection for *in vitro* dry matter digestibility and forage yield in switchgrass. Crop Sci. 32:253-258.

Hopkins, A. A., K. P. Vogel, K. J. Moore, K. D. Johnson, and I. T. Carlson. 1994. Genetic differences between elite switchgrass strains for agronomic, biofuel, and forage quality traits. Crop Sci. (In review).

Hoveland, C. S. 1993. Importance and economic significance of the *Acremonium* endophytes to performance of animals and grass plant. Agric. Ecosystems Environ. 44:3-12.

Hovin, A. W. 1980. Cool-season grasses. p. 285-298. *In* Walter R. Fehr and Henry H. Hadley (ed.) Hybridization of crop plants. ASA-CSSA. Madison, WI.

Howarth, R. E., K.-J. Cheng, W. Majak, and J. W. Costerton. 1986. Ruminant bloat. p. 516-527. *In* L. P. Milligan, W. L. Grovum, and A. Dobson (ed.) Control of digestion and metabolism in ruminants. Proc. Sixth Int. Symp. Ruminant Physiol., Banff, Canada. 10-14 Sept. 1984. Prentice Hall, Englewood Cliffs, NJ.

Howarth, R. E., and B. P. Goplen. 1983. Improvement of forage quality through production management and plant breeding. Can. J. Plant Sci. 63:895-902.

Iiyama, T. B., T. B. T. Lam, P. J. Meikle, K. Ng, D. I. Rhodes, and B. A. Stone. 1993. Cell wall biosynthesis and its regulation. p. 621-684. *In* H.G. Jung, D. R. Buxton, R. D. Hatfield, and J. Ralph (ed.) Forage cell wall structure and digestibility. ASA-CSSA-SSSA. Madison, WI.

Jones, D. A. 1981. Cyanide and coevolution. p. 509-516. *In* B. Vennesland, E. E. Conn, C. J. Knowles, J. Westley, and F. Wissing (ed.) Cyanide in Biology. Academic Press, New York.

Kaeppler, H. F., D. A. Somers, H. W. Rines, and A. F. Cockburn. 1992. Silicon carbide fiber-mediated stable transformation of plant cells. Theor. App. Genet. 84:560-566.

Kalton, R. R., J. Shields, and P. Richardson. 1989a. Registration of 'Palaton' reed canarygrass. Crop Sci. 29:1327.

Kalton, R. R., P. Richardson, and J. Shields. 1989b. Registration of 'Venture' reed canarygrass. Crop Sci. 29:1327-1328.

Kamstra, L. D., J. C. Ross, and D. C. Ronning. 1973. *In vivo* and *in vitro* relationships in evaluating digestibility of selected smooth bromegrass synthetics. Crop Sci. 13:575-576.

Karn, J. F., and J. M. Krupinsky. 1983. Chemical composition of intermediate wheatgrass affected by foliar diseases and stem smut. Phytopathology 73:1152-1155.

Kemp, A., and M. L. t'Hart. 1957. Grass tetany in grazing milking cows. Neth. J. Agric. Sci. 5:4-17.

Kneebone, William R. 1960. Grass breeding and livestock production. Econ. Bot. 14:300-315.

Kuc, J., and O. E. Nelson. 1964. The abnormal lignins produced by the brown midrib mutants of maize. I. The brown-midrib-1 mutation. Arch. Biochem. Biophys. 105:103-113.

Lamb, J. F. S., K. P. Vogel, and P. E. Reece. 1984. Genotype and genotype x environment interaction effects on forage yield and quality of crested wheatgrass. Crop Sci. 24:559-564.

Lee, Myoung Hoon, and J. L. Brewbaker. 1984. Effects of brown midrib-3 on yields and yield components of maize. Crop Sci. 24:105-108.

Lush, J. W., P. K. Karau, D. O. Balogu, and L. M. Gourley. 1984. Brown midrib sorghum or corn silage for milk production. J. Dairy Sci. 67:1739-1744.

Marten, G. C., R. M. Jordan, and A. W. Hovin. 1981. Improved lamb performance associated with breeding for alkaloid reduction in reed canarygrass. Crop Sci. 21:295-298.

Marten, Gordon C. 1989. Breeding forage grasses to maximize animal performance. p. 71-104. *In* D. A. Sleper, K. H. Asay, and J. F. Pedersen (ed.) Contributions from breeding forage and turf grasses. CSSA Special Pub. Num. 15. CSSA. Madison, WI.

McElroy, A. R., and B. R. Christie. 1986. Variation in digestibility decline with advance in maturity among timothy (*Phleum pratense* L.) genotypes. Can. J. Plant Sci. 66:323-328.

McIntyre, C. Lynne, John M. Manners, John R. Wilson, Heather Way, and Donovan Sharp. 1993a. Genetic engineering of pasture legumes and grasses for reduced lignin content and increased digestibility. *In* Proc. XVII International Grassland Congr., Palmerston North, New Zealand and Rockhampton, Queensland, Australia. 8-28 Feb. 1993. (In press).

McIntyre, C. Lynne, Sharon Abraham, Heather M. Bettenay, Ruth Sandeman, Christine Hayes, Donovan Sharp, Adrian Elliot, John M. Manners, and John Watson. 1993b. Improving pasture digestibility: Low lignin forages. *In* Proc. XVII International Grassland Congr., Palmerston North, New Zealand and Rockhampton, Queensland, Australia. 8-28 Feb. 1993. (In press).

Mertens, D. R. 1993. Kinetics of cell wall digestion and passage in ruminants. p. 535-570. *In* H. G. Jung, D. R. Buxton, R. D. Hatfield, and J. Ralph (ed.) Forage cell wall structure and digestibility. ASA-CSSA-SSSA. Madison, WI.

Molyneux, Russell J., and Michael H. Ralphs. 1992. Plant toxins and palatability to herbivores. J. Range Manage. 45:13-18.

Moore, K. J., L. E. Moser, K. P. Vogel, S. S. Waller, B. E. Johnson, and J. F. Pedersen. 1991. Describing and quantifying growth stages of perennial forage grasses. Agron. J. 83:1073-1077.

Moore, K. J., K. P. Vogel, T. J. Klopfenstein, R. A. Masters, and B. A. Anderson. 1994. Evaluation of four stains of intermediate wheatgrass under grazing. Agron. J. (In review).

Moore, K. J., K. P. Vogel, A. A. Hopkins, J. F. Pedersen, and L. E. Moser. 1993. Improving the digestibility of warm-season perennial grasses. *In* Proc. XVII International Grassland Congr., Palmerston North, New Zealand and Rockhampton, Queensland, Australia. 8-28 Feb. 1993. (In press).

Mosely, G., and D. H. Baker. 1991. The efficacy of a high magnesium grass cultivar in controlling hypomagnesaemia in grazing animals. Grass Forage Sci. 46:375-380.

Pedersen, J. F., H. J. Gorz, F. A. Haskins, and W. M. Ross. 1982. Variability of quality and agronomic traits in forage sorghum. Crop Sci. 22:853-856.

Pedersen, J. F., G. D. Lacefield, and D. M. Ball. 1990. A review of the agronomic characteristics of endophyte-free and endophyte-infected tall fescue. Appl. Agric. Res. 5:188-194.

Poethig, R. Scott. 1990. Phase change and the regulation of shoot morphogenesis in plants. Science 250:923-930.

Roth, L. S., G. C. Marten, W. A. Compton, and D. D. Stuthman. 1970. Genetic variation of quality traits in maize (*Zea mays* L.) forage. Crop Sci. 10:365-367.

Schmidt, S. P., and T. G. Osborn. 1993. Effects of endophyte-infected tall fescue on animal performance. Agric. Ecosystems Environ. 44:233-262.

Schroeder, H. E., M. R. I. Khan, W. R. Knibb, D. Spencer, and T. J. V. Higgins. 1991. Expression of a chicken ovalbumin gene in three lucerne cultivars. Austr. J. Plant Physiol. 18:495-505.

Shelby, R. A., and L. W. T. Dalrymple. 1987. Incidence and distribution of the tall fescue endophyte in the United States. Plant Dis. 71:783-786.

Siegel, M. R., G. C. M. Latch, and M. C. Johnson. 1987. Fungal endophytes of grasses. Ann. Rev. Phytopathol. 25:293-315.

Sleper, D. A., K. P. Vogel, K. H. Asay, and H. F. Mayland. 1989. Using plant breeding and genetics to overcome the incidence of grass tetany. J. Anim. Sci. 67:3456-3462.

Smolenski, Stanislaus J., Douglas A. Kinghorn, and Manuel F. Balandrin. 1981. Toxic constituents of legume forage plants. Econ. Bot. 35:321-355.

Studemann, J. A., and C. S. Hoveland. 1988. Fescue endophyte: History and impact on animal agriculture. J. Prod. Agric. 1:39-44.

Taylor, Norman L. 1980. Clovers. p. 261-272. *In* Walter R. Fehr and Henry H. Hadley (ed.) Hybridization of crop plants. ASA-CSSA. Madison, WI.

Tilley, J. A., and R. A. Terry. 1963. A two-stage technique of the *in vitro* digestion of forage crops. J. Br. Grassl. Soc. 18:104-111.

Verma, D. P. S., A. J. Delauney, and T. Nguyen. 1987. A strategy towards antisense regulation of plant gene expression. p. 155-162. *In* George Broening, John Harada, Tsune Kosuge, and Alexander Hollaender (ed.) Tailoring genes for crop improvement. Plenum Press, New York.

Vogel, K. P., R. Britton, H. J. Gorz, and F. A. Haskins. 1984. *In vitro* and *in vivo* analyses of hays of switchgrass strains selected for high and low IVDMD. Crop Sci. 24:977-980.

Vogel, K. P., B. C. Gabrielsen, J. K. Ward, B. E. Anderson, H. F. Mayland, and R. A. Masters. 1993a. Forage quality, mineral constituents, and performance of beef yearlings grazing two crested wheatgrasses. Agron. J. 85:584-590.

Vogel, K. P., H. J. Gorz, and F. A. Haskins. 1981a. Heritability estimates of forage yield, *in vitro* dry matter digestibility, crude protein, and heading date in indiangrass. Crop Sci. 21:35-38.

Vogel, K. P., H. J. Gorz, and F. A. Haskins. 1981b. Divergent selection for *in vitro* dry matter digestibility in switchgrass. Crop Sci. 21:39-41.

Vogel, K. P., F. A. Haskins, H. J. Gorz, B. A. Anderson, and J. K. Ward. 1991. Registration of 'Trailblazer' switchgrass. Crop Sci. 31:1388.

Vogel, K. P., H. F. Mayland, P. E. Reece, and J. F. S. Lamb. 1989. Genetic variability for mineral element concentration of crested wheatgrass. Crop Sci. 29:1146-1150.

Vogel, K. P., and K. J. Moore. 1993. Quantifying economic value of forage breeding programs. p. 201-205. Proc. American Forage and Grassland Council. Georgetown, TX.

Vogel, K. P., P. E. Reece, and J. F. S. Lamb. 1986. Genotype and genotype x environment interaction effects for forage yield and quality of intermediate wheatgrass. Crop Sci. 26:653-658.

Vogel, K. P., P. E. Reece, and J. T. Nichols. 1993b. Genotype and genotype x environment interaction effects on forage yield and quality of intermediate wheatgrass in swards. Crop Sci. 33:37-41.

Vogel, K. P., and J. F. Pedersen. 1993. Breeding systems for cross-pollinated perennial grasses. Plant Breeding Rev. 11:251-274.

Voight, P. W. 1971. Registration of Morpa weeping lovegrass. Crop Sci.11: 312-313.

Voight, P. W. 1975. Improving palatability of range plants. *In* Robert S. Campbell and Carlton Herbel (ed.) Improved range plants. Range Symposium Series No. 1. Soc. Range Manage. Denver, CO.

Voight, P. W., W. R. Kneebone, E. H. McIlvain, M. C. Shoop, and J. E. Webster. 1970. Palatability, chemical composition, and animal gains of weeping lovegrass, *Eragrositis curvula* (Schrad.) Nees. Agron. J. 62:673-676.

Ward, M. G., J. K. Ward, B. E. Anderson, and K. P. Vogel. 1989. Grazing selectivity and in vivo digestibility of switchgrass strains selected for differing digestibility. J. Anim. Sci. 67:1418-1424.

Wheeler, J. L., and J. L. Corbett. 1989. Criteria for breeding forages of improved feeding value: Results of a Delphi survey. Grass Forage Sci. 44:77-83.

White, J. F., Jr. 1987. Widespread distribution of endophytes in the Poaceae. Plant Disease 71:340-342.

Woodhead, S., and E. A. Bernays. 1978. The chemical basis of resistance of *Sorghum bicolor* to attack by *Locusta migratoria*. Entomol. Exp. Appl. 24:123-144.

CHAPTER 23

MODIFICATION OF FORAGE QUALITY AFTER HARVEST

L. L. Berger, G. C. Fahey, Jr.,
L. D. Bourquin, and E. C. Titgemeyer

INTRODUCTION

The goal of most forage production systems is to produce the highest quality forage possible. However, there are many production systems where forage quality cannot be optimized. In these situations, post-harvest treatment may be the only opportunity to improve forage quality. For example, approximately 1 kg of residue is produced for each kg of grain harvested. This ratio of grain:residue translates into an excess of 400 million tonnes of crop residue produced each year in the USA and 330 million tonnes in Europe and the Soviet Union (Agricultural Statistics, 1988; Sundstøl, 1988). Corn stover (*Zea mays* L.), wheat straw (*Triticum aestivum* L.), and soybean residues (*Glycine max* Merrill) account for greater than 75% of the total USA farm residue supply. The annual crops of corn, wheat, and soybeans produce about 161, 102, and 102 million tonnes of residue, respectively, and most of this material is left in the field after grain harvest. Not all of this residue should or could be harvested for animal feed; nevertheless, the potential supply is staggering.

As the demand for direct human use of cereal grains increases, the role of crop residues and by-product feedstuffs will become increasingly important. Greenhalgh (1984) identified four reasons for the resurgence in interest in the feeding of agricultural residues: 1) new feeding strategies for ruminants which involve fibrous residues; 2) more precise knowledge of animal nutrient requirements, resulting in the identification of situations where less than maximal nutrition is required; 3) better supplementation regimens which optimize ruminal digestion and postruminal utilization of nutrients; and 4) new methods for improving the nutritive value of fibrous feedstuffs.

To improve the nutritive value of agricultural residues for livestock, some form of processing is generally required. The purpose of processing is to increase acceptability of high fiber feeds to the animal, thus increasing daily feed intake, and to enhance rate and(or) extent of digestion, thus increasing

---

L.L. Berger and G.C. Fahey, Jr., Dep. of Animal Sciences, Univ. of Illinois, 1207 W. Gregory, Urbana, IL 61801; L.D. Bourquin, Dep. of Food Sci. and Human Nutrition, Michigan State Univ., East Lansing, MI 48824; E.C. Titgemeyer, Dep. of Animal Science and Industry, Kansas State Univ., Manhattan, KS 66506.

nutrient availability (Nicholson, 1981). Today, even in the developed countries, there is renewed appreciation of the value of high fiber feeds, largely because of the introduction of methods (physical-mechanical, chemical, microbiological) for improving their nutritional value. The processed material must compete both nutritionally and economically with conventional animal feeds to command a place of importance in animal production systems.

Ultimately, it is the cost of the treatment process that will determine if a treated fibrous residue will become part of a diet formulation. A problem common to all treatments is that the cost often exceeds the value of the end-product (Satter, 1983); further, the treated product may be of low to moderate value because of the protein and(or) energy supplementation that may be required to achieve a completely balanced diet. This precludes large expenditures for upgrading the intake potential and(or) increasing the digestibility of a given agricultural residue. In the end, the livestock producer must decide on the economics of processing fibrous and by-product feedstuffs available in his area. If, as in many parts of the world, fibrous and by-product feedstuffs are the major or perhaps the only feedstuffs available, then the goal of a feeding program must be to maximize microbial digestion and utilization of ingested material. Extensive processing then may be quite practical. On the other hand, if high quality feedstuffs as well as lower quality fibrous and by-product feeds are available, use of elaborate processing may be more difficult to justify (Walker, 1984).

This chapter will provide information on a number of different types of roughage processing techniques. Emphasis will be placed on those perceived to be most important to present-day and future agricultural practices. Discussion will be directed towards answering two key questions: 1) How do these modifications improve roughage utilization by ruminants?, and 2) What is the potential for further improvement and economic competitiveness in the future?

PHYSICAL TREATMENTS

Numerous physical processing techniques have been used, with varying degrees of success, to enhance the utilization of fibrous feedstuffs by ruminants. The more common methods - grinding and pelleting, steam treatment, and mechanical separation of plant parts - will be reviewed briefly with regard to their effects on forage and roughage utilization. Although many recent advances in this field have been made in the treatment of wood residues, this area of research generally will not be addressed. Likewise, pretreatments designed to enhance utilization of lignocellulosics for bioconversion procedures, such as ethanol or single cell protein production, have contributed significantly to our understanding of physical processing methods, but are beyond the scope of this review.

Grinding and Pelleting

Undoubtedly the most studied physical treatment for enhancing the utilization of fibrous feedstuffs by ruminants is grinding and pelleting. Grinding decreases particle size, increases surface area, and increases the bulk density of leaf and stem fractions of forages (Laredo and Minson, 1975). Ground roughages often are further processed by pelleting or cubing before being fed. Benefits derived from pelleting include a further increase in bulk density, decreased dustiness, and increased ease of handling. Numerous reviews have been published on various aspects of feeding ground and pelleted forages (Minson, 1963; Beardsley, 1964; Moore, 1964; Meyer et al., 1965; Osbourn et al., 1976; Thomson and Beever, 1980).

The most consistent animal responses to grinding and pelleting of forages are increases in feed intake and gain and improvement of feed efficiency. Beardsley (1964) used data compiled from several studies with steers and estimated that grinding and pelleting long hay increased feed intake 25%, increased daily gain 98%, and reduced the feed:gain ratio 36%. Minson (1963) concluded that improvements in feed intake and liveweight gain were inversely related to unprocessed forage quality; thus, greater improvements occur when roughages of poor initial quality are treated. However, significant increases in voluntary intake of processed poor quality roughages often occur only in conjunction with nitrogen (N) supplementation of the diet (Campling and Freer, 1966; Weston, 1967). Greenhalgh and Reid (1973) demonstrated that intake responses to grinding were greater for sheep (45%) than cattle (11%) and greater in young animals (38%) than in mature ones (17%). Intake of ground roughages generally increases with decreasing particle size and is not maximized until mean particle size is less than 1 mm for most forages (Osbourn et al., 1976). Less extreme processing of forages, such as chopping or wafering, does not result in consistent improvements in animal performance although these may result in reduced diet wastage and sorting (Beardsley, 1964). Meyer et al. (1959) concluded that fine grinding was the major factor causing the increased consumption of ground and pelleted hay, while pelleting increased acceptability of some diets by reducing dustiness.

Digestibility of ground and pelleted roughages generally is depressed relative to that of the parent material fed in either the long or chopped forms. The average depression in dry matter (DM) digestibility in 21 studies evaluated by Minson (1963) was 33 g kg^{-1}. Thomson and Beever (1980) estimated somewhat larger digestibility depressions, and generalized these as greater for grasses (up to 150 g kg^{-1}) than legumes (30 to 60 g kg^{-1}). Digestibility depressions due to processing increase in magnitude as feed intake increases (Blaxter et al., 1956). Decreases in diet digestibility often are more than offset by increased gross energy (GE) intakes, such that digestible energy (DE) intake is increased (Minson, 1963). This effect is illustrated by results of Heaney et al. (1963) where sheep consumed chopped or pelleted orchardgrass (*Dactylis glomerata* L.), timothy (*Phleum pratense* L.), or alfalfa (*Medicago sativa* L.) at three levels of maturity. Increases in GE and DE intakes as a result of forage processing were greater as forage maturity advanced, reflecting the increased intake response as forage quality decreased. This experiment also demonstrated the relatively larger depressions in energy digestibility for grasses than for legumes.

The decrease in digestibility of ground and pelleted forages is primarily due to reduced fiber digestion (Beardsley, 1964). While *in vitro* studies demonstrate that fine milling of forage increases cell wall degradability (Dehority and Johnson, 1961), *in vivo* studies generally show a reduction in cell wall digestion (Thomson and Beever, 1980). This effect can be attributed to the shorter gastrointestinal tract residence time for ground than for long or chopped roughages (Minson, 1963; Alwash and Thomas, 1974). The relationship between digesta transit time and total tract digestibility is evident from the results of Blaxter et al. (1956). The increased flow of undigested structural carbohydrate to the small intestine can account for 65 to 75% of the increase in energy flowing out of the reticulorumen of ruminants consuming ground pelleted forages when compared to unprocessed forages (Thomson and Beever, 1980).

Fine grinding and pelleting of forages dramatically reduces the time that ruminants spend eating and ruminating (Weston and Hogan, 1967; Osuji et al., 1975) and, consequently, saliva production is reduced significantly (Osuji et al., 1975; Thomson and Beever, 1980). As a result of the decreased buffering capacity, animals consuming processed forages often have significantly depressed ruminal pH (Moore, 1964). Low ruminal pH and increased rate of digesta passage result in decreased cell wall carbohydrate fermentation in animals consuming processed forages. Relative to parent material, the rate of fermentation of processed forages is increased (Moore, 1964). However, total production of volatile fatty acids (VFA) as a proportion of GE is depressed significantly as a consequence of reduced extent of ruminal fermentation (Osbourn et al., 1976). In addition, reductions in the acetate:propionate ratio often result with forage processing (Moore, 1964). This has been implicated as a reason for the reduction in milk fat percentage often observed when lactating dairy cattle consume processed forages (Shaver et al., 1986; Woodford and Murphy, 1988).

Grinding and pelleting forages also affect ruminal N metabolism. Ruminal degradation of forage protein often is decreased by processing, an effect attributed to the heating that forages undergo during grinding and pelleting (Thomson and Beever, 1980). Thus, escape of dietary protein to the small intestine generally is increased by forage processing (Thomson and Beever, 1980). This effect is demonstrated by results of Beever et al. (1981) who fed sheep equal amounts of DM from Italian ryegrass (*Lolium multiflorum* Lam.) or timothy that was either chopped or ground and pelleted. Amino acid intakes were slightly lower for sheep consuming ground and pelleted diets. This was attributed both to leaf loss during forage processing and to destruction of amino acids by heat generated during grinding and pelleting. Lysine content of both pelleted forages was especially depressed, indicating that formation of Maillard complexes probably occurred (Beever et al., 1981). Ruminal degradability of forage protein decreased an average of 220 g kg^{-1} by processing, leading to a significant increase in undegraded feed protein reaching the small intestine. Efficiency of microbial protein synthesis was not influenced by forage processing in this study. However, reductions in ruminal organic matter (OM) digestion led to a slight reduction in net microbial protein synthesis. This depression was more than offset by increased intestinal flow of undegraded feed protein; thus, total amino acids reaching the small intestine increased an average of 16 g d^{-1} in this study. This increase occurred despite a 7% mean reduction in apparent GE digestibility for the two processed forages (Beever et al., 1981). Although processing increases the escape of dietary protein to the small intestine, it should be recognized that intestinal availability of this additional protein may be reduced if heat damage is excessive (Thomson and Beever, 1980).

Rode et al. (1985) found that grinding and pelleting alfalfa hay led to a 15% increase in efficiency of microbial protein synthesis in lactating dairy cows consuming 800 g kg^{-1} alfalfa diets *ad libitum*, even though feed intake was similar among diets. Cattle consuming pelleted hay had a faster ruminal particulate dilution rate and a slower ruminal fluid dilution rate than cattle consuming long alfalfa; thus, microbial efficiency was positively correlated with particulate dilution rate and inversely correlated with fluid dilution rate (Rode et al., 1985). Bergen et al. (1982) concluded that efficiency of microbial protein synthesis was positively affected by factors such as forage processing that increased ruminal particulate dilution rate. Increased feed intake also would be expected to result in increased particulate dilution rate and, thus, increase microbial efficiency as well (Bergen et al., 1982). An increase in feed intake as a result of forage processing might be expected to further enhance microbial efficiency, provided ruminally degradable N is not limiting.

As noted previously, digestibility depressions resulting from forage processing are generally larger for grasses than legumes. Because most of the depression in OM digestion is due to reduced cell wall fermentation, the digestibility of grasses could be more affected due to their greater content of structural carbohydrates relative to legumes (Thomson and Beever, 1980). Another potential factor is that grasses usually contain greater amounts of water-soluble carbohydrates (WSC) than legumes. Presumably, ruminal fermentation of this fraction would lead to greater pH depressions for pelleted grasses, while the high buffering capacity of legumes would help minimize pH depressions (Osbourn et al., 1976). Indeed, Beever et al. (1981) found that ruminal digestion of Italian ryegrass (220 g kg^{-1} WSC) was depressed considerably more than that of timothy (55 g kg^{-1} WSC).

Although DE and metabolizable energy (ME) values for processed forages are generally lower than that of the parent material, a consistent finding is that the net energy (NE) of processed forages is increased relative to the parent material (Osbourn et al., 1976). The increase in processed forage NE content is greatest for poor quality forages (Osbourn et al., 1976). Research by Osuji et al. (1975) and Webster et al. (1975) indicated that reductions in energy expenditure due to decreased eating and ruminating time were considerably smaller than the observed reduction in heat increment for processed forages. It was concluded that reduced heat energy losses were largely attributable to reduced losses in gut and tissue metabolism (Webster et al., 1975), implying that metabolism of the end-products of digestion was more efficient for ground and pelleted forages.

Grinding and pelleting roughages does not enhance structural carbohydrate utilization by ruminants. Increases in animal performance primarily are consequences of increased DE intake. Although utilization of low quality roughages is enhanced by grinding and pelleting, the processed material still has a relatively poor NE content. Grinding and pelleting in conjunction with chemical treatment provides a means to further enhance the feeding value of such low quality roughages. For example, NaOH treatment combined with grinding and pelleting of cornstalks increased animal performance more than either treatment alone (NRC, 1983), indicating a potential for combinations of physical and chemical treatments.

The economic feasibility of grinding and(or) pelleting forage is determined by the cost of alternative sources of DE and the class of livestock being fed. Current costs for grinding forages are in the range of US $15 to $25 per tonne, with an additional expense of approximately US $10 per tonne for pelleting. If grains or byproducts are relatively cheap, it is improbable that the costs of grinding and pelleting forages can be justified, especially for use in maintenance diets. However, with expensive grains or in diets for high producing ruminants, grinding and pelleting forages can give a good return on investment.

Steam Treatment

Improving the utilization of lignocellulosic materials by steam pressure treatment has received considerable attention in recent years. Although much of the research on steam treatment concerns its use as a pretreatment for lignocellulose bioconversion processes such as ethanol production, several studies have investigated its utility for improving quality of ruminant feedstuffs. Treatment conditions vary considerably in the published literature, but most employ pressures in the range of 5 to 40 kg cm^{-2} for durations of less than 5 min. One variation on the basic procedure is the steam explosion process in which the pressurized material is quickly exposed to atmospheric pressure after the desired reaction time. Under such conditions, water inside the substrate vaporizes rapidly and further disintegrates the substrate (Grous et al., 1986).

Although steam pressure treatment involves physical disruption of the cell wall structure, its mechanism of action is largely hydrolytic in nature. Acetyl esters of hemicelluloses are cleaved, thus releasing a considerable amount of acetic acid that lowers substrate pH and assists in the hydrolysis (Bender et al., 1970). Other effects include production of furfurals and phenolic derivatives and solubilization of much of the hemicellulosic fraction, some of which may be destroyed under severe treatment conditions (Walker, 1984). Depending on treatment conditions, up to 200 g kg^{-1} of substrate DM may be lost (Walker, 1984). Cellulose is not solubilized to any significant degree, but steam explosion treatment has been demonstrated to significantly increase pore volume available to soluble probes approximating the size of cellulase enzymes (Grous et al., 1986). In some experiments, acid detergent insoluble nitrogen

(ADIN) and lignin concentrations increased in steam-treated substrates, indicating that non-enzymatic browning reactions had occurred (Oji and Mowat, 1978, 1979).

A series of lamb feeding studies confirmed the potential for steam treatment to improve utilization of fibrous residues. Klopfenstein and Bolsen (1971) included untreated, 17.5 kg cm^{-2}, or 28 kg cm^{-2} for 50 s steam-treated corn cobs in lamb diets at 500 g kg^{-1} of diet DM. Dry matter digestibilities were increased from 583 g kg^{-1} for diets containing untreated cobs to 660 and 633 g kg^{-1} for diets containing 17.5 and 28 kg cm^{-2}-treated cobs. Umunna et al. (1972) fed similarly treated corn cobs (17.5 kg cm^{-2}) at 500 g kg^{-1} of diet DM and found increased average daily gains (183 vs. 104 g d^{-1}) and decreased feed:gain ratios (6.7 vs. 9.0 kg kg^{-1}) for lambs consuming treated cobs. In a subsequent trial, corn cobs (untreated, 14 kg cm^{-2}, 17.5 kg cm^{-2}) were incorporated at 700 g kg^{-1} of diet DM for lambs. Average daily gains (g d^{-1}) and feed:gain ratios (kg kg^{-1}) for lambs consuming untreated, 14 kg cm^{-2} and 17.5 kg cm^{-2}-treated corn cobs were 56, 12.1; 100, 10.4; and 150, 7.9, respectively (Umunna et al., 1972). Dry matter digestibility of 700 g kg^{-1} wheat straw diets fed to lambs was improved from 535 to 663 g kg^{-1} by steam treatment (21 kg cm^{-2}, 50 s; Umunna and Klopfenstein, 1972). A subsequent growth study showed a 50% improvement in daily gains and a 36% improvement in feed efficiency of lambs consuming treated wheat straw-containing diets (Umunna and Klopfenstein, 1972). Klopfenstein et al. (1974) found improvements in lamb daily gains (186 vs. 87 g d^{-1}) and feed:gain ratios (6.2 vs. 8.5 kg kg^{-1}) when steam-treated corn cobs (17.5 kg cm^{-2}, 50 s) replaced untreated cobs (790 g kg^{-1} of diet DM).

Garrett et al. (1981) obtained less promising results when rice *(Oryza sativa)* straw was steam-treated (28 kg cm^{-2}) for 20 or 90 s and incorporated into diets (650 g kg^{-1} of DM) for lambs. Daily gains and feed intakes were similar for untreated and 20 s treated material, but were significantly depressed for 90 s treated straw. Organic matter and cellulose digestibilities of both treated straw diets were approximately 40 g kg^{-1} lower than for untreated rice straw diets, while N digestibility was depressed an average of 150 g kg^{-1} for treated straw diets.

Material in previously discussed experiments was all treated in batch-type systems, but continuous-flow steam treatment equipment has been developed (Bender, 1979) and is currently being used (Stake Technology Ltd., Norval, Ontario, Canada) for treating aspen *(Populus tremuloides)* wood and various other materials for cattle feed (Walker, 1984). The patent describes the system as using steam pressure of at least 14 kg cm^{-2} for at least 15 s to treat material, after which time it is rapidly expelled and cooled (Bender, 1979). The end product is approximately 500 g kg^{-1} water with a pH of 3.5 to 3.8 (Walker, 1984).

Several studies have been conducted to evaluate a variety of materials treated by this continuous-flow system. Oji and Mowat (1978) compared steam-treated corn stover (16.2 kg cm^{-2}, 15 min) to untreated material in 870 g kg^{-1} corn stover diets fed to lambs. Treatment increased feed intake 55% and increased digestibilities of OM, GE, and cellulose 8, 10, and 16%, respectively. Oji and Mowat (1979) subsequently examined corn stover treated at the same pressure but for only 4 min. Intakes of 784 g kg^{-1} corn stover diets by sheep were similar for treated and untreated material, but digestibilities of OM and cellulose were improved 15 and 37%, respectively, for diets containing steam-treated corn stover. Horton et al. (1991) examined the feeding value of steam-treated bagasse (20 kg cm^{-2}, 2 min) for beef cattle. When bagasse was incorporated into diets at approximately 600 g kg^{-1} of DM, feed intake was 18% greater with treated than with untreated bagasse. Nutrient digestibilities were similar for treated and untreated bagasse diets, but steers consuming treated material had greatly improved N retention (42 vs. 16 g d^{-1}).

Potential improvements will probably result from combining steam and chemical treatments. Because steam treatment is mainly hydrolytic in effect, using oxidative chemicals in combination with the steam may have a synergistic effect.

Process costs for bagasse treatment by continuous-flow steam treatment were reported in 1983 to be US $30 to $40 per tonne DM (Satter, 1983). This did not include the cost of drying and pelleting the treated material, if desired. Although the reported processing cost is considerable, it may be possible to improve the cost effectiveness of steam treatment. Using waste heat or low quality steam as byproducts from manufacturing processes could reduce costs. Another alternative is to utilize a portion of the biomass supply as a fuel source. Such a system would be especially suited for treating residual material that is accumulated in a central location, as is the case for bagasse and oat hulls.

Mechanical Separation of Plant Parts

The primary aim of this type of physical processing is the separation of plant materials into discrete fractions of differing quality that are better adapted for specific uses. End products may be either higher (i.e., leaves) or lower (i.e., stems) in quality than the original material. Separation processing in its simplest form involves leaf/stem fractionation of hays and crop residues. More involved processes include whole crop cereal harvesting and leaf protein concentrate isolation from fresh forages.

Owing to their relative ease of separation, fractionation of leaf and stem portions of hays and crop residues has been the most thoroughly investigated method of separation processing. Large scale leaf/stem fractionation of dried forage generally involves the use of controlled velocity air currents in closed systems to separate light (leaf) from heavy (stem) particles (Mowat and Wilton, 1984). The process is well established for alfalfa processing (Kohler et al., 1972).

Several studies have demonstrated that the feeding value for ruminants is greater for leaf fractions than for stems. Mowat et al. (1965a) studied the *in vitro* digestibility of leaf and stem fractions isolated from timothy, smooth bromegrass (*Bromus inermis* Leyss), orchardgrass, and alfalfa harvested at 12 stages of maturity. At early stages of maturity, leaf and stem fractions of all substrates had roughly similar IVDMD values. With advancing maturity,

IVDMD of leaf and stem fractions were similar for timothy and bromegrass, whereas stem IVDMD was considerably less than leaf IVDMD for alfalfa and orchardgrass. Crude protein content of leaves was 1.5 to 3.0 times that of stems. Percentage leaf in substrates decreased with advancing maturity and was highly correlated to IVDMD, although there were exceptions. Orchardgrass had the highest percentage leaf and the lowest IVDMD of any substrate at advanced stages of maturity (Mowat et al., 1965b).

Laredo and Minson (1973) fractionated five tropical grasses at three stages of maturity into predominantly leaf and stem fractions using a gravity separator and examined their intake and digestibility by sheep. Averaged across substrates, leaf fractions contained 33% more N and 6, 11, and 26% less NDF, ADF, and lignin than stems. Average intakes of leaf fractions were 46% greater than for stems (57.8 vs. 39.6 g $kg^{-0.75}$). Total tract DM digestibility of leaf fractions was lower than for stems (526 vs. 558 g kg^{-1}) due to shorter mean reticulorumen particulate residence time for leaf diets than stems (23.5 vs. 31.8 h). However, digestible DM intake was 38% greater for leaves than for stems (30.4 vs. 22.1 g $kg^{-0.75}$). In a subsequent experiment, Laredo and Minson (1975) again found that voluntary intake of leaf fractions by sheep from three tropical grasses was greater than for stems (40.3 vs. 30.0 g $kg^{-0.75}$). Grinding and pelleting increased voluntary intake of leaf and stem fractions 88 and 60% to 75.9 and 48.0 g $kg^{-0.75}$, respectively. Total tract DM digestibilities were not different for leaf and stem fractions even though mean ruminal particulate retention time for sheep consuming leaf diets was shorter than for sheep consuming stems.

Separation of chopped whole-crop cereals into four fractions was achieved by techniques similar to those for leaf/stem separation. Fractions were: 1) grain; 2) a mixture of light and broken grain, nodes, and some heavily lignified straw; 3) "heavy" straw; and 4) "light" straw (chaff, leaf, and weeds; Mowat and Wilton, 1984). Rexen (1978) developed techniques to separate straw internodes from other straw components (nodes, chaff, leaves). The internode fraction was more suitable for paper production than intact straw, whereas the remaining material was found to be superior to whole straw in feeding value for ruminants (Rexen, 1978).

Studies also have been conducted on fractional separation of crop residues. Leask and Daynard (1973) separated corn stover into leaf, husk, and stalk fractions and found that IVDMD of husk and leaf fractions was greater than intact stover IVDMD, whereas isolated stalks had considerably lower IVDMD values than stover. Intact stover IVDMD declined at a rate of 15 g kg^{-1} wk^{-1} following grain physiological maturity, indicating that stover harvesting and separation should be conducted soon after grain maturity to maximize its nutritive value (Leask and Daynard, 1973). Separated chaff is of superior feeding value relative to intact straw. Coxworth et al. (1981) found greater DE intakes by sheep consuming wheat chaff than by those consuming wheat straw (202 vs. 152 kcal $kg^{-0.75}$). When both wheat chaff and straw were ammoniated, the difference in DE intake was maintained (294 vs. 233 kcal $kg^{-0.75}$).

Isolation of leaf protein concentrate from forages involves the fractionation of fresh forage into a liquid rich in protein, sugars, and minerals and a fibrous residue having a reduced protein content relative to the parent forage (Houseman and Connell, 1976). Leaf protein concentrate is isolated from the liquid fraction and is used for monogastric animal and human consumption, whereas the fibrous residue may be fed to ruminants fresh or further processed by ensiling, dehydrating, or leaf/stem separation (Kohler et al., 1972). Utilization of various products of fresh forage separation by ruminants has been reviewed (Houseman and Connell, 1976; Kohler et al., 1979).

Potential future improvements in mechanical separation lie in the prospect of being able to fractionate a forage into individual components to the extent that the value-added nature of each fraction justifies the additional expense. A separation procedure yielding highly refined fractions will probably require both mechanical and chemical phases.

Although economic evaluations of mechanical separation procedures are generally lacking, a number of features of the processes are quite expensive. As with most physical treatment procedures, transporting lignocellulosic material to a central location for processing is a significant expense. Dehydration of fresh forages prior to leaf/stem separation also is energy intensive; thus, improvements in product quality must be significant to offset processing costs. This type of improvement generally is not possible with most products, although a significant amount of alfalfa is used for manufacturing dehydrated alfalfa pellets and alfalfa meal. Perhaps one of the most viable uses of separation processing would be whole crop harvesting of cereals for use on the farm where the crop was grown (Wilton, 1978; Wilton et al., 1980). Such a procedure would reduce costs associated with transporting materials to processing plants, but might require more capital investment by individual producers than is economically feasible.

CHEMICAL TREATMENTS

Numerous chemical treatment methods have been developed for forages and crop residues in the past century. The two major treatment categories involve hydrolytic and oxidative agents. Common hydrolytic agents are NaOH, other alkali metal hydroxides, NH_3, and urea. Common oxidants are ozone, SO_2, and various other delignifying agents such as chlorite, peracetic acid, permanganate, etc. Combinations of hydrolytic agents and oxidants also have been studied. Each of these treatment processes exerts distinct effects on plant cell wall structure and composition.

Items that must be considered when selecting fibrous feedstuffs as candidates for chemical treatment include stage of plant maturity and plant family (monocotyledon vs. dicotyledon). With regard to stage of maturity, the generalization can be made that chemical treatment is most beneficial for treating mature, lignified substrates. No benefit of chemical treatment is realized on the cell soluble components of the plant; indeed, a negative effect has been demonstrated (Atwell et al., 1991a; Cameron et al., 1991a).

Therefore, it is advisable that the substrate contain a high fiber content. Fibrous feedstuffs having considerable cell contents even at an advanced stage of maturity generally are not good candidates for chemical treatment.

With regard to plant family, the cellulose component of monocots and dicots generally is similar. Crystallinity and degree of polymerization increase as plant maturity advances. Monocots have a much greater concentration of esterified hydroxycinnamic acids (p-coumaric acid [PCA] and ferulic acid [FA]) than dicots. A potentially large number of polysaccharide-lignin associations can occur with monocots whereas with dicots, polysaccharides and lignin reside in more discrete compartments. These differences in fiber chemistry cause monocots to be more responsive to chemical treatment than dicots.

Hydrolytic Treatments

Hydrolytic treatments have been used to improve the feeding value of roughages in Europe since the late nineteenth century (Sundstøl, 1988). Today the most widely used chemical is NaOH. Other hydroxides such as KOH and $Ca(OH)_2$ have been researched, but are more expensive or less effective than NaOH. The phenomenon most widely associated with the alkali treatment of roughages is the partial solubilization of hemicelluloses, lignin, and silica and the hydrolysis of uronic and acetic acid esters (Rexen and Thomsen, 1976; Jackson, 1977; Klopfenstein, 1978; Chesson, 1981; Chesson et al., 1983; Sundstøl, 1988). A disruption of intermolecular hydrogen bonding in cellulose also is believed to result from treatment with alkali.

The increased rate of hydration of alkali-treated roughages also may contribute to the improvement observed in digestibility and intake. Hydration of plant fiber is necessary prior to microbial degradation in the rumen. Allen and Mertens (1988) identified hydration as an important factor affecting lag time for ruminal fiber digestion. The reduced time necessary for hydration of plant fiber would presumably reduce the time necessary for bacterial colonization of plant fiber and may be one factor contributing to the more extensive colonization of treated grasses observed by Latham et al. (1979).

Early work involved a wet method of application, termed the Beckmann method. Cereal straws were soaked in a dilute solution of NaOH over a 3-d period and then washed to remove the residual chemical. This method increased the *in vivo* OM digestibility of rye (*Secale cereale* L.) straw from 460 to 760 g kg^{-1} (Sundstøl, 1988) and barley (*Hordeum vulgare*) straw from 520 to 760 g kg^{-1} (Wanapat et al., 1985). There were two major drawbacks with this process: first, the waste water was contaminated with residual NaOH and posed a threat to the environment; second, because the treated materials were washed prior to feeding, the material solubilized by the treatment (approximately 150 to 200 g kg^{-1} of the original DM) was lost. These concerns resulted in development of dry treatment processes (Rexen and Thomsen, 1976). Several variations of the dry procedure exist, depending upon the desired form of treated feed (i.e., pellets, briquettes, etc.). The basic differences between the wet and dry methods are that the roughage is sprayed with a solution of NaOH and the treated material is not washed prior to feeding. Increases in OM digestibility using this method are somewhat lower than those observed using the wet method. Wanapat et al. (1985) compared effects of several alkali treatment methods on *in vivo* OM digestibility of barley straw. Organic matter digestibilities were 524, 678, and 757 g kg^{-1} for untreated, dry-treated, and wet-treated barley straw, respectively.

Across 24 studies with monocotyledonous crop residues, DM intake was improved by 22% as a result of NaOH treatment. Two studies with dicots indicated only a 6% improvement in DM intake due to NaOH treatment. In all studies, chemical treatment was effective and treated material was fed at levels exceeding 600 g kg^{-1} of diet DM. These results are encouraging in view of the fact that other studies showed NaOH-treated straw to be poorly accepted by animals when fed alone (Rexen and Bach Knudsen, 1984). The remaining dietary components also have a bearing on how the total diet will be accepted. Results of 32 studies with NaOH-treated crop residues (monocots) showed a 30% improvement in DM digestibility while 3 studies with dicots indicated a 17% improvement in digestibility.

Other chemicals such as $Ca(OH)_2$ and KOH have been researched as possible alkaline hydrolytic agents (Owen et al., 1984). Both $Ca(OH)_2$ and KOH are attractive as treatments as they would provide Ca and K to the animal. Calcium and potassium hydroxide also would supply soil nutrients. However, calcium hydroxide by itself is generally ineffective in improving digestibility (Bass et al., 1982) and KOH is too expensive to be economically feasible (Owen et al., 1984).

Future improvements are most likely to come from combining a physical process like pelleting with the application of NaOH. As understanding of fiber structure increases, it may be possible to develop catalysts that could enhance the saponification reaction, resulting in a more efficient treatment process.

Treatment with NaOH costs US $25 to $35 per tonne. This includes grinding and storage, but not substrate costs. Treatment with other hydroxides is either too expensive or ineffective for widespread use.

If the manure from ruminants fed NaOH-treated forages is put back on the same land from which the forages were harvested, Na accumulation in the soil should not become a problem. However, if manure application is more concentrated, then the risk of Na accumulation in the soil or contamination of run-off is a potential environmental problem. Proper management of the high-sodium manure can alleviate these concerns.

Ammonia and Urea

Currently in the US, the most widely used method of alkali treatment is ammoniation (Sundstøl and Coxworth, 1984; Males, 1987; Mason et al., 1988). Anhydrous NH_3, NH_4OH, thermoammoniation, and use of urea as a source of NH_3 all have been investigated. The mechanism of action of ammoniation is assumed to be similar to that described for NaOH. However, several features distinguish treatment of roughages with NH_3 from treatment with NaOH. Increases in digestibility of roughages are generally not as dramatic as those observed with NaOH. Males (1987) stated that the average improvement in

IVDMD of NaOH-treated straws was 160 g kg^{-1} higher than with NH$_3$-treated straws. In addition to increasing structural carbohydrate digestibility, NH$_3$ treatment is an effective means of decreasing the amount of supplemental N needed in diets containing high levels of treated residues (Sundstøl and Coxworth, 1984). Ammoniation also has been an effective means of increasing the intake of low quality roughages such as corn stover (Morris and Mowat, 1980), wheat straw (Saenger et al., 1983), and mature tall fescue (*Festuca arundinacea*; Chestnut et al., 1988). Amount of ammonia (optimal level = 20 to 30 kg tonne^{-1}), temperature, length of treatment time, water content (optimal level = 293 kg tonne^{-1}), and type and quality of material being treated are factors affecting the efficacy of the process (Sundstøl and Coxworth, 1984).

An OECD (Organization for Economic Cooperation and Development) collaborative study was conducted with thermoammoniated barley straw (30 kg NH$_3$ tonne straw DM^{-1}; treated at 90°C for 16 h). Both untreated and treated materials were distributed to several research institutions for detailed studies on its utilization. Graham and Aman (1984) found few differences in neutral sugar and uronic acid contents of the straws but noted that acetyl groups, PCA, and FA concentrations were reduced markedly by treatment. Neutral detergent fiber content was reduced from 842 g kg^{-1} for untreated barley straw to 785 g kg^{-1} for treated straw. Hemicellulose concentration decreased from 319 to 255 g kg^{-1} as a result of treatment. Cellulose, ADL, and ADIN were unaffected by treatment (Van Soest et al., 1984). For sheep fed *ad libitum*, digestibilities of DM, OM, and ADF increased 19, 16, and 25% as a result of NH$_3$ treatment.

Urea as a source of NH$_3$ for roughage treatment has been used for some time and application of this chemical has developed in two primary directions: 1) in industrial processing of roughage combined with grinding and pelleting, and 2) at the farm scale by mixing solutions of urea with roughage and leaving it for some time to allow the urease enzyme to decompose urea and form NH$_3$ (Sundstøl and Coxworth, 1984). In some cases, the roughage contains sufficient enzyme. In other cases, a source of urease such as jackbean (*Canavalia ensiformis*) meal must be added.

When urea is used, NH$_3$ is released only after urea has been mixed with the roughage in the confines of a pelleter, silo, or some other structure. This is a much safer method than those requiring handling of anhydrous or aqueous NH$_3$. However, results with urea have been variable. Williams and Innes (1983) showed that the roughage water content was of major importance for the decomposition of urea added to the roughage. Other factors inhibiting the breakdown of urea are low temperature and low urease activity.

Twenty-one studies using NH$_3$-treated crop residues indicated an average increase in DM intake of 22%. Six studies using NH$_3$-treated grasses showed a 14% increase in intake while 1 study with ammoniated cotton plant (*Gossypium hirsutum* L.) showed a 4% increase in DM intake. With regard to digestibility, 32 studies on NH$_3$-treated crop residues indicated a 15% increase in DM digestibility as a result of NH$_3$ treatment, 10 studies with grasses showed a 16% increase, and 1 study using the cotton plant indicated a 16% increase in digestion. In 5 studies where urea was used as the NH$_3$ source and DM intake was measured, treatment increased intake an average of 13%. In 7 studies reviewed, urea treatment resulted in a 23% increase in DM digestion.

Another treatment process, Ammonia Fiber Explosion (AFEX) has been developed (Dale, 1983) and tested to a limited extent. Ammonia fiber explosion treatment of coastal bermudagrass [*Cynodon dactylon* (L.) Pers.] removed the waxy coating on the surface of the samples, but caused no physical disruption when treatment intensity was low (Turner et al., 1990). However, as the severity of treatment increased, exposure of internal structures of the substrates occurred. Changes in structure correlated with increased IVDMD (680 vs. 850 g kg^{-1} for control vs. treated coastal bermudagrass, respectively; Hagevoort et al., 1990). Bagasse was tested and IVDMD values increased from 300 to 550 g kg^{-1} with level of treatment intensity. No *in vivo* studies have been reported to date.

Ammoniation costs are usually between US $20 to $25 per tonne. This is cheaper than NaOH treatment because grinding of the forage is not required and storage costs basically involve the costs of the plastic. In addition, the NH_3 has some value as a crude protein source, even though approximately two-thirds of what is originally applied is lost when the stack is aerated prior to feeding.

Environmental problems associated with ammoniation are minimal and are more often related to disposal of the plastic covering the stack than to ammonia.

Oxidative Treatments

Oxidative treatment reagents actively attack and degrade a major proportion of cell wall lignin (Chang and Allen, 1971). Ozone and hydrogen peroxide also have been shown to randomly cleave glycosidic linkages of cell wall polysaccharides (Nevell, 1985). Sulfur dioxide oxidation results in considerable solubilization of polysaccharides, suggesting a similar effect (Miron and Ben-Ghedalia, 1982). The net effect of treatment with oxidizing agents is a significant reduction in lignin content and an increase in soluble carbohydrate concentration (Ben-Ghedalia et al., 1982, 1983). Various peroxides, chlorine-containing compounds, and ozone represent some of the major oxidative treatments investigated (Chandra and Jackson, 1971; Klopfenstein et al., 1972; Yu et al., 1975; Ben-Ghedalia and Miron, 1981b; Ben-Ghedalia et al., 1982, 1983; Ben-Ghedalia and Shefet, 1983). All of these compounds have been employed by the wood pulping industry for a number of years as delignifying agents (Chang and Allen, 1971).

Peroxide oxidation is dependent upon pH of the medium. While the exact mechanism of hydrogen peroxide (H_2O_2) oxidation of lignin is not completely understood, one proposed mechanism is that the hydroperoxide anion (HO_2^-), formed when H_2O_2 dissociates (pH 11.6), attacks the phenylpropane units of lignin at electron deficient loci. However, alkaline hydrogen peroxide (AHP) oxidation of lignin is quite limited in scope as compared to ozone and chlorine compound oxidation. Degradation by AHP generally is limited to specific side groups, whereas other oxidants are more likely to degrade aromatic rings.

Chandra and Jackson (1971) and Klopfenstein et al. (1972) investigated the use of peroxides for improving the nutritive value of low quality roughages. Chandra and Jackson (1971) treated ground corn cobs with solutions containing various concentrations of H_2O_2. They noted a linear decrease in lignin content and a linear increase in *in situ* DM disappearance (ISDMD) with increasing level of H_2O_2 treatment. Klopfenstein et al. (1972) investigated the efficacy of both sodium peroxide (Na_2O_2) and combinations of Na_2O_2 and H_2O_2. Acid detergent lignin of alfalfa was not lowered by either treatment; however, ADL concentration of corn cobs was lowered. Sodium hydroxide also was included as a treatment in this study. The authors noted that due to the higher cost of the peroxides and the fact that similar results were obtained when either NaOH or the peroxides were used, NaOH was a more practical treatment.

Comprehensive studies to evaluate the merits of chlorine-containing compounds for increasing the nutritive value of roughages were conducted by Yu et al. (1975). Treatment with chlorine compounds often increases IVDMD but reduces voluntary intake by as much as 50% compared to the untreated feed (Yu et al., 1975; Ford et al., 1987). In light of results collected to date, chlorine compounds appear impractical despite their delignifying capabilities (Chandra and Jackson, 1971; Yu et al., 1975).

Ozone has been investigated extensively by Ben-Ghedalia and colleagues as a means of improving the feeding value of roughages (Ben-Ghedalia and Miron, 1981b; Ben-Ghedalia et al., 1982, 1983; Ben-Ghedalia and Shefet, 1983). Treatment of wheat straw with ozone resulted in a 20 and 63% reduction in lignin and hemicellulose concentrations, respectively. *In vitro* OM digestibility was increased from 440 to 670 g kg^{-1} by ozone. Ozonolysis also has been shown to improve the digestibility of cotton straw (Ben-Ghedalia et al., 1983), grape (*Vitis* sp.) branches (Ben-Ghedalia, 1982), and mesquite (*Proscopisis* sp.; Bryant et al., 1984). Ben-Ghedalia and Shefet (1983) showed that ozone treatment makes cell wall polysaccharides more available for digestion by ruminal microorganisms when compared to untreated or NaOH-treated straw in experiments using cannulated sheep. In an *in vivo* study, Ben-Ghedalia et al. (1983) examined effects of feeding sheep (45 to 50 kg) ozonated cotton straw at 500 g kg^{-1} of diet DM. Ruminal VFA concentration, pH, NH_3-N, and liquid dilution rate were similar for lambs fed diets containing either the untreated or the ozone-treated cotton straw.

At present, ozonation holds little promise as a practical means of producing large quantities of treated material for animal production. The level of technology required and the fact that ozone at ground level is a pollutant argue against its use. However, useful information concerning the relative effectiveness of oxidative as opposed to hydrolytic treatment of dicotyledonous residues has been gained. The lignin in dicots is apparently a much greater constraint to cell wall digestion than alkali-labile phenolic material. This explains the greater success of oxidative delignifying agents as opposed to alkaline hydrolytic agents in improving utilization of dicots.

Another oxidant positively affecting digestion is SO_2. A 50 g kg^{-1} SO_2 treatment converted wheat straw into a highly digestible residue (Ben-Ghedalia and Miron, 1981b). Sulfur dioxide disrupts lignin-hemicellulose bonds and solubilizes a portion of the plant lignin. Ben-Ghedalia and Miron (1981b) reported that SO_2 treatment of wheat straw solubilized all of the straw hemicelluloses and converted them to cell solubles, and decreased the lignin content (63.9 vs. 73.3 g kg^{-1}) when compared to the untreated straw; NaOH (50 g kg^{-1}) treatment had little effect on either fraction. *In vitro* OM digestibility was increased by 80% with the SO_2 treatment, whereas NaOH treatment improved IVOMD by 50% compared to the untreated wheat straw.

Ben-Ghedalia and Miron (1984b) conducted an experiment to measure the digestibility of SO_2-treated wheat straw by sheep. This study was of particular importance to determine the potential of SO_2-treated wheat straw for ruminant diets as it frequently has been found that improvements in *in vitro* digestibilities due to chemical treatment often overestimate treatment effects on *in vivo* digestibilities (Berger et al., 1979). The SO_2 treatment of wheat straw increased OM digestibility from 463 to 652 g kg^{-1}, the effect being lower than the 80% increase found *in vitro* (Ben-Ghedalia and Miron, 1981b). The phenomenon of lower *in vivo* values compared to *in vitro* values apparently is not confined to NaOH-treated residues, but is a characteristic of other chemical agents applied to cereal residues.

In a growth experiment (Ben-Ghedalia and Miron, 1987), lambs were fed two diets containing untreated or SO_2-treated (35 g kg^{-1}) wheat straw plus poultry litter at a ratio of 1:1, and a concentrate mixture (corn, soybean meal). The amount of wheat straw plus poultry litter in the complete diets was increased during the 95 d of the study from 500 to 700 g kg^{-1}. The lambs did not respond negatively to the above dietary changes. The average liveweight gain of the SO_2-treated wheat straw-fed lambs was 330 g d^{-1} and that of the untreated wheat straw-fed lambs, 267 g d^{-1}. The increased performance of animals fed the SO_2-treated wheat straw diet was due to a 13% higher feed intake (1,565 vs. 1,385 g d^{-1}) and an 8.5% improvement in feed efficiency.

In a second lamb growth study (Ben-Ghedalia et al., 1988), a high energy concentrate diet (barley, corn, soybean meal) was compared against diets in which 600 g kg^{-1} of the concentrate was replaced with a combination (1:1) of either untreated wheat straw and poultry litter or SO_2 (35 g kg^{-1})-treated wheat straw and poultry litter. Diets were fed *ad libitum* for the 114 d growth experiment. Gains were 266, 356, and 363 g d^{-1} for the untreated straw, SO_2-treated straw, and concentrate diets, respectively.

These results indicate that SO_2-treated residues may be incorporated into the diets of ruminants where a high level of productivity is desired. However, SO_2-treated wheat straw has been fed only to sheep and further research with larger ruminants is necessary to determine its potential in ruminant production systems.

Insufficient data are available to determine the practical potential of using SO_2 to treat crop residues. Initial projections suggest that the costs would be greater than NaOH or NH_3 treatment. However, it has been the only treatment method to date to markedly improve the digestibility of dicots.

Consequently, in certain situations, it may be the most efficacious treatment available.

Mixtures of Hydrolytic Agents and Oxidative Agents

Less research is available on the efficacy of using mixtures of chemicals to improve the nutritive value of lignocellulosics. Streeter and Horn (1982) conducted a study to examine the effects of NH_3-peracetic acid treatment of wheat straw on cell wall composition and IVDMD. Peracetic acid has been used successfully by wood chemists to delignify wood in the preparation of holocellulose. These workers reasoned that improvements in roughage digestibility by alkaline hydrolysis could be further enhanced by subsequent treatment with an oxidizing agent which would affect lignin. The sequential addition of both NH_3 (0 to 105 g kg^{-1} straw DM) and peracetic acid (0 to 120 g kg^{-1} straw DM) resulted in linear increases in digestibility. An 85% improvement in IVDMD compared to control (630 vs. 340 g kg^{-1}) occurred when combinations of the highest concentrations of both chemicals were applied. Peracetic acid significantly decreased lignin concentration while NH_3 had no effect. Hemicellulose content was reduced significantly by NH_3; however, peracetic acid had no effect on this constituent. Cellulose content of straw was unaffected by either chemical. The authors noted that peracetic acid is probably too expensive and dangerous to be considered a viable treatment option. However, the important conclusion of the study was that the combined effects of alkaline hydrolysis and oxidative delignification were additive. These promising results indicated that further research should be conducted to determine more economical and safe methods of employing this concept.

A hydrolytic agent-oxidant mixture with potential as a lignocellulosic pretreatment is AHP (Kerley et al., 1985). Dilute, alkaline solutions of H_2O_2 solubilize approximately 500 g kg^{-1} of the lignin and many of the hemicelluloses originally present in a variety of lignocellulosics. The original AHP treatment process consisted of suspending material in water containing 10 g L^{-1} H_2O_2 (1 g substrate 50 ml^{-1} solution). Sodium hydroxide is added to bring the pH to 11.5, and the mixture is stirred gently at room temperature. After several hours, the insoluble residue is collected, washed repeatedly with water until the filtrate pH is neutral, and dried in a vacuum oven at 50°C. The cellulose-rich residue which remains after treatment with AHP is extremely susceptible to *Trichoderma reesii* cellulase, with cellulose to glucose conversion efficiencies close to 100% for wheat straw and corn stover. In combined saccharification/fermentation experiments using cellulase and *Saccharomyces cerevisiae*, ethanol yields from these crop residues exceeded 90% of the theoretical yield, indicating that lignin degradation products generated by this pretreatment were not particularly inhibitory to microbial growth. A marked increase in hydration capacity of the cellulose-rich residues also occurred as a result of AHP treatment (Gould, 1984; Gould and Freer, 1984).

Sheep studies were conducted comparing the intake potential and nutrient digestibilities of diets containing varying levels of untreated and AHP-treated wheat straw (AHP-WS; Kerley et al., 1985, 1986). Results of these studies showed significant advantage to feeding AHP-WS compared to untreated wheat straw. Lambs consuming diets containing 720 g kg^{-1} AHP-WS on a DM basis consumed over 100% more DM than did those consuming control diets. Dry matter, NDF, and ADF digestibilities were greater for lambs consuming treated wheat straw than for those consuming untreated wheat straw. The DE and ME contents of diets also increased when lambs consumed treated straw.

Results using mature sheep were equally dramatic. Feed intakes were held constant at 1,200 g d^{-1} so that digestibility comparisons could be made among treatments without the confounding effects of variable feed intakes. Dry matter digestibilities were approximately 25 percentage units greater for treated straw diets compared to their controls. Neutral detergent fiber and cellulose digestibilities were 30 to 40 percentage units greater for sheep fed diets containing low and high levels of treated straw when compared with controls.

The AHP process described above is currently in use by a number of industrial organizations who manufacture fiber sources for humans. However, the cost of processing is too great for it to be considered a means of treatment of lignocellulosics for livestock feed. Also, the original AHP treatment process suffers from some of the same drawbacks encountered with the Beckmann NaOH process. It is a wet treatment which results in a considerable loss of solubilized hemicelluloses (Gould and Freer, 1984). Additionally, concerns about pollution exist as the solution in which the substrate is soaked contains a significant amount of residual Na from the alkali used in the process. Consequently, a dry method for applying the AHP treatment process was developed (Cameron et al., 1990). This involves a mixer-sprayer system which delivers 50 g kg^{-1} NaOH and 20 g kg^{-1} H$_2$O$_2$ to the substrate DM. Water is added such that the final product is approximately 650 g DM kg^{-1}. The dry AHP treatment process is superior to a dry NaOH treatment as evidenced by the fact that growing wethers consumed 16 g kg^{-1} more digestible DM when fed AHP-WS compared to animals fed NaOH-treated wheat straw.

Much research has been conducted to determine the efficacy of including AHP-treated substrates (dry method) in ruminant diets. Initially, several studies were conducted with sheep (Atwell et al., 1991a; Willms et al., 1991b). When a diet containing 650 g kg^{-1} AHP-WS was fed at restricted intake (2.2% of body weight), lambs digested 770 g kg^{-1} of the dietary OM. When diets containing 800 g kg^{-1} AHP-WS were fed, sheep ate 3.2% of their body weight and digested 700 g kg^{-1} of dietary OM and 700 g kg^{-1} of dietary NDF.

Based on the results of the sheep experiments, steer growing and finishing trials were conducted. In the growing experiment (Willms et al., 1991a), steers were fed 660 g kg^{-1} roughage diets comparing AHP-WS, a 1:1 mixture of AHP-WS and corn silage, and corn silage. Dry matter intakes (kg d^{-1}), average daily gain (kg) and gain/feed (kg kg^{-1}) for the AHP-WS, AHP-WS:corn silage mixture, and corn silage were: 7.5, 1.08, and 0.14; 8.2, 1.33, and 0.16; and 8.0, 1.56 and 0.19, respectively. Even though steers fed AHP-WS gained slower and less efficiently than those fed corn silage, the actual gains were 235% of projected gains based on the NE values of untreated WS (NRC, 1984).

In a steer finishing experiment (Willms et al., 1991a), AHP-WS was compared to alfalfa hay and corn silage at 100 and 200 g kg^{-1} of dietary DM. There were no differences in performance due to roughage source. Carcass measurements were similar among treatments.

In another experiment (Berger and Fahey, unpublished data), Angus heifers were fed either AHP-treated oat hulls (AHP-OH) or dehydrated alfalfa pellet diets in an intake and digestibility study. Treated oat hulls comprised approximately 700 g kg^{-1} of the oat hull pellet. Other ingredients were soybean meal, urea, and minerals. Heifers consumed 11.2 and 8.9 kg d^{-1} and digested 737 and 686 g kg^{-1} of the AHP-OH and alfalfa pellet DM, respectively. Due to the greater intakes and digestibilities, heifers fed AHP-OH consumed 368 g kg^{-1} more digestible DM than those fed alfalfa.

Five experiments (Cameron et al., 1990, 1991a, b, c ; Atwell et al., 1991b) were conducted to determine the effects of varying amounts of AHP-treated lignocellulosics in the diet on feed intake, nutrient digestion, ruminal characteristics, and performance responses by dairy cows and heifers. As observed for sheep and beef cattle, AHP treatment of wheat straw or oat hulls allowed for extensive and rapid fermentation of fiber in the gastrointestinal tract of dairy cattle. Dry matter intakes and milk yields for mid-lactation dairy cows consuming AHP-WS (20.7 and 25.4 kg d^{-1}, respectively)- and AHP-OH (27.0 and 33.9 kg d^{-1}, respectively)-based diets, demonstrates the potential of AHP-treated roughages for dairy production.

Another approach was taken by Adebowale et al. (1989) who sprayed untreated wheat straw, corn stover, and corn cobs with either 10, 50, or 100 g H_2O_2 kg^{-1} substrate with subsequent ammoniation of all materials (30 kg anhydrous NH_3 tonne^{-1}; 90°C; 23 h). Ammonia treatment increased potential degradability of wheat straw from 599 to 722 g kg^{-1} and corn stover from 677 to 741 g kg^{-1} but had little effect on corn cobs which increased only from 580 to 596 g kg^{-1}. Further improvements in degradability occurred as a result of AHP treatment. Degradability of wheat straw increased to 802 g kg^{-1} for the 10 g kg^{-1} AHP treatment. Values increased to 815 g kg^{-1} for corn stover and 722 g kg^{-1} for corn cobs. For all substrates, rate of digestion increased also. These results demonstrated that H_2O_2 treatment increased fermentability of structural carbohydrates when subsequently treated with gaseous NH_3.

Substituting NH_3 or $Ca(OH)_2$ for a portion of the sodium hydroxide has the potential to reduce costs and make the AHP process more environmentally friendly. Treating residues like oat hulls or bagasse which already are collected at one site offers the greatest potential for practical application of the technology. To produce a value-added product, these treated residues could be mixed with other ingredients, pelleted or cubed, and sold in niche markets.

The dry AHP treatment process costs US $35 to $40 per tonne. Although this is more expensive than some other treatment procedures, the final product also has a greater nutritional value. When substrate costs are added to treatment costs, total costs may not be competitive with other sources of DE available in North America. However, in other parts of the world where cereal grains are more expensive, this process may be economically competitive.

Environmental concerns center around the disposal of manure that is high in sodium. As noted previously, if the manure is spread on the land from which the residue is harvested, there should be little concern. However, if more concentrated application of the manure is required, then the effects of sodium on soil fertility could be a concern.

MICROBIAL/ENZYMATIC TREATMENTS

Biological treatment of roughages offers the advantage of fewer chemicals and lower energy inputs for treatment than required for chemical or physical manipulation of feeds. The clear disadvantages are the much longer times required for biological treatments compared to chemical/physical treatments and, in the case of fungal inoculation, the greater loss of substrate upon incubation.

Biological treatment of fibrous feeds has advantages which include: 1) an improvement in digestibility (rate or extent) of the lignocellulose; 2) an improvement in feed quality as reflected by increased protein content and(or) acceptability; 3) destruction of harmful compounds present in feedstuffs; and 4) especially in the case of silage, improved storage characteristics (discussed below). The effects of white-rot fungi (WRF) on delignification and digestibility of lignocellulosics and enzymatic treatment of forages to improve digestibility will be the topics discussed in this section.

White-rot Fungi for Treatment of Lignocellulosic Material

Use of WRF to increase digestibility of lignocellulosics has been studied extensively, yet the literature reports details of many failures. Because many factors (e.g., environment, selection of fungal species, inoculum size, length of incubation) may limit the success of WRF treatment, we will focus on the successes which researchers have had in improving feed quality rather than on the failures.

One of the disadvantages of using WRF to improve roughage digestibility is the loss of OM associated with fungal growth. Even in cases where digestibility is radically improved, the total yield of digestible OM may be diminished. Thus, Reid (1989) suggested that the digestibility of WRF-colonized feedstuffs should be expressed as a percentage of the original substrate. From an economic standpoint, the exchange of a substrate for a lesser amount of a more digestible product may or may not be of value. Because intake is a critical factor limiting forage utilization and is generally positively related to degradability, an equal amount of digestible OM can generally be considered more valuable if it is present as a more energy-dense product. Therefore, digestibilities presented for biologically treated products will be expressed as a percentage of the treated product and not as a percentage of the original substrate.

The ideal organism for treatment of lignocellulosics to improve digestibility would be capable of 1) degrading structures that limit fermentation (e.g., lignin) while not removing structures which can be effectively utilized by the animal (e.g., carbohydrates); 2) growing rapidly so as to minimize treatment time; and 3) competing effectively against other microorganisms which are present. Of all biological species, WRF appear to degrade lignin most rapidly and extensively. White-rot fungal species include those which degrade all wood components at similar rates and those which preferentially degrade lignin and hemicelluloses (Blanchette, 1984; Blanchette et al., 1985). Those species which preferentially degrade lignin will be the most useful for increasing the digestibilities of feedstuffs.

Growth and lignin degradation by WRF are oxygen-requiring activities. In early stages of fungal colonization, the WRF rapidly remove soluble carbohydrates present in the substrate. Following depletion of the readily available carbohydrate, lignin-degrading enzymes appear and lignin degradation commences. Lignin is usually degraded by WRF only during secondary metabolism (metabolic events which are activated only after primary growth is complete); often this is triggered by nutrient (usually N) limitation. However, lignin is not a growth substrate for WRF and, therefore, the degradation of lignin must be linked to degradation of an alternate carbon source (e.g., carbohydrate). Thus, even under ideal conditions, some carbohydrates will be lost from the substrate. Those WRF which do not degrade cellulose will be most useful for delignifying feeds. After extensive colonization and delignification, WRF begin to degrade carbohydrate from the delignified cell wall. Clearly, this is not ideal as it will be associated with excessive losses of OM. At some point prior to extensive carbohydrate loss, the product will be of maximal digestibility, but will still retain most of the available carbohydrate. Thus, timing of the harvest of the product is extremely important.

Various substrates have been successfully treated with WRF under laboratory conditions such that a partially delignified product with improved digestibility was obtained. Bagasse was inoculated using four different cultures of WRF, and *Pleurotus sajorcaju* was found to selectively delignify the product and increase IVDMD from 300 to 490 g kg^{-1} (Kewalramani et al., 1988). This occurred after 40 d of incubation and was associated with a 170 g kg^{-1} loss of weight and a 260 g kg^{-1} loss of lignin. Rolz et al. (1986) tested 12 species of WRF on bagasse. *Bondarzewia berkeleyi* was most effective; it removed 370 to 640 g kg^{-1} of the lignin and improved enzymatic degradability from 130 to 190 g kg^{-1} in control bagasse to 350 to 430 g kg^{-1} for inoculated bagasse. The extensive delignification also was associated with large DM losses ranging from 360 to 420 g kg^{-1} of initial weight.

Yadav (1987) optimized conditions for treatment of wheat straw with *Coprinus* sp. and was able to increase IVDMD from 400 to 650 g kg^{-1}. About one-half of the lignin and one-fourth of the DM was removed by the WRF after 21 d of incubation. In that experiment, the WRF incubations were scaled up to 50 kg batches with only modest losses in efficiency.

Ruminal fermentability of rice husk has increased from 190 to 330 g kg^{-1} following incubation for 35 d with a WRF *Pleurotus ostreatus*; the treated product contained a lignin content of 124 g kg^{-1} compared to 210 g kg^{-1} for the intact rice husks (Beg et al., 1986).

The conversion of poplar (*Populus tremuloides* Michx.) into a digestible feedstuff by WRF was investigated by Reade and McQueen (1983). Incubation with *Ganoderma applanatum*, *Phanerochaete chrysosporium*, and *Polyporus versicolor* resulted in a 250 to 300 g kg^{-1} loss of lignin and a 150 to 200 g kg^{-1} loss of DM after a 3 to 4 wk fermentation and increased ruminal digestibility to 600 to 650 g kg^{-1} (control poplar = 300 g kg^{-1}). *Polyporus anceps* resulted in a product with a greater digestibility (720 g kg^{-1}), greater lignin loss (430 g kg^{-1}) and only 180 g kg^{-1} DM loss, but required 8 wk to reach these levels.

Zadrazil (1980) noted the importance of matching substrate with inoculum. *Stropharia rugosounnulata* was most effective for increasing the digestibility of rape (*Brassica campestris* L.; from 340 to 820 g kg^{-1}) while *Pleurotus cornucopiae* was most effective for sunflower (*Helianthus annuus*; from 410 to 640 g kg^{-1}). None of the species tested by Zadrazil (1980) was effective in improving the digestibility of rice husks.

In vivo digestibility experiments with WRF-treated forages are scarce. Rai et al. (1989) conducted two experiments in which rice straw was inoculated with *Coprinus fimetarius* 386 and incubated for either 2 or 4 wk before being fed to goats. The lignin content of the straw increased and the cellulose content decreased by the fungal treatment, indicating an ineffective treatment. In both experiments, fungal treatment led to decreased cellulose digestibility (11 and 6%). Despite reduced cellulose digestibility, intake of the fungally-treated straw was higher such that digestible nutrient intake was as high or higher than that of uninoculated straw (digestible OM intake was unaffected in one experiment and increased by 56% in the other).

Calzada et al. (1987) and Bakshi et al. (1985) fed wheat straw residue which remained following harvest of edible mushrooms (*Pleurotus* sp.). Calzada observed similar intakes (0.39 and 0.42 kg d^{-1}) and digestibilities (550 and 580 g kg^{-1} of OM) by lambs fed untreated and spent wheat straw. Bakshi et al. (1985) found that intake was similar (8.76 and 8.54 kg d^{-1}), but digestibility was lower for diets containing spent straw (490 g kg^{-1}) than for those containing untreated straw (620 g kg^{-1}) when fed to buffaloes (*Bubalus bubalis*). In neither of these studies was the lignin content of the spent straw lower than that of the control straw.

Future improvements in the use of fungi to enhance the nutritive value of low quality roughage requires the identification of fungal strains that can partially delignify the substrate, without destroying the potentially digestible carbohydrates. This would allow increased substrate recovery which is necessary to make the process economically viable. For example, traditional cross-breeding has produced cellulase-less mutants of *P. chrysosporium* which are better able to delignify wood (Johnsrud, 1988).

Reid (1989) estimated that the costs of biological delignification of wood by SSF with WRF would be in the range of US $20 tonne^{-1}, exclusive of the cost of the wood; this requires amortization of the cost of facilities over a large number of batches. The cost of the wood and the protein supplementation required to formulate a nutritionally balanced diet were estimated to be the most expensive inputs into the system. Currently, there is an absence of data showing the costs of a successful fungal treatment procedure for low quality roughages.

Enzymatic Treatment of Lignocellulosic Material

Enzymes have been used to treat fibrous feeds in an effort to improve their quality. Silage pH can be decreased more rapidly, thereby leading to improved ensiling characteristics, if cellulases are added to forages prior to storage (Autrey et al., 1975; Jorgensen and Cowan, 1989; van der Meer and Ketelaar, 1989). This effect is a result of increasing the availability of rapidly fermentable sugars by hydrolysis of cellulose (Henderson and McDonald, 1977).

Cellulases also have been added to forages in order to improve their digestibility. Daniels and Hashim (1977) observed that a fungal cellulase added to a rice hull-based diet improved the *in vivo* digestibility of DM, energy, and crude protein. Vanbelle and Bertin (1989) used several commercial cellulase enzymes (which also contained significant hemicellulolytic activity) and observed increases in the *in vitro* digestibility of capim (*Panicum maximum* Jacq.; from 298 to 402 g kg^{-1}) and of alfalfa (from 648 to 765 g kg^{-1}). However, Morrison (1988, 1991) treated barley straw, ryegrass, and alfalfa with either cellulase or hemicellulase and observed decreases in digestibilities *in vitro*, presumably because the sugars which were solubilized by the enzymatic treatment and removed prior to evaluating the product were from the more readily degradable fraction of the cell wall. The extent of rice straw degradability was not affected by treating it with a polysaccharidase containing a broad spectrum of activities (Nakashima et al., 1988). However, the soluble fraction was increased from 150 to 210 g kg^{-1} of DM and the rate of fermentation was increased from 5 to 8 h^{-1}. Nakashima and Orskov (1989) also reported increased solubility of barley straw treated with the polysaccharidase. The enzyme was found to be more active on fiber of leaf blades and leaf sheaths than nodes or internodes. Enzyme treatment of barley straw following pre-treatment with NaOH or alkaline H_2O_2 did not affect the extent of degradability, but similar to its action on untreated straw, increased the soluble fraction.

The combination of chemical and enzymatic treatment of forage was tested by Ben-Ghedalia and Marcipar (1979) who observed small increases in digestibility of wheat straw *in vitro* when it was treated with cellulases from *Aspergillus niger* or *Trichoderma viride* following treatment of the forage with 50 g kg^{-1} NaOH or NH_3 gas. Ben-Ghedalia and Miron (1981a) found that following chemical treatment of wheat straw, cellulases increased the rate of fermentation although the extent (at 48 h) was relatively unchanged. Sulfur dioxide was found to be a better pretreatment than either ozone or NaOH.

Khazaal et al. (1990) treated barley straw with a ligninase enzyme from *P. chrysosporium* and measured the effects on composition and degradability. Modest reductions in the lignin content of the straw were observed (from 116 to 105 g kg^{-1} of OM). However, no significant improvements in digestibility were observed. The lack of response to the isolated enzyme may be because lignin degradation is a free radical reaction which may require other enzymatic means to shift the equilibrium of the reactions involving degradation and repolymerization. Ander (1990) suggests that cellobiose:quinone oxidoreductase should be added to enzyme mixtures designed to degrade lignin in order to shift the equilibrium away from repolymerization and toward degradation.

Biotechnology can be expected to rapidly expand our knowledge of microbial enzymes. Research is currently underway to create strains of *P. chrysosporium* which produce large amounts of ligninases (Holzbaur et al., 1991). Normally, ligninase enzymes are induced only during secondary metabolism, but biotechnology may allow the creation of strains of WRF which produce greater amounts of ligninase during primary growth. If these enzymes could be harvested and applied to lignocellulosics, a faster and more extensive biological delignification may be possible.

The ligninase genes of WRF have been cloned into *E. coli*. Unfortunately, the enzyme as produced by *E. coli* often lacks the heme group and is insoluble, thus requiring costly downstream processing prior to use (Holzbauer et al., 1991). The enzyme used by Nakashima and Ørskov (1989) was reported to cost US $65 kg^{-1}, and this was typically used at a level of 5 g kg^{-1}, resulting in a cost per tonne of US $325, clearly an uneconomical situation. Obviously, advances in biotechnology and increased production of efficacious enzymes would be expected to lower the cost of enzymatic treatments.

Much research remains to be done on the microbiological front to make this treatment method useful and economical. However, positive responses to delignification with WRF suggest that much potential exists for this environmentally safe treatment process. In the future, biotechnological advances can be expected to improve the economic outlook for biological treatment of lignocellulosic residues.

ADDITIVES AND PRESERVATIVES

Hays

The efficiency of conserving forage as hay is limited by losses in yield and nutrients while the crop dries to approximately 200 g kg^{-1} moisture. Reduction in feeding value of hay increases when its moisture exceeds 200 g kg^{-1} and ambient temperature exceeds 40°C (Van Soest, 1965). Feeding value reductions of 250 g kg^{-1} or more have occurred when storing moist hay.

Propionic acid has been the most common acid used to preserve moist hay because of its fungistatic properties. Lacey and Lord (1977) showed that with uniform distribution, mold growth could be prevented by a ratio of propionic acid to water in the hay of 1.25:100. This is equivalent to 30 to 40 g kg^{-1} propionic acid on hay containing 250 to 300 g kg^{-1} moisture. However, because uniform distribution is difficult and propionic acid is quite volatile under field conditions, 10 g kg^{-1} propionic acid often is required for mold prevention in hay containing 300 g kg^{-1} moisture. Atwal and Heslop (1987) reported that treating alfalfa hay containing 236 g kg^{-1} moisture with 30 g kg^{-1} propionic acid resulted in significantly lower NDF concentration and higher energy digestibility than similar hays with 3 g kg^{-1} propionic acid. In this study, after 36 wk in storage, the 3 g kg^{-1} propionic acid-treated alfalfa had DM, crude protein, and energy digestibilities 37, 76, and 40 g kg^{-1}, respectively, lower than control hay baled at 17.8 g kg^{-1} moisture. In contrast, Baron and Greer (1988) reported that hay harvested at 250 or 350 g kg^{-1} moisture and treated with commercial propionic acid-based preservatives had similar in vitro OM digestibilities to the same hay harvested and stored at 150 g kg^{-1} moisture. Many factors such as harvesting conditions of the dry hay, storage conditions, density of the bale, ambient temperature, relative humidity, uniformity of acid application, and length of storage affect the response to propionic acid treatment.

The cost effectiveness of propionic acid application is determined primarily by the cost of the acid relative to the value of the hay. A 10 percentage unit reduction in storage losses of hay harvested between 200 and 300 g kg^{-1} moisture is common with 10 g kg^{-1} propionic acid treatment (Jorgensen et al., 1973). The value of the extra hay should be counted against the treatment cost. There also may be situations where propionic acid treatment would allow the salvaging of a high-quality hay already dried to less than 300 g kg^{-1} moisture and threatened by an impending rain. However, under rain-free drying conditions, the increased feeding value and yield of harvesting high-moisture hay (> 200 g kg^{-1} moisture) compared to allowing the hay to dry will not pay for the propionic acid treatment.

Anhydrous ammonia has been used as a hay preservative. Its desirable properties include being an effective fungicide, a weak alkali that can improve fiber digestibility, and a source of nonprotein N that can be utilized by ruminants. The application of 10 g kg^{-1} anhydrous ammonia (DM basis) to conventional rectangular bales of alfalfa or brome hays at 300 g kg^{-1} moisture (Johnson et al., 1981; Thorlacius and Robertson, 1984) and alfalfa or mixed hay at 350 g kg^{-1} moisture (Knapp et al., 1975; Thorlacius and Robertson, 1984) suppressed mold growth, reduced storage losses, and improved digestibility. Atwal et al. (1986) reported that application of 10 g kg^{-1} anhydrous ammonia to large round bales of alfalfa containing approximately 300 g kg^{-1} moisture reduced DM losses during storage and improved crude protein and energy digestibility. However, when the ammoniated hay was uncovered and ground, molding occurred within 4 wk even though the ambient temperature averaged only 6°C. This suggests that once the ammoniated high-moisture hay is uncovered, it needs to be fed within a few weeks to avoid molding. Consequently, depending on how rapidly the hay will be fed, it may be more efficacious to ammoniate several small stacks instead of one large stack.

Although using 10 g kg^{-1} ammonia on high quality hay as a preservative has not caused animal health problems, treating with higher levels of ammonia has resulted in significant animal losses. The toxicity problem has been termed bovine hysteria, crazy cattle disease, or bovine bonkers. The toxin not only affects the animal consuming the ammoniated forage, but is passed through the milk to the young (Weiss et al., 1986). The ammoniated hays that have caused this disease include alfalfa, bromegrass, fescue, orchardgrass, bermudagrass, wheat, barley, sudangrass, and sorghum. The three most common conditions that have been present when bovine hysteria has occurred are: 1) treating high quality forages, 2) uneven distribution of the ammonia, or 3) applying 30 g kg^{-1} or more ammonia on a DM basis.

The active compound has not been clearly identified, but is believed to be an alkaloid, possibly an indole (Weiss et al., 1986). It is likely that the presence of soluble carbohydrates is required to form the toxic compound. Similar problems have been observed in cattle and sheep fed ammoniated molasses. The low level of soluble carbohydrates in crop residues probably explains why toxicity problems have not occurred when treating these forages.

Although ammonia can be an excellent preservative of high-moisture hays, at the producer level there can be significant risks involved. First, producers must uniformly apply ammonia at approximately 10 g kg^{-1} of DM. Secondly, the ammoniated hay should not be fed alone. Thirdly, feeding the ammoniated hay to lactating cattle and sheep should be avoided as the offspring are often more susceptible than are the mothers. Finally, ammoniated high quality hay should not be fed to lactating dairy cattle.

Urea also has been used as a high-moisture hay preservative. Its main advantage over ammonia is that it is easier to apply uniformly at the time of baling. In addition, it is less corrosive and easier to work with than ammonia. However, urea has not always been effective as a hay preservative. Ghate and Bilanski (1979) treated alfalfa of varying moisture (300, 400, and 500 g kg^{-1}) with three concentrations of urea (17.5, 35, and 53 g kg^{-1} of DM) and found little benefit compared to untreated hay. In contrast, Alhadhrami et al. (1989) showed that adding 20 or 40 g kg^{-1} urea to alfalfa hay containing approximately 25 g kg^{-1} moisture improved in vitro DM digestibility, reduced heating, and prevented mold growth. The 40 g kg^{-1} urea treatment was more effective than the 20 g kg^{-1} treatment.

The limitations of using urea as a hay preservative relates to the fact that urease must be present either from the plant or microbial populations for it to have fungicidal activity. Secondly, to get effective preservation, urea concentrations in excess of 20 g kg^{-1} of DM are required. This level of urea limits the use of the treated hays because most producers prefer to feed 10 g kg^{-1} or less urea in the final diet. Finally, the ammonia released from the urea could cause the formation of toxic compound(s) that result in bovine hysteria.

One alternative to chemical preservation of high-moisture hays is to use bacterial inoculants. If homofermentative lactic acid-producing bacteria would grow in moist hay, they may prevent the growth of spore-forming organisms and *clostridia* until normal dehydration could stabilize the hay. Nelson et al. (1989b) harvested alfalfa as large round bales at 643, 734, or 847 g kg^{-1} DM with or without an inoculant. Inoculation aided in the preservation of alfalfa harvested at 734 g kg^{-1} DM, but provided no benefit with the 643 or 847 g kg^{-1} DM bales. In a second trial, alfalfa was harvested as small rectangular bales at 566, 735, or 836 g kg^{-1} DM (Nelson et al., 1989a). An inoculant was applied to half of the bales harvested at 566 and 735 g kg^{-1} DM. Inoculation had no effect on DM intake or digestibility when the hay treatments were fed to lambs.

Preservation, as measured by temperature and crude protein recovery, was improved by inoculating the 566 g kg^{-1} DM hay, but had the opposite effect on the 735 g kg^{-1} DM hay. These data are interpreted to suggest that many variables can affect the efficacy of a bacterial inoculant for hay preservation. Additional research is needed to determine if bacterial inoculants can be a reliable method of improving hay preservation.

Use of drying agents can be a method of improving nutritional value of hays. Reduced drying times mean less opportunity for rain damage and the nutrient loss which accompanies it. Animal performance responses to hays treated with drying agents have been inconsistent. Ollerman et al. (1989) reported that dairy cows fed alfalfa hay treated with a commercial potassium carbonate-sodium carbonate-based drying agent increased DM intake and milk yield compared to cows fed control hay. Digestibilities of DM, NDF, and ADF also increased when treated hay was fed ad libitum to dairy cows (Hong et al., 1988). In contrast, Ziemer et al. (1991) reported that feeding alfalfa hay treated with a commercial potassium carbonate-sodium carbonate based drying agent did not affect DM intake, milk yield, or nutrient digestibility in midlactation cows compared to untreated hay. In all three trials, drying agents had little effect on nutrient composition of the hay. Additional research is needed to determine the conditions under which drying agents may be cost effective.

Silages

The use of additives and preservatives to enhance the fermentation and nutritive value of silages has become a common practice within the last ten years. Based on their desired effect, Muck and Bolsen (1991) categorized these products as 1) stimulants, 2) inhibitors, or 3) nutrient sources. Stimulants increase lactic acid production by introducing lactic acid bacteria or substrates which yield lactic acid. Inhibitors reduce bacterial fermentation by reducing pH or partially sterilizing the forage to slow the fermentation or retard aerobic deterioration. Nutrient sources are added at ensiling to improve the feeding value of the resulting silage.

Bacterial inoculants are the most common stimulant being added to silages. These lactic acid bacteria are usually homofermentative in that they produce only lactic acid from the fermentation of six carbon sugars. Increased lactic acid production usually results in a more rapid decline in pH which decreases proteolysis and subsequent generation of ammonia.

Muck and Bolsen (1991) summarized over 40 different trials comparing the performance of beef and dairy cattle fed control and inoculated silages. In 40 trials where DM intake was measured, approximately one-third of the trials showed increased (averaging 12%) intakes, as a result of inoculating the forage at ensiling. Significant improvements (averaging 5%) in milk production occurred in 8 of 20 trials as a result of feeding inoculated silage. In 9 of 35 comparisons, daily gains were improved an average of 6% for cattle fed the treated silage. Feed efficiency, expressed as milk or gain per tonne of crop ensiled, also was improved an average of 7% in 9 of 24 studies. Although the responses observed from inoculation are not consistent, significant improvements in animal performance occur. Most of the improvements were associated with improved DM digestibility (Muck and Bolsen, 1991). In general, responses were greater with alfalfa or grass silages than with corn or sorghum silages. Consequently, type of forage being ensiled, class of ruminant being fed, costs of inoculation, and required response to break even are all factors to consider in determining the cost effectiveness of inoculating silages.

Enzyme addition, alone or in combination with inoculants, is a relatively new method of improving the nutritive value of ensiled forages. The focus of most recent research has been to partially degrade the fiber components during ensiling with the goal of increasing both DM intake and rate of fiber utilization. A secondary purpose has been to degrade fiber to individual sugars, which could then be used as substrate for fermentation. The enzymes being added have included cellulases, xylanases, cellobiases, amylases, pectinases, proteases, and glucose oxidase.

Responses to enzyme additions have been inconsistent. Fiber concentrations were reduced in only 10 of 19 studies reviewed by Muck and Bolsen (1991). The most consistent factor determining whether enzyme additions reduced fiber concentration was forage species. Grasses were twice as likely to respond than were legumes (primarily alfalfa). This should not be surprising in that many of the original products were developed in Europe for use on grass silages. Secondly, the fiber structure in grasses may be more susceptible to enzymatic attack than that in legumes.

In general, extent of DM or fiber digestibility has not been improved by enzyme additions. However, rate of DM or fiber digestion was improved in 5 of 6 trials where it was measured (Muck and Bolsen, 1991). These data are interpreted to suggest that enzyme additions do not increase the potentially digestible fiber, but may increase the susceptibility of the potentially digestible fiber to bacterial attack in the rumen.

Performance responses to enzyme additions have been strongly related to type of forage treated. For example, Stokes (1992) reported that the enzyme treatment of a timothy-alfalfa mixture fed to lactating Holsteins improved DM intake 9.6% and milk production 2.4%. In contrast, Jaster and Moore (1988) found no change in DM intake or milk production as a result of feeding enzyme-treated alfalfa haylage to lactating dairy cows. These trials are comparable because the type of enzymes added, diet formulations, stage of production, and general management of the cattle were similar. In general, 1 out of 5 trials has shown a significant animal response to enzyme treatment.

In many research trials, bacterial inoculants have been used in combination with enzymes. The theory has been that enzymatic degradation of fiber would provide more sugars to the desirable bacteria, resulting in an enhanced fermentation, and thus improving the nutritive value and preservation characteristics simultaneously. In practice, it appears that matching the enzymes and inoculants so that they complement each other is not an easy task. For example, Stokes (1992) reported that a cellulase-based enzyme mixture improved the fermentation quality and nutritional value of a grass-legume mixture. Inoculation with lactic acid bacteria also improved the efficiency of silage fermentation. However, when the enzymes and inoculant were combined, their actions were not additive or synergistic but tended to be antagonistic. Clearly, these data show that complementarity should not be assumed when combining effective inoculants and enzyme mixtures.

Acids, such as propionic, acetic, lactic, benzoic, and hydrochloric, are a group of silage additives which are commonly used in certain parts of the world, but not in North America. Acids are most beneficial when the concentration of sugar in the forage is below that required for an optimal fermentation. A rapid decrease in pH is needed to prevent the growth of potentially pathogenic microorganisms such as *Clostridia* and *Listeria*. Crops that benefit most from acid addition are usually high in moisture, inadequate in sugar, and have a high buffering capacity. Legumes such as alfalfa, clover, or trefoil are good examples of forages which could benefit from acid additions.

One of the consistent benefits from acid addition is increased protein preservation. For example, Glenn and Waldo (1986) reported that the percentage of the total N as soluble NPN decreased from 657 to 414 g kg^{-1} as a result of treating alfalfa silage with formic acid. The immediate decline in pH, following acid addition, reduced proteolytic activity during the first hours of fermentation when it is normally most rapid. Nagel and Broderick (1992) reported that treating alfalfa (350 g kg^{-1} DM) with 2.8 g of formic acid 100 g^{-1} of DM at ensiling increased the nonammonia N flow to the small intestine of lactating Holstein cows from 185 to 306 g d^{-1}. Milk production was increased from 29.2 to 32.6 kg d^{-1} by the formic acid treatment. Increased DM intakes and N retention have been reported in sheep and dairy heifers fed formic acid-treated alfalfa silage. Similar results have been obtained in Northern Europe where formic acid is commonly added to direct-cut silage.

One disadvantage of formic acid treatment is that it tends to reduce aerobic stability of the silage. This occurs because acid treatment reduces the acid concentration in the final silage and leaves more sugars, which yeast and molds prefer compared to fermentation acids. Other disadvantages include the corrosive nature of formic acid, the requirement for specialized equipment, and cost.

Propionic, and to a lesser extent acetic acid, promote aerobic stability but do not dramatically reduce pH at the time of their addition. Propionic acid is especially inhibitory to yeast and molds and thus will decrease secondary fermentations and improve bunk life. The other acids have been tested primarily in a laboratory setting, and thus insufficient data are available to determine if they have practical value under field conditions.

Ammonia and urea have been added at ensiling to increase the crude protein concentration, decrease proteolysis, and improve bunk life by reducing yeast and mold levels. Urease in the ensiling mass converts urea to ammonia which serves as a N source for bacterial growth. Ammonia, and to a lesser extent urea, increase the pH of the forage initially and usually result in greater concentrations of fermentation endproducts in NPN-treated silages. The NPN treatment of silage usually shifts the fermentation towards more acetic and less lactic acid (Kung et al., 1989).

In a summary of 13 trials published between 1985 and 1990, Muck and Bolsen (1991) reported that DM or OM digestibility was improved in 10 trials by ammonia addition. In general, ammonia was more effective at improving digestibility than urea when added on an equal N basis. The ammonia additions varied from approximately 10 to 30 g kg^{-1} of the forage DM.

One disadvantage of adding NPN at ensiling is that DM recoveries often are reduced. If the silage is wetter than 700 g kg^{-1} moisture, NPN additions favor *Clostridial* growth which reduce DM recovery. For example, Glenn (1990) reported that adding 8.2 g kg^{-1} ammonia on a DM basis to direct-cut alfalfa silage (202 g kg^{-1} DM) decreased DM recovery from 79 to 71 g kg^{-1}. In drier silages, the fermentation shift from lactic acid to acetic acid as a result of NPN addition also will increase DM loss.

Urea is frequently added to silage crops that are relatively high in energy but deficient in protein. Corn silage is the most common example. Ely (1978) summarized the early research and found small (2 to 3%), but consistent, improvements in gain, feed efficiency, and milk production by adding urea at ensiling compared to adding an equal amount of urea at feeding. Urea has the advantage of being easier to handle than ammonia, but is usually more expensive per unit of N.

In summary, NPN additions at ensiling are the cheapest means of increasing N content, improving DM digestibility, and enhancing aerobic stability of the resulting silage. However, DM recoveries are reduced and performance responses are small. Consequently, the economic merit of NPN additions is determined primarily by the savings in protein supplement costs.

CONCLUSION

At the present time, a combination of some physical-mechanical treatment (such as grinding) with a chemical treatment (such as ammoniation) is probably most cost-effective in upgrading the nutritive value of most major fibrous and by-product feedstuffs. In general, different types of effects of physical vs. chemical treatment on digestion of cell wall carbohydrates occur. Physical treatments generally provide a greater surface area for attack by enzymes, whereas chemical treatments usually will increase rate of fiber digestion. With regard to chemical treatment, the "ideal" substrate is one which is at an advanced stage of maturity and highly lignified. Monocots respond to both hydrolytic and oxidative treatments while dicots respond better to oxidative treatment than hydrolytic treatment. Hydrolytic agents disrupt hemicellulose:lignin interactions and have a swelling effect. Oxidants degrade part of the polyphenolic matrix and randomly cleave polysaccharides. Mixtures of hydrolytic and oxidative agents have an additive effect on digestion phenomena important to ruminant livestock. Fungi have great potential as an environmentally friendly method of increasing structural carbohydrate digestibility. Use of molecular biology techniques may be an opportunity to develop more rapid and efficient strains of fungi. Molecular biology also may allow the large scale production of enzymes targeted to attack specific fiber moieties. Considerable research probably will be required before either fungi or enzymes are used on a practical basis. Additives and preservatives can enhance the nutritional value of forages by allowing more of the nutrients harvested to be available for absorption and subsequent utilization by the animal.

REFERENCES

Adebowale, E. A., E. R. Orskov, and P. M. Hotten. 1989. Rumen degradation of straw. 8. Effect of alkaline hydrogen peroxide on degradation of straw using either sodium hydroxide or gaseous ammonia as source of alkali. Anim. Prod. 48:553-559.

Agricultural Statistics. 1988. U. S. Department of Agriculture, Washington, DC.

Alhadhrami, G., J. T. Huber, G. E. Higginbotham, and J. M. Harper. 1989. Nutritive value of high moisture alfalfa hay preserved with urea. J. Dairy Sci. 72:972-979.

Allen, M. S., and D. R. Mertens. 1988. Evaluating constraints on fiber digestion by rumen microbes. J. Nutr. 118:261-270.

Alwash, A. H., and P. C. Thomas. 1974. Effect of the size of hay particles on digestion in the sheep. J. Sci. Food Agric. 25:139-147.

Ander, P. 1990. The use of white-rot fungi and their enzymes for biopulping and biobleaching. p. 287-295. In M. P. Coughlan and M. T. A. Collaco (ed.) Advances in biological treatment of lignocellulosic materials. Elsevier Applied Science, London.

Atwal, A. S., and L. C. Heslop. 1987. Effectiveness of propionic acid for preserving alfalfa hay in large round bales. Can. J. Anim. Sci. 67:75-82.

Atwal, A. S., L. C. Heslop, and K. Lievers. 1986. Effectiveness of anhydrous ammonia as a preservative for high-moisture alfalfa hay in large round bales. Can. J. Anim. Sci. 66:743-753.

Atwell, D. G., N. R. Merchen, E. H. Jaster, G. C. Fahey, Jr., and L. L. Berger. 1991a. Site and extent of nutrient digestion by sheep fed alkaline hydrogen peroxide-treated wheat straw-alfalfa hay combinations at restricted intakes. J. Anim. Sci. 69:1697-1706.

Atwell, D. G., N. R. Merchen, E. H. Jaster, G. C. Fahey, Jr., L. L. Berger, E. C. Titgemeyer, and L. D. Bourquin. 1991b. Intake, digestibility, and in situ digestion kinetics of treated wheat straw and alfalfa mixtures fed to Holstein heifers. J. Dairy Sci. 74:3524-3534.

Autrey, K. M., T. A. McCaskey, and J. A. Little. 1975. Cellulose digestibility of fibrous materials treated with *Trichoderma viride* cellulase. J. Dairy Sci. 58:67-71.

Bakshi, M. P. S., V. K. Gupta, and P. N. Langar. 1985. Acceptability and nutritive evaluation of *Pleurotus* harvested spent wheat straw in buffaloes. Agric. Wastes 13:51-57.

Baron, V. S., and G. G. Greer. 1988. Comparison of six commercial hay preservatives under simulated storage conditions. Can. J. Anim. Sci. 68:1195-1207.

Bass, J. M., J. J. Parkins, and G. Fishwick. 1982. The effect of calcium hydroxide on the digestibility of chopped oat straw supplemented with a solution containing urea, calcium, phosphorus, sodium, trace elements and vitamins. Anim. Feed Sci. Technol. 7:93-100.

Beardsley, D. W. 1964. Symposium on forage utilization: Nutritive value of forage as affected by physical form. Part II. Beef cattle and sheep studies. J. Anim. Sci. 23:239-245.

Beever, D. E., D. F. Osbourn, S. B. Cammell, and R. A. Terry. 1981. The effect of grinding and pelleting on the digestion of Italian ryegrass and timothy by sheep. Br. J. Nutr. 46:357-370.

Beg, S., S. I. Zafar, and F. H. Shah. 1986. Rice husk biodegradation by *Pleurotus ostreatus* to produce a ruminant feed. Agric. Wastes 17:15-21.

Bender, R. 1979. Method of treating lignocellulose materials to produce ruminant feed. U. S. Patent 4,136,207.

Bender, F., D. P. Heaney, and A. Bowden. 1970. Potential of steamed wood as a feed for ruminants. Forest Prod. J. 20:36-41.

Ben-Ghedalia, D. 1982. Degradation of cell-wall monosaccharides of chemically treated grape branches by rumen microorganisms. Eur. J. Appl. Microbiol. Biotechnol. 16:142-145.

Ben-Ghedalia, D., and A. Marcipar. 1979. The effect of chemical pretreatments and subsequent enzymatic treatments on the organic matter digestibility *in vitro* of wheat straw. Nutr. Rep. Int. 19:499-505.

Ben-Ghedalia, D., and J. Miron. 1981a. The effect of combined chemical and enzyme treatments on the saccharification and *in vitro* digestion rate of wheat straw. Biotechnol. Bioeng. 23:823-831.

Ben-Ghedalia, D., and J. J. Miron. 1981b. Effects of sodium hydroxide, ozone, and sulphur dioxide on the composition and *in vitro* digestibility of wheat straw. J. Sci. Food Agric. 32:224-228.

Ben-Ghedalia, D., and J. Miron. 1984. The digestibility of wheat straw treated with sulphur dioxide. J. Agric. Sci. (Camb.) 102:517-520.

Ben-Ghedalia, D., and J. Miron. 1987. Intensive growth of lambs on sulphur dioxide-treated straw diets. Nutr. Rep. Int. 35:1129-1135.

Ben-Ghedalia, D., J. Miron, Y. Est, and E. Yosef. 1988. SO_2 treatment for converting straw into a concentrate-like feed: A growth study with lambs. Anim. Feed Sci. Technol. 19:219-229.

Ben-Ghedalia, D., and G. Shefet. 1983. Chemical treatments for increasing the digestibility of cotton straw. II. Effect of ozone and sodium hydroxide treatments on the digestibility of cell wall monosaccharides. J. Agric. Sci. (Camb.) 100:401-406.

Ben-Ghedalia, D., G. Shefet, and Y. Dror. 1983. Chemical treatments for increasing the digestibility of cotton straw. 1. Effect of ozone and sodium hydroxide treatments on rumen metabolism and on the digestibility of cell walls and organic matter. J. Agric. Sci. (Camb.) 100:393-400.

Ben-Ghedalia, D., G. Shefet, J. Miron, and Y. Dror. 1982. Effect of ozone and sodium hydroxide treatments on some chemical characteristics of cotton straw. J. Sci. Food Agric. 33:1213-1218.

Bergen, W. G., D. B. Bates, D. E. Johnson, J. C. Waller, and J. R. Black. 1982. Ruminal microbial protein synthesis and efficiency. p. 99-112. *In* F. N. Owens (ed.) Protein requirements for cattle: Symposium. MP-109. Oklahoma State University Press, Stillwater.

Berger, L., T. Klopfenstein, and R. Britton. 1979. Effect of sodium hydroxide on efficiency of rumen digestion. J. Anim. Sci. 49:1317-1323.

Blanchette, R. A. 1984. Screening wood decayed by white rot fungi for preferential lignin degradation. Appl. Environ. Microbiol. 48:647-653.

Blanchette, R. A., L. Otjen, M. J. Effland, and W. E. Eslyn. 1985. Changes in structural and chemical components of wood delignified by fungi. Wood Sci. Technol. 19:35-46.

Blaxter, K. L., N. McC. Graham, and F. W. Wainman. 1956. Some observations on the digestibility of food by sheep, and on related problems. Br. J. Nutr. 10:69-91.

Bryant, E. C., T. Mills, J. S. Pitts, M. Carrigan, and E. P. Wiggers. 1984. Ozone treated mesquite for supplementing steers in west Texas. J. Range Manage. 37:420-422.

Calzada, J. F., L. F. Franco, M. C. de Arriola, C. Rolz, and M. A. Ortiz. 1987. Acceptability, body weight changes and digestibility of spent wheat straw after harvesting of *Pleurotus sajor-caju*. Biol. Wastes 22:303-309.

Cameron, M. G., G. C. Fahey, Jr., J. H. Clark, N. R. Merchen, and L. L. Berger. 1990. Effects of feeding alkaline hydrogen peroxide-treated wheat straw-based diets on digestion and production by dairy cows. J. Dairy Sci. 73:3544-3554.

Cameron, M. G., G. C. Fahey, Jr., J. H. Clark, N. R. Merchen, and L. L. Berger. 1991a. Effects of feeding alkaline hydrogen peroxide-treated wheat straw-based diets on intake, digestion, ruminal fermentation, and production responses by mid-lactation dairy cows. J. Anim. Sci. 69:1775-1787.

Cameron, M. G., M. R. Cameron, G. C. Fahey, Jr., J. H. Clark, L. L. Berger, and N. R. Merchen. 1991b. Effects of treating oat hulls with alkaline hydrogen peroxide on intake and digestion by mid-lactation dairy cows. J. Dairy Sci. 74:177-189.

Cameron, M. G., J. D. Cremin, Jr., G. C. Fahey, Jr., J. H. Clark, L. L. Berger, and N. R. Merchen. 1991c. Chemically treated oat hulls in diets for dairy heifers and wethers: Effects on intake and digestion. J. Dairy Sci. 74:190-201.

Campling, R. C., and M. Freer. 1966. Factors affecting the voluntary intake of food by cows. 8. Experiments with ground, pelleted roughages. Br. J. Nutr. 20:229-244.

Chandra, S., and M. G. Jackson. 1971. A study of various chemical treatments to remove lignin from coarse roughages and increase their digestibility. J. Agric. Sci. (Camb.) 77:11-17.

Chang, H.-M., and G. G. Allan. 1971. Oxidation. p. 433-485. *In* K. V. Sarkanen and C. H. Ludwig (ed.) Lignins: Occurrence, formation, structure and reactions. Wiley-Interscience, New York.

Chestnut, A. B., L. L. Berger, and G. C. Fahey, Jr. 1988. Effects of conservation methods and anhydrous ammonia or urea treatments on composition and digestion of tall fescue. J. Anim. Sci. 66:2044-2056.

Chesson, A. 1981. Effects of sodium hydroxide on cereal straws in relation to the enhanced degradation of structural polysaccharides by rumen microorganisms. J. Sci. Food Agric. 32:745-758.

Chesson, A., A. H. Gordon, and J. A. Lomax. 1983. Substituent groups linked by alkali-labile bonds to arabinose and xylose residues of legume, grass and cereal straw cell walls and their fate during digestion by rumen microorganisms. J. Sci. Food Agric. 34:1330-1340.

Coxworth, E., J. Kernan, J. Knipfel, O. Thorlacius, and L. Crowle. 1981. Review: Crop residues and forages in western Canada; potential for feed use either with or without chemical or physical processing. Agric. Environm. 6:245-256.

Dale, B. E. 1983. Biomass refining: Protein and ethanol from alfalfa. Ind. Eng. Chem. Prod. Res. Dev. 22:466-472.

Daniels, L. B., and R. B. Hashim. 1977. Evaluation of fungal cellulases in rice hull based diets for ruminants. J. Dairy Sci. 60:1563-1567.

Dehority, B. A., and R. R. Johnson. 1961. Effect of particle size upon the *in vitro* cellulose digestibility of forages by rumen bacteria. J. Dairy Sci. 44:2242-2249.

Ely, L. O. 1978. The use of feedstuffs in silage production. *In* Fermentation of Silage - A Review. M.E. McCullough (ed.). National Feed Ingredients Association, Washington, DC.

Ford, C. W., R. Elliott, and P. J. Maynard. 1987. The effect of chlorite delignification on digestibility of some grass forages and on intake and rumen microbial activity in sheep fed barley straw. J. Agric. Sci. (Camb.) 108:129-136.

Garrett, W. N., H. G. Walker, Jr., G. O. Kohler, M. R. Hart, and R. P. Graham. 1981. Steam treatment of crop residues for increased ruminant digestibility. II. Lamb feeding studies. J. Anim. Sci. 51:409-413.

Ghate, S. R., and W. K. Bilanski. 1979. Treating high-moisture alfalfa with urea. Trans. ASAE 23:504-508.

Glenn, B. P. 1990. Effects of dry matter concentration and ammonia treatment of alfalfa silage on digestion and metabolism by heifers. J. Dairy Sci. 73:1081-1090.

Glenn, B. P., and D.R. Waldo. 1986. Alfalfa and orchardgrass silages treated with formaldehyde and formic acid or anhydrous ammonia for heifers. J. Dairy Sci. 69:1317-1328.

Gould, J. M. 1984. Alkaline peroxide delignification of agricultural residues to enhance enzymatic saccharification. Biotechnol. Bioeng. 26:46-52.

Gould, J. M., and S. N. Freer. 1984. High-efficiency ethanol production from lignocellulosic residues pretreated with alkaline H_2O_2. Biotechnol. Bioeng. 26:628-631.

Graham, H., and P. Aman. 1984. A comparison between degradation *in vitro* and *in sacco* of untreated and ammonia-treated barley straw. Anim. Feed Sci. Technol. 10:199-211.

Greenhalgh, J. F. D. 1984. Preface. p. i-ii. *In* F. Sundstøl and E. Owen (ed.) Straw and other fibrous by-products as feed. Elsevier Science Publishers, Amsterdam, The Netherlands.

Greenhalgh, J. F. D., and G. W. Reid. 1973. Long- and short-term effects on intake of pelleting a roughage for sheep. Anim. Prod. 19:77-86.

Grous, W. R., A. O. Converse, and H. E. Grethlein. 1986. Effect of steam explosion pretreatment on pore size and enzymatic hydrolysis of poplar. Enzyme Microbiol. Technol. 8:274-280.

Hagevoort, G. R., F. M. Byers, M. T. Holtzapple, J. H. Jun, L. W. Greene, and G. E. Carstens. 1990. Enhancing the nutritive value of forages with an Ammonia Fiber Explosion (AFEX) technique. J. Anim. Sci. (Suppl. 1):584 (Abstr.)

Heaney, D. P., W. J. Pigden, D. J. Minson, and G. I. Pritchard. 1963. Effect of pelleting on energy intake of sheep from forages cut at three stages of maturity. J. Anim. Sci. 22:752-757.

Henderson, A. R., and P. McDonald. 1977. The effect of cellulase preparations on the chemical changes during the ensilage of grass in laboratory silos. J. Sci. Food Agric. 28:486-490.

Holzbaur, E. L. F., A. Andrawis, and M. Tien. 1991. Molecular biology of lignin peroxidases from *Phanerochaete chrysosporium*. p. 197-223. *In* S. A. Leong and R. M. Berka (ed.) Molecular industrial mycology. Systems and applications for filamentous fungi. Marcel Dekker, Inc., New York.

Hong, B. J., G. A. Broderick, and R. P. Walgenbach. 1988. Effect of chemical conditioning of alfalfa on drying rate and nutrient digestion in ruminants. J. Dairy Sci. 71:1851-1859.

Horton, G. M. J., F. M. Pate, and W. D. Pitman. 1991. The effects of steam-pressure treatment, pelleting and ammoniation on the feeding value of sugarcane bagasse for cattle. Can. J. Anim. Sci. 71:79-86.

Houseman, R. A., and J. Connell. 1976. The utilization of the products of green-chop fractionation by pigs and ruminants. Proc. Nutr. Soc. 35:213-220.

Jackson, M. G. 1977. Review article: The alkali treatment of straws. Anim. Feed Sci. Technol. 2:105-130.

Jaster, E. H., and K. J. Moore. 1988. Fermentation characteristics and feeding value of enzyme-treated alfalfa haylage. J. Dairy Sci. 71:705-711.

Johnson, T. R., J. W. Thomas, K. R. Aherns, Z. R. Helsel, and C. M. Hansen. 1981. Ammonia treatment of wet baled hay. J. Anim. Sci. 53(Suppl. 1):275 (Abstr.)

Johnsrud, S. C. 1988. Selection and screening of white-rot fungi for delignification and upgrading of lignocellulosic materials. p. 50-55. In F. Zadrazil and P. Reiniger (ed.) Treatment of lignocellulosics with white rot fungi. Elsevier Applied Science, London.

Jorgensen, N. A., G. P. Barrington, and P. R. Fritschel. 1973. Preservation of high moisture baled hay by chemical treatment. J. Dairy Sci. 56:677-678 (Abstr.)

Jorgensen, O. B., and D. Cowan. 1989. Use of enzymes in feed and in ensiling. p. 347-355. In M. P. Coughlan (ed.) Enzyme systems for lignocellulose degradation. Elsevier Applied Science, London.

Kerley, M. S., G. C. Fahey, Jr., L. L. Berger, J. M. Gould, and F. L. Baker. 1985. Alkaline hydrogen peroxide treatment unlocks energy in agricultural by-products. Science 230:820-822.

Kerley, M. S., G. C. Fahey, Jr., L. L. Berger, N. R. Merchen, and J. M. Gould. 1986. Effects of alkaline hydrogen peroxide treatment of wheat straw on site and extent of digestion in sheep. J. Anim. Sci. 63:868-878.

Kewalramani, N., D. N. Kamra, D. Lall, and N. N. Pathak. 1988. Bioconversion of sugarcane bagasse with white rot fungi. Biotechnol. Letters 10:369-372.

Khazaal, K. A., E. Owen, A. P. Dodson, P. Harvey, and J. Palmer. 1990. A preliminary study of the treatment of barley straw withligninase enzyme: Effect on in vitro digestibility and chemical composition. Biol. Wastes 33:53-62.

Klopfenstein, T. J. 1978. Chemical treatment of crop residues. J. Anim. Sci. 46:841-848.

Klopfenstein, T. J., and K. K. Bolsen. 1971. High temperature pressure treated crop residues. J. Anim. Sci. 33:290 (Abstr.)

Klopfenstein, T.J., R.P. Graham, H.G. Walker, Jr., and G.O. Kohler. 1974. Chemicals with pressured treated cobs. J. Anim. Sci. 39:243 (Abstr.)

Klopfenstein, T. J., V. E. Krause, M. J. Jones, and W. Woods. 1972. Chemical treatment of low quality roughages. J. Anim. Sci. 35:418-422.

Knapp, W. R., D. A. Holt, and V. L. Lechtenberg. 1975. Hay preservation and preservation and quality improvement by anhydrous ammonia treatment. Agron. J. 67:766-769.

Kohler, G. O., E. M. Bickoff, and W. M. Beeson. 1972. Processed products for feed and food industries. p. 659-676. In C. H. Hanson (ed.) Alfalfa science and technology. ASA, Madison, WI.

Kohler, G. O., H. G. Walker, Jr., and D. D. Kuzmicky. 1979. Processing and use of crop residues including alfalfa press cake. Fed. Proc. 38:1934-1938.

Kung, L., Jr., W. M. Craig, and L. D. Satter. 1989. Ammonia-treated alfalfa silage for lactating dairy cows. J. Dairy Sci. 72:2565-2572.

Lacey, J., and K. A. Lord. 1977. Methods of testing chemical additives to prevent molding of hay. Ann. Appl. Biol. 87:327-335.

Laredo, M. A., and D. J. Minson. 1973. The voluntary intake, digestibility, and retention time by sheep of leaf and stem fractions of five grasses. Aust. J. Agric. Res. 24:875-888.

Laredo, M. A., and D. J. Minson. 1975. The effect of pelleting on the voluntary intake and digestibility of leaf and stem fractions of three grasses. Br. J. Nutr. 33:159-170.

Latham, M. J., D. G. Hobbs, and P. J. Harris. 1979. Adhesion of rumen bacteria to alkali-treated plant stems. Ann. Rech. Vet. 10:244-245.

Leask, W. C., and T. B. Daynard. 1973. Dry matter yield, *in vitro* digestibility, percent protein, and moisture of corn stover following grain maturity. Can. J. Plant Sci. 53:515-522.

Males, J. R. 1987. Optimizing the utilization of cereal crop residues for beef cattle. J. Anim. Sci. 65:1124-1130.

Mason, V. C., R. D. Hartley, A. S. Keene, and J. M. Cobby. 1988. The effect of ammoniation on the nutritive value of wheat, barley and oat straws. I. Changes in chemical composition in relation to digestibility *in vitro* and cell wall degradability. Anim. Feed Sci. Technol. 19:159-171.

Meyer, J. H., R. Kromann, and W. N. Garrett. 1965. Digestion (influence of roughage preparation). p. 262-271. *In* R. W. Dougherty, R. S. Allen, W. Burroughs, N. L. Jacobson, and A. D. McGilliard (ed.) Physiology of digestion in the ruminant. Butterworths, Washington, DC.

Meyer, J. H., W. C. Weir, J. B. Dobie, and J. L. Hull. 1959. Influence of the method of preparation on the feeding value of alfalfa hay. J. Anim. Sci. 18:976-982.

Minson, D. J. 1963. The effect of pelleting and wafering on the feeding value of roughage - A review. J. Br. Grassl. Soc. 18:39-44.

Miron, J., and D. Ben-Ghedalia. 1982. Effect of hydrolyzing and oxidizing agents on the composition and degradation of wheat straw monosaccharides. European J. Appl. Microbiol. Biotechnol. 15:83-87.

Moore, L. A. 1964. Symposium on forage utilization: Nutritive value of forage as affected by physical form. Part I. General principles involved with ruminants and effect of feeding pelleted or wafered forage to dairy cattle. J. Anim. Sci. 23:230-238.

Morris, P. J., and D. N. Mowat. 1980. Nutritive value of ground and/or ammoniated corn stover. Can. J. Anim. Sci. 60:327-336.

Morrison, I. M. 1988. Influence of chemical and biological pretreatments on the degradation of lignocellulosic material by biological systems. J. Sci. Food Agric. 42:295-304.

Morrison, I. M. 1991. Changes in the biodegradability of ryegrass and legume fibres by chemical and biological pretreatments. J. Sci. Food Agric. 54:521-533.

Mowat, D. N., R. S. Fulkerson, W. E. Tossell, and J. E. Winch. 1965a. The *in vitro* digestibility and protein content of leaf and stem portions of forages. Can. J. Plant Sci. 45:321-331.

Mowat, D. N., R. S. Fulkerson, W. E. Tossell, and J. E. Winch. 1965b. The *in vitro* dry matter digestibility of several species and varieties and their plant parts with advancing stages of maturity. p. 801-806. *In* Proc. IX International Grassland Congress, São Paulo, Brazil.

Mowat, D. N., and B. Wilton. 1984. Whole crop harvesting, separation and utilization. p. 293-304. *In* F. Sundstøl and E. Owen (ed.) Straw and other fibrous by-products as feed. Elsevier Science Publishers, Amsterdam, The Netherlands.

Muck, R. E., and K. K. Bolsen. 1991. Silage preservation and silage additive products. p. 105-126. *In* K.K. Bolsen (ed.) Hay and silage management. National Feed Ingredients Association. Washington, DC.

Nagel, S. A., and G. A. Broderick. 1992. Effect of formic acid or formaldehyde treatment of alfalfa silage on nutrient utilization by dairy cows. J. Dairy Sci. 75:140-154.

Nakashima, Y., E. R. Orskov, P. M. Hotten, K. Ambo, and Y. Takase. 1988. Rumen degradation of straw. 6. Effect of polysaccharidase enzymes on degradation characteristics of ensiled rice straw. Anim. Prod. 47:421-427.

Nakashima, Y., and E. R. Orskov. 1989. Rumen degradation of straw. 7. Effects of chemical pre-treatment and addition of propionic acid on degradation characteristics of botanical fractions of barley straw treated with a cellulase preparation. Anim. Prod. 48:543-551.

Nelson, M. L., D. M. Headley, and J. A. Loesche. 1989a. Control of fermentation in high-moisture baled alfalfa by inoculation with lactic acid-producing bacteria. II. Small rectangular bales. J. Anim. Sci. 67:1586-1592.

Nelson, M. L., T. J. Klopfenstein, and R. A. Britton. 1989b. Control of fermentation in high-moisture baled alfalfa by inoculation with lactic acid-producing bacteria. I. Large round bales. J. Anim. Sci. 67:1577-1585.

Nevell, T. P. 1985. Oxidation of cellulose. p. 243-265. *In* T. P. Nevell and S. H. Zeronian (ed.) Cellulose chemistry and its applications. Ellis Horwood Ltd., Chichester.

Nicholson, J. W. G. 1981. Nutrition and feeding aspects of the utilization of processed lignocellulosic waste materials by animals. Agric. Environ. 6:205-228.

NRC. 1983. Underutilized resources as animal feedstuffs. National Academy Press, Washington, DC.

Oellermann, S. O., M. J. Arambel, and J. L. Walters. 1989. Effect of chemical drying agents on alfalfa hay and milk production response when fed to dairy cows in early lactation. J. Dairy Sci. 72:501-504.

Oji, U. I., and D. N. Mowat. 1978. Nutritive value of steam-treated corn stover. Can. J. Anim. Sci. 58:177-181.

Oji, U. I., and D. N. Mowat. 1979. Nutritive value of thermoammoniated and steam-treated maize stover. I. Intake, digestibility and nitrogen retention. Anim. Feed Sci. Technol. 4:177-186.

Osbourn, D. F., D. E. Beever, and D. J. Thomson. 1976. The influence of physical processing on the intake, digestion and utilization of dried herbage. Proc. Nutr. Soc. 35:191-200.

Osuji, P. O., J. G. Gordon, and A. J. F. Webster. 1975. Energy exchanges associated with eating and rumination in sheep given grass diets of different physical forms. Br. J. Nutr. 34:59-71.

Owen, E., T. Klopfenstein, and N. A. Urio. 1984. Treatment with other chemicals. p. 248-275. In F. Sundstøl and E. Owen (ed.) Straw and other fibrous by-products as feed. Elsevier Science Publishers, Amsterdam, The Netherlands.

Reade, A. E., and R. E. McQueen. 1983. Investigation of white-rot fungi for the conversion of poplar into a potential feedstuff for ruminants. Can. J. Microbiol. 29:457-463.

Reid, I. D. 1989. Solid-state fermentations for biological delignification. Enzyme Microbiol. Technol. 11:786-803.

Rexen, F. P. 1978. Straw utilization in Denmark. p. 55-58. In Report on Straw Utilization Conference. Agricultural Development and Advisory Service. Ministry of Agriculture, Fisheries and Food. Oxford.

Rexen, F. P., and K. E. Bach Knudsen. 1984. Industrial-scale dry treatment with sodium hydroxide. p. 127-161. In F. Sundstøl and E. Owen (ed.) Straw and other fibrous by-products as feed. Elsevier Science Publishers, Amsterdam, The Netherlands.

Rexen, F., and K. V. Thomsen. 1976. The effect on digestibility of a new technique for alkali treatment of straw. Anim. Feed Sci. Technol. 1:73-83.

Rode, L. M., D. C. Weakley, and L. D. Satter. 1985. Effect of forage amount and particle size in diets of lactating dairy cows on site of digestion and microbial protein synthesis. Can. J. Anim. Sci. 65:101-111.

Rolz, C., R. de Leon, M. C. de Arriola, and S. de Cabrera. 1986. Biodelignification of lemon grass and citronella bagasse by white-rot fungi. Appl. Environ. Microbiol. 52:607-611.

Saenger, P. F., R. P. Lemenager, and K. S. Hendrix. 1983. Effects of anhydrous ammonia treatment of wheat straw upon *in vitro* digestion, performance and intake by beef cattle. J. Anim. Sci. 56:15-20.

Satter, L. D. 1983. Comparison of procedures for treating lignocellulose to improve its nutrient value. p. 213-228. *In* Nuclear techniques for assessing and improving ruminant feeds, International Atomic Energy Agency, Vienna.

Shaver, R. D., A. J. Nytes, L. D. Satter, and N. A. Jorgensen. 1986. Influence of amount of feed intake and forage physical form on digestion and passage of prebloom alfalfa hay in dairy cows. J. Dairy Sci. 69:1545-1559.

Stokes, M. R. 1992. Effects of an enzyme mixture, an inoculant, and their interaction on silage fermentation and dairy production. J. Dairy Sci. 75:764-773.

Streeter, C. L., and G. W. Horn. 1982. Effect of treatment of wheat straw with ammonia and peracetic acid on digestibility *in vitro* and cell wall composition. Anim. Feed Sci. Technol. 7:325-329.

Sundstøl, F., and E. M. Coxworth. 1984. Ammonia treatment. p. 196-247. *In* F. Sundstøl and E. Owen (ed.) Straw and other fibrous by-products as feed. Elsevier Science Publishers, Amsterdam, The Netherlands.

Sundstøl, F. 1988. Straw and other fibrous by-products. Livestock Prod. Sci. 19:137-158.

Thomson, D. J., and D. E. Beever. 1980. The effect of conservation and processing on the digestion of forages by ruminants. p. 291-308. *In* Y. Ruckebusch and P. Thivend (ed.) Digestive physiology and metabolism in ruminants. MTP Press Ltd., Lancaster, England.

Thorlacius, S. O., and J. A. Robertson. 1984. Effectiveness of anhydrous ammonia as a preservative for high-moisture hay. Can. J. Anim. Sci. 64:867-880.

Turner, N. D., C. M. McDonough, F. M. Byers, M. T. Holtzapple, B. E. Dale, J. H. Jun, and L. W. Greene. 1990. Disruption of forage structure with an ammonia fiber explosion process. Proc. Western Section ASAS 41:494-497.

Umunna, N. N., and T. Klopfenstein. 1972. Response of lambs fed pressure treated wheat straw. J. Anim. Sci. 35:1136 (Abstr.)

Umunna, N. N., T. Klopfenstein, and K. Bolsen. 1972. Response of lambs fed pressure treated corn cobs. J. Anim. Sci. 35:277-278 (Abstr.)

Vanbelle, M., and G. Bertin. 1989. Screening of fungal cellulolytic preparations for application in ensiling processes. p. 357-369. *In* M. P. Coughlan (ed.) Enzyme systems for lignocellulose degradation. Elsevier Applied Science, London.

Van der Meer, J. M., and R. Ketelaar. 1989. Evaluation of enzymatically changed feeds. p. 383-396. *In* M. P. Coughlan (ed.) Enzyme systems for lignocellulose degradation. Elsevier Applied Science, London.

Van Soest, P. J. 1965. Use of detergents in analysis of fibrous feeds. III. Study of effects of heating and drying on yield of fibre and lignin in forages. J. Assoc. Off. Agric. Chem. 48:785-790.

Van Soest, P. J., A. Mascarehnas-Ferreira, and R. D. Hartley. 1984. Chemical properties of fibre in relation to nutritive quality of ammonia-treated forages. Anim. Feed Sci. Technol. 10:155-164.

Walker, H. G. 1984. Physical treatment. p. 79-105. *In* F. Sundstøl and E. Owen (ed.) Straw and other fibrous by-products as feed. Elsevier Science Publishers, Amsterdam, The Netherlands.

Wanapat, J. M., F. Sundstøl, and T. H. Garmo. 1985. A comparison of alkali treatment methods to improve the nutritive value of straw. I. Digestibility and metabolizability. Anim. Feed Sci. Technol. 12:295-309.

Webster, A. J. F., P. O. Osuji, F. White, and J. F. Ingram. 1975. The influence of food intake on portal blood flow and heat production in the digestive tract of sheep. Br. J. Nutr. 34:125-139.

Weiss, W. P., H. R. Conrad, C. M. Martin, R. F. Cross, and W. L. Shockey. 1986. Etiology of ammoniated hay toxicosis. J. Anim. Sci. 63:525-532.

Weston, R. H. 1967. Factors limiting the intake of feed by sheep. II. Studies with wheaten hay. Aust. J. Agric. Res. 18:983-1002.

Weston, R. H., and J. P. Hogan. 1967. The digestion of chopped and ground roughages by sheep. I. The movement of digesta through the stomach. Aust. J. Agric. Res. 18:789-801.

Williams, P. E. V., and G. M. Innes. 1983. Factors affecting the hydrolysis of urea when applied to barley straw. Br. Soc. Anim. Prod., Abstract 36.

Willms, C. L., L. L. Berger, N. R. Merchen, and G. C. Fahey, Jr. 1991a. Utilization of alkaline hydrogen peroxide-treated wheat straw in cattle growing and finishing diets. J. Anim. Sci. 69:3917-3924.

Willms, C. L., L.L. Berger, N. R. Merchen, and G. C. Fahey, Jr. 1991b. Effects of supplemental protein source and level of urea on intestinal amino acid supply and feedlot performance of lambs fed diets based on alkaline hydrogen peroxide-treated wheat straw. J. Anim. Sci. 69:4925-4938.

Wilton, B. 1978. Whole crop cereals: Harvesting, drying and separation. Agric. Eng. 33:4-5.

Wilton, B., F. Amini, and I. Randjbar. 1980. Whole crop cereals: A low-cost approach. Agric. Eng. 35:7-10.

Woodford, S. T., and M. R. Murphy. 1988. Effect of forage physical form on chewing activity, dry matter intake, and rumen function of dairy cows in early lactation. J. Dairy Sci. 71:674-686.

Yadav, J. S. 1987. Influence of nutritional supplementation on solid-substrate fermentation of wheat straw with an alkaliphilic white-rot fungus (*Coprinus* sp.). Appl. Microbiol. Biotechnol. 26:474-478.

Yu, Y., J. W. Thomas, and R. S. Emery. 1975. Estimated nutritive value of treated forages for ruminants. J. Anim. Sci. 41:1742-1751.

Zadrazil, F. 1980. Conversion of different plant waste into feed by *Basidiomycetes*. Eur. J. Appl. Microbiol. 9:243-248.

Ziemer, C. J., A. J. Heinrichs, C.J. Canale, and G. A. Varga. 1991. Chemical drying agents for alfalfa hay: Effect on nutrient digestibility and lactational performance. J. Dairy Sci. 74:2674-2680.

CHAPTER 24

FORAGE QUALITY INDICES: DEVELOPMENT AND APPLICATION

John E. Moore

INTRODUCTION

Crampton (1957) demonstrated the importance of voluntary intake as a component of forage quality indices; he began that paper with these words: "The work thus far reported, aimed at establishing some index of the overall feeding value of forage, has been disappointing." At the first National Conference on Forage Quality Evaluation and Utilization, Heaney (1970) discussed quantitative forage quality indices and emphasized that such indices must include both voluntary intake and digestibility. Today, the disappointment is that in spite of many advances in understanding the biological determinants of forage quality, there has been only limited use of forage quality indices in practical feeding systems.

Definition of Forage Quality

A major limitation to practical application of forage quality information is the lack of a uniform quantitative definition or expression of forage quality. Mott (1959) suggested that differences in forage quality are expressed best as differences in animal performance (e.g., daily gain or milk production) under the conditions that 1) animals used to compare forages have a potential for production and are uniform among treatments, 2) forages are available in quantities adequate for maximum intake, and 3) no supplemental energy and protein are provided. Most livestock producers understand that differences in forage quality should mean differences in animal performance.

Animal performance is a sound theoretical definition of forage quality and can be useful for relative comparison among forages fed to, or grazed by, growing or lactating animals. To be useful in livestock feeding, however, forage quality information must be available before feeding. Furthermore, such information must be expressed in terms that can be used to predict animal performance when forage is fed alone, and to formulate supplements or mixed forage-concentrate diets to meet requirements for a targeted rate of animal production. In order to accomplish these objectives, a given forage must be assigned an absolute quality value (i.e., index number). The definition of forage quality in an absolute sense requires careful consideration of the many other factors that affect animal performance in forage-based livestock systems (Table 1).

Animal Science Dep., Univ. of Florida, Gainesville, FL 32611-0900

Table 1. Factors affecting daily production per animal in a forage-based ruminant production system (adapted from Moore, 1980).

Forage Factors	Non-Forage Factors
A. Characteristics (quality-related)	A. Animal Potential (appetite, efficiency)

1. Chemical
    Nutrients
        protein
        minerals
    Cell wall
        concentration
        composition
    Antiquality components
    Palatability factors
	1. Animal
	    Genotype
	    Age
	    Sex
	    Physiological state

2. Physical
    Solubility
    Degradability
	2. Previous treatment
	    Feeding rate
	    Diseases
	    Parasites

3. Structural
    Microanatomy
    Morphology
        leaf:stem ratio
        growth habit
    Canopy structure
        bulk density
        grass:legume ratio
        living:dead tissue ratio
	3. Climate
	    Temperature
	    Humidity
	    Solar radiation
	    Precipitation
	
	4. Implants

B. Quantity (forage per animal)	B. Supplemental Feeds (restricted intake)

1. In confinement
    Daily excess
    Supplement effects
	1. Improve forage utilization
	    Protein
	    Minerals
	    Vitamins
	    Additives

2. On pasture
    Forage yield
    Grazing pressure
    Supplement effects
	2. Provide digestible energy
	    Grains
	    Molasses
	    Byproducts
	
	3. Change forage intake
	    Increase
	    Decrease (substitution)

In forage research, quantities of forage frequently are inadequate for direct measurement of animal gain or milk production. In some cases, animal responses such as voluntary intake and nutrient digestibility are measured in short-term trials, often with sheep. In other cases, only small samples are available, and intake and digestibility must be predicted from laboratory analyses. With proper conversion formulas, measures and predictions of intake and digestibility may be used to provide adequate quantitative estimates of animal performance (i.e., forage quality). To improve accuracy, conversion formulas should include partial efficiency of digestible nutrient utilization as well as intake and digestibility (Raymond, 1969).

The term "forage nutritive value" often is confused with "forage quality." In his classic paper on forage quality indices mentioned above, Crampton (1957) chose "Intake and Nutritive Value" for the running head, and showed clearly that forage quality is the combination of voluntary intake and nutritive value. Reid and Klopfenstein (1983) noted that during the 1960's researchers became aware of the "need to consider intake, digestibility, (and) efficiency as inter-related but, possibly, independent components" of forage quality. At the first National Conference, Mott and Moore (1970) contrasted quality, intake, and nutritive value, and suggested that nutritive value is determined by 1) nutrient concentration, 2) nutrient digestibility, and 3) nature of digestion end-products. Nutritive value should refer to inherent characteristics of consumed forage which determine its digestible energy (DE) concentration and partial efficiency of DE utilization. Nutritive value information has been used for many years in livestock feeding, and when a forage is fed in restricted amounts, its nutritive value is of primary importance. A discussion of the nutritive value of a forage should not, however, include any reference to voluntary intake except the possible effect of intake on digestibility.

Index vs. Predictor of Forage Quality

An "index" of the quality of a given forage is a single number which represents the combination of its potential voluntary intake and nutritive value. Heaney (1970) stated that "To be effective, any index of feeding value must have certain inherent characteristics such as: it must be capable of being measured with some degree of precision, the measured value obtained from a small group of animals under controlled conditions must be applicable to a more general livestock population, and it must be related to the production achieved when livestock are fed the evaluated feedstuff." In addition, the ideal index should be independent of forage and animal types. The theoretical basis of a forage quality index should be either observed animal performance (e.g., daily gain or milk production), or voluntary intake and nutritive value. Such an index applies only when animals are given free-choice access to the forage and when forage is the sole source of energy and protein.

At the very least, a forage quality index must be useful for making relative comparisons among forages differing in genotype and season of growth, as well as maturity. At best, the forage quality index would be used in livestock feeding, and would be expressed in, or could be converted to, units of energy requirements and(or) animal performance. If the index is to be used in livestock feeding, it must 1) provide the quantitative information necessary to predict potential animal performance when the forage is fed alone, and 2) provide the basis for formulation of supplements or mixed diets to meet production targets.

A "predictor" of forage quality or nutritive value is a quality-related characteristic of forage (Table 1) which can be measured by traditional laboratory analyses or estimated by Near Infrared Reflectance Spectroscopy (NIRS). Such characteristics should be used only when it has been established that they are themselves, or are correlated highly with, actual determinants of forage quality, voluntary intake, or nutritive value. Predictors of forage quality or nutritive value must be used in forage agronomy research and extension forage testing because only small quantities of each forage are available. These quality-related characteristics are not really measures or indices of forage quality or nutritive value, but they may be used successfully to compare forages or predict forage quality and nutritive value. One or more predictors may be used in equations which convert laboratory data to estimates of voluntary intake, nutritive value, or an index of quality. Predicted intake and nutritive value could be used to calculate "predicted quality indices" if proper equations have been developed and validated.

The objectives of the remainder of this chapter are to discuss 1) the historical development of quantitative indices of forage quality, and 2) the practical application of forage quality indices, including their prediction from laboratory analyses.

DEVELOPMENT OF FORAGE QUALITY INDICES

Nutritive Value Index (1960)

Based on the observation of Crampton (1957) that voluntary dry matter (DM) intake was an essential component of an overall index of forage quality, Crampton et al. (1960) developed the Nutritive Value Index (NVI) calculated as follows:

$$NVI = \text{Relative Intake} * (\text{Gross Energy [GE] Digestibility, \%})/100 \quad [1]$$

where:

$$\text{Relative Intake} = \frac{\text{Observed DM Intake} * 100}{\text{Standard DM Intake}} \quad [2]$$

The DM intakes were expressed as grams per day per kg metabolic weight [MW = (live body weight in kg)$^{0.75}$]. The NVI equations may be rearranged as follows:

$$NVI = \frac{\text{Observed DM Intake} * \text{GE Digestibility, \%}}{\text{Standard DM Intake}} \quad [3]$$

The standard DM intake was a constant chosen to be 80 g/MW for sheep; this was the average intake by sheep of an "early-cut, chopped, dehydrated legume hay." The daily intake of this type of forage was greater generally than that of other dry forages in their studies. For cattle, the standard DM intake was 140 g/MW (Crampton et al., 1962) because it was assumed that both sheep and cattle would consume the standard forage at the rate of 3% of live body weight. Because the Relative Intake of the standard forage was 100 and its GE digestibility was 70%, its NVI was 70. The NVI of most other forages was less than 70. The NVI system included equations to predict NVI from 12-h in vitro cellulose digestion for both chopped and ground forages (Donefer et al., 1960, 1962).

Even though Crampton et al. (1960) used the words "nutritive value" in their index, it is clear that NVI is a "forage quality" index. The correlation (r) between NVI and daily gains by lambs was 0.88 for a combination of chopped and ground forages (Lloyd et al., 1960). Although sound theoretically, application of NVI is limited to relative comparisons among forages because NVI is not expressed in units of nutrient requirements or animal performance. Crampton et al. (1962) provided equations for converting predicted NVI to a daily "yield" of DE assuming a constant GE concentration in forages of 4.4 kcal/g DM.

Digestible Energy Intake (1966)

Heaney et al. (1966) and Heaney (1970) suggested the use of daily DE intake (DEI, kcal/MW) as a forage quality index. An equation for calculating DEI is as follows:

$$\text{DEI} = \text{Observed DM Intake} * \text{DE Concentration} \qquad [4]$$

Observed DM intake was expressed as g/MW, and DE concentration as kcal/g DM. The DEI equation (Eq. [4]) differs from the rearranged NVI equation (Eq. [3]) in that, for DEI, DE concentration was used in place of GE digestibility, and there was no DM intake constant. The same in vivo data are required for calculation of both NVI and DEI: voluntary DM intake, fecal DM excretion, and GE concentrations in forage and feces.

Both NVI and DEI provided nearly equivalent statistical comparisons among 16 timothy (*Phleum pratense*) hays fed to lambs (Heaney et al., 1966). Advantages of DEI over NVI are 1) DEI is expressed in units of energy requirements, 2) calculation of DEI is simpler, 3) DEI does not require a standard forage, and 4) use of DE concentration rather than GE digestibility accounts for variation in forage organic matter (OM) concentration due to variations in forage ash and soil contamination (Heaney et al., 1966).

Relative Feed Value (1976)

In 1972, the American Forage and Grassland Council (AFGC) formed a Hay Marketing Task Force, and a Forage Analysis Subcommittee. The subcommittee stimulated the application of NIRS technology to forage analysis (Norris et al.,

1976) and developed a forage quality index called Relative Feed Value (RFV; Rohweder et al., 1976a, 1976b, 1978). The RFV was calculated as follows:

$$\text{RFV} = \frac{\text{Observed DM Intake} * \text{Digestible DM, \%}}{\text{Base Digestible DM Intake}} \qquad [5]$$

Intakes were expressed as g/MW. Digestible DM was the same as DM digestibility. The first base digestible DM intake was 40 g/MW for sheep (Rohweder et al., 1978). To use RFV for cattle, the base digestible DM intake was set to 69 g/MW (Rohweder et al., 1981). Currently, the RFV equation for dairy cattle expresses intake as a percentage of live body weight (BW), and uses a base digestible DM intake of 1.29% of BW (personal communication, D. Undersander, University of Wisconsin, Madison). The base digestible DM intake was chosen so that RFV would be equal to 100 for a legume hay just at the point of full bloom. In trials with sheep, RFV ranged from 54 for a mature tropical grass to 167 for an immature legume (Moore, 1978).

The RFV equation (Eq. [5]) is similar to the rearranged NVI equation (Eq. [3]) except that, for RFV, digestible DM is substituted for GE digestibility, and a constant digestible DM intake is substituted for a constant total DM intake. The in vivo data required for calculation of RFV are very simple: voluntary DM intake and fecal DM excretion. When digestibilities of DM and GE are correlated highly, then NVI, DEI, and RFV are correlated highly, also. When forages vary in concentrations of fat, protein, forage ash, or soil contamination, however, the correlation between digestibilities of DM and GE may be decreased.

A major concern of the AFGC Forage Analysis Subcommittee was the development of equations to predict RFV from laboratory and NIRS analysis. These equations, presented in the publications cited above, have been revised as additional data became available. Further discussion of the prediction of forage quality indices is presented in a later section of this chapter.

Fill Unit and Feed Unit (1978)

Researchers of the Institut National de la Recherche Agronomique (INRA, France) developed the Fill Unit for expressing and predicting the voluntary intake of forage-based diets (Jarrige et al., 1986; Jarrige, 1989). The Fill Unit was a companion to the Feed Unit used for many years to describe the Net Energy (NE) concentrations of feeds. In the context of forage quality, Fill Unit relates to voluntary intake and Feed Unit to nutritive value. Both Fill Units and Feed Units were expressed in terms of reference feeds. The Fill Unit (per kg DM) was calculated as follows:

$$\text{Fill Unit} = \frac{\text{Reference DM Intake}}{\text{Observed DM Intake}} \qquad [6]$$

The DM intakes were expressed as g/MW. The reference DM intakes by "standard" animals were 75 g/MW for sheep, 95 g/MW for growing cattle, and 140 g/MW for lactating dairy cows. These rates of intake were expected for "an average pasture grass cut at the grazing stage of the first growth" which, by definition, contains 1 Fill Unit per kg of DM. The expected DM intakes of the "reference" forage for calculating Fill Unit were very similar to those of the "standard" forage used to calculate Relative Intake and NVI (75 or 80 g/MW for sheep, and 140 g/MW for cattle). Fill Units (Eq. [6]) are related inversely to Relative Intake (Eq. [2]).

Even though different reference DM intake constants were used for calculating Fill Units for different types of animals, a given non-reference forage had three Fill Units, one each for sheep, growing cattle, and lactating cattle. The ranges in Fill Units from leafy legumes to cereal straw were 0.75 to 2.5 for sheep, 0.8 to 1.8 for growing cattle, and 0.87 to 1.6 for lactating cows. Although forages were ranked in a similar order by each of the three Fill Units, the range was much wider for sheep than for cattle. This discrepancy shows that sheep intake values must be converted to cattle values by use of regression equations rather than by constants. Equations for converting DM intake by sheep to that by growing and lactating cattle were given by Jarrige (1989).

The Feed Unit was the ratio of the NE concentration of a feed to the NE concentration of "standard" barley grain for a particular type of production. One Feed Unit corresponded to the amount of NE supplied by 1 kg of standard barley fed above maintenance. The Feed Unit was calculated as follows:

$$\text{Feed Unit} = \frac{\text{ME Concentration} * \text{Partial Efficiency}}{\text{NE Concentration of Standard Barley}} \quad [7]$$

Energy concentrations were expressed as kcal/kg, and partial efficiencies were expressed as decimal fractions. Partial efficiencies were calculated from ME concentration, and represented the efficiency of conversion of ME to NE for milk or meat production. The NE concentrations of standard barley were 1700 kcal/kg for milk production and 1820 kcal/kg for meat production. The ranges in Feed Units from maize to cereal straw were 1.1 to 0.45 for milk production, and 1.15 to 0.34 for meat production.

The Fill Unit/Feed Unit system is a practical livestock feeding system which considers variation in both forage intake and nutritive value. Because Feed Unit is a measure of NE, the system does consider the efficiency of utilization of digestion end-products. There seems to have been no attempt to combine Fill Unit (i.e., voluntary intake) with Feed Unit (i.e., nutritive value) to give an overall index of forage quality.

Quality Index (1982)

The Florida Extension Forage Testing Program has reported Quality Index (QI) along with Crude Protein (CP) and Total Digestible Nutrients (TDN) concentrations since 1982 (Moore et al., 1984). For extension purposes, QI was defined as voluntary TDN intake as a multiple of the TDN requirement for maintenance. To facilitate calculation of TDN, the assumption was made that TDN concentration was numerically equal to digestible OM (DOM) concentration; this is true if digestible ether extract is negligible as it is in most forages. The first equation used to calculate QI was as follows (Moore, 1978):

$$QI = \frac{\text{Observed DM Intake} * (\text{Digestible OM}, \%)/100}{\text{TDN Requirement for Maintenance}} \qquad [8]$$

Observed DM intake and TDN requirement for maintenance were expressed as g/MW. The TDN requirement for maintenance used initially was that for sheep (29 g/MW; NRC, 1975). The QI equation is similar to the rearranged NVI equation (Eq. [3]) except that, for QI, DOM concentration is substituted for GE digestibility, and a constant TDN intake is substituted for a constant total DM intake. The QI and RFV equations (Eq. [8] and [5]) are similar in that DOM and digestible DM are related closely. The major difference between QI and RFV is that, for QI, the reference base is a defined animal requirement for energy rather than the quality of a forage chosen arbitrarily. The base QI was set to 1 rather than 100 in order to avoid confusion of QI with RFV.

The maintenance TDN requirement was selected as a reference base for QI so that QI could be used both for relative comparisons among forages and for predicting animal performance in mathematical models (Spreen et al., 1985). When QI is less than 1, weight loss would be expected; such a forage would be of low quality. When QI equals 1.0, growing animals fed that forage would neither gain nor lose weight. When QI equals 1.8, growing cattle would gain approximately 0.6 kg/d, and lactating cows would produce approximately 10 kg/d, assuming no weight change (NRC, 1971, 1976). This rate of animal performance would be expected with forages of medium quality. In trials with sheep, QI ranged from 0.74 for a mature tropical grass to 2.30 for an immature legume (Moore, 1978).

The in vivo data required for calculation of QI are the same as that for RFV except that OM concentration of forage and feces must be determined. One reason for using OM rather than DM was the variable sand contamination of most forages in Florida. If samples are analyzed for DM in porcelain crucibles, OM can be determined easily by ashing the dry residue in a muffle furnace. This procedure requires only one weighing in addition to those required for DM determination.

The calculation of QI has been revised (Moore et al., 1990) to include partial efficiency of TDN utilization based on NRC (1984) equations. When dietary nutrients are in balance, partial efficiencies for both maintenance and gain are a function of ME concentration (Raymond, 1969; NRC, 1984; Jarrige, 1989; Minson, 1990). In the revised approach, DE concentration was calculated from DOM and CP concentrations using a local regression equation, and ME was calculated as DE * 0.82 (NRC, 1984). Concentrations of NE for maintenance (NE_m) and NE for gain (NE_g) were calculated from ME concentration using NRC (1984) cubic equations. Efficiency constants of ME utilization were calculated for maintenance ($k_m = NE_m/ME$) and gain ($k_g = NE_g/ME$). There is a wide range in efficiency of ME utilization for both maintenance and gain (Figure 1). When concentration of ME is low, as in many subtropical grasses, the efficiency constants are quite low.

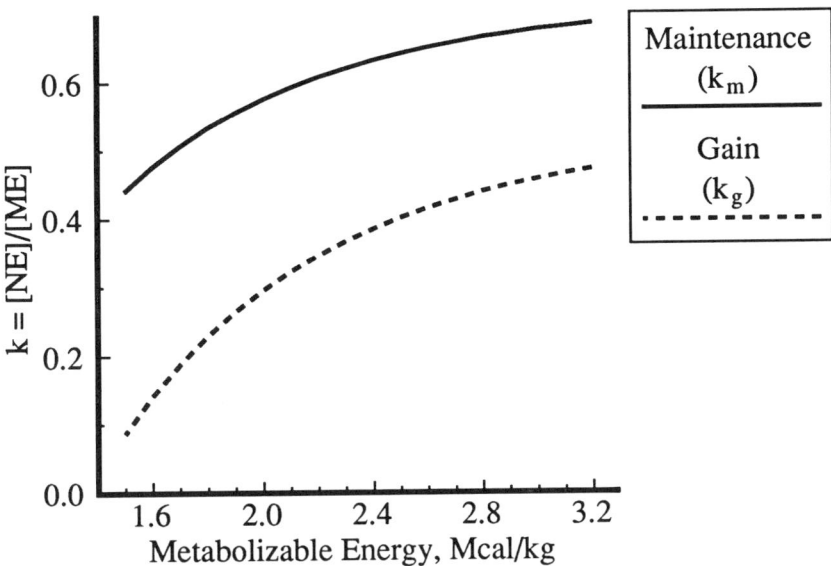

Figure 1. Relationship between the partial efficiency (k) of metabolizable energy (ME) utilization and ME concentration of feeds; NE = net energy for maintenance or gain calculated from ME using NRC (1984) equations.

To determine the effect that differences in efficiency constants would have on predicted gains, a hypothetical data set was constructed with variation in both ME intake and ME concentration. The NE_m requirement was assumed to be 0.077 Mcal/MW (NRC, 1984), and k_m was used to partition total ME intake into that used for maintenance (ME_m) and gain (ME_g). Intake of NE_g was calculated by multiplying ME_g intake by k_g, and gains were calculated from NE_g intake using the equation for medium frame, non-implanted steers (NRC, 1984). Because k_m varied with ME concentration, the maintenance ME requirement was not constant (Figure 2). Further, because k_g varied with ME concentration, rates of gain were dependent on both ME intake and ME concentration. The relationships between gain and NVI, DEI, or RFV are the same as that between gain and ME intake, because none of those indices are adjusted for a maintenance requirement, and their calculations do not include differences in efficiency.

When ME intake was expressed as a multiple of the ME_m requirement most, but not all, of the variation was removed from the relationship between gain and ME intake in the hypothetical data set (Figure 3). Preston (1988) developed a similar procedure for predicting gain of growing-finishing cattle from intake and nutritive value data. Equations were derived from NRC (1984) tabular data to predict gain from DM intake, expressed as a multiple of the amount of DM required for maintenance, and the NE_m concentration of feed. The range of ME concentrations used in developing the equations was from 2.0 to 2.7 Mcal/kg, a range over which k_m and k_g values were high and reasonably constant (Figure 1). Specific equations were calculated for different classes of cattle. The technique was

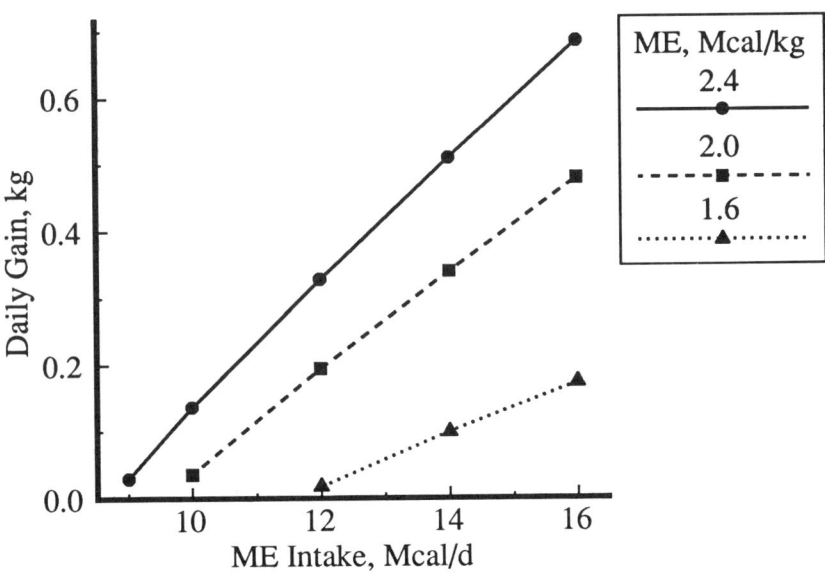

Figure 2. Relationship between expected daily gain and metabolizable energy (ME) intake for a hypothetical data set including three diets having ME concentrations typical of the range expected in forages; gains based on NRC (1984) equations.

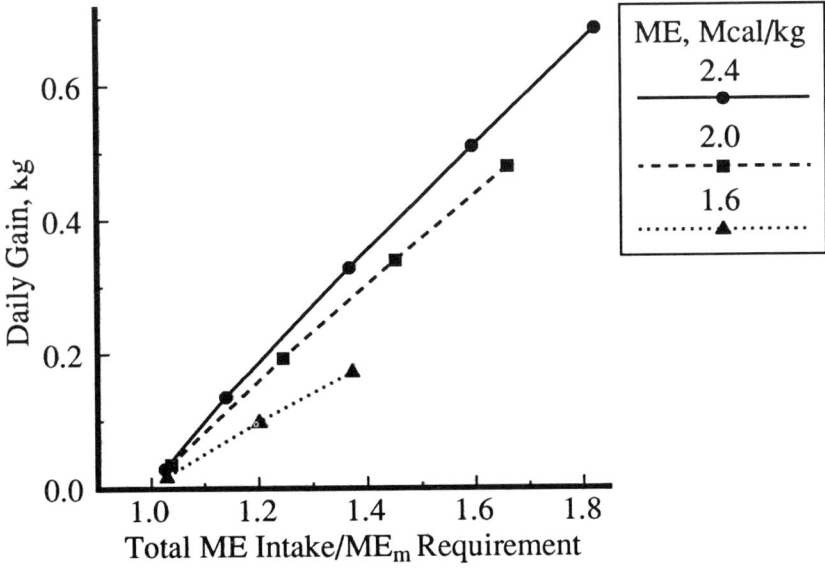

Figure 3. Relationship between expected daily gain and metabolizable energy (ME) intake as a multiple of the ME requirement for maintenance (ME_m), using the hypothetical data set described in Figure 2.

used to predict gains by cattle grazing small plots where intakes were calculated from forage disappearance during short grazing periods, and ME was calculated from in vitro fermentation (Mowrey et al., 1992). No overall index of forage quality was calculated, however.

In order to remove the remainder of the variation seen in Figure 3 between gains and the ratio of total ME intake to ME_m requirement, a base forage was included in the calculation of QI (Moore et al., 1990). In the revised model, intakes of ME_m and ME_g were adjusted to those of a base forage having the same NE_m and NE_g intakes as those of the forage being evaluated. Base forage ME_m and ME_g intakes were calculated using k_m and k_g of the base forage (11% CP and 57% TDN; 2.1 Mcal ME/kg). Total ME intake of the base forage was the sum of its ME_m and ME_g intakes. The revised calculation of QI was as follows:

$$QI = \frac{\text{Base Total ME Intake}}{\text{Base ME Requirement for Maintenance}} \quad [9]$$

The intake and requirement for ME were expressed as Mcal/MW. When QI was calculated as described above for the hypothetical data set, the relationship between QI and gain was not affected by ME concentration (Figure 4). In effect, QI is an expression of voluntary NE_g intake of a forage independent of its ME concentration. Further, the relationship between gain and QI is nearly independent of body weight.

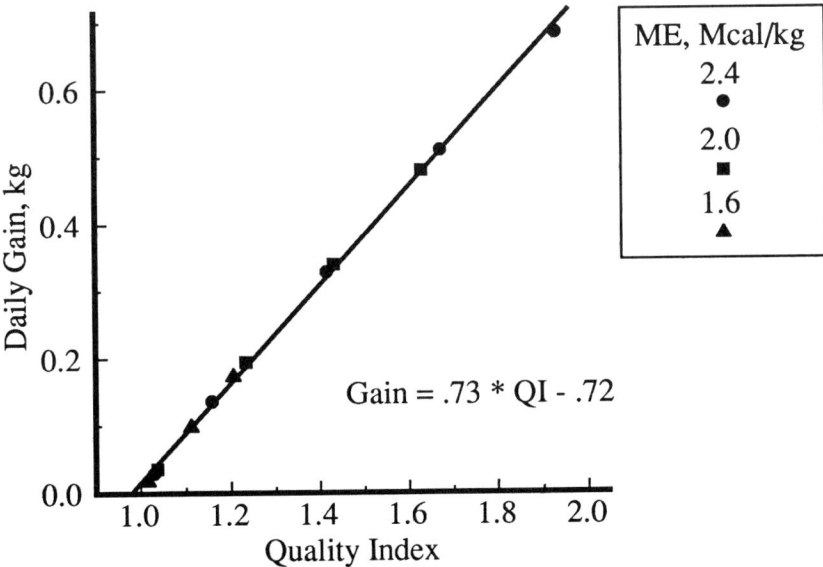

Figure 4. Relationship between expected daily gain and Quality Index using the hypothetical data set described in Figure 2.

The revised QI and gain model was tested on 11 bermudagrass (*Cynodon dactylon*) hays fed to growing steers in four experiments (Figure 5). The QI of each hay was calculated from voluntary intake of DM by steers, digestibility of OM by sheep, and OM and CP concentrations. Observed gains were similar to gains predicted from QI, except for one hay for which gain was much higher than expected. It was concluded that gains by growing cattle can be predicted accurately from QI, that QI can be predicted accurately from voluntary DM intake and ME concentration, and that ME concentration can be predicted accurately from DOM and CP concentrations. It was concluded, also, that the NRC (1984) system is applicable to forages of low ME concentration.

As an index of forage quality, QI incorporates both voluntary intake and nutritive value, and nutritive value includes both ME concentration and efficiency of utilization of ME for maintenance and gain. Therefore, QI is useful both as an overall index of forage quality for making relative comparisons among forages, and as an input for livestock production models which predict animal performance or formulate mixed diets. Because TDN and ME are related closely, the working definition of QI for discussion with livestock producers remains "voluntary intake of TDN as a multiple of the maintenance TDN requirement."

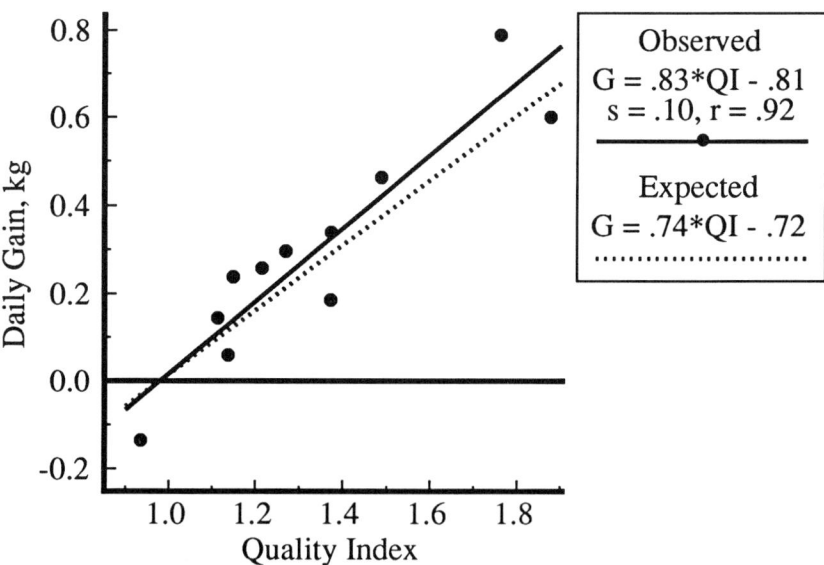

Figure 5. Relationship between observed and expected daily gain and Quality Index calculated from intake and digestibility of 11 bermudagrass hays fed to steers (Moore and Kunkle, 1993).

APPLICATION OF FORAGE QUALITY INDICES

Forage quality information obtained before feeding may have practical application in two areas: 1) forage marketing, and 2) livestock production models. At present, RFV, Fill Unit, and QI are used in one or both of these areas. It is essential that predictions of forage quality indices have acceptable accuracy and precision if they are to be of value to forage-livestock producers.

Forage Marketing

If forage quality information is used in forage marketing, it is because there is a price differential among forages related to their quality and(or) nutritive value. Moore (1978) suggested that producers will invest in a high quality forage if 1) the forage is a large part of the total diet, 2) animal productivity is increased by an increase in forage quality, and 3) the increased animal productivity is more profitable. These conditions exist often in milk production, beef cattle backgrounding, and heifer development, but not in finishing beef cattle or maintaining beef cow herds.

In forage marketing, there is a need for some uniform language to facilitate communication among buyers and sellers (Petritz, 1989a). Many different approaches to evaluating or "grading" forages have been taken, and a confusing array of terms are in use. Petritz (1989a) challenged "the U.S. hay industry to develop one descriptive system for hay." Such a system is needed, of course, but it must take into consideration the multiple uses of forages in livestock feeding.

For many years, nutritive value information, such as CP, TDN, and NE for lactation, has been used to compare the relative monetary value of forages. Although nutritive value is very useful, it does not represent total forage quality because voluntary intake is not represented. Each of the indices NVI, DEI, RFV, and QI could function as a "one number index" that reflects relative differences in potential forage quality when the forage is offered alone and free-choice. Both NVI and RFV have as their point of reference a particular forage to which producers may relate in a general sense. No one species or maturity of forage is, however, of the same quality every year or in every location. A limitation to the use of DEI as a practical forage quality index is that the units are not understood readily and do not permit easy reference to either a standard forage or rate of animal performance. Neither NVI, DEI, nor RFV adjust intake to the maintenance requirement or include the efficiency of utilization of digestion end-products as does QI. Also, only QI is expressed in terms of animal requirements and can be related easily to rates of animal performance when forage is offered alone and free-choice.

The RFV has received much attention in the USA and is used extensively in several states. Although it cannot be used directly in diet formulation, RFV has been used very effectively in educating livestock producers about variations in forage quality and helping them to produce higher quality forage. The Forage Analysis Subcommittee of AFGC used RFV to establish a system of hay grades for marketing grasses and legumes (Rohweder et al., 1978, 1981). Grades were based on ranges of RFV's and were intended to group forages having similar stages of maturity. In an evaluation of the AFGC hay grading system, Turnbull et al. (1982)

fed lactating cows diets containing alfalfa (*Medicago sativa*) hays varying in RFV's from 95 to 148 which were in grades from 1 to 4. They found that RFV's, but not grades, were useful in predicting milk production. Hays with RFV's at the opposite ends of the RFV range within a specific grade were more different in terms of animal performance than were hays of similar RFV but in adjacent grades. In effect, the use of forage grades truncates index numbers and diminishes the usefulness of the quality information available in the index itself.

A forage quality index is an important, but not the only, item of information needed in forage marketing. In many cases, the forage will not be offered free-choice, and only nutritive value information is needed. The price of a forage may depend upon whether it is to be used as a source of protein, energy, or both (Petritz, 1989b). Often, the forage may be either supplemented with concentrates or be part of a total mixed ration (TMR). Also, associative effects between forages and concentrates may mask differences in potential forage quality. Monetary value of forage must reflect its intended use in a feeding program. There is no reason to buy the highest priced forage if it will not increase the profitability of animal production.

Livestock Production Models

The market value of a forage can be determined only when all factors affecting its use, and the use of other alternatives, are considered in relation to requirements of animals (Petritz, 1989b). The ultimate value of forage quality indices, therefore, may be as components of livestock production models. Models can predict animal performance and economic return by interrelating nutrient requirements of animals, forage quality and nutritive value, concentrate nutritive value, associative effects among feeds, and factors affecting animal potential. To be of value to livestock producers, these models must be interactive and allow input of site-specific information obtained easily from producers and forage testing programs. Now that inexpensive high-speed computers are readily available, a new age of practical forage evaluation should begin.

For an index of forage quality to be successful in livestock production models, the index must be on the same basis as, or easily converted to, nutrient requirements. The Fill Unit/Feed Unit system, and QI along with ME (or TDN) concentration, provide quantitative forage quality and nutritive value information that can be used to describe the energy contribution of forages when fed alone. Incorporation of this information in comprehensive production models involves consideration of non-forage factors that influence forage intake and partial efficiency of nutrient utilization.

Types of Models

There are two types of production models which differ in the way forage is utilized; one applies when forage is offered or grazed free-choice and a limited amount of concentrate supplement is fed, and the other applies when forage is included in a TMR at a predetermined percentage of the total diet. A few examples are presented below.

Free-choice forage intake models The INRA system of using Fill Units to quantify intake potential and Feed Unit to quantify nutritive value is an effective way of including forage quality information in models of livestock feeding systems (Jarrige, 1989). In order to calculate intake of a given forage by a given animal using the Fill Unit/Feed Unit system, the Intake Capacity (IC) of animals was expressed in terms of Fill Units per animal per day and calculated as follows:

$$IC = \frac{(MW) * (Observed\ DM\ Intake) * (Fill\ Unit/kg\ DM)}{1000} \quad [10]$$

Observed DM intake was expressed as g/MW. For a given forage the product of its observed DM intake and Fill Unit/kg DM was equal to the reference DM intake. Therefore, IC was independent of forage, but varied with type and weight of animal. The IC's of "standard" sheep, growing cattle, and lactating cattle were 1.62, 8.5, and 17.0 Fill Units, respectively. The expected voluntary DM intake (VDMI) of a given forage by a given animal was calculated as follows:

$$VDMI\ (kg/d) = \frac{IC,\ Fill\ Units\ per\ animal\ per\ day}{Fill\ Units/kg\ forage\ DM} \quad [11]$$

Extensive tables of Fill Unit, Feed Unit, and Intake Capacity data were constructed from results of numerous feeding and digestion trials (Jarrige, 1989). These characteristics of forages and animals were used to predict animal performance and formulate mixed diets to meet requirements for targeted rates of animal production.

Spreen et al. (1985) used QI and TDN in a bioeconomic simulation model that predicted gains and breakeven prices of growing cattle on forage-based diets. Their model was based on the NRC (1976) NE system. It was assumed that maintenance TDN requirement was a function of body weight but independent of forage. Intake of forage DM when fed alone was predicted as follows:

$$DM\ Intake = \frac{QI * TDN\ Requirement\ for\ Maintenance}{(TDN\ Concentration,\ \%)/100} \quad [12]$$

Both DM intake and TDN requirement for maintenance were expressed as g/MW. Differences among feeds in efficiency of TDN utilization were taken into consideration by converting TDN concentrations to NE_m and NE_g concentrations using specific equations for different classes of feeds; these equations were derived from data in feed composition tables (NRC, 1976). Intake of DM was partitioned into that for maintenance and gain, and NE_g intake was calculated and used to predict gain. As in the Fill Unit system (Jarrige, 1989), Spreen et al. (1985) provided tabular data on QI and TDN concentrations for several forages over their growing season. The model was designed to facilitate entry of site-specific information from the keyboard in answer to instructions that appeared on the monitor.

The model of Spreen et al. (1985) has been updated to use the revised QI calculation and NRC (1984) equations. The QI system has the flexibility to use other equations if needed. For example, adjustment of energy efficiency constants for specific forages can be made, as suggested by Minson (1981), if such adjustment is justified. Also, the QI system can use the equation for predicting gain from NE_g intake which is most appropriate for a given class of animals, and it allows adjustments for factors affecting the potential of animals. It is very important to have an accurate estimate of the NE_m requirement for specific types of animals and feeding situations, such as grazing.

Total Mixed Ration Models Mertens (1987) described an approach to modelling intake and digestibility of TMR's based on diet characteristics known to influence intake and digestibility by lactating dairy cows. Three intake control mechanisms were considered: 1) physical limits dependent on the animals fill capacity and the filling effect of the diet, 2) physiological limits depending on the animals energy requirement for maximum animal production and the available energy value of the diet, and 3) psychogenic limits influenced by external factors which may reduce intake below the potential intake as determined by either physical or physiological limits. In this model, neutral detergent fiber (NDF) was considered to be a major determinant of intake because of its role in determining both the fill effect and the energy availability of diets. More details on this model were presented by D.R. Mertens in Chapter 11.

The Cornell Net Carbohydrate and Protein System (Fox et al., 1992; Russell et al., 1992; Sniffen et al., 1992) is an example of a very comprehensive model with much practical potential for both growing and lactating animals. The Cornell model includes consideration of diet composition, feed additives, nutrient availability, rumen fermentation, animal characteristics, and environmental factors that influence partial efficiency of energy and protein utilization. It should be possible to incorporate forage quality information into this model in the form of forage quality indices such as QI. Also, certain components of the Cornell model could be used in other, simpler models to increase the validity of animal performance predictions based on indices such as QI.

Adjustments for Differences in Animal Potential

Many animal-related factors influence the performance of forage-fed animals (Table 1). A major concern is differences among animals in voluntary forage intake when forage is offered free-choice. Also, consideration must be given to differences among animals in partial efficiency of nutrient utilization. Factors such as animal species (sheep vs. cattle), body weight, genetic potential, compensatory gain, lactation, pregnancy, nutrient deficiencies, toxicities, additives, and implants must be taken into account if forage quality indices are to be applied accurately. The INRA system (Jarrige, 1989) includes several equations for converting sheep intake data to cattle intake data. The NRC (1987) published a number of equations for predicting DM intake of farm livestock. Forbes (1986) reviewed thoroughly the many factors affecting voluntary intake and described relationships mathematically. The Cornell model (Fox et al., 1992) provides adjustments for several environmental conditions which influence animal potential and voluntary intake.

Interactions with Concentrates

When forages are offered free-choice and supplemental concentrates are offered in restricted amounts, forage intake may either increase, decrease or not change (Horn and McCollum, 1987; Moore, 1992). In general, forage intake will be decreased by concentrates when forage quality is high, other nutrients are in balance with energy, and concentrates are fed in large amounts. On the other hand, small amounts of concentrates may increase voluntary forage intake when forages are of low quality, especially when they have high ratios of TDN to CP. In addition, there may be "associative effects" such that the digestibility of the total diet is not the same as that calculated from the weighted intakes and digestibilities of the ingredients. Associative effects may change the partial efficiency of ME utilization, also. For example, at equivalent DM intakes, daily gains were increased by substituting ruminally undegradable protein for urea in complete diets for growing cattle (Goedeken et al., 1987).

When forage intake decreases due to concentrate feeding, this effect is called substitution of forage with concentrate. "Substitution rate" (SR) is the decrease in forage intake per unit of concentrate fed. As a result of substitution, animal response to supplemental energy concentrates will be less than expected based upon the intake of the forage fed alone and the amount of concentrate being fed. Spreen et al. (1985) and INRA (Jarrige, 1989) included SR in their models.

Accurate predictions of supplementation effects on forage intake and total diet digestibility are required if forage quality indices are to be used in livestock production models. The changes in forage intake, digestibility, and animal performance due to restricted feeding of protein and energy supplements may be affected by many factors, including but not limited to 1) forage characteristics such as potential intake when fed alone, energy digestibility, and protein concentration, and 2) supplement characteristics such as feeding rate, DE concentration and source (starch vs. digestible fiber), and protein concentration and type (ruminally degradable vs. undegradable). The change in voluntary forage intake due to supplementation with high starch concentrates may be predicted from voluntary intake of forage when fed alone, TDN:CP ratio of the forage, forage CP and ME concentrations, supplement feeding rate, and supplement CP and ME concentrations (Moore, 1992; Moore and Brant, 1992; Brant, 1993).

A special case of substitution and associative effects is the use of TMR's, especially for lactating dairy cows. Such diets may include several forages and combinations of concentrates (both energy and protein). Because TMR's often are based on high quality forages and include high percentages of concentrates, intake and digestibility of the forage component is very likely less than that expected when forage is fed alone. In such cases, complex models such as the Cornell model may be necessary to predict animal performance accurately, but they must consider the effects of concentrates on forage intake and digestibility.

Limits to Using Indices on Pasture

Under grazing conditions, ingestive behavior (i.e., bite weight, bite rate, and grazing time) may be a more important determinant of forage intake than are physical or metabolic controls (Hodgson, 1982). This subject was covered thoroughly by J. Hodgson, D.A. Clark, and R.J. Mitchell in Chapter 19. The behavioral mechanism may control intake on pasture when bite weight is limited by low forage availability, low bulk density, or low acceptability due to lack of palatability or presence of dead herbage in the grazed horizon. In such cases, grazing animals cannot compensate for low bite weight by increasing grazing time and bite rate, and indices of forage quality will not apply because intake will not be limited by characteristics of consumed forage. For example, Duble et al. (1971) demonstrated that quality-related forage characteristics were related to animal performance only when available forage was above a minimum threshold. Information on nutritive value would be useful in this situation, however, along with knowledge of factors affecting bite weight.

Prediction of Quality Indices and Nutritive Value

The laboratory analysis of forage samples, i.e., "forage testing," has a long history around the world. Many different laboratory procedures and approaches for the prediction of intake and digestibility have been investigated. At the first National Conference, the subject was reviewed thoroughly by several speakers. Weir (1970) began his paper with the following words: "A test of nutritive value to be applicable to routine evaluation of forages ideally should have the following characteristics:
 1. Provide a reasonably accurate prediction of nutritive value.
 2. Require a small sample of the forage under test (one kilogram or less).
 3. Simple enough to permit rapid evaluation with a minimum of equipment in commercial laboratories."

These requirements are still valid today and much effort has been spent in attempts to achieve them. As described in the preceding chapters, much research has been done since the first National Conference to elucidate the determinants of forage nutritive value and intake. D.J. Minson and J.R. Wilson in Chapter 13, and W.P. Weiss in Chapter 16, reviewed the many methods which have been tested for prediction of intake and digestibility. Little attention has been given, however, to incorporating this new information into practical forage testing programs.

Both public and private laboratories are involved in extension or advisory programs, and there has been concern about the lack of uniformity among laboratories in methods and results (Barnes, 1970; Tietz, 1990, 1993). In an attempt to improve uniformity among forage testing laboratories, AFGC sponsored the National Forage Testing Association (NFTA), a non-profit corporation consisting of researchers, extension specialists, and commercial forage testing laboratories. In 1993, 96 laboratories in the United States and Canada were certified by NFTA (Anonymous, 1993). Certification is awarded on the basis of the analysis of alfalfa samples for DM, CP, and acid detergent fiber (ADF). A sample is sent to each laboratory four times a year and, to be certified, results on three samples must be within a specified range of accuracy.

Even though laboratories may have consistent results on standard samples, there are many other factors that influence the applicability of forage testing results. It is imperative that the sample submitted to the laboratory be representative of the forage being offered. Even if the sample represents the forage being offered, however, it may not represent that being consumed because of selection of plant parts. Frequently, forages are fed with supplements or used as components of complete diets. Further, animals being fed the forages differ in their appetite and production potential. Because of the heterogeneity of forages, differences in forage usage, and variation among animals, it is remarkable that the chemical composition of a gram of forage ever has any relationship to the performance of animals.

Most forage testing laboratories convert one or more of the analytical results to an estimate of forage nutritive value or an index of quality. These conversion formulas are regression equations, and they vary both in the predictor (i.e., independent variable) and in the coefficients (i.e., intercept and slope). In many cases, different equations are used for different classes of feeds and forages. Different laboratories may use different equations for the same combination of independent and dependent variables. The source of the equations is often unknown. Each equation was developed on a different data base, and it may be applicable to only a narrow range of forages. In some cases, equations are shared without concern as to whether the equation is valid in another location.

The fact that so many different equations are in use illustrates their empirical origin. Not only do regression equations differ in terms of intercept and slope, there is wide variation in the fit of the regression equation to the data (e.g., correlation coefficient or root mean square error). At the first National Conference, Mott and Moore (1970) asked: ". . . why are the indicators or the multitude of other components which are now determined on forages, under some circumstances good estimators of forage nutritive value and at other times not?" They answered as follows: ". . . the measurement of a single constituent of the forage works quite well for predicting DDM (digestible DM) if the investigator restricts the number of kinds of forages to a few closely related species, to a fairly narrow spectrum of management systems and to similar environmental conditions." Seventeen years later, Mertens (1987) described the limitations of empirical prediction equations as follows: ". . . regression coefficients are a function of the specific data set used to derive the equation and they predict well only when the situation to be predicted is similar to the original data set, i.e., both predicted and original data sets belong to the same population. It is a statistical limitation that the farther the unknown situation is from the mean conditions of the original population, the less accurately it will be predicted." After discussing this point in Chapter 16, W.P. Weiss concluded that ". . . each laboratory should generate its own equations."

The discrepancies in relationships between laboratory analyses and forage nutritive value or quality suggest that a given empirical prediction equation might be inappropriate for a given application because it is either imprecise, inaccurate, or both. Equations may lack precision and(or) accuracy for one or more of the following reasons:

1. Laboratory methods were based on empirical rather than on rational relationships between forage characteristics and voluntary intake or digestibility.
2. Equations developed on small sets of samples did not include enough variability to be applicable to a larger set.

3. Equations developed on one set of samples were not applicable to another because the determinants of nutritive value and intake were not the same in both sets.
4. Variations occurred among experiments in types of animal, rate and frequency of feeding, preparation of forage, and length of preliminary and collection periods.
5. The sample of forage analyzed did not represent that consumed during the animal trial.
6. Intake of experimental forages was not determined by forage characteristics.

When properly validated and limited in application, models based on empirical prediction equations may be used quite successfully. The use of ADF and NDF to predict RFV is a very common practice and has been quite useful in midwestern dairy applications. Lippke and Herd (1990) developed a simple model for predicting intake and gain or milk production from CP and ADF. Analysis of forage for CP, NDF, and in vitro OM digestion was shown to be valid in the QI system for predicting gains by cattle fed bermudagrass hay (Figure 6). Even though such models may be valid for the setting in which they were developed, they should not be applied in another setting without being validated there.

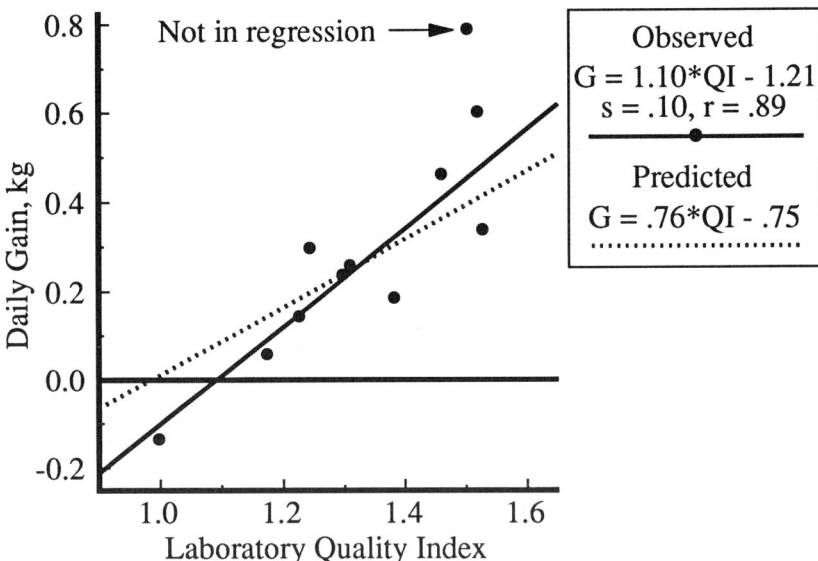

Figure 6. Relationship between observed and expected daily gain and Quality Index calculated from crude protein, neutral detergent fiber, and in vitro organic matter digestibility of 11 bermudagrass hays fed to steers (Moore and Kunkle, 1993).

Toward a Rational Future in Forage Testing

The widely divergent empirical relationships between forage composition and animal responses raise two major questions about the biological relationships between quality-related characteristics of forages and their voluntary intake and nutritive value:
1. Are the characteristics of forage which determine voluntary intake or nutritive value consistent among all forage genotypes, maturities, management practices, and growth environments?
2. Are the predictors determined by common laboratory procedures measurements of, or related to, those characteristics of forage which determine voluntary intake or nutritive value?

It is imperative that these questions be considered, if not answered, in order to improve the accuracy of the forage quality and nutritive value information which will be used in extension-oriented, livestock production models. The following discussion is intended to set the stage for the future by reiterating some fundamental principles.

Prediction of Voluntary Intake

At the first National Conference, Waldo (1970) demonstrated the complex nature of the factors controlling voluntary intake of forages and mixed diets. More recently, Forbes (1986) presented a thorough coverage of this subject for all farm animals. The complex plant-animal interactions which determine the intake of forage were discussed by D.R. Mertens in Chapter 11, D.J. Minson and J.R. Wilson in Chapter 13, and A.W. Illius and M.S. Allen in Chapter 21. In this discussion, the effects of ingestive behavior on intake of pasture by grazing animals will not be considered.

The classical theories of physical and physiological intake control mechanisms for ruminants were developed using varying mixtures of forages and concentrates. For example, Conrad et al. (1964) studied a total of 134 diets, of which 20 were legume and(or) grass, 20 were alfalfa and corn (*Zea mays*) silage mixtures, and the other 94 included concentrates up to 46% of the diet. The range in DM digestibility was from 53 to 80%, and was influenced primarily by dietary concentrate proportion. In that study, and others like it, digestibility was correlated with intake; the correlation was positive when DE intake was less than maximum, and negative when DE requirement was met. Among forages fed alone, however, the correlation between voluntary intake and digestibility is generally positive but often very low (Moore and Mott, 1973; Minson, 1990; Ketelaars and Tolkamp, 1992a). Also, the performance of animals consuming forage-only diets is seldom equal to that achieved when concentrate diets are fed; this suggests that the DE requirement for maximum production may seldom be a factor controlling voluntary intake of forages fed alone. Theories and concepts based on studies of mixed diets may not, therefore, be appropriate for prediction of voluntary intake of forages fed alone. Only when forage is fed alone and free-choice can determinants of forage quality be expressed, and forage quality indices be determined, without being influenced by interactions with concentrates.

Conventional wisdom says that rumen fill and retention time are the major determinants of forage intake, and that NDF is the analysis of choice for predicting intake. Analysis for NDF is used widely as the single predictor of intake in some forage testing programs. Although there is no doubt that NDF is an important characteristic of forages, the empirical relationships between intake and NDF are quite variable, and correlations are often very low (Moore and Mott, 1973; Abrams et al., 1981; Moore et al., 1981; Minson, 1990). Mertens (1987) found that NDF intake of lactating dairy cattle was 1.25% of body weight when they consumed TMR's that produced maximum daily milk yields. D.R. Mertens, in Chapter 11, defined that rate of NDF intake as the "intake constraint." The NDF intake constraint cannot be expected to apply, however, when forages are fed alone. For example, growing Holstein steers fed a bermudagrass hay consumed NDF at the rate of 2.2% of body weight (Hall et al., 1990). Digestibility of NDF in that hay was 62%. In TMR's, intake of forage and, thus, NDF likely will be depressed due to substitution effects, as described above.

Although NDF concentration per se has questionable value as a single predictor of voluntary intake when forage is fed alone, it must be considered as the basis for dividing forage into two fractions, cell walls and cell contents, each of which "behave" differently in the gastrointestinal tract and, thus, influence intake. Perhaps there are fractions of NDF that behave differently from each other, and behave differently in one forage than in another. Such differences may be reflected in variations in degradation and passage rates among forages. Flores et al. (1993) suggested that the greater quality of 'Mott' elephantgrass (*Pennisetum purpureum*) compared to 'Pensacola' bahiagrass (*Paspalum notatum*) was due, in part, to a smaller proportion of sclerenchyma fibers in leaf blades of elephantgrass. Study of the distribution of NDF in different forage tissues may provide important new concepts for use in understanding forage characteristics which are determinants of voluntary intake. D.J. Minson and J.R. Wilson reviewed this subject in Chapter 13.

Recently, doubts have been raised about the conventional theories of intake control. Grovum (1987) suggested that "The simplistic view that only physical factors limit the intakes of roughage diets and that only chemical or physiological factors limit intakes of concentrate diets should be abandoned." Forbes (1988) stated that "We can no longer consider the various theories of intake control as alternatives but rather as complementary and contributing to a multifactorial control system. No single factor is essential for normal feed intake and many manipulations which stay within the physiological range have effects which can only be picked up by close attention to details of feeding behaviour." After concluding that conventional theories of intake control were inadequate (Ketelaars and Tolkamp, 1992a), Tolkamp and Ketelaars (1992) suggested that voluntary intake by ruminants is controlled by a mechanism which maximizes the efficiency of oxygen utilization. Ketelaars and Tolkamp (1992b) suggested that ". . . intracellular pH . . . appears to be a good candidate to sense and control the intensity of feeding behavior." D.R. Mertens, in Chapter 11, compared the intake prediction model of Tolkamp and Ketelaars (1992) to that of Mertens (1987). Both models gave similar predictions of intake when NDF concentration was less than about 65% of DM. When NDF was greater than 65%, as it is in many forages fed alone, predicted intake was lower for the Mertens (1987) model than for that of

Tolkamp and Ketelaars (1992). In Chapter 11, D.R. Mertens pointed out the importance of energy demand on intake; he asked "Does intake determine animal performance (intake as an input) or does animal performance determine intake (intake as a response)?" It is clear that there are many determinants of forage intake, and that they interact in a complex manner.

Because of the complexities of forage composition, structure, and degradation, and of voluntary forage intake control, it is naive to think that any one laboratory procedure will be satisfactory as a universal predictor of voluntary intake of forages fed alone. Prior to the first National Conference, many chemical, solubility and physical procedures were being investigated for the development of predictors of intake, digestibility, or both (Barnes, 1973). Since then, there has been much research directed toward understanding intake regulation but little effort directed toward routine intake prediction. Perhaps consideration should once again be given to developing more comprehensive laboratory analysis systems which include measures of forage characteristics that affect forage particle degradation and passage from the rumen and reticulum, and which consider the metabolic effects of digestion end-products. Such systems might include quantitative analysis of physical or anatomical characteristics of forages, as well as chemical composition and end-products of in vitro digestion. Incorporating such information in routine forage testing will require some innovative approaches in the laboratory. Until such truly rational prediction systems are developed and made practical, however, carefully validated empirical equations will have to be used in routine forage testing programs.

Prediction of Nutritive Value

In a classic paper which should be studied by all those interested in forage testing, Crampton and Maynard (1938) made the following statements: ". . . any relation crude fiber may have to the digestibility of a feed may be in part fortuitous. . . . the digestibility of the dietary carbohydrate does not follow its partition into crude fiber and nitrogen-free extract with any marked certainty, especially in the case of roughages, . . . Furthermore, it is logical to believe that if a division of the carbohydrate fraction could be made into parts which were either biological or chemical units, the usefulness of the feeding stuffs analysis in predicting probable feeding value would be enhanced. The problem of such a partition resolves itself largely into consideration of the chief constituents of the 'cell wall' carbohydrates - cellulose, hemicellulose and lignin. . . . it may not follow that, other factors constant, a decrease in total digestibility of the feed will reflect an increase in total lignin, inasmuch as the manner of its deposition in the plant may be a factor of importance in this respect." These statements have been challenges to researchers for 56 years.

In another classic paper, Lucas and Smart (1959) described the concept of a "nutritive entity" which had constant true digestibility in all forages. Based on this concept, Van Soest (1967) developed the NDF analysis to estimate total cell walls, and showed that cell contents (neutral detergent solubles) met the criteria of a nutritive entity, but NDF did not because of its highly variable digestibility. At the first National Conference, Mott and Moore (1970) and Van Soest (1970) discussed the importance of the nutritive entity concept. Mott and Moore (1970) suggested that: "When making decisions with respect to components to be determined in the laboratory, the question should be asked, 'does this component meet the requirements of a nutritive entity?' If it does not then its usefulness for estimating forage nutritive value cannot be expected to be very great."

The summative equation for predicting digestible DM concentration of forages proposed by Van Soest (1967) was a very rational approach based on the nutritive entity concept. Analysis for NDF estimated the cell contents which was shown to be a nutritive entity and have a metabolic fecal excretion. Digestible NDF was determined from NDF concentration and an empirical prediction of NDF digestibility from ADF and lignin. Weiss et al. (1992) developed a modified summative equation for predicting TDN of both forages and concentrates from chemical composition. Summative equations have a rational or theoretical base, but still include some empirical components. This subject was reviewed by W.P. Weiss in Chapter 16. When empirical relationships are validated for a given situation, their inclusion in an equation having a rational foundation is an acceptable procedure. The importance of validating empirical relationships was demonstrated by Duble et al. (1971) who found that the summative equation of Van Soest (1967) was not acceptable for tropical grasses. This discrepancy was probably because the empirical prediction of NDF digestibility was not valid for the forages being tested.

No doubt the reason that in vitro digestion with ruminal microorganisms or purified enzymes is still used so widely is that it measures at least two nutritive entities: cell contents and digestible cell walls. Also, when DM or OM digestion is determined in vitro, the growth of microorganisms contributes a fraction similar to the in vivo metabolic fecal excretion. No other single laboratory procedure or combination of procedures predicts digestibility as well. Modification of this procedure to incorporate consideration of passage should improve predictions of digestibility (Allen and Mertens, 1988).

Factors affecting nutritive value other than energy digestibility must be given more attention in the future. The concentration and ruminal degradability of protein, digestible protein to DE ratio, anti-quality components, mineral interactions, and perhaps many other factors may affect the partial efficiency of DE utilization. Such factors must be included in forage testing programs if they are suspected to be important in practical situations.

Role of NIRS in Forage Testing

The development of NIRS technology made possible the rapid, routine forage testing program described by Weir (1970). This technique is being used widely and successfully to provide estimates of CP, ADF, NDF, and in vitro digestion. Without NIRS, or some similar technology, it would be impossible to consider the expansion of forage testing to include a larger number of more complex analyses. The technique does not substitute, however, for further research on determinants of forage nutritive value and quality. As discussed above, careful consideration must be given to analysis of forages for other, more complex characteristics that may prove to be valid predictors of voluntary intake and nutritive value. When these characteristics are known, at least for a given type of forage, the availability of NIRS makes it feasible to analyze forages more rapidly, and at less cost. It has been suggested that NIRS be used to predict forage quality and nutritive value directly, but it would be very difficult to collect the large number of samples required for accurate calibration of the instrument.

There are two major considerations for development of expanded application of NIRS in forage testing:

1. Development and validation of predictors and prediction models must be based on research with animals where forage voluntary intake and digestibility are measured in vivo.
2. Calibration of NIRS instruments for analysis of predictors must be based on samples similar to those used in the animal trials mentioned above.

Although the NIRS analysis is rapid, the research required for an effective NIRS-based forage testing program is long-term. Neither the development of new predictors nor the calibration of NIRS instruments can be accomplished with just a few samples. The use of NIRS analysis in concert with practical livestock production models has the potential of revolutionizing forage testing and making it possible to utilize quantitative forage quality information more effectively in ruminant production.

CONCLUSIONS

Indices of forage quality have been based on the hypothesis that the quality of a forage is a function of its voluntary intake and nutritive value. Several indices have been developed and they have potential practical application in both marketing of harvested forage and as inputs for livestock production models. Indices which are based on animal requirements should be more closely related to animal performance than are those based on the intake of a standard forage. Including partial efficiency along with digestibility in the nutritive value component adds some degree of accuracy to an index.

Accurate forage quality indices will be related to animal performance under the conditions that 1) animals used to compare forages have uniform potential for production, 2) forages are available in quantities adequate for maximum intake, and 3) no supplemental energy and protein are provided. Because many factors other than forage quality affect animal performance, it is necessary to develop practical computer models of livestock production that are site-specific and user-friendly. Such models must incorporate forage quantity, animal potential, and supplemental feeds, as well as quantitative information on forage quality and nutritive value. For such models to have practical application, routine forage testing procedures must provide the necessary forage quality information rapidly and in units of animal requirements.

Many forage characteristics are known to be determinants of forage intake, digestibility, and partial efficiency of energy utilization. It is not surprising, therefore, that empirical prediction equations based on simple laboratory predictors have limited application outside the database on which they were developed. As discussed in previous chapters, much is known about the basic biological determinants of forage quality. The challenge for the immediate future is to incorporate this basic knowledge into routine forage testing programs. It may be time to reinvestigate some older procedures, and develop some new ones, that have potential for practical application as predictors of forage intake and nutritive value.

Finally, it is appropriate to repeat the challenge presented by R.L. Reid who stated at the end of Chapter 1: "It is hoped that simplified models, or model components, will be developed to effect further improvements in feeding and grazing management at the farm level." To meet his challenge, the following must occur:

1. Fundamental understanding of the biological determinants of forage intake and nutritive value must be translated into routine analysis using NIRS or a similar technology.
2. The NIRS output must be converted to accurate quantitative predictions of forage quality indices and expressions of nutritive value of forage fed alone and free-choice.
3. Forage quality indices and expressions of nutritive value must be inputs for comprehensive, site-specific computer models that integrate all known factors affecting the performance of animals in a particular forage-based livestock production enterprise.

Forage evaluation and utilization research between now and the third National Conference (2019?) should focus on these goals.

REFERENCES

Abrams, S.M., H. Hartadi, C.M. Chaves, J.E. Moore, and W.R. Ocumpaugh. 1981. Relationship of forage-evaluation techniques to the intake and digestibility of tropical grasses. p. 508-511. *In* J.A. Smith and V.W. Hays (ed.) Proc. XIV Int. Grassl. Congr., Lexington, KY. 15-24 June 1981. Westview Press, Boulder, CO.

Anonymous. 1993. NFTA certifies 96 forage-test labs. Hay and Forage Grower (April):18-21.

Allen, M.S., and D.R. Mertens. 1988. Evaluating constraints on fiber digestion by rumen microorganisms. J. Nutr. 118:261-270.

Barnes, R.F. 1970. Collaborative research with the two-stage in vitro rumen fermentation technique. p. N1-N20. *In* R.F. Barnes, D.C. Clanton, C.H. Gordon, T.J. Klopfenstein, and D.R. Waldo (ed.) Proc. Natl. Conf. Forage Qual. Eval. Util., Lincoln, NE. 3-4 Sept. 1969. Nebraska Center for Continuing Education, Lincoln.

Barnes, R.F. 1973. Laboratory methods of evaluating feeding value of herbage. p. 179-214. *In* G.W. Butler and R. W. Bailey (ed.) Chemistry and biochemistry of herbage, Vol. 3. Academic Press, New York.

Brant, M.H. 1993. Predicting the effects of supplementation of forage diets on forage intake and total diet metabolizable energy concentration. M.S. thesis. Univ. of Florida, Gainesville.

Conrad, H.R., A.D. Pratt, and J.W. Hibbs. 1964. Regulation of feed intake in dairy cows. I. Change in importance of physical and physiological factors with increasing digestibility. J. Dairy Sci. 47:54-62.

Crampton, E.W. 1957. Interrelations between digestible nutrient and energy content, voluntary dry matter intake, and the overall feeding value of forages. J. Anim. Sci. 16:546-552.

Crampton, E.W., and L.A. Maynard. 1938. The relation of cellulose and lignin content to the nutritive value of animal feeds. J. Nutr. 15:383-394.

Crampton, E.W., E. Donefer, and L.E. Lloyd. 1960. A nutritive value index for forages. J. Anim. Sci. 19:538-544.

Crampton, E.W., E. Donefer, and L.E. Lloyd. 1962. Caloric equivalent of the nutritive value index. J. Anim. Sci. 21:628-632.

Donefer, E., E.W. Crampton, and L.E. Lloyd. 1960. Prediction of the nutritive value index of a forage from in vitro rumen fermentation data. J. Anim. Sci. 19:545-552.

Donefer, E., L.E. Lloyd, and E.W. Crampton. 1962. Prediction of the nutritive value index of forages fed chopped or ground using an in vitro rumen fermentation method. J. Anim. Sci. 21:815-818.

Duble, R.L., J.A. Lancaster, and E.C. Holt. 1971. Forage characteristics limiting animal performance on warm-season perennial grasses. Agron. J. 63:795-798.

Flores, J.A., J.E. Moore, and L.E. Sollenberger. 1993. Determinants of forage quality in Pensacola bahiagrass and Mott elephantgrass. J. Anim. Sci. 71:1606-1614.

Forbes, J.M. 1986. The voluntary food intake of farm animals. Butterworths, Boston.

Forbes, J.M. 1988. Metabolic aspects of the regulation of voluntary food intake and appetite. Nutr. Res. Rev. 1:145-168.

Fox, D.G., C.J. Sniffen, J.D. O'Connor, J.B. Russell, and P.J. Van Soest. 1992. A net carbohydrate and protein system for evaluating cattle diets: III. Cattle requirements and diet adequacy. J. Anim. Sci. 70:3578-3596.

Goedeken, F., T. Klopfenstein, R. Stock, and R. Britton. 1987. Hydrolyzed feather meal as a protein source for growing ruminants. p. 50-51. *In* Beef Cattle Report. MP-52. Univ. of Nebraska-Lincoln.

Grovum, W.L. 1987. A new look at what is controlling food intake. p. 1-39. *In* F.N. Owens (ed.) Symposium Proc.: Feed Intake by Beef Cattle, Stillwater, OK. 20-22 Nov. 1986. Anim. Sci. Dep., Oklahoma State Univ., Stillwater.

Hall, K.L., A.L. Goetsch, and L.A. Forster, Jr. 1990. Effects of buffer or dl-methionine with different amounts of supplemental corn on feed intake and nutrient digestion by Holstein steers consuming bermudagrass hay. J. Anim. Sci. 68:1674-1682.

Heaney, D.P. 1970. Voluntary intake as a component of an index to forage quality. p. C1-C10. *In* R.F. Barnes, D.C. Clanton, C.H. Gordon, T.J. Klopfenstein, and D.R. Waldo (ed.) Proc. Natl. Conf. Forage Qual. Eval. Util., Lincoln, NE. 3-4 Sept. 1969. Nebraska Center for Continuing Education, Lincoln.

Heaney, D.P., W.J. Pigden, and G.I. Pritchard. 1966. Comparative energy availability for lambs of four timothy varieties at progressive growth stages. J. Anim. Sci. 25:142-149.

Hodgson, J. 1982. Ingestive behaviour. p. 113-138. *In* J.D. Leaver (ed.) Forage intake handbook. Br. Grassl. Soc., Hurley, UK.

Horn, G.W., and F.T. McCollum. 1987. Energy supplementation of grazing ruminants. p. 125-134. *In* M.B. Judkins, D.C. Clanton, M.K. Petersen, and J.D. Wallace (ed.) Proc. Grazing Livestock Nutr. Conf., Jackson, WY. 23-24 July 1987. Univ. of Wyoming, Laramie.

Jarrige, R. (ed.) 1989. Ruminant nutrition: Recommended feed allowances and feed tables. John Libbey Eurotext, Paris, France.

Jarrige, R., C. Demarquilly, J.P. Dulphy, A. Hoden, J. Robelin, C. Beranger, Y. Geay, M. Journet, C. Malterre, D. Micol, and M. Petit. 1986. The INRA "fill unit" system for predicting the voluntary intake of forage-based diets in ruminants: A review. J. Anim. Sci. 63:1737-1758.

Ketelaars, J.J.M.H, and B.J. Tolkamp. 1992a. Toward a new theory of feed intake regulation in ruminants 1. Causes of differences in voluntary feed intake: critique of current views. Livest. Prod. Sci. 30:269-296.

Ketelaars, J.J.M.H, and B.J. Tolkamp. 1992b. Toward a new theory of feed intake regulation in ruminants 3. Optimum feed intake: in search of a physiological background. Livest. Prod. Sci. 31:235-258.

Lippke, H., and D.B. Herd. 1990. FORAGVAL: Intake and gain by cattle estimated from acid detergent fiber and crude protein of the forage diet. Texas Agric. Exp. Stn. MP - 1708. Texas A&M Univ., College Station.

Lloyd, L.E., E.W. Crampton, E. Donefer, and S.E. Beacom. 1960. The effect of chopping versus grinding on the nutritive value index of early versus late cut red clover and timothy hays. J. Anim. Sci. 19:859-866.

Lucas, H.L., Jr., and W.W.G. Smart, Jr. 1959. Chemical composition and the digestibility of forages. p. 23-26. *In* Proc. 16th S. Pasture Forage Crop Imp. Conf., State College, MS.

Mertens, D.R. 1987. Predicting intake and digestibility using mathematical models of ruminal function. J. Anim. Sci. 64:1548-1558.

Minson, D.J. 1981. A flexible, two-component feeding system derived from Blaxter's three-component metabolizable-energy system. Animal Feed Sci. Tech. 6:223-234.

Minson, D.J. 1990. Forage in ruminant nutrition. Academic Press, San Diego, CA.

Moore, J.E. 1978. Forage quality and animal performance. p. 27-34. *In* Proc. Forage Grassl. Conf., Raleigh, NC. 13-15 Feb. 1978. Am. Forage Grassl. Counc., Lexington, KY.

Moore, J.E. 1980. Crop quality and utilization: Forage crops. p. 61-91. *In* C.S. Hoveland (ed.) Crop quality, storage, and utilization. ASA, CSSA, Madison, WI.

Moore, J.E. 1992. Matching protein and energy supplements to forage quality. p. 31-44. *In* B. Haskins and B. Harris (ed.) Proc. 3rd Annual Florida Rumin. Nutr. Symp., Gainesville, FL. 23-24 Jan. 1992. Univ. of Florida, Gainesville.

Moore, J.E., and M.H. Brant. 1992. Meeting energy requirements of beef cattle with forages and concentrates. J. Anim. Sci. 70(Suppl. 1):181 (Abstr.).

Moore, J.E., and W.E. Kunkle. 1993. Predicting TDN and Quality Index of bermudagrass using routine laboratory procedures. J. Anim. Sci. 71(Suppl. 1):6 (Abstr.).

Moore, J.E., and G.O. Mott. 1973. Structural inhibitors of quality in tropical grasses. p. 53-98. *In* A.G. Matches (ed.) Anti-quality components of forages. CSSA Spec. Publ. 4, CSSA, Madison, WI.

Moore, J.E., J.C. Burns, A.C. Linnerud, and R.J. Monroe. 1981. Relationships between the properties of southern forages and animal response. p. 19-36. *In* Proc. 37th S. Pasture Forage Crop Imp. Conf., Nashville, TN. 19-22 May 1980. Science and Education Administration, USDA, New Orleans, LA.

Moore, J.E., W.E. Kunkle, K.A. Bjorndal, R.S. Sand, C.G. Chambliss, and P. Mislevy. 1984. Extension forage testing program utilizing near infrared reflectance spectroscopy. p. 41-52. *In* Proc. Forage Grassl. Conf., Houston, TX. 23-26 Jan. 1984. Am. Forage Grassl. Counc., Lexington, KY.

Moore, J.E., W.E. Kunkle, S.C. Denham, and R.D. Allshouse. 1990. Quality index: Expression of forage quality and predictor of animal performance. J. Anim. Sci. 68(Suppl. 1):572 (Abstr.).

Mott, G.O. 1959. Symposium on forage quality. IV. Animal variation and measurement of forage quality. Agron. J. 51:223-226.

Mott, G.O., and J.E. Moore. 1970. Forage evaluation techniques in perspective. p. L1-L10. *In* R.F. Barnes, D.C. Clanton, C.H. Gordon, T.J. Klopfenstein, and D.R. Waldo (ed.) Proc. Natl. Conf. Forage Qual. Eval. Util., Lincoln, NE. 3-4 Sept. 1969. Nebraska Center for Continuing Education, Lincoln.

Mowrey, D.P., A.G. Matches, and R.L. Preston. 1992. Technical note: Utilization of sainfoin by grazing steers and a method for predicting daily gain from small-plot grazing data. J. Anim. Sci. 70:2262-2266.

Norris, K.H., R.F. Barnes, J.E. Moore, and J.S. Shenk. 1976. Predicting forage quality by infrared reflectance spectroscopy. J. Anim. Sci. 43:889-897.

NRC. 1971. Nutrient requirements of dairy cattle (4th Ed.). National Academy Press, Washington, DC.

NRC. 1975. Nutrient requirements of sheep (5th Ed.). National Academy Press, Washington, DC.

NRC. 1976. Nutrient requirements of beef cattle (5th Ed.). National Academy Press, Washington, DC.

NRC. 1984. Nutrient requirements of beef cattle (6th Ed.). National Academy Press, Washington, DC.

NRC. 1987. Predicting feed intake of food-producing animals. National Academy Press, Washington, DC.

Petritz, D. 1989a. Confusing hay terms don't make the grade. Hay and Forage Grower (March):38.

Petritz, D. 1989b. Use an accurate method for pricing alfalfa hay. Hay and Forage Grower (April):22.

Preston, R.L. 1988. A simplified procedure for estimating gain and feed efficiency in growing-finishing cattle. p. 41-43. *In* Anim. Sci. Res. Rept., Texas Tech. Univ. Agric. Sci. Tech. Rept. T-5-251.

Raymond, W.F. 1969. The nutritive value of forage crops. Adv. Agron. 21:1-108.

Reid, R.L., and T.J. Klopfenstein. 1983. Forages and crop residues: Quality evaluation and systems of utilization. J. Anim. Sci. 57(Suppl. 2):534-562.

Rohweder, D.A., R.F. Barnes, and N. Jorgensen. 1976a. A standardized approach to establish market value for hay. p. 112. *In* Agronomy Abstracts. ASA, Madison, WI.

Rohweder, D.A., R.F. Barnes, and N. Jorgensen. 1978. Proposed hay grading standards based on laboratory analyses for evaluating quality. J. Anim. Sci. 47:747-759.

Rohweder, D.A., N. Jorgensen, and R.F. Barnes. 1976b. Using chemical analyses to provide guidelines in evaluating forages and establishing hay standards. Feedstuffs 48(47):22.

Rohweder, D.A., N. Jorgensen, and R.F. Barnes. 1981. Proposed hay-grading standards based on laboratory analyses for evaluating quality. p. 534-538. *In* J.A. Smith and V.W. Hays (ed.) Proc. XIV Int. Grassl. Congr., Lexington, KY. 15-24 June 1981. Westview Press, Boulder, CO.

Russell, J.B., J.D. O'Connor, D.G. Fox, P.J. Van Soest, and C.J. Sniffen. 1992. A net carbohydrate and protein system for evaluating cattle diets: I. Ruminal fermentation. J. Anim. Sci. 70:3551-3561.

Sniffen, C.J., J.D. O'Connor, P.J. Van Soest, D.G. Fox, and J.B. Russell. 1992. A net carbohydrate and protein system for evaluating cattle diets: II. Carbohydrate and protein availability. J. Anim. Sci. 70:3562-3577.

Spreen, T.H., J.A. Ross, J.W. Pheasant, J.E. Moore, and W.E. Kunkle. 1985. A simulation model for backgrounding feeder cattle in Florida. Florida Agric. Exp. Stn. Bull. 850 (Tech.).

Tietz, N. 1990. Establishing a stand. Fluctuating forage tests erode farmer confidence. Hay and Forage Grower (January):4.

Tietz, N. 1993. Grower tests forage labs. Hay and Forage Grower (March):14-17.

Tolkamp, B.J., and J.J.M.H Ketelaars. 1992. Toward a new theory of feed intake regulation in ruminants 2. Costs and benefits of feed consumption: an optimization approach. Livest. Prod. Sci. 3:297-317.

Turnbull, G.W., D.W. Claypool, and E.G. Dudley. 1982. Performance of lactating cows fed alfalfa hays graded by relative feed value system. J. Dairy Sci. 65:1205-1211.

Van Soest, P.J. 1967. Development of a comprehensive system of feed analyses and its application to forages. J. Anim. Sci. 26:119-127.

Van Soest, P.J. 1970. The chemical basis for the nutritive evaluation of forages. p. U1-U19. *In* R.F. Barnes, D.C. Clanton, C.H. Gordon, T.J. Klopfenstein, and D.R. Waldo (ed.) Proc. Natl. Conf. Forage Qual. Eval. Util., Lincoln, NE. 3-4 Sept. 1969. Nebraska Center for Continuing Education, Lincoln.

Waldo, D.R. 1970. Factors influencing the voluntary intake of forage. p. E1-E22. *In* R.F. Barnes, D.C. Clanton, C.H. Gordon, T.J. Klopfenstein, and D.R. Waldo (ed.) Proc. Natl. Conf. Forage Qual. Eval. Util., Lincoln, NE. 3-4 Sept. 1969. Nebraska Center for Continuing Education, Lincoln.

Weir, W.C. 1970. Basis for use of simplified techniques to evaluate alfalfa for hay and brush for browse in California. p. R1-R6. *In* R.F. Barnes, D.C. Clanton, C.H. Gordon, T.J. Klopfenstein, and D.R. Waldo (ed.) Proc. Natl. Conf. Forage Qual. Eval. Util., Lincoln, NE. 3-4 Sept. 1969. Nebraska Center for Continuing Education, Lincoln.

Weiss, W.P., H.R. Conrad, and N.R. St. Pierre. 1992. A theoretically-based model for predicting total digestible nutrient values of forages and concentrates. Anim. Feed Sci. Tech. 39:95-110.